GATE PREP SERIES

GATE

Graduate Aptitude Test in Engineering

2022

AF651183

26 Years

Chapter-Wise Solved Papers

(1996-2021)

Civil Engineering

G. K. PUBLICATIONS (P) LTD.

Title : GATE 2022 : Civil Engineering 26 Years Chapter-wise Solved Papers (1996-2021)

Language : English

Editor's Name : Vinit Garg

Typeset & Published by :

Career Launcher Infrastructure (P) Ltd.

A-45, Mohan Cooperative Industrial Area, Near Mohan Estate Metro Station, New Delhi - 110044

Marketed by :

G.K. Publications (P) Ltd.

Plot No. 9A, Sector-27A, Mathura Road, Faridabad, Haryana-121003

ISBN : **978-93-90820-89-4**

Printer's Details : Made in India, New Delhi.

GATE
Civil Engineering

Contents

Solved Papers (Chapter-Wise)

Verbal Ability

Numerical Ability

Engineering Mathematics

Technical Section

GATE

Graduate Aptitude Test in Engineering

IIT Institutes

GATE 2021 will be conducted by
Indian Institute of Technology, Mumbai

Indian Institute of Technology, Kharagpur

GATE 2020 conducted by
Indian Institute of Technology, Delhi

Indian Institute of Technology, Chennai

Indian Institute of Technology, Guwahati

Indian Institute of Technology, Roorkee

Indian Institute of Technology, Kanpur

Preface

The Graduate Aptitude Test in Engineering (GATE) is an online exam conducted by the IITs for admissions to PG courses in IITs, IISc Bangalore, NITs and many state run universities as well as private universities. Also there are more than 37 PSUs that use GATE score for recruitments. A large number of corporates are also using GATE score as a tool to screen students for placements.

GK Publications is well known as the "publisher of choice" to students preparing for GATE and other technical examinations in the country. We published the first set of books in 1994 when GATE exam, both objective and conventional, was conducted in the paper and pencil environment, and used as a check point for entry to post graduate courses in IITs and IISCs. At that time, students had little access to technology and relied mainly on instructor led learning followed by practice with books available for these examinations.

A lot has changed since then!

Today, GATE is conducted in an online only mode with multiple choice and numerical based questions. The score is valid for three years and is used not only for post graduate courses but is also used by major PSUs for recruitment. Today's students have easy access to technology and the concept of a monologue within the classroom has changed to dialogue where students come prepared with concepts and then discuss topics. They learn a lot of things on the go with their mobile devices and practice for mock tests online.

We, as a leading publisher of GATE books, have also embraced change. Today, our books are no more guides and papers only but come with a fully supported mobile app and a web portal. The mobile app provides access to video lectures, short tests and regular updates about the exam. The web portal in additional to what is available on the app provides full length mock tests to mimic the actual exam and help you gauge your level of preparedness. The combination of practice content in print, video lectures, and short and full length tests on mobile and web makes this product a complete courseware for GATE preparation.

This book includes previous years GATE questions along with detailed solution of each question for better understanding. It will help the GATE aspirants to know an idea about the pattern of questions asked in GATE examination.

We also know that improvement is a never ending process and hence we welcome your suggestions and feedback or spelling and technical errors if any. Please write to us at gkp@gkpublications.com

We hope that our small effort will help you prepare well for the examination.

We wish you all the best!

GK Publications Pvt. Ltd.

About GATE

The Graduate Aptitude Test in Engineering (GATE) conducted by IISc and IITs has emerged as one of the bench mark tests for engineering and science aptitude in facilitating admissions for higher education (M.Tech./Ph.D.) in IITs, IISc and various other Institutes/Universities/Laboratories in India. With the standard and high quality of the GATE examination in 27 disciplines of engineering and science, Humanities and Social Sciences subjects, it identifies the candidate's understanding of a subject and aptitude and eligibility for higher studies. During the last few years, GATE score is also being used as one of the criteria for recruitment in Government Organizations such as Cabinet Secretariat, and National/State Public Sector Undertakings in India. Because of the importance of the GATE examination, the number of candidates taking up GATE exams has increased tremendously. GATE exams are conducted by the IITs and IISc as a computer based test having multiple choice questions and numerical answer type questions. The questions are mostly fundamental, concept based and thought provoking. From 2017 onwards GATE Exam is being held in Bangladesh, Ethiopia, Nepal, Singapore, Sri Lanka and United Arab Emirates. An Institute with various nationalities in its campus widens the horizons of an academic environment. A foreign student brings with him/her a great diversity, culture and wisdom to share. Many GATE qualified candidates are paid scholarships/assistantship, especially funded by Ministry of Human Resources Development, Government of India and by other Ministries. Indian Institute of Technology Bombay is the Organizing Institute for GATE 2021.

Why GATE?

Admission to Post Graduate and Doctoral Programmes

Admission to postgraduate programmes with MHRD and some other government scholarships/ assistantships in engineering colleges/institutes is open to those who qualify through GATE. GATE qualified candidates with Bachelor's degree in Engineering/Technology/Architecture or Master's degree in any branch of Science/Mathematics/Statistics/Computer Applications are eligible for admission to Master/Doctoral programmes in Engineering/Technology/Architecture as well as for Doctoral programmes in relevant branches of Science with MHRD or other government scholarships/ assistantships. Candidates with Master's degree in Engineering/Technology/Architecture may seek admission to relevant Ph.D programmes with scholarship/assistantship without appearing in the GATE examination.

Financial Assistance

A valid GATE score is essential for obtaining financial assistance during Master's programs and direct Doctoral programs in Engineering/Technology/Architecture, and Doctoral programs in relevant branches of Science in Institutes supported by the MHRD or other Government agencies. As per the directives of the MHRD, the following procedure is to be adopted for admission to the post-graduate programs (Master's and Doctoral) with MHRD scholarship/assistantship. Depending upon the norms adopted by a specific institute or department of the Institute, a candidate may be admitted directly into a course based on his/her performance in GATE only **or** based on his/her performance in GATE **and** an admission test/interview conducted by the department to which he/she has applied **and/or** the candidate's academic record. If the candidate is to be selected through test/interview for post-graduate programs, a minimum of 70% weightage will be given to the performance in GATE and the remaining 30% weightage will be given to the candidate's performance in test/interview and/or academic record, as per MHRD guidelines. The admitting institutes could however prescribe a minimum passing percentage of marks in the test/interview. Some colleges/institutes specify GATE qualification as the mandatory requirement even for admission without MHRD scholarship/assistantship.

To avail of the financial assistance (scholarship), the candidate must first secure admission to a program in these Institutes, by a procedure that could vary from institute to institute. Qualification in GATE is also a minimum requirement to apply for various fellowships awarded by many Government organizations. Candidates are advised to seek complete details of admission procedures and availability of MHRD scholarship/assistantship from the concerned admitting institution. The criteria for postgraduate admission with scholarship/assistantship could be different for different institutions. The management of the post-graduate scholarship/assistantship is also the responsibility of the admitting institution. Similarly, reservation of seats under different categories is as per the policies and norms prevailing at the admitting institution and Government of India rules. *GATE offices will usually not entertain any enquiry about admission, reservation of seats and / or award of scholarship / assistantship.*

PSU Recruitments

As many as 37 PSUs are using GATE score for recruitment. It is likely that more number of PSUs may start doing so by next year. Below is the list of PSUs:

MDL, BPCL, GAIL, NLC LTD, CEL, Indian Oil, HPCL, NBPC, NECC, BHEL, WBSEDCL, NTPC, ONGC, Oil India, Power Grid, Cabinet Secretariat, Govt. of India, BAARC, NFL, IPR, PSPCL, PSTCL, DRDO, OPGC Ltd., THDC India Ltd., BBNL, RITES, IRCON, GHECL, NHAI, KRIBHCO, Mumbai Railway Vikas Corporation Ltd. (MRVC Ltd.), National Textile Corporation, Coal India Ltd., BNPM, AAI, NALCO, EdCIL India.

Important :

1. Admissions in IITs/IISc or other Institutes for M.Tech./Ph.D. through GATE scores shall be advertised separately by the Institutes and GATE does not take the responsibility of admissions.
2. Cabinet Secretariat has decided to recruit officers for the post of Senior Field Officer (Tele) (From GATE papers of EC, CS, PH), Senior Research Officer (Crypto) (From GATE papers of EC, CS, MA), Senior Research Officer (S&T) (From GATE papers EC, CS, CY, PH, AE, BT) in the Telecommunication Cadre, Cryptographic Cadre and Science & Technology Unit respectively of Cabinet Secretariat. The details of the scheme of recruitment shall be published in National Newspaper/Employment News by the concerned authority.
3. Some PSUs in India have expressed their interest to utilize GATE scores for their recruitment purpose. The Organizations who intend to utilize GATE scores shall make separate advertisement for this purpose in Newspapers etc.

Who Can Appear for GATE?

Eligibility for GATE

Before starting the application process, the candidate must ensure that he/she meets the eligibility criteria of GATE given in Table.

Eligibility Criteria for GATE 2021

Degree/Program	Qualifying Degree/Examination	Description of Eligible Candidates	Expected Year of Completion
B.E. / B.Tech. / B.Pharm.	Bachelor's degree in Engineering / Technology (4 years after 10+2 or 3 years after B.Sc. / Diploma in Engineering / Technology)	Currently in the 3rd year or higher grade or already completed	2022
B. Arch.	Bachelor's degree of Architecture (5- year course) / Naval Architecture (4- year course) / Planning (4- year course)	Currently in the 3rd year or higher grade or already completed	2023 (for 5-year program), 2022 (for 4-year program)
B.Sc. (Research) / B.S.	Bachelor's degree in Science (Post-Diploma/4 years after 10+2)	Currently in the 3rd year or higher grade or already completed	2022
Pharm. D. (after 10+2)	6 years degree program, consisting of internship or residency training, during third year onwards	Currently in the 3rd/4th/5th/6th year or already completed	2024
M.B.B.S.	Degree holders of M.B.B.S. and those who are in the 5th/6th/7th semester or higher semester of such programme.	5th, 6th, 7th or higher semester or already completed	2022
M. Sc. / M.A. / MCA or equivalent	Master's degree in any branch of Arts/Science/Mathematics/Statistics/ Computer Applications or equivalent	Currently in the first year or higher or already Completed	2022
Int. M.E./ M.Tech. (Post-B.Sc.)	Post-B.Sc Integrated Master's degree programs in Engineering/ Technology (4-year program)	Currently in the 1st / 2nd/3rd/4th year or already completed	2024
Int. M.E./ M.Tech. or Dual Degree (after Diploma or 10+2)	Integrated Master's degree program or Dual Degree program in Engineering/Technology (5-year program)	Currently in the 3rd /4th/5th year or alreadycompleted	2023
B.Sc. / B.A. / B.Com.	Bachelor degree in any branch of Science / Arts / Commerce (3 years program)	Currently in the 3rd year or already completed	2021
Int. M.Sc. / Int. B.S. / M.S.	Integrated M.Sc. or 5-year integrated B.S.-M.S. program	Currently in the 3rd year or higher or already completed	2022
Professional Society Examinations (equivalent to B.E. / B.Tech. / B.Arch.)	B.E./B.Tech./B.Arch. equivalent examinations of Professional Societies, recognized by MoE/UPSC/AICTE (e.g. AMIE by Institution of Engineers-India, AMICE by the Institute of Civil Engineers-India and so on)	Completed Section A or equivalent of such professional courses	NA

In case a candidate has passed one of the qualifying examinations as mentioned above in 2020 or earlier, the candidate has to submit the degree certificate / provisional certificate / course completion certificate / professional certificate / membership certificate issued by the society or institute. In case, the candidate is expected to complete one of the qualifying criteria in 2021 or later as mentioned above, he/she has to submit a certificate from Principal or a copy of marks card for section A of AMIE.

Certificate From Principal

Candidates who have to submit a certificate from their college Principal have to obtain one from his/her institution beforehand and upload the same during the online submission of the application form.

Candidates With Backlogs

Candidates, who have appeared in the final semester/year exam in 2021, but with a backlog (arrears/failed subjects) in any of the papers in their qualifying degree should upload a copy of any one of the mark sheets of the final year,

OR

obtain a declaration from their Principal along with the signature and seal beforehand and upload the same during the online submission of the application form.

GATE Structure

Structure of GATE

GATE 2021 will be conducted on 27 subjects (papers). Table below shows the list of papers and paper codes for GATE 2021. A candidate is allowed to appear in ANY ONE or UP TO TWO papers of the GATE examination. However, note that the combination of TWO papers in which a candidate can appear MUST be from the pre-defined list as given in Table. Also note that for a paper running in multiple sessions, a candidate will be mapped to appear for the examination in one of the sessions ONLY.

List of GATE Papers and Corresponding Codes

GATE Paper	Code	GATE Paper	Code
Aerospace Engineering	AE	Instrumentation Engineering	IN
Agricultural Engineering	AG	Mathematics	MA
Architecture and Planning	AR	Mechanical Engineering	ME
Bio-medical Engineering	BM	Mining Engineering	MN
Biotechnology	BT	Metallurgical Engineering	MT
Civil Engineering	CE	Petroleum Engineering	PE
Chemical Engineering	CH	Physics	PH
Computer Science and Information Technology	CS	Production and Industrial Engineering	PI
Chemistry	CY	Statistics	ST
Electronics and Communication Engineering	EC	Textile Engineering and Fibre Science	TF
Electrical Engineering	EE	Engineering Sciences	XE*
Environmental Science & Engineering	ES	Humanities & Social Sciences	XH**
Ecology and Evolution	EY	Life Sciences	XL***
Geology and Geophysics	GG		

Note: *Environmental Science and Engineering (ES) and Humanities and Social Sciences (XH) are two new papers introduced in GATE 2021.*

*XE Paper Sections	Code	**XH Paper Sections	Code	***XL Paper Sections	Code
Engineering Mathematics (Compulsory) (15 marks)	A	Reasoning and Comprehension (Compulsory) (25 marks)	B1	Chemistry (Compulsory) (25 marks)	P
Any TWO optional Sections (2x35 = 70 marks)		Any ONE optional Section (60 marks)		Any TWO optional Sections (2x30 = 60 marks)	
Fluid Mechanics	B	Economics	C1	Biochemistry	Q
Materials Science	C	English	C2	Botany	R
Solid Mechanics	D	Linguistics	C3	Microbiology	S
Thermodynamics	E	Philosophy	C4	Zoology	T
Polymer Science and Engineering	F	Psychology	C5	Food Technology	U
Food Technology	G	Sociology	C6		
Atmospheric and Oceanic Sciences	H				

*XE (Engineering Sciences), **XH (Humanities & Social Sciences), ***XL (Life Sciences), papers are of general nature and will be comprised of Sections listed in the above table

Note: *Each subject/paper is of total 100 marks. General Aptitude (GA) section of 15 marks is common for all papers. Hence remaining 85 marks are for the respective subject/paper code.*

Combination of Two Papers Allowed to Appear in GATE 2021 (subject to availability of infrastructure and schedule)

Code of The First (Primary) Paper	Codes of Papers Allowed as The Second Paper
AE	XE
AG	ES
AR	CE
BM	BT / XL
BT	BM / XL
CE	AR / ES
CH	CY / PE / XE
CS	MA
CY	CH / XL
EC	IN / PH
EE	IN
ES	AG / CE
EY	XL
GG	MN / PE / PH
IN	EC / EE / PH
MA	CS / PH / ST
ME	XE
MN	GG / XE
MT	PH / XE
PE	CH / GG / XE
PH	EC / GG / IN / MA / MT / ST
PI	XE
ST	MA / PH
TF	XE
XE	AE / CH / ME / MN / MT / PE / PI / TF
XH	----
XL	BM / BT / CY / EY

General Aptitude Questions

All the papers will have a few questions that test the General Aptitude (Language and Analytical Skills), apart from the core subject of the paper.

Duration and Examination Type

All the papers of the GATE 2021 examination will be for 3 hours duration and they consist of 65 questions for a total of 100 marks. Since the examination is an ONLINE computer based test (CBT), at the end of the stipulated time (3-hours), the computer screen will automatically close the examination inhibiting any further action.

Candidates will be permitted to occupy their allotted seats 40 minutes before the scheduled start of the examination. Candidates can login and start reading the instructions 20 minutes before the start of examination. The late login time (if any) recorded by the computer system MUST NOT be beyond 30 minutes from the actual starting time of the examination. Under NO circumstances, will a candidate be permitted to login after 30 minutes from the actual examination starting time. Candidates will NOT be permitted to leave the examination hall before the end of the examination.

Pattern of Question Papers

GATE 2021 may contain questions of THREE different types in all the papers:

(i) **Multiple Choice Questions (MCQ)** carrying 1 or 2 marks each, in all the papers and sections. These questions are objective in nature, and each will have choice of four answers, out of which ONLY ONE choice is correct.

Negative Marking for Wrong Answers: For a wrong answer chosen in a MCQ, there will be negative marking. For 1-mark MCQ, 1/3 mark will be deducted for a wrong answer. Likewise, for 2-mark MCQ, 2/3 mark will be deducted for a wrong answer.

(ii) **Multiple Select Questions (MSQ)** carrying 1 or 2 marks each in all the papers and sections. These questions are objective in nature, and each will have choice of four answers, out of which ONE or MORE than ONE choice(s) is / are correct.

Note: There is NO negative marking for a wrong answer in MSQ questions. However, there is NO partial credit for choosing partially correct combinations of choices or any single wrong choice.

(iii) **Numerical Answer Type (NAT)** Questions carrying 1 or 2 marks each in most of the papers and sections. For these questions, the answer is a signed real number, which needs to be entered by the candidate using the virtual numeric keypad on the monitor (keyboard of the computer will be disabled). No choices will be shown for these types of questions. The answer can be a number such as 10 or –10 (an integer only). The answer may be in decimals as well, for example, 10.1 (one decimal) or 10.01 (two decimals) or –10.001 (three decimals). These questions will be mentioned with, up to which decimal places, the candidates need to present the answer. Also, for some NAT type problems an appropriate range will be considered while evaluating these questions so that the candidate is not unduly penalized due to the usual round-off errors. Candidates are advised to do the rounding off at the end of the calculation (not in between steps). Wherever required and possible, it is better to give NAT answer up to a maximum of three decimal places.

Note: *There is NO negative marking for a wrong answer in NAT questions.*
Also, there is NO partial credit in NAT questions.

Marking Scheme – Distribution of Marks and Questions

General Aptitude (GA) Questions

In all papers, GA questions carry a total of 15 marks. The GA section includes 5 questions carrying 1-mark each (sub-total 5 marks) and 5 questions carrying 2-marks each (sub-total 10 marks).

Question Papers other than GG, XE, XH and XL

These papers would contain 25 questions carrying 1-mark each (sub-total 25 marks) and 30 questions carrying 2-marks each (sub-total 60 marks) consisting of some MCQ type questions, while the remaining may be MSQ and / or NAT questions.

GG (Geology and Geophysics) Paper

Apart from the General Aptitude (GA) section, the GG question paper consists of two parts: Part A and Part B. Part A is compulsory for all the candidates. Part B contains two sections: Section 1 (Geology) and Section 2 (Geophysics). Candidates will have to attempt questions in Part A and questions in either Section 1 or Section 2 of Part B. The choice of Section 1 OR Section 2 of Part B has to be made at the time of online registration in GOAPS. At the examination hall, candidate cannot request for change of section.

Part A consists of 25 questions carrying 1-mark each (sub-total 25 marks and some of these will be MSQ and/or numerical answer type questions while remaining questions will be MCQ type). Either section of Part B (Section 1 and Section 2) consists of 30 questions carrying 2-marks each (sub-total 60 marks and some of these will be MSQ and/or numerical answer type questions while remaining questions will be MCQ type).

XE Paper (Engineering Sciences)

A candidate appearing in the XE paper has to answer the following:

- GA – General Aptitude carrying a total of 15 marks.
- Section A – Engineering Mathematics (Compulsory): This section contains 11 questions carrying a total of 15 marks: 7 questions carrying 1-mark each (sub-total 7 marks), and 4 questions carrying 2-marks each (sub-total 8 marks). Some questions will be of numerical answer type while remaining questions will be MCQ type.
- Any two of XE Sections B to H: The choice of two sections from B to H can be made during the examination after viewing the questions. Only TWO optional sections can be answered at a time. A candidate wishing to change midway of the examination to another optional section must first choose to deselect one of the previously chosen optional sections (B to H). Each of the optional sections of the XE paper (Sections B through H) contains 22 questions carrying a total of 35 marks: 9 questions carrying 1-mark each (sub-total 9 marks) and 13 questions carrying 2-marks each (sub-total 26 marks). Some questions will be of MSQ and/or numerical answer type while remaining questions will be MCQ type.

XH Paper (Humanities and Social Sciences)

A candidate appearing in the XH paper has to answer the following:

- **GA – General Aptitude** carrying a total of 15 marks.
- **Section B1 – Reasoning and Comprehension (Compulsory):** This section contains 15 questions carrying a total of 25 marks: 5 questions carrying 1-mark each (sub-total 5 marks) and 10 questions carrying 2-marks each (sub-total 20 marks). Some questions will be of MSQ and/or numerical answer type while remaining questions will be MCQ type.
- **Any ONE of XH Sections C1 to C6:** The ONE choice of section from C1 to C6 has to be made at the time of online registration in GOAPS. At the examination hall, candidate cannot request for change of section. Each of the optional sections of the XH paper (Sections C1 through C6) contains 40 questions carrying a total of 60 marks: 20 questions carrying 1-mark each (sub-total 20 marks) and 20 questions carrying 2-marks each (sub-total 40 marks). Some questions will be of MSQ and/or numerical answer type while remaining questions will be MCQ type.

XL Paper (Life Sciences)

A candidate appearing in the XL paper has to answer the following:

- **GA – General Aptitude** carrying a total of 15 marks.
- **Section P–Chemistry (Compulsory):** This section contains 15 questions carrying a total of 25 marks: 5 questions carrying 1-mark each (sub-total 5 marks) and 10 questions carrying 2-marks each (sub-total 20 marks). Some questions will be of MSQ and/or numerical answer type while remaining questions will be MCQ type.
- **Any two of XL Sections Q to U:** The choice of two sections from Q to U can be made during the examination after viewing the questions. Only TWO optional sections can be answered at a time. A candidate wishing to change midway of the examination to another optional section must first choose to deselect one of the previously chosen optional sections (Q to U). Each of the optional sections of the XL paper (Sections Q through U) contains 20 questions carrying a total of 30 marks: 10 questions carrying 1-mark each (sub-total 10 marks) and 10 questions carrying 2-marks each (sub-total 20 marks). Some questions will be of MSQ and/or numerical answer type while remaining questions will be MCQ type.

GATE Score

After the evaluation of the answers, the actual (raw) marks obtained by a candidate will be considered for computing the GATE Score. For multi-session papers (subjects), raw marks obtained by the candidates in different sessions will be converted to Normalized marks for that particular subject. Thus, raw marks (for single session papers) or normalized marks (for multi-session papers) will be used for computing the GATE Score, based on the qualifying marks.

Calculation of Normalized Marks for Multi-Session Papers

In GATE 2021 examination, some papers may be conducted in multi-sessions. Hence, for these papers, a suitable normalization is applied to take into account any variation in the difficulty levels of the question papers across different sessions. The normalization is done based on the fundamental assumption that "in all multi-session GATE papers, the distribution of abilities of candidates is the same across all the sessions". This assumption is justified since the number of candidates appearing in multi-session papers in GATE 2021 is large and the procedure for allocation of session to candidates is random. Further, it is also ensured that for the same multi-session paper, the number of candidates allotted in each session is of the same order of magnitude.

Based on the above, and considering various normalization methods, the committee arrived at the following formula for calculating the normalized marks for the multi-session papers.

Nsormalization mark of jth candidate in the ith session $\widehat{M_{ij}}$ is given by

$$\widehat{M_{ij}} = \frac{\bar{M}_t^g - M_q^g}{\bar{M}_{ti} - M_{iq}}(M_{ij} - M_{iq}) + M_q^g$$

where

M_{ij} : is the actual marks obtained by the jth candidate in i^{th} session

$\bar{M}_t^g$: is the average marks of the top 0.1% of the candidates considering all sessions

M_q^g : is the sum of mean and standard deviation marks of the candidates in the paper considering all sessions

$\bar{M}_{ti}$: is the average marks of the top 0.1% of the candidates in the ith session

M_{iq} : is the sum of the mean marks and standard deviation of the ith session

Calculation of GATE Score for All Papers

For all papers for which there is only one session, actual marks obtained by the candidates will be used for calculating the GATE 2021 Score. For papers in multi-sessions, normalized marks will be calculated corresponding to the raw marks obtained by a candidate and the GATE 2021 Score will be calculated based on the normalized marks.

The GATE 2021 score will be computed using the formula given below.

$$\text{GATE Score} = S_q + (S_t - S_q)\frac{(M - M_q)}{(\bar{M}_t - M_q)}$$

where

M : marks obtained by the candidate (actual marks for single session papers and normalized marks for multi-session papers)

M_q : is the qualifying marks for general category candidate in the paper

$\bar{M}_t$: is the mean of marks of top 0.1% or top 10 (whichever is larger) of the candidates who appeared in the paper (in case of multi-session papers including all sessions)

S_q : 350, is the score assigned to M_q

S_t : 900, is the score assigned to $\bar{M}_t$

In the GATE 2021 the qualifying marks (M_q) for general category student in each subject will be 25 marks (out of 100) or $\mu + \sigma$, whichever is larger. Here μ is the mean and σ is the standard deviation of marks of all the candidates who appeared in the paper.

After the declaration of results, GATE Scorecards can be downloaded by the GATE qualified candidates ONLY.

The GATE 2021 Committee has the authority to decide the qualifying mark/score for each GATE paper. In case of any claim or dispute with respect to GATE 2021 examination or score, the Courts and Tribunals in Mumbai alone will have the exclusive jurisdiction to entertain and settle them.

GATE Syllabus

Section 1: Engineering Mathematics

Linear Algebra: Matrix algebra; Systems of linear equations; Eigen values and Eigen vectors.

Calculus: Functions of single variable; Limit, continuity and differentiability; Mean value theorems, local maxima and minima; Taylor series; Evaluation of definite and indefinite integrals, application of definite integral to obtain area and volume; Partial derivatives; Total derivative; Gradient, Divergence and Curl, Vector identities; Directional derivatives; Line, Surface and Volume integrals.

Ordinary Differential Equation (ODE): First order (linear and non-linear) equations; higher order linear equations with constant coefficients; Euler-Cauchy equations; initial and boundary value problems.

Partial Differential Equation (PDE): Fourier series; separation of variables; solutions of one- dimensional diffusion equation; first and second order one-dimensional wave equation and two-dimensional Laplace equation.

Probability and Statistics: Sampling theorems; Conditional probability; Descriptive statistics – Mean, median, mode and standard deviation; Random Variables – Discrete and Continuous, Poisson and Normal Distribution; Linear regression.

Numerical Methods: Error analysis. Numerical solutions of linear and non-linear algebraic equations; Newton's and Lagrange polynomials; numerical differentiation; Integration by trapezoidal and Simpson's rule; Single and multi-step methods for first order differential equations.

Section 2: Structural Engineering

Engineering Mechanics: System of forces, free-body diagrams, equilibrium equations; Internal forces in structures; Frictions and its applications; Centre of mass; Free Vibrations of undamped SDOF system.

Solid Mechanics: Bending moment and shear force in statically determinate beams; Simple stress and strain relationships; Simple bending theory, flexural and shear stresses, shear centre; Uniform torsion, Transformation of stress; buckling of column, combined and direct bending stresses.

Structural Analysis: Statically determinate and indeterminate structures by force/ energy methods; Method of superposition; Analysis of trusses, arches, beams, cables and frames; Displacement methods: Slope deflection and moment distribution methods; Influence lines; Stiffness and flexibility methods of structural analysis.

Construction Materials and Management

Construction Materials: Structural Steel – Composition, material properties and behaviour; Concrete - Constituents, mix design, short-term and long-term properties. Construction.

Management: Types of construction projects; Project planning and network analysis - PERT and CPM; Cost estimation.

Concrete Structures: Working stress and Limit state design concepts; Design of beams, slabs, columns; Bond and development length; Prestressed concrete beams.

Steel Structures: Working stress and Limit state design concepts; Design of tension and compression members, beams and beam- columns, column bases; Connections - simple and eccentric, beam-column connections, plate girders and trusses; Concept of plastic analysis -beams and frames.

Section 3: Geotechnical Engineering

Soil Mechanics: Three-phase system and phase relationships, index properties; Unified and Indian standard soil classification system; Permeability - one dimensional flow, Seepage through soils – two - dimensional flow, flow nets, uplift pressure, piping, capillarity, seepage force; Principle of effective stress and quicksand condition; Compaction of soils; One- dimensional consolidation, time rate of consolidation; Shear Strength, Mohr's circle, effective and total shear strength parameters, Stress-Strain characteristics of clays and sand; Stress paths.

Foundation Engineering: Sub-surface investigations - Drilling bore holes, sampling, plate load test, standard penetration and cone penetration tests; Earth pressure theories - Rankine and Coulomb; Stability of slopes – Finite and infinite slopes, Bishop's method; Stress distribution in soils – Boussinesq's theory; Pressure bulbs,

Shallow foundations – Terzaghi's and Meyerhoff's bearing capacity theories, effect of water table; Combined footing and raft foundation; Contact pressure; Settlement analysis in sands and clays; Deep foundations – dynamic and static formulae, Axial load capacity of piles in sands and clays, pile load test, pile under lateral loading, pile group efficiency, negative skin friction.

Section 4: Water Resources Engineering

Fluid Mechanics: Properties of fluids, fluid statics; Continuity, momentum and energy equations and their applications; Potential flow, Laminar and turbulent flow; Flow in pipes, pipe networks; Concept of boundary layer and its growth; Concept of lift and drag.

Hydraulics: Forces on immersed bodies; Flow measurement in channels and pipes; Dimensional analysis and hydraulic similitude; Channel Hydraulics - Energy-depth relationships, specific energy, critical flow, hydraulic jump, uniform flow, gradually varied flow and water surface profiles.

Hydrology: Hydrologic cycle, precipitation, evaporation, evapo-transpiration, watershed, infiltration, unit hydrographs, hydrograph analysis, reservoir capacity, flood estimation and routing, surface run-off models, ground water hydrology - steady state well hydraulics and aquifers; Application of Darcy's Law.

Irrigation: Types of irrigation systems and methods; Crop water requirements - Duty, delta, evapo-transpiration; Gravity Dams and Spillways; Lined and unlined canals, Design of weirs on permeable foundation; cross drainage structures.

Section 5: Environmental Engineering

Water and Waste Water Quality and Treatment: Basics of water quality standards – Physical, chemical and biological parameters; Water quality index; Unit processes and operations; Water requirement; Water distribution system; Drinking water treatment.

Sewerage system design, quantity of domestic wastewater, primary and secondary treatment. Effluent discharge standards; Sludge disposal; Reuse of treated sewage for different applications.

Air Pollution: Types of pollutants, their sources and impacts, air pollution control, air quality standards, Air quality Index and limits.

Municipal Solid Wastes: Characteristics, generation, collection and transportation of solid wastes, engineered systems for solid waste management (reuse/ recycle, energy recovery, treatment and disposal).

Section 6: Transportation Engineering

Transportation Infrastructure: Geometric design of highways - cross-sectional elements, sight distances, horizontal and vertical alignments.

Geometric design of railway Track – Speed and Cant.

Concept of airport runway length, calculations and corrections; taxiway and exit taxiway design.

Highway Pavements: Highway materials - desirable properties and tests; Desirable properties of bituminous paving mixes; Design factors for flexible and rigid pavements; Design of flexible and rigid pavement using IRC codes

Traffic Engineering: Traffic studies on flow and speed, peak hour factor, accident study, statistical analysis of traffic data; Microscopic and macroscopic parameters of traffic flow, fundamental relationships; Traffic signs; Signal design by Webster's method; Types of intersections; Highway capacity.

Section 7: Geomatics Engineering

Principles of surveying; Errors and their adjustment; Maps - scale, coordinate system; Distance and angle measurement - Levelling and trigonometric levelling; Traversing and triangulation survey; Total station; Horizontal and vertical curves.

Photogrammetry and Remote Sensing - Scale, flying height; Basics of remote sensing and GIS.

Chapter-Wise Analysis

GATE PAPERS (Civil Engineering)

Subject	2011	2012	2013	2014	2015	2016	2017	2018	2019	2020	2021
1. Engineering Mathematics											
1 Mark Question	3	3	2	3	3	4	6	3	3	5	5
2 Marks Question	5	5	4	4	5	5	4	4	4	4	4
Total Marks	**13**	**13**	**10**	**11**	**13**	**14**	**14**	**11**	**11**	**13**	**13**
2. Engineering Mechanics											
1 Mark Question						1	1	0	2	1	0
2 Marks Question	–	–	–	–	–	0	1	2	0	0	1
Total Marks						**1**	**3**	**4**	**2**	**1**	**2**
3. Solid Mechanics											
1 Mark Question	1	2	0	2	2	0	2	2	2	0	2
2 Marks Question	0	0	2	2	1	2	3	1	1	1	2
Total Marks	**1**	**2**	**4**	**6**	**4**	**4**	**8**	**4**	**4**	**2**	**6**
4. Structural Analysis											
1 Mark Question	0	1	2	2	2	1	0	1	0	1	1
2 Marks Question	4	3	3	3	1	4	2	3	3	3	1
Total Marks	**8**	**7**	**8**	**8**	**4**	**9**	**4**	**7**	**6**	**7**	**3**
5. Constructional Material & Management											
1 Mark Question						2	0	0	0	2	1
2 Marks Question	–	–	–	–	–	1	1	0	1	0	2
Total Marks						**4**	**2**	**0**	**2**	**2**	**5**
6. Concrete Structure											
1 Mark Question	5	3	5	4	4	1	1	4	3	0	1
2 Marks Question	2	3	1	3	3	0	1	3	4	2	1
Total Marks	**9**	**9**	**7**	**10**	**10**	**1**	**3**	**10**	**11**	**4**	**3**
7. Steel Structure											
1 Mark Question	0	1	3	0	0	0	0	1	1	1	0
2 Marks Question	1	1	2	3	2	2	1	1	1	1	1
Total Marks	**2**	**3**	**7**	**6**	**4**	**4**	**2**	**3**	**3**	**3**	**2**
8 Soil Mechanics and Foundation Engineering											
1 Mark Question	5	5	3	4	3	5	5	5	4	4	3
2 Marks Question	5	5	6	3	6	6	5	5	5	5	5
Total Marks	**15**	**15**	**15**	**10**	**15**	**17**	**15**	**15**	**14**	**14**	**13**

Subject	2011	2012	2013	2014	2015	2016	2017	2018	2019	2020	2021
9. Fluid Mechanics and Hydraulics											
1 Mark Question	1	3	3	2	3	1	2	4	2	2	2
2 Marks Question	0	2	1	2	3	3	3	2	3	2	3
Total Marks	**1**	**7**	**5**	**6**	**9**	**7**	**8**	**8**	**8**	**6**	**8**
10. Hydrology & Irrigation											
1 Mark Question	3	2	2	2	0	0	1	1	1	3	2
2 Marks Question	5	3	4	3	3	1	3	3	3	2	2
Total Marks	**13**	**8**	**10**	**8**	**6**	**2**	**7**	**7**	**7**	**7**	**6**
11. Environmental Engineering											
1 Mark Question	3	2	2	3	3	4	3	2	2	3	3
2 Marks Question	4	4	4	3	2	3	3	2	3	4	4
Total Marks	**11**	**10**	**10**	**9**	**7**	**10**	**9**	**6**	**8**	**11**	**11**
12. Transportation Engineering											
1 Mark Question	3	3	3	2	2	3	2	1	2	3	4
2 Marks Question	3	3	2	3	3	2	1	2	1	4	3
Total Marks	**9**	**9**	**7**	**8**	**8**	**6**	**4**	**5**	**4**	**11**	**10**
13. Surveying											
1 Mark Question	1	0	0	1	3	3	2	1	3	0	1 (MSQ)
2 Marks Question	1	1	1	1	1	1	2	2	1	2	1
Total Marks	**3**	**2**	**2**	**3**	**5**	**5**	**6**	**5**	**5**	**4**	**3**
14. General Aptitude											
1 Mark Question	5	5	5	5	5	5	5	5	5	5	5
2 Marks Question	5	5	5	5	5	5	5	5	5	5	5
Total Marks	**15**	**15**	**15**	**15**	**15**	**15**	**15**	**15**	**15**	**15**	**15**

Verbal Ability

CHAPTER 1

English Grammar

MCQ TYPE QUESTIONS

2016

1. Computers were invented for performing only high-end useful computations. However, it is no understatement that they have taken over our world today. The internet, for example, is ubiquitous. Many believe that the internet itself is an unintended consequence of the original invention. With the advent of mobile computing on our phones, a whole new dimension is now enabled. One is left wondering if all these developments are good or, more importantly, required.

Which of the statement(s) below is/are logically valid and can be inferred from the above paragraph?

(*i*) The author believes that computers are not good for us.

(*ii*) Mobile computers and the internet are both intended inventions

(*a*) (*i*) only (*b*) (*ii*) only

(*c*) both (*i*) and (*ii*) (*d*) neither (*i*) nor (*ii*)

2. All hill-stations have a lake. Ooty has two lakes. Which of the statement(s) below is/are logically valid and can be inferred from the above sentences?

(*i*) Ooty is not a hill-station.

(*ii*) No hill-station can have more than one lake.

(*a*) (*i*) only (*b*) (*ii*) only

(*c*) both (*i*) and (*ii*) (*d*) neither (*i*) nor (*ii*)

3. Identify the correct spelling out of the given options :

(*a*) Managable (*b*) Manageable

(*c*) Mangaeble (*d*) Managible

4. Choose the statement(s) where the underlined word is used correctly :

(*i*) A prone is a dried plum.

(*ii*) He was lying prone on the floor.

(*iii*) People who eat a lot of fat are prone to heart disease.

(*a*) (*i*) and (*iii*) only (*b*) (*iii*) only

(*c*) (*i*) and (*ii*) only (*d*) (*ii*) and (*iii*) only

5. The policeman asked the victim of a theft, "What did you ?"

(*a*) loose (*b*) lose

(*c*) loss (*d*) louse

6. In a world filled with uncertainty, he was glad to have many good friends. He had always assisted them in times of need and was confident that they would reciprocate. However, the events of the last week proved him wrong.

Which of the following inference(s) is/are logically valid and can be inferred from the above passage?

(*i*) His friends were always asking him to help them.

(*ii*) He felt that when in need of help, his friends would let him down.

(*iii*) He was sure that his friends would help him when in need.

(*iv*) His friends did not help him last week.

(*a*) (*i*) and (*ii*) (*b*) (*iii*) and (*iv*)

(*c*) (*iii*) only (*d*) (*iv*) only

7. An apple costs Rs. 10. An onion costs Rs. 8.

Select the most suitable sentence with respect to grammar and usage.

(*a*) The price of an apple is greater than an onion.

(*b*) The price of an apple is more than onion.

(*c*) The price of an apple is greater than that of an onion.

(*d*) Apples are more costlier than onions.

8. He <u>turned a deaf ear</u> to my request. What does the underlined phrasal verb mean?

(*a*) ignored (*b*) appreciated

(*c*) twisted (*d*) returned

2015

9. Choose the statement where underlined word is used correctly

(*a*) The minister <u>insured</u> the victims that everything would be all right.

(*b*) He <u>ensured</u> that the company will not have to bear any loss.

(*c*) The actor got himself <u>ensured</u> against any accident.

(*d*) The teacher <u>insured</u> students of good results

10. The following question presents a sentence, part of which is underlined. Beneath the sentence you find four ways of phrasing the underline part. Following the requirements of the standard written English, select the answer that produces the most effective sentence.

Tuberculosis, together with its effects, ranks one of the leading causes of death in India.

(*a*) ranks as one of the leading causes of death
(*b*) rank as one of the leading causes of death
(*c*) has the rank of one of the leading causes of death
(*d*) are one of the leading causes of death

11. Choose the statement where underlined word is used correctly.

(*a*) When the teacher eludes to different authors, he is being elusive
(*b*) When the thief keeps eluding the police, he is being elusive
(*c*) Matters that are difficult to understand, identify or remember are allusive
(*d*) Mirages can be allusive, but a better way to express them is illusory

12. Fill in the blank with the correct idiom/phrase.

That boy from the town was a __________ in the sleepy village.

(*a*) Dog out of herd (*b*) Sheep from the heap
(*c*) Fish out of water (*d*) Bird from the flock

13. Select the appropriate option in place of underlined part of the sentence.

Increased productivity necessary reflects greater efforts made by the employees.

(*a*) Increase in productivity necessary
(*b*) Increase productivity is necessary
(*c*) Increase in productivity necessarily
(*d*) No improvement required

14. Ram and Shyam shared a secret and promised to each other that it would remain between them. Ram expressed himself in one of the following ways as given in the choices below. Identify the correct way as per standard English.

(*a*) It would remain between you and me
(*b*) It would remain between me and you
(*c*) It would remain between you and I
(*d*) It would remain with me

15. What is the adverb for the given word below? Misogynous

(*a*) Misogynousness (*b*) Misogynity
(*c*) Misogynously (*d*) Misogynous

16. In the following sentence certain parts are underlined and marked P, Q and R. One of the parts may contain certain error or may not be acceptable in standard written communication. Select the part containing an error. Choose D as your Answer: if there is no error.

The student corrected all the errors (P) that the instructor marked (Q) on the answer book (R)

(*a*) P (*b*) Q
(*c*) R (*d*) No Error

17. Choose the statement where underlined word is used correctly.

(*a*) The industrialist had a personnel jet.
(*b*) I write my experience in my personnel diary.
(*c*) All personnel are being given the day off.
(*d*) Being religious is a personnel aspect.

18. Out of the following four sentences, select the most suitable sentence with respect to grammer and usage:

(*a*) Since the report lacked needed information, it was of no use to them.
(*b*) The report was useless to them because there were no needed information in it.
(*c*) Since the report did not contain the needed information, it was not real useful to them.
(*d*) Since the report lacked needed information, it would not had been useful to them.

19. Select the alternative meaning of the underlined part of the sentence.

The chain snatchers took to their heels when the police party arrived.

(*a*) took shelter in a thick jungle
(*b*) open indiscriminate fire
(*c*) took to flight
(*d*) unconditionally surrendered

2014

20. Which of the following options is the closest in meaning to the word underlined in the sentence below?

In a democracy, everybody has the freedom to disagree with the government.

(*a*) dissent (*b*) descent
(*c*) decent (*d*) decadent

21. 'Advice' is ________.

(*a*) a verb (*b*) a noun
(*c*) an adjective (*d*) both a verb and a noun

22. Which of the options given below best completes the following sentence?

She will feel much better if she ________.

(*a*) will get some rest

(*b*) gets some rest

(*c*) will be getting some rest

(*d*) is getting some rest

23. While trying to collect an envelope
I

from under the table, Mr. X fell down and
II III

was losing consciousness.
IV

Which one of the above underlined parts of the sentence is NOT appropriate?

(*a*) I (*b*) II

(*c*) III (*d*) IV

24. Which of the following options is the closest in meaning to the phrase underlined in the sentence below?

It is fascinating to see life forms cope with varied environmental conditions.

(*a*) Adopt to (*b*) Adapt to

(*c*) Adept in (*d*) Accept with

25. In a press meet on the recent scam, the minister said, "The buck stops here". What did the minister convey by the statement?

(*a*) He wants all the money

(*b*) He will return the money

(*c*) He will assume final responsibility

(*d*) He will resist all enquiries

2013

26. The professor ordered to the students to go
I II III

out of the class.
IV

Which of the above underlined parts of the sentence is grammatically incorrect?

(*a*) I (*b*) II

(*c*) III (*d*) IV

27. Choose the grammatically INCORRECT sentence:

(*a*) He is of Asian origin.

(*b*) They belonged to Africa.

(*c*) She is an European.

(*d*) They migrated from India to Australia.

28. Choose the grammatically CORRECT sentence:

(*a*) Two and two add four.

(*b*) Two and two become four.

(*c*) Two and two are four.

(*d*) Two and two make four.

29. All engineering students
I

should learn mechanics,
II

mathematics and
III

how to do computation.
IV

Which of the above underlined parts of the sentence is not appropriate?

(*a*) I (*b*) II

(*c*) III (*d*) IV

2012

30. Choose the most appropriate alternative from the options given below to complete the following sentence :

Suresh's dog is the one _____ was hurt in the stampede.

(*a*) that (*b*) which

(*c*) who (*d*) whom

31. Choose the grammatically **INCORRECT** sentence :

(*a*) They gave us the money back less the service charges of Three Hundred rupees.

(*b*) This country's expenditure is not less than that of Bangladesh.

(*c*) The committee initially asked for a funding of Fifty Lakh rupeesr, but later settled for a lesser sum.

(*d*) This country's expenditure on educational reforms is very less.

32. One of the parts (A, B, C, D) in the sentence given below contains an ERROR. Which one of the following is **INCORRECT** ?

I requested that he should be given the driving test today instead of tomorrow.

(*a*) requested that

(*b*) should be given

(*c*) the driving test

(*d*) instead of tomorrow

ANSWERS

MCQ Type Questions

1. (*d*)	**2.** (*d*)	**3.** (*b*)	**4.** (*d*)	**5.** (*b*)	**6.** (*b*)	**7.** (*c*)	**8.** (*a*)	**9.** (*b*)	**10.** (*a*)
11. (*b*)	**12.** (*c*)	**13.** (*c*)	**14.** (*a*)	**15.** (*c*)	**16.** (*b*)	**17.** (*c*)	**18.** (*a*)	**19.** (*c*)	**20.** (*a*)
21. (*b*)	**22.** (*b*)	**23.** (*d*)	**24.** (*b*)	**25.** (*c*)	**26.** (*b*)	**27.** (*c*)	**28.** (*d*)	**29.** (*d*)	**30.** (*a*)
31. (*d*)	**32.** (*b*)								

EXPLANATIONS

MCQ TYPE QUESTIONS

4. 'lying prone' means lie down flat. 'Prone to' means vulnerable to.

5. Lose (verb)

9. *Insured*–the person, group, or organization whose life or property is covered by an insurance policy.
Ensured–to secure or guarantee

11. Elusive: Difficult to answer.

12. From the statement, it appears that boy found it tough to adapt to a very different situation.

16. The is not required in 'Q'.

30. The correct answer is 'that'. In the given sentence 'was hurt in the stampede' determines which dog belongs to Suresh.

■■

2 CHAPTER Sentence Completion

MCQ TYPE QUESTIONS

2016

1. The man who is now Municipal Commissioner worked as ________________.

(a) the security guard at a university
(b) a security guard at the university
(c) a security guard at university
(d) the security guard at the university

2. Nobody knows how the Indian cricket team is going to <u>cope with</u> the difficult and seamer-friendly wickets in Australia.

Choose the option which is closest in meaning to the underlined phrase in the above sentence.

(a) put up with (b) put in with
(c) put down to (d) put up against

3. The chairman requested the aggrieved shareholders to ________ him.

(a) bare with (b) bore with
(c) bear with (d) bare

4. The unruly crowd demanded that the accused be ________ without trial.

(a) hanged (b) hanging
(c) hankering (d) hung

5. Which of the following is CORRECT with respect to grammar and usage?

Mount Everest is __________.

(a) the highest peak in the world
(b) highest peak in the world
(c) one of highest peak in the world
(d) one of the highest peak in the world

6. Despite the new medicine's ____________ in treating diabetes, it is not __________widely.

(a) effectiveness --- prescribed
(b) availability --- used
(c) prescription --- available
(d) acceptance --- proscribed

7. The students _____ the teacher on teachers' day for twenty years of dedicated teaching.

(a) facilitated (b) felicitated
(c) fantasized (d) facillitated

8. After India's cricket world cup victory in 1985, Shrotria who was playing both tennis and cricket till then, decided to concentrate only on cricket. And <u>the rest is history</u>.

What does the underlined phrase mean in this context?

(a) history will rest in peace
(b) rest is recorded in history books
(c) rest is well known
(d) rest is archaic

9. If I were you, I _________ that laptop. It's much too expensive.

(a) won't buy (b) shan't buy
(c) wouldn't buy (d) would buy

10. Choose the most appropriate set of words from the options given below to complete the following sentence.

_________ is a will, _________ is a way.

(a) Wear, there, their (b) Were, their, there
(c) Where, there, there (d) Where, their, their

2015

11. Choose the most appropriate word from the options given below to complete the following sentence. The official answered _______ that the complaints of the citizen would be looked into.

(a) respectably (b) respectfully
(c) reputably (d) respectively

12. Choose the most appropriate word from the options given below to complete the following sentence.

The principal presented the chief guest with a _________, as token of appreciation.

(a) momento (b) memento
(c) momentum (d) moment

13. Choose the appropriate word/phrase, out of the four options given below, to complete the following sentence : Frogs _______.

(a) Croak (b) Roar
(c) Hiss (d) Patter

14. Extreme focus on syllabus and studying for tests has become such a dominant concern of Indian students that they close their minds to anything ______ to the requirements of the exam

(a) related (b) extraneous
(c) outside (d) useful

15. The Tamil version of _____ John Abraham-starrer Madras café _____ cleared by the Censor Board with no cuts last week but the film's distributors _____ no takers among the exhibitors for a release in Tamil Nadu _____ this Friday.
(*a*) Mr., was, found, on (*b*) a, was, found, at
(*c*) the, was, found, on (*d*) a, being, find at

16. Choose the appropriate word/phase, out of the four options given below, to complete the following sentence:

Apparent lifelessness _____ dormant life.

(*a*) harbours (*b*) lead to
(*c*) supports (*d*) affects

17. Choose the correct verb to fill in the blank below :

Let us _____________.

(*a*) Introvert (*b*) Alternate
(*c*) Atheist (*d*) Altruist

18. Choose the most appropriate word from the options given below to complete the following sentence :

If the Athlete had wanted to come first in the race, he __________ several hours everyday

(*a*) should practice (*b*) should have practiced
(*c*) practiced (*d*) should be practicing

19. Choose the appropriate word/phrase, out of the four options given below, to complete the following sentence:

Dhoni, as well as the other team members of Indian team, _____ present on the occasion.

(*a*) were (*b*) was
(*c*) has (*d*) have

20. We ________ our friend's birthday and we _______ how to make it up to him.
(*a*) completely forgot — don't just known
(*b*) forget completely — don't just know
(*c*) completely forget —just don't know
(*d*) forgot completely —just don't know

21. Which of the following options is the closest in meaning to the sentence below?

She enjoyed herself immensely at the party.

(*a*) She had a terrible time at the party
(*b*) She had a horrible time at the party
(*c*) She had a terrific time at the party
(*d*) She had a terrifying time at the party

22. Didn't you buy ______ when you went shopping?
(*a*) any paper
(*b*) much paper
(*c*) no paper
(*d*) a few paper

2014

23. Choose the most appropriate word from the options given below to complete the following sentence:

A person suffering from Alzheimer's disease ________ short-term memory loss.

(*a*) experienced (*b*) has experienced
(*c*) is experiencing (*d*) experiences

24. Choose the most appropriate word from the options given below to complete the following sentence:

__________ is the key to their happiness; they are satisfied with what they have.

(*a*) Contentment (*b*) Ambition
(*c*) Perseverance (*d*) Hunger

25. Choose the most appropriate word from the options given below to complete the following sentence:

One of his biggest _____ was his ability to forgive.

(*a*) vice (*b*) virtues
(*c*) choices (*d*) strength

26. After the discussion, Tom said to me, 'Please revert!' He expects me to ________.
(*a*) retract (*b*) get back to him
(*c*) move in reverse (*d*) retreat

27. The value of one U.S. Dollar is 65 Indian Rupees today, compared to 60 last year. The Indian Rupee has ________.
(*a*) Depressed (*b*) Depreciated
(*c*) Appreciated (*d*) Stabilized

28. Choose the most appropriate word from the options given below to complete the following sentence:

Communication and interpersonal skills are ______ important in their own ways.

(*a*) each (*b*) both
(*c*) all (*d*) either

29. Choose the most appropriate pair of words from the options given below to complete the following sentence:

She could not ________ the thought of ________ the election to her bitter rival.

(*a*) bear, loosing (*b*) bare, loosing
(*c*) bear, losing (*d*) bare, losing

30. Choose the most appropriate phrase from the options given below to complete the following sentence:

The aircraft ________ take off as soon as its flight plan was filed.

(*a*) is allowed to (*b*) will be allowed to
(*c*) was allowed to (*d*) has been allowed to

31. Choose the most appropriate word from the options given below to complete the following sentence:

Many ancient cultures attributed disease to supernatural causes. However, modern science has largely helped ______ such notions.

(*a*) impel (*b*) dispel
(*c*) propel (*d*) repel

32. If she __________ how to calibrate the instrument, she _______ done the experiment.

(*a*) knows, will have
(*b*) knew, had
(*c*) had known, could have
(*d*) should have known, would have

33. Who ______ was coming to see us this evening?

(*a*) you said (*b*) did you say
(*c*) did you say that (*d*) had you said

34. Choose the most appropriate word from the options given below to complete the following sentence:

He could not understand the judges awarding her the first prize, because he thought that her performance was quite ____________.

(*a*) Superb (*b*) Medium
(*c*) Mediocre (*d*) Exhilarating

2013

35. Friendship, no matter how ________ it is, has its limitations.

(*a*) cordial (*b*) intimate
(*c*) secret (*d*) pleasant

36. Were you a bird, you_____ in the sky.

(*a*) would fly (*b*) shall fly
(*c*) should fly (*d*) shall have flown

37. Complete the sentence:

Dare __________ mistakes.

(*a*) commit (*b*) to commit
(*c*) committed (*d*) committing

38. The Headmaster ______ to speak to you.

Which of the following options is incorrect to complete the above sentence?

(*a*) is wanting (*b*) wants
(*c*) want (*d*) was wanting

2012

39. Choose the most appropriate alternative from the options given below to complete the following sentence :

Despite several ______ the mission succeeded in its attempt to resolve the conflict.

(*a*) attempts (*b*) setbacks
(*c*) meetings (*d*) delegations

40. Choose the most appropriate alternative from the options given below to complete the following sentence :

If the tired soldier wanted to lie down, he ___ the mattress out on the balcony.

(*a*) should take (*b*) shall take
(*c*) should have taken (*d*) will have taken

41. Choose the most appropriate word from the options given below to complete the following sentence :

Given the seriousness of the situation that he had to face, his ___ was impressive.

(*a*) beggary (*b*) nomenclature
(*c*) jealousy (*d*) nonchalance

2011

42. Choose the most appropriate word from the options given below to complete the following sentence:

If you are trying to make a strong impression on your audience, you cannot do so by being understated, tentative or _______ .

(*a*) hyperbolic (*b*) restrained
(*c*) argumentative (*d*) indifferent

43. Choose the most appropriate word(s) from the options given below to complete the following sentence.

I contemplated _____ Singapore for my vacation but decided against it.

(*a*) to visit (*b*) having to visit
(*c*) visiting (*d*) for a visit

44. Choose the most appropriate word from the options given below to complete the following sentence:

Under ethical guidelines recently adopted by the Indian Medical Association, human genes are to be manipulated only to correct diseases for which_______treatments are unsatisfactory.

(*a*) similar (*b*) most
(*c*) uncommon (*d*) available

45. Choose the most appropriate word from the options given below to complete the following sentence:

It was her view that the country's problems had been_________by foreign technocrats, so that to invite them to come back would be counter-productive.

(*a*) identified
(*b*) ascertained
(*c*) exacerbated
(*d*) analysed

2010

46. Choose the most appropriate word from the options given below to complete the following sentence:
If we manage to ________ our natural resources, we would leave a better planet for our children.
(*a*) uphold (*b*) restrain
(*c*) cherish (*d*) conserve

47. Choose the most appropriate word from the options given below to complete the following sentence:
His rather casual remarks on politics ________ his lack of seriousness about the subject.
(*a*) masked (*b*) belied
(*c*) betrayed (*d*) suppressed

ANSWERS

MCQ Type Questions

1. (*b*)	**2.** (*a*)	**3.** (*c*)	**4.** (*a*)	**5.** (*a*)	**6.** (*a*)	**7.** (*b*)	**8.** (*c*)	**9.** (*c*)	**10.** (*c*)
11. (*b*)	**12.** (*b*)	**13.** (*a*)	**14.** (*b*)	**15.** (*c*)	**16.** (*a*)	**17.** (*b*)	**18.** (*b*)	**19.** (*b*)	**20.** (*c*)
21. (*c*)	**22.** (*a*)	**23.** (*d*)	**24.** (*a*)	**25.** (*b*)	**26.** (*b*)	**27.** (*b*)	**28.** (*b*)	**29.** (*c*)	**30.** (*c*)
31. (*b*)	**32.** (*c*)	**33.** (*b*)	**34.** (*c*)	**35.** (*b*)	**36.** (*a*)	**37.** (*b*)	**38.** (*c*)	**39.** (*b*)	**40.** (*a*)
41. (*d*)	**42.** (*b*)	**43.** (*c*)	**44.** (*d*)	**45.** (*c*)	**46.** (*a*)	**47.** (*c*)			

EXPLANATIONS

MCQ TYPE QUESTIONS

2. From the given underlined phrase <u>copewith</u> means put up with.

4. 'hanged' means death by hanging 'hung' is used only with things and not with people.

6. Here 'effectiveness' is noun and 'prescribed' is verb. So these words are apt and befitting with the given word 'medicine'.

12. The principal presented the chief guest with a memento, as token of appreciation.

13. Frogs make 'croak' sound.

14. Extraneous -irrelevant or unrelated to the subject being dealt with.

15. John-Abraham starrer Madras Cafe talks about the movie not the person, so Mr. is ruled out. 'Find no takers' is not the correct phrase. At this Friday is incorrect.

So, option (*c*) is correct.

16. Apparent: looks like; dormant: hidden
Harbour: give shelter; Effect (verb): results in

18. For condition regarding something which already happened, should have practiced is the correct choice.

23. The correct word to fill in the blank is experiences. This is because the sentence is in present tense and the sense is in present indefinite.

24. Contentment is the right word which is appropriate to the meaning of the sentence. Contentment is the sign of true happiness as it gives a sense of fulfillment.

39. The word despite tells us that mission succeeded even though there were problems. Hence the correct answer is Setbacks.

42. Tone of the sentence clearly indicates a word that is similar to understated is needed for the blank. Alternatively, word should be antonym of strong (fail to make strong impression). Therefore, best choice is restrained which means controlled / reserved / timid.

43. Contemplate is a transitive verb and hence is followed by a gerund Hence the correct usage of contemplate is verb + ing form.

44. The context seeks to take a deviation only when existing / present / current / alternative treatments are unsatisfactory. So word for the blank should be a close synonym of existing / present / current / alternative. Available is the closest of all.

45. Clues: foreign technocrats did something negatively to the problems – so it is counter-productive to invite them. All other options are non-negative. Best choice is exacerbated which means aggravated or worsened.

47. Betrayed, **means** 'showed' or revealed.

■■

3
CHAPTER

Synonyms

MCQ TYPE QUESTIONS

2016

1. The Buddha said, "Holding on to anger is like <u>grasping</u> a hot coal with the intent of throwing it at someone else; you are the one who gets burnt." Select the word below which is closest in meaning to the word underlined above.

(*a*) burning (*b*) igniting
(*c*) clutching (*d*) flinging

2015

2. The word similar in meaning to 'dreary' is

(*a*) cheerful (*b*) dreamy
(*c*) hard (*d*) dismal

3. Choose the word most similar in meaning to the given word :

Educe

(*a*) Exert (*b*) Educate
(*c*) Extract (*d*) Extend

4. Choose the most suitable one word substitute for the following expression

Connotation of a road or way

(*a*) Pertinacious (*b*) Viaticum
(*c*) Clandestine (*d*) Ravenous

5. Choose the word most similar in meaning to the given word:

Awkward

(*a*) Inept (*b*) Graceful
(*c*) Suitable (*d*) Dreadful

6. A generic term that includes various items of clothing such as a skirt, a pair of trousers and a shirt as

(*a*) fabric
(*b*) textile
(*c*) fibre
(*d*) apparel

7. Which of the following combinations is incorrect?

(*a*) Acquiescence – Submission
(*b*) Wheedle – Roundabout
(*c*) Flippancy – Lightness
(*d*) Profligate – Extravagant

2014

8. A student is required to demonstrate a high level of <u>comprehension</u> of the subject, especially in the social sciences.

The word closest in meaning to <u>comprehension</u> is

(*a*) understanding (*b*) meaning
(*c*) concentration (*d*) stability

9. While receiving the award, the scientist said, "I feel vindicated". Which of the following is closest in meaning to the word 'vindicated'?

(*a*) punished (*b*) substantiated
(*c*) appreciated (*d*) chastened

10. Choose the word that is opposite in meaning to the word "coherent".

(*a*) sticky (*b*) well-connected
(*c*) rambling (*d*) friendly

11. Match the columns:

	Column-1		**Column-2**
1.	eradicate	P.	misrepresent
2.	distort	Q.	soak completely
3.	saturate	R.	use
4.	utilize	S.	destroy utterly

(*a*) 1 : S, 2 : P, 3 : Q, 4 : R
(*b*) 1 : P, 2 : Q, 3 : R, 4 : S
(*c*) 1 : Q, 2 : R, 3 : S, 4 : P
(*d*) 1 : S, 2 : P, 3 : R, 4 : Q

2013

12. Which of the following options is the closest in meaning to the word given below:

Primeval

(*a*) Modern (*b*) Historic
(*c*) Primitive (*d*) Antique

13. Which one of the following options is the closest in meaning to the word given below?

Nadir

(*a*) Highest (*b*) Lowest
(*c*) Medium (*d*) Integration

14. They were requested not to **quarrel** with others. Which one of the following options is the closest in meaning to the word **quarrel**?

(*a*) make out (*b*) call out
(*c*) dig out (*d*) fall out

2012

15. Which one of the following options is the closest in meaning to the word given below?

Mitigate

(*a*) Diminish (*b*) Divulge

(*c*) Dedicate (*d*) Denote

16. Which one of the following options is the closest in meaning to the word given below ?

Latitude

(*a*) Eligibility (*b*) Freedom

(*c*) Coercion (*d*) Meticulousness

2011

17. Which of the following options is the closest in the meaning to the word below :

Inexplicable

(*a*) Incomprehensible (*b*) Indelible

(*c*) Inextricable (*d*) Infallible

2010

18. Which of the following options is the closest in meaning to the word below:

Circuitous

(*a*) cyclic (*b*) indirect

(*c*) confusing (*d*) crooked

ANSWERS

MCQ Type Questions

1. (*c*)	**2.** (*d*)	**3.** (*c*)	**4.** (*b*)	**5.** (*a*)	**6.** (*d*)	**7.** (*b*)	**8.** (*a*)	**9.** (*b*)	**10.** (*c*)
11. (*a*)	**12.** (*c*)	**13.** (*b*)	**14.** (*b*)	**15.** (*a*)	**16.** (*b*)	**17.** (*a*)	**18.** (*b*)		

EXPLANATIONS

MCQ TYPE QUESTIONS

1. The meaning of underlined word grasping means clutching (or holding something tightly).

2. *dreary* - depressingly dull and bleak or repetitive.

3. The word similar in meaning to Educe is Extract.

13. Given: A word "Nadir"

To find: Closest meaning word to 'Nadir' in the given options.

Analysis: The dictionary meaning of 'Nadir' is 'Lowest' which is potion (*b*)

Source: Google

Given: A sequence 44, 42, 40,......

To find: Maximum sum of the sequence

Analysis: To find the maximum sum, we have to ignore all the negative terms (they will only decrease the sum). Hence, we terminate the sequence at 0.

Now, the sequence is 44, 42,......,2, 0.

This sequence is an Arithmetic progression with common difference –2. Now, the problem reduces to finding the sum of this A.P.

To find the sum of A.P., we have to find the number of terms in A.P.

Let n be the number of terms

a be the first term

l be the last term

d be the common difference.

$n = ?$

$a = 44, \quad l = 0, \quad d = -2$

Now, $l = a + (n - 1)d$

15. 'Mitigate' means to make something less harmful, serious, intense etc. Hence correct answer is 'Diminis'

17. Inexplicable means not explicable; that cannot be explained, understood, or accounted for. So the best synonym here is incomprehensible.

18. Circuitous means round about or not direct

So circuitous : indirect.

4 CHAPTER

Antonyms

MCQ TYPE QUESTIONS

2015

1. Which word is not a synonym for the word *vernacular*?
   (*a*) regional
   (*b*) indigeneous
   (*c*) indigent
   (*d*) colloquial

2011

2. Choose the word from the options given below that is most nearly opposite in meaning to the given word:
   Amalgamate
   (*a*) merge
   (*b*) split
   (*c*) collect
   (*d*) separate
3. Choose the word from the options given below that is most nearly **opposite** in meaning to the given word :
   Frequency
   (*a*) periodicity
   (*b*) rarity
   (*c*) gradualness
   (*d*) persistency

ANSWERS

MCQ Type Questions

1. (*c*) **2.** (*b*) **3.** (*b*)

EXPLANATIONS

MCQ TYPE QUESTIONS

1. *Vernacular* - expressed or written in the native language of a place

   Indigent - deficient in what is requisite
2. Amalgamate means combine or unite to form one organization or structure. So the best option here is split. Separate on the other hand, although a close synonym, it is too general to be the best antonym in the given question while Merge is the synonym; Collect is not related.
3. Best antonym is rarity which means shortage or scarcity.

5 CHAPTER

Reasoning Ability

MCQ TYPE QUESTIONS

2016

1. Find the odd one in the following group of words.
mock, deride, praise, jeer

(*a*) mock (*b*) deride
(*c*) praise (*d*) jeer

2. Pick the odd one from the following options.

(*a*) CADBE (*b*) JHKIL
(*c*) XVYWZ (*d*) ONPMQ

3. Pick the odd one out in the following :
13, 23, 33, 43, 53

(*a*) 23 (*b*) 33
(*c*) 43 (*d*) 53

4. R2D2 is a robot. R2D2 can repair aeroplanes. No other robot can repair aeroplanes. Which of the following can be logically inferred from the above statements?

(*a*) R2D2 is a robot which can only repair aeroplanes.
(*b*) R2D2 is the only robot which can repair aeroplanes.
(*c*) R2D2 is a robot which can repair only aeroplanes.
(*d*) Only R2D2 is a robot.

5. A poll of students appearing for masters in engineering indicated that 60% of the students believed that mechanical engineering is a profession unsuitable for women. A research study on women with masters or higher degrees in mechanical engineering found that 99% of such women were successful in their professions.

Which of the following can be logically inferred from the above paragraph?

(*a*) Many students have misconceptions regarding various engineering disciplines.
(*b*) Men with advanced degrees in mechanical engineering believe women are well suited to be mechanical engineers.
(*c*) Mechanical engineering is a profession well suited for women with masters or higher degrees in mechanical engineering.
(*d*) The number of women pursuing higher degrees in mechanical engineering is small.

6. Sourya committee had proposed the establishment of Sourya Institutes of Technology (SITs) in line with Indian Institutes of Technology (IITs) to cater to the technological and industrial needs of a developing country.

Which of the following can be logically inferred from the above sentence?

Based on the proposal,

(*i*) In the initial years, SIT students will get degrees from IIT.
(*ii*) SITs will have a distinct national objective.
(*iii*) SIT like institutions can only be established in consulation with IIT.
(*iv*) SITs will serve technological needs of a developing country.

(*a*) (*iii*) and (*iv*) only (*b*) (*i*) and (*iv*) only
(*c*) (*ii*) and (*iv*) only (*d*) (*ii*) and (*iii*) only

7. Fact : If it rains, then the field is wet.
Read the following statements:

(*i*) It rains (*ii*) The field is not wet
(*iii*) The field is wet (*iv*) It did not rain

Which one of the options given below is NOT logically possible, based on the given fact?

(*a*) If (*iii*), then (*iv*) (*b*) If (*i*), then (*iii*)
(*c*) If (*i*), then (*ii*) (*d*) If (*ii*), then (*iv*)

8. Students taking an exam are divided into two groups, **P** and **Q** such that each group has the same number of students. The performance of each of the students in a test was evaluated out of 200 marks. It was observed that the mean of group **P** was 105, while that of group **Q** was 85. The standard deviation of group **P** was 25, while that of group **Q** was 5. Assuming that the marks were distributed on a normal distribution, which of the following statements will have the highest probability of being **TRUE**?

(*a*) No student in group **Q** scored less marks than any student in group **P**.
(*b*) No student in group **P** scored less marks than any student in group **Q**.
(*c*) Most students of group **Q** scored marks in a narrower range than students in group **P**.
(*d*) The median of the marks of group **P** is 100.

9. A smart city integrates all modes of transport, uses clean energy and promotes sustainable use of resources. It also uses technology to ensure safety and security of the city, something which critics argue, will lead to a surveillance state.

Which of the following can be logically inferred from the above paragraph?

(*i*) All smart cities encourage the formation of surveillance states.

(*ii*) Surveillance is an integral part of a smart city.

(*iii*) Sustainability and surveillance go hand in hand in a smart city.

(*iv*) There is a perception that smart cities promote surveillance.

(*a*) (*i*) and (*iv*) only (*b*) (*ii*) and (*iii*) only

(*c*) (*iv*) only (*d*) (*i*) only

10. Find the missing sequence in the letter series.

B, FH, LNP, _ _ _ _.

(*a*) SUWY (*b*) TUVW

(*c*) TVXZ (*d*) TWXZ

11. Michael lives 10 km away from where I live. Ahmed lives 5 km away and Susan lives 7 km away from where I live. Arun is farther away than Ahmed but closer than Susan from where I live. From the information provided here, what is one possible distance (in km) at which I live from Arun's place?

(*a*) 3.00 (*b*) 4.99

(*c*) 6.02 (*d*) 7.01

12. Leela is older than her cousin Pavithra. Pavithra's brother Shiva is older than Leela. When Pavithra and Shiva are visiting Leela, all three like to play chess. Pavithra wins more often than Leela does.

Which one of the following statements must be TRUE based on the above?

(*a*) When Shiva plays chess with Leela and Pavithra, he often loses.

(*b*) Leela is the oldest of the three.

(*c*) Shiva is a better chess player than Pavithra.

(*d*) Pavithra is the youngest of the three.

13. Based on the given statements, select the appropriate option with respect to grammar and usage.

Statements :

I. The height of Mr. X is 6 feet.

II. The height of Mr. Y is 5 feet.

(*a*) Mr. X is longer than Mr. Y.

(*b*) Mr. X is more elongated than Mr. Y

(*c*) Mr. X is taller than Mr. Y

(*d*) Mr. X is lengthier than Mr. Y

14. Social science disciplines were in existence in an amorphous form until the colonial period when they were institutionalized. In varying degrees, they were intended to further the colonial interest. In the time of globalization and the economic rise of postcolonial countries like India, conventional ways of knowledge production have become obsolete.

Which of the following can be logically inferred from the above statements?

I. Social science disciplines have become obsolete.

II. Social science disciplines had a pre-colonial origin.

III. Social science disciplines always promote colonialism.

IV. Social science must maintain disciplinary boundaries.

(*a*) II only (*b*) I and III only

(*c*) II and IV only (*d*) III and IV only

15. Two and a quarter hours back, when seen in a mirror, the reflection of a wall clock without number markings seemed to show 1 : 30. What is the actual current time shown by the clock?

(*a*) 8:15 (*b*) 11:15

(*c*) 12:15 (*d*) 12:45

16. M and N start from the same location. M travels 10 km East and then 10 km North-East. N travels 5 km South and then 4 km South-East. What is the shortest distance (in km) between M and N at the end of their travel?

(*a*) 18.60 (*b*) 22.50

(*c*) 20.61 (*d*) 25.00

17. M has a son Q and a daughter R. He has no other children. E is the mother of P and daughter-inlaw of M. How is P related to M?

(*a*) P is the son-in-law of M.

(*b*) P is the grandchild of M.

(*c*) P is the daughter-in-law of M.

(*d*) P is the grandfather of M.

18. The number that least fits this set: (324, 441, 97 and 64) is ______.

(*a*) 324

(*b*) 441

(*c*) 97

(*d*) 64

19. The overwhelming number of people infected with rabies in India has been flagged by the World Health Organization as a source of concern. It is estimated that inoculating 70% of pets and stray dogs against rabies can lead to a significant reduction in the number of people infected with rabies.

Which of the following can be logically inferred from the above sentences ?

(*a*) The number of people in India infected with rabies is high

(*b*) The number of people in other parts of the world who are infected with rabies is low

(*c*) Rabies can be eradicated in India by vaccinating 70% of stray dogs

(d) Stray dogs are the main source of rabies worldwide

20. A flat is shared by four first year undergraduate students. They agreed to allow the oldest of them to enjoy some extra space in the flat. Manu is two months older than Sravan, who is three months younger than Trideep. Pavan is one month older than Sravan. Who should occupy the extra space in the flat ?

(*a*) Manu (*b*) Sravan

(*c*) Trideep (*d*) Pavan

21. Today, we consider Ashoka as a great ruler because of the copious evidence he left behind in the form of stone carved edicts. Historians tend to correlate greatness of a king at his time with the availability of evidence today. Which of the following can be logically inferred from the above sentences?

(*a*) Emperors who do not leave significant sculpted evidence are completely forgotten.

(*b*) Ashoka produced stone carved edicts to ensure that later historians will respect him.

(*c*) Statues of kings are a reminder of their greatness.

(*d*) A king's greatness, as we know him today, is interpreted by historians.

22. Fact 1 : Humans are mammals.

Fact 2 : Some humans are engineers.

Fact 3 : Engineers build houses.

If the above statements are facts, which of the following can be logically inferred ?

I. All mammals build houses.

II. Engineers are mammals.

III. Some humans are not engineers.

(*a*) II only (*b*) III only

(*c*) I, II and III (*d*) I only

2015

23. Mr. Vivek walks 6 meters North-East, then turns and walks 6 meters South- East, both at 60 degrees to East. He further moves 2 meters South and 4 meters West. What is the straight distance in meters between the point he started from and the point he finally reached?

(*a*) $2\sqrt{2}$ (*b*) 2

(*c*) $\sqrt{2}$ (*d*) $1/\sqrt{2}$

24. There are 16 teachers who can teach Thermodynamics (TD), 11 who can teach Electrical Sciences (ES), and 5 who can teach both TD and Engineering Mechanics (EM). There are a total of 40 teachers, 6 cannot teach any of the three subjects, i.e. EM, ES or TD. 6 can teach only ES. 4 can teach all three subjects, i.e. EM, ES and TD. 4 can teach ES and TD. How many can teach both ES and EM but not TD?

(*a*) 1 (*b*) 2

(*c*) 3 (*d*) 4

25. The given question is followed by two statements: select the most appropriate option that solves the question

Capacity of a solution tank A is 70% of the capacity of tank B. How many gallons of solution are in tank A and tank B?

Statements:

I. Tank A is 80% full and tank B is 40% full

II. Tank A if full contains 14,000 gallons of solution

(*a*) Statement I alone is sufficient

(*b*) Statement II alone is sufficient

(*c*) Either statement I or II alone is sufficient

(*d*) Both the statements I and II together are sufficient

26. Operators $\square, \diamond$ and $\rightarrow$ are defined by

$$a \square b = \frac{a-b}{a+b};\ a \diamond b = \frac{a+b}{a-b};\ a \rightarrow b = ab.$$

Find the value $(66 \square 6) \rightarrow (66 \diamond 6)$

(*a*) –2 (*b*) –1

(*c*) 1 (*d*) 2

27. Humpty Dumpty sits on a wall every day while having lunch. The wall sometimes breaks. A person sitting on the wall falls if the wall breaks.

Which one of the statements below is logically valid and can be inferred from the above sentences?

(*a*) Humpty Dumpty always falls while having lunch

(*b*) Humpty Dumpty does not fall sometimes while having lunch

(*c*) Humpty Dumpty never falls during dinner

(*d*) When Humpty Dumpty does not sit on the wall, the wall does not break

28. If ROAD is written as URDG, then SWAN should be written as :

(*a*) VXDQ (*b*) VZDQ

(*c*) VZDP (*d*) UXDQ

29. Select the pair that best expresses a relationship similar to that expressed in the pair :

Children : Pediatrician

(*a*) Adult : Orthopaedist

(*b*) Females : Gynaecologist

(*c*) Kidney : Nephrologist

(*d*) Skin : Dermatologist

30. The head of a newly formed government desires to appoint five of the six selected members P,Q,R,S,T and U to portfolios of Home, Power, Defense, Telecom, and Finance. U does not want any portfolio if S gets one of the five. R wants either Home or Finance or no portfolio. Q says that if S gets either Power or Telecom, then she must get the other one. T insists on a portfolio if P gets one.

Which is the valid distribution of portfolios?

(*a*) P-Home, Q-Power, R-Defense, S-Telecom, T-Finance

(*b*) R-Home, S-Power, P-Defense, Q-Telecom, T-Finance

(*c*) P-Home, Q-Power, T-Defense, S-Telecom, U-Finance

(*d*) Q-Home, U-Power, T-Defense, R-Telecom, P-Finance

31. Most experts feel that in spite of possessing all the technical skills required to be a batsman of the highest order, he is unlikely to be so due to lack of requisite temperament. He was guilty of throwing away his wicket several times after working hard to lay a strong foundation. His critics pointed out that until he addressed this problem success at the highest level will continue to elude him.

Which of the statement (s) below is/are logically valid and can be inferred from the above passage?

(*i*) He was already a successful batsman at the highest level

(*ii*) He has to improve his temperament in order to become a great batsman

(*iii*) He failed to make many of his good starts count

(*iv*) Improving his technical skills will guarantee success

(*a*) (*iii*) and (*iv*) (*b*) (*ii*) and (*iii*)

(*c*) (*i*), (*ii*) and (*iii*) (*d*) (*ii*) only

32. Alexander turned his attention towards India, since he had conquered Persia.

Which one of the statements below is logically valid and can inferred from the above sentence?

(*a*) Alexander would not have turned his attention towards India had he not conquered Persia.

(*b*) Alexander was not ready to rest on his laurels, and wanted to march to India

(*c*) Alexander was completely in control of his army and could command it to move towards India

(*d*) Since Alexander's kingdom extended to Indian borders after the conquest of Persia, he was keen to move further

33. Tanya is older than Enc.

Cliff is older than Tanya.

Eric is older than Cliff.

If the first two statements are true, then the third statement is

(*a*) True (*b*) False

(*c*) Uncertain (*d*) Data insufficient

34. Given below are two statements followed by two conclusions. Assuming these statements to be true, decide which one logically follows:

Statements:

I. No manager is a leader.

II. All leaders are executive.

Conclusions:

I. No manager is a executive.

II. All executive is a manager.

(*a*) Only conclusion I follows.

(*b*) Only conclusion II follows.

(*c*) Neither conclusion I nor II follows.

(*d*) Both conclusion I and II follow.

35. Find the missing sequence in the letter series below :

A, CD, GHI, _?___, UVWXY

(*a*) LMN (*b*) MNO

(*c*) MNOP (*d*) NOPQ

36. Ms. X will be in Bagdogra from 1/5/14 to 20/5/14 and from 22/5/14 to 31/5/14. On the morning of 21/5/14, she will reach Kochi via Mumbai.

Which one of the statements below is logically valid and can be inferred from the above sentences?

(*a*) Ms. X will be in Kochi for only one day only in May

(*b*) Ms. X will be in Kochi for only one day in May

(*c*) Ms. X will be only in Kochi for one day in May

(*d*) Only Ms. X will be in Kochi for one day in May

37. Given below are two statements followed by two conclusions. Assuming these statements to be true, decide which one logically follows.

Statement:

I. All film stars are playback singers.

II. All film directors are film stars.

Conclusions:

I. All film directors are playback singers.

II. Some film stars are film directors.

(*a*) Only conclusion I follows

(*b*) Only conclusion II follows

(*c*) Neither conclusion I nor II follows

(*d*) Both conclusions I and II follow

38. Lamenting the gradual sidelining of the arts in school curricula, a group of prominent artists wrote to the Chief Minister last year, asking him to allocate more funds to support arts education in schools. However, no such increase has been announced in this year's Budget. The artists expressed their deep anguish at their request not being approved, but many of them remain optimistic about funding in the future.

Which of the statement(s) below is/are logically valid and can be inferred from the above statements?

i. The artists expected funding for the arts to increase this year.

ii. The Chief Minister was receptive to the idea of increasing funding for the arts.

iii. The Chief Minister is a prominent artist.

iv. Schools are giving less importance to arts education nowadays.

(*a*) *iii* and *iv* (*b*) *i* and *iv*

(*c*) *i*, *ii* and *iv* (*d*) *i* and *iii*

39. Four branches of a company are located at M, N, O, and P. M is north of N at a distance of 4 km; P is south of O at a distance of 2 km; N is southeast of O by 1 km. What is the distance between M and P in km?

(*a*) 5.34 (*b*) 6.74

(*c*) 28.5 (*d*) 45.49

40. The given statement is followed by some courses of action. Assuming the statement to be true, decide the correct option.

Statement:

There has been a significant drop in the water level in the lakes supplying water to the city.

Course of action:

I. The water supply authority should impose a partial cut in supply to tackle the situation.

II. The government should appeal to all the residents through mass media for minimal use of water.

III. The government should ban the water supply in lower areas.

(*a*) Statements I and II follow.

(*b*) Statements I and III follow.

(*c*) Statements II and III follow.

(*d*) All statements follow.

2014

41. In a group of four children, Som is younger to Riaz. Shiv is elder to Ansu. Ansu is youngest in the group. Which of the following statements is/are required to find the eldest child in the group?

Statements

1. Shiv is younger to Riaz.

2. Shiv is elder to Som.

(*a*) Statement 1 by itself determines the eldest child.

(*b*) Statement 2 by itself determines the eldest child.

(*c*) Statements 1 and 2 are both required to determine the eldest child.

(*d*) Statements 1 and 2 are not sufficient to determine the eldest child.

42. X is 1 km northeast of Y. Y is 1 km southeast of Z. W is 1 km west of Z. P is 1 km south of W. Q is 1 km east of P. What is the distance between X and Q in km?

(*a*) 1 (*b*) $\sqrt{2}$

(*c*) $\sqrt{3}$ (*d*) 2

43. Rajan was not happy that Sajan decided to do the project on his own. On observing his unhappiness, Sajan explained to Rajan that he preferred to work independently.

Which one of the statements below is logically valid and can be inferred from the above sentences?

(*a*) Rajan has decided to work only in a group.

(*b*) Rajan and Sajan were formed into a group against their wishes.

(*c*) Sajan had decided to give in to Rajan's request to work with him.

(*d*) Rajan had believed that Sajan and he would be working together.

44. Find the odd one in the following group: ALRVX, EPVZB, ITZDF, OYEIK

(*a*) ALRVX

(*b*) EPVZB

(*c*) ITZDF

(*d*) OYEIK

45. Anuj, Bhola, Chandan, Dilip, Eswar and Faisal live on different floors in a six-storeyed building (the ground floor is numbered 1, the floor above it 2, and so on). Anuj lives on an even-numbered floor. Bhola does not live on an odd numbered floor. Chandan does not live on any of the floors below Faisal's floor. Dilip does not live on floor number 2. Eswar does not live on a floor immediately above or immediately below Bhola. Faisal lives three floors above Dilip. Which of the following floor-person combinations is correct?

	Anuj	*Bhola*	*Chandan*	*Dilip*	*Eswar*	*Faisal*
(*a*)	6	2	5	1	3	4
(*b*)	2	6	5	1	3	4
(*c*)	4	2	6	3	1	5
(*d*)	2	4	6	1	3	5

46. Find the next term in the sequence: 13M, 17Q, 19S, ________

(*a*) 21W (*b*) 21V

(*c*) 23W (*d*) 23V

47. If 'KCLFTSB' stands for 'best of luck' and 'SHSWDG' stands for 'good wishes', which of the following indicates 'ace the exam'?

(*a*) MCHTX (*b*) MXHTC

(*c*) XMHCT (*d*) XMHTC

48. "India is a country of rich heritage and cultural diversity."

Which one of the following facts best supports the claim made in the above sentence?

(*a*) India is a union of 28 states and 7 union territories.

(*b*) India has a population of over 1.1 billion.

(*c*) India is home to 22 official languages and thousands of dialects.

(*d*) The Indian cricket team draws players from over ten states.

49. In which of the following options will the expression P < M be definitely true?

(*a*) M < R > P > S (*b*) M > S < P < F

(*c*) Q < M < F = P (*d*) P = A < R < M

50. Find the next term in the sequence:

7G, 11K, 13M, ________

(*a*) 15Q

(*b*) 17Q

(*c*) 15P

(*d*) 17P

51. The multi-level <u>hierarchical</u> pie chart shows the population of animals in a reserve forest. The correct conclusions from <u>this</u> information are:

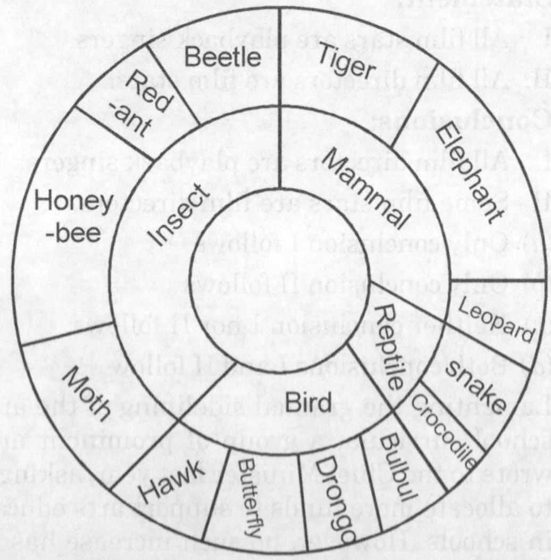

(*i*) Butterflies are birds

(*ii*) There are more tigers in this forest than red ants

(*iii*) All reptiles in this forest are either snakes or crocodiles

(*iv*) Elephants are the largest mammals in this forest

(*a*) (*i*) and (*ii*) only

(*b*) (*i*), (*ii*), (*iii*) and (*iv*)

(*c*) (*i*), (*iii*) and (*iv*) only

(*d*) (*i*), (*ii*) and (*iii*) only

52. Find the odd one in the following group

Q,W,Z,B B,H,K,M W,C,G,J M,S,V,X

(*a*) Q, W, Z, B (*b*) B, H, K, M

(*c*) W, C, G, J (*d*) M, S, V, X

53. Lights of four colors (red, blue, green, yellow) are hung on a ladder. On every step of the ladder there are two lights. If one of the lights is red, the other light on that step will always be blue. If one of the lights on a step is green, the other light on that step will always be yellow. Which of the following statements is not necessarily correct?

(*a*) The number of red lights is equal to the number of blue lights

(*b*) The number of green lights is equal to the number of yellow lights

(*c*) The sum of the red and green lights is equal to the sum of the yellow and blue lights

(*d*) The sum of the red and blue lights is equal to the sum of the green and yellow lights

54. Read the statements:
All women are entrepreneurs.
Some women are doctors.
Which of the following conclusions can be logically inferred from the above statements?
(*a*) All women are doctors
(*b*) All doctors are entrepreneurs
(*c*) All entrepreneurs are women
(*d*) Some entrepreneurs are doctors

55. Find the odd one from the following group:
(*a*) W,E,K,O (*b*) I,Q,W,A
(*c*) F,N,T,X, (*d*) N,V,B,D

56. For submitting tax returns, all resident males with annual income below ₹ 10 lakh should fill up Form P and all resident females with income below ₹ 8 lakh should fill up Form Q. All people with incomes above ₹ 10 lakh should fill up Form R, except non residents with income above ₹ 15 lakhs, who should fill up Form S. All others should fill Form T. An example of a person who should fill Form T is
(*a*) a resident male with annual income ₹ 9 lakh
(*b*) a resident female with annual income ₹ 9 lakh
(*c*) a non-resident male with annual income ₹ 16 lakh
(*d*) a non-resident female with annual income ₹ 16 lakh

57. Which number does not belong in the series below?
2, 5, 10, 17, 26, 37, 50, 64
(*a*) 17 (*b*) 37
(*c*) 64 (*d*) 26

58. A dance programme is scheduled for 10.00 a.m. Some students are participating in the programme and they need to come an hour earlier than the start of the event. These students should be accompanied by a parent. Other students and parents should come in time for the programme. The instruction you think that is appropriate for this is
(*a*) Students should come at 9.00 a.m. and parents should come at 10.00 a.m.
(*b*) Participating students should come at 9.00 a.m. accompanied by a parent, and other parents and students should come by 10.00 a.m.
(*c*) Students who are not participating should come by 10.00 a.m. and they should not bring their parents. Participating students should come at 9.00 a.m.
(*d*) Participating students should come before 9.00 a.m. Parents who accompany them should come at 9.00 a.m. All others should come at 10.00 a.m.

59. The number of people diagnosed with dengue fever (contracted from the bite of a mosquito) in north India is twice the number diagnosed last year. Municipal authorities have concluded that measures to control the mosquito population have failed in this region.
Which one of the following statements, if true, does not contradict this conclusion?
(*a*) A high proportion of the affected population has returned from neighbouring countries where dengue is prevalent
(*b*) More cases of dengue are now reported because of an increase in the Municipal Office's administrative efficiency
(*c*) Many more cases of dengue are being diagnosed this year since the introduction of a new and effective diagnostic test
(*d*) The number of people with malarial fever (also contracted from mosquito bites) has increased this year

60. Geneticists say that they are very close to confirming the genetic roots of psychiatric illnesses such as depression and schizophrenia, and consequently, that doctors will be able to eradicate these diseases through early identification and gene therapy.
On which of the following assumptions does the statement above rely?
(*a*) Strategies are now available for eliminating psychiatric illnesses
(*b*) Certain psychiatric illnesses have a genetic basis
(*c*) All human diseases can be traced back to genes and how they are expressed
(*d*) In the future, genetics will become the only relevant field for identifying psychiatric illnesses

2013

61. Select the pair that best expresses a relationship similar to that expressed in the pair:
Medicine: Health
(*a*) Science : Experiment
(*b*) Wealth : Peace
(*c*) Education : Knowledge
(*d*) Money : Happiness

62. Abhishek is elder to Savar.
Savar is younger to Anshul.
Which of the given conclusions is logically valid and is inferred from the above statements?
(*a*) Abhishek is elder to Anshul
(*b*) Anshul is elder to Abhishek
(*c*) Abhishek and Anshul are of the same age
(*d*) No conclusion follows

63. Complete the sentence:
Universalism is to particularism as diffuseness is to ______
(*a*) specificity
(*b*) neutrality
(*c*) generality
(*d*) adaptation

64. After several defeats in wars, Robert Bruce went in exile and wanted to commit suicide. Just before committing suicide, he came across a spider attempting tirelessly to have its net. Time and again, the spider failed but that did not deter it to retrain from making attempts. Such attempts by the spider made Bruce curious. Thus, Bruce started observing the near-impossible goal of the spider to have the net. Ultimately, the spider succeeded in having its net despite several failures. Such act of the spider encouraged Bruce not to commit suicide. And then, Bruce went back again and won many a battle, and the rest is history.
Which one of the following assertions is best supported by the above information?
(*a*) Failure is the pillar of success.
(*b*) Honesty is the best policy.
(*c*) Life begins and ends with adventures.
(*d*) No adversity justifies giving up hope.

65. Statement: You can always give me a ring whenever you need.
Which one of the following is the best inference from the above statement?
(*a*) Because I have a nice caller tune.
(*b*) Because I have a better telephone facility.
(*c*) Because a friend in need is a friend indeed.
(*d*) Because you need not pay towards the telephone bills when you give me a ring.

66. Statement: There were different streams of freedom movements in colonial India carried out by the moderates, liberals, radicals, socialists, and so on.
Which one of the following is the best inference from the above statement?
(*a*) The emergence of nationalism in colonial India led to our Independence.
(*b*) Nationalism in India emerged in the context of colonialism.
(*c*) Nationalism in India is homogeneous.
(*d*) Nationalism in India is heterogeneous.

67. Select the pair that best expresses a relationship similar to that expressed in the pair :
water : pipe : :
(*a*) cart : road
(*b*) electricity : wire
(*c*) sea : beach
(*d*) music: instrument

68. All professors are researchers
Some scientists are professors
Which of the given conclusions is logically valid and is inferred from the above arguments :
(*a*) All scientists are researchers
(*b*) All professors are scientists
(*c*) Some researchers are scientists
(*d*) No conclusion follows

2012

69. Wanted Temporary, Part-time persons for the post of Field Interviewer to conduct personal interviews to collect and collate economic data. Requirements: High School-pass, must be available for Day, Evening and Saturday work. Transportation paid, expenses reimbursed.
Which one of the following is the best inference from the above advertisement?
(*a*) Gender-discriminatory
(*b*) Xenophobic
(*c*) Not designed to make the post attractive
(*d*) Not gender-discriminatory

70. Given the sequence of terms, AD CG FK JP, the next term is
(*a*) OV (*b*) OW
(*c*) PV (*d*) PW

2011

71. A transporter receives the same number of orders each day. Currently, he has some pending orders (backlog) to be shipped. If he uses 7 trucks, then at the end of the 4th day he can clear all the orders. Alternatively, if he uses only 3 trucks, then all the orders are cleared at the end of the 10th day. What is the minimum number of trucks required so that there will be no pending order at the end of the 5th day?
(*a*) 4
(*b*) 5
(*c*) 6
(*d*) 7

72. Few school curricula include a unit on how to deal with bereavement and grief, and yet all students at some point in their lives suffer from losses through death and parting.
Based on the above passage which topic would not be included in a unit on bereavement?

(*a*) How to write a letter of condolence

(*b*) What emotional stages are passed through in the healing process

(*c*) What the leading causes of death are

(*d*) How to give support to a grieving friend

73. The question below consists of a pair of related words followed by four pairs of words. Select the pair that best expresses the relation in the original pair :

Gladiator : Arena

(*a*) dancer : stage

(*b*) commuter : train

(*c*) teacher : classroom

(*d*) lawyer : courtroom

74. The horse has played a little known but very important role in the field of medicine. Horses were injected with toxins of diseases until their blood built up immunities. Then a serum was made from their blood. Serums to fight with diphtheria and tetanus were developed this way.
It can be inferred from the passage, that horses were

(*a*) given immunity of diseases

(*b*) generally quite immune to diseases

(*c*) given medicines to fight toxins

(*d*) given diphtheria and tetanus serums

2010

75. The question below consists of a pair of related words followed by four pairs of words. Select the pair that best expresses the relation in the original pair.

Unemployed : Worker

(*a*) fallow : land (*b*) unaware : sleeper

(*c*) wit : jester (*d*) renovated : house

76. Hari (H), Gita (G), Irfan (I) and Saira (S) are siblings (i.e. brothers and sisters). All were born on 1st January. The age difference between any two successive siblings (that is born one after another) is less than 3 years. Given the following facts:

I. Hari's age + Gita age > Irfan's age + Saira's age

II. The age difference between Gita and Saira is 1 year. However, Gita is not the oldest and Saira is not the youngest.

III. There are no twins.

In what order were they born (oldest first)?

(*a*) HSIG

(*b*) SGHI

(*c*) IGSH

(*d*) IHSG

77. Modern warfare has changed from large scale clashes of armies to suppression of civilian populations. Chemical agents that do their work silently appear to be suited to such warfare; and regretfully, there exist people in military establishments who think that chemical agents are useful tools for their cause.

Which of the following statements best sums up the meaning of the above passage:

(*a*) Modern warfare has resulted in civil strife

(*b*) Chemical agents are useful in modern warfare

(*c*) Use of chemical agents in warfare would be undesirable

(*d*) People in military establishments like to use chemical agents in war

78. If 137 + 276 = 435 how much is 731 + 672?

(*a*) 534 (*b*) 1403

(*c*) 1623 (*d*) 1513

NUMERICAL TYPE QUESTIONS

2015

1. Fill in the missing value

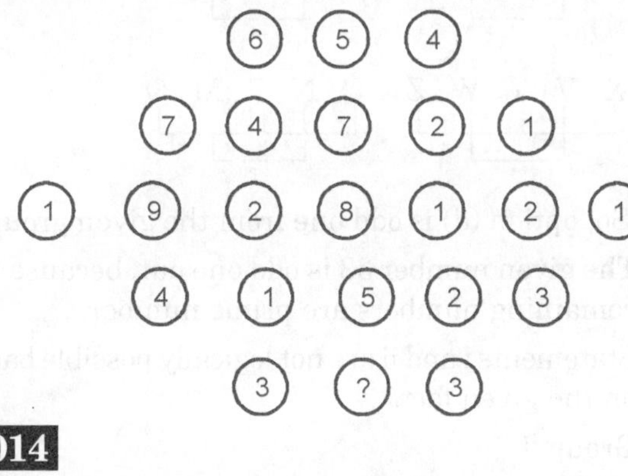

2014

2. The next term in the series 81, 54, 36, 24,.... is ________

3. Fill in the missing number in the series.

2 3 6 15 ___ 157.5 630

4. What is the next number in the series?

12 35 81 173 357 ____

ANSWERS

MCQ Type Questions

1. (*c*)	**2.** (*d*)	**3.** (*b*)	**4.** (*b*)	**5.** (*a*)	**6.** (*c*)	**7.** (*c*)	**8.** (*c*)	**9.** (*b*)	**10.** (*c*)
11. (*c*)	**12.** (*d*)	**13.** (*c*)	**14.** (*a*)	**15.** (*d*)	**16.** (*c*)	**17.** (*b*)	**18.** (*c*)	**19.** (*a*)	**20.** (*c*)
21. (*d*)	**22.** (*b*)	**23.** (*a*)	**24.** (*a*)	**25.** (*d*)	**26.** (*c*)	**27.** (*b*)	**28.** (*b*)	**29.** (*b*)	**30.** (*b*)
31. (*b*)	**32.** (*a*)	**33.** (*b*)	**34.** (*c*)	**35.** (*c*)	**36.** (*b*)	**37.** (*d*)	**38.** (*b*)	**39.** (*a*)	**40.** (*a*)
41. (*a*)	**42.** (*c*)	**43.** (*d*)	**44.** (*d*)	**45.** (*b*)	**46.** (*c*)	**47.** (*b*)	**48.** (*c*)	**49.** (*d*)	**50.** (*b*)
51. (*d*)	**52.** (*c*)	**53.** (*d*)	**54.** (*d*)	**55.** (*d*)	**56.** (*b*)	**57.** (*c*)	**58.** (*b*)	**59.** (*d*)	**60.** (*b*)
61. (*c*)	**62.** (*d*)	**63.** (*a*)	**64.** (*d*)	**65.** (*c*)	**66.** (*d*)	**67.** (*b*)	**68.** (*c*)	**69.** (*d*)	**70.** (*a*)
71. (*c*)	**72.** (*c*)	**73.** (*d*)	**74.** (*b*)	**75.** (*d*)	**76.** (*b*)	**77.** (*c*)	**78.** (*c*)		

Numerical Type Questions

1. (3) **2.** (16) **3.** (45) **4.** (725 to 725)

EXPLANATIONS

MCQ TYPE QUESTIONS

1. Given words 'mock, deride and jeer' are synonyms which means mockery. So, the given word praise is odd one.

2.

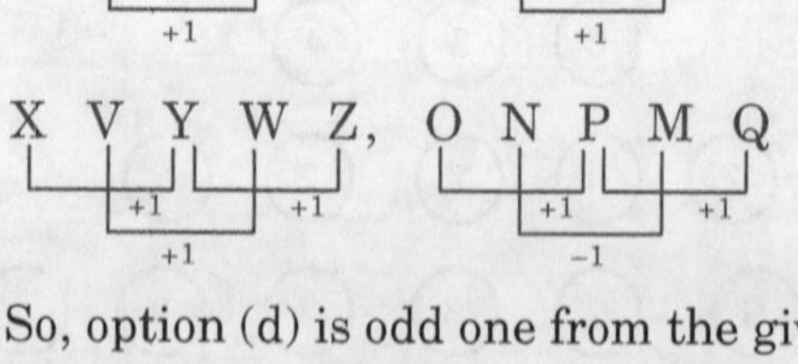

So, option (d) is odd one from the given group.

3. The given number 33 is odd one out, because the remaining numbers are prime number.

7. Statements i and ii are not logically possible based on the given fact.

8. Group 'P'

Mean (μ) = 105

Standard deviation (σ_1) = 25

$\Pr(\mu - \sigma \le x \le \mu + \sigma) = 0.6827$

$\therefore$ 68% within one standard deviation

$$\mu_1 - \sigma_1 = 105 - 25 = 80$$

$$\mu_1 + \sigma_1 = 105 + 25 = 130$$

$\therefore$ range = 80 to 130

Distribution of P:

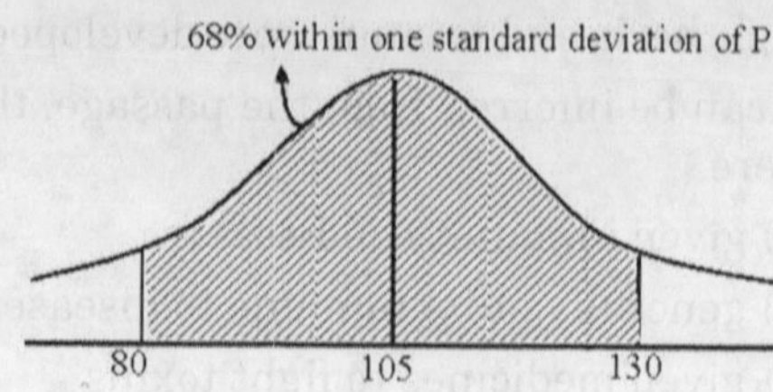

Group 'Q'

Mean (μ_2) = 85

Standard deviation (σ_2) = 5

$\Pr(\mu - \sigma \le x \le \mu + \sigma) = 0.6827$

$\therefore$ 68% within one standard deviation

$$\mu_2 - \sigma_2 = 85 - 5 = 80$$

$$\mu_2 + \sigma_2 = 85 + 5 = 90$$

$\therefore$ Range of Q in one standard deviation is 80 to 90

Distribution of 'Q'

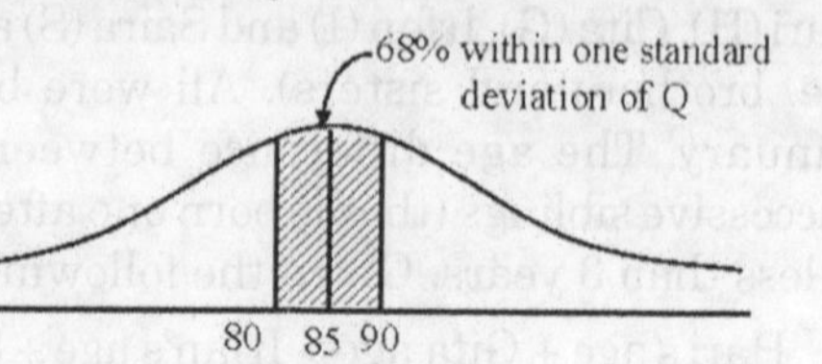

68% within one standard deviation of Q is narrower

$\therefore$ 68% within one standard deviation of Q means most students of group Q.

$\therefore$ Most students of group 'Q' scored marks in a narrower range than students in group 'P'

10. The following letter series is in the order of even letters series

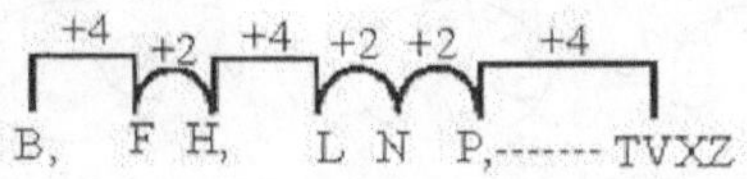

11. From given data, the following diagram can be possible.

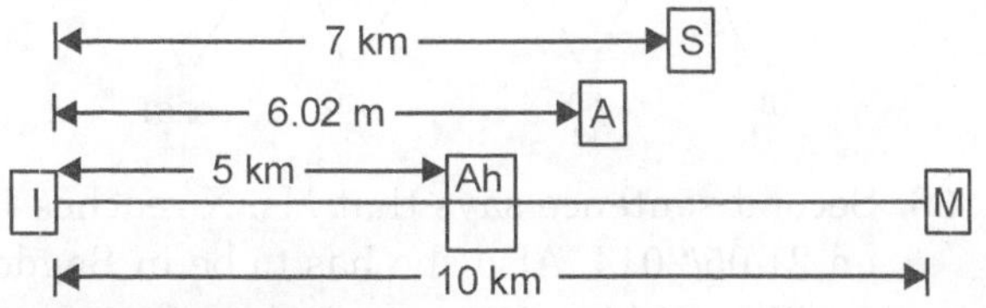

Here S → Susan

A → Arun

Ah → Ahmed

M → Michael

From the above diagram, Arun lives farthest away than Ahmed means more than 5 km but closer than Susan means less than 7 km, from the given alternatives.

So, only option 'c' is possible.

12. From the given question two statements will be followed.

For statment I

Arrange the given data according to their ages

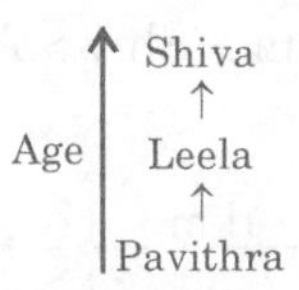

For statement II

Arrange the given data according to their winning

So, from statement I and II, it is clear that only option (*d*) is possible (i.e., statement I).

13.

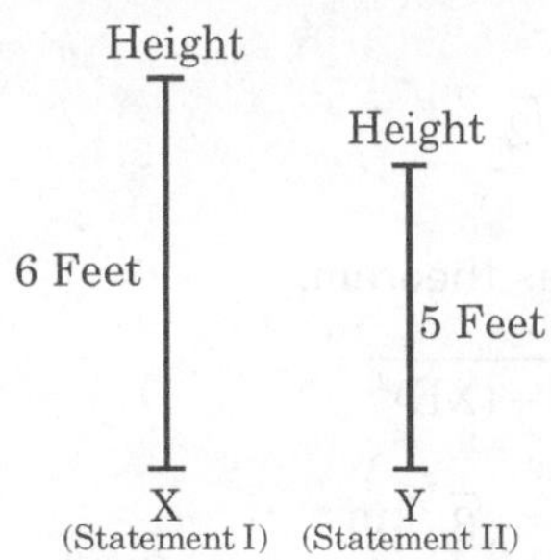

Hence from the given figure Mr. X is taller than Mr. Y by 1 foot.

14. Social science disciplines had a pre-colonial origin.

15.

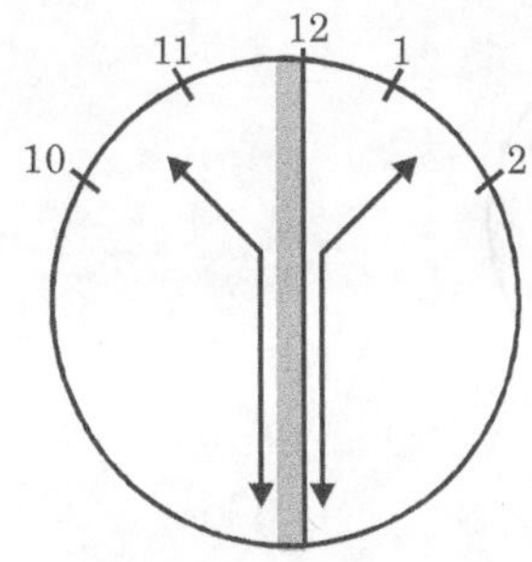

Mirror image of 1 : 20 is 10 : 30

10 : 30 was the time two and quarter hour back so time now will be 12 : 45

16.

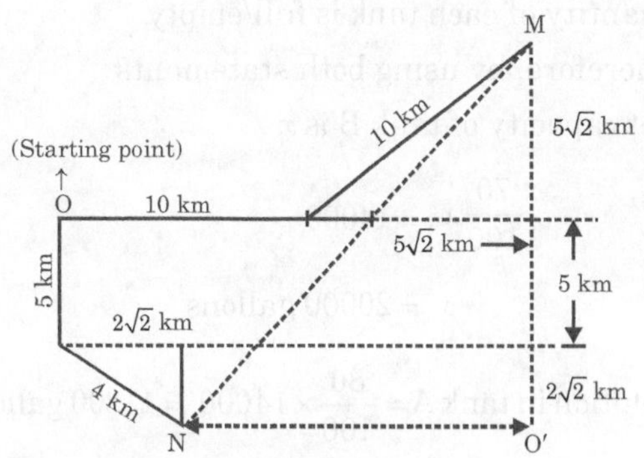

From the given figure

$$MN = \sqrt{(O'M)^2 + (O'N)^2}$$

$$O'M = 5\sqrt{2} + 5 + 2\sqrt{2} = 5 + 7\sqrt{2}$$

$$O'N = 10 + 5\sqrt{2} - 2\sqrt{2} = 10 + 3\sqrt{2}$$

$$MN = \sqrt{(5+7\sqrt{2})^2 + (10+3\sqrt{2})^2}$$

$$= \sqrt{25 + 98 + 70\sqrt{2} + 100 + 18 + 60\sqrt{2})}$$

$$\approx 20.61 \text{ km}$$

17.

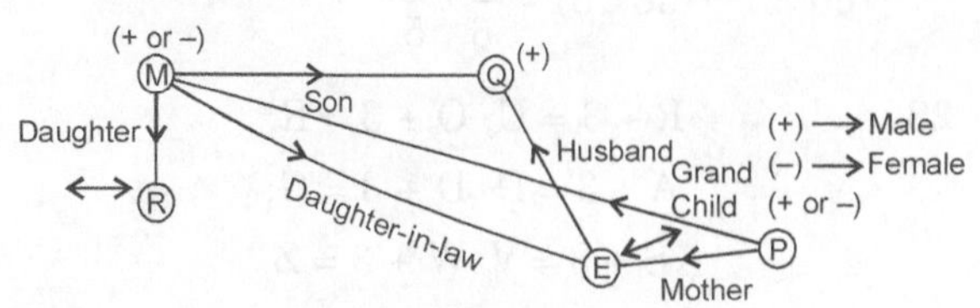

So, from the above relation diagram it is clear that P is the grandchild of M.

20. Manu age = sravan age + 2 months

Manu age = Trideep age – 3 months

Pavan age = Sravan's age + 1 month

From the above statement

Trideep age > Man > Pavan > Sravan

Hence, Trideep can occupy the extra space in the flat.

24.

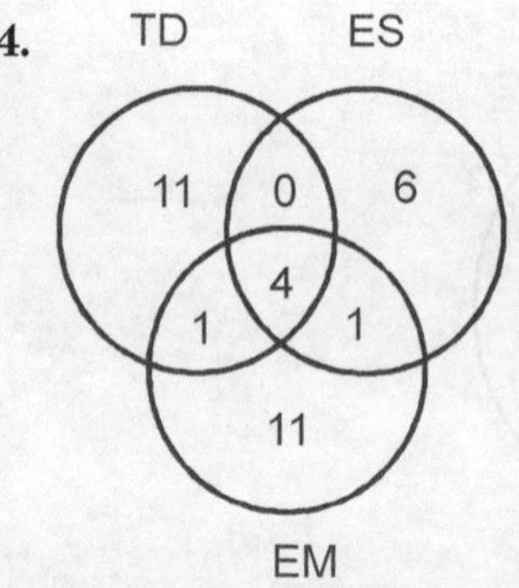

25. Statement-I can be used to solve the question if capacity of both tanks is already known

Statement-II can be used if it is known what quantity of each tank is full/empty.

Therefore, by using both statements

Let capacity of tank B is x

$$\frac{70}{100}x = 14000$$

$$\Rightarrow \quad x = 20000 \text{ gallons}$$

$$\text{Solution in tank A} = \frac{80}{100} \times 14000 = 11200 \text{ gallons}$$

$$\text{Solution in tank B} = \frac{40}{100} \times 20000 = 8000 \text{ gallons}$$

$\therefore$ Total solution = 11200 + 8000 = 19200 gallons

26. $66\square 6 = \frac{66-6}{66+6} = \frac{60}{72} = \frac{5}{6}$

$66\diamond 6 = \frac{66+6}{66-6} = \frac{72}{60} = \frac{6}{5}$

$(66\square 6) \rightarrow (66\diamond 6) = \frac{5}{6} \times \frac{6}{5} = 1$

28. R + 3 = U, O + 3 = R,
A + 3 = D, D + 3 = G;
S + 3 = V, W + 3 = Z,
A + 3 = D, N + 3 = Q

29. Community of people : Doctor

30. Since U does not want any portfolio, (*c*) and (*d*) are ruled out. R wants Home, or Finance or No portfolio, (*a*) is not valid.

Hence option (*b*) is correct.

34. S + 1: S + 2 :

Therefore concluding diagram can be

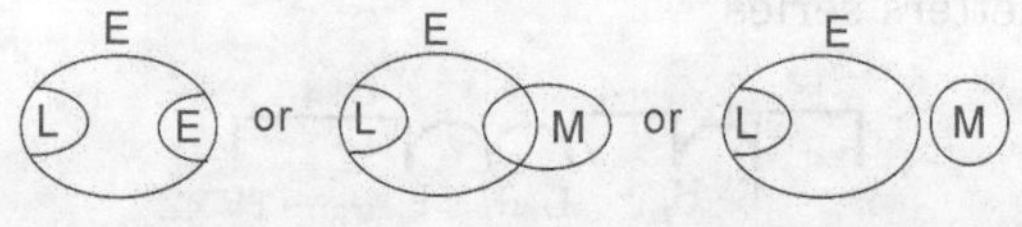

35.

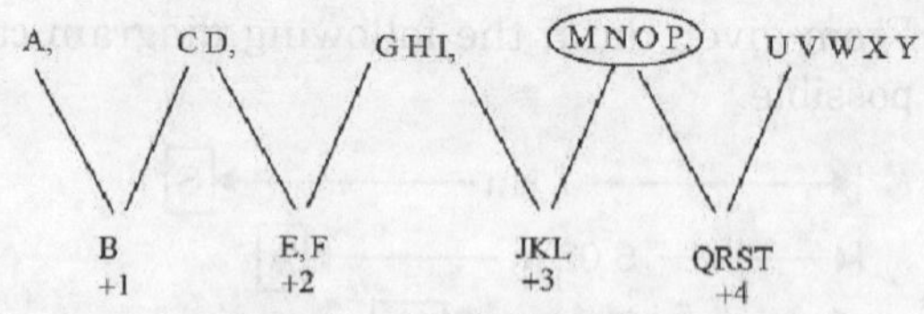

36. Second sentence says that Ms. X reaches Kochi on 21/05/2014. Also she has to be in Bagdogora on 22/05/2014.

$\therefore$ She stays in Kochi for only one day in may.

39.

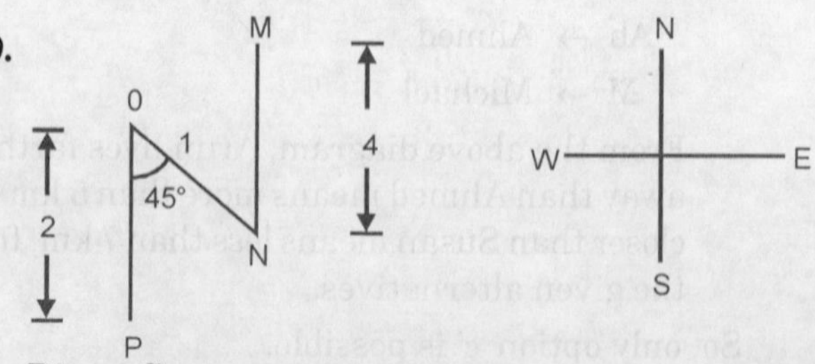

41. Riaz > Som ...(*i*)
Shiv > Ansu ...(*ii*)

Ansu is youngest in the group

$\therefore$ From equations (*i*) and (*ii*)

$\therefore$ Riaz > Som > Shiv > Ansu

42.

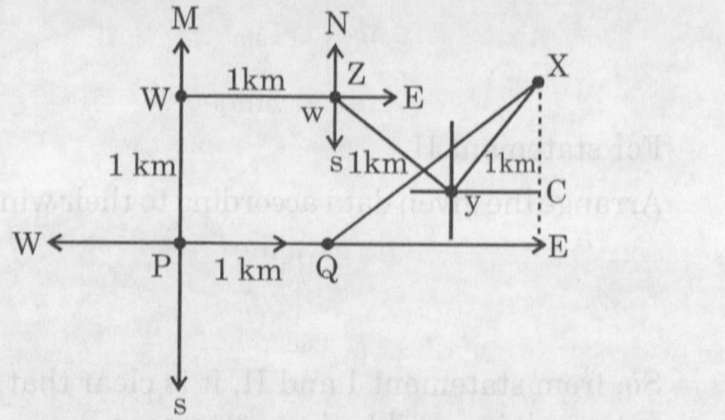

Here XE = 1 km

$$QE = \frac{2}{\sqrt{2}} = \sqrt{2}$$

$\therefore$ using Pythagoras theorem,

$$XQ = \sqrt{(QE)^2 + (XE)^2}$$

$$= \sqrt{2+1} = \sqrt{3} \text{ km}$$

$\therefore$ Statement I itself determines the eldest child which statement 2 is incorrect as Ansu is youngest in the group is not satisfied by statement 2

Hence option (*a*) is correct

44. ALRVX → only one vowel
EPVZB → only one vowel
ITZDF → only one vowel
OYEIK → three vowels

45. (*a*) Anuj: Even numbered floor (2, 4, 6)
(*b*) Bhola: Even numbered floor (2, 4, 6)
(*c*) Chandan lives on the floor above that of Faisal.
(*d*) Dilip: not on 2nd floor.
(*e*) Eswar: does not live immediately above or immediately below Bhola
From the options its clear, that only option (*b*) satisfies condition (*e*).
So, correct Ans is (*b*).

47. KCLFTSB: BST-Best, F-Of, LCK-Luck (Reverse order)
SHSWDG: GD-Good, WSHS-Wishes (Reverse order)
Similarly "ace the Exam'- C-Ace, T-The, XM-Exam

48. Diversity is shown in terms of difference language.

51. It is not mentioned that elephant is the largest animal.

52.

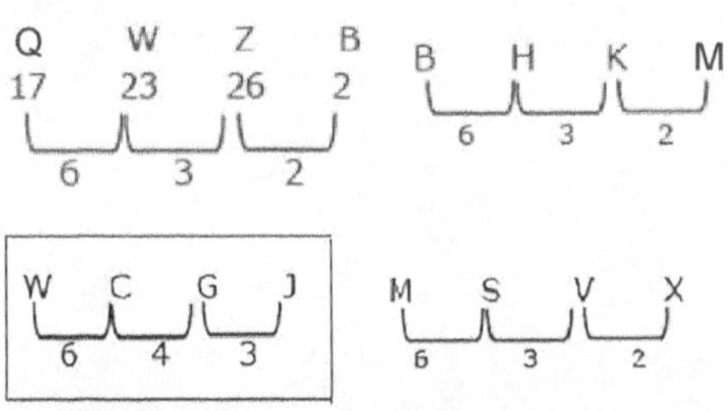

55.

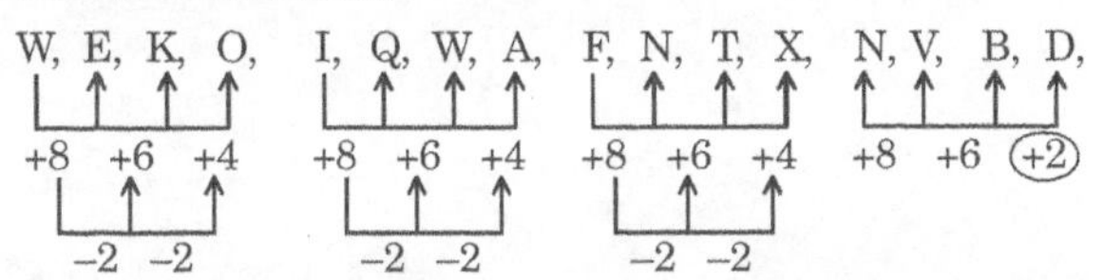

Hence the odd one from the following group is N, V, B, D.

56. By reading the instructions, it is clear that person who fills form T is a resident female with annual income of ₹ 9 Lakh.

59. Among given option, option (*d*) *i.e.* the number of people with malarial fever has increased this year do not contradic the conclusion.

70.

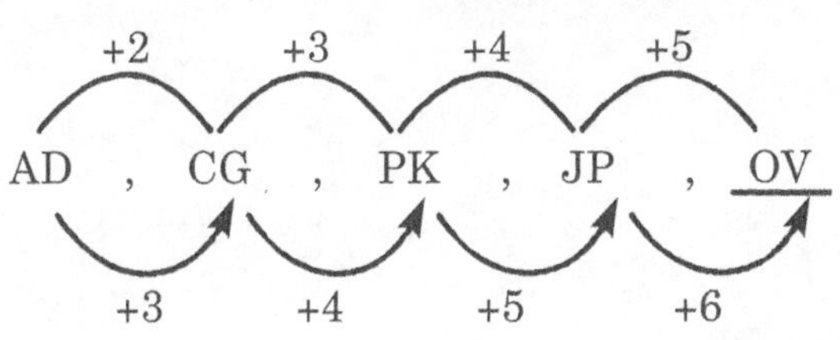

71. Let each truck carry 100 units.

$$2800 = 4n + e$$

where n = normal

$$3000 = 10n + e$$

where e = excess/pending

$$\therefore \quad n = \frac{100}{3}, \quad e = \frac{8000}{3}$$

$$5 \text{ days} \Rightarrow 500\,x = \frac{5100}{3} + \frac{8000}{3}$$

$$\Rightarrow \quad 500x = \frac{8500}{3} = 17$$

$$\Rightarrow \quad x > 5$$

Minimum possible = 6

72. The given passage clearly deals with how to deal with bereavement and grief and so after the tragedy occurs and not about precautions. Therefore, irrespective of the causes of death, a school student rarely gets into details of causes—which is beyond the scope of the context. Rest all are important in dealing with grief.

73. Given relationship is worker: workplace. A gladiator is
(*i*) a person, usually a professional combatant trained to entertain the public by engaging in mortal combat with another person or a wild.
(*ii*) A person engaged in a controversy or debate, especially in public.

74. From the passage it cannot be inferred that horses are given immunity as in option (*a*), since the aim is to develop medicine and in turn immunize humans. Option (*b*) is correct since it is given that horses develop immunity after some time. Refer "until their blood built up immunities". Even option(*c*) is invalid since medicine is not built till immunity is developed in the horses. Option (*d*) is incorrect since specific examples are cited to illustrate and this cannot capture the essence.

76.

$$H + G > I + S \quad ...(1)$$

and $$G - S = 1 \quad ...(2)$$

G is not oldest, S is not youngest

$$\therefore \quad H + 1 > I$$

Irfan older than Hari
Gita older than Sarita
From given option SGHI

77. Use of chemical agents in warfare would be undesirable

78.

$$\begin{array}{r l} 137- & \overline{001}\ \ \overline{011}\ \ \overline{111} \\ +276- & 010\ \ 111\ \ 110 \\ \hline & 100\ \ 011\ \ 101 = 435 \end{array}$$

Octal number system

$$\begin{array}{r l} \therefore\quad 731- & 111\ \ 011\ \ 001 \\ 672- & 110\ \ 111\ \ 110 \\ \hline & 100\ \ 010\ \ 011 \end{array}$$

Hence 1623

NUMERICAL TYPE QUESTIONS

1. Middle number is the average of number on both sides

$\therefore$ Average of 3 and 3 is $\frac{3+3}{2} = \frac{6}{2} = 3$

2. $81 - 54 = 27;\ 27 \times \frac{2}{3} = 18$

$54 - 36 = 18;\ 18 \times \frac{2}{3} = 12$

$36 - 24 = 12;\ 12 \times \frac{2}{3} = 8$

$\therefore\quad 24 - 8 = 16$

3. 2 3 6 15 [45] 157.5 630

1.5 2 2.5 3 3.5 4

$\frac{\text{2nd number}}{\text{1st number}}$ is in increasing order as shown above

4. The given series is

12, 35, 81, 173, 357, ...

The given series follows the following pattern

$12 \times 2 + 11 = 35$

$35 \times 2 + 11 = 81$

$81 \times 2 + 11 = 173$

$173 \times 2 + 11 = 357$

$357 \times 2 + 11 = 725$

Numerical Ability

1
CHAPTER

Numbers and Algebra

MCQ TYPE QUESTIONS

2016

1. In a quadratic function, the value of the product of the roots (α, β) is 4. Find the value of

$$\frac{\alpha^n + \beta^n}{\alpha^{-n} + \beta^{-n}}$$

(*a*) n^4 (*b*) 4^n
(*c*) 2^{2n-1} (*d*) 4^{n-1}

2. Among 150 faculty members in an institute, 55 are connected with each other through Facebook® and 85 are connected through WhatsApp®. 30 faculty members do not have Facebook® or WhatsApp® accounts. The number of faculty members connected only through Facebook® accounts is ________.

(*a*) 35 (*b*) 45
(*c*) 65 (*d*) 90

3. If $|9y - 6| = 3$, then $y^2 - \frac{4y}{3}$ is ________.

(*a*) 0 (*b*) $+\frac{1}{3}$
(*c*) $-\frac{1}{3}$ (*d*) Undefined

4. If $q^{-a} = \frac{1}{r}$ and $r^{-b} = \frac{1}{s}$ and $s^{-c} = \frac{1}{q}$, the value of *abc* is .

(*a*) $(rqs)^{-1}$ (*b*) 0
(*c*) 1 (*d*) r^+q^+s

5. The sum of the digits of a two digit number is 12. If the new number formed by reversing the digits is greater than the original number by 54, find the original number.

(*a*) 39 (*b*) 57
(*c*) 66 (*d*) 93

6. Two finance companies, P and Q, declared fixed annual rates of interest on the amounts invested with them. The rates of interest offered by these companies may differ from year to year. Year-wise annual rates of interest offered by these companies are shown by the line graph provided below.

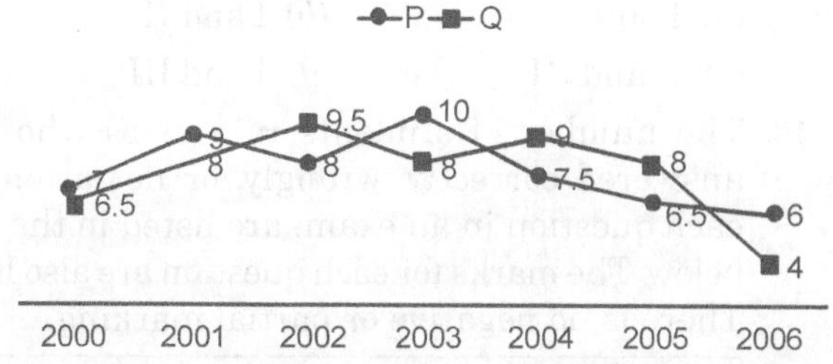

If the amounts invested in the companies, P and Q, in 2006 are in the ratio 8 : 9, then the amounts received after one year as interests from companies P and Q would be in the ratio:

(*a*) 2 : 3 (*b*) 3 : 4
(*c*) 6 : 7 (*d*) 4 : 3

2015

7. Read the following table giving sales data of five types of batteries for years 2006 to 2012

Year	Type I	Type I	Type I	Type I	Type I
2006	75	144	114	102	108
2007	90	126	102	84	126
2008	96	114	75	105	135
2009	105	90	150	90	75
2010	90	75	135	75	90
2011	105	60	165	45	120
2012	115	85	160	100	145

Out of the following, which type of battery achieved highest growth between the years 2006 and 2012?

(*a*) Type V (*b*) Type III
(*c*) Type II (*d*) Type I

8. If $a^2 + b^2 + c^2 = 1$, then $ab + bc + ac$ lies in the interval

(*a*) $\left[1, \frac{2}{3}\right]$

(*b*) $\left[\frac{-1}{2}, 1\right]$

(*c*) $\left[-1, \frac{1}{2}\right]$

(*d*) [2,–4]

9. If the list of letters, P, R, S, T, U is an arithmetic sequence, which of the following are also in arithmetic sequence?

I. 2P, 2R, 2S, 2T, 2U

II. P–3, R–3, S–3, T–3, U–3

III. P^2, R^2, S^2, T^2, U^2

(*a*) I only (*b*) I and II

(*c*) II and III (*d*) I and III

10. The number of students in a class who have answered correctly, wrongly, or not attempted each question in an exam, are listed in the table below. The marks for each question are also listed. There is no negative or partial marking.

Q. No.	Marks	Answered Correctly	Answered Wrongly	Not Attempted
1	2	21	17	6
2	3	15	27	2
3	1	11	29	4
4	2	23	18	3
5	5	31	12	1

What is the average of the marks obtained by the class in the examination?

(*a*) 2.290 (*b*) 2.970

(*c*) 6.795 (*d*) 8.795

2014

11. The monthly rainfall chart on 50 years of rainfall in Agra is shown in the following figure. Which of the following are true? (k percentile is the value such that k percent of the data fall below that value)

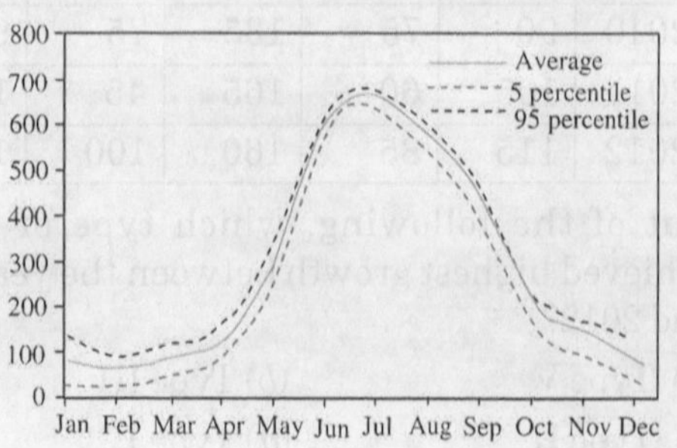

(*i*) On average, it rains more in July than in December

(*ii*) Every year, the amount of rainfall in August is more than that in January

(*iii*) July rainfall can be estimated with better confidence than February rainfall

(*iv*) In August, there is at least 500 mm of rainfall

(*a*) (*i*) and (*ii*)

(*b*) (*i*) and (*iii*)

(*c*) (*ii*) and (*iii*)

(*d*) (*iii*) and (*iv*)

12. The total exports and revenues from the exports of a country are given in the two charts shown below. The pie chart for exports shows the quantity of each item exported as a percentage of the total quantity of exports. The pie chart for the revenues shows the percentage of the total revenue generated through export of each item. The total quantity of exports of all the items is 500 thousand tonnes and the total revenues are 250 crore rupees. Which item among the following has generated the maximum revenue per kg?

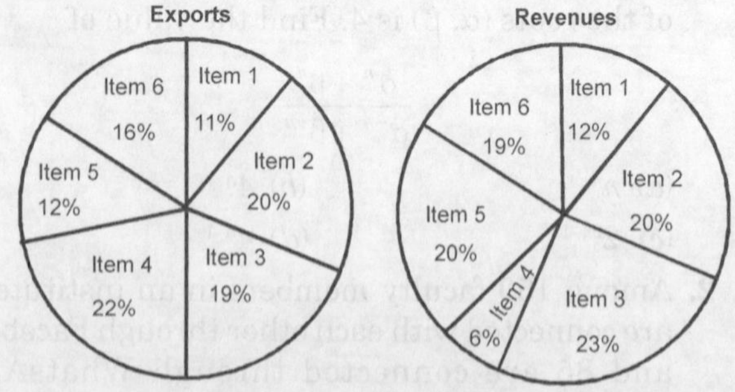

(*a*) Item 2 (*b*) Item 3

(*c*) Item 6 (*d*) Item 5

13. The statistics of runs scored in a series by four batsmen are provided in the following table. Who is the most consistent batsman of these four?

Batsman	Average	Standard deviation
K	31.2	5.21
L	46.0	6.35
M	54.4	6.22
N	17.9	5.90

(*a*) K (*b*) L

(*c*) M (*d*) N

14. The exports and imports (in crores of ₹) of a country from 2000 to 2007 are given in the following bar chart. If the trade deficit is defined as excess of imports over exports, in which year is the trade deficit 1/5th of the exports?

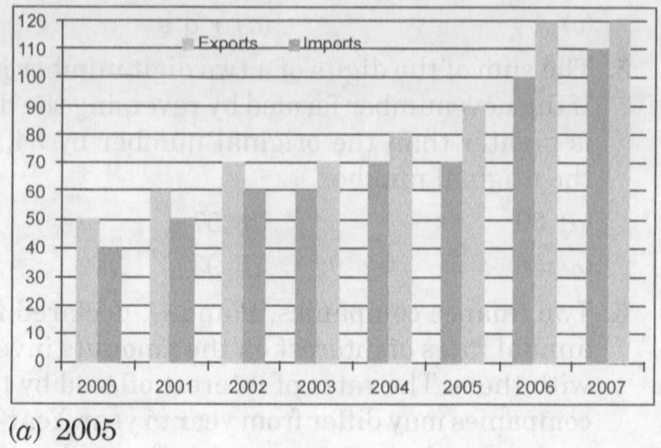

(*a*) 2005

(*b*) 2004

(*c*) 2007

(*d*) 2006

15. The table below has question-wise data on the performance of students in an examination. The marks for each question are also listed. There is no negative or partial marking in the examination.

Q.No.	Marks	Answered Correctly	Answered Wrongly	Not Attempted
1	2	21	17	6
2	3	15	27	2
3	2	23	18	3

What is the average of the marks obtained by the class in the examination?

(*a*) 1.34 (*b*) 1.74
(*c*) 3.02 (*d*) 3.91

16. The ratio of male to female students in a college for five years is plotted in the following line graph. If the number of female students in 2011 and 2012 is equal, what is the ratio of male students in 2012 to male students in 2011?

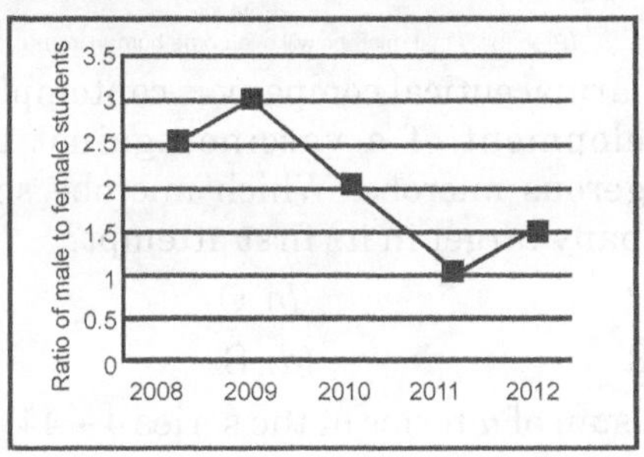

(*a*) 1 : 1 (*b*) 2 : 1
(*c*) 1.5 : 1 (*d*) 2.5 : 1

17. What is the average of all multiples of 10 from 2 to 198?

(*a*) 90 (*b*) 100
(*c*) 110 (*d*) 120

18. The value of $\sqrt{12+\sqrt{12+\sqrt{12+...}}}$ is

(*a*) 3.464 (*b*) 3.932
(*c*) 4.000 (*d*) 4.444

19. If x is real and $|x^2 - 2x + 3| = 11$, then possible values of $|-x^3 + x^2 - x|$ include

(*a*) 2, 4 (*b*) 2, 14
(*c*) 4, 52 (*d*) 14, 52

2013

20. A number is as much greater than 75 as it is smaller than 117. The number is

(*a*) 91
(*b*) 93
(*c*) 89
(*d*) 96

21. Following table provides figures (in rupees) on annual expenditure of a firm for two years - 2010 and 2011.

Category	2010	2011
Raw material	5200	6240
Power & fuel	7000	9450
Salary & wages	9000	12600
Plant & machinery	20000	25000
Advertising	15000	19500
Research & Development	22000	26400

In 2011, which of the following two categories have registered increase by same percentage?

(*a*) Raw material and Salary & wages
(*b*) Salary & wages and Advertising
(*c*) Power & fuel and Advertising
(*d*) Raw material and Research & Development

22. What will be the maximum sum of 44, 42, 40,... ?

(*a*) 502 (*b*) 504
(*c*) 506 (*d*) 500

23. The current erection cost of a structure is ₹ 13,200. If the labour wages per day increase by 1/5 of the current wages and the working hours decrease by 1/24 of the current period, then the new cost of erection in ₹ is

(*a*) 16,500 (*b*) 15,180
(*c*) 11,000 (*d*) 10,120

24. Find the sum of the expression

$$\frac{1}{\sqrt{1}+\sqrt{2}}+\frac{1}{\sqrt{2}+\sqrt{3}}+\frac{1}{\sqrt{3}+\sqrt{4}}+.....+\frac{1}{\sqrt{80}+\sqrt{81}}$$

(*a*) 7 (*b*) 8
(*c*) 9 (*d*) 10

25. In the summer of 2012, in New Delhi, the mean temperature of Monday to Wednesday was 41°C and of Tuesday to Thursday was 43°C. If the temperature on Thursday was 15% higher than that of Monday, then the temperature in °C on Thursday was

(*a*) 40 (*b*) 43
(*c*) 46 (*d*) 49

26. The set of values of p for which the roots of the equation $3x^2 + 2x + p(p - 1) = 0$ are of opposite sign is

(*a*) $(-\infty, 0)$ (*b*) $(0, 1)$
(*c*) $(1, \infty)$ (*d*) $(0, \infty)$

27. Find the sum to n terms of the series 10 + 84 + 734 + ...

(*a*) $\frac{9(9^n+1)}{10}+1$ (*b*) $\frac{9(9^n-1)}{8}+1$

(*c*) $\frac{9(9^n-1)}{8}+n$ (*d*) $\frac{9(9^n-1)}{8}+n^2$

28. If $3 \le X \le 5$ and $8 \le Y \le 11$ then which of the following options is TRUE?

(a) $\frac{3}{5} \le \frac{X}{Y} \le \frac{8}{5}$ (b) $\frac{3}{11} \le \frac{X}{Y} \le \frac{5}{8}$

(c) $\frac{3}{11} \le \frac{X}{Y} \le \frac{8}{5}$ (d) $\frac{3}{5} \le \frac{X}{Y} \le \frac{8}{11}$

29. Following table gives data on tourists from different countries visiting India in the year 2011.

Country	Number of Tourists
USA	2000
England	3500
Germany	1200
Italy	1100
Japan	2400
Australia	2300
France	1000

Which two countries contributed to the one third of the total number of tourists who visited India in 2011?

(a) USA and Japan

(b) USA and Australia

(c) England and France

(d) Japan and Australia

2012

30. If $(1.001)^{1259} = 3.52$ and $(1.001)^{2062} = 7.85$, then $(1.001)^{3321} =$

(a) 2.23 (b) 4.33

(c) 11.37 (d) 27.64

31. Raju has 14 currency notes in his pocket consisting of only ₹ 20 notes and ₹ 10 notes. The total money value of the notes is ₹ 230. The number of ₹ 10 notes that Raju has is

(a) 5 (b) 6

(c) 9 (d) 10

32. The data given in the following table summarizes the monthly budget of an average household.

Category	Amount (Rs.)
Food	4000
Clothing	1200
Rent	2000
Savings	1500
Other expenses	1800

The approximate percentage of the monthly budget **NOT** spent on savings is

(a) 10% (b) 14%

(c) 81% (d) 86%

2011

33. P, Q, R and S are four types of dangerous microbes recently found in a human habitat. The area of each circle with its diameter printed in brackets represents the growth of a single microbe surviving human immunity system within 24 hours of entering the body. The danger to human beings varies proportionately with the toxicity, potency and growth attributed to a microbe shown in the figure below :

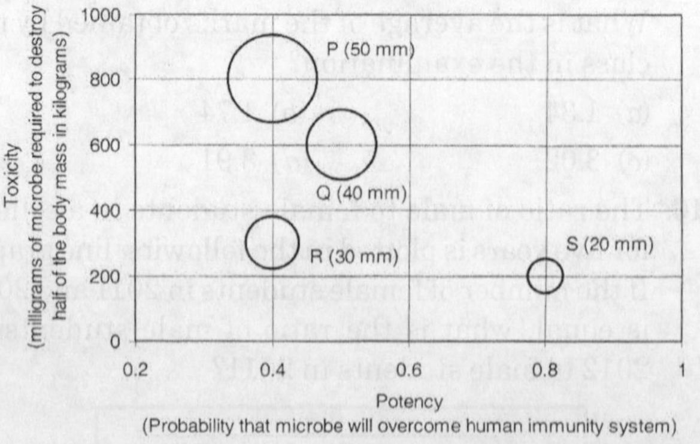

A pharmaceutical company is contemplating the development of a vaccine against the most dangerous microbe. Which microbe should the company target in its first attempt?

(a) P (b) Q

(c) R (d) S

34. The sum of n terms of the series 4 + 44 + 444 + ... is

(a) $(4/81)[10^{n+1} - 9n - 1]$

(b) $(4/81)[10^{n-1} - 9n - 1]$

(c) $(4/81)[10^{n+1} - 9n - 10]$

(d) $(4/81)[10^{n} - 9n - 10]$

35. Given that $f(y) = |y|/y$, and q is any non-zero real number, the value of $|f(q) - f(-q)|$ is

(a) 0 (b) –1

(c) 1 (d) 2

NUMERICAL TYPE QUESTIONS

2016

1. The following graph represents the installed capacity for cement production (in tonnes) and the actual production (in tonnes) of nine cement plants of a cement company. Capacity utilization of a plant is defined as ratio of actual production of cement to installed capacity. A plant with installed capacity of at least 200 tonnes is called a large plant and a plant with lesser capacity is called a small plant. The difference between total production of large plants and small plants, in tonnes is _______.

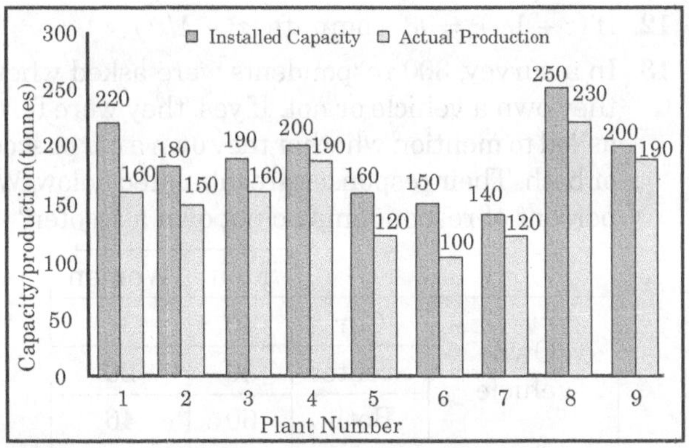

2. The numeral in the units position of $211^{870} + 146^{127} \times 3^{424}$ is ________.

2015

3. The exports and imports (in crores of Rs.) of a country from the year 2000 to 2007 are given in the following bar chart. In which year is the combined percentage increase in imports and exports the highest ?

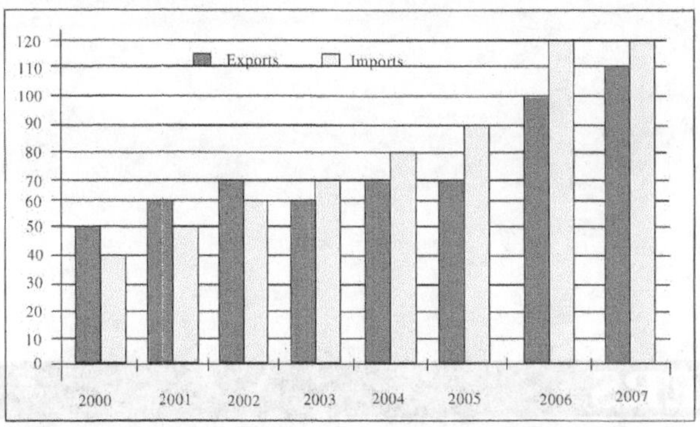

4. The pie chart below has the breakup of the number of students from different departments in an engineering college for the year 2012. The proportion of male to female students in each department is 5 : 4. There are 40 males in Electrical Engineering. What is the difference between numbers of female students in the Civil department and the female students in the Mechanical department?

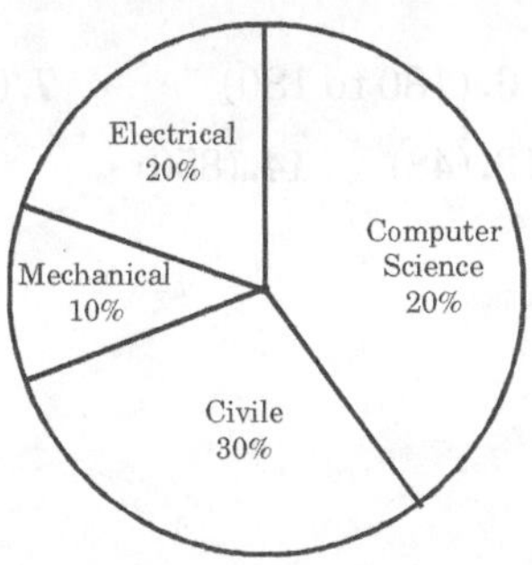

2014

5. A foundry has a fixed daily cost of ₹ 50,000 whenever it operates and variable cost of ₹ 800Q, where Q is the daily production in tonnes. What is the cost of production in ₹ per tonne for a daily production of 100 tonnes?

6. The smallest angle of a triangle is equal to two thirds of the smallest angle of a quadrilateral. The ratio between the angles of the quadrilateral is 3 : 4 : 5 : 6. The largest angle of the triangle is twice its smallest angle. What is the sum, in degrees, of the second largest angle of the triangle and the largest angle of the quadrilateral?

7. In a sequence of 12 consecutive odd numbers, the sum of the first 5 numbers is 425. What is the sum of the last 5 numbers in the sequence?

8. A firm producing air purifiers sold 200 units in 2012. The following pie chart presents the share of raw material, labour, energy, plant & machinery, and transportation costs in the total manufacturing cost of the firm in 2012. The expenditure on labour in 2012 is ₹ 4,50,000. In 2013, the raw material expenses increased by 30% and all other expenses increased by 20%. What is the percentage increase in total cost for the company in 2013?

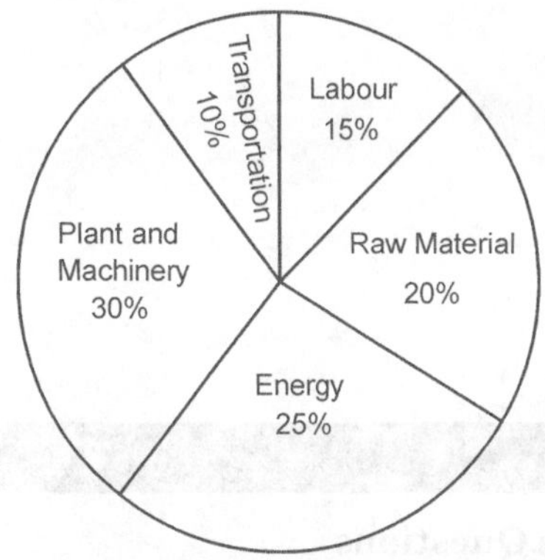

9. A firm producing air purifiers sold 200 units in 2012. The following pie chart presents the share of raw material, labour, energy, plant & machinery, and transportation costs in the total manufacturing cost of the firm in 2012. The expenditure on labour in 2012 is ₹ 4,50,000. In 2013, the raw material expenses increased by 30% and all other expenses increased by 20%. If the company registered a profit of ₹ 10 lakhs in 2012, at what price (in ₹) was each air purifier sold?

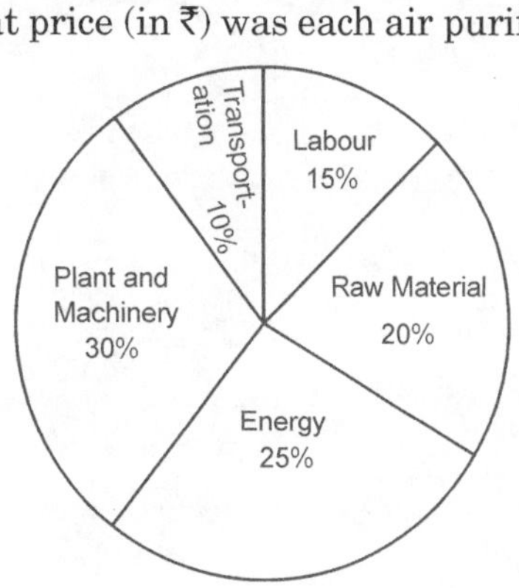

10. The sum of eight consecutive odd numbers is 656. The average of four consecutive even numbers is 87. What is the sum of the smallest odd number and second largest even number?

11. The ratio of male to female students in a college for five years is plotted in the following line graph. If the number of female students doubled in 2009, by what percent did the number of male students increase in 2009?

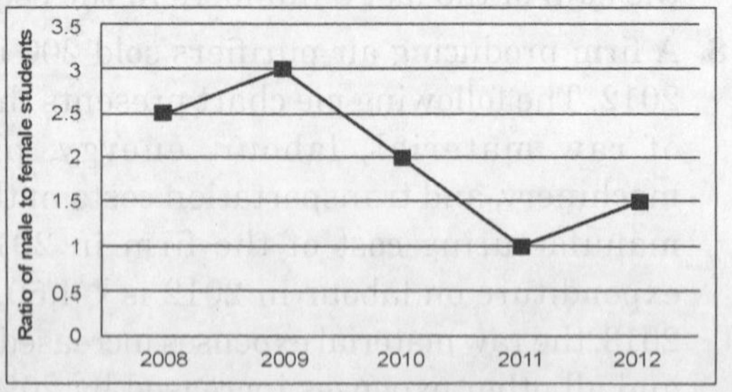

12. If $(z + 1/z)^2 = 98$, compute $(z^2 + 1/z^2)$

13. In a survey, 300 respondents were asked whether they own a vehicle or not. If yes, they were further asked to mention whether they own a car or scooter or both. Their responses are tabulated below. What percent of respondents do not own a scooter?

		Men	Women
Own vehicle	Car	40	34
	Scooter	30	20
	Both	60	46
Do not own vehicle		20	50

14. Round–trip tickets to a tourist destination are eligible for a discount of 10% on the total fare. In addition, groups of 4 or more get a discount of 5% on the total fare. If the one way single person fare is ₹ 100, a group of 5 tourists purchasing round–trip tickets will be charged ₹ ________.

ANSWERS

MCQ Type Questions

1. (*b*)	**2.** (*a*)	**3.** (*c*)	**4.** (*c*)	**5.** (*a*)	**6.** (*d*)	**7.** (*d*)	**8.** (*b*)	**9.** (*b*)	**10.** (*c*)
11. (*b*)	**12.** (*d*)	**13.** (*a*)	**14.** (*d*)	**15.** (*c*)	**16.** (*c*)	**17.** (*b*)	**18.** (*c*)	**19.** (*d*)	**20.** (*d*)
21. (*d*)	**22.** (*c*)	**23.** (*a*)	**24.** (*b*)	**25.** (*c*)	**26.** (*b*)	**27.** (*d*)	**28.** (*b*)	**29.** (*c*)	**30.** (*d*)
31. (*a*)	**32.** (*d*)	**33.** (*d*)	**34.** (*c*)	**35.** (*d*)					

Numerical Type Questions

1. (120) **2.** (7) **3.** (2006) **4.** (32) **5.** (1300 to 1300) **6.** (180 to 180) **7.** (495) **8.** (22)
9. (20,000) **10.** (163) **11.** (140 to 140) **12.** (96) **13.** (48) **14.** (850)

EXPLANATIONS

MCQ TYPE QUESTIONS

1. $\dfrac{\alpha^n+\beta^n}{\alpha^{-n}+\beta^{-n}} = \dfrac{\alpha^n+\beta^n}{\dfrac{1}{\alpha^n}+\dfrac{1}{\beta^n}} = \dfrac{\alpha^n+\beta^n}{\dfrac{\beta^n+\alpha^n}{\alpha^n\times\beta^n}} = \alpha^n\times\beta^n$

$\alpha^n\times\beta^n = (\alpha\times\beta)^n = (4)^n$

So, the value of $\dfrac{\alpha^n+\beta^n}{\alpha^{-n}+\beta^{-n}}$ is 4^n.

2. From question

Total number of faculty members = 150

The number of faculty members having facebook account = (FB) = 55

The number of faculty members having whatsapp = (W) = 85

The number of faculty members do not have face book or Whats App accounts = 30

The number of faculty members having any account = 150 – 30 = 120

The number of faculty members having both the accounts = (FB + W) – 120 = (55 + 85) – 120 = 20

$\therefore$ The number of faculty members connected only through Facebook accounts = 55 – 20 = 35

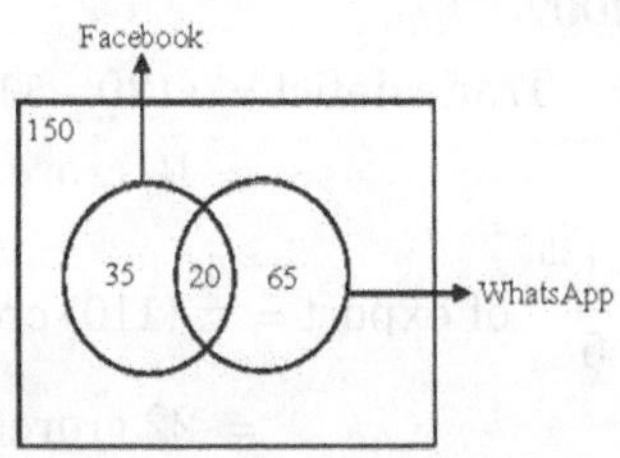

So the number of faculty members connected only through facebook accounts is 35.

3. Given, $|9y-6| = 3$

It has two values either

$$(9y-6) = 3 \quad \text{...}(i)$$

Or $\quad -(9y-6) = 3$

$$9y-6 = -3 \quad \text{...}(ii)$$

From equation (*i*),

$$9y = 9$$

$$\therefore \quad y = 1$$

Or from equation (*ii*),

$$9y = 3$$

$$\therefore \quad y = \frac{1}{3}$$

Now $\quad y^2 = \dfrac{4y}{3}$

put $\quad y = 1,$

we get $\quad = 1-\dfrac{4}{3} = -\dfrac{1}{3}$

and also put $\quad y = \dfrac{1}{3}$

we get, $\quad \dfrac{1}{9}-\dfrac{4}{9} = \dfrac{-3}{9} = \dfrac{-1}{3}$

So, the required value is $\dfrac{-1}{3}$.

4. Given $\quad q^{-a} = \dfrac{1}{r}$

$$\Rightarrow \quad \frac{1}{q^a} = \frac{1}{r}$$

$$\Rightarrow \quad q^a = r \quad \text{...}(i)$$

$$r^{-b} = \frac{1}{s}$$

$$\Rightarrow \quad \frac{1}{r^b} = \frac{1}{s}$$

$$\Rightarrow \quad s = r^b \quad \text{...}(ii)$$

$$s^{-c} = \frac{1}{q}$$

$$\Rightarrow \quad \frac{1}{s^c} = \frac{1}{q}$$

$$\Rightarrow \quad s^c = q \quad \text{...}(iii)$$

From equation (*i*),

$$q^a = r$$

$$(s^c)^a = r \quad \text{(from eq. } (iii))$$

$$S^{ac} = r$$

$$S^{ac} = (s)^{\frac{1}{b}} \quad \text{(from eq. } (ii))$$

$$S^{abc} = S = S^1$$

$$\therefore \quad abc = 1$$

So the value of *abc* is 1

5. Let the two digit number be xy

Now sum of the digit of a two digit number is 12

$$\therefore \quad x+y = 12 \quad \text{...}(i)$$

Now according to question,

$$(10y+x)-(10x+y) = 54$$

$$\therefore \quad 9y-9x = 54$$

$$\therefore \quad y-x = 6 \quad \text{...}(ii)$$

Now solving equation (*i*) and (*ii*) we get,

$$x = 3 \text{ and } y = 9$$

Hence the original number is 39.

$\therefore$ Option (*a*) is correct.

6. Amount invested by Company P in 2006 = ₹ 8x

and amount invested by Company Q in 2006 = ₹ $9x$

Interest from Company P

$$= \frac{\frac{₹\, 8x \times 6}{100}}{\frac{₹\, 9x \times 4}{100}}$$

$$= \frac{4}{3}$$

∴ Option (d) is correct.

7. Type-I achieved a growth of 53% in the period which is higher than any other type of battery

10. Total number of students in class = 44

Average of marks,

$$= \frac{(2\times 21)+(3\times 15)+(1\times 11)+(2\times 23)+(5\times 31)}{42}$$

$$= \frac{42+45+11+46+155}{44} = \frac{299}{44} = 6.795$$

∴ Option (c) is correct.

11. In the question the monthly average rainfall chart for 50 years has been given.

Let us check the options.

(i) On average, it rains more in July than in December ⇒ correct.

(ii) Every year, the amount of rainfall in August is more than that in January.

⇒ may not be correct because average rainfall is given in the question.

(iii) July rainfall can be estimated with better confidence than February rainfall.

⇒ From chart it is clear the gap between 5 percentile and 95 percentile from average is higher in February than that in July ⇒ correct.

(iv) In August at least 500 mm rainfall

⇒ May not be correct, because its 50 year average.

So correct option (b) (i) and (iii).

12. Item : 2

$$\frac{\frac{20}{100}\times 250\times 10^7}{\frac{20}{100}\times 500\times 10^3}$$

$0.5\times 10^4 = 5\times 10^3$ $\boxed{1}$ = Item 2

Item : 3

$$\frac{23\times 250\times 10^7}{19\times 500\times 10^3}$$

1.2 = Item 3

Item : 6

$\frac{19}{16} = 1.18 =$ Item 6

Item : 5

$\frac{20}{12} = \frac{5}{3} = 1.6 \Rightarrow$ $\boxed{1.6 = \text{Item 5}}$

14. For 2005,

trade defict = (90 – 70) crores = 20 crores

Now $\frac{1}{5}^{th}$ of export = $\frac{1}{5}$(70) crores

= 14 crores ≠ 20 crores.

Hence option (a) is wrong.

For 2004,

Trade defict =(80 – 70) crores = 10 crores

Now $\frac{1}{5}^{th}$ of export = $\frac{1}{5}$(70) crores

= 14 crores ≠ 20 crores.

Hence option (b) is wrong.

For 2007

Trade defict = (120 – 110) crores

= 10 crores

Now $\frac{1}{5}^{th}$ of export = $\frac{1}{5}$(110) crores

= 22 crores ≠ 10 crores.

Hence option (c) is wrong.

For 2006,

Trade defict = (120 – 100) crores

= 20 crores

Now $\frac{1}{5}^{th}$ of export = $\frac{1}{5}$(100) crores

= 20 crores = Trade deficts

Hence option (d) is correct.

15. Total question

44 × 2 = 88

44 × 3 = 132

$\frac{144}{132} = \frac{88}{308}$

Total marks obtained = (21×2) + (15×3) + (23×2)

= 133

Total Number of students = 44

$$\text{Average} = \frac{133}{44} = 3.02$$

24. Given: The expression

$$S = \frac{1}{\sqrt{1}+\sqrt{2}} + \frac{1}{\sqrt{2}+\sqrt{3}} + ... + \frac{1}{\sqrt{80}+\sqrt{81}}$$

To find: Sum of the expression.

Analysis: Each term of the expression is having irrational denominator. One can also see, that both the elements in the denominator of a term differ by only 1.

This gives an idea that on rationalising the denominator, each denominator will become 1, so the sum will be easier than earlier.

Rationalising the denominator,

$$S = \frac{\sqrt{2}-\sqrt{1}}{(\sqrt{2})^2-(\sqrt{1})^2} + \frac{\sqrt{3}-\sqrt{2}}{(\sqrt{3})^2-(\sqrt{2})^2} + ... + \frac{\sqrt{81}-\sqrt{80}}{(\sqrt{81})^2-(\sqrt{80})^2}$$

$$= \frac{\sqrt{2}-\sqrt{1}}{1} + \frac{\sqrt{3}-\sqrt{2}}{3-2} + ... + \frac{\sqrt{81}-\sqrt{80}}{81-80}$$

$$= \sqrt{2}-\sqrt{1}+\sqrt{3}-\sqrt{2}+....+\sqrt{81}-\sqrt{80}$$

Now, one can see that the terms will cancel out.

Each term except $\sqrt{1}$ & $\sqrt{81}$

Occurs twice, once as positive and once a negative. So, after cancelling out, we are left with the following:

$$S = \sqrt{81}-\sqrt{1} = 9 - 1 = 8$$

So, the answer is (*b*).

25. $\frac{\text{Mon + Tues + Wed.}}{3} = 41$

Mon + Tues + Wed. = 123 ...(1)

$\frac{\text{Tues + Wed + Thurs.}}{3} = 43°$

Tue + Wed + Thu. = 129° ...(2)

(2) – (1)

Tues + Wed + Thu – (Mon + Tues. + Wed.)

= 129 – 123 = 6°

Thu. – Mon. = 6° ⇒ $\frac{115x}{100} - x = 6°$

Thus. = Mon × $\frac{115}{100} = \frac{15x}{100} = 6°$

Mon = x $x = 40°$

Thurs = $\frac{115x}{100}$

∴ Thurs = $\frac{115 \times 40}{100} = 46°$

30. Let $1.001 = x$

$x^{1259} = 3.52$, and $x^{2062} = 7.85$

∴ $x^{3321} = x^{1259} . x^{2062}$

$= 3.52 \times 7.85$

$= 27.64$

31. Let number of ₹ 20 notes be x and ₹ 10 notes be y

∴ $20x + 10y = 230$ ...(1)

and $x + y = 14$ ...(2)

Solving equations (1) and (2), we have

$x = 9$ and $y = 5$

Hence numbers of 10 rupee notes are 5.

32. Total budget = 10,500

Expenditure other than savings = 9000

Hence approximate percentage of monthly budget

$$= \frac{9000}{10500} = 86\%$$

33. By observation of the table, we can say S

	P	Q	R	S
Requirement	800	600	300	200
Potency	0.4	0.5	0.4	0.8

34. Let S = 4 (1 + 11 + 111 + ...)

$$= \frac{4}{9}(9 + 99 + 999 + ...)$$

$$= \frac{4}{9}\{(10-1) + (10^2-1) + (10^3-1) + ...\}$$

$$= \frac{4}{9}\{(10 + 10^2 + ... + 10^n) - n\}$$

$$= \frac{4}{9}\left\{10\frac{(10^n-1)}{9} - n\right\}$$

$$= \frac{4}{81}\{10^{n+1} - 9n - 10\}$$

35. Given : $f(y) = \frac{|y|}{y}$

$$\Rightarrow \quad f(q) = \frac{|q|}{q}$$

$$f(-q) = \frac{|-q|}{-q}$$

$$= \frac{-|q|}{q}$$

$$|f(q) - f(q)| = \frac{|q|}{q} + \frac{|q|}{q}$$

$$= \frac{2|q|}{q} = 2$$

NUMERICAL TYPE QUESTIONS

1. From the given graph large plants which are having installed capacity of at least 200 tonnes are 1, 4, 8 and 9.

So, the total production of large plants

$$= 160 + 190 + 230 + 190$$
$$= 770$$

Now the remaining plants with installed capacity is less then 200 tonnes are 2, 3, 5, 6 and 7.

So, total production of small plants

$$= 150 + 160 + 120 + 100 + 120$$
$$= 650$$

Then difference = 770 – 650

$$= 120$$

2. Unit digit of $211^{870} + 146^{127} \times 3^{424}$ is $1 + 6 \times 1 = 7$
3. Increase in exports in 2006

$$= \frac{100-70}{70} = 42.8\%$$

Increase in imports in 2006

$$= \frac{120-90}{90} = 33.3\%$$

which is more than any other year.

4. According to question,

There are 40 males in Electrical Engg.

∴ Total number of females in Electrical Engg.

$$= 72 - 40 = 32$$

∴ Total number of students in Electrical Engg. = 72

⇒ 20% = 72

∴ 30% = Total number of students in Civil Engg.

∴ Total number of students in Civil Engg.

$$= \frac{30 \times 72}{20} = 36 \times 3 = 108$$

∴ Number of female students in Civil Engg.

$$= \frac{4 \times 108}{9} = 4 \times 12 = 48$$

Now, total number of students in Mechanical Engg.

$$= \frac{10 \times 72}{20} = 36$$

∴ Number of female students in Mechanical Engg.

$$= \frac{4 \times 36}{9} = 16$$

∴ Difference between female students in Civil and Mechanical departments = 48 – 16 = 32

5. Fixed cost = ₹ 50,000

Variable cost = ₹ 800Q

Q = daily production in tones

For 100 tonnes of production daily, total cost of production = 50,000 + 800 × 100 = 130,000

So, cost of production per tonne of daily production

$$= \frac{130,000}{100} = ₹\ 1300.$$

6. Let the angles of quadrilateral are $3x$, $4x$, $5x$, $6x$

So, $3x + 4x + 5x + 6x = 360$

$$x = 20$$

Smallest angle of quadrilateral = 3 × 20 = 60°

Smallest angle of triangle = $\frac{2}{3} \times 60° = 40°$

Largest angle of triangle = 2 × 40° = 60°

Three angles of triangle are 40°, 60°, 80°

Largest angle of quadrilateral is 120°

Sum (2nd largest angle of triangle + largest angle of quadrilateral) = 60° + 120° =180°.

7. 8th observation is 7 × 2 = 14 more than 1st observation

9th observation is 14 more than 2nd observation

10th observation is 14 more than 3rd observation

11th observation is 14 more than 4th observation

12th observation is 14 more than 5th observation

Total 14 × 5 =70

Sum of the first five numbers = 425

Sum of last five numbers = 495

8.

	2012	2013
Transport (10%)	300,000	360,000
Labour (15%)	450,000	540,000
Raw material (20%)	750,000	780,000
Energy (25%)	750,000	900,000
Plant and Machinery (30%)	900,000	1,080,000
Total	**3,000,000**	**3,660,000**

Percentage increase in total cost = 22%

9. Total expenditure = $\frac{15}{100}\ x = 4,50,000$

$$x = 3 \times 10^6$$

Profit = 10 lakhs

So, total selling price = 40,00,000 ...(1)

Total purifies = 200 ...(2)

S.P. of each purifier = (1)/(2) = 20,000

10. Eight consecutive odd number = 656

$a - 6, a - 1, a - 2, a, a + 2, a + 4, a + 6$

$$a + 8 = 656$$
$$a = 81$$

Smallest $m = 75$ ...(1)

Average consecutive even numbers

$$\Rightarrow \frac{a-2+a+a+2+a+4}{4} = 87$$

$$\Rightarrow a = 86$$

Second largest number = 88 ...(2)

Adding equation (1) and (2),

$$75 + 88 = 163$$

11. Ratio of male to female student in 2008 = 5 : 2

In 2009, let the male student be $5x$ and female student be $2x$

Now if number of female students double in 2009

$\therefore$ Number of female student = $4x$

$\therefore$ Number of male student = $12x$

Gender	2008	2009
Male	$5x$	$12x$
Female	$2x$	$4x$

% of male student increase in 2009

$$= \left[\left(\frac{12x-5x}{5x}\right) \times 100\right]\%$$

$$= \left[\frac{7}{5} \times 100\right]\%$$

$$= 140\%$$

12. Expanding

$$\left(z + \frac{1}{z}\right)^2 = 98$$

$$\therefore \quad z^2 + \frac{1}{z^2} + 2(z)\left(\frac{1}{z}\right) = 98$$

$$\therefore \quad z^2 + \frac{1}{z^2} = 96$$

13. Total respondents = 300

Now men respondents who do not have scooter

$= 40 + 20 = 60$

And women respondents who do not have scooter

$= 34 + 50 = 84$

$\therefore$ Total respondents who do not have scooter

$= 60 + 84 = 144$

$$\therefore \text{Required percent} = \left(\frac{144}{300} \times 100\right)\% = 48\%$$

14. One way single person fare = ₹ 100

$\therefore$ Two way fare for single person = ₹ 200

$\therefore$ For 5 persons two way fare = ₹ 1000

Now, total discount = (10 + 5)% = 15%

$$\text{Discount amount} = ₹ \left[\frac{15}{100} \times 1000\right] = ₹\ 150$$

Amount to be paid = ₹(1000 – 150) = ₹ 850

■■

2

CHAPTER

Percentage and Its Applications

MCQ TYPE QUESTIONS

2016

1. In a huge pile of apples and oranges, both ripe and unripe mixed together, 15% are unripe fruits. Of the unripe fruits, 45% are apples. Of the ripe ones, 66% are oranges. If the pile contains a total of 5692000 fruits, how many of them are apples?

(*a*) 2029198 (*b*) 2467482
(*c*) 2789080 (*d*) 3577422

2. A person moving through a tuberculosis prone zone has a 50% probability of becoming infected. However, only 30% of infected people develop the disease. What percentage of people moving through a tuberculosis prone zone remains infected but does not show symptoms of disease?

(*a*) 15 (*b*) 33
(*c*) 35 (*d*) 37

3. (x % of y) + (y % of x) is equivalent to _______.

(*a*) 2 % of xy (*b*) 2 % of (xy/100)
(*c*) xy % of 100 (*d*) 100 % of xy

2014

4. The population of a new city is 5 million and is growing at 20% annually. How many years would it take to double at this growth rate?

(*a*) 3 – 4 years (*b*) 4 – 5 years
(*c*) 5 – 6 years (*d*) 6 – 7 years

5. One percent of the people of country X are taller than 6 ft. Two percent of the people of country Y are taller than 6 ft. There are thrice as many people in country X as in country Y. Taking both countries together, what is the percentage of people taller than 6 ft?

(*a*) 3.0 (*b*) 2.5
(*c*) 1.5 (*d*) 1.25

6. Industrial consumption of power doubled from 2000-2001 to 2010-2011. Find the annual rate of increase in percent assuming it to be uniform over the years.

(*a*) 5.6 (*b*) 7.2
(*c*) 10.0 (*d*) 12.2

7. The Gross Domestic Product (GDP) in Rupees grew at 7% during 2012-2013. For international comparison, the GDP is compared in US Dollars (USD) after conversion based on the market exchange rate. During the period 2012-2013 the exchange rate for the USD increased from ₹ 50/ USD to ₹ 60/ USD. India's GDP in USD during the period 2012-2013

(*a*) increased by 5% (*b*) decreased by 13%
(*c*) decreased by 20% (*d*) decreased by 11%

2013

8. A firm is selling its product at ₹ 60 per unit. The total cost of production is ₹ 100 and firm is earning total profit of ₹ 500. Later, the total cost increased by 30%. By what percentage the price should be increased to maintained the same profit level.

(*a*) 5 (*b*) 10
(*c*) 15 (*d*) 30

2012

9. An automobile plant contracted to buy shock absorbers from two suppliers X and Y. X supplies 60% and Y supplies 40% of the shock absorbers. All shock absorbers are subjected to a quality test. The ones that pass the quality test are considered reliable. Of X's shock absorbers, 96% are reliable. Of Y's shock absorbers, 72% are reliable.
The probability that a randomly chosen shock absorber, which is found to be reliable, is made by Y is

(*a*) 0.288 (*b*) 0.334
(*c*) 0.667 (*d*) 0.720

2011

10. There are two candidates P and Q in an election. During the campaign, 40% of the voters promised to vote for P, and rest for Q. However, on the day of election 15% of the voters went back on their promise to vote for P and instead voted for Q. 25% of the voters went back on their promise to vote for Q and instead voted for P. Suppose, P lost by 2 votes, then what was the total number of voters?

(*a*) 100 (*b*) 110
(*c*) 90 (*d*) 95

NUMERICAL TYPE QUESTION

2015

1. From a circular sheet of paper of radius 30 cm, a sector of 10% area is removed. If the remaining part is used to make a conical surface, then the ratio of the radius and height of the cone is ____.

ANSWERS

MCQ Type Questions

1. (*a*)	**2.** (*c*)	**3.** (*a*)	**4.** (*a*)	**5.** (*d*)	**6.** (*b*)	**7.** (*d*)	**8.** (*a*)	**9.** (*b*)	**10.** (*a*)

Numerical Type Question

1. (2.06)

EXPLANATIONS

MCQ TYPE QUESTIONS

1. Given,

Total no. of fruits = 5692000

Unripe type of apples = 45% of 15% of 5692000

$$= \frac{45}{100} \times \frac{15}{100} \times 5692000$$

$$= 384210$$

Ripe type of apples $= (100 - 66)\% \times (100 - 15)\% \times 5692000$

$$= \frac{34}{100} \times \frac{85}{100} \times 5692000$$

$$= 1644988$$

$\therefore$ Total no. of apples = 384210 + 1644988

= 2029198

2. The percentage of people moving through a tuberculosis prone zone remains infected but does not show symptoms of disease.

$$= (100 - 30)\% \times 50\% = \frac{70}{100} \times \frac{50}{100} = \frac{35}{100} = 35\%$$

3. (x% of y) + (y% of x)

$$\frac{x}{100} \times y + \frac{y}{100} \times x = \frac{2xy}{100} = 2\% \text{ of } xy$$

$\therefore$ Option (*a*) is correct.

4. $$A = P\left[1 + \frac{r}{100}\right]^t$$

$$10 \text{ million} = 5 \text{ million}\left[1 + \frac{20}{100}\right]^t$$

$$2 = \left[1 + \frac{1}{5}\right]^t$$

$$2 = \left(\frac{6}{5}\right)^t = \log 2 = t \log\left(\frac{6}{5}\right)$$

$$0.301 = t\, 0.0791$$

$$t = \frac{0.301}{0.0791} = 3.8 \text{ years}$$

$\therefore$ Hence around 3–4 years would it take to double the growth rate

5. Let number of people in country y = 100

So, number of people in country x = 300

Total number of people taller than 6ft in both the

$$\text{countries} = 300 \times \frac{1}{100} + 100 \times \frac{2}{100} = 5$$

% of people taller than 6ft in both the countries

$$= \frac{5}{400} \times 100 = 1.25\%.$$

7. Per ₹ 100 final value ₹ 107

$\Rightarrow$ per $\frac{100}{50}$ Dollars final value $\frac{107}{60}$

for 100 dollars ?

$$= \frac{100 \times 50}{100} \times \frac{107}{60} = 89.16$$

Descreased by 11%

9.

	X	Y
Supply	60%	40%
Reliable	96%	72%
Overall	0.576	0.288

$$\therefore \quad P(x) = \frac{0.288}{0.576 + 0.288} = 0.334$$

10.

P	Q
40%	60%
–6%	+6%
+15%	–15%
49%	51%

$\therefore \quad 2\% = 2 \quad \Rightarrow \quad 100\% = 100$

NUMERICAL TYPE QUESTION

1. 90% of area of sheet = Cross sectional area of cone

$\Rightarrow 0.9 \times \pi \times 30 \times 30 = \pi \times r_1 \times 30$

($\because$ Slant height of cone (l) = 30)

$\Rightarrow 27 \text{ cm} = r_1$

$\therefore$ Height of the cone $= \sqrt{30^2 - 27^2} = 13.08$ cm

Hence ratio of radius and height is 2.06.

3 CHAPTER

Time and Work

MCQ TYPE QUESTIONS

2016

1. P, Q, R and S are working on a project. Q can finish the task in 25 days, working alone for 12 hours a day. R can finish the task in 50 days, working alone for 12 hours per day. Q worked 12 hours a day but took sick leave in the beginning for two days. R worked 18 hours a day on all days. What is the ratio of work done by Q and R after 7 days from the start of the project?

(*a*) 10:11 (*b*) 11:10
(*c*) 20:21 (*d*) 21:20

2. S, M, E and F are working in shifts in a team to finish a project. M works with twice the efficiency of others but for half as many days as E worked. S and M have 6 hour shifts in a day whereas E and F have 12 hours shifts. What is the ratio of contribution of M to contribution of E in the project?

(*a*) 1 : 1 (*b*) 1 : 2
(*c*) 1 : 4 (*d*) 2 : 1

3. It takes 10 s and 15 s, respectively, for two trains travelling at different constant speeds to completely pass a telegraph post. The length of the first train is 120 m and that of the second train is 150 m. The magnitude of the difference in the speeds of the two trains (in m/s) is ______.

(*a*) 2.0 (*b*) 10.0
(*c*) 12.0 (*d*) 22.0

4. The velocity V of a vehicle along a straight line is measured in m/s and plotted as shown with respect to time in seconds. At the end of the 7 seconds, how much will the odometer reading increase by (in m) ?

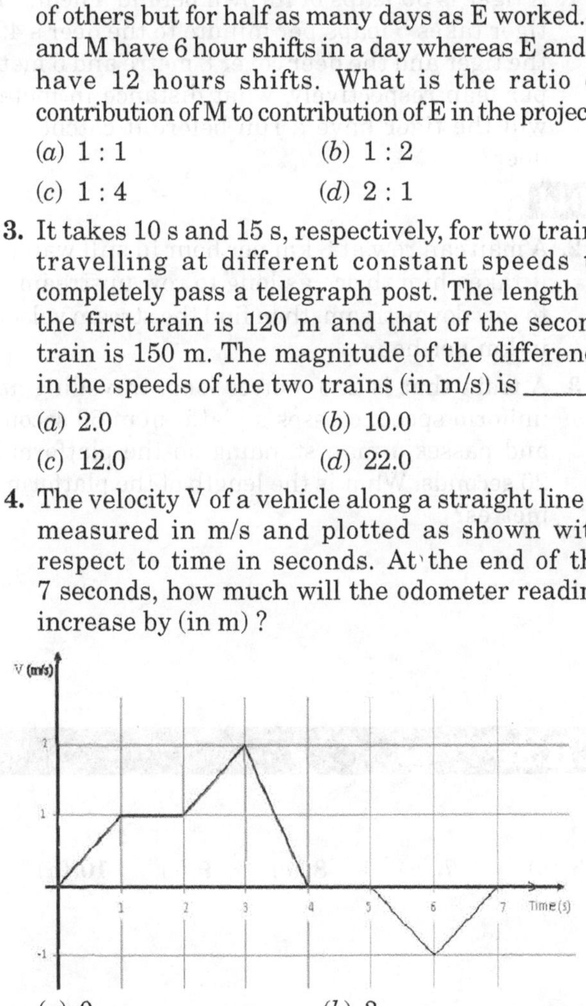

(*a*) 0 (*b*) 3
(*c*) 4 (*d*) 5

5. Ananth takes 6 hours and Bharath takes 4 hours to read a book. Both started reading copies of the book at the same time. After how many hours is the number of pages **to be** read by Ananth, twice that **to be** read by Bharath? Assume Ananth and Bharath read all the pages with constant pace.

(*a*) 1 (*b*) 2
(*c*) 3 (*d*) 4

2015

6. An electric bus has onboard instruments that report the total electricity consumed since the start of the trip as well as the total distance covered. During a single day of operation, the bus travels on stretches *M*, *N*, *O* and *P*, in that order. The cumulative distances travelled and the corresponding electricity consumption are shown in the table below.

Stretch	Comulative distance(km)	Electricity used (kWh)
M	20	12
N	45	25
O	75	45
P	100	57

The stretch where the electricity consumption per km is minimum is

(*a*) M
(*b*) N
(*c*) O
(*d*) P

2014

7. It takes 30 minutes to empty a half-full tank by draining it at a constant rate. It is decided to simultaneously pump water into the half-full tank while draining it. What is the rate at which water has to be pumped in so that it gets fully filled in 10 minutes?

(*a*) 4 times the draining rate
(*b*) 3 times the draining rate
(*c*) 2.5 times the draining rate
(*d*) 2 times the draining rate

2013

8. A tourist covers half of his journey by train at 60 km/h, half of the remainder by bus at 30 km/h and the rest by cycle at 10 km/h. The average speed of the tourist in km/h during his entire journey is

(*a*) 36 (*b*) 30
(*c*) 24 (*d*) 18

9. A car travels 8 km in the first quarter of an hour, 6 km in the second quarter and 16 km in the third quarter. The average speed of the car in km per hour over the entire journey is

(*a*) 30 (*b*) 36
(*c*) 40 (*d*) 24

2011

10. The fuel consumed by a motorcycle during a journey while traveling at various speeds is indicated in the graph below.

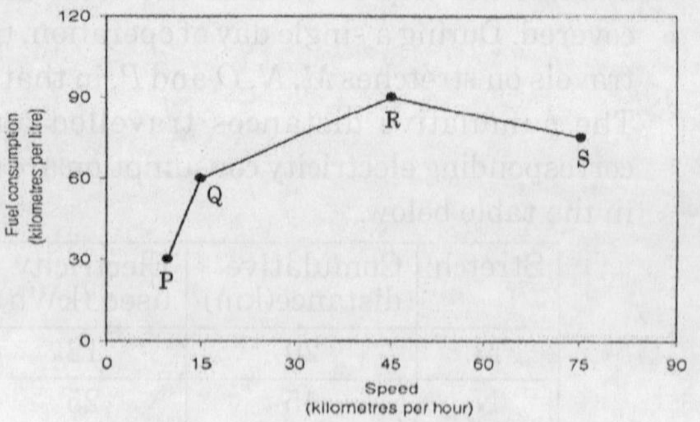

The distances covered during four laps of the journey are listed in the table below

Lap	Distance (kilometers)	Average speed (kilometers per hour)
P	15	15
Q	75	45
R	40	75
S	10	10

From the given data, we can conclude that the fuel consumed per kilometre was least during the lap

(*a*) P
(*b*) Q
(*c*) R
(*d*) S

2010

11. 5 skilled workers can build a wall in 20 days; 8 semi-skilled workers can build a wall in 25 days; 10 unskilled workers can build a wall in 30 days. If a team has 2 skilled, 6 semi-skilled and 5 unskilled workers, how long will it take to build the wall?

(*a*) 20 days
(*b*) 18 days
(*c*) 16 days
(*d*) 15 days

NUMERICAL TYPE QUESTIONS

2015

1. A tiger is 50 leaps of its own behind a deer. The tiger takes 5 leaps per minute to the deer's 4. If the tiger and the deer cover 8 metre and 5 metre per leap respectively, what distance in meters will the tiger have a run before it catches the deer?

2014

2. A man can row at 8 km per hour in still water. If it takes him thrice as long to row upstream, as to row downstream, then find the stream velocity in km per hour.

3. A train that is 280 metres long, travelling at a uniform speed, crosses a platform in 60 seconds and passes a man standing on the platform in 20 seconds. What is the length of the platform in metres?

ANSWERS

MCQ Type Questions

1. (*c*) **2.** (*b*) **3.** (*a*) **4.** (*d*) **5.** (*c*) **6.** (*b*) **7.** (*a*) **8.** (*c*) **9.** (*c*) **10.** (*a*)
11. (*d*)

Numerical Type Questions

1. (800) **2.** (4) **3.** (560 to 560)

EXPLANATIONS

MCQ TYPE QUESTIONS

1. Q can finish 1 day's 1 hour's work

$$= \frac{1}{25 \times 12} = \frac{1}{300}$$

R can finish 1 day's 1 hour's work

$$= \frac{1}{50 \times 12}$$

$$= \frac{1}{600}$$

Now Q working hours

$$= (7 - 2) \times 12$$

$$= 60 \text{ hr}$$

$\therefore$ R working hours = 7 × 18 = 126

After 7 days, the ratio of work done by Q & R is

$$Q : R$$

$$\frac{60}{300} : \frac{126}{600}$$

$$Q : R = 120 : 126$$

$$= 20 : 21$$

2. M is twice as efficient as E but worked for half as many days. So in this case they will do equal work if their shifts had same timings. But M's shift is for 6 hours, while E's shift for 12 hours. Hence, E will do twice the work as M.

Ratio of contribution of M : E in work is 1 : 2.

3. Here the speed of first train (S_1) = $\frac{120}{10}$ = 12 m/s

and the speed of second train (S_2) = $\frac{150}{15}$ = 10 m/s

$\therefore$ The magnitude of difference in the speeds of two trains = $(S_1 - S_2)$ = (12 – 10) m/s = 2 m/s

4. From the given figure.

The odometer reading increases from starting point to end point.

Odometer reading = Area of the given diagram

Area of the velocity and time graph per second

1st sec $\Rightarrow$ triangle

$$= \frac{1}{2} \times 1 \times 1 = \frac{1}{2}$$

2nd sec $\Rightarrow$ square

$$= 1 \times 1 = 1$$

3rd sec $\Rightarrow$ square + triangle

$$= 1 \times 1 + \frac{1}{2} \times 1 \times 1 = 1\frac{1}{2}$$

4th sec $\Rightarrow$ triangle

$$= \frac{1}{2} \times 1 \times 2 = 1$$

5th sec $\Rightarrow$ straight line

$$= 0$$

6th sec $\Rightarrow$ triangle

$$= \frac{1}{2} \times 1 \times 1 = \frac{1}{2}$$

7th sec $\Rightarrow$ triangle

$$= \frac{1}{2} \times 1 \times 1 = \frac{1}{2}$$

Total Odometer reading at 7 seconds

$$= \left(\frac{1}{2} + 1 + 1\frac{1}{2} + 1 + 0 + \frac{1}{2} + \frac{1}{2}\right) \text{m}$$

$$= 5 \text{ m}$$

So, the odometer reading is increased by 5 m.

6. For $M \Rightarrow \frac{12}{20} = 0.6$

$$N \Rightarrow \frac{25}{45} = 0.55$$

$$O \Rightarrow \frac{45}{75} = 0.6$$

$$P \Rightarrow \frac{57}{100} = 0.57$$

7. $V_{half} = 30(s)$ drawing rate = s

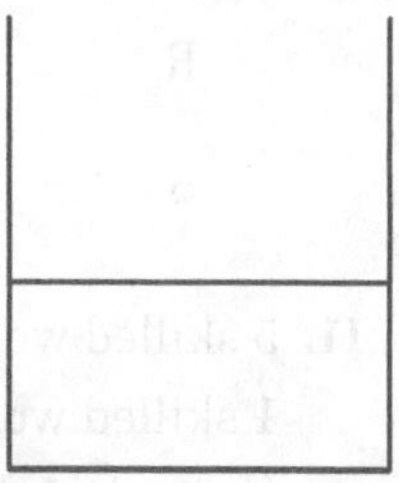

Total volume = 60 S tank

$(s^1)(10) - (s)10 = 30s$

$s^1(s) - s = 3s$

$s1 = 4s$

$s^1 = 4$ drawing rate

8. Given: A tourist covers half the journey at 60 km/h, one fourth at 30 km/h and the remaining one fourth at 10 km/h

To find: Avearge speed of the whole journey.

Analysis: One approach to this is to find individual time consumed in each of the three parts and use that value to get total time, and find average speed.

So,

Lets assume total length of journey = x km

$\therefore$ First phase at 60 km/h

$$\text{Time } t_1 = \frac{\text{Distance}}{\text{Speed}} = \frac{x}{2}\frac{1}{60} = \frac{x}{120} h$$

Second phase at 30 km/h

$$\text{Distance} = \frac{x}{4}\text{km}$$

$$\text{Time } t_2 = \frac{x}{4.30} = \frac{x}{120}$$

Third phase at 10 km/h

$$\text{Distance} = \frac{x}{4} \text{ km}$$

$$\text{Time } t_3 = \frac{x}{4.10} = \frac{x}{40}h$$

$$\therefore \quad \text{Total time} = \frac{x}{120} + \frac{x}{120} + \frac{x}{40} = \frac{x}{40}\left(\frac{1}{3}+\frac{1}{3}+1\right)$$

$$= \frac{x}{40}\frac{(1+1+3)}{3} = \frac{5x}{120}$$

$$\therefore \text{Average speed} = \frac{\text{Total Distance}}{\text{Total time}}$$

$$= \frac{x \cdot 120}{5x} = 24 \text{ km/}h$$

Hence, the answer is (c)

9. $\text{Average Speed} = \dfrac{\text{Total distance}}{\text{total time taken}}$

$$\text{Total Distance} = \frac{8+6+16}{15+15+15} = \frac{30 \times 60}{45}$$

= 40km/h.

10.	Fuel consumption	Actual
P	60 km/l	$\frac{15}{60} = \frac{1}{4}l$
Q	90 km/l	$\frac{75}{90} = \frac{5}{6}l$
R	75 km/l	$\frac{40}{75} = \frac{8}{15}l$
S	30 km/l	$\frac{10}{30} = \frac{1}{3}l$

11. 5 skilled workers build wall in 20 days

1 skilled worker build wall in 20 × 5 days

Hence in 1 day, part of work done by skilled work

$$= \frac{1}{100}$$

Similarly in 1 day part of work done by semi-skilled workers

$$= \frac{1}{25 \times 8}$$

and in 1 day part of work done by un-skilled worker

$$= \frac{1}{30 \times 10}$$

So part of work done in 1 day by 2 skilled, 6 semi-skilled and 5 unskilled

$$= \frac{2}{100} + \frac{6}{200} + \frac{5}{300} = \frac{1}{15}$$

So work done by given workers in days = 15

NUMERICAL TYPE QUESTIONS

1. Tiger – 1 leap ⇒ 8 meter

Speed = 5 leap/hr = 40m/min

Deer → 1 leap = 5 meter

speed = 4hr = 20m/min

Let at time 't' the tiger catches the deer.

∴ Distance travelled by deer + initial distance between them

50 × 8 ⇒ 400m = distance covered by tiger.

⇒ 40 × t = 400 + 20t

$$\Rightarrow t = \frac{400}{200} = 20 \text{ min}$$

⇒ total distance ⇒ 400 + 20 × t = 800 m

2. Speed of man = 8;

Left distance = d

$$\text{Time taken} = \frac{d}{8}$$

Upstream: Speed of stream = s

⇒ speed upstream = S' = (8 – s)

$$t' = \left(\frac{d}{8-s}\right)$$

Downstream:

$$\text{Given speed downstream} = t'' = \frac{d}{8+s}$$

$$\Rightarrow 3t' = t'' \Rightarrow \frac{3d}{8-s} = \frac{d}{8+s}$$

$$\Rightarrow \frac{3d}{8-s} = \frac{d}{8+s}$$

$$\Rightarrow \quad s = 4 \text{ km/hr}$$

3. Let the length of platform = x.m

and train length = 280 m (given)

According to Question

$$\frac{x+280}{60} = \frac{280}{20}$$

$$\therefore \quad x = 560 \text{ m.}$$

16. Take number of female students in 2011 = 100

$\therefore$ Number of male in 2011 = 100

No. of female in 2012 = 100

No. of male in 2012 = 150

$$\text{Ratio} = \frac{150}{100}$$

17. All multiples of 10 from 2 to 198

$= 10, 20, \ldots\ldots\ldots\ldots 190$

Now $a_n = a + (n-1)d$

$$190 = 10 + (n-1) \times 10$$

$$180 = (n-1) \times 10$$

$$18 = n - 1$$

$$\therefore \quad n = 19$$

Hence total number of multiples are 19.

Now $S_n = \frac{n}{2}[2a + (n-1)d]$

$$= \frac{19}{2}[20 + (18)(10)]$$

$$= \frac{19}{2}[20 + 180] = 1900$$

Now average of all multiples of 10 from 2 to198

$$= \frac{10 + 20 + \ldots + 190}{19}$$

$$\therefore \quad = \frac{1900}{19} = 100$$

18. $\sqrt{12 + \sqrt{12 + \sqrt{12 + }}}\ldots$

For a particular type of question

i.e. $\sqrt{a + \sqrt{a + \sqrt{a + }}}\ldots$ we find a two consecutive number whose product is a. The greater number among the two consecutive number is the answer.

Here $a = 12$

i.e. $12 = 3 \times 4$
↓
greater number

Hence $\sqrt{12 + \sqrt{12 + \sqrt{12 + }}}\ldots = 4$

19. Here it is given that x is real

Now $|x^2 - 2x + 3| = 11$

$\Rightarrow$ $x^2 - 2x + 3 = 11$ or $x^2 - 2x + 3 = -11$

For $x^2 - 2x + 3 = 11 \Rightarrow x^2 - 2x - 8 = 0$

$$b^2 - 4ac = 4 - 4(1)(-8) = 36 > 0$$

Hence roots are real.

and for $x^2 - 2x + 14 = 0$

$$b^2 - 4ac = 4 - 4(1)(14) < 0$$

Hence roots are imaginary

$\therefore$ The value of x is the root of $x^2 - 2x - 8 = 0$

Now $x = \frac{+2 \pm \sqrt{4 - 4(1)(-8)}}{2}$

$$= \frac{2 \pm \sqrt{36}}{2}$$

$$= \frac{2 \pm 6}{2}$$

$\therefore$ $x = 4, -2.$

Now for $x = 4$,

$$|-x^3 + x^2 - x| = |-64 + 16 - 4| = 52.$$

and for $x = -2$

$$|-x^3 + x^2 - x| = |+8 + 4 + 2| = 14$$

Hence possible values of $|-x^3 + x^2 - x| = 14, 52$

22. $0 = 44 + (n-1)(-2)$

$$n - 1 = 22$$

$$n = 23$$

Sum $= \frac{n}{2}(a + l)$

$$= \frac{23}{2}(44 + 0)$$

$$= 23 \times 22 = 506$$

Hence (c) is the answer

23. Let 'd' be the # of days required

Let 'h' be the daily working hours.

$\therefore$ dh can be taken as a measure of the total work be done in order to erect the structure.

Now working hours per day $h' = \frac{23}{24}h$

$\therefore$ # of days $d' = \frac{dh}{h'} = \frac{dh}{\left(\frac{23}{24}h\right)} = \frac{24}{23}d$

Let c = daily wages earlier

$$\therefore \quad c = \frac{13,200}{d}$$

Now daily wages $c' = \frac{6}{5}c$

$\therefore$ Total new cost $= c' \times d'$

$$= \frac{6}{5} \times \frac{13200}{d} \times \frac{24}{23}d \approx 16,500$$

4

CHAPTER

Ratio, Proportion and Mixtures

MCQ TYPE QUESTIONS

2015

1. A cube of side 3 units is formed using a set of smaller cubes of side 1 unit. Find the proportion of the number of faces of the smaller cubes visible to those which are NOT visible.

(*a*) 1 : 4

(*b*) 1 : 3

(*c*) 1 : 2

(*d*) 2 : 3

2011

2. A container originally contains 10 litres of pure spirit. From this container 1 litre of spirit is replaced with 1 litre of water. Subsequently, 1 litre of the mixture is again replaced with 1 litre of water and this process is repeated one more time. How much spirit is now left in the container?

(*a*) 7.58 litres

(*b*) 7.84 litres

(*c*) 7 litres

(*d*) 7.29 litres

ANSWERS

MCQ Type Questions

1. (*c*) **2.** (*d*)

EXPLANATIONS

MCQ TYPE QUESTIONS

1.

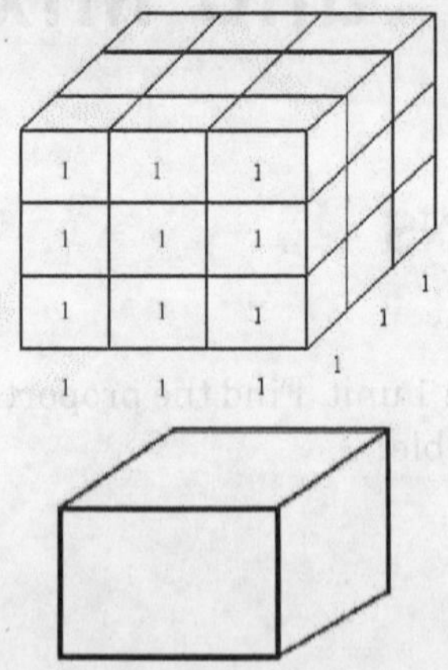

Number of faces per cube = 6

Total number of cubes = 9 × 3 = 27

$\therefore$ Total number of faces = 27 × 6 = 162

$\therefore$ Total number of non visible faces = 162 – 54 = 108

$$\therefore \frac{\text{Number of visible faces}}{\text{Number of non visible faces}} = \frac{54}{108} = \frac{1}{2}$$

2. $10\left(\frac{10-1}{10}\right)^3 = 10\left(\frac{9}{10}\right)^3 = \frac{729}{1000}$

$$\therefore \frac{729}{1000} \times 1 = 7.29 \text{ litres}$$

5 CHAPTER Permutations and Combinations & Probability

MCQ TYPE QUESTIONS

2016

1. Shaquille O' Neal is a 60% career free throw shooter, meaning that he successfully makes 60 free throws out of 100 attempts on average. What is the probability that he will successfully make exactly 6 free throws in 10 attempts?

(*a*) 0.2508 (*b*) 0.2816

(*c*) 0.2934 (*d*) 0.6000

2015

2. Four cards are randomly selected from a pack of 52 cards. If the first two cards are kings, what is the probability that the third card is a king?

(*a*) 4/52 (*b*) 2/50

(*c*) 1/52 × (1/52) (*d*) 1/52 × (1/52) × (1/50)

3. Five teams have to compete in a league, with every team playing every other team exactly once, before going to the next round. How many matches will have to be held complete the league round of matches?

(*a*) 20 (*b*) 10

(*c*) 8 (*d*) 5

4. Right triangle PQR is to be constructed in the xy – plane so that the right angle is at P and line PR is parallel to the-axis. The x and y coordinates of P, Q, and R are to be integers that satisfy the inequalities: $-4 \le x \le 5$ and $6 \le y \le 16$. How many different triangles could be constructed with these properties?

(*a*) 110 (*b*) 1,100

(*c*) 9,900 (*d*) 10,000

5. A coin is tossed thrice. Let X be the event that head occurs in each of the first two tosses. Let Y be the event that a tail occurs on the third toss. Let Z be the event that two tails occurs in three tosses. Based on the above information, which one of the following statements is TRUE?

(*a*) X and Y are not independent

(*b*) Y and Z are dependent

(*c*) Y and Z are independent

(*d*) X and Z independent

6. Ram and Ramesh appeared in an interview for two vacancies in the same department. The probability of Ram's selection is 1/6 and that of Ramesh is 1/8. What is the probability that only one of them will be selected?

(*a*) $\frac{47}{48}$ (*b*) $\frac{1}{4}$

(*c*) $\frac{13}{48}$ (*d*) $\frac{35}{48}$

7. Given set A = {2, 3, 4, 5} and Set B = {11, 12, 13, 14, 15}, two numbers are randomly selected, one from each set. What is probability that the sum of the two numbers equals 16 ?

(*a*) 0.20 (*b*) 0.25

(*c*) 0.30 (*d*) 0.33

8. The probabilities that a student passes in Mathematics, Physics and Chemistry are m, p and c respectively. Of these subjects, the student has 75% chance of passing in at least one, a 50% chance of passing in at least two and a 40% chance of passing in exactly two. Following relations are drawn in $m, p.c$:

I. $p + m + c = \frac{27}{20}$ II. $p + m + c = \frac{13}{20}$

III. $(p) \times (m) \times (c) = \frac{1}{10}$

(*a*) Only relation I is true.

(*b*) Only relation II is true.

(*c*) Relations II and III are true.

(*d*) Relations I and III are true.

2014

9. A five digit number is formed using the digits 1,3,5,7 and 9 without repeating any of them. What is the sum of all such possible five digit numbers?

(*a*) 6666660 (*b*) 6666600

(*c*) 6666666 (*d*) 6666606

10. You are given three coins: one has heads on both faces, the second has tails on both faces, and the third has a head on one face and a tail on the other. You choose a coin at random and toss it, and it comes up heads. The probability that the other face is tails is

(*a*) 1/4 (*b*) 1/3

(*c*) 1/2 (*d*) 2/3

2013

11. Out of all the 2-digit integers between 1 and 100, a 2-digit number has to be selected at random. What is the probability that the selected number is not divisible by 7?
   (*a*) 13/90 (*b*) 12/90
   (*c*) 78/90 (*d*) 77/90

12. What is the chance that a leap year, selected at random, will contain 53 Saturdays?
   (*a*) 2/7 (*b*) 3/7
   (*c*) 1/7 (*d*) 5/7

13. In a factory, two machines M1 and M2 manufacture 60% and 40% of the autocomponents respectively. Out of the total production, 2% of M1 and 3% of M2 are found to be defective. If a randomly drawn autocomponent from the combined lot is found defective, what is the probability that it was manufactured by M2?
   (*a*) 0.35 (*b*) 0.45
   (*c*) 0.5 (*d*) 0.4

2012

14. A and B are friends. They decide to meet between 1 PM and 2 PM on a given day. There is a condition that whoever arrives first will not wait for the other for more than 15 minutes. The probability that they will meet on that day is
   (*a*) $\frac{1}{4}$ (*b*) $\frac{1}{16}$
   (*c*) $\frac{7}{16}$ (*d*) $\frac{9}{16}$

2010

15. Given digits 2, 2, 3, 3, 3, 4, 4, 4, 4 how many distinct 4 digit numbers greater than 3000 can be formed?
   (*a*) 50 (*b*) 51
   (*c*) 52 (*d*) 54

NUMERICAL TYPE QUESTIONS

2015

1. How many four digit numbers can be formed with the 10 digits 0, 1, 2, ... 9 if no number can start with 0 and if repetitions are not allowed?

2014

2. In any given year, the probability of an earthquake greater than Magnitude 6 occurring in the Garhwal Himalayas is 0.04. The average time between successive occurrences of such earthquakes is _____ years.

3. 10% of the population in a town is HIV$^+$. A new diagnostic kit for HIV detection is available; this kit correctly identifies HIV$^+$ individuals 95% of the time, and HIV$^-$ individuals 89% of the time. A particular patient is tested using this kit and is found to be positive. The probability that the individual is actually positive is _____

4. A batch of one hundred bulbs is inspected by testing four randomly chosen bulbs. The batch is rejected if even one of the bulbs is defective. A batch typically has five defective bulbs. The probability that the current batch is accepted is _____

ANSWERS

MCQ Type Questions

1. (*a*)	**2.** (*b*)	**3.** (*b*)	**4.** (*c*)	**5.** (*d*)	**6.** (*b*)	**7.** (*a*)	**8.** (*d*)	**9.** (*b*)	**10.** (*b*)
11. (*d*)	**12.** (*a*)	**13.** (*c*)	**14.** (*c*)	**15.** (*b*)					

Numerical Type Questions

1. 4536 **2.** (25 to 25) **3.** (0.47 to 0.48) **4.** (0.8145)

EXPLANATIONS

MCQ TYPE QUESTIONS

1. The question clearly explains that shaquille makes 60 free throws out of 100

Hence, Probability of free throw = $\frac{60}{100} = 0.6$

And probability of NOT free throw

$= 1 - 0.6 = 0.4$

So required probability of exactly 6 throws in 10 attempts will be given by

$^{10}C_6\, 0.6^6 \times 0.4^4 = 0.2508$

2. There are 4 kings in a pack of 52 cards.

If 2 cards are selected and both are kings, remaining cards will be 50 out of which 2 will be kings.

3. For a match to be played, we need 2 teams

No. of matches = No. of ways of selections 2 teams out of 5

$= {}^5C_2 = 10$

5. $x = \{HHT, HHH\}$

y depends on x

$z = \{TTH, TTT\}$

$\therefore$ '*d*' is the correct choice.

6. P (Ram) = 1/6; p(Ramesh) = 1/8

p(only at) = p (Ram) × p (not ramesh) + p(Ramesh) × p(n_0 × R_{am})

$= \frac{1}{6} + \frac{7}{8} + \frac{1}{8} \times \frac{5}{6}$

$\Rightarrow \frac{12}{40} = \frac{1}{4}$

7. Here the given sets are

$A = \{2, 3, 4, 5\}$

and $B = \{11, 12, 13, 14, 15\}$

desired outcome = {(2, 14), (3, 13), (4, 12), (5, 11)}

$\therefore \quad n(E) = 4$

and total outcome = 4 × 5 = 20

$\therefore \quad n(s) = 20$

$\therefore$ Probability of an event

$= \frac{n(E)}{n(s)}$

$= \frac{4}{20} = \frac{1}{5} = 0.2$

9. The digit in unit place is selected in 4! Ways

The digit in tens place is selected in 4! Ways

The digit in hundreds place is selected in 4! Ways

The digit in thousands place is selected in 4! Ways

The digit in ten thousands place is selected in 4! Ways

Sum of all values for 1

$4! \times 1 \times (10^0 + 10^1 + 10^2 + 10^3 + 10^4)$

$= 4! \times 11111 \times 1$

Similarly for '3' $4! \times (11111) \times 3$

Similarly for '5' $4! \times (11111) \times 5$

Similarly for '7' $4! \times (11111) \times 7$

Similarly for '9' $4! \times (11111) \times 9$

$\therefore$ sum of all such numbers

$4! \times (11111) \times (1 + 3 + 5 + 7 + 9)$

$= 24 \times (11111) \times 25$

$= 6666600$

10. The probability that the other face is tail is $\frac{1}{3}$.

11. Given: All the 2-digit numbers

To find: Probability of selecting a number at random not divisible by 7

Analysis: Total number of 2-digit numbers

10, 11, ..., 99 → 90 numbers

Unfavourable numbers *i.e.* numbers divisible by 7 : 14, 21, ..., 98

Number of unfavourable numbers = 13

This can be hand counted or we can use the following properties:

Last term = First time + (no. – 1) Diff

98 = 14 + (no. – 1) 7

$\Rightarrow$ no. = 13

$\Rightarrow$ Probability = $\frac{\text{Total} - \text{unfavourable}}{\text{Total}}$

$= \frac{90 - 13}{90} = 77/90$

Hence, the solution is (*d*)

12. at a leap year

52 weeks, and 2 extra day they are (Mon, Tues) (Tues, Wed) (Wed Thu.) (Thus. Fri) (Fri. Sat) (Sat. Sun) (Sun Mond.)

$n(s) = 7$

$n(E) = 2 \quad P(E) = \frac{2}{7}$

14.

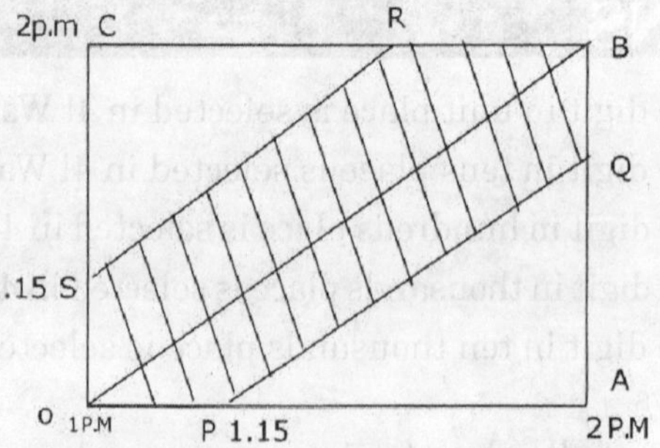

OB is the line when both A and B arrive at same time.

Total sample sapce = 60 × 60 = 3600

Favourable cases = Area of OABC – 2(Area of SRC)

$$= 3600 - 2 \times \left(\frac{1}{2} \times 45 \times 45\right)$$

$$= 1575$$

$$\therefore \text{ Required probability } = \frac{1575}{3600} = \frac{7}{16}$$

15. 2, 2, 3, 3, 4, 4, 4, 4 $\quad x > 3000$

322 — 3, 4
323 — 2, 3, 4
324 — 2, 3, 4
332 — 2, 3, 4
333 — 2, 4
334 — 2, 3, 4
342 — 2, 3, 4
343 — 2, 3, 4
344 — 2, 3, 4
422 — 3, 4
423 — 2, 3, 4
424 — 2, 3, 4
432 — 2, 3, 4
433 — 2, 3, 4
434 — 2, 3, 4
442 — 2, 3, 4
443 — 2, 3, 4
444 — 2, 3, 4

NUMERICAL TYPE QUESTIONS

1. In thousands place, 9 digits except 0 can be placed

In hundreds place, 9 digits can be placed (including 0, excluding the one used in thousands place)

In tens place, 8 digits can be placed (excluding the ones used in thousands and hundreds place)

In ones place, 7 digits can be placed (excluding the one used in thousands, hundreds and tens place)

Total number of combinations

$$= 9 \times 9 \times 8 \times 7 = 4536$$

2. Given, in one year, probability of an earthquake > 6 magnitude = 0.04

So, average time between successive earthquakes = 1/0.04 = 25 years

3. Given, 10% of the population is HIV^+, so probability = 0.1 for HIV^+ and 0.9 for HIV^-

The diagnostic kit identifies → 0.95 HIV^+ and 0.89 HIV^- correctly

So, the required probability that the person is actually HIV^+ = 0.95/2 = 0.475

4. Probability for one bulb to be non-defective is $\frac{95}{100}$

∴ Probabilities that none of the bulbs is defectives $\left(\frac{95}{100}\right)^4 = 0.8145$

■■

6
CHAPTER

Miscellaneous

MCQ TYPE QUESTIONS

2016

1. In a 2 × 4 rectangle grid shown below, each cell is a rectangle. How many rectangles can be observed in the grid?

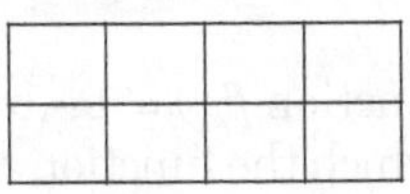

(*a*) 21 (*b*) 27

(*c*) 30 (*d*) 36

2.

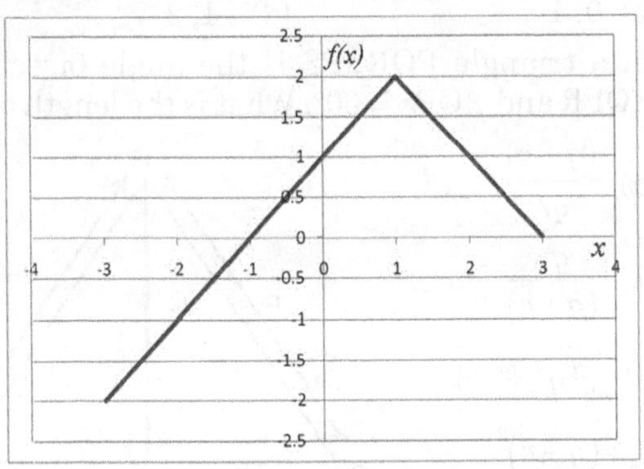

Choose the correct expression for $f(x)$ given in the graph.

(*a*) $f(x) = 1 - |x - 1|$ (*b*) $f(x) = 1 + |x - 1|$

(*c*) $f(x) = 2 - |x - 1|$ (*d*) $f(x) = 2 + |x - 1|$

3. The volume of a sphere of diameter 1 unit is ________ than the volume of a cube of side 1 unit.

(*a*) least (*b*) less

(*c*) lesser (*d*) low

4. A window is made up of a square portion and an equilateral triangle portion above it. The base of the triangular portion coincides with the upper side of the square. If the perimeter of the window is 6 m, the area of the window in m² is ________.

(*a*) 1.43 (*b*) 2.06

(*c*) 2.68 (*d*) 2.88

5. The binary operation □ is defined as $a \square b = ab + (a + b)$, where a and b are any two real numbers. The value of the identity element of this operation, defined as the number x such that $a \square x = a$, for any a, is ______.

(*a*) 0 (*b*) 1

(*c*) 2 (*d*) 10

6. Which of the following curves represents the function $y = \ln\left(\left|e^{[|\sin(|x|)|]}\right|\right)$ for $|x| < 2\pi$?

Here, x represents the abscissa and y represents the ordinate.

(*a*)

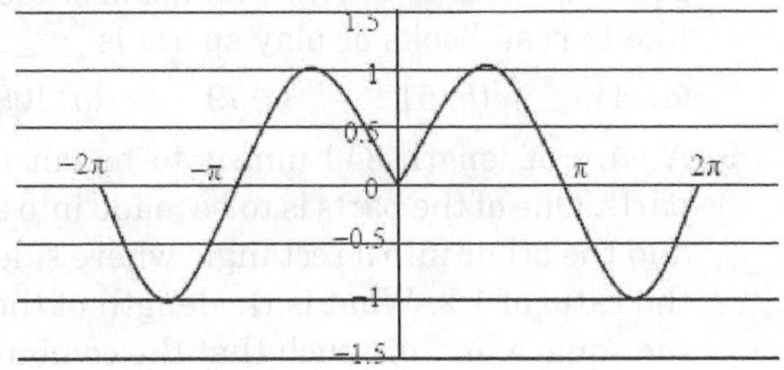

(*b*)

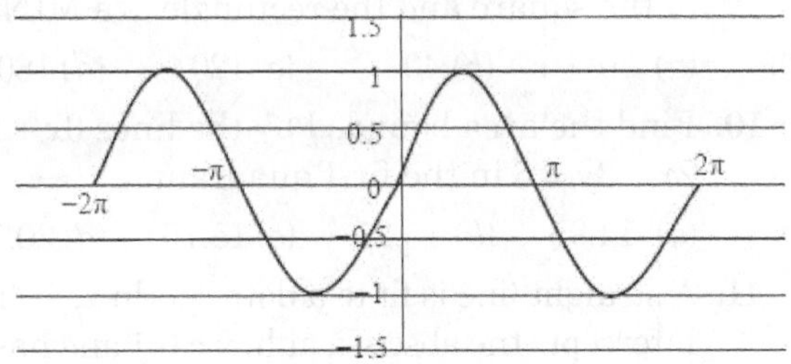

(*c*)

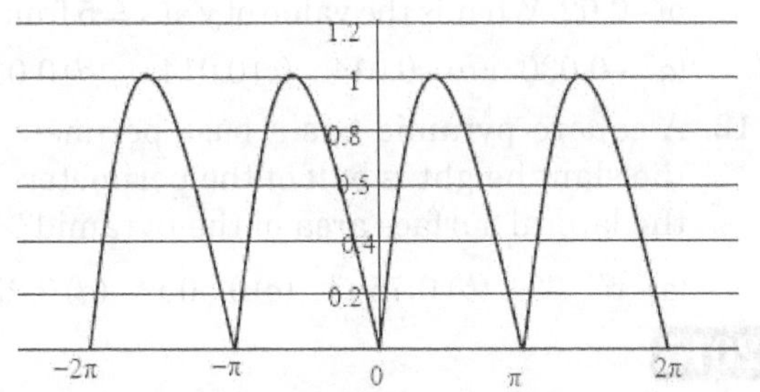

(*d*)

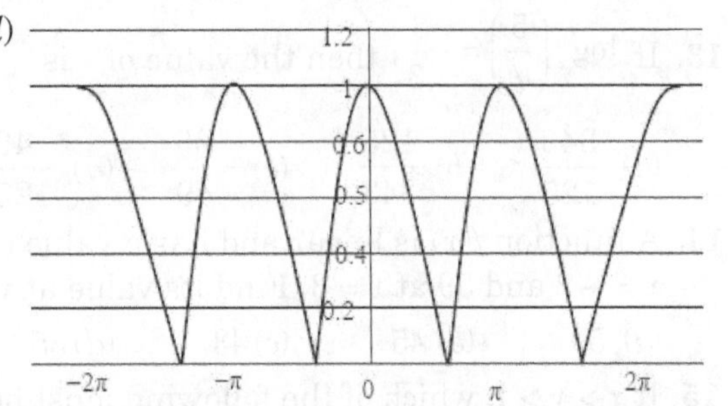

7. Given (9 inches)$^{1/2}$ = (0.25 yards)$^{1/2}$, which one of the following statements is TRUE?

(*a*) 3 inches = 0.5 yards

(*b*) 9 inches = 1.5 yards

(*c*) 9 inches = 0.25 yards

(*d*) 81 inches = 0.0625 yards

8. The Venn diagram shows the preference of the student population for leisure activities.

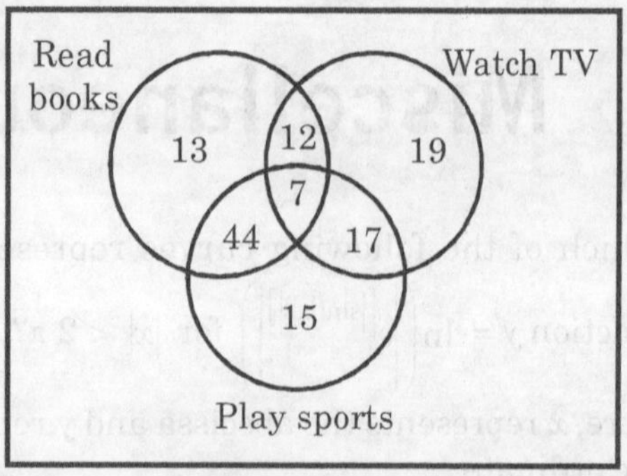

From the data given, the number of students who like to read books or play sports is ____.

(*a*) 44 (*b*) 51 (*c*) 79 (*d*) 108

9. A wire of length 340 mm is to be cut into two parts. One of the parts is to be made into a square and the other into a rectangle where sides are in the ratio of 1:2. What is the length of the side of the square (in mm) such that the combined area of the square and the rectangle is a MINIMUM?

(*a*) 30 (*b*) 40 (*c*) 120 (*d*) 180

10. Find the area bounded by the lines $3x + 2y = 14$, $2x - 3y = 5$ in the first quadrant

(*a*) 14.95 (*b*) 15.25 (*c*) 15.70 (*d*) 20.35

11. A straight line is fit to a data set ($\ln x, y$). This line intercepts the abscissa at $\ln x = 0.1$ and has a slope of -0.02. What is the value of y at $x = 5$ from the fit?

(*a*) -0.030 (*b*) -0.014 (*c*) 0.014 (*d*) 0.030

12. A square pyramid has a base perimeter x, and the slant height is half of the perimeter. What is the lateral surface area of the pyramid?

(*a*) x^2 (*b*) $0.75\,x^2$ (*c*) $0.50\,x^2$ (*d*) $0.25\,x^2$

2015

13. If $\log_x\left(\frac{5}{7}\right) = -\frac{1}{3}$, then the value of x is

(*a*) $\frac{343}{125}$ (*b*) $\frac{125}{343}$ (*c*) $-\frac{25}{49}$ (*d*) $-\frac{49}{25}$

14. A function $f(x)$ is linear and has a value of 29 at $x = -2$ and 39 at $x = 3$. Find its value at $x = 5$.

(*a*) 59 (*b*) 45 (*c*) 43 (*d*) 35

15. If $x > y > 1$, which of the following must be true?

(*i*) $\ln x > \ln y$ (*ii*) $e^x > e^y$

(*iii*) $y^x > x^y$ (*iv*) $\cos x > \cos y$

(*a*) *i* and *ii* (*b*) *i* and *iii*

(*c*) *iii* and *iv* (*d*) *ii* and *iv*

16. log tan1° + log tan2° +-------+ log tan89° is ____.

(*a*) 1 (*b*) $1/\sqrt{2}$ (*c*) 0 (*d*) -1

17. Based on the given statements, select the most appropriate option to solve the given question.

What will be the total weight of 10 poles each of same weight?

Statements:

(I) One fourth of the weight of a pole is 5 kg

(II) The total weight of these poles is 160 kg more than the total weight of two poles.

(*a*) Statement II alone is not sufficient

(*b*) Statement II alone is not sufficient

(*c*) Either I or II alone is sufficient

(*d*) Both statements I and II together are not sufficient.

18. Consider a function $f(x) = 1 - |x|$ on $-1 \le x \le 1$. The value of x at which the function attains a maximum, and the maximum value of function are:

(*a*) 0, -1 (*b*) -1, 0

(*c*) 0, 1 (*d*) -1, 2

19. In a triangle PQR, PS is the angle bisector of ∠QPR and ∠QPS = 60°. What is the length of PS?

(*a*) $\frac{(q+r)}{qr}$

(*b*) $\frac{qr}{(q+r)}$

(*c*) $\sqrt{(q^2+r^2)}$

(*d*) $\frac{(q+r)^2}{qr}$

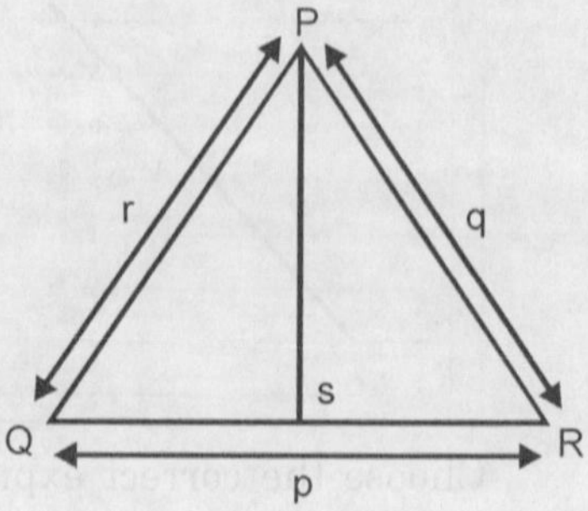

20. Based on the given statements, select the most appropriate option to solve the given question.

If two floors in a certain building are 9 feet apart, how many steps are there in a set of stairs that extends from the first floor to the second floor of the building?

Statements:

I. Each step is ¾ foot high.

II. Each step is 1 foot wide.

(*a*) Statement I alone is sufficient, but statement II alone is not sufficient.

(*b*) Statement II alone is sufficient, but statementI alone is not sufficient.

(*c*) Both statements together are sufficient, but neither statement alone is sufficient.

(*d*) Statement I and II together are not sufficient.

2014

21. If $y = 5x^2 + 3$, then the tangent at $x = 0$, $y = 3$

(*a*) passes through $x = 0$, $y = 0$

(*b*) has a slope of $+1$

(*c*) is parallel to the x-axis

(*d*) has a slope of -1

22. Let $f(x, y) = x^n y^m = P$. If x is doubled and y is halved, the new value of f is

(a) 2^{n-m} P (b) 2^{m-n} P
(c) $2(n - m)$ P (d) $2(m - n)$ P

23. A regular die has six sides with numbers 1 to 6 marked on its sides. If a very large number of throws show the following frequencies of occurrence: $1 \to 0.167$; $2 \to 0.167$; $3 \to 0.152$; $4 \to 0.166$; $5 \to 0.168$; $6 \to 0.180$. We call this die

(a) irregular (b) biased
(c) Gaussian (d) insufficient

24. Consider the equation : $(7526)_8 - (Y)_8 = (4364)_8$, where $(X)_N$ stands for X to the base N. Find Y.

(a) 1634 (b) 1737 (c) 3142 (d) 3162

25. At what time between 6 *a.m.* and 7 *a.m.* will the minute hand and hour hand of a clock make an angle closest to 60°?

(a) 6 : 22 *a.m.* (b) 6 : 27 *a.m.*
(c) 6 : 38 *a.m.* (d) 6 : 45 *a.m.*

26. The roots of $ax^2 + bx + c = 0$ are real and positive a, b and c are real. Then $ax^2 + b|x| + c = 0$ has

(a) No roots (b) 2 real roots
(c) 3 real roots (d) 4 real roots

2012

27. The cost function for a product in a firm is given by $5q^2$, where q is the amount of production. The firm can sell the product at a market price of ₹ 50 per unit. The number of units to be produced by the firm such that the profit is maximized is

(a) 5 (b) 10 (c) 15 (d) 25

28. Which of the following assertions are **CORRECT ?**

P : Adding 7 to each entry in a list adds 7 to the mean of the list

Q : Adding 7 to each entry in a list adds 7 to the standard deviation of the list

R : Doubling each entry in a list doubles the mean of the list

S : Doubling each entry in a list leaves the standard deviation of the list unchanged

(a) P, Q (b) Q, R (c) P, R (d) R, S

29. A political party orders an arch for the entrance to the ground in which the annual convention is being held. The profile of the arch follows the equation $y = 2x - 0.1x^2$ where y is the height of the arch in meters. The maximum possible height of the arch is

(a) 8 meters (b) 10 meters
(c) 12 meters (d) 14 meters

30. There are eight bags of rice looking alike, seven of which have equal weight and one is slightly heavier. The weighing balance is of unlimited capacity. Using this balance, the minimum number of weighings required to identify the heavier bag is

(a) 2 (b) 3 (c) 4 (d) 8

2011

31. If log (P) = (1/2) Log (Q) = (1/3) Log (R), then which of the following options is TRUE?

(a) $P^2 = Q^3 R^2$ (b) $Q^2 = PR$
(c) $Q^2 = R^3 P$ (d) $R = P^2 Q^2$

32. The variable cost (V) of manufacturing a product varies according to the equation V = 4q, where q is the quanity produced. The fixed cost (F) of production of same product reduces with q according to the equation $F = 100/q$. How many units should be produced to minimize the total cost (V + F)?

(a) 5 (b) 4 (c) 7 (d) 6

33. Three friends R, S and T shared toffee from a bowl. R took 1/3rd of the toffees, but returned four to the bowl. S took 1/4th of what was left but returned three toffees to the bowl. T took half of the remainder but returned two back into the bowl. If the bowl had 17 toffees left, how many toffees were originally there in the bowl?

(a) 38 (b) 31 (c) 48 (d) 41

NUMERICAL TYPE QUESTIONS

2015

1. In the given figure angle Q is a right angle, PS : QS = 3:1, RT : QT = 5:2 and PU : UR = 1:1. If area of triangle QTS is 20 cm², then the area of triangle PQR in cm² is ______.

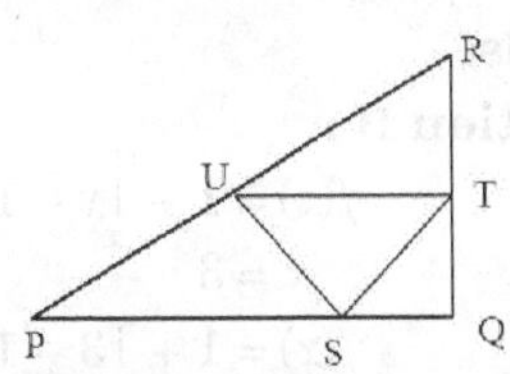

2. If p, q, r, s are distinct integers such that:

$f(p, q, r, s) = \max(p, q, r, s)$

$g(p, q, r, s) = \min(p, q, r, s)$

$h(p, q, r, s)$ = remainder of $(p \times q)/(r \times s)$ if $(p \times q) > (r \times s)$ or remainder of $(r \times s)/(p \times q)$ if $(r \times s) > (p \times q)$

Also a function $fgh(p, q, r, s) = f(p, q, r, s) \times g(p, q, r, s) \times h(p, q, r, s)$

Also the same operations are valid with two variable function of the form $f(p, q)$.

What is the value of $fg(h(2, 5, 7, 3), 4, 6, 8$

2014

3. When a point inside of a tetrahedron (a solid with four triangular surfaces) is connected by straight lines to its corners, how many (new) internal planes are created with these lines? ________.

ANSWERS

MCQ Type Questions

1. (*c*)	**2.** (*c*)	**3.** (*c*)	**4.** (*b*)	**5.** (*a*)	**6.** (*c*)	**7.** (*c*)	**8.** (*d*)	**9.** (*b*)	**10.** (*b*)
11. (*a*)	**12.** (*d*)	**13.** (*a*)	**14.** (*c*)	**15.** (*a*)	**16.** (*c*)	**17.** (*c*)	**18.** (*c*)	**19.** (*b*)	**20.** (*a*)
21. (*c*)	**22.** (*a*)	**23.** (*b*)	**24.** (*c*)	**25.** (*a*)	**26.** (*d*)	**27.** (*a*)	**28.** (*c*)	**29.** (*b*)	**30.** (*a*)
31. (*b*)	**32.** (*a*)	**33.** (*c*)							

Numerical Type Questions

1. 280 **2.** (8) **3.** (6)

EXPLANATIONS

MCQ TYPE QUESTIONS

1. In the given 2 × 4 rectangle grid, the following type of rectangles are present.

One figured rectangles = 8

Two figured rectangles = 10

Three figured rectangles = 4

Four figured rectangles = 5

Six figured rectangles = 2

Eight figured rectangles = 1

∴ Total number of rectangles

$= 8 + 10 + 4 + 5 + 2 + 1 = 30$

So the number of rectangles observed in the given grid is 30.

2. From the given graph function $f(x)$ must be equals to zero at $x = 3$.

From Option A :

$f(x) = 1 - |x - 1|$

at $x = 3$

$f(x) = 1 - |3 - 1| = 1 - 2 = -1$

So, it is false

From Option B :

$f(x) = 1 + |x - 1|$

at $x = 3$

$f(x) = 1 + |3 - 1| = 1 + 2 = 3$

So, it is false

From Option C :

$f(x) = 2 - |x - 1|$

at $x = 3$

$f(x) = 2 - |3 - 1| = 2 - 2 = 0$

So, it is true.

From Option D :

$f(x) = 2 + |x - 1|$

at $x = 3$

$f(x) = 2 - |3 - 1| = 2 + 2 = 4$

So, it is false.

Hence the correct expression for $f(x)$ given in the graph is $2 - |x - 1|$.

3. 'Lesser than' is apt because the sentence should be in comparative degree.

4. The possible window will be as follows :

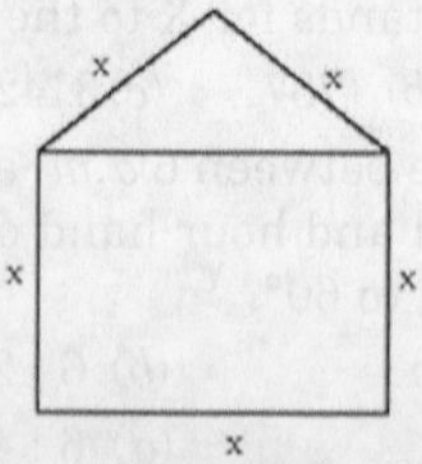

Perimeter of window = $5x = 6$m

$x = 1.2$ m

So area of window = Area of triangle + Area of square

$$= \frac{\sqrt{3}}{4}x^2 + x^2$$

$$= \frac{\sqrt{3}}{4}(1.2)^2 + (1.2)^2 = 2.06 \text{ m}^2$$

5. The binary operation □ is defined

$\Rightarrow a \,\square\, b = ab + (a + b)$

$a \,\square\, x = a$

∴ From the equation '*b*' is the variable

Option A: $x = 0$

$a \,\square\, o = a \times 0 + (a + 0) = 0 + a = a$

Option B: $x = 1$

$a \,\square\, 1 = a \times 1 + (a + 1) = a + a + 1 = 2a + 1$

Option C: $x = 2$

$a \,\square\, 2 \Rightarrow a \times 2 + (a + 2) = 2a + a + 2 = 3a + 2$

Option D: $x = 10$

$a \,\square\, 10 \Rightarrow a \times 10 + (a + 10) = 10a + a + 10$

$= 11a + 10$

∴ Option 'A' only True.

7. $(9 \text{ inches})^{1/2} = (0.25 \text{ yards})^{1/2}$,

Solving we get 9 inch = 0.25 yards

(since 1 inch = 0.028 yard)

8. Given Venn diagram is

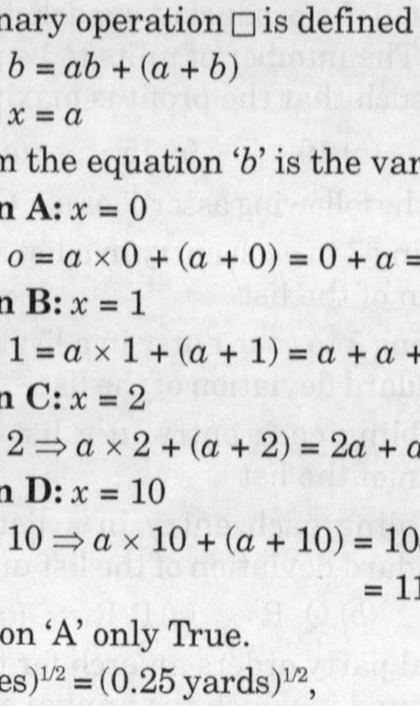

The number of students who like to read books or play sports

$$= 13 + 12 + 44 + 7 + 15 + 17 = 108$$

10. $A = \left[\frac{14}{3}, 0\right]$

$B = [0, 7]$,

$C = \left[\frac{5}{2}, 0\right]$

$D = \left[0, \frac{-5}{3}\right]$

$E = [4, 1]$,

$F = [0, 1]$

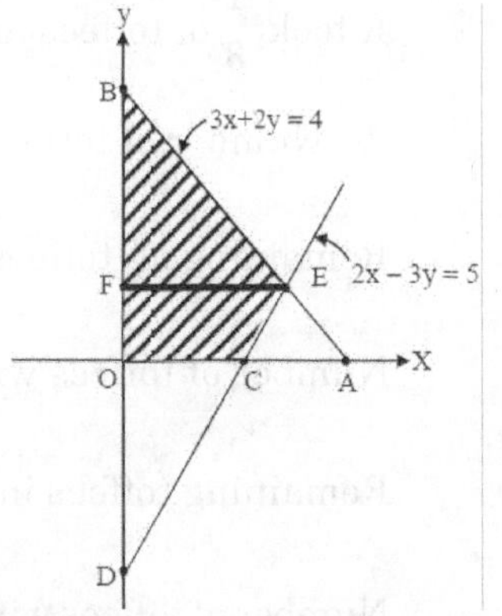

Required area = Area of Δ BFE + Area of quadrilateral FEOC

$$= \frac{1}{2} \times 4 \times 6 + \frac{1}{2} \times (4 + 2.5) \times 1$$

$$= 2 \times 6 + \frac{1}{2} \times 6.5$$

$$= 12 + 3.25 = 15.25 \text{ sq. units}$$

So, the area bounded by the given lines is 15.25 sq. units.

11. The equation of a straight line is

$$y = mx + c$$

$$m = \text{slope} = -0.02$$

set $(\log x, y)$

Now, the line intercepts the abscissa at

$$\log x = 0.1$$

So, at abscissa $y = 0$

$\therefore$ $y = mx + c$

$0 = -0.02 \times 0.1 + c$

$\therefore$ $c = 0.002$

$y = mx + c$

$y = -0.02 \times \log x + c$

At $x = 5$

$\therefore$ $y = -0.02 \times \log 5 + 0.002 = -0.030$

So, the value of y is -0.030.

13. $\frac{5}{7} = x^{-1/3} \Rightarrow \frac{7}{5} = x^{1/3} \Rightarrow \left(\frac{7}{5}\right)^3 = x \Rightarrow x = \frac{343}{125}$

14. $f(x) = 2x + 33 = 2(5) + 33 = 43$

15. For whole numbers, greater the value greater will be its log.

Same logic for power of e.

16. $\log \tan 1° + \log \tan 89° = \log(\tan 1° \times \tan 89°)$

$= \log(\tan 1° \times \cot 1°)$

$= \log 1 = 0$

Using the same logic total sum is '0'.

18. The given function is

$$f(x) = 1 - |x| \text{ on } -1 \le x \le 1$$

In order to find maximum and minimum value of function in interval [–1, 1]

At $x = -1$, $f(-1) = 1 - |-1| = 1 - 1 = 0$

At $x = -0.5$, $f(-0.5) = 1 - |-0.5| = 1 - 0.5 = 0.5$

At $x = 0$, $f(0) = 1 - |0| = 1$

At $x = 0.5$, $f(0.5) = 1 - |0.5| = 0.5$

At $x = 1$, $f(1) = 1 - |1| = 1 - 1 = 0$

$\therefore$ Hence, maximum value occurs at $x = 0$ and its maximum value is 1.

$\therefore$ Option (c) is correct.

21. $y = 5x^2 + 3$, $\frac{dy}{dx} = 10x$

Slope of tangent $= \left(\frac{dy}{dx}\right)_{x=0, y=3} = 10 \times 0 = 0$

Slope = 0

$\Rightarrow$ tangent is parallel to x-axis.

22.

$$P' = 2^n X^n \left(\frac{1}{2}\right)^m y^m$$

$$= 2^{n-m} X^n Y^m = 2^{n-m} P$$

23. For a very large number of throws, the frequency should be same for unbiased throw. As it not same, then the die is baised.

24. $(7526)_8 - (y)_8 = (4364)_8$

$\Rightarrow y_8 = (7526)_8 - (4364)_8$

$$\begin{array}{cccc} & 4 & (8+2=10) & \\ 7 & \not{5} & 2 & 6 \\ 4 & 3 & 6 & 4 \\ \hline 3 & 1 & 4 & 2 \end{array}$$

When we have base 8, we borrow 8 instead of 10 as done in normal subtraction.

25.

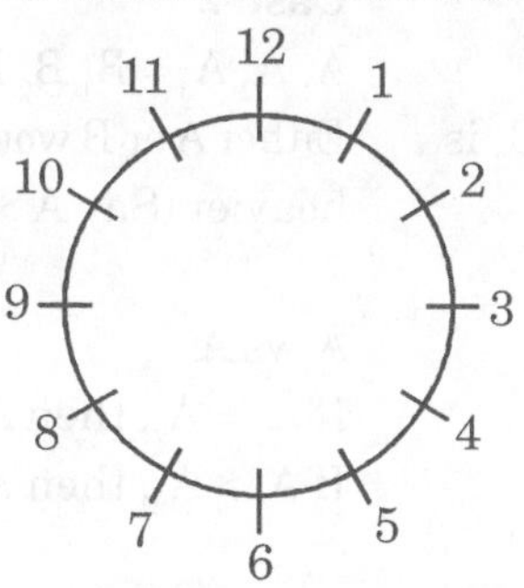

From option

At 6 a.m.

Both hand – 180° aparts,

At 6 : 20 a.m. – both hand apart

$\Rightarrow$ $180° + 10° - 120° = 70°$

At 6 : 22 a.m. – both hand apart

$\Rightarrow$ $180° + 12° - 132° = 60°$ (closest)

26. $ax^2 + bx + c = 0$

Roots are real and positive

$\therefore \quad b^2 - 4ac > 0$

By the above condition we get two positive real roots

Now, $ax^2 + b\,|x| + c = 0$

This can be written as;

$ax^2 + bx + c$

$\Delta = b^2 - 4ac > 0$

Now for $ax^2 - bx + c$

$(-b)^2 - 4ac$

$\Rightarrow \quad b^2 - 4ac$

Again $b^2 - 4ac > 0$

So again we get real roots

Thus we have 4 real roots of $ax^2 + b\,|x| + c = 0$

27. $P = 50q - 5q^2$

$$\frac{dp}{dq} = 50 - 10q;\ \frac{d^2p}{dq^2} < 0$$

$\therefore$ p is maximum at $50 - 10q = 0$ or, $q = 5$

Else check with options

28. P and R always holds true

Else consider a sample set {1, 2, 3, 4} and check accordingly

29. $y = 2x - 0.1x^2;\quad \frac{dy}{dx} = 2 - 0.2x;\quad \frac{d^2y}{dx^2} < 0$

$\therefore$ y maximises at $2 - 0.2x = 0$

$\Rightarrow x = 10$

$\therefore$ $y = 20 - 10 = 10$m

30. Let us categorize bags in three groups as

$A_1 A_2 A_3 \quad B_1 B_2 B_3 \quad C_1 C_2$

1st weighing A vs B

Case-1	**Case-2**
$A_1 A_2 A_3 = B_1 B_2 B_3$	$A_1 A_2 A_3 \neq B_1 B_2 B_3$
Then either C_1 or C_2 is heavier	Either A or B would be heavier (Say A > B)
2nd weighing	
C_1 vs C_2	A_1 vs A_2
If $C_1 > C_2$, then C_1	If $A_1 = A_2$, then A_3
If $C_1 < C_2$, then C_2	If $A_1 > A_2$, then A_1
If $A_1 < A_2$, then A_2	

31. $\log P = \frac{1}{2} \log Q$

$= \frac{1}{3} \log (R) = k$

$\therefore \quad P = b^k,\ Q = b^{2k},\ R = b^{3k}$

Hence, $Q^2 = b^{4k} = b^{3k}\, b^k = PR$

32. Checking with all options in formula: (4q+100/q) i.e. (V+F). Option (*a*) gives the minimum cost.

33. Let total number of toffees in bowl be x

R took $\frac{1}{3}$ of toffees and returned 4 to the bowl

$\therefore$ Number of toffees with R = $\frac{1}{3}x - 4$

Remaining of toffees in bowl = $\frac{2}{3}x + 4$

Number of toffees with S = $\frac{1}{4}\left(\frac{2}{3}x+4\right) - 3$

Remaining toffees in bowl = $\frac{3}{4}\left(\frac{2}{3}x+4\right) + 4$

Number of toffees with T = $\frac{1}{2}\left(\frac{3}{4}\left(\frac{2}{3}x+4\right)+4\right)+2$

Remaining toffees in bowl

$$= \frac{1}{2}\left[\frac{3}{4}\left(\frac{2}{3}x+4\right)+4\right]+2$$

Given : $\frac{1}{2}\left[\frac{3}{4}\left(\frac{2}{3}x+4\right)+4\right]+2 = 17$

$\Rightarrow \quad \frac{3}{4}\left(\frac{2}{3}x+4\right) = 27$

$\Rightarrow \quad x = 48$

NUMERICAL TYPE QUESTIONS

1. Let area of triangle PQR be 'A'

$$\frac{SQ}{PQ} = \frac{1}{1+3} = \frac{1}{4}$$

$$\frac{QT}{QR} = \frac{2}{2+5} = \frac{2}{7}$$

$\therefore$ Area of $\Delta QTS = \frac{1}{2} \times SQ \times QT$

$$= \frac{1}{2} \times \left(\frac{1}{4}PQ\right) \times \left(\frac{2}{7}QR\right)$$

$$= \frac{1}{4} \times \frac{2}{7} \times \left(\frac{1}{2} \times PQ \times QR\right)$$

$$= \frac{1}{14} \times \text{Area of } \Delta PQR$$

Given 20 cm² = $\frac{1}{14} \times A$

$\therefore \quad A = 14\ \ 20 = 280$ cm²

2. $fg(h(2, 5, 7, 3), 4, 6, 8) = fg(1, 4, 6, 8)$

$= f(1, 4, 6, 8) \times g(1, 4, 6, 8)$

$= 8 \times 1 = 8$

Engineering Mathematics

1

CHAPTER

Linear Algebra

MCQ TYPE QUESTIONS

2016

1. If the entries in each column of a square matrix M add up to 1, then an eigenvalue of M is

(a) 4 (b) 3

(c) 2 (d) 1

2. Consider the following linear system.

$x + 2y - 3z = a;$ $2x + 3y + 3z = b$

$5x + 9y - 6z = c$

This system is consistent if a, b and c satisfy the equation

(a) $7a - b - c = 0$ (b) $3a + b - c = 0$

(c) $3a - b + c = 0$ (d) $7a - b + c = 0$

2015

3. The smallest and largest Eigen values of the following matrix are :

$$\begin{bmatrix} 3 & -2 & 2 \\ 4 & -4 & 6 \\ 2 & -3 & 5 \end{bmatrix}$$

(a) 1.5 and 2.5 (b) 0.5 and 2.5

(c) 1.0 and 3.0 (d) 1.0 and 2.0

4. Let $A = [\alpha_{ij}], 1 \le i, j \le n$ with $n \ge 3$ and $a_{ij} = i.j.$ The rank of A is

(a) 0 (b) 1

(c) $n - 1$ (d) n

5. The two Eigen values of the matrix $\begin{bmatrix} 2 & 1 \\ 1 & p \end{bmatrix}$ have a ratio of 3 : 1 for $p = 2$. What is another value of p for which the Eigen values have the same ratio of 3 : 1?

(a) –2 (b) 1

(c) 7/3 (d) 14/3

2014

6. The sum of Eigen values of the matrix, [M] is

$$\text{where } [M] = \begin{bmatrix} 215 & 650 & 795 \\ 655 & 150 & 835 \\ 485 & 355 & 550 \end{bmatrix}$$

(a) 915 (b) 1355

(c) 1640 (d) 2180

2012

7. The eigenvalues of matrix $\begin{bmatrix} 9 & 5 \\ 5 & 8 \end{bmatrix}$ are

(a) – 2.42 and 6.86 (b) 3.48 and 13.53

(c) 4.70 and 6.86 (d) 6.86 and 9.50

2011

8. [A] is a square matrix which is neither symmetric nor skew-symmetric and $[A]^T$ is its transpose. The sum and difference of these matrices are defined as $[S] = [A] + [A]^T$ and $[D] = [A] - [A]^T$, respectively. Which of the following statements is true?.

(a) Both [S] and [D] are symmetric

(b) Both [S] and [D] are skew-symmetric

(c) [S] is skew-symmetric and [D] is symmetric

(d) [S] is symmetric and [D] is skew-symmetric

2010

9. The inverse of the matrix $\begin{bmatrix} 3+2i & i \\ -i & 3-2i \end{bmatrix}$ is

(a) $\dfrac{1}{12}\begin{bmatrix} 3+2i & i \\ i & 3-2i \end{bmatrix}$ (b) $\dfrac{1}{12}\begin{bmatrix} 3-2i & -i \\ i & 3+2i \end{bmatrix}$

(c) $\dfrac{1}{14}\begin{bmatrix} 3+2i & -i \\ i & 3-2i \end{bmatrix}$ (d) $\dfrac{1}{14}\begin{bmatrix} 3-2i & -i \\ i & 3+2i \end{bmatrix}$

2009

10. A square matrix B is skew–symmetric if

(a) $B^T = -B$ (b) $B^T = B$

(c) $B^{-1} = B$ (d) $B^{-1} = B^T$

2008

11. The product of matrices $(PQ)^{-1}P$ is

(a) P^{-1} (b) Q^{-1}

(c) $P^{-1} Q^{-1} P$ (d) PQP^{-1}

12. Eigen values of the matrix $[P] = \begin{bmatrix} 4 & 5 \\ 2 & -5 \end{bmatrix}$ are

(a) –7 and 8 (b) –6 and 5

(c) 3 and 4 (d) 1 and 2

13. The following simultaneous equations

$$x + y + z = 3$$
$$x + 2y + 3z = 4$$
$$x + 4y + kz = 6$$

will NOT have a unique solution for k equal to

(a) 0 (b) 5

(c) 6 (d) 7

2007

14. The minimum and the maximum eigen values of the matrix $\begin{bmatrix} 1 & 1 & 3 \\ 1 & 5 & 1 \\ 3 & 1 & 1 \end{bmatrix}$ are −2 and 6, respectively. What is the other eigen value ?

(*a*) 5 (*b*) 3

(*c*) 1 (*d*) −1

15. For what values of α and β the following simultaneous equations have an infinite number of solutions ?

$x + y + z = 5;$

$x + 3y + 3z = 9;$

$x + 2y + \alpha z = \beta$

(*a*) 2, 7 (*b*) 3, 8

(*c*) 8, 3 (*d*) 7, 2

16. The inverse of the 2 × 2 matrix $\begin{bmatrix} 1 & 2 \\ 5 & 7 \end{bmatrix}$ is

(*a*) $\frac{1}{3}\begin{bmatrix} -7 & 2 \\ 5 & -1 \end{bmatrix}$ (*b*) $\frac{1}{3}\begin{bmatrix} 7 & 2 \\ 5 & 1 \end{bmatrix}$

(*c*) $\frac{1}{3}\begin{bmatrix} 7 & -2 \\ -5 & 1 \end{bmatrix}$ (*d*) $\frac{1}{3}\begin{bmatrix} -7 & -2 \\ -5 & -1 \end{bmatrix}$

2006

17. For a given matrix A = $\begin{bmatrix} 2 & -2 & 3 \\ -2 & -1 & 6 \\ 1 & 2 & 0 \end{bmatrix}$, one of the eigenvalues is 3. The other two eigen values are

(*a*) 2, – 5 (*b*) 3, – 5

(*c*) 2, 5 (*d*) 4 – 5

NUMERICAL TYPE QUESTIONS

2016

1. The spot speeds (expressed in km/hr) observed at a road section are 66, 62, 45, 79, 32, 51, 56, 60, 53, and 49. The median speed (expressed in km/hr) is ________

(*Note : answer with one decimal accuracy*)

2015

2. For what value of *p* the following set of equations will have no solution ?

$$2x + 3y = 5 \ ; \quad 3x + py = 10$$

2014

3. Given the matrices J = $\begin{bmatrix} 3 & 2 & 1 \\ 2 & 4 & 2 \\ 1 & 2 & 6 \end{bmatrix}$ and K = $\begin{bmatrix} 1 \\ 2 \\ -1 \end{bmatrix}$, the product K^TJK is ________.

4. The determinant of matrix $\begin{bmatrix} 0 & 1 & 2 & 3 \\ 1 & 0 & 3 & 0 \\ 2 & 3 & 0 & 1 \\ 3 & 0 & 1 & 2 \end{bmatrix}$ is ________

5. The rank of the matrix $\begin{bmatrix} 6 & 0 & 4 & 4 \\ -2 & 14 & 8 & 18 \\ 14 & -14 & 0 & -10 \end{bmatrix}$ is ________

2013

6. What is the minimum number of multiplications involved in computing the matrix product PQR? Matrix P has 4 rows and 2 columns, matrix Q has 2 rows and 4 columns, and matrix R has 4 rows and 1 column. ________

ANSWERS

MCQ Type Questions

1. (All options are correct) **2.** (*b*) **3.** (*d*) **4.** (*b*) **5.** (*d*) **6.** (*a*) **7.** (*b*) **8.** (*d*)
9. (*b*) **10.** (*a*) **11.** (*b*) **12.** (*b*) **13.** (*d*) **14.** (*b*) **15.** (*a*) **16.** (*a*) **17.** (*b*)

Numerical Type Questions

1. (54.5) **2.** (4.5) **3.** (23 to 23) **4.** (88 to 88) **5.** (2.0 to 2.0) **6.** (16)

EXPLANATIONS

MCQ TYPE QUESTIONS

3. The given matrix is

$$A = \begin{bmatrix} 3 & -2 & 2 \\ 4 & -4 & 6 \\ 2 & -3 & 5 \end{bmatrix}$$

Inorder to find Eigen value of above matrix

$$|A - \lambda I| = 0$$

$$\therefore \begin{vmatrix} 3-\lambda & -2 & 2 \\ 4 & -4-\lambda & 6 \\ 2 & -3 & 5-\lambda \end{vmatrix} = 0$$

$$\therefore (3-\lambda)\left[(-4-\lambda)(5-\lambda)+18\right]+2\left[20-4\lambda-12\right] +2\left[-12+8+2\lambda\right]=0$$

$$(3-\lambda)\left[-20+4\lambda-5\lambda+\lambda^2+18\right] + 2\left[8-4\lambda\right] +2\left[-4+2\lambda\right] = 0$$

$$\therefore (3-\lambda)(\lambda^2-\lambda-2) +16-8\lambda-8+4\lambda = 0$$

$$3\lambda^2-3\lambda-6-\lambda^3+\lambda^2+2\lambda+16-8\lambda-8+4\lambda = 0$$

$$\therefore -\lambda^3+4\lambda^2-5\lambda+2 = 0$$

$$\therefore \lambda^3-4\lambda^2+5\lambda-2 = 0$$

$$\lambda_1 = 2 \text{ and } \lambda_2 = 1$$

Thus smallest eigen value of matrix is 1 and largest eigen value of matrix is 2.

$\therefore$ Option (*d*) is correct

7. $\begin{vmatrix} 9-\lambda & 5 \\ 5 & 8-\lambda \end{vmatrix} = 0$

$\Rightarrow (9-\lambda)(8-\lambda) = 25 \quad \Rightarrow \lambda = 3.48, 13.53$

8. $[A] = \begin{bmatrix} a_{11} & a_{12} \\ a_{21} & a_{22} \end{bmatrix}; \quad [A]^T = \begin{bmatrix} a_{11} & a_{21} \\ a_{12} & a_{22} \end{bmatrix}$

$$[S] = \begin{bmatrix} 2a_{11} & a_{12}+a_{21} \\ a_{21}+a_{12} & 2a_{22} \end{bmatrix}$$

$$[D] = \begin{bmatrix} 0 & a_{12}-a_{21} \\ a_{21}-a_{12} & 0 \end{bmatrix}$$

For symmetric matrix,

$$[S] = [S]^T, [D] = [D]^T$$

For skew-symmetric matrix,

$$[S] = -[S]^T, [D] = -[D]^T$$

As it can be seen that both matrix [S] and [D] are symmetric only.

9. Given : $A = \begin{bmatrix} 3+2i & i \\ -i & 3-2i \end{bmatrix}$

Now $A^{-1} = \dfrac{[\text{adj}.A]^T}{|A|}$

where $[\text{adj } A] = \begin{bmatrix} 3-2i & +i \\ -i & 3+2i \end{bmatrix}$

$\Rightarrow [\text{adj } A]^T = \begin{bmatrix} 3-2i & -i \\ i & 3+2i \end{bmatrix}$

and $|A| = (3+2i)(3-2i) - (i)(-i)$

$= 9 - 4i^2 + i^2 = 9 + 4 - 1 = 12$

$\therefore A^{-1} = \dfrac{1}{12}\begin{bmatrix} 3-2i & -i \\ i & 3+2i \end{bmatrix}$

11. $(PQ)^{-1} P = Q^{-1} P^{-1} P$ .... [since $(PQ)^{-1} = Q^{-1} P^{-1}$]

$= Q^{-1}$ ... [since $P^{-1} P = I$ (Unit Matrix)]

12. Given : $[P] = \begin{bmatrix} 4 & 5 \\ 2 & -5 \end{bmatrix}$

To find Eigen values

$$\begin{vmatrix} 4-\lambda & 5 \\ 2 & -5-\lambda \end{vmatrix} = 0$$

$\Rightarrow (4-\lambda)(-5-\lambda) - 10 = 0$

$\Rightarrow \lambda^2 + \lambda - 30 = 0$

Solving, we get

$\lambda = -6, 5$ (eigen values)

13. The system will not have a unique solution if

$$\Delta = 0$$

i.e. $\Delta I = \begin{vmatrix} 1 & 1 & 1 \\ 1 & 2 & 3 \\ 1 & 4 & k \end{vmatrix} = 0$

$\Rightarrow 1(2k - 12) - 1(k - 3) + 1(4 - 2) = 0$

$\Rightarrow k = 7$

14. $A = \begin{bmatrix} 1 & 1 & 3 \\ 1 & 5 & 1 \\ 3 & 1 & 1 \end{bmatrix}$

Eigen values are given by

$$|A - \lambda I| = 0$$

$$\Rightarrow \begin{bmatrix} 1-\lambda & 1 & 3 \\ 1 & 5-\lambda & 1 \\ 3 & 1 & 1-\lambda \end{bmatrix} = 0$$

Apply $C_1 \to C_1 - C_3$

$$\begin{vmatrix} -2-\lambda & 1 & 3 \\ 0 & 5-\lambda & 1 \\ 2+\lambda & 1 & 1-\lambda \end{vmatrix} = 0$$

$$\Rightarrow (2+\lambda)\begin{vmatrix} -1 & 1 & 3 \\ 0 & 5-\lambda & 1 \\ 1 & 1 & 1-\lambda \end{vmatrix} = 0$$

$\Rightarrow (2+\lambda)\left[-1\{(5-\lambda)(1-\lambda)-1\}+1\{1-3(5-\lambda)\}\right]=0$

$\Rightarrow (2+\lambda)\left[5-6\lambda+\lambda^2-1-1+15-3\lambda\right]=0$

$\Rightarrow (2+\lambda)(\lambda^2-9\lambda+18)=0$

$\Rightarrow (2+\lambda)(\lambda-6)(\lambda-3)=0$

$\Rightarrow \lambda=-2, 3, 6$

16. *Given* : Matrix, $A=\begin{bmatrix}1 & 2\\ 5 & 7\end{bmatrix}$

co-factors are

$a_{11}=7 \qquad a_{21}=-2$

$a_{12}=-5 \qquad a_{22}=1$

$\therefore$ Matrix by co-factors $=\begin{bmatrix}7 & -5\\ -2 & 1\end{bmatrix}$

$\therefore$ Adjoint A $=\begin{bmatrix}7 & -2\\ -5 & 1\end{bmatrix}$

Value of determinant of matrix A,

$|A| = 1\times7-2\times5=-3$

$A^{-1}=\dfrac{\text{Adjoint A}}{|A|}$

$=\dfrac{1}{-3}\begin{bmatrix}7 & -2\\ -5 & 1\end{bmatrix}=\dfrac{1}{3}\begin{bmatrix}-7 & 2\\ 5 & -1\end{bmatrix}$

17. *Given* : $A=\begin{bmatrix}2 & -2 & 3\\ -2 & -1 & 6\\ 1 & 2 & 0\end{bmatrix}$

For eigen values

$|A-\lambda I|=0$

$\therefore \begin{vmatrix}2-\lambda & -2 & 3\\ -2 & -1-\lambda & 6\\ 1 & 2 & -\lambda\end{vmatrix}=0$

$\Rightarrow \lambda^3-\lambda^2-21\lambda+45=0$

$\therefore \lambda_1+\lambda_2+\lambda_3=-\left(-\dfrac{1}{1}\right)=1$ ...(*i*)

and $\lambda_1\lambda_2\lambda_3=-45$ ...(*ii*)

Given : $\lambda_1=3$

$\therefore \lambda_2+\lambda_3=-2$

and $\lambda_2\lambda_3=-15$

Solving, we get

$\lambda_2=3$

and $\lambda_3=-5$

NUMERICAL TYPE QUESTIONS

1. Median speed calculation = ~~32~~, ~~45~~, ~~49~~, ~~51~~, 53, 56 ~~60~~, ~~62~~, ~~66~~, ~~79~~

Median speed $=\dfrac{53+56}{2}=54.5$ km/hr

2. The given equation are

$2x+3y=5$ and $3x+Py=10$

$\therefore \begin{bmatrix}2 & 3\\ 3 & P\end{bmatrix}\begin{bmatrix}x\\ y\end{bmatrix}=\begin{bmatrix}5\\ 10\end{bmatrix}$

Now comparing with A X = B

Argment matrix $[A/B]=\begin{bmatrix}2 & 3 & 5\\ 3 & P & 10\end{bmatrix}$

Now applying $R_2\to 2R_2-3R_1$,

we get $\begin{bmatrix}2 & 3 & 5\\ 0 & 2P-9 & 5\end{bmatrix}$

If $\int(A/B)\neq\int(A)$, then system will have no solution

$\therefore 2P-9=0$

$\therefore P=\dfrac{9}{2}=4.5$

3. $K^T JK=\begin{bmatrix}1 & 2 & -1\end{bmatrix}\begin{bmatrix}3 & 2 & 1\\ 2 & 4 & 2\\ 1 & 2 & 6\end{bmatrix}\begin{bmatrix}1\\ 2\\ -1\end{bmatrix}=\begin{bmatrix}6 & 8 & -1\end{bmatrix}\begin{bmatrix}1\\ 2\\ -1\end{bmatrix}$

$=23$

4. $A=\begin{bmatrix}0 & 1 & 2 & 3\\ 1 & 0 & 3 & 0\\ 2 & 3 & 0 & 1\\ 3 & 0 & 1 & 2\end{bmatrix}$

$\Delta=-1\begin{vmatrix}1 & 3 & 0\\ 2 & 0 & 1\\ 3 & 1 & 2\end{vmatrix}+2\begin{vmatrix}1 & 0 & 0\\ 2 & 3 & 1\\ 3 & 0 & 2\end{vmatrix}-3\begin{vmatrix}1 & 0 & 3\\ 2 & 3 & 0\\ 3 & 0 & 1\end{vmatrix}$

$=-1[1(-1)-3(4-3)]+2[1(6)]-3[1(3)+3(-9)]$

$=-1[-1-3]+12-3[3-27]$

$=4+12-3[-24]=4+12+72=88.$

5. The given matrix is

$A=\begin{bmatrix}6 & 0 & 4 & 4\\ -2 & 14 & 8 & 18\\ 14 & -14 & 0 & -10\end{bmatrix}$

The matrix is of order 3×4

Hence $\rho(A)\le 3$

$\begin{vmatrix}6 & 0 & 4\\ -2 & 14 & 8\\ 14 & -14 & 0\end{vmatrix}=6(0+112)+4(28-196)$

$=672-672=0$

and $\begin{vmatrix}0 & 4 & 4\\ 14 & 8 & 18\\ -14 & 0 & -10\end{vmatrix}=-4(-140+252)+4(112)$

$=-4(112)+4(112)=0$

Hence $\rho(A)\le 2$.

Now $\begin{vmatrix}6 & 0\\ -2 & 14\end{vmatrix}=84\neq0$

Hence $\rho(A)=2$

$\therefore$ Rank of given matrix is 2.

2 CHAPTER

Calculus

MCQ TYPE QUESTIONS

2016

1. The value of $\int_0^{\infty} \frac{1}{1+x^2} dx + \int_0^{\infty} \frac{\sin x}{x} dx$ is

(a) $\frac{\pi}{2}$ (b) π

(c) $\frac{3\pi}{2}$ (d) 1

2. The area of the region bounded by the parabola $y = x^2 + 1$ and the straight line $x + y = 3$ is

(a) $\frac{59}{6}$ (b) $\frac{9}{2}$

(c) $\frac{10}{3}$ (d) $\frac{7}{6}$

3. The magnitudes of vectors **P**, **Q** and **R** are 100 kN, 250 kN and 150 kN, respectively as shown in the figure.

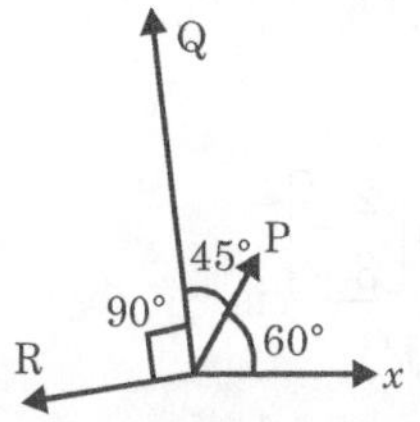

The respective values of the magnitude (in kN) and the direction (with respect to the x-axis) of the resultant vector are

(a) 290.9 and 96.0° (b) 368.1 and 94.7°

(c) 330.4 and 118.9° (d) 400.1 and 113.5°

4. The optimum value of the function

$$f(x) = x^2 - 4x + 2 \text{ is}$$

(a) 2 (maximum) (b) 2 (minimum)

(c) – 2 (maximum) (d) – 2 (minimum)

5. The Fourier series of the function,

$$f(x) = 0, \quad -\pi < x \le 0$$
$$= \pi - x, \quad 0 < x < \pi$$

in the interval $[-\pi, \pi]$ is

$$f(x) = \frac{\pi}{4} + \frac{2}{\pi}\left[\frac{\cos x}{1^2} + \frac{\cos 3x}{3^2} + ...\right] + \left[\frac{\sin x}{1} + \frac{\sin 2x}{2} + \frac{\sin 3x}{3} + ...\right].$$

The convergence of the above Fourier series at $x = 0$ gives

(a) $\sum_{n=1}^{\infty} \frac{1}{n^2} = \frac{\pi^2}{6}$ (b) $\sum_{n=1}^{\infty} \frac{(-1)^{n+1}}{n^2} = \frac{\pi^2}{12}$

(c) $\sum_{n=1}^{\infty} \frac{1}{(2n-1)^2} = \frac{\pi^2}{8}$ (d) $\sum_{n=1}^{\infty} \frac{(-1)^{n+1}}{2n-1} = \frac{\pi}{4}$

6. What is the value of $\lim_{\substack{x\to 0 \\ y \to 0}} \frac{xy}{x^2+y^2}$?

(a) 1 (b) – 1

(c) 0 (d) Limit does not exist

7. The angle of intersection of the curves $x^2 = 4y$ and $y^2 = 4x$ at point (0, 0) is

(a) 0° (b) 30°

(c) 45° (d) 90°

8. The quadratic approximation of $f(x) = x^3 - 3x^2 - 5$ at the point $x = 0$ is

(a) $3x^2 - 6x - 5$ (b) $-3x^2 - 5$

(c) $-3x^2 + 6x - 5$ (d) $3x^2 - 5$

2015

9. Given $i = \sqrt{-1}$, the value of the definite integral,

$$I = \int_0^{\pi/2} \frac{\cos x + i \sin x}{\cos x - i \sin x} dx \text{ is :}$$

(a) 1 (b) –1

(c) i (d) $-i$

10. $\lim_{n\to\infty} \left(1 + \frac{1}{x}\right)^{2x}$ is equal to

(a) e^{-2} (b) e

(c) 1 (d) e^2

11. While minimizing the function $f(x)$, necessary and sufficient conditions for a point, x_0 to be a minima are:

(a) $f'(x_0) > 0$ and $f''(x_0) = 0$

(b) $f'(x_0) < 0$ and $f''(x_0) = 0$

(c) $f'(x_0) = 0$ and $f''(x_0) < 0$

(d) $f'(x_0) = 0$ and $f''(x_0) > 0$

2014

12. $\lim_{x\to\infty} \left(\frac{x + \sin x}{x}\right)$ equals to

(a) $-\infty$ (b) 0

(c) 1 (d) ∞

13. With reference to the conventional Cartesian (x, y) coordinate system, the vertices of a triangle have the following coordinates : $(x_1, y_1) = (1, 0)$; $(x_2, y_2) = (2, 2)$; and $(x_3, y_3) = (4, 3)$. The area of the triangle is equal to

(a) $\frac{3}{2}$ (b) $\frac{3}{4}$

(c) $\frac{4}{5}$ (d) $\frac{5}{2}$

14. If $\{x\}$ is a continuous, real valued random variable defined over the interval $(-\infty, +\infty)$ and its occurrence is defined by the density function given as: $f(x) = \frac{1}{\sqrt{2\pi} \times b} e^{-\frac{1}{2}\left(\frac{x-a}{b}\right)^2}$ where 'a' and 'b' are the statistical attributes of the random variable $\{x\}$. The value of the integral $\int_{-\infty}^{a} \frac{1}{\sqrt{2\pi} \times b} e^{-\frac{1}{2}\left(\frac{x-a}{b}\right)^2} dx$ is

(a) 1 (b) 0.5

(c) π (d) $\frac{\pi}{2}$

15. The expression $\lim_{\alpha \to 0} \frac{x^\alpha - 1}{\alpha}$ is equal to

(a) $\log x$ (b) 0

(c) $x \log x$ (d) ∞

2013

16. The solution for $\int_0^{\pi/4} (\cos^4 3\theta \sin^3 6\theta d\theta$ is

(a) 0 (b) $\frac{1}{15}$

(c) 1 (d) $\frac{8}{3}$

2012

17. The infinite series $1 + x + \frac{x^2}{2!} + \frac{x^3}{3!} + \frac{x}{3!} + \frac{x^4}{4!} + ...$ corresponds to

(a) $\sec x$ (b) e^x

(c) $\cos x$ (d) $1 + \sin^2 x$

18. For the parallelogram OPQR shown in the sketch, $\overline{OP} = a\hat{i} + b\hat{j}$ and $\overline{OR} = c\hat{i} + d\hat{j}$. The area of the parallelogram is

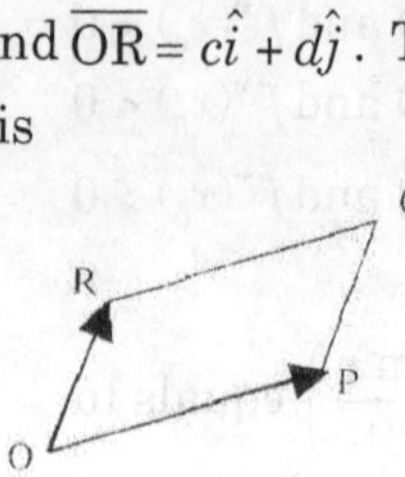

(a) $a d - b c$ (b) $a c + b d$

(c) $a d + b c$ (d) $a b - c d$

2011

19. What should be the value of λ such that the function defined below is continuous at $x = \pi/2$?

$$f(x) = \begin{cases} \dfrac{\lambda \cos x}{\frac{\pi}{2} - x} & \text{if} \neq \pi/2 \\ 1 & \text{if } x = \pi/2 \end{cases}$$

(a) 0 (b) $2/\pi$

(c) 1 (d) $\pi/2$

20. What is the value of the definite integral, $\int_0^a \frac{\sqrt{x}}{\sqrt{x} + \sqrt{a - x}} dx$?

(a) 0 (b) $a/2$

(c) a (d) $2a$

21. If $\vec{a}$ and $\vec{b}$ are two arbitrary vectors with magnitudes a and b, respectively, $|\vec{a} \times \vec{b}|^2$ will be equal to

(a) $a^2b^2 - (\vec{a}.\vec{b})^2$

(b) $ab - \vec{a} \cdot \vec{b}$

(c) $a^2b^2 + (\vec{a}.\vec{b})^2$

(d) $ab + \vec{a} \cdot \vec{b}$

2010

22. The $\lim_{x \to 0} \frac{\sin\left[\frac{2}{3}x\right]}{x}$ is

(a) $\frac{2}{3}$ (b) 1

(c) $\frac{3}{2}$ (d) ∞

23. A parabolic cable is held between two supports at the same level. The horizontal span between the supports is L. The sag at the mid–span is h. The equation of the parabola is

$$y = 4h\frac{x^2}{L^2},$$

where x is horizontal coordinate and y is vertical coordinate with the origin at the centre of the cable.

The expression for the total length of the cable is

(a) $\int_0^L \sqrt{1 + 64\frac{h^2x^2}{L^4}} dx$ (b) $\int_0^{L/2} \sqrt{1 + 64\frac{h^3x^2}{L^4}} dx$

(c) $\int_0^{L/2} \sqrt{1 + 64\frac{h^2x^2}{L^4}} dx$ (d) $2\int_0^{L/2} \sqrt{1 + 64\frac{h^2x^2}{L^4}} dx$

24. Given a function

$$f(x, y) = 4x^2 + 6y^2 - 8x - 4y + 8$$

The optimal value of f (x, y)

(a) is a minimum equal to $\frac{10}{3}$

(b) is a maximum equal to $\frac{10}{3}$

(c) is a minimum equal to $\frac{8}{3}$

(d) is a maximum equal to $\frac{8}{3}$

2009

25. For a scalar function $f(x, y, z) = x^2 + 3y^2 + 2z^2$, gradient at the point P (1, 2, –1) is

(a) $2\vec{i} + 6\vec{i} + 4\vec{k}$ (b) $2\vec{i} + 12\vec{j} - 4\vec{k}$

(c) $2\vec{i} + 12\vec{j} + 4\vec{k}$ (d) $\sqrt{56}$

26. For a scalar function

$$f(x, y, z) = x^2 + 3y^2 + 2z^2,$$

the directional derivative at the point P (1, 2, – 1) in the direction of a vector $\vec{i} - \vec{j} + 2\vec{k}$ is

(a) – 18 (b) $-3\sqrt{6}$

(c) $3\sqrt{6}$ (d) 18

2008

27. The equation $k_x \frac{\partial^2 h}{\partial x^2} + k_z \frac{\partial^2 h}{\partial z^2} = 0$ can be transformed to $\frac{\partial^2 h}{\partial x_1^2} + \frac{\partial^2 h}{\partial z^2} = 0$ by substituting

(a) $x_1 = x\frac{k_z}{k_x}$ (b) $x_1 = x\frac{k_x}{k_z}$

(c) $x_1 = x\sqrt{\frac{k_x}{k_z}}$ (d) $x_1 = x\sqrt{\frac{k_z}{k_x}}$

28. The value of $\int_0^3 \int_0^x (6 - x - y)\, dx\, dy$ is

(a) 13.5 (b) 27.0

(c) 40.5 (d) 54.0

29. The inner (dot) product of two vectors $\vec{P}$ and $\vec{Q}$ is zero. Angle between the two vectors is

(a) 0° (b) 30°

(c) 90° (d) 120°

2007

30. A velocity vector is given as

$$\vec{V} = 5xy\,\vec{i} + 2y^2\,\vec{j} + 3yz^2\,\vec{k}.$$

The divergence of this velocity vector at (1, 1, 1) is

(a) 9 (b) 10

(c) 14 (d) 15

31. Evaluate $\int_0^\infty \frac{\sin t}{t} dt$

(a) π (b) $\frac{\pi}{2}$

(c) $\frac{\pi}{4}$ (d) $\frac{\pi}{3}$

32. Potential function φ is given as

$$\phi = x^2 - y^2.$$

What will be the stream function (ψ) with. the condition ψ = 0 at x = y = 0?

(a) $2xy$ (b) $x^2 + y^2$

(c) $x^2 - y^2$ (d) $2x^2y^2$

2006

33. The directional derivative of

$$f(x, y, z) = 2x^2 + 3y^2 + z^2$$

at the point P : (2, 1, 3) in the direction of the vector a = i –2 k is

(a) – 2.785 (b) – 2.145

(c) – 1.789 (d) 1.000

34. What is the area common to the circles $r = a$ and $r = 2a \cos \theta$?

(a) 0.524 a^2

(b) 0.614 a^2

(c) 1.047 a^2

(d) 1.228 a^2

NUMERICAL TYPE QUESTIONS

2016

1. The area between the parabola $x^2 = 8y$ and the straight line $y = 8$ is ________.

2015

2. The directional derivative of the field $u(x, y, z) = x^2 - 3yz$ in the direction of the vector $(\hat{i} + \hat{j} - 2\hat{k})$ at point (2, – 1, 4) is _______.

ANSWERS

MCQ Type Questions

1. (b)	**2.** (b)	**3.** (c)	**4.** (d)	**5.** (c)	**6.** (d)	**7.** (d)	**8.** (b)	**9.** (c)	**10.** (d)
11. (d)	**12.** (c)	**13.** (a)	**14.** (b)	**15.** (a)	**16.** (b)	**17.** (b)	**18.** (a)	**18.** (c)	**20.** (b)
21. (a)	**22.** (a)	**23.** (d)	**24.** (a)	**25.** (b)	**26.** (b)	**27.** (d)	**28.** (a)	**29.** (c)	**30.** (d)
31. (b)	**32.** (a)	**33.** (c)	**34.** (d)						

Numerical Type Questions

1. (85.33) **2.** (– 5.72)

EXPLANATIONS

MCQ TYPE QUESTIONS

3. $R_X = \Sigma F_X = 100 \cos 60 - 250 \sin 15 - 150 \cos 15$
$= -159.59$ kN

$R_Y = \Sigma F_Y = 100 \sin 60 + 250 \cos 15 - 150 \sin 15$
$= 289.261$ kN

$$R = \sqrt{R_X^2 + R_Y^2} = \sqrt{159.59^2 + 289.261^2} = 330.36 \text{ kN}$$

$$\tan\theta = \frac{289.261}{159.59} = 1.815$$

$\theta = 61.1°$

Angle with respect to X – axis = 118.9°

8. R tan 30→ = 0.577 R

12.
$$\lim_{x\to\infty}\left(\frac{x+\sin x}{x}\right) = \lim_{x\to\infty}\left(1+\frac{\sin x}{x}\right)$$
$= 1 + 0 = 1.$

13. Area triangle
$$= \frac{1}{2}\begin{vmatrix} x_1 & x_2 & x_3 \\ y_1 & y_2 & y_3 \\ 1 & 1 & 1 \end{vmatrix}$$
$$= \frac{1}{2}\begin{vmatrix} 1 & 2 & 4 \\ 0 & 2 & 3 \\ 1 & 1 & 1 \end{vmatrix}$$
$$= \frac{1}{2}\,|1(2-3) - 2(0-3) + 4(0-2)|$$
$$= \frac{1}{2}|-3| = \frac{3}{2}$$

15.
$$\lim_{\alpha\to 0}\frac{x^\alpha - 1}{\alpha}\left[\frac{0}{0}\text{ form}\right]$$

Hence applying L Hospital rule

Let $y = x^\alpha$

$\therefore \quad \log y = \alpha \log x$

$$\frac{1}{y}\frac{dy}{dx} = \frac{\alpha}{x} + \log x\frac{d\alpha}{dx}$$

$$\therefore \quad \frac{dy}{dx} = x^\alpha\left[\frac{\alpha}{x} + \log x\frac{d\alpha}{dx}\right]$$

Now applying L Hospital rule

$$\lim_{\alpha\to 0}\frac{x^\alpha\left[\dfrac{\alpha}{x} + \log x\dfrac{d\alpha}{dx}\right]}{\dfrac{d\alpha}{dx}} = \log x$$

17. e^x

Put x = 2 in given equation which comes to about 7 and $e^2 = 7$

19. For a function $f(x)$ to be continuous, at $x = a$

$$\operatorname*{Lt}_{x\to a} f(x) = f(a)$$

$$\Rightarrow \quad \operatorname*{Lt}_{x\to\frac{\pi}{2}} \frac{\lambda\cos x}{\frac{\pi}{2} - x} = 1$$

$$\Rightarrow \quad \operatorname*{Lt}_{x\to\frac{\pi}{2}} \frac{(-\lambda\sin x)}{-1} = 1$$

$$\Rightarrow \quad \lambda\cdot\sin\frac{\pi}{2} = 1$$

$$\Rightarrow \quad \lambda = 1$$

20. *Given :*
$$I = \int_0^a \frac{\sqrt{x}}{\sqrt{x}+\sqrt{a-x}}\cdot dx \qquad ...(i)$$

Putting $a - x = t$, we get

$$I = -\int_a^0 \frac{\sqrt{a-t}}{\sqrt{a-t}+\sqrt{t}}\cdot dt$$
$$= \int_0^a \frac{\sqrt{a-t}\cdot dt}{\sqrt{a-t}+\sqrt{t}}$$
$$= \int_0^a \frac{\sqrt{a-x}}{\sqrt{a-x}+\sqrt{x}}\cdot dx \qquad ...(ii)$$

Adding equations (i) and (ii), we have

$$2I = \int_0^a \frac{\sqrt{x}+\sqrt{a-x}}{\sqrt{x}+\sqrt{a-x}}\cdot dx = \int_0^a dx = a$$

$$\Rightarrow \quad I = \frac{a}{2}$$

21. $\left|\bar{a}\times\bar{b}\right|^2 = (ab\sin\theta)^2 = (ab)^2(1-\cos^2\theta)$

$\Rightarrow \quad a^2b^2 - \left|\bar{a}\cdot\bar{b}\right|^2$

22. *Given :* $\lim\limits_{x\to 0}\dfrac{\sin\left(\frac{2}{3}x\right)}{x}$

It is 0|0 form, so apply L-Hospital's rule

$$\lim_{x\to 0}\frac{\frac{2}{3}\cos\left(\frac{2}{3}x\right)}{1}$$

Taking limit, we get

$$\frac{2}{3}\cos 0^\circ = \frac{2}{3}$$

23. Given : $y = 4h\dfrac{x^2}{L^2}$

$$\therefore \quad \frac{dy}{dx} = \frac{8hx}{L^2}$$

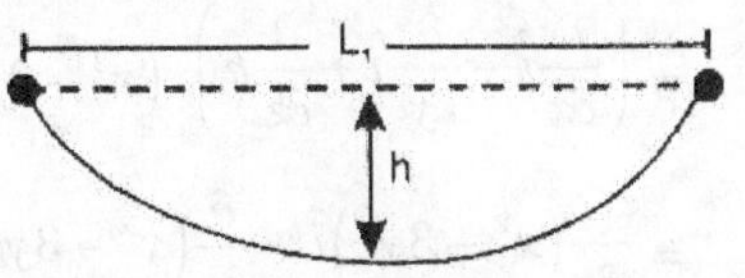

Now, $$\frac{ds}{dx} = \sqrt{1+\left(\frac{dy}{dx}\right)^2}$$

$$\therefore \quad s = 2\int_0^{L/2} 1+\left(\frac{dy}{dx}\right)^2 dx$$

$$= 2\int_0^{L/2}\sqrt{\left(1+64\frac{h^2x^2}{L^4}\right)}dx$$

24. Given : $f(x, y) = 4x^2 + 6y^2 - 8x - 4y + 8$

$$\therefore \quad \frac{\partial f(x,y)}{\partial x} = 8x - 8$$

Now $$\frac{\partial f(x,y)}{\partial x} = 0$$

$\Rightarrow \quad 8x - 8 = 0$

$\Rightarrow \quad x = 1$

and $$\frac{\partial^2 f(x,y)}{\partial x^2} = 8 \ (+\text{ ve})$$

So, at $x = 1$, $f(x, y)$ is minimum.

$$\therefore \quad \frac{\partial f(x,y)}{\partial y} = 12y - 4$$

$$= 0$$

$$\Rightarrow \quad y = \frac{1}{3}$$

and $$\frac{\partial^2 +(x,y)}{\partial y^2} = 12 \ (+\text{ ve})$$

So, at $y = \dfrac{1}{3}$, $f(x, y)$ is minimum

Putting the value of $x = 1$, and $y = \dfrac{1}{3}$.

we get

$$f(x, y) = 4\times 1^2 + 6\times\frac{1}{9} - 8 - 4\times\frac{1}{3} + 8$$

$$= \frac{10}{3} \text{ (minimum)}$$

27. Substituting $x_t = x\sqrt{\dfrac{k_z}{k_x}}$

we get $$\frac{\partial^2 h}{\partial xt^2} + \frac{\partial^2 h}{\partial z^2} = 0$$

28. $$\int_0^3\int_0^x (6 - x - y)\,dx\,dy$$

$$= \int_0^3\left(6y - xy - \frac{y^2}{2}\right)_0^x dx$$

$$= \int_0^3\left(6x - x^2 - \frac{x^2}{2}\right)dx$$

$$= \left(3x^2 - \frac{x^3}{3} - \frac{x^3}{6}\right)_0^6$$

$= 13.5$

29. When two vectors are perpendicular to each other, their dot multiplication is zero

i.e. $\vec{P}.\vec{Q} = 0$

and $\vec{P}$ and $\vec{Q}$ are perpendicular.

30. Div. $$\vec{V} = \frac{\partial V}{\partial x}\hat{i} + \frac{\partial V}{\partial y}\hat{j} + \frac{\partial V}{\partial z}\hat{k}$$

$$\vec{V} = 5xy\,\vec{i} + 2y^2\,\vec{j} + 3yz^2\,\vec{k}$$

$$\frac{\partial V}{\partial x}\hat{i} = 5y + 0 + 0 = 5y$$

$$\frac{\partial v}{\partial y}\hat{j} = O + 4y + O = 4y$$

$$\frac{\partial V}{\partial z}\hat{k} = 0 + 0 + 6yz = 6yz$$

Div. $\vec{V} = 5y + 4y + 6yz$

At (1, 1, 1), div. $\vec{V} = 5 + 4 + 6$

$= \mathbf{15}$

31. $$L\left(\frac{\sin at}{t}\right) = \cot^{-1}\left(\frac{s}{a}\right)$$

$$\int_0^\infty e^{-st}\frac{\sin(at)}{t}dt = \cot^{-1}\left(\frac{s}{a}\right)$$

Put $a = 1$ and $s = 0$

$$\therefore \quad \int_0^\infty \frac{\sin t}{t}\, dt = \cot^{-1}(0) = \frac{\pi}{2}$$

32. Potential function , $\phi = x^2 - y^2$

$$\therefore \quad \frac{\partial \phi}{\partial x} = 2x = U$$

and $$\frac{\partial \psi}{\partial y} = U = 2x$$

$$\frac{\partial \psi}{\partial y} = 2x$$

Integrating, $\int \partial\psi = \int 2x \partial y$

$$\therefore \quad \psi = 2xy + c$$

where, ψ = stream function.

At $x = y = 0$, $\psi = 0$

$\Rightarrow$ $c = 0$

$$\therefore \quad \psi = 2xy$$

33. Given : $f(x, y, z) = 2x^2 + 3y^2 + z^2$

$$\text{grad } f = \frac{\partial f}{\partial x}\hat{i} + \frac{\partial f}{\partial y}\hat{j} + \frac{\partial f}{\partial z}\hat{k}$$

$$= 4x\,\hat{i} + 6y\frac{PI^3}{3EI} + 2z\,\hat{k}$$

At point (2, 1, 3), grad $f = 8i + 6j + 6k$

$\therefore$ Unit vector in direction $\hat{a}$

$$= \frac{i - 2k}{\sqrt{1+4}} = \frac{i-2k}{\sqrt{5}}$$

$\therefore$ Directional derivative

$$= \text{grad} \,.\, \hat{a}$$

$$= (8\hat{i} + 6\hat{j} + 6\hat{k}) \,.\, \left(\frac{i - 2\hat{k}}{\sqrt{5}}\right)$$

$$= \frac{8}{\sqrt{5}} - \frac{12}{\sqrt{5}}$$

$$= -1.789$$

34. At $r = a$, $x^2 + y^2 = a^2$

At $r = 2a\cos\theta$,

$$x^2 + y^2 - 2ax = 0$$

$$\Rightarrow \quad (x-a)^2 + y^2 = a^2$$

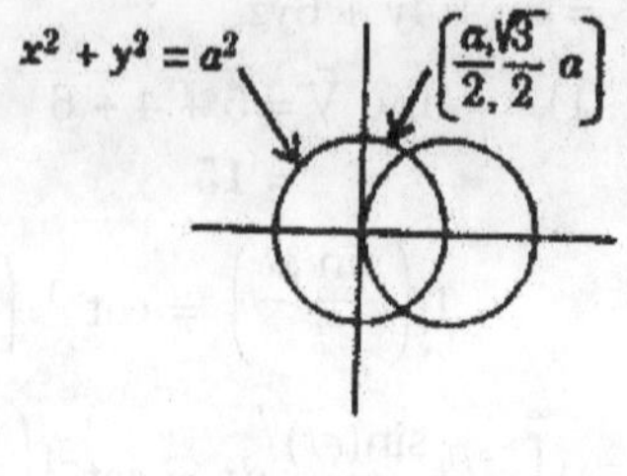

Hence required area

$$= 2\left[\int_0^{a/2} \sqrt{a^2 - (x-a)^2}\,dx + \int_{a/2}^{a} \sqrt{a^2 - x^2}\,dx\right]$$

$$= 2\left[\frac{\pi}{3} - \frac{\sqrt{3}}{4}\right] a^2$$

$$= \left(\frac{2\pi}{3} - \frac{\sqrt{3}}{2}\right) a^2$$

$$= 1.228\, a^2$$

NUMERICAL TYPE QUESTION

2. Here $u(x, y, z) = x^2 - 3yz$

Now directional derivative of u

$$\therefore \quad \nabla u = \left(\frac{\partial}{\partial x}\hat{i} + \frac{\partial}{\partial y}\hat{j} + \frac{\partial}{\partial z}\hat{k}\right) u$$

$$= \left(\frac{\partial}{\partial x}\hat{i} + \frac{\partial}{\partial y}\hat{j} + \frac{\partial}{\partial z}\hat{k}\right)(x^2 - 3yz)$$

$$= \frac{\partial}{\partial x}(x^2 - 3yz)\hat{i} + \frac{\partial}{\partial y}(x^2 - 3yz)\hat{j} + \frac{\partial}{\partial z}(x^2 - 3yz)\hat{k}$$

$$= 2x\hat{i} - 3z\hat{j} - 3y\hat{k}$$

Now directional derivative of u in direction of vector $\hat{i} + \hat{j} - 2\hat{k}$

i.e. $$\nabla u . \hat{a} = \frac{(2x\hat{i} - 3z\hat{j} - 3y\hat{k}).(\hat{i} + \hat{j} - 2\hat{k})}{\sqrt{1+1+4}}$$

$$= \frac{2x - 3z + 6y}{\sqrt{6}}$$

Now $$\nabla u . \hat{a}\big|_{(2,-1,4)} = \frac{2(2) - 3(4) + 6(-1)}{\sqrt{6}}$$

$$= \frac{4 - 12 - 6}{\sqrt{6}}$$

$$= \frac{-14}{\sqrt{6}} = -\frac{7\sqrt{6}}{3}$$

■■

3 CHAPTER

Ordinary and Partial Differential Equations

MCQ TYPE QUESTIONS

2016

1. The type of partial differential equation

$$\frac{\partial^2 P}{\partial x^2} + \frac{\partial^2 P}{\partial y^2} + 3\frac{\partial^2 P}{\partial x \partial y} + 2\frac{\partial P}{\partial x} - \frac{\partial P}{\partial y} = 0 \text{ is}$$

(a) elliptic (b) parabolic

(c) hyperbolic (d) none of these

2. The solution of the partial differential equation $\frac{\partial u}{\partial t} = \alpha \frac{\partial^2 u}{\partial x^2}$ is of the form

(a) $C\cos(kt)\left[C_1 e^{(\sqrt{k/\alpha})x} + C_2 e^{-(\sqrt{k/\alpha})x}\right]$

(b) $Ce^{kt}\left[C_1 e^{(\sqrt{k/\alpha})x} + C_2 e^{-(\sqrt{k/\alpha})x}\right]$

(c) $Ce^{kt}\left[C_1 \cos(\sqrt{k/\alpha})x + C_2 \sin(-\sqrt{k/\alpha})x\right]$

(d) $C\sin(kt)\left[C_1 \cos(\sqrt{k/\alpha})x + C_2 \sin(-\sqrt{k/\alpha})x\right]$

3. The respective expressions for complimentary function and particular integral part of the solution of the differential equation $\frac{d^4 y}{dx^4} + 3\frac{d^2 y}{dx^2} = 108x^2$ are

(a) $\left[c_1 + c_2 x + c_3 \sin\sqrt{3}x + c_4 \cos\sqrt{3}x\right]$ and $[3x^4 - 12x^2 + c]$

(b) $\left[c_2 x + c_3 \sin\sqrt{3}x + c_4 \cos\sqrt{3}x\right]$ and $[5x^4 - 12x^2 + c]$

(c) $\left[c_1 + c_3 \sin\sqrt{3}x + c_4 \cos\sqrt{3}x\right]$ and $[3x^4 - 12x^2 + c]$

(d) $\left[c_1 + c_2 x + c_3 \sin\sqrt{3}x + c_4 \cos\sqrt{3}x\right]$ and $[5x^4 - 12x^2 + c]$

2015

4. Consider the following differential equation :

$$x(ydx + xdy)\cos\frac{y}{x} = y(xdy - ydx)\sin\frac{y}{x}$$

Which of the following is the solution of the above equation (c is an arbitrary constant) ?

(a) $\frac{x}{y}\cos\frac{y}{x} = c$ (b) $\frac{x}{y}\sin\frac{y}{x} = c$

(c) $xy\cos\frac{y}{x} = c$ (d) $xy\sin\frac{y}{x} = c$

2014

5. The integrating factor for the differential equation $\frac{dp}{dt} + k_2 P = k_1 L_0 e^{-k_1 t}$ is

(a) $e^{-k_1 t}$ (b) $e^{-k_2 t}$

(c) $e^{k_1 t}$ (d) $e^{k_2 t}$

2012

6. The solution of the ordinary differential equation $\frac{dy}{dx} + 2y = 0$ for the boundary condition, $y = 5$ at $x = 1$ is

(a) $y = e^{-2x}$ (b) $y = 2e^{-2x}$

(c) $y = 10.95^{e-2y}$ (d) $y = 36.95e^{-2x}$

2011

7. The solution of the differential equation $\frac{dy}{dx} + \frac{y}{x} = x$ with the condition that $y = 1$ at $x = 1$, is

(a) $y = \frac{2}{3x^2} + \frac{x}{3}$ (b) $y = \frac{x}{2} + \frac{1}{2x}$

(c) $y = \frac{2}{3} + \frac{x}{3}$ (d) $y = \frac{2}{3x} + \frac{x^2}{3}$

2010

8. The order and degree of the differential equation

$$\frac{d^3 y}{dx^3} + 4\sqrt{\left(\frac{dy}{dx}\right)^3 + y^2} = 0 \text{ are respectively}$$

(a) 3 and 2

(b) 2 and 3

(c) 3 and 3

(d) 3 and 1

9. The solution to the ordinary differential equation

$$\frac{d^2 y}{dx^2} + \frac{dy}{dx} - 6y = 0 \text{ is}$$

(a) $y = c_1 e^{3x} + c_2 e^{-2x}$

(b) $y = c_1 e^{3x} + c_2 e^{2x}$

(c) $y = c_1 e^{-3x} + c_2 e^{2x}$

(d) $y = c_1 e^{-3x} + c_2 e^{-2x}$

10. The partial differential equation that can be formed from $z = ax + by + ab$ has the form (with $P = \frac{\partial z}{\partial x}$ and $q\frac{\partial z}{\partial y}$)

(a) $z = px + qy$ (b) $z = px + pq$

(c) $z = px + qy + pq$ (d) $z = qx + pq$

2009

11. Solution of the differential equation $3y\frac{dy}{dx} + 2x = 0$ represents a family of

(a) ellipses (b) circles

(c) parabolas (d) hyperbolas

12. Laplace transform for the following function is $f(x) = \cosh(ax)$

(a) $\frac{a}{s^2 - a^2}$ (b) $\frac{s}{s^2 - a^2}$

(c) $\frac{a}{s^2 + a^2}$ (d) $\frac{s}{s^2 + a^2}$

2008

13. The general solution of $\frac{d^2y}{dx^2} + y = 0$ is

(a) $y = P\cos x + Q\sin x$

(b) $y = P\cos x$

(c) $y = P\sin x$

(d) $y = P\sin^2 x$

14. Solution of $\frac{dy}{dx} = -\frac{x}{y}$ at x = 1 and $y = \sqrt{3}$ is

(a) $x - y^2 = -2$

(b) $x + y^2 = 4$

(c) $x^2 - y^2 = -2$

(d) $x^2 + y^2 = 4$

2007

15. The degree of the differential equation $\frac{d^2x}{dt^2} + 2x^3 = 0$ is

(a) 0 (b) 1

(c) 2 (d) 3

16. The solution for the differential equation $\frac{dy}{dx} = x^2y$ with the condition that $y = 1$ at $x = 0$ is

(a) $y = e^{\frac{1}{2x}}$ (b) $\ell n(y) = \frac{x^3}{3} + 4$

(c) $\ell n(y) = \frac{x^2}{2}$ (d) $y = e^{\frac{x^3}{3}}$

2006

17. The solution of the differential equation, $x^2\frac{dy}{dx} + 2xy - x + 1 = 0$, at $x = 1$, $y = 0$ is

(a) $\frac{1}{2} - \frac{1}{x} + \frac{1}{2x^2}$ (b) $\frac{1}{2} - \frac{1}{x} - \frac{1}{2x^2}$

(c) $\frac{1}{2} + \frac{1}{x} + \frac{1}{2x^2}$ (d) $-\frac{1}{2} + \frac{1}{x} + \frac{1}{2x^2}$

NUMERICAL TYPE QUESTION

2015

1. Consider the following second order linear differential equation

$$\frac{d^2y}{dx^2} = -12x^2 + 24x - 20$$

The boundary conditions are at $x = 0, y = 5$ and at $x = 2, y = 21$

The value of at $x = 1$ is ________.

ANSWERS

MCQ Type Questions

1. (c) **2.** (c) **3.** (a) **4.** (c) **5.** (d) **6.** (d) **7.** (d) **8.** (a) **9.** (c) **10.** (c)

11. (a) **12.** (a) **13.** (a) **14.** (d) **15.** (b) **16.** (d) **17.** (a)

Numerical Type Question

1. (18)

EXPLANATIONS

MCQ TYPE QUESTIONS

4. The differential equation

$$x(ydx + xdy)\cos\frac{y}{x} = y(xdy - ydx)\sin\frac{y}{x}$$

$$\therefore\ x(ydx+xdy)\cos\frac{y}{x} + \left(\sin\frac{y}{x}\right)y(xdy - ydx) = 0$$

$$\therefore\ (ydx+xdy)\cos\left(\frac{y}{x}\right) + \left(-\sin\frac{y}{x}\right)y\left(\frac{xdy-ydx}{x}\right) = 0$$

$$\therefore\ (ydx + xdy)\cos\left(\frac{y}{x}\right) + (xy)\left(-\sin\frac{y}{x}\right)\left(\frac{xdy-ydx}{x^2}\right) = 0$$

Now by intergrating we get

$$xy\cos\left(\frac{y}{x}\right) = c$$

5. The integrating factor for the differential equation

$$\frac{dp}{dt} + K_2 p = K_1 L_0 e^{-K_1 t}$$

is $e^{K_2 t}$

6. *Given :* $\frac{dy}{dx} + 2y = 0$

and at $x = 1, y = 5$

Differential equation is,

$$(D + 2)y = 0$$

$\Rightarrow$ $D = -2$

It's solution is $y = C_1 e^{-2x}$

At $x = 1$,

$$5 = c_1 e^{-2}$$

$\Rightarrow$ $5e^2 = c_1 = 36.95$

Hence required solution is

$$y = 3695e^{-2x}$$

7. $$\frac{dy}{dx} + \frac{y}{x} = x$$

Putting $v = y^{-1}$

$\Rightarrow$ $dv = -y^{-2}\,dy = -v^2 \cdot dy$

$$\therefore\ \frac{-v(v^{-2}dv)}{dx} + \frac{y}{x} = x$$

$$\Rightarrow\ -(v^{-1}x)\frac{dv}{dx} + y = x^2$$

$$\Rightarrow\ -(v^{-1}x)\frac{dv}{dx} + v^{-1} = x^2$$

$$\Rightarrow\ +\int d(v^{-1}x) = \int x^2 dx$$

$$\Rightarrow\ v^{-1}x = \frac{x^3}{3} + c$$

$$\Rightarrow\ y = \frac{x^2}{3} + \frac{c}{3x}$$

At $x = 1, y = 1$

$$\therefore\ 1 = \frac{1}{3} + \frac{c}{3}$$

$\Rightarrow$ $c = 2$

Hence $$y = \frac{x^2}{3} + \frac{2}{3x}$$

8. Given : $$\frac{d^3y}{dx^3} + 4\sqrt{\left(\frac{dy}{dx}\right)^3 + y^2} = 0$$

$$\Rightarrow\ \frac{d^3y}{dx^3} = -4\sqrt{\left(\frac{dy}{dx}\right)^3 + y^2}$$

$$\Rightarrow\ \left(\frac{d^3y}{dx^3}\right)^2 = 16\left[\left(\frac{dy}{dx}\right)^3 + y^2\right]$$

$$\Rightarrow\ \left(\frac{d^3y}{dx^3}\right)^2 - 16\left(\frac{dy}{dx}\right)^3 - 16y^2 = 0$$

So, order = 3 and degree = 2

9. Given : $$\frac{d^2y}{dx^2} + \frac{dy}{dx} - 6y = 0$$

$\therefore$ $(D^2 + D - 6)y = 0$

It's auxiliary equation is

$$D^2 + D - 6 = 0$$

$\Rightarrow$ $(D + 3)(D - 2) = 0$

$\Rightarrow$ $D = -3, 2$

Hence solution is,

$$y = c_1 e^{-3x} + c_2 e^{2x}$$

10. Given : $z = ax + by + ab$ ...(i)

$$p = \frac{\partial z}{\partial x} = a$$

and $$q = \frac{\partial z}{\partial y} = b$$

Putting values of a and b in equation (i), we get

$$z = px + qy + pq$$

13. $\frac{d^2y}{dx^2} + y = 0$

$\Rightarrow$ $(D^2 + 1)\, y = 0$ $\quad \ldots \left(D = \frac{d}{dx}\right)$

Auxiliary equation is,

$D^2 + 1 = 0$

$\therefore$ $D = \pm i$

So its solution is,

$y = C_1 e^{ix} + C_2 e^{-ix}$

Using Euler's theorem,

$y = C_1 e^{\circ} (\cos x + i \sin x) + C_2 e^{\circ} (\cos x - i \sin x)$

$y = (C_1 + C_2) \cos x + i (C_1 - C_2) \sin x \quad \ldots (i)$

Now let $C_1 + C_2 = P$ and $i (C_1 - C_2) = Q$

Hence from equation (i)

$y = P \cos x + Q \sin x$

14. Given : $\frac{dy}{dx} = \frac{-x}{y}$

$\Rightarrow$ $y \,.\, dy = -x\, dx$

Integrating, $\frac{y^2}{2} = \frac{-x^2}{2} + c$

But at $x = 1, y = \sqrt{3}$

$\Rightarrow$ $c = 2$

Hence solution is,

$\frac{y^2}{2} = \frac{-x^2}{2} + 2$

$\Rightarrow$ $y^2 + x^2 = 4$

15. Degree of differential equation $\frac{d^2x}{dt^2} + 2x^3 = 0$ is 1.

16. Given : $\frac{dy}{dx} = x^2 y$

Integrating $\int \frac{dy}{y} = \int x^2 dx$

$\Rightarrow$ $\log y = \frac{x^3}{3} C$

At $x = 0, y = 1$

$\Rightarrow$ $c = 0$

$\therefore$ $\log y = \frac{x^3}{3}$

$\Rightarrow$ $y = e^{x^3/3}$

17. Given, differential equation is

$\frac{dy}{dx} + \left(\frac{2}{x}\right) y = \frac{1}{x} - \frac{1}{x^2}$

$\therefore$ Integrating factor (I. F.)

$= e^{\int \left(\frac{2}{x}\right) dx}$

$= x^2$

Its solution is,

$x^2 y = \int x^2 \left(\frac{1}{x} - \frac{1}{x^2}\right) dx$

$= \int (x - 1)\, dx$

$= \frac{x^2}{2} - x + c$

From end condition : $c = \frac{1}{2}$

$\therefore$ $x^2 y = \frac{x^2}{2} - x + \frac{1}{2}$

$\Rightarrow$ $y = \frac{1}{2} - \frac{1}{x} + \frac{1}{2x^2}$

4 CHAPTER Probability and Statistics

MCQ TYPE QUESTIONS

2016

1. Type II error in hypothesis testing is

(*a*) acceptance of the null hypothesis when it is false and should be rejected

(*b*) rejection of the null hypothesis when it is true and should be accepted

(*c*) rejection of the null hypothesis when it is false and should be rejected

(*d*) acceptance of the null hypothesis when it is true and should be accepted

2. Probability density function of a random variable X is given below

$$f(x) = \begin{cases} 0.25 & \text{if } 1 \le x \le 5 \\ 0 & \text{otherwise} \end{cases}$$

$P(X \le 4)$ is

(*a*) $\frac{3}{4}$ (*b*) $\frac{1}{2}$

(*c*) $\frac{1}{4}$ (*d*) $\frac{1}{8}$

3. X and Y are two random independent events. It is known that $P(X) = 0.40$ and $P(X \cup Y^C) = 0.7$. Which one of the following is the value of $P(X \cup Y)$?

(*a*) 0.7 (*b*) 0.5

(*c*) 0.4 (*d*) 0.3

4. If $f(x)$ and $g(x)$ are two probability density functions,

$$f(x) = \begin{cases} \frac{x}{a} + 1 & : -a \le x < 0 \\ -\frac{x}{a} + 1 & : 0 \le x \le a \\ 0 & : \text{otherwise} \end{cases}$$

$$g(x) = \begin{cases} -\frac{x}{a} & : -a \le x < 0 \\ \frac{x}{a} & : 0 \le x \le a \\ 0 & : \text{otherwise} \end{cases}$$

Which one of the following statements is true?

(*a*) Mean of $f(x)$ and $g(x)$ are same; Variance of $f(x)$ and $g(x)$ are same

(*b*) Mean of $f(x)$ and $g(x)$ are same; Variance of $f(x)$ and $g(x)$ are different

(*c*) Mean of $f(x)$ and $g(x)$ are different; Variance of $f(x)$ and $g(x)$ are same

(*d*) Mean of $f(x)$ and $g(x)$ are different; Variance of $f(x)$ and $g(x)$ are different

2014

5. A fair (unbiased) coin was tossed four times in succession and resulted in the following outcomes: (i) Head, (ii) Head, (iii) Head, (iv) Head. The probability of obtaining a 'Tail' when the coin is tossed again is

(*a*) 0

(*b*) $\frac{1}{2}$

(*c*) $\frac{4}{5}$

(*d*) $\frac{1}{5}$

2012

6. The annual precipitation data of a city is normally distributed with mean and standard deviation as 1000 mm and 200 mm, respectively. The probability that the annual precipitation will be more than 1200 mm is

(*a*) < 50% (*b*) 50%

(*c*) 75% (*d*) 100%

7. In an experiment, positive and negative values are equally likely to occur. The probability of obtaining at mosi one negative value in five trials is

(*a*) $\frac{1}{32}$

(*b*) $\frac{2}{32}$

(*c*) $\frac{3}{32}$

(*d*) $\frac{6}{32}$

2011

8. There are two containers, with one containing 4 Red and 3 Green balls and the other containing 3 Blue and 4 Green balls. One ball is drawn at random from each container. The probability that one of the balls is Red and the other is Blue will be

(a) 1/7 (b) 9/49

(c) 12/49 (d) 3/7

2010

9. Two coins are simultaneously tossed. The probability of two heads simultaneously appearing is

(a) $\frac{1}{8}$ (b) $\frac{1}{6}$

(c) $\frac{1}{4}$ (d) $\frac{1}{2}$

2009

10. In the solution of the following set of linear equations by Gauss elimination using partial pivoting

$$5x + y + 2z = 34;$$

$$4y - 3z = 12;$$

and $$10x - 2y + z = -4;$$

the pivots for elimination of x and y are

(a) 10 and 4

(b) 10 and 2

(c) 5 and 4

(d) 5 and – 4

11. The standard normal probability function can be approximated as

$$Fx_N = \frac{1}{1+\exp\left(-1.7255x_N|x_N|^{0.12}\right)}$$

where x_N = standard normal deviate.

If mean and standard deviation of annual precipitation are 102 cm and 27 cm respectively, then probability that the annual precipitation will be between 90 cm and 102 cm is

(a) 66.7 %

(b) 20.0 %

(c) 33.3 %

(d) 16.7 %

2008

12. If probability density function of a random variable X is

$f(x) = x^2$ for $-1 \le x \le 1$; and

= 0 for any other value of x

then, percentage probability $P\left(-\frac{1}{3} \le x \le \frac{1}{3}\right)$ is

(a) 0.247

(b) 2.47

(c) 24.7

(d) 247

13. A person on a trip has a choice between private car and public transport. The probability of using a private car is 0.45. While using the public transport, further choices available are bus and metro, out of which probability of commuting by a bus is 0.55. In such a situation, the probability (rounded up to two decimals) of using a car, bus and metro, respectively would be

(a) 0.45, 0.30 and 0.25

(b) 0.45, 0.25 and 0.30

(c) 0.45, 0.55 and 0.00

(d) 0.45, 0.35 and 0.20

2006

14. A class of first year B. Tech students is composed of four batches A, B, C and D, each consisting of 30 students. It is found that the sessional marks of students in Engineering Drawing in batch C have a mean of 6.6 and standard deviation of 2.3. The mean and standard deviation of the marks for the entire class are 5.5 and 4.2, respectively. It is decided by the course instructor to normalize the marks of the students of all batches to have the same mean and standard deviation as that of the entire class. Due to this, the marks of a student in batch C are changed from 8.5 to

(a) 6.0

(b) 7.0

(c) 8.0

(d) 9.0

15. There are 25 calculators in a box. Two of them are defective. Suppose 5 calculators are randomly picked for inspection (*i.e.* each has the same chance of being selected), what is the probability that only one of the defective calculators will be included in the inspection ?

(*a*) $\frac{1}{2}$

(*b*) $\frac{1}{3}$

(*c*) $\frac{1}{4}$

(*d*) $\frac{1}{5}$

NUMERICAL TYPE QUESTIONS

2015

1. Consider the following probability mass function (p.m.f.) of a random variable X :

$$p(x, q) = \begin{cases} q & \text{if } X = 0 \\ 1-q & \text{if } X = 1 \\ 0 & \text{otherwise} \end{cases}$$

If $q = 0.4$, the variance of X is _______ .

2. For probability density function of a random variable, x is

$$f(x) = \frac{x}{4}(4 - x^2) \text{ for } 0 \le x \le 2$$

$$= 0 \text{ otherwise}$$

The mean $_z$ of the random varialbe is _____

ANSWERS

MCQ Type Questions

1. (*a*) **2.** (*a*) **3.** (*a*) **4.** (*b*) **5.** (*b*) **6.** (*a*) **7.** (*d*) **8.** (*c*) **9.** (*c*) **10.** (*a*)
11. (*b*) **12.** (*b*) **13.** (*) **14.** (*d*) **15.** (*b*)

Numerical Type Questions

1. (0.24) **2.** (1.0667)

EXPLANATIONS

MCQ TYPE QUESTIONS

5. The probability of getting 'tail' when the coin is tossed again is $\frac{1}{2}$

6. $1 - P(7) = 1 - \left(\frac{1200-1000}{200}\right)$

$= 1 - P\left(\frac{1}{T}\right)$

$= 0.159 < 50\%$

7. $p = q = \frac{1}{2}$

Probability $= {}^{n}C_r \; p^r q^{n-r}$

$= {}^{5}C_0 \left(\frac{1}{2}\right)^0 \left(\frac{1}{2}\right)^5 + {}^{5}C_1 \left(\frac{1}{2}\right)^1 \left(\frac{1}{2}\right)^4$

$= \frac{6}{32}$

8. $P(AB) = P(A) \cdot P(B)$

$= \frac{4}{7} \times \frac{3}{7} = \frac{12}{49}$

9. Probability $= \frac{\text{Number of outcomes favourable}}{\text{Total number of possible outcomes}}$

Total number of possible outcomes

(HH, TT, HT, TH) = 4

Number of outcomes favourable (HH) = 1

$\therefore$ Required probability $= \frac{1}{4}$

12. $P\left(-\frac{1}{3} \le x \le \frac{1}{3}\right) = \int_{-\frac{1}{3}}^{\frac{1}{3}} x^2\, dx$

$= \left[\frac{x^3}{3}\right]_{-1/3}^{1/3}$

$= 0.0247 = 2.47\ \%$

15. Required probability

$= \frac{{}^{23}C_4 \cdot {}^{2}C_1}{{}^{25}C_5} = \frac{1}{3}$

NUMERICAL TYPE QUESTION

1. According to question

x	0	1
$P(x,q)$	q	$1-q$

Now $q = 0.4$ (given)

$\therefore$

x	0	1
$P(x,q)$	0.4	$(1-0.4) = 0.6$

Now variance of x i.e $v(x) = E(x^2) - [E(x^2)]^2$

Now $E(x) = \sum x_i P(x,q) = 0 \times 0.4 + 1 \times 0.6$

$= 0 + 0.6 = 0.6$

and $E(x^2) = \sum x_i^2 P(x,q)$

$= (0)^2 \times 0.4 + (1)^2 \times 0.6$

$= 0 + 0.6 = 0.6$

$\therefore$ $V(x) = E(x) - [E(x^2)]^2$

$= 0.6 - (0.6)^2 = 0.6 - 0.36 = 0.24$

5

CHAPTER

Numerical Methods

MCQ TYPE QUESTIONS

2015

1. The integral $\int_{x_1}^{x_2} x^2 dx$ with $x_2 > x_1 > 0$ is evaluated analytically as well as numerically using a single application of the trapezoidal rule. If I is the exact value of the integral obtained analytically and J is the approximate value obtained using the trapezoidal rule, which of the following statements is correct about their relationship ?

(a) J > 1

(b) J < 1

(c) J = 1

(d) Insufficient data to determine the relationship

2012

2. The estimate of $\int_{0.5}^{1.5} \frac{dx}{x}$ obtained using Simpson's rule with three-point function evaluation exceeds the exact value by

(a) 0.235 (b) 0.068

(c) 0.024 (d) 0.012

3. The error in $\left.\frac{d}{dx}f(x)\right|_{x=x_0}$ for a continuous function estimated with $h = 0.03$ using the central difference formula $\left.\frac{d}{dx}f(x)\right|_{x=x_0} \approx \frac{f(x_0+h)-f(x_0-h)}{2h}$, is 2×10^{-3}. The values of x_0 and $f(x_0)$ are 19.78 and 500.01, respectively. The corresponding error in the central difference estimate for $h = 0.02$ is approximately

(a) 1.3×10^{-4}

(b) 3.0×10^{-4}

(c) 4.5×10^{-4}

(d) 9.0×10^{-4}

2011

4. The square root of a number N is to be obtained by applying the Newton Raphson iterations to the equation $x^2 - N = 0$. If i denotes the iteration index, the correct iterative scheme will be

(a) $x_{i+1} = \frac{1}{2}\left(x_i + \frac{N}{x_i}\right)$ (b) $x_{i+1} = \frac{1}{2}\left(x_i^2 + \frac{N}{x_i^2}\right)$

(c) $x_{i+1} = \frac{1}{2}\left(x_i + \frac{N^2}{x_i}\right)$ (d) $x_{i+1} = \frac{1}{2}\left(x_i - \frac{N}{x_i}\right)$

2010

5. The table below gives values of a function F(x) obtained for values of x at intervals of 0.25.

x	0	0.25	0.5	0.75	1.0
F(x)	1	0.9412	0.8	0.64	0.50

The value of the integral of the function between the limits 0 to 1 using Simpson's rule is

(a) 0.7854 (b) 2.3562

(c) 3.1416 (d) 7.5000

2007

6. A body originally at 60°C cools down to 40°C in 15 minutes when kept in air at a temperature of 25°C. What will be the temperature of the body at the end of 30 minutes ?

(a) 35.2°C (b) 31.5°C

(c) 28.7°C (d) 15°C

7. The following equation needs to be numerically solved using the Newton–Raphson method.

$$x^3 + 4x - 9 = 0$$

The iterative equation for this purpose is (k indicates the iteration level)

(a) $x_{k+1} = \frac{2x_k^3 + 9}{3x_k^2 + 4}$ (b) $x_{k+1} = \frac{3x_k^2 + 4}{2x_k^2 + 9}$

(c) $x_{k+1} = x_k - 3x_k^2 + 4$ (d) $x_{k+1} = \frac{4x_k^2 + 3}{9x_k^2 + 2}$

8. If one root of the equation

$$x^3 - 10x^2 + 31x - 30 = 0$$

is 5, then other two roots are

(a) 2 and 3 (b) 2 and 4

(c) 3 and 4 (d) –2 and –3

2006

9. A 2nd degree polynomial, $f(x)$, has values of 1, 4 and 15 at $x = 0$, 1, and 2, respectively. The integral $\int_0^2 f(x)\,dx$ is to be estimated by applying the trapezoidal rule to this data. What is the error (defined as "true value — approximate value") in the estimate ?

(*a*) $-\frac{4}{3}$ (*b*) $-\frac{2}{3}$

(*c*) 0 (*d*) $\frac{2}{3}$

NUMERICAL TYPE QUESTIONS

2016

1. Newton-Raphson method is to be used to find root of equation $3x - e^x + \sin x = 0$. If the initial trial value for the root is taken as 0.333, the next approximation for the root would be ________ (*note: answer up to three decimal*)

2015

2. The quadratic equation $x^2 - 4x + 4 = 0$ is to be solved numerically, starting with the initial guess $x_0 = 3$. The Newton-Raphson method is applied once to get a new estimate and then the Secant method is applied once using the initial guess and this new estimate. The estimated value of the root after the application of the Secant method is ________.

3. In Newton-Raphson iterative method, the initial guess value (x_{ini}) is considered as zero while finding the roots of the equation: $f(x) = -2 + 6x - 4x^2 + 0.5x^3$. The correction, Δx, to be added to x_{ini} in the first iteration is ________.

4. For step-size $\Delta x = 0.4$, the value of following integral using Simpson 's 1/3 rule is ________.

$$\int_0^{0.8} \left(0.2 + 25x - 200x^2 + 675x^3 - 900x^4 + 400x^5\right) dx$$

2013

5. There is no value of x that can simultaneously satisfy both the given equations. Therefore, find the 'least squares error' solution to the two equations, i.e., find the value of x that minimizes the sum of squares of the errors in the two equations. ________

$$2x = 3$$

$$4x = 1$$

6. Find the magnitude of the error (correct to two decimal places) in the estimation of following integral using Simpson's $\frac{1}{3}$ Rule. Take the step length as 1. ________

$$\int_0^4 (x^4 + 10)\,dx$$

7. Find the value of λ such that the function $f(x)$ is a valid probability density function.

$$f(x) = \begin{cases} \lambda(x-1)(2-x), & 1 \le x \le 2 \\ 0, & \text{otherwise} \end{cases}$$

ANSWERS

MCQ Type Questions

1. (*a*) **2.** (*d*) **3.** (*d*) **4.** (*a*) **5.** (*a*) **6.** (*b*) **7.** (*a*) **8.** (*a*) **9.** (*a*)

Numerical Type Questions

1. (0.3601) **2.** 2.32 to 2.34 **3.** 0.3333 **4.** 1.36 to 1.37 **5.** (0.5) **6.** (0.52 to 0.55)

7. (6)

EXPLANATIONS

MCQ TYPE QUESTIONS

1. By trapezoidal rule,

Approximated value of $\int_a^b f(x)dx$ is always greater than analytical value

$\therefore \quad J > I$

Here J is approximate value and I is analytical value.

2. Exact value of $\int_{0.5}^{1.5} \frac{1}{x}dx = 1.0986$

Using Simpson's rule in three-point form,

$$h = \frac{b-a}{2} = \frac{1.5-0.5}{2} = 0.5$$

So,

x	0.5	1	1.5
y	2	1	0.67

$$\int \frac{1}{x}dx = \frac{h}{3}[y_0 + y_n + 4_{y1}]$$

$$= \frac{0.5}{3}[2 + 0.67 + 4 \times 3] = 1.1116$$

It exceeds exects by 0.012 (approx)

4.
$$x_{i+1} = x_i - \frac{f(x_i)}{f'(x_i)}$$

$$f(x) = x^2 - N$$

$\Rightarrow \quad f'(x_i) = 2x_i,$

and $\quad f(x_i) = x_i^2 - N$

$$\therefore \quad x_{i+1} = x_i - \frac{x_i^2 - N}{2x_i}$$

$$= \frac{1}{2}\left[x_i + \frac{N}{x_i}\right]$$

5. By Simpson's rule,

$$\int_h^b f(x)dx \approx \frac{h}{3}(f_0 + 4f_1 + 2f_2 + 4f_3 + ... + 2f_{2m-2} + 4f_{2m-1} + f_{2m})$$

Here, $h = 0.25$,

$f_0 = 1$,

$f_1 = 0.9412$,

$f_2 = 0.8$,

$f_3 = 0.64$,

$f_4 = 0.50$

$$\therefore \int_h^6 f(x)dx \approx \frac{0.25}{3}(1 + 4 \times 0.9412 + 2 \times 0.8 + 4 \times 0.64 + 0.5$$

$$= 0.7854$$

Alternately

Given : $h = 0.25$.

By Simpson's rule

$$\int_{x_0} f(x)dx$$

$$= \frac{h}{3}[y_o + y_n + 4(y_1 + y_3 + y_{n-1}) + 2(y_2 + y_4 + ...y_{n-2}]$$

$$= \frac{0.25}{3}[1 + 0.5 + 4(0.9412 + 0.64) + 2(0.8)]$$

$$= 0.7854$$

6. According to Stiffen law of cooling

$$\frac{d\theta}{dt} = k(\theta - \theta_0)$$

Case I : $\theta = \frac{60° + 40°}{2}$ and $\theta_0 = 25°$

$$\frac{60° - 40}{15 - 0} = k\left[\frac{60 + 40}{2} - 25\right] \quad ...(i)$$

Case II :

$$\frac{40 - x}{30 - 15} = k\left[\frac{40 + x}{2} - 25\right] \quad ...(ii)$$

Divide equation (*i*) by (*ii*), we get

$$\frac{20}{15} \times \frac{15}{40 - x} = \frac{k(50 - 25)}{k\left(\frac{40 + x}{2} - 25\right)}$$

$$\Rightarrow \quad 25(40 - x) = 20\left[\frac{40 + x}{2} - 25\right]$$

$$\Rightarrow \quad 1000 - 25x = 400 + 10x - 500$$

$$\Rightarrow \quad 35x = 1100$$

$$\Rightarrow \quad x = \frac{1100}{35}$$

$$= 31.43°C$$

7.
$$f(x) = x^3 + 4x - 9 = 0$$

According to Newton's–Raphson's method.

$$x_{k+1} = x_k - \frac{f(x)}{f'(x)}$$

$$\Rightarrow \quad x_{k+1} = x_k - \frac{x_k^3 + 4x_k - 9}{3x_k^2 + 4}$$

$$\Rightarrow \quad x_{k+1} = \frac{3x_k^3 + 4x_k - x_k^3 - 4x_k + 9}{3x_k^2 + 4}$$

$$\Rightarrow \quad 2x_{k+1} = \frac{2x_k^3 + 9}{3x_k^2 + 4}$$

8. *Given* : $x^3 - 10x^2 + 31x - 30 = 0$

Its one root is 5, hence

$$(x-5)(x^2-5x+6)=0$$

$$\Rightarrow \quad (x-5)(x^2-3x-2x+6)=0$$

$$\Rightarrow \quad (x-5)(x-3)(x-2)=0$$

$$\Rightarrow \quad x = 5, 3, 2$$

NUMERICAL TYPE QUESTION

2. Here $f(x) = x^2 - 4x + 4 = 0$

$$\therefore \quad f'(x) = \frac{d}{dx}(x^2-4x+4) = 2x-4$$

Now initial condition is at $x_0 = 3$

By Newton Rapson Method

$$x_1 = x_0 - \frac{f(x_0)}{f'(x_0)}$$

Now $f(x_0) = f(3) = (3)2 - 4(3) + 4$

$= 9 - 12 + 4 = 1$

and $f'(x_0) = f'(3) = 2(3) - 4 = 2$

$$\therefore \quad x_1 = 3 - \frac{1}{2} = \frac{5}{2}$$

Now for Secant Method

$$x_0 = \frac{5}{2} \text{ and } x_1 = 3$$

$\therefore$ Secant Method

$$x_2 = x_1 - \frac{(x_1-x_0)}{f(x_1)-f(x_0)} \times f(x_1)$$

Now $f(x_0) = f\left(\frac{5}{2}\right) = \left(\frac{5}{2}\right)^2 - 4\left(\frac{5}{2}\right) + 4$

$$= \frac{25}{4} - 10 + 4 = \frac{25}{4} - 6 = \frac{1}{4}$$

and $f(x_1) = f(3) = 1$

$$\therefore \quad x_2 = x_1 - \frac{(x_1-x_0)}{f(x_1)-f(x_0)} \times f(x_1)$$

$$= 3 - \frac{\left(3-\frac{5}{2}\right)}{1-\frac{1}{4}} \times 1$$

$$= 3 - \frac{\left(\frac{1}{2}\right)}{\frac{3}{4}}$$

$$= 3 - \frac{4}{6} = \frac{12}{6} = 2$$

Thus estimated value of root after the application of secant method is 1.33.

Technical Section

1 CHAPTER Engineering Mechanics

MCQ TYPE QUESTIONS

2016

1. The magnitudes of vectors **P**, **Q** and **R** are 100 kN, 250 kN and 150 kN, respectively as shown in the figure.

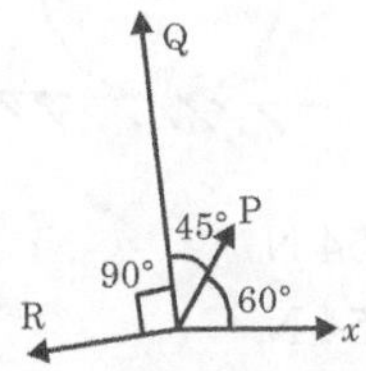

The respective values of the magnitude (in kN) and the direction (with respect to the x-axis) of the resultant vector are

(*a*) 290.9 and 96.0° (*b*) 368.1 and 94.7°
(*c*) 330.4 and 118.9° (*d*) 400.1 and 113.5°

2. A rigid member ACB is shown in the figure. The member is supported at A and B by pinned and guided roller supports, respectively. A force **P** acts at C as shown. Let R_{Ah} and R_{Bh} be the horizontal reactions at supports A and B, respectively, and R_{Av} be the vertical reaction at support A. Selfweight of the member may be ignored.

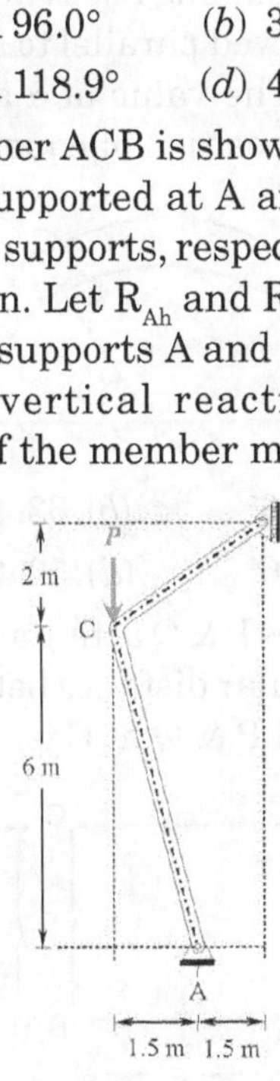

Which one of the following sets gives the correct magnitudes of R_{Av}, R_{Bh} and R_{Ah} ?

(*a*) $R_{AV} = 0; R_{Bh} = \frac{1}{3}P$; and $R_{Ah} = \frac{2}{3}P$

(*b*) $R_{AV} = 0; R_{Bh} = \frac{2}{3}P$; and $R_{Ah} = \frac{1}{3}P$

(*c*) $R_{AV} = P; R_{Bh} = \frac{3}{8}P$; and $R_{Ah} = \frac{1.5}{8}P$

(*d*) $R_{AV} = P; R_{Bh} = \frac{1.5}{8}P$; and $R_{Ah} = \frac{1.5}{8}P$

3. An assembly made of a rigid arm A-B-C hinged at end A and supported by an elastic rope C-D at end C is shown in the figure. The members may be assumed to be weightless and the lengths of the respective members are as shown in the figure.

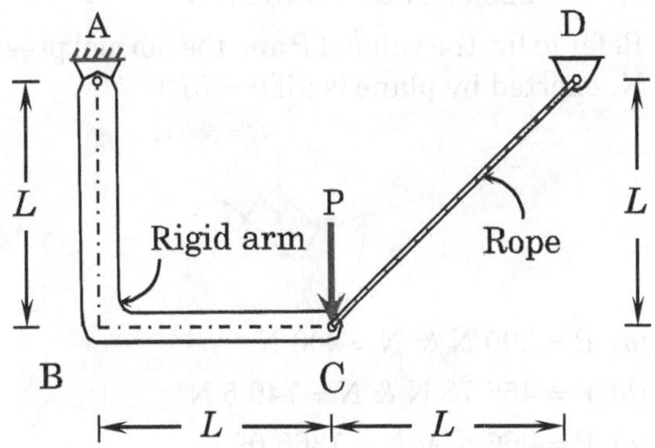

Under the action of a concentrated load P at C as shown, the magnitude of tension developed in the rope is

(*a*) $\frac{3P}{\sqrt{2}}$

(*b*) $\frac{P}{\sqrt{2}}$

(*c*) $\frac{3P}{8}$

(*d*) $\sqrt{2}$ P

4. An elastic isotropic body is in a hydrostatic state of stress as shown in the figure. For no change in the volume to occur, what should be its Poisson's ratio?

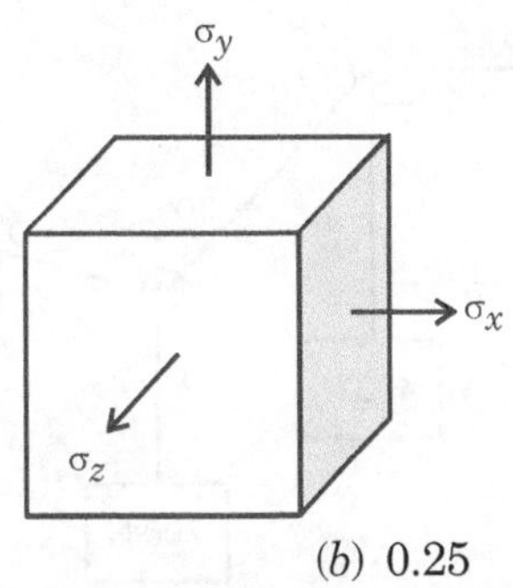

(*a*) 0.00 (*b*) 0.25
(*c*) 0.50 (*d*) 1.00

2015

5. Refer to fig. the value of P & N exerted on cylinder by road & inclination where weight of cylinder is = 700N.

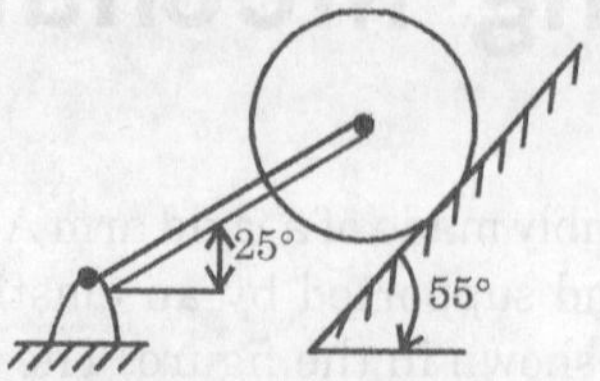

(*a*) P = 732.5 N & N = 732.5 N

(*b*) P = 662.05 N & N = 732.5 N

(*c*) P = 662.05 N & N = 222.85 N

(*d*) P = 222.85 N & N = 732.5 N

6. Refer to fig. the value of P and the normal pressure N, exerted by plane is :(if θ = 75°)

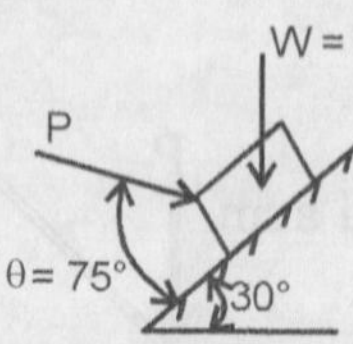

(*a*) P = 500 N & N = 900 N

(*b*) P = 456.78 N & N = 740.5 N

(*c*) P = 966 N & N = 1366.09 N

(*d*) P = 900 N & N = 500 N

7. Refer to fig., the value of P & F is:

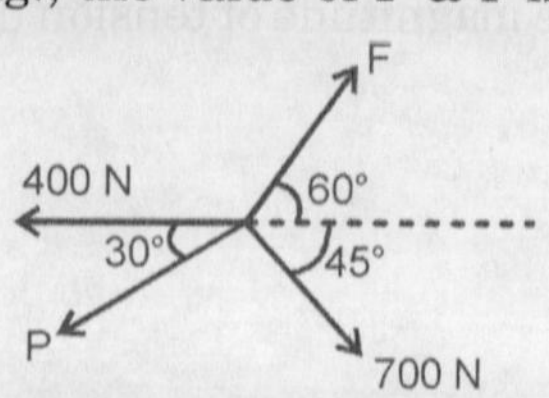

(*a*) P = 120.79 N & F = 397.20 N

(*b*) P = 250.05 N & F = 397.20 N

(*c*) P = 397.20 N & F = 250.05 N

(*d*) P = 120.79 N & F = 250.05 N

8. The system of knotted cords is shown in fig., the value of tensile force in each cord A, B, C, D respectively is:

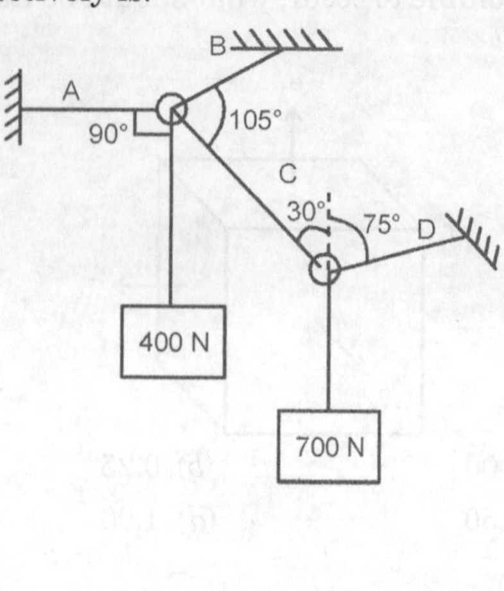

(*a*) 2000.05, 1919.52, 1000.00, 785.55 N

(*b*) 2111.65, 1857.66, 998.25, 571.37 N

(*c*) 2314.4, 1919.52, 1105.17, 571.37 N

(*d*) None of the above

9. Refer to fig., cords are loop around a small spacer separating two cylinders each weighing 500 N and pass over a frictionless pulley 700 N and 300 N, the value of θ and normal pressure N between cylinder and the smooth horizontal surface.

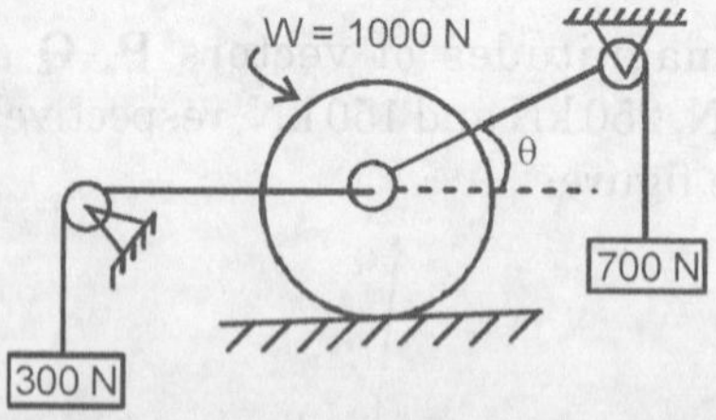

(*a*) 64.6°, 367.54 N

(*b*) 34.6°, 367.54 N

(*c*) 64.6°, 360.54 N

(*d*) 34.6°, 350.50 N

10. Two cylinder weighing 200 N and 400 N connected by rigid rod curved parallel to smooth cylindrical surface (fig.). The value of α and β respectively which defines position of equation is :

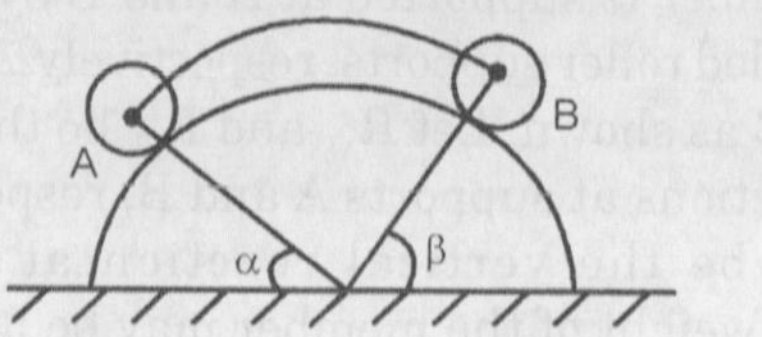

(*a*) 55.43°, 26.56° (*b*) 63.43°, 26.56°

(*c*) 63.43°, 30.56° (*d*) 50.43°, 20.56°

11. ABCD has forces P & Q acting along its sides (Fig.). The perpendicular distance between resultant of P & Q at A and P & Q at C is:

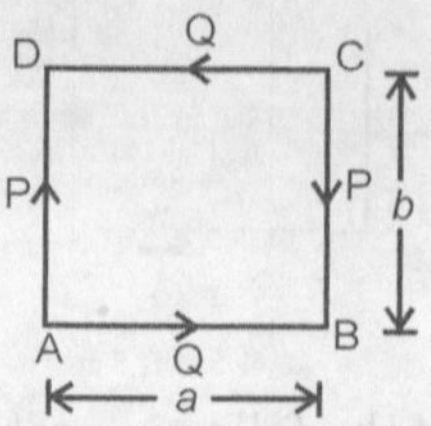

(*a*) $\dfrac{P.a - Q.b}{\sqrt{P^2 + Q^2}}$

(*b*) $\dfrac{Q.b - P.a}{\sqrt{P^2 + Q^2}}$

(*c*) $\dfrac{Q.a - P.b}{\sqrt{P^2 + Q^2}}$

(*d*) $\dfrac{a.b}{\sqrt{P^2 + Q^2}}$

12. A man of weight, W = 850 N holds one end of rope that passes over a pulley vertically above his head and to the other end of which is attached a weight Q = 635 N. Find the force with which man's feet press against the floor. (Fig.)

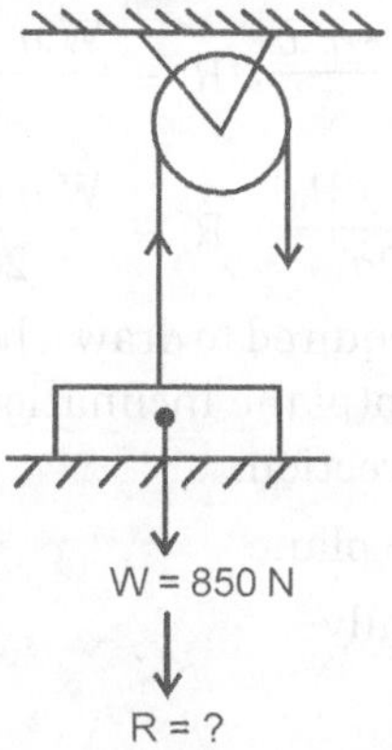

(*a*) 215 N upwards

(*b*) 300 N downwards

(*c*) 215 N downwards

(*d*) 300N upwards

13. A bar AB of weight Q and length $2l$ rests on a very small frictionless roller at α and against a smooth vertical wall at A, the value of 'α' that the bar must make with the horizontal in equilibrium. (Fig.)

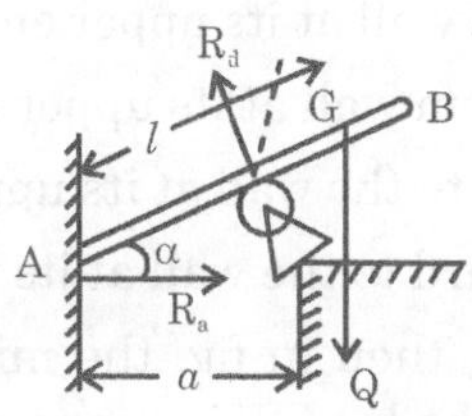

(*a*) $\cos^{-1}\sqrt[3]{\ell}$ (*b*) $\cos^{-1}\sqrt[3]{\dfrac{a}{\ell}}$

(*c*) $\cos^{-1}\sqrt[3]{\dfrac{\ell}{a}}$ (*d*) $\cos^{-1}\sqrt[3]{a}$

14. The beam AB supports a load which varies with an intensity of 440 N/m to 660 N/m, (fig.) the magnitude and position of the resultant load respectively are :

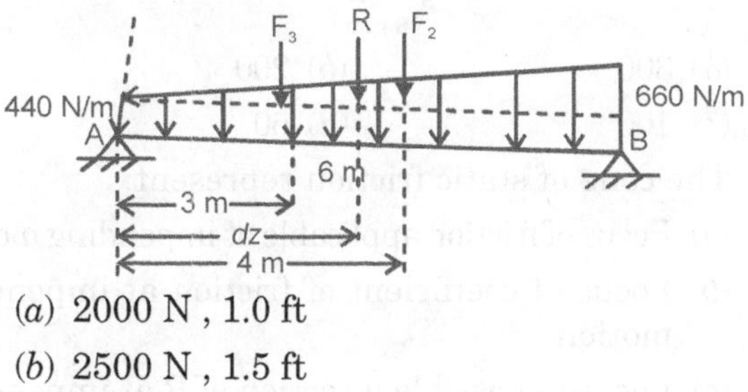

(*a*) 2000 N , 1.0 ft

(*b*) 2500 N , 1.5 ft

(*c*) 2988 N , 2.5 ft

(*d*) 3960 N , 3.33 ft

15. A 16ft wing of an airplane is subjected to a lift which varies from zero at the tip to 400N per ft at the fuselage according to $w = 90x^{1/2}$ N per ft, where x is measured from tip, the value of resultant force acting on wing and its location from wing tip respectively are:

(*a*) 3840 N , 9.6 ft

(*b*) 3000 N , 9.6 ft

(*c*) 3550 N , 3.6 ft

(*d*) 3840 N , 4.6 ft

16. The effect of certain non-concurrent force system is defined by ΣF_x = + 80 kN, ΣF_y = –40 kN and ΣM_o = 300 kN-m counter clockwise, the value of the point at which the resultant intersects the x-axis and y-axis respectively is:

(*a*) $a_x = 7.5\ m,\ a_y = 4.75\ m$

(*b*) $a_x = 7.5\ m,\ a_y = 3.75\ m$

(*c*) $a_x = 2.5\ m,\ a_y = 3.75\ m$

(*d*) $a_x = 6.5\ m,\ a_y = 2.75\ m$

17. A cantilever beam is built into wall of 3.5 ft. thickness so that it rest against A & B, the beam is 15 ft. long & weighs 150 N per ft., (fig.) the value of reactions at A & B respectively are :

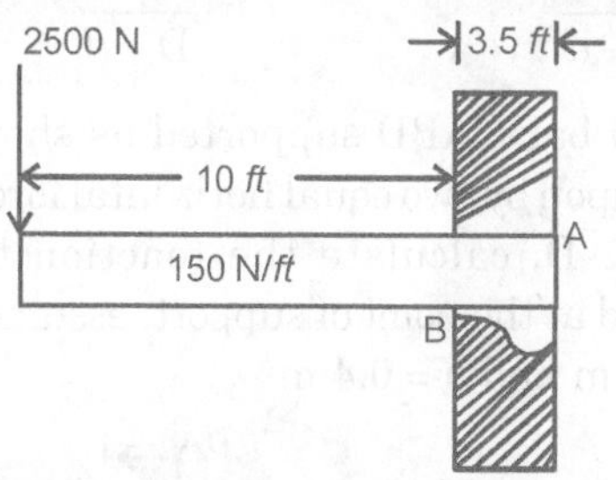

(*a*) 14550 N, 2769.90 N

(*b*) 16550 N, 23835.9 N

(*c*) 15000 N, 2500 N

(*d*) None of the above

18. In fig., the upper beam is supported at D and a roller at C which separates the upper and lower beam, the value of reactions A, B, C & D respectively are:-

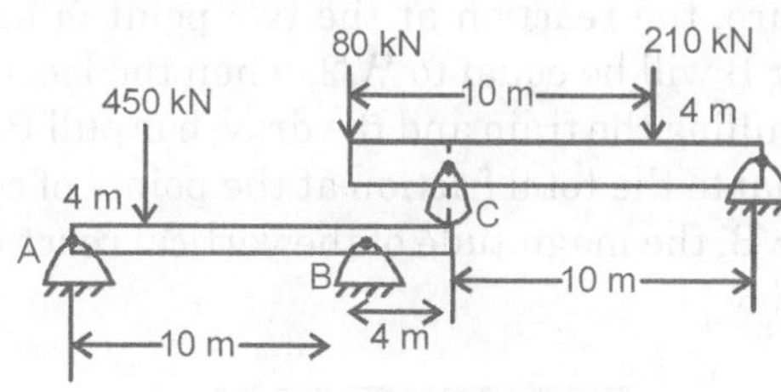

(*a*) 191.6 N, 454.4N, 196N, 94N

(*b*) 291.6 N, 454.4N, 196N, 94N

(*c*) 191.6 N, 354.4N, 200N, 94N

(*d*) 191.6 N, 454.4N, 200N, 90N

19. The differential chain hoist consists of two concentric pulleys rigidly fastened together, the pulleys from two sprockets for an endless chain looped over them in two loops, in one loop is mounted a movable pulley supporting a load W, neglecting friction, the value of maximum load W that can just be raised by a pulley P supplied, (figure)

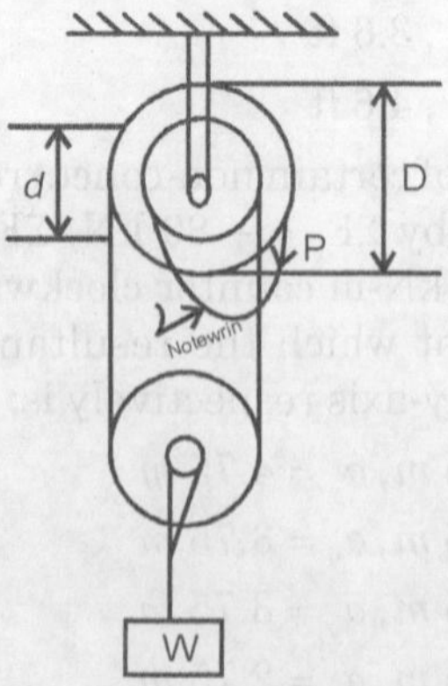

(*a*) $\frac{2D}{D-d}$ (*b*) $\frac{2P}{D-d}$

(*c*) $\frac{2PD}{D-d}$ (*d*) $\frac{PD}{D-d}$

20. A rigid bar CABD supported as shown in fig. is acted upon by two equal horizontal forces P applied at C & D, calculate the reaction that will be induced at the point of support, assume l = 1.5 m, a = 0.65 m and b = 0.4 m

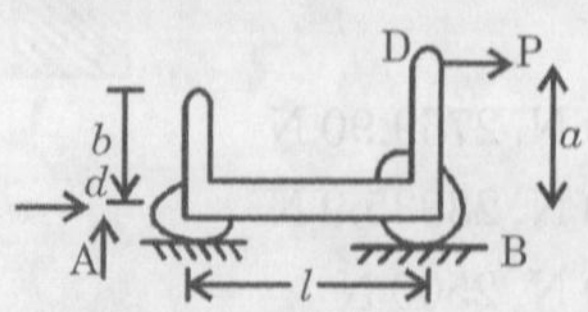

(*a*) P, 0.166 P

(*b*) 0.166 P, P

(*c*) 0.166 P, 0.166 P

(*d*) 0.166 P, 0

21. Owing to weight W of the locomotive shown in figure, the reaction at the two point of support A & B will be equal to W/2, when the locomotive is pulling the train and the draw bar pull P is just equal to the total friction at the points of contact A & B, the magnitude of the vertical reactions R_a and R_b.

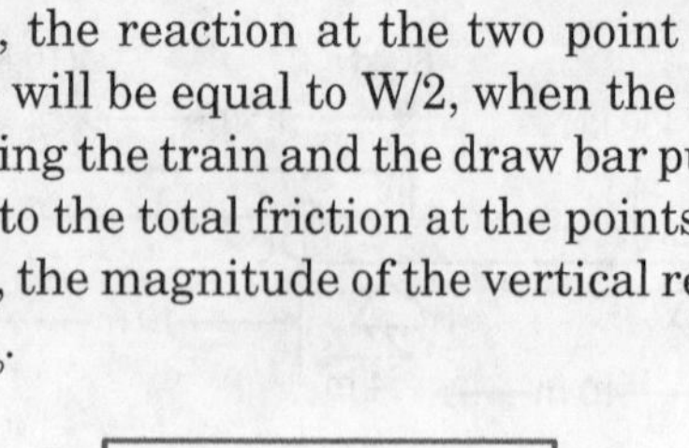

(*a*) $R_a = \frac{W.a - P.b}{2a}$, $R_b = \frac{W.a + P.b}{2a}$

(*b*) $R_a = \frac{W.a + P.b}{2a}$, $R_b = \frac{W.a + P.b}{2a}$

(*c*) $R_a = \frac{W.a - P.b}{2a}$, $R_b = \frac{W.a - P.b}{2a}$

(*d*) $R_a = \frac{W.a + P.b}{2a}$, $R_b = \frac{W.a - P.b}{2a}$

22. Least force required to draw a body up the inclined plane is Wsin(plane inclination + friction angle) applied in direction:

(*a*) Along the plane

(*b*) Horizontally

(*c*) Vertically

(*d*) At an angle equal to angle of friction to the inclined plane.

23. The incorrect statement for dry friction is:

(*a*) Proportional to normal load between surfaces

(*b*) Dependent on materials of contact surface

(*c*) Directly proportional to velocity of sliding

(*d*) Independent of area of contact surface

24. On ladder resting on ground and learning against smooth wall, force of friction will be:

(*a*) Towards wall at its upper end

(*b*) Away from wall at its upper end

(*c*) Upward to the wall at its upper end

(*d*) Downward to the wall at its upper end

25. If μ_s = 0.25, then in fig. the minimum value of x at which the load W is supported is:

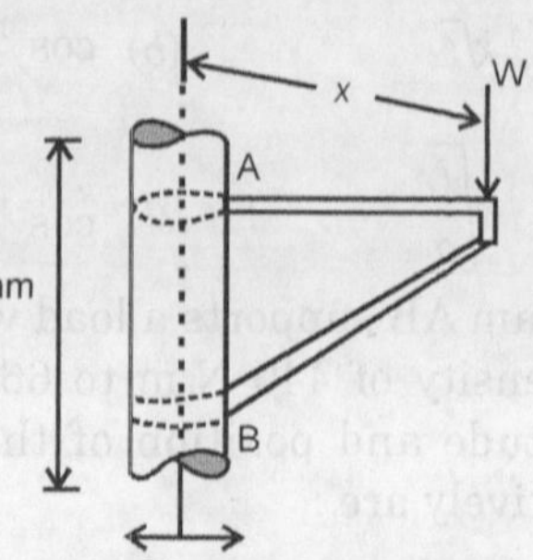

(*a*) 300 (*b*) 200

(*c*) 100 (*d*) 50

26. The cone of static friction represent:

(*a*) Locus of friction applicable at impending motion

(*b*) Locus of coefficient of friction at impending motion

(*c*) Locus of possible direction of R at impending motion

(*d*) Locus of projections of area at impending motion

27. Refer to fig., the maximum height at which horizontal force can be applied to slide the block without toppling :

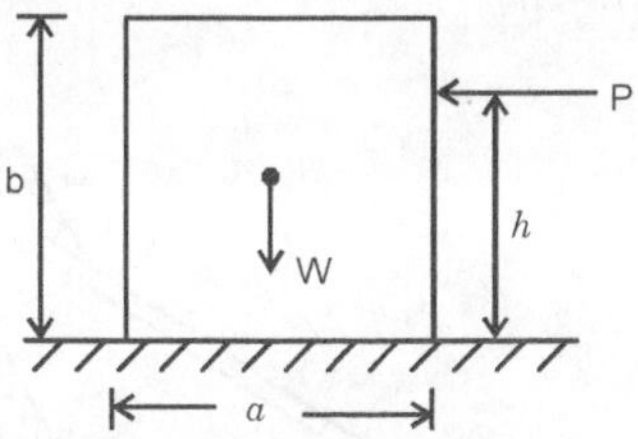

(a) $h = a - 2$ (b) $h = a(2\ )$

(c) $h = a + 2$ (d) $h = a/2$

28. A body of weight W is placed on rough inclined plane, the force applied parallel to the plane so that body moves upwards is:

(a) $P = \dfrac{W}{\cos\alpha}\sin(\alpha - \phi)$

(b) $P = \dfrac{W}{\sin\alpha}\cos(\alpha - \phi)$

(c) $P = \dfrac{W}{\cos\alpha}\sin(\alpha + \phi)$

(d) $P = \dfrac{W}{\sin\alpha}\cos(\alpha - \phi)$

29. Refer to fig., the inclination and magnitude of minimum force required to set the block into impending motion:

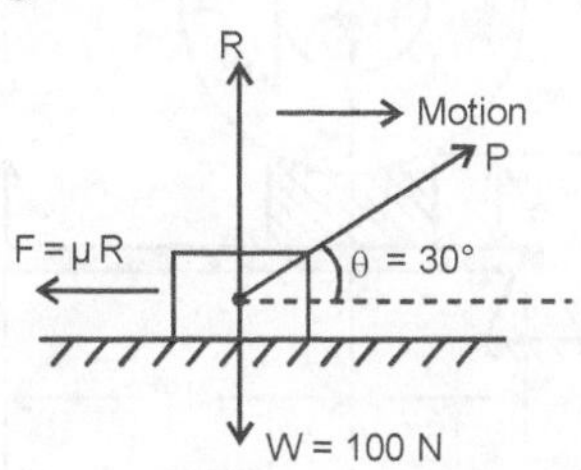

W = 100N , θ = 30°,   = 0.3

(a) $\alpha = 20.3°$ and $P_{min} = 30.5$ N

(b) $\alpha = 15.5°$ and $P_{min} = 45.3$ N

(c) $\alpha = 16.7°$ and $P_{min} = 28.7$ N

(d) $\alpha = 11.3°$ and $P_{min} = 25.23$ N

30. In fig., the flat belt connect pulley A and moter B, with $\mu_s = 0.25$, $\mu_k = 0.20$ and $T_{max} = 300$N. the value of largest torque is:

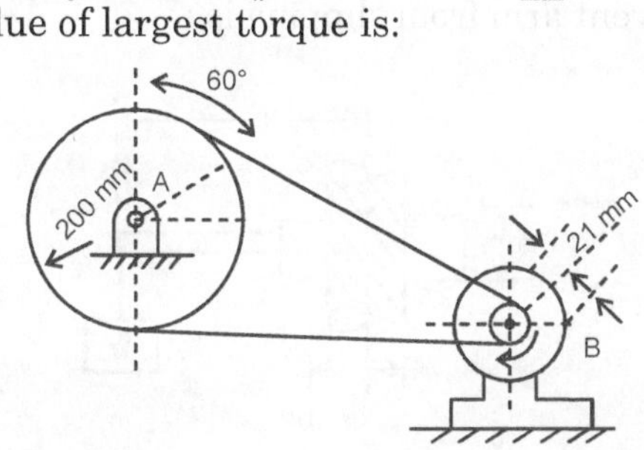

(a) 1552.5 N-m (b) 300 N-m

(c) 1777.2 N-m (d) 244.54 N-m

31. In case of collar bearing, friction is developed between

(a) Full circular areas

(b) Over ring shaped areas when shaft end is hollow

(c) Two ring shaped areas in contact

(d) Unpredictable

32. The journal bearings provide

(a) Axial support to shaft and axles

(b) Lateral support to rotate shaft

(c) No support over shaftes

(d) None

33. The circle of friction is produced when line of action is

(a) Tangent to the circle

(b) Perpendicular to the circle

(c) Parallel to the circle

(d) None of the above

34. The screw will unwind under the load to maintain equilibrium if,

(a) $\phi_s < \theta$

(b) $\phi_s > \theta$

(c) $\phi_s = \theta$

(d) ϕ_s does not get effected

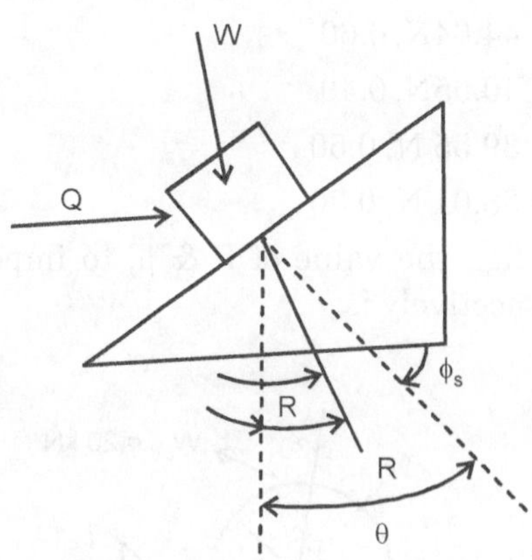

35. Considering a block resting on a board and subjected to no other force than W & R; the conditions represent motion is impending is:-

(a) $\theta = 0$ (b) $\theta > \phi_s$

(c) $\theta = \phi_s$ (d) $\theta < \phi_s$

36. Which of the following will represent that force acting on block is unbalanced:

(a) $\theta \neq \phi_s$ (b) $\theta > \phi_s$

(c) $\theta = \phi_s$ (d) $\theta < \phi_s$

37. Considering fig., having $\mu_1 = 0.30, \mu_2 = 0.60$; the value of least force P, required to move the block is :

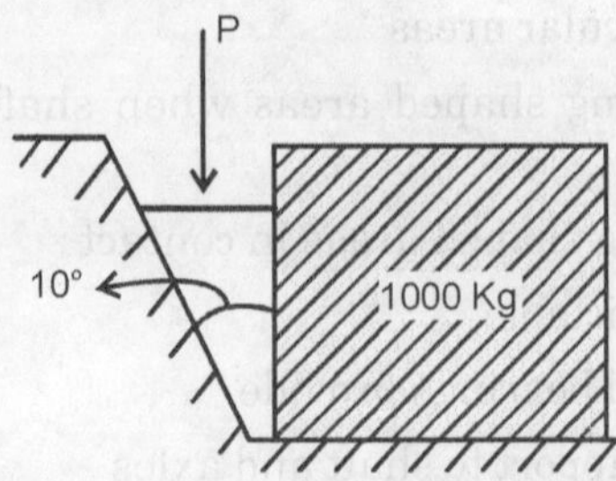

(*a*) 6831.10 N (*b*) 4972.37 N
(*c*) 6831.1 N (*d*) 500 N

38. Reference to fig., the block is tied up by horizontal string having tension of 20N, find the value of frictional force

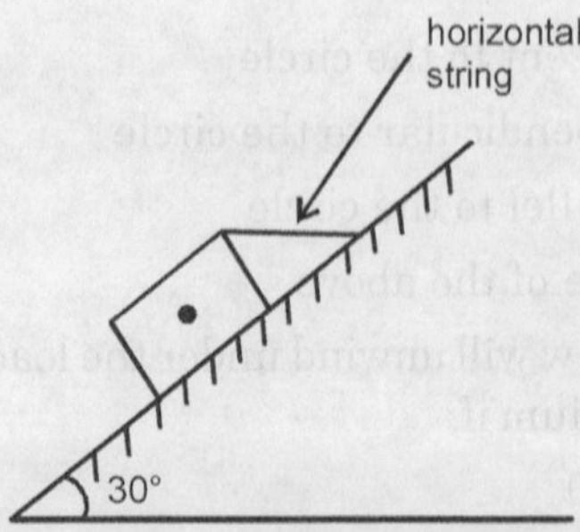

(*a*) 2.67 N (*b*) 2.0 N
(*c*) 3.67 N (*d*) 3 N

39. In the above question, find the value of normal reaction and coefficient of friction respectively are:

(*a*) 44.64N, 0.60
(*b*) 40.05N, 0.40
(*c*) 39.55 N, 0.50
(*d*) 38.01 N, 0.30

40. In fig., the value of P & μ, to impend slipping respectively is:

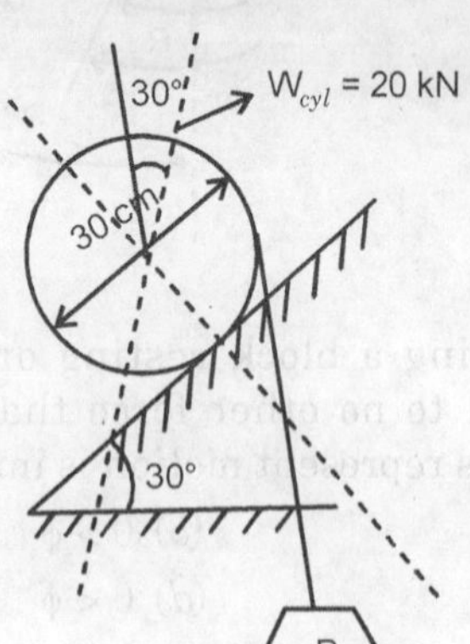

(*a*) 30 N, 0.06 (*b*) 20N, 0.06
(*c*) 20N, 0.577 (*d*) 30N, 0.577

41. Refer to Fig.; A & B are connected with strut, if $\mu = 0.25$ & $W_B = 2700$N, the minimum value of W_A to prevent motion is:

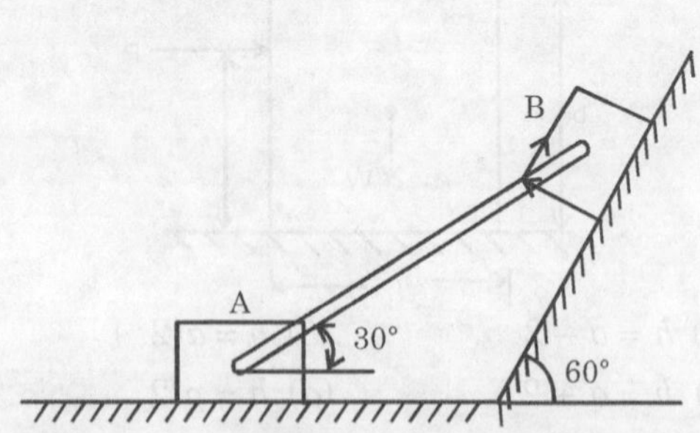

(*a*) 3857.66 N (*b*) 5984.20 N
(*c*) 4522.89 N (*d*) 3526.74 N

42. The minimum value of θ so that ladder can be placed against wall to prevent slipping is if $\mu = 0.2$

(*a*) 30.38°
(*b*) 29.38°
(*c*) 67.38°
(*d*) 38.38°

43. The value of P in Figure is ($\mu = 0.2$)

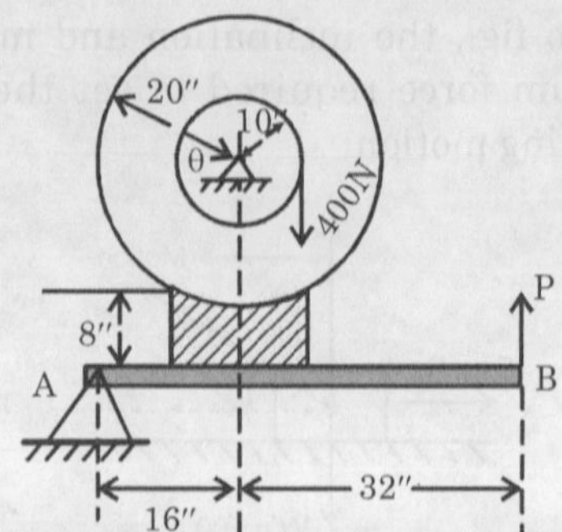

(*a*) 200lb (*b*) 300lb
(*c*) 400lb (*d*) 500lb

44. A horizontal arm having bushing of 20mm is slipped over a 20mm diameter vertical rod as shown in Figure. The value of μ is 0.25, then the minimum length 'L' at which 'W' can be placed to prevent arm from slipping is :

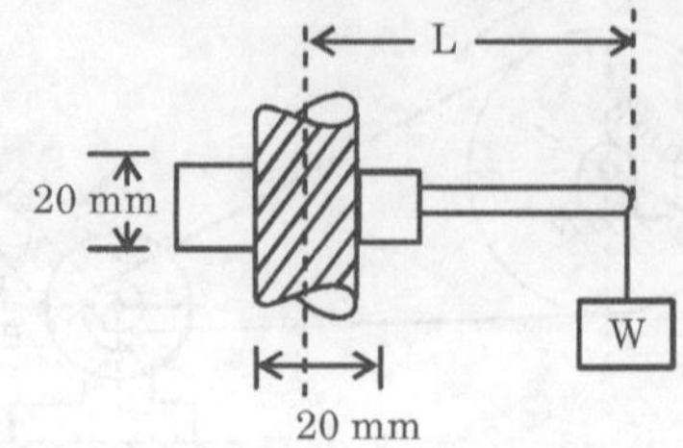

(*a*) 20mm (*b*) 30mm
(*c*) 40mm (*d*) 50mm

45. Refer to Fig., the maximum value of angle 'α' at which slipping occurs is:

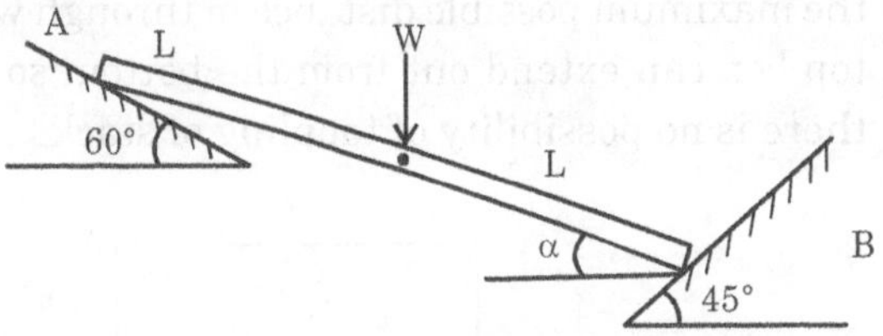

(*a*) 36.21° (*b*) 45.31°

(*c*) 56.50° (*d*) 38.75°

46. A wedge is used to split logs. If ϕ is the angle of friction between the wedge and log, the maximum value of angle of the wedge so that it will remain embedded in the log:

(*a*) ϕ (*b*) 2ϕ

(*c*) 3ϕ (*d*) 0

47. A particle starts with Vo and acceleration-velocity relation is prescribed as a = – kv , the expression that prescribes the displacement-time relation for the particle is:

(*a*) $x = \frac{1}{k}(1 - e^{-kt})$ (*b*) $x = \frac{v_0}{k}(1 - e^{-kt})$

(*c*) $x = \frac{1}{k}(v_0 - e^{-kt})$ (*d*) $x = \frac{v_0}{k}(e^{-kt})$

48. An electric train runs between two station which are 3.5 km apart and takes 8minutes from start to stop. The train has a constant running speed of 40 km/hr between the end of acceleration and beginning of retardation. If the acceleration retardation are both uniform and numerically equal to each other, the value of both respectively are:

(*a*) 940.233 m/sec² (*b*) 942.157 m/sec²

(*c*) 941.176 m/sec² (*d*) 652.368 m/sec²

49. Two stones are dropped from rest from two points both 250 m above the ground. One of the stone during its downward motion, hit a fixed inclined plane located at 85 m distance from ground. The plane is inclined at an angle to the horizontal, as a result of this impact the direction of velocity of stone becomes horizontal. (Fig.)The value of ratio of times of fall of two stones to reach the ground, is

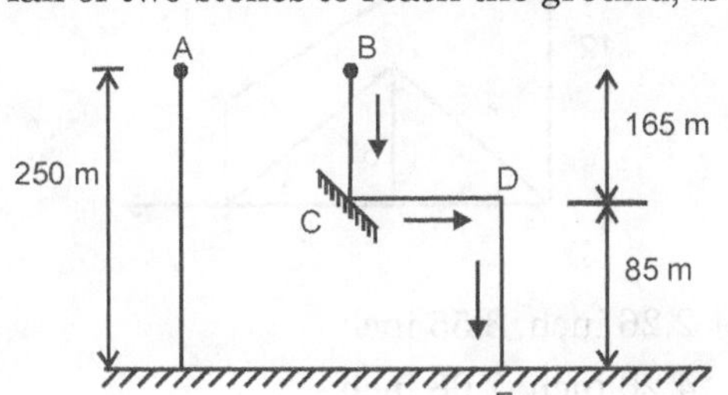

(*a*) 1.39 sec (*b*) 2 sec

(*c*) 4 sec (*d*) 3.3 sec

50. A Partical moves in X-Y plane and its coordinates are defined by relation $x = -3\sin t$ and $y = 2 - 3\cos t$

(*i*) the equation of path along which the particle moves is-

(*a*) $x^2 + y^2 = 3^2$

(*b*) $x^2 + (2 - y)^2 = 2^2$

(*c*) $x^2 + (2 - y)^2 = 3^2$

(*d*) none of the above

(*ii*) the equation for position vector, velocity vector and acceleration vector for the particle at the instant when the motion starts respectively is:

(*a*) $\vec{r} = -\hat{j};\ \vec{v} = -\hat{i};\ \vec{a} = \hat{j}$

(*b*) $\vec{r} = -1\ \hat{j};\ \vec{v} = -3\hat{i};\ \vec{a} = 3\hat{j}$

(*c*) $\vec{r} = \hat{j};\ \vec{v} = 3\hat{i};\ \vec{a} = 3\hat{j}$

(*d*) $\vec{r} = \hat{j};\ \vec{v} = -\hat{i};\ \vec{a} = 3\hat{j}$

51. A balloon of weight W is falling vertically downward with constant acceleration, *a*. The amount of ballast, Q must be thrown out in order to give balloon an equal upward acceleration is (Fig.)

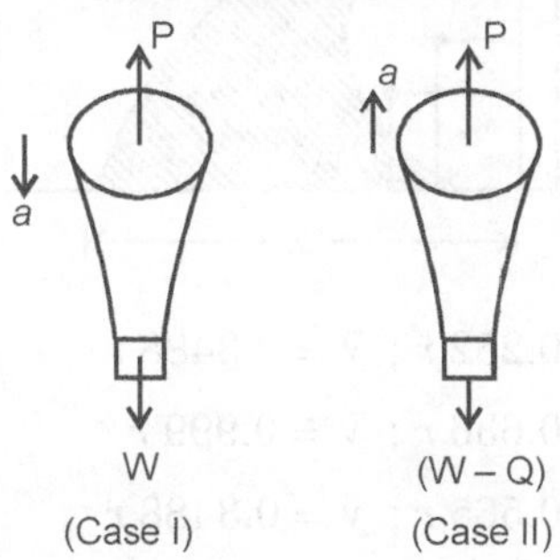

(*a*) $\frac{W}{(g+a)}$ (*b*) $\frac{2W}{(g+a)}$

(*c*) $\frac{2Wa}{(g+a)}$ (*d*) $\frac{Wa}{(g+a)}$

52. A cylinder of diameter 6 m rolls without slipping along a horizontal surface PQ, if its centre has a uniform velocity of 40 m/s, (Fig.) the velocity of B & D lying on rim of cylinder respectively is

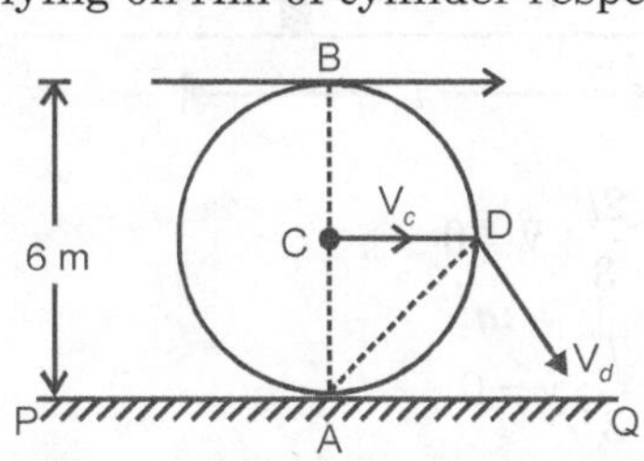

(*a*) 200m/s and 100 m/s

(*b*) 240m/s and 150 m/s

(*c*) 240m/s and 163 m/s

(*d*) 200m/s and 163 m/s

53. A racing car travels around a circular track of 400 m radius with a speed of 484 km/hr, what angle α should the floor of the track make with the horizontal in order to safe guard against sliding:-

(*a*) 88.7°

(*b*) 80°

(*c*) 76.5°

(*d*) 68.9°

54. A 59 kg of woman and a 73 kg man sits on a see saw 3.5 m long where is the centre of mass?

(*a*) 2.55m

(*b*) 1.75 m

(*c*) 1.94cm

(*d*) None of the above

55. The centroid of shaded area in Figure created by cutting a semicircle of diameter *r* from a quarter circle of radius *r* :-

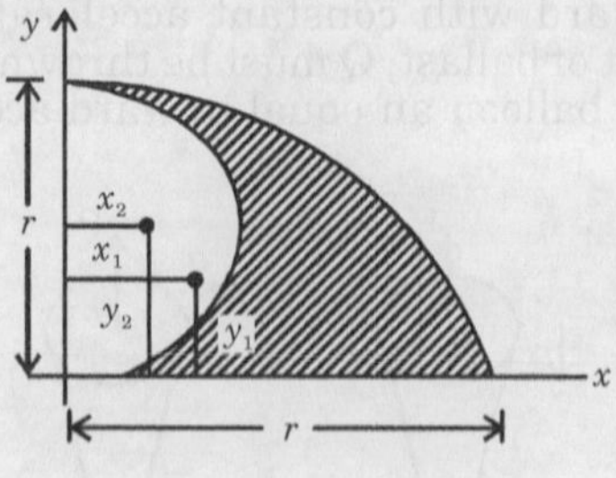

(*a*) $\bar{x} = 0.252\,r$; $\bar{y} = 0.3488\,r$

(*b*) $\bar{x} = 0.636\,r$; $\bar{y} = 0.999\,r$

(*c*) $\bar{x} = 0.565\,r$; $\bar{y} = 0.3488\,r$

(*d*) $\bar{x} = 0.636\,r$; $\bar{y} = 0.3488\,r$

56. The density at a point of cylinder rod (Fig.) varies with the first power of the distance of the point from one end of the rod. Locate the centre of mass.

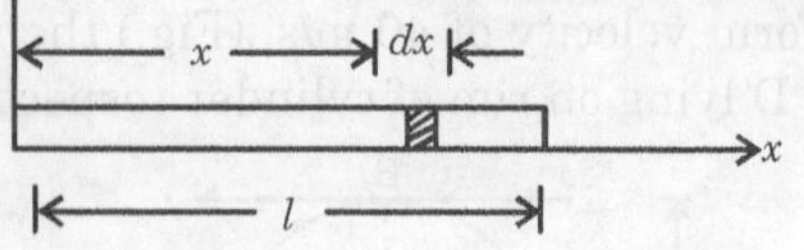

(*a*) $\bar{x} = \frac{2l}{3}; \bar{y} = 0$

(*b*) $\bar{x} = \frac{l}{3}; \bar{y} = 0$

(*c*) $\bar{x} = \frac{2}{3}; \bar{y} = 0$

(*d*) $\bar{x} = \frac{l}{2}; \bar{y} = 0$

57. Three identical boxes, each having length *l* and weight *w* are placed as shown in Fig., find out the maximum possible distance m through which top box can extend out from the bottom so that there is no possibility of toppling of stack.

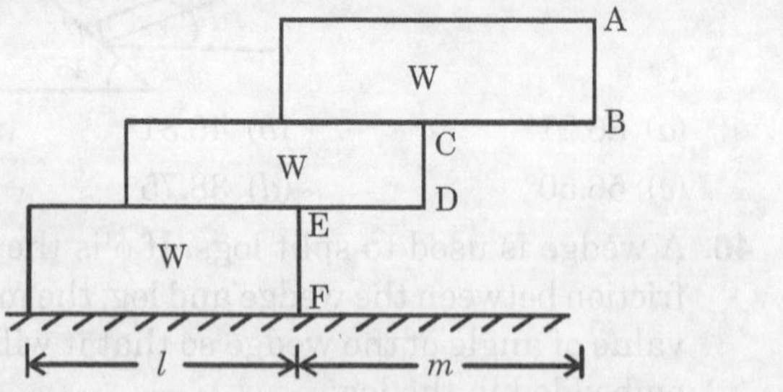

(*a*) $\frac{l}{3}$ (*b*) $\frac{3l}{4}$

(*c*) $\frac{l}{4}$ (*d*) Zero

58. The coordinate of the centroid of the area w.r.t the axes indicated in the Figure, are :

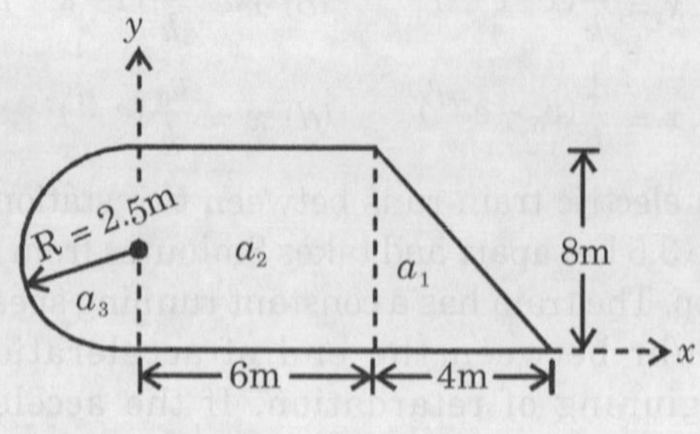

(*a*) 4.93 m, 3.03 m

(*b*) 5.51 m, 2.55 m

(*c*) 0,0

(*d*) 3.03 m, 5.51 m

59. Determine the centroid of the line that form the boundary of the shaded area as shown in the Figure.

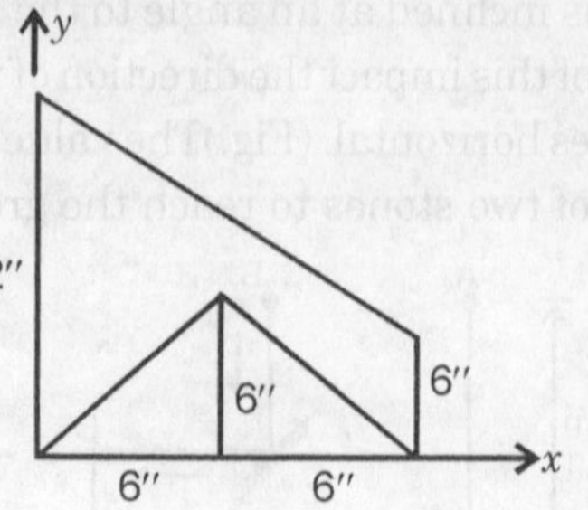

(*a*) 2.26 inch, 3.55 inch

(*b*) 4.26 inch, 2.08 inch

(*c*) 5.26 inch, 5.408 inch

(*d*) 5.26 inch, 4.55 inch

60. The coordinate of centroid of area as shown in figure is:

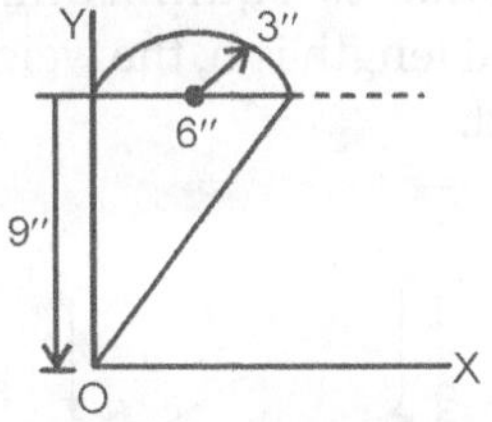

(*a*) 2.34 inch, 7.47 inch (*b*) 3.33 inch, 5.88 inch

(*c*) 0, 7.47 inch (*d*) 0, 0

61. A cylinder homogeneous wire of uniform cross-section is bent into shape shown in the figure, determine the coordinates of the centroid

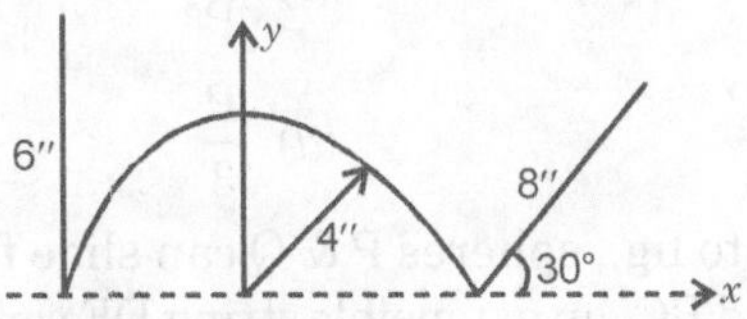

(*a*) 2.55 inch, 2.48 inch

(*b*) 1.34 inch, 2.48 inch

(*c*) 1.34 inch, 2.55 inch

(*d*) 1.34 inch, 0

62. Determine the centroid of following figure as shown (units are in m):-

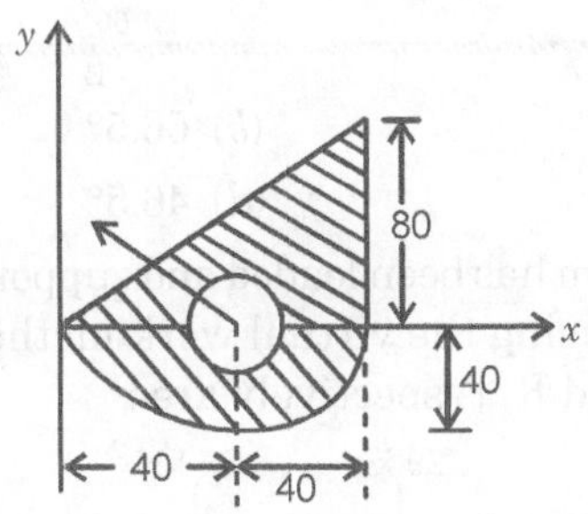

(*a*) 49.57, 9.58

(*b*) 32.75, 7.88

(*c*) 25, 10

(*d*) 0, 0

63. Determine the centroid of following figure:-

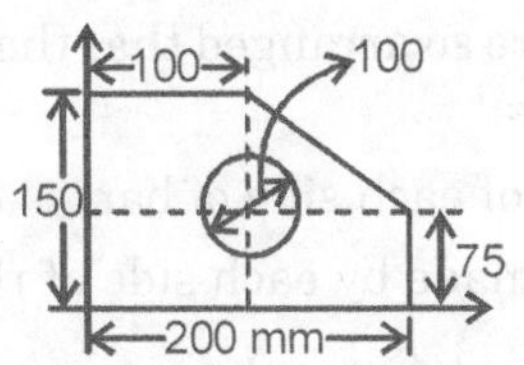

(*a*) 95.5 mm, 56.45 mm

(*b*) 80.25 mm, 40.00 mm

(*c*) 86.4 mm, 64.8 mm

(*d*) 86.4 mm, 29.23 mm

64. Reference to figure, calculate the value of spring constant is so that maximum deflection is limited δ, the block is held by spring and a bullet is fired with velocity v and mass M:

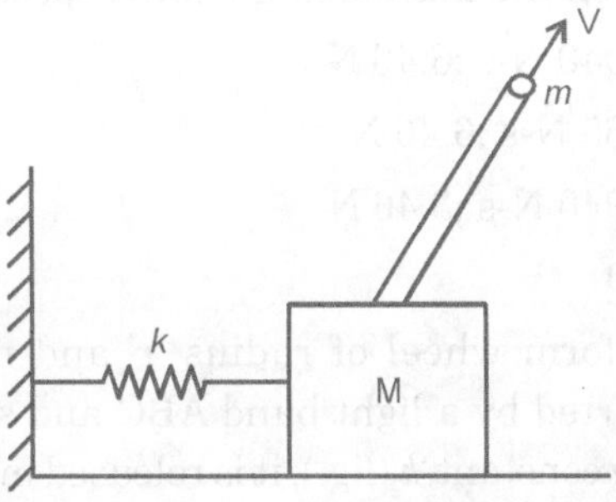

(*a*) $\dfrac{Mv}{\delta^2}$ (*b*) $\dfrac{Mv^2}{\delta^2}$

(*c*) $\dfrac{Mv^2}{\delta}$ (*d*) $\dfrac{v^2}{\delta^2}$

65. Reference to the figure, a mass m is suddenly put on the spring, the deflection in spring is:-

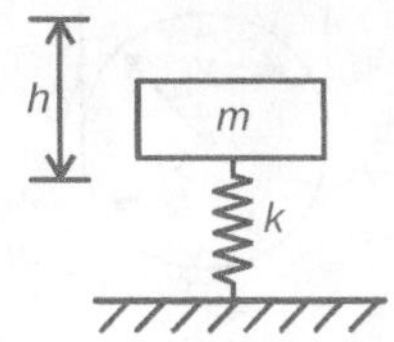

(*a*) $\dfrac{4mg}{k}$ (*b*) $\dfrac{2mgh}{k}$

(*c*) $\dfrac{2mg}{k}$ (*d*) $\dfrac{mg}{k}$

66. A particle of mass m tied at the end of an extensible string is rotated at *w* rad/s. the string is pulled through a slot at the centre of table top at the speed V(fig.), the speed of the particle when it reaches (*r*/2) from the centre is :

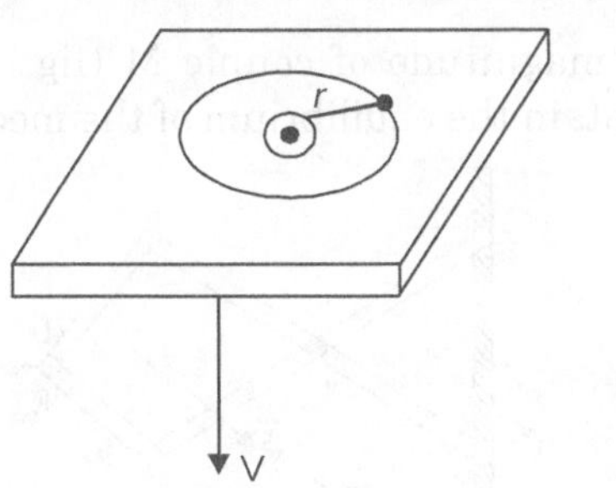

(*a*) $\sqrt{4w^2r^2}$

(*b*) $\sqrt{v^2 + 4w^2r^2}$

(*c*) $\sqrt{v + 4w^2r^2}$

(*d*) 0

67. A glass marble whose height is 0.6 N g = falls from a height of 16 m and rebounds to a height of 10, the marble and floor remains in contact for 0.3 second, the value of impulse and average force between the marble and floor respectively is :

(*a*) 1.940 N-s ;6.46 N

(*b*) 2.65 N-s ;6.46 N

(*c*) 1.940 N-s ;3.46 N

(*d*) 0;0

68. A uniform wheel of radius '*r*' and mass '*m*' is supported by a light band ABC and spring with spring constant '*k*'(fig.), it is released initially from a unstretched position of spring, the value of angular displacement for equilibrium position, assuming no slippage between disc and band during motion is:

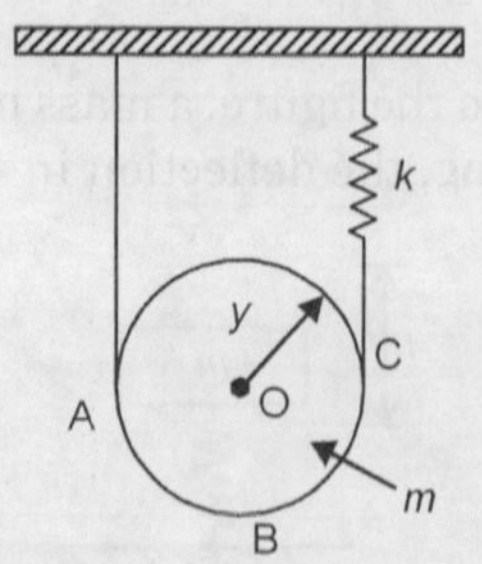

(*a*) $\theta_e = \dfrac{mg}{2Kr}$

(*b*) $\theta_e = \dfrac{mg}{4Kr}$

(*c*) $\theta_e = \dfrac{mg}{Kr}$

(*d*) $\theta_e = \dfrac{mg}{6\,Kr}$

69. The magnitude of couple M (fig.) required to maintain the equilibrium of the mechanism is:

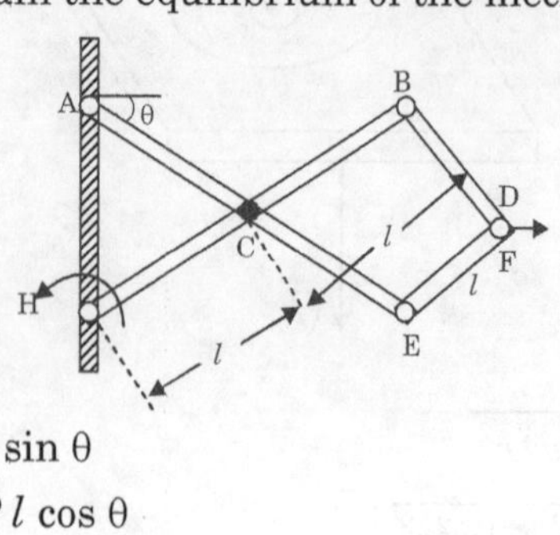

(*a*) P l sin θ

(*b*) 3 P l cos θ

(*c*) 3 P l sin θ

(*d*) P l cos θ

70. What is the value of θ and tension in spring when it corresponds to equilibrium position (fig.)? Unstretched length is h, the weight of mechanism is neglected.

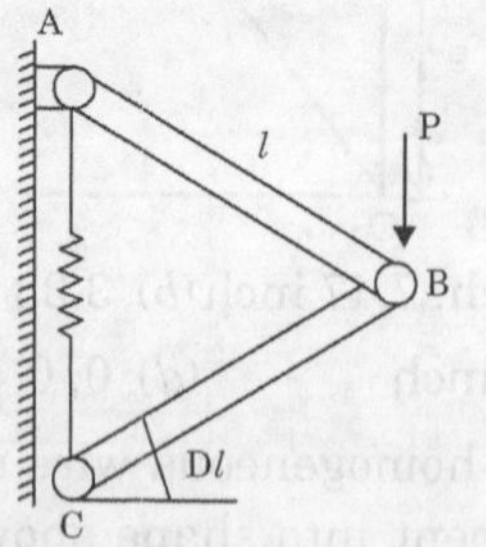

(*a*) $\dfrac{P}{2}$ (*b*) $\dfrac{2}{P}$

(*c*) $\dfrac{3P}{2}$ (*d*) $\dfrac{P}{3}$

71. Refer to fig., spheres P & Q can slide freely along AC and BC, an extensible string DE connects these spheres. The value of θ which defines the position of equilibrium is :

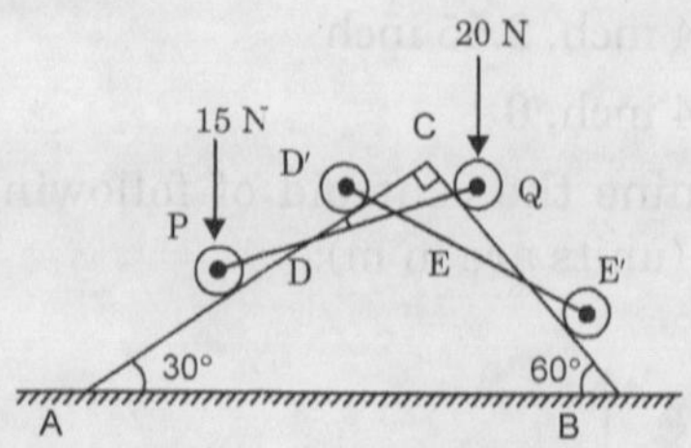

(*a*) 26.5° (*b*) 66.5°

(*c*) 36.5° (*d*) 46.5°

72. The beam has been loaded and supported as shown in fig., using the virtual work method, the value of R_a and R_b respectively are :

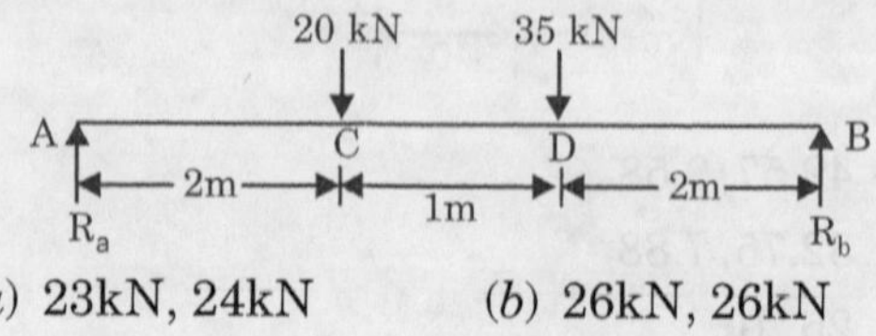

(*a*) 23kN, 24kN (*b*) 26kN, 26kN

(*c*) 26kN, 29kN (*d*) 29kN, 26kN

73. What is the relation between active forces P and Q (fig.) for equilibrium of system of forces on bars, the bars are so arranged that they form identical rhombuses.

l = length of each side of bar

θ = angle made by each side of rhombus

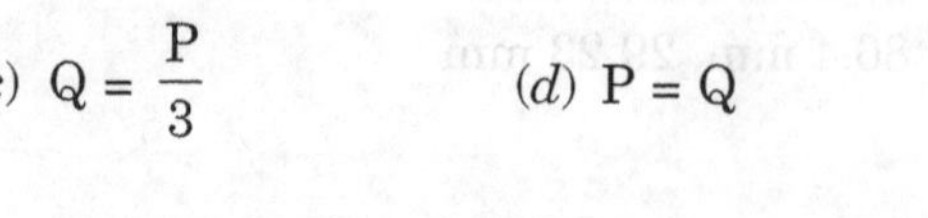

(*a*) P = 3Q (*b*) P = 2Q

(*c*) $Q = \dfrac{P}{3}$ (*d*) P = Q

74. Refer to figure using virtual work, the value of reaction at A from the truss is :

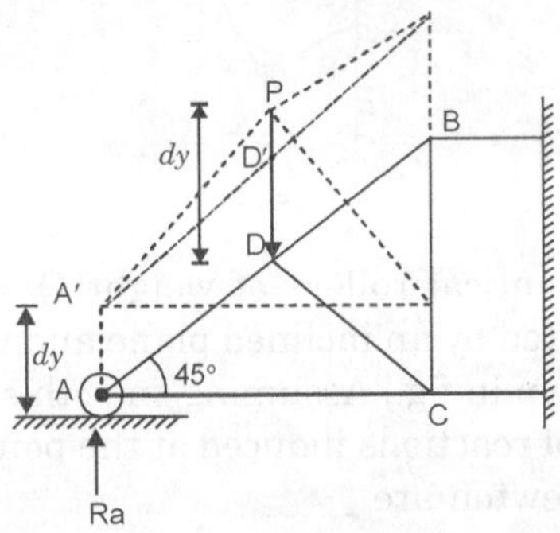

(a) 4P (b) 3P

(c) 2P (d) P

75. A uniform rod of length '2 *l*' rest in equilibrium against a smooth vertical wall and smooth peg 'P', the inclination of rod with wall is : (fig.)

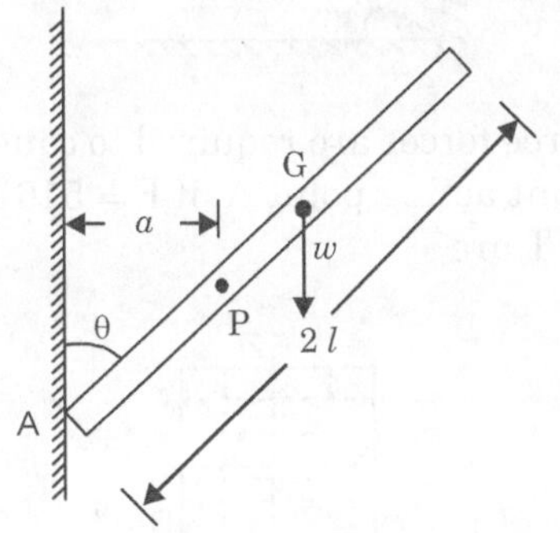

(a) $\theta = \sin^{-1}\left(\frac{a}{l}\right)^{\frac{1}{3}}$

(b) $\theta = \sin^{-1}\left(\frac{l}{a}\right)^{\frac{1}{3}}$

(c) $\theta = \sin^{-1}\left(\frac{2a}{l}\right)^{\frac{1}{2}}$

(d) $\theta = \sin^{-1}\left(\frac{2l}{a}\right)^{\frac{1}{2}}$

76. A hexagonal frame (fig.) ABCDE made of six bars of equal length and weight, one bar is fixed and system lies in vertical plane. The mid-points of two upper non- horizontal bars are connected by PQ. Using virtual work, the value of tension in string is :

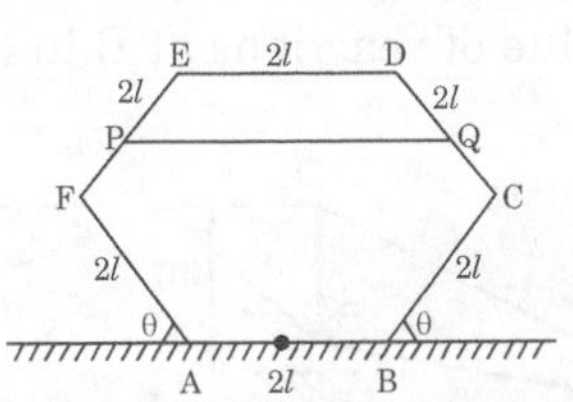

(a) T = 3W cot θ (b) T = 6W cot θ

(c) T = 3W tan θ (d) T = 6W tan θ

77. A regular hexagonal frame ABCDEF (fig.), is made of six uniform identical bars 'W'. The frame rests in a vertical plane and the bar AB is fixed in horizontal position. The mid points of AB and DE are connected by light string. The value of tension in string is :

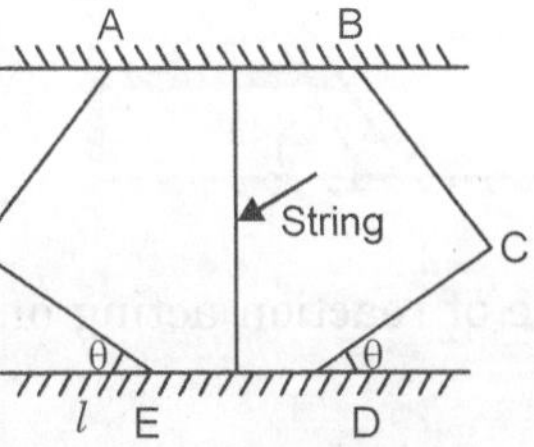

(a) W (b) 2W

(c) 3W (d) 4W

78. The value of F_{DH} for hydraulic lift as shown in fig. is :

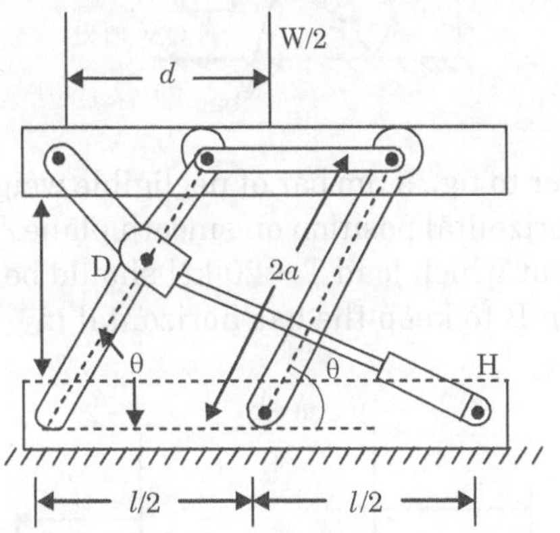

(a) $\frac{W.S}{L}\cot\theta$

(b) $\frac{W.S}{L}\tan\theta$

(c) $\frac{W}{L}\cot\theta$

(d) $\frac{W}{S}\cot\theta$

NUMERICAL TYPE QUESTIONS

1. Refer to fig., the value of tension 'T' in (N) in the cable is ______

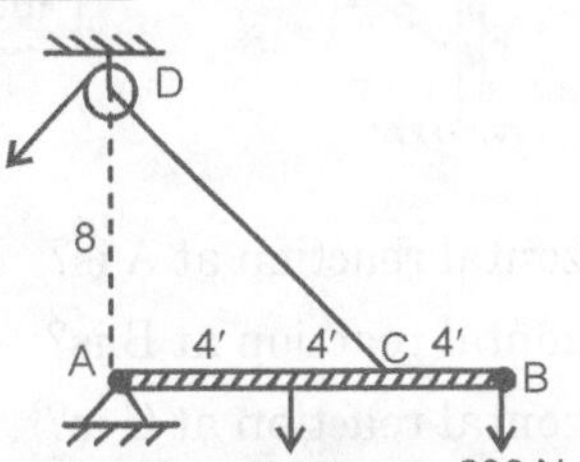

2. Refer to the figure, the rod weights 550 N and has centre of gravity at G. The value of tension in the cable is ______ kN

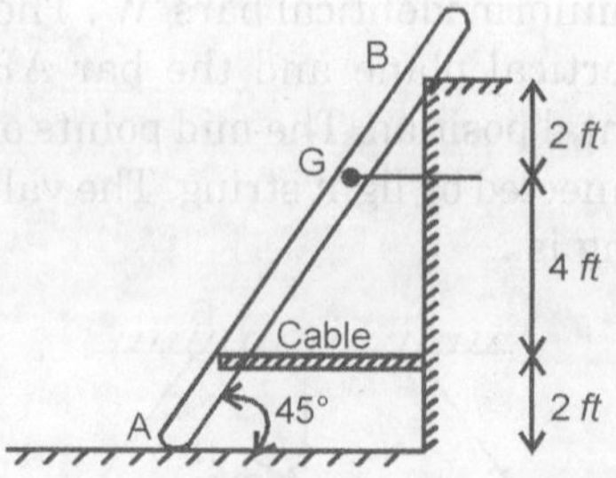

3. The value of reaction acting on ball in figure is ______

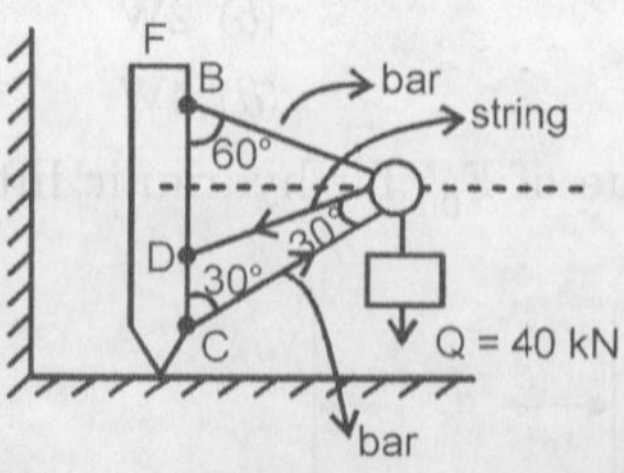

4. Refer to fig., a 5m bar of negligible weight rest in a horizontal position on smooth plane. The value of x at which load T = 20 kN should be placed at from B to keep the bar horizontal in______ m

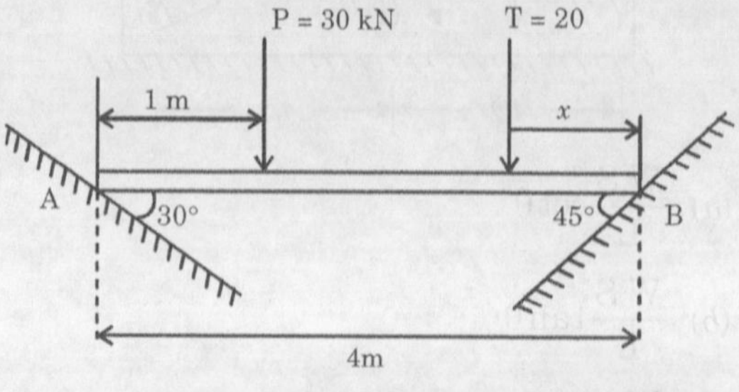

5. Refer to figure the frame supported at pivots A & B. Each member weighs 10 kN/m. The value of :

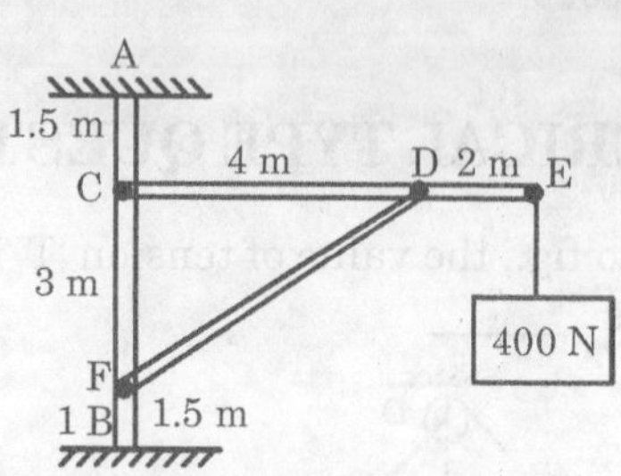

(*i*) Horizontal reaction at A is?

(*ii*) Horizontal reaction at B is?

(*iii*) Horizontal reaction at C is?

6. An electric light fixture of weight Q = 188N is supported as shown in fig., value of tensile forces in Newton S_1 and S_2 in the wires BA and BC, if there angles of inclination are given is:

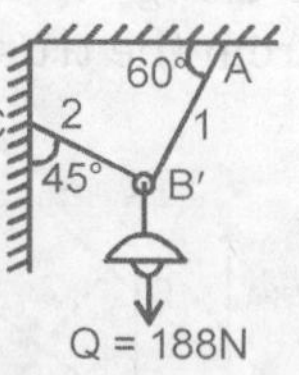

7. Two identical rollers of weight Q = 650 N are supported by an inclined plane and vertical wall as shown in fig.. Assuming smooth surfaces, the value of reactions induced at the points A, B and C in Newton are ______

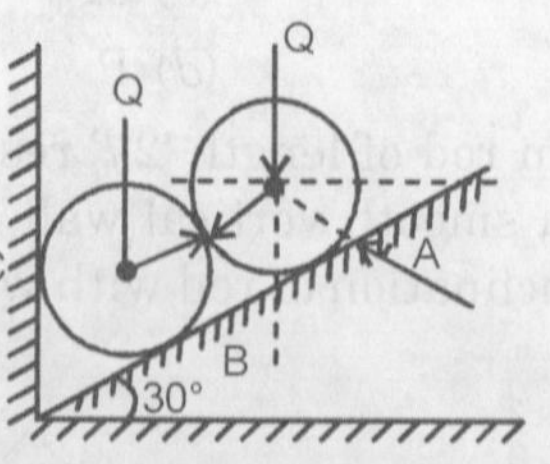

8. The three forces are required to cause horizontal resultant acting point A, if F = 516 N, the value of P & T are

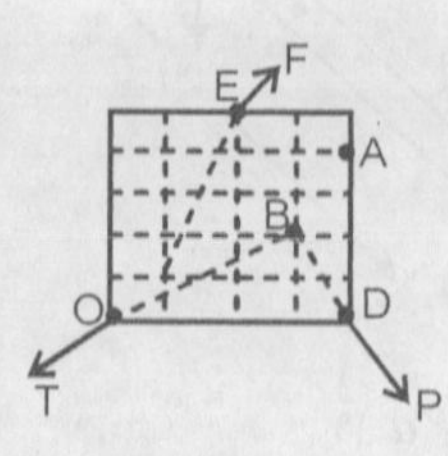

9. The fink truss is shown in fig., supported by roller at A and hinge at B, the given loads are normal to inclined member, the value of reactions at A & B in Newron are :

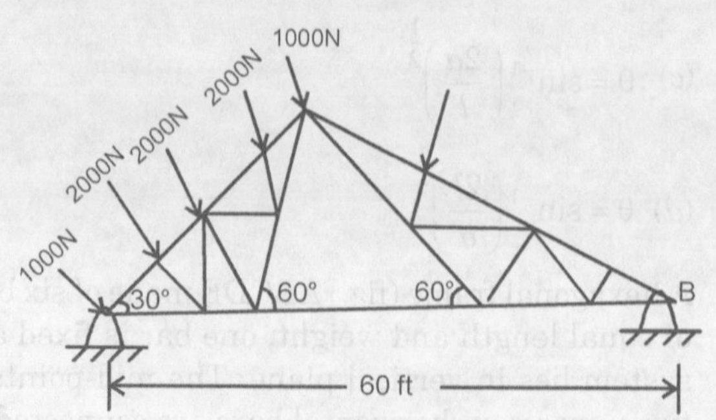

10. The truss shown in fig. is supported by a hinge at A and a roller at B, a load of 40 kN is applied at C, the value of reactions at B in (kN) is______

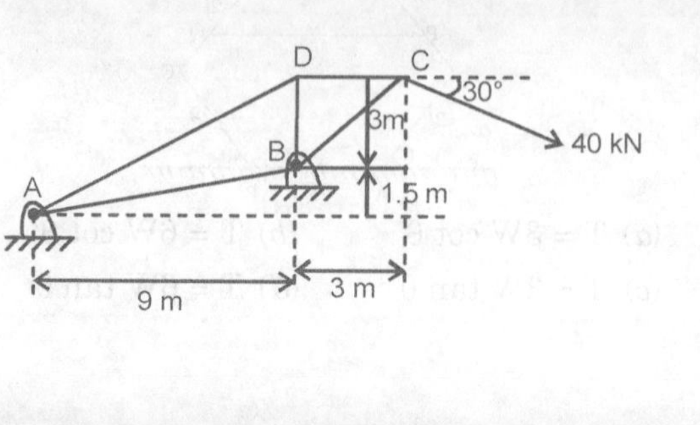

11. A wheel of 10" radius carries a load of 1500 N (Figure)

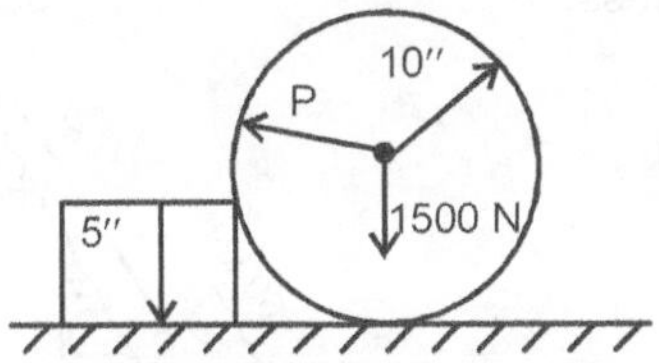

(*i*) The value of horizontal force P in newton applied at center which is necessary to start the wheel over a 5" block and reactions at the block

(*ii*) If force P may be inclined at an angle with the horizontal, the minimum value of P in newton to start the wheel over the block, angle P and reactions at block

12. The two 12 ft beam are to be moved horizontally with each other and load P shifted to a new position on CD so that all three reactions are equal

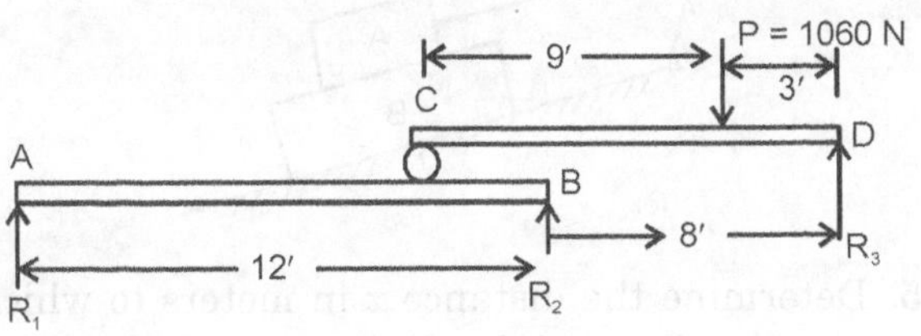

(*i*) How far in ft apart will R_2 and R_3 then be?

(*ii*) how far P in ft be from D

13. The four wheels of the locomotive produce vertical forces on the horizontal girder AB, the reactions R_a & R_b in (kN) at the supports are, if the loads P = 120 kN and Q = 85 kN.

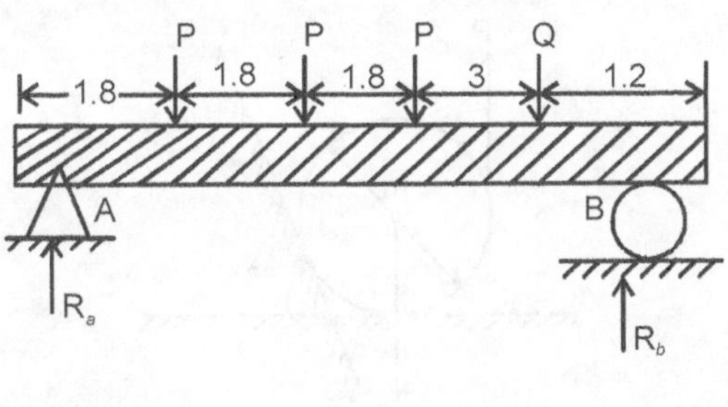

14. A prismatic bar AB of weight Q = 58.5kN is supported by two vertical wires at its end carries at D, a load P = 95 kN as shown in figure, the forces S_a & S_b in KN.in two wires are

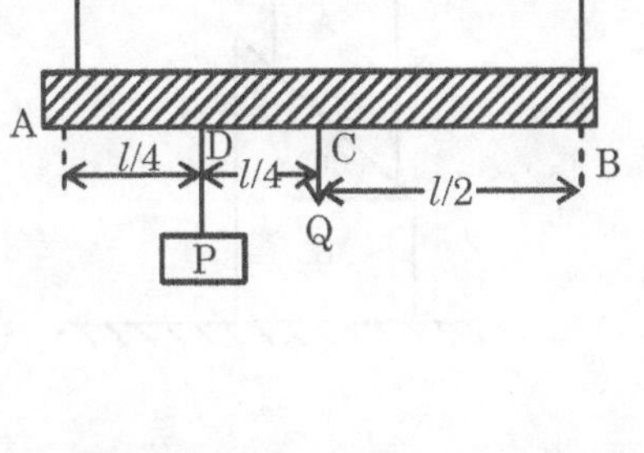

15. A steel cylinder 1.5 m in diameter and weighing 800 N is placed between two cross arms which make an angel of 60° with each other and are pinned at P as shown in figure. A rope attached to the arms near their upper end prevent from slipping, the tension in rope is _______ N

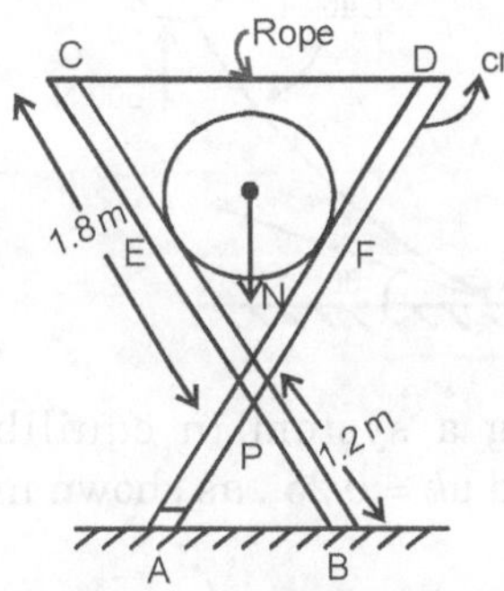

16. A rigid link ABC supported by cylinder at A & C, is loaded by a force 35 kN as shown in figure each cylinder has a diameter of 30 cm & AB = BC = 40 cm, neglecting friction and self weight of link, estimate the support reactions at D, E & F in kN

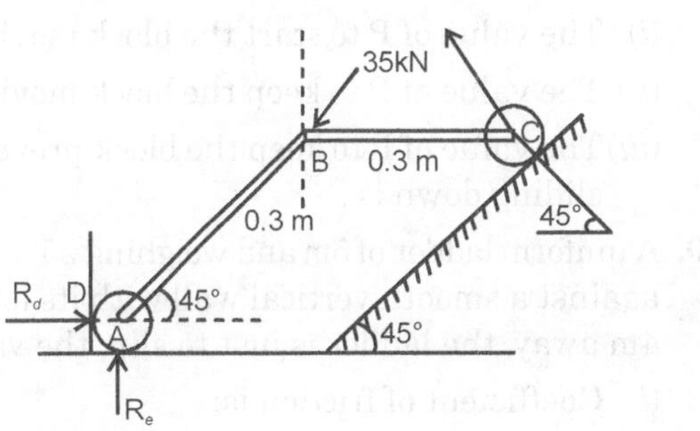

17. A square board ABCD is fixed in vertical plane, two pulleys P & Q, on the board carry loads of 35kN & 60 kN suspended by strings as shown in figure the coordinates of centre of pulleys P(1,4)m & Q(4,1) m respectively. The magnitude of the resultant is _______ N

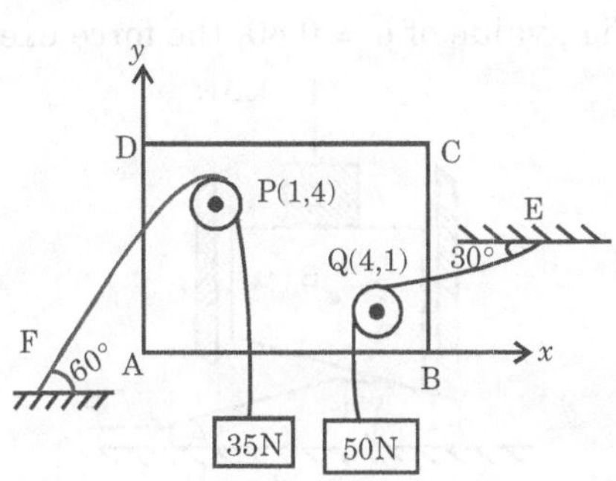

18. A balloon exerts an upward force of 600 N on its ground cable at the attachment point as shown, a horizontal wind causes shift in position shown fig., if cable weighs 8 N/m, the maximum tension in the cable is, if length is 50 m ______ (N)

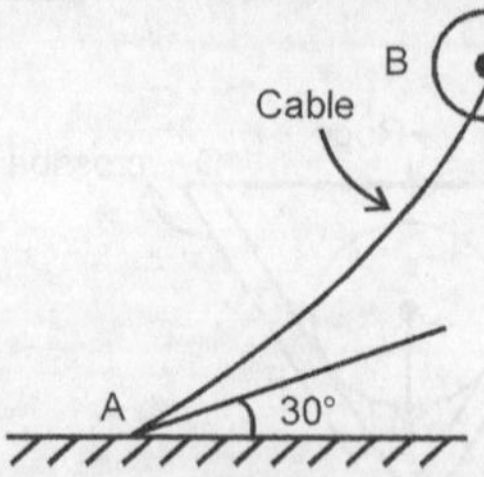

19. Considering a system in equilibrium with $\mu s = 0.35$ and $\mu k = 0.25$, as shown in Fig. :

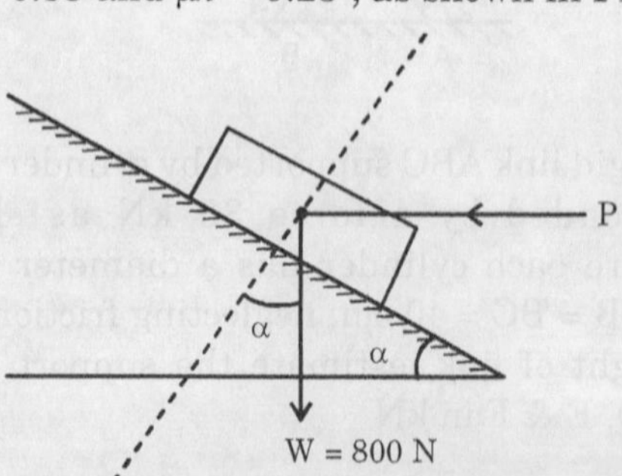

(*i*) The value of P to start the block moving up is

(*ii*) The value of P to keep the block moving up is

(*iii*) The value of P to keep the block prevent from sliding down is :

20. A uniform ladder of 5m and weighing 25N is placed against a smooth vertical wall with its lower end 4m away, the ladder is just to slip, the value of

(*i*) Coefficient of friction is:

(*ii*) Frictional force acting at the point of contact between ladder and floor:

21. A clamp of double threaded screw $d = 10$ mm and $\mu s = 0.67$, if $T = 50$ N-m is applied in tightening the clamp, the value of

(*i*) Force exerted on wood is

(*ii*) Torque required to loosen the clamp

22. In Fig., value of $\mu_s = 0.40$, the force exerted to,

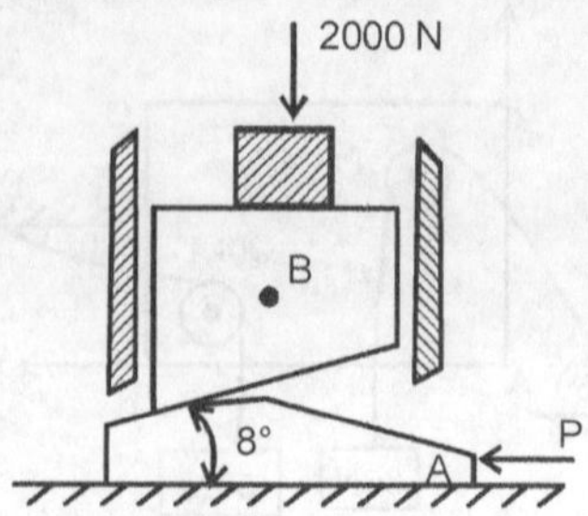

(*i*) Raise the block is

(*ii*) Lower the block is

23. In Fig., the value of W to start the movement of block to right is? where $\mu = 0.10$ & pulleys are frictionless.

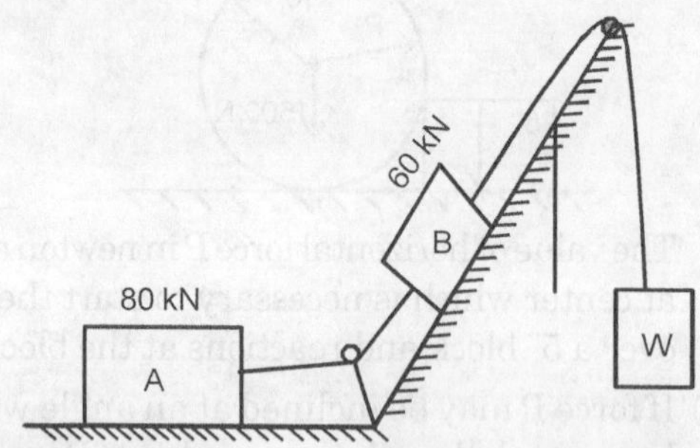

24. Refer to figure the value of θ is such that at inclination, the motion of B impends, where $\mu = 0.2$, $W_A = 150$ N and $W_B = 250$ N.

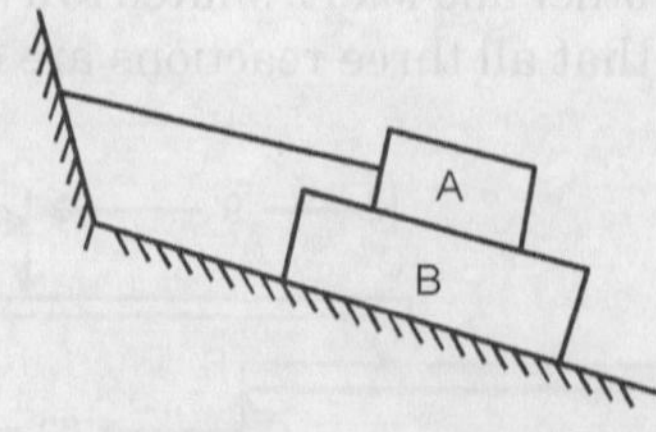

25. Determine the distance x in meters to which a man having weight 72 kg can climb without causing a 6m ladder to slip at its lower end. The top of the 18kg ladder has a small roller, and at the ground the coefficient of static friction is tan 15°.

26. In figure, calculate the value of P to start the turning of cylinder, if $\mu = 0.30$.

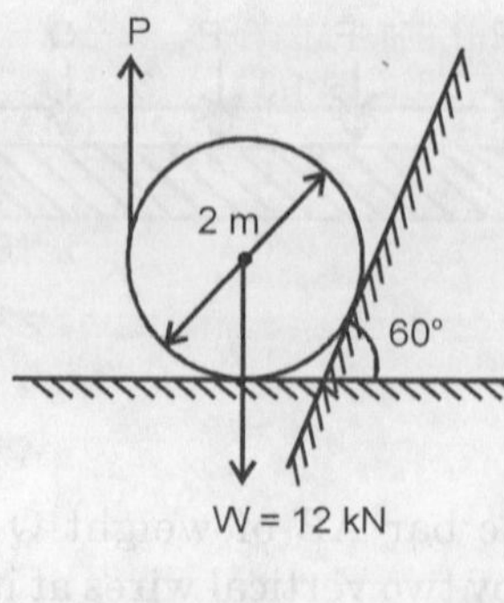

27. In figure the minimum weight of B that will keep it at rest is; given:

$W_A = 200$N, $\mu = 15°$.

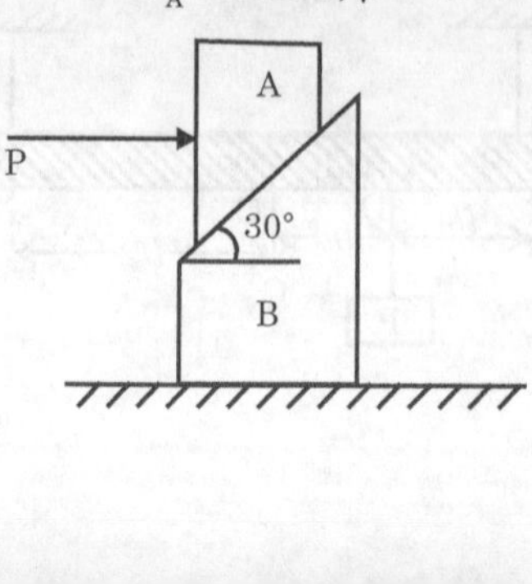

28. In figure calculate the value of P to be applied so that wedges to start impending motion under the block. Angle of friction is 10°.

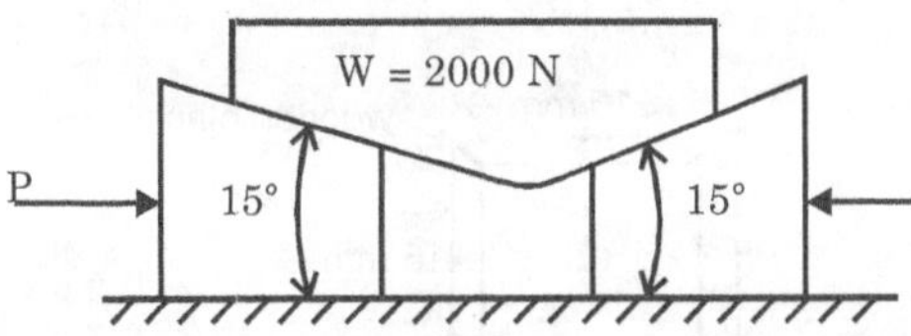

29. To adjust the vertical position of column under a vertical load of 200N, two 5° wedges are used (figure). The value of P to start the wedges is? Angle of friction at all contact surface is 25°.

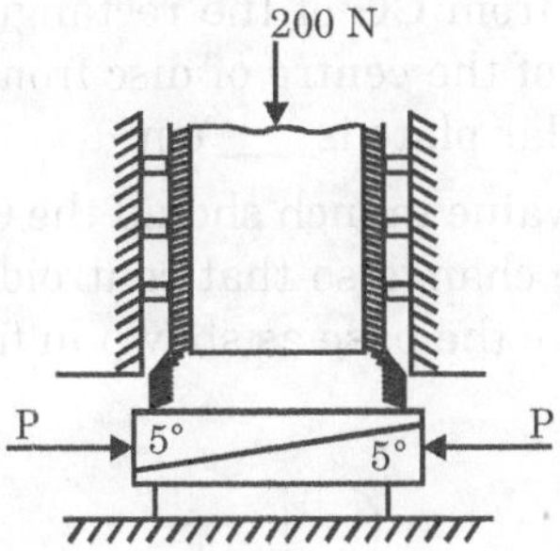

30. A car is moving in a straight line starts from rest, accelerates uniformly at 15 m/sec^2 for first 3 secs then travel at uniform speed for the next 2 secs and finally deaccelerates uniformly at 10m/sec^2 for next 4 secs, the value of distance travelled by the car during the given time period upto 1 decimal place in meters is ______

31. A stone is thrown vertically upwards with a velocity of 25 m/s from the top of a tower 30 m high, (figure) the value of:

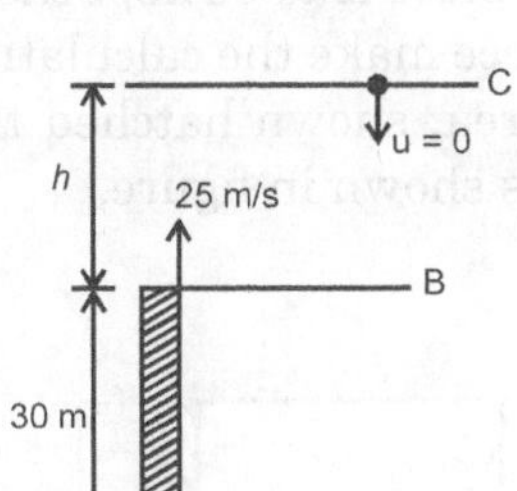

(*i*) maximum height in metre to which stone will rise in the flight

(*ii*) velocity of stone in m/s during its downward travel at a point in the same level as the point of projection.

(*iii*) the time require for the stone to reach the ground in second.

32. A weight; w = 4500 N is supported in a vertical plane by string and pulleys arranged as shown. If the free end A of the string is pulled vertically downward with constant acceleration, the tension in string in N is______ (a = 28 m/s^2) (figure)

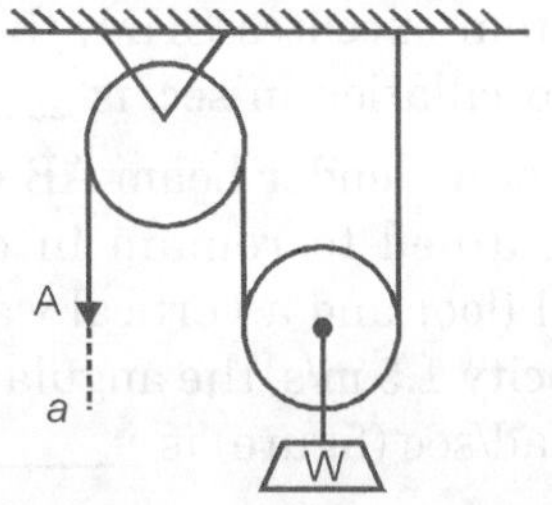

33. A bullet leaves the muzzle of a gun with velocity v = 800 m/s,. Assuming constant acceleration from barrel to muzzle, the time 't' occupied by the bullet to travelling the gun barrel which is 5 mm diameter is______sec

34. On a bumpy road, the motorist applies breaks and the speed of the vehicle reduces uniformly from 70 km/h at A to 50 km/h at B. The points A and B are 150 m a part, if the vehicle has a resultant acceleration of 2.5 m/s^2 at point A,

(*i*) the value of radius of curvature of vehicle in metre at this point is ______

(*ii*) the acceleration of vehicle in (m/s^2) at point B where the road has a radius of curvature to 90 m is ______

35. A projectile is aimed at a mark on the horizontal plane through the point of projection and falls 15 m short when the angle of projection is 15° while it overshoots the mark by 35 m when inclination is 70°, neglecting air resistance

The value of correct distance of the target is ______m

36. Two guns are pointed at each other one upward at an angle of 30° and other at same angle of depression, the nuzzle of gun are 80 m apart, of the guns are with the velocities of 450m/s upward and 350 m/s downward, determine

(*i*) The shots will meet in ____ sec

(*ii*) The shots will meet at ____ m

37. A motor cyclist wants to jump over a ditch as shown in figure

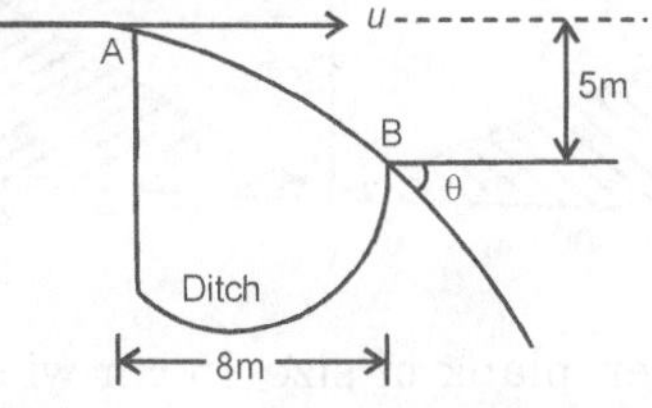

The minimum horizontal velocity of cyclist at A in (m/s) is ______

38. A helical spring of negligible mass is found to extend 0.20 mm under a mass of 0.5 kg, this spring is made to carry a mass of 50 kg at its lower end, subsequently the spring and mass systems is displaced vertically through 150 mm an released, if the system executes SHM, the value of time period of oscillation in sec. is _____

39. The ends of a slender beam AB of length 3.5 m are constrained to remain in contact with a horizontal floor and a vertical wall, if the end A has a velocity 1.5 m/s, the angular velocity of the beam in rad/sec (figure) is _____

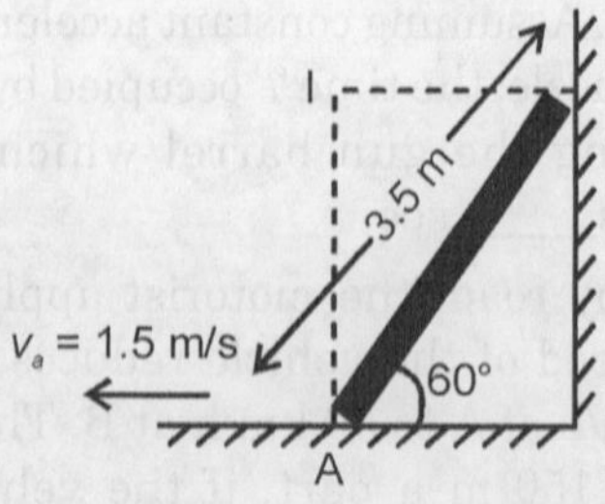

40. A triangular plate in the form of an isosceles triangle ABC has base BC = 10 cm and altitude = 12 cm, (fig.) from this plate a portion in the shape of an isosceles triangle OBC is removed if O is the mid- point of the altitude of triangle ABC then the distance of CG of remainder section from the base is_____ cm

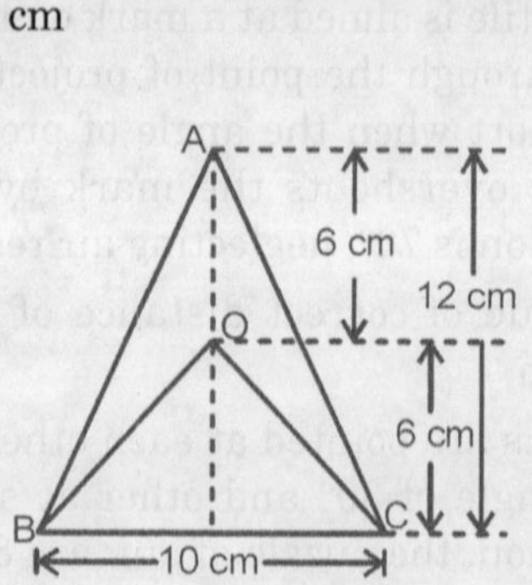

41. The centroid of shaded area in fig. is to be required on *y*-axis, the distance '*b*' in (mm) that will fulfil this requirement is _____

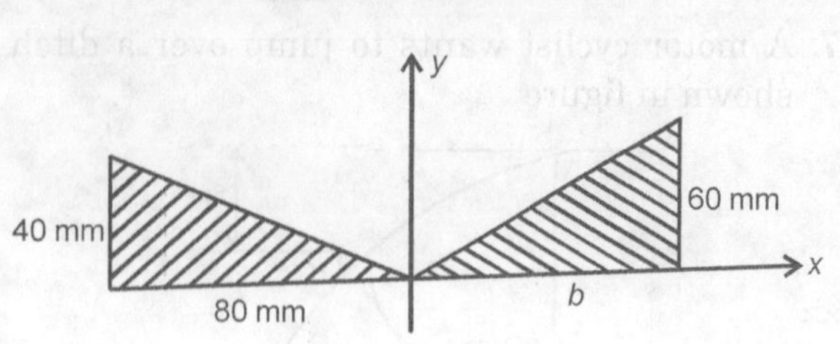

42. A wooden plank of size 15 cm width and 1 cm thick is fixed on the top of a semi-circular steel plate of radius 5 cm and thickness 1 cm (fig.), the maximum height h of the wooden plank so that CG of composite body lies in junction of two is _____ mm

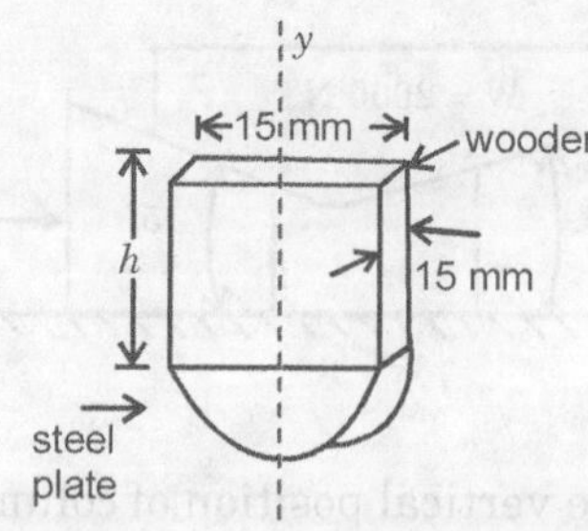

43. From a rectangular plate whose cross section is 9 cm × 4 cm as shown in figure, a circular disc of 3 cm^2 in area is removed. If CG of the remainder is 1 mm from CG of the rectangular plate, the distance of the centre of disc from the centre of rectangular plate is_____ cm

44. To what value in inch should the 6 inch width of flange be change so that centroid of area is 2.5 inch above the base as shown in fig.,

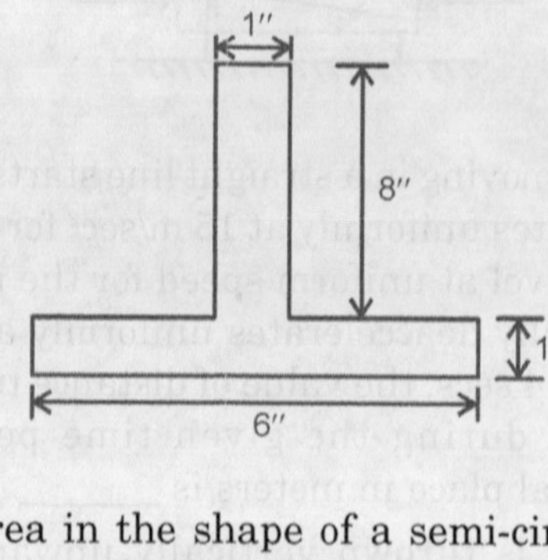

45. An area in the shape of a semi-circle of 10 cm radius has been purchased from square plate of 50 cm × 50 × 106 cm in size, if the centroid of the remainder piece is to be kept the same as for the original piece make the calculation for the width *w* of the area, shown hatched that needs to be removed as shown in figure,

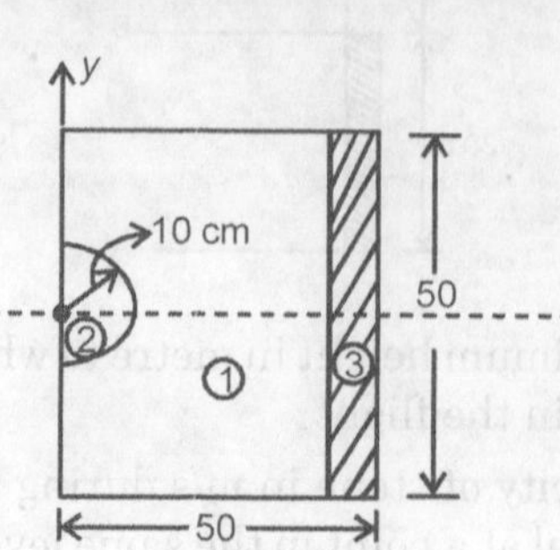

46. A solid circular metal ring has a rectangular cross section of size 8 cm × 6 cm, if its external and internal radii are 12 cm and 10 cm respectively, calculate the total surface area of ring in cm^2

47. Three objects are rolling a straight line, a 5 kg mass is going 3m/s to the right, 4 kg mass is going 2m/s to the left and 8 kg mass is travelling 8 m/s to the right. How fast in (m/s) is each going relative to the centre of mass?

48. Determine the centroid of composite figure 4.20 about *x*-*y* coordinate, take $x = 40$ mm

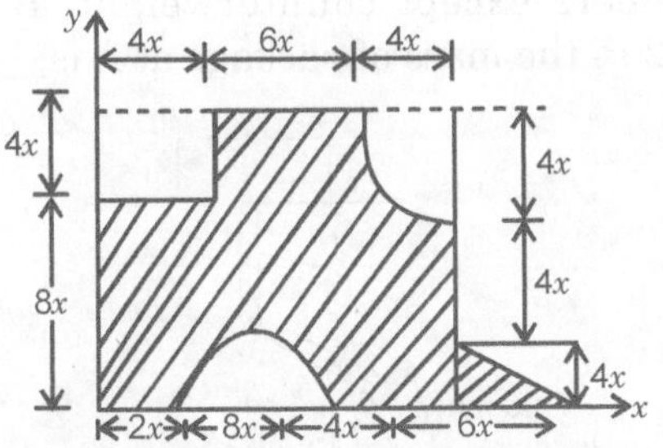

49. A man *f* weight 800 N stands in a boat so that he is 5 m from a pier on shoe, he walks 2.5 m in the boat towards the pier and then stops. How far in metre from pier man will be at the end of the time.

50. A locomotive weight 600 N has a velocity of 20 kmph and locks into a freight car of weight 86 kN that is at rest on a track after coupling the value of velocity *v* in m/s for the entire system to be continued to move

51. A 775.4 N man sits in a 445.75 N and fire a rifle bullet horizontally. The value of velocity '*v*' in (m/s) with the canon will move after the shot. The rifle has a nuzzle velocity 550 m/s and the weight of bullet is 0.30 N

52. A wooden block of 32.25 N rest on a smooth horizontal surface. A revolver bullet weighing 0.20 N is hot horizontally into the side if the block, attains velocity of 3.5 m/s, the value of nuzzle velocity is _____m/s

53. A shell of mass 80 kg is fired horizontally with a velocity of 300 m/s by a gun of 4500 kg mass, the value of

(*i*) The velocity in m/s with which the gun recoils

(*ii*) The velocity force required to stop the gun in 1.5 m distance_______ in (kN)

(*iii*) The time required to stop the gun in sec is ____

54. A hammer of mass of 300 kg is made to fall freely from height $h = 4$ m on the head of pile of 2500 kg mass. The pile is driven 6 cm in to the ground in one below, (fig.) the value of common velocity of pile and hammer after the impact is_____ m/s

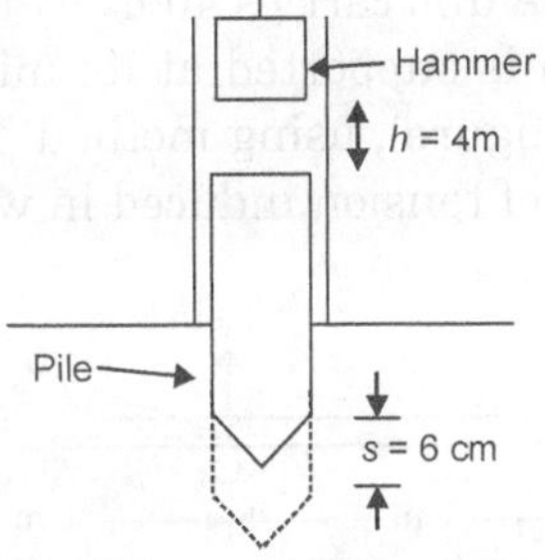

55. A machine gun fires 200 bullets per minute, each bullet has a mass of 75 gm and firing velocity is 800 m/s, presuming the firing to be continuous make calculation for the average reactionin newton of gun against the support is ________

56. Two identical rods ABC & DBE (fig) are connected by a pin at B and by spring CE. Knowing that spring is 40mm long when unstretched and spring constant is 0.8 N-mm, the value of *x* in mm corresponding to equilibrium when a 24 N load is applied at E is_____

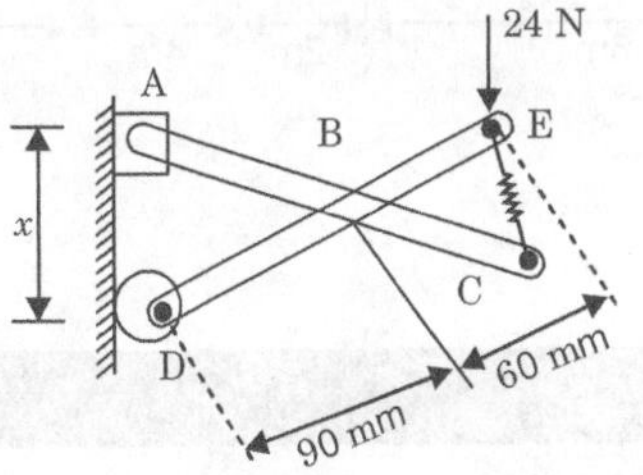

57. Refer to fig., the force in N in top members DE by method of virtual work is: (the length of all members are equal).

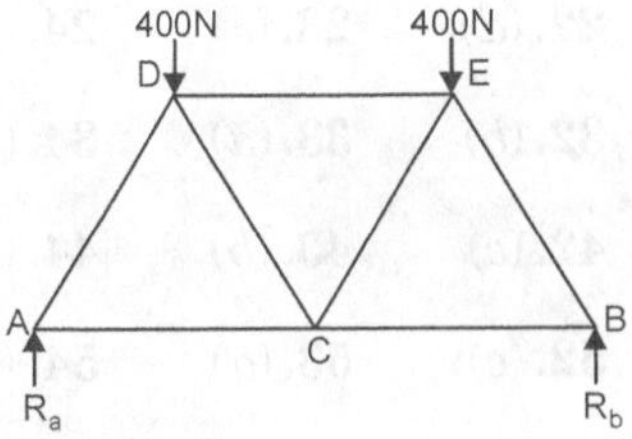

58. A weight N = 2500N is to be raised by a system of pulleys, using principle of virtual work, the value of 'P' (figure) in newton which can hold the system in equilibrium is

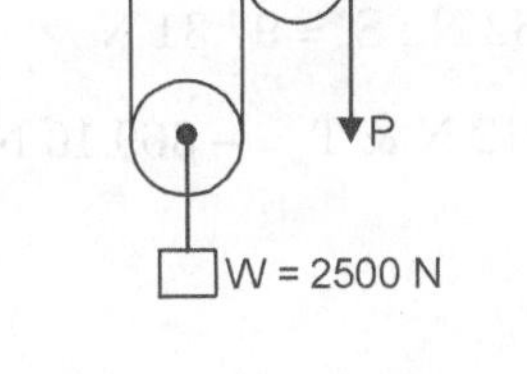

59. A beam of length 2m and negligible weight is hinged at A and carries 400N weight at free end. The beam is supported at its mid point C by a wire CD (figure), using method of virtual work, the value of tension induced in wire is_____N

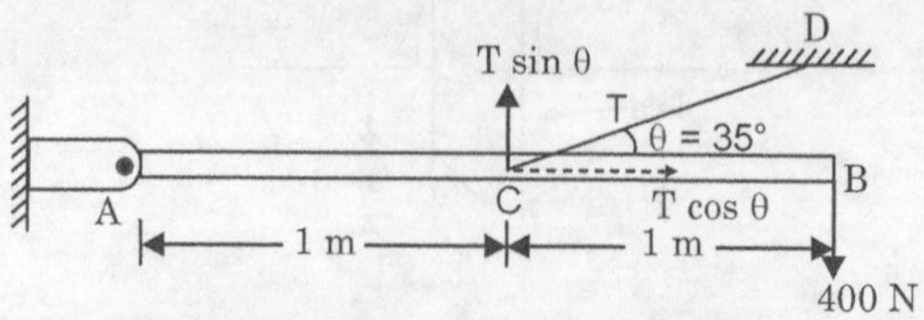

60. Two beams (figure), AC and CD of lengths 10 and 11 m are hinged at C, supported on rollers at A and D and hinged support B, 8m away from A. the beam is acted upon by load 850N at E. By using the principle of virtual work the value of :

(*i*) Reaction at hinge point is ____kN

(*ii*) Reaction at B is _____ kN

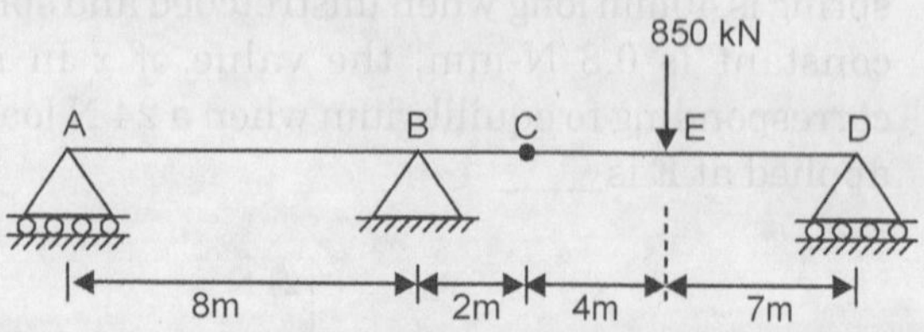

61. A uniform ladder of 400N rests against smooth wall and rough horizontal floor making angle of 75°, using virtual work the value of frictional force between foot of ladder and horizontal floor is____N

62. The mechanism is used to weigh. In fig., a package placed at A causes the weighted pointer to rotate through the angle, α, neglecting the weights of members except counterweight, B = 4kg, if α = 20°, the mass of package at A is____ kg

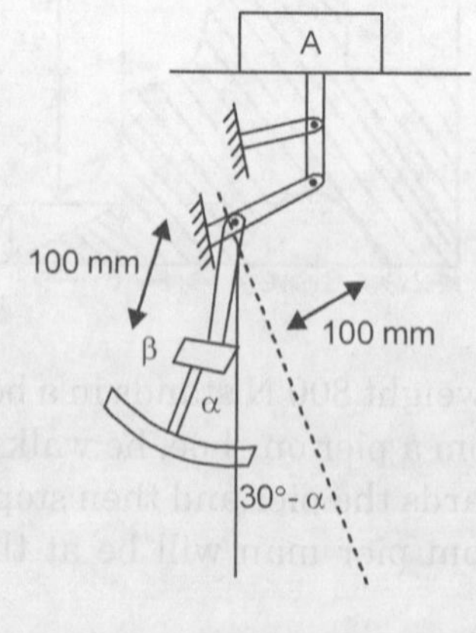

ANSWERS

MCQ Type Questions

1. (*c*)	**2.** (*d*)	**3.** (*b*)	**4.** (*c*)	**5.** (*b*)	**6.** (*c*)	**7.** (*a*)	**8.** (*c*)	**9.** (*a*)	**10.** (*b*)
11. (*a*)	**12.** (*c*)	**13.** (*b*)	**14.** (*d*)	**15.** (*a*)	**16.** (*b*)	**17.** (*b*)	**18.** (*a*)	**19.** (*c*)	**20.** (*c*)
21. (*a*)	**22.** (*d*)	**23.** (*c*)	**24.** (*c*)	**25.** (*a*)	**26.** (*c*)	**27.** (*d*)	**28.** (*c*)	**29.** (*c*)	**30.** (*d*)
31. (*c*)	**32.** (*b*)	**33.** (*a*)	**34.** (*a*)	**35.** (*c*)	**36.** (*b*)	**37.** (*b*)	**38.** (*a*)	**39.** (*a*)	**40.** (*c*)
41. (*b*)	**42.** (*c*)	**43.** (*b*)	**44.** (*d*)	**45.** (*a*)	**46.** (*b*)	**47.** (*b*)	**48.** (*c*)	**49.** (*a*)	**50.** (*b*)
51. (*c*)	**52.** (*c*)	**53.** (*a*)	**54.** (*c*)	**55.** (*d*)	**56.** (*a*)	**57.** (*b*)	**58.** (*a*)	**59.** (*c*)	**60.** (*a*)
61. (*b*)	**62.** (*a*)	**63.** (*c*)	**64.** (*b*)	**65.** (*c*)	**66.** (*b*)	**67.** (*a*)	**68.** (*b*)	**69.** (*c*)	**70.** (*a*)
71. (*b*)	**72.** (*c*)	**73.** (*c*)	**74.** (*d*)	**75.** (*a*)	**76.** (*b*)	**77.** (*c*)	**78.** (*a*)		

Numerical Type Questions

1. 1077.01

2. 235.7

3. 69.28

4. 1.83

5. i) 423.33kN ; ii) 423.33kN ; iii) 485

6. $S_1 = 137.62$ N ; $S_2 = 97.31$ N

7. $R_a = 562.9$ N, $R_b = 938.19$ N, $R_c = 750.55$

8. P = 777.012 N & $T_y = -369.16$ N

9. $R_A = 2309.40$ N & $R_B = 4618.8$ N

10. 42.13

11. *i)* 3000N; *ii)* 750 N

12. i) 3.96 ft; ii) 8.03 ft

13. 235.625 ,209.375 **14.** 100.5, 53 **15.** 373.95

16. 31.5,103.80, 19.806 **17.** 101.49 N

18. 692.82 N **19.** (i) 780.41 N; (ii) 648.67 N; (iii) 80 N

20. (i) 0.67; (ii) 16.67 N

21. (i) 11.481 kN; (ii) 28.68 kN-m **22.** i) 2989.5 N, 2523.3; ii) 1875.22 N, 1175.6

23. 53.196 kN **24.** 23.71° **25.** 3.15m

26. 5.569 kN **27.** 546.4 N **28.** 642.63

29. 208.73 N

30. 257.5 m **31.** i) 56.25m ; ii) 25m/s ; iii) 6.54sec

32. 5461.01 N **33.** 0.0021 sec

34. i) 156.16m; ii) 2.27m/s^2 **35.** i) 66.41m; ii) 20.11° **36.** (i) 0.1 sec; (ii) 17.45 m

37. (i) 7.92 m/s; (ii) 9.904 m/s at 51.4° **38.** 0.283 sec **39.** 0.495 rad/sec

40. 6 **41.** 65.32mm **42.** 40.82mm

43. – 1.1 **44.** 10 inch **45.** 2.76cm

46. 6635.04 **47.** – 1.18 m/s, – 6.18 m/s, 3.82 m/s

48. 326.404, 219.124 **49.** 3.642m **50.** 9.718

51. 0.135 m/s **52.** 567.87 m/s

53. (*i*) – 5.33m/s ; (*ii*) 42.61kN (*iii*) 0.125 sec **54.** 0.9482

55. 133.33 N **56.** 71.25 **57.** 346.42

58. 1250 N **59.** 1394.75

60. (*i*) 540.90 kN and (*ii*) 676.12 kN **61.** 53.5 **62.** 0.1361 kg

EXPLANATIONS

MCQ TYPE QUESTIONS

1. $R_X = \Sigma F_X = 100\cos 60 - 250\sin 15 - 150\cos 15$

$= -159.59$ kN

$R_Y = \Sigma F_Y = 100\sin 60 + 250\cos 15 - 150\sin 15$

$= 289.261$ kN

$$R = \sqrt{R_X^2 + R_Y^2} = \sqrt{159.59^2 + 289.261^2} = 330.36 \text{ kN}$$

$$\tan\theta = \frac{289.261}{159.59} = 1.815$$

$\theta = 61.1°$

Angle with respect to X – axis = 118.9°

2. $\Sigma F_Y = 0 \quad R_{AV} = P \quad \Sigma M_B = 0$

$R_{AV} \times 1.5 - P \times 3 + R_{Ah} \times 8 = 0$

$$R_{Ah} = \frac{1.5P}{8}$$

$\Sigma F_X = 0$

$$R_{Ah} = R_{Bh} = \frac{1.5P}{8}$$

3.

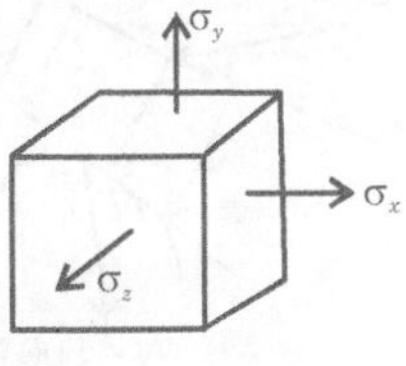

$$\epsilon_x = \frac{\sigma_x}{E} - \mu\frac{\sigma_y}{E} - \mu\frac{\sigma_z}{E}$$

$$\epsilon_y = \frac{\sigma_y}{E} - \mu\frac{\sigma_x}{E} - \mu\frac{\sigma_z}{E}$$

$$\epsilon_x = \frac{\sigma_z}{E} - \mu\frac{\sigma_x}{E} - \mu\frac{\sigma_y}{E}$$

$$\epsilon_V = \frac{\Delta V}{V} = \frac{\Delta X}{X} + \frac{\Delta Y}{Y} + \frac{\Delta Z}{Z}$$

$$\frac{\Delta V}{V} = \epsilon_x + \epsilon_y + \epsilon_z$$

$$= \frac{(1-2\mu)}{E}\left(\sigma_x + \sigma_y + \sigma_z\right)$$

Therefore, $\epsilon_V = 0$, either $(1 - 2\mu) = 0$

Or, $\sigma_x + \sigma_y + \sigma_z = 0$

Therefore, $1 - 2\mu = 0$

$$\mu = \frac{1}{2} = 0.5$$

5. F.B.D. is shown in figure

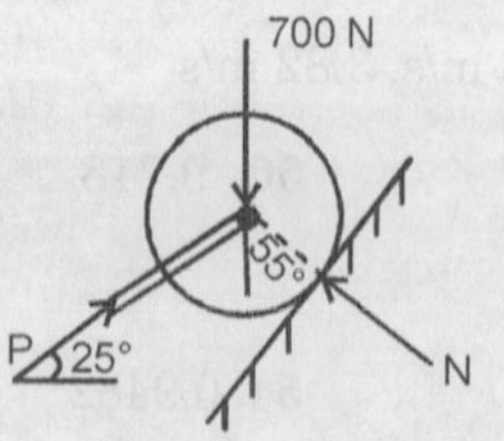

$$\Sigma F_H = 0$$

$$P\cos 25° = N\sin 55°$$

$$\Rightarrow \quad P = 0.9038\ N \qquad \text{...(1)}$$

$$\Sigma F_V = 0$$

$$\Rightarrow P\sin 25° + N\cos 55° = 700\ N \qquad \text{....(2)}$$

Substituting equation (1) in (2)

$$(0.9038)\sin 25° + N\cos 55° = 700$$

$$\Rightarrow \quad 0.9556\ N = 700$$

$$\Rightarrow \quad N = 732.5\ N$$

and $\quad P = 662.05\ N$

6. F.B.D. is shown in figure

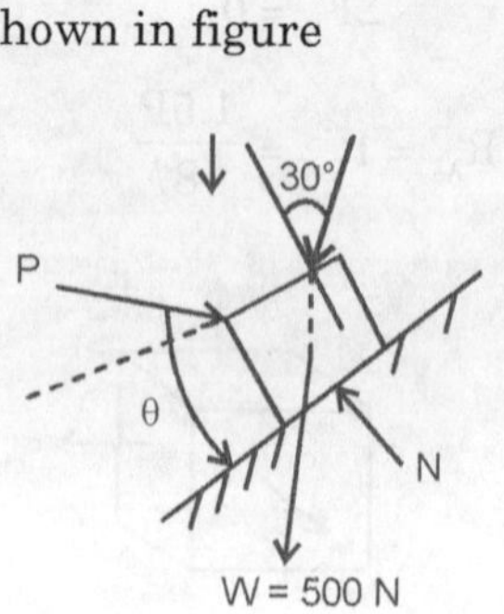

$$\Sigma F_X = 0$$

$$\Rightarrow \quad P\cos\theta = W\sin 30°$$

$$\Rightarrow \quad P\cos 75° = W\sin 30°$$

$$\Rightarrow \quad P = \frac{(W\sin 30°)}{\cos 75°}$$

$$\Rightarrow \quad P = 966\ N$$

$$\Sigma F_y = 0$$

$$\Rightarrow \quad N = P\sin\theta + W\cos 30°$$

$$= 966\sin 75° + W\cos 30°$$

$$\Rightarrow \quad N = 1366.09\ N$$

7.

$$\Sigma F_H = 0$$

$$\Rightarrow F\cos 60° + 700\cos 45° - 400 - P\cos 30° = 0$$

$$\Rightarrow F\cos 60° + 494.97 - 400 - P\cos 30° = 0$$

$$\Rightarrow \quad F = 189.4 + 1.732P \qquad \text{...(1)}$$

$$\Sigma F_V = 0$$

$$F\sin 60° = 400\sin 45° + P\sin 30°$$

$$\Rightarrow (189.4 + 1.732\ P)\sin 60°$$

$$= 400\sin 45° + P\sin 30°$$

$$\Rightarrow \quad 164.02 + 1.499\ P = 282.8 + 0.5\ P$$

$$\Rightarrow \quad 0.999\ P = 118.78$$

$$\Rightarrow \quad P = 120.79\ N$$

and $\quad F = 397.20\ N$

8. FBD for 700 N load is shown in figure (*a*)

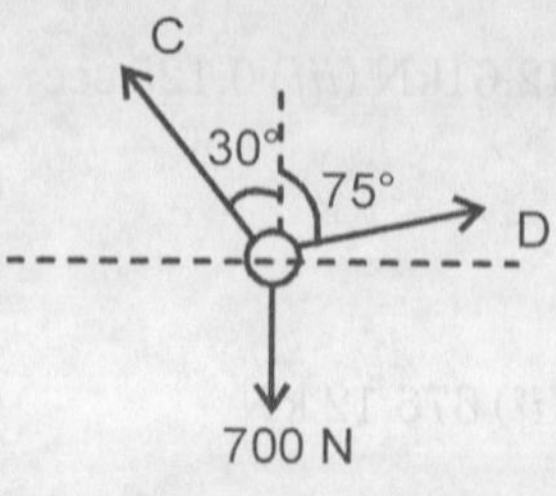

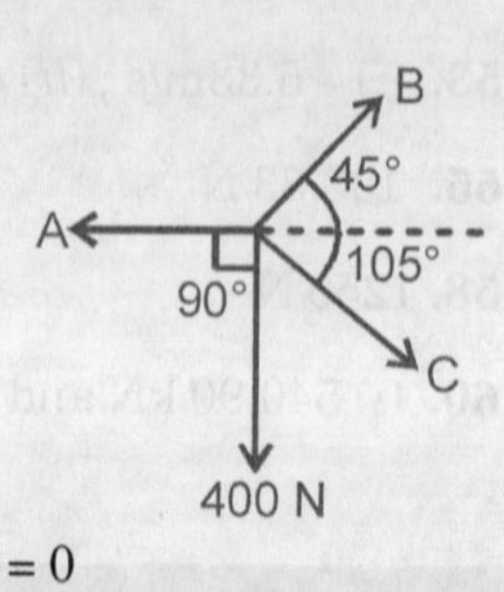

$$\Sigma F_H = 0$$

$$D\sin 75° = C\sin 30°$$

$$\Rightarrow \quad D = 0.517\ C$$

$$\Sigma F_V = 0$$

$$D\cos 75° + C\sin 30° = 700$$

$$\Rightarrow \quad 0.258\ D + 0.5\ C = 700$$

$$\therefore \quad C = 1105.17\ N$$

and $\quad D = 571.37\ N$

FBD for 400N load is shown in Fig.1.7 (*b*)

$$\Sigma F_V = 0$$

$$B\sin 75° = 400 + C\cos 30°$$

$$\Rightarrow \quad B = 1919.52N$$

$$\Sigma F_H = 0$$

$$A = B\cos 45° + C\sin 30°$$

$$\Rightarrow \quad A = 2314.4\ N$$

9. F.B.D. is shown in figure

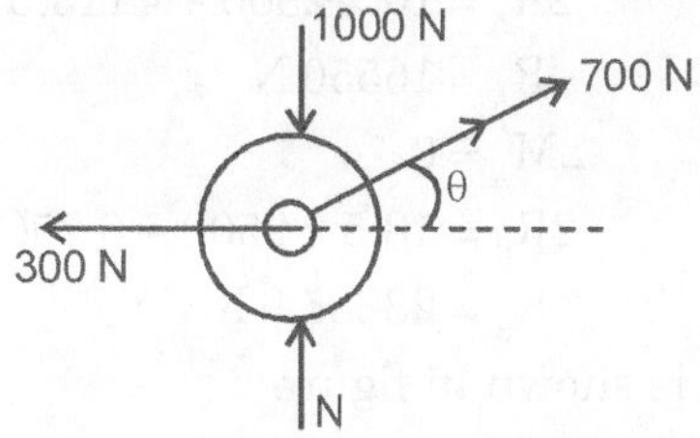

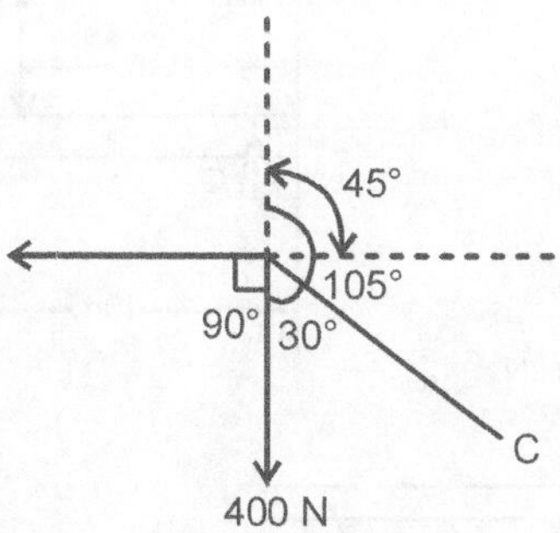

$$\Sigma F_H = 0$$

$$700 \cos\theta = 300$$

$$\Rightarrow \quad \theta = 64.6°$$

$$\Sigma F_V = 0$$

$$N + 700 \sin\theta = 1000$$

$$\Rightarrow \quad N = 367.54 \text{ N}$$

10. F.B.D. is shown in figure

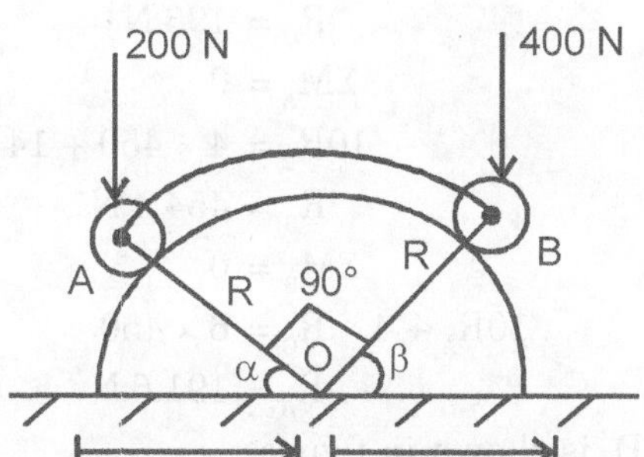

$$\Sigma M_O = 0$$

$$200 R \cos\alpha = 400 R \cos\beta$$

$$\cos\alpha = 2\cos\beta$$

$$\because \quad \alpha + \beta = 90°$$

$$\Rightarrow \quad \alpha = 90° - \beta$$

$$\cos(90° - \beta) = 2\cos\beta$$

$$\sin\beta = 2\cos\beta$$

$$\Rightarrow \quad \tan\beta = 2$$

$$\Rightarrow \quad \beta = \tan^{-1}2 = 63.43°$$

$$\therefore \quad \alpha = 26.56°$$

11. F.B.D. is shown in figure

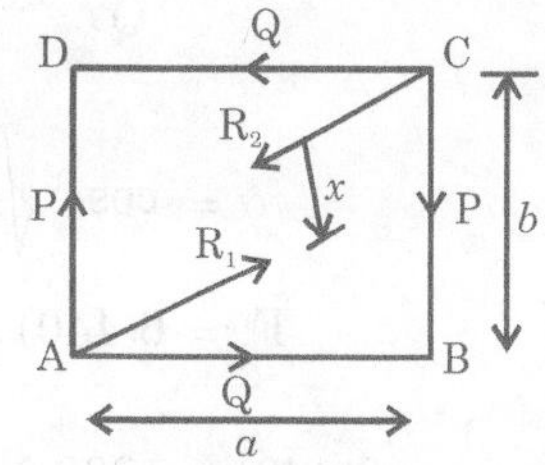

$$R_1 = \sqrt{P^2 + Q^2} \quad \text{at A}$$

$$R_2 = \sqrt{P^2 + Q^2} \quad \text{at C}$$

R_1 and R_2 constitutes couple

$$\therefore \quad M_{couple} = \sqrt{P^2 + Q^2}\,.x$$

Net moment of couple formed by

P and Q = $P \cdot a - Q \cdot b$

$$\therefore \quad \sqrt{P^2 + Q^2}\,.x = P.a - Q.b$$

$$\therefore \quad x = \frac{P.a - Q.b}{\sqrt{P^2 + Q^2}}$$

12. Tension in the string S is equal to the load to it, Q = 635 (figure)

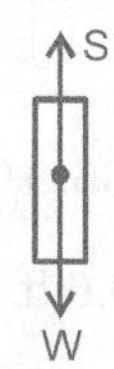

Now, applying parallelogram law, Resultant force,

$$R = \sqrt{W^2 + S^2 + 2WS\cos 180°}$$

$$= \sqrt{W^2 + S^2 - 2WS}$$

$$= \sqrt{(W - S)^2} = (W - S)$$

$$\Rightarrow \quad R = 850 - 635 = 215 \text{ N}$$

Reaction on the man's feet = 215 N downwards.

13.

$$\Sigma y = 0$$

$$R_d \cos\alpha = Q$$

Now taking moments about A,

$$\frac{R_d a}{\cos\alpha} - Q.\ell.\cos\alpha = 0$$

$$\Rightarrow \quad \frac{Q.a}{\cos^2\alpha} - Q.\ell.\cos\alpha = 0$$

$$\Rightarrow \qquad \cos^3 \alpha = \frac{Q.a}{Q\ell}$$

$$\Rightarrow \qquad \alpha = \cos^{-1} \sqrt[3]{\frac{a}{\ell}}$$

14. $$F_1 = 6(440) = 2640 \text{ N} - \text{m}$$

$$F_2 = \frac{1}{2} \times 6(440) = 1230 \text{ N} - \text{m}$$

$$R = F_1 + F_2 = 3960 \text{ N}$$

$$R.d = 3F_1 + 4F_2$$

$$\Rightarrow \qquad 3960d = 3(2640) + 4(1320)$$

$$d = 3.33\text{m}$$

15. $$R = \int_0^{16} y.dx = 90\int_0^{16} \sqrt{x}.dx$$

$$= 90\left[\frac{x^{3/2}}{3/2}\right]_0^{16} = 3840 \text{ N}$$

$$R.d = \int_0^{16} x(y.dx) = 90\int_0^{16} x\sqrt{x}.dx$$

$$\Rightarrow \qquad 3840.d = 90x\left[\frac{x^{5/2}}{5/2}\right]_0^{16}$$

$$\Rightarrow \qquad 3840.d = 36\left[16^{5/2}\right]_0^{16}$$

$$d = 9.6 \text{ ft}$$

16. From figure

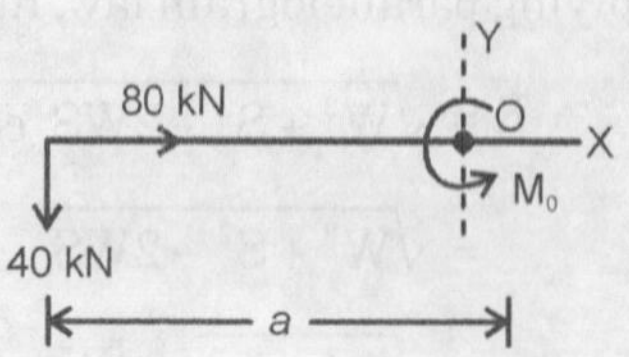

$$M_o = 300 \text{ kN-}m$$

$$\Rightarrow \qquad 40.a_x = 300$$

$$\Rightarrow \qquad a_x = 7.5 \ m$$

$$M_o = 300 \text{ kN-}m$$

$$\Rightarrow \qquad 80.a_y = 300$$

$$\Rightarrow \qquad a_y = 3.75 \ m$$

17. F.B.D. is shown in figure

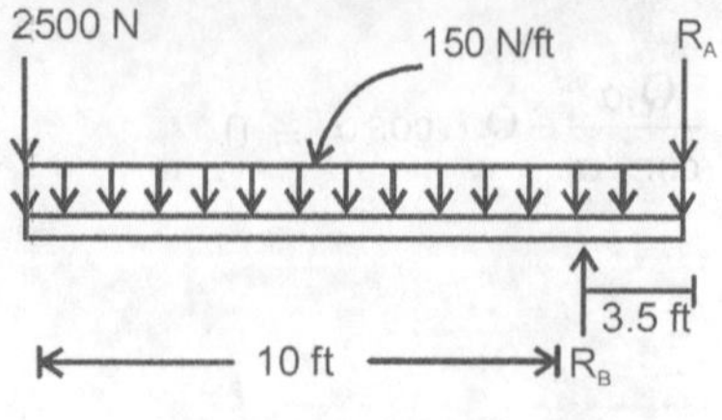

$$\Sigma M_B = 0$$

$$2R_A = 10 \times 2500 + 4\,(13.5 \times 150)$$

$$\Rightarrow \qquad R_A = 16550 \text{ N}$$

$$\Sigma M_A = 0$$

$$2R_B = 13.5 \times 2500 + 6.75(13.75 \times 150)$$

$$\Rightarrow \qquad R_B = 23835.9 \text{ N}$$

18. F.B.D. is shown in figure

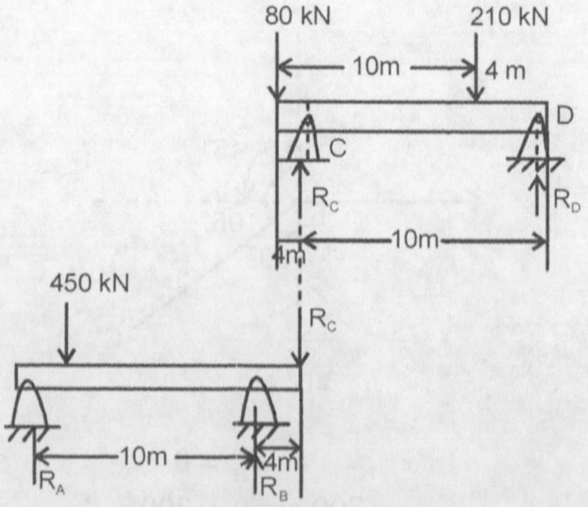

$$\Sigma M_C = 0$$

$$10R_D + 4 \times 80 = 6 \times 210$$

$$\Rightarrow \qquad R_D = 94 \text{ N}$$

$$\Sigma M_D = 0$$

$$10R_C = 14 \times 80 + 4 \times 210$$

$$\Rightarrow \qquad R_C = 196 \text{ N}$$

$$\Sigma M_A = 0$$

$$10R_B = 4 \times 450 + 14 \times R_C$$

$$\Rightarrow \qquad R_B = 454.4 \text{ N}$$

$$\Sigma M_B = 0$$

$$10R_A + 4 \times R_C = 6 \times 450$$

$$\Rightarrow \qquad R_A = 191.6 \text{ N}$$

19. F.B.D. is shown in figure

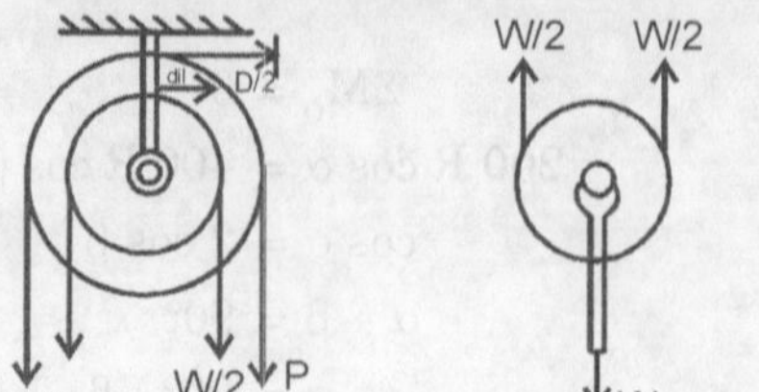

$$\Sigma M_O = 0$$

$$\frac{1}{2}W.\left(\frac{1}{2}\right)D = P\left(\frac{1}{2}\right)D + \left(\frac{1}{2}\right)W\left(\frac{1}{2}\right)d$$

$$\frac{1}{4}W.D = P\left(\frac{1}{2}\right)D + \left(\frac{1}{4}\right)Wd$$

$$\frac{1}{4}\text{W.D} - \left(\frac{1}{4}\right)\text{W}d = \text{P}\left(\frac{1}{2}\right)\text{D}$$

$$\left(\frac{1}{4}\right)\text{W}(\text{D} - d) = \frac{1}{2}\text{P.D}$$

$$\text{W} = \frac{\frac{1}{2}\text{P.D}}{\frac{1}{4}(\text{D} - d)} = \frac{2\text{PD}}{\text{D} - d}$$

20. F.B.D. is shown in figure

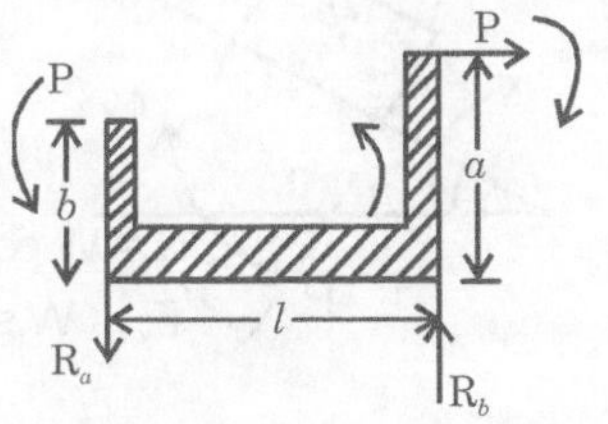

$$\Sigma\text{V} = 0$$
$$\text{R}_a = \text{R}_b$$
$$\Sigma\text{M} = 0$$
$$\text{R}_a = \text{R}_b$$
$$\text{R}_a \times l + \text{P} \times b = \text{P} \times a$$
$$\text{R}_b = \text{P}\frac{0.65 - 0.4}{1.5} = 0.166\text{P}$$
$$\text{R}_a = 0.166\text{P}$$

21.
$$\Sigma\text{V} = 0$$
$$\text{R}_a + \text{R}_b = \text{W}$$
$$\Sigma\text{M}_\text{B} = 0$$
$$\text{R}_a \times 2a + \text{P} \times b = \text{W} \times a$$
$$\text{R}_a = \frac{\text{W}.a - \text{P}.b}{2a}$$
$$\text{R}_b = \frac{\text{W}.a + \text{P}.b}{2a}$$

22. $\text{P} = \text{W}\sin(\alpha + \Phi)$

23. $\text{F} = \mu_s \text{N}$

24.

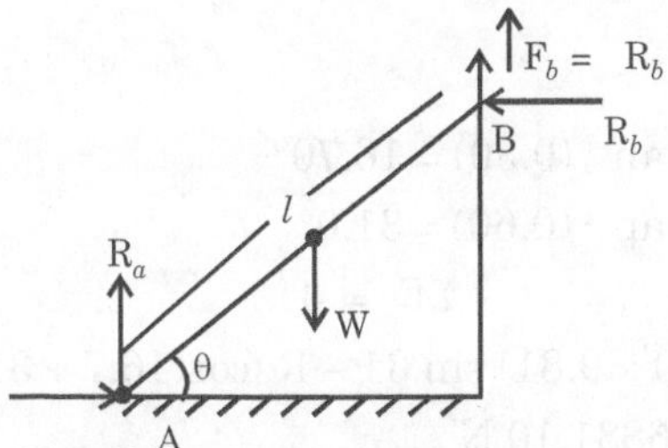

25. FBD of fig. is shown in figure

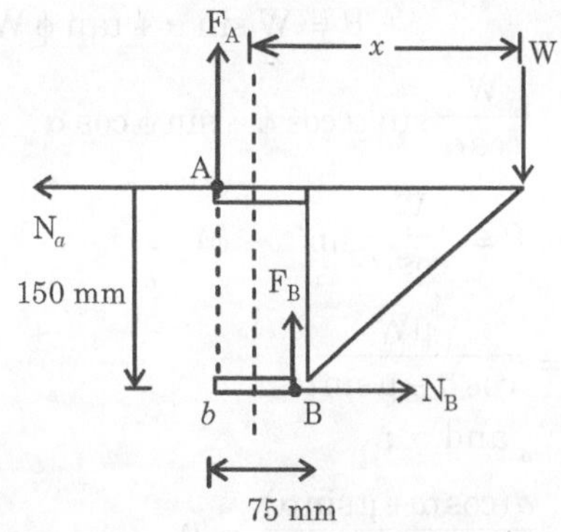

$$\Sigma\text{F}_x = 0$$
$$\text{N}_\text{A} = \text{N}_\text{B}$$
$$\Sigma\text{F}_x = 0$$
$$\Rightarrow 0.25(\text{N}_\text{A} + \text{N}_\text{B}) = \text{W}$$
$$\Rightarrow \text{N}_\text{A} = \text{N}_\text{B} = 2\text{W}$$
$$\Sigma\text{M}_\text{B} = 0$$
$$\Rightarrow \text{N}_\text{A} \times 150 - \text{F}_\text{A} \times 75 - \text{W}(x - 37.5) = 0$$
$$\Rightarrow (2\text{W}) \times 150 - (0.25\text{N}_\text{A}) \times 75 - \text{W}.x + 37.5\text{W} = 0$$
$$\Rightarrow x = 300 \text{ mm}$$

26.

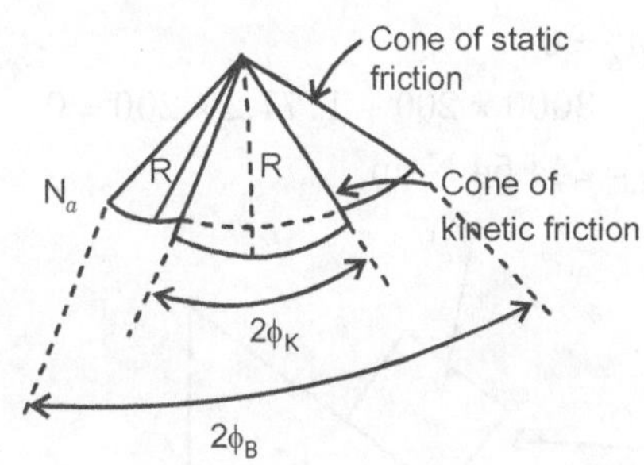

27.
$$\tan\theta = \frac{a/2}{h} = \frac{a}{2h}$$
$$\Rightarrow \quad h = \frac{a}{2\tan\theta}$$

Now $\mu = \tan\theta$

$$\therefore \quad h = \frac{a}{2\mu}$$

28. Resolving forces in figure

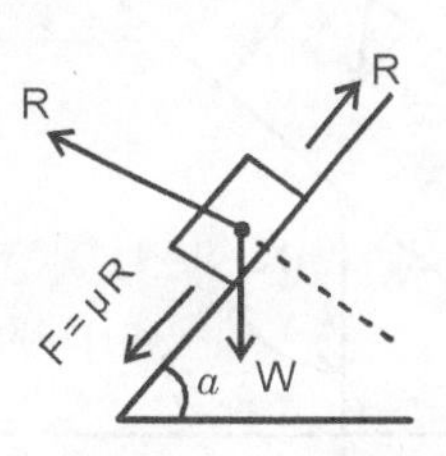

$$P = W\sin\alpha + \mu W\cos\alpha$$

Also, $\mu = \tan\phi$

$$P = W\sin\alpha + \tan\phi\, W\cos\alpha$$

$$\Rightarrow P = \frac{W}{\cos\alpha}\sin\alpha\cos\phi + \sin\phi\cos\alpha$$

Hence, $P = \dfrac{W}{\cos\alpha}\sin(\alpha+\phi)$

29. $P = \dfrac{\mu W}{\cos\alpha + \mu\sin\alpha}$

For P_{min} and α :

$$\frac{d(\cos\alpha + \mu\sin\alpha)}{dx} = 0$$

$$\Rightarrow \alpha = 16.7°$$

And $P_{min} = \dfrac{0.3\times 100}{\cos\alpha + \mu\sin\alpha}$

$$\Rightarrow P_{min} = 28.73 \text{ N}$$

30. At B :

$$\frac{T_1}{T_2}e^{\mu_s\theta} = e^{0.25(2/3\pi)} = 1.688$$

$T_2 = 300$ N and $T_1 = 1777.2$N

At A;

$\Sigma M_A = 0$

$\Rightarrow M_B - 3000\times 200 + 1777.2\times 200 = 0$

$\Rightarrow M_B = 244.54$ N–m

34.

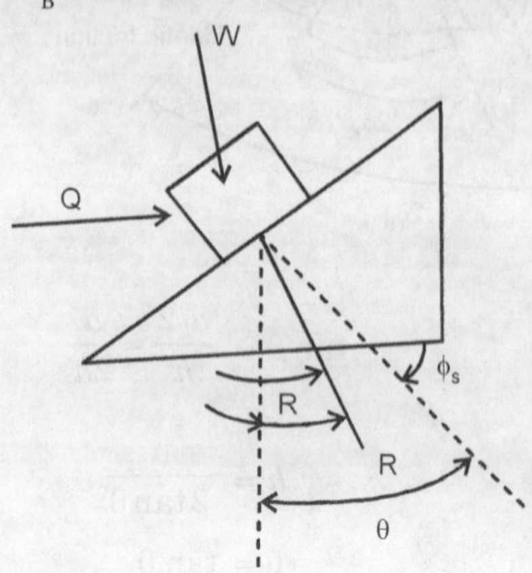

35. From figure

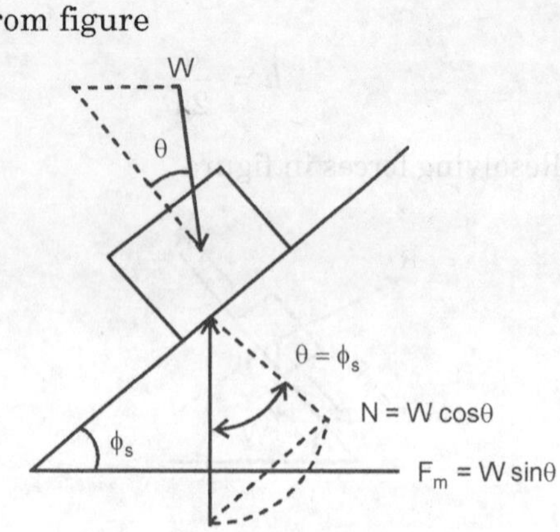

$$N = W\cos\theta$$
$$F = W\cos\theta$$

$$\tan\phi = \frac{F}{N} = \tan\theta = \tan\phi_s = \mu_s$$

$\therefore$ $F = F_m = \mu_s N$

$\phi = \phi_s \rightarrow$ impending motion

36. In figure

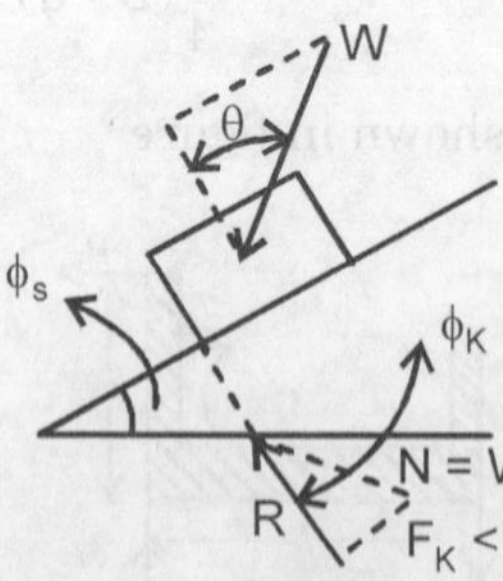

$N = W\cos\theta$

$F_{max} = F_m = \mu_s N = N\tan\phi < N\tan\theta$

$F_{max} < W$ sinh (Motion starts)

$F = F_k = \mu_k N$

$\phi = \phi_k < \phi_s < \theta$

which shows R is not vertical

$\therefore$ force acting are unbalanced.

37. From figure

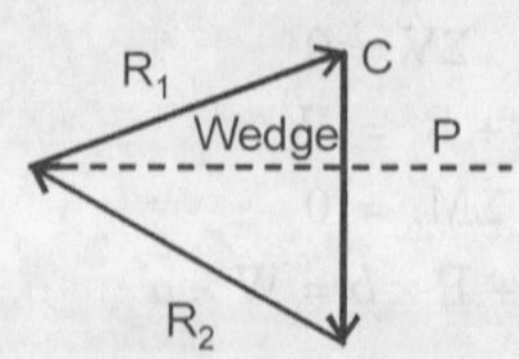

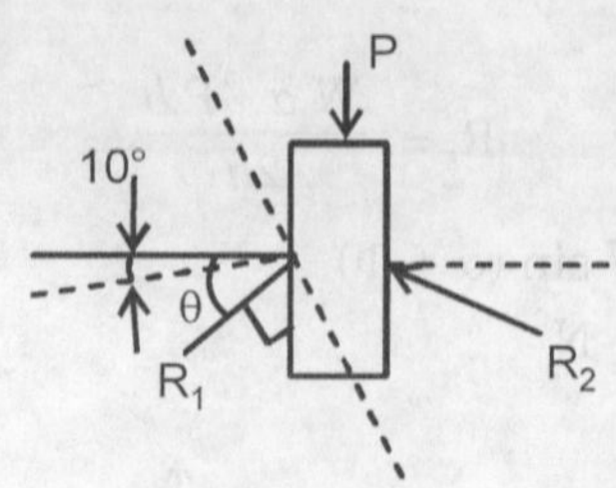

$\phi_1 = \tan^{-1}(0.30) = 16.70°$

$\phi_2 = \tan^{-1}(0.60) = 31.0°$

$$\Sigma F_a = 0$$

$\Rightarrow (1000\times 9.81)\sin 31 - R_2\cos(16.7+31) = 0$

$\Rightarrow R_2 = 6831.10$ N

$$\Sigma F_a = 0$$

$\Rightarrow R_2\cos[90 - (2\phi_1 + 8°] - P\cos[\phi_1 + 8°] = 0$

$\Rightarrow$ 6831.1 cos [90 – 2 × 16.70 + 8°]
– P [cos(16.70 + 8°)] = 0

$\Rightarrow$ 4517.4 – P [0.9085] = 0

$\Rightarrow$ P = 4972.37 N

38. The F.B.D. of figure is shown in figure

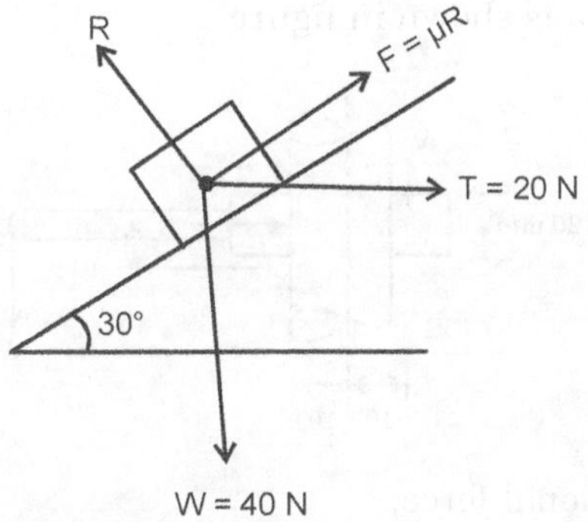

$$\Sigma F_x = 0$$

$\Rightarrow$ $\mu R - W \sin 30° + T \cos 30° = 0$

$\Rightarrow$ $\mu R = -T \cos 30° + W \sin 30°$

$= -(20) \cos 30° + 40 \sin 30°$

$\Rightarrow$ $\mu R = 2.67$ N = F

39. Normal reaction: R :

$R - W \cos 30° - T \sin 30° = 0$

$R = W \cos 30° + T \sin 30°$

$R = 44.64$ N

Coefficient of friction:

$F = \mu R$

$\Rightarrow \mu = \frac{F}{R} = 0.60$

40. The F.B.D. is shown in figure

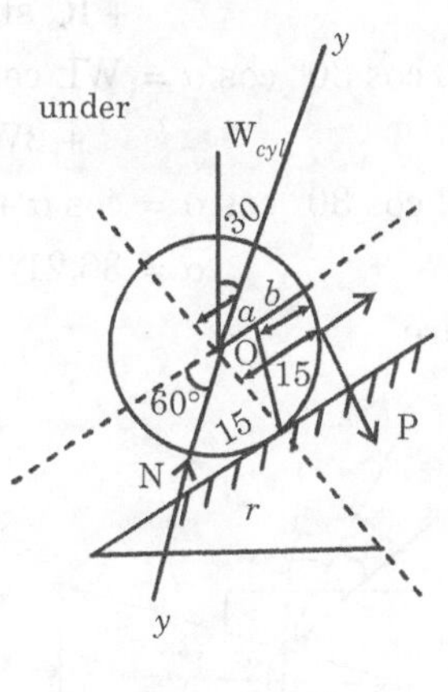

$a = 15 \cos 60°$

$a = 7.5$cm

$b = 15 - a = 7.5$cm

$\Sigma M_A = 0$

$P.b = W_{cyl}.a$

$\Rightarrow \quad P = W_{cyl} \cdot \frac{a}{b} = 20 \times \frac{7.5}{7.5} = 20$kN

$\Sigma M_o = 0$

$15f = 15P$

$f = P = 20$kN

$\Sigma F_y = 0$

$N = 20 \cos 30° + P \cos 30°$

$= 10 + 20 \cos 30° = 34.64$kN

$\therefore \quad f = \mu N$

$\mu = \frac{f}{N} = \frac{20}{34.64}$

$\mu = 0.577$

41. F.B.D. is shown in figure

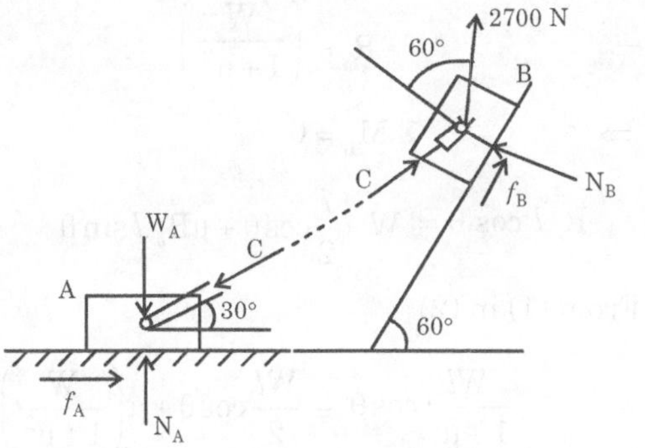

For B, $\Sigma F_y = 0$

$N_B = 2700 \cos 60° + C \sin 30°$

$= 1350 + 0.5C$

$f_B = \mu N_B = \mu(1350 + 0.5C)$

$= 0.25(1350 + 0.5C)$

$\Rightarrow \quad f_B = 337.5 + 0.125\,C$

Also, $\Sigma F_X = 0$

$f_B + C \cos 30° = 2700 \sin 60°$

$f_B = 2700 \sin 60° - C \cos 30°$

$\Rightarrow$ $337.5 + 0.125C = 2700 \sin 60° - C \cos 30°$

$\Rightarrow$ $337.5 + 0.125C = 2338.2 - 0.866\,C$

$\Rightarrow$ $C = 2018.89$ N

For A ; $\Sigma F_Y = 0$

$\Rightarrow$ $N_A = W_A + C \sin 30°$

$\Rightarrow$ $N_A = W_A + 1009.44$

$\therefore$ $f_A = \mu\, N_A$

$f_A = 0.25\,(W_A + 1009.44)$

$\because$ $f_A = C \cos 30°$ $\quad (\Sigma F_X = 0)$

$\therefore$ $C \cos 30° = 0.25(W_A + 1009.44)$

$\Rightarrow$ $2018.89 \cos 30° = 0.25\,W_A + 252.36$

$\Rightarrow$ $W_A = 5984.20$ N

42. In figure

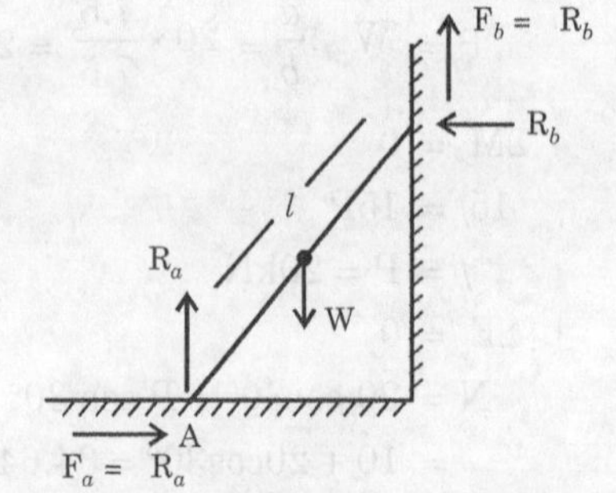

$$\Sigma F_X = 0$$

and $$\Sigma F_X = 0$$

$$R_b = \mu R_a$$

and $$R_a + \mu R_b = W$$

$\therefore$ $$R_a + \mu(\mu R_a) = W$$

$\Rightarrow$ $$R_a = \left(\frac{W}{1+\mu^2}\right) \quad ...(1)$$

$\Rightarrow$ $$\Sigma M_B = 0$$

$$R_a.l \cos\theta = W \times \frac{l}{2}\cos\theta + \mu R_a.l \sin\theta \quad ...(2)$$

From (1) in (2)

$$\frac{Wl}{1+\mu^2}\cos\theta = \frac{Wl}{2}\cos\theta + \mu\left(\frac{W}{1+\mu^2}\right)l\sin\theta$$

$$\left(\frac{\mu}{1+\mu^2}\right)\sin\theta = \left(\frac{\mu}{1-\mu^2} - \frac{1}{2}\right)\cos\theta$$

$\Rightarrow$ $$\tan\theta = \left(\frac{1-\mu^2}{2\mu}\right) = \frac{1-(0.2)^2}{2\times 0.2}$$

$$\theta = 67.38°$$

43. F.B.D. of the figure is shown in figure

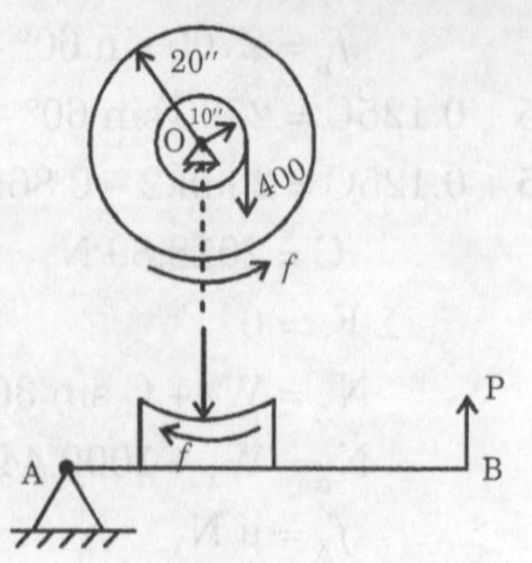

$$\Sigma M_o = 0$$

$$20 f = 10(400)$$

$$f = 200\, lb$$

$$f = \mu N$$

$$200 = 0.20N$$

$$N = 1000\, lb$$

$$\Sigma M_A = 0$$

$$48P + 8f = 16\, N$$

$$48P + 8(200) = 16(1000)$$

$\Rightarrow$ $$P = 300\, lb$$

44. F.B.D. is shown in figure

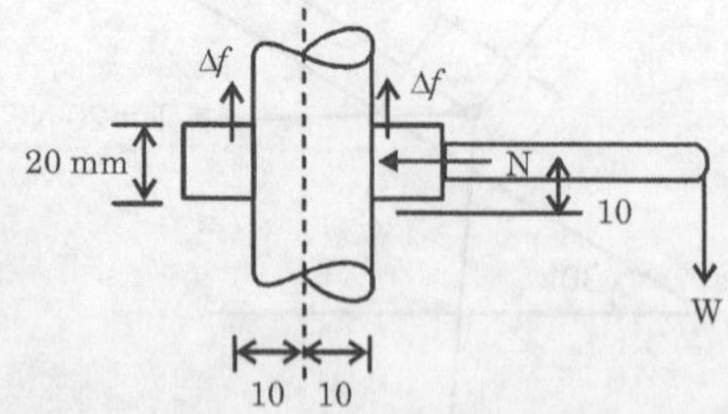

Frictional force,

$$f = \mu N = 0.20\,(0.10\, WL)$$

$$= 0.02\, WL$$

$$\Sigma F_V = 0$$

$$f = W$$

$$0.02\, WL = W$$

$$L = 50mm$$

45. F.B.D. is shown in figure

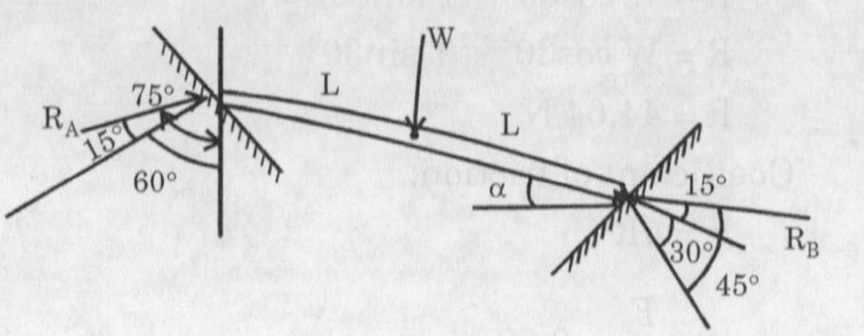

$$\Sigma M_A = 0$$

$\Rightarrow$ $$(R_B \cos 30°)(2L\cos\alpha) = WL\cos\alpha + R_B \sin 30°\,(2L\sin\alpha)$$

$\Rightarrow$ $$2WL\cos 30°\cos\alpha = WL\cos\alpha + 2WL\sin 30°\sin\alpha$$

$\Rightarrow$ $$2\cos 30°\cos\alpha = \cos\alpha + \sin 30°\sin\alpha$$

$\Rightarrow$ $$\alpha = 36.21°$$

46. From figure

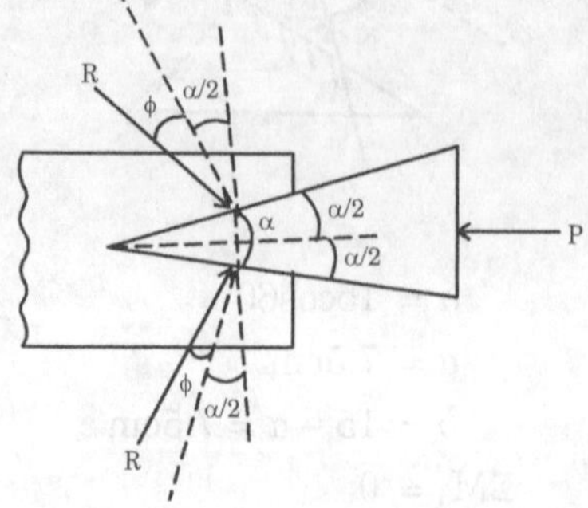

$$\Sigma F_H = 0$$

$$P = 2R \sin\left(\phi + \frac{\alpha}{2}\right)$$

$$\frac{2P}{2\alpha} = \frac{R}{2}\cos\alpha\left(\phi + \frac{\alpha}{2}\right) = 0 \text{ (for P to be maximum)}$$

$$\Rightarrow \cos\left(\phi + \frac{\alpha}{2}\right) = 0$$

$$\Rightarrow \phi + \frac{\alpha}{2} = 90°$$

$$\Rightarrow 2\phi + \alpha = 180°$$

$$\Rightarrow \alpha = 180° - 2\phi$$

$$\therefore \sin\alpha = \sin(180 - 2\phi)$$

$$\Rightarrow \sin\alpha = \sin 180 \cos 2\phi - \cos 180 \sin 2\phi$$

$$\Rightarrow \sin\alpha = -(-1)\sin 2\phi$$

$$\Rightarrow \alpha = 2\phi$$

47. $a = -kv;$

$$\frac{dv}{dt} = -kv$$

$$\int_{v_0}^{v} \frac{dv}{v} = -\int_0^t k\,dt$$

or $\log_e\left(\frac{v}{v_0}\right) = -kt$

$$\therefore \quad v = v_0 e^{-kt}$$

$$\because \quad v = \frac{dx}{dt};$$

$$\therefore \quad \frac{dx}{dt} = v_0 e^{-kt}$$

$$\int_0^x dx = \int_0^t v_0 e^{-kt}\,dt$$

$$x = \frac{v_0}{k}(1 - e^{-kt})$$

48. Refer to figure

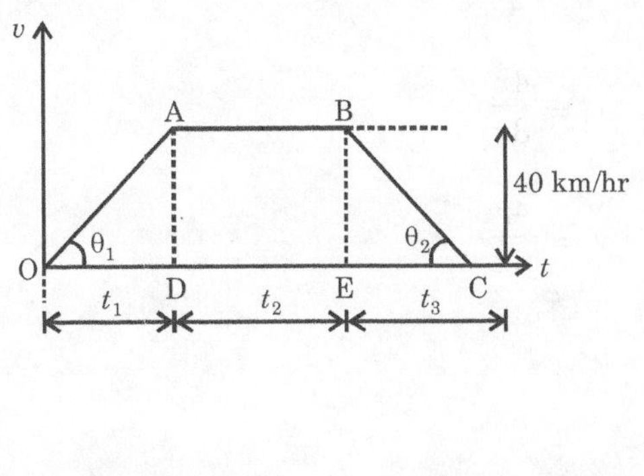

Acceleration and retardation are numerically equal to each other,

$$\therefore \quad \tan\theta_1 = \tan\theta_2$$

$$\Rightarrow \quad \theta_1 = \theta_2$$

$$\Delta OAD \sim \Delta BCE$$

$$\therefore \quad OD = EC$$

$$\Rightarrow \quad t_1 = t_3$$

$$\therefore \quad s = s_1 + s_2 + s_3$$

$$= 20t_1 + 40t_2 + 20t_3$$

$$\Rightarrow \quad 3.5 = 40t_1 + 40t_2$$

$$\Rightarrow \quad t_1 + t_2 = 0.0875 \quad ...(1)$$

$$\Rightarrow \quad t_1 + t_2 + t_3 = \frac{8}{60} = 0.13 \text{ hr}$$

$$\Rightarrow \quad 2t_1 + t_2 = 0.13 \quad ...(2)$$

From equation 1 and 2,

$$\Rightarrow \quad t_1 = 0.0425 \text{ hr}$$

And $t_2 = 0.045$ hr

$$a = \frac{V_a - V_o}{t_1} = \frac{40 - 0}{0.0425}$$

$$= 941.176 \text{ ms}^2$$

49. For stone A:

$u = 0$, $s = 250$ m

t = time taken to reach ground

$$\therefore \quad s = ut + \frac{1}{2}gt^2$$

$$\Rightarrow \quad 250 = 0 \times t + \frac{1}{2} \times 9.81 \times t^2$$

$$\Rightarrow \quad t = 7.139 \text{ sec.}$$

For stone B:

$u = 0$; $s = 165$m

$$s = ut + \frac{1}{2}gt^2$$

$$\Rightarrow \quad 165 = 0 + \frac{1}{2}gt^2$$

$$\Rightarrow \quad t = 5.79 \text{ sec (from B to C)}$$

∵ The direction of the stone B become horizontal

∴ initial velocity will be zero for remaining distance.

$$85 = 0 \times t_1 + \frac{1}{2} \times 9.81 \times t_2^2$$

$$\Rightarrow \quad t_2 = 4.16 \text{ sec.}$$

∴ time taken for B to reach ground:-

$$\Rightarrow \quad t_1 + t_2 = 9.95 \text{ sec}$$

$\therefore$ ratio of time taken by B & A to reach ground

$$= \frac{9.95}{7.13} = 1.39 \text{ sec}$$

50. (*i*) $x = -3 \sin t \Rightarrow \sin t = -\frac{x}{3}$

$$y = 2 - 3\cos t \Rightarrow \cos t = \frac{(2-y)}{3}$$

$$\therefore \quad \sin^2 t + \cos^2 t = \left(-\frac{x}{3}\right)^2 + \left(\frac{2-y}{3}\right)^2$$

$$\Rightarrow \quad 1 = \frac{x^2}{3^2} + \frac{(2-y)^2}{3^2}$$

$$\Rightarrow \quad x^2 + (2-y)^2 = 3^2$$

(*ii*) $\vec{r} = x\hat{i} + y\hat{j}$

$$\Rightarrow \vec{r} = (-3\sin t)\,\hat{i} + (2\text{-}3\cos t)\,\hat{j}$$

Velocity vector,

$$\vec{v} = \frac{dr}{dt} = \frac{d}{dt}\,[(-3\sin t)\,\hat{i} + (2-3\cos t)\,\hat{j}\,]$$

$$= (-3\cos t)\,\hat{i} + (3\sin t)\,\hat{j}$$

Acceleration vector,

$$\vec{a} = \frac{dv}{dt} = \frac{d}{dt}\,[(-3\cos t)\,\hat{i} + (3\sin t)\,\hat{j}\,]$$

$$= (3\sin t)\,\hat{i} + (3\cos t)\,\hat{j}$$

At the instant of start, at $t = 0$

$$\therefore \quad \vec{r} = (\text{-}3\sin 0)\,\hat{i} + (2 - 3\cos 0)\,\hat{j} = -1\,\hat{j}$$

$$\vec{v} = (-3\cos 0)\,\hat{i} + (3\sin 0)\,\hat{j} = -3\,\hat{i}$$

$$\vec{a} = (3\sin 0)\,\hat{i} + (3\cos 0)\,\hat{j} = 3\,\hat{j}$$

51. Case 1 : When balloon is falling-

$$\frac{W}{g}a = W - P \qquad \text{...(eq.1)}$$

Case 2 : To balance the balloon

$$\frac{W-Q}{g}a = P - (W - Q) \qquad \text{...(eq.2)}$$

Adding equation 1 and equation 2

$$\Rightarrow \quad \frac{W}{g}a - W + P + \frac{W-Q}{g}a - (P - (W - Q)) = 0$$

$$\Rightarrow \quad \frac{Wa + (W-Q)a}{g} = W - P + P - (W - Q)$$

$$\Rightarrow 2Wa = Qg + Qa$$

$$\Rightarrow Q = \frac{2Wa}{(g+a)}$$

52. A is the instantaneous centre of the cylinder,

$$V_c = \omega.AC$$
$$V_B = \omega.AB$$
$$V_D = \omega.AD$$

$$\omega_{cyl} = \frac{v_c}{AC} = \frac{40}{\frac{6}{2}} = 13.3\,\frac{\text{rad}}{\text{sec}}$$

$$AD = \sqrt{DC^2 + AC^2} = \sqrt{3^2 + 3^2} = 4.23\,\frac{\text{m}}{\text{s}}$$

$$V_D = \omega.AD = \omega \times 4.24 = 163\,\frac{\text{m}}{\text{s}}$$

$$V_B = \omega.AB = \omega \times 6 = 240\,\frac{\text{m}}{\text{s}}$$

53. $v = 48.4$ km/hr $= 134.4$ m/s

$r = 400$ m

$$\text{angle of breaking, } \tan\alpha = \frac{v^2}{r_s}$$

$$a = \tan^{-1}\frac{134.41^2}{400} = 88.7°$$

54.

$$m_1 = 59 \text{ kg} \quad x_1 = 0 \text{ m}$$
$$m_2 = 73 \text{ kg} \quad x_2 = 3.5 \text{ m}$$

$$x_{cm} = \frac{m_1x_1 + m_2x_2}{m_1 + m_2}$$

$$x_{cm} = \frac{59\times 0 + 73\times 3.5}{59 + 71}$$

$$x_{cm} = 1.94 \text{ m}$$

55. For the quarter circle,

$$A_1 = \frac{1}{4}\times\pi\times r^2$$

$$x_1 = \frac{4r}{3\pi}$$

$$y_1 = \frac{4r}{3\pi}$$

For semicircle

$$A_2 = \left(\frac{1}{4}\times\pi\times r^2\right)\times\frac{1}{2}$$

$$x_2 = \frac{4\left(\frac{1}{2}\right)r}{3\pi} = \frac{2r}{3\pi}$$

$$y_2 = \frac{r}{2}$$

For the shaded area

$$A = A_1 - A_2$$

$$= \frac{1}{4}\times\pi\times r^2 - \frac{1}{8}\times\pi\times r^2$$

$$= \frac{1}{8}\times\pi\times r^2$$

Therefore, $A\bar{x} = \Sigma ax$

$$\frac{1}{8}\times\pi\times r^2\bar{x} = \frac{1}{4}\times\pi\times r^2\times\frac{4r}{3\pi} - \frac{1}{8}\times\pi\times r^2\times\frac{2r}{3\pi}$$

$$\frac{1}{8}\times\pi\times r^2\bar{x} = \frac{1}{8}r^3$$

$$\bar{x} = \frac{2r}{\pi} = 0.636\text{ r}$$

$$A\bar{y} = \Sigma ay$$

$$\frac{1}{8}\times\pi\times r^2\bar{y} = \frac{1}{4}\times\pi\times r^2\times\frac{4r}{3\pi} - \frac{1}{8}\times\pi\times r^2\times\frac{r}{2}$$

$$\frac{1}{8}\times\pi\times r^2\bar{y} = 0.137\, r^3$$

$$\bar{y} = 0.3488\, r$$

56. The centre of mass of a wire, is

$$\bar{x} = \frac{\Sigma\, x.dm}{\Sigma\, dm},$$

$$\bar{y} = \frac{\Sigma\, y.dm}{\Sigma\, dm},$$

Let x-axis is so chosen that it passes through centre of cylinder rod, therefore $\bar{y} = 0$

Let density at the distance x from left end be $\rho = kx.dx$

$$\Sigma\, x.dm = \int_0^l kx^2 dx = \frac{kl^3}{3}$$

$$\Sigma\, dm = \int_0^l kx\, dx = \frac{kl^3}{3}$$

$$\Sigma\, x.dm = \int_0^l kx\, dx = \frac{kl^3}{3}$$

$$\bar{x} = \frac{2l}{3}$$

57. F.B.D. is shown in figure

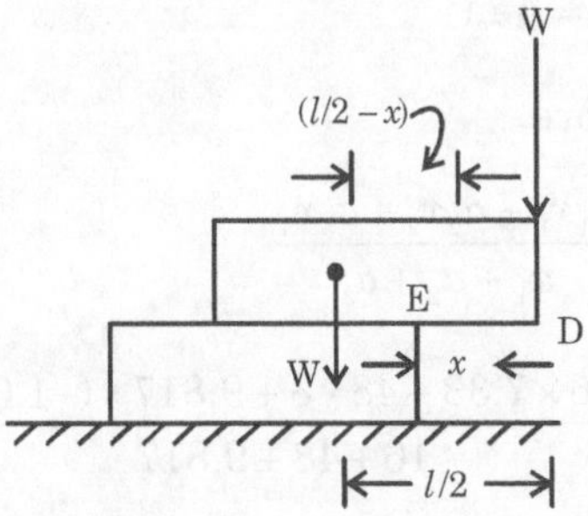

To avoid toppling over the top of the box, its centre of gravity is with the outer edge of CD of middle box.

The maximum permissible overhang of the box

$$= \frac{l}{2}$$

Therefore distance, CB $= \frac{l}{2}$

Under equation Conditions

$$\Sigma M_e = 0$$

$$W.x - W\left(\frac{l}{2} - x\right) = 0$$

Or, $$x = \frac{l}{2} - x$$

$$x = \frac{l}{4}$$

Therefore, $$m = \frac{l}{2} + \frac{l}{4}$$

$$= \frac{3l}{4}$$

58. Triangular segment:Rectangular Segment:

$A_1 = \frac{1}{2}\times 4\times 8 = 16\text{m}^2$ $\quad a_2 = 6 \times 8 = 48\text{ m}^2$

$x_1 = 6 + \frac{1}{3}\times 4 = 7.33\text{ m}$ $\quad x_2 = \frac{6}{3} = 3\text{ m}$

$y_1 = \frac{8}{3} = 2.66$ m $\quad y_2 = \frac{8}{2} = 4$ m

Semi-circular segment:-

$a_3 = \frac{1}{2}\pi.(2.5)^2 = 9.817\ m^2$

$x_3 = -\frac{4R}{3\pi} = -\frac{4(2.5)}{3\pi} = 1.06m$

$y_3 = \frac{8}{2} = 4$ m

Therefore,

$$\bar{x} = \frac{a_1x_1 + a_2x_2 + a_3x_3}{a_1 + a_2 + a_3}$$

$$= \frac{16\times 7.33 + 48\times 3 + 9.817\times(-1.06)}{16 + 48 + 9.817}$$

= 4.93 m

$$\bar{y} = \frac{a_1y_1 + a_2y_2 + a_3y_3}{a_1 + a_2 + a_3}$$

$$= \frac{16\times 2.66 + 48\times 4 + 9.817\times(-1.06)}{16 + 48 + 9.817}$$

= 3.03 m

59. From figure

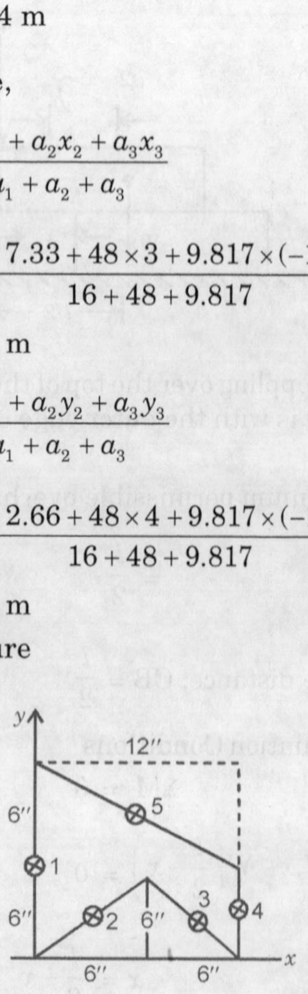

$L_1 = 12$ inch

$L_2 = \sqrt{6^2 + 6^2} = 6\sqrt{2}$ inch

$x_1 = 0, y_1 = 0$

$x_2 = \frac{1}{2}(6) = 3$ inch

$y_2 = \frac{1}{2}(6) = 3$ inch

$L_3 = \sqrt{6^2 + 6^2} = 6\sqrt{2}$ inch

$L_4 = 6$ inch

$x_3 = 6 + \frac{1}{2}(6) = 9$ inch

$x_4 = 12$ inch

$y_3 = \frac{1}{2}(6) = 3$ inch

$y_4 = 3$ inch

$L_5 = \sqrt{12^2 + 6^2} = 6\sqrt{5}$ inch

$x_5 = \frac{1}{2}(12) = 6$ inch

$y_5 = 6 + \frac{1}{2}(6) = 9$ inch

$L = L_1 + L_2 + L_3 + L_4 + L_5$

$= 12 + 6\sqrt{2} + 6\sqrt{2} + 6 + 6\sqrt{5}$

$L = 48.387$ inch

$\sum \bar{x} = \sum lx$

$\Rightarrow 48.387\,\bar{x} = 12(0) + 6\sqrt{2}(3) + 6\sqrt{2}(9)$

$+ 6(12) + 6\sqrt{5}(6)$

$\Rightarrow \bar{x} = 5.26$ inch

$\sum \bar{y} = \sum ly$

$\Rightarrow 48.387\,\bar{y} = 12(6) + 6\sqrt{2}(3) + 6\sqrt{2}(3)$

$+ 6(3) + 6\sqrt{5}(9)$

$\Rightarrow \bar{y} = 5.408$ inch

60. $A_1 = \frac{1}{2}\times 6\times 9 = 27\ inch^2$

$x_1 = \frac{1}{3}\times 6 = 2$ inch

$y_1 = \frac{2}{3}\times 9 = 6$ inch

$A_1 = \frac{1}{2}\times\pi\times 3^2 = 14.14$ inch

$x_2 = r = 3$ inch

$y_2 = 9 + \frac{4\times 3}{3\pi}$

$= 10.27$ inch

$A_1 + A_2 = A$

$= 27 + 14.14 = 41.14\ inch^2$

$A\bar{x} = \sum ax$

$41.14\,\bar{x} = 27 \times 2 + 14.14 \times 3$

$\Rightarrow \quad A\bar{y} = \sum ay$

$\Rightarrow \quad 41.14\bar{y} = 27 \times 6 + 14.14 \times 10.27$

$\Rightarrow \quad \bar{y} = 7.47$ inch

61. F.B.D. is shown in figure

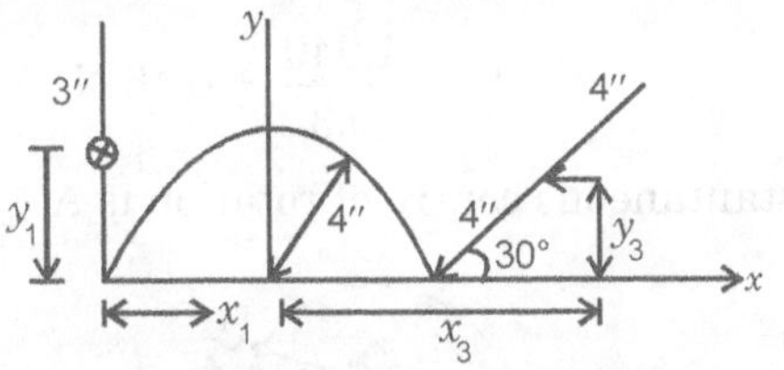

$L_1 = 6$ inch

$L_2 = \theta \times 4 = 12.5664$ inch

$X_1 = 4$ inch

$X_2 = 0$

$Y_1 = 3$ inch

$Y_2 = \frac{2 \times 4}{\pi} = 2.5465$ inch

$L_3 = 8$ inch

$X_2 = 4 + 4\cos 30° = 7.4641$ inch

$Y_3 = 4\sin 30° = 2$ inch

$L\bar{x} = \sum lx$

$\Rightarrow 25.5664\bar{x} = 6(-4) - 12.5664(0) + 8 \times (7.46411)$

$\Rightarrow \bar{x} = 1.34$ inch

$L\bar{y} = \sum ly$

$\Rightarrow 25.5664\bar{y} = 6 \times 3 + 12.5664(2.5664) + 8 \times (2)$

$\Rightarrow \bar{y} = 2.48$ inch

62. A_1 = area of triangle

$= \frac{1}{2} \times 80 \times 80 = 3200 \text{ m}^2$

A_2 = area of semi-circle

$= \frac{\pi d^2}{8} - \frac{\pi R^2}{2} = \frac{\pi 40^2}{8} - \frac{\pi 20^2}{2}$

$= 2513.274 \text{ m}^2$

A_3 = area of circle

$= \frac{\pi d^2}{2} = 1256.64 \text{ m}^2$

$$\bar{x} = \frac{a_1x_1 + a_2x_2 + a_3x_3}{a_1 + a_2 + a_3}$$

$$= \frac{3200 \times \left(2 \times \frac{80}{3}\right) + 2513.274 \times (40) - 1256.64 \times 40}{3200 + 2513.274 - 1256.64}$$

$= 49.57$ m

$$\bar{y} = \frac{a_1y_1 + a_2y_2 + a_3y_3}{a_1 + a_2 + a_3}$$

$$= \frac{3200 \times \left(\frac{80}{3}\right) + 2513.274 \times (-16.97) - 1256.64 \times 0}{3200 + 2513.274 - 1256.64}$$

$= 9.58$ m

63. A_1 = area of rectangle

$= 200 \times 150 = 30000 \text{ mm}^2$

A_2 = area of triangle

$= \frac{1}{2} \times b \times h = \frac{\frac{1}{2} \times 200 \times 150}{2} = 3750 \text{ mm}^2$

A_3 = area of circle $= \pi\, 50^2$

$= 7853.98 \text{ mm}^2$

$$\bar{x} = \frac{a_1x_1 + a_2x_2 + a_3x_3}{a_1 + a_2 + a_3}$$

$$= \frac{30000 \times 100 - 3750 \times 166.67 - 7853.98 \times 100}{30000 - 37550 - 7853.98}$$

$= 86.4$ mm

$$\bar{y} = \frac{a_1y_1 + a_2y_2 + a_3y_3}{a_1 + a_2 + a_3}$$

$$= \frac{30000 \times 75 - 3750 \times 125 - 7853.98 \times 75}{30000 - 37550 - 7853.98}$$

$= 64.8$ mm

64. $m.v_x = -$ M.V

$\frac{1}{2}k.\delta^2 = \frac{1}{2}\text{M}.v^2$

The negative direction of velocity shows that bullet direction and direction of M will be opposite.

The K.E. will be stored as PE.

Therefore,

$$\text{M}.v^2 = \phi.k.\delta^2$$

$$k = \frac{\text{M}v^2}{\delta^2}$$

65. According to conservation of energy,

Change in PE = change in KE

$$mg(h + \delta) = \frac{1}{2}k\delta^2$$

$$h = 0$$

Therefore,

$$mg(\delta) = \frac{1}{2}k\delta^2$$

$$\delta = \frac{2mg}{k}$$

66. Moment of resultant force at the centre is 0

i.e., $r \times F = 0$

Therefore, according to conservation of angular momentum H remains constant,

$$r \times mv = H = \text{constant}$$

$$r \times mv = r \times m(v_r\hat{e}_r + v_\theta\hat{e}_\theta)$$

Since v_r & v_θ coincide, therefore,

$$r \times m(v_\theta\hat{e}_\theta) = H$$

And $v_\theta = rw$

Therefore, $r^2w = H = \text{constant}$

$$\Rightarrow \quad r_1^2w_1 = r_2^2 = w_2$$

$$\Rightarrow \quad w_2 = \left(\frac{r_1}{r_2}\right)^2 w_1$$

Since, $r_1 = r$

and $r_2 = \frac{r}{2}$

and $w_1 = w$

Therefore,

$$\Rightarrow \quad w_2 = \left(\frac{r}{\frac{r}{2}}\right)^2 w = 4w$$

Therefore, resultant speed of the particle is,

$$\sqrt{v^2 + 4w^2r^2}$$

67. Velocity with which marble strikes the floor

$$= \sqrt{2gh_1} = \sqrt{2 \times 9.81 \times 16}$$

$$= -17.71 \text{ m/s}$$

downward direction

Velocity of rebounding $= \sqrt{2gh_2} = \sqrt{2 \times 9.81 \times 10}$

$$= 14.01 \text{ m/s}$$

upward direction

Therefore, Impulse,

$$H = m(v_2 - v_1)$$

$$= \frac{0.6}{9.81}(14.01 - 17.71)$$

$$H = 1.940 \text{ N-s}$$

Also $F.t = 1.940$

$$\Rightarrow \quad F = \frac{1.940}{0.3} = 6.46 \text{ N}$$

68. Instantaneous centre of rotation is A.

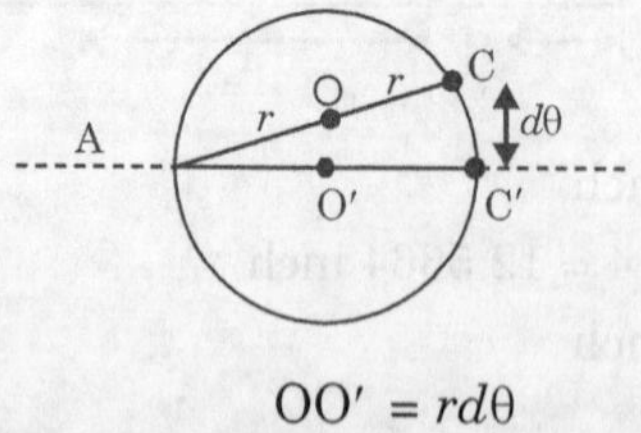

$$OO' = rd\theta$$

$$CC' = 2\,rd\theta$$

Potential Energy of system is function of θ,

$$V = V_g + V_e$$

$$= -mgr\theta + \frac{1}{2}k(2r\theta)^2$$

For equilibrium; $\frac{dV}{d\theta} = 0$

$$-mgr + k(2r\theta)\cdot(2r) = 0$$

$$\theta_e = \frac{mg}{4\,Kr}$$

69. F.B.D. is shown in figure

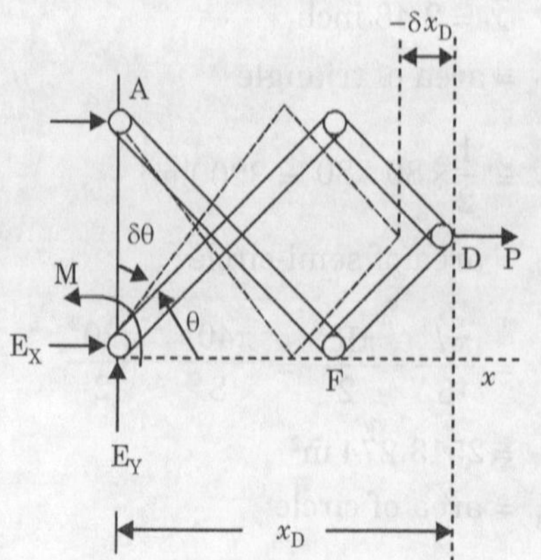

Applying a positive virtual increment to θ at E,

$$\delta U = 0$$

$$= \delta U_M + \delta U_P$$

$$= M\,\delta\theta + P\,\delta x_D \qquad \text{...(1)}$$

δθ along positive θ increase with δθ → + δθ

M and δθ have same sense → + Mδθ

For positive $\delta\theta$, x_D will decrease by

$$\delta x_D \rightarrow -\delta x_D$$

P and $-\delta x_D$ have opposite sence

$$-P(-\delta x_D) \rightarrow +P\,\delta x_D$$

$$x_D = 3\,l\cos\theta$$

$$\delta x_D = -3l\sin\theta\,\delta\theta \quad ...(2)$$

$\therefore$ (1) & (2) gives

$$0 = M\,\delta\theta + P(-3\,l\sin\theta\,\delta\theta)$$

$$\Rightarrow M = 3\,Pl\sin\theta$$

70. F.B.D. is shown in figure

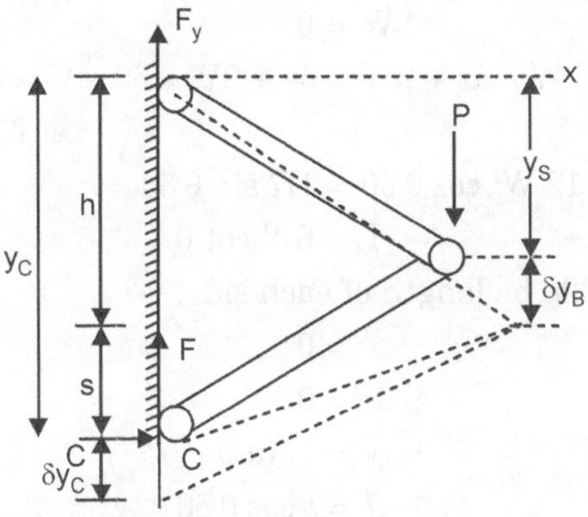

Applying a positive virtual increment to θ at A,

$$\delta U = \delta U_B + \delta U_F = 0$$

$$0 = P\delta y_B - F\delta y_C \quad ...(1)$$

y_B increases with $\delta y_B \rightarrow +\delta y_B$

P and δy_B have same sense $\rightarrow +\delta y_B$

For positive $\delta\theta$, y_c will increase by $\delta y_C \rightarrow +\delta y_C$

F and δy_C have opposite sense $\rightarrow -F(\delta y_C) \rightarrow -F\,\delta y_C$

$\therefore$ $y_B = l\sin\theta$

& $y_C = 2\,l\sin\theta$

$$F = k_s$$

$$\Rightarrow \delta y_B = l\cos\theta\,\delta\theta$$

$$\Rightarrow \delta y_C = 2\,l\cos\theta\,\delta\theta$$

$$F = k(y_C - h)$$

$$= k(2\,l\sin\theta - h) \quad ...(2)$$

$$\Rightarrow 0 = P(l\cos\theta\,\delta\theta) - k(2\,l\sin\theta - h)\cdot(2\,l\cos\theta\,\delta\theta)$$

$$\Rightarrow \sin\theta = \frac{P + 2kh}{4\,kl}$$

Now,

$$F = k\left[2l\left(\frac{P+2kh}{4\,kl}\right) - h\right]$$

$$F = \frac{P}{2}$$

71. $DE = l\cos\theta$

and $CE = l\sin\theta$

When a virtual displacement, $d\theta$ is given to string DE,

$$\delta_{DE} = -l\sin\theta\,\delta\theta$$

$$\delta_{CE} = l\cos\theta\,\delta\theta$$

The resolved parts of weight P & Q along plane are 15.sin30° and 20.sin60°

$$U_D = (15\sin 30°)\cdot\delta_{DE}$$

$$= 15\sin 30°.(-l\sin\theta\,\delta\theta)$$

$$= -7.5\,l\sin\theta\,\delta\theta$$

$$U_C = (20\sin 60°)\cdot\delta_{CE}$$

$$= 20\sin 60°.(l\cos\theta\,\delta\theta)$$

$$= 17.3\,l\cos\theta\,\delta\theta$$

$$\therefore U_{total} = U_D + U_E$$

$$= 0 \text{ (virtual work)}$$

$$\Rightarrow -7.5\,l\sin\theta\,\delta\theta = -17.3\,l\cos\theta\,\delta\theta$$

$$\Rightarrow \frac{\sin\theta}{\cos\theta} = \frac{17.3}{7.5}$$

$$\Rightarrow \theta = 66.5°$$

72. F.B.D. is shown in figure

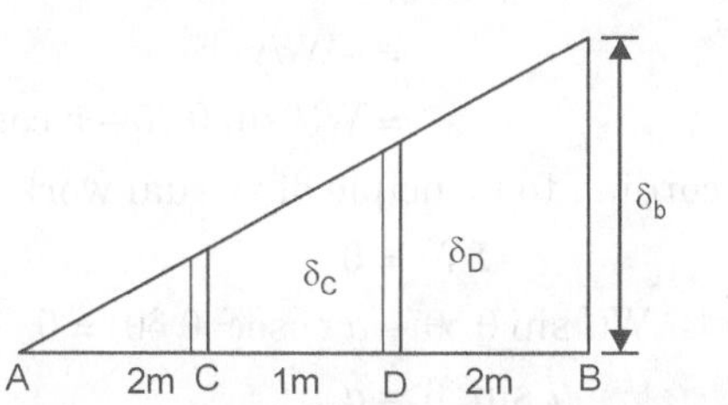

$\delta_b \rightarrow$ vertical virtual displacement

From the triangle similarity,

$$\frac{\delta_C}{2} = \frac{\delta_D}{3} = \frac{\delta_b}{5}$$

$$\Rightarrow \delta_c = 0.456\,\delta_b$$

$$\Rightarrow \delta_D = 0.6\,\delta_b$$

$$\therefore \text{Total work} = R_b \times \delta_b - 35 \times \delta_d - 20 \times \delta_c + R_a \times 0 = 0$$

$$\Rightarrow R_b\cdot\delta_b - 21\,\delta_b - 8\delta_b = 0$$

$$\Rightarrow R_b = 29\text{ kN}$$

$$R_a + R_b = 20 + 35$$

$$\Rightarrow R_a = 26\text{ kN}$$

73. Distance of P from

$$A = 6\,l\cos\theta$$

Distance of Q from

$$A = 2\,l\cos\theta$$

Virtual displacement of P is B – B′

$\therefore \quad B - B' = dx_1$

$= -6\,l \sin\theta\, d\theta$

Virtual displacement of Q is C – C′

$\therefore \quad C - C' = dx_2$

$= -2\,l \sin\theta\, d\theta$

According to principle of virtual work :

$\Sigma W = 0$

$P(-6\,l \sin\theta\, d\theta) - Q(2\,l \sin\theta\, d\theta) = 0$

$\Rightarrow \quad P = \dfrac{Q}{3}$

74. Let virtual displacement in truss be '*dy*'

$\therefore \quad \Sigma W = 0$

$R_a \times AA' = P \times DD'$

$\Rightarrow \quad R_a = P \qquad (\because AA' = DD' = dy)$

75. Vertical height of G above the peg 'P' is '*y*'

$y = AG\cos\theta - a\cot\theta$

$= l\cos\theta - a\cot\theta$

Let system rotate through 'δθ'.

$\therefore \quad dy = -\,l\sin\theta\,\delta\theta + a\,\text{cosec}^2\theta\,\delta\theta$

Work done by weight 'W' of rod

$= -Wdy$

$= W(l\sin\theta\,\delta\theta - a\,\text{cosec}^2\theta\,\delta\theta)$

According to principle of virtual work :

$\Sigma W = 0$

$\Rightarrow \quad W(l\sin\theta\,\delta\theta - a\,\text{cosec}^2\theta\,\delta\theta) = 0$

$\Rightarrow \quad l\sin^3\theta = a$

$\Rightarrow \quad \theta = \sin^{-1}\left(\dfrac{a}{l}\right)^{\frac{1}{3}}$

76. F.B.D. is shown in figure

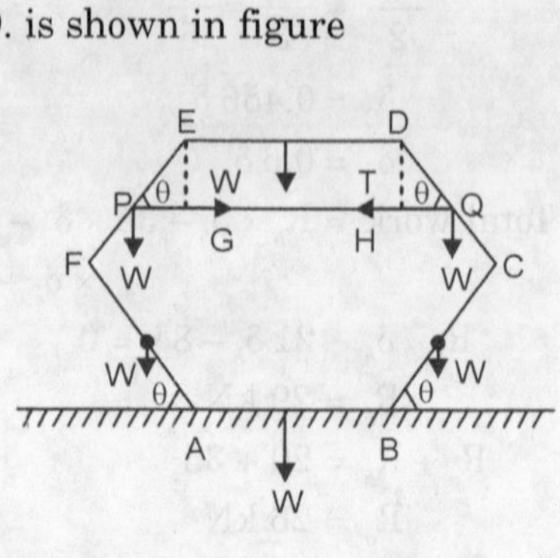

Let $2l$ be length of each side.

$L_{PQ} = PG + GH + HQ$

$= l\cos\theta + 2l + a\cos\theta$

$= 2l + 2l\cos\theta$

$\therefore \quad \delta_{PQ} = -\,2l\sin\theta\,\delta\theta$

Let x, y, z and h be the height of mid-points of AB, BC, AF, CD, FE and ED measured above AB.

$x = 0\,;$

$\delta x = 0$

$y = l\sin\theta\,;$

$\delta y = l\cos\theta\,\delta\theta$

$z = 3l\sin\theta\,;$

$\delta z = 3l\cos\theta\,\delta\theta$

$h = 4l\sin\theta\,;$

$\delta h = 4l\cos\theta\,\delta\theta$

By principle of virtual work :

$\Sigma W = 0$

$\Rightarrow \quad W \times \delta x + 2W \times \delta y + 2W \times \delta z + W \times \delta h + T \times \delta_{PQ} = 0$

$\Rightarrow \quad 12\,Wl\cos\theta\,\delta\theta = 2Tl\sin\theta\,\delta\theta$

$\Rightarrow \quad T = 6W\cot\theta$

77. let '$2l$' be length of each side.

$x = 0\,;$

$\delta x = 0$

$y = l\sin\theta\,;$

$\delta y = l\cos\theta\,\delta\theta$

$z = 3l\sin\theta\,;$

$\delta z = 3l\cos\theta\,\delta\theta$

$h = 4l\sin\theta\,;$

$\delta h = 4l\cos\theta\,\delta\theta$

By principle of virtual work :

$\Sigma W = 0$

$\Rightarrow \quad W \times 0 + 2W \times l\cos\theta\,\delta\theta + 2W(3l\cos\theta\,\delta\theta) + W(4l\cos\theta\,\delta\theta) - T(4l\cos\theta\,\delta\theta) = 0$

$\Rightarrow \quad T = 3W$

78. F.B.D. is shown in figure

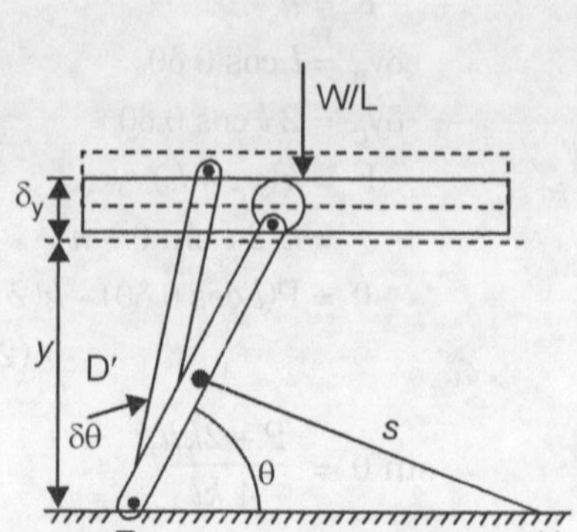

$y = 2a\sin\theta\,;$

$\delta y = 2a\cos\theta\,\delta\theta$

$s^2 = a^2 + L^2 - 2aL\cos\theta$

$2s\,\delta s = -\,2aL\,(-\sin\theta)\,\delta\theta$

$$\Rightarrow \qquad \delta s = \frac{a L \sin\theta}{s}\cdot\delta\theta$$

By principle of virtual work :

$$\Sigma W = 0$$

$$-\frac{1}{2}W\delta y + F_{DH}\delta s = 0$$

$$\Rightarrow \qquad F_{DH} = \frac{W.S}{L}\cot\theta$$

NUMERICAL TYPE QUESTIONS

1. The F.B.D. is shown in figure

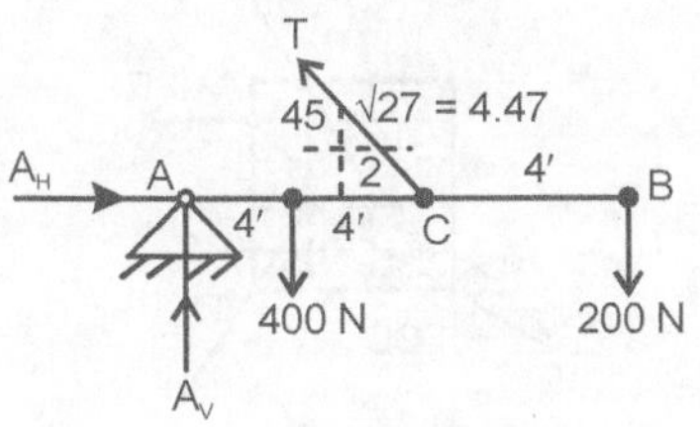

$$\Sigma M_A = 0$$

$$4\times\left(\frac{5}{5.385}T\right) = 4(400) + 12(200)$$

$$3.714\,T = 1600 + 2400$$

$$T = 1077.01\text{ N}$$

2. F.B.D. is shown in figure

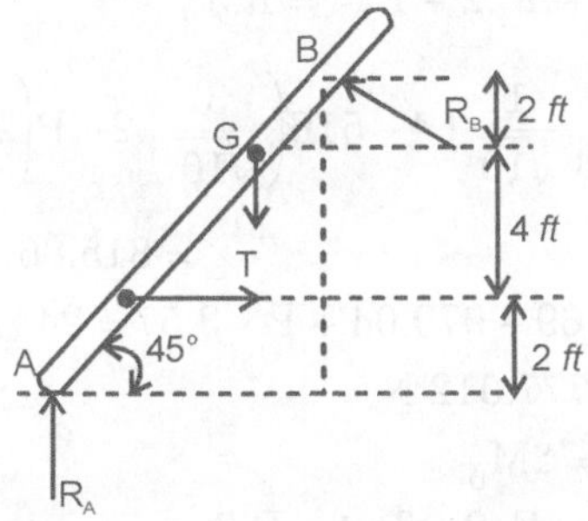

$$AB = \sqrt{8^2+8^2} = 8\sqrt{2}\text{ m}$$

$$\Sigma F_H = 0$$

$$T = R_B\cos 45°$$

$$\Sigma M_A = 0$$

$$2T + 6(550) = 8\sqrt{2}\,R_B$$

$$T + 1650 = 4\sqrt{2}\,R_B$$

$$R_B\cos 45° + 1650 = 4\sqrt{2}\,R_B$$

$$R_B = 333.34\text{ kN}$$

$$\therefore \qquad T = 333.34\cos 45°$$

$$T = 235.70\text{ kN}$$

3. F.B.D. is shown in figure

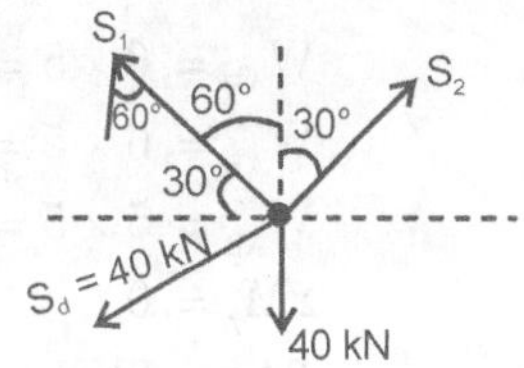

$$\Sigma X = 0$$

$$S_1\cos 35° + 40\sin 60° = S_2\sin 30°$$

$$\Rightarrow \qquad \frac{\sqrt{3}}{2}S_1 + 40\frac{\sqrt{3}}{2} = \frac{S_2}{2}$$

$$\Rightarrow \qquad S_2 = \sqrt{3}\,S_1 + 40\sqrt{3} \qquad ...(1)$$

$$\Sigma Y = 0$$

$$S_1\sin 30° + S_2\cos 30° = S_d\cos 60° + 40$$

$$\Rightarrow \qquad \frac{S_1}{2} + S_2\frac{\sqrt{3}}{2} = \frac{40}{2} + 40$$

$$\Rightarrow \qquad 120 = S_1 + \sqrt{3}\,S_2 \qquad ...(2)$$

from (1) and (2)

$$\Rightarrow \qquad S_1 + 3S_1 + 120 = 120 \Rightarrow S_1 = 0\text{ kN}$$

$$\therefore \qquad S_2 = 69.28\text{ kN}$$

4. F.B.D. is shown in figure

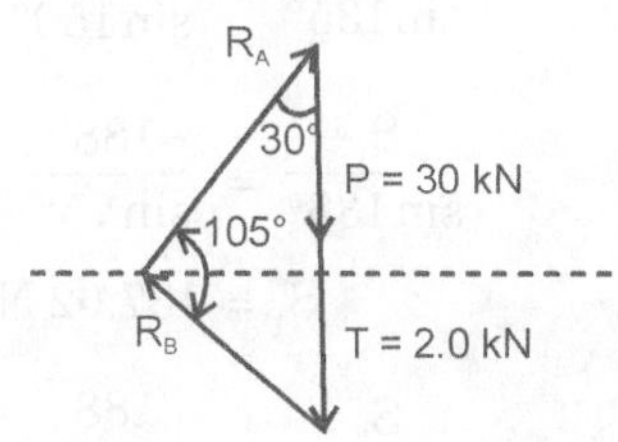

$$\frac{R_A}{\sin 45°} = \frac{20+30}{\sin 105°}$$

$$\Rightarrow \qquad R_A = 36.60\text{ kN}$$

$$\Sigma M_B = 0$$

$$4(R_A\cos 30°) = 30\times 3 + 20x$$

$$\Rightarrow \qquad 4\times 31.6 = 90 + 20x$$

$$\Rightarrow \qquad x = 1.83\text{ m}$$

5. The F.B.D. is shown in figure

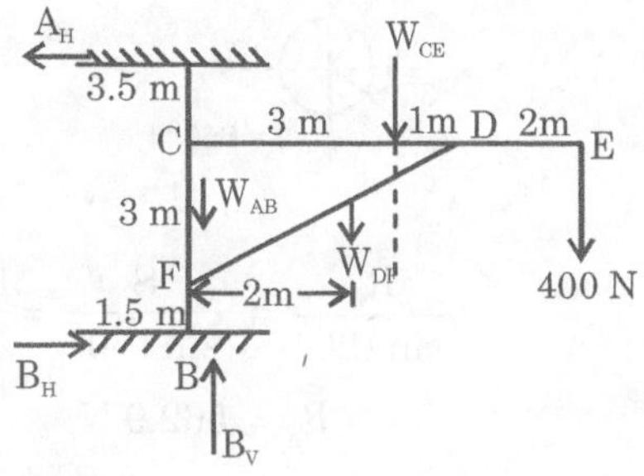

$$L_{DF} = \sqrt{4^2 + 3^2} = 5\text{m}$$

$$W_{AB} = 6 \times 5 = 30 \text{ kN}$$

$$W_{AB} = 6 \times 5 = 30 \text{ kN}$$

$$W_{AB} = 5 \times 5 = 25 \text{ kN}$$

(*i*) $\Sigma M_A = 0$

$\Rightarrow$ $6A_H = 3W_{CE} + 2W_{DF} + 6(400)$

$\Rightarrow$ $6A_H = 3 \times 30 + 2 \times 25 + 6(400)$

$\Rightarrow$ $A_H = 423.33 \text{ kN}$

(*ii*) $\Sigma F_H = 0$

$B_H = A_H$

$\Rightarrow$ $B_H = 423.33 \text{ kN}$

(iii) $\Sigma F_V = 0$

$B_V = W_{AB} + W_{CE} + W_{DF} + 400$

$\Rightarrow$ $B_V = 485 \text{ kN}$

6. F.B.D. is shown in figure

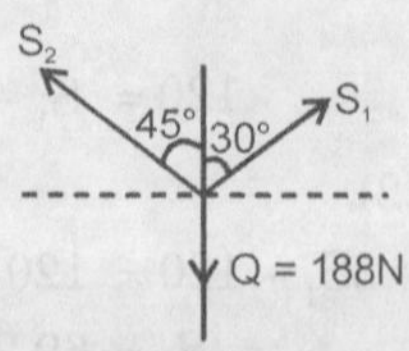

$$\frac{S_1}{\sin 135°} = \frac{S_2}{\sin 150°} = \frac{S_3}{\sin 75°}$$

$$\Rightarrow \frac{S_1}{\sin 135°} = \frac{188}{\sin 75°}$$

$\Rightarrow$ $S_1 = 137.62 \text{ N}$

$$\Rightarrow \frac{S_2}{\sin 150°} = \frac{188}{\sin 75°}$$

$\Rightarrow$ $S_2 = 97.31 \text{ N}$

7. F.B.D. is shown in figure

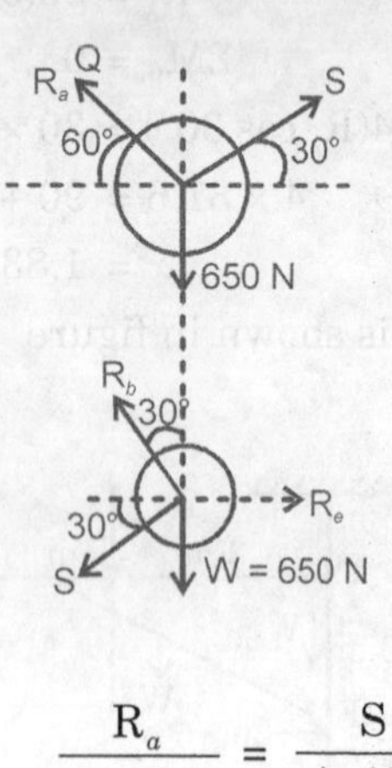

$$\frac{R_a}{\sin 120°} = \frac{S}{\sin 150°} = \frac{650}{\sin 90°}$$

$\Rightarrow$ $R_a = 562.9 \text{ N}$

$\Rightarrow$ $S = 325 \text{ N}$

$\Sigma Y = 0$

$R_b \cos 60° = 650 + S \sin 30°$

$$\Rightarrow R_b \frac{\sqrt{3}}{2} = 650 + S \times \frac{1}{2}$$

$\Rightarrow$ $R_b = 938.19 \text{ N}$

$\Sigma X = 0$

$R_c = R_b \sin 30° + S \sin 30°$

$R_c = 750.55 \text{ N}$

8. F.B.D. is shown in figure

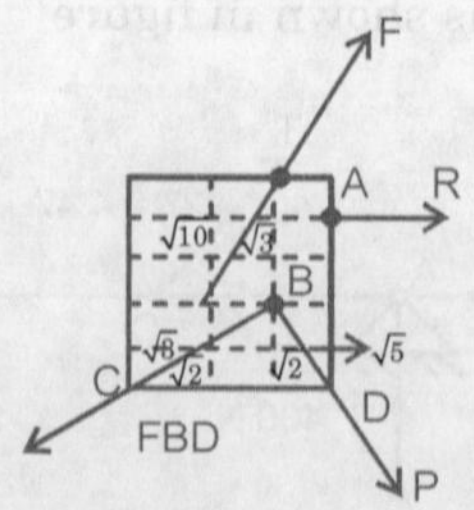

$M_R = \Sigma M_B$

$R.1 = F_x.2 + F_y.1$

$$\Rightarrow R = 516\left(\frac{1}{\sqrt{10}}\right).2 + 516\left(\frac{3}{\sqrt{10}}\right).1$$

$R = 326.34 + 489.52 = 815.86 \text{ N}$

$M_R = \Sigma M_C$

$F_x.4 - F_y.2 + P_y.4 = R.3$

$$\Rightarrow 516\left(\frac{1}{\sqrt{10}}\right).4 - 516\left(\frac{3}{\sqrt{10}}\right).2 + P\left(\frac{2}{\sqrt{5}}\right).4 = 815.86 \times 3$$

$\Rightarrow 652.69 - 979.04 + P \times 3.57 = 2447.58$

$\Rightarrow P = 777.012 \text{ N}$

$M_R = \Sigma M_D$

$F_x.4 - F_y.2 - T_y.4 = R.3$

$$\Rightarrow 516\left(\frac{1}{\sqrt{10}}\right).4 + 516\left(\frac{3}{\sqrt{10}}\right).2 - T_y\left(\frac{2}{\sqrt{3}}\right).4 = 815.86 \times 3$$

$\Rightarrow 652.69 + 979.041 - T_y \times 2.21 = 2447.58$

$T_y = -369.16 \text{ N}$

9. F.B.D. is shown in figure

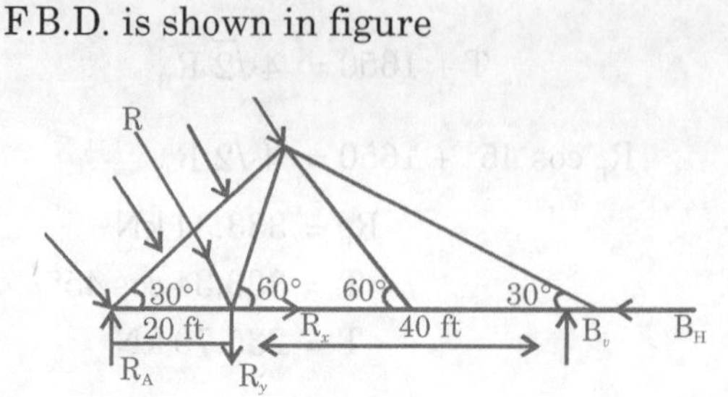

$R = 2 \times 1000 + 3 \times 2000 = 8000$ N

$R_x = R \sin 30° = 8000 \sin 30° = 4000$ N

$R_y = R \cos 30° = 8000 \cos 30° = 6928.20$ N

$$\Sigma M_D = 0$$

$$60R_A = 20R_y$$

$$60R_A = 20 \times 6928.20$$

$$R_A = 2309.40 \text{ N}$$

$$R_B = \sqrt{B_H^2 + B_V^2}$$

$$= \sqrt{4000^2 + 2309.4^2}$$

$$= 4618.8 \text{ N}$$

$$\tan \theta_{Bx} = \frac{B_v}{B_H} = \frac{2309.40}{4000}$$

$$\theta_{Bx} = 30°$$

10. F.B.D. is shown in figure

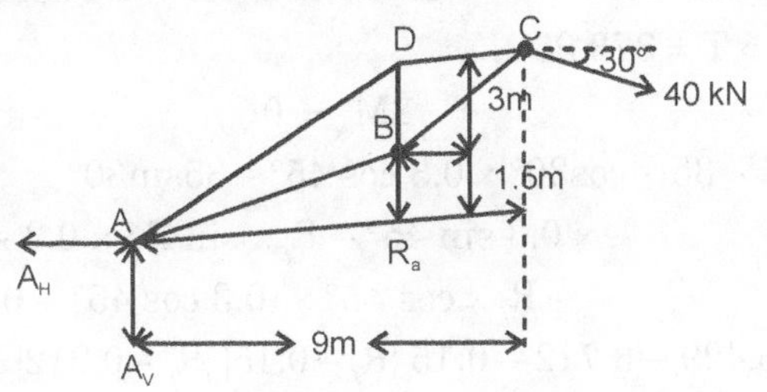

$$\Sigma M_A = 0$$

$$9R_B = 3 + 1.5(40 \cos 30°) + (9 + 3)\, 40 \sin 30°$$

$$\Sigma F_H = 0$$

$$A_H = 40 \cos 30° = 34.64 \text{ kN}$$

$$\Sigma M_B = 0$$

$$9A_V = 1.5A_H + 3 \times 40 \cos 30° + 3 \times 40 \sin 30°$$

$$A_V = 23.98 \text{ kN}$$

$$R_B = \sqrt{A_H^2 + A_V^2}$$

$$= \sqrt{34.64^2 + 23.98^2}$$

$$= 42.13 \text{ kN}$$

Therefore, $\tan \theta_{Bx} = \frac{A_v}{A_H} = \frac{23.98}{34.64}$

$$\theta = 35°$$

11.

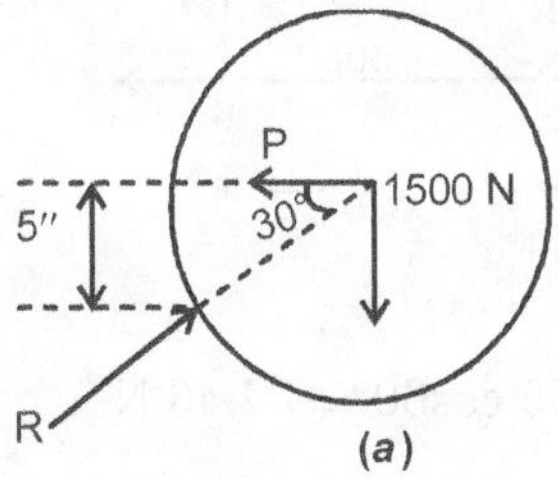

(*a*)

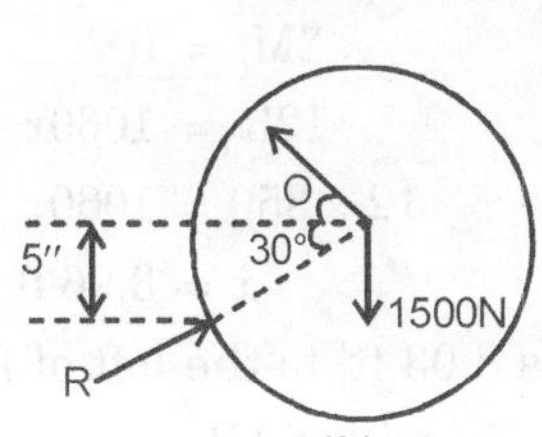

(*b*)

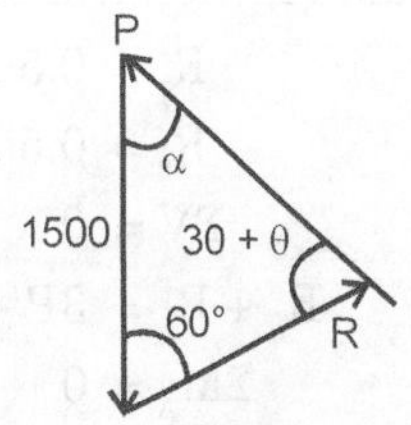

(*c*)

(*i*) F.B.D. is shown in figure (*a*)

$$\tan 30° = \frac{1500}{P} \Rightarrow P = 2598.07 \text{ N}$$

$$\sin 30° = \frac{1500}{R} \Rightarrow R = 3000 \text{ N}$$

(*ii*) F.B.D. is shown in figure (*b*),

$$\frac{P}{\sin 60°} = \frac{1500}{\sin(30° + \theta)}$$

$$\Rightarrow \quad P = \frac{1500 \sin 60°}{\sin(30° + \theta)}$$

$$\frac{dP}{d\theta} = \frac{1500 \sin 60° \cos(30° + \theta)}{\sin^2(30° + \theta)} = 0$$

$$-1500 \sin 60° \cos(30° + \theta) = 0$$

$$\cos(30° + \theta) = 0$$

$$\theta = 60°$$

$$P_{min} = \frac{1500 \cos 60°}{\sin(30° + 60°)}$$

$$= 1299.03 \text{ N}$$

$$\alpha = 180° - 60° - (30° + \theta)$$

$$\alpha = 30°$$

$$\frac{R}{\sin \alpha} = \frac{1500}{\sin(30° + \theta)}$$

$$= \frac{1500}{\sin(30° + 60°)}$$

$$R = 750 \text{ N}$$

12.

$$R_3 = 0.5\, R_C$$

$$R_3 = 350 \text{ N}$$

$$\Sigma F_V = 0$$

$$R_C + R_3 = P$$

$$R_C + 0.5\, R_C = 1060$$

$$R_C = 700 \text{ N}$$

$$\Sigma M_C = 0$$
$$12R_3 = 1060x$$
$$12 \times 350 = 1060x$$
$$x = 3.96 \text{ ft.}$$

Hence, P is 8.03 ft. to the left of D

Since, R_D is in mid of AB

Therefore, equal reaction will be produced R_1 & R_2

Also, $R_1 = 0.5 \times 700 = 350$ N

$R_2 = 0.5 \times 700 = 350$ N

13. $\Sigma V = 0$

$$R_a + R_b = 3P + Q$$
$$\Sigma M_A = 0$$

$R_b \times 9.6 = 120 \times 1.8 + 120 \times 3.6 + 10 \times 5.4 + 85 \times 8.4$

$R_b = 209.375$ kN

$R_a = 235.625$ kN

14. F.B.D. is shown in figure

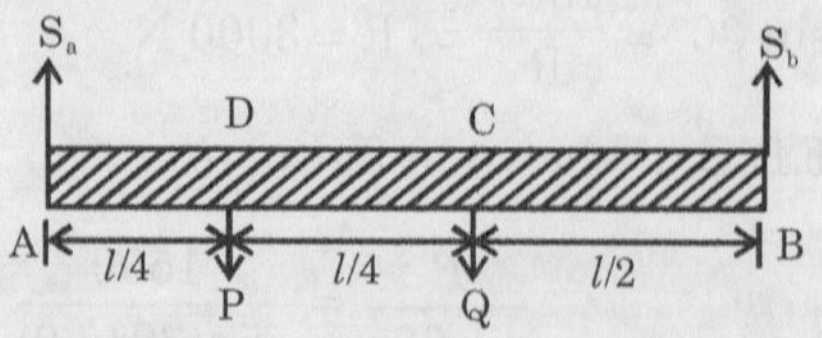

$$\Sigma V = 0$$
$$S_a + S_b = P + Q$$
$$S_a + S_b = 95 + 58.5 = 153.5\text{kN}$$
$$\Sigma M_A = 0$$
$$S_b \times l = P \times \left(\frac{l}{4}\right) + Q \times \left(\frac{l}{2}\right)$$
$$S_b = \left(\frac{P}{4}\right) + \left(\frac{Q}{2}\right)$$
$$= \frac{95}{4} + \frac{58.5}{2}$$
$$S_b = 53 \text{ kN}$$
$$S_a = 100.5 \text{ kN}$$

15. F.B.D. is shown in figure

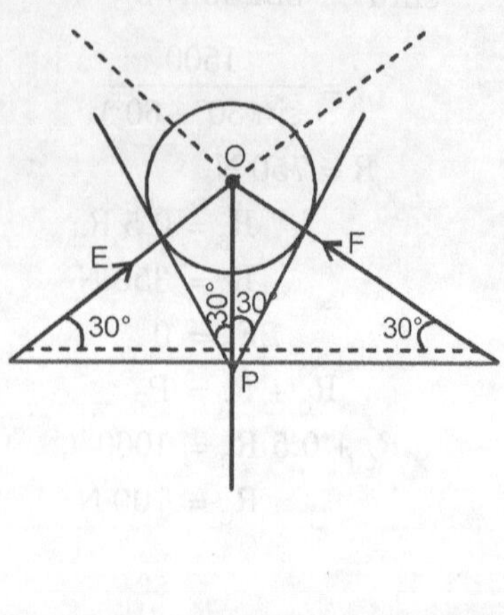

R_e, R_f and W meet at a point, applying Lami's theorem,

$$\frac{R_e}{\sin 120°} = \frac{R_f}{\sin 120°} = \frac{W}{\sin 120°}$$

$\Rightarrow$ $R_e = R_f = W = 500$ N

Considering equation of arm BC and taking moments at P,

$$\Sigma M_P = 0$$

$\Rightarrow$ $T \times 1.8 \cos 30° - R_e \times PE - R_b \times 1.2 \sin 30° = 0$

In triangle OEP,

$$\frac{OE}{PE} = \tan 30°$$

$\Rightarrow$ $PE = \frac{OE}{\tan 30°} = \frac{0.5}{0.577}$

$= 0.866$ m

$\Rightarrow T \times 1.8 \cos 30° - R_e \times 0.866 - R_b \times 1.2 \sin 30° = 0$

$\Rightarrow T = 373.95$ N

16. $\Sigma M_A = 0$

$\Rightarrow$ $35 \times \cos 30° \times 0.3 \cos 45° - 35 \sin 30° \times 0.3 \sin 45° - R_f \times \sin 45° \times 0.3 \sin 45° - R_f \times \cos 45° \times (0.3 \cos 45° + 0.3) = 0$

$6.429 - 3.712 - 0.15 |R_f - 0.15| |R_f - 0.212| R_f = 0$

$10.141 = 0.512R_f$

$\Rightarrow$ $R_f = 19.806$ kN

$\Sigma F_x = 0$

$\Rightarrow$ $R_d - 35 \sin 30° - R_f \sin 45° = 0$

$\Rightarrow$ $R_d - 35 \sin 30° - 19.806 \sin 45° = 0$

$\Rightarrow$ $R_d = 31.5$ kN

$\Sigma F_y = 0$

$\Rightarrow$ $R_e - 35 \cos 30° - R_e \cos 45° = 0$

$\Rightarrow$ $0.292R_e = 30.310$

$\Rightarrow$ $R_e = 103.80$ kN

17. F.B.D. is shown in figure

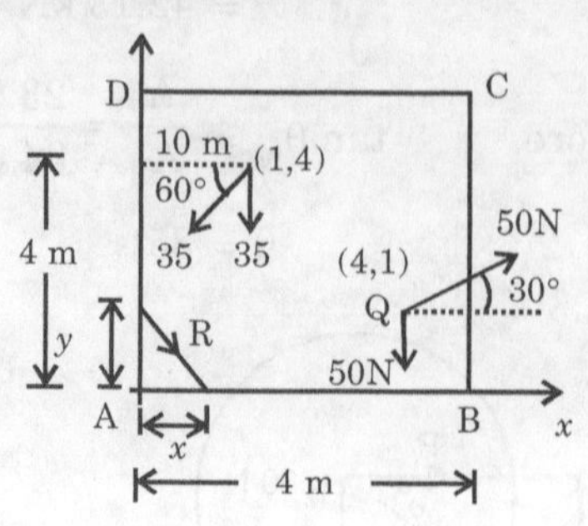

$$\Sigma F_x = 0$$

$\Rightarrow$ $60 \cos 30° - 35 \cos 60° = 34.46$ N

$$\Sigma F_y = 0$$

$\Rightarrow$ 60 sin30° – 35 sin60° – 60 – 35 = – 95.31 N

$$R = \sqrt{\Sigma F_x^2 + \Sigma F_y^2}$$

$$= \Sigma\, 34.46^2 + \Sigma\,(-95.31)^2$$

$$= 101.49 \text{ N}$$

$$\alpha = \tan^{-1}\frac{\Sigma F_x}{\Sigma F_y} = 70.02°$$

18. F.B.D. is shown in figure

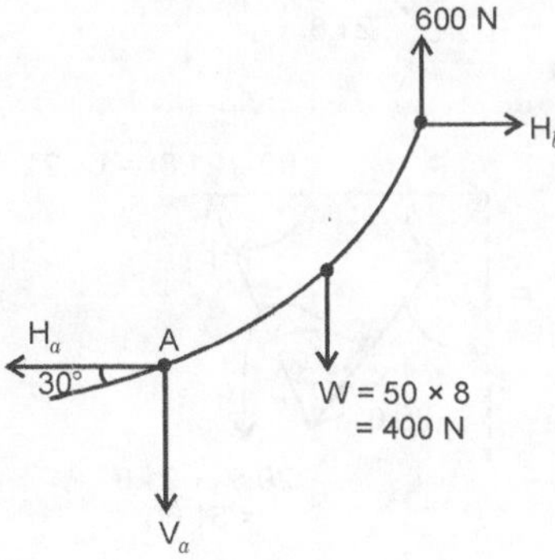

$\Sigma F_x = 0 \;\Rightarrow\; H_b - H_a = 0$

$\Sigma F_y = 0 \;\Rightarrow\; 600 + V_a - 400 = 0$

$\Rightarrow V_a = -200$ N

$$\frac{V_a}{H_a} = \tan 30°$$

$H_a = 346.4$ N

$$T_{max} = \sqrt{400^2 + 346.4^2}$$

$= 692.82$ N

19. (*i*) $\mu_s = \tan\phi_s$

$\Rightarrow$ $\phi_s = \tan^{-1}\mu_s$

$= \tan^{-1} 0.35$

$= 19.29°$

$P = W\tan(\phi + \alpha)$

$= W\tan(19.29° + 25°)$

$= W\tan(44.29°)$

$= 800\tan(44.29°)$

$= 780.41$ N

(*ii*) $\mu_k = \tan\phi_k$

$\Rightarrow$ $\phi_k = \tan^{-1}\mu_k$

$= \tan^{-1} 0.25$

$= 14.03°$

$P = W\tan(\phi + \alpha)$

$= W\tan(14.03° + 25°)$

$= W\tan(44.29°)$

$= 800\tan(39.03°)$

$= 648.67$ N

(*iii*) for block move down

$P = W\tan(\phi - \alpha)$

to prevent from sliding $P < W\tan(\phi - \alpha)$

$\Rightarrow$ $P < 800\tan(14.03° - 25°)$

$\Rightarrow$ $P < 800\tan(-5.71)$

$\Rightarrow$ $P = 80$N

20. In figure

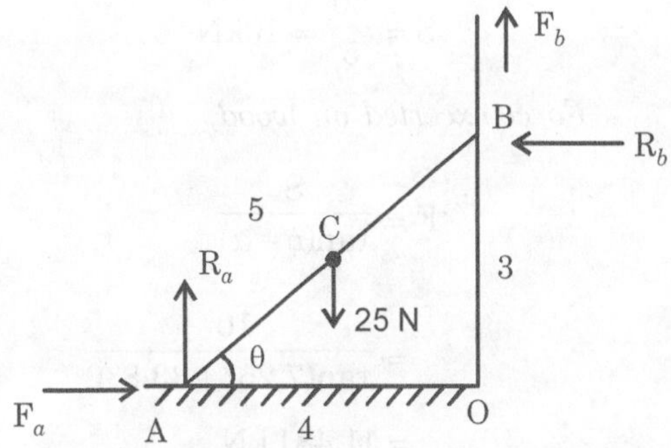

(*i*) $\Sigma F_y = 0 \Rightarrow R_a - 25\text{N} \Rightarrow R_a = 25$N

$\Sigma F_y = 0 \Rightarrow R_b = \mu R_a$

$M_A = 0$

$R_b \times OB = 25 \times AC\cos\theta$

Where, $OB = \sqrt{5^2 + 4^2} = 3$m

$\therefore$ $\sin\theta = \frac{3}{5}$ & $\cos\theta = \frac{4}{5}$

$\Rightarrow$ $$R_b = 25 \times \frac{AC}{OB}\cos\theta$$

$$= 25 \times \frac{5}{2} \times \frac{1}{3} \times \frac{4}{5}$$

$R_b = 16.67$ N

$\Rightarrow$ $$\mu = \frac{R_b}{R_a} = \frac{16.67}{25} = 0.67$$

(*ii*) $F_a = \mu R_a = 0.67 \times 25 = 16.67$ N

21. In Figure

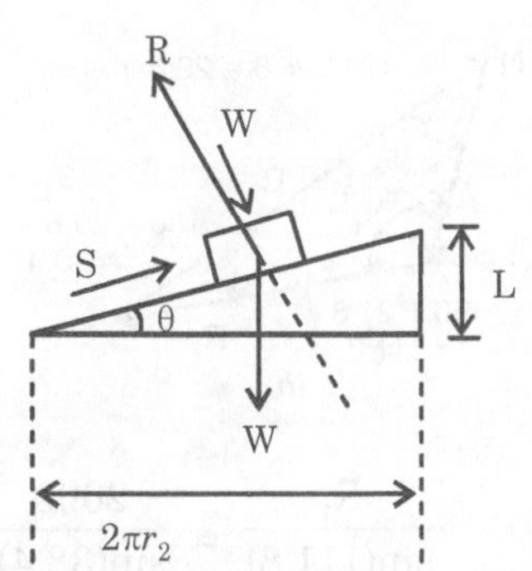

(*i*) $\alpha = 2p = 4$ mm

where p is number of thread

$$\tan\theta = L/2\pi r = 4/2\pi(5) = 0.1273$$

$$\Rightarrow \quad \theta = 7.25°$$

$$\mu_s = 0.67$$

$$\therefore \quad \phi_s = \tan^{-1}\mu_s = 33.8°$$

$$\Rightarrow \quad T = Sr$$

$$\Rightarrow \quad S = \frac{50}{5} = 10\text{kN}$$

∴ *Force excerted on wood,*

$$F = \frac{S}{\tan(\phi+\alpha)}$$

$$= \frac{10}{\tan(7.25° + 33.8°)}$$

$$= 11.481 \text{ kN}$$

(*ii*) for self-locking, $\phi_s > \theta$

$$S = F\tan(\phi_s - \theta)$$

$$= 11.48\tan(33.8° - 7.25)$$

$$= 5.73 \text{ kN}$$

$$\therefore \quad T = S.r = 5.73 \times r$$

$$= 28.68 \text{ kN–m}$$

22. $\phi_s = \tan^{-1}\mu_s = 21.8°$

(*i*) For raising the block

F.B.D of B is shown in figure(*a*) :

From figure (*b*)

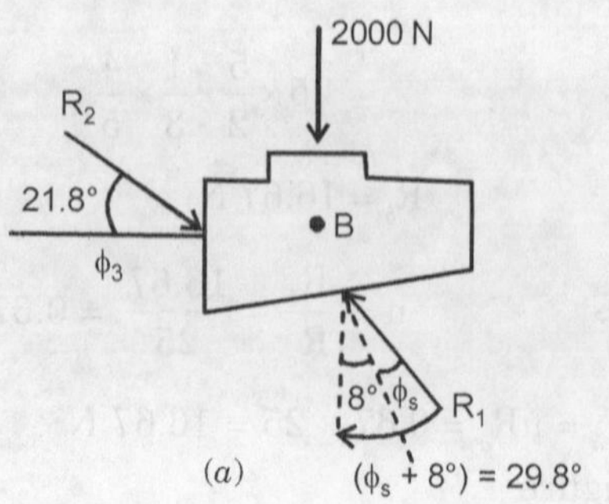

(*a*)

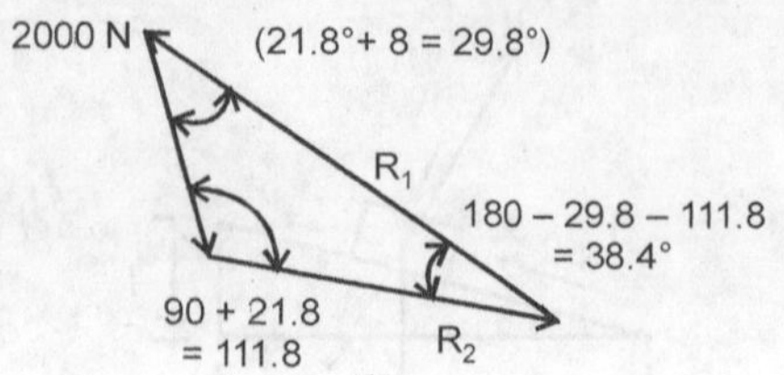

(*b*)

$$\frac{R_1}{\sin(111.8)} = \frac{2000}{\sin(38.4)}$$

$$\Rightarrow \quad R_1 = 2989.5 \text{ N}$$

F.B.D of A is shown in figure :

From figure.

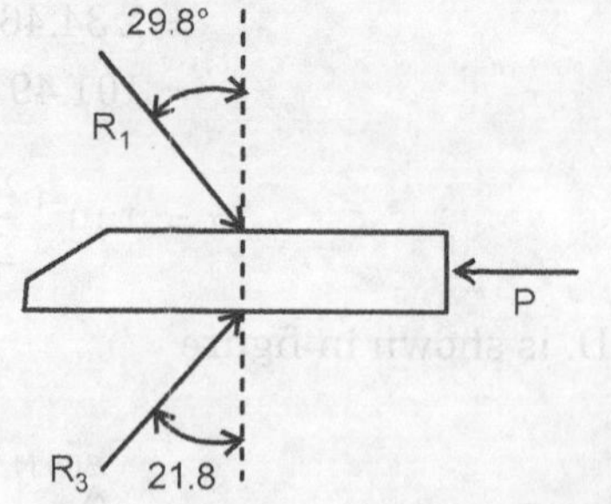

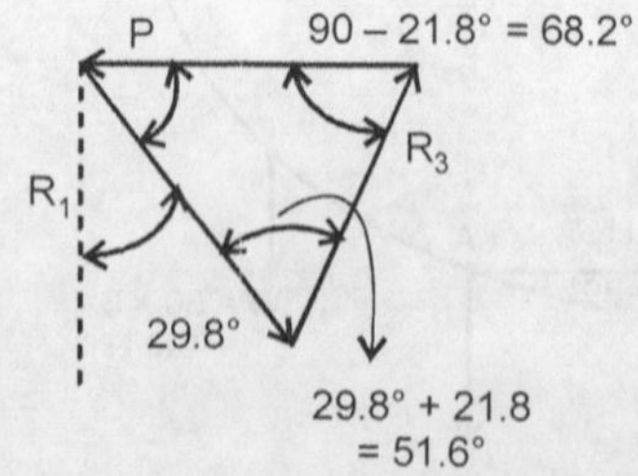

$$\frac{P}{\sin(51.6)} = \frac{R_1}{\sin(68.2)}$$

$$\Rightarrow \quad R_1 = 2523.3 \text{ N}$$

(*ii*) For lowering the block

F.B.D of B is shown in figure:

From figure

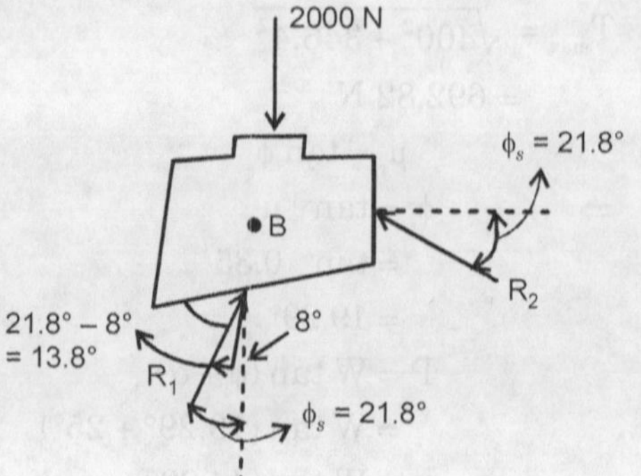

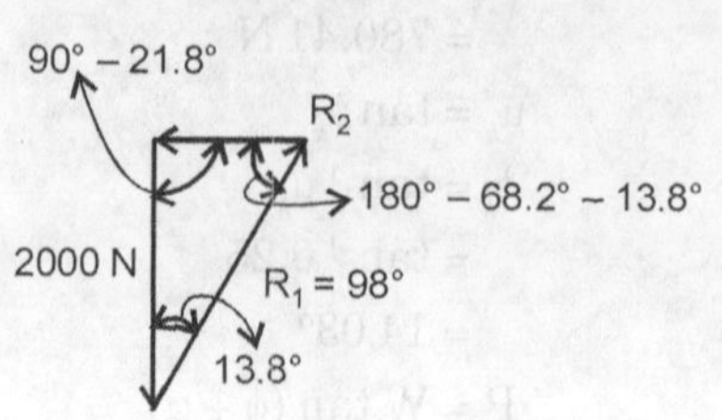

$$\frac{R_1}{\sin(6.8)} = \frac{2000}{\sin(98)}$$

$$\Rightarrow \quad R_1 = 1875.22 \text{ N}$$

F.B.D of A is shown in figure :

From figure.

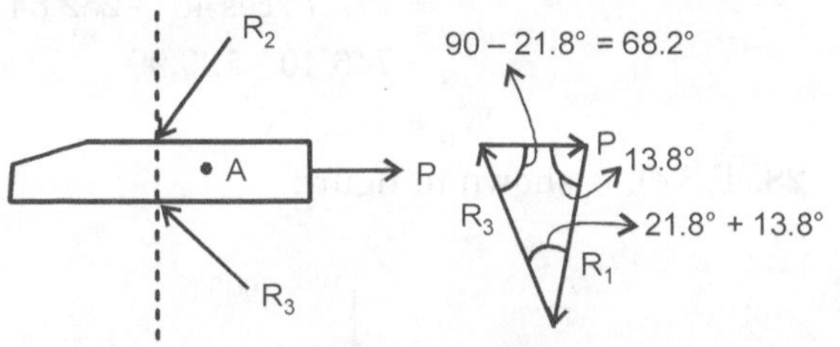

$$\frac{P}{\sin(35.6)} = \frac{R_1}{\sin(68.2)}$$

$$\Rightarrow \quad R_1 = 1175.6 \text{ N}$$

23. The F.B.D. is shown in figure

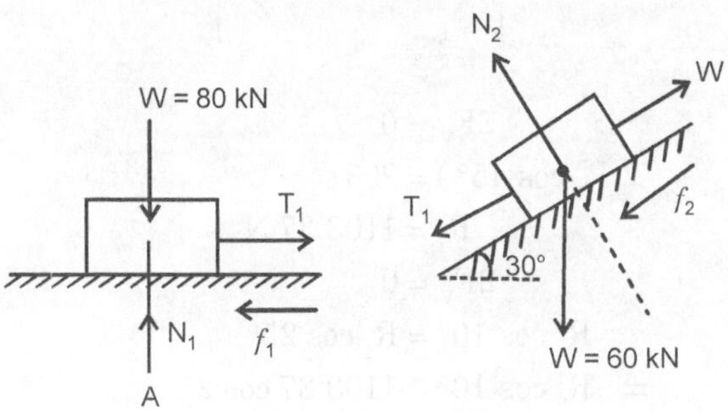

$$N_1 = W = 80 \text{ kN}$$

$$f_1 = \mu N_1 = 0.10 \times 80$$

$$\Rightarrow \quad f_1 = 8 \text{ kN} = T_1$$

$$N_2 = W \cos 30° = 60 \cos 30°$$

$$N_2 = 51.96 \text{ kN}$$

$$\therefore \quad f_2 = \mu N_2 = 5.196 \text{ N}$$

$$\therefore \quad W = 80 \sin 30° + T_1 + f_2$$

$$W = 53.196 \text{ kN}$$

24. The F.B.D. is shown in figure

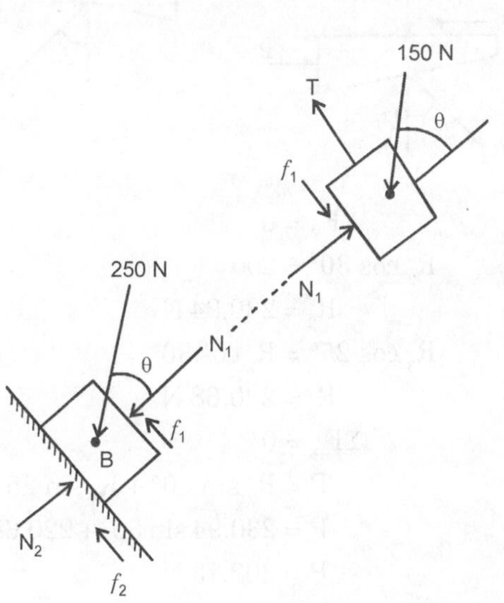

$$N_1 = 150 \cos\theta$$

$$f_1 = 0.2\,(150 \cos\theta)$$

$$= 30 \cos\theta \text{ N}$$

$$N_2 = N_1 + 250 \cos\theta$$

$$= 150 \cos\theta + 250 \cos\theta$$

$$= 400 \cos\theta$$

$$\therefore \quad f_2 = 0.2\,(400 \cos\theta)$$

$$f_1 + f_2 = 250 \sin\theta$$

$$30 \cos\theta + 80 \cos\theta = 250 \sin\theta$$

$$\Rightarrow \quad 110 \cos\theta = 250 \sin\theta$$

$$\Rightarrow \quad \tan\theta = \frac{110}{250}$$

$$\theta = 23.71°$$

25. F.B.D. is shown in figure;

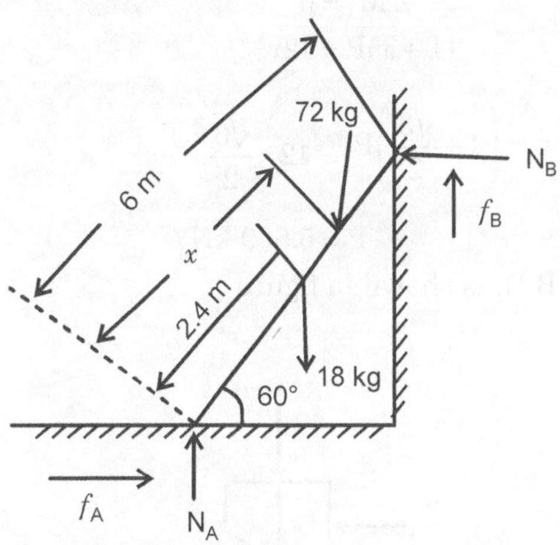

$$\Sigma F_V = 0$$

$$N_A + f_B = 18 + 72$$

$$N_A = 90 - f_B = 90 - \mu N_B$$

$$\Rightarrow \quad N_A = 90 - \tan 15°\, N_B$$

$$\Sigma F_H = 0$$

$$N_A \tan 15° = N_B$$

$$(90 - \tan 15°\, N_B) \tan 15° = N_B$$

$$\Rightarrow \quad 90 \tan 15° - N_B \tan^2 15° = N_B$$

$$\Rightarrow \quad N_A = \frac{90 \tan 15°}{1 + \tan^2 15°}$$

$$N_B = 22.5 \text{ kg}$$

$$\therefore \quad f_B = \mu N_B = 22.5 \tan 15°$$

$$\Sigma M_B = 0$$

$$N_B\,(6 \sin 60°) + f_B\,(6 \cos 60°)$$

$$= 18(2.4 \cos 60°) + 72(x \cos 60°)$$

$$\Rightarrow \quad 6(22.5) \tan 60°) + 6(6.03) = 43.2 + 72x$$

$$\Rightarrow \quad x = 3.15 \text{m}$$

26. F.B.D. is shown in figure

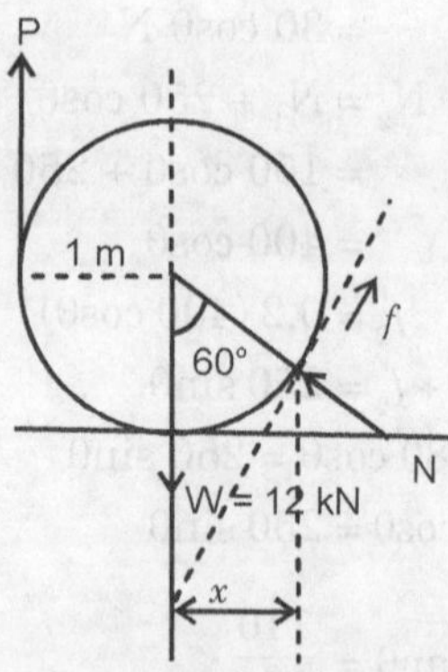

$$x = (1) \sin 60° = \frac{\sqrt{3}}{2} m$$

$$\Sigma M_A = 0$$

$$(1 + x)P = 12x$$

$$\left(1+\frac{\sqrt{3}}{2}\right)P = 12\times\frac{\sqrt{3}}{2}$$

$$P = 5.569 \text{ kN}$$

27. F.B.D. is shown in figure

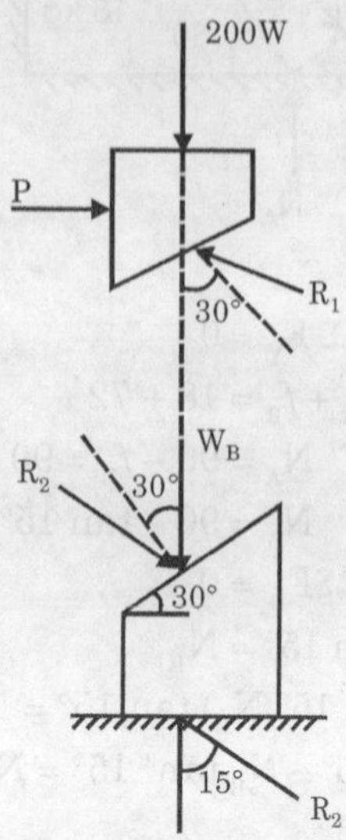

$$\Sigma F_V = 0$$

$$R_1 \cos 45° = 200$$

$$\Rightarrow \quad R_1 = 282.84 \text{ N}$$

$$\Sigma F_H = 0$$

$$R_2 \sin 15° = R_1 \sin 45°$$

$$\Rightarrow \quad R_2 = (R_1 \sin 45°)/\sin 15°$$

$$= 772.74 \text{ N}$$

$$\Sigma F_V = 0$$

$$W_B + R_1 \cos 45° = R_2 \cos 45°$$

$$\Rightarrow \quad W_B = R_2 \cos 45° - R_1 \cos 45°$$

$$= 772.74 \cos 45° - 282.84 \cos 45°$$

$$= 746.40 - 199.99$$

$$\Rightarrow \quad W_B = 546.4 \text{ N}$$

28. F.B.D. is shown in figure

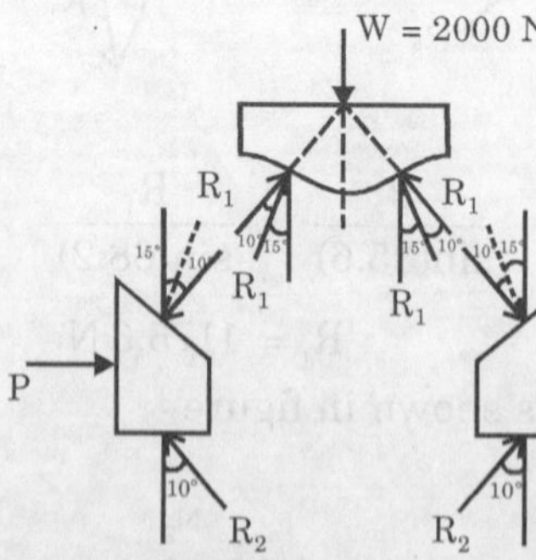

$$\Sigma F_V = 0$$

$$2(R_1 \cos 45°) = 200$$

$$R_1 = 1103.37 \text{ N}$$

$$\Sigma F_V = 0$$

$$R_2 \cos 10° = R_1 \cos 25°$$

$$\Rightarrow R_2 \cos 10° = 1103.37 \cos 25°$$

$$R_2 = 1015.42 \text{ N}$$

$$\Sigma F_H = 0$$

$$P = R_1 \sin 25° + R_2 \sin 10°$$

$$= 1103.37 \sin 25° + 1015.42 \sin 10°$$

$$P = 642.63 \text{ N}$$

29. FBD is shown in figure:

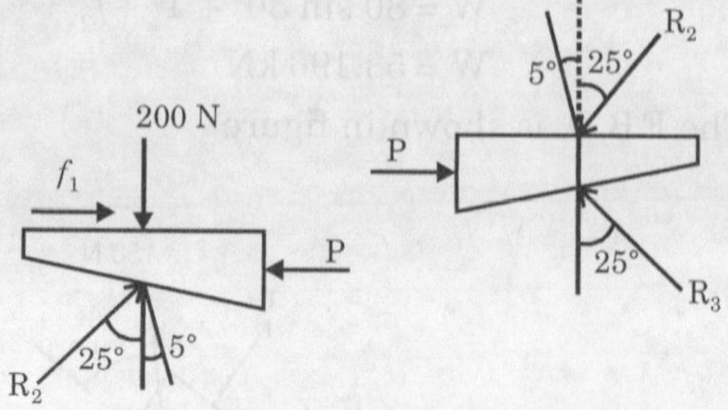

$$\Sigma F_V = 0$$

$$R_2 \cos 30° = 200$$

$$\Rightarrow \quad R_2 = 230.94 \text{ N}$$

$$R_3 \cos 25° = R_2 \cos 30°$$

$$\Rightarrow \quad R_3 = 220.68 \text{ N}$$

$$\Sigma F_H = 0$$

$$P = R_2 \sin 30° + R_3 \sin 25°$$

$$\Rightarrow \quad P = 230.94 \sin 30° + 220.68 \sin 25°$$

$$\Rightarrow \quad P = 208.73 \text{ N}$$

30. Refer figure

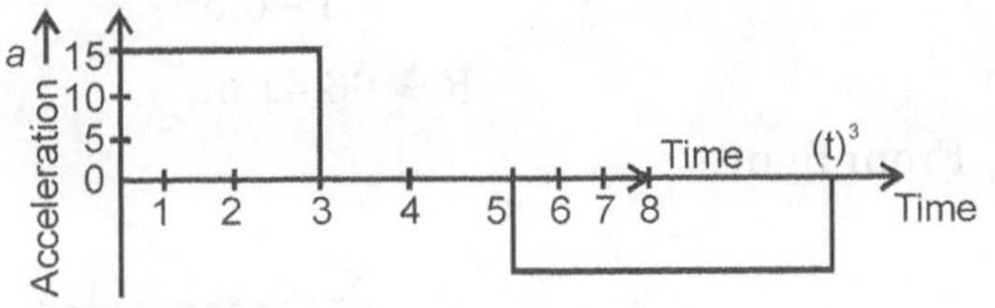

Using relation, V = $u + at$

Velocity after 3 secs = 0 + 15 × 3 = 45 m/s

Velocity after 4 secs = 45 + (–10) × 4 = 5 m/s

Hence, distance travelled by car

$$= \left(\frac{1}{2}\times 3\times 45\right)+(45\times 2)+\left(\frac{45+5}{2}\right)\times 4$$

= 257.5 m

31. (*i*) From B to C

$v = u + at$

$u = 25, v = 0, a = -g = -10$

$\therefore\ 0 = 25 - 10t$

$\Rightarrow\ t = 2.5$ sec

and $v^2 - u^2 = 2as$

$0 - 25^2 = 2 \times (-10) \times h$

$\Rightarrow\ h = 31.25$ m

$\therefore$ total height from ground

= 25 + 31.25 = 56.25 m

(*ii*) From C to B

$v_b = u + gt$

Where v_b is velocity of ball

$\Rightarrow\ v_b = 0 + 10 \times 2.5$

$\Rightarrow\ v_b = 25$ m/s

Apparently the velocity with which stone reaches B is same with which it has thrown upwards.

(*iii*) From B to A

$$h = ut + \frac{1}{2}gt^2$$

$$\Rightarrow\ 31.25 = 25 \times t + \frac{1}{2} \times 10 \times t^2$$

$$\Rightarrow\ 31.25 \times 2 = 2 \times 25 \times t + \frac{1}{2} \times 10 \times t^2$$

$\Rightarrow\ 10t^2 + 25t - 62.5 = 0$

$\Rightarrow\ t = 1.54$ sec

$\therefore$ total time during which stone is in flight

$= t_{B\to C} + t_{C\to B} + t_{B\to A}$

= 2.5 + 2.5 + 1.54

t = 6.54 sec

32. Differential equation of motion is

$$2s - w = \frac{w}{g}\times\frac{a}{2}$$

$$\Rightarrow\ 2s = w + \frac{wa}{2g}$$

$$\Rightarrow\ 2s = w\left(1+\frac{a}{2g}\right)$$

$$\Rightarrow\ s = \frac{w}{2}\left(1+\frac{a}{2g}\right)$$

$$\Rightarrow\ s = \frac{4500}{2}\left(1+\frac{28}{2\times 9.81}\right)$$

$\Rightarrow\ s = 5461.01$ N

33. Initial velocity of bullet, $u = 0$

Final velocity of bullet, $v = 800$ m/s

Total distance, $s = 0.85$ m

$\because\ v^2 - u^2 = 2as$

$\Rightarrow\ v^2 = 2as - 0$

$\Rightarrow\ v^2 = 2as$

$$\Rightarrow\ a = \frac{v^2}{2s}$$

$$\Rightarrow\ a = \frac{800^2}{2\times 0.85}$$

$\Rightarrow\ a = 376450.58$ m/s²

$\because\ v = u + at$

$\Rightarrow\ 800 = 376470.58 \times t$

$\Rightarrow\ t = 0.0021$ sec

34. (*i*) v_a = 70 km/hr = 19.44 m/s

v_b = 50 km/hr = 13.88 m/s

$v^2 - u^2 = 2as$

$$a_t = \frac{13.88^2 - 19.44^2}{2\times 150}$$

$= -0.617\text{ms}^{-2}$

at A : $a = 2.5\ \text{ms}^{-2}$

$a_t = -0.617\ \text{ms}^{-2}$

$$a_n = \sqrt{a^2 + a_t^2}$$

$$= \sqrt{2.5^2 + (-0.617)^2} = 2.42\ \text{ms}^{-2}$$

$$a_n = \frac{v^2}{r}$$

$$\Rightarrow \quad r = \frac{v^2}{a_n} = \frac{19.44^2}{2.42} = 156.16 \text{ m}$$

$$(ii) \; a_n = \frac{v^2}{r} = \frac{13.88^2}{90} = 156.16 \text{ ms}^{-2}$$

$$a_t = -0.617 \text{ ms}^{-2}$$

$$a_n = \sqrt{a^2 + a_t^2}$$

$$= \sqrt{156.16^2 + (-0.617)^2}$$

$$= 2.27 \text{ ms}^{-2}$$

35. From figure

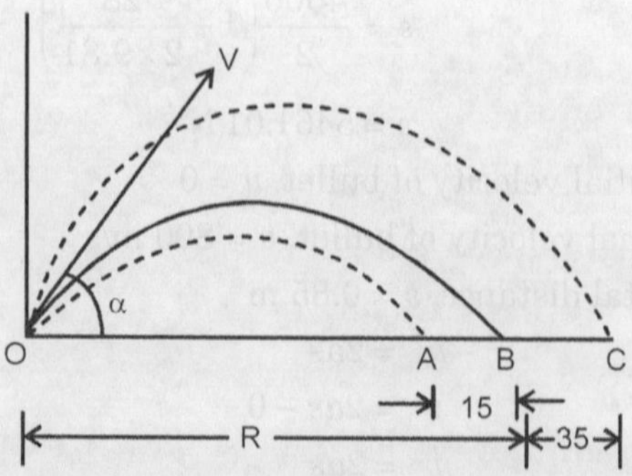

Let range, OB = R

For α = 25°

Range OA = R – 15

$$\text{Therefore,} \quad R - 15 = \frac{v^2 \sin 2\alpha}{g}$$

$$= \frac{v^2 \sin 30^\circ}{g}$$

$$R - 15 = \frac{0.5v^2}{g} \quad \text{...(1)}$$

For α = 40°

Range OC = R + 35

$$= \frac{v^2 \sin^2(40^\circ)}{g}$$

$$= \frac{v^2 \sin^2 40^\circ}{g}$$

$$= \frac{0.985\, v^2}{g} \quad \text{...(2)}$$

$$\frac{\text{equation 1}}{\text{equation 2}} = \frac{R-15}{R+35} = 0.507$$

$$R - 15 = 0.507\,R + 17.745$$

$$R = \frac{17.745 + 15}{1 - 0.507}$$

$$\Rightarrow \quad R = 66.41 \text{ m}$$

36. From figure

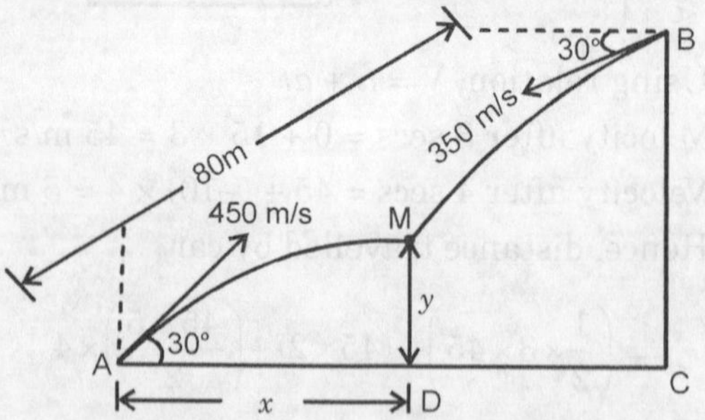

(*i*) horizontal distance between A & B,

AC = 80 cos 30° = 69.28 m

Horizontal distance covered by shot fired from A in time t

x = AD = 450 cos 30° × t = 389.71 t ...(1)

Horizontal distance covered by shot fired from B in time t

$$DC = 69.28 - t$$
$$= 350 \cos 30^\circ \times t$$
$$= 303.10\, t \quad \text{...(2)}$$

Adding equation 1 & 2,

$$AD + DC = AC$$
$$389.71\, t + 303.1\, t = 69.28$$
$$t = 0.1 \text{ sec}$$

$$(ii) \quad x = v_x.t = v \cos\alpha.t = 450 \cos 30^\circ \times t$$

$$\Rightarrow \quad x = 38.97 \text{ m}$$

$$y = v \sin\alpha.t - \frac{1}{2}gt^2$$

$$= 350 \sin 30^\circ \times 0.1 - \frac{1}{2} \times 9.81 \times 0.1^2$$

$$\Rightarrow \quad y = 17.45 \text{ m}$$

coordinates of meeting point is M (38.97, 17.45) m

37. From figure

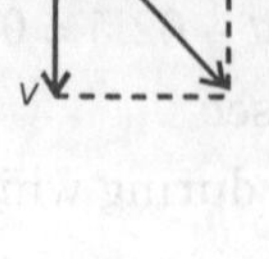

(*i*) $h = ut$

$\Rightarrow 8 = ut$

$$5 = \frac{1}{2}gt^2 = \frac{1}{2} \times 9.81 \times \left(\frac{8}{u}\right)^2$$

$$10u^2 = g \times 64$$
$$\Rightarrow u = 7.92 \text{m/s}$$
$$u_A = 7.92 \text{ m/s}$$

38. $k = \dfrac{0.5 \times 9.81}{0.20} = 24.52$ N/m

$$T = 2\pi\sqrt{\frac{m}{k}} = 2\pi\sqrt{\frac{50}{24.52 \times 10^3}} = 0.283 \text{ sec}$$

39. Instantaneous centre of AB is the point of intersection of perpendicular to V_a & V_b

IA = AB sin 60° = 3.5 sin 60° = 3.03 m

IB = AB cos 60 ° = 3.5 cos 60° = 1.75 m

Therefore $\omega_{ab} = \dfrac{v_a}{IA} = 0.495$ rad/sec

40. $A_1 = \dfrac{1}{2} \times b \times h$

$$= \frac{1}{2} \times 10 \times 12 = 60 \text{ cm}^2$$

$$Y_1 = \frac{h}{3} = \frac{12}{3} = 4 \text{ cm from BC}$$

$$A_1 = \frac{1}{2} \times 10 \times 6 = 30 \text{ cm}^2$$

$$Y_2 = \frac{h}{3} = \frac{6}{3} = 2 \text{ cm from BC}$$

$$y = \frac{A_1 y_1 + A_2 y_2}{A_1 + A_2}$$

$$= \frac{60 \times 4 + (-30) \times 2}{60 + (-30)} = 6 \text{ cm}$$

41. $A_1 = \dfrac{1}{2} \times 80 \times 40 = 1600 \text{ mm}^2$

$$x_1 = -\frac{2}{3} \times 80 = -\frac{160}{3} \text{ mm}$$

$$A_1 = \frac{1}{2} \times 60 \times 40 = 30\,b \text{ mm}^2$$

$$x_2 = -\frac{2}{3} xb = -\frac{2}{3} b$$

Since centroid is to be on y axis, therefore $x = 0$

$$A\bar{x} = \sum ax$$

$$0 = 1600\left(-\frac{160}{3}\right) + 30b\left(\frac{2}{3}\right)b$$

$$\frac{2 \times 30}{3} r^2 = \left(\frac{160 \times 1600}{3}\right)$$

$$b = 65.32 \text{ mm}$$

42.

$$m_w = \rho_w.(h.t.b)$$
$$= \rho_w.(h.15.150)$$
$$= 2250\, h\, \rho_w$$

$$m_s = \rho_s.\frac{\pi r^2}{2} t$$

$$= \rho_s.\frac{\pi(50)^2 \times 15}{2}$$

$$= 58904.8\, \rho_s$$

With the coordinate axes,

$$y_w = \frac{h}{2}$$

and $\quad y_s = -\dfrac{4r}{3\pi}$

$$= -\frac{4 \times 50}{3\pi}$$

$$= 21.22 \text{ mm}$$

Negative sign shows that the centroid for steel plate lies below the x-axis

Therefore,

$$\bar{y} = \frac{m_w y_w + m_s y_s}{m_w + m_s}$$

$$= \frac{1500 \times h\rho_w \times \frac{h}{2} + 58904.8\rho_s(-21.22)}{1500.h.\rho_w + 58904.8\rho_s}$$

Since CG of system passes through x-axis, therefore $\bar{y} = 0$

Therefore,

$$1500 \times h\rho_w + \frac{h}{2} + 58904.8\rho_s(-21.22) = 0$$

$$\rho_s = 10\rho_w$$

$$1500\,\frac{h^2}{2}\rho_w + 58904.8\rho_s.(-21.22) = 0$$

$$\frac{h^2}{2} = \frac{58904.8 \times 21.22}{1500}$$

$$h = 40.82 \text{ mm}$$

43. From figure, Taking moment of areas about G,

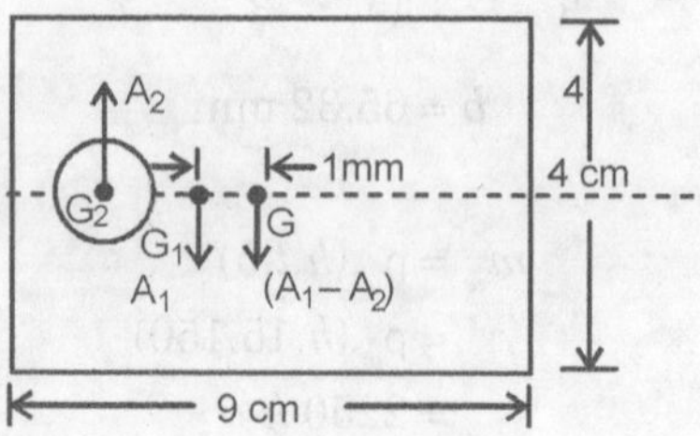

$A_1 = A_2 \times G_1.G_2 + (A_1 - A_2)\, G_1.G$

$(9 \times 4) \times 0 = 3 \times G_1.G_2 + (36 - 3) \times 0.1$

$$G_1.G_2 = -\frac{33 \times 0.1}{3} = -1.1 \text{ cm}$$

Negative sign shows that G_2 is left of G_1

44. From figure,

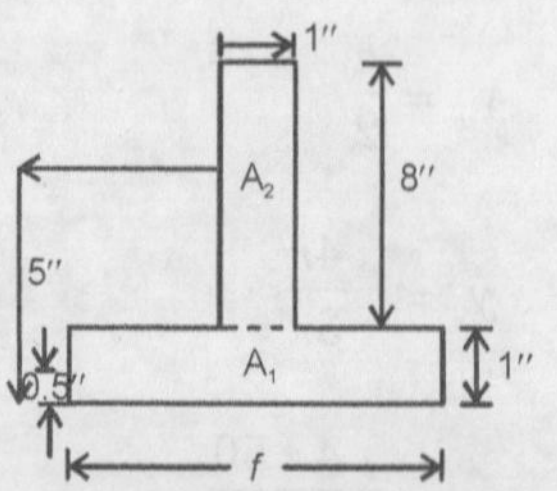

$A_1 = 1.(f) = f \text{ inch}^2$

$A_2 = 8.(f) = 8 \text{ inch}^2$

$y_1 = 0.5$ inch

$$y_1 = 1 + \frac{1}{2} = 5 \text{ inch}$$

$A = A_1 + A_2 = f + 8$

$A\bar{y} = \sum ay$

$\Rightarrow (f + 8)(2.5) = 0.5f + 8 \times 5$

$\Rightarrow 2.5f + 20 = 0.5f + 40$

$\Rightarrow 2f = 20$

$\Rightarrow f = 10$ inch

45. Square 50 × 50 in size

$A_1 = 50 \times 50 = 2500 \text{ cm}^2$

Centroid is at (25, 25)

Semi-circular area (removed),

$$A_2 = -\frac{\pi}{2} \times 10^2 = -157 \text{ cm}^2$$

Centroid is at $\left(\frac{4 \times 10}{3\pi}, 25\right)$

Rectangular area (removed)

$A_3 = -50 \times W = 50 \text{ W cm}^2$

Centroid is at $\left(\frac{50\text{ W}}{2}, 25\right)$

$A\bar{x} = A_1 x_1 + A_2 x_2 + A_3 x_3$

$(2500 - 157 - 50\text{ W}) \times 25$

$$= 2500 \times 25 - 157 \times \frac{4 \times 10}{3\pi} - 50\text{W}\left(50 - \frac{\text{W}}{2}\right)$$

$25\text{W}^2 - 1250\text{W} + 3258.33 = 0$

W = 47.24 and 2.76 cm

W = 47.24 is not possible

Therefore, W = 2.76 cm

46. According to 1st theorem of Pappus- Guldinus,

Surface area generated

= length of revolving shape × distance travelled

= (circumference of rectangular cross section) $\times 2\pi\bar{y}$

$$= 2(8 + 6) \times 2\pi \frac{12 + 10}{2} = 6635.04 \text{ cm}^2$$

47. $$v_{cm} = \frac{m_1 v_1 + m_2 v_2 + m_3 v_3}{m_1 + m_2 + m_3}$$

$$v_{cm} = \frac{5 \times 3 + 4 \times (-2) + 8 \times 8}{17} = 4.18 \text{ m/s}$$

Relative speed of other mass,

$v_5 = 3 - 4.18 = -1.18$ m/s

$v_4 = -2 - 4.18 = -6.18$ m/s

$v_8 = 8 - 4.18 = 3.82$ m/s

48. A_1 = area of rectangle

$= (12x).(14x) = 168x^2$

A_2 = area of rectangle to be subtracted

$= (4x).(4x) = 16x^2$

A_3 = area of semi-circle to be subtracted

$$= \frac{\pi R^2}{2} = \frac{\pi(4x)^2}{2} = 25.13x^2$$

A_4 = area of quarter circle to be subtracted

$$= \frac{\pi R^2}{2} = \frac{\pi(4x)^2}{4} = 12.56x^2$$

A_4 = area of triangle

$$\frac{1}{2} \times (6x) \times (4x) = 12x^2$$

$$\bar{x} = \frac{a_1 x_1 + a_2 x_2 + a_3 x_3}{a_1 + a_2 + a_3}$$

$$= \frac{168x^2 \times 280 - 16x^2 \times 80 - 25.13x^2 \times 240 - 12.56x^2(492.09) + 12x^2 \times 640}{168x^2 - 16x^2 - 25.13x^2 - 12.56x^2 + 12x^2}$$

= 326.404 mm

$$\bar{y} = \frac{a_1y_1 + a_2y_2 + a_3y_3}{a_1 + a_2 + a_3}$$

$$= \frac{168x^2 \times 280 - 16x^2 \times 400 - 25.13x^2 \times 67.906 - 12.56x^2(412.093) + 12x^2 \times 53.33}{168x^2 - 16x^2 - 25.13x^2 - 12.56x^2 + 12x^2}$$

= 219.124 mm

49. From figure

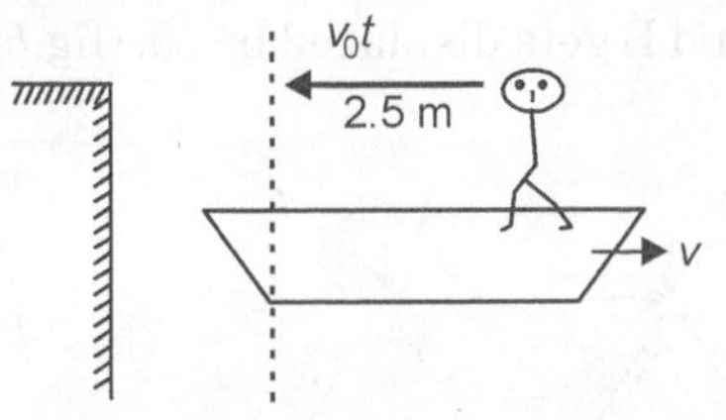

weight of man, W_1 = 800 N

Weight of man, W_2 = 950 N

Let v_0 is initial velocity of man and t is the time

Then $v_0t = x$

$v_0t = 2.5$ m

$v_0 = (2.5/t)$ m/s

According to conservation of momentum,

$$w_1v_0 = (w_1 + w_2).v$$

$$v = \frac{w_1v_0}{w_1 + w_2}$$

Distance covered by boat,

$$s = v.t = \left(\frac{w_1v_0}{w_1 + w_2}\right)\cdot t$$

$$s = \frac{(800 \times 2.5)\cdot t}{t(800 + 950)}$$

$\Rightarrow$ $s = 1.142\ m$

Posi tion of man from pier

$= 5 + s - x$

$= 5 + 1.142 - 2.5$

$= 3.642$ m

50. From fig., by Conservation of momentum :

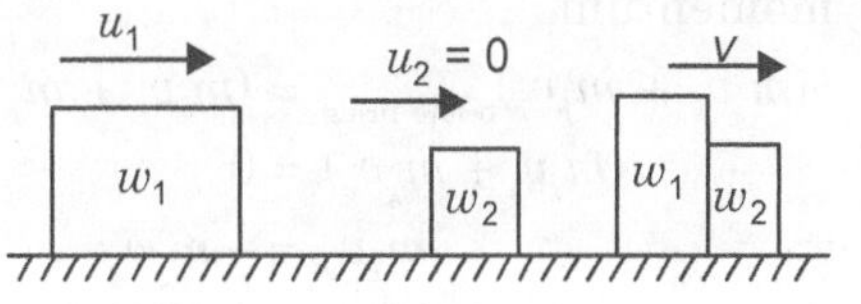

$$w_1u_1 + w_2u_2 = (w_1 + w_2)\cdot v$$

$$\Rightarrow \quad v = \frac{w_1u_1 + w_2u_2}{w_1 + w_2}$$

$$= \frac{\frac{600 \times 20 \times 5}{18}}{600 + 86} = 9.718\frac{m}{s}$$

51. Weight of man W_1 = 775.4 N

Weight of canon, W_2 = 445.75 N

Weight of bullet, W_3 = 0.30 N

Velocity of nuzzle, u = 550 m/s

V = final velocity of canon

According to conservation of momentum,

$$w_3.u = (w_1 + w_2)\cdot v$$

$$v = \frac{W_3 \cdot u}{W_1 + W_2}$$

$$= \frac{0.30 \times 550}{77.55 + 445.75}$$

$\Rightarrow$ $v = 0.135$ m/s

52. Weight of wooden block,

W_1 = 32.25 N

Weight of bullet,

W_2 = 0.20 N

According to conservation of momentum,

$$W_2.u = (W_1 + W_2)\cdot v$$

$$u = \left(\frac{W_1 + W_2}{W_2}\right)\cdot v$$

$$= \frac{(32.25 + 0.20)}{0.20} \times 3.5$$

$u = 567.87$ m/s

53. From figure

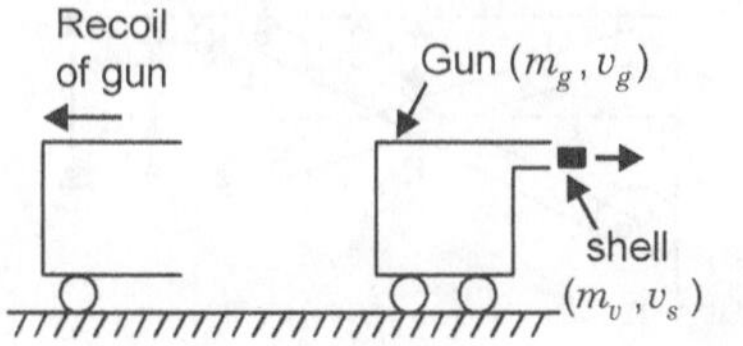

(*i*) According to the principle of conservation of momentum,

$$(m_s v_s + m_g v_g)_{\text{before firing}} = (m_s v_s + m_g v_g)_{\text{after firing}}$$

$$(m_s v_s + m_g v_g) = 0$$

$$\Rightarrow \quad m_s v_s = -m_g v_g$$

$$80 \times 300 = -4500 \times v_g$$

$$v_g = -5.33 \text{ m/s}$$

(*ii*) $v^2 - u^2 = 2as$

$$\Rightarrow \quad a = \frac{v^2 - u^2}{2s} = \frac{0^2 - 5.33^2}{2 \times 1.5}$$

$$= -9.469 \text{ ms}^{-2}, \text{ retardation}$$

$$F = ma$$

$$= 4500 \times 9.469$$

$$= 42613.35 \text{ N} = 42.61 \text{ kN}$$

(*iii*) $v = u + at$

$$t = \frac{v - u}{a} = \frac{0 - (-5.33)}{42.61}$$

$$= 0.125 \text{ sec}$$

54. $v^2 - u^2 = 2gh$

$$v^2 - 0^2 = 2 \times 9.81 \times 4$$

$$= 8.85 \text{ m/s}$$

According to conservation of momentum

$$mv = (m + M) \cdot v$$

$$\Rightarrow \quad 300 \times 8.85 = (300 + 2500) \cdot v$$

$$\Rightarrow \quad v = 0.9482 \text{ m/s}$$

55. Time taken in firing one bullet,

$$t = \frac{60}{200}$$

$$= 0.3 \text{ sec}$$

For impulse-momentum reaction,

$$F \cdot t = m(V - u)$$

$$F \cdot 0.3 = \frac{50}{1000}(0 - 800)$$

$$\Rightarrow \quad F = -133.33 \text{ N}$$

56. F.B.D. is shown in figure

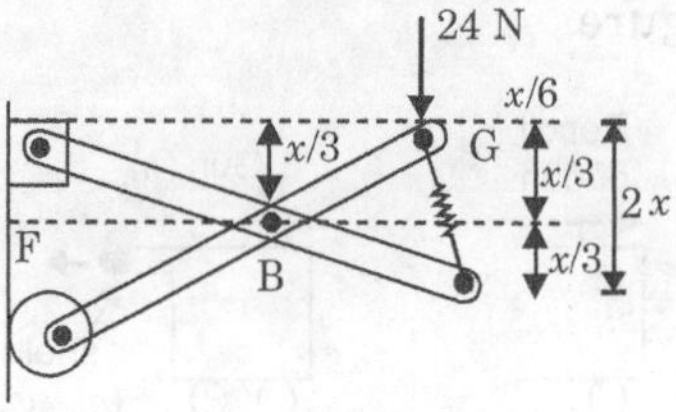

Deformation of spring,

$$S = EC - 40 \text{ mm}$$

$$= \frac{2x}{3} - 40 \text{ mm}$$

$$V = \frac{1}{2}ks^2 - 24\frac{x}{6}$$

$$= \frac{1}{2}(0.8)\left(\frac{2x}{3} - 40\right)^2 - 4x$$

For equilibrium;

$$\frac{dV}{dx} = 0$$

$$\frac{dV}{dx} = (0.8)\left(\frac{2x}{3} - 40\right)\left(\frac{2}{3}\right) - 4 = 0$$

$$x = 71.25 \text{ mm}$$

57. The member DE is cut to make the system movable and introducing an axial force P (fig.*a*), B and E gets displaced by δθ. (fig.*b*).

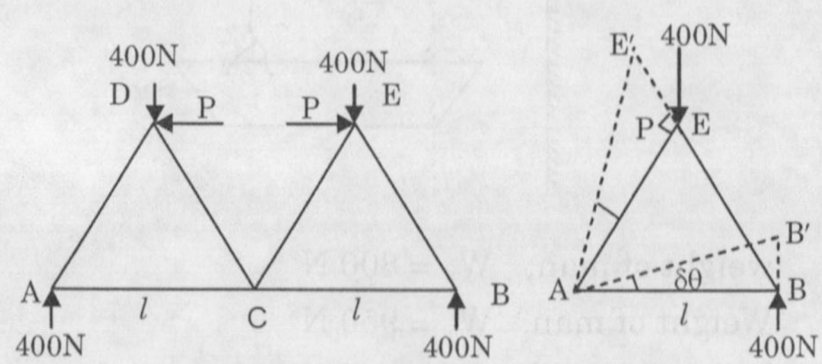

(*a*) (*b*)

Vertical displacement of

$$B = BB' = l\,\delta\theta.$$

$$\therefore \quad U_B = R_b \cdot l\,\delta\theta = 400\,l\,\delta\theta$$

The horizontal and vertical displacement at E are,

$$EE' \sin 60° = l\,\delta\theta \times 0.866$$

$$EE' \cos 60° = l\,\delta\theta \times 0.5$$

$$\text{Work done by P} = -P\,l\,\delta\theta \sin 60°$$

$$= -Pl\,\delta\theta \times 0.866$$

Work done by 400N load at

$$E = -400\,l\,\delta\theta \times 0.5$$

According to virtual work principle :

$$400\,l\,\delta\theta - 0.866\ Pl\,\delta\theta - 100\,l\,\delta\theta = 0$$

$$\Rightarrow \quad 400 - 0.866P - 100 = 0$$

$$\Rightarrow \quad P = 346.42 \text{ N}$$

58. $\delta_b \rightarrow$ vertical virtual downward displacement of force 'P'.

Work done by $P = -P \times \delta_y$

$$\text{Work done by load} = W \times \frac{\delta_y}{2}$$

$$\therefore \quad \text{Total work} = -P\delta_y + W \cdot \frac{\delta_y}{2}$$

As per virtual work, work done by all forces must be zero.

$$\therefore \quad -P\delta_y + W\cdot\frac{\delta_y}{2} = 0$$

$$\Rightarrow \quad P = \frac{W}{2}$$

$$\Rightarrow \quad P = 1250 \text{ N}$$

59. Let virtual displacement be y_0 at B, then corresponding displacement at C;

$$y_c = y_b \times \frac{AC}{AB} = y_b \times \frac{1}{2}$$

$$= 0 = 0.5\, y_b$$

$$T_V = T \sin\theta\ ;$$

$$T_H = T\cos\theta$$

$$W_B = -W \times y_b$$

$$= -400\, y_b$$

Work done due to

$$T_V = T\sin\theta \times y_c$$

$$= T\sin\theta\,(0.5\, y_b)$$

Applying virtual work,

$$\Sigma W = 0$$

$$T\sin\theta\,(0.5\, y_b) - 400\, y_b = 0$$

$$\Rightarrow \quad T = \frac{400 y_b}{\sin\theta(0.5 y_b)} = 1394.75 \text{ N}$$

60. F.B.D. is shown in figure.

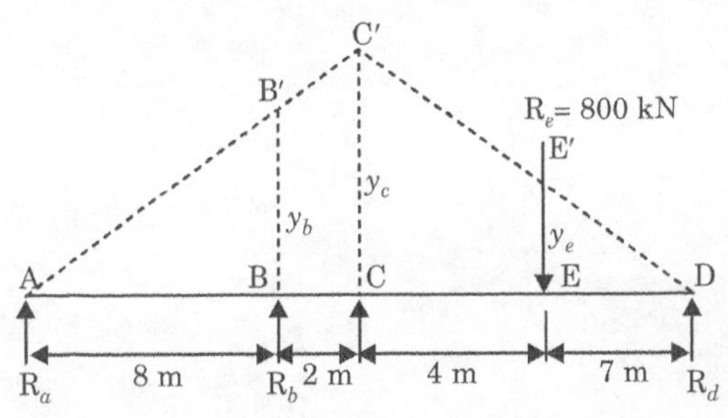

By similarity of triangles :

$$y_b = y_c \times \frac{AB}{AC} = y_c \times \frac{8}{10}$$

$$y_e = y_c \times \frac{ED}{CD} = y_c \times \frac{7}{11}$$

(*i*) Work done at $\quad R_c = R_c \times y_c$

Work done at $\quad R_e = -R_e \times y_e$

According to principle of virtual work :

$$R_c \times y_c - R_e \times y_e = 0$$

$$\Rightarrow \quad R_c \times y_c - 850 \times y_c \times \frac{7}{11} = 0$$

$$\Rightarrow \quad R_c = 540.90 \text{ kN}$$

(*ii*) Work done at $\quad R_b = R_b \times y_b$

Work done at $\quad R_c = -R_c \times y_c$

According to principle of virtual work :

$$\Sigma W = 0$$

$$R_b \times y_b - R_c \times y_c = 0$$

$$\Rightarrow \quad R_b \times \frac{8}{10} y_c - R_c \times y_c = 0$$

$$\Rightarrow \quad R_b = 676.12 \text{ kN}$$

61. In figure, Height of G above base line AC,

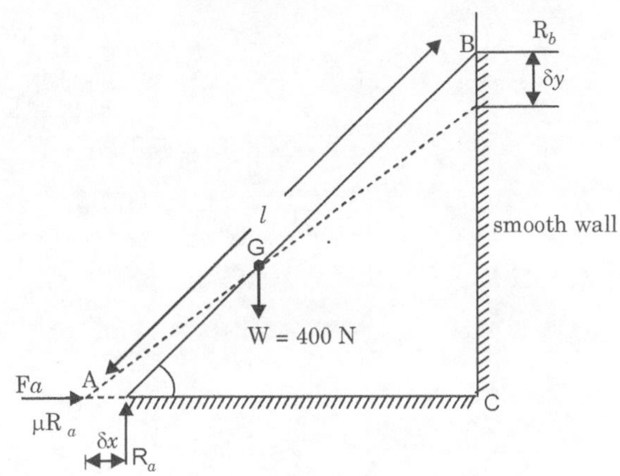

$$y = AG\sin\theta = \frac{l}{2}\sin\theta$$

$$x = AC = l\cos\theta$$

$$\therefore \quad \delta_x = -l\sin\theta\,\delta\theta$$

$$\text{and} \quad \delta_y = \frac{l}{2}\cos\theta\,\delta\theta$$

Work done at A by frictional force

$$= F_a \times \delta_x$$

Work done at A by weight of ladder

$$= W \times \delta_y$$

According to principle of virtual work :

$$\Sigma W = 0$$

$$F_a \times \delta_x + W \times \delta_y = 0$$

$$F_a \times (-l\sin\theta\,\delta\theta) + 400\left(\frac{l}{2}\cos\theta\,\delta\theta\right) = 0$$

$$\Rightarrow \quad -2F_a\sin\theta + 400\cos\theta = 0$$

$$\Rightarrow \quad -F_a = -\frac{400}{2}\frac{\cos\theta}{\sin\theta}$$

$$\Rightarrow \quad F_a = 53.5 \text{ N}$$

62. F.B.D. is shown in figure

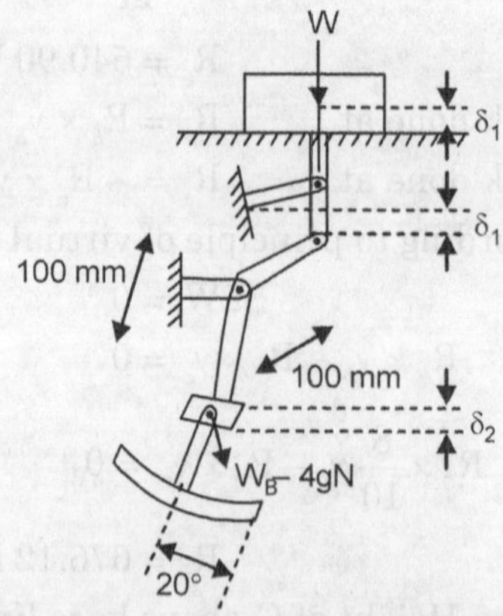

B rotates counterclock wise by $\delta\theta$.

$$\delta_2 = (100 \text{ mm})(\cos(20° - \delta\theta) - \cos 20°)$$
$$= 100 \sin 20° \,\delta\theta$$
$$\delta_1 = (100 \text{ mm})(\sin(10° - \delta\theta) - \sin 10°)$$
$$= 100 \cos 10° \,\delta\theta$$

By principle of virtual work :

$$\Sigma W = 0$$
$$4g.\,\delta_2 - W\delta_1 = 0$$
$$\Rightarrow \quad W = 13.6140\,g$$
$$\therefore \quad W = m_A g.$$
$$\Rightarrow \quad m_A = 0.1361 \text{ kg}$$

■■

2 CHAPTER

Solid Mechanics

MCQ TYPE QUESTIONS

2016

1. A 3 m long simply supported beam of uniform cross section is subjected to a uniformly distributed load of w = 20 kN/m in the central 1 m as shown in the figure.

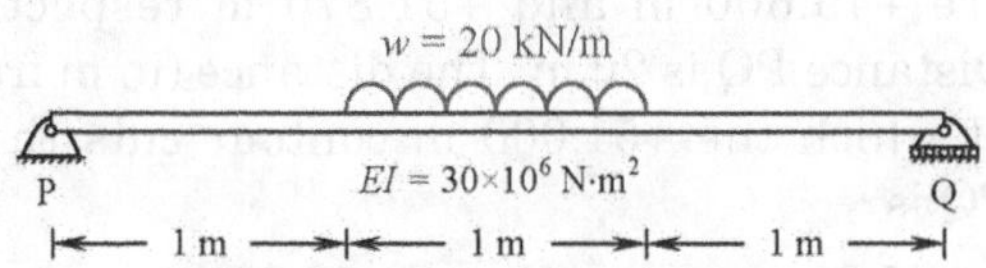

If the flexural rigidity (EI) of the beam is 30×10^6 N-m², the maximum slope (expressed in radians) of the deformed beam is

(*a*) 0.681×10^{-7} (*b*) 0.943×10^{-7}

(c) 4.310×10^{-7} (*d*) 5.910×10^{-7}

2. Two beams PQ (fixed at P and with a roller support at Q, as shown in Figure I, which allows vertical movement) and XZ (with a hinge at Y) are shown in the Figures I and II respectively. The spans of PQ and XZ are L and 2L respectively. Both the beams are under the action of uniformly distributed load (W) and have the same flexural stiffness, EI (where, E and I respectively denote modulus of elasticity and moment of inertia about axis of bending). Let the maximum deflection and maximum rotation be δ_{max1} and θ_{max1}, respectively, in the case of beam PQ and the corresponding quantities for the beam XZ be δ_{max2} and θ_{max2}, respectively.

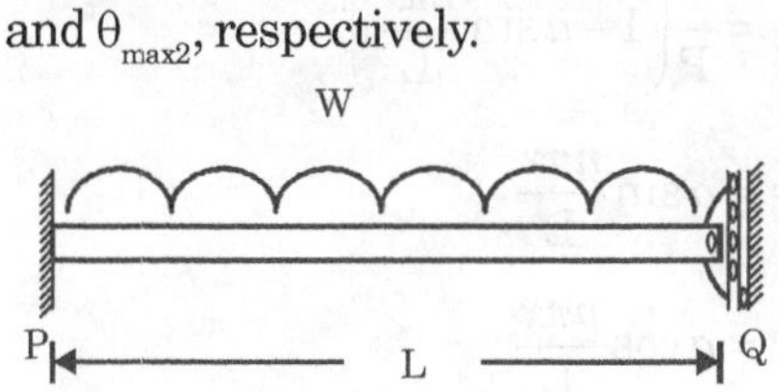

Figure I

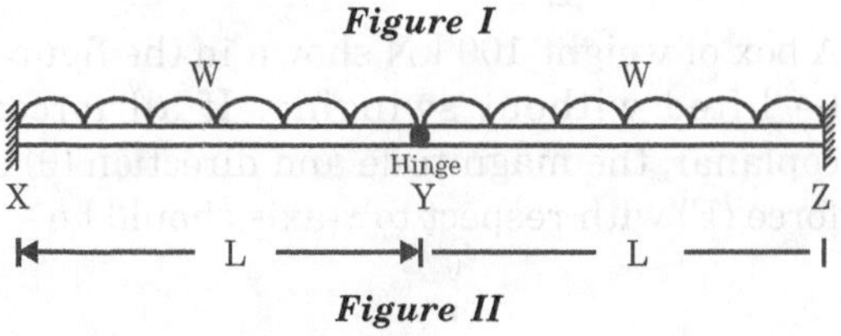

Figure II

Which one of the following relationships is true?

(*a*) $\delta_{max1} \neq \delta_{max2}$ and $\theta_{max1} \neq \theta_{max2}$

(*b*) $\delta_{max1} = \delta_{max2}$ and $\theta_{max1} \neq \theta_{max2}$

(*c*) $\delta_{max1} \neq \delta_{max2}$ and $\theta_{max1} = \theta_{max2}$

(*d*) $\delta_{max1} = \delta_{max2}$ and $\theta_{max1} = \theta_{max2}$

3. For the stress state (in MPa) shown in the figure, the major principal stress is 10 MPa.

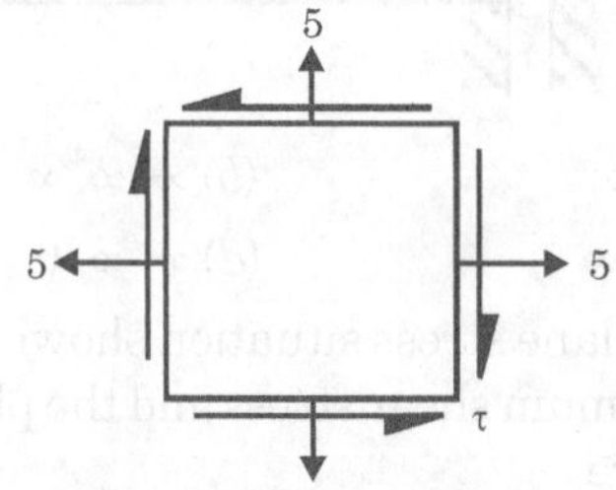

The shear stress τ is

(*a*) 10.0 MPa (*b*) 5.0 MPa

(c) 2.5 MPa (*d*) 0.0 MPa

4. Consider the structural system shown in the figure under the action of weight W. All the joints are hinged. The properties of the members in terms of length (L), area (A) and the modulus of elasticity (E) are also given in the figure. Let L, A and E be 1 m, 0.05 m² and 30×10^6 N/m², respectively, and W be 100 kN.

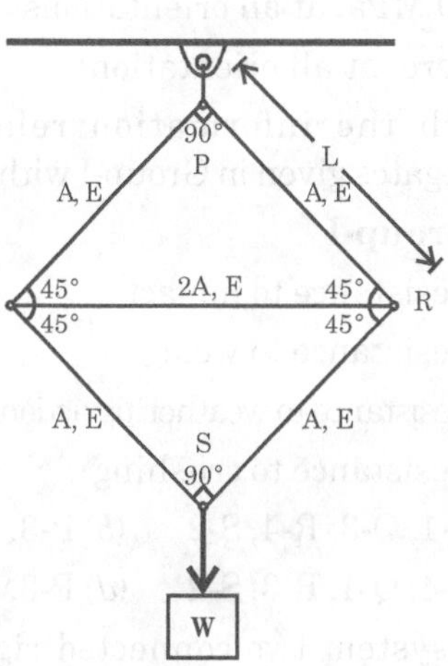

Which one of the following sets gives the correct values of the force, stress and change in length of the horizontal member QR?

(*a*) Compressive force = 25 kN; Stress = 250 kN/m²; Shortening = 0.0118 m

(*b*) Compressive force = 14.14 kN; Stress = 141.4 kN/m²; Extension = 0.0118 m

(c) Compressive force = 100 kN; Stress = 1000 kN/m²; Shortening = 0.0417 m

(*d*) Compressive force = 100 kN; Stress = 1000 kN/m²; Extension = 0.0417 m

2015

5. A guided support as shown in the figure below is represented by three springs (horizontal, vertical and rotational) with stiffness k_x, k_y and k_θ respectively. The limiting values of k_x, k_y and k_θ are

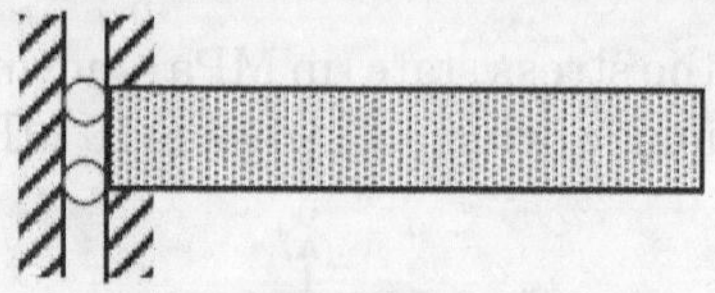

(a) ∞ 0, ∞ (b) ∞, ∞, ∞

(c) 0, ∞, ∞ (d) ∞, ∞, 0

6. For the plane stress situation shown in the figure, the maximum shear stress and the plane on which it acts are

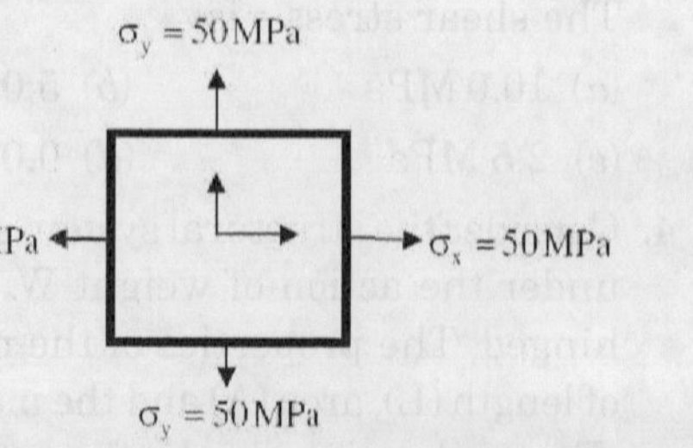

(a) –50 MPa, on a plane 45° clockwise *w.r.t.* *x*-axis

(b) –50 MPa, on a plane 45° anti-clockwise *w.r.t.* *x*-axis

(c) 50 MPa, at all orientations

(d) Zero, at all orientations

7. Match the information related to test on aggregates given in Group-I with that in Group-II.

Group-I	Group-II
P. Resistance to impact	1. Hardness
Q. Resistance to wear	2. Strength
R. Resistance to weathering action	3. Toughness
S. Resistance to crushing	4. Soundness

(a) P-1, Q-3, R-4, S-2 (b) P-3, Q-1, R-4, S-2

(c) P-4, Q-1, R-3, S-2 (d) P-3, Q-4, R-2, S-1

8. In a system two connected rigid bars AC and BC are of identical length, L with pin supports at A and B. The bars are interconnected at C by a frictionless hinge. The rotation of the hinge is restrained by a rotational spring of stiffness, k. The system initially assumes a straight line configuration, ACB. Assuming both the bars as weightless, the rotation at supports, A and B, due to a transverse load, P applied at C is

(a) $\dfrac{PL}{4k}$ (b) $\dfrac{PL}{2k}$

(c) $\dfrac{P}{4k}$ (d) $\dfrac{Pk}{4L}$

2014

9. The possible location of shear centre of the channel section, shown below, is

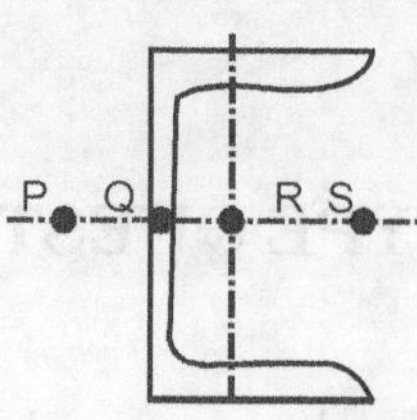

(a) P (b) Q

(c) R (d) S

10. The Reduced Levels (RLs) of the points P and Q are +49.600 m and +51.870 m respectively. Distance PQ is 20 m. The distance (in m from P) at which the +51.000 m contour cuts the line PQ is

(a) 15.00 (b) 12.33

(c) 3.52 (d) 2.27

11. If the following equation establishes equilibrium in slightly bent position, the mid-span deflection of a member shown in the figure is

$$\frac{d^2y}{dx^2} + \frac{P}{EI}y = 0$$

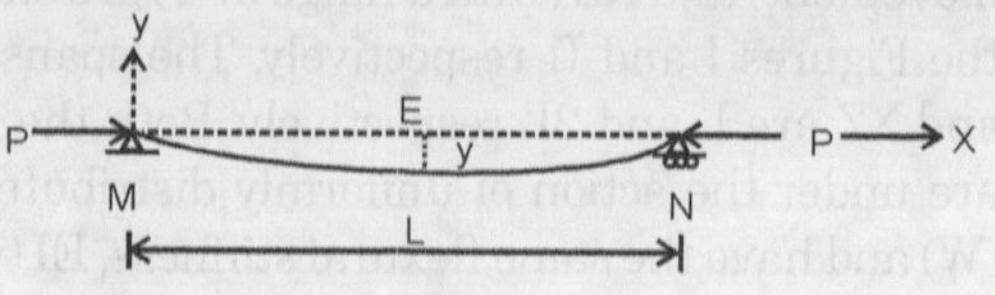

If *a* is amplitude constant for *y*, then

(a) $y = \dfrac{1}{P}\left(1 - a\cos\dfrac{2\pi x}{L}\right)$

(b) $y = \dfrac{1}{P}\left(1 - a\sin\dfrac{2\pi x}{L}\right)$

(c) $y = a\sin\dfrac{n\pi x}{L}$

(d) $y = a\cos\dfrac{n\pi x}{L}$

12. A box of weight 100 kN shown in the figure is to be lifted without swinging. If all forces are coplanar, the magnitude and direction (θ) of the force (F) with respect to *x*-axis should be

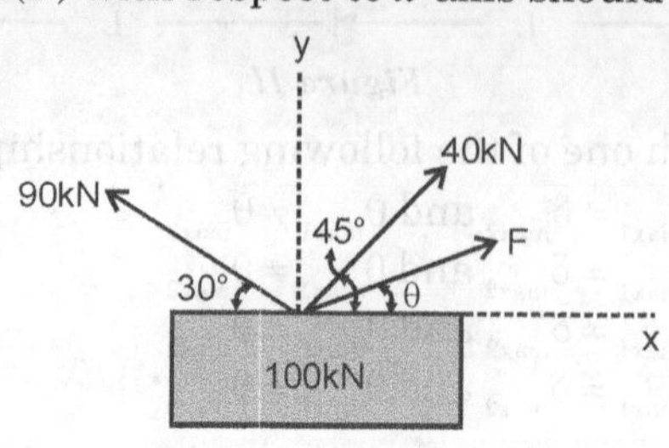

(*a*) F = 56.389 kN and θ = 28.28°

(*b*) F = –56.389 kN and θ = –28.28°

(*c*) F = 9.055 kN and θ = 1.414°

(*d*) F = –9.055 kN and θ = –1.414°

13. The first moment of area about the axis of bending for a beam cross-section is

(*a*) moment of inertia

(*b*) section modulus

(*c*) shape factor

(*d*) polar moment of inertia

14. The values of axial stress (σ) in kN/m², bending moment (M) in kNm, and shear force (V) in kN acting at point P for arrangement shown in the figure are respectively

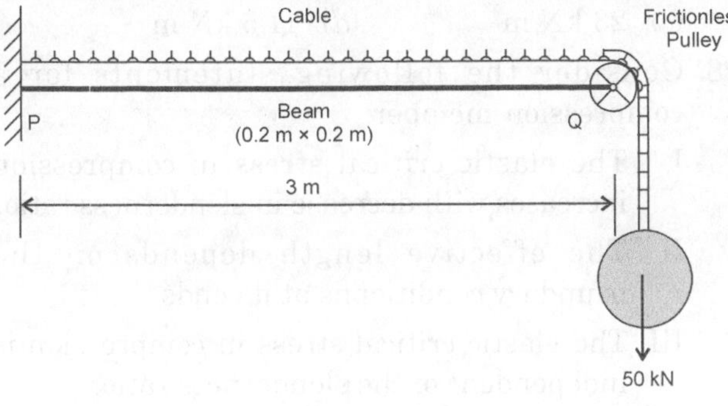

(*a*) 1000, 75 and 25

(*b*) 1250, 150 and 50

(*c*) 1500, 225 and 75

(*d*) 1750, 300 and 100

15. The tension (in kN) in a 10 m long cable, shown in the figure, neglecting its self-weight is

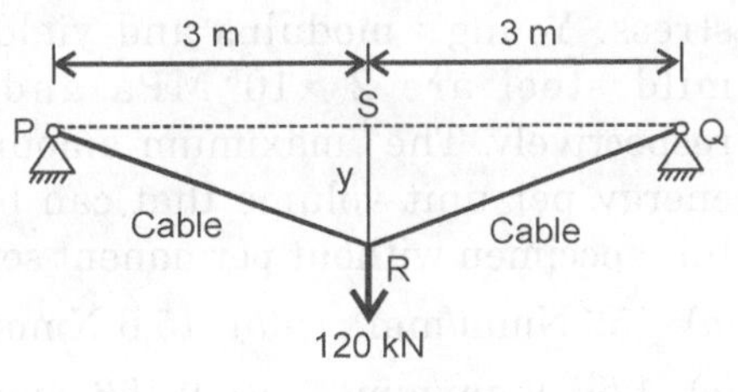

(*a*) 120 (*b*) 75

(*c*) 60 (*d*) 45

2013

16. Beam PQRS has internal hinges in spans PQ and RS as shown. The beam may be subjected to a moving distributed vertical load of maximum intensity 4 kN/m of any length anywhere on the beam. The maximum absolute value of the shear force (in kN) that can occur due to this loading just to the right of support Q shall be

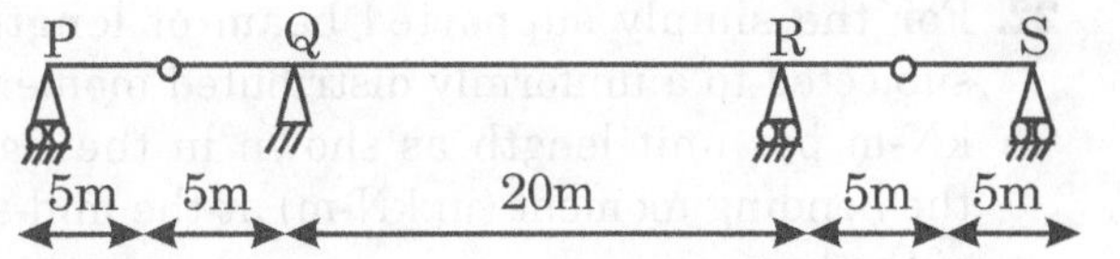

(*a*) 30 (*b*) 40

(*c*) 45 (*d*) 55

2012

17. The Poisson's ratio is defined as

(*a*) $\left|\frac{\text{axial stress}}{\text{lateral stress}}\right|$ (*b*) $\left|\frac{\text{lateral strain}}{\text{axial strain}}\right|$

(*c*) $\left|\frac{\text{lateral stress}}{\text{axial stress}}\right|$ (*d*) $\left|\frac{\text{axial strain}}{\text{lateral strain}}\right|$

18. The ratio of the theoretical critical buckling load for a column with fixed ends to that of another column with the same dimensions and material, but with pinned ends, is equal to

(*a*) 0.5 (*b*) 1.0

(*c*) 2.0 (*d*) 4.0

2011

19. For the fillet weld of size '*s*' shown in the adjoining figure the effective throat thickness is

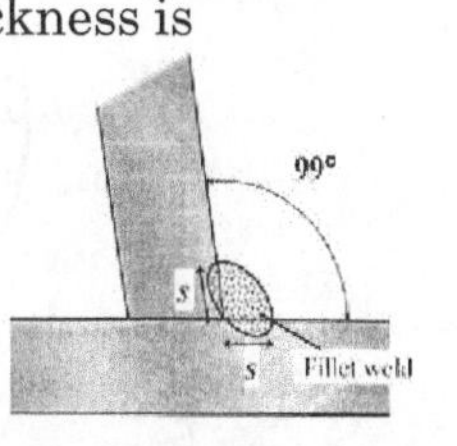

(*a*) 0.61 *s*

(*b*) 0.65 *s*

(*c*) 0.70 *s*

(*d*) 0.75 *s*

2010

20. Two people weighing W each are sitting on a plank of length L floating on water at $\frac{L}{4}$ from either end. Neglecting weight of the plank, bending moment at the centre of the plank is

(*a*) $\frac{WL}{8}$ (*b*) $\frac{WL}{16}$

(*c*) $\frac{WL}{32}$ (*d*) zero

21. For the truss shown in the figure, the force in the member QR is

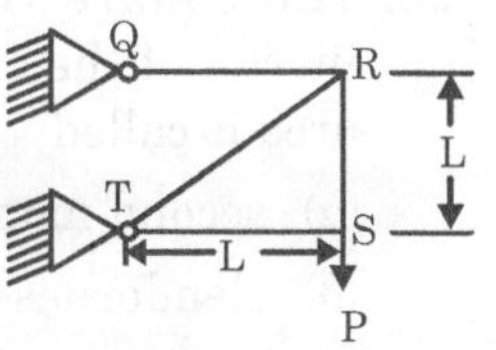

(*a*) zero

(*b*) $\frac{P}{\sqrt{2}}$

(*c*) P

(*d*) $\sqrt{2}$P

22. For the simply supported beam of length L, subjected to a uniformly distributed moment M kN-m per unit length as shown in the figure, the bending moment (in kN-m) at the mid-span of the beam is

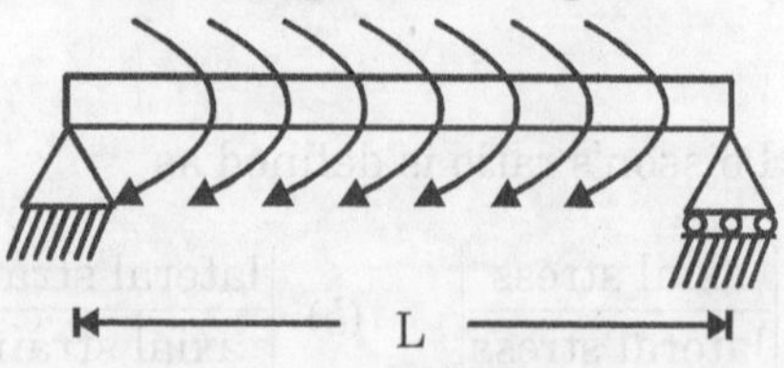

(*a*) zero (*b*) M

(*c*) ML (*d*) $\frac{M}{L}$

23. A disc of radius *r* has a hole of radius $\frac{r}{2}$ cut-out as shown. The centroid of the remaining disc (shaded portion) at a radial distance from the centre "O" is

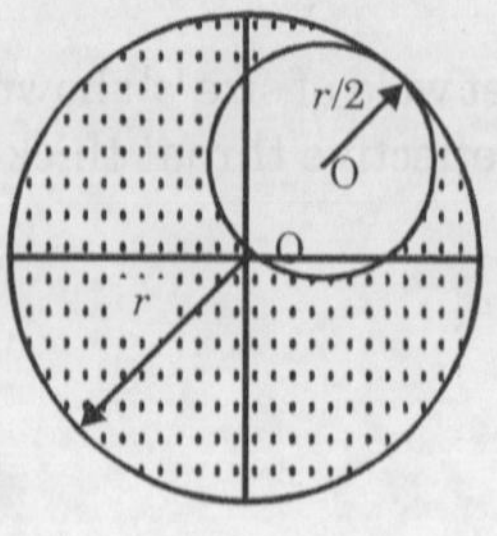

(*a*) $\frac{r}{2}$ (*b*) $\frac{r}{3}$

(*c*) $\frac{r}{6}$ (*d*) $\frac{r}{8}$

2009

24. A thin walled cylindrical pressure vessel having a radius of 0.5 m and wall thickness of 25 mm is subjected to an internal pressure of 700 kPa. The Hoop stress developed is

(*a*) 14 MPa

(*b*) 1.4 MPa

(*c*) 0.14 MPa

(*d*) 0.014 MPa

25. The square root of the ratio of moment of inertia of the cross section to its cross sectional area is called

(*a*) second moment of area

(*b*) slenderness ratio

(*c*) section modulus

(*d*) radius of gyration

26. Consider the following statements:

I. On a principal plane, only normal stress acts.

II. On a principal plane, both normal and shear stresses act.

III. On a principal plane, only shear stress acts.

IV. Isotropic state of stress is independent of frame of reference.

The TRUE statements are

(*a*) I and IV (*b*) II

(*c*) II and IV (*d*) II and III

27. A hollow circular shaft has an outer diameter of 100 mm and a wall thickness of 25 mm. The allowable shear stress in the shaft is 125 MPa. The maximum torque the shaft can transmit is

(*a*) 46 kN m (*b*) 24.5 kN m

(*c*) 23 kN m (*d*) 11.5 kN m

28. Consider the following statements for a compression member:

I. The elastic critical stress in compression increases with decrease in slenderness ratio.

II. The effective length depends on the boundary conditiopns at its ends.

III. The elastic critical stress in compression is independent of the slenderness ratio.

IV. The ratio of the effective length to its radius of gyration is called as slenderness ratio.

The TRUE statements are

(*a*) II and III (*b*) III and IV

(*c*) II, II and IV (*d*) I, II and IV

2008

29. A mild steel specimen is under uni-axial tensile stress. Young's modulus and yield stress for mild steel are 2×10^5 MPa and 250 MPa respectively. The maximum amount of strain energy per unit volume that can be stored in this specimen without permanent set is

(*a*) 156 Nmm/mm^3 (*b*) 15.6 Nmm/mm^3

(*c*) 1.56 Nmm/mm^3 (*d*) 0.156 Nmm/mm^3

30. The maximum shear stress in a solid shaft of circular cross-section having diameter '*d*' subjected to a torque T is 'τ'. If troque is increased by four times and diameter of the shaft is increased by two times, then maximum shear stress in the shaft will be

(*a*) 2τ (*b*) τ

(*c*) $\frac{\tau}{2}$ (*d*) $\frac{\tau}{4}$

2007

31. An axially loaded bar is subjected to a normal stress of 173 MPa. The shear stress in the bar is

(*a*) 75 MPa (*b*) 86.5 MPa
(*c*) 100 MPa (*d*) 122.3 MPa

32. A steel column, pinned at both ends, has a buckling load of 200 kN. If the column is restrained against lateral movement at its mid-height, its buckling load will be

(*a*) 200 kN (*b*) 283 kN
(*c*) 400 kN (*d*) 800 kN

33. The stiffness coefficient k_{ij} indicates

(*a*) force at *i* due to a unit deformation at *j*
(*b*) deformation at *j* due to a unit force at *i*
(*c*) deformation at *i* due to a unit force at *j*
(*d*) force at *j* due to a unit deformation at *i*

34. For an isotropic material, the relationship between the Young's modulus (E), shear modulus (G) and Poisson's ratio (μ) is given by

(*a*) $G = \frac{E}{2(1+\mu)}$ (*b*) $E = \frac{G}{2(1+\mu)}$

(*c*) $G = \frac{E}{(1+2\mu)}$ (*d*) $G = \frac{E}{2(1-\mu)}$

35. A metal bar of length 100 mm is inserted between two rigid supports and its temperature is increased by 10°C. If coefficient of thermal expansion is 12×10^{-6} per °C and the Young's modulus is 2×10^5 MPa, then stress in the bar is

(*a*) zero (*b*) 12 MPa
(*c*) 24 MPa (*d*) 2400 MPa

36. A rigid bar is suspended by three rods made of the same material as shown in the figure. The area and length of the central rod are 3A and L, respectively while that of the two outer rods are 2A and 2L, respectively. If a downward force of 50 kN is applied to the rigid bar, then forces in the central and each of the outer rods will be

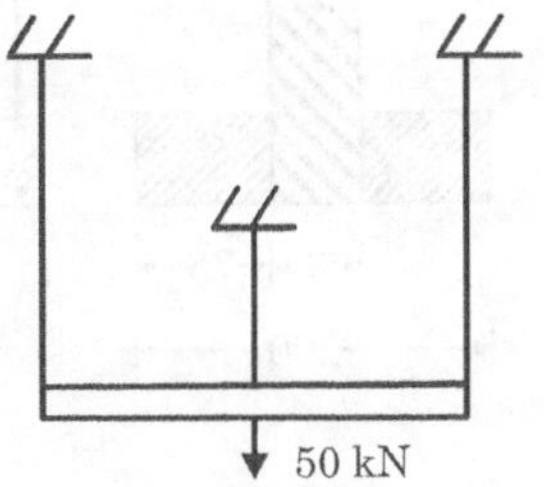

(*a*) 16.67 kN each (*b*) 30 kN and 15 kN
(*c*) 30 kN and 10 kN (*d*) 21.4 kN and 14.3 kN

37. The maximum and minimum shear stresses in a hollow circular shaft of outer diameter 20 mm and thickness 2 mm, subjected to a torque of 92.7 N.m will be

(*a*) 59 MPa and 47.2 MPa
(*b*) 100 MPa and 80 MPa
(*c*) 118 MPa and 160 MPa
(*d*) 200 MPa and 160 MPa

38. The shear stress at the neutral axis in a beam of triangular section with a base of 40 mm and height 20 mm, subjected to a shear force of 3 kN is

(*a*) 3 MPa
(*b*) 6 MPa
(*c*) 10 MPa
(*d*) 20 MPa

2006

39. The necessary and sufficient condition for a surface to be called as a free surface is

(*a*) no stress should be acting on it
(*b*) tensile stress acting on it must be zero
(*c*) shear stress acting on it must be zero
(*d*) no point on it should be under any stress

40. Mohr's circle for the state of stress defined by

$\begin{bmatrix} 30 & 0 \\ 0 & 30 \end{bmatrix}$ MPa is a circle with

(*a*) center at (0, 0) and radius 30 MPa
(*b*) center at (0, 0) and radius 60 MPa
(*c*) center at (30, 0) and radius 30 MPa
(*d*) center at (30, 0) and zero radius

41. The buckling load P = P_{cr} for the column AB in the figure, as K_T approaches infinity, become $\alpha \frac{\pi^2 EI}{L^2}$, where α is equal to

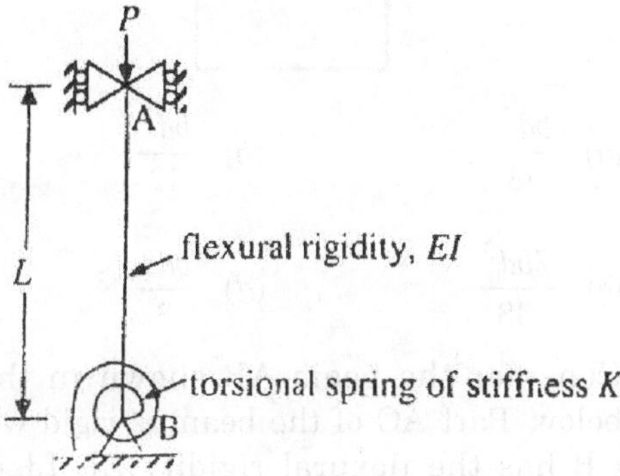

(*a*) 0.25 (*b*) 1.00
(*c*) 2.05 (*d*) 4.00

42. A long shaft of diameter d is subjected to twisting moment T at its ends. The maximum normal stress acting at its cross-section is equal to

(a) zero

(b) $\frac{16T}{\pi d^3}$

(c) $\frac{32T}{\pi d^3}$

(d) $\frac{64T}{\pi d^3}$

43. A simply supported beam AB has the bending moment diagram as shown in the following figure.

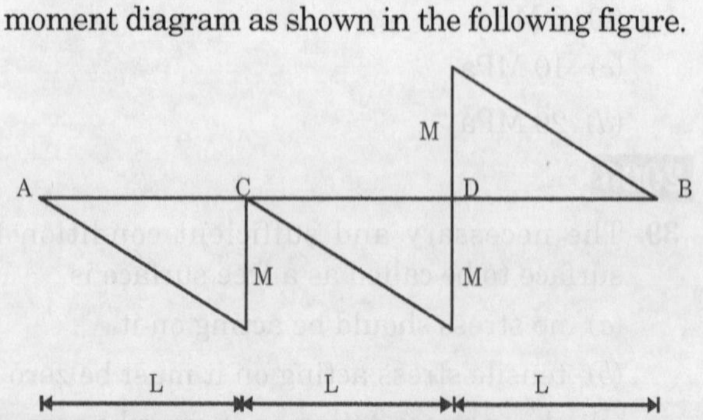

The beam is possibly under the action of following loads:

(a) Couples of M at C and 2M at D

(b) Couples of 2M at C and M at D

(c) Concentrated loads of M/L at C and 2M/L at D

(d) Concentrated load of M/L at C and couple of 2M at D

44. For the section shown below, second moment of the area about an axis $d/4$ distance above the bottom of the area is

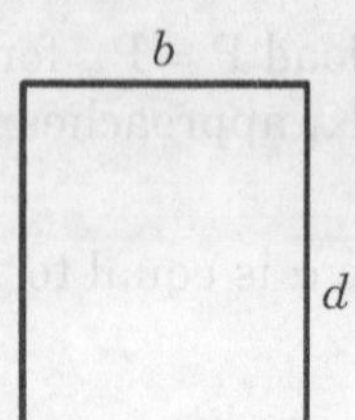

(a) $\frac{bd^3}{48}$ (b) $\frac{bd^3}{12}$

(c) $\frac{7bd^3}{48}$ (d) $\frac{bd^3}{3}$

45. Consider the beam AB shown in the figure below. Part AC of the beam is rigid while Part CB has the flexural rigidity EI. Identify the correct combination of deflection at end B and bending moment at end A, respectively.

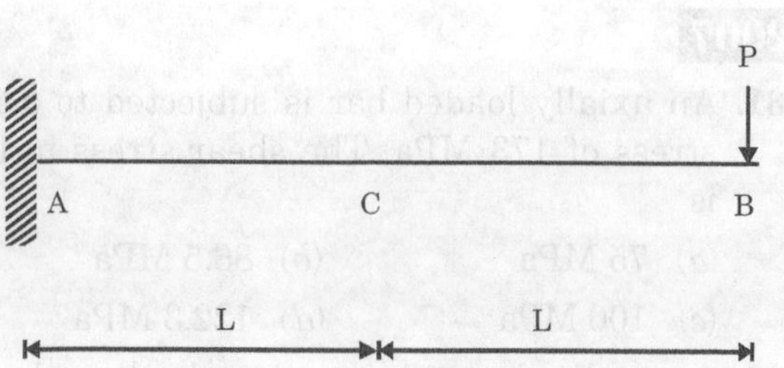

(a) $\frac{PL^3}{3EI}$, 2PL (b) $\frac{PL^3}{3EI}$, PL

(c) $\frac{8PL^3}{3EI}$, 2PL (d) $\frac{8PL^3}{3EI}$, PL

46. A beam with the cross-section given below is subjected to a positive bending moment (causing compression at the top) of 16 kN-m acting around the horizontal axis. The tensile force acting on the hatched area of the cross-section is

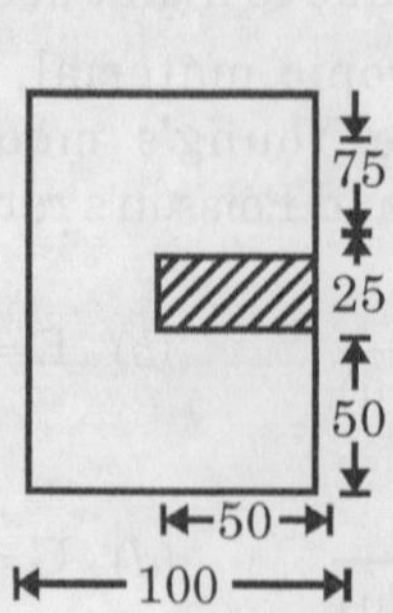

(a) zero (b) 5.9 kN

(c) 8.9 kN (d) 17.8 kN

47. T-section of a beam is formed by gluing wooden planks as shown in the figure below. If this beam transmits a constant vertical shear force of 3000 N, the glue at any of the four joints will be subjected to a shear force (in kN per meter length) of

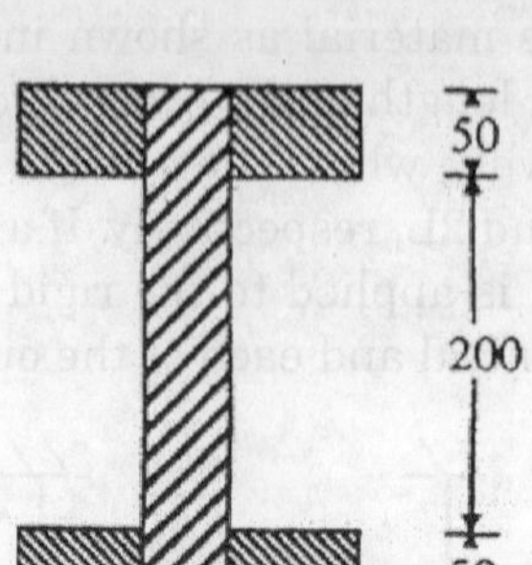

(a) 3.0 (b) 4.0

(c) 8.0 (d) 10.7

48. If a beam of rectangular cross-section is subjected to a vertical shear force V, then shear force carried by the upper one-third of the cross-section is

(*a*) zero (*b*) $\frac{7V}{27}$

(*c*) $\frac{8V}{27}$ (*d*) $\frac{V}{3}$

49. A thin-walled long cylindrical tank of inside radius *r* is subjected simultaneously to internal gas pressure p and axial compressive force F at its ends. In order to produce '*pure shear*' state of stress in the wall of the cylinder, F should be equal to

(*a*) πpr^2 (*b*) $2\pi pr^2$

(*c*) $3\pi pr^2$ (*d*) $4\pi pr^2$

2005

50. The symmetry of stress tensor at a point in the body under equilibrium is obtained from

(*a*) conservation of mass

(*b*) force equilibrium equations

(*c*) moment equilibrium equations

(*d*) conservation of energy

51. The components of strain tensor at a point in the plane strain case can be obtained by measuring longitudinal strain along

(*a*) any two arbitrary directions

(*b*) any three arbitrary directions

(*c*) two mutually orthogonal directions

(*d*) any arbitrary direction

52. A cantilever beam of length l, width b and depth d is loaded with a concentrated vertical load at the tip. If yielding starts at a load P, then collapse load shall be

(*a*) 2.0 P (*b*) 1.5 P

(*c*) 1.2 P (*d*) P

53. A circular shaft shown in the figure is subjected to torsion T at two points A and B. The torsional rigidity of portions CA and BD is GJ_1 and that of portion AB is GJ_2. The rotations of shaft at points A and B are θ_1 and θ_2. The rotation θ_1 is

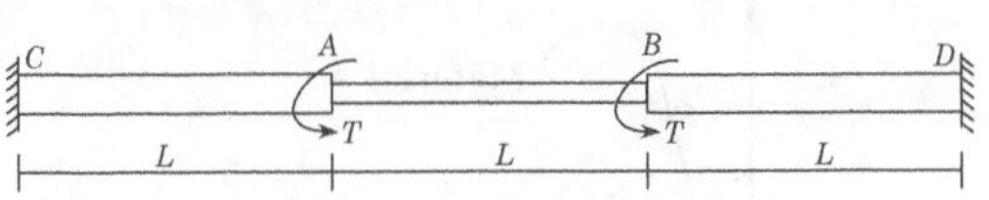

(*a*) $\frac{TL}{GJ_1 + GJ_2}$ (*b*) $\frac{TL}{GJ_1}$

(*c*) $\frac{TL}{GJ_2}$ (*d*) $\frac{TL}{GJ_1 - GJ_2}$

54. If principal stresses ina two-dimensional case are – 10 MPa and 20 MPa respectively, then maximum shear stress at the pont is

(*a*) 10 MPa (*b*) 15 MPa

(*c*) 20 MPa (*d*) 30 MPa

55. The bending moment diagram for a beam is given below :

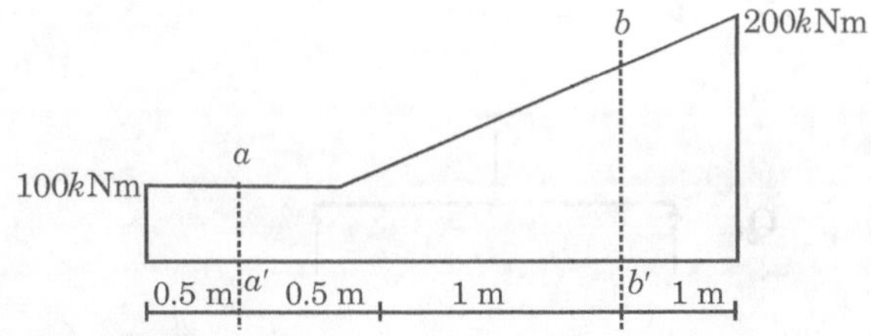

The shear force at sections aa′ and bb′ respectively are of the magnitude

(*a*) 100 kN, 150 kN (*b*) zero, 100 kN

(*c*) zero, 50 kN (*d*) 100 kN, 100 kN

2003

56. A bar of varying square cross-section is loaded symmetrically as shown in the figure. Loads shown are placed on one of the axes of symmetry of cross-section. Ignoring self weight, the maximum tensile stress in N/mm^2 anywhere is

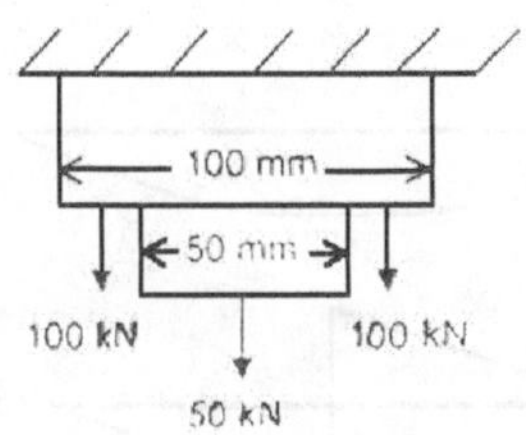

(*a*) 16.0 (*b*) 20.0

(*c*) 25.0 (*d*) 30.0

57. A curved member with a straight vertical leg is carrying a vertical load at Z, as shown in the figure. The stress resultants in the XY segment are

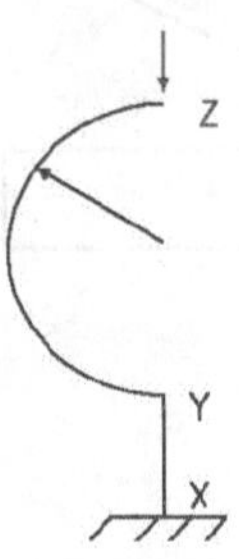

(*a*) bending moment, shear force and axial force

(*b*) bending moment and axial force only

(*c*) bending moment and shear force only

(*d*) axial force only

58. Group 1 shows different loads acting on a beam and Group 2 shows different bending moment distributions. Match the load with the corresponding bending moment diagram.

Group 1

P. 

Q. 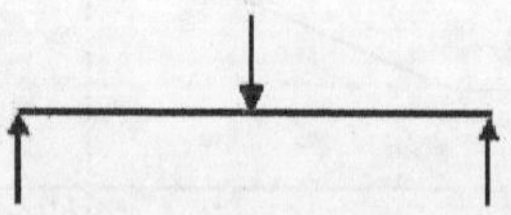

R. 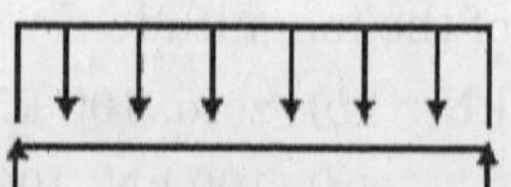

S. 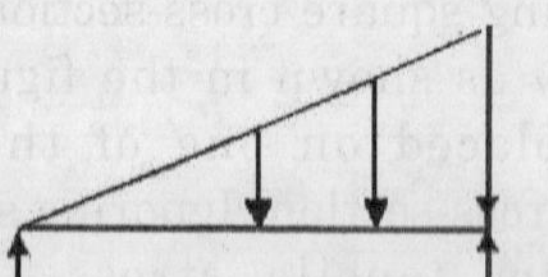

Group 2

1. 

2. 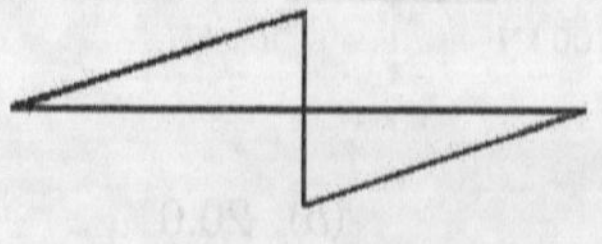

3. 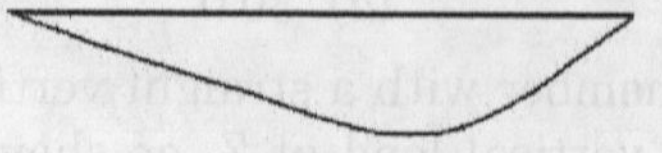

4. 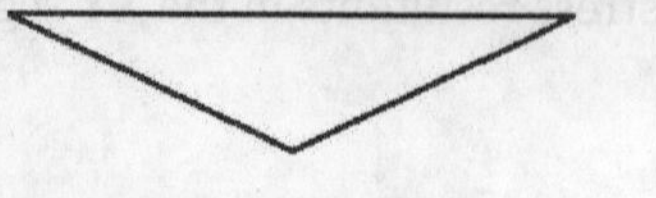

5. 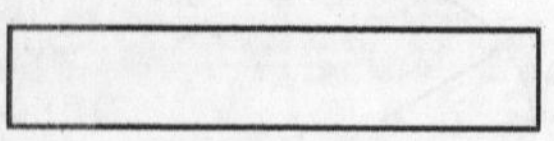

Codes :

	P	Q	R	S
(*a*)	4	2	1	3
(*b*)	5	4	1	3
(*c*)	2	5	3	1
(*d*)	2	4	1	3

2002

59. For the loading given in the figure below, two statements (I and II) are made.

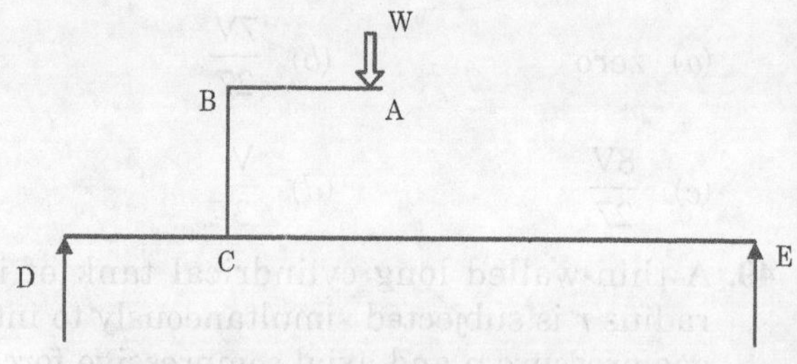

I. Member AB carries shear force and bending moment

II. Member BC carries axial load and shear force.

Which of the following is true ?

(*a*) Statement I is true but II is false

(*b*) Statement I is false but II is true

(*c*) Both statements I and II are true

(*d*) Both statements I and II are false

60. The shear modulus (G), modulus of elasticity (E) and the Poisson's ratio (v) of a material are related as,

(*a*) $G = \frac{E}{[2(1+v)]}$ (*b*) $E = \frac{G}{[2(1+v)]}$

(*c*) $G = \frac{E}{[2(1-v)]}$ (*d*) $G = \frac{E}{[2(v-1)]}$

2000

61. Pick the incorrect statement from the following four statements :

(*a*) On the plane which carries maximum normal stress, the shear stress is zero.

(*b*) Principal planes are mutually orthogonal.

(*c*) On the plane which carries maximum shear stress, the normal stress is zero.

(*d*) The principal stress axes and principal strain axes coincide for an isotropic material.

62. The stress-strain diagram for two materials A and B is shown below :

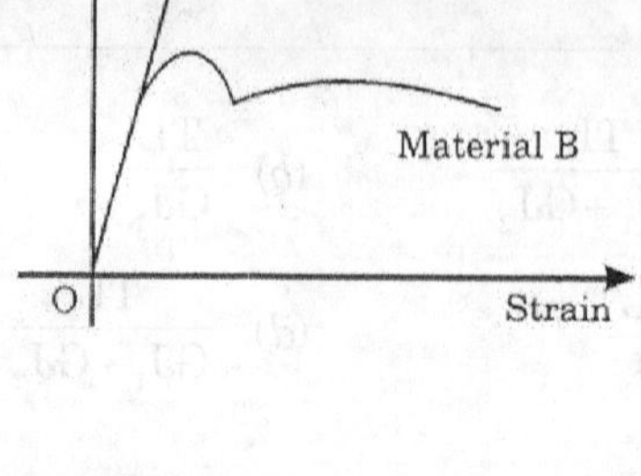

The following statements are made based on this diagram :

I. Material A is more brittle than material B.

II. The ultimate strength of material B is more than that of A.

With reference to the above statements, which of the following applies ?

(*a*) Both the statements are false.

(*b*) Both the statements are true.

(*c*) I is true but II is false.

(*d*) I is false but II is true.

63. Four column of the same material and having identical geometric properties are supported in different ways as shown below :

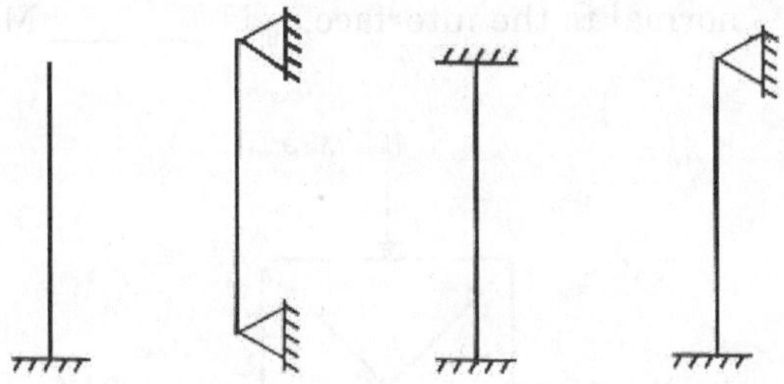

I II III IV

It is required to order these four beams in the increasing order of their respective first buckling loads. The correct order is given by

(*a*) I, II, III, IV (*b*) III, IV, I, II

(*c*) II, I, IV, III (*d*) I, II, IV, III

64. The four cross sections shown below are required to be ordered in the increasing order of their respective shape factors.

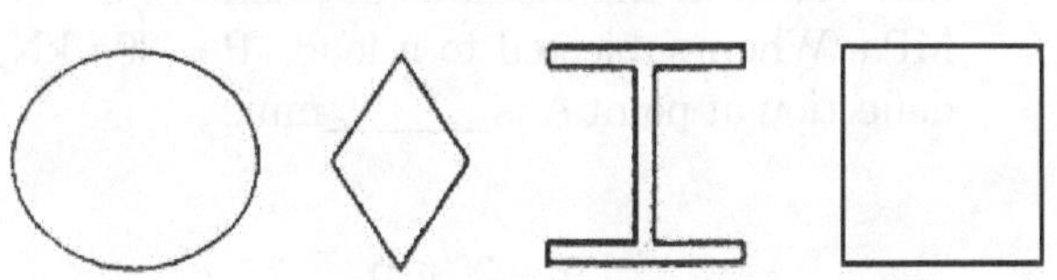

I II III IV

Which of the following order is correct ?

(*a*) III, I, IV, II (*b*) I, II, III, IV

(*c*) III, IV, I, II (*d*) III, IV, II, I

65. A simply supported beam with an overhang is traversed by a unit concentrated moment from the left to the right as shown below.

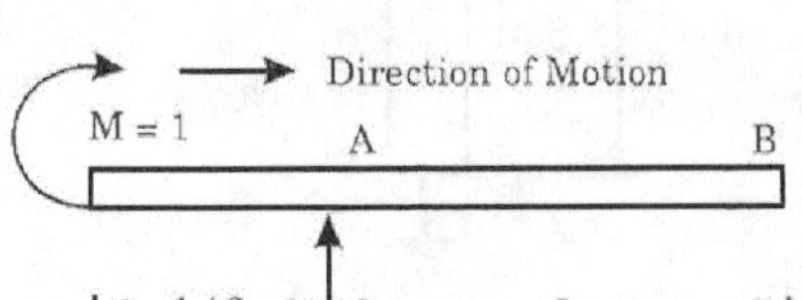

The influence line for reaction at B is given by

(*a*) 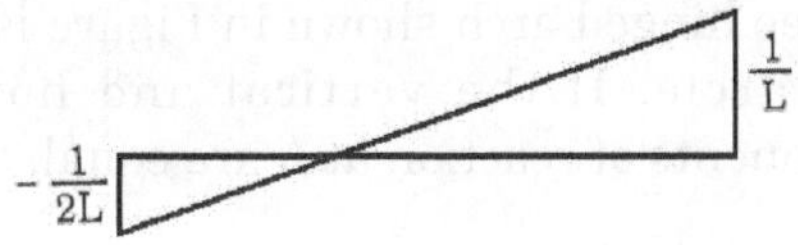

(*b*) 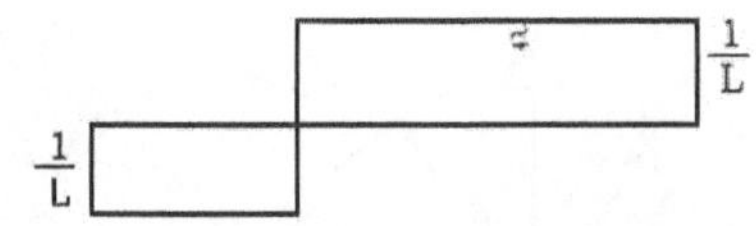

(*c*) 

(*d*) zero everywhere

1999

66. Two perpendicular axes x and y of a section are called principal axes when

(*a*) Moments of inertia about the axes are equal ($I_x = I_y$)

(*b*) Product moment of inertia (I_{xy}) is zero

(*c*) Product moments of inertia ($I_x \times I_y$) is zero

(*d*) Moment of inertia about one of the axes is greater than the other

67. In section, shear centre is a point through which, if the resultant load passes, the section will not be subjected to any

(*a*) Bending (*b*) Tension

(*c*) Compression (*d*) Torsion

68. The slope of the elastic curve at the free end of a cantilever beam of span L, and with flexural rigidity EI, subjected to uniformly distributed load of intensity w is

(*a*) $\frac{wL^3}{6EL}$

(*b*) $\frac{wL^3}{3EL}$

(*c*) $\frac{wL^4}{8EL}$

(*d*) $\frac{wL^3}{2EL}$

69. If an element of a stressed body is in a state of pure shear with a magnitude of 80N/mm², the magnitude of maximum principal stress at that location is

(*a*) 80 N/mm² (*b*) 113.14 N/mm²

(*c*) 120 N/mm² (*d*) 56.57 N/mm²

1998

70. A three hinged arch shown in Figure is quarter of a circle. If the vertical and horizontal components of reaction at A are equal, the value of θ is

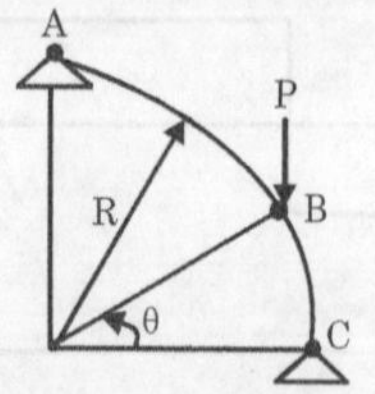

(a) 60° (b) 45°

(c) 30° (d) None in (0°, 90°)

71. A cantilever beam is shown in the Figure. The moment to be applied at free end for zero vertical deflection at that point is

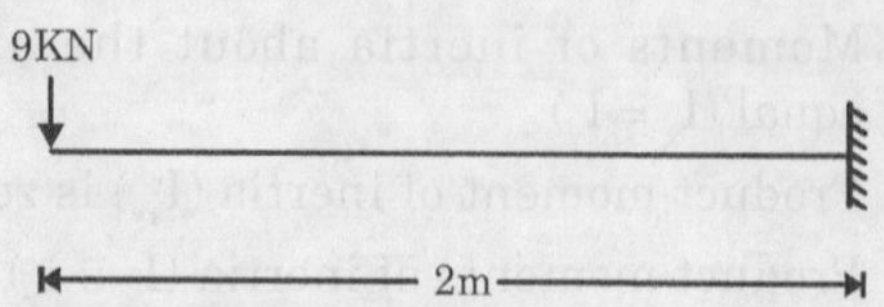

(a) 9 kN.m clockwise

(b) 9 kN.m anti-clockwise

(c) 12kN.m clockwise

(d) 12 kN.m anti-clockwise

1997

72. The force in the member DE of the truss shown in the figure is

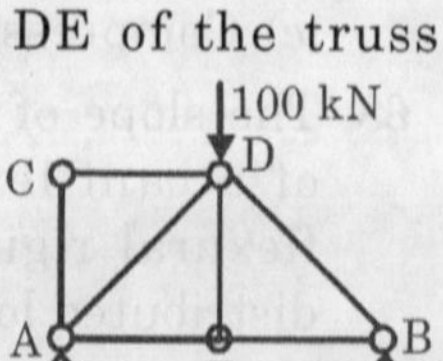

(a) 100.0 kN

(b) zero

(c) 35.5 kN

(d) 25.0 kN

73. A propped cantilever beam of span L, is loaded with uniformly distributed load of intensity w/ unit length, all through the span. Bending moment at the fixed end is

(a) $\frac{wL^2}{8}$ (b) $\frac{wL^2}{2}$

(c) $\frac{wL^2}{12}$ (d) $\frac{wL^2}{24}$

74. A cantilever beam of span 'L' is loaded with a concentrated load 'P' at the free end. Deflection of the beam at the free end is

(a) $\frac{PL^3}{48\,EI}$ (b) $\frac{5\,PL^3}{384\,EI}$

(c) $\frac{PL^3}{3\,EI}$ (d) $\frac{PL^3}{6\,EI}$

1996

75. Tangential or shear stress on an oblique plane at an angle ϕ to the cross-section of the bar subjected to uniaxial stress σ_x is equal to

(a) $\frac{\sigma_x}{2}\sin 2\phi$

(b) $\sigma_x \sin 2\phi$

(c) $\sigma_x \cos 2\phi$

(d) $\sigma_x \cos\phi$

NUMERICAL TYPE QUESTIONS

2015

1. Two triangular wedges are glued together as shown in the following figure. The stress acting normal to the interface, σ_n is ________ MPa

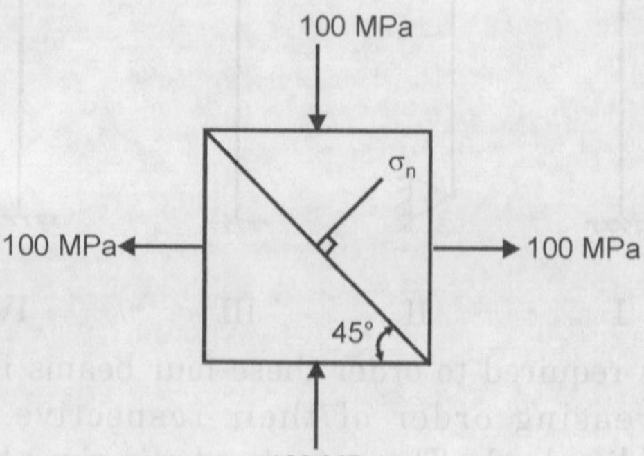

2. A tapered circular rod of diameter varying from 20 mm to 10 mm is connected to another uniform circular rod of diameter 10 mm as shown in the following figure. Both bars are made of same material with the modulus of elasticity, E = 2 ×10⁵ MPa. When subjected to a load P = 30π kN, the deflection at point A is ________ mm.

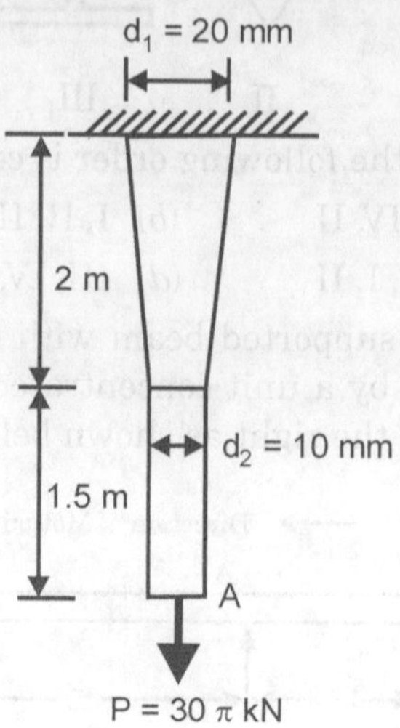

3. A horizontal beam ABC is loaded as shown in the figure below. The distance of the point of contraflexure from end A (in m) is ________.

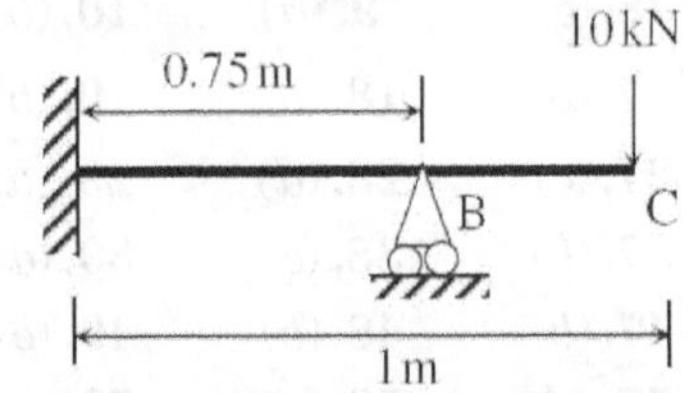

4. A steel strip of length, L = 200 mm is fixed at end A and rests at B on a vertical spring of stiffness, k = 2 N/mm. The steel strip is 5 mm wide and 10 mm thick. A vertical load, P = 50 N is applied at B, as shown in the figure. Considering E = 200 GPa, the force (in N) developed in the spring is ______.

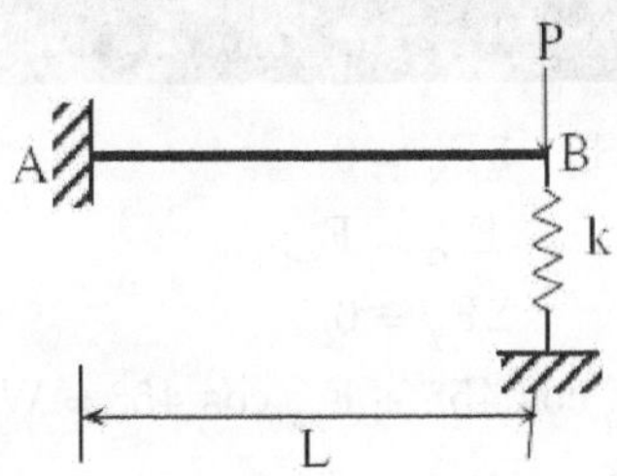

5. A simply supported beam AB of span, L = 24 m is subjected to two wheel loads acting at a distance, d = 5 m apart as shown in the figure below. Each wheel transmits a load, P = 3 kN and may occupy any position along the beam. If the beam is an I-section having section modulus, S = 16.2 cm^3, the maximum bending stress (in GPa) due to the wheel loads is __________

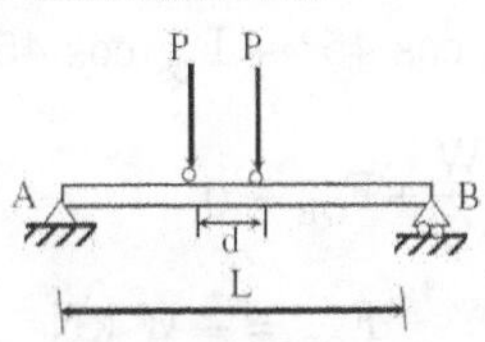

2014

6. For the cantilever beam of span 3 m (shown below), a concentrated load of 20 kN applied at the free end causes a vertical displacement of 2 mm at a section located at a distance of 1 m from the fixed end. If a concentrated vertically downward load of 10 kN is applied at the section located at a distance of 1 m from the fixed end (with no other load on the beam), the maximum vertical displacement in the same beam (in mm) is________ .

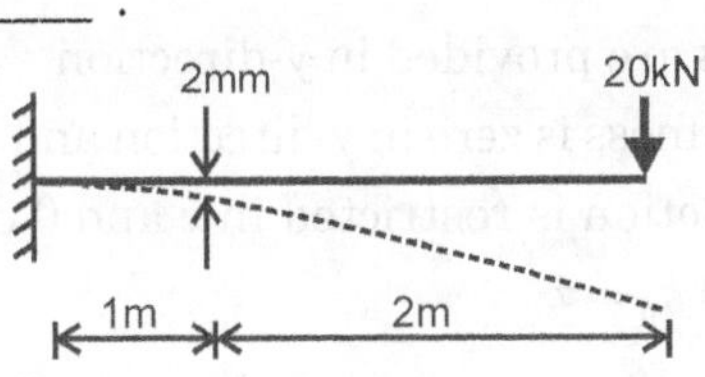

7. Polar moment of inertia (I_p), in cm^4, of a rectangular section having width, b = 2 cm and depth, d = 6 cm is ________

8. The beam of an overall depth 250 mm (shown below) is used in a building subjected to two different thermal environments. The temperatures at the top and bottom surfaces of the beam are 36°C and 72°C respectively. Considering coefficient of thermal expansion (α) as 1.50 × 10^{-5} per °C, the vertical deflection of the beam (in mm) at its mid-span due to temperature gradient is ________

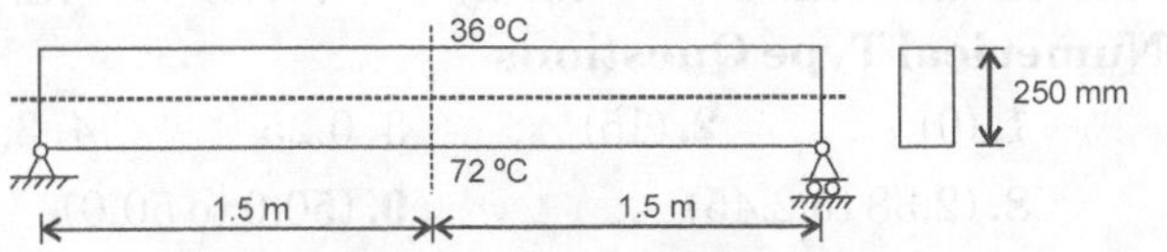

9. The axial load (in kN) in the member PQ for the arrangement/assembly shown in the figure given below is ________

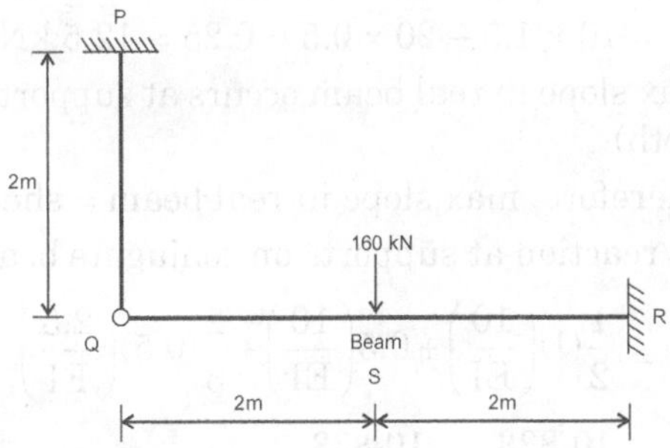

10. For the state of stresses (in MPa) shown in the figure below, the maximum shear stress (in MPa) is ________

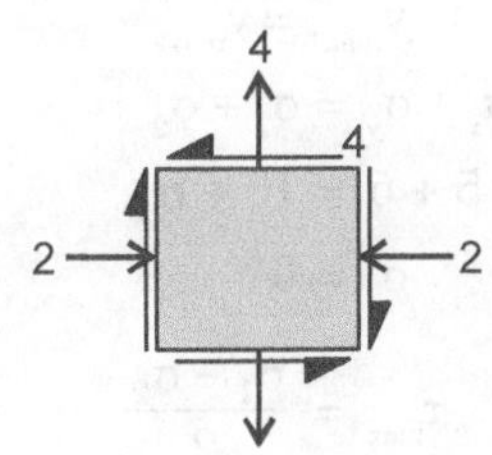

2013

11. a rectangular concrete beam 250 mm wide and 600 mm deep is pre-stressed by means of 16 high tensile wires, each of 7 mm diameter, located at 200 mm from the bottom face of the beam at a given section. If the effective pre-stress in the wires is 700 MPa, what is the maximum sagging bending moment (in kNm) (correct to 1-decimal place) due to live load that this section of the beam can withstand without causing tensile stress at the bottom face of the beam? Neglect the effect of dead load of beam. ______

ANSWERS

MCQ Type Questions

1. (No options are correct)

2. (*d*)	**3.** (*b*)	**4.** (*c*)	**5.** (*a*)	**6.** (*d*)	**7.** (*b*)	**8.** (*a*)	**9.** (*a*)	**10.** (*b*)	
11. (*c*)	**12.** (*a*)	**13.** (*b*)	**14.** (*b*)	**15.** (*b*)	**16.** (*c*)	**17.** (*b*)	**18.** (*d*)	**19.** (*b*)	**20.** (*d*)
21. (*c*)	**22.** (*a*)	**23.** (*c*)	**24.** (*a*)	**25.** (*d*)	**26.** (*a*)	**27.** (*c*)	**28.** (*d*)	**29.** (*d*)	**30.** (*c*)
31. (*b*)	**32.** (*d*)	**33.** (*a*)	**34.** (*a*)	**35.** (*c*)	**36.** (*c*)	**37.** (*b*)	**38.** (*c*)	**39.** (*a*)	**40.** (*d*)
41. (*d*)	**42.** (*b*)	**43.** (*a*)	**44.** (*c*)	**45.** (*a*)	**46.** (*c*)	**47.** (*b*)	**48.** (*b*)	**49.** (*c*)	**50.** (*c*)
51. (*b*)	**52.** (*b*)	**53.** (*a*)	**54.** (*b*)	**55.** (*c*)	**56.** (*c*)	**57.** (*d*)	**58.** (*d*)	**59.** (*c*)	**60.** (*c*)
61. (*c*)	**62.** (*c*)	**63.** (*d*)	**64.** (*c*)	**65.** (*c*)	**66.** (*c*)	**67.** (*d*)	**68.** (*a*)	**69.** (*a*)	**70.** (*b*)
71. (*d*)	**72.** (*b*)	**73.** (*a*)	**74.** (*c*)	**75.** (*a*)					

Numerical Type Questions

1. (0) **2.** (15) **3.** 0.25 **4.** 3.2 **5.** 1.783 **6.** (1 to 1) **7.** (40.00 to 40.00)

8. (2.38 to 2.45) **9.** (50.0 to 50.0) **10.** (5.0 to 5.0) **11.** (85.5 to 86.5)

EXPLANATIONS

MCQ TYPE QUESTIONS

1. Max BM at the centre of beam

$$= 10 \times 1.5 - 20 \times 0.5 \times 0.25 = 12.5 \text{ kN-m}$$

Max slope in real beam occurs at supports P & Q (both)

Therefore, max slope in real beam = shear force

Or reaction at supports on conjugate beam

$$= \frac{1}{2}(1)\left(\frac{10}{EI}\right) + 0.5\left(\frac{10}{EI}\right) + \frac{2}{3}(0.5)\left(\frac{2.5}{EI}\right)$$

$$= \frac{10.833}{EI} = \frac{10.833}{30 \times 1000} = 3.61 \times 10^{-4} \text{ radian}$$

2. Figure I is the FBD of left part of Fig – II

Therefore, $\theta_{max1} = \theta_{max2}$

$$y_{max1} = y_{max2}$$

3.

$$\sigma_x + \sigma_y = \sigma_1 + \sigma_2$$

$$5 + 5 = 10 + \sigma_2$$

$$\sigma_2 = 0$$

$$\tau_{max} = \frac{\sigma_1 - \sigma_2}{2}$$

$$= \frac{10 - 0}{2} = 5 \text{ MPa}$$

4.

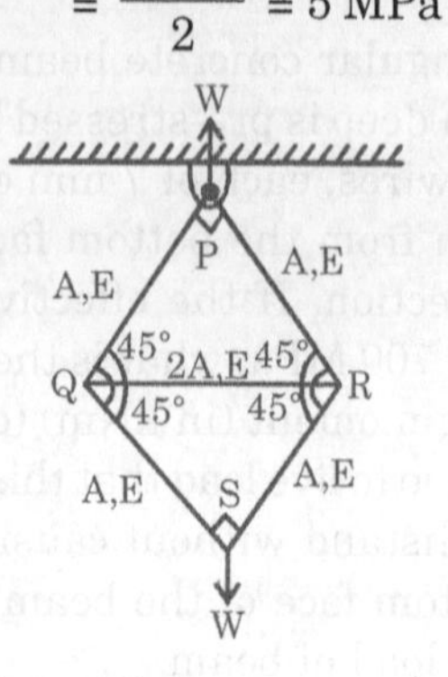

At joint S, $\Sigma F_X = 0$

$$F_{SQ} = F_{SR}$$

$$\Sigma F_Y = 0$$

$$F_{SQ} \cos 45° + F_{SR} \cos 45° = W$$

$$F_{SQ} + F_{SR} = W\sqrt{2}$$

$$F_{SQ} = F_{SR} = W\sqrt{2}$$

Similarly at point P we get,

$$F_{SQ} = F_{SR} = W\sqrt{2}$$

Now at joint Q,

$$\Sigma F_X = 0$$

$$F_{QP} \cos 45° + F_{QS} \cos 45° + F_{QR} = 0$$

$$\frac{W}{2} + \frac{W}{2} + F_{QR} = 0$$

$$F_{QR} = -W \text{ kN}$$

$$= -100 \text{ kN}$$

Stress in $Q_R = \frac{F_{QR}}{2 \times 0.05}$

$$= 1000 \text{ kN}$$

Shortening due to compressive stress

$$= \frac{100 \times \sqrt{2} \times 1}{2 \times 0.05 \times 30 \times 1000}$$

$$= 0.0471 \text{ m}$$

5. Rollers are provided in *y*-direction

$\therefore$ Stiffness is zero in *y*-direction and hence, $k_y = 0$

Also motion is restricted in *x* and θ direction

$\therefore$ $k_x = k_\theta = \infty$

6. 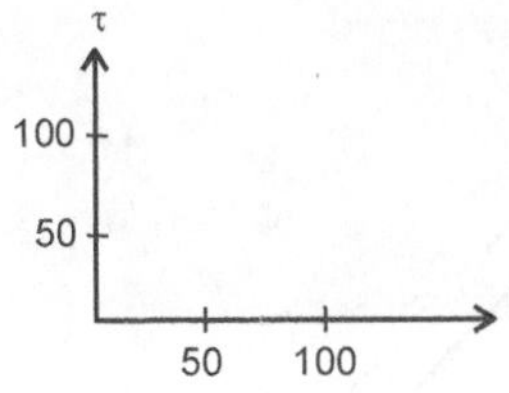

$$\sigma_x = \sigma_y = 50 \text{ MPa}$$

∴ Mohr's circle is a point

So, maximum shear stress is zero at all orientations.

7. Toughness – Resistance to impact
Hardness – Resistance to wear
Soundness – Resistance to weathering
Strength – Resistance to crushing

8. 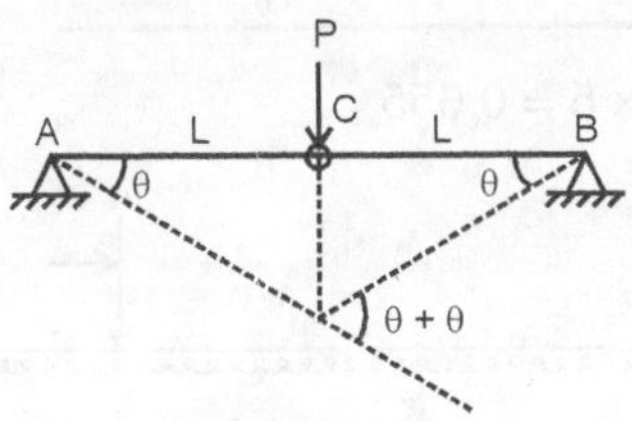

Overturning Moment = PL

Restoring moment = $k\ 4\theta$

$$\therefore \quad PL = 4k\theta$$

$$\theta = \frac{PL}{4k}$$

9. 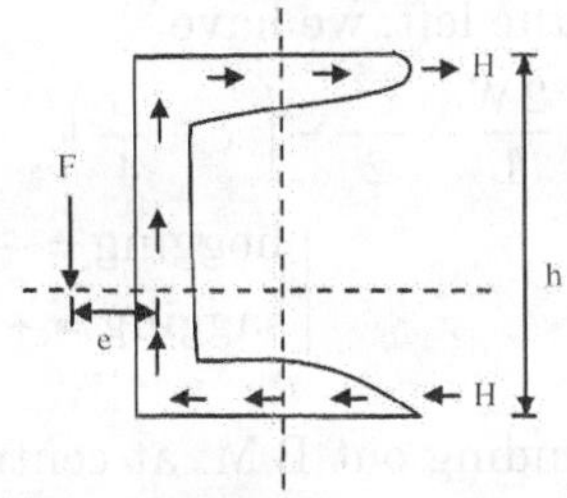

$$F.e = H \times h$$

$$\Rightarrow F = \frac{H \times h}{e}$$

Only possible location is P

10. 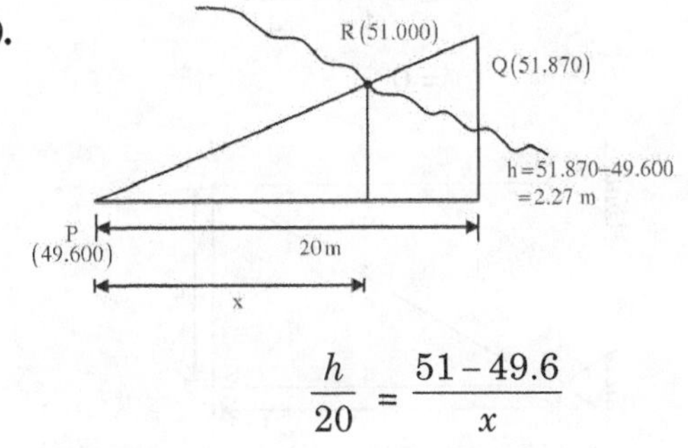

$$\frac{h}{20} = \frac{51 - 49.6}{x}$$

$$\Rightarrow \quad x = \frac{1.4 \times 20}{2.27} = 12.33 \text{ m}$$

11. 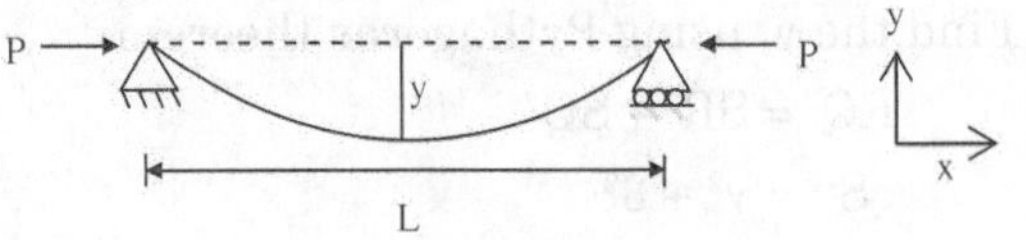

$$\frac{d^2y}{dx^2} = \frac{-P}{EI}.y$$

$$= -K^2.\ y$$

Solution of this differential equation is

$$y = a \sin kx + b \cos kx$$

at $\quad x = 0, y = 0$

$\Rightarrow \quad b = 0$

at $\quad x = L, y = 0$

$\Rightarrow \quad 0 = a \sin KL$

$\Rightarrow \quad KL = n\pi$

$\Rightarrow \quad K = \dfrac{n\pi}{L}$

$\therefore \quad y = a \sin \dfrac{n\pi}{L} x$

12. For no swinging, $\Sigma F_H = 0$

$\Rightarrow \quad 90 \cos 30° = F\cos = 40 \cos 45°$

$\Rightarrow \quad F \cos \theta = 49.658 \quad ...(i)$

Also, $\quad \Sigma F_V = 0$

$\Rightarrow \quad 100 = 90 \sin 30° + 40 \sin 45° + F \sin \theta$

$\Rightarrow \quad F \sin \theta = 26.715 \quad ...(ii)$

Solving (*i*) and (*ii*)

$$F = 56.389 \text{ kN},$$

$$\theta = 28.28°$$

13. First moment of area about the axis of bending for a beam cross-section ⇒ section modulus.
Moment of Inertia ⇒ second moment of area.
Polar moment of inertia $\boxed{I_{zz} = I_{xx} + I_{yy}}$ the correct choice is (*b*)

14. Shear force at P = 50 kN(↑) because the fig. shows a cantilever whoes free end is carrying a point load of 50 kN(↓).
Bending moment at P = 50 × 3 = 150 kN.m
So option (*b*) satisfy these answer.

15. Total length of cable is 10 m.

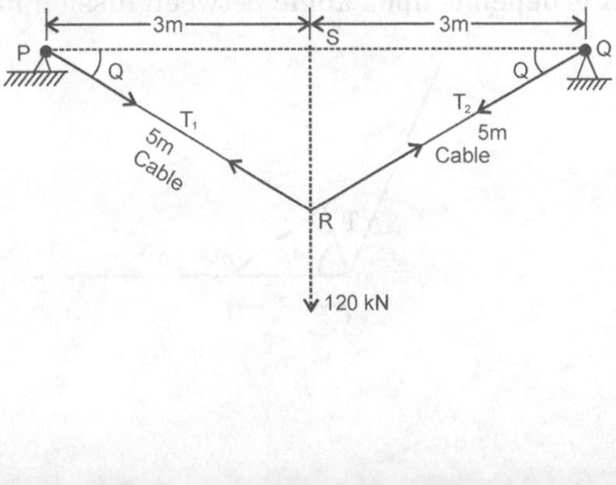

Find the y, using Pythagoras theorem

$$RQ^2 = SR^2 + SQ^2$$

$$S^{-2} = y^2 + 3^2$$

$$y = 4\text{m}$$

Then resolving the forces.

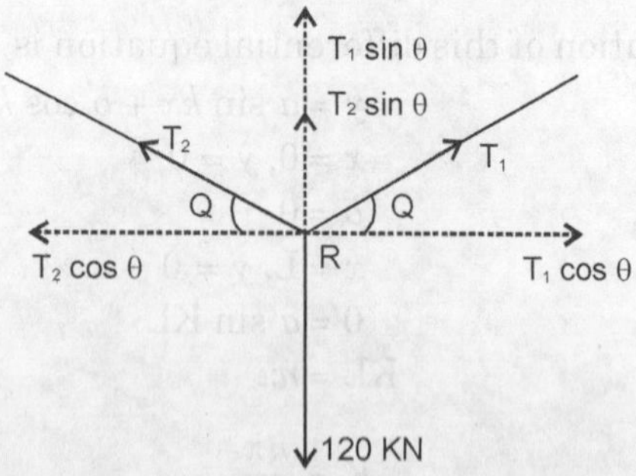

$$\Sigma x = 0$$

$$T_1 = T_2$$

$$\Sigma y = 0$$

$$T_1 \sin\theta + T_1 \sin\theta = 120$$

$$2\,T_1 = \frac{120}{\sin\theta}$$

$$\sin\theta = \frac{SR}{RQ} = \frac{4}{5}$$

$$2\,T_1 = \frac{120}{\sin\theta} \Rightarrow T_1 = \frac{60}{4/5} = 75$$

17. Poisson ratio, $\mu = \dfrac{\dfrac{dy}{y}}{\dfrac{dx}{x}}$

18. For fixed end column, $P_{C_1} = \dfrac{4\pi^2 EI}{L^2}$

For pinned end column, $P_{C_2} = \dfrac{\pi^2 EI}{L^2}$

$$\therefore \quad \frac{P_{C_1}}{P_{C_2}} = 4$$

19. Throat thinkness T = K × fillest size

K is depends upon angle between fussion forces.

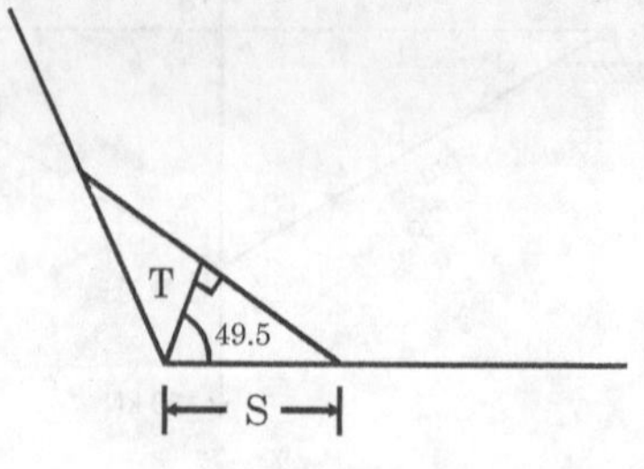

⇒

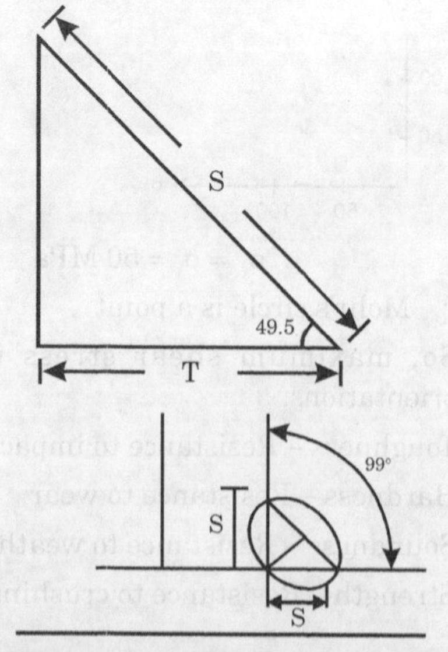

$$\therefore \quad T = \cos 49.5 \times 5 = 0.655$$

20.

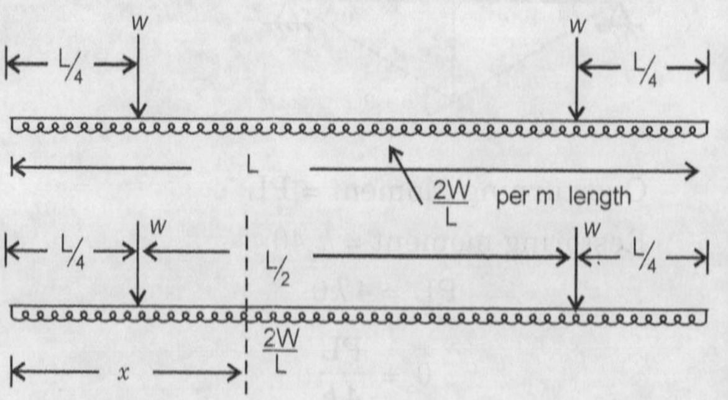

Take a section from the left, we have

$$M_x = \frac{2W}{L} \cdot \frac{x^2}{2} - \left(x - \frac{L}{4}\right)$$

$$\begin{bmatrix} \text{hogging} = -\text{ve} \\ \text{sagging} = +\text{ve} \end{bmatrix}$$

Putting $x = \dfrac{L}{2}$ for finding out B.M. at centre of plank

$$M_{L/2} = \frac{2W}{L} \cdot \frac{L^2}{4 \times 2} - W\left(\frac{L}{2} - \frac{L}{4}\right)$$

$$= \frac{WL}{4} - \frac{WL}{4}$$

$$= 0$$

21.

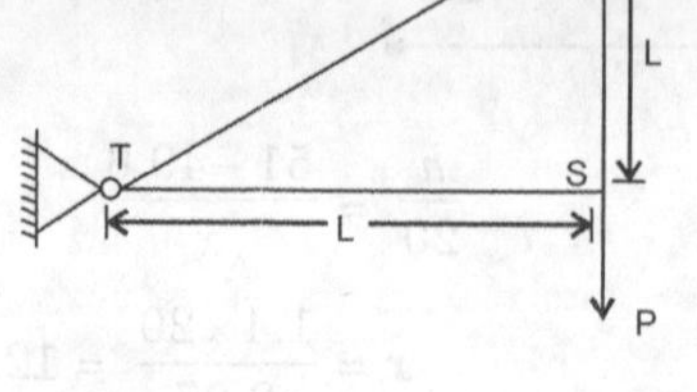

Considering joint S

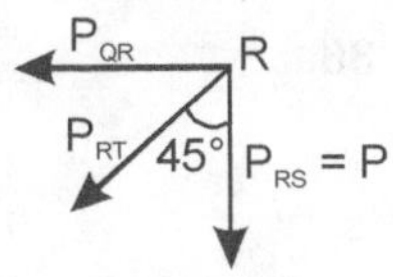

Force in the member RS = P

Considering joint R

Resolving horizontally and vertically, we have

$$\Sigma V = 0$$

$$\therefore \quad P_{RT} \cos 45° + P = 0$$

$$\Rightarrow \quad P_{RT} = -\sqrt{2}P$$

$$\Sigma H = 0$$

$$\therefore \quad P_{QR} + P_{RT} \sin 45° = 0$$

$$\Rightarrow \quad P_{QR} = -\sqrt{2}P \times \frac{1}{\sqrt{2}} = -P$$

$\therefore$ Force in member QR = P(compresion)

22.

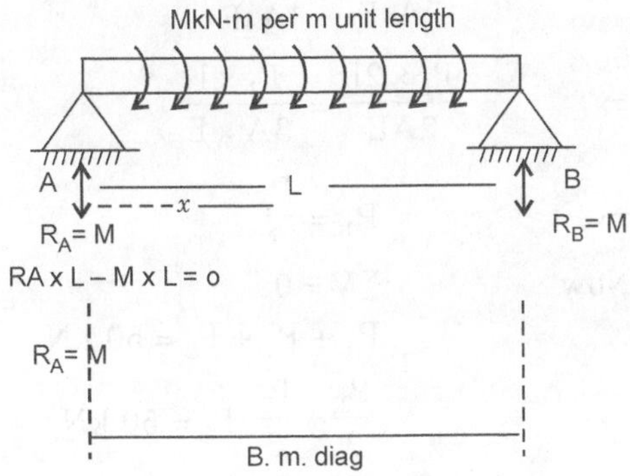

Take a section at a distance x from left support

$$M_x = R_A \times x - M \times x$$
$$= M \times x - M\,x$$
$$= 0$$

23.

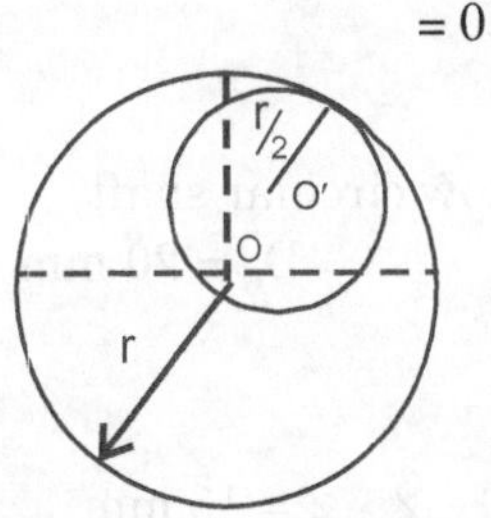

Area of circular disc = πr^2

$$\text{Area of hole} = \pi\left(\frac{r}{2}\right)^2 = \frac{\pi r^2}{4}$$

$$\therefore \quad \text{Remaining area} = \frac{3\pi r^2}{4}$$

Taking moment of the area about 'O' and dividing remaining area, to find the centroid

$$\therefore \quad \bar{x} = \frac{\frac{\pi r^2}{4} \times \frac{r}{2} - \pi r^2 \times 0}{\frac{3}{4} \times \pi r^2} = \frac{r}{6}$$

24. Hoop stress, $p = \frac{P.r}{t}$

$$= \frac{700 \times 10^3 \times 0.5}{0.025}$$
$$= 14 \times 10^6 \text{ N/m}^2$$
$$= 14 \text{ MPa}$$

29. Maximum strain energy per unit value

$$U_v = \frac{p^2}{2E}$$

where, E = modulus of elasticity

p = stress in the bar (yield stress)

$$\therefore \quad U_V = \frac{(250)^2}{2 \times 2 \times 10^5}$$
$$= 0.156 \text{ N.mm / mm}^3$$

30. By Torsion formula,

$$\frac{f_s}{\left(d/2\right)} = \frac{T}{J}$$

where, T = torsional moment

J = polar moment of inertia

f_s = shear stress = τ

d = diameter of shaft.

For solid circular section,

$$J = \frac{\pi}{32}d^4$$

$$\therefore \quad f_{s1} = \frac{T}{\frac{\pi}{32}d^4} \cdot \frac{d}{2}$$

$$\text{and} \quad f_s = \frac{16T}{\pi d^3} = \tau$$

Now, $T_1 = 4T$ and $d_1 = 2d$

$$\therefore \quad \tau_1 = \frac{16 \times 4T}{\pi.(2d)^3} = \frac{4}{8}\frac{16T}{\pi d^3}$$

$$\Rightarrow \quad \tau_1 = \frac{1}{2}.\tau$$

31. Shear stress in bar, $q = \frac{P}{2} \sin 2\theta$

For maximum shear, $\theta = 45°$

$$\therefore \quad q = \frac{173}{2} \sin 2 \times 45°$$
$$= 86.5 \text{ MPa.}$$

32. Buckling load, $P_E = \dfrac{\pi^2 EI}{l^2}$

Case I :

Effective length, $l = L$

$$\therefore \quad 200 = \frac{\pi^2 EI}{L^2} \quad ...(i)$$

Case II :

Effective length, $l = \dfrac{L}{2}$

$$P_E = \frac{4\pi^2 EI}{L^2} \quad ...(ii)$$

Divide equation (*ii*) by (*i*), we get

$$\frac{P_E}{200} = \frac{4\pi^2 EI}{L^2} \times \frac{L^2}{\pi^2 EI}$$

$$\therefore \quad P_E = \mathbf{800\ kN}$$

33. $P_i = k_{ij}\Delta_j$

where, P_i = force at point i

Δ_j = deformation at j

k_{ij} = stiffness coefficient

k_{ij} is defined as the force at i due to unit deformation at j

34. Relation between E, G, μ is

$$E = 2G(1 + \mu)$$

$$\therefore \quad G = \frac{E}{2(1+\mu)}$$

where, E = Young's Modulus

G = Shear modulus

$\mu = \dfrac{1}{m}$

= Poisson's ratio

35. *Given :*

Length of bar, L = 100 mm

Temperature increase, $\Delta t = 10°C$

Coefficient of thermal expansion,

$\alpha = 12 \times 10^{-6}$ per °C

$E = 2 \times 10^5$ MPa

$$\therefore \text{ Stress in bar, } P = \frac{\Delta E}{L}$$

$$= \frac{L\alpha t.E}{L}$$

$$= \alpha t E$$

$$= 12 \times 10^{-6} \times 2 \times 10^{-5} \times 10$$

$$= \mathbf{24\ MPa}$$

36.

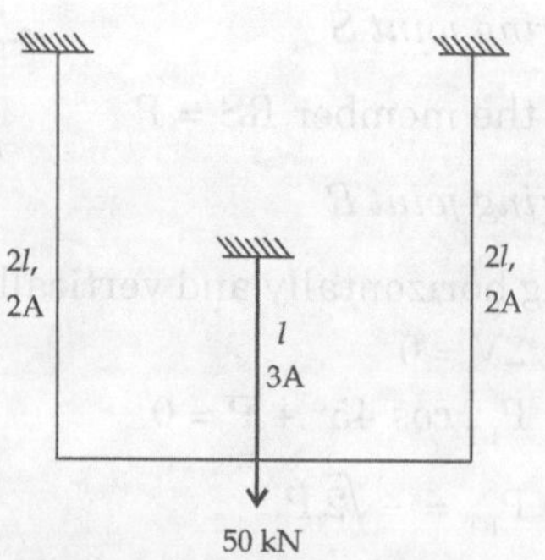

Let force in the outer rod = P_1

and force in the inner rod = P_2

Now, Deformation of outer rod = deformation of inner rod

i.e. $\Delta_1 = \Delta_2$

$$\Rightarrow \quad \frac{P_1 L_1}{A_1 E} = \frac{P_2 L_2}{A_2 E}$$

$$\Rightarrow \quad \frac{P_1 \times 2l}{2AE} = \frac{P_2 \times l}{3A \times E}$$

$$\Rightarrow \quad P_1 = \frac{P_2}{3}$$

Now $\Sigma V = 0$

$$\therefore \quad P_1 + P_1 + P_2 = 50 \text{ kN}$$

$$\Rightarrow \quad \frac{P_2}{3} + \frac{P_2}{3} + P_2 = 50 \text{ kN}$$

$$\Rightarrow \quad P_2 = \frac{50 \times 3}{5} = 30 \text{ kN}$$

$$\therefore \quad P_1 = \frac{30}{3} = 10 \text{ N}$$

37. *Given :*

Outer diameter of hollow circular shaft, $D_0 = 20$ mm

Thickness = 2mm

Torque, T = 92 NN-m.

Inner diameter, $d = 20 - 2 \times 2 = 16$ mm

∴ Maximum shear stress,

$$f = \frac{16\ TD_0}{\pi(D_0^4 - d^4)} = \frac{16 \times 92.7 \times 10^3 \times 20}{\pi \times (20^4 - 16^4)}$$

$$= 99.95 \approx 100 \text{ N/mm}^2$$

$$= \mathbf{100\ MPa}$$

38. Shear force, F = 3kN.

$$I = \frac{1}{36} bh^3$$

$$= \frac{1}{36} \times 40 \times 20^3$$

$$= 8888.88 \text{ mm}^4$$

Shear force at x distance from the apex

$$q = \frac{E}{3I}\, x\,(h - x)$$

$$= \frac{3 \times 1000}{3 \times 8888.88} \times \frac{2}{3} \times 20\left(20 - \frac{2}{3} \times 20\right)$$

$$= 10 \text{ MPa} = \mathbf{10\ N/mm^2}$$

39. No stress should be acting on it.

40. Given : $I = \begin{bmatrix} 30 & 0 \\ 0 & 30 \end{bmatrix}$

where, $p_1 = 30$ MPa,

$p_2 = 30$ MPa

and $q = 0$

Therefore diagram is

$\therefore$ Centre $\equiv (30, 0)$

Radius = 0 m.

41. Since, column B likes both end fixed, therefore

$$l_{eff} = \frac{L}{2}$$

$$\therefore \quad P = P_{cr} = \frac{\pi^2 EI}{\left(\frac{L}{2}\right)^2} = \frac{4\pi^2 EI}{L^2}$$

$$\Rightarrow \quad \alpha = 4$$

42. We know, $\dfrac{T}{J} = \dfrac{f}{Y}$

$$\Rightarrow \quad \frac{T}{\frac{\pi}{32}d^4} = \frac{f}{\frac{d}{2}}$$

$$\Rightarrow \quad \frac{16T}{\pi d^3} = f$$

43.

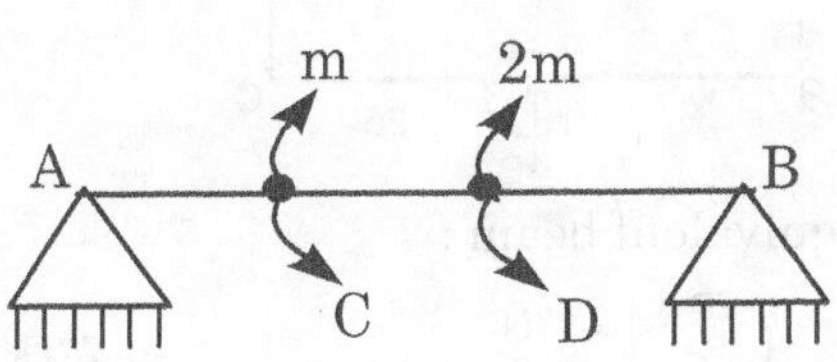

44. $I_{xx} = I_{cc} + A\bar{y}^2$

$$= \frac{1}{12}bd^3 + bd\left(\frac{d}{4}\right)^2 = \frac{7bd^3}{48}$$

45. Moment, M = P × 2L = 2PL

Member AC is rigid, so end C behave like fixed end

$\therefore$ Deflection at C = $\dfrac{PL^3}{3EI}$

(By double Integration)

46. Given : M = 16×10^6 N-mm

$$I = \frac{1}{12} \times 100 \times (150)^3$$

$$= 28125000 \text{ mm}^2$$

$$\therefore \quad f_{max} = \frac{M}{I}\, y_{max} = \frac{16 \times 10^6}{28125000} \times 75$$

$$= 42.67 \text{ N/mm}^2$$

$\therefore$ Tensile force,

$$P_T = \frac{f_{max}}{y_{max}} A\bar{y} = \frac{42.67}{75} \times (25 \times 50) \times \frac{25}{2}$$

$$= 8.9 \text{ kN}$$

47. $q = \dfrac{PL}{4k}\dfrac{F}{I_b}(A\bar{y})$

53. Rotation, $\theta = \dfrac{TL}{GJ_1 + GJ_2}$

55. Shear force at section aa′ = 0

....(because at aa′, BMD = constant)

$\therefore$ Shear force at section bb′

$$= \frac{200 - 100}{2} \times 1 = 50 \text{ kN}$$

56.

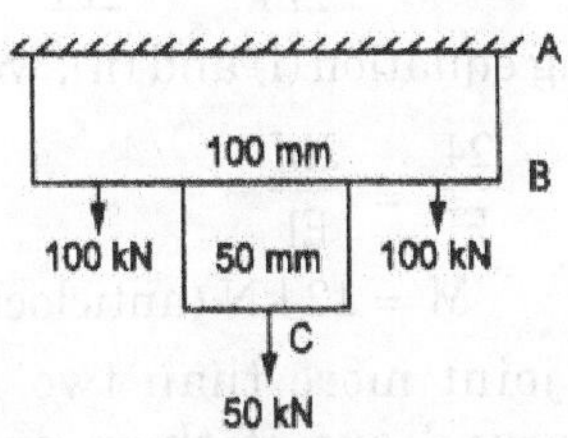

F.B.D. of AB

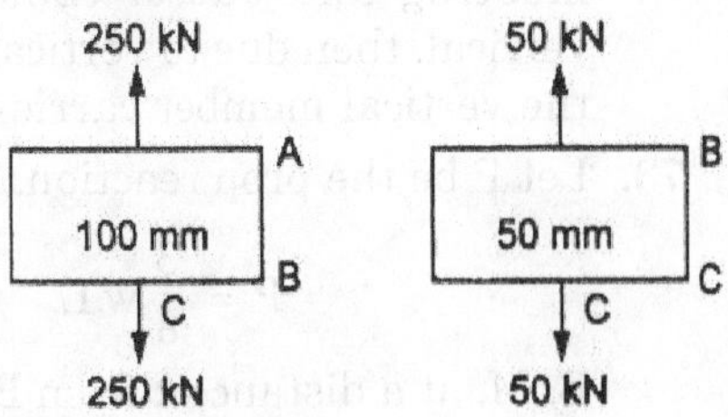

$$\therefore \quad \text{Stress in AB} = \frac{250 \times 10^3}{100 \times 100} = 25 \text{ N/mm}^2$$

and stress in BC = $\dfrac{50 \times 10^3}{50 \times 50} = 20$ N/mm^2

So, maximum tensile stress = 25 N/mm^2

57. From the figure, it is clear that in XY only axial stress is present.

64.

S.N.	Shape	Shape factor
I.	Circle	1.67
II	Diamand	2.0
III.	I - section	As per dimentions but less than 1.5
IV.	Rectangle	1.5

65. Influence Line Diagram = $M - R_A L$.

It is irrespective of x

70. Values of horizontal and vertical components of reaction at A are equal at θ = 45 Since the vertical and horizontal components are given by sin θ and cos θ respectively, and sin θ = cos θ at 45 , therefore the components are equal

71. Deflection at free end of cantilever

$$y = \frac{wL^3}{3EI}$$

$$= \frac{9 \times 2^3}{3EI} = \frac{24}{EI} \quad .....(i)$$

Let M be the anticlockwise moment applied at B such that, $y_B = 0$

Deflection at B due to moment

$$M = \frac{ML^2}{2EI} = \frac{M \times 4}{2EI} = \frac{2M}{EI} \quad ....(ii)$$

Equating equation (i) and (ii), we get

$$\frac{24}{EI} = \frac{2M}{EI}$$

or, $M = 12$ kN (anticlock wise)

72. If at a joint more than two members are meeting and out of them one member is vertical, then due to vertical load at that joint, the vertical member carries a zero force.

73. Let P be the prop reaction, then

$$P = \frac{3}{8} w.L.$$

B. M. at a distance x from B is given by,

$$M_x = E.I \frac{d^2y}{dx^2}$$

$$= -P.x + \frac{w.x^2}{2}$$

$$= \frac{3}{8} w.L.x + \frac{w.x^2}{2}$$

Now, at x = 0, $M_B = 0$

and at $x = L$, $M_A = \frac{3}{8} \cdot \frac{w.L^2}{2} + \frac{w.L^2}{2} = \frac{w.L^2}{8}$

74. The load 'P' is acting at the free end. The B.M. diagram is a triangle having a maximum ordinate P.L. at end A.

$$\sum_B^A (A\bar{X}) = \frac{1}{2}.P.L.L\left(\frac{2}{3}L\right) = \frac{PL^3}{3}$$

$$\therefore \quad y_B^A = \frac{1}{El} \cdot \frac{PL^3}{3} = \frac{PL^3}{3El}$$

$\therefore$ Deflection at free end = $\frac{PL^3}{3El}$

NUMERICAL TYPE QUESTIONS

1.
$$\sigma_n = \frac{(\sigma_x + \sigma_y)}{2} + \frac{(\sigma_x - \sigma_y)}{2} \cos 2\theta$$

$$= \frac{(100 - 100)}{2} + \frac{(100 + 100)}{2} \cos 90° = 0$$

2.

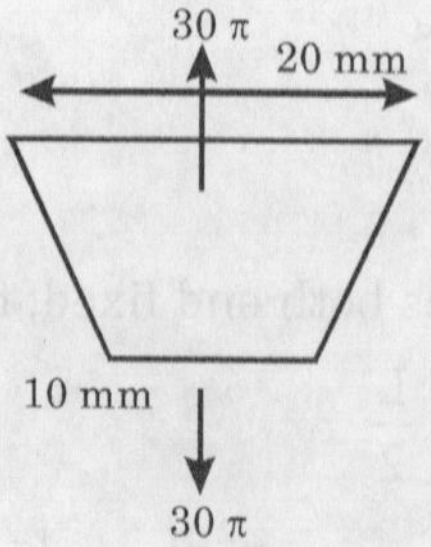

$$\Delta_1 = \frac{4P.L}{\pi d_1 d_2 \times E}$$

$$\Delta_1 = \frac{4 \times 30\pi \times 2 \times 10^6}{\pi \times 20 \times 10 \times 2 \times 10^6} = 6\text{mm}$$

$$\Delta_2 = \frac{P.L}{A \times E}$$

$$\Delta_1 = \frac{4 \times 30\pi \times 1.5 \times 10^6}{\pi \times 10 \times 10 \times 2 \times 10^6} = 9 \text{ mm}$$

$$\Delta = \Delta_1 + \Delta_2 = 15 \text{ mm}$$

3.

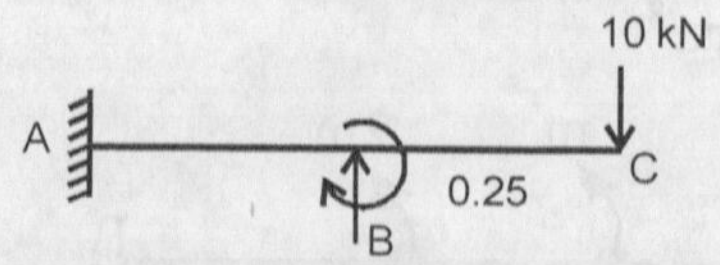

Equivalent beam :

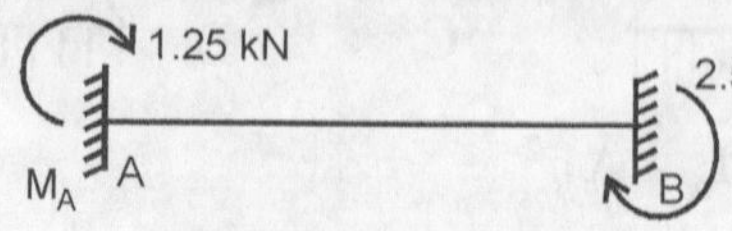

$$\varepsilon M_A = 0$$

$$\Rightarrow 10 \times 1 - R_B \times 0.75 + 1.25 = 0$$

$$R_B = 15 \text{ kN}$$

$$\therefore \quad R_A = -5 \text{ kN}$$

Point of contraflexure

$$M_A + R_A \times x = 0$$

$$1.25 - 5 \times x = 0$$

$$x = 0.25\ m$$

4. $$\delta = \frac{PL^3}{3EI}$$

Force developed in spring

$$= k \times \delta = k \times \frac{PL^3}{3EI}$$

$$= \frac{2 \times 50 \times (200)^3}{3 \times 200 \times 10^3 \times \frac{5 \times 10^3}{12}} = 3.2$$

5.

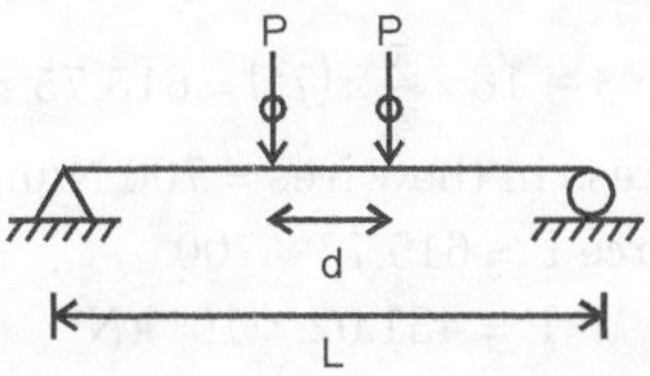

L = 24 m

d = 5 m

ρ = 3 kN

Z = 16.2 cm^3

Max. bending moment

For max. B.M., the resultant of the load system and point load should be equidistant from the centre

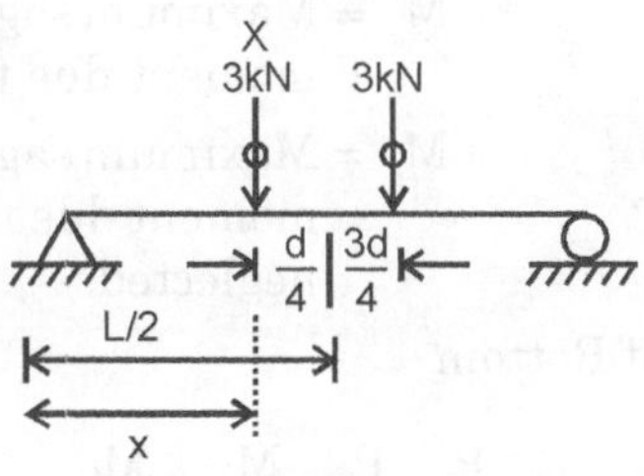

Considering the maximum. BM, occurs below the X - load

$\therefore \quad SM_A = 0$

$R_B \times 24 - (3\text{ kN}) \times 15.75 - (3\text{ kN} \times 10.75) = 0$

$R_B = 3.3125$ kN

and $\quad R_A = 6 - R_B$

$= 2.6875$ kN

$\therefore \quad M_x = R_A \times x$

$= 2.6875 \times (12 - 1.25)$

$= 2.6875 \times 10.75$

$= 28.89$ kN

$$\therefore \quad \sigma = \frac{M}{Z} = \frac{28.89 \times 10^6 \times 10^3}{16.2}$$

$= 1783.33 \times 10^6$ N/m^2

$= 1.783$ GPa

6.

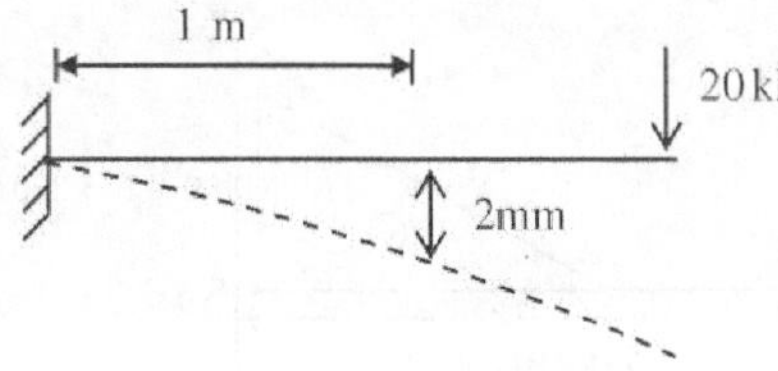

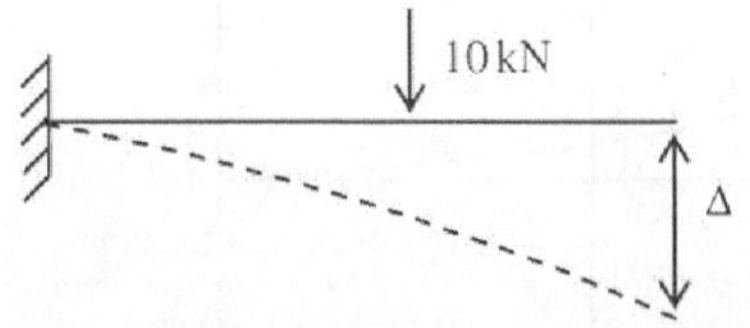

By Bettis theorem,

$20 \times \Delta = 10 \times 2 \Rightarrow \Delta = 1$ mm.

7. Polar moment of Inertia in cm^4 of rectangular section $b = 2$ cm $d = 6$ cm

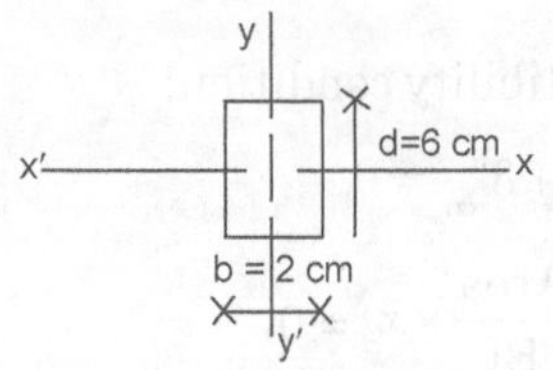

$$I_P \Rightarrow I_{xx} + I_{yy}$$

$$\Rightarrow \frac{bd^3}{12} + \frac{db^3}{12} = \frac{2 \times 6^3}{12} + \frac{6 \times 2^3}{12} \Rightarrow 40 \text{ cm}^4$$

9. The Axial load (in kN) in the member PQ for the arrangement/assembly shown in fig.

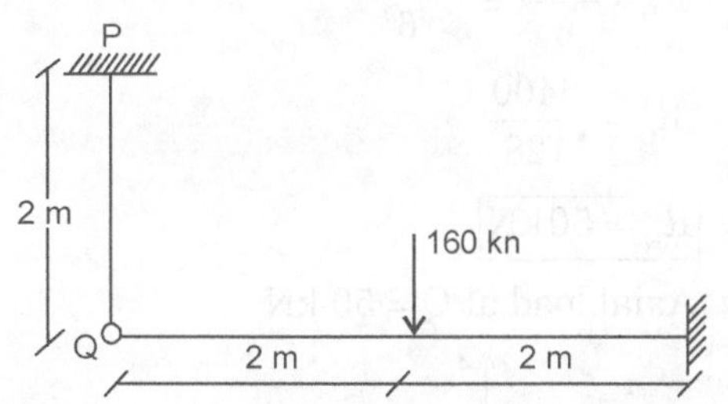

Hinge at Q (Solve by Area moment theorem)

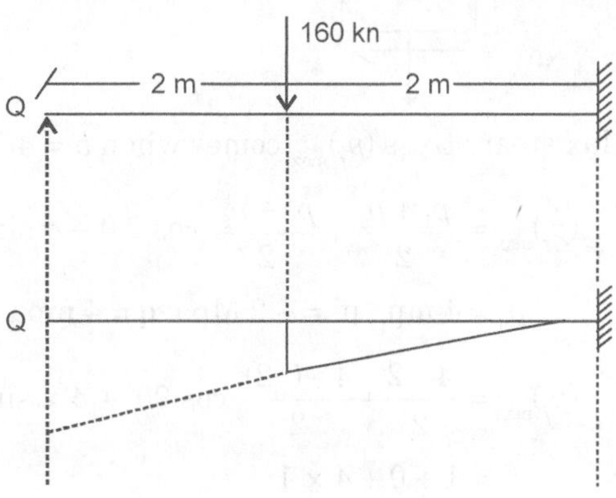

1st case remove Pin support at Q draw the bending moment diagram

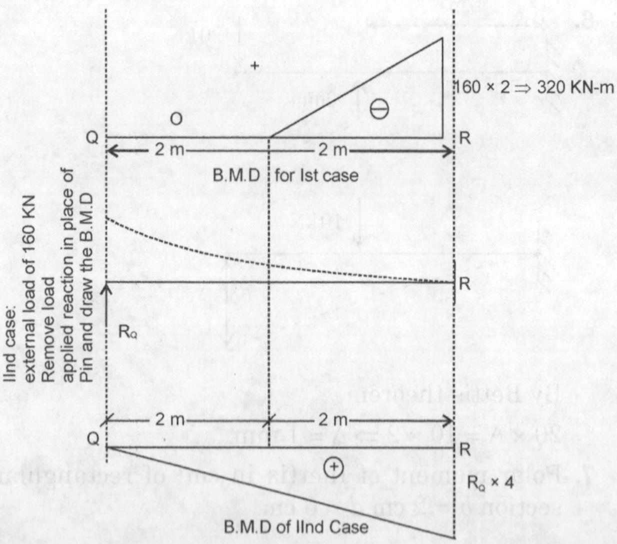

Compatibility condition:

$$\delta_{Q_1}^{R} + \delta_{Q_2}^{R} = 0$$

$$\sum_{Q}^{R} \frac{\text{Area}}{\text{EI}} \times \bar{x} = 0$$

$$\sum_{Q_1}^{R} \frac{\text{Area}}{\text{EI}} \times \bar{x} + \sum_{Q_2}^{R} \frac{\text{Area}}{\text{EI}} \times \bar{x} = 0$$

$$\left\{0 + \frac{1}{2} \times \frac{-320 \times 2}{\text{EI}} \times \left[2 + \frac{2}{3} \times 2\right]\right\} + \frac{1}{2} \times \frac{(R_Q \times 4) \times 4}{\text{EI}} \times \frac{2}{3} \times 4 = 0$$

$$R_Q \times \frac{128}{6} = \frac{6400}{6}$$

$$R_Q = \frac{6400}{128}$$

$$\boxed{R_Q = 50\,\text{kN}}$$

Axial load at Q = 50 kN

10.

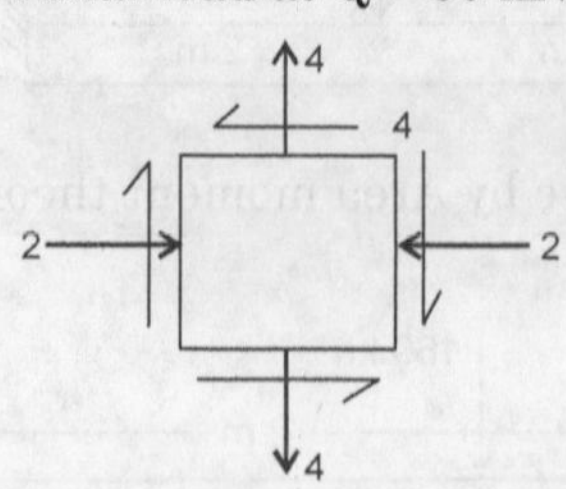

Max stear stress $(p_f)_{max}$ comes when b = 45°.

$$(p_f)_{max} = \frac{p_1 + p_2}{2} + \frac{p_1 + p_2}{2} \cos 2\theta + q \sin 2\theta$$

$$p_1 = 4 \text{ mp}_4,\ p_2 = -2 \text{ Mpa},\ q = 4 \text{ mpa}$$

$$(p_f)_{max} = \frac{4-2}{2} + \frac{4-(-2)}{2} \cos 90° + 4 \times \sin 90°$$

$$= 1 + 0 + 4 \times 1$$

$$(p_f)_{max} = 5 \text{ Mpa}$$

11.

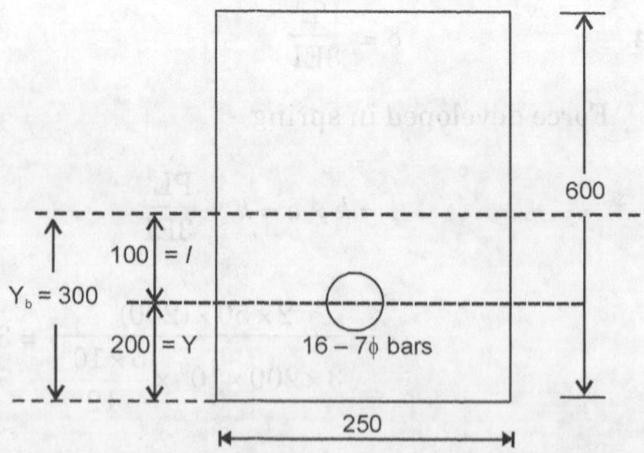

Area a steel wires = $16 \times \frac{\pi}{4} \times (7^2) = 615.75 \text{ mm}^2$

Effective pre-stress in the wires = 700 N/mm²

Pre-stressing force P = 615.75 × 700

P = 431.02 × 10³ kN

Area of cross-section

$$A = 250 \times 600 = 150 \times 10^3 \text{ mm}^2$$

Second moment of Inertia

$$I = \frac{250 \times (600)^3}{12} = 4.5 \times 10^9 \text{mm}^4$$

$$\text{Section modulus } (Z_b) = \frac{I}{y_b} = \frac{4.5 \times 10^9}{300} = 15 \times 10^6 \text{ mm}^3$$

M_q = Maximum sagging bending moment due to live load.

M_g = Maximum sagging bending moment due to deal load is neglected.

Stress of Bottom

$$f_b = \frac{P}{A} + \frac{P_l}{Z_b} - \frac{Mg}{Z_b} - \frac{Mq}{Z_b}$$

f_b = 0 because section of beam can with stand without causing tensile stress at bottom face.

M_g = 0 because it is neglected.

$$0 = \frac{431.02 \times 10^3}{150 \times 10^3} + \frac{431.02 \times 10^3 \times 100}{15 \times 10^6} - \frac{Mq}{15 \times 10^6}$$

$$\frac{Mq}{15 \times 10^6} = 2.87 + 2.87$$

Mq = 86.152 × 10⁶ N-M

Mq = 86.1 kN-M (One decimel place)

■■

3 CHAPTER

Structural Analysis

MCQ TYPE QUESTIONS

2016

1. Consider the plane truss with load P as shown in the figure. Let the horizontal and vertical reactions at the joint B be H_B and V_B, respectively and V_C be the vertical reaction at the joint C.

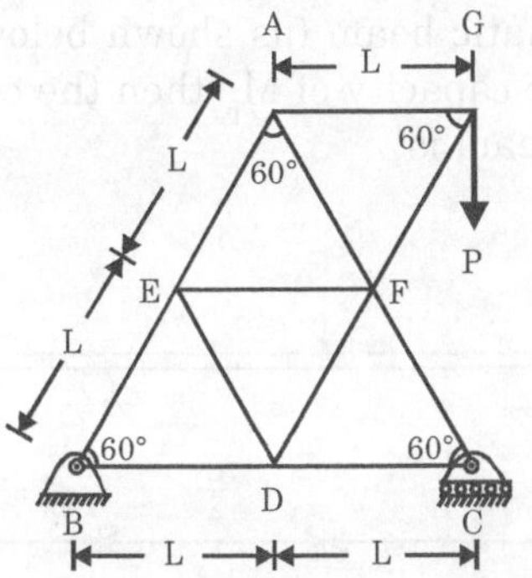

Which one of the following sets gives the correct values of V_B, H_B and V_C?

(*a*) $V_B = 0$; $H_B = 0$; $V_C = P$

(*b*) $V_B = P/2$; $H_B = 0$; $V_C = P/2$

(*c*) $V_B = P/2$; $H_B = P$ (sin 60°); $V_C = P/2$

(*d*) $V_B = P$; $H_B = P$ (cos 60°); $V_C = 0$

2. A plane truss with applied loads is shown in the figure.

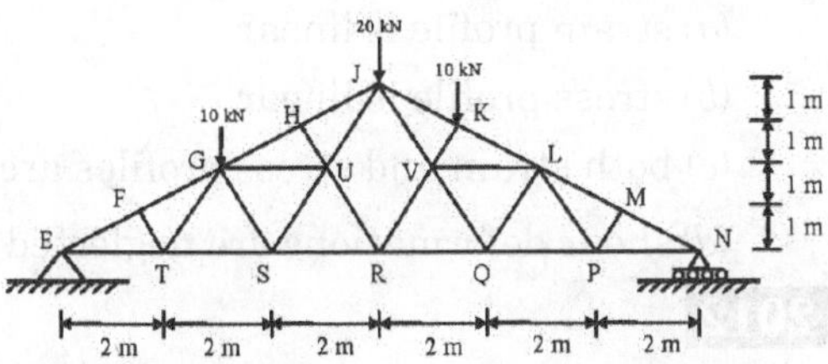

The members which do not carry any force are

(*a*) FT, TG, HU, MP, PL

(*b*) ET, GS, UR, VR, QL

(*c*) FT, GS, HU, MP, QL

(*d*) MP, PL, HU, FT, UR

3. The kinematic indeterminacy of the plane truss shown in the figure is

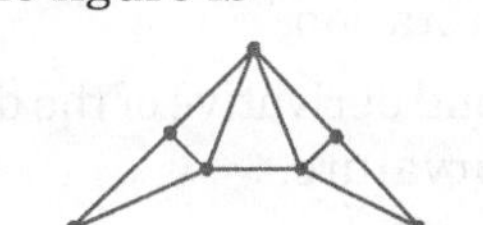

(*a*) 11 (*b*) 8

(*c*) 3 (*d*) 0

4. The portal frame shown in the figure is subjected to a uniformly distributed vertical load *w* (per unit length).

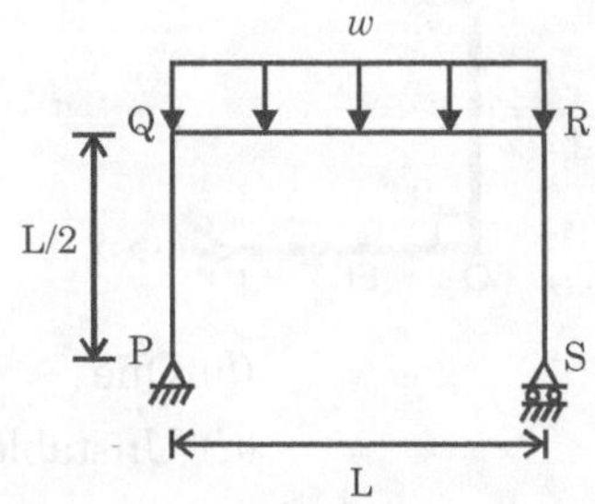

The bending moment in the beam at the joint 'Q' is

(*a*) zero (*b*) $\frac{wL^2}{24}$ (hogging)

(*c*) $\frac{wL^2}{12}$ (hogging) (*d*) $\frac{wL^2}{8}$ (sagging)

2015

5. For the beam shown below, the stiffness coefficient K_{22}, can be written as

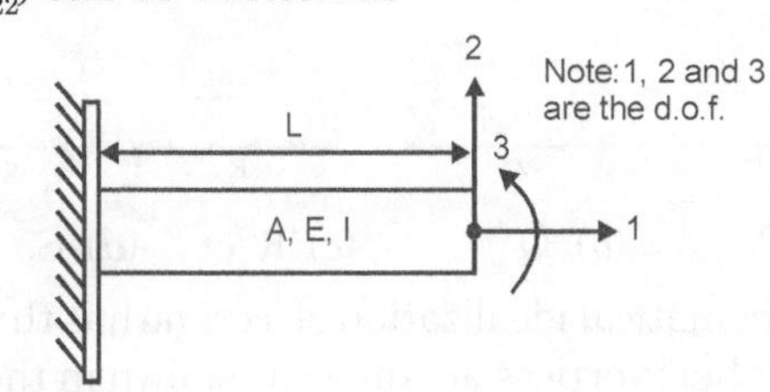

(*a*) $\frac{6EI}{L^2}$ (*b*) $\frac{12EI}{L^3}$

(*c*) $\frac{3EI}{L}$ (*d*) $\frac{EI}{6L^2}$

6. A fixed end beam is subjected to a load, W at 1/3rd span from the left support as shown in the figure. The collapse load of the beam is

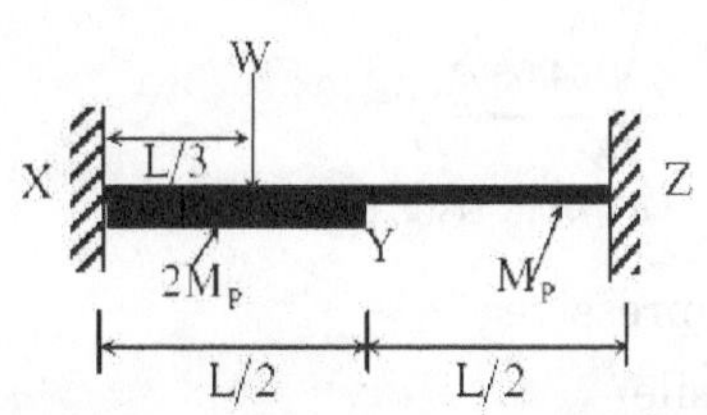

(*a*) 16.5 M_p/L (*b*) 15.5 M_p/L

(*c*) 15.0 M_p/L (*d*) 16.0 M_p/L

2014

7. The degree of static indeterminacy of a rigid jointed frame PQR supported as shown in the figure is

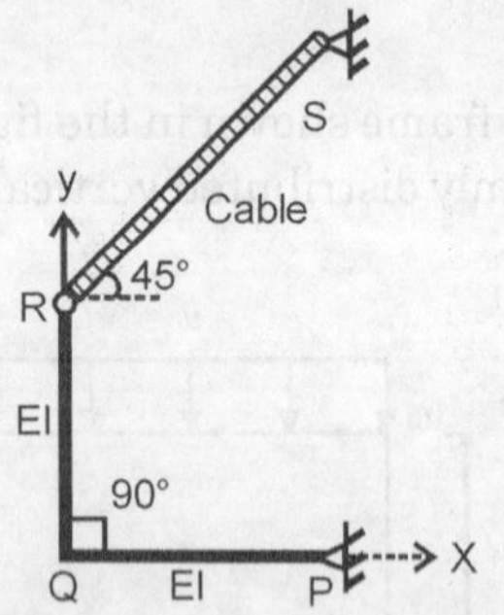

(*a*) Zero (*b*) One

(*c*) Two (*d*) Unstable

8. In a beam of length L, four possible influence line diagrams for shear force at a section located at a distance of $\frac{L}{4}$ from the left end support (marked as P, Q, R and S) are shown below. The correct influence line diagram is

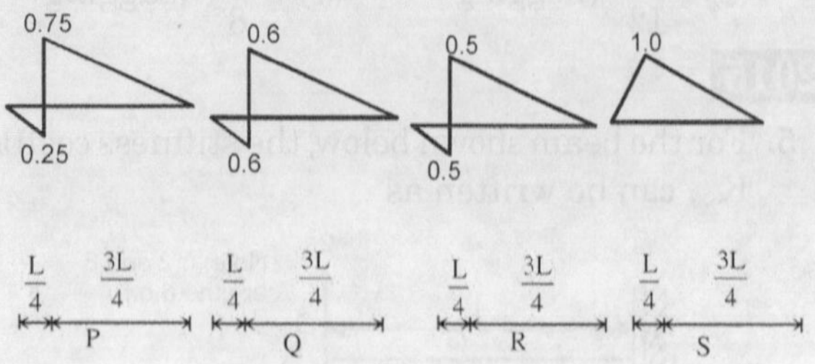

(*a*) P (*b*) Q (*c*) R (*d*) S

9. Mathematical idealization of a crane has three bars with their vertices arranged as shown in the figure with a load of 80 kN hanging vertically. The coordinates of the vertices are given in parentheses. The force in the member QR, F_{QR} will be

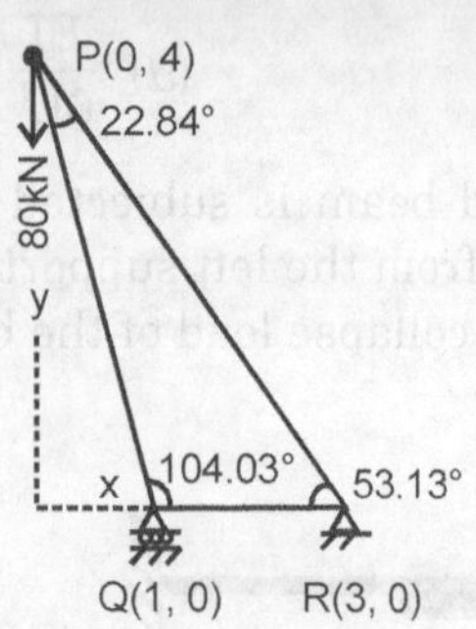

(*a*) 30 kN Compressive

(*b*) 30 kN Tensile

(*c*) 50 kN Compressive

(*d*) 50 kN Tensile

10. Considering the symmetry of a rigid frame as shown below, the magnitude of the bending moment (in kNm) at P (preferably using the moment distribution method) is

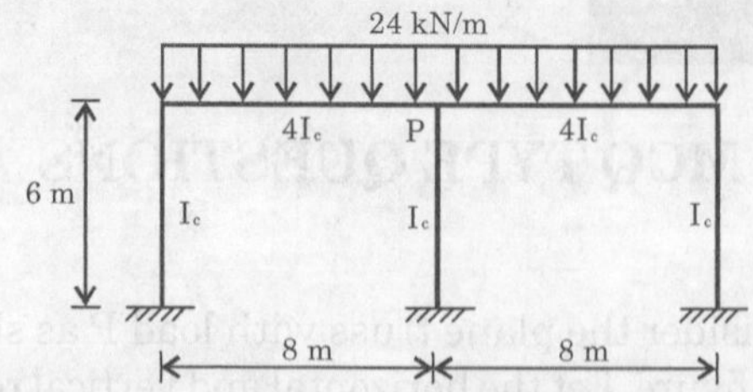

(*a*) 170 (*b*) 172

(*c*) 176 (*d*) 178

11. A prismatic beam (as shown below) has plastic moment capacity of M_p, then the collapse load P of the beam is

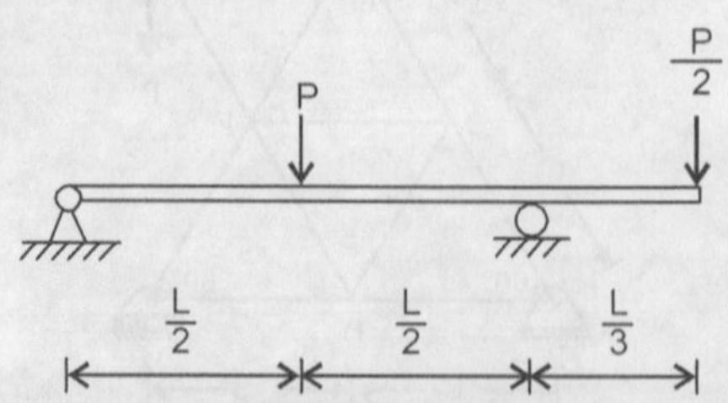

(*a*) $\frac{2M_p}{L}$ (*b*) $\frac{4M_p}{L}$

(*c*) $\frac{6M_p}{L}$ (*d*) $\frac{8M_p}{L}$

2013

12. The 'plane section remains plane' assumption in bending theory implies:

(*a*) strain profile is linear

(*b*) stress profile is linear

(*c*) both strain and stress profiles are linear

(*d*) shear deformations are neglected

2012

13. The following statements are related to bending of beams:

I. The slope of the bending moment diagram is equal to the shear force.

II. The slope of the shear force diagram is equal to the load intensity

III. The slope of the curvature is equal to the flexural rotation

IV. The second derivative of the deflection is equal to the curvature.

The only **FALSE** statement is

(*a*) I (*b*) II

(*c*) III (*d*) IV

14. A simply supported beam is subjected to a uniformly distributed load of intensity w per unit length, on half of the span from one end. The length of the span and the flexural stiffness are denoted as l and EI respectively. The deflection at mid-span of the beam is

(*a*) $\frac{5}{6144}\frac{wl^4}{EI}$ (*b*) $\frac{5}{768}\frac{wl^4}{EI}$

(*c*) $\frac{5}{384}\frac{wl^4}{EI}$ (*d*) $\frac{5}{192}\frac{wl^4}{EI}$

15. The sketch shows a column with a pin at the base and rollers at the top. It is subjected to an axial force P and a moment M at mid-height. The reaction(s) at R is/are

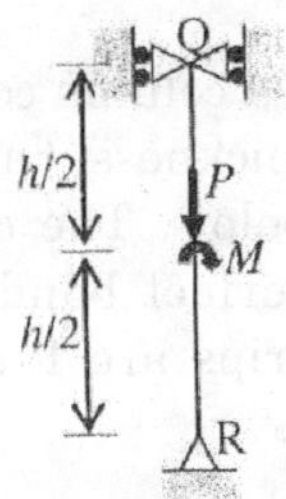

(*a*) a vertical force equal to P

(*b*) a vertical force equal to P/2

(*c*) a vertical force equal to P and a horizontal force equal to M/h

(*d*) a vertical force equal to P/2 and a horizontal force equal to M/h

16. A symmetric frame PQR consists of two inclined members PQ and QR, connected at 'Q' with a rigid joint, and hinged at 'P' and 'R'. The horizontal length PR is l. If a weight W is suspended at 'Q', the bending moment at 'Q' is

(*a*) $\frac{Wl}{2}$ (*b*) $\frac{Wl}{4}$

(*c*) $\frac{Wl}{8}$ (*d*) zero

2011

17. The value of W that results in the collapse of the beam shown in the adjoining figure and having a plastic moment capacity of M_P is

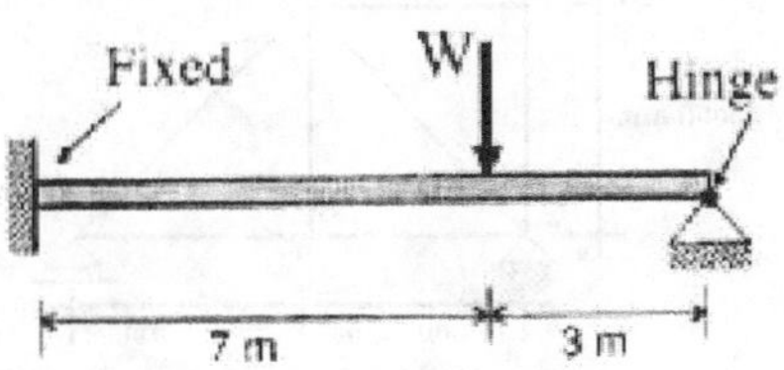

(*a*) (4/21) M_P (*b*) (3/10) M_P
(*c*) (7/21) M_P (*d*) (13/21) M_P

18. For the cantilever bracket, PQRS, loaded as shown in the adjoining figure (PQ = RS = L, and, QR = 2L), which of the following statements is **FALSE**?

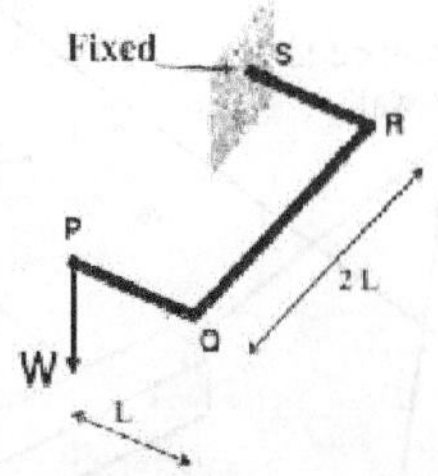

(*a*) The portion RS has a constant twisting moment with a value of 2WL.

(*b*) The portion QR has a varying twisting moment with a maximum value of WL.

(*c*) The portion P has a varying bending moment with a maximum value of WL

(*d*) The portion PQ has no twisting moment.

2010

19. The number of independent elastic constants for a linear elastic isotropic and homogeneous material is

(*a*) 4 (*b*) 3 (*c*) 2 (*d*) 1

20. A three hinged parabolic arch having a span of 25 m and a rise of 5 m carries a point load of 10 kN at quarter span from the left end as shown in the figure. The resultant reaction at the left support and its inclination with the horizontal are respectively

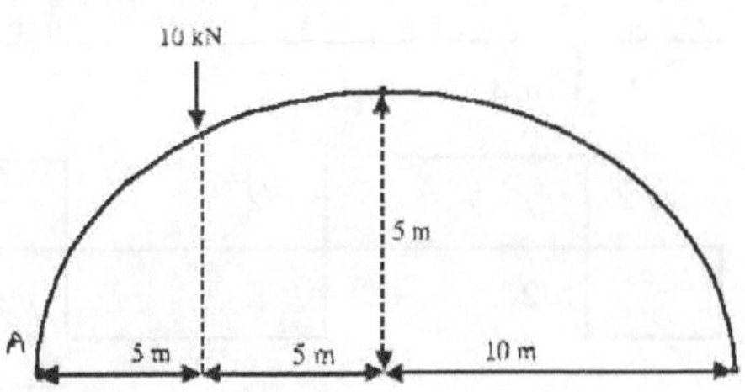

(*a*) 9.01 kN and 56.31° (*b*) 9.01 kN and 33.69°
(*c*) 7.50 kN and 56.31° (*d*) 2.50 kN and 33.69°

2009

21. The point within the cross sectional plane of a beam through which the resultant of the external loading on the beam has to pass through to ensure pure bending without twisting of the cross-section of the beam is called

(*a*) moment centre

(*b*) centroid

(*c*) shear centre

(*d*) elastic center

22. The degree of static indeterminacy of a rigidly jointed frame in a horizontal plane and subjected to vertical loads only, as shown in figure below, is

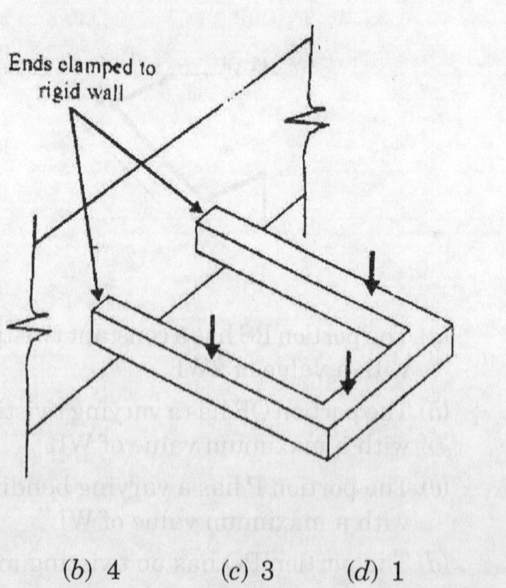

(*a*) 6 (*b*) 4 (*c*) 3 (*d*) 1

23. Group I gives the shear force diagrams and Group II gives the diagrams of beams with supports and loading. Match the Group I with Group II

Group I

P.
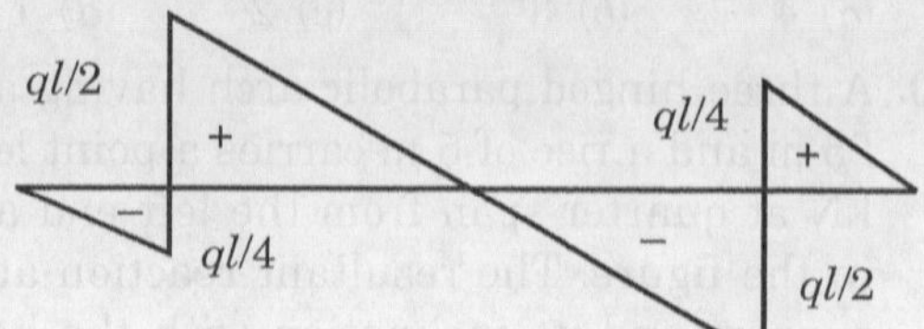

Q.
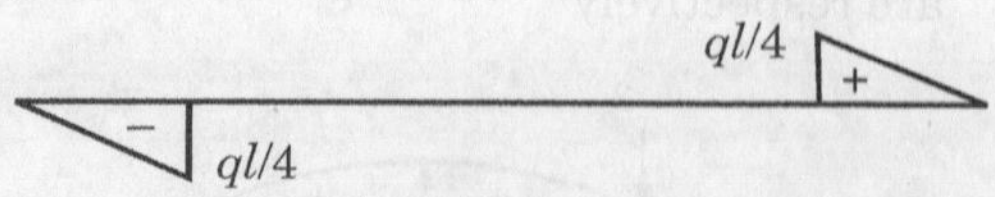

R.
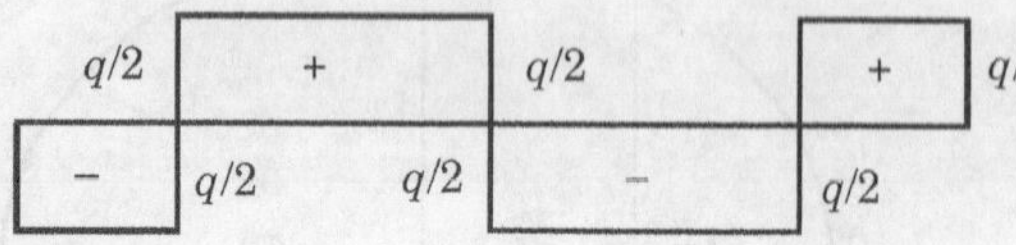

S.
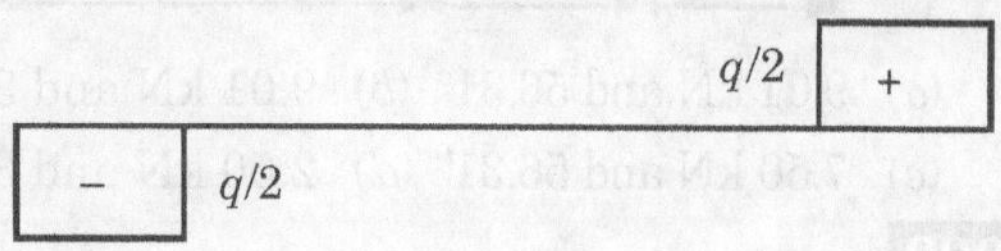

Group II

1.

2.
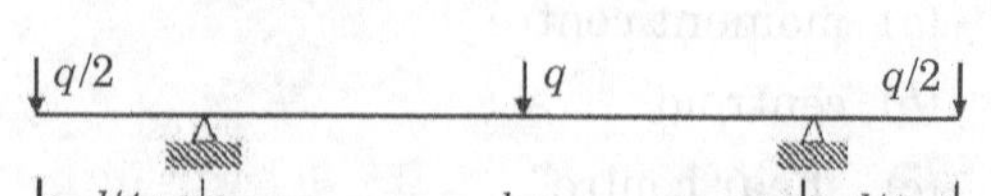

3.
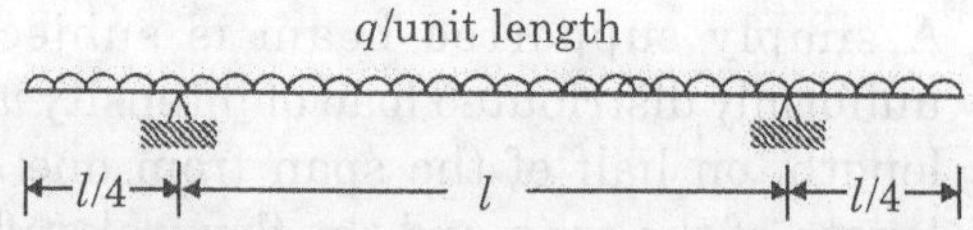

4.
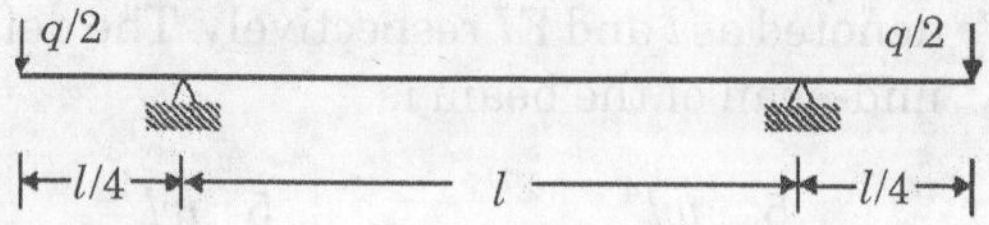

Codes :

	P	Q	R	S
(*a*)	3	1	2	4
(*b*)	3	4	2	1
(*c*)	2	1	4	3
(*d*)	3	4	3	4

2008

24. Cross-section of a column consisting of two steel strips, each of thickness t and width b as shown in the figure below. The critical loads of the column with perfect bond and without bond between the strips are P and P_o respectively. The ratio P/P_o is

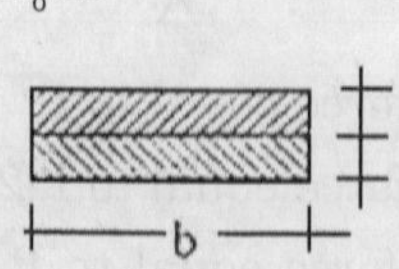

(*a*) 2 (*b*) 4 (*c*) 6 (*d*) 8

25. The degree of static indeterminacy of the rigid frame having two internal hinges as shown in the figure below, is

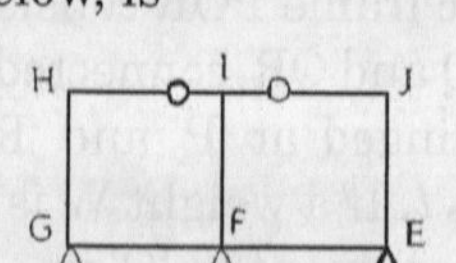

(*a*) 8 (*b*) 7
(*c*) 6 (*d*) 5

26. The members EJ and IJ of a steel truss shown in the figure below are subjected to a temperature rise of 30°C. The coefficient of thermal expansion of steel is 0.000012 per°C per unit length. The displacement(mm) of joint E relative to joint H along the direction HE of the truss, is

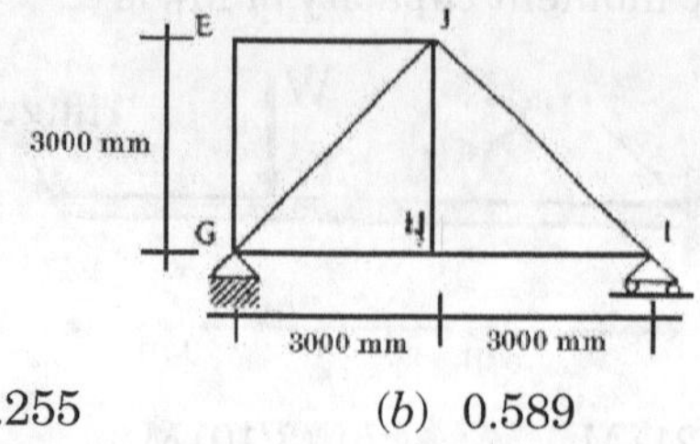

(*a*) 0.255 (*b*) 0.589
(*c*) 0.764 (*d*) 1.026

27. The span(s) to be loaded uniformly for maximum positive (upward) reaction at support P, as shown in the figure below, is (are)

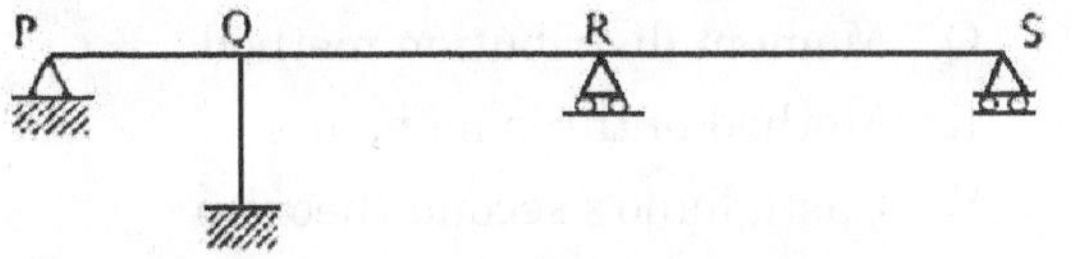

(a) PQ only (b) PQ and QR

(c) QR and RS (d) PQ and RS

28. A vertical rod PQ of length L is fixed at its top end P and has a flange fixed to the bottom end Q. A weight W is dropped vertically from a height h (<L) on to the flange. The axial stress in the rod can be reduced by

(a) increasing the length of the rod

(b) decreasing the length of the rod

(c) decreasing the area of cross-section of the rod

(d) increasing the modulus of elasticity of the material

29. The maximum tensile stress at the section X-X shown in the figure below is

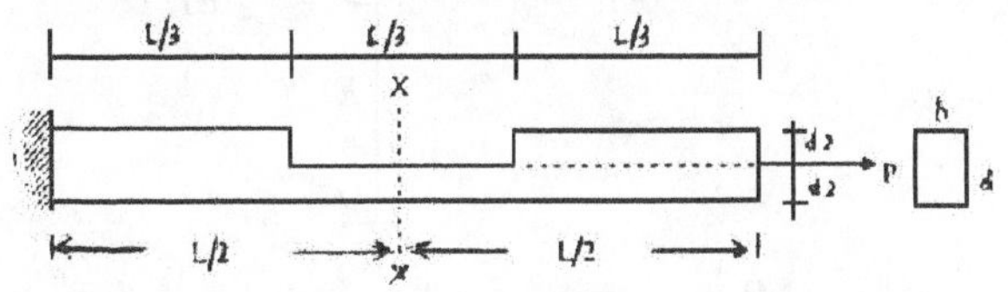

(a) $\frac{8P}{bd}$ (b) $\frac{6P}{bd}$

(c) $\frac{4P}{bd}$ (d) $\frac{2P}{bd}$

30. The stepped cantilever is subjected to moments, M as shown in the figure below. The vertical deflection at the free end (neglecting the self weight) is

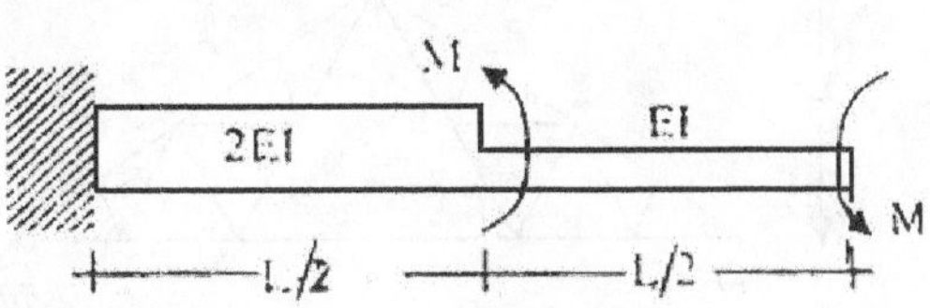

(a) $\frac{ML^2}{8EI}$ (b) $\frac{ML^2}{4EI}$

(c) $\frac{ML^2}{2EI}$ (d) Zero

2007

31. U_1 and U_2 are the strain energies stored in a prismatic bar due to axial tensile forces P_1 and P_2, respectively. The strain energy U stored in the same bar due to combined action of P_1 and P_2 will be

(a) $U = U_1 + U_2$ (b) $U = U_1U_2$

(c) $U < U_1 + U_2$ (d) $U > U_1 + U_2$

32. The right triangular truss is made of members having equal cross sectional area of 1550 mm^2 and Young's modulus of 2×10^5 MPa. The horizontal deflection of the joint Q is

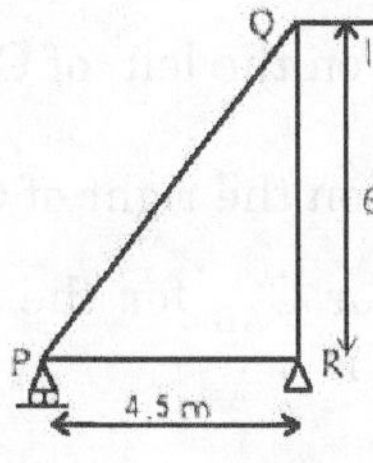

(a) 2.47 mm (b) 10.25 mm

(c) 14.31 mm (d) 15.68 mm

33. The influence line diagram (ILD) shown is for the member

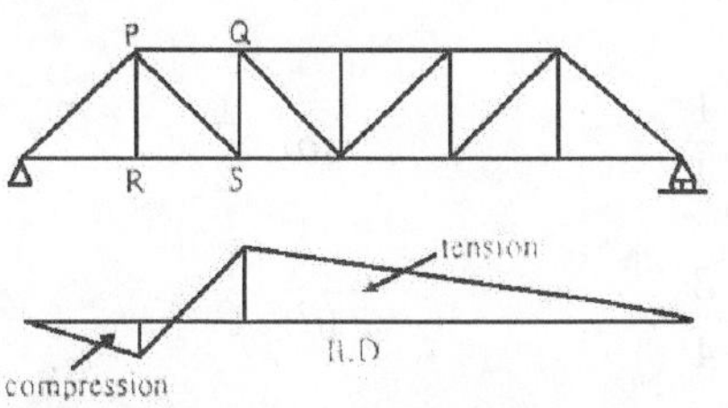

(a) PS (b) RS

(c) PQ (d) QS

2006

34. Vertical reaction developed at B in the frame below due to the applied load of 100 kN (with 150, 000 mm^2 cross-sectional area and 3.125×10^9 mm^4 moment of inertia for both members) is

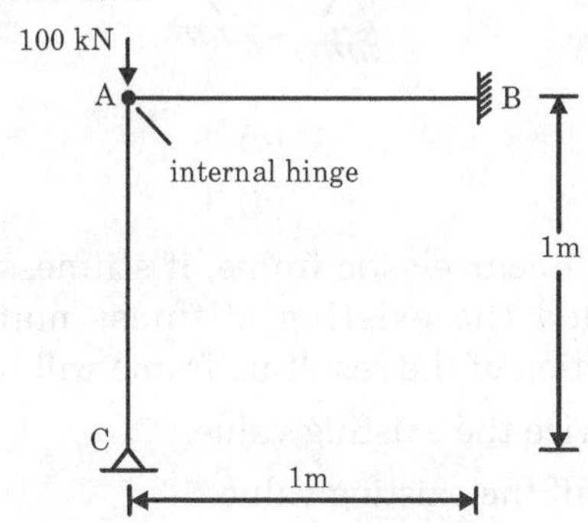

(a) 5.9 kN (b) 302 kN

(c) 66.3 kN (d) 94.1 kN

35. Consider the beam ABCD and the influence line as shown below. The influence line pertains to

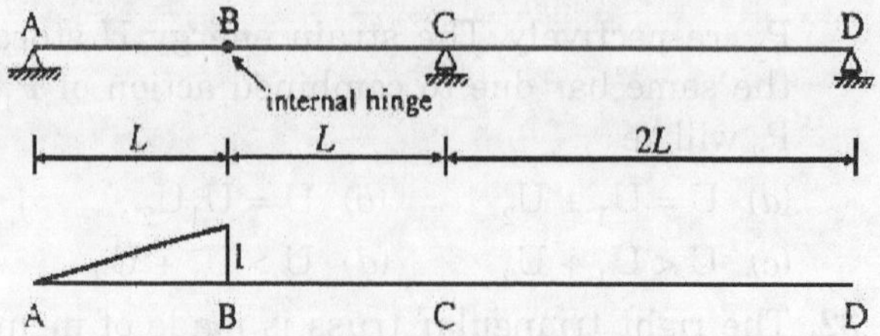

(*a*) reaction at A, R_A

(*b*) shear force at B, V_B

(*c*) shear force on the left of C, V_C^-

(*d*) shear force on the right of C, V_C^+

36. Carry-over factor C_{AB} for the beam shown in the figure below is

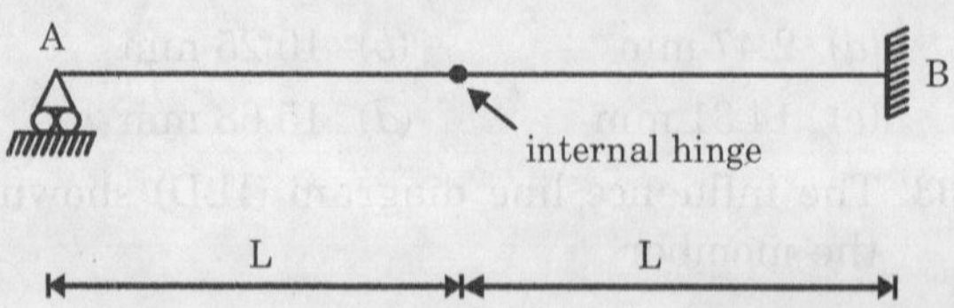

(*a*) $\frac{1}{4}$ (*b*) $\frac{1}{2}$

(*c*) $\frac{3}{4}$ (*d*) 1

2005

37. Considering beam as axially rigid, the degree of freedom of a plane frame shown below is

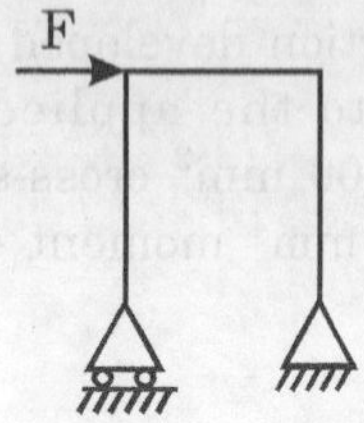

(*a*) 9 (*b*) 8

(*c*) 7 (*d*) 6

38. For a linear elastic frame, if stiffness matrix is doubled the existing stiffness matrix, then deflection of the resulting frame will be

(*a*) twice the existing value

(*b*) half the existing value

(*c*) the same as existing value

(*d*) indeterminate value

39. Match the following :

Group 1

P. Slope deflection method

Q. Moment distribution method

R. Method of three moments

S. Castigliano's second theorem

Group 2

1. Force method

2. Displacement method

Codes :

	P	**Q**	**R**	**S**
(*a*)	1	2	1	2
(*b*)	1	1	2	2
(*c*)	2	2	1	1
(*d*)	2	1	2	1

40. All members of the frame shown below have the same flexural rigidity EI and length L. If a moment M is applied at joint B, then rotation of the joint is

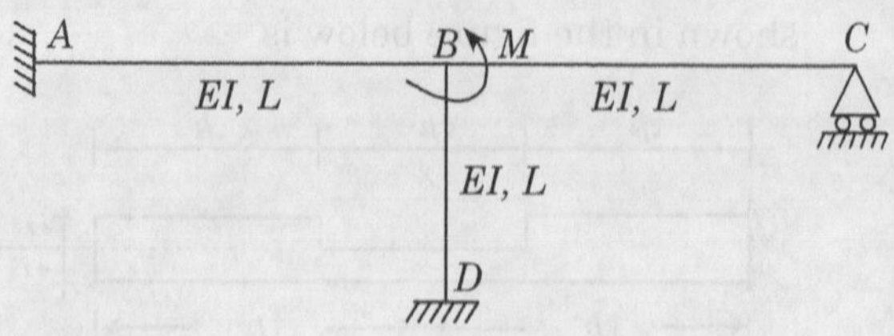

(*a*) $\frac{ML}{12EI}$ (*b*) $\frac{ML}{11EI}$

(*c*) $\frac{ML}{8EI}$ (*d*) $\frac{ML}{7EI}$

Linked Answer Q. 41-42

A truss is shown in the figure. Members are to equal cross section A and same modulus of elasticity E. A vertical force P is applied at point C.

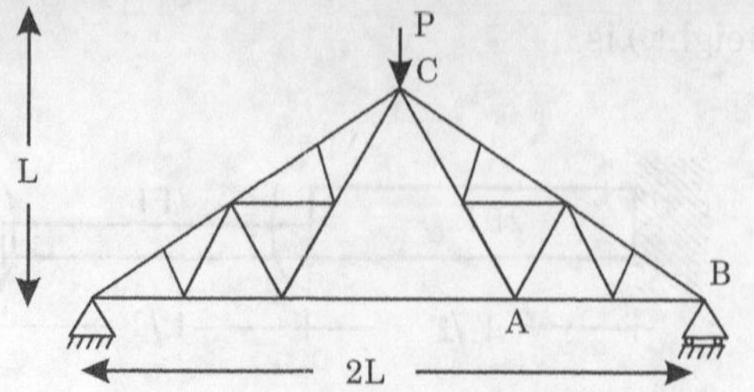

41. Force in the member AB of the truss is

(*a*) $\frac{P}{\sqrt{2}}$ (*b*) $\frac{P}{\sqrt{3}}$

(*c*) $\frac{P}{2}$ (*d*) P

42. Deflection of the point C is

(a) $\frac{(2\sqrt{2}+1)}{2}\frac{PL}{EA}$ (b) $\sqrt{2}\frac{PL}{EA}$

(c) $(2\sqrt{2}+1)\frac{PL}{EA}$ (d) $(\sqrt{2}+1)\frac{PL}{EA}$

2004

43. For the plane truss shown in the figure, the number of zero force members for the given loading is

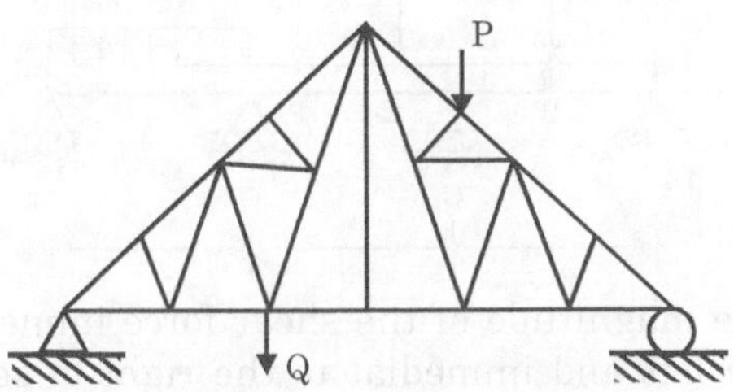

(a) 4 (b) 8

(c) 11 (d) 13

44. The unit load method used in structural analysis is

(a) applicable only to statically indeterminate structures

(b) another name for stiffness method

(c) an extension of Maxwell's reciprocal theorem

(d) derived from Castigliano's theorem

45. For linear elastic systems, the type of isplacement function for the strain energy is

(a) linear (b) quadratic

(c) cubic (d) quartic

46. For a linear elastic structural system, inimization of potential energy yields

(a) compatibility conditions

(b) constitutive relations

(c) equilibrium equations

(d) strain-displacement relations

47. For the plane frame with an overhang as shown below, assuming negligible axial deformation, the degree of static indeterminacy, d, and the degree of kinematic indeterminacy, k, are

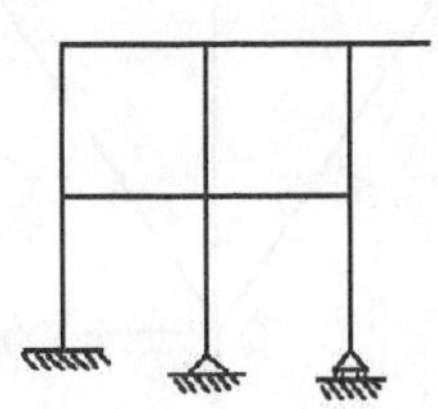

(a) $d = 3$ and $k = 10$ (b) $d = 3$ and $k = 13$

(c) $d = 9$ and $k = 10$ (d) $d = 9$ and $k = 13$

48. A homogeneous, simply supported prismatic beam of width B, depth D and span L is subjected to a concentrated load of magnitude P. The load can be placed anywhere along the span of the beam. The maximum flexural stress developed in beam is

(a) $\frac{2}{3}\frac{PL}{BD^2}$

(b) $\frac{3}{4}\frac{PL}{BD^2}$

(c) $\frac{4}{3}\frac{PL}{BD^2}$

(d) $\frac{3}{2}\frac{PL}{BD^2}$

49. For the linear elastic beam shown in the figure, the flexural rigidity, EI, is 781250 kN-m^2. When w = 10 kN/m, the vertical reaction R_A at A is 50 kN. The value of R_A for w = 100 kN/m is

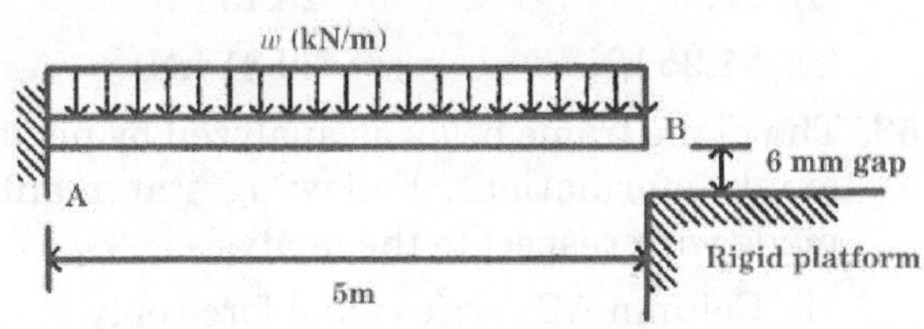

(a) 500 kN (b) 425 kN

(c) 250 kN (d) 75 kN

50. In a two dimensional stress analysis, the state of stress at a point is shown below.

If s = 120 MPa and t = 70 MPa, then σ_x and σ_y are respectively

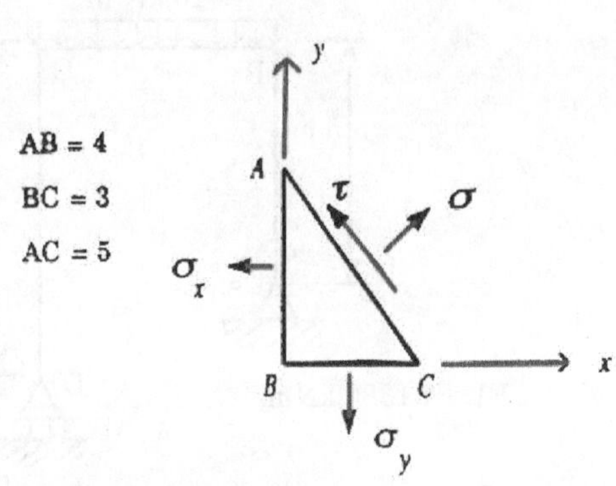

(a) 26.7 MPa and 172.5 MPa

(b) 54 MPa and 128 MPa

(c) 67.5 MPa and 213.3 MPa

(d) 16 MPa and 138 MPa

51. A circular solid shaft of span L = 5 m is fixed at one end and free at the other end. A twisting moment T = 100 kN-m is applied at the free end. The torsional rigidity GJ is 50000 kN-m²/rad. Following statements are made for this shaft.

I. Maximum rotation is 0.01 rad

II. Torsional strain energy is 1 kN-m

With reference to the above statements, which of the following applies?

(*a*) Both statements are true
(*b*) Statement I is true but II is false
(*c*) Statement II is true but I is false
(*d*) Both the statements are false

52. A three-hinged parabolic arch ABC has a span of 20 m and a central rise of 4 m. The arch has hinges at the ends at the centre. A train of two point loads of 20 kN and 10 kN, 5 m apart, crosses this arch from left to right, with 20 kN load leading. The maximum thrust induced at the supports is

(*a*) 25.00 kN (*b*) 28.13 kN
(*c*) 31.25 kN (*d*) 32.81 kN

53. The plane frame below is analyzed by neglecting axial deformations. Following statements are made with respect to the analysis.

I. Column AB carries axial force only

II. Vertical deflection at the center of beam BC is 1 mm

With reference to the above statements, which of the following applies?

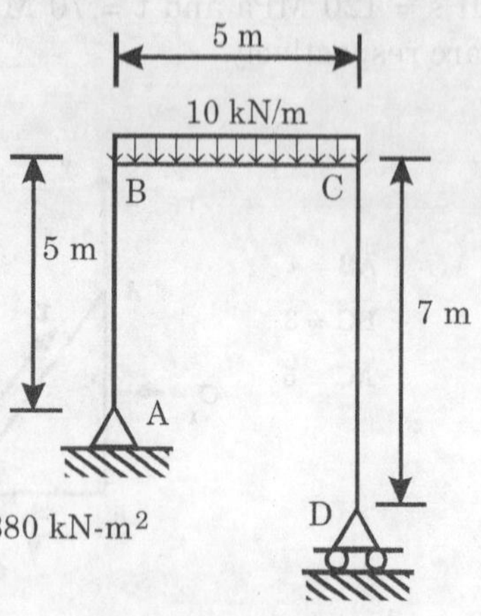

(*a*) Both the statements are true
(*b*) Statement I is true but II is false
(*c*) Statement II is true but I is false
(*d*) Both the statements are false

Common Data Q. 54-55

A three-span continuous beam has an internal hinge at B. Section B is at the mid-span of AC. Section E is at the mid-span of CG. The 20 kN load is applied at section B whereas 10 kN loads are applied at sections D and F as shown in the figure. Span GH is subjected to uniformly distributed load of magnitude 5 kN/m. For the loading shown, shear force immediate to the right of section E is 9.84 kN upwards and the sagging moment at section E is 10.31 kN-m.

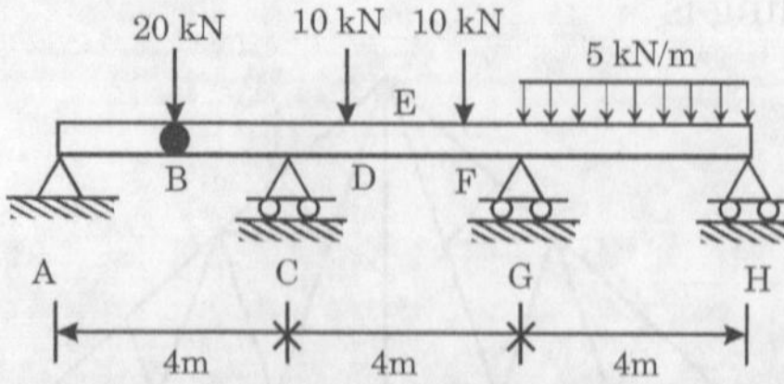

54. The magnitude of the shear force immediate to the left and immediat to the right of section B are, respectively

(*a*) 0 and 20 kN (*b*) 10 kN and 10 kN
(*c*) 20 kN and 0 (*d*) 9.84 kN and 10.16 kN

55. The vertical reaction at support H is

(*a*) 15 kN upward (*b*) 9.84 kN upward
(*c*) 15 kN downward (*d*) 9.84 kN downward

2003

56. Muller Breslau principle in structural analysis is used for

(*a*) drawing influence line diagram for any force function
(*b*) writing virtual work equation
(*c*) super-position of load effects
(*d*) none of these

57. In a redundant joint model, three bar members are pin connected at Q as shown in the figure. Under some load placed at Q, elongation of the members MQ and OQ are found to be 48 mm and 35 mm respectively. Then horizontal displacement *u* and the vertical displacement v of the node Q, in mm, will be respectively

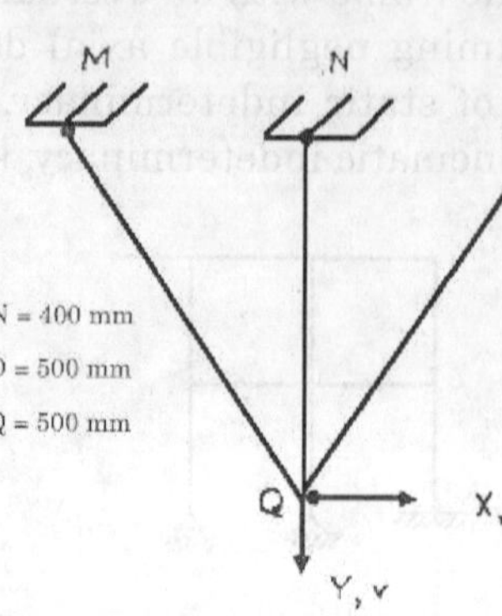

(*a*) –6.64 and 56.14 (*b*) 6.64 and 56.14
(*c*) 0.0 and 59.41 (*d*) 59.41 and 0.0

58. A "H" shaped frame of uniform flexural rigidity EI is loaded as shown in the figure. The relative outward displacement between points K and O is

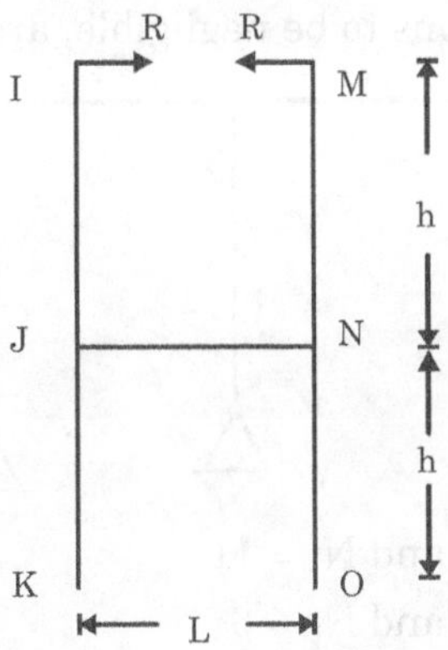

(a) $\frac{RLh^2}{EI}$ (b) $\frac{RL^2h}{EI}$

(c) $\frac{RLh^2}{3EI}$ (d) $\frac{RL^2h}{3EI}$

59. A simply supported beam of uniform rectangular cross-section of width b and depth h is subjected to linear temperature gradient, 0° at the top and T° at the bottom, as shown in the figure. The coefficient of linear expansion of the beam material is α. The resulting vertical deflection at the mid-span of the beam is

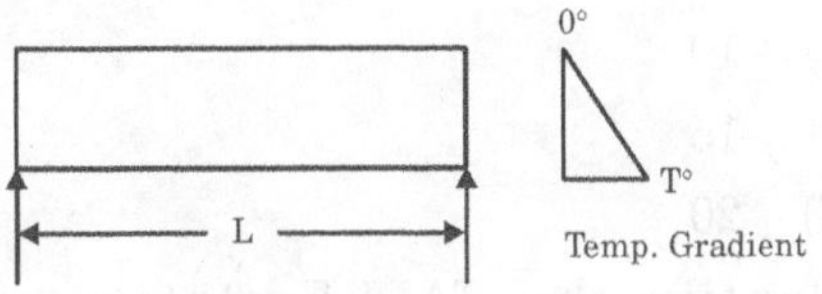

(a) $\frac{\alpha Th^2}{8L}$ upward (b) $\frac{\alpha TL^2}{8h}$ upward

(c) $\frac{\alpha Th^2}{8L}$ downward (d) $\frac{\alpha TL^2}{8h}$ downward

60. A truss, as shown in the figure, is carrying 180 kN load at node L_2. The force in the diagonal member M_2U_4 will be

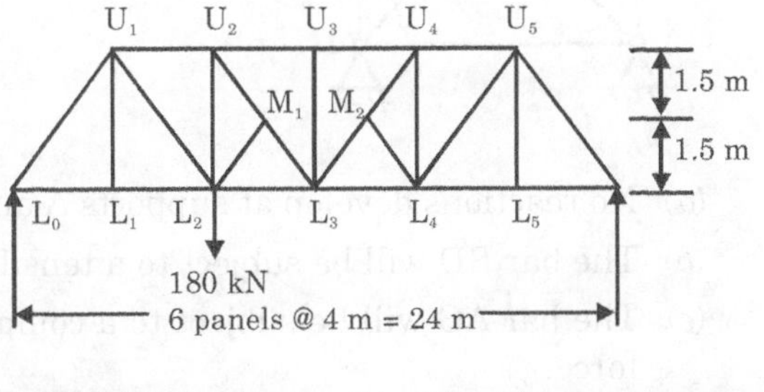

(a) 100 kN tension (b) 100 kN compression

(c) 80 kN tension (d) 80 kN compression

61. A steel portal frame has dimensions, plastic moment capacities and applied loads as shown in the figure. The vertical load is always twice of the horizontal load. The collapse load P required for the development of a beam mechanism is

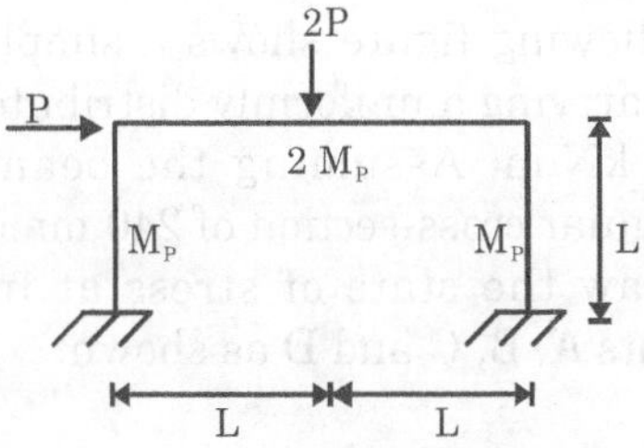

(a) $\frac{3M_p}{L}$ (b) $\frac{4M_p}{L}$

(c) $\frac{6M_p}{L}$ (d) $\frac{8M_p}{L}$

2002

62. In the propped cantilever beam carrying a uniformly distributed load of w N/m, shown in the following figure, the reaction at the support B is

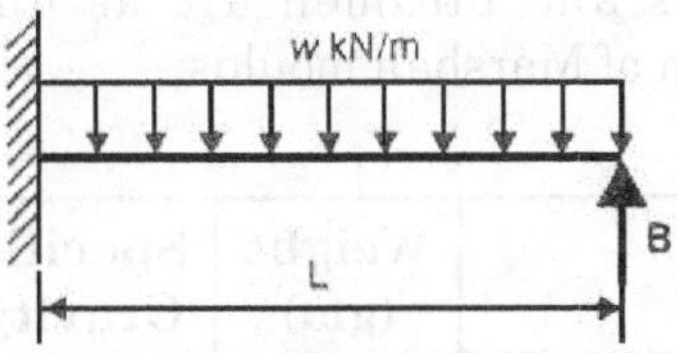

(a) $\frac{5}{8}wL.$ (b) $\frac{3}{8}wL$

(c) $\frac{1}{2}wL$ (d) $\frac{3}{4}wL$

63. A schematic representation of a P_u-M_u interaction diagram for the design of reinforced concrete columns is given in the following figure. Based on the given diagram, answer the following questions:

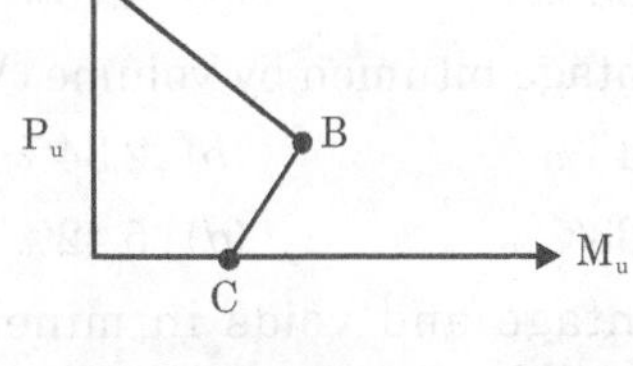

(a) What do the points A and C physically signify?

(b) What is the basic difference between the portions AB and BC?

(c) In the region BC, why does the moment capacity of the column increase even as the axial load is also being increased?

(d) Design codes often require the designer to ensure adequate strength for a minimum eccentricity. How is such a provision incorporated into the interaction diagram?

64. The following figure shows a simply supported beam carrying a uniformly distributed load (udl) of 10 kN/m. Assuming the beam to have a rectangular cross-section of 240 mm(b) × 400mm (h), draw the state of stress at infinitesimal elements A, B, C and D as shown.

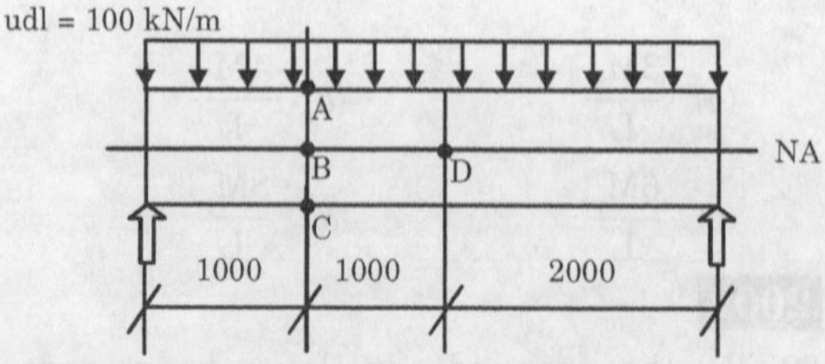

Common Data Q. 65-67

The specific gravities and weight proportions for aggregates and bitumen are as under for the preparation of Marshall moulds:

	Weight (gm)	Specific Gravity
Aggregate1	825	2.63
Aggregate2	1200	2.51
Aggregate 3	325	2.46
Aggregate 4	150	2.43
Bitumen	100	1.05

The volume and weight of one Marshall mould was found to be 475 cc and 1100 gm. Assuming absorption of bitumen in aggregates is zero.

65. Percentage air voids (V_v) will be

(a) 1.5% (b) 2.34%

(c) 3.33% (d) 4.76%

66. Percentage bitumen by volume (V_b) will be

(a) 1.43% (b) 2.34%

(c) 3.33% (d) 5.42%

67. Percentage and voids in mineral aggregates (V_{MA}) will be

(a) 1.43% (b) 2.34%

(c) 3.33% (d) 4.76%

2001

68. The degree of static indeterminacy, N_s, and the degree of kinematic indeterminacy, N_k, for the plane frame shown below, assuming axial deformations to be negligible, are given by

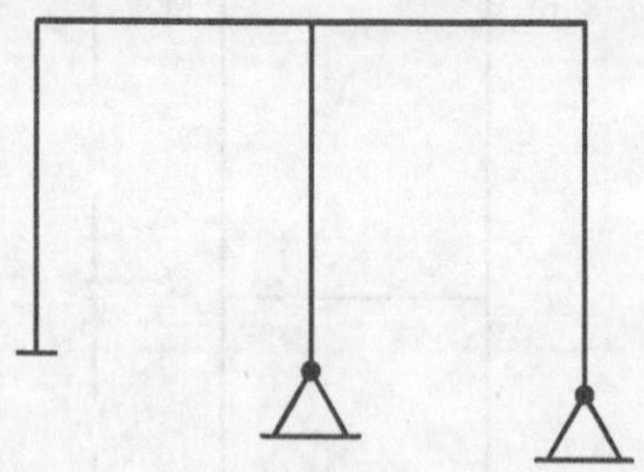

(a) $N_s = 6$ and $N_k = 11$

(b) $N_s = 6$ and $N_k = 6$

(c) $N_s = 4$ and $N_k = 6$

(d) $N_s = 4$ and $N_k = 4$

69. The bending moment (in kNm units) at the mid-span location X in the beam with overhangs shown below is equal to

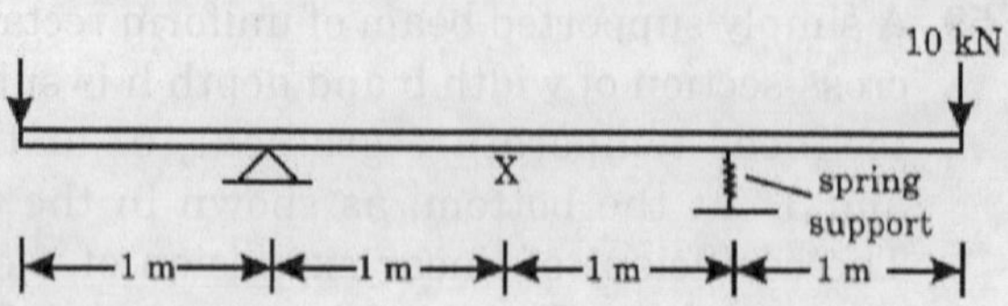

(a) 0

(b) –10

(c) –15

(d) –20

70. Identify the FALSE statement from the following, pertaining to the effects due to a temperature rise DT in the bar BD alone in the plane truss shown below.

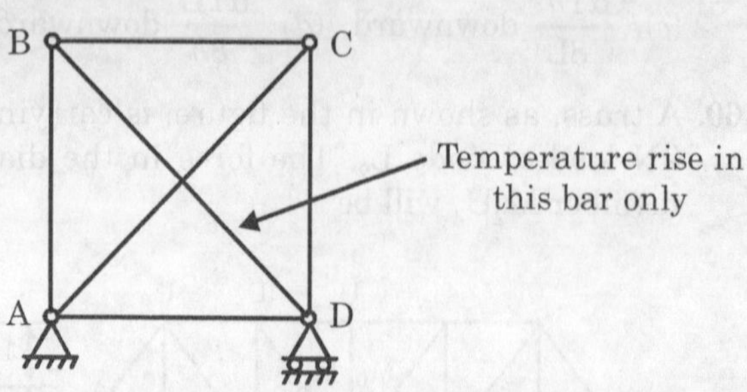

(a) No reactions develop at supports A and D.

(b) The bar BD will be subject to a tensile force.

(c) The bar AC will be subject to a compressive force.

(d) The bar BC will be subject to a tensile force.

71. Identify the correct deflection diagram corresponding to the loading in the plane frame shown below.

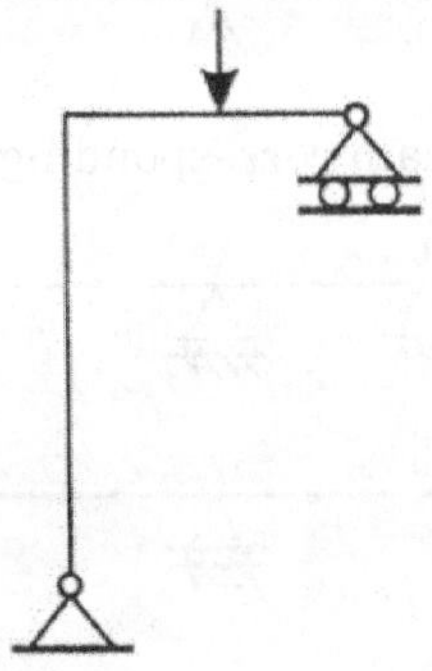

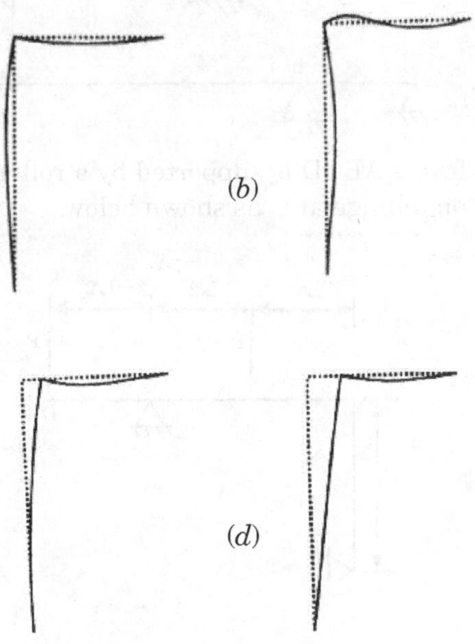

72. Identify the FALSE statement from the following, pertaining to the methods of structural analysis.

(*a*) Influence lines for stress resultants in beams can be drawn using Muller Breslau's Principle.

(*b*) The Moment Distribution Method is a force method of analysis, not a displacement method.

(*c*) The Principle of Virtual Displacements can be used to establish a condition of equilibrium.

(*d*) The Substitute Frame Method is not applicable to frames subjects to significant sidesway.

73. The frame below shows three beam elements OA, OB and OC, with identical length L and flexural rigidity E*l*, subject to an external moment M applied at the rigid joint O. The correct set of bending moments {M_{OA}, M_{OB}, M_{OC}} that develop at O in the three beam elements OA, OB and OC respectively, is

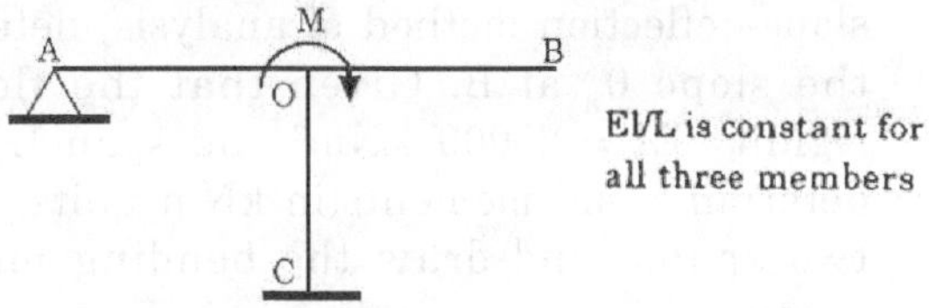

(*a*) $\frac{3M}{8}, \frac{M}{8}, \frac{4M}{8}$

(*b*) $\frac{3M}{11}, \frac{4M}{11}, \frac{4M}{11}$

(*c*) $\frac{M}{3}, \frac{M}{3}, \frac{M}{3}$

(*d*) $\frac{3M}{7}, 0, \frac{4M}{7}$

74. Identify from the following, correct value of the bending moment M_A (in kNm units) at the fixed end A in the statically determinate beam shown below (with internal hinges at B and D), when a uniformly distributed load of 10 kN/m is placed on all spans.

(*Hint :* Drawing influence line for M_A or applying the Principle of Virtual Displacements makes the solution easy).

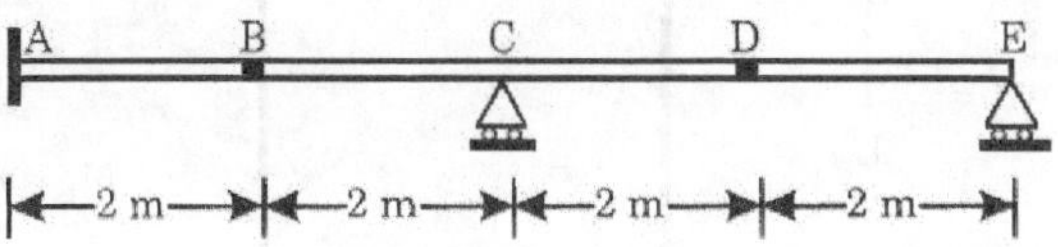

(*a*) –80 (*b*) –40

(*c*) 0 (*d*) +40

75. The end moment (in kNm) units) developed in the roof level beams in the laterally loaded frame shown below (with all columns having identical cross-sections), according to the Cantilever Method of simplified analysis, is

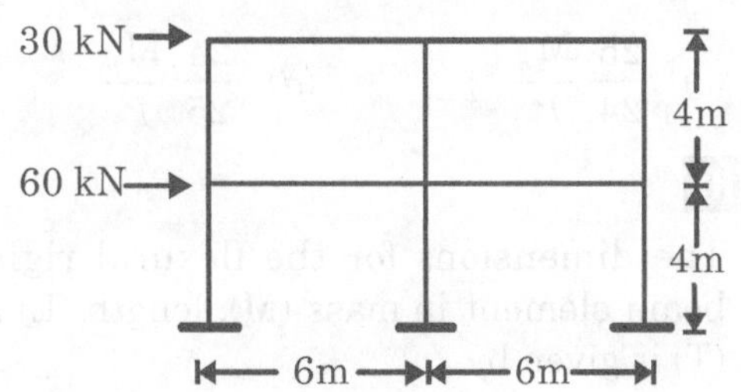

(*a*) 7.5 (*b*) 15

(*c*) 20 (*d*) 30

76. The two-span continuous beam shown below is subject to a clockwise rotational slip $\theta_A = 0.004$ radian at the fixed end A. Applying the slope-deflection method of analysis, determine the slope θ_B at B. Given that the flexural rigidity EI = 25000 kNm² and span L = 5m, determine end moments(in kNm units) in the two spans, and draw the bending moment diagram.

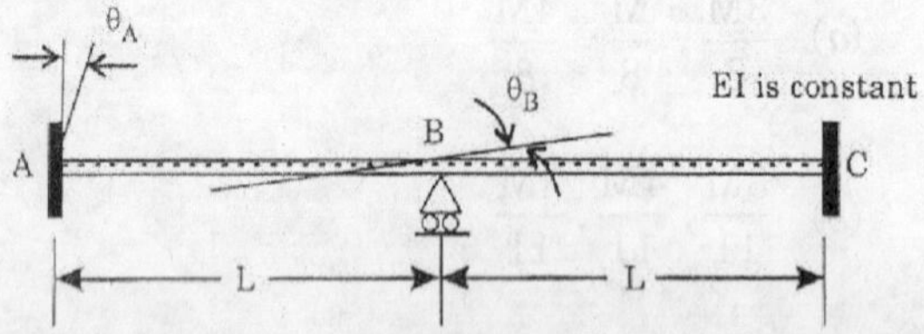

77. With reference to the plane frame (portal with overhanging beam) shown below, sketch four possible failure mechanisms, clearly marking the plastic hinge locations and mode of failure in each case. For the case of the 'combined machanism', the collapse load W in terms of the plastic moment capacity M_p (assumed to be constant at all sections) and the dimension L will be

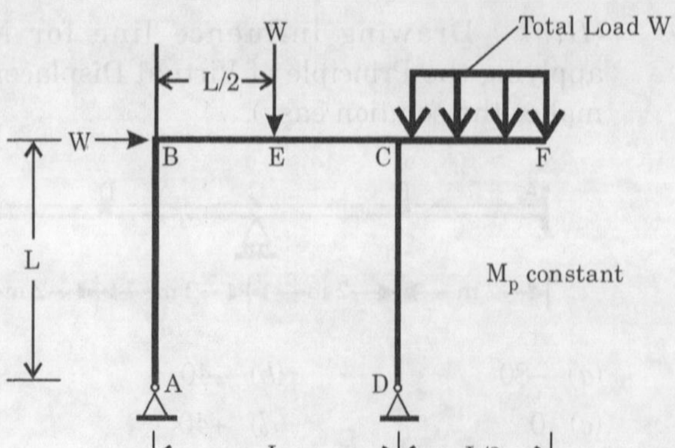

(a) $\frac{25}{24}\frac{M_p}{l}$ (b) $\frac{24}{25}\frac{M_p}{l}$

(c) $\frac{25}{24}\frac{M_p}{l^2}$ (d) $\frac{24}{25}\frac{M_p}{l^2}$

2000

78. The dimensions for the flexural rigidity of a beam element in mass (M), length (L) and time (T) is given by

(a) MT^{-2} (b) $ML^3 T^{-2}$

(c) $ML^{-1} T^{-2}$ (d) $ML^{-1} T^2$

79. A two span beam with an internal hinge is shown below.

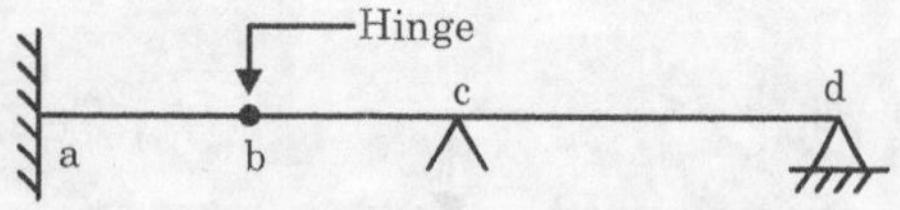

Conjugate beam corresponding to this beam is

(a) 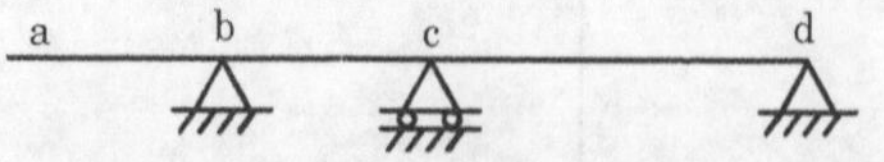

(b) 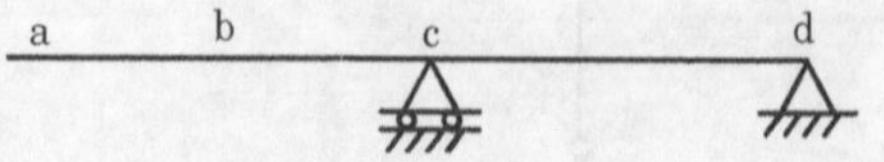

(c) 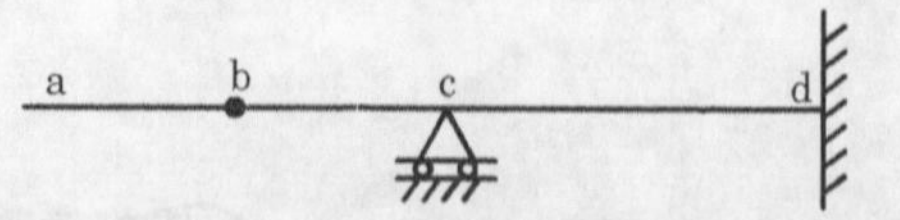

(d) 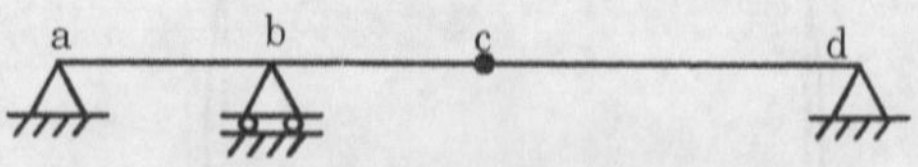

80. A frame ABCD is supported by a roller at A and is on a hinge at C as shown below.

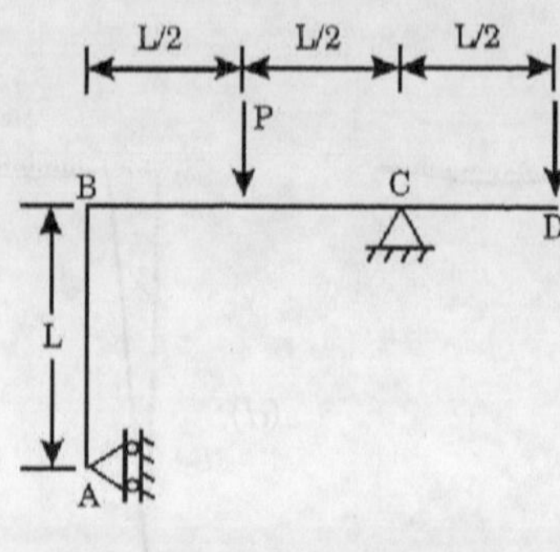

The reaction at the roller end A is given by

(a) P (b) 2 P

(c) $\frac{P}{2}$ (d) zero

81. The following two statements are made with reference to the planar truss shown below.

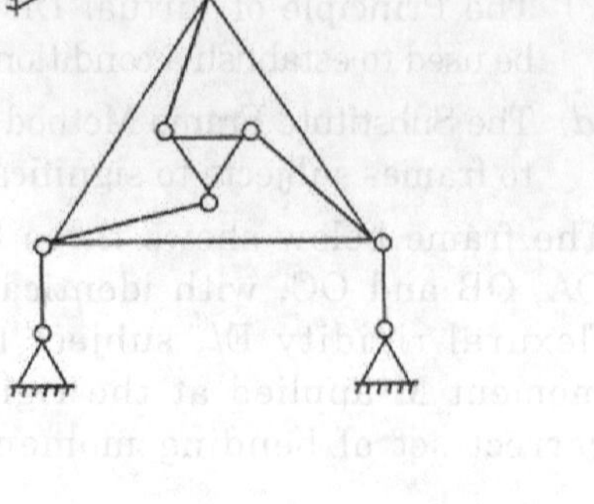

I. The truss is statically determinate

II. The truss is kinematically determinate

With reference to the above statements, which of the following applies ?

(*a*) Both statements are true.

(*b*) Both statements are false.

(*c*) II is true but I false.

(*d*) I is true but II is false.

82. A cantilever beam of length L and a cross section with shape factor f supports a concentrated load P as shown below.

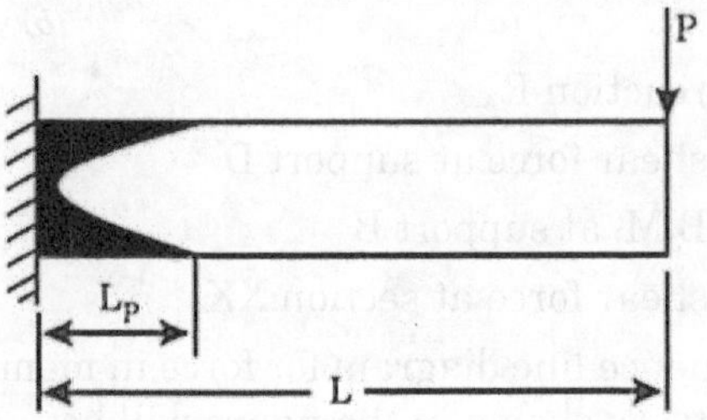

Length L_p of the plastic zone, when the maximum bending moment equals the plastic moment M_p, is given by

(*a*) $\frac{L_p}{L} = \frac{1}{f}$ (*b*) $\frac{L_p}{L} = L(1-f)$

(*c*) $\frac{L_p}{L} = 1 - \frac{1}{\sqrt{f}}$ (*d*) $\frac{L_p}{L} = 1 - \frac{1}{f}$

83. For the structure shown below, the vertical deflection at point A is given by

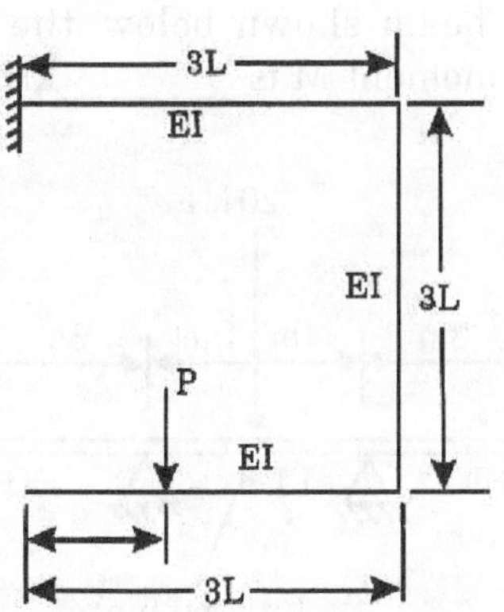

(*a*) $\frac{PL^3}{81EI}$ (*b*) $\frac{2PL^3}{81EI}$

(*c*) Zero (*d*) $\frac{PL^3}{72EI}$

1999

84. For a fixed beam with span L, having plastic moment capacity of M_p, the ultimate central concentrated load will be

(*a*) $\frac{4M_p}{L}$ (*b*) $\frac{M_p}{8L}$

(*c*) $\frac{6M_p}{L}$ (*d*) $\frac{8M_p}{L}$

1998

85. A propped cantilever beam is shown in figure. The plastic moment capacity of the beam is M_0. The collapse load P is

(*a*) $4M_0/L$

(*b*) $6\,M_0/L$

(*c*) $8\,M_0/L$

(*d*) $12\,M_0/L$

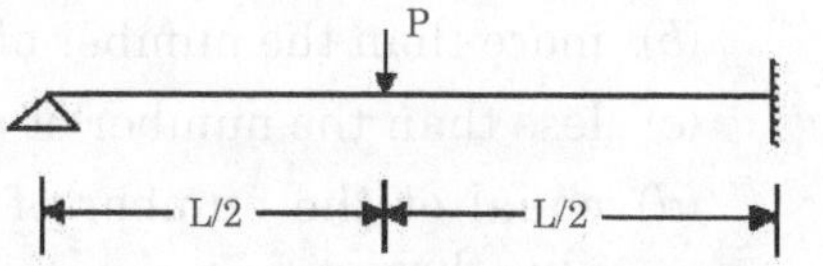

86. The strain energy stored in member AB of the pin-jointed truss is shown in figure when E and A are same for all members, is

(*a*) $\frac{2P^2L}{AE}$

(*b*) $\frac{P^2L}{AE}$

(*c*) $\frac{P^2L}{2AE}$

(*d*) Zero

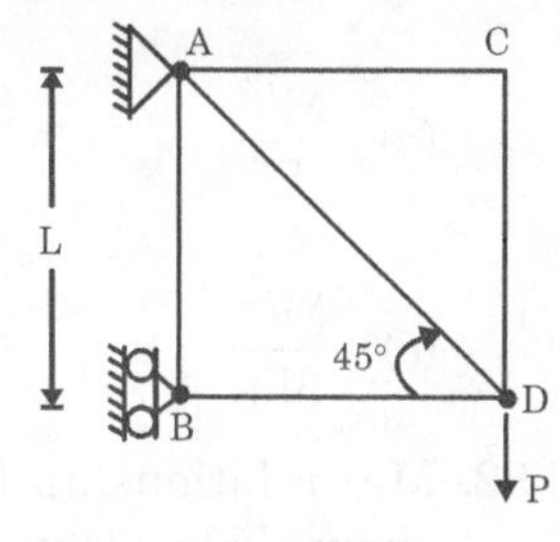

87. The stiffness matrix of a beam element is given as $(2EI/L)\begin{bmatrix} 2 & -1 \\ 1 & 2 \end{bmatrix}$. Then the flexibility matrix is

(*a*) $\left(\frac{L}{2EI}\right)\begin{bmatrix} 2 & 1 \\ 1 & 2 \end{bmatrix}$ (*b*) $\left(\frac{L}{6EI}\right)\begin{bmatrix} 1 & -2 \\ -2 & 1 \end{bmatrix}$

(*c*) $\left(\frac{L}{3EI}\right)\begin{bmatrix} 2 & -1 \\ -1 & 2 \end{bmatrix}$ (*d*) $\left(\frac{L}{5EI}\right)\begin{bmatrix} 2 & -1 \\ -1 & 2 \end{bmatrix}$

1997

88. The degree of kinematic indeterminacy of the rigid frame with clamped ends at A and D shown in the figure is,

(*a*) 4

(*b*) 3

(*c*) 2

(*d*) zero

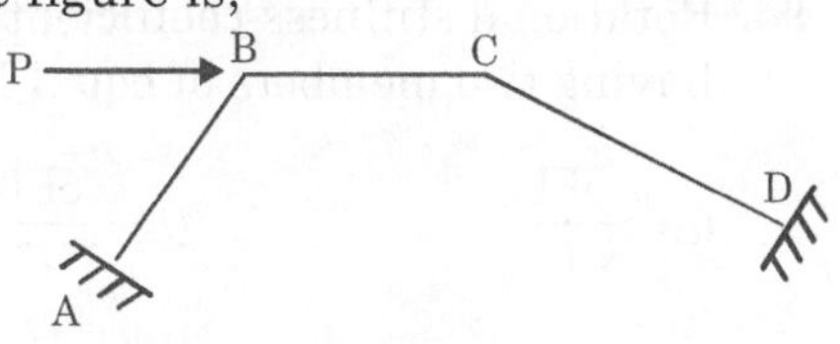

89. For the frame shown in the figure, the maximum bending moment in the column is

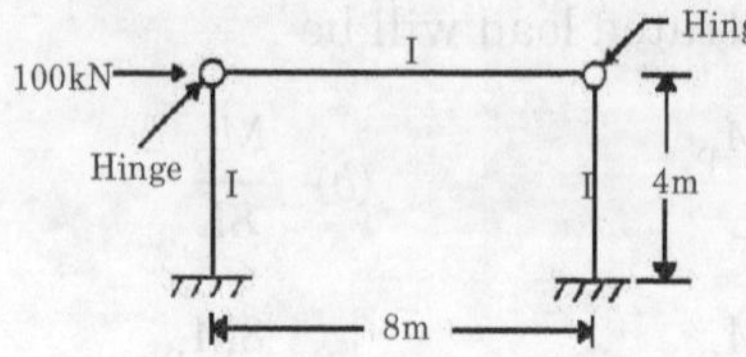

(*a*) zero (*b*) 400 kN m

(*c*) 100 kN m (*d*) 200 kN m

90. The order or the flexibility matrix for a structure is

(*a*) equal to the number of redundant forces

(*b*) more than the number of redundant forces

(*c*) less than the number of redundent forces

(*d*) equal ot the number of redundant forces plus three

1996

91. The deflection of a cantilever beam at free end *b* applied with a moment M at the same point is

(*a*) $\frac{Ml^2}{EI}$ (*b*) $\frac{Ml^2}{2EI}$

(*c*) $\frac{Ml^2}{3EI}$ (*d*) $\frac{Ml^2}{4EI}$

92. M-θ relationship for a simply supported beam shown below is given by

(*a*) $\frac{Ml}{EI} = 2\theta$ (*b*) $\frac{Ml}{EI} = 3\theta$

(*c*) $\frac{Ml}{EI} = 4\theta$ (*d*) $\frac{Ml}{EI} = 6\theta$

93. A two-hinged semi-circular arch of radius 'R' carries a concentrated load 'W' at the crown. The horizontal thrust is

(*a*) $\frac{W}{2\pi}$ (*b*) $\frac{W}{\pi}$

(*c*) $\frac{2W}{3\pi}$ (*d*) $\frac{4W}{3\pi}$

94. Rotational stiffness-coefficient, K_{11} for the frame having two members of equal EI/*i* is given by.

(*a*) $\frac{5EI}{l}$ (*b*) $\frac{6EI}{l}$

(*c*) $\frac{7EI}{l}$ (*d*) $\frac{8EI}{l}$

95. Horizontal stiffness coefficient, K_{11} of bar *ab* is given by,

(*a*) $\frac{AE}{l\sqrt{2}}$ (*b*) $\frac{AE}{2l}$

(*c*) $\frac{AE}{l}$ (*d*) $\frac{2AE}{l}$

96. For beam shown in Fig. (*a*), an influence line diagram is shown in Fig. (*b*). This refers to

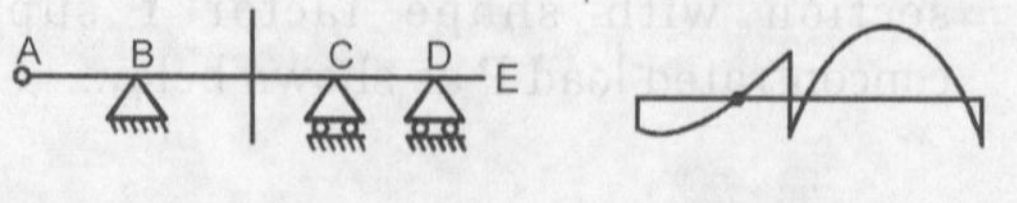

(*a*) (*b*)

(*a*) reaction R_A

(*b*) shear force at support D

(*c*) B.M. at support B

(*d*) shear force at section XX

97. Influence line diagram for force in member XY of the truss shown in the figure will be

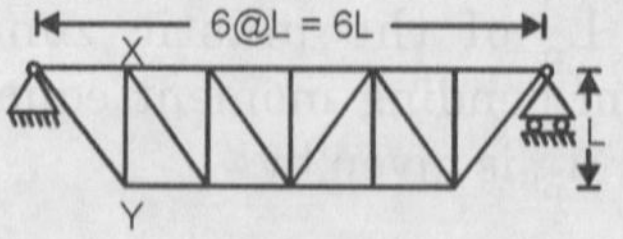

(*a*) 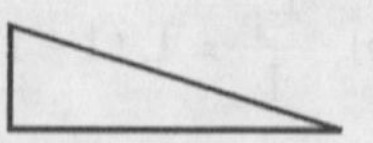 (*b*) 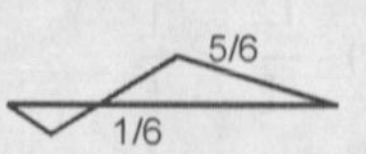

(*c*) 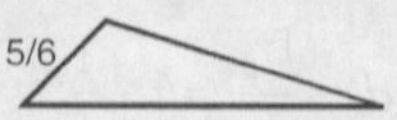

(*d*) 

NUMERICAL TYPE QUESTIONS

2015

1. For the beam shown below, the value of the support moment M is ________ kN-m.

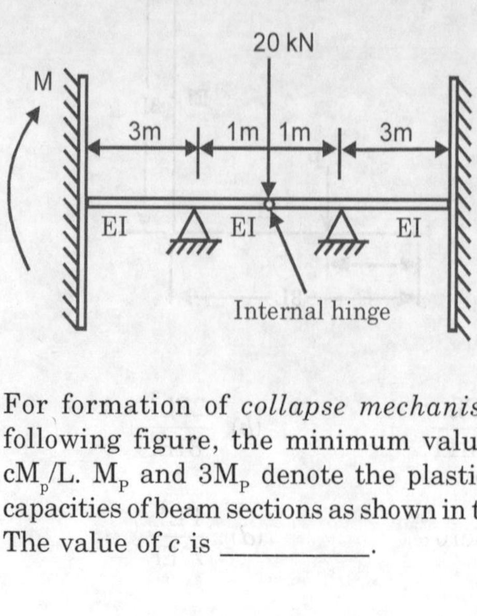

2. For formation of *collapse mechanism* in the following figure, the minimum value of P_u is cM_p/L. M_P and $3M_P$ denote the plastic moment capacities of beam sections as shown in this figure. The value of *c* is ________.

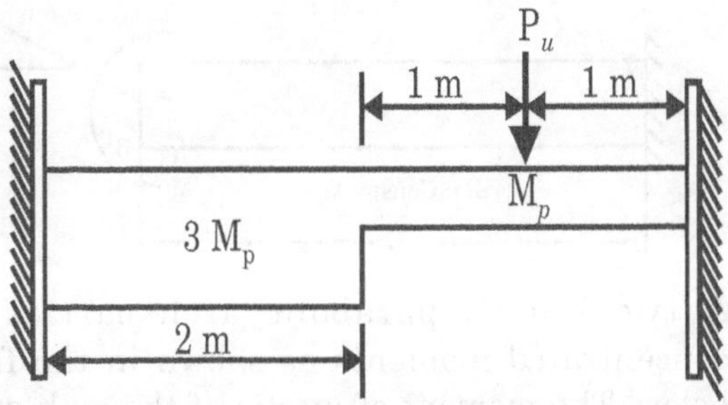

3. For the 2D truss with the applied loads shown below, the strain energy in the member XY is ____ kN-m. For member XY, assume AE = 30 kN, where A is cross-section area and E is the modulus of elasticity.

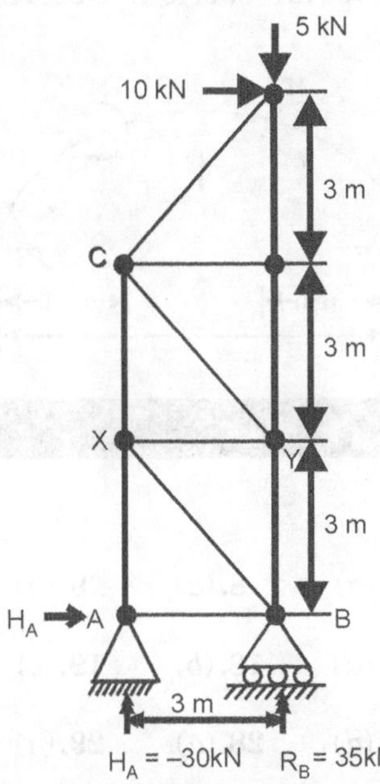

4. Two beams are connected by a linear spring as shown in the following figure. For a load P as shown in the figure, the percentage of the applied load P carried by the spring is ________

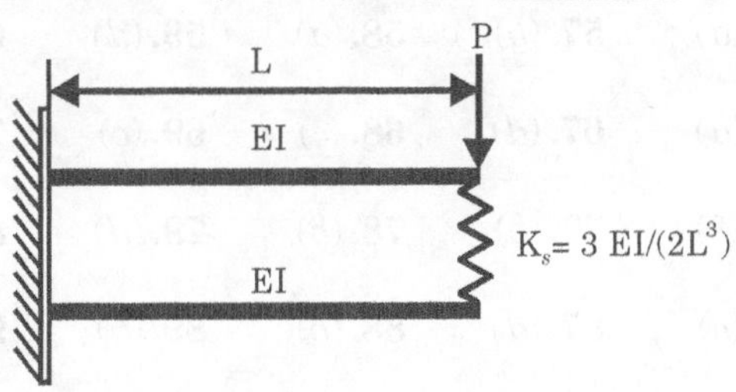

2014

5. For the truss shown below, the member PQ is short by 3 mm. The magnitude of vertical displacement of joint R(in mm) is

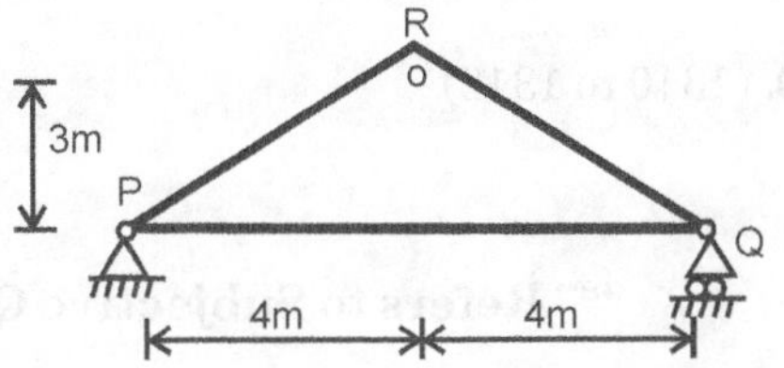

6. The static indeterminacy of the two-span continuous beam with an internal hinge, shown below, is ________

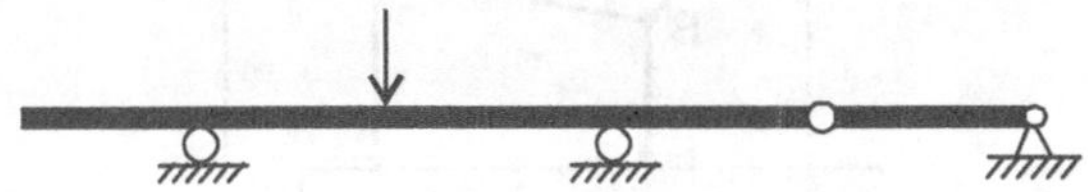

2013

7. A symmetric I-section(with width of each flange = 50 mm, thickness of each flange = 10 mm, depth of web = 100 *mm*, and thickness of web = 10 mm) of steel is subjected to a shear force of 100 kN. Find the magnitude of the shear stress (in N/mm²) in the web at its junction with the top flange. ________

8. All members in the rigid-jointed frame shown are prismatic and have the same flexural stiffness EI Find the magnitude of the bending moment at Q(in kNm) due to the given loading. ________

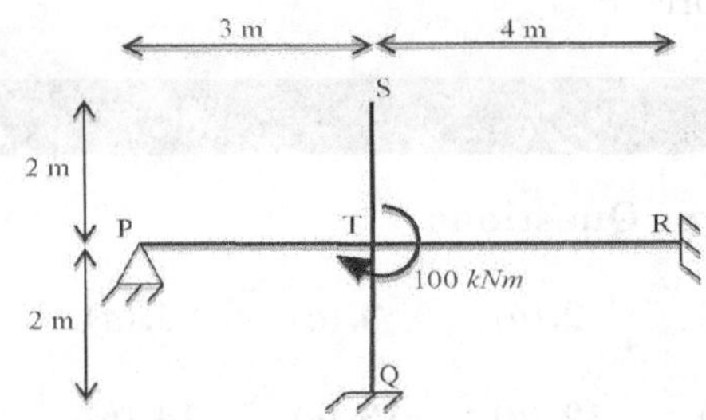

9. A uniform beam (EI = constant) PQ in the form of a quarter-circle of radius R is fixed at end P and free at the end Q, where a load W is applied as shown. The vertical downward displacement, δ_q, at the loaded point Q is given by: $\delta_q = \beta\left(\frac{WR^3}{EI}\right)$.

Find the value of β(correct to 4-decimal places).

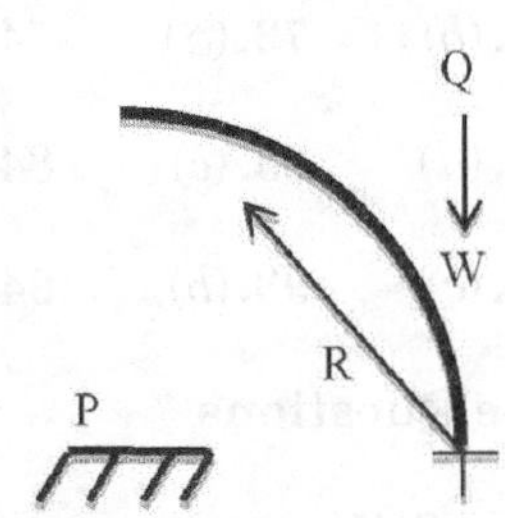

10. A uniform beam weighing 1800 N is supported at E and F by cable ABCD. Determine the tension (in N) in segment AB of this cable (correct of 1-decimal place). Assume the cables ABCD, BE and CF to be weightless ______

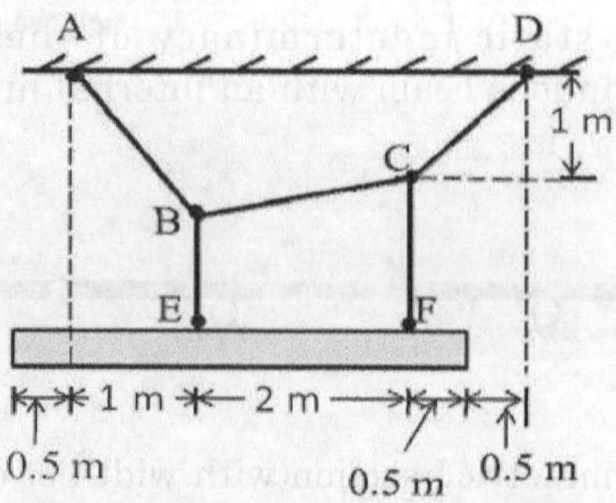

2000

11. A cantilever beam AB carries a concentrated force P and a moment $M = \frac{PL}{3}$ at its tip as shown below.

Show using Castigliano's theorem that, if angle of inclination α of the line of action of the force P is such that $\sin \alpha = \frac{1}{2}$, then displacement of the point B, due to bending will be in the direction of force P.

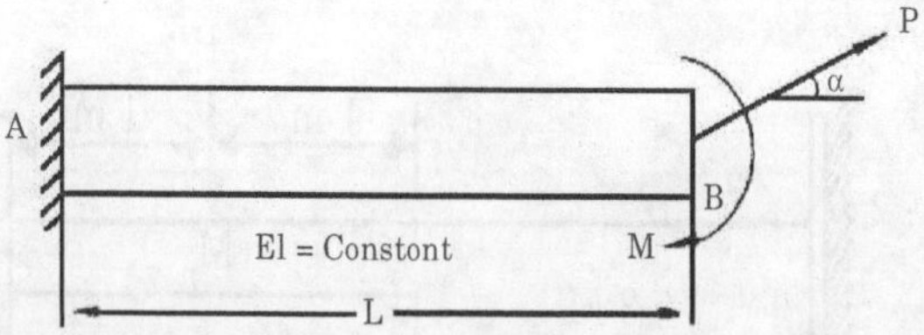

12. A two hinged parabolic arch carries two concentrated moments as shown in the figure below. The moment of inertia of the arch at any particular cross section is equal to the moment of interia at the crown multiplied by the secant of the angle θ, where θ is the angle between horizontal and the tangent to the arch axis at that particular section. Determine support reactions.

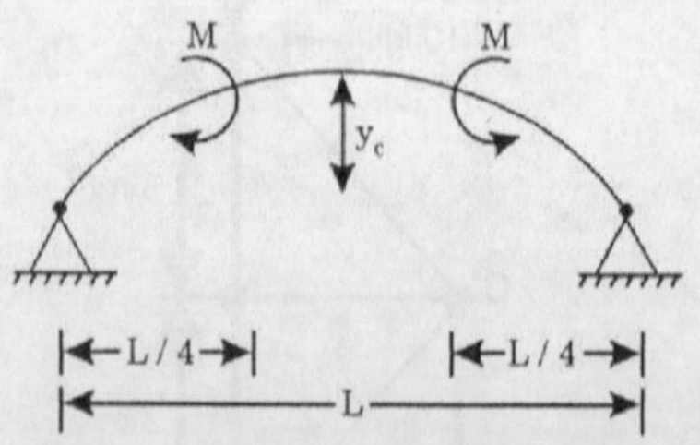

ANSWERS

MCQ Type Questions

1. (a)	**2.** (a)	**3.** (a)	**4.** (a)	**5.** (b)	**6.** (c)	**7.** (a)	**8.** (a)	**9.** (a)	**10.** (c)
11. (c)	**12.** (a)	**13.** (c)	**14.** (b)	**15.** (c)	**16.** (d)	**17.** (d)	**18.** (b)	**19.** (c)	**20.** (a)
21. (c)	**22.** (c)	**23.** (a)	**24.** (b)	**25** (d)	**26.** (c)	**27.** (d)	**28.** (a)	**29.** (a)	**30.** (c)
31. (d)	**32.** (d)	**33.** (a)	**34.** (a)	**35.** (b)	**36.** (d)	**37.** (d)	**38.** (c)	**39.** (c)	**40.** (b)
41. (c)	**42.** (c)	**43.** (a)	**44.** (a)	**45.** (a)	**46.** (c)	**47.** (d)	**48.** (d)	**49.** (a)	**50.** (c)
51. (b)	**52.** (c)	**53.** (a)	**54.** (a)	**55.** (b)	**56.** (a)	**57.** (b)	**58.** (a)	**59.** (d)	**60.** (a)
61. (a)	**62.** (b)	**63.** (*)	**64.** (*)	**65.** (c)	**66.** (a)	**67.** (d)	**68.** (c)	**69.** (c)	**70.** (b)
71. (d)	**72.** (b)	**73.** (d)	**74.** (c)	**75.** (b)	**76.** (*)	**77.** (b)	**78.** (b)	**79.** (d)	**80.** (d)
81. (d)	**82.** (d)	**83.** (c)	**84.** (d)	**85.** (b)	**86.** (c)	**87.** (d)	**88.** (b)	**89.** (b)	**90.** (a)
91. (b)	**92.** (c)	**93.** (b)	**94.** (c)	**95.** (a)	**96.** (c)	**97.** (b)			

Numerical Type Questions

1. (5) **2.** (13.3) **3.** (5) **4.** (25) **5.** (1.0 to 2.5) **6.** (0 to 0)

7. (70 to 72) **8.** (25) **9.** (0.785 to 0.786) **10.** (1310 to 1313)

11. (*) **12.** (*)

"*" Refers to Subjective Question.

EXPLANATIONS

MCQ TYPE QUESTIONS

1. The force P is passing through support at C

$\therefore\ V_C = P$

and other two reactions become zero

$\therefore\ V_B = 0, H_B = 0$

2. Taking FBD of $jt_F = F_{FT} = 0$

Taking FBD of $jt_T = F_{TG} = 0$

Taking FBD of $jt_H = F_{HU} = 0$

Taking FBD of $jt_M = F_{MP} = 0$

Taking FBD of $jt_P = F_{PL} = 0$

3. Kinematic indeterminacy

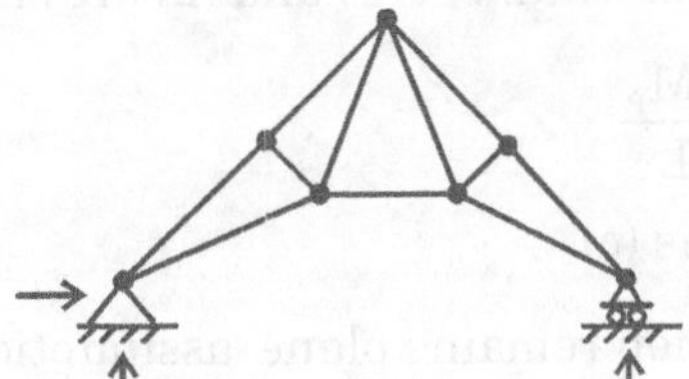

No. of joints = 7

No. of unrestrained displacement possible

$= 7 \times 2 = 14$

No. of restraints = 2 + 1 = 3

Therefore, degree of freedom = 14 – 3 = 11

4.

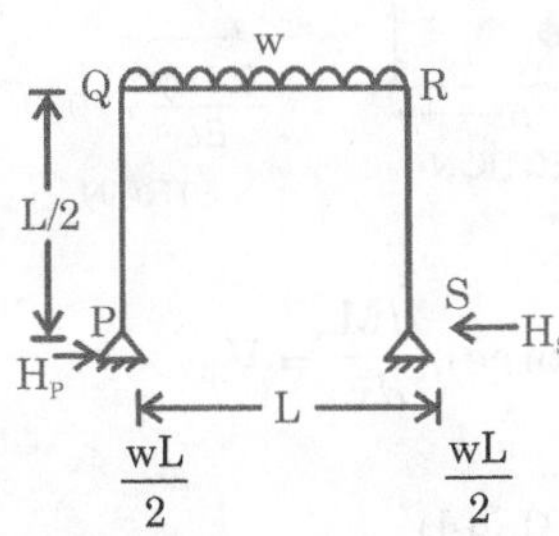

$$\Sigma F_X = 0$$

$$H_P - H_S = 0$$

Therefore,

$$H_P = H_S = 0$$

$$\Sigma M_Q = \frac{H_P \times l}{2} = 0$$

Bending moment at Q = 0

5.

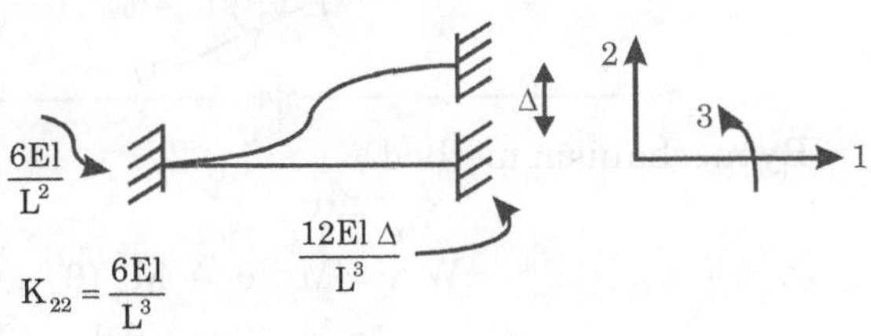

$$K_{22} = \frac{6EI}{L^3}$$

6.

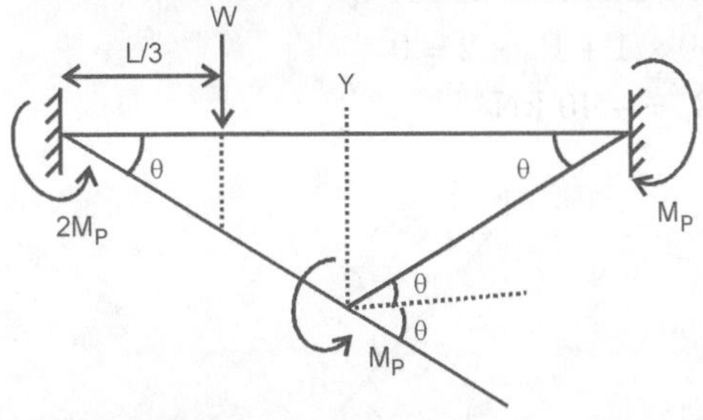

No. of hinges required = $D_s + 1 = 2 + 1 = 3$

External workdone = Internal workdone

$$W \cdot \frac{L}{3} \cdot \theta = 2M_P\theta + M_P(2\theta) + M_P\theta$$

$$\frac{WL\theta}{3} = 5M_P\theta$$

$$\Rightarrow \qquad W = \frac{15M_P}{L}$$

7. $D_s = D_{sc} + D_{si} = (r_e - 3) + 3c - r_r$

$= (4 - 3) + 3 \times 0 - 1 = 0$

8.

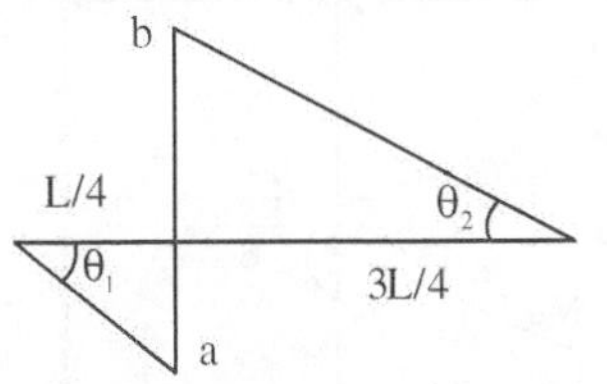

$\theta_1 = \theta_2$

So, $\quad \dfrac{a}{L/4} = \dfrac{b}{3L/4}$

$\Rightarrow \quad b = 3a$

Also $\quad a + b = 1$

$\Rightarrow \quad a = 2/4 = 0.25$

$\Rightarrow \quad b = 3/4 = 0.75$

9.

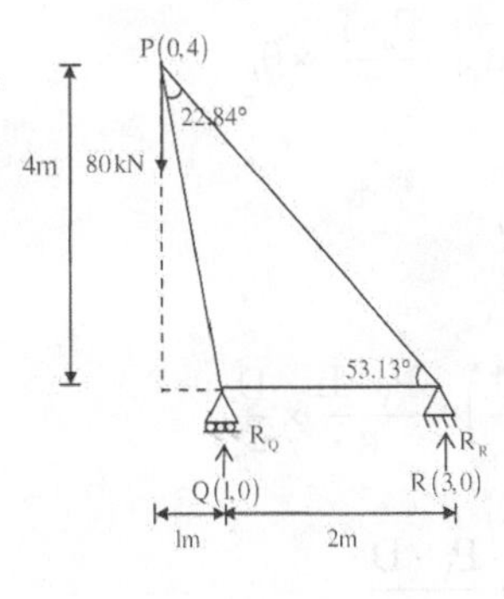

$\Sigma F_V = 0$

$\Rightarrow R_Q + R_R = 80 \quad ...(i)$

Taking moment about Q = 0

$\Rightarrow 80 \times 1 + R_R \times 2 = 0$

$\Rightarrow R_R = -40$ kN

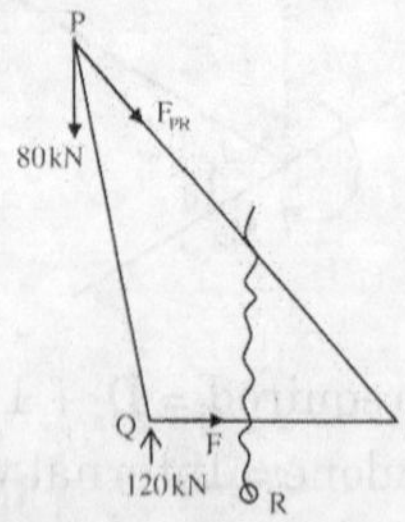

From (i), we get $R_Q = 120$ kN

Taking moment about P = 0

$F_{QR} \times 4 + 120 \times 1 = 0$

$\Rightarrow F_{QR} = -30$kN (–) means indicates compressive

11.

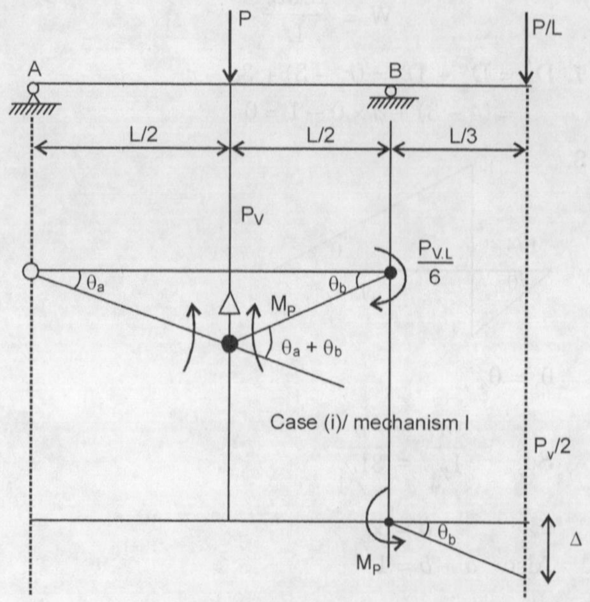

Case-II/Mechanism 2

By load Minimisation technique $\boxed{W_e = W_i}$

for case (i) Mechanism

$$P_V \times D = M_P(\theta_a + \theta_b) + \frac{P_v \cdot L}{6} \times \theta_b$$

$$\theta_a = \theta_b = \frac{D}{L/2}$$

$$P_V \times D = M_P\left(\frac{D}{L/2} + \frac{D}{L/2}\right) + \frac{P_V \cdot L}{6} \times \frac{D}{L/2}$$

$$P_V \times D = M_P\left(\frac{4D}{L}\right) + \frac{P_V \cdot D}{3}$$

$$P_V \times D - \frac{P_V D}{3} = M_P \cdot \frac{4D}{L}$$

$$\frac{2}{3} P_V \cdot D = M_P \cdot \frac{4D}{L}$$

$$P_V = \frac{6M_P}{L}$$

For case 2 / Mechanism 2

$$\frac{P_V}{2} \cdot D = M_P \cdot \theta_b = M_P \frac{D}{L/3}$$

$$P_V = \frac{6M_P}{L}$$

Loads occur in Mechanism (1) and (2) are same.

So $\quad P_V = \dfrac{6M_P}{L}$

Correct choice is (c)

12. The 'Plane section remains plane' assumption in bending theory implies strain profile is linear. See in figure

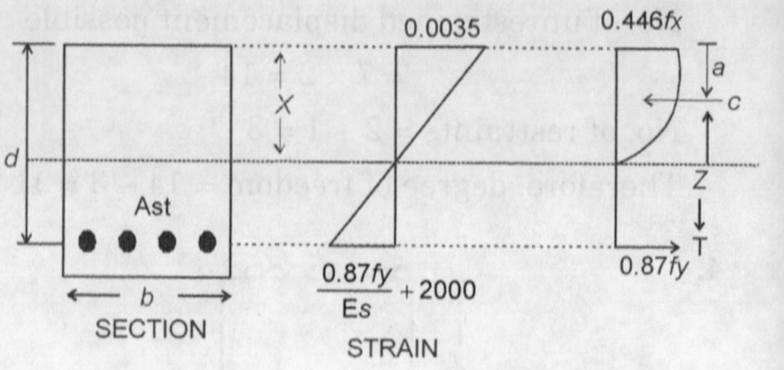

13. (Shear force), $\dfrac{dM}{dx} = V$

$$\frac{dv}{dx} = w \text{ (load)}$$

$$\frac{d^2y}{dx^2} = \psi \text{ (curvature)}$$

17.

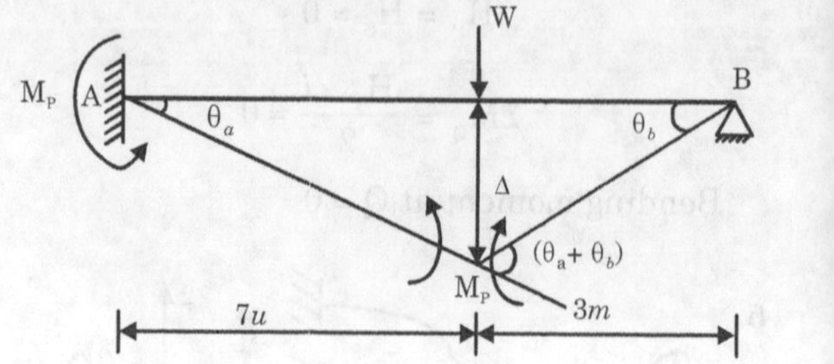

By mechanism method

$$W_e = W_i$$

$$\therefore \quad W.\Delta = M_p.\theta_a + M_p(\theta_a + \theta_b)$$

$$= M_p(2\theta_a + \theta_b)$$

$$\Rightarrow \quad W. \cancel{\Delta} = M_p \left(2 \times \frac{\cancel{\Delta}}{7} + \frac{\cancel{\Delta}}{3}\right)$$

$$\Rightarrow \quad W = \frac{13}{21} Mp$$

19. Equation of elastic constants is

$$E = 2N\left(1 + \frac{1}{m}\right) = 3K\left(1 - \frac{2}{m}\right)$$

All the elastic constants can be found if any 2 of them are known.

Hence number of independent elastic constants are 2.

20.

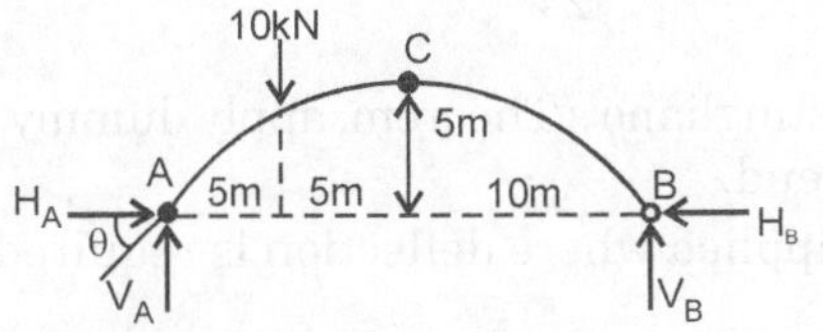

Taking moment at B

From $\quad \Sigma M_B = 0$

$$V_A \times 20 = 10 \times 15$$

$$\Rightarrow \quad V_A = 7.5 \text{ kN}$$

Taking moment at C

From $\quad \Sigma M_C = 0$

$$H_A \times 5 - V_A \times 10 + 10 \times 5 = 0$$

$$\Rightarrow \quad H_A \times 5 - 7.5 \times 10 + 50 = 0$$

$$\Rightarrow \quad H_A = 5 \text{ kN}$$

$$\therefore \quad \text{Resultant} = \sqrt{(7.5^2 + 5^2)} = 9.01 \text{ kN}$$

Angle with horizontal is given by

$$\tan\theta = \frac{V_A}{H_A} = \frac{7.5}{5} = 1.5$$

$$\Rightarrow \quad \theta = 36.31°$$

22. Here, m = 3, r = 6, j = 4

$$\therefore \quad S_0 = (3m + r) - 3j$$

$$= (3 \times 3 + 6) - 3 \times 4 = 3$$

24.

$$P_{cr} = \frac{\pi^2 EI}{L^2}$$

Having all other properties constant

Load when bond is perfect,

$$P_1 = \left\{\frac{1}{12}.b(2t)^3\right\} \frac{\pi^2 E}{L^2}$$

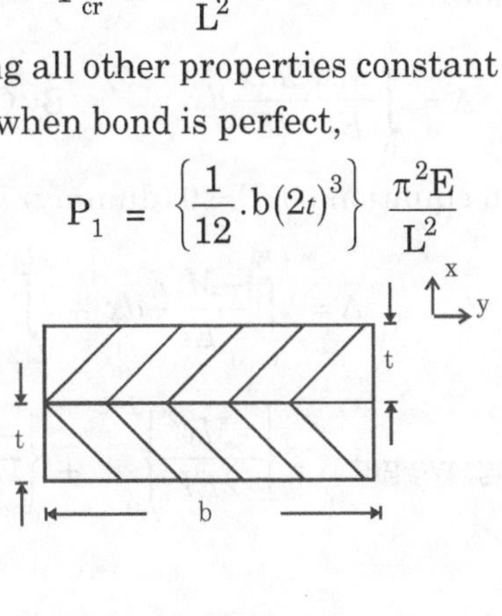

Load without bond,

$$P_2 = \frac{\pi^2 E}{L^2} \left\{2 \times \frac{1}{12}.bt^3\right\}$$

$$\therefore \quad \frac{P_1}{P_2} = \frac{b(2t)^3}{2.bt^3} = 4$$

25. If m = number of members

r_r = number of internal reactios component released

r_e = number of external reaction components.

j = number of joints

j' = number of releases treated as joints.

then static indeterminacy in given by

$$I_s = 3m - 3(j + j') - r_r + r_e$$

$$= 3 \times 9 - 3(6 + 2) - 2 + 4 = 5$$

26.

$$L_{IJ} = \sqrt{3000^2 + 3000^2} = 4242.6$$

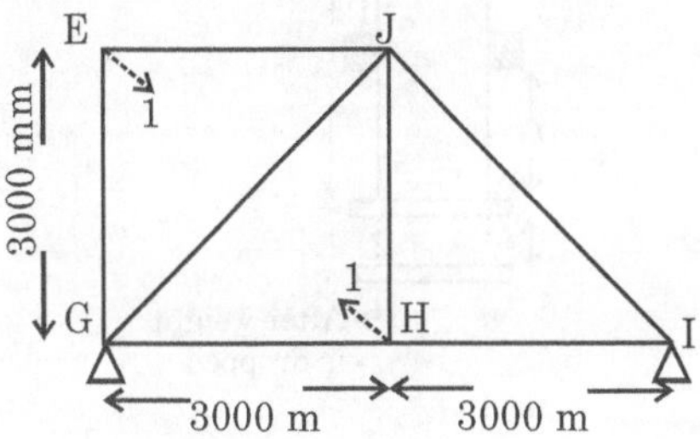

To find relative displacement of joints E and H, apply unit loads at each joint as shown in the figure.

Member	Temp. Rise	Lαt = dL	U	U.dL
EJ	30°	1.08	–0.707	–0.764
EG	0		–0.707	0
GH	0		–0.707	0
HJ	0		–0.707	0
HI	0		0	0
GJ	0		1.0	0
JI	30°	1.591	0	0

$$\Sigma U \times dL = -0.764 \text{ mm}$$

– ve sign indicates the joints go away from each other.

27. Using Muller Breslaw Principle, given unit displacement to Reaction at P then

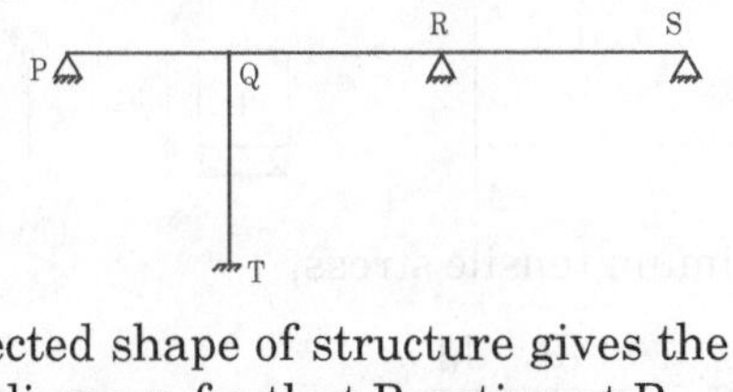

deflected shape of structure gives the Influence line diagram for that Reaction at P.

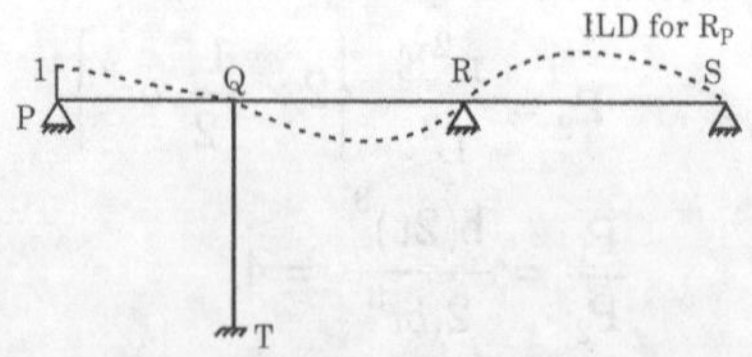

Now when the positive influence line area loaded, then the reaction will be maximum

So when PQ and RS are loaded, the Reaction will be maximum

28. Work done by falling weight,

$$W = p\,(h + \Delta) = p\left(h + \frac{p.L}{E}\right)$$

where, p = axial stress, L = length of bar.

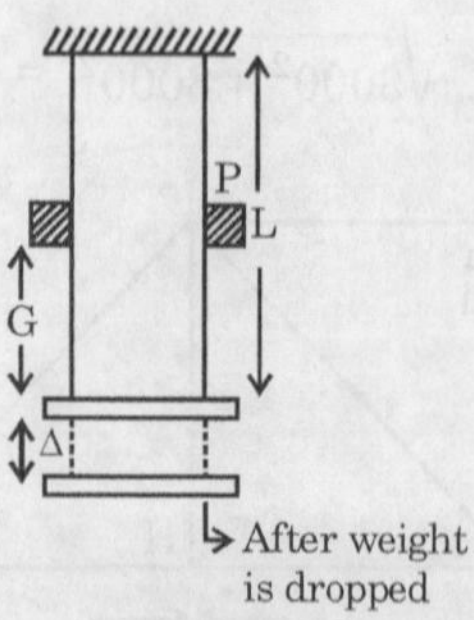

Work stored in bar, $U = \frac{1}{2}R.\Delta = \frac{1}{2}\,p.A.\frac{p.L}{E}$

$$U = \frac{1}{2}\frac{p^2.AL}{E}$$

Now $\qquad W = u$

Solving we get, $\qquad p = \frac{P}{A}\left(1 + \sqrt{1 + \frac{2AEh}{PL}}\right)$

From above relation, preduced as L increases.

29. This is a case of combined direct and bending stress

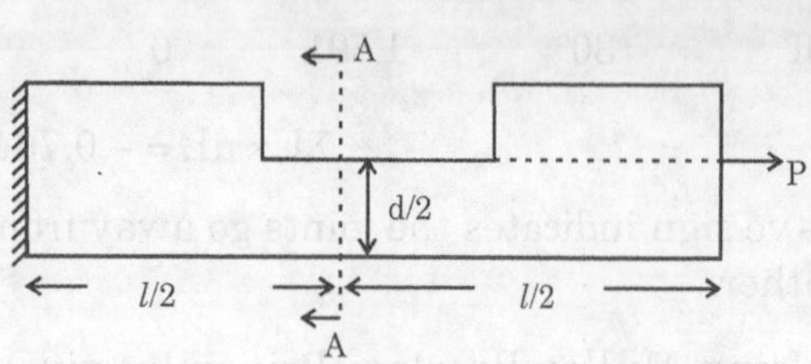

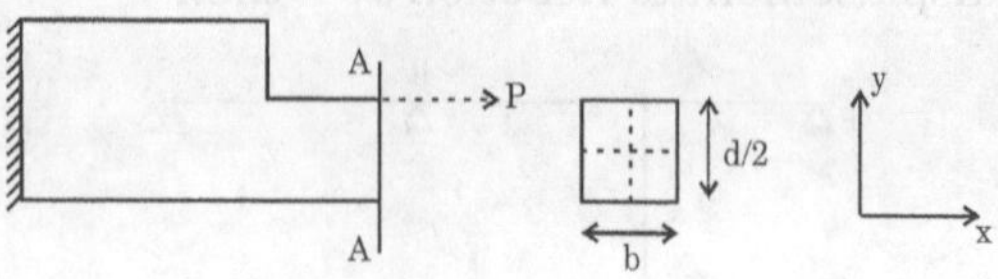

Maximum tensile stress,

$$\sigma_{max} = \frac{P}{A} + \frac{M}{Z}$$

where, A = area of crossection at

$$x - x = b \times \frac{d}{2}$$

$$Z = \frac{I_x}{y} = \frac{\frac{1}{12}.b\left(\frac{d}{2}\right)^3}{\left(\frac{d}{4}\right)} = \frac{1}{24}bd^2$$

$$M = P \times \frac{d}{4} = \frac{Pd}{4}$$

$$\therefore \quad \sigma_{max} = \frac{P}{bd/2} + \frac{Pd/4}{\frac{1}{24}bd^2} = \frac{2P}{bd} + \frac{6P}{bd} = \frac{8P}{bd}$$

30. Using Castingliano's Theorem, apply dummy load F at free end.

(Load is applied where deflection is required)

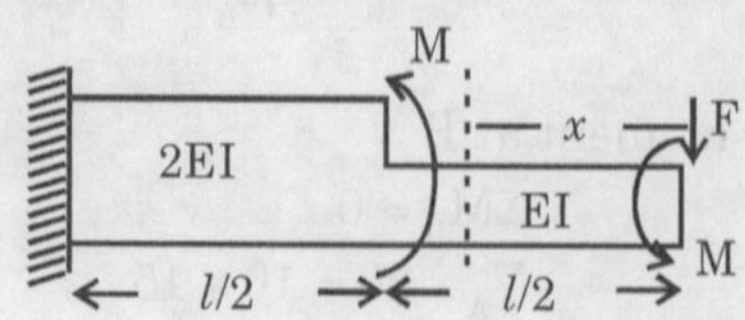

1. From $x = 0$ to $\frac{L}{2}$:

   Bending moment, $M_x = F.x - M$

   $$\therefore \quad \frac{\partial M_x}{\partial F} = x$$

2. From $x = \frac{L}{2}$ to l,

   Bending moment, $M_x = F.x - 2M$

   $$\therefore \quad \frac{\partial M_x}{\partial F} = x$$

$\therefore$ Deflection,

$$\Delta = \int_0^{L/2} \frac{(F.x - M)x\,dx}{EI} + \int_{L/2}^{L} \frac{(F.x - 2M)x\,dx}{2EI} \qquad ...(i)$$

In general,

$$\Delta = \int_0^L \frac{Mx}{EI}\frac{\partial Mx}{\partial F}dx \qquad \text{...(By Castigliano)}$$

Now in equation (*i*), F = 0 (dummy)

$$\therefore \quad \Delta = \int_0^{L/2} \frac{-M.x}{EI}dx + \int_{L/2}^{L} \frac{-2M.x}{2EI}dx$$

Solving, we get $\Delta = \left\{\frac{-Mx^2}{2EI}\right\}_0^{L/2} + \left\{\frac{-Mx^2}{2EI}\right\}_{L/2}^{L}$

$$\Rightarrow \qquad \Delta = \frac{ML^2}{2EI}$$

–ve sign indicates the deflection is in upward direction.

$$\therefore \qquad \Delta = \frac{ML^2}{2EI} \uparrow$$

31. Strain energy stored in a prismatic bar due to axial load P_1

$$U_1 = \frac{P_1^2 L}{2AE}$$

Stored energy stored in a prismatic bar due to load P_2

$$U_2 = \frac{P_2^2 L}{2AE}$$

$$U = \frac{(P_1 + P_2)^2 L}{2AE}$$

$$= \frac{(P_1^2 + P_2^2 + 2P_1P_2)L}{2AE}$$

$$= \frac{P_1^2 L}{2AE} + \frac{P_2^2 L}{2AE} + \frac{P_1P_2L}{AE}$$

$$\therefore \qquad U = U_1 + U_2 + \frac{P_1P_2L}{AE}$$

$$\therefore \qquad U > U_1 + U_2$$

32. Here, $A = 1550 \text{ mm}^2$

$E = 2 \times 10^5 \text{ N/mm}^2$

$PQ = \sqrt{(6)^2 + 4.5^2}$

$= 7.5$ m

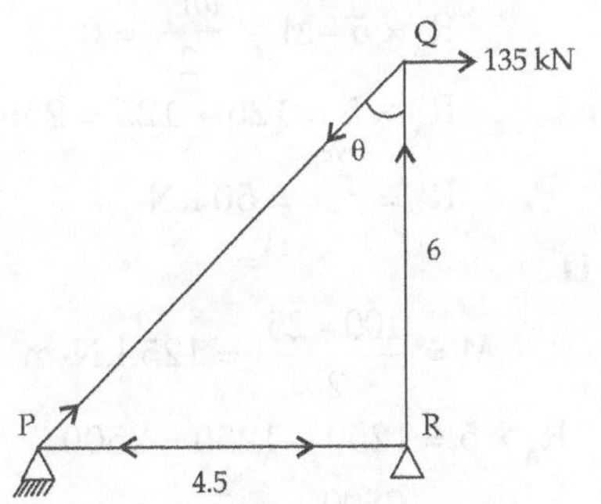

$$\therefore \qquad \sin\theta = \frac{4.5}{7.5} = 0.6$$

$$\text{and} \qquad \cos\theta = \frac{6}{7.5} = 0.8$$

Member	Length (l)	Force due to External load (P)	Force due to Unit load (u)	$P_u l$
PQ	7.5	225	1.67	2818.125
QR	6	–180	–1.33	1436.4
PR	4.5	–135	–1	607.5

$$\therefore \qquad \Sigma P_u l = 4862.025$$

Horizontal deflection of Q,

$$\Sigma H = \sum_{1}^{n} \frac{PuL}{AE}$$

$$= \frac{1}{AE} \Sigma PuL$$

$$= \frac{1}{1550 \times 2 \times 105} \times 4862.025$$

$$= 1.568 \times 10^{-5} \text{ m}$$

$$= \mathbf{15.68 \text{ mm}}$$

33.

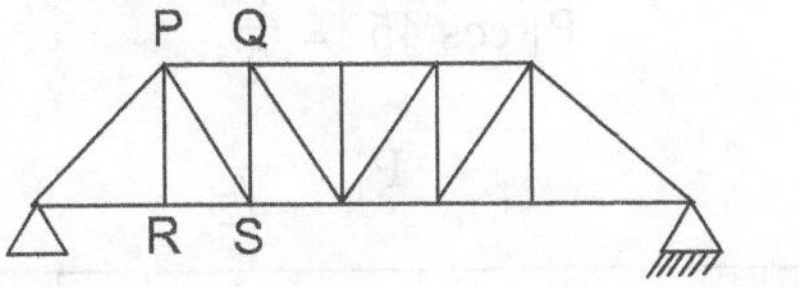

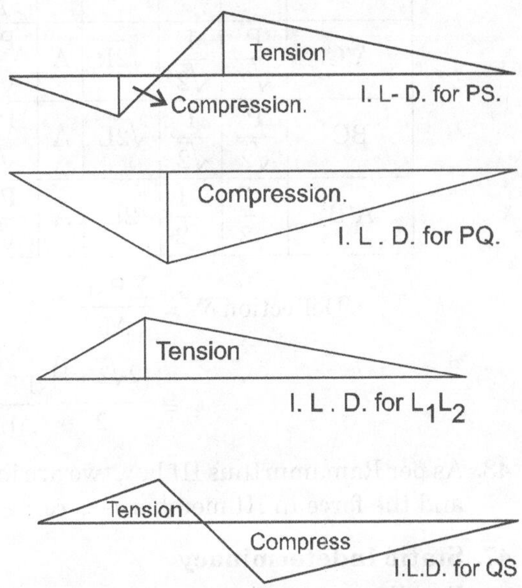

35. Given influence line diagram shows when unit load move from A to B, V_B change from 0 to 1.

37. Let j = number of joints

Degree of freedom of given frame

$= 3j$ – (number of constraints + number of members)

$= 3 \times 4 - (3 + 3)$

$= 6$

40. M is applied at B.

$$M = \frac{4EI\theta_1}{L} + \frac{4EI\theta_2}{L} + \frac{3EI\theta_3}{L}$$

Rotation at the joint $= \theta_1 + \theta_2 + \theta_3 = \theta$

$$\therefore \qquad \theta = \frac{ML}{11EI}$$

41. *Taking a section at A – B.*

$$P_{CB}\cos\theta = P_{AB}$$

But symmetry vertical reaction of B = $\frac{P}{2}$

$$P_{CB}\sin\theta = \frac{P}{2}$$

$$\therefore \quad P_{AB} = P_{CB}\cos\theta = P_{CB}\sin\theta = \frac{P}{2} \text{ (as } \theta = 45°)$$

42. Given : $\theta = 45°$

$$\therefore \quad P_1 \cos 45° = \frac{P}{2}$$

$$\Rightarrow \quad P_1 = \frac{P}{\sqrt{2}}$$

Member	P	k	L	A	$\frac{PkL}{A}$
A′C	$\frac{P}{\sqrt{2}}$	$\frac{1}{\sqrt{2}}$	$\sqrt{2}L$	A	$\frac{PL}{\sqrt{2}}$
BC	$\frac{P}{\sqrt{2}}$	$\frac{1}{\sqrt{2}}$	$\sqrt{2}L$	A	$\frac{PL}{\sqrt{2}}$
A′B	$-\frac{P}{2}$	$-\frac{1}{2}$	2L	A	$\frac{PL}{\sqrt{2}}$

$$\text{Deflection } \delta V = \frac{\Sigma PkL}{AE} = \frac{(2\sqrt{2}+1)}{2}\frac{PL}{AE}$$

43. As per Ramamurthus III law, two are in same line and the force in III member is zero.

47. Static Indeterminacy

D_3 = External indeterminacy + Internal indeterminacy

$= D_{se} + D_{si}$

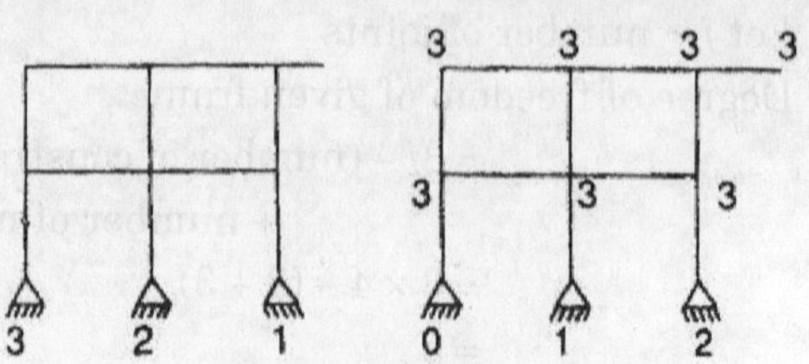

where, $D_{se} = R - 3 = 6 - 3 = 3$

$D_{si} = 3\,C = 3$ (closed boxes) = 3(2) = 6

$D_s = 3 + 6 = 9$

Kinematic Indeterminacy

Number of free rotations (horizontal, vertical rotational)

Neglecting axial deformations,

total free rotations = 24 – 11 = 13

d = 9

K = 13

48.

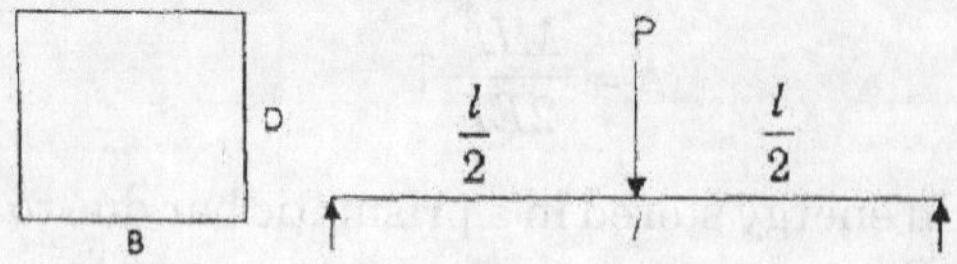

$$M = fZ$$

$$\therefore \quad \frac{PL}{4} = f_{max.}\left(\frac{BD^2}{6}\right)$$

$$\Rightarrow \quad f_{max} = \frac{3}{2}\frac{PL}{BD^2}$$

When the load is at centre, maximum B.M occurs.

49.

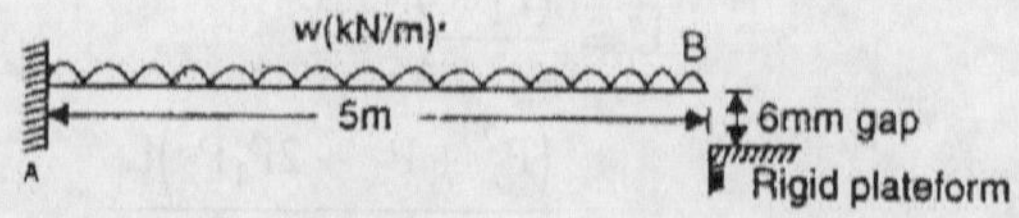

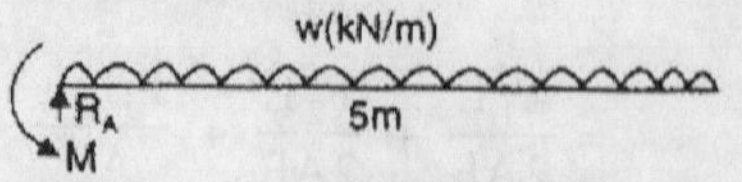

Case I.

$$M = \frac{wL^2}{2} = \frac{10 \times 25}{2} = 125 \text{ kN-m}$$

Now, $R_A \times 5 - M - \frac{wL^2}{2} = 0$

$\therefore \quad R_A \times 5 = 125 + 125 = 250$ kN-m

$\Rightarrow \quad R_A = \frac{25}{5} = 50$ kN

Case II.

$$M = \frac{100 \times 25}{2} = 125 \text{ kN-m}$$

$\therefore \quad R_A \times 5 = 1250 + 1250 = 2500$

$\Rightarrow \quad R_A = \frac{2500}{5} = 500$ kN.

50.

$$\sigma = \frac{\sigma_x - \sigma_y}{2}\cos 2\theta + \frac{\sigma_x - \sigma_y}{2}$$

$$\tau = -\left(\frac{\sigma_x - \sigma_y}{2}\right)\sin 2\theta$$

$$\tau_0 = -(\sigma_x - \sigma_y)\sin\theta\cos\theta$$

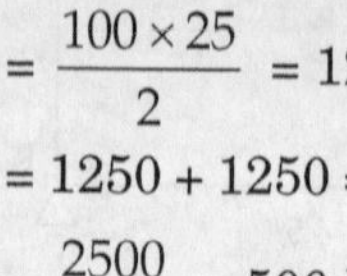

Here, '–' ve because opposite direction

$$\sin\theta = \frac{3}{5}$$

$$\cos\theta = \frac{4}{5}$$

$$\therefore \quad \tau_0 = -(\sigma_x - \sigma_y) \times \frac{3}{5} \times \frac{4}{5}$$

$$\Rightarrow \quad \sigma_y - \sigma_x = -145.83 \quad \text{....}(i)$$

Again, $$120 = \frac{\sigma_x - \sigma_y}{2}\cos 2\theta + \frac{\sigma_x - \sigma_y}{2}$$

$$\therefore \quad 240 = (\sigma_x - \sigma_y)(2\cos^2\theta - 1) + (\sigma_x + \sigma_y)$$

$$= (\sigma_x + \sigma_y)\left(2 \times \frac{16}{25} - 1\right) + (\sigma_x + \sigma_y)$$

$$\Rightarrow \quad 1.28\sigma_x + 0.72\sigma_y = 240 \quad \text{...}(ii)$$

Solving equations (*i*) and (*ii*), we get

$\sigma_x = 67.5$ MPa,

and $\sigma_y = 213.3$ MPa

51. Given : L = 5 mts

T = 100kN–*m*

GJ = 5000kN-m²/rad

Now, $$\frac{T}{\tau} = \frac{f_s}{R} = \frac{G\theta}{l}$$

$$= \frac{Tl}{GJ} = \frac{100 \times 5}{50000} = 0.01 \text{ radians}$$

Torsional strain energy

$$= \frac{T^2 l}{2GJ} = \frac{(100)^2 \times 5}{2 \times 50000} = 0.5 \text{ kN.m}$$

So statement I is correct; and statement II is false.

52. Maximum thrust when 20 kN load on crown.

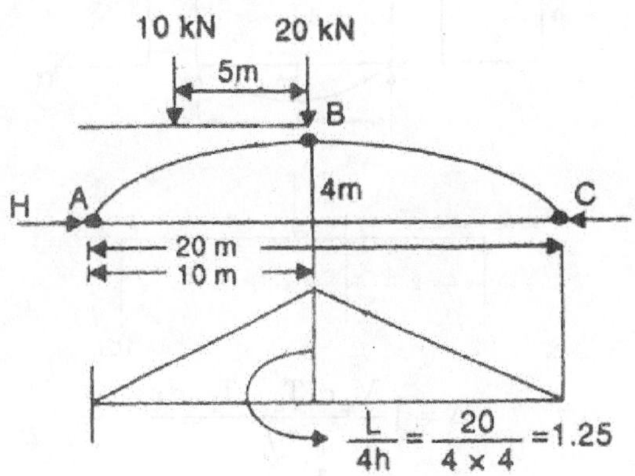

$$\therefore \quad H_{max} = 20 \times 1.25 + 10 \times 1.25 \times \frac{5}{10}$$

$$= 31.25 \text{ kN}$$

53. Vertical deflection at the centre of the span

$$= \frac{5}{384}\frac{wL^4}{EI}$$

$$= \frac{5}{384} \times \frac{10 \times 5^4}{81380}$$

$$= 1 \text{ mm.}$$

54. Idealize the structure

Shear force immediate left and right = 0 and 20 kN

55. *Given:*

Shear force immediate to the right of section E = 9.84 kN (↑)

Sagging moment at section E = 10.31 kN-m

$$\therefore \quad R_H + R_G - 10 - 5 \times 4 = -9.84$$

$$\Rightarrow \quad R_H + R_G = -9.84 + 30$$

$$\Rightarrow \quad R_H + R_G = 20.16 \quad \text{...}(i)$$

and $$R_H \times 6 + R_G \times 2 - 10 \times 1 - 5 \times 4 \times 4 = -10.31$$

$$\Rightarrow \quad 3R_H + R_G = 39.845 \quad \text{...}(ii)$$

Solving equations (*i*) and (*ii*), we get

$R_H = 9.84$kN (upward).

56. According to Muller-Breslau principle, the influence line for any stress-function of a structure, such as S.F., B.M. or any reactive force or moment is given by deflected curve of a structure obtained by imposing a unit distortion in the direction of the stress function.

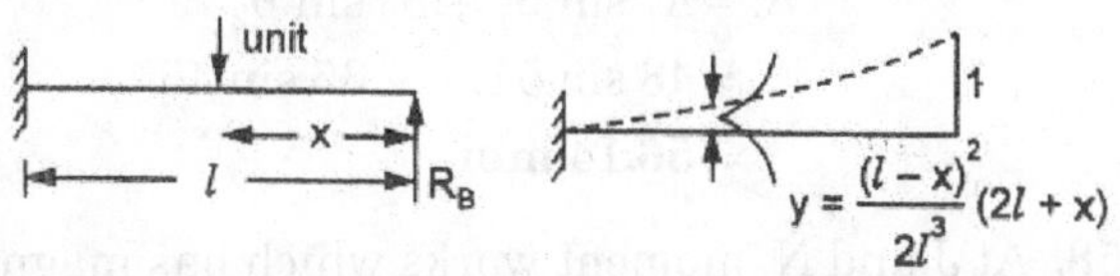

To draw influence line for R_B , a displacement equal to 1 is given in the direction of R_B.

By principle,

$$R_B \times 1 - 1 \times y = 0$$

$$\therefore \quad R_B = y$$

57.

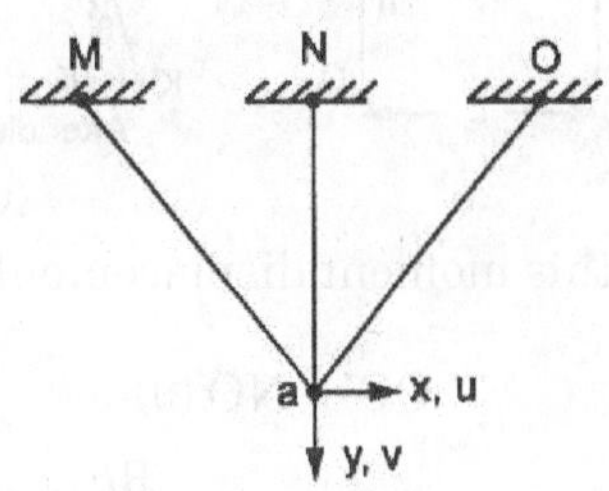

Given : Δ MQ = 48 mm

Δ OQ = 35 mm

Clearly, position of bar members becomes

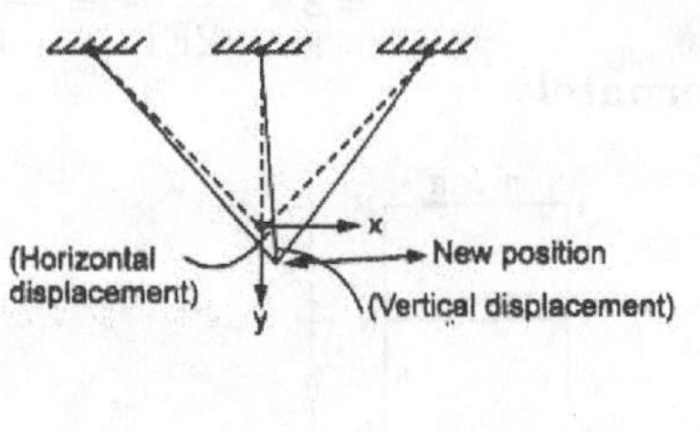

Both the displacements take place positive values. So, we have only one option, *i.e.* 6.64 and 56.14.

To find Vertical displacement

A force F is applied at Q as shown

$$\Delta_y = \frac{\partial U}{\partial F} = \frac{1}{EI}\int M\frac{\partial M}{\partial F}dS$$

To find horizontal displacement

A force F is applied at Q as shown

$$\Delta_x = \frac{\partial U}{\partial F} = \frac{1}{EI}\int M\frac{\partial M}{\partial F}dS$$

$\therefore \quad \Delta_x = 6.64$ mm

and $\quad \Delta_y = 56.14$ mm

Alternately

$$\Delta_u = \Delta_1 \cos\theta_1 - \Delta_2 \cos\theta_2$$

Here, $\tan\theta_2 = \frac{500}{500} = 1$

$\Rightarrow \quad \theta_2 = 45°$

and $\quad \tan\theta_1 = \frac{500}{400} = 1.25$

$\Rightarrow \quad \theta_1 = 51.3°$

$\therefore \quad \Delta_u = 4P\cos 51.3° - 35\cos 45°$

$= 6.64$ mm

$\Delta_v = \Delta_1 \sin\theta_1 + \Delta_2 \sin\theta_2$

$= 48 \sin 51.3° + 35 \sin 45°$

$=$ **56.14 mm**

58. At J and N, moment works which has magnitude Rh.

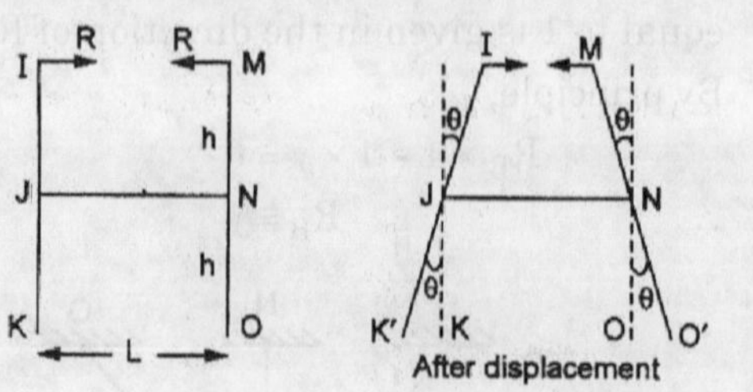

Due to this moment displacement becomes $\frac{RhL}{2EI}$

Now $\quad OO' = NO(\theta)$

$$= h\theta = \frac{RhL}{2EI} \times h = \frac{Rh^2L}{2EI}$$

Now total displacement between K and O

$$= 2 \times \frac{Rh^2L}{2EI} = \frac{RLh^2}{EI}$$

Alternately

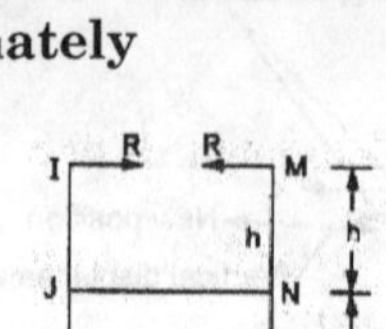

$$\theta = \frac{RhL}{2EI}$$

$\therefore \quad \delta = \theta h$

$$= \frac{RLh^2}{2EI}$$

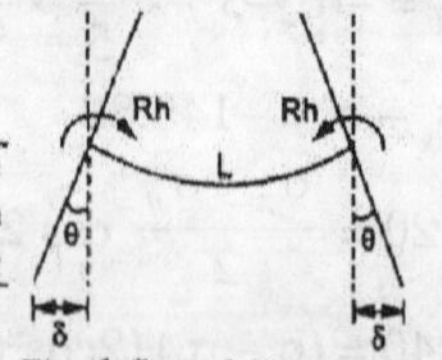

Fig. deflected diagram

Hence relative displacement between K and O

$$= \frac{RLh^2}{EI}$$

59.

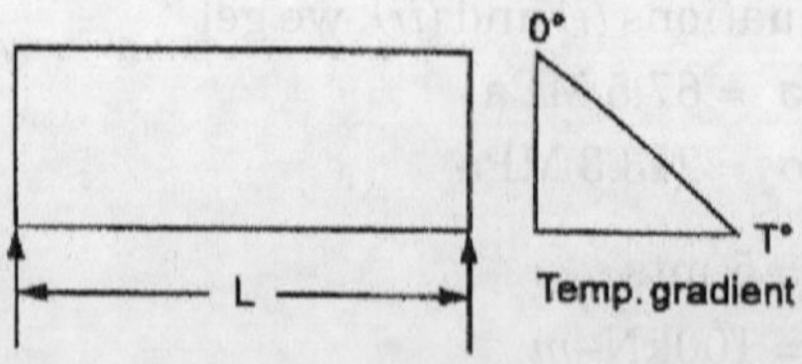

After expansion of beam

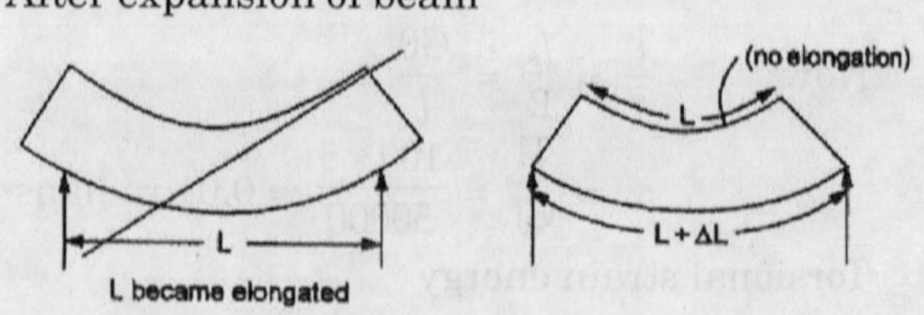

$$\delta = \frac{\alpha TL^2}{8h} \text{ (downward)}$$

Alternately

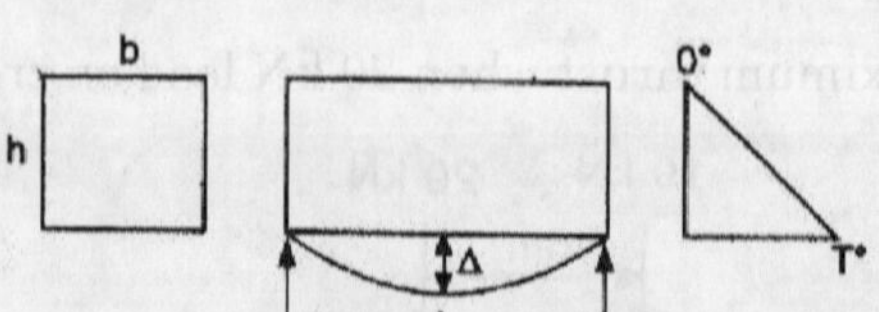

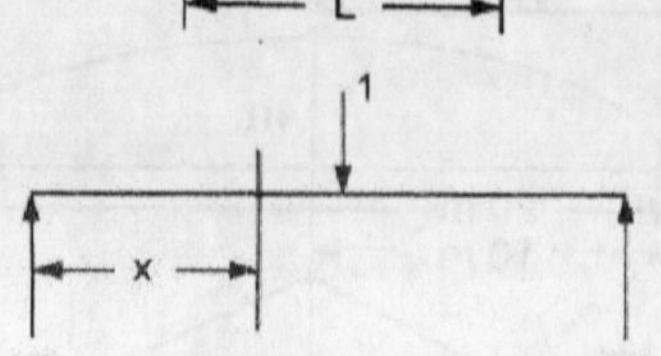

$$\Delta = \int \frac{M_u \alpha (T_2 - T_1)\,dx}{h}$$

$$= 2\int_0^{l/2} \frac{\frac{1}{2}x.\alpha\,(T-0)\,dx}{h}$$

$$= 2 \times \frac{1}{2}\frac{\alpha\,T}{h}\left(\frac{x^2}{2}\right)_0^{l/2}$$

$$\Delta = \frac{\alpha TL^2}{8h} \text{ (downward)}$$

60.

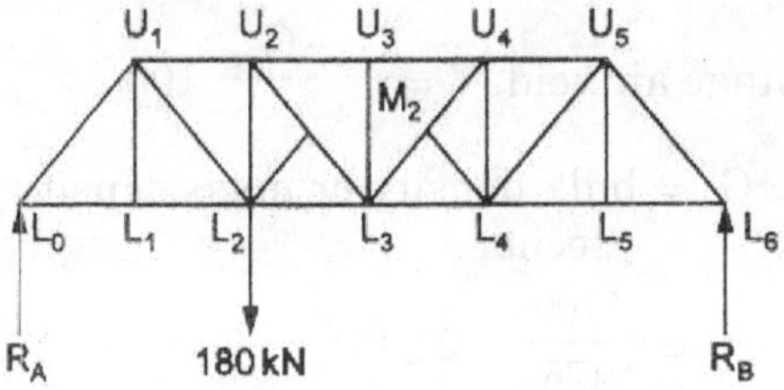

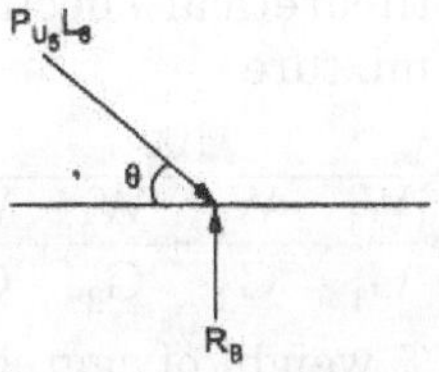

After solving, $R_A = 120$ kN

$R_B = 60$ kN

Now, $P_{U_5L_6} \sin\theta = R_B$

$\Rightarrow \quad P_{U_5L_6} = \dfrac{R_B}{\sin\theta} = 100$ kN

$\therefore \quad P_{U_5L_6} = P_{U_5L_6} = 100$ kN

and same is the force in U_4M_2 and is tension.

Alternately

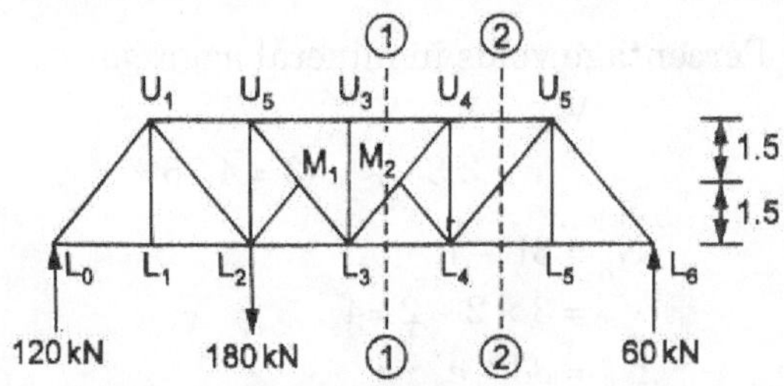

From method of section 1 – 1

For L.H.S. $\Sigma V = 0$

$\therefore \quad F_{M_2L_2} = \dfrac{120 \times 5}{3}$

$= 200$ (compression)

Moment about $L_3 = 0$

$\therefore \quad F_{U_3U_4} = 240$ (compression)

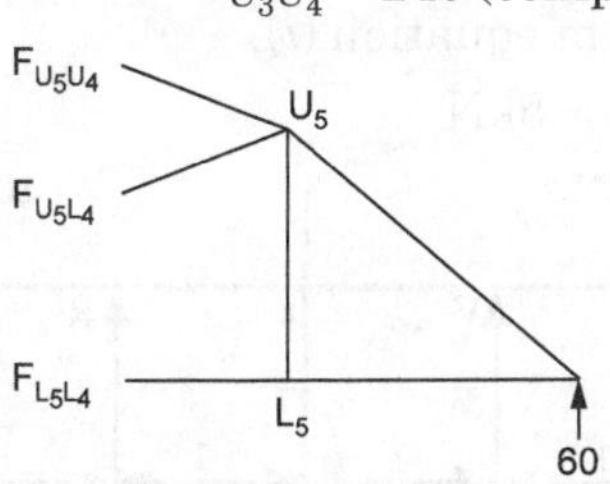

From method of section about 2 – 2

$\Sigma V = 0$

$\therefore \quad F_{U_5L_4} = 100$ (tension)

Now, about $\Sigma M\ L_5 = 0$

$\therefore \quad F_{U_5L_4} = 160$ (compression)

Considering equilibrium of joint U_4

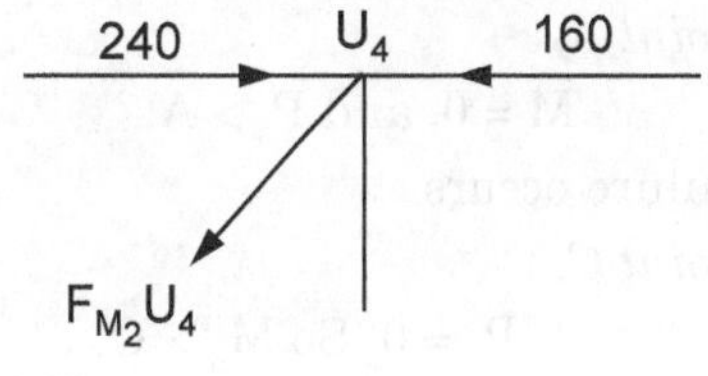

From $\Sigma H = 0$

$$F_{M_2U_4} \times \frac{4}{5} = 240 - 160$$

$$\therefore \quad F_{M_2U_4} = \frac{80 \times 5}{4}$$

$= 100$ kN (tension)

61. (*a*) **Beam mechanism :**

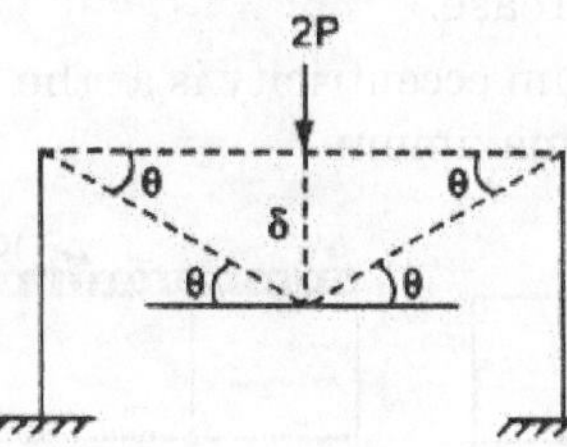

$$2M_P\theta + 2M_P\theta = 2PL\theta$$

$$\therefore \quad L = \frac{2M_P}{L}$$

(*b*) **Combined mechanism :**

Plastic hinges are formed at A,C,D and under the load

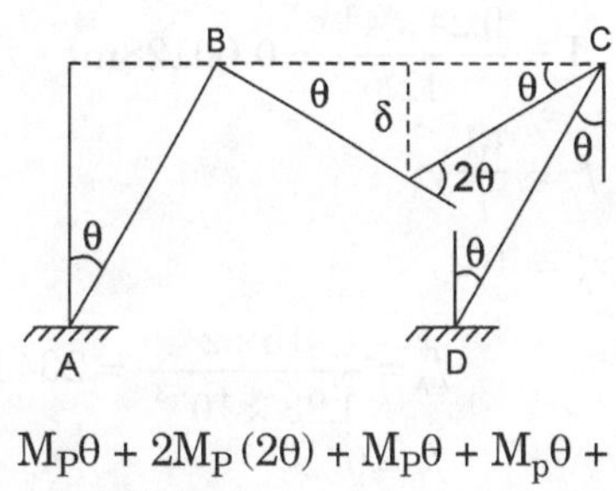

$$\therefore \quad M_P\theta + 2M_P(2\theta) + M_P\theta + M_p\theta + M_P\theta = 2PL\theta$$

$$\Rightarrow \quad P_u = \frac{3M_P}{L}$$

Alternately

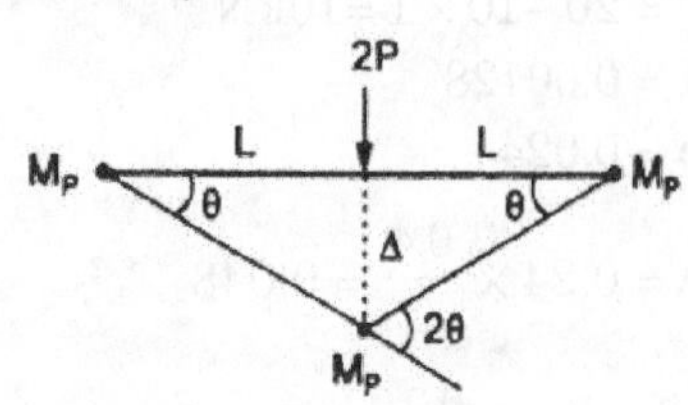

$$2P \times \Delta = M_P\theta + M_P\theta + 2M_P \times 2\theta$$

$$\Rightarrow \quad 2P\Delta = 6M_P\theta$$

$$\Rightarrow \quad 2P_u \times \frac{\theta L}{L\theta} = 6M_P\theta$$

$$\Rightarrow \quad P_u = \frac{6M_P}{2L} = \mathbf{\frac{3M_P}{L}}$$

63. (*a*) *At point A,*

$M = 0$, and $P_u > A$;

so failure occurs.

At point C,

$P_u = 0$, So $M_u > C$,

so failure occure

(*b*) Portion AB is *compression member* and Portion BC is called *tension zone.*

(*c*) In region BC region of tension failure at C, ultimate moment is M_u while the ultimate load is 0.

As load is of tension in nature, it produces negative moment and the net moment will be less than M_u, Hence moment carrying capacity will increase.

(*d*) Minimum eccentricity is at the point A, so the load is maximum.

64. 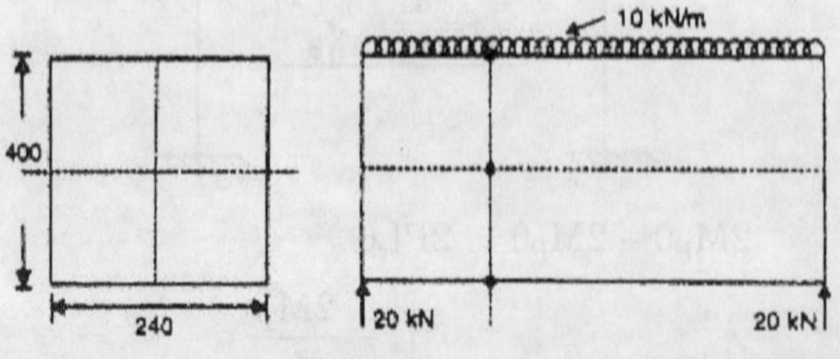

We Know, $f = \frac{M}{I} Y$

where, $M = 20 \times 1 - 10 \times 1 \times \frac{1}{2} = 15$ KN-m

$$I = \frac{0.24 \times 4^3}{12} = 0.00128 m^3$$

Now, $f_A = \frac{M}{I} y_A$

where, $y_A = 0.2$

$$f_A = \frac{15 \times .2}{1.28 \times 10^{-3}} = 2343.75 kN/m^2$$

$$f_B = 0$$

and $f_C = -\frac{15 \times 0.2}{1.28 \times 10^{-3}} = -2343.75 kN/m^3$

Now, $F_x = 20 - 10 \times 1 = 10kN$

$I = 0.00128$

$b = 0.024$

$$A = 0.24 \times \frac{0.4}{2} = 0.048$$

$$\bar{y} = 0.24 \times \frac{0.2}{2} = 0.24$$

$$\therefore \quad \tau = \frac{F_0}{I.b} A\bar{y}$$

$$= \frac{10}{0.00128 \times .24} \times 0.048 \times 0.024$$

$= 37.5$ kN/m

65. Percentage air void, $V_v = \frac{G_t - G_m}{G_m} 100$

where, G_m = bulk density or mass density of the specimen

$$= \frac{1100}{475} = 2.315$$

and G_t = theoretical specific gravity of the mixture

$$= \frac{100}{\frac{W_1}{G_1} + \frac{W_2}{G_2} + \frac{W_3}{G_3} + \frac{W_4}{G_4}},$$

where, W = % weight of aggregate

$$\Rightarrow \quad G_t = \frac{100}{\frac{8.25}{2.63} + \frac{12}{2.51} + \frac{3.25}{2.46} + \frac{1.5}{2.43}}$$

$$= \frac{100}{3.14 + 4.78 + 1.32 + 0.62} = 10.14$$

$$\therefore \quad V_v = \frac{10.14 - 2.34}{2.34} = 3.33\%$$

66. Percentage bitumen by vacum,

$$V_b = G_m \frac{W_4}{G_4} = 2.35 \times \frac{1.5}{2.43} = 1.43\%$$

67. Percentage voids in mineral aggregate,

$$V_{MA} = V_v + V_b$$

$$= 3.33 + 1.43 = 4.76\%$$

68. $N_S = 3L - R$

$= 3 \times 2 - 2 = 4$

$N_K = 3J - e$

$= 3 \times 6 - (3 + 2 + 2)$

$= 6$

69. $R + V = 30$ ...(*i*)

Now, $V \times 2 - 20 \times 3 + 10 \times 1 = 0$

$$\Rightarrow \quad V = \frac{60 - 10}{2} = 25kN$$

Hence from equation (*i*),

$R = 5kN$

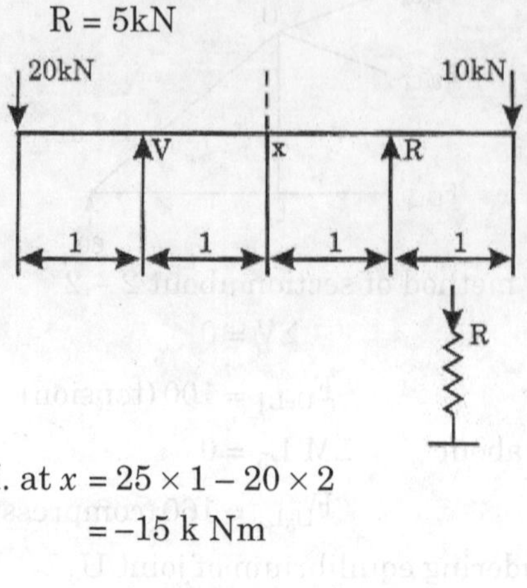

B. M. at $x = 25 \times 1 - 20 \times 2$

$= -15$ k Nm

73. OA $-\frac{3}{4}\frac{I}{L}$

OB $\rightarrow 0$

OC $= \frac{I}{L}$

Total stiffness $= \frac{7}{4}\frac{I}{L}$

Stiffness Factors : $\frac{3M}{7}, 0, \frac{4M}{7}$

74.

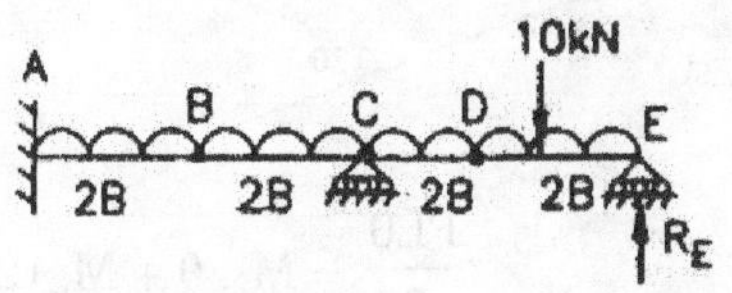

Here, $R_E = 10$ kN

Now, $R_E \times 6 + R_C \times 2 = 10 \times 6 \times 3$

$$\Rightarrow R_C = \frac{180-60}{2} = 60kN$$

$$M_A = 10\times 8 + 60 \times 4 - 10 \times 8 \times 4$$
$$= 80 + 240 - 320 = 0$$

76.

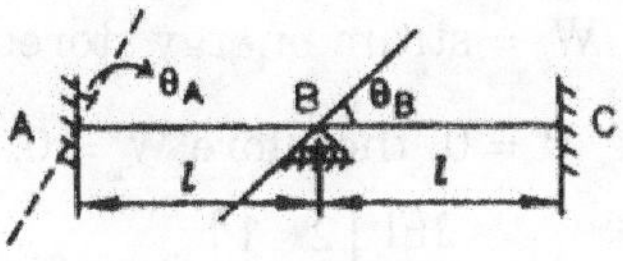

From sloped deflection equation

$$M_{AB} = \frac{2EI}{L}(2\theta_A+\theta_B)$$

$$M_{BA} = \frac{2EI}{L}(\theta_A+2\theta_B)$$

$$M_{BC} = \frac{2EI}{L}(2\theta_B+0)$$

$$M_{CB} = \frac{2EI}{L}(\theta_B)$$

At joint B, $M_{BA}+M_{BC} = 0$

$$\Rightarrow \frac{2EI}{L}[\theta_A+2\theta_B+2\theta_B] = 0$$

$$\Rightarrow 4\theta_B+\theta_A = 0$$

$$\Rightarrow \theta_B = -\frac{\theta_A}{4}$$

$$= -\frac{0.004}{4} \text{ radians}$$

$$= -0.001 \text{ radians (anticlockwise)}$$

$$M_{AB} = \frac{2\times 25,000}{5}(2\,0\,004 - 0\,001)$$
$$= 70kNm$$

$$M_{BA} = \frac{50,000}{5}(0.004 - 0.002)$$
$$= 20\,kNm$$

$$M_{BC} = 10,000(-2\times 0.001)$$
$$= -20kNm$$

$$M_{CB} = 10,000\times -0.001 = -10kNm$$

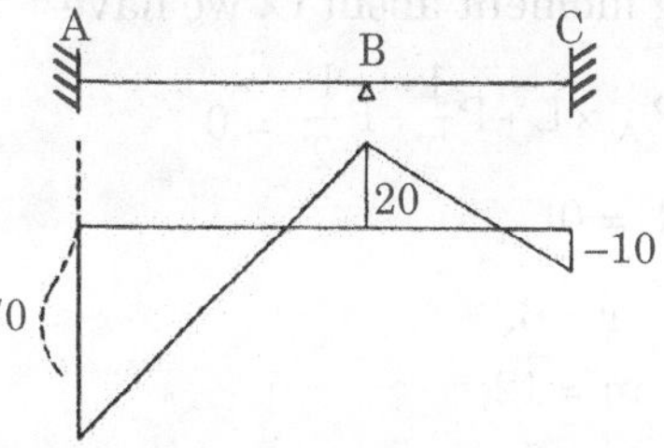

77. Sway mechanism (hinge form at P and Q)

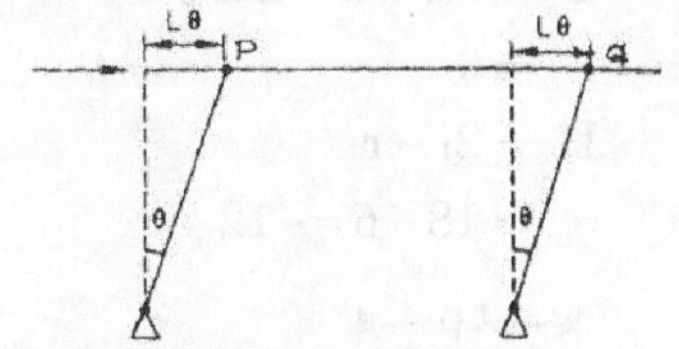

Beam mechanism (hinge at P, Q and R)

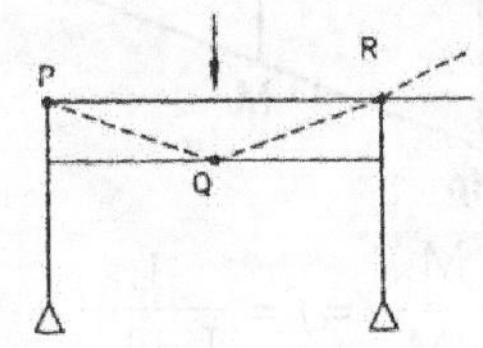

Combined mechanism hinge at P and Q.

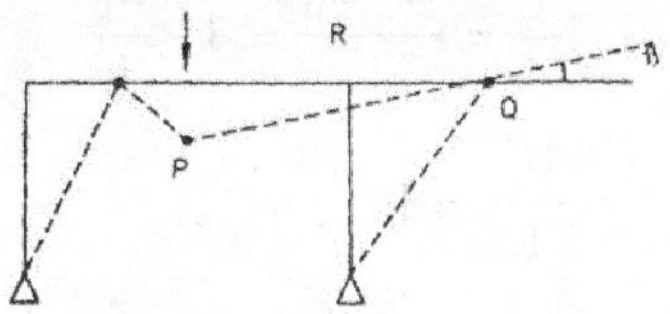

Equating virtual work done, we have

$$w\frac{l}{2}\theta + w\frac{l}{3}\theta = 4M_P\theta$$

Let UDL acting as a concertrated load at $\frac{2}{3}\left(\frac{L}{2}\right) = \frac{L}{3}$ from end.

$$\therefore \quad wl\left(\frac{\theta}{2}+\frac{\theta}{3}\right) = 4M_P\theta$$

$$\Rightarrow \quad wl\left(\frac{3+2}{6}\right)\theta = 4M_p\theta$$

$$\Rightarrow \quad wl \times \frac{5}{6} = 4\,M_p$$

$$\therefore \quad \text{Collapse load, } w = \frac{24}{25}\frac{M_p}{l}$$

78. $E.I = ML^{-1}T^{-2} \times L^4 = ML^3T^{-2}$

79.

In real beam	In conjugate beam
Fix-end	Free end
Internal hinge	Roller support
Roller support	Internal hinge
Hinge	Hinge

80. The frame is statically determinate.
Taking moment about C, we have

$$R_A \times L + P\frac{1}{2} - P\frac{L}{2} = 0$$

$\Rightarrow \quad R_A = 0$

81. Here, $r = 6$,
$m = 12$,
$j = 9$

Then $D_s = 12 + 6 - 2 \times 9$
$= 0$
$D_k = 2j - r$
$= 18 - 6 = -12$

82.

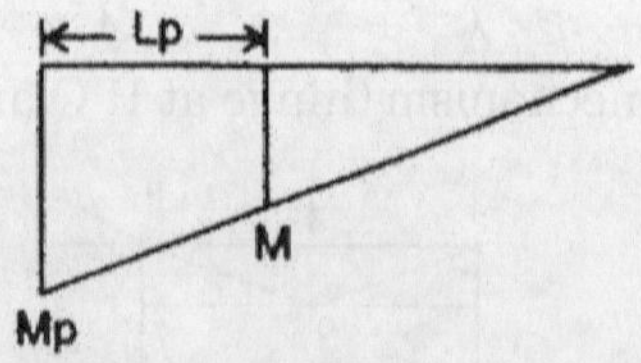

$$\frac{M_P}{M} = f = \frac{L}{L - L_p}$$

$$\Rightarrow \quad \frac{1}{f} = \frac{L - L_P}{L} = 1 - \frac{L_P}{L}$$

$$\Rightarrow \quad \frac{L_P}{L} = 1 - \frac{1}{f}$$

83. Consider free body diagram.

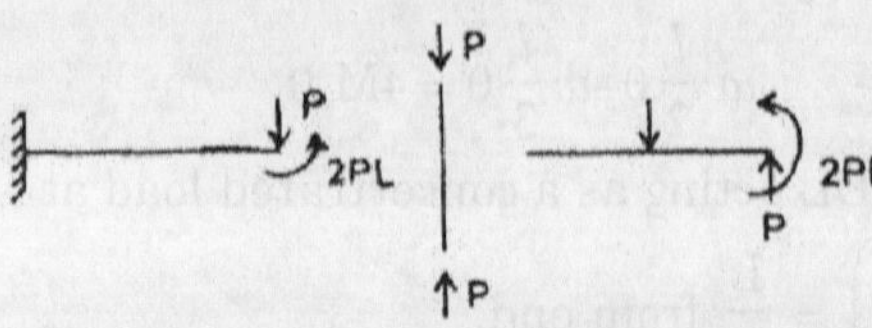

$$\text{Deflection at } A = \frac{PL^3}{3EI} - \frac{ML^2}{2EI}$$

$$= \frac{P.(3L)^3}{3EI} - \frac{2PL.(3L)^2}{2EI}$$

$$= \frac{P \times 27L^3}{3EI} - \frac{2PL.9L^2}{2EI} = 0$$

84. Consider a beam with span L, having plastic moment capacity of M_P
Let C is the collapse load.
Considering Bending Moment diagram, from geometry of collapse, we have

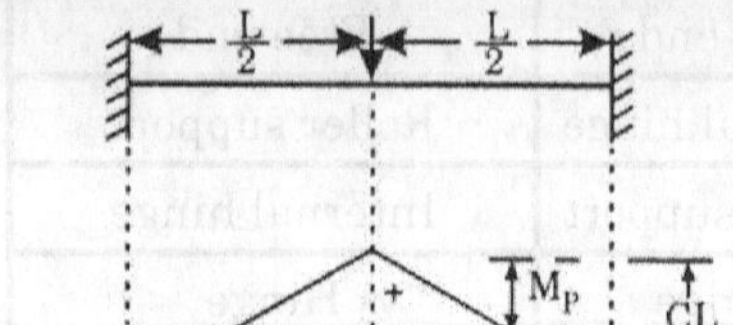

$$M_P + M_P = \frac{Cl}{4}$$

$$\therefore \quad \text{Collapse load, } C = \frac{8M_P}{L}$$

85. For kinematic equilibirium,

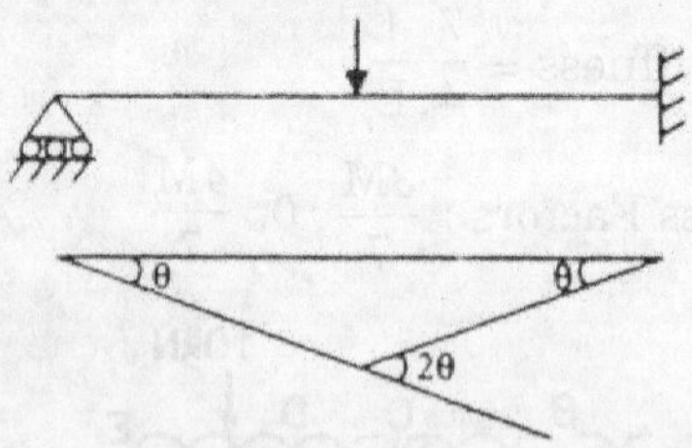

$$\frac{PL\theta}{2} = M_0.\,\theta + M_0\,(2\theta)$$

$$\text{or} \quad P = \frac{6M_0}{L}$$

86. Since the force in the member AB is zero, therefore the strain energy stored in the member AB will be zero.

Let W_i = strain energy stored $= \dfrac{P^2L}{2EI}$

Since $P = 0$, therefore $W_i = 0$.

87. If $A = \dfrac{2EI}{L}\begin{bmatrix} 2 & 1 \\ 1 & 2 \end{bmatrix}$ is stiffness matrix, then

$$|A| = \begin{bmatrix} 2 & 1 \\ 1 & 2 \end{bmatrix} = [4 - 1] = 3$$

$$\text{The matrix of cofactors} = \begin{bmatrix} 2 & -1 \\ -1 & 2 \end{bmatrix}$$

$$\text{Transpose of matrix, } A^T = \begin{bmatrix} 2 & -1 \\ 1 & 2 \end{bmatrix}$$

$$\text{Inverse of matrix} = \frac{A^T}{|A|} = \frac{L}{2EI}\begin{bmatrix} 2 & -1 \\ -1 & 2 \end{bmatrix} \times \frac{1}{3}$$

$$= \frac{L}{6EI}\begin{bmatrix} 2 & -1 \\ -1 & 2 \end{bmatrix}$$

= Flexibility matrix

88. Since the support A and D are fixed, there can not be displacements of those joints. However joints B and C each have three degrees of freedom. Thus the kinemetic indeterminacy of structure is 6. Since axial deformatons can be neglected for each column and horizontal displacement of B and C would be equal. Thus the given frame is kinematically indeterminate to three degrees, i.e. one each at B and C and one horizontal displacement.

89. The maximum bending moment in the column will be at fixed end and is equal to $100 \times 4 = 400$ kNm

NUMERICAL TYPE QUESTIONS

1.

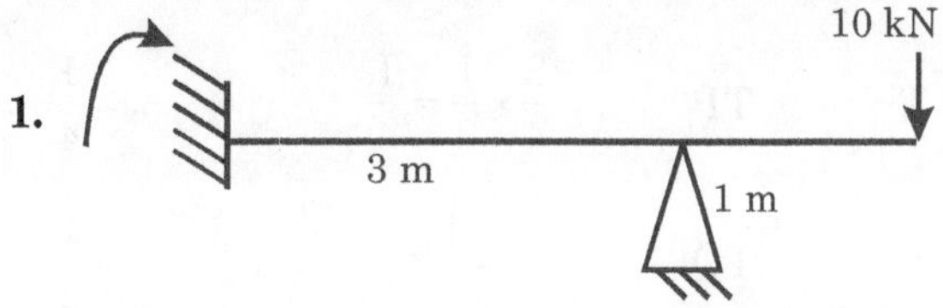

$$\frac{10\times 1}{2} = 5\text{kN}-\text{m}$$

2. $P_{u,\,min} = \dfrac{CM_P}{L}$

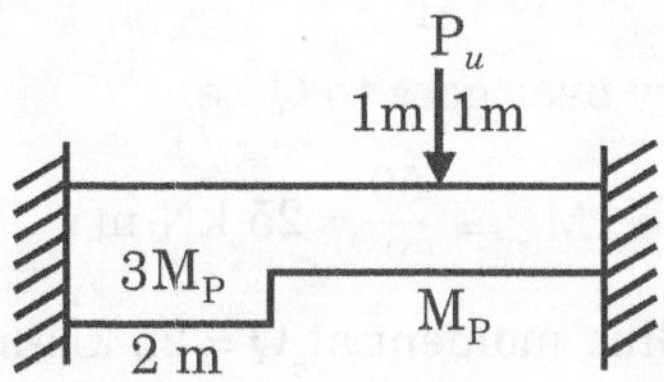

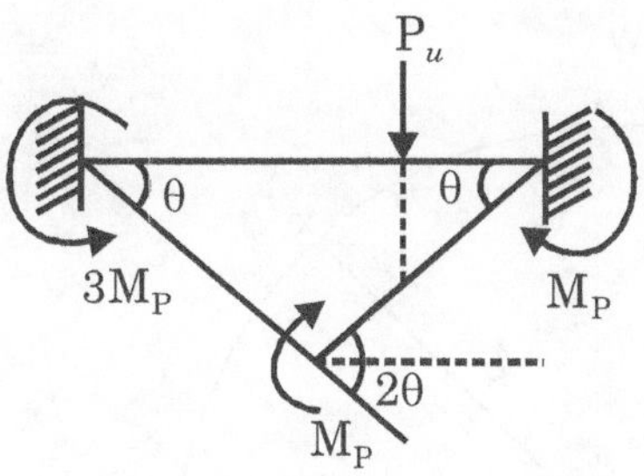

I[st] Possibility,

$$\therefore \quad P_u \times \frac{L}{4}\theta = (3M_P + 2M_P + M_P)\theta$$

$$P_u = \frac{6M_P}{L}\times 4 = \frac{24M_P}{L}$$

IInd Possibility,

$$\therefore \quad P_u.\frac{3L}{4}\theta = 3M_P\theta + M_P(\theta+\alpha) + M_P(\alpha)$$

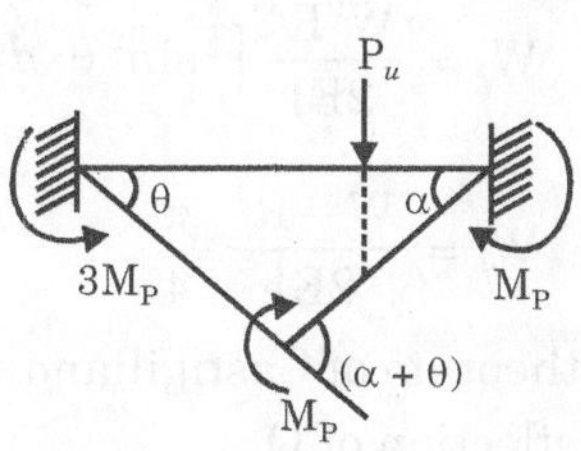

Also, $\quad 3\theta = \alpha$

$$\therefore \; P_u \times \frac{3L}{4}\theta = 3M_P\theta + M_P(\theta+3\theta) + M_P(3\theta)$$

$$\therefore \quad P_u\frac{L}{4} = \frac{10M_P\theta}{3\theta}$$

$$\Rightarrow \quad P_u\frac{L}{4} = \frac{10}{3}M_P$$

$$\Rightarrow \quad P_u = \frac{40}{3}\frac{M_P}{L}$$

$$P_{u\,min} = \frac{40}{3}\frac{M_P}{L}$$

$$\therefore \quad C = \frac{40}{3} = 13.33$$

3. $R_A \times 3 + 10 \times 9 = 0$

$$\Rightarrow \quad R_A = -30\text{ N}$$

$$R_C = 35\text{ kN}$$

$$F_x = 10\text{ kN}$$

$$U = \frac{F^2\times L}{2A_E} = \frac{10\times 10\times 3}{2\times 30}$$

$$= 5\text{ kN}-\text{m}$$

5. Since deflection at R is required. So, let us apply a virtual unit load at point R in upwards direction.

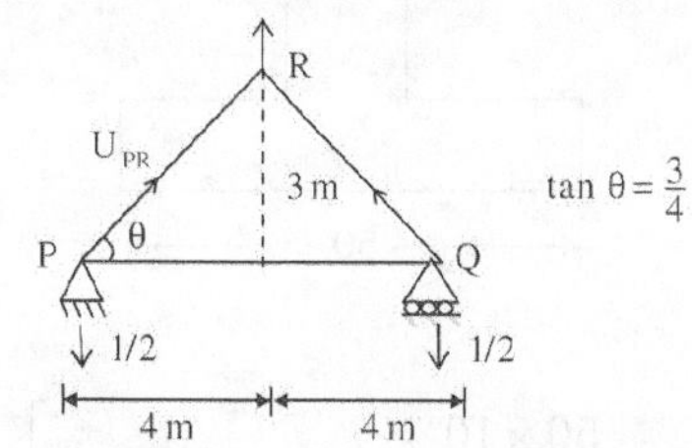

At point P $\Sigma F_V = 0$

$$\Rightarrow \quad U_{PR}\sin\theta = \frac{1}{2}$$

$$\Sigma F_H = 0$$

$$\Rightarrow \quad U_{PR}\cos\theta + U_{PQ} = 0$$

$$\Rightarrow \quad U_{PQ} = -U_{PR}\cos\theta$$

$$= -\frac{1}{2\sin\theta}.\cos\theta$$

$$= -\frac{1}{2}+\frac{1}{\tan\theta}$$

$$= -\frac{1}{2}\times\frac{4}{3} = {}^{-2}\!/_{3}$$

$$\delta_R = U_{PQ}\times\lambda_{PQ}$$

$$= \frac{-2}{3}\times(-3) \;[\because \text{PQ is 3 mm short}] = 2\text{ mm}$$

So, deflection at R = 2 mm (upwards)

6. Static Indeterminacy (SI)

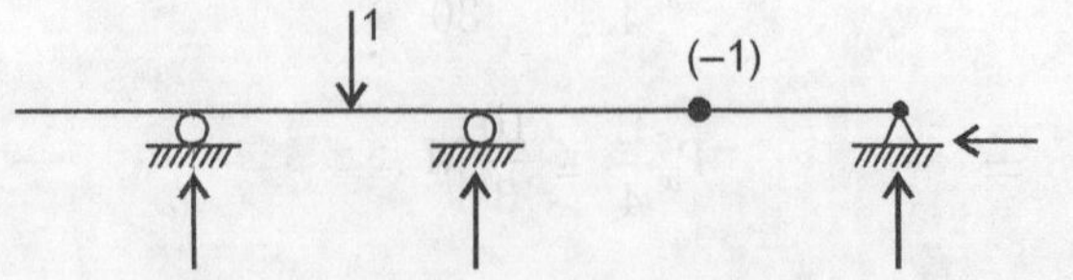

(1 + 1 + 2)   – 1
(2 roller + 1 hinge) (Internal hinge)

= 3

Static Indeterminacy

= No. of Unknowns – Equation of equilibrium

= 3 – 3

= 0

7. Symetrical I-section

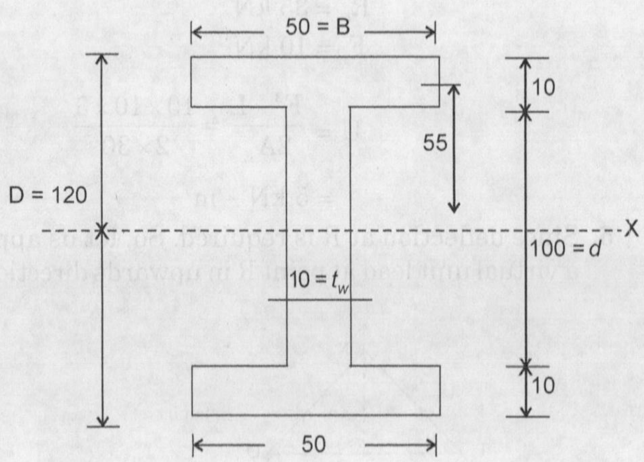

$$I_{xx} = 2 \times \left[\frac{50 \times 10^3}{12} + (50 \times 10) \times 55^2\right] + \frac{10 \times 100^3}{12}$$

$$I_{xx} = 3.86 \times 10^6 \text{ mm}^4$$

Magnitude of shear stress in the web at its junction with top of flange.

$$q_w = \frac{B}{t_w} q_t$$

$$q_t = \frac{F}{8I}(D^2 - d^2)$$

$$= \frac{100 \times 10^3}{8 \times 3.86 \times 10^6} \times ((120)^2 - (100)^2)$$

$$q_t = 14.22 \text{ N/mm}^2$$

$$\therefore q_w = \frac{50}{10} \times 14.22 = 71.1 \text{ N/mm}^2$$

8. As per the statement the moment of 100 kN-m is divided in a ratio as per D.F of member TR, TP and TQ

So calculate the distribution FACTOR.

JOINT	MEMBER	Relative Stiffness (k)	Σk	D.F. $= \frac{k}{\Sigma k}$
T	TP	$\frac{3}{4} \times \frac{I}{3} = \frac{I}{4}$		$-\frac{1}{4}$
	TQ	$-\frac{I}{2}$	I	$-\frac{1}{2}$
	TR	$-\frac{I}{4}$		$-\frac{1}{4}$

So moment comes in TQ i.e.

$$M_{TQ} = \frac{1}{2} \times 100 = 50 \text{ kN-m}$$

And the carry over goes to Q i.e.

$$M_{QT} = \frac{50}{2} = 25 \text{ kN-m}$$

$\because$ The bending moment at Q = 25 kN-m

9.

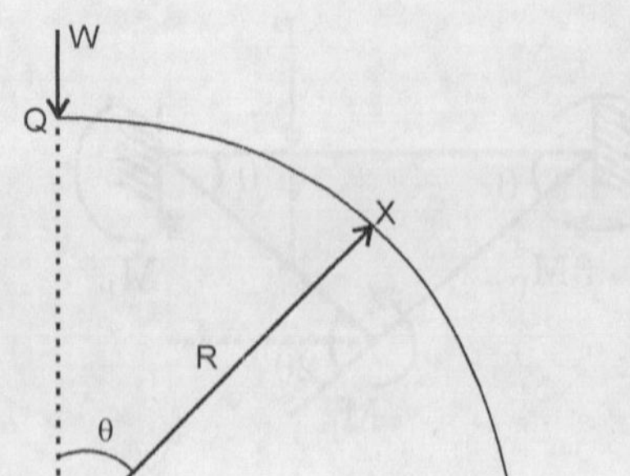

The bending moment at any section X, the radius vector OX making an angle θ with the verticle is given by

$$M = -WR \sin\theta$$

$\therefore$ Strain energy stored

$$W_i = \int \frac{M^2 ds}{2EI} = \int_0^{\frac{\pi}{2}} \frac{W^2R^2 \sin^2\theta}{2EI} \cdot R.d\theta$$

$$W_i = \frac{W^2R^3}{2EI} \int_0^{\frac{\pi}{2}} \sin^2\theta \cdot d\theta$$

$$W_i = \frac{W^2 \cdot R^3}{2EI} \cdot \frac{\pi}{4}$$

Apply Ist theorem of Castigiliano

Vertical deflection of Q,

$$S_q = \frac{\partial W_i}{\partial W} = \frac{2WR^3}{2EI} \cdot \frac{\pi}{4}$$

$$S_q = \frac{\pi}{4}\left(\frac{WR^3}{EI}\right)$$

So, the value of $\beta = 0.7854$

10.

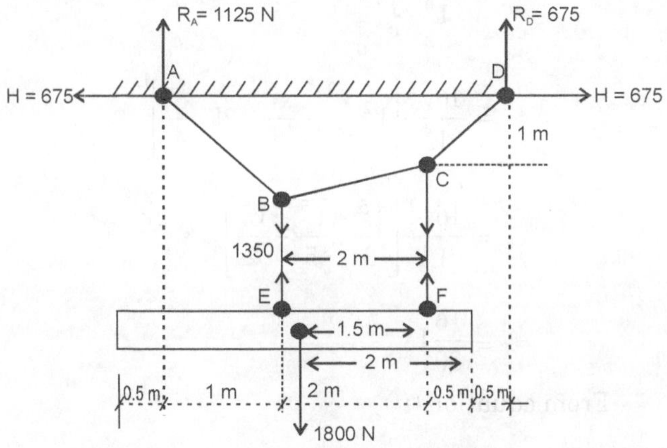

Force in cable BE and CF due 1800 N load

$$\Sigma Y = 0 \qquad F_{BE} + F_{CF} = 1800 \qquad ...(i)$$

$$\Sigma M_C = 0 \qquad F_{BE} \times 2 = 1800 \times 1.5$$

$$F_{BE} = 1350 \text{ N}$$

$$F_{CF} = 1800 - 1350 = 450 \text{ N}$$

After that the question will solve as like cable suspension bridge problem.

Taking moment about C, considering only right hand side portion

$$R_D \times 1.0 = H \times 1.0$$

$$R_D = H$$

Taking moment about C, considering only left hand side portion

$$R_A \times 3.0 = H \times 1.0 + 1350 \times 2.0$$

$$R_A = \frac{H}{3} + 900$$

But, $\quad R_A + R_B = 1800$ N

$$\frac{H}{3} + 900 + H = 1800$$

$$\frac{4}{3}H = 1800 - 900$$

$$= 900$$

$$H = 675 \text{ N}$$

$$R_A = \frac{675}{3} + 900 = 1125 \text{ N}$$

$$R_D = 675 \text{ N}$$

then the tension in cable AB = T_{AB}

$$T_{AB} = \sqrt{(675)^2 + (1125)^2}$$

$$= 1311.9 \text{ N}$$

11.

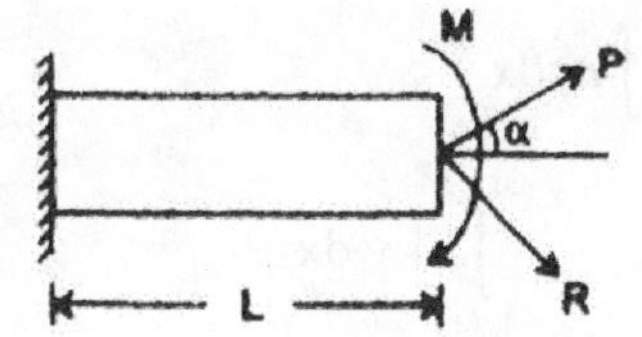

Given load is along P, so deflection perpendicular to P is zero.

By Castiglino's first theorem

$$\int_0^L \frac{M}{EI} \cdot \frac{\partial M_x}{\partial R} . dx = 0 \qquad ...(i)$$

where, R = reaction perpendicular to P.

Moment in X-direction

$$M_x = P \sin \alpha . x - R \cos \alpha . x - M = 0$$

$$\therefore \qquad \frac{\partial M_x}{\partial R} = -\cos \alpha.$$

From equation (*i*)

$$\int_0^L [-Px^2 \sin \alpha \cos \alpha + R \cos^2 \alpha . x + M.x \cos \alpha]\, dx = 0$$

Putting R = 0 and integrating, we get

$$-\frac{P}{2} \sin 2\alpha . L^3 + M.\frac{L^2}{2} \cos \alpha = 0$$

Putting $M = \frac{P.L}{3}$, we get

$$\frac{P.L^3}{6} \cos \alpha = \frac{P}{6} L^3 \sin 2\alpha$$

$$\Rightarrow \qquad \cos \alpha = 2 \sin \alpha \cos \alpha$$

$$\therefore \qquad \sin \alpha = \frac{1}{2}$$

12.

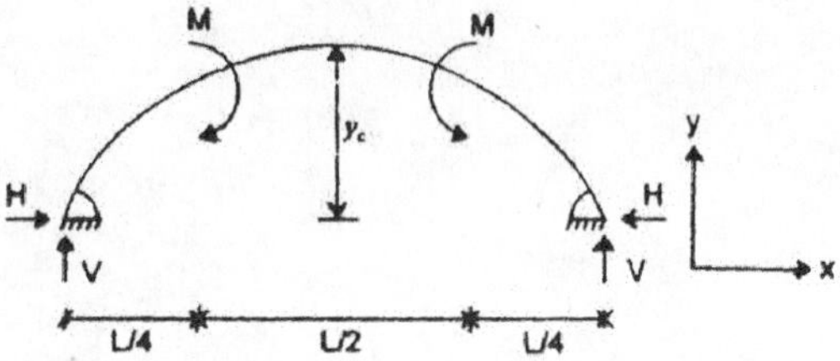

At any section distant x

$$y = \frac{4}{L^2} y_c x(L-x) \qquad ...(i)$$

and horizontal reaction

$$H = -\frac{\int \mu y dx}{\int y^2 dx} \qquad ...(ii)$$

Here, $\int \mu y dx$

$$= \int_{L/4}^{3L/4} M.y\, dx$$

$$= \frac{4y_c}{L^2}.M \int_{L/4}^{3L/4} (Lx - x^2)dx$$

$$= \frac{4y_c}{L^2}.M\left[\frac{1}{2}x^2 - \frac{x^3}{3}\right]_{L/4}^{3L/4}$$

$$= \frac{4y_c}{L^2} M\left[\left(\frac{9}{32}L^3 - \frac{27}{192}L^3\right) - \left(\frac{L^3}{32} - \frac{L^3}{64}\right)\right]$$

$$= \frac{11}{24} M.y_c L$$

and $\int_o^L y^2 dx$

$$= \int_o^L \left[\frac{4y_c}{L_2}x(L-x)\right]^2 dx$$

$$= \frac{16y_c^2}{L^4}\int_0^L x^2 (L-x)^2 dx$$

$$= \frac{16y_c^2}{L^4}\int_0^L (x^2L^2 + x^4 - 2Lx^3)\, dx$$

$$= \frac{16y_c^2}{L^4}\left[L^2\frac{x^3}{3} + \frac{x^5}{5} - 2L.\frac{x^4}{4}\right]_o^L$$

$$= \frac{16y_c^2}{L^4}\left[\frac{L^5}{3} + \frac{L^5}{5} - \frac{L^5}{2}\right]$$

$$= \frac{16}{30}Ly_c^2$$

From equation (*ii*)

$$H = -\left(\frac{11}{24}My_c .L\right) \times \left(\frac{30}{16}.\frac{1}{y_c^2 L}\right)$$

$$= \frac{55}{64}\frac{M}{y_c}$$

and from equation (*i*),

$$V = 0$$

4 CHAPTER Construction Materials and Management

MCQ TYPE QUESTIONS

2016

1. A construction project consists of twelve activities. The estimated duration (in days) required to complete each of the activities along with the corresponding network diagram is shown below.

	Activity	Duration (days)
A	Inauguration	1
B	Foundation work	7
C	Structural construction-1	30
D	Structural construction-2	30
E	Brick masonry work	25
F	Plastering	7
G	Flooring	25
H	Electrification	7
I	Plumbing	7
J	Wood work	7
K	Coloring	3
L	Handing over function	1

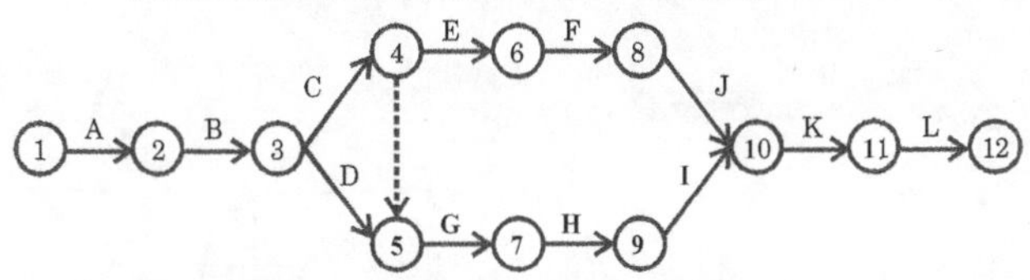

Total floats (in days) for the activities 5-7 and 11-12 for the project are, respectively,

(*a*) 25 and 1 (*b*) 1 and 1

(*c*) 0 and 0 (*d*) 81 and 0

2. The activity-on-arrow network of activities for a construction project is shown in the figure. The durations (expressed in days) of the activities are mentioned below the arrows.

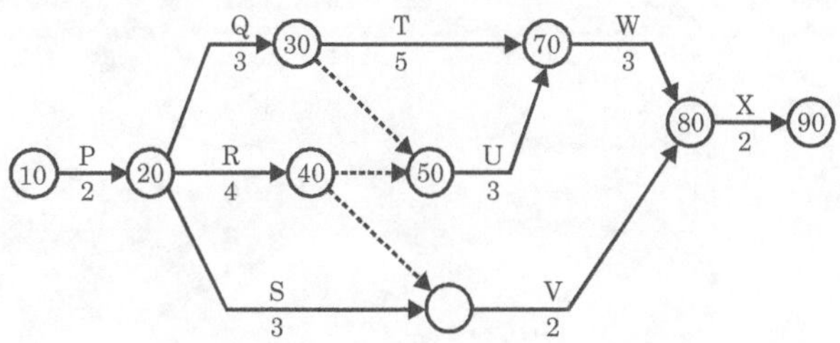

The critical duration for this construction project is

(*a*) 13 days

(*b*) 14 days

(*c*) 15 days

(*d*) 16 days

NUMERICAL TYPE QUESTION

2016

1. The Optimistic Time (O), Most likely Time (M) and Pessimistic Time (P) (in days) of the activities in the critical path are given below in the format O-M-P.

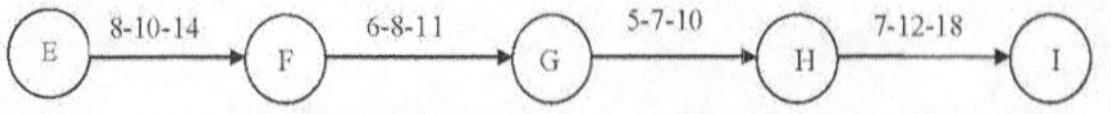

The expected completion time (in days) of the project is ________.

ANSWERS

MCQ Type Questions

1. (*c*) **2.** (*c*)

Numerical Type Question

1. (38)

EXPLANATIONS

MCQ TYPE QUESTION

1.

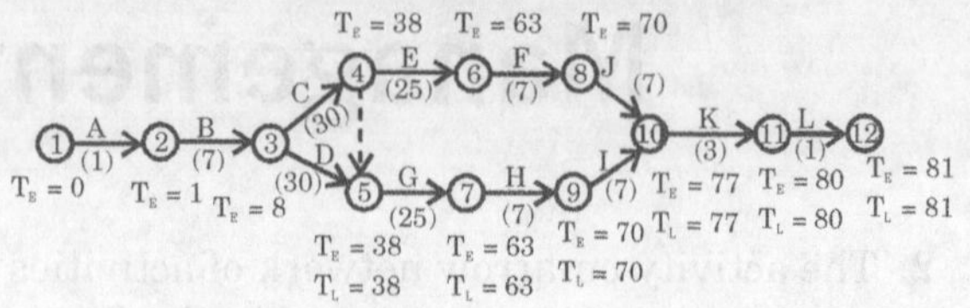

Activity (5)–(7), $F_T = (63 - 38) - 25 = 0$

(11)–(12), $F_T = (81 - 80) - 1 = 0$

NUMERICAL TYPE QUESTION

1. Activity — Expected duration,

$$t_E = \frac{t_o + 4t_m + t_p}{6}$$

E – F $\quad \dfrac{8 + 4x10 + 14}{6} = 10.33$

F – G $\quad \dfrac{6 + 4x8 + 11}{6} = 8.16$

G – H $\quad \dfrac{5 + 4x7 + 10}{6} = 7.16$

H – I $\quad \dfrac{7 + 4x12 + 18}{6} = 12.16$

Expected completion time

5 CHAPTER

Concrete Structures

MCQ TYPE QUESTIONS

2016

1. In shear design of an RC beam, other than the allowable shear strength of concrete (τ_c), there is also an additional check suggested in IS 456-2000 with respect to the maximum permissible shear stress ($\tau_{c\ max}$). The check for $\tau_{c\ max}$ is required to take care of

(*a*) additional shear resistance from reinforcing steel

(*b*) additional shear stress that comes from accidental loading

(*c*) possibility of failure of concrete by diagonal tension

(*d*) possibility of crushing of concrete by diagonal compression

2. A reinforced concrete (RC) beam with width of 250 mm and effective depth of 400 mm is reinforced with Fe415 steel. As per the provisions of IS 456-2000, the minimum and maximum amount of tensile reinforcement (expressed in mm^2) for the section are, respectively

(*a*) 250 and 3500

(*b*) 205 and 4000

(*c*) 270 and 2000

(*d*) 300 and 2500

3. As per IS 456-2000 for the design of reinforced concrete beam, the maximum allowable shear stress ($\tau_{c\ max}$) depends on the

(*a*) grade of concrete and grade of steel

(*b*) grade of concrete only

(*c*) grade of steel only

(*d*) grade of concrete and percentage of reinforcement

4. As per Indian standards for bricks, minimum acceptable compressive strength of any class of burnt clay bricks in dry state is

(*a*) 10.0 MPa

(*b*) 7.5 MPa

(*c*) 5.0 MPa

(*d*) 3.5 MPa

2015

5. Consider the following statements for air-entrained concrete :

(*i*) Air-entrainment reduces the water demand for a given level of workability

(*ii*) Use of air-entrained concrete is required in environments where cyclic freezing and thawing is expected.

Which of the following is TRUE?

(*a*) Both (*i*) and (*ii*) are True

(*b*) Both (*i*) and (*ii*) are False

(*c*) (*i*) is True and (*ii*) is False

(*d*) (*i*) is False and (*ii*) is True

6. Workability of concrete can be measured using slump,compaction factor and Vebe time. Consider the following statements for workability of concrete :

(*i*) As the slump increases, the Vebe time increases

(*ii*) As the slump increases, the compaction factor increases

Which of the following is TRUE ?

(*a*) Both (*i*) and (*ii*) are True

(*b*) Both (*i*) and (*ii*) are False

(*c*) (*i*) is True and (*ii*) is False

(*d*) (*i*) is False and (*ii*) is True

7. Consider the singly reinforced beam shown in the figure below :

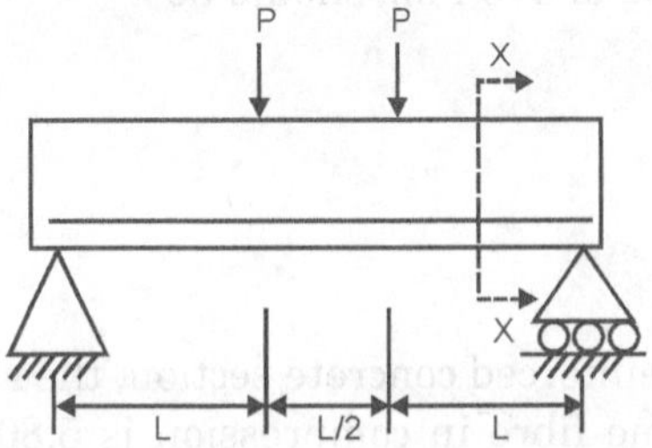

At cross-section XX, which of the following statements is TRUE at the limit state?

(*a*) The variation of stress is linear and that of strain is non-linear

(*b*) The variation of strain is linear and that of stress is non-linear

(*c*) The variation of both stress and strain is linear

(*d*) The variation of both stress and strain is non-linear

8. A column of size 450 mm × 600 mm has unsupported length of 3.0 m and is braced against side sway in both directions. According to IS 456 : 2000, the minimum eccentricities (in mm) with respect to major and minor principle axes are

(*a*) 20.0 and 20.0 (*b*) 26.0 and 21.0
(*c*) 26.0 and 20.0 (*d*) 21.0 and 15.0

9. In a pre-stressed concrete beam section shown in the figure, the net loss is 10% and the final prestressing force applied at X is 750 kN. The initial fiber stresses (in N/mm²) at the top and bottom of the beam were:

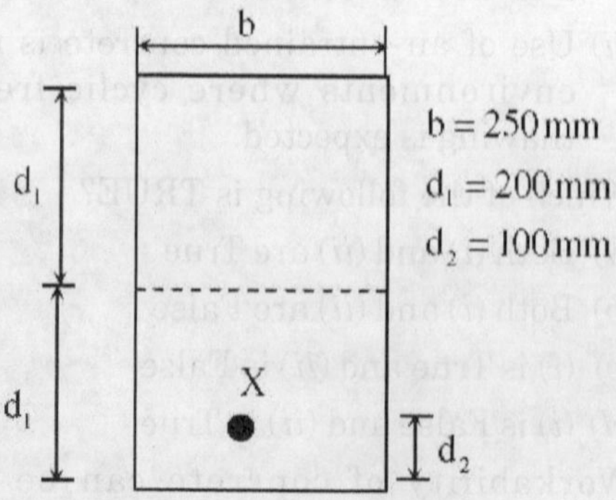

(*a*) 4.166 and 20.833 (*b*) –4.166 and –20.833
(*c*) 4.166 and –20.833 (*d*) –4.166 and 20.833

2014

10. A rectangular beam of width (*b*) 230 mm and effective depth (*d*) 450 mm is reinforced with four bars of 12 mm diameter. The grade of concrete is M20 and grade of steel if Fe500. Given that for M20 grade of concrete the ultimate shear strength, τ_{uc} = 0.36 N/mm² for steel percentage, p = 0.25, and τ_{uc} = 0.48 N/mm² for p = 0.50. For a factored shear force of 45kN, the diameter (in mm) of Fe500 steel two legged stirrups to be used at spacing of 375 mm, should be

(*a*) 8
(*b*) 10
(*c*) 12
(*d*) 16

11. In a reinforced concrete section, the stress at the extreme fibre in compression is 5.80 MPa. The depth of neutral axis in the section is 58 mm and the grade of concrete is M25. Assuming linear elastic behavior of the concrete, the effective curvature of the section (in per mm) is

(*a*) 2.0×10^{-6}
(*b*) 3.0×10^{-6}
(*c*) 4.0×10^{-6}
(*d*) 5.0×10^{-6}

12. Group I contains representative stress-strain curves as shown in the figure, while **Group II** gives the list of materials. Match the stress-strain curves with the corresponding materials.

Group I	**Group II**
P. Curye J	1. Cement paste
Q. Curve K	2. Coarse aggregate
R. Curve L	3. Concrete

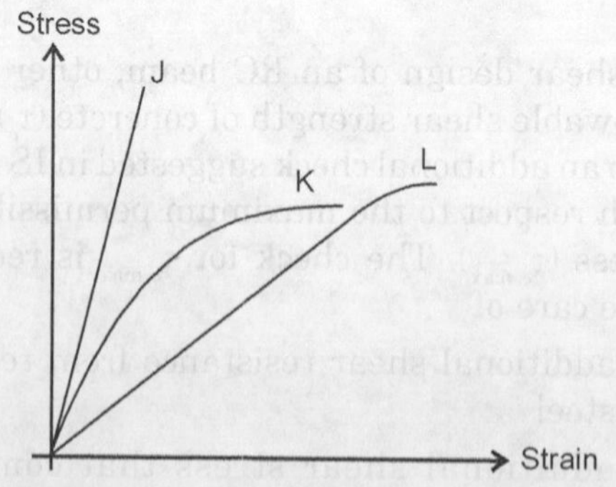

Codes :

	P	Q	R
(*a*)	1	3	2
(*b*)	2	3	1
(*c*)	3	1	2
(*d*)	3	2	1

13. The target mean strength f_{cm} for concrete mix design obtained from the characteristic strength f_{ck} and standard deviation s, as defined in IS:456-2000, is

(*a*) $f_{ck} + 1.35\sigma$ (*b*) $f_{ck} + 1.45\sigma$
(*c*) $f_{ck} + 1.55\sigma$ (*d*) $f_{ck} + 1.65\sigma$

14. The modulus of elasticity, $E = 5000\sqrt{f_{ck}}$ where f_{ck} is the characteristic compressive strength of concrete, specified in IS:456-2000 is based on

(*a*) tangent modulus (*b*) initial tangent modulus
(*c*) secant modulus (*d*) chord modulus

2013

15. Maximum possible value of Compacting Factor for fresh (green) concrete is

(*a*) 0.5 (*b*) 1.0
(*c*) 1.5 (*d*) 2.0

16. The creep strains are

(*a*) caused due to dead loads only
(*b*) caused due to live loads only
(*c*) caused due to cyclic loads only
(*d*) independent of loads

17. Four columns of a building are to be located within a plot size of 10 m × 10 m. The expected load on each column is 4000 kN. Allowable bearing capacity of the soil deposit is 100 kN/m². The type of foundation best suited is

(*a*) isolated footing (*b*) raft foundation
(*c*) pile foundation (*d*) combined footing

18. Select the strength parameter of concrete used in design of plain jointed cement concrete pavements from the following choices:

(*a*) Tensile strength (*b*) Compressive strength
(*c*) Flexural strength (*d*) Shear strength

19. Following bearings are observed while traversing with a compass.

Line	Fore Bearing	Back Bearing
AB	126°45′	308°00′
BC	49°15′	227°30′
CD	340°30′	161°45′
DE	258°30′	78°30′
EA	212°30′	31°45′

After applying the correction due to local attraction, the corrected fore bearing of line BC will be

(*a*) 48°15′ (*b*) 50°15′
(*c*) 49°45′ (*d*) 48°45′

2012

20. If a small concrete cube is submerged deep in still water in such a way that the pressure exerted on all faces of the cube is *p*, then the maximum shear stress developed inside the cube is

(*a*) 0 (*b*) $\frac{p}{2}$
(*c*) *p* (*d*) 2*p*

21. As per IS 456:2000, in the Limit State Design of a flexural member, the strain in reinforcing bars under tension at ultimate state should not be less than

(*a*) $\frac{f_y}{E_s}$ (*b*) $\frac{f_y}{E_s} + 0.002$
(*c*) $\frac{f_y}{L15\ E_s}$ (*d*) $\frac{f_y}{L15\ E_y} + 0.002$

22. Which one of the following is categorised as a long-term loss of prestress in a prestressed concrete member?

(*a*) Loss due to elastic shortening
(*b*) Loss due to friction
(*c*) Loss due to relaxation of strands
(*d*) Loss duo to anchorage slip

23. A concrete beam prestressed with a parabolic tendon is shown in the sketch. The eccentricity of the tendon is measured from the centroid of the cross-section. The applied prestressing force at service is 1620 kN. The uniformly distributed load of 45 kN/m includes the self-weight.

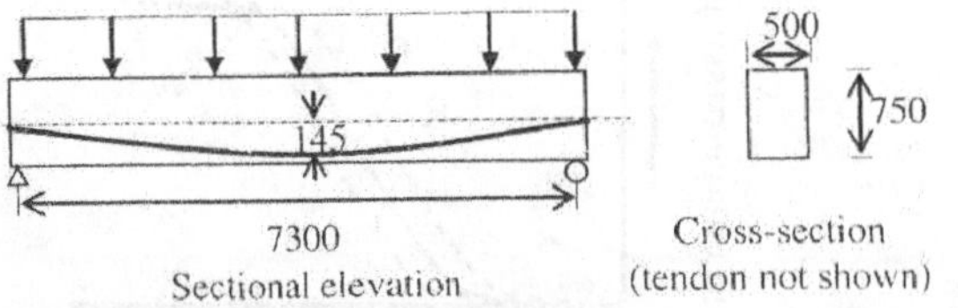

All dimensions are in mm

The stress (in N/mm²) in the bottom fibre at mid-span is

(*a*) tensile 2.90
(*b*) compressive 2.90
(*c*) tensile 4.32
(*d*) compressive 4.32

2011

24. A 16 mm thick plate measuring 650 mm× 420 mm is used as a base plate for an ISHB 300 column subjected to a factored axial compressive load of 2000 kN. As per IS 456-2000, the minimum grade of concrete that should be used below the base plate for safely carrying the load is

(*a*) M 15 (*b*) M 20
(*c*) M 30 (*d*) M 40

25. Consider a reinforcing bar embedded in concrete. In a marine environment this bar undergoes uniform corrosion, which leads to the deposition of corrosion products on its surface and an increase in the apparent volume of the bar. This subjects the surrounding concrete to expansive pressure. As a result, corrosion induced cracks appear at the surface of concrete. Which of the following statements is TRUE?

(*a*) Corrosion cause circumferential tensile stresses in concrete and the cracks will be parallel to the corroded reinforcing bar.
(*b*) Corrosion causes radial tensile stesses in concrete and the cracks will be parallel to the corroded reinforcing bar.
(*c*) Corrosion causes circumferential tensile stresses in concrete and the cracks will be perpendicular to the direction of the corroded reinforcing bar.
(*d*) Corrosion cause radial tensile stresses in concrete and the cracks will be perpendicular to the direction of the corroded reinforcing bar.

26. The results for sieve analysis carried out for three types of sand, P, Q and R, are given in the adjoining figure. If the fineness modulus values of the three sands are given as FM_P, FM_Q and FM_R, it can be stated that

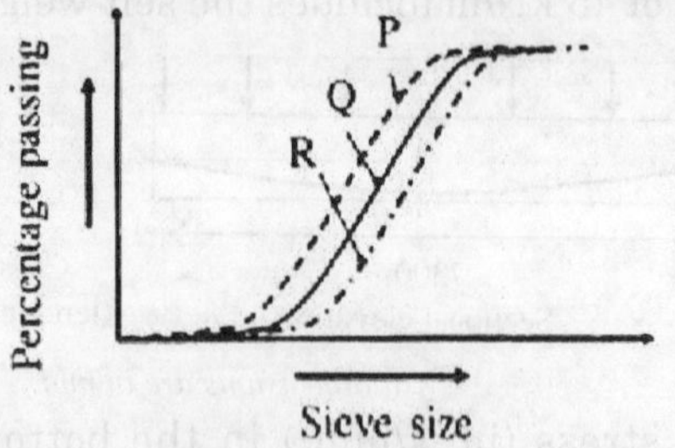

(a) $FM_Q = \sqrt{FM_P \times FM_R}$

(b) $FM_Q = 0.5\ (FM_P + FM_R)$

(c) $FM_P > FM_Q > FM_R$

(d) $FM_P < FM_Q < FM_R$

27. The cross-section of a thermo-mechanically treated (TMT) reinforcing bar has

(a) soft ferrite-pearlite throughout.

(b) hard martensite throughout.

(c) a soft ferrite-pearlite core with a hard martensitic rim.

(d) a hard martensitic core with a soft pearlite-bainitic rim.

28. Consider a simply supported beam with a uniformly distributed load having a neutral axis (NA) as shown. For points P (on the neutral axis) and Q (at the bottom of the beam) the state of stress is best represented by which of the following pairs?

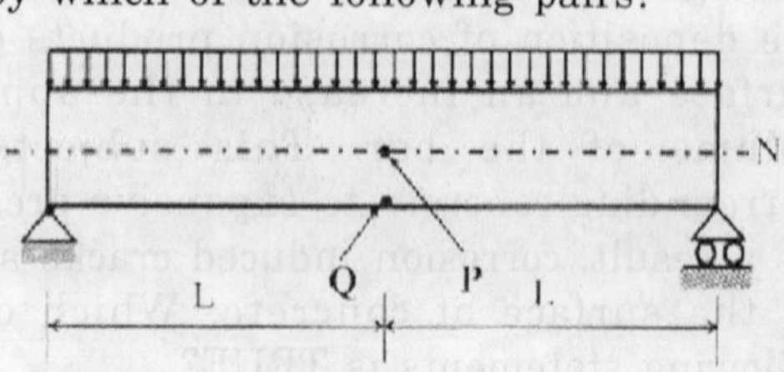

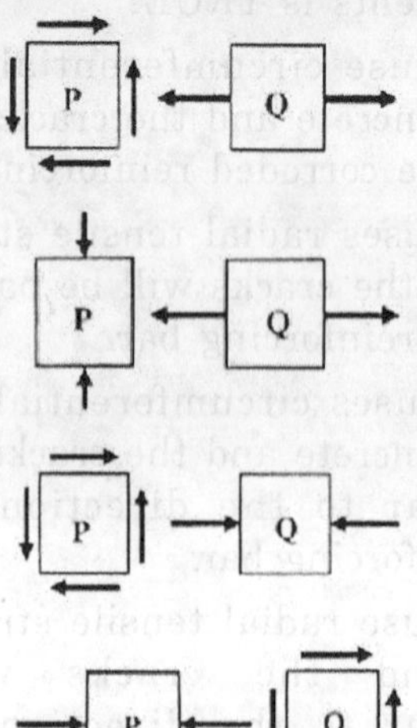

29. Consider a bar of diameter 'D' embedded in a large concrete block as shown in the adjoining figure, with a pull out force P being applied. Let σ_b and σ_{st} be the tensile strength (between the bar and concrete) and the tensile strength of the bar, respectively. If the block is held in position and it is assumed that the material of the block dies not fail, which of the following options represents the maximum value of P?

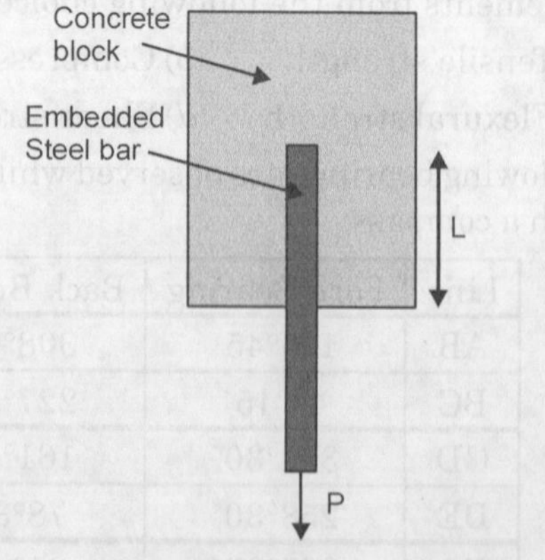

(a) Maximum of $\left(\frac{\pi}{4}D^2\sigma_b\right)$ and $(\pi DL\sigma_{st})$

(b) Maximum of $\left(\frac{\pi}{4}D^2\sigma_{st}\right)$ and $(\pi DL\sigma_b)$

(c) Minimum of $\left(\frac{\pi}{4}D^2\sigma_{st}\right)$ and $(\pi DL\sigma_b)$

(d) Minimum of $\left(\frac{\pi}{4}D^2\sigma_b\right)$ and $(\pi DL\sigma_{st})$

30. Consider two RCC beams, P and Q, each having the section 400 mm × 750 mm (effective depth, d = 750 mm) made with concrete having a τ_{max} = 2.1 N/mm². For the reinforcement provided and the grade of concrete used, it may be assumed that the τ_c = 0.75 N/mm². The design shear in beam P is 400 kN and in beam Q is 750 kN. Considering the provisions of IS 456 - 2000, which of the following statements is TRUE?

(a) Shear reinforcement should be designed for 175 kN for beam P and the section for beam Q should be revised.

(b) Nominal shear reinforcement is required for beam P and the shear reinforcement should be designed for 120 kN for beam Q.

(c) Shear reinforcement should be designed for 175 kN for beam P and the shear reinforcement should be designed for 525 kN for beam Q.

(d) The sections for both beams P and Q need to be revised.

2010

31. As per Indian standard code of practice for prestressed concrete (IS : 1343 - 1980) the minimum grades of concrete to be used for post-tensioned and pre-tensioned structural elements are respectively

(*a*) M20 for both (*b*) M 40 and M 30

(*c*) M 15 and M20 (*d*) M 30 and M 40

2009

32. The modulus of rupture of concrete in terms of its characteristic cube compressive strength (f_{ck}) in MPa according to Is 456:2000 is

(*a*) $5000 f_{ck}$ (*b*) $0.7 f_{ck}$

(*c*) $5000 \sqrt{f_{ck}}$ (*d*) $0.7 \sqrt{f_{ck}}$

33. For limit state of collapse, the partial safety factors remommended by IS 456:2000 for estimating the design strength of concrete and reiforcing steel are respectively

(*a*) 1.15 and 1.5

(*b*) 1.0 and 1.0

(*c*) 1.5 and 1.15

(*d*) 1.5 and 1.0

34. A rectangular concrete beam of width 120 mm and depth 200 mm is prestressed by pretensioning to a force of 150 kN at an eccentricity of 20 mm. The cross sectional area of the prestressing steel is 187.5 mm^2. Take modulus of elasticity of steel and concrete as 2.1×10^5 MPa and 3.0×10^{-4} MPa repectively. The percentage loss of stress in the prestressing steel due to elastic deformation of concrete is

(*a*) 8.75 (*b*) 6.125

(*c*) 4.81 (*d*) 2.19

35. Column I gives a list of test methods for evaluating properties of concrete and Column II gives the list of properties.

Column I

P. Resonant frequency test

Q. Rebound hammer test

R. Split cylinder test

S. Compacting factor test

Column II

1. Tensile strength

2. Dynamic modulus of elasticity

3. Workability

4. Compressive strength

The correct match of the test with the property is

Codes :

	P	Q	R	S
(*a*)	2	4	1	3
(*b*)	2	1	4	3
(*c*)	2	4	3	1
(*d*)	4	3	1	2

2008

36. A reinforced concrete structure has to be constructed along a sea coast. The minimum grade of concrete to be used as per IS :456-2000 is

(*a*) M 15 (*b*) M 20

(*c*) M 25 (*d*) M 30

37. Un-factored maximum bending moments at a section of a reinforced concrete beam resulting from a frame analysis are 50, 80, 120 and 180 kNm under dead, live, wind and earthquake loads respectively. The design moment (kNm) as per IS: 456-2000 for the limit state of collapse (flexure) is

(*a*) 195 (*b*) 350

(*c*) 345 (*d*) 372

38. A reinforced concrete column contains longitudinal steel equal to 1 percent of net cross-sectional area of the column. Assume modular ratio as 10. The loads carried (using the elastic theory) by the longitudinal steel and the net area of concrete, are P_s and P_c respectively. The ratio P_s/P_c expressed as percent is

(*a*) 0.1 (*b*) 1

(*c*) 1.1 (*d*) 10

39. A pre-tensioned concrete member of section 200 mm× 250 mm contains tendons of area 500 mm^2 at centre of gravity of the section. The prestress in the tendons is 1000N/mm^2. Assuming modular ratio as 10, the stress (N/mm^2) in concrete is

(*a*) 11 (*b*) 9

(*c*) 7 (*d*) 5

2007

40. Consider the following statements :

I. The compressive strength of concrete decreases with increase in water-cement ratio of the concrete mix.

II. Water is added to the concrete mix for hydration of cement and workability

III. Creep and shrinkage of concrete are independent of the water-cement ratio in the concrete mix.

The true statements are

(*a*) I and II (*b*) I, II and III

(*c*) II and III (*d*) only II

41. The percentage loss of prestress due to anchorage slip of 3 mm in a concrete beam of length 30 m which is post-tensioned by a tendon with an initial stress of 1200 N/mm^2 and modulus of elasticity equal to 2.1×10^5 N/mm^2 is

(a) 0.0175 (b) 0.175

(c) 1.75 (d) 17.5

42. A concrete beam of rectangular cross-section of size 120 mm (width) and 200 mm (depth) is prestressed by a straight tendon to an effective force of 150 kN at an eccentricity of 20 mm (below the centroidal axis in the depth direction). The stresses at the top and bottom fibres of the section are

(a) 2.5 N/mm^2 (compression), 10N/mm^2 (compression)

(b) 10 N/mm^2 (tension), 2.5 N/mm^2 (compression)

(c) 3.75 N/mm^2 (tension), 3.75 N/mm^2 (compression)

(d) 2.75 N/mm^2 (compression), 3.75 N/mm^2 (compression)

43. Consider the following statements :

I. Modulus of elasticity of concrete increases with increase in compressive strength of concrete.

II. Brittleness of concrete increases with decrease in compressive strength of concrete.

III. Shear strength of concrete increases with increase in compressive strength of concrete.

The **TRUE** statements are

(a) II and III (b) I, II and III

(c) I and II (d) I and III

2006

44. If characteristic strength of concrete f_{ck} is defined as the strength below which not more than 50% of the test results are expected to fall, then expression for f_{ck} in terms of mean strength f_m and standard deviation S would be

(a) $f_m - 0.1645S$ (b) $f_m - 1.645S$

(c) f_m (d) $f_m + 1.645S$

45. Assuming concrete below the neutral axis to be cracked, the shear stress across the depth of a singly-reinforced rectangular beam section

(a) increases parabolically to the neutral axis and then drops suddenly to zero value

(b) increases parabolically to the neutral axis and then remains constant over the remaining depth

(c) increases linearly to the neutral axis and then remains constant up to the tension steel

(d) increases parabolically to the neutral axis and then remains constant up to the tension steel

46. As per IS: 456-2000, consider the following statements

I. The modular ratio considered in the working stress method depends on the type of steel used

II. There is an upper limit on the nominal shear stress in beams (even withshear reinforcement) due to the possibility of crushing of concrete in diagonal compression.

III. A rectangular slab whose length is equal to its width may not be a two-way slab for some support conditions.

The TRUE statements are

(a) only I and II

(b) only II and III

(c) only I and III

(d) I, II and III

47. Consider the following statements:

I. The width-to-thickness ratio limitations on the plate elements under compression in steel members are imposed by IS: 800-1984 in order to avoid fabrication difficulties.

II. In a doubly reinforced concrete beam, the strain in compressive reinforcement is higher than the strain in the adjoining concrete.

III. If a cantilever I-section supports slab construction all along its length with sufficient friction between them, the permissible bending stress in compression will be the same as that in tension.

The TRUE statements are

(a) only I and II (b) only II and III

(c) only I and III (d) I, II and III

2005

48. IS : 1343 - 1980 limits the minimum characteristic strength of prestressed concrete for post tensioned works and pretension work as

(a) 25 MPa, 30 MPa respectively

(b) 25 MPa, 35 MPa respectively

(c) 30 MPa, 35 MPa respectivly

(d) 30 MPa, 40 MPa respectively

49. The permissible stress in axial tension s_{st} in steel member on the net effective area of the section shall not exceed which of the following value (f_y is the yield stress)

(a) $0.80f_y$

(b) $0.75f_y$

(c) $0.60f_y$

(d) $0.50f_y$

50. The partial factor of safety for concrete as per IS : 456-2000 is

(a) 1.50 (b) 1.15

(c) 0.87 (d) 0.446

51. A concrete beam of rectangular cross section of 200 mm × 400 mm is prestressed with a force 400 kN at eccentricity 100 mm. The maximum compressive stress in the concrete is

(*a*) 12.5 N/mm^2 (*b*) 7.5 N/mm^2

(*c*) 5.0 N/mm^2 (*d*) 2.5 N/mm^2

52. Which one of the following is NOT correct for steel sections as per IS : 800-1984?

(*a*) The maximum bending stress in tension or in compression in extreme fibre calculated on the effective section of a beam shall not exceed0.66 f_y.

(*b*) The bearing stress in any part of a beam when calculated on the net area shall not exceed 0.75 f_y.

(*c*) The direct stress in compression on the gross sectional area of axial loaded compression member shall not exceed 0.6 f_y.

(*d*) None of these.

53. The flexural strength of M 30 concrete as per IS : 456-2000 is

(*a*) 3.83 MPa (*b*) 5.47 MPa

(*c*) 21.23 MPa (*d*) 30.0 MPa

54. In a random sampling procedure for cube strength of concrete, one sample consists of X number of specimens. These specimens are tested at 28 days and average strength of these X specimens is considered as test result of the sample, provided the individual variation in the strength of specimens is not more than ± Y percent of the average strength. The values of X and Y as per IS : 456-2000 are

(*a*) 4 and 10 respectively

(*b*) 3 and 10 respectively

(*c*) 4 and 15 respectively

(*d*) 3 and 15 respectively

55. A rectangular column section of 250 mm × 400 mm is reinforced with five steel bars of grade Fe 500, each of 20 mm diameter. Concrete mix is M 30. Axial load on the column section with minimum eccentricity as per IS : 456-2000 using limit state method can be applied upto

(*a*) 1707.37 (*b*) 1805.30

(*c*) 1806.40 (*d*) 1903.7

Linked Answer Q. 56 - 57

Assume straight line instead of parabola for stress-strain curve of concrete as follows and partial factor of safety as 1.0.

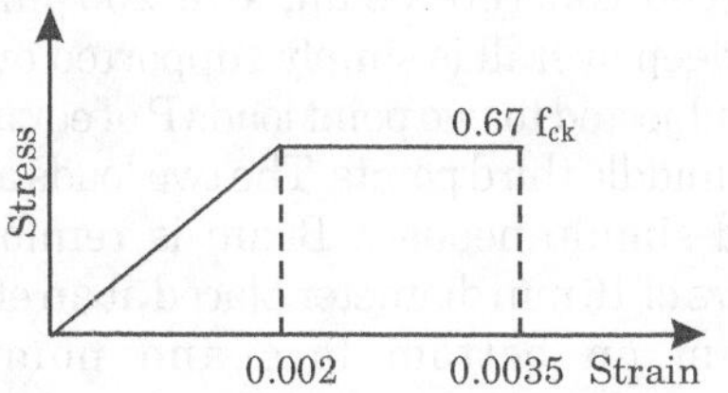

A rectangular under-reinforced concrete section of 300 mm width and 500 mm effective depth is reinforced with 3 bars of grade Fe 415, each of 16 mm diameter. Concrete mix is M 20.

56. The depth of the neutral axis from the compression fibre is

(*a*) 76 mm (*b*) 81 mm

(*c*) 87 mm (*d*) 100 mm

57. The depth of the neutral axis obtained as per IS : 456–2000 differs from the depth of neutral axis obtained in the above question

(*a*) 15 mm (*b*) 20 mm

(*c*) 25 mm (*d*) 32 mm

2004

58. In the limit state design method of concrete structures, the recommended partial material safety factor (γ_m) for steel according to IS: 456-2000 is

(*a*) 1.5 (*b*) 1.15

(*c*) 1.00 (*d*) 0.87

59. For avoiding the limit state of collapse, the safety of R.C. structures is checked for appropriate combinations of dead load (DL), imposed load or live load (IL), wind load (WL) and earthquake load (EL). Which of the following load combinations is NOT considered?

(*a*) 0.9 DL + 1.5 WL

(*b*) 1.5 DL + 1.5 WL

(*c*) 1.5 DL + 1.5 WL + 1.5 EL

(*d*) 1.2 DL + 1.2 IL + 1.2 WL

60. An R.C. short column with 300 mm x 300 mm square cross-section is made of M20 grade concrete and has 4 numbers, 20 mm diameter longitudinal bars of Fe 415 steel. It is under the action of a concentric axial compressive load. Ignoring reduction in the area of concrete due to steel bars, the ultimate axial load carrying capacity of the column is

(*a*) 1659 kN

(*b*) 1548 kN

(*c*) 1198 kN

(*d*) 1069 kN

61. An R.C. square footing of side length 2 m and uniform effective depth 200 mm is provided for a 300 mm × 300 mm column. The line of action of the vertical compressive load passes through the centroid of the footing as well as of the column. If magnitude of the load is 320 kN, then nominal transverse (one way) shear stress in the footing is

(*a*) 0.26 N/mm^2

(*b*) 0.30 N/mm^2

(*c*) 0.34 N/mm^2

(*d*) 0.75 N/mm^2

62. A simply supported prestressed concrete beam is 6 m long and 300 mm wide. Its gross depth is 600 mm. It is prestressed by horizontal cable tendons at a uniform eccentricity of 100 mm. The prestressing tensile force in the cable tendons is 1000 kN. Neglect the self weight of beam. The maximum normal compressive stress in the beam at transfer is

(*a*) Zero (*b*) 5.55 N/mm^2

(*c*) 11.11 N/mm^2 (*d*) 15.68 N/mm^2

Common Data Q. 63 - 64

At the limit state of collapse, an R.C. beam is subjected to flexural moment 200 kN-m, shear force 20 kN and torque 9 kN-m. The beam is 300 mm wide and has a gross depth of 425 mm, with an effective cover of 25 mm. The equivalent nominal shear stress (τ_{ve}) as calculated by using the design code turns out to be lesser than the design shear strength (τ_c) of the concrete.

63. The equivalent shear force (V_c) is

(*a*) 20 kN (*b*) 54 kN

(*c*) 56 kN (*d*) 68 kN

64. The equivalent flexural moment (M_{el}) for designing the longitudinal tension steel is

(*a*) 187 kN-m (*b*) 200 kN-m

(*c*) 209 kN-m (*d*) 213 kN-m

2003

65. The effective length of a column in a reinforced concrete building frame, as per IS : 456-2000, is independent of the

(*a*) frame type *i.e.*, braced (no sway) or un-braced (with sway)

(*b*) span of the beam

(*c*) height of the column

(*d*) loads acting on the frame

66. The working stress method of design specifies the value of modular ratio, $m = \dfrac{280}{3\sigma_{cbc}}$,

where σ_{cbc} is allowable stress in bending compression in concrete.

To what extent does the above value of m make any allowance for the creep of concrete?

(*a*) No compensation

(*b*) Full compensation

(*c*) Partial compensation

(*d*) The two are unrelated

67. Maximum strains in an extreme fibre in concrete and in the tension reinforcement (Fe-415 grade and E_s = 200 kN/mm^2) in a balanced section at limit state of flexure are respectively

(*a*) 0.0035 and 0.0038 (*b*) 0.002 and 0.0018

(*c*) 0.0035 and 0.0041 (*d*) 0.002 and 0.0031

68. Top ring beam of an Intze tank carries a hoop tension of 120 kN. The beam cross-section is 250 mm wide and 400 mm deep and it is reinforced with 4 bars of 20 mm diameter of Fe 415 grade. Modular ratio of the concrete is 10. The tensile stress in N/mm^2 in the concrete is

(*a*) 1.02 (*b*) 1.07

(*c*) 1.20 (*d*) 1.32

69. The state of two dimensional stresses acting on a concrete lamina consists of a direct tensile stress, σ_x = 1.5 N/mm^2, and shear stress τ = 1.20 N/mm^2, which cause cracking of concrete. Then the tensile strength of the concrete in N/mm^2 is

(*a*) 1.50 (*b*) 2.08

(*c*) 2.17 (*d*) 2.29

70. Group I contains some properties of concrete/ cement and Group 2 contains list of some tests on concrete/cement. Match the property with the corresponding test.

Group I	Group II
P. Workability of concrete	**1.** Cylinder splitting test
Q. Direct tensile strength of concrete	**2.** Vee-Bee test
R. Bond between concrete and steel	**3.** Surface area test
S. Fineness of cement	**4.** Fineness modulus test
	5. Pull out test.

Codes :

	P	Q	R	S
(*a*)	2	1	5	3
(*b*)	4	5	1	3
(*c*)	2	1	5	4
(*d*)	2	5	1	4

71. A concrete column carries an axial load of 450 kN and a bending moment of 60 kN m at its base. An isolated footing of size 2m by 3m, with 3m side along the plane of the bending moment, is provided under the column. Centres of gravity of column and footing coincide. The net maximum and the minimum pressures in kN/m^2 on soil under the footing are respectively.

(*a*) 95 and 55 (*b*) 95 and 75

(*c*) 75 and 55 (*d*) 75 and 75

Common Data Q. 72 – 73

A reinforced concrete beam, size 200 mm wide and 300 mm deep overall is simply supported over a span of 3m. It is subjected to two point loads P of equal magnitude placed at middle third points. The two loads are gradually increased simultaneously. Beam is reinforced with 2 HYSD bars of 16 mm diameter placed at an effective cover of 40 mm on bottom face and nominal shear reinforcement. The characteristic compressive strength and the bending tensile strength of the concrete are 20.0 N/mm^2 and 2.2N/mm^2 respectively.

72. Ignoring the presence of tension reinforcement, what is the value of load P in kN when the first flexure crack will develop in the beam?

(*a*) 4.5
(*b*) 5.0
(*c*) 6.6
(*d*) 7.5

73. The theoretical failure load of the beam for attainment of limit state of collapse in flexure is

(*a*) 23.7 kN
(*b*) 25.6 kN
(*c*) 28.7 kN
(*d*) 31.6 kN

2002

74. Read the following two statements

I. Maximum strain in concrete at the outermost compression fibre is taken to be 0.0035 in bending

II. The maximum compressive strain in concrete in axial compression is taken as 0.002.

Keeping the provisions of IS 456-2000 on limit state design in mind, which of the following is true?

(*a*) Statement I is true but II is false
(*b*) Statement I is false but II is true
(*c*) Both statements I and II are true
(*d*) Both statement I and II are false

75. As per the provisions of IS 456-2000, the (short term) modulus of elasticity of M25 grade concrete (in N/mm^2) can be assumed to be

(*a*) 25000 (*b*) 28500
(*c*) 30000 (*d*) 36000

76. As per the provisions of IS 456-2000, in the limit state method for design of beams, the limiting value of the depth of neutral axis in a reinforced concrete beam of effective depth'd' is given as

(*a*) 0.53d
(*b*) 0.48d
(*c*) 0.46d
(*d*) any of the above depending on the different grades of steel

Common Data Q. 77

A 10 m long prestressing bed is used to cast 4 (pretensioned) prestressed concrete beams of 2.3 m each. A schematic representation of the bed is given in the following figure. The continuous prestressing reinforcement is pulled at the end 'Y' of the bed through a distance of 20 mm to introduce the required 'prestress', before the concrete is cast. After the concrete has hardened, the prestressing reinforcement is cut at points A, B, C, D and E.

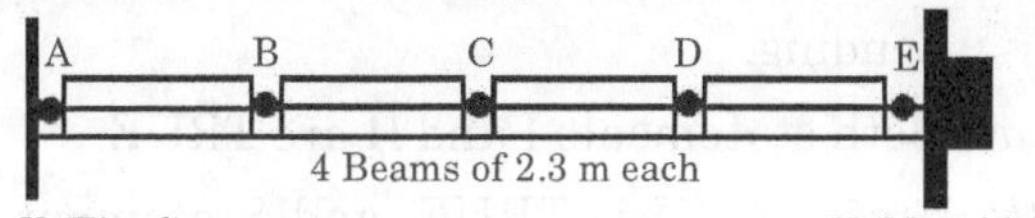

77. Asssuming that the prestress is introduced without eccentricity, what is the loss in prestress on account of elastic deformation of concrete. Assume E_s = 200,000N/mm^2, E_c = 20,000 N/mm^2, Area of prestressing reinforcement is 500mm^2, size of beams = 200 mm (b) × 400 mm(h).

(*a*) 15 N/m^2 (*b*) 20 N/m^2
(*c*) 25 N/m^2 (*d*) 35 N/m^2

78. Consider the T-section shown in the following figure made up of two plates100 mm 10 mm

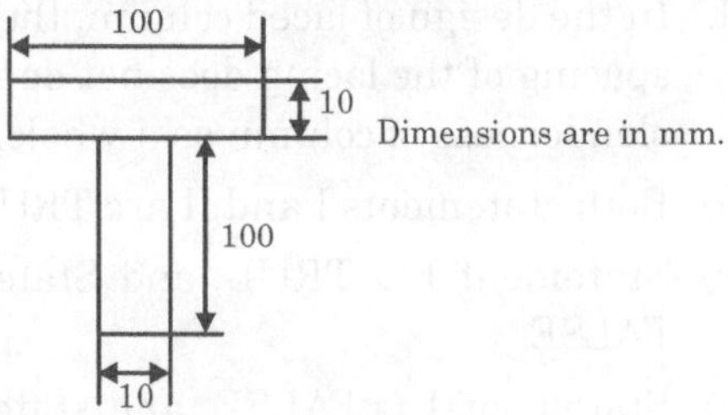

What is the shape factor ?

(*a*) 1.95 (*b*) 2.65
(*c*) 3.95 (*d*) 4.35

79. Give reasons for the following in not more than 20 words.

(*a*) A maximum permissible distance between ties in reinforced concrete columns is usually specified.
(*b*) A concrete mix is targeted to give higher compressive strength than the required characteristic strength.
(*c*) In the limit state design of reinforced concrete beams, it is a requirement that the maximum strain in the tension reinforcement in the section at failure is not less than a given value.
(*d*) In the case of slabs running over supports, reinforcement needs to be provided on the top in the neighbourhood of the supports.
(*e*) The load carrying capacity of an RC column with appropriate helical reinforcement can be taken to be slightly higher than that having lateral ties.

2001

80. Consider the following two statements related to reinforced concrete design, and identify whether they are TRUE or FALSE.

I. Curtailment of bars in the flexural tension zone in beams reduces the shear strength at the cut-off locations.

II. When a rectangular column section is subject to biaxially eccentric compression, the neutral axis will be parallel to the resultant axis of bending.

(*a*) Both statements I and II are TRUE.

(*b*) Statement I is TRUE, and Statement II is FALSE.

(*c*) Statement I is FALSE, and statement II is TRUE.

(*d*) Both Statements I and II are FALSE.

81. Consider the following two statements related to structural steel design, and identify whether they are TRUE or FALSE.

I. The Euler buckling load of a slender steel column depends on the yield strength of steel.

II. In the design of laced column, the maximum spacing of the lacing does not depend on the slenderness of column as a whole.

(*a*) Both statements I and II are TRUE.

(*b*) Statement I is TRUE, and Statement II is FALSE.

(*c*) Statement I is FALSE, and statement II is TRUE.

(*d*) Both Statements I and II are FALSE.

Linked Answer Q. 82-83

The plan of a reinforced concrete column section, and the distribution of strains at the ultimate limit state are shown below. The concrete is of M20 grade and the steel of Fe 250 grade. Also sketched below, for convenience, are the concrete compression stress block and the design stress-strain curve for Fe 250, with all notations as per IS 456. Ignore the reduction in concrete area due to the embedded steel.

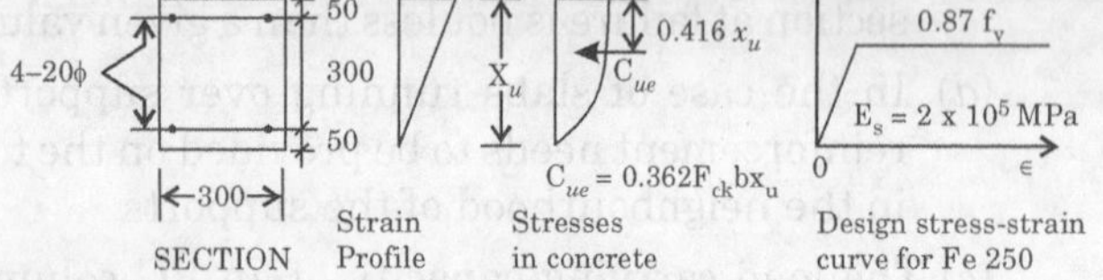

82. Determine ultimate axial compression capacity P_u (in kN units).

(*a*) 377.4 (*b*) 477.4

(*c*) 577.4 (*d*) 877.4

83. Determine corresponding eccentricity e (in mm units) of loading , with respect to the centroidal axis at the ultimate limit state.

(*a*) 30 (*b*) 40

(*c*) 50 (*d*) 60

2000

84. The following two statements are made with reference to a simply supported under - reinforced RCC beam :

I. Failure takes place by crushing of concrete before the steel has yielded.

II. The neutral axis moves up as the load is increased.

With reference to the above statements, which of the following applies ?

(*a*) Both the statements are false.

(*b*) I is true but II is false.

(*c*) Both the statements are ture.

(*d*) I is false but II is true.

85. A simply supported rectangular prestressed concrete beam 200 mm wide and 400 mm deep has an effective span of 12 m. The prestressing cable has a triangular profile with zero eccentricity at ends and 70 mm at the midspan as shown in the figure below. The effective prestress is 800 kN after all losses. What is the value of a point load, the beam can support at the midspan if the pressure line passes through the upper kern of the section? The weight density of the material of the beam can be taken to be 25 kN/m³.

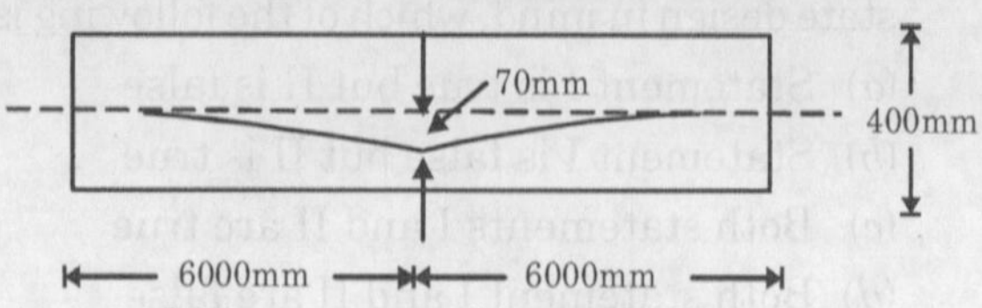

(*a*) 2.14 kN (*b*) 3.14 kN

(*c*) 4.14 kN (*d*) 5.14 kN

1999

86. In reinforced concrete, pedestal is defined as compression member, whose effective length does not exceed its dimension by

(*a*) 12 times (*b*) 3 times

(*c*) 16 times (*d*) 8 times

87. The characteristic strength of concrete is defined as that compressive strength below which not more than

(*a*) 10% of results fall (*b*) 5% of results fall

(*c*) 2% of results fall (*d*) None of these

1998

88. A reinforced concrete wall carrying vertical loads is generally designed as per recommendations given for columns. The ratio of minimum reinforcements in the vertical and horizontal directions is

(*a*) 2 : 1 (*b*) 5 : 3

(*c*) 1 : 1 (*d*) 3 : 5

1997

89. IS 459 - 1978 recommends to provide certain minimum steel in a RCC beam

(*a*) to ensure compression failure

(*b*) to avoid rupture of steel in case a flexural failure occurs

(*c*) to hold the stirrup steel in position

(*d*) to provide enough ductility to the beam

90. The permissible bending tensile stress in concrete for the vertical wall of an R.C. water tank made of M 25 concrete is

(*a*) 8.5 N/mm^2

(*b*) 6.0 N/mm^2

(*c*) 2.5 N/mm^2

(*d*) 1.8 N/mm^2

91. Permissible bending tensile stress in high yield strength deformed bars of grade 415 N/mm^2 in a beam is

(*a*) 190 N/mm^2

(*b*) 230 N/mm^2

(*c*) 140 N/mm^2

(*d*) None of these

92. A prestressed concrete rectangular beam of size 300 mm × 900 mm is prestressed with an intial prestressing force of 700 kN at an eccentricity of 350 mm at midspan. Stress at top of the due to prestress alone, in N/mm^2 is

(*a*) – 3.46 (tension)

(*b*) 2.59 (compression)

(*c*) zero

(*d*) 8.64 (compression)

1996

93. If ϕ = Nominal diameter of reinforcing bar, f_s = Compressive stress in the bar and f_{bd} = design bond stress of concrete, the anchorage length, L_a of straight bar in compression is equal to

(*a*) $L_a = \dfrac{\phi f_s}{f_{bd}}$ (*b*) $L_a = \dfrac{\phi f_s}{2f_{bd}}$

(*c*) $L_a = \dfrac{\phi f_s}{\pi f_{bd}}$ (*d*) $L_a = \dfrac{\phi f_s}{4f_{bd}}$

NUMERICAL TYPE QUESTIONS

2016

1. For M25 concrete with creep coefficient of 1.5, the long-term static modulus of elasticity (expressed in MPa) as per the provisions of IS:456-2000 is ________.

2. A haunched (varying depth) reinforced concrete beam is simply supported at both ends, as shown in the figure. The beam is subjected to a uniformly distributed factored load of intensity 10 kN/m.

The design shear force (expressed in kN) at the section X-X of the beam is ______

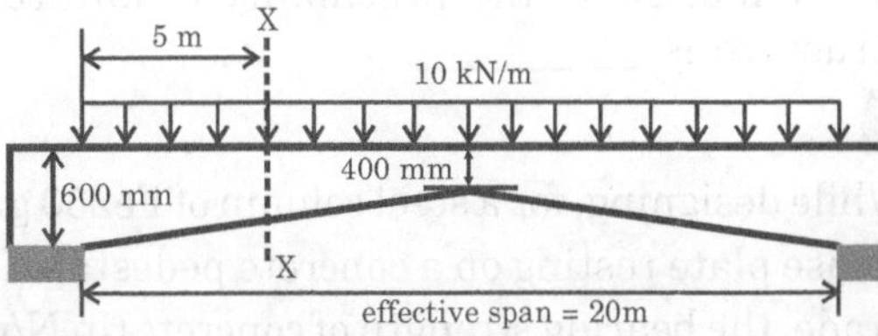

3. A 450 mm long plain concrete prism is subjected to the concentrated vertical loads as shown in the figure. Cross section of the prism is given as 150 mm × 150 mm. Considering linear stress distribution across the cross-section, the modulus of rupture (expressed in MPa) is ______

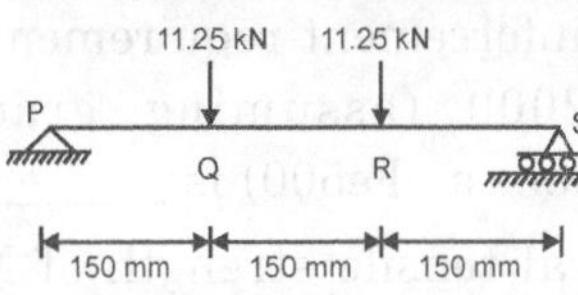

2015

4. The development length of a deformed reinforcement bar can be expressed as $\left(\dfrac{1}{k}\right)\left(\dfrac{\phi\sigma_s}{\tau_{bd}}\right)$. From the IS : 456–2000, the value of k can be calculated as ________.

5. The composition of an air-entrained concrete is given below :

Water	: 184 kg/m^3
Ordinary Portland Cement (OPC)	: 368 kg/m^3
Sand	: 606 kg/m^3
Coarse aggregate	: 1155 kg/m^3

Assume the specific gravity of OPC, sand and coarse aggregate to be 3.14, 2.67 and 2.74, respectively. The air content is ______ liters/m^3.

6. Consider the singly reinforced beam section given below (left figure). The stress block parameters for the cross-section from IS : 456-2000 are also given below (right figure). The moment of resistance for the given section by the limit state method is ________ kN-m.

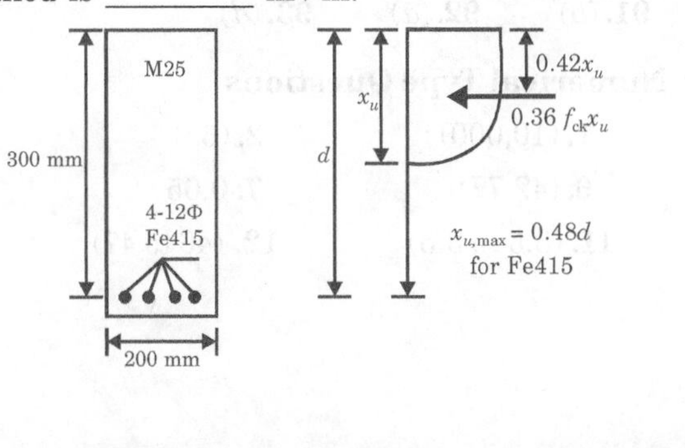

7. A simply supported reinforced concrete beam of length 10 m sags while undergoing shrinkage. Assuming a uniform curvature of 0.004 m^{-1} along the span, the maximum deflection (in m) of the beam at mid-span is ______.

8. According to the concept of Limit State Design as per IS456: 2000, the probability of failure of a structure is ______.

2014

9. While designing, for a steel column of Fe250 grade, a base plate resting on a concrete pedestal of M20 grade, the bearing strength of concrete (in N/mm²) in limit state method of design as per IS: 456-2000 is ______.

10. For a beam of cross-section, width = 230 mm and effective depth = 500 mm, the number of rebars of 12 mm diameter required to satisfy minimum tension reinforcement requirement specified by IS: 456-2000 (assuming grade of steel reinforcement as Fe500) is ______.

11. The flexural tensile strength of M25 grade of concrete, in N/mm², per IS:456-2000 is ______

2013

12. As per IS 456:2000 for M20 grade concrete and plain barsin tension the design bond stress τ_{bd} = 1.2 MPa Further, IS 456:2000 permits this design bond stress value to be increased by 60% for HSD bars. The stress in the HSDreinforcing steel barsin tension, α_s = 360 MPa. Find the required development length, L_d, for HSD barsin terms of the bar diameter, ϕ. ______

2000

13. A continuous beam 250 mm × 450 mm carries 6 numbers of 12 mm diameter longitudinal bars as shown. The factored shear force at the point of inflection is 200 kN. Check if the beam is safe in bond. Assume M15 mix with σ_{ck} = 15 N/mm² and mild steel with σ_y = 250 N/mn². A clear cover of 25 mm can be assumed. The design bond stress for mild steel bars in M15 concrete is specified to be 1.0 N/mm².

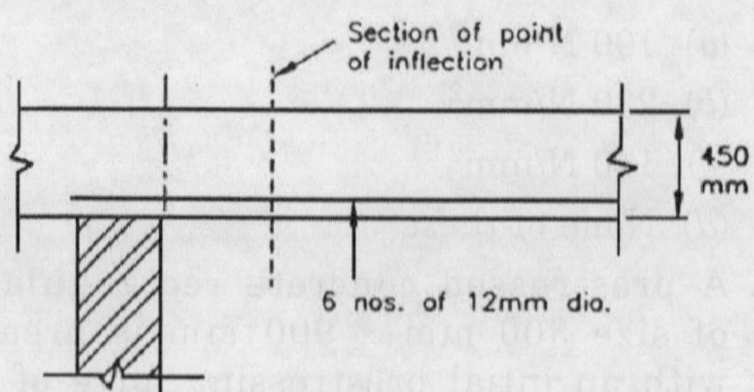

ANSWERS

MCQ Type Questions

1. (*d*)	**2.** (*b*)	**3.** (*b*)	**4.** (*d*)	**5.** (*a*)	**6.** (*d*)	**7.** (*b*)	**8.** (*b*)	**9.** (*d*)	**10.** (*a*)
11. (*c*)	**12.** (*b*)	**13.** (*d*)	**14.** (*b*)	**15.** (*b*)	**16.** (*a*)	**17.** (*c*)	**18.** (*c*)	**19.** (*d*)	**20.** (*a*)
21. (*d*)	**22.** (*c*)	**23.** (*b*)	**24.** (*b*)	**25.** (*c*)	**26.** (*a*)	**27.** (*c*)	**28.** (*a*)	**29.** (*c*)	**30.** (*a*)
31. (*d*)	**32.** (*d*)	**33.** (*c*)	**34.** (*b*)	**35.** (*a*)	**36.** (*d*)	**37.** (*d*)	**38.** (*d*)	**39.** (*b*)	**40.** (*a*)
41. (*c*)	**42.** (*a*)	**43.** (*b*)	**44.** (*c*)	**45.** (*d*)	**46.** (*b*)	**47.** (*a*)	**48.** (*d*)	**49.** (*c*)	**50.** (*a*)
51. (*a*)	**52.** (*d*)	**53.** (*a*)	**54.** (*d*)	**55.** (*a*)	**56.** (*d*)	**57.** (*c*)	**58.** (*b*)	**59.** (*c*)	**60.** (*d*)
61. (*a*)	**62.** (*c*)	**63.** (*d*)	**64.** (*d*)	**65.** (*b*)	**66.** (*c*)	**67.** (*a*)	**68.** (*b*)	**69.** (*c*)	**70.** (*c*)
71. (*a*)	**72.** (*c*)	**73.** (*d*)	**74.** (*c*)	**75.** (*a*)	**76.** (*d*)	**77.** (*c*)	**78.** (*c*)	**79.** (*)	**80.** (*b*)
81. (*d*)	**82.** (*c*)	**83.** (*d*)	**84.** (*d*)	**85.** (*c*)	**86.** (*b*)	**87.** (*b*)	**88.** (*a*)	**89.** (*b*)	**90.** (*d*)
91. (*b*)	**92.** (*a*)	**93.** (*d*)							

Numerical Type Questions

1. (10,000)	**2.** (65)	**3.** (12)	**4.** (6.4)	**5.** (51)
6. (42.77)	**7.** 0.05	**8.** 0.0975	**9.** (9 to 9)	**10.** (2 to 2)
11. (3.5 to 3.5)	**12.** (46 to 47)	**13.** (*)		

"*" Refers to Subjective Question.

EXPLANATIONS

MCQ TYPE QUESTIONS

1. If $\tau_v > \tau_c$ max, diagonal compression failure occurs in concrete

2. Minimum % Steel

$$\frac{(A_{st})_{min}}{b.d} = \frac{0.85}{f_u}$$

$$(A_{st})_{min} = \frac{0.85}{415} \times 250 \times 400$$

$$= 205 \text{ mm}^2$$

Max. Reinforcement cannot be calculated as total depth is not given,

$$(A_{st})_{max} = 4\%A_g = 4\%(b \times D)$$

Infact there is no need to calculate$(A_{st})_{max}$ as per the options given

3.

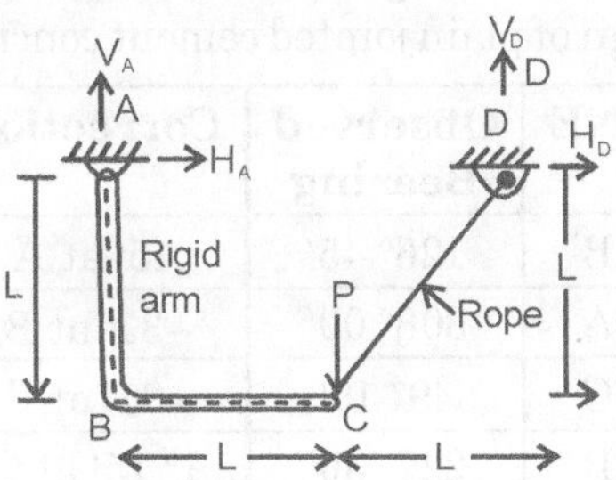

$$\Sigma M_A = 0$$

$$V_D \times 2l - P.l = 0$$

$$V_D = \frac{P}{2}$$

Also, $\Sigma Fy = 0$

$$V_A = P - V_D = P - \frac{P}{2} = \frac{P}{2}$$

$$\Sigma F_x = 0$$

$$H_A + H_D = 0 \quad \text{...(1)}$$

$$\Sigma M_C = 0$$

$$V_A . l + H_A . l = 0$$

$$H_A = -V_A = -\frac{P}{2} \quad \text{...(2)}$$

By 1 & 2,

Component of force 'P' in the direction of rope

$$= \frac{P}{\sqrt{2}}$$

Therefore, tension in rope $= \dfrac{P}{\sqrt{2}}$

4. As per BIS : 1077-1957 & 1910 code specification, minimum compressive strength or crushing strength of bricks should be 3.5 N/mm²

5. The primary purpose of air entrainment is to increase the durability of the hardened concrete, especially in climates subject to freeze thaw; secondary purpose is to increase workability as the bubbles contribute to workability by acting as a sort of lubricant for all the aggregates.

6. As the slump increases, the Vebe time decreases.

7.

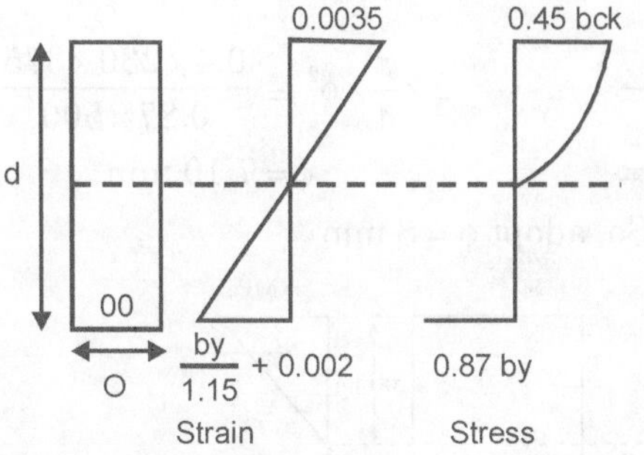

8. Minimum eccentricity,

$$\left.\begin{aligned} e_{min} &= \frac{L}{500} + \frac{D}{30} \\ &\text{or} \\ e_{min} &= 20 \text{ mm} \end{aligned}\right\} \text{whichever is less}$$

$$e_{xx} = \frac{3000}{500} + \frac{600}{30} = 26 \text{ mm}$$

$$e_{yy} = \frac{300}{500} + \frac{450}{30} = 21 \text{ mm}$$

9. $\rho_i = 1.1 \times 750$ kN

$$\sigma_i = \frac{\rho}{A} \pm \frac{\rho_e}{I} y$$

$$= \frac{1.1 \times 750 \times 10^3}{250 \times 400} \pm \frac{1.1 \times 750 \times 10^3 \times 100 \times 200}{\frac{250 \times (400)^3}{12}}$$

$$= 8.33 \pm 12.5$$

$$-4.166 \text{ and } 20.833$$

10. $\tau_v = \dfrac{V_U}{bd} = \dfrac{45 \times 1000}{230 \times 450} = 0.434 \text{ N/mm}^2$

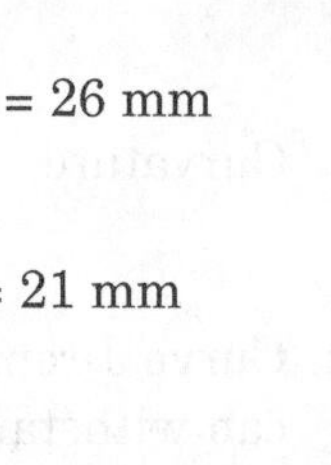

% tensile reinforcement (p)

$$= \frac{4 \times \frac{\pi}{4} \times (12)^2}{230 \times 450} \times 100$$

$$= 0.437\%$$

$$\tau_c = 0.36 + \frac{0.12}{0.25} \times (0.437 - 0.25)$$

$$= 0.45 \text{ N/mm}^2$$

$$\tau_v = \tau_c$$

So, minimum shear reinforcement is required

Minimum shear reinforcement

$$\frac{A_{sv}}{b \times S_v} = \frac{0.4}{0.87 f_y}$$

$$\Rightarrow \quad A_{sv} = \frac{0.4 \times b \times S_v}{0.87 \times f_y}$$

$$\Rightarrow \quad 2 \times \frac{\pi}{4} \times \phi^2 = \frac{0.4 \times 230 \times 375}{0.87 \times 500}$$

$$\Rightarrow \quad \phi = 7.10 \text{ mm}$$

So, adopt $\phi = 8$ mm

11. 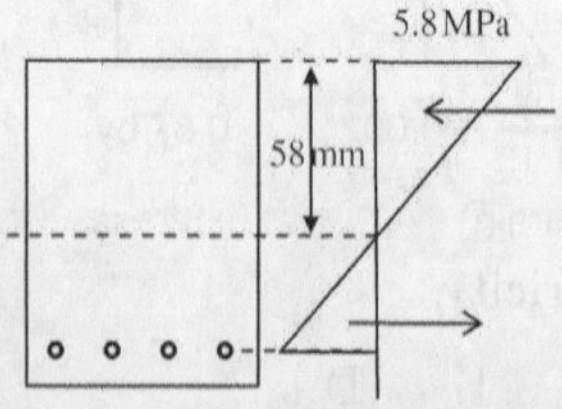

$$E = 5000\sqrt{f_{ck}}$$
$$= 5000 \times \sqrt{25}$$
$$= 25000 \text{ N/mm}^2$$

$$\frac{M}{I} = \frac{f}{y} = \frac{E}{R}$$

Curvature, $\frac{1}{R} = \frac{f}{yE} = \frac{5.8}{58 \times 25000}$

$$= 4 \times 10^{-6} \text{ per mm}$$

12. Curve J-represent corse aggregate because they can withstand more stress without showing any strain/less strain compare to others.

Curve k ⇒ represent concrete because among the rest two choice concrete can take more stress and showing less strain than cement paste.

Curve L ⇒ represent cement paste because it is showing less resistance. And the stress is proportional to strain in it.

Correct choice is (*b*)

13. Target mean strength (f_{cm}) ⇒ f_{ck} + 1.65 σ

Correct chose is (*d*)

(As per IS : 456.2000)

14. Initial tangent modulus.

15. – Compacting factor measures the compactability of concrete which is an important aspect of workability by measuring the amount of compaction achieved for a given amount of work.

– The compacting factor test has been held to be more accurate than slump test.

– The range of compacting factor 0.65 to 0.95 for extremely low to high workability.

16. The creep strains are eased due to dead loads only.

→ Creep strains comes due to age of loading so only dead load alive from begining to end of structure life.

17. Expected load coming on each column = 4000 kN

No. of Column = 4

Total load coming from all column = 4 × 4000
= 16000 kN

In raft footing

Pressure coming to soil $= \frac{P}{A} = \frac{16000}{10 \times 10}$

$= 160 \text{ kN/m}^2 > 100 \text{ kN/m}^2$

Therefore, pressure acting downwards to soil is more than bearing capacity of soil, so the raft is not provided this condition.

So go for pile foundation.

18. Flexural strength parameter of concrete used in design of plain jointed cement concrete pavements.

19.

LINE	Observed Bearing	Correction	Correct Bearing
AB	126° 45′	45′ at A	127° 30′
BA	308° 00′	–31′ at B	308° 30′
BC	49° 15′	–31′ at B	48° 45′
CB	227° 30′	1° 15′ at C	228° 45′
CD	340° 30′	1° 15′ at C	341° 45′
DC	161° 45′	0 at D	161° 45′
DE	258° 30′	0 at D	258° 30′
ED	78° 30′	0 at E	78° 30′
EA	212° 30′	0 at E	212° 30′
AE	31° 45′	45′ at A	32° 30′

The difference bewteen back and fore bearing of the line DC is exactly 180°. Hence Both the station D and E are free from local attraction and all other bearing measured at these station are also correct.

After calculation the force bearing of line BC

= 48°45′.

20. $\sigma_x = \sigma_y = \sigma_z = P$

and $\tau_{xy} = 0$ (only normal forces act over it)

So, $\tau_0 = \frac{\sigma_x - \sigma_y}{2} \sin 2\theta - \tau_{xy} \cos 2\theta = 0$

$$\tau_{max} = \sqrt{\left(\frac{\sigma_x - \sigma_y}{2}\right)^2 + \tau_{xy}^2} = 0$$

21. From IS 456 : 2000

22. Loss due to relaxation, shrinkage, and creep are time dependent loss

23. $A = 375 \times 10^3 \text{ mm}^2$

$$\frac{I}{y} = z = \frac{bd^2}{6} = 46875 \times 10^3 \text{ mm}^3$$

Moment due to UDL

$$= \frac{wl^2}{8} = 299.76 \times 10^6 \text{Nmm}$$

So, stress at bottom at mid – span

$$= \frac{-P}{A} - \frac{P_e}{z} + \frac{M_0}{2}$$

$= -2.90 \text{ N/mm}^2$ (compressive)

24. Permissible stress,

$$\sigma_{cbc} = \frac{\text{Load}}{\text{Area of base plate}}$$

$$= \frac{P}{\text{Area of base plate}}$$

Factored load, $P_{vz} = 2000$ kH

$$\text{Load, } P = \frac{2000}{1.5}$$

$= 1333.33$ kN

$$\therefore \quad \sigma_{cbc} = \frac{1333.33 \times 10^3}{650 \times 420}$$

$= 4.88$

$\approx 5 \text{ N/mm}^2$

So from IS: 456 – 2000, minimum grade of concrete that should be used for carrying the safe load is M15.

31. Minimum grade of Conerete

For post-tensioned structure : M 30

For pre-tensioned structure : M 40

Because in pre-stressing in pre-tensioned structure, the prestress is transfer through bond action, so concrete must be better than post-tensioned concrete.

Hence M30 and M40.

36. According to clause 8.2.2.1 and Refering Table 3 as well as Table 5, of IS code, the minimum grade of concrete for reinforced members constructed along sea cost is M 30.

38.

$$\frac{P_S}{P_C} = \frac{\sigma_S A_S}{\sigma_C A_C}$$

where, σ_S = stress in steel = $m\sigma_C$

σ_C = stress in concrete

A_S = area of steel

= 1% of $A_C = 0.01 A_C$

$$\therefore \quad \frac{P_S}{P_C} = \frac{m\sigma_C}{\sigma_C} \times 0.01 \frac{A_C}{A_C}$$

$= 10 \times 0.01 = 0.1 = 10\%$

40. *True statements are*

I. Compressive strength of concrete decreases with increase in water-cement ratio.

II. Water is added to the concrete mix for hydration of cement and workability.

41. *Given :*

Length of concrete beam = 30 m

Anchorage slip = 3 mm

Initial stress, $f_s = 1200 \text{ N/mm}^2$

$E = 2 \times 1 \times 10^5 \text{ N/mm}^2$

$$\text{Loss of strain} = \frac{3}{30 \times 1000} = 1 \times 10^{-4}$$

$$E_s = \frac{f_s}{E} = \frac{1200}{2.1 \times 10^5} = 5.71 \times 10^{-4}$$

$$\therefore \quad \text{Percentage loss} = \frac{1 \times 10^{-4}}{5.71 \times 10^{-4}} \times 100$$

$= 1.75$ %

42. Given :

Size of beam = 120mm × 200 mm

$$I = \frac{1}{12} bd^3$$

$$= \frac{1}{12} \times 120 \times (200)^3$$

$= 40 \times 10^4 \text{ mm}^4$

Prestressed force, $f = 150$ kN

$$\therefore \quad F_l = \frac{I}{A}$$

$$= \frac{150 \times 1000}{120 \times 200}$$

$= 6.25 \text{ N/mm}^2$

$$F_b = \frac{f.e.y}{I}$$

$$= \frac{150 \times 10^3 \times 100}{40 \times 10^4} = 3.75 \text{ N/mm}^2$$

Stress at top extreme fiber,

$F_t = F_l + F_b$

$= 6.25 - 3.75$

$= 2.50 \text{ N/mm}^2$ (compression)

Stress at bottom fiber,

$F_b = F_l - F_b$

$= 6.25 + 3.75$

$= 10 \text{ N/mm}^2$ (compression)

43. All the three statements are true

44. From figure we can say that when strength below which not more than 50%, then

$$f_{ck} = f_m$$

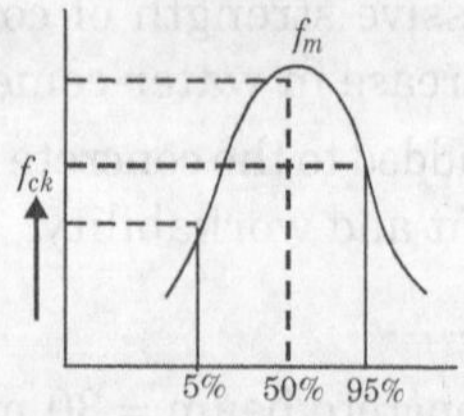

45. For rectangular section, shear stress increases parabolically.

51.

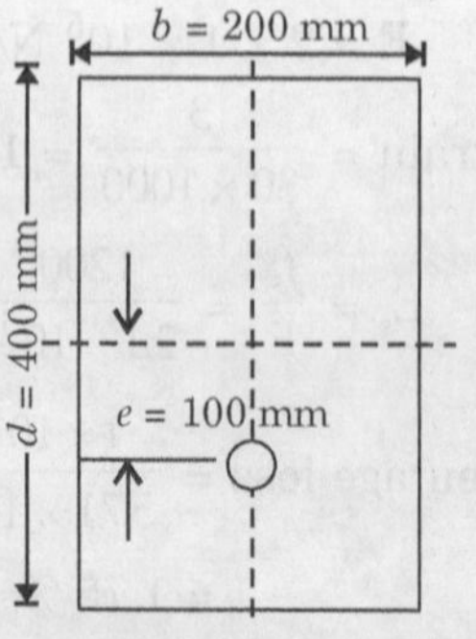

Given : b = 200 mm,

d = 400 mm

P = 400 kN,

e = 100 mm

Maximum compressive strength in concrete

$$= \frac{P}{A} + \frac{Pe}{Z}$$

where, Z = section modulus

$$= \frac{bd^2}{6}$$

∴ Maximum compressive strength in concrete

$$= \frac{400 \times 1000}{(200 \times 400)} + \frac{400 \times 1000 \times 100 \times 6}{200 \times 400^2}$$

$$= 12.5 \text{ N/mm}^2$$

53. Flexural strength of concrete = $0.7\sqrt{f_{ck}}$

Since, given M 30, hence

$$f_{ck} = 30 \text{ N/mm}^2$$

∴Flexural strength of concrete

$$= 0.7\sqrt{30}$$

$$= 3.83 \text{ MPa}$$

54. According to IS 456 – 2000, for random sampling number of specimens taken as 3.

Average strength variation 15% is permitted.

X = 3; Y = 15

55.

$$A = 250 \times 400 = 10^5 \text{ mm}^2$$

$$A_{sc} = 5 \times \frac{\pi}{4}(20)^2 = 1570.8 \text{ mm}^2$$

$$\therefore \quad A_c = A - A_s = 98429.2 \text{ mm}^2$$

Axial load,

$$P_n = 0.4 f_{ck} A_c + 0.67 f_y \cdot A_{sc}$$

$$= 0.4 \times 30 \times 98429.2 + 0.67 \times 500 \times 1570.8$$

$$= 1707.37 \text{ kN}$$

56. Here, $p = \frac{A_{st}}{bd}$

$$= \frac{3 \times \frac{\pi}{4} \times 16^2}{300 \times 500}$$

$$= 0.004021$$

$$\therefore \quad \frac{x_u}{d} = 2.417 \frac{p f_y}{f_{ck}}$$

$$= 2.417 \times 0.004021 \times \frac{415}{20}$$

$$= 0.20$$

$$\Rightarrow \quad x_u = \mathbf{100.8 \text{ mm}}$$

57. By codal provision, depth of N.A. = 25 mm

59. Partial safety factors for load under limit stage of collapse.

Load Combination	DL	LL	WL
DL + LL	1.5	1.5	—
DL + WL	1.5	—	1.5
DL + WL	0.9		1.5
DL + LL + WL	1.2	1.2	1.2

60. Neglecting area of steel,

$$A_C = A$$

$$\therefore \quad P_u = 0.4 f_{ck} A_c + 0.67 f_y A_{sc}$$

$$= 0.4 \times 20(300)^2 + 0.67 \times 415\ (1256.637)$$

$$= 1069 \text{ kN}$$

61. *R.C. square footing*

Length = 2 m

⇒ B = 2000

Effective depth, d = 200 mm

P = 320 kN

b = 300 mm

= size of column

We know, $F = q\left[\frac{B-b}{2} - d\right] B$

where , q = Net upward pressure

$$= \frac{\text{load on the column}}{\text{footing area}}$$

$$= \frac{320 \times 10^3}{(2000)^2}$$

$$= 0.08$$

Net effective shear force

$$F = 0.08\left[\frac{2000-300}{2} - 200\right]2000$$

$$= 104000 \text{ N}$$

Nominal transverse (one way) shear

$$= \frac{F}{Bd}$$

$$= \frac{104000}{2000 \times 200}$$

$$= 0.26 \text{ N/mm}^2$$

62. Maximum normal compressive stress

$$= \frac{P}{A} + \frac{M}{Z}$$

where, $M = Pe$

$$= (10^3 \times 1000) \times 100$$

$$= 10^8 \text{ Nmm}$$

$$Z = \frac{I}{y}$$

$$= \frac{300 \times 600^3}{12 \times 300}$$

$$= 18 \quad 10^6 \text{ mm}^3$$

$$\therefore \quad \frac{M}{Z} = 5.555 \text{ N/mm}^2$$

$$\frac{P}{A} = \frac{10^3 \times 1000}{300 \times 600}$$

$$= 5.55 \text{ N/mm}^2$$

Maximum normal compressive stres

$$= 5.55 + 5.55$$

$$= 11.11 \text{ N/mm}^2$$

63. Equivalent shear force,

$$V_e = V + 1.6\frac{T}{b}$$

$$= 20 + 1.6 \times \frac{9}{0.3}$$

$$= 68 \text{ kN.}$$

64. Equivalent flexural moment,

$$M_{e1} = M_v + M_T$$

$$= 200 + \frac{9}{1.7}\left(1+\frac{D}{b}\right)$$

$$= 212.79 \text{ kN-m}$$

$$\approx 213 \text{ kN-m}$$

65. Effective length of a column depends upon the conditions of its end. It may be obtained from the ratio of effective length to unsupported length (l_{eff}/l).

66. Given : $m = \dfrac{280}{(3\sigma_{cbc})}$

The expression given for m partially takes into account long-term effects such as creep. Therefore this m is not the same as the modular ratio derived based on the value of

$$E_C = 5700\sqrt{f_{ck}}$$

where f_{ck} is characteristic cube strength of concrete in N/mm².

67. Maximum strain in concrete at the outermost compression fibre is taken as 0.0035 in bending. Maximum strain in the tension reinforcement in the section at failure shall not be less than maximum strain :

Here, $E_S = 2 \times 10^5$ N/mm²

and $f_y = 415$

$$\therefore \quad \text{Maximum strain} = \frac{f_y}{1.15\,E_S} + 0.002$$

$$= \frac{415}{1.15 \times 200 \times 10^3} + 0.002$$

$$= 0.0038$$

(in tension reinforcement)

68. Tensil stress in concrete is given by

$$\sigma_{T\,[\text{in concrete}]} = \frac{P}{b \times d + (m-1)A_{st}}$$

Here, $P = 120$ kN

$b = 250$ mm

$d = 400$ mm

$m = 10$

$$A_{st} = 4 \times \frac{\pi}{4}(20)^2$$

$$= \pi \times 400$$

$$= 1256.56 \text{ mm}^2$$

∴ Tensile stress in concrete

$$= \frac{120 \times 10^3}{250 \times 400 + (10-1) \times 1256.56}$$

$$= \mathbf{1.07\ N/mm^2}$$

69. Given : $\sigma_x = 1.5$ N/mm²

$\tau = 1.2$ N/mm²

Principal tensile stresses are :

$$\sigma_1 = \frac{\sigma_x + \sigma_y}{2} \pm \sqrt{\left(\frac{\sigma_x - \sigma_y}{2}\right)^2 + (\tau_{xy})^2}$$

$$= \frac{1.5}{2} \pm \sqrt{\left(\frac{1.5}{2}\right)^2 + (1.2)^2}$$

$= 2.165$

$\approx$ **2.17 N/mm²**

71. Maximum and Minimum stresses are given by

$$\sigma_{max,\ min} = \frac{P}{A}\left[1 \pm \frac{\sigma_e}{b}\right]$$

Here, e = eccentricity

$$= \frac{M}{P} = \frac{60}{400} = 0.133$$

$b = 3m$

$A = 2 \times 3 = \sigma m^2$

$P = 400$ kN

$$\therefore \quad \sigma_{max,\ min} = \frac{450}{2 \times 3}\left[1 \pm \frac{6 \times 0.133}{3}\right]$$

$= 75\,[1 \pm 0.266] = 75 \pm 20$

$\therefore \quad \sigma_{max} = 95$ kN/m²

and $\quad \sigma_{min} = 75$ kN/m²

72. For HYSD bar

$\sigma_{cbc} = 5$

$\sigma_{st} = 230$

$b = 200$ mm

$d = 300 - 40 = 260$ mm

$m = 19$

$$\therefore \quad \frac{n_c}{d} = \frac{m\sigma_{cbc}}{m\sigma_{cbc} + \sigma_{st}} = K = 0.292$$

$\Rightarrow \quad n_c = 75.92$ mm

$$\text{Again,} \quad \frac{bn^2}{2} = m\,A_s + (d - n)$$

$$\text{where,} \quad A_{st} = 2 \times \frac{\pi}{4}(10)^2\ (10^2$$

$= 402.12$mm²

$\therefore \quad n = 107.8$ mm

Hence $\quad n > n_c$

$$\therefore \quad M = \frac{1}{2}\sigma_{cbc}\,bn\left(d - \frac{n}{3}\right)$$

$= 12.077$ kN–m

Again, size = (200 × 300) mm²

Self weight = 1.44 kN/m

$= 0.2 \times 0.3 \times 24000$

$$\therefore \quad M = \frac{1.44 \times 3^2}{8} = 1.62 \text{ kN/m}$$

Let $y_f = 1.5$

Total M = 1.5 (1.62 + P)

$= 12.077$ kN-m

$\therefore \quad P = 6.45 \approx$ **6.6 kN**

73.

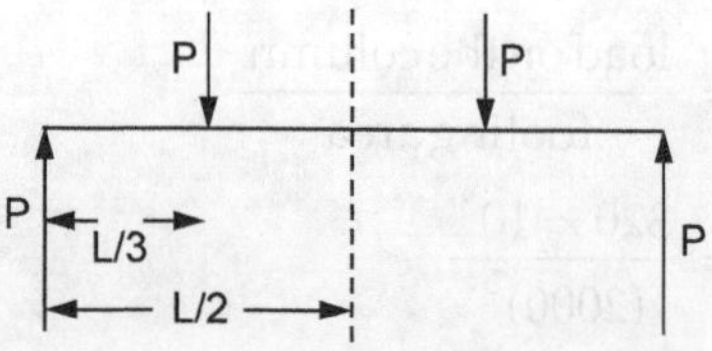

$$\text{Maximum } M = \frac{PL}{3}$$

$$= \frac{P \times 3}{3} = P \text{ N-m}$$

Given : $f_{ck} = 20$ N/mm²

$b = 200$ mm

$d = 260$ mm

$$\text{Then,} \quad \frac{(x_u)_{max}}{d} = \frac{700}{1100 + 0.87 f_y}$$

$$= \frac{700}{1100 + 0.87 \times 415}$$

$= 0.48$

$\Rightarrow \quad (x_u)_{max} = 0.48 \times 260$

$= 124.8$ mm

$$\text{Again,} \quad \frac{x_u}{d} = \frac{0.87 f_y A_{st}}{0.37 f_{ck}\, bd}$$

$= 0.377$

$\Rightarrow \quad x_u = 98$ mm

$\therefore \quad (x_u)_{max} > x_u$

$$\therefore \quad M_u = 0.87 f_y A_{st}\, d\left[1 - \frac{A_{st}}{bd}\frac{f_y}{f_{ck}}\right]$$

$$= 0.87 \times 415 \times 402.12 \times 260 \left[1 - \frac{402.12}{200 \times 260} \times \frac{415}{20}\right]$$

$= 31.69$ kN-m

P = **31.69 kN**

77. $$\alpha_e = \frac{E_S}{E_C} = \frac{200000}{20000} = 10$$

Stress in concrete at the steel

$$f_c = \frac{P}{A}$$

$$\text{Strain} = \frac{\Delta l}{e} = \frac{20}{10 \times 10000}$$

$\therefore \quad$ Stress = E_C × Strain

$$= \frac{200000 \times 20}{10000}$$

$= 400$ N/m²

P = Stress Area

$= 400 \times 500$

$= 200$kN

$$f_c = \frac{200 \times 100}{200 \times 400} = 2.5 \text{ N/mm}^2$$

Elastic Loss = $f_c \alpha$

= 10×2.5

= $25 N/m^2$

78.

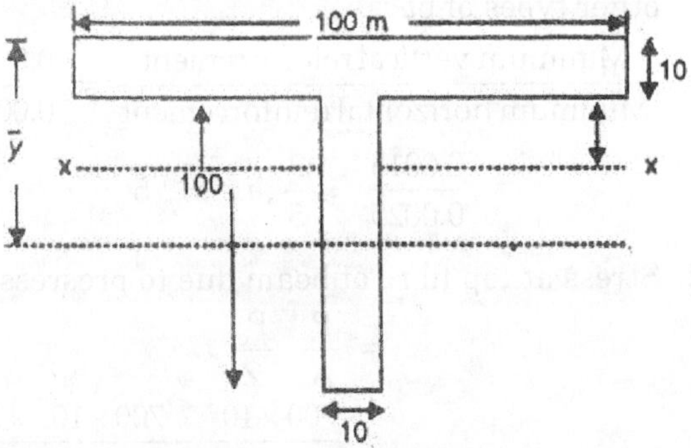

Distance of N.A from top,

$$\bar{y} = \frac{100 \times 10 \times 5 + 100 \times 10 \times (50+10)}{100 \times 10 + 100 \times 10}$$

= 32.5 mm

$$I_{xx} = \frac{100 \times 10^3}{12} + 100 \times 10 \times (32.5-5)^2 + \frac{10 \times 100^3}{12} + 10 \times 100 \times (50-32.5)^2$$

= 8333.33 + 7562.50 + 8333.33 + 306250

= $107.92 \times 10^4 mm^4$

Section modulus,

$$Z_x = \frac{I_{max}}{y_{min}}$$

$$= \frac{107.92 \times 10^4}{(110-32.5)}$$

= $13.924 \times 10^3 mm^3$

Let section XX be the equal area at a distance y, from top.

∴ Total area of section, A = 100 × 10 + 100 × 10 = $2000 m^2$

∴ $100 \times 10 + (x_1 - 10) \times 10 = \frac{2000}{2}$

⇒ $1000 + (x_1 - 10) \times 10 = 1000$

⇒ $100 + x_1 - 10 = 100$

∴ $x_1 = 10$

Now, $Z_p = \frac{A}{2}\left(\overline{y_1} + \overline{y_2}\right)$

where,

$$y_1 = \frac{100 \times 10 \times (10-5) + (10-10) \times (10-10)^2}{100 \times 10 + 10 \times (10-10)}$$

= 5mm

$$y_2 = \frac{(100+10)-10}{2}$$

= 50 mm

Plastic modulus, $Z_p = 1000(y_1 - y_2)$

= 1000(5+50)

= $55 \times 10^3 mm^3$

∴ Shape factor = $\frac{Z_p}{Z_{xx}}$

$$= \frac{55000}{13.92 \times 10^3}$$

= 3.95

79. (*a*) Distance between tie-in reinforcement concrete colume is spacified 5 mmϕ@ 190 mm pitch.

(*b*) A concrete mix is targeted to give high compressive stress due to compaction and compressive characters of mixture and binding material.

(*c*) Because compressive force exerted on the slab is distributed on support through the rein forcement.

(*d*) Because in the helical reinforcement, number of bars is generally greater than lateral tie and spaced slightly close together.

82. $T_u = 0.87 f_y A_{st}$

$= 0.87 \times 250 \times \frac{\pi}{4}(20)^2 \times 4$

= 573318 N

$P_u = 0.362 f_{ck}\, x_{ub} + f_{sc} A_{sc}$

where, f_{sc} is stress in the compression reinforcement to be read from stress-strain curve

$$\in_{sc} = \frac{0.0035(x_u - d')}{x_u}$$

Assume $\frac{x_u}{d} = 0.4$

⇒ $x_u = 0.4 \times 350 = 140$

∴ $\in_{sc} = \frac{0.0035(140-50)}{140}$

= 0.00225

For strain greater than 0.2%

$f_{sc} = 0.87 f_y$

= 0.87 × 250

= $217.5 N/m^2$

∴ $P_u = 0.362 \times 20 \times 140 \times 300 + 217.5 \times \frac{\pi}{4}(20)^2 \times 4$

= 577398.56 N

= T_u (adopt)

83. Here, Centroid = $\frac{400}{2}$ = 200 mm

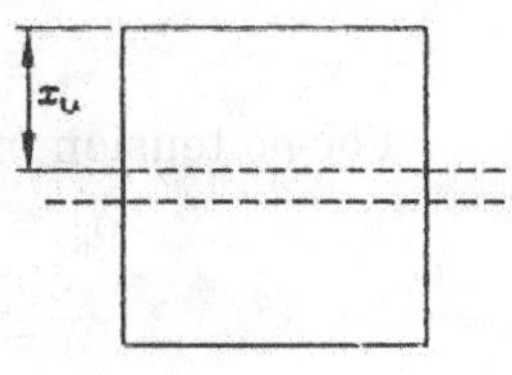

x_u = 140 mm

∴ $e = \text{Centroid} - x_u$

= 200 − 140

= 60mm

85. 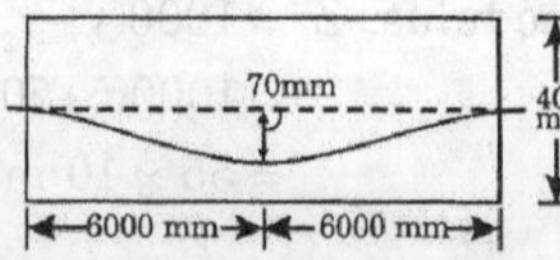

Given : b = 200 mm,

d = 400 mm,

l = 12 m

Effective prestress, $\lambda P_i = 800$ kN

$\gamma_c = 25$ kN/m³

Area of beam, A = bd

$= 200 \times 400 = 80 \times 10^3$ mm³

$$I = \frac{1}{12} bd^3$$

$$= \frac{1}{12} 200 \times (400)^3$$

$= 1.067 \times 10^9$ mm⁴

At mid span, e = 70 mm

$\Rightarrow \quad y_t = 270$ mm

and $\quad y_b = 130$ mm

$$\therefore \quad Z_t = \frac{1.067 \times 10^9}{270}$$

$= 3.951 \times 10^6$ mm³

$$Z_b = \frac{1.067 \times 10^9}{130}$$

$= 8.205 \times 10^6$ mm³

Hence at service load

$$\sigma_{bot} = -\frac{\lambda P_i}{A} - \frac{\lambda P_{i.e}}{Z_b} + \frac{M_d}{Z_b} + \frac{M_l}{Zb} \quad ...(i)$$

Now, self weight of beam,

$w = A\gamma_c$

$= 0.2 \times 0.4 \times 25$ kN/m²

$= 2$ kN/m²

So dead load moment,

$$M_d = \frac{wl^2}{8} = 36 \text{ kN-m.}$$

Live load moment, $M_l = \frac{Wl}{4} = 3W$

Putting values in equation (*i*), we get

$$\sigma_{bot} = -\frac{800 \times 10^3}{80 \times 10^3} - \frac{800 \times 10^3 \times 70}{8.25 \times 10^6} + \frac{36 \times 10^6}{8.206 \times 10^6} + 3W$$

$= -10 - 6.825 + 4.388 + 3W$

$= -12.437 + 3W$

For no tension in beam,

$\sigma_{bot} = 0$

$$\therefore \quad W = \frac{12.437}{3}$$

$= 4.14$ kN.

88. As per I.S. code, the minimum reinforcement in the vertical and horizontal directions is 0.0012 and 0.0020 for deformed bars up to bar diameter 16 mm respectively, while it is 0.0015 and 0.0025 for other types of bars

$$\frac{\text{Minimum vertical reinforcement}}{\text{Minimum horizontal reinforcement}} = \frac{0.0012}{0.0020} \text{ or}$$

$$\frac{0.0015}{0.0025} = \frac{3}{5}, \textit{ i.e. } 3 : 5$$

92. Stress at top fibre of beam due to presress force

$$= \frac{P}{A} - \frac{P_c}{Z}$$

$$= \frac{700 \times 10^3}{300 \times 900} - \frac{700 \times 10^3 \times 350}{\frac{1}{6} \times 300 \times 900^2}$$

$= 2.592 - 6.046.$

$= -3.457$

≈ -3.46 N/mm²(tensile)

NUMERICAL TYPE QUESTIONS

1. Long term modulus of concrete,

$$E_{ce} = \frac{E_C}{1+\theta} = \frac{5000\sqrt{25}}{1+1.5} = 10000 \text{ MPa}$$

2. 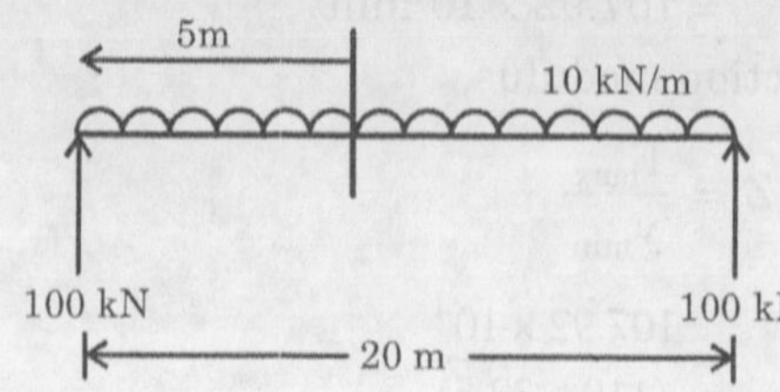

Cross sectional area at section X-X of the beam is

$$\text{Depth (m)} = 600 - \frac{600-400}{10} \times 5$$

$= 600 - 100 = 500$ m

Shear force at section X-X

$= 100 - (10 \times 5)$

$= 50$ kN

Moment at section X-X

$$= M_u = 100 \times 5 - 10 \times 5 \times \left(\frac{5}{2}\right)$$

$= 500 - 125$

$= 375$ kN-m

Design shear force at section X-X = V_u

$$= V + \frac{M_u}{d} \tan\beta$$

$$= 50 + \left(\frac{\frac{375}{0.5} \times 200}{10000}\right) = 65 \text{ kN}$$

3.

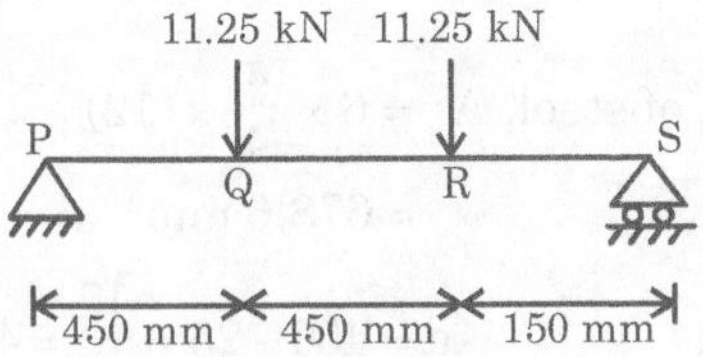

Compressive strength of prism

$= 150 \text{ mm} \times 150 \text{ mm}$

Maximum moment

$= 11.25 \times (350 + 75) - (11.25 \times 75)$

$= 11.25\,(225 \times 150)$

$= 1687.5$ kN-mm

Section modulus $= \dfrac{I}{y}$

$$= \frac{\dfrac{b.h^3}{36} \times 3}{2h}$$

$$= \frac{bh^2}{24} = \frac{150 \times 150^2}{24}$$

Modulus of rupture $= \dfrac{M}{2}$

$$= \frac{1.6875 \times 10^6}{150 \times 150^2} \times 24$$

$$= 120 \frac{N}{mm^2} = 12 \text{ MPa}$$

4. $L_d = \dfrac{\phi\sigma_s}{4\tau_{bd}}$ But for deformed bars τ_{bd} is increased by 60%

So, $L_d = \dfrac{\phi\sigma_s}{4 \times 1.6 \times \tau_{bd}}$, $L = \dfrac{\phi\tau_s}{6.4\tau_{bd}}$

So, $k = 6.4$

5. $\dfrac{M_c}{\rho_c} + \dfrac{M_s}{\rho_s} + \dfrac{M_a}{\rho_a} + V_w + V_a = 1$

$$\Rightarrow \frac{368}{3.14 \times 1000} + \frac{606}{2.67 \times 1000} + \frac{1155}{2.74 \times 1000} + \frac{184}{1000} + V_v = 1$$

$\Rightarrow 0.117 + 0.227 + 0.421 + 0.184 + V_v = 1$

$\Rightarrow V_v = 0.051 = 0.051 \times 1000 = 51$

6. $x_u = \dfrac{0.87\, f_y\, A_{st}}{0.36\, f_{ck}\, B}$

$A_{st} = 4 \times \dfrac{\pi}{4} \times (12)^2$

$= 452.38 \text{ mm}^2$

$$x_u = \frac{0.87 \times 415 \times 452.38}{0.36 \times 25 \times 200} = 90.739$$

$x_{u,\max} = 0.48\, d = 300 \times 0.48 = 144$

$x_u < x_u$ lim

This is under reinforced section

$MR = 0.87\, f_y\, A_{st}\,(d - 0.42\, x_u)$

$$= 0.87 \times 415 \times 452.38\,(300 - 0.42 \times 90.74) \frac{1}{10^6}$$

$= 42.77$ kN/m

7. Considering the curved beam as a part of a gaint circle

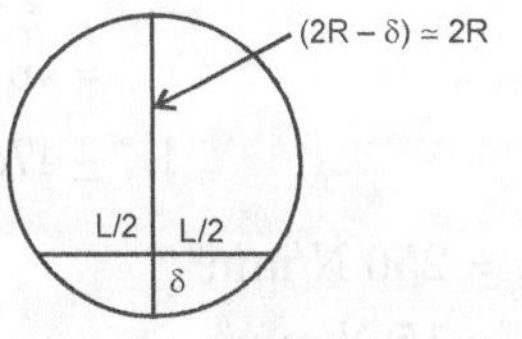

$l = 10$ m, $\dfrac{1}{R} = 0.004 \text{ m}^{-1}$

$\therefore \quad \dfrac{L}{2} \times \dfrac{L}{2} = \delta \times 2R$

$$\delta = \frac{L^2}{8R} = \frac{100 \times 0.004}{8}$$

$= 0.05$

8. As we consider f_{ck} as the value of compressive strength for a sample below which not more than 5% specimen are expected to fail.

$\therefore$ chances of no failure

$= 100 - 5 = 95\%$

So, probability of failure

$= [1 - \rho^2]$

$= (1 - 0.95)^2$

$= 0.0975$

9. Permissible bearing strength

$= 0.45\, f_{ck}$

$= 0.45 \times 20$

$= 9 \text{N/mm}^2$

10. $\dfrac{(A_{st})_{min}}{bd} = \dfrac{0.85}{f_y}$

$\Rightarrow \quad (A_{st})_{min} = \dfrac{0.85}{500} \times 230 \times 500$

$= 195.5 \text{ mm}^2$

$n \times \dfrac{\pi}{4} d^2 = 195.5$

$\Rightarrow \quad n = 1.73$

So, take $n = 2$

11. Flexural tensile strength $\Rightarrow 0.7\sqrt{f_{ck}}$

here $f_{ck} = 25$ (characteristic compressive strength of concrete).

$\therefore$ Flexural tensile strenght $= 0.7\sqrt{25}$

$\Rightarrow 3.5 \text{ N/mm}^2$

12.

$$\sigma_s = 360 \text{ MPa}$$

$$\text{design bend stress } (\tau_{bd}) = 1.6 \times 1.2$$

$$= 1.92 \text{ for HSD bars}$$

$$\text{Development Length } L_d = \frac{\phi\sigma_s}{4\tau_{bd}} z$$

$$= \frac{\phi \times 360}{4 \times 1.92}$$

$$= 46.875\,\phi$$

$$\therefore \quad L_d \simeq 47\phi$$

13. Given : $f_y = 250 \text{ N/mm}^2$

$f_{ck} = 15 \text{ N/mm}^2$

Cover = 25 mm.

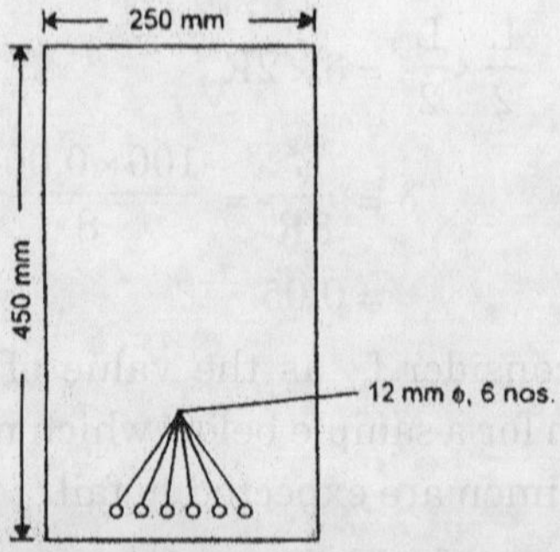

Bending Moment,

$$M = 0.87 f_y A_{st}\left(d - \frac{f_y}{f_{ck}} \cdot \frac{A_{st}}{b}\right)$$

$$\text{Area of steel, } A_{st} = 6 \times \frac{\pi}{4} \times (12)^2$$

$$= 678.6 \text{ mm}^2$$

Also, $d = 450 - 25 - \frac{12}{2} = 419\text{mm}.$

$$\therefore \quad M = 0.87 \times 250 \times 678.6\left(419 - \frac{250}{15} \times \frac{678.6}{250}\right)$$

$$= 55.16 \text{ kN–m}.$$

Then development length,

$$L_d = \frac{\phi\sigma_s}{4\tau_{bd}}$$

$$= \frac{\phi \times 0.87 \times 250}{4 \times 1}$$

$$= 54.375\phi$$

Also development length,

$$L_d < \frac{M_1}{V} + L_o$$

But $L_o = 0$

$$\therefore \quad L_d < \frac{55.16}{200}$$

$$\Rightarrow \quad \phi < \frac{55.16}{200} \times \frac{1}{54.375}$$

Since $\phi < 5$, hence beam is not same in bond.

6 CHAPTER

Steel Structures

MCQ TYPE QUESTIONS

2016

1. A propped cantilever of span L carries a vertical concentrated load at the mid-span. If the plastic moment capacity of the section is M_p, the magnitude of the collapse load is

(a) $\frac{8M_p}{L}$ (b) $\frac{6M_p}{L}$

(c) $\frac{4M_p}{L}$ (d) $\frac{2M_p}{L}$

2. Two plates are connected by fillet welds of size 10 mm and subjected to tension, as shown in the figure. The thickness of each plate is 12 mm. The yield stress and the ultimate tensile stress of steel are 250 MPa and 410 MPa, respectively. The welding is done in the workshop (γ_{mw} = 1.25).

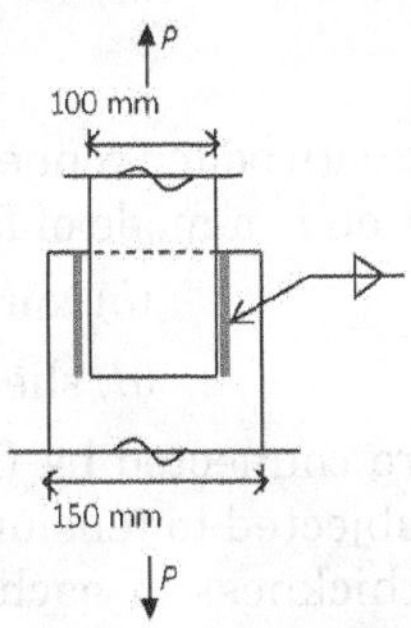

As per the Limit State Method of IS 800: 2007, the minimum length (rounded off to the nearest higher multiple of 5 mm) of each weld to transmit a force P equal to 270 kN (factored) is

(a) 90 mm (b) 105 mm

(c) 110 mm (d) 115 mm

3. Two bolted plates under tension with alternative arrangement of bolt holes are shown in figures 1 and 2. The hole diameter, pitch, and gauge length are d, p and g, respectively.

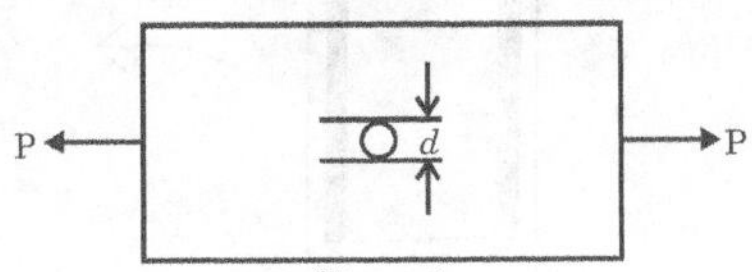

Figure 1

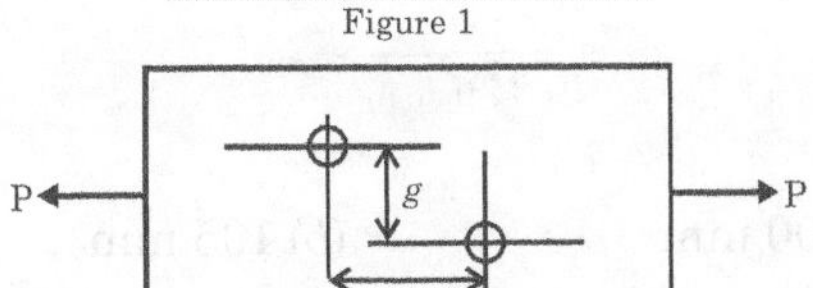

Figure 2

Which one of the following conditions must be ensured to have higher net tensile capacity of configuration shown in Figure 2 than that shown in Figure 1 ?

(a) $p^2 > 2gd$ (b) $p^2 < \sqrt{4gd}$

(c) $p^2 > 4gd$ (d) $p > 4gd$

4. A fixed-end beam is subjected to a concentrated load (P) as shown in the figure. The beam has two different segments having different plastic moment capacities (M_p, $2M_p$) as shown.

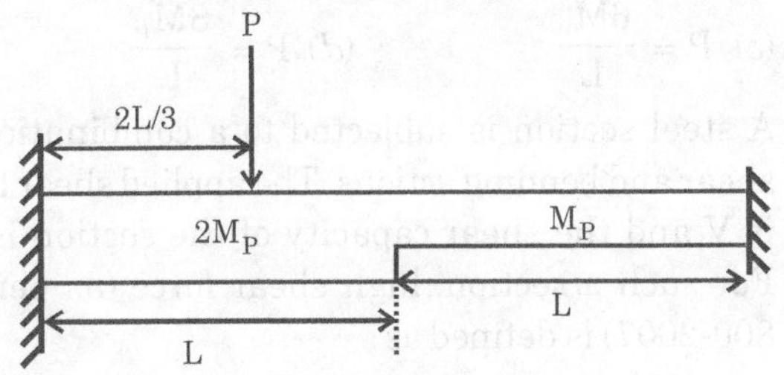

The minimum value of load (P) at which the beam would collapse (ultimate load) is

(a) $\frac{7.5\ M_P}{L}$ (b) $\frac{5.0\ M_P}{L}$

(c) $\frac{4.5\ M_P}{L}$ (d) $\frac{2.5\ M_P}{L}$

2014

5. Match the information given in Group-I with those in Group-II.

Group-I	Group-II
P. Factor to decrease ultimate strength to design strength	1. Upper bound on ultimate load
Q. Factor to increase working load to ultimate load for design	2. Lower bound on ultimate load
R. Statical method of ultimate load analysis	3. Material partial safety factor
S. Kinematical mechanism method of ultimate load analysis	4. Load factor

	P	Q	R	S
(a)	1	2	3	4
(b)	2	1	4	3
(c)	3	4	2	1
(d)	4	3	2	1

6. The ultimate collapse load (P) in terms of plastic moment M_p by kinematic approach for a propped cantilever of length L with P acting at its mid-span as shown in the figure, would be

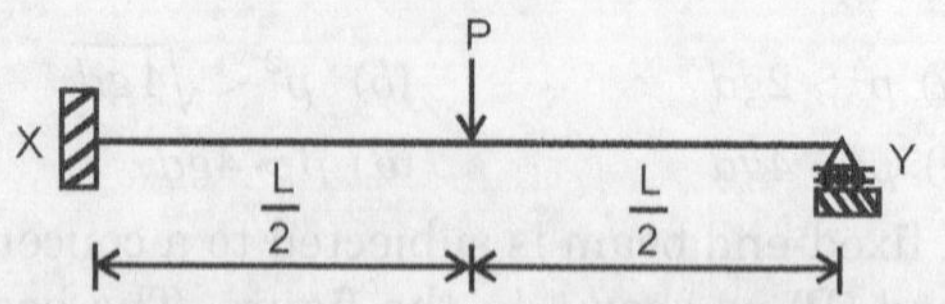

(a) $P = \frac{2M_P}{L}$ (b) $P = \frac{4M_P}{L}$

(c) $P = \frac{6M_P}{L}$ (d) $P = \frac{8M_P}{L}$

7. A steel section is subjected to a combination of shear and bending actions. The applied shear force is V and the shear capacity of the section is V_s. For such a section, high shear force (as per IS: 800-2007) is defined as

(a) $V > 0.6\ V_s$

(b) $V > 0.7\ V_s$

(c) $V > 0.8\ V_s$

(d) $V > 0.9\ V_s$

8. The tension and shear force (both in kN) in each bolt of the joint, as shown below, respectively are

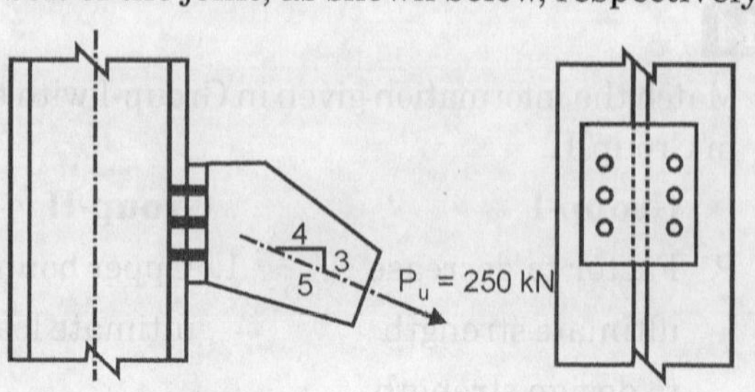

(a) 30.33 and 20.00 (b) 30.33 and 25.00

(c) 33.33 and 20.00 (d) 33.33 and 25.00

9. A steel member 'M' has reversal of stress due to live loads, whereas another member 'N' has reversal of stress due to wind load. As per IS 800:2007, the maximum slenderness ratio permitted is

(a) less for member 'M' than that of member 'N'

(b) more for member 'M' than for member 'N'

(c) same for both the members

(d) not specified in the Code

10. Prying forces are

(a) shearing forces on the bolts because of the joints

(b) tensile forces due to the flexibility of connected parts

(c) bending forces on the bolts because of the joints

(d) forces due the friction between connected parts

2013

11. As per IS 800:2007, the cross-section in which the extreme fiber can reach the yield stress, but cannot develop the plastic moment of resistance due to failure by local buckling is classified as

(a) plastic section

(b) compact section

(c) semi-compact section

(d) slender section

12. Two steel columns P (length L and yield strength f_y = 250 MPa) and Q (length 2L and yield strength f_y = 500 MPa) have the same cross-sections and end-conditions. The ratio of buckling load of column P to that of column Q is

(a) 0.5 (b) 1.0

(c) 2.0 (d) 4.0

2012

13. In a steel plate with bolted connections, the rupture of the net section is a mode of failure under

(a) tension (b) compression

(c) flexure (d) shear

14. Two plates are connected by fillet welds of size 10 mm and subjected to tension, as shown in the sketch. The thickness of each plate is 12 mm. The yield stress and the ultimate tensile stress of steel are 250 MPa and 410 MPa respectively. The welding is done in the workshop (γ_{mw} = 1.25). As per the Limit State Method of IS 800 : 2007, the minimum length (rounded off to the nearest higher multiple of 5 mm) of each weld to transmit a force P equal to 270 kN is

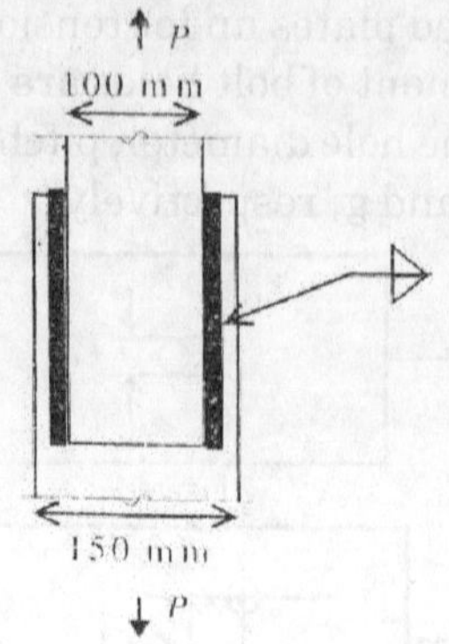

(a) 100 mm (b) 105 mm

(c) 110 mm (d) 115 mm

2011

15. The adjoining figure shows a schematic representation of a steel plate girder to be used as a simply supported beam with a concentrated load. For stiffeners, PQ (running along the beam axis) and RS (running between the top and bottom flanges) which of the following pairs of statements will be TRUE?

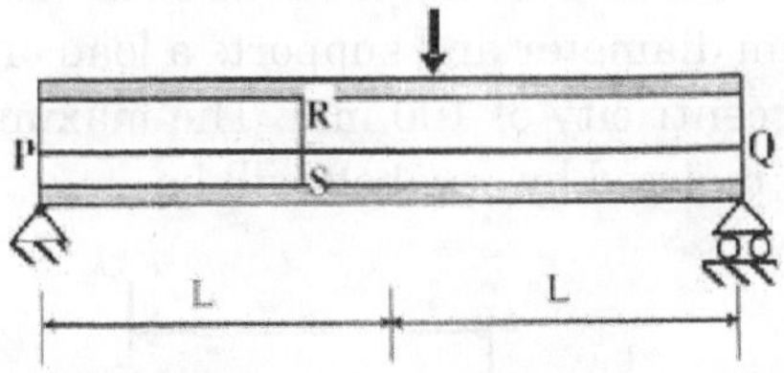

(a) (i) RS should be provided under the concentrated load only.

(ii) PQ should be placed in the tension side of the flange.

(b) (i) RS helps to prevent local buckling of the web.

(ii) PQ should be placed in the compression side of the flange.

(c) (i) RS should be provided at supports.

(ii) PQ should be placed along the neutral axis.

(d) (i) RS should be provided away from points of action of concentrated loads.

(ii) PQ should be provided on the compression side of the flange.

2010

16. The effective length of a column of length L fixed against rotation and translation at one end and free at the other end is

(a) 0.5 L

(b) 0.7 L

(c) 1.414 L

(d) 2L

17. A solid circular shaft of diameter d and length L is fixed at one end free at the other end. A torque T is applied at the free end. The shear modulus of the material is G. The angle of twist at the free end is

(a) $\frac{16TL}{\pi d^4 G}$ (b) $\frac{32TL}{\pi d^4 G}$

(c) $\frac{64TL}{\pi d^4 G}$ (d) $\frac{128TL}{\pi d^4 G}$

18. A double cover butt riveted joint is used to connect two flat of 200 mm width and 14 mm thickness as shown in the figure. There are twelve power driven rivets of 20 mm diameter at a pitch of 50 mm in both directions on either side of the plate. Two cover plates of 10 mm thickness are used. the capacity of the joint in tension considering bearing and shear ONLY, with permissible bearing and shear stresses as 300 MPa and 100 MPa respectively is

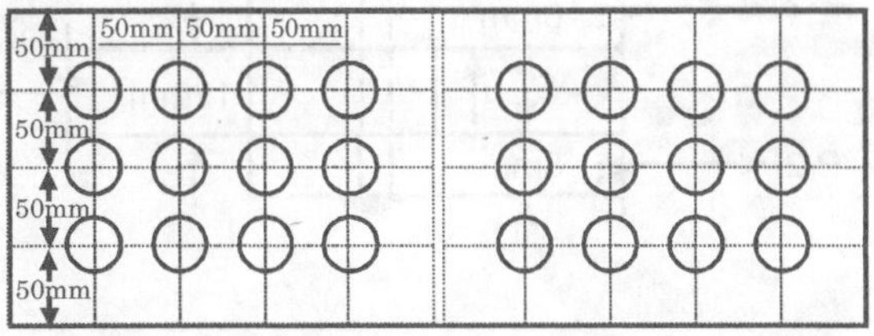

(a) 1083.6 kN

(b) 871.32 kN

(c) 541.8 kN

(d) 433.7 kN

19. Two plates, subjected to direct ension, each of 10 mm thickness and having widths of 100 mm and 175 mm, respectively are to be fillet welded with an overlap of 200 mm. Given that the permissible weld stress is 110 MPa and the permissible stress in steel is 150 MPa, the length of the weld required using the maximum permissible weld size as per IS : 800-1984 is

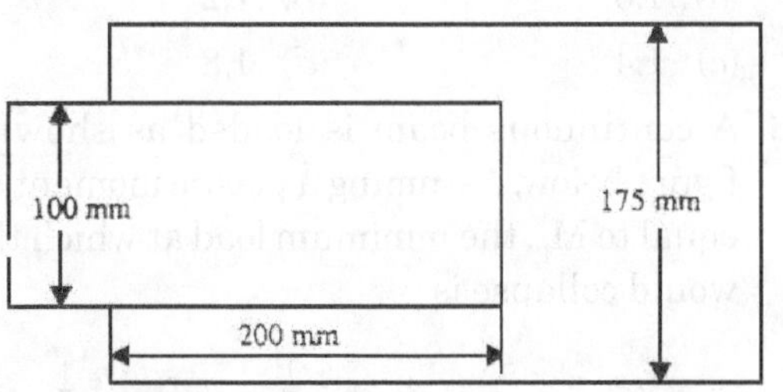

(a) 245.3 mm

(b) 229.2 mm

(c) 205.5 mm

(d) 194.8 mm

2009

20. In the theory of plastic bending of beams, the ratio of plastic moment to yield moment is called

(a) shape factor

(b) plastic section modulus

(c) modulus of resilience

(d) rigidity modulus

21. A 12 mm thick plate is conected to two 8 mm thick plates, on either side through a 16 mm diameter power driven field rivet as shown in the figure below. Assuming permissible shear stress as 90 MPa and permissible bearing stress as 270 MPa in the rivet, the rivet value of the joint is

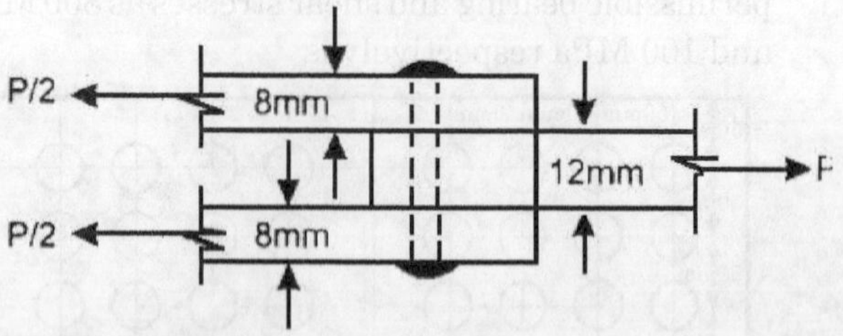

(*a*) 56.70 kN

(*b*) 43.29 kN

(*c*) 36.19 kN

(*d*) 21.65 kN

2008

22. Rivets and bolts subjected to both shear stress ($\tau_{vf, cal}$) and axial tensile stress ($\sigma_{tf, cal}$) shall be so proportioned that the stresses do not exceed the respective allowable stresses τ_{vf} and σ_{tf} and the value of $\left(\frac{\tau_{vf,\,cal}}{\tau_{vf}} + \frac{\sigma_{tf,\,cal}}{\sigma_{tf}}\right)$ does not exceed

(*a*) 1.0 (*b*) 1.2

(*c*) 1.4 (*d*) 1.8

23. A continuous beam is loaded as shown in the figure below. Assuming a plastic moment capacity equal to M_p, the minimum load at which the beam would collapse is

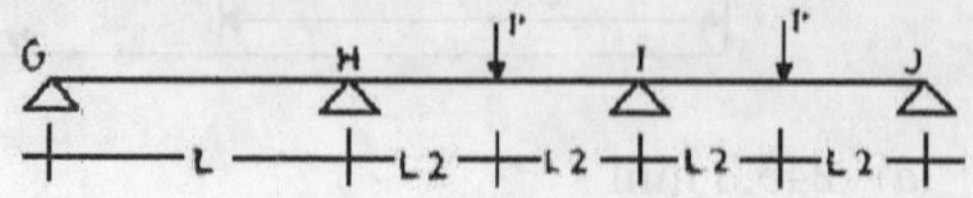

(*a*) $\frac{4M_p}{L}$ (*b*) $\frac{6M_p}{L}$

(*c*) $\frac{8M_p}{L}$ (*d*) $\frac{10M_p}{L}$

2007

24. A steel flat of rectangular section of size 70 × 6 mm is connected to a gusset plate by three bolts each having a shear capacity of 15 kN in holes having diameter 11.5 mm. If allowable tensile stress in the flat is 150 MPa, maximum tension that can be applied to the flat is

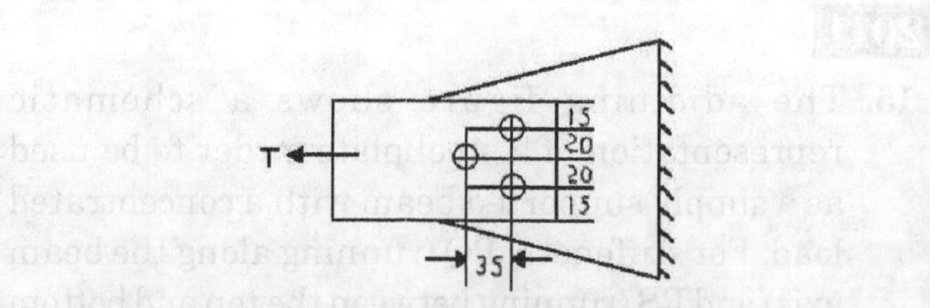

(*a*) 42.3 kN (*b*) 52.65 kN

(*c*) 59.5 kN (*d*) 63.0 kN

25. A bracket connection is made with four bolts of 10 mm diameter and supports a load of 10 kN at an eccentricity of 100 mm. The maximum force to be resisted by any bolt will be

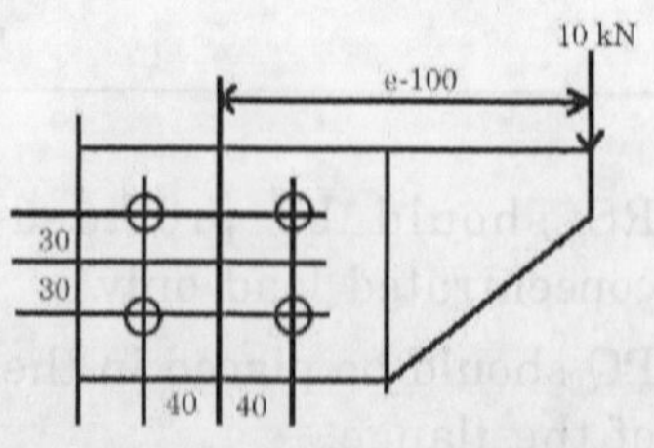

(*a*) 5 kN (*b*) 6.5 kN

(*c*) 6.8 kN (*d*) 7.16 kN

26. The plastic collapse load W_p for the propped cantilever supporting two point loads as shown in figure in terms of plastic moment capacity, M_p is given by

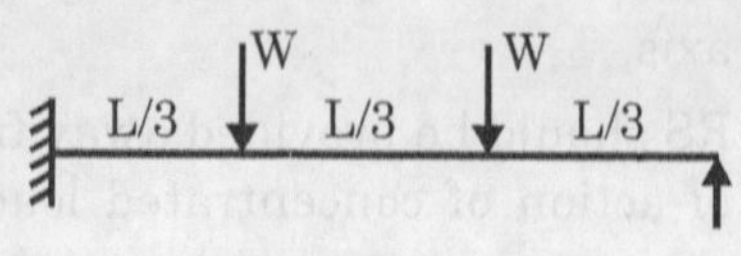

(*a*) $\frac{3M_p}{L}$ (*b*) $\frac{4M_p}{L}$

(*c*) $\frac{5M_p}{L}$ (*d*) $\frac{6M_p}{L}$

2006

27. In the design of welded tension members, consider the following statements :

I. The entire cross-sectional area of the connected leg is assumed to contribute to the effective area in case of angles.

II. Two angles back-to-back and tack-welded as per the codal requirements may be assumed to behave as a tee section.

III. A check on slenderness ratio may be necessary in some cases.

The TRUE statements are

(*a*) only I and II (*b*) only II and III

(*c*) only I and III (*d*) I, II and III

28. Consider the following statements

I. Effective length of a battened column is usually increased to account for the additional load on battens due to the lateral expansion of columns.

II. As per IS: 800-1984, permissible stress in bending compression depends on both Euler buckling stress and the yield stress of steel.

III. As per IS: 800-1984, the effective length of a column effectively held in position at both ends but not restrained against rotation, is taken to be greater than that in the ideal end conditions.

The TRUE statements are

(*a*) only I and II

(*b*) only II and III

(*c*) only I and III

(*d*) I, II and III

29. When triangular section of a beam as shown below becomes a plastic hinge, then compressive force acting on the section (with σ_y denoting the yield stress) becomes

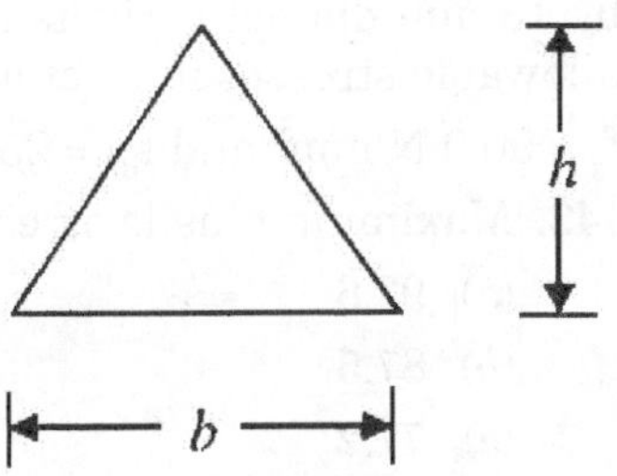

(*a*) $\dfrac{bh\sigma_y}{4}$

(*b*) $\dfrac{2bh\sigma_y}{9}$

(*c*) $\dfrac{bh\sigma_y}{2}$

(*d*) $\dfrac{bh\sigma_y}{3}$

2005

30. A fillet-welded joint of 6mm size is shown in the figure. The welded surfaces meet at 60-90 degree and permissible stress in the fillet weld is 108 MPa. The safe load that can be transmitted by the joint is

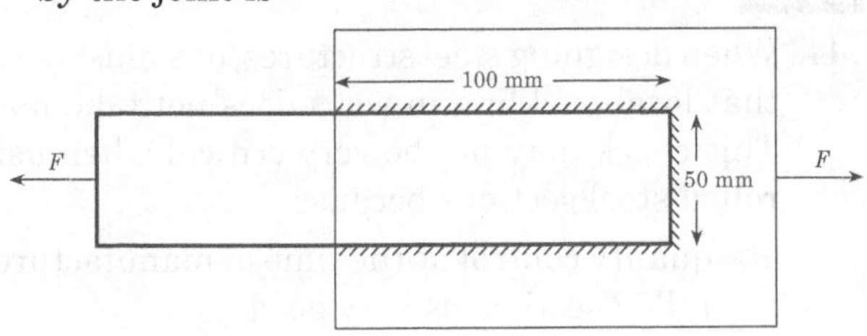

(*a*) 162.7 kN

(*b*) 151.6 kN

(*c*) 113.4 kN

(*d*) 109.5 kN

2004

31. Rivet value is defined as

(*a*) lesser of the bearing strength of rivet and the shearing strength of the rivet

(*b*) lesser of the bearing strength of rivet and the tearing strength of thinner plate

(*c*) greater of the bearing strength of rivet and the shearing strength of the rivet

(*d*) lesser of the shearing strength of the rivet and the tearing strength of thinner plate

32. In a plate girder , the web plate is connected to the flange plates by fillet welding. The size of the fillet welds is designed to safely resist

(*a*) bending stresses in the flanges

(*b*) vertical shear force at the section

(*c*) horizontal shear force between the flanges and the web plate

(*d*) forces causing buckling in the web

33. A moment M of magnitude 50 kN-m is transmitted to a column flange through a bracket by using four 20 mm diameter rivets as shown in the figure.

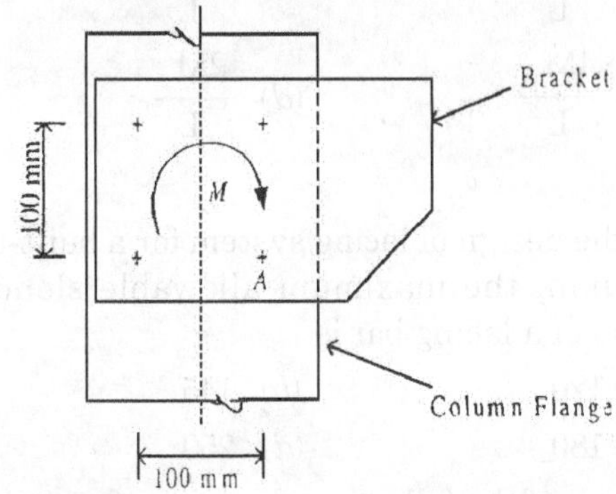

The shear force induced in the rivet A is

(*a*) 250 kN (*b*) 175.8 kN

(*c*) 125 kN (*d*) 88.4 kN

34. Two equal angles ISA 100 mm × 100 mm of thickness 10 mm are placed back-to-back and connected to the either side of a gusset plate through a single row of 16 mm diameter rivets in double shear. The effective areas of the connected and unconnected legs of each of these angles are 775 mm^2 and 950 mm^2, respectively. If these angles are NOT tackriveted, then net effective area of this pair of angles is

(*a*) 3650 mm^2 (*b*) 3450 mm^2

(*c*) 3076 mm^2 (*d*) 2899 mm^2

35. A strut in a steel truss is composed of two equal angles ISA 150 mm × 150 mm of thickness 100 mm connected back-to-back to the same side of a gusset plate. The cross sectional area of each angle is 2921 mm^2 and moment of inertia ($I_{xx} = I_{yy}$) is 6335000 mm^4. The distance of the centroid of the angle from its surface ($C_x = C_y$) is 40.8 mm. The minimum radius of gyration of the strut is

(*a*) 93.2 mm (*b*) 62.7 mm

(*c*) 46.6 mm (*d*) 29.8 mm

36. A square steel slab base of are 1 m^2 is provided for a column made of two rolled channel sections. The 300 mm × 300 mm column carries an axial compressive load of 2000 kN. The line of action of the load passes through the centroid of the column section as well as of the slab base. The permissible bending stress in the slab base is 185 MPa. The required minimum thickness of the slab base is

(*a*) 110 mm (*b*) 89 mm

(*c*) 63 mm (*d*) 55 mm

37. A propped cantilever of span L is carrying a vertical concentrated load acting at midspan. The plastic moment of the section of M_p. The magnitude of the collapse load is

(*a*) $\frac{8M_p}{L}$ (*b*) $\frac{6M_p}{L}$

(*c*) $\frac{4M_p}{L}$ (*d*) $\frac{2M_p}{L}$

2003

38. In the design of lacing system for a built-up steel column, the maximum allowable slenderness ratio of a lacing bar is

(*a*) 120 (*b*) 145

(*c*) 180 (*d*) 250

39. Which of the following elements of a pitched roof industrial steel building primarily resists lateral load parallel to the ridge ?

(*a*) bracings (*b*) purlins

(*c*) truss (*d*) columns

40. The stiffness K of a beam deflecting in a symmetric mode, as shown in the figure, is

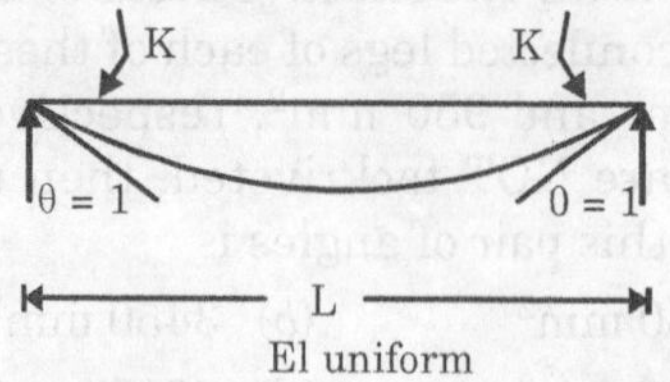

(*a*) $\frac{EI}{L}$ (*b*) $\frac{2EI}{L}$

(*c*) $\frac{4EI}{L}$ (*d*) $\frac{6EI}{L}$

41. Group I contains some elements in design of a simply supported plate girder and Group 2 gives some qualitative locations on the girder. Match the items of two lists as per good design practice and relevant codal provisions.

Group I	Group II
P. Flange splice	**1.** at supports (minimum)
Q. Web splice	**2.** away from centre of span
R. Bearing stiffeners	**3.** away from support
S. Horizontal stiffener	**4.** in the middle of span
	5. longitudinally somewhere in the compression flange.

Codes :

	P	Q	R	S
(*a*)	2	3	1	5
(*b*)	4	2	1	3
(*c*)	3	4	2	1
(*d*)	1	5	2	3

Common Data Q. 42-43

A truss tie consisting of 2 ISA 75 × 75 × 8 mm carries a pull of 150 kN. At ends the two angles are connected, one each on either side of a 10 mm thick gusset plate, by 18 mm diameter rivets arranged in one row. The allowable stresses in rivet are

$f_s = 90.0$ N/mm² and $f_{br} = 250$ N/mm².

42. Maximum tensile stress in the tie in N/mm² is

(*a*) 93.6

(*b*) 87.5

(*c*) 77.2

(*d*) 66.0

43. Minimum number of rivets required at each end is

(*a*) 2

(*b*) 3

(*c*) 4

(*d*) 5

2002

44. When designing steel structures, one must ensure that local buckling in webs does not take place. This check may not be very critical when using rolled steel sections because

(*a*) quality control at the time of manufacture of rolled sections is very good

(*b*) web depths available are small

(*c*) web stiffeners are in-built in rolled sections

(*d*) depth to thickness ratios (of the web) are appropriately adjusted

45. An ISMB 500 is used as a beam in a multi-storey construction. From the viewpoint of structural design, it can be considered to be *laterally restrained*' when

(*a*) tension flange is 'laterally restrained'

(*b*) compression flange is 'laterally restrained'

(*c*) web is adequately stiffened

(*d*) conditions in (A) and (C) are met

46. A steel beam (with a constant EI, and span L) is fixed at both ends and carries a uniformly distributed load (w kN/m), which is gradually increased till the beam reaches the stage of plastic collapse (refer to the following figure). Assuming 'B' to be at mid-span, which of the following is true ?

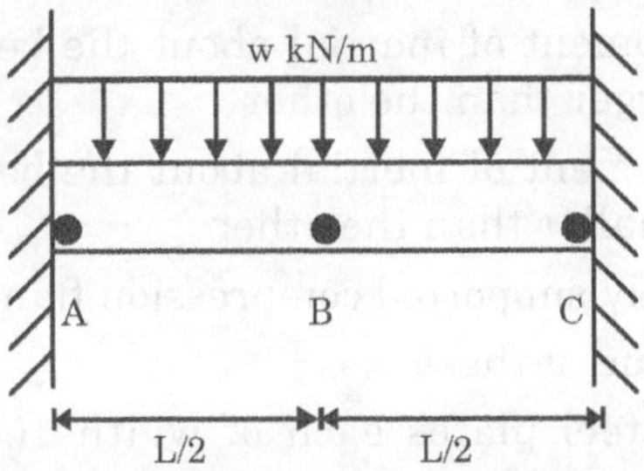

(*a*) Hinges are formed at A, B and C together

(*b*) Hinges are formed at B and then at A and C together

(*c*) Hinges are formed at A and C together and then at B

(*d*) Hinges are formed at A and C only

47. ISA 100 × 100 × 10mm (cross sectional area = 1908 mm²) serves as tensile member. This angle is welded to a gusset plate along A and B appropriately as shown. Assuming yield strength of the steel to be 260 N/mm², tensile strength of this member can be taken to be approximately

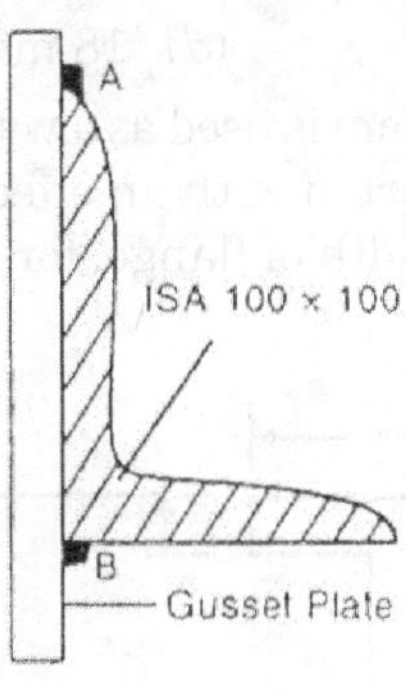

(*a*) 500kN

(*b*) 300kN

(*c*) 225kN

(*d*) 375kN

48. ISA 100 × 100 × 10 mm (cross sectional area = 1908 mm²) is welded along A and B (Refer to figure for above question), such that lengths of the weld along A and B are l_1 and l_2, respectively. Which of the following is a possibly acceptable combination of l_1 and l_2?

(*a*) l_1 = 60 mm and l_2 = 150 mm

(*b*) l_1 = 150mm and l_2 = 60 mm

(*c*) l_1 = 150 mm and l_2 = 150 mm

(*d*) Any of the above, depending on the size of the weld

49. Give reasons for the following in not more than 20 words.

(*a*) A maximum permissible distance between lacings and battens in steel columns is usually specified.

(*b*) It is sometimes preferable to have unequal flange angles with the longer legs horizontal in a plate girder.

(*c*) If two channels sections need to be used as a steel column, they may be connected *face-to face* rather than *back to back*.

(*d*) It is sometimes preferred to have a small gap between the web and the flange plate in a plate girder.

(*e*) A maximum permissible outstand may be specified for flanges in built-up sections.

2001

50. Identify the most efficient but joint (with double cover plates) for a plate in tension from the patterns (plan views) shown below, each comprising 6 identical bolts with the same pitch and gauge.

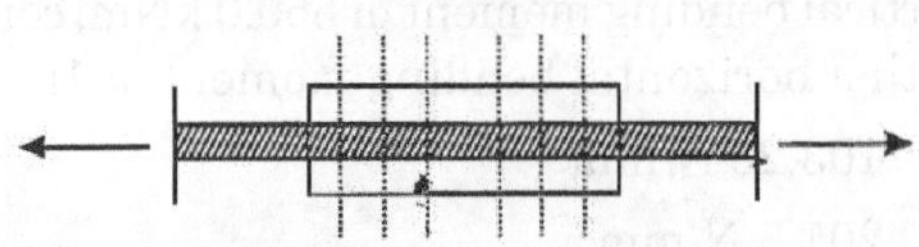

COMMON ELEVATION

(all plates have same thickness)

(*a*) 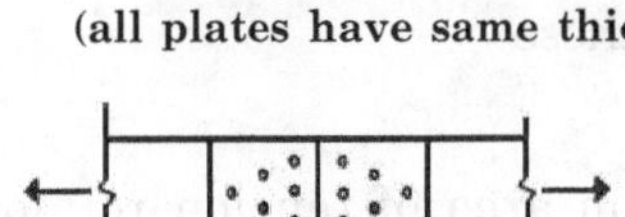

(*b*) 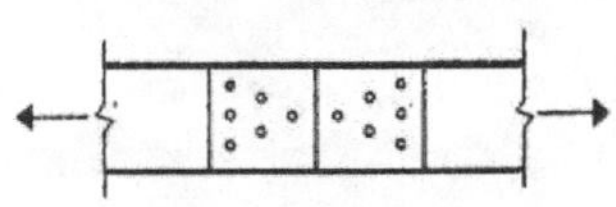

(*c*) 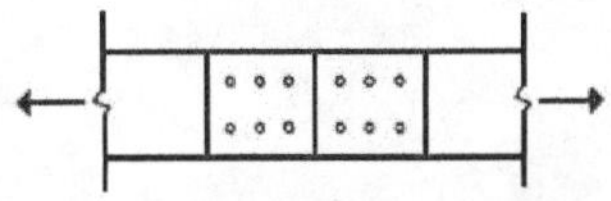

(*d*) 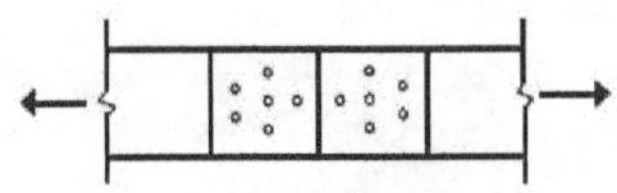

Common Data Q. 51-53

The relevant cross-sectional details of a compound beam comprising a symmetric I-section and a channel section (with welded connections), proposed for a steel gantry girder, are given below (all dimensions in mm).

For computing $I_{yy,eff}$ include full contribution of the channel section, but only top flange of the I-section.

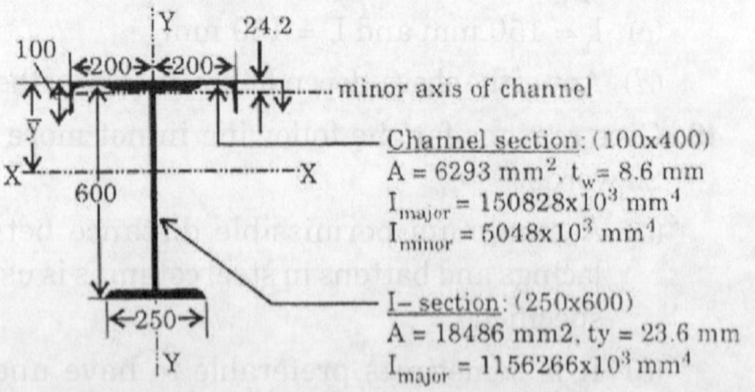

51. Determine depth of the centroidal axis $\overline{y}$

(*a*) 136.8 mm

(*b*) 236.4 mm

(*c*) 190.5 mm

(*d*) 325.0 mm

52. Second moments of area, I_{xx} and $I_{yy,eff}$ of the compound section

(*a*) 5×10^8 mm⁴, 1.21×10^8 mm⁴

(*b*) 10×10^8 mm⁴, 1.38×10^8 mm⁴

(*c*) 15×10^8 mm⁴, 1.81×10^8 mm⁴

(*d*) None of these

53. Determine maximum compressive stress that developes at a top corner location on account of a vertical bending moment of 550.0 kNm, combined with a horizontal bending moment of 15.0 kNm.

(*a*) 103.25 N/mm²

(*b*) 201.5 N/mm²

(*c*) 62.57 N/mm²

(*d*) None of these

1999

54. The minimum area of tension reinforcement in a beam shall be greater than

(*a*) $\frac{0.85 \text{ bd}}{f_y}$

(*b*) $\frac{0.87 f_y}{\text{bd}}$

(*c*) 0.04 bd

(*d*) $\frac{0.4 \text{ bd}}{y}$

55. Maximum strain at the level of compression steel for a rectangular section having effective cover to compression steel as d' and neutral axis depth from compression face x_u is

(*a*) $0.0035\left(1-\frac{d'}{x_u}\right)$ (*b*) $0.002\left(1-\frac{d'}{x_u}\right)$

(*c*) $0.0035\left(1-\frac{x_u}{d'}\right)$ (*d*) $0.002\left(1-\frac{x_u}{d'}\right)$

56. A steel beam supporting loads from the floor slab as well as from wall is termed as

(*a*) Stringer beam (*b*) Lintel beam

(*c*) Spandrel beam (*d*) Header beam

57. The problem of lateral buckling can arise only in those steel beams which have

(*a*) moment of inertial about the bending axis larger than the other

(*b*) moment of inertial about the bending axis smaller than the other

(*c*) fully supported compression flange

(*d*) none of these

58. Two steel plates each of width 150 mm and thickness 10 mm are connected with three 20 mm diameter rivets placed in a zig - zag pattern. The pitch of the rivets is 75 mm and gauge is 60 mm. If the allowable tensile stress is 150 MPs, the maximum tensile force that the joint can withstand is

(*a*) 195.66 kN (*b*) 195.00 kN

(*c*) 192.75 kN (*d*) 225.00 kN

1998

59. The maximum permissible deflection for a gantry gride, spanning over 6m, on which an EOT (electric overhead travelling) crane of capacity 200 k.N is operating, is

(*a*) 8 mm (*b*) 10mm

(*c*) 12 mm (*d*) 18 mm

60. An isolated T beam is used as a walkway. The beam is simply supported with an effective span of 6m. The effective width of flange, for the cross-section shown in Figure, is

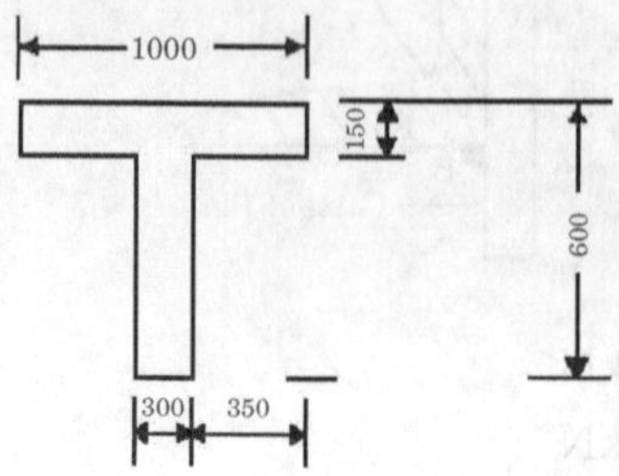

(*a*) 900 mm (*b*) 1000 mm

(*c*) 1259 mm (*d*) 2200 mm

61. The plastic modulus of a section is 4.8×10^{-4} m³. The shape factor is 1.2. The plastic moment capacity of the section is 120 kN.m. The yield stress of the material is
(a) 100MPa (b) 240 MPa
(c) 250 MPa (d) 300MPa

1997

62. Maximum size of a fillet weld for a plate of square edge is
(a) 1.5 mm less than the thickness of the plate
(b) one half of the thickness of the plate
(c) thickness of the plate itself
(d) 1.5 mm more than the thickness of the plate

63. Factor of safety adopted by IS : 800 - 1984 while arriving at the permissible stress in axial compression is
(a) 2.00
(b) 1.00
(c) 1.67
(d) 1.50

64. Effective length of a rafter member between two nodes at a distance L, perpendicular to the plane of the truss is
(a) 2.00 L
(b) 0.85
(c) 1.50
(d) 1.00 L

65. Allowable average shear stress in an unstiffened web for beams made of steel of grade 250 N/mm² is
(a) 250 N/mm²
(b) 165 N/mm²
(c) 150 N/mm²
(d) 100 N/mm²

1996

66. Generally the maximum deflection/span ratio of a steel member should bot exceed (a)
(a) $\frac{1}{750}$
(b) $\frac{1}{500}$
(c) $\frac{1}{325}$
(d) $\frac{1}{250}$

67. The effective length of a circular electric pole of length L and constant diameters erected on ground is,
(a) 0.80 L
(b) 1.20 L
(c) 1.50 L
(d) 2.00 L

68. In IS : 800 (1984), the permissible compressive stress in column is based on,
(a) Euler formula
(b) Secant formula
(c) Rankine – Gordan formula
(d) Rankine – Merchant fourmula

69. Generally (fatigue life of welded steel structure/ fatigue life of riveted steel structure) ratio is
(a) smaller than 1
(b) equal to 1
(c) greater than 1
(d) greater than 2.1

NUMERICAL TYPE QUESTIONS

2016

1. The semi-compact section of a laterally unsupported steel beam has an elastic section modulus, plastic section modulus and design bending compressive stress of 500 cm³, 650 cm³ and 200 MPa, respectively. The design flexural capacity (expressed in kNm) of the section is________.

2014

2. A bracket plate connected to a column flange transmits a load of 100 kN as shown in the following figure. The maximum force for which the bolts should be designed is ________ kN.

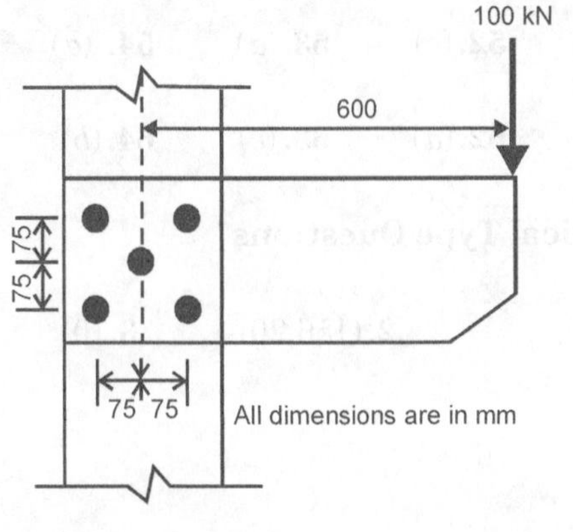

2013

3. The pin-jointed 2-D truss is loaded with a horizontal force of 15 kN at joint S and another 15 kN vertical force at joint U, as shown. Find the force in member RS (in kN) and report your answer taking tension as positive and compression as negative. ______

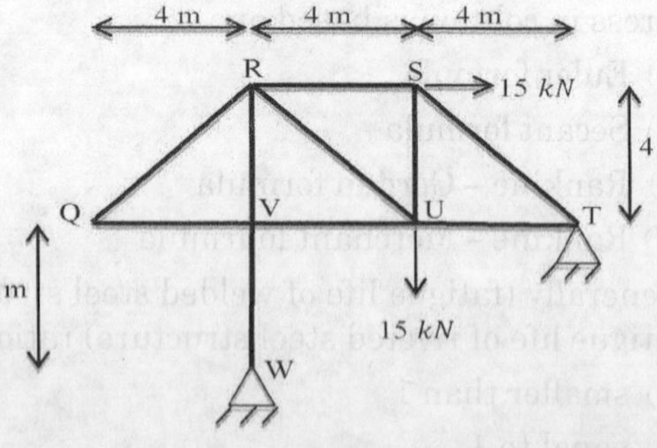

2000

4. A 12mm bracket plate is connected to a column flange as shown in the figure below. The bracket transmits a load of W = 200 kN to the column flange. A 10mm fillet weld is provided along AB, BC and CD. If e = 350 mm, b = 200 mm and d = 650 mm, verify if size of the weld provided is adequate. Allowable shearing stress in the fillet weld can be taken to be 108 MPa.

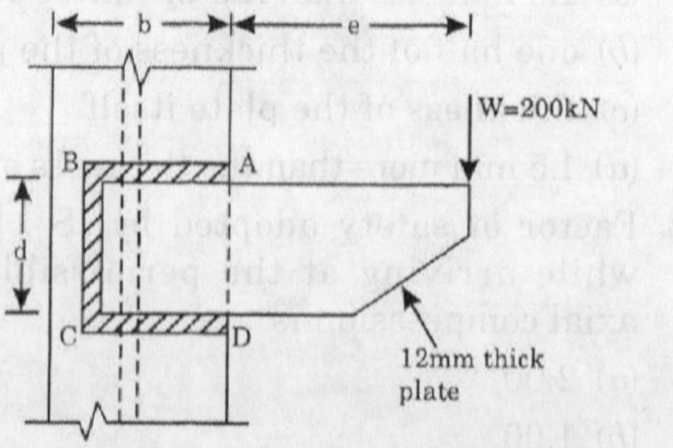

ANSWERS

MCQ Type Questions

1. (*b*)	**2.** (*b*)	**3.** (*c*)	**4.** (*a*)	**5.** (*d*)	**6.** (*c*)	**7.** (*a*)	**8.** (*d*)	**9.** (*a*)	**10.** (*b*)
11. (*c*)	**12.** (*d*)	**13.** (*a*)	**14.** (*b*)	**15.** (*b*)	**16.** (*d*)	**17.** (*b*)	**18.** (*b*)	**19.** (*b*)	**20.** (*a*)
21. (*b*)	**22.** (*c*)	**23.** (*b*)	**24.** (*c*)	**25.** (*d*)	**26.** (*d*)	**27.** (*b*)	**28.** (*a*)	**29.** (*a*)	**30.** (*c*)
31. (*a*)	**32.** (*c*)	**33.** (*b*)	**34.** (*b*)	**35.** (*c*)	**36.** (*d*)	**37.** (*b*)	**38.** (*b*)	**39.** (*a*)	**40.** (*b*)
41. (*a*)	**42.** (*a*)	**43.** (*c*)	**44.** (*d*)	**45.** (*b*)	**46.** (*b*)	**47.** (*a*)	**48.** (*b*)	**49.** (*)	**50.** (*a*)
51. (*b*)	**52.** (*c*)	**53.** (*a*)	**54.** (*a*)	**55.** (*a*)	**56.** (*a*)	**57.** (*b*)	**58.** (*c*)	**59.** (*a*)	**60.** (*a*)
61. (*c*)	**62.** (*a*)	**63.** (*c*)	**64.** (*b*)	**65.** (*d*)	**66.** (*c*)	**67.** (*b*)	**68.** (*d*)	**69.** (*c*)	

Numerical Type Questions

1. (100) **2.** (156.20) **3.** (0) **4.** (*)

"*" Refers to Subjective Question

EXPLANATIONS

MCQ TYPE QUESTIONS

1. $$W_C \times \delta = 3M_P.\theta$$

$$? = \frac{l}{2}.\theta$$

$$W_C \times \frac{l}{2}.\theta = 3M_P.\theta$$

$$W_C = \frac{6M_P}{l}$$

2. $$F_u = 410 \text{ N/mm}^2;$$

$$f_y = 250 \text{ N/mm}^2;$$

$$\gamma_{mw} = 1.25;$$

$$S = 10 \text{ mm};$$

factored load P = 270kN

Assuming length of each side to be l_j & effective length offillet weld $L_w = 2\,l_j$

Effective Throat Thickness,

$$t_t = K \times S = 0.7 \times 10 = 7 \text{ mm}$$

Equating design axial load (P) to the design shear strength of fillet weld (P_{dw})

$$P = P_{dw} = L_w.t_t.\frac{f_u}{\sqrt{3}\gamma_{mw}}$$

$$270 \times 10^3 = L_w \times 7 \times \frac{410}{\sqrt{3} \times 1.25}$$

$$= \left(2 \times l_j \times 7 \times \frac{410}{\sqrt{3} \times 1.25}\right)$$

$$L_j = 101.83 \text{ mm} = 105 \text{ mm}$$

3. Condition is : The loss due to second bolt is less than the gain in strength due to staggering of the second bolt

$$\frac{p^2}{4g} > d,$$

i.e., $$p^2 > 4gd$$

4. Number of hinges for the mechanism

$$= D_s + 1 = 2 + 1 = 3$$

Case I : When hinges are at A, B & C

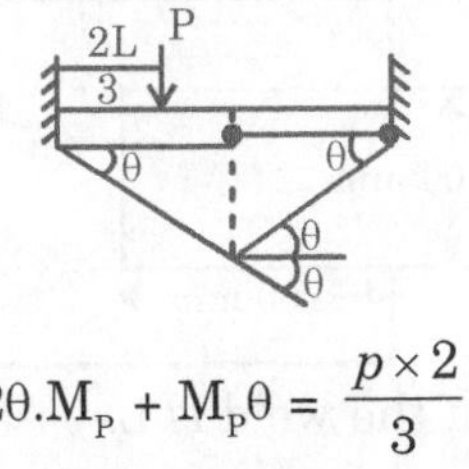

$$2M_P\theta + 2\theta.M_P + M_P\theta = \frac{p \times 2}{3}L.\theta$$

$$5M_P\theta = \frac{2}{3}P.L.\theta$$

$$P = \frac{\frac{15}{2}M_P}{L}$$

$$= 7.5\frac{M_P}{L}$$

Case II : Hinges are formed at A, B & D

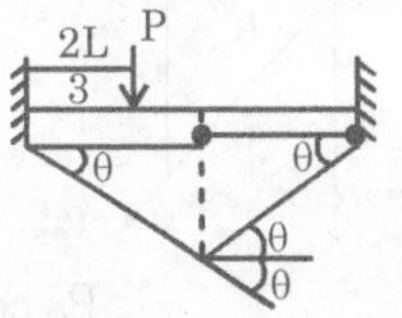

$$2M_P\alpha + M_P(\alpha + \beta) + M_P\beta = \frac{2P}{3} \times L.\alpha$$

$$5.5\frac{M_P}{L} \times \frac{3}{2} = P$$

$$P = \frac{8.25.\,M_P}{L}$$

Therefore, $$\frac{2}{3}\alpha = \frac{4}{3}\beta.L$$

$$\alpha = 2.\beta$$

Collapse load is $7.5\frac{M_P}{L}$

5. Static method → Upper bound method of ultimate load analysis

Kinematic method → Lower bound on ultimate load

$Q_{Design} = Q_w \times$ load factor

$M_0 = M_u \times \gamma_m$, γ_m = material partial safetyfactor

6. $D_s = 1$

So, no. of plastic hinges = $D_s + 1 = 2$

External work done = $P.\Delta = P.\left(\frac{L}{2}.\theta\right)$

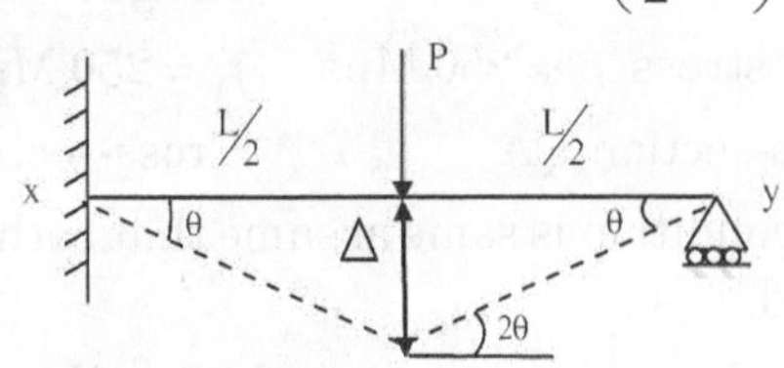

Internal work done

$$M_p.\theta + M_p.2\theta = 3M_p.\theta$$

By principal of virtual work = $P.\frac{L}{2}.\theta = 3M_p.\theta$

$$\Rightarrow \quad P = \frac{6M_p}{L}$$

7. As per clause 9.2.1 (IS: 800-2007) for combined shear and bending: Factored value of applied shear force is greater than or equal to shear strength for high shear.

i.e., $V > 0.6V_s$

8. $\sin\theta = \frac{3}{5}$, $\cos\theta = \frac{4}{5}$

$$P_U \cos\theta = \frac{4}{5}.P_U$$

$$P_U \sin\theta = \frac{3}{5}.P_U$$

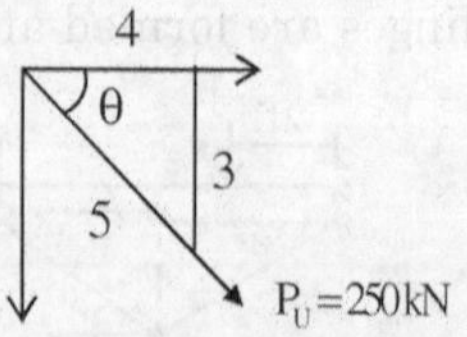

$$\text{Tension in each bolt} = \frac{P_U \cos\theta}{6} = \frac{4P_U}{5\times 6}$$

$$= \frac{4}{30}\times 250 = 33.33 \text{ kN}$$

$$\text{Shear in each bolt} = \frac{P_U \sin\theta}{6} = \frac{3}{5}\times\frac{P_U}{6}$$

$$= \frac{3}{5\times 6}\times 250 = 25 \text{ kN}$$

9. According to IS : 800 – 1984

Under Clause 3.7 and in Table 3.1,

Maximum slenderness ratio in M due to live loads = 180

And Maximum slenderness ratio in N due to windload = 350

Hence, λ for M < λ for N.

10. Prying forces are tensile forces due to the flexibility of connected parts.

12. There are two column P and Q

Column 'P'	**Column 'Q'**
Length = L	Length = 2L
Yield stress f_Y = 250 Mpa	f_Y = 250 Mpa
Cross-section = A	Cross-section = A

End condition is same assume that both ends are pissed.

Left = L   Left = 2L

$$P_{buckling} = \frac{\pi^2 EI}{l_{eff}^2} \qquad Q_{buckling} = \frac{\pi^2 EI}{l_{eff}^2}$$

$$P_{buckling} = \frac{\pi^2 EI}{L^2} \qquad Q_{buckling} = \frac{\pi^2 EI}{(2L)^2}$$

$$\text{Ratio of } \frac{P_{buckling}}{Q_{buckling}} = \frac{\frac{\pi^2 EI}{L^2}}{\frac{\pi^2 EI}{4L^2}} = \frac{1}{1}\times\frac{4}{1}$$

$$\frac{P_{buckling}}{Q_{buckling}} = 4$$

16.

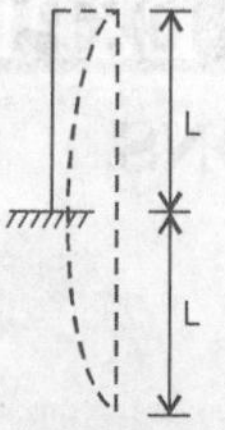

Effective length = 2 L

17. We know, $\frac{T}{I_P} = \frac{G\theta}{L}$

where, I_P = polar moment of inertia = $\frac{\pi}{32}d^4$

$$\therefore \quad \theta = \frac{TL}{I_P G} = \frac{32TL}{\pi d^4 G}$$

18. Gross diameter of rivet, d = 20 + 1.5 = 21.5 mm

There are two rivets per pitch length.

Since each rivet is in double sheal, therefore strength of rivet in double shear

$$= 2\times\left(2\times\frac{\pi}{4}\times 21.5^2\times 100\right)$$

$$= 145.220 \text{ kN}$$

Strength of rivet in bearing

$$= 2\times 21.5\times 14\times 300 = 180.6 \text{ kN}$$

∴ Rivet value = 145.220 kN

Since joint is strengthened by 6 rivets, therefore capacity of the joint = 145.22 × 6 = 871.32 kN

Alternately

Strength of rivet in shear,

$$P_S = 2\times\frac{\pi}{4}\times d^2\times\tau$$

$$= 2\times\frac{\pi}{4}\times 21.5^2\times 100 = 72.61 \text{ kN}$$

Strength of joint in bearing,

$$P_b = d.tp_f$$

$$= 21.5\times 14\times 300 = 90.3 \text{ kN}$$

∴ Rivet value, R_v = 72.61 kN

∴ Capacity of joint = NR_V = 72.61 × 12

= 871.32 kN

19.

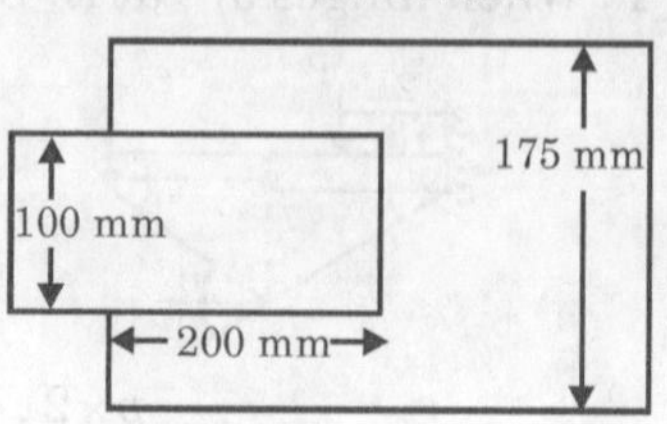

Let length of the weld is L.

Size of weld = Thickness of thinner plate – 1.5

=10 – 1.5 = 8.5 mm

Throat thickness, t = 0.7 × 8.5 = 5.95 mm

The joint will be designed on the basis of strength of small plate.

$\therefore$ Total load taken by small plate

$$= 150 \times 100 \times 10 = 150 \text{ kN}$$

Allowable load per line

$$= 5.95 \times 110 = 654.5 \text{ N}$$

$$\therefore \quad L = \frac{150 \times 10^3}{654.5} = 229.2 \text{ mm}$$

21. Here, $d = 16 + 1.5 = 17.5$

$$\text{Shear strength} = \frac{\pi}{4} \times d^2 \times 2 \times 90$$

$$= \frac{\pi}{4} \times (17.5)^2 \times 2 \times 90 = 43.29 \text{ kN}$$

Bearing strength $= p \times d \times T$

$$= 270 \times 17.5 \times 12 = 56.7 \text{ kN}$$

So, minimum strength = 43.29

Rivet value = 43.29 kN

22. Refer IS 800 : 1984 code for General steel construction clause 8.9.4.5

$$\frac{\tau_{vf,cal}}{\tau_{vf}} + \frac{\sigma_{tf,cal}}{\sigma_{tf}} \not> 1.4$$

23.

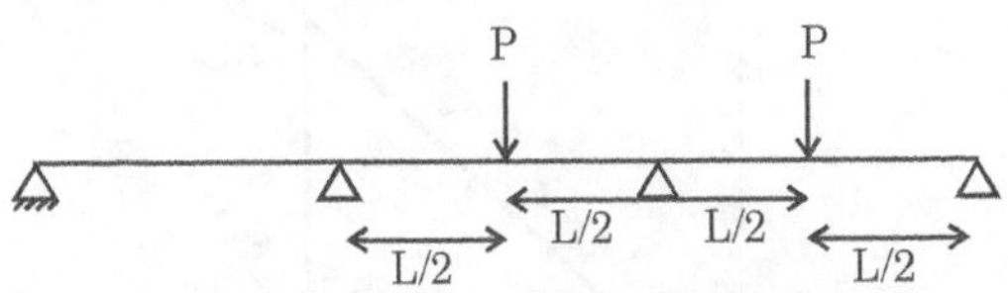

There will be two possible collapse mechanisms

(i)

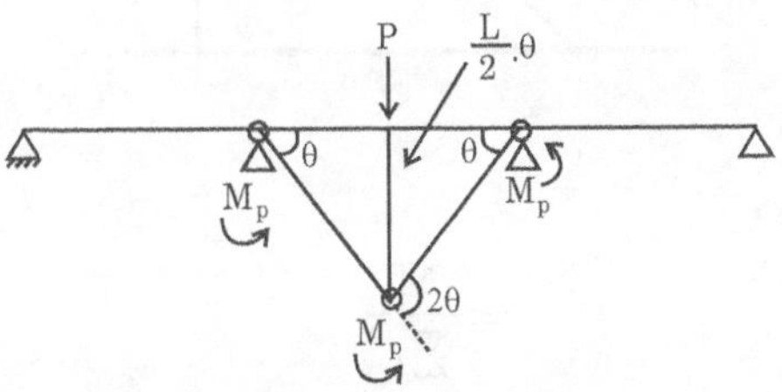

Here, $M_p.\theta + M_p.2\theta + M_p\theta = p.\frac{L}{2}.\theta$

$$\Rightarrow \quad \frac{8M_p}{L} = p$$

(ii)

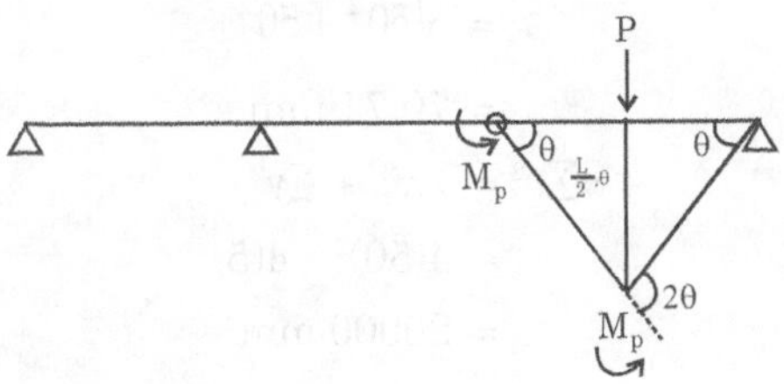

Here, $M_p.\theta + M_p.2\theta = p.\frac{L}{2}.\theta$

$$\Rightarrow \quad p = \frac{6M_p}{L}$$

The lower of two callapse values will be the collapse load.

$$\therefore \quad p = \frac{6Mp}{L}$$

24. Size of steel flat = 70 × 6mm

Diameter of bolts, $t = 11.5$ mm

Net minimum area will be at the setion $a\ b\ c\ d\ e$

$s = 35,$

$g = 20$

$n = 3$

$m = 2$

But $\quad A_{net} = t\left(L - nd + \frac{ms^2}{4g}\right)$

$$\therefore \quad A_{net} = 6\left[70 - 3 \times 11.5 + \frac{2 \times 35^2}{4 \times 20}\right]$$

$$= 396.75 \text{ mm}^2$$

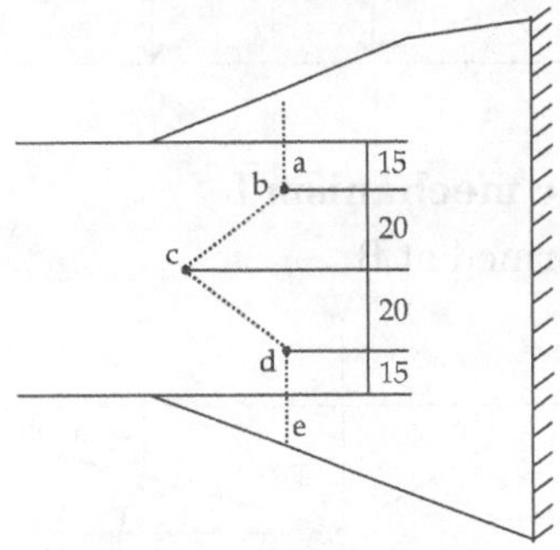

Strength of plate = 150 N/mm^2

Maximum tension in flat, $P = A_{net} \times \sigma_{at}$

$$P = 796.75 \times 150$$

Where, $\sigma_{at} = 150$ N/mm^2 = 59.5 kN

25. Diameter of bolt, $d = 10$ mm

Load, P = 10 kN

$e = 100$ mm

Number of bolts, $n = 4$

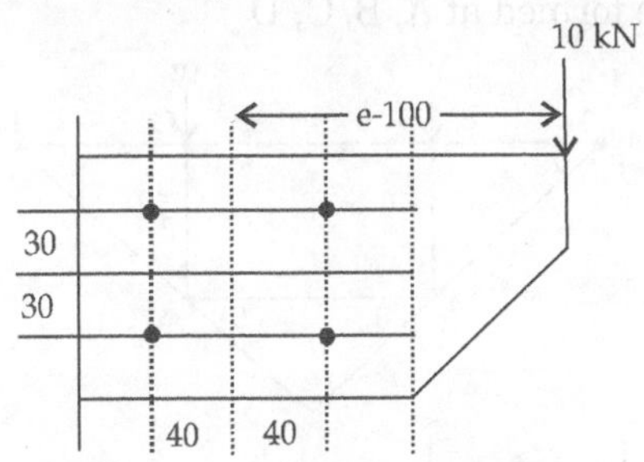

Bending moment, M = P.e

$$= 10 \times 10^3 \times 100$$

$$= 10 \times 10^5 \text{ N-mm}$$

$$F_1 = \frac{P}{n} = \frac{10 \times 100}{n} = 2.5 \text{ kN}$$

$$F_2 = \frac{M.r}{\sum r^2}$$

$$r = \sqrt{(30)^2 + (40)^2} = 50 \text{ mm}$$

$$\therefore \quad \sum r^2 = 4 \times (50)^2 = 10,000 \text{ mm}$$

$$\therefore \quad F_2 = \frac{10 \times 10^5 \times 50}{10,000} = 5000 = 5 \text{ kN.}$$

$$\cos \alpha = \frac{30}{50} = 0.6,$$

$$\sin \alpha = 0.8$$

Maximum load resisted by bolt,

$$R = \sqrt{(F_1 + F_2 \sin \alpha)^2 + (F_2 \cos \alpha)^2}$$

$$= \sqrt{(2.5 + (5 \times 0.81))^2 + (5 \times 0.6)^2} = 7.16 \text{ kN}$$

26.

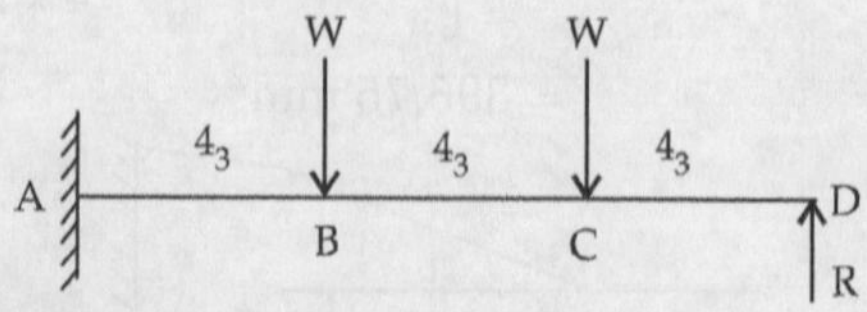

Collapse mechanism I

Hinge formed at B

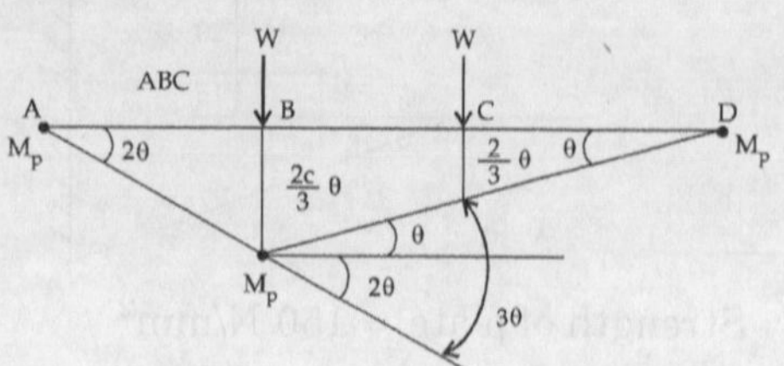

$$W_p \times \frac{2L}{3} \theta + W_p \times \frac{L}{3} \theta = M_p . 2\theta + M_p 3\theta + M_p \theta$$

$$\Rightarrow \quad W_p . L . \theta = \sum M_p . \theta$$

$$\Rightarrow \quad W_p = \frac{\sum M_p}{L}$$

Collapse mechanism II

Hinge formed at A, B, C, D

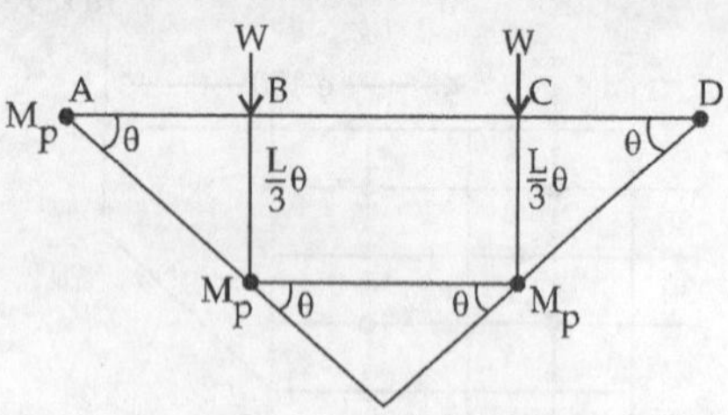

$$W_p \times \frac{L}{3} \theta + W_p \frac{L}{3} \theta = M_P \theta + M_p \theta + M_p \theta + M_p \theta$$

$$\Rightarrow \quad \frac{2L}{3} W_p \theta = 4 M_p \theta$$

$$\Rightarrow \quad W_p = \frac{6M_p}{L}$$

29. Compressive force $= \frac{1}{2} \times$ area of triangle $\times \sigma_y$

$$= \frac{1}{2} \times \frac{1}{2} \times b \times h \times \sigma_y = \frac{bh\sigma_y}{4}$$

30.

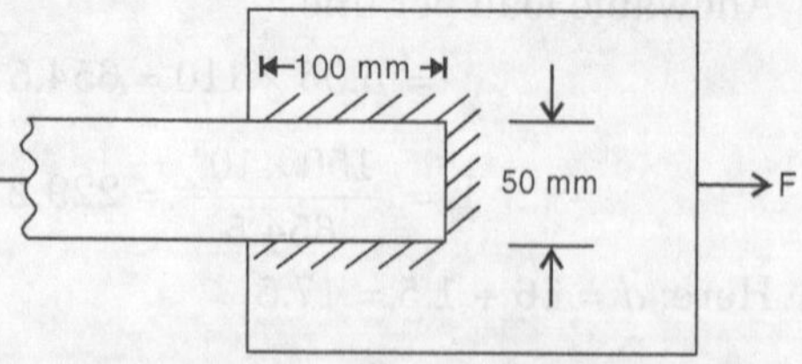

$$F = l \times t \times P_q$$

where, t = throat thickness

l = effective length = $2 \times 100 + 50 = 250$ mm

P_d = permissible stress

= 108 MPa (Given)

F = Safe load

S = size of fillet weld

$$\therefore \quad F = 0.7 l \times S \times P_q = 0.7 \times 250 \times 6 \times 108 = 113.4 \text{ kN}$$

33. Shear force induced in the rivet

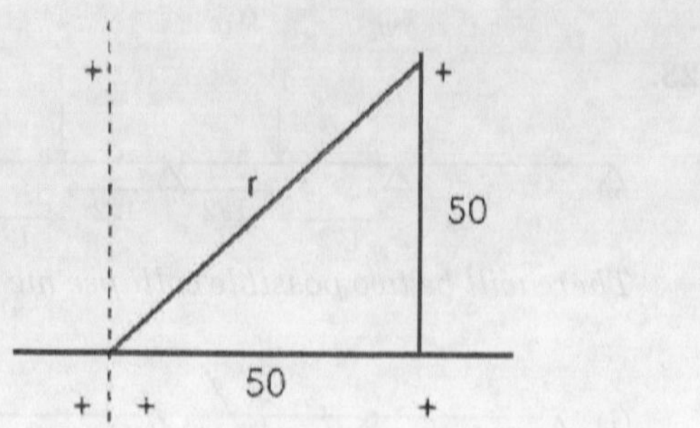

$$= \frac{M_r}{\sum r^2}$$

$$= \frac{50 \times 10^6 \times 70.71}{50 \times 10^6}$$

$$= 176.775 \text{ kNm}$$

$$r = \sqrt{50^2 + 50^2}$$

$$= 70.71 \text{ mm}$$

$$\sum r^2 = \sum x^2 + \sum y^2$$

$$= 4(50)^2 + 4(5)^2$$

$$= 20000 \text{ mm}^2$$

$$M = 50 \times 10^6 \text{ Nmm.}$$

34. Area of coneected leg = 775 mm²

Unconnected leg = 950 mm²

Net effective area = $2 \times 775 + 2 \times 950$

= 3450 mm²

35.

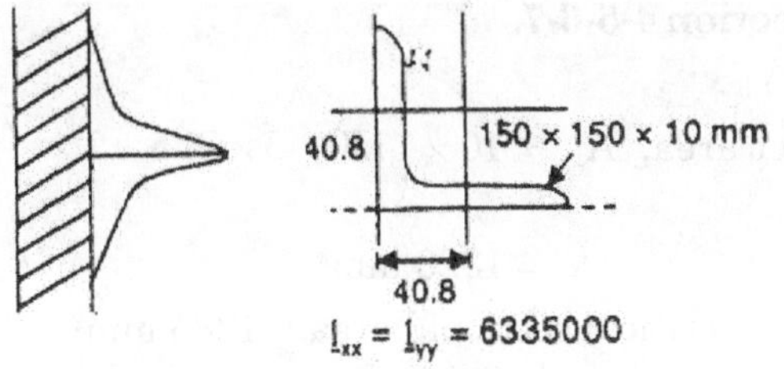

$$r = \sqrt{\frac{2I_{xx}}{2A}} = \sqrt{\frac{I_{xx}}{A}}$$

$$= \sqrt{\frac{6335000}{2921}} = 46.57 \text{ mm}$$

37.

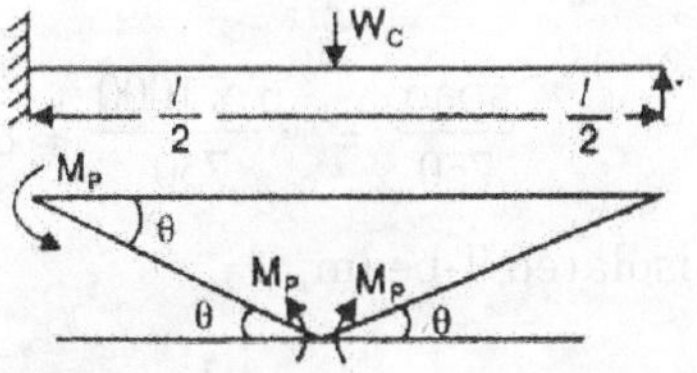

Internal workdone = External workdone

$$\therefore \quad 3M_p\theta = W_c\delta = W_c\left(\frac{1}{2}\theta\right)$$

$$\Rightarrow \quad W_c = \frac{6M_p}{L}$$

38. According to steel code, stenderness ratio 'λ' of the lacing bars for compression member shall not exceed 145.

39. According to steel code, the main element of a pitched roof industrial steel building primarily resists lateral load parallel to ridge in bracings.

40. $$\theta_A = \theta_B = \frac{ML}{2EI}$$

Here, $\theta_A = \theta_B = 1$

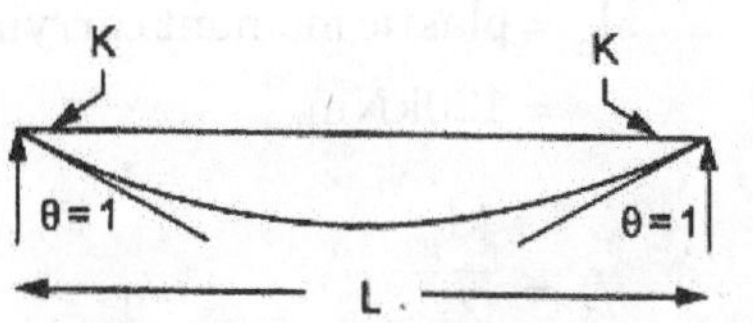

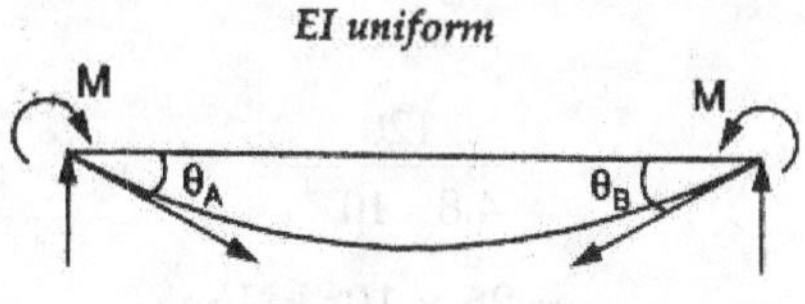

$$\therefore \quad \frac{ML}{2EI} = 1$$

$$\Rightarrow \quad M = \frac{2EI}{L} = K$$

42. Area of truss tie = (75 – 4) × 2 × 8

$= 1136 \text{ mm}^2$

Total area = 2 × 1136 – 4 D*t*

$= 1602.56 \text{ mm}^2$

$$\therefore \quad \text{Maximum tensile stress} = \frac{P}{1602.56}$$

$$= \frac{150 \times 10^3}{1602.56}$$

$$= \mathbf{93.6\ N/mm^2}$$

43. Shearing force = $2 \times \frac{\pi}{4} \times (19.5)^2 \times 90 = 53.6$ N

Bearing force = D × *t* × 250 = 48.75 N

$$\therefore \quad \text{Number of rivets} = \frac{150}{48.75}$$

$= 3.076$

$= \mathbf{4\ (say)}$

49. (*a*) Minimum slenderness ratio of the component member $\frac{C}{r_{yy}}$ is not greater than 50 or 0.7 times the slenderness ratio of the member.

(*b*) To resist horizontal shear force and addition vertical load from the flange to the web., load on the top flange is W per unit length.

(*c*) Channel placed face to face provided larger value of radius of gyration as compared to back to back. Then face to face are ideal section.

(*d*) Due to heavy load on the section preventing buckling, the small gap is provided.

(*e*) Maximum permissible outstand is $\frac{1440t}{\sqrt{fy}}$ subject to minimum of 90*t*, and $\frac{560t}{\sqrt{fy}}$ subjected to maximum of 35*t* for weld plate.

51-52.

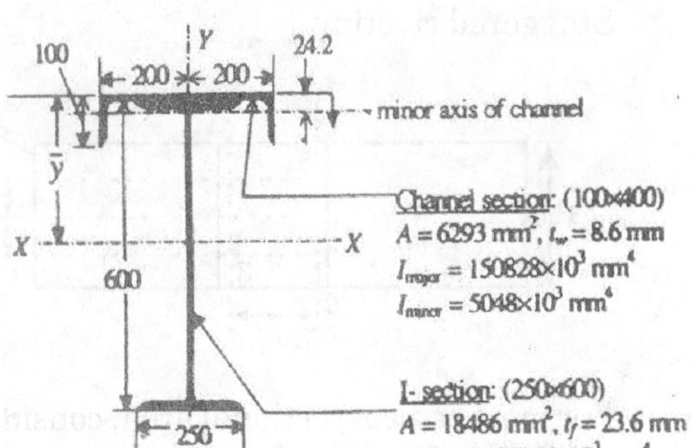

$$\bar{y} = \frac{6293 \times 24.2 + 18486 \times (300 + 8.6)}{(6293 + 18486)}$$

$= 236.40 \text{mm}$

$$I_{xx} = 1156266 \times 103 + 18486(300 - 236.40 - 8.6)^2$$
$$+ 5048 \times 10^3 + 6293(236.40 - 24.2)^2$$

$= 15 \times 10^8 \text{ mm}^4$

$$I_{yy} = 150828 \times 10^3 + \frac{23.6 \times 250^3}{12}$$

$$= 1.81 \times 10^8 \text{ mm}^4$$

53. Maximum compressive stress at top

$$\sigma = \frac{M_v \overline{Y}}{I_{xx}} + \frac{M_h \overline{X}}{I_{yy}}$$

$$= \frac{550 \times 10^6 \times 236.40}{15 \times 10^8} + \frac{15 \times 10^6 \times 200}{1.81 \times 10^8}$$

$$= 86.68 + 16.57$$

$$= 103.25 \text{ N/mm}^2$$

55. Ultimate strength in compression, $\varepsilon_{cu} = 0.0035$

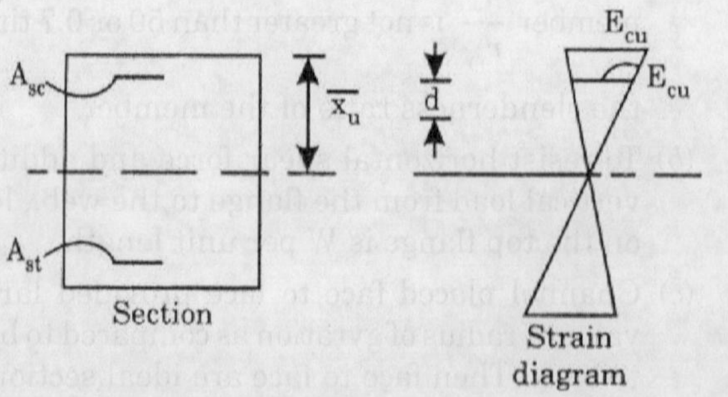

Let Maximum strain at the level of compression steel is ε_{cc}

By geometry, $\dfrac{\varepsilon_{cc}}{x_u - d'} = \dfrac{\varepsilon_{cu}}{x_u} = 0.0035\left(1 - \dfrac{d'}{x_u}\right)$

58. Given, diameter of rivet = 20 mm

$\therefore$ Gross dimeter = 20 + 1.5 = 21.5 mm

Pitch = 75 cm Gauge = 60 mm

Staggered riveting

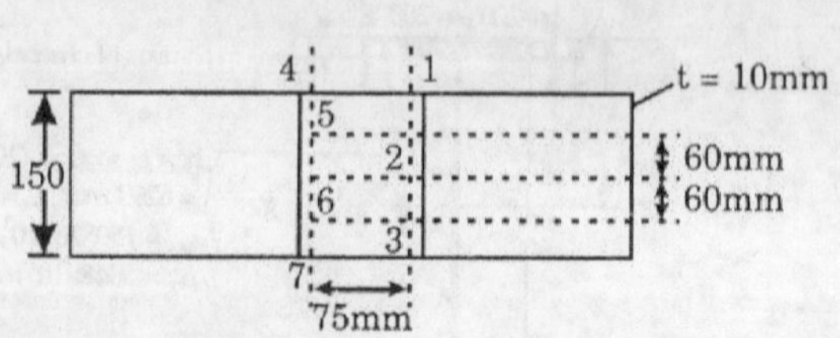

For most critical sectional area, consider different section as follows:

Section 1-2-3.

$$\text{Net area, } A_{net} = t \times \left(b - n \times d + 1 \times \frac{s^2}{2g}\right)$$

$$= 10(150 - 1 \times 21.5 + 0) = 1285 \text{ mm}^2$$

Section 4-5-6-7.

$$\text{Net area, } A_{net} = 10 \times \left(150 - 3 \times 21.5 + 2 \times \frac{7.5^2}{4 \times 60}\right)$$

$$= 1323 \text{ mm}^2$$

Most critical sectional area = 1285 mm²

Maximum tensile force at that joint = 150 × 1285 = 192.75 kN

59. Maximum permissible deflection as per I.S code 800, 1984 for overhead travelling and electically operated crane up to capacity of 500 kN is given by

$$\frac{\text{span}}{750} = \frac{6 \times 1000}{750} = 8 \text{ mm.}$$

60. For isolated T-beam,

$$b_f = \frac{I_0}{(I_0 / b) + 4} + b_w$$

where $I_0 = 6\text{m}$,

$b = 1000$ mm

and $b_w = 300$ mm

$$\therefore \quad b_f = \frac{6 \times 1000}{(6000/1000) + 4}$$

$$= 300 = 900 \text{ mm}$$

61. $M_p = Z_p \cdot f_y$

where, f_y = yield stress

Z_p = plastic stress modulus

$= 4.8 \times 10^{-4}\ \text{m}^3$

M_p = plastic moment carrying capacity

= 120kNm

$$\therefore \quad f_y = \frac{M_p}{Z_p}$$

$$= \frac{120}{4.8 \times 10^{-4}}$$

$$= 25 \times 10^4 \text{ kN/m}^2$$

$$= \frac{25 \times 10^4 \times 10^3}{10^6}$$

$$= 250 \text{ N/ mm}^2 \text{ or MPA}$$

65. As per I.S. code 800 : 1984, allowable average shear stress for unstiffened web of steel

$$= 0.4 f_y = 0.4 \times 250 = 100 \text{ N/mm}^2$$

NUMERICAL TYPE QUESTIONS

1. Elastic section modulus of the section

$$Z_e = 500 \text{ cm}^3 = 500 \times 10^3 \text{ mm}^3$$

Plastic section modulus of the section

$$Z_p = 650 \text{ cm}^3 = 650 \times 10^3 \text{ mm}^3$$

Design bending compressive stress f_{cd} = 200 Mpa

The bending strength of laterally unsupported beam is given by

$$M_d = \beta_b . Z_p . f_{cd}$$

β_b = 1.0 (For plastic and compact sections)

= Z_e/Z_p (for semi compact sections)

Z_e = Elastic section modulus of steel beam

The design flexural capacity of the section (M_d)

$$M_d = Z_e f_{cd} = 500 \times 10^3 \times 200$$

$$= 100 \times 10^6 \text{ N-mm} = 100 \text{ kN-m}$$

2. $$F_D = \frac{P}{n} = \frac{100}{5} = 20\text{kN}$$

$$F_t = \frac{(P.d)r}{\sum r^2} = \frac{100 \times 600 \times 75\sqrt{2}}{4 \times (75\sqrt{2})^2} = 141.42 \text{ kN}$$

$$\theta = 45°$$

$$\cos\theta = \frac{1}{\sqrt{2}}$$

$$F_R = \sqrt{F_D^2 + F_t^2 + 2 \times F_D \times F_t \cos\theta}$$

$$F_R = \sqrt{(20)^2 + (141.42)^2 + 2 \times 20 \times 141.42 \frac{1}{\sqrt{2}}}$$

$$= 156.20 \text{ kN}$$

3. First of all find the reaction at support R_T and R_W

$$R_W + R_T = 15 \quad ...(i)$$

$$\Sigma M_W = 0$$

$$R_T \times 8 = 15 \times 4 + 15 \times 8$$

$$R_T = \frac{180}{8} = 22.5 \text{ kN}(\uparrow) \text{upward direction}$$

$$R_W = 15 - 22.5$$

$$= -7.5 \text{ kN } (\downarrow) \text{ downward direction}$$

Consider the joint T

(assume initially tension) tension +ve and compression –ve

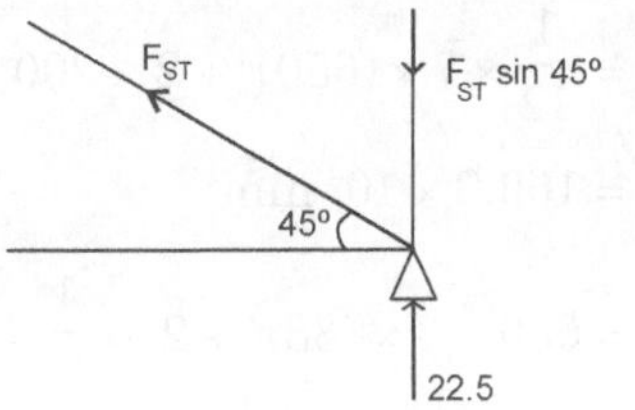

$$\Sigma Y = 0$$

$$F_{ST} \sin 45° + 22.5 = 0$$

$$F_{ST} = \frac{22.5}{\sin 45°} = -31.819 \text{ kN}$$

$$F_{ST} = 31.819 \text{ (comp.)}$$

Consider the Joint S

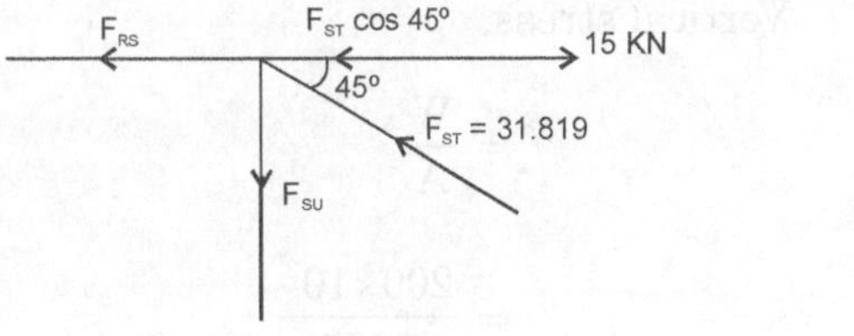

$$\Sigma X = 0$$

$$F_{RS} + F_{ST} \cos 45° = 15$$

$$F_{RS} + 31.819 \times \cos 45° = 15$$

$$F_{RS} = 15 - 22.5 = -7.5 \text{ (comp.)}$$

∵ Force in member RS = 7.5 (comp.)

4. *Given* : Filet size = 10 mm

Throat thickness, t = 7 mm

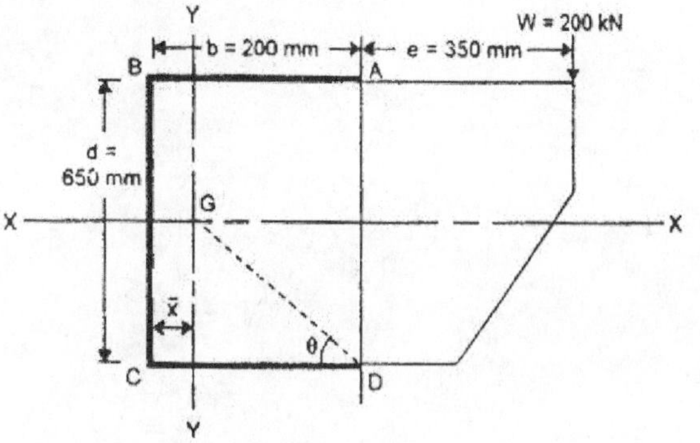

Distance of centroid of weld area from BC

$$\bar{x} = \frac{2 \times 200 \times t \times 100}{2 \times 200 \times t + 650t}$$

$$\therefore \quad t = \frac{28 \times 10^4}{7350}$$

$$= 30 \text{ mm}$$

Eccentricity of load

$$= 350 + 200 - 38$$

$$= 512 \text{ mm}$$

$$I_{xx} = \frac{1}{12} \times 7 \times (650)^3 + 2 \times 200 \times 7 \times (200)^2$$
$$= 160.2 \times 10^6 \text{ mm}^4$$

$$I_{yy} = 650 \times 7 \times (38)^2 + 2 \times \frac{1}{12} \times 7$$
$$\times (200)^3 + 2 \times 200 \times 7\,(100 - 38)^2$$
$$= 26.667 \times 10^6 \text{ mm}^4$$

Now, $J = I_{xx} + I_{yy}$
$$= 160.2 \times 10^6 + 26.667 \times 10^6$$
$$= 186.87 \times 10^6 \text{ mm}^4$$

Area, $A = 2 \times 200 \times 7 + 650 \times 7$
$$= 7359 \text{ mm}^2$$

Vertical stress,

$$f_a = \frac{W}{A}$$
$$= \frac{200 \times 10^3}{7335}$$
$$= 27.21 \text{ N/mm}^2$$

Maximum stress due to torsion will occur at A or D.

Length of radius vector for A or D

$$r = \sqrt{(650/2)^2 + (200-38)^2}$$
$$= 363.14 \text{ mm}$$

∴ Maximum shear stress due to torsion,

$$f_t = \frac{T}{J}.r$$
$$= \frac{200 \times 10^3 \times 512}{186.87 \times 10^6} \times 363.14$$
$$= 199 \text{ N/mm}^2$$

$$\cos\theta = \frac{(200-38)}{363.14}$$
$$= 0.446$$

∴ Resultant stress,

$$f_r = \sqrt{f_a^2 + f_1^2 + 2f_a . f_t \cos\theta}$$
$$= \sqrt{(27.21)^2 + (199)^2 + 2 \times 27.21 \times 199 \times 0.446}$$
$$= 212.5 \text{ N/mm}^2$$

Allowable shear stress = 108 N/mm²

Since $f_r > 108$, hence size of fillet is not adequate

■■

7 CHAPTER

Soil Mechanics and Foundation Engineering

MCQ TYPE QUESTIONS

2016

1. In the consolidated undrained triaxial test on a saturated soil sample, the pore water pressure is zero

(*a*) during shearing stage only

(*b*) at the end of consolidation stage only

(*c*) both at the end of consolidation and during shearing stages

(*d*) under none of the above conditions

2. A fine grained soil is found to be plastic in the water content range of 26-48%. As per Indian Standard Classification System, the soil is classified as

(*a*) CL (*b*) CH

(*c*) CL-ML (*d*) CI

3. A vertical cut is to be made in a soil mass having cohesion *c*, angle of internal friction φ, and unit weight γ. Considering K_a and K_p as the coefficients of active and passive earth pressures, respectively, the maximum depth of unsupported excavation is

(*a*) $\dfrac{4c}{\gamma\sqrt{K_p}}$ (*b*) $\dfrac{2c\sqrt{K_p}}{\gamma}$

(*c*) $\dfrac{4c\sqrt{K_a}}{\gamma}$ (*d*) $\dfrac{4c}{\gamma\sqrt{K_a}}$

4. A homogeneous gravity retaining wall supporting a cohesionless backfill is shown in the figure. The lateral active earth pressure at the bottom of the wall is 40 kPa.

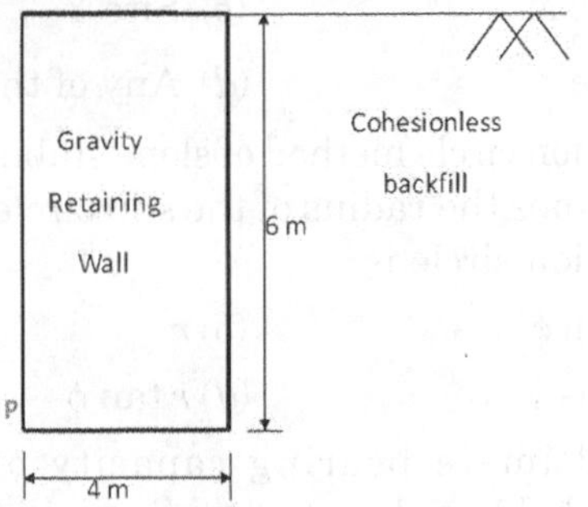

The minimum weight of the wall (expressed in kN per m length) required to prevent it from overturning about its toe (Point P) is

(*a*) 120 (*b*) 180

(*c*) 240 (*d*) 360

5. Seepage is occurring through a porous media shown in the figure. The hydraulic conductivity values (k_1, k_2, k_3) are in m/day.

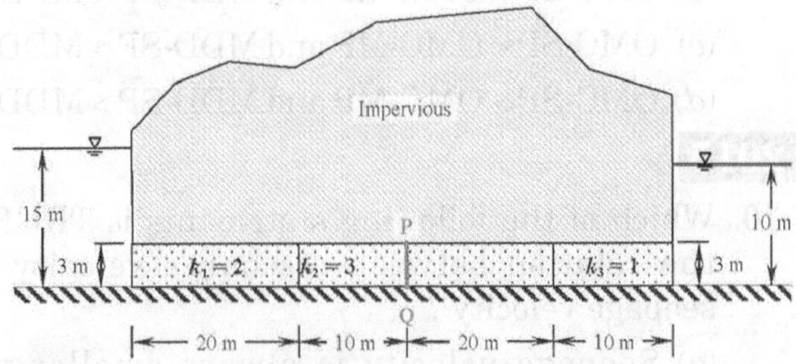

The seepage discharge (m³/day per m) through the porous media at section PQ is

(*a*) $\dfrac{7}{12}$ (*b*) $\dfrac{1}{2}$

(*c*) $\dfrac{9}{16}$ (*d*) $\dfrac{3}{4}$

6. A strip footing is resting on the surface of a purely clayey soil deposit. If the width of the footing is doubled, the ultimate bearing capacity of the soil

(*a*) becomes double

(*b*) becomes half

(*c*) becomes four-times

(*d*) remains the same

7. The relationship between the specific gravity of sand (G) and the hydraulic gradient (*i*) to initiate quick condition in the sand layer having porosity of 30% is

(*a*) G = 0.7 *i* + 1

(*b*) G = 1.43 *i* – 1

(*c*) G = 1.43 *i* + 1

(*d*) G = 0.7 *i* – 1

8. The results of a consolidation test on an undisturbed soil, sampled at a depth of 10 m below the ground level are as follows :

Saturated unit weight : 16 kN/m³

Pre-consolidation pressure : 90 kPa

The water table was encountered at the ground level. Assuming the unit weight of water as 10 kN/m³, the over-consolidation ratio of the soil is

(*a*) 0.67 (*b*) 1.50

(*c*) 1.77 (*d*) 2.00

9. OMC-SP and MDD-SP denote the optimum moisture content and maximum dry density obtained from standard Proctor compaction test, respectively. OMC-MP and MDD-MP denote the optimum moisture content and maximum dry density obtained from the modified Proctor compaction test, respectively. Which one of the following is correct?
   (*a*) OMC-SP < OMC-MP and MDD-SP < MDD-MP
   (*b*) OMC-SP > OMC-MP and MDD-SP < MDD-MP
   (*c*) OMC-SP < OMC-MP and MDD-SP > MDD-MP
   (*d*) OMC-SP > OMC-MP and MDD-SP > MDD-MP

2015

10. Which of the following statements is TRUE for the relation between discharge velocity and seepage velocity ?
    (*a*) Seepage velocity is always smaller than discharge velocity
    (*b*) Seepage velocity can never be smaller than discharge velocity
    (*c*) Seepage velocity is equal to the discharge velocity
    (*d*) No relation between seepage velocity and discharge velocity can be established

11. Which of the following statements is NOT correct ?
    (*a*) Loose sand exhibits contractive behavior upon shearing
    (*b*) Dense sand when sheared under undrained condition, may lead to generation of negative pore pressure
    (*c*) Black cotton soil exhibits expansive behavior
    (*d*) Liquefaction is the phenomenon where cohesionless soil near the downstream side of dams or sheet-piles loses its shear strength due to high upward hydraulic gradient

12. A fine-grained soil has 60% (by weight) silt content. The soil behaves as *semi-solid* when water content is between 15% and 28%. The soil behaves *fluid-like* when the water content is more than 40%. The *'Activity'* of the soil is
    (*a*) 3.33 (*b*) 0.42
    (*c*) 0.30 (*d*) 0.20

13. In an unconsolidated undrained triaxial test, it is observed that an increase in cell pressure from 150 kPa to 250 kPa leads to a pore pressure increase of 80 kPa. It is further observed that, an increase of 50 kPa in deviatoric stress results in an increase of 25 kPa in the pore pressure. The value of *Skempton's pore pressure parameter B* is.
    (*a*) 0.5 (*b*) 0.625
    (*c*) 0.8 (*d*) 1.0

14. Which of the following statements is TRUE for degree of disturbance of collected soil sample ?
    (*a*) Thinner the sampler wall, lower the degree of disturbance of collected soil sample
    (*b*) Thicker the sampler wall, lower the degree of disturbance of collected soil sample
    (*c*) Thickness of the sampler wall and the degree of disturbance of collected soil sample are unrelated
    (*d*) The degree of disturbance of collected soil sample is proportional to the inner diameter of sampling tube

15. An earth embankment is to be constructed with compacted cohesionless soil. The volume of the embankment is 5000 m^3 and the target dry unit weight is 16.2 kN/m^3. Three nearby sites (see figure below) have been identified from where the required soil can be transported to the construction site. The void ratios (e) of different sites are shown in the figure. Assume the specific gravity of soil to be 2.7 for all three sites. If the cost of transportation per km is twice the cost of excavation per m^3 of borrow pits, which site would you choose as the most economic solution? (Use unit weight of water = 10 kN/m^3)

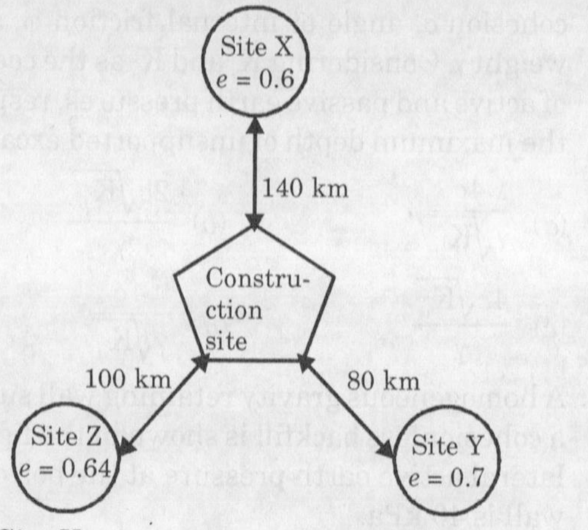

    (*a*) Site X (*b*) Site Y
    (*c*) Site Z (*d*) Any of the sites

16. In friction circle method of slope stability analysis, if r defines the radius of the slip circle, the radius of friction circle is
    (*a*) $r \sin \phi$ (*b*) r
    (*c*) $r \cos \phi$ (*d*) $r \tan \phi$

17. Net ultimate bearing capacity of a footing embedded in a clay stratum
    (*a*) increases with depth of footing only
    (*b*) increases with size of footing only
    (*c*) increases with depth and size of footing
    (*d*) is independent of depth and size of footing

18. In the water content of a fully saturated soil mass is 100% the void ratio of the sample is

(*a*) Less than specific gravity of soil

(*b*) equal to specific gravity of soil

(*c*) greater than specific gravity of soil

(*d*) independent of specific gravity of soil

19. The relationship between porosity (η), specific yield (S_y) and specific retention (S_r) of an unconfined aquifer is

(*a*) $S_y + S_r = \eta$ (*b*) $S_y + \eta = S_r$

(*c*) $S_r + \eta = S_y$ (*d*) $S_y + S_r + \eta = 1$

20. Surcharge loading required to placed on the horizontal backfill of a smooth retaining vertical wall so as to completely eliminate tensile crack is :

(*a*) 2 *c* (*b*) $2\,ck_a$

(*c*) $2c\sqrt{k_a}$ (*d*) $2c/\sqrt{k_a}$

21. A 6 m high retaining wall having a smooth vertical back face retains a layered horizontal backfill. Top 3 *m* thick layer of the backfill is sand having an angle of internal friction, φ = 30° while the bottom layer is 3 *m* thick clay with cohesion, *c* = 20 kPa. Assume unit weight for both sand and clay as 18 kN/m³. The total active earth pressure per unit length of the wall (in kN/m) is:

(*a*) 150 (*b*) 216

(*c*) 156 (*d*) 196

22. Stress path equation for tri-axial test upon application of deviatoric stress is, $q = 10\sqrt{3} + 0.5p$. The respective values of cohesion, *c* (in kPa) and angle of internal friction, φ are :

(*a*) 20 and 20° (*b*) 20 and 30°

(*c*) 30 and 30° (*d*) 30 and 20°

2014

23. The degree of disturbance of the sample collected by the sampler is expressed by a term called the "area ratio". If the outer diameter and inner diameter of the sampler are D_o and D_i respectively, the area ratio is given by

(*a*) $\dfrac{D_o^2 - D_i^2}{D_i^2}$

(*b*) $\dfrac{D_i^2 - D_o^2}{D_i^2}$

(*c*) $\dfrac{D_o^2 - D_i^2}{D_o^2}$

(*d*) $\dfrac{D_i^2 - D_o^2}{D_o^2}$

24. For a saturated cohesive soil, triaxial test yields the angle of internal friction (φ) as zero. The conducted test is

(*a*) Consolidated Drained (CD) test

(*b*) Consolidated Undrained (CU) test

(*c*) Unconfined Compression (UC) test

(*d*) Unconsolidated Undrained (UU) test

25. The action of negative skin friction on the pile is to

(*a*) increase the ultimate load on the pile

(*b*) reduce the allowable load on the pile

(*c*) maintain the working load on the pile

(*d*) reduce the settlement of the pile

26. A long slope is formed in a soil with shear strength parameters: *c*′ = 0 and φ′ = 34°. A firm stratum lies below the slope and it is assumed that the water table may occasionally rise to the surface, with seepage taking place parallel to the slope. Use γ_{sat} = 18 kN/m³ and γ_w = 10 kN/m³. The maximum slope angle (in degrees) to ensure a factor of safety of 1.5, assuming a potential failure surface parallel to the slope, would be

(*a*) 45.3 (*b*) 44.7

(*c*) 12.3 (*d*) 11.3

27. Group I contains representative load-settlement curves for different modes of bearing capacity failures of sandy soil. Group II enlists the various failure characteristics. Match the load-settlement curves with the corresponding failure characteristics.

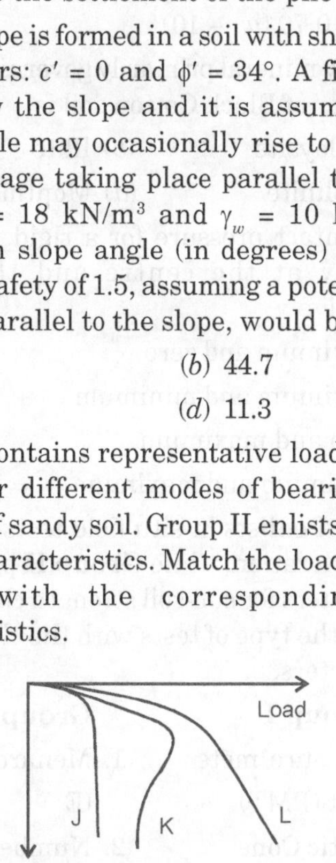

Group I	Group II
P. Curve J	1. No apparent heaving of soil around the footing
Q. Curve K	2. Rankine's passive zone develops imperfectly
R. Curve L	3. Well defined slip surface extends to ground surface

	P	Q	R
(*a*)	1	3	2
(*b*)	3	2	1
(*c*)	3	1	2
(*d*)	1	2	3

28. A given cohesionless soil has e_{max} = 0.85 and e_{min} = 0.50. In the field, the soil is compacted to a mass density of 1800 kg/m³ at a water content of 8%. Take the mass density of water as 1000kg/m³ and G_s as 2.7. The relative density (in %) of the soil is

(*a*) 56.43 (*b*) 60.25

(*c*) 62.87 (*d*) 65.71

29. As per Indian Standard Soil Classification System (IS: 1498-1970), an expression for **A-line** is

(*a*) $I_P = 0.73\ (w_L - 20)$

(*b*) $I_P = 0.70\ (w_L - 20)$

(*c*) $I_P = 0.73\ (w_L - 10)$

(*d*) $I_P = 0.70\ (w_L - 10)$

30. The clay mineral primarily governing the swelling behavior of Black Cotton soil is

(*a*) Halloysite (*b*) Illite

(*c*) Kaolinite (*d*) Montmorillonite

31. The contact pressure for a rigid footing resting on clay at the centre and the edges are respectively

(*a*) maximum and zero

(*b*) maximum and minimum

(*c*) zero and maximum

(*d*) minimum and maximum

32. Group I enlists in-situ field tests carried out for soil exploration, while **Group II** provides a list of parameters for sub-soil strength characterization. Match the type of tests with the characterization parameters.

Group I	Group II
P. Pressure meter Test (PMT)	1. Menard's modulus (E_m)
Q. Static Cone Penetration Test (SCPT)	2. Number of blows (N)
R. Standard Penetration Test (SPT)	3. Skin resistance (f_c)
S. Vane Shear Test (VST)	4. Undrained cohesion (c_u)

Codes :

	P	Q	R	S
(*a*)	1	3	2	4
(*b*)	1	2	3	4
(*c*)	2	3	4	1
(*d*)	4	1	2	3

33. A circular raft foundation of 20 m diameter and 1.6 m thick is provided for a tank that applies a bearing pressure of 110 kPa on sandy soil with Young's modulus, E'_s = 30 MPa and Poisson's ratio, υ_s = 0.3. The raft is made of concrete (E_c = 30 GPa and υ_c = 0.15). Considering the raft as rigid, the elastic settlement (in mm) is

(*a*) 50.96 (*b*) 53.36

(*c*) 63.72 (*d*) 66.71

2013

34. In its natural condition, a soil sample has a mass of 1.980 kg and a volume of 0.001 m³. After being completely dried in an oven, the mass of the sample is 1.800 kg. Specific gravity G is 2.7. Unit weight of water is 10 kN/m³. The degree of saturation of the soil is

(*a*) 0.65 (*b*) 0.70

(*c*) 0.54 (*d*) 0.61

35. The ratio N_f/N_d is known as shape factor, where N_f is the number of flow lines and N_d is the number of equipotential drops. Flow net is always drawn with a constant b/a ratio, where b and a are distances between two consecutive flow lines and equipotential lines, respectively. Assuming that b/a ratio remains the same, the shape factor of a flow net will change if the

(*a*) upstream and downstream heads are interchanged

(*b*) soil in the flow space is changed

(*c*) dimensions of the flow space are changed

(*d*) head difference causing the flow is changed

36. Following statementsare made on compacted soils, wherein DS stands for the soils compacted on dry side of optimum moisture content and WS stands for the soils compacted on wet side of optimum moisture content. Identify the *incorrect* statement.

(*a*) Soil structure is flocculated on DS and dispersed on WS.

(*b*) Construction pore water pressure is low on DS and high on WS.

(*c*) On drying, shrinkage is high on DS and low on WS.

(*d*) On access to water, swelling is high on DS and low on WS.

37. The soil profile below a lake with water level at elevation = 0 m and lake bottom at elevation = – 10 m is shown in the figure, where k is the permeability coefficient. A piezometer (stand pipe) installed in the sand layer shows a reading of +10 m elevation. Assume that the piezometric head is uniform in the sand layer. The quantity of water (in m^3/s) flowing into the lake from the sand layer through the silt layer per unit area of the lake bed is

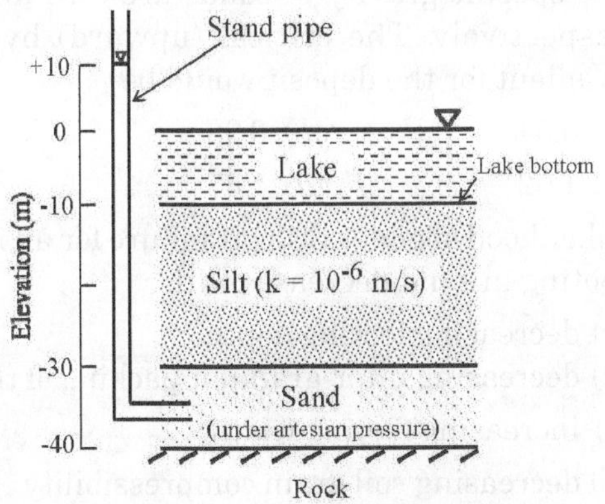

(*a*) 1.5×10^{-6} (*b*) 2.0×10^{-6}

(*c*) 1.0×10^{-6} (*d*) 0.5×10^{-6}

38. The soil profile above the rock surface for a 25° infinite slope is shown in the figure, where s_u is the undrained shear strength and γ_t is total unit weight. The slip will occur at a depth of

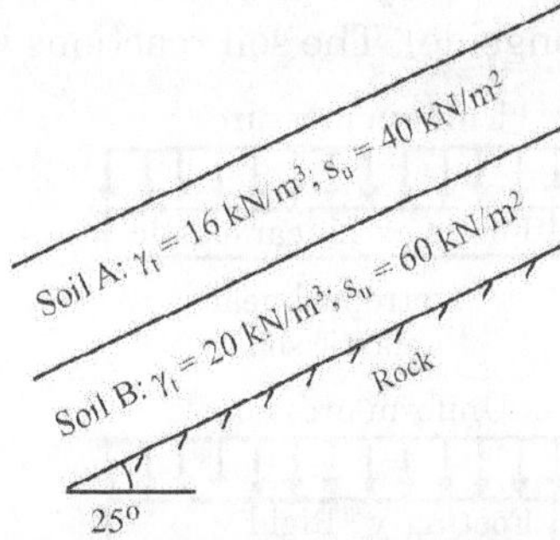

(*a*) 8.83 m (*b*) 9.79 m

(*c*) 7.83 m (*d*) 6.53 m

39. Two different soil types (Soil 1 and Soil 2) are used as backfill behind a retaining wall as shown in the figure, where γ_t is total unit weight, and c' and ϕ' are effective cohesion and effective angle of shearing resistance. The resultant active earth force per unit length (in kN/m) acting on the wall is

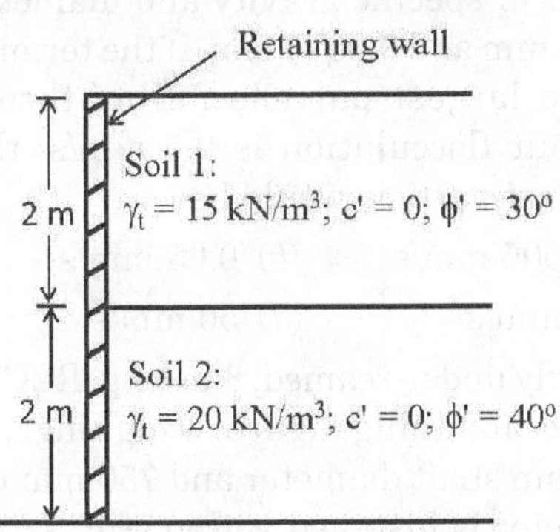

(*a*) 31.7 (*b*) 35.2

(*c*) 51.8 (*d*) 57.0

40. The percent voids in mineral aggregate (VMA) and percent air voids (V_v) in a compacted cylindrical bituminous mix specimen are 15 and 4.5 respectively. The percent voids filled with bitumen (VFB) for this specimen is

(*a*) 24 (*b*) 30

(*c*) 54 (*d*) 70

2012

41. The effective stress friction angle of a saturated, cohesionless soil is 38°. The ratio of shear stress to normal effective stress on the failure plane is

(*a*) 0.781 (*b*) 0.616

(*c*) 0.488 (*d*) 0.438

42. Two series of compaction tests were performed in the laboratory on an inorganic clayey soil employing two different levels of compaction energy per unit volume of soil. With regard to the above tests, the following two statements are made.

I. The optimum moisture content is expected to be more for the tests with higher energy.

II. The maximum dry density is expected to be more for the tests with higher energy.

The **CORRECT** option evaluating the above statements is

(*a*) Only I is TRUE

(*b*) Only II is TRUE

(*c*) Both I and II are TRUE

(*d*) Neither I nor II is TRUE

43. As per the Indian Standard soil classification system, a sample of silty clay with liquid limit of 40% and plasticity index of 28% is classified as

(*a*) CH (*b*) CI

(*c*) CL (*d*) CL-ML

44. A smooth rigid retaining wall moves as shown in the sketch causing the backfill material to fail. The backfill material is homogeneous and isotropic, and obeys the Mohr-Coulomb failure criterion. The major principal stress is

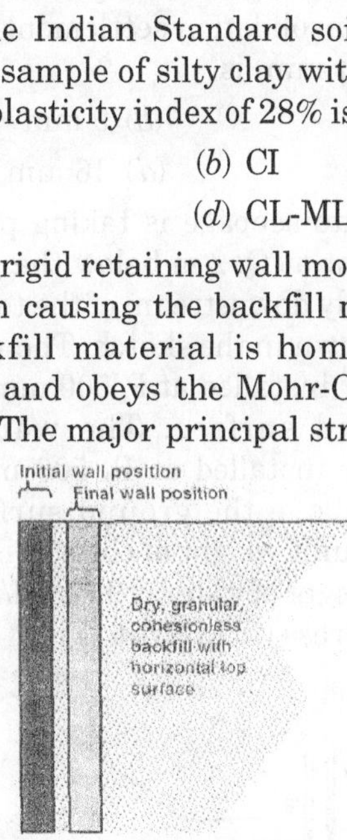

(*a*) parallel to the wall face and acting downwards

(*b*) normal to the wall face

(*c*) oblique to the wall face acting downwards

(*d*) oblique to the wall face acting upwards

45. An embankment is to be constructed with a granular soil (bulk unit weight = 20 kN/m^3) on a saturated clayey silt deposit (undrained shear strength = 25 kPa). Assuming undrained general shear failure and bearing capacity factor of 5.7, the maximum height (in m) of the embankment at the point of failure is

(*a*) 7.1 (*b*) 5.0
(*c*) 4.5 (*d*) 2.5

46. Two soil specimens with identical geometric dimensions were subjected to falling head permeability tests in the laboratory under identical conditions. The fall of water head was measured after an identical time interval. The ratio of initial to final water heads for the test involving the first specimen was 1.25. If the coefficient of permeability of the second specimen is 5-times that of the first, the ratio of initial to final water heads in the test involving the second specimen is

(*a*) 3.05 (*b*) 3.80
(*c*) 4.00 (*d*) 6.25

47. A layer of normally consolidated, saturated silty clay of 1 m thickness is subjected to one dimensional consolidation under a pressure increment of 20 kPa. The properties of the soil are: specific gravity = 2.7, natural moisture content = 45%, compression index = 0.45, and recompression index = 0.05. The initial average effective stress within the layer is 100 kPa. Assuming Terzaghi's theory to be applicable, the primary consolidation settlement (rounded off to the nearest mm) is

(*a*) 2 mm (*b*) 9 mm
(*c*) 14 mm (*d*) 16 mm

48. Steady state seepage is taking place through a soil element at Q, 2 m below the ground surface immediately downstream of the toe of an earthen dam as shown in the sketch. The water level in a piezometer installed at P, 500 mm above Q, is at the ground surface. The water level in a piezometer installed at R, 500 mm below Q, is 100 mm above the ground surface. The bulk saturated unit weight of the soil is 18 kN/m^3 and the unit weight of water is 9.81 kN/m^3. The vertical effective stress (in kPa) at Q is

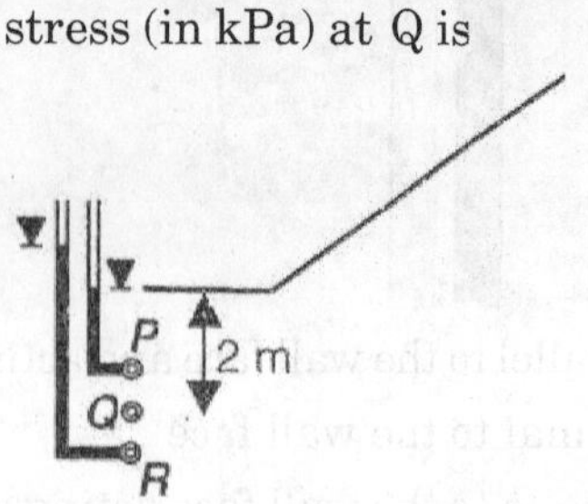

(*a*) 14.42 (*b*) 15.89
(*c*) 16.38 (*d*) 18.34

2011

49. For a saturated sand deposit, the void ratio and the specific gravity of solids are 0.70 and 2.67, respectively. The critical (upward) hydraulic gradient for the deposit would be

(*a*) 0.54 (*b*) 0.98
(*c*) 1.02 (*d*) 1.87

50. Likelihood of general shear failure for an isolated footing in sand decreases with

(*a*) decreasing footing depth
(*b*) decreasing inter-granular packing of the sand
(*c*) increasing footing width
(*d*) decreasing soil grain compressibility

51. For a sample of dry, cohesionless soil with friction angle, ϕ, the failure plane will be inclined to the major principal plane by an angle equal to

(*a*) ϕ (*b*) 45°
(*c*) $45° - \phi/2$ (*d*) $45° + \phi/2$

52. Two geometrically identical isolated footings, X (linear elastic) and y (rigid), are loaded identically (shown alongside). The soil reactions will

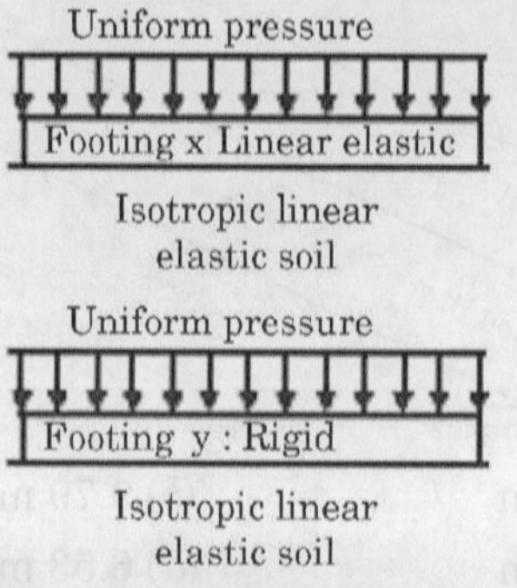

(*a*) be uniformly distributed for Y but not for X
(*b*) be uniformly distributed for X but not for Y
(*c*) be uniformly distributed for both X and Y
(*d*) not be uniformly distributed for both X and Y

53. A soil is composed of solid spherical grains of identical specific gravity and diameter between 0.075 mm and 0.0075 mm. If the terminal velocity of the largest particle falling through water without flocculation is 0.5 mm/s, that for the smallest particle would be

(*a*) 0.005 mm/s (*b*) 0.05 mm/s
(*c*) 5 mm/s (*d*) 50 mm/s

54. A singly under-reamed, 8-m long, RCC pile (shown in the adjoining figure) weighing 20 kN with 350 mm shaft diameter and 750 mm under-ream diameter is installed within stiff, saturated silty clay (undrained shear strength is 50 kPa, adhesion fator is 0.3, and the applicable bearing capacity factor is 9) to counteract the impact of soil swelling on a structure constructed above. neglecting suction and the contribution of the under-ream to the adhesive shaft capacity, what would be the estimated ultimate tensile capacity (rounded off to the nearest integer value of kN) of the pile?

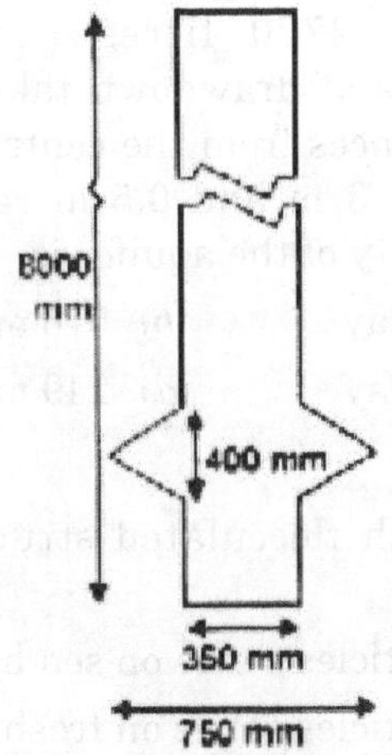

(*a*) 132 kN (*b*) 156 kN
(*c*) 287 kN (*d*) 301 kN

55. Identical surcharge are placed at ground surface at sites X and Y, with soil conditions shown alongside and water table at ground surface. The silty clay layers at X and Y are identical. The thin sand layer at Y is continuous and free-draining with a very larger discharge capacity. If primary consolidation at X is estimated to complete in 36 months, what would be the corresponding time for completion of primary consolidation at Y?

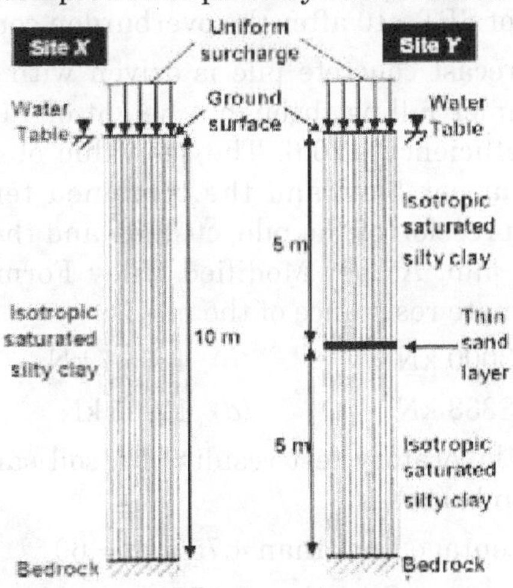

(*a*) 2.25 months (*b*) 4.5 months
(*c*) 9 months (*d*) 36 months

56. A field vane testing instrument (shown alongside) was inserted completely into a deposit of soft, saturated silty clay with the vane rod vertical such that the top of the blades were 500 mm below the ground surface. Upon application of a rapidly increasing torque about the vane rod, the soil was found to fail when the torque reached 4.6 Nm. Assuming mobilization of undrained shear strengh on all failure surfaces to be uniform and the resistance mobilized on the surface of the vane rod to be negligible, what would be the peak undrained shear strength (rounded off to the nearest integer value of kPa) of the soil?

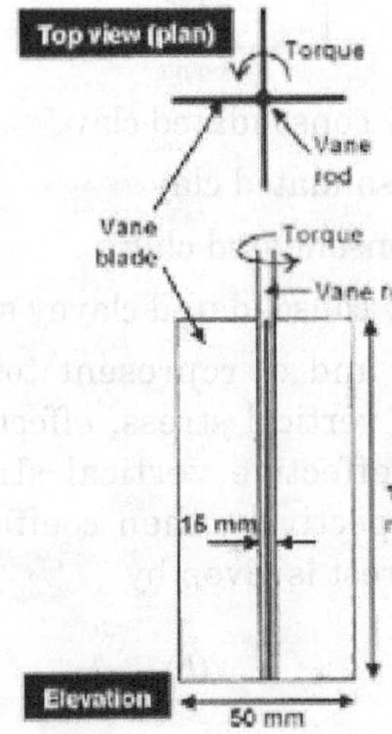

(*a*) 5 kPa (*b*) 10 kPa
(*c*) 15 kPa (*d*) 20 kPa

2010

57. In a compaction test, G, *w*, S and *e* represent the specific gravity, water content, degree of saturation and void ratio of the soil sample, respectively. If γ_w represents unit weight of water and γ_d represents dry unit weight of the soil, then equation of zero air voids line is

(*a*) $\gamma_d = \dfrac{G\gamma_w}{1+Se}$

(*b*) $\gamma_d = \dfrac{G\gamma_w}{1+Gw}$

(*c*) $\gamma_d = \dfrac{Gw}{e+\gamma_w S}$

(*d*) $\gamma_d = \dfrac{Gw}{1+Se}$

58. A fine grained soil has liquid limit of 60 and plastic limit of 20. As per the plasticity chart, according to IS classification, the soil is represented by the letter symbols

(*a*) CL (*b*) CI
(*c*) CH (*d*) CL-ML

59. Quick sand condition occurs when

(*a*) void ratio of the soil becomes 1.0
(*b*) upward seepage pressure in soil becomes zero
(*c*) upward seepage pressure in soil becomes equal to the saturated unit weight of the soil
(*d*) upward seepage pressure in soil becomes equal to the submerged unit weight of the soil.

60. The e-log p curve shown in the figure is representative of

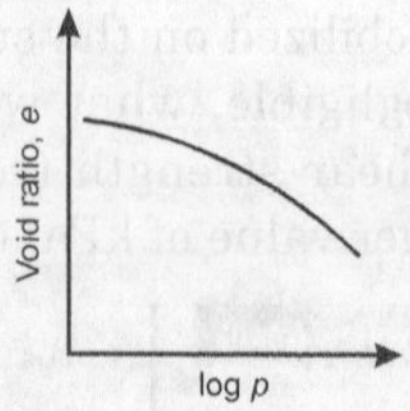

(a) normally consolidated clay
(b) over consolidated clay
(c) under consolidated clay
(d) normally consolidated clayey sand

61. If σ_h, σ_v, σ_h and σ_v represent total horizontal stress, total vertical stress, effective horizontal stress and effective vertical stress on a soil element respectively, then coefficient of earth pressure at rest is given by

(a) $\frac{\sigma_h}{\sigma_v}$ (b) $\frac{\sigma'_h}{\sigma'_v}$
(c) $\frac{\sigma_v}{\sigma_h}$ (d) $\frac{\sigma'_v}{\sigma'_h}$

62. The vertical stress at point P_1 due to the point load Q on the ground surface as shown in figure is σ_z. According to Boussinesq's equation, the vertical stress at point P_2 shown in figure will be

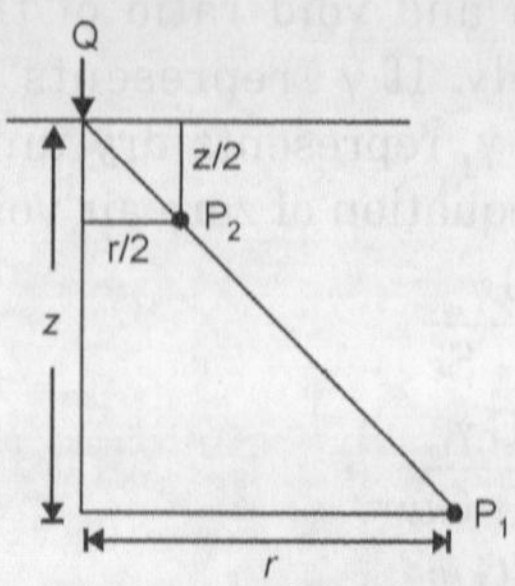

(a) $\frac{\sigma_z}{2}$ (b) σ_z
(c) $2\sigma_z$ (d) $4\sigma_z$

63. An open ended steel barrel of 1 m height and 1 m diameter is filled with saturated fine sand having coefficient of permability of 10^{-2} m/s. The barrel stands on a saturated bed of gravel. The time required for the water level in the barrel to drop by 0.75 m is

(a) 58.9s (b) 75s
(c) 100s (d) 150s

64. The ultimate load capacity of a 10 m long concrete pile of square cross section 500 mm × 500 mm driven into a homogeneous clay layer having undrained cohesion value of 40 kPa is 700 kN. If cross section of the pile is reduced to 250 mm × 2500 mm and length of the pile is increased to 20 m, then ultimate load capacity will be

(a) 350 kN (b) 632.5 kN
(c) 722.5 kN (d) 1400 kN

65. A well of diameter 20 cm fully penetrates a confined aquifer. After a long period of pumping at rate of 2720 litres per minute, the observations of drawdown taken at 10 m and 100 m distances from the centre of the well are found to be 3 m and 0.5 m respectively. The transmissivity of the aquifer is

(a) 676 m^2/day (b) 576 m^2/day
(c) 526 m^2/day (d) 249 m^2/day

2009

66. Deposit with flocculated structure is formed when

(a) clay particles settle on sea bed
(b) clay particles settle on fresh water lake bed
(c) sand particles settle on river bed
(d) sand particles settle on sea bed

67. Dilatancy correction is required when a strata is

(a) cohesive and saturated and also has N value of SPT >15
(b) saturated silt/fine sand and N value of SPT < 10 after the overburden correction
(c) saturated silt/fine sand and N value of SPT > 15 after the overburden correction
(d) coarse sand under dry condition and N value of SPT <10 after the overburden correction

68. A precast concrete pile is driven with a 50 kN hammer falling through a height of 1.0 m with an efficiency of 0.6. The set value observed is 4 mm per blow and the combined temporary compression of the pile, cushion and the ground is 6 mm. As per Modified Hiley Formula, the ultimate resistance of the pile is

(a) 3000 kN (b) 4285.7 kN
(c) 8333 kN (d) 11905 kN

69. The laboratory test results of a soil sample are given below:

Percentage finer than 4.75 mm = 60

Percentage finer than 0.075 mm = 30

Liquid Limit = 35 %

Plastic Limit = 27 %

The soil classification is

(a) GM
(b) SM
(c) GC
(d) ML–MI

70. A plate load test is carried out on a 300 mm × 300 mm plate placed at 2 m below the ground level to determine the bearing capacity of a 2 m × 2 m footing placed at same depth of 2 m on a homogeneous sand deposit extending 10 m below ground. The ground water table is 3 m below the ground level. Which of the following factors *does not* require a correction to the bearing capacity determined based on the load test ?

(*a*) Absence of the overburden pressure during the test

(*b*) Size of the plate is much smaller than the footing size

(*c*) Influence of the ground water table

(*d*) Settlement is recorded only over a limited period of one or two days

2008

71. Group symbols assigned to silty sand and clayey sand are respectively

(*a*) SS and CS

(*b*) SM and CS

(*c*) SM and SC

(*d*) MS and CS

72. When a retaining wall moves away from the backfill, then pressure exerted on the wall is termed as

(*a*) passive earth pressure

(*b*) swelling pressure

(*c*) pore pressure

(*d*) active earth pressure

73. Compaction by vibratory roller is the best method of compaction in case of

(*a*) moist silty sand

(*b*) well graded dry sand

(*c*) clay of medium compressibility

(*d*) silt of high compressibility

74. The liquid limit(LL), plastic limit (PL) and shrinkage limit (SL) of a cohesive soil satisfy the relation

(*a*) LL >PL < SL

(*b*) LL > PL > SL

(*c*) LL < PL < SL

(*d*) LL < PL > SL

75. A footing 2 m × 1 m exerts a uniform pressure of 150 kN/m^2 on the soil. Assuming a load dispersion of 2 vertical to 1 horizontal, the average vertical stress (kN/m^2) at 1.0 m below the footing is

(*a*) 50 (*b*) 75

(*c*) 80 (*d*) 100

76. A direct shear test was conducted on a cohesionless soil (c = 0) specimen under a normal stress of 200 kN/m^2. The specimen failed at a shear stress of 100kN/m^2. The angle of internal friction of the soil (degrees) is

(*a*) 26.6 (*b*) 29.5

(*c*) 30.0 (*d*) 32.6

77. A pile of 0.50 m diameter and of length 10 m is embedded in a deposit of clay. The undrained strength parameters of the clay are cohesion = 60 kN/m^2 and the angle in internal friction = 0. The skin friction capacity (kN) of the pile for an adhesion factor of 0.6, is

(*a*) 671 (*b*) 565

(*c*) 283 (*d*) 106

78. A saturated clay stratum draining both at the top and bottom undergoes 50 percent consolidation in 16 years under an applied load. If an additional drainage layer were present at the middle of the clay stratum, 50 percent consolidation would occur in

(*a*) 2 years (*b*) 4 years

(*c*) 8 years (*d*) 16 years

79. A test plate 30 cm× 30 cm resting on a sand deposit settles by 10 mm under a certain loading intensity. A footing 150 cm× 200 cm resting on the same sand deposit and loaded to the same load intensity settles by

(*a*) 2.0 mm

(*b*) 27.8 mm

(*c*) 3.02 mm

(*d*) 50.0 mm

80. A volume of 3.0 × 10^6 m^3 of groundwater was pumped out from an unconfined aquifer, uniformly from an area of 5 km^2. The pumping lowered the water table from initial level of 102 m to 99 m. The specifc yield of the aquifer is

(*a*) 0.20 (*b*) 0.30

(*c*) 0.40 (*d*) 0.50

2007

81. A clay soil sample is tested in a triaxial apparatus in consolidated-drained conditions at a cell pressure of 100 kN/m^2. What will be the pore water pressure at a deviator stress of 40 kN/m^2?

(*a*) 0 kN/m^2

(*b*) 20 kN/m^2

(*c*) 40 kN/m^2

(*d*) 60 kN/m^2

82. The number of blows observed in a Standard Penetration Test (SPT) for different penetration depths are given as follows

Penetration of sampler	Number of blows
0 – 150 mm	6
150 – 300 mm	8
300 – 450 mm	10

The observed N value is

(*a*) 8 (*b*) 14

(*c*) 18 (*d*) 24

83. The vertical stress at some depth below the corner of a 2m × 3m rectangular footing due to a certain load intensity is 100 kN/m^2. What will be the vertical stress in kN/m^2 below the centre of a 4m × 6m rectangular footing at the same depth and same load intensity ?

(*a*) 25 (*b*) 100

(*c*) 200 (*d*) 400

84. Sieve analysis on a dry soil sample of mass 1000 g showed that 980 g and 270 g of soil pass through 4.75 mm and 0.075 mm sieve, respectively. The liquid limit and plastic limits of the soil fraction passing through 425μ sieves are 40% and 18% respectively. The soil may be classified as

(*a*) SC (*b*) MI

(*c*) CI (*d*) SM

85. The water content of a saturated soil and the specific gravity of soil solids were found to be 30% and 2.70, respectively. Assuming the unit weight of water to be 10 kN/m^3, the saturated unit weight (kN/m^3) and the void ratio of the soil are

(*a*) 19.4, 0.81

(*b*) 18.5, 0.30

(*c*) 19.4, 0.45

(*d*) 18.5, 0.45

86. The factor of safety of an infinite soil slope shown in the figure having the properties

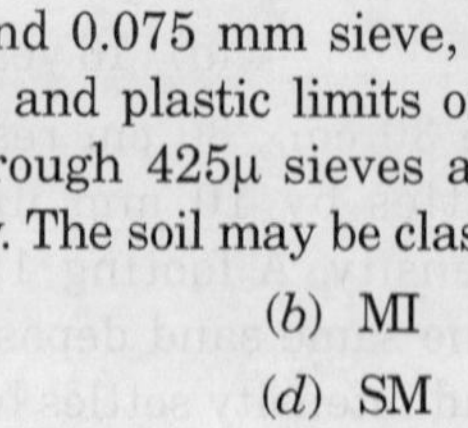

$c = 0$, $\phi = 35°$, $\gamma_{dry} = 16$ kN/m^3

and $\gamma_{sat} = 20$ kN/m^3

is approximately equal to

(*a*) 0.70 (*b*) 0.80

(*c*) 1.00 (*d*) 1.20

87. Match the following groups

Group-I	Group-II
P. Constant head permeability test	1. Pile foundations
Q. Consolidation test	2. Specific gravity
R. Pycnometer test	3. Clay soil
S. Negative skin friction	4. Sand

Codes:

	P	Q	R	S
(*a*)	4	3	2	1
(*b*)	4	2	3	1
(*c*)	3	4	2	1
(*d*)	4	1	2	3

88. The bearing capacity of a rectangular footing of plan dimensions 1.5 m×3 m resting on the surface of a sand deposit was estimated as 600 kN/m^2 when the water table is far below the base of the footing. The bearing capacities in kN/m^2 when the water level rises to depths of 3m, 1.5 m and 0.5 m below the base of the footing are

(*a*) 600, 600, 400

(*b*) 600, 450, 350

(*c*) 600, 500, 250

(*d*) 600, 400, 250

89. What is the ultimate capacity in kN of the pile group shown in the figure assuming the group to fail as a single block ?

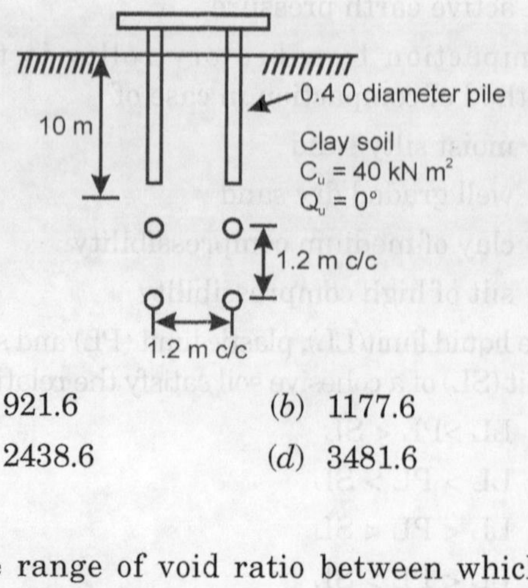

(*a*) 921.6 (*b*) 1177.6

(*c*) 2438.6 (*d*) 3481.6

2006

90. The range of void ratio between which quick sand condition occurs in cohesionless granular soil deposits is

(*a*) 0.4 – 0.5

(*b*) 0.6 – 0.7

(*c*) 0.8 – 0.9

(*d*) 1.0 – 1.1

91. Figure given below shows a smooth vertical gravity retaining wall with cohesionless soil backfill having an angle of internal friction ϕ. In the graphical representation of Rankine's active earth pressure for the retaining wall shown in figure, length OP represents

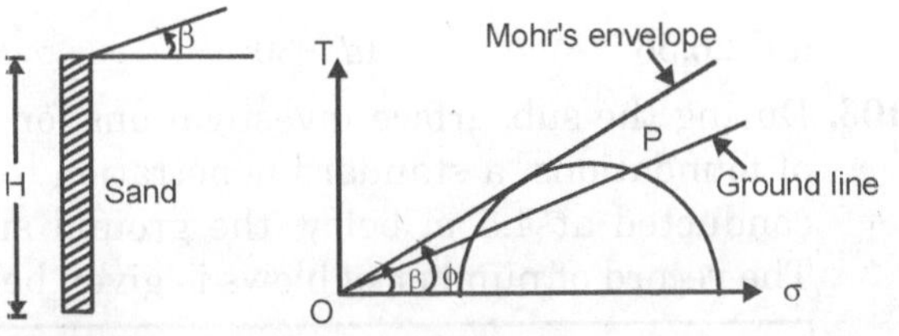

(*a*) vertical stress at the base

(*b*) vertical stress at a height H/3 from the base

(*c*) lateral earth pressure at the base

(*d*) lateral earth pressure at a height H/3 from the base

92. Which of following statement is NOT true in the context of capillary pressure in soils?

(*a*) Water is under tension in capillary zone

(*b*) Pore water pressure is negative in capillary zone

(*c*) Effective stress increases due to capillary pressure

(*d*) Capillary pressure is more in coarse grained soils

93. To provide safety against piping failure, with a factor of safety of 5, what should be the maximum permissible exit gradient for soil with specific gravity of 2.5 and porosity of 0.35?

(*a*) 0.155 (*b*) 0.167

(*c*) 0.195 (*d*) 0.213

94. List I below gives the possible types of failure for a finite soil slope and List II gives the reasons for these different types of failure. Match the items in List I with the items in List II

List I	**List II**
P. Base failure	1. Soils above and below the toe have same strength
Q. Face failure	2. Soil above the toe is comparatively weaker
R. Toe failure	3. Soil above the toe is comparatively stronger

Codes:

	P	**Q**	**R**
(*a*)	1	2	3
(*b*)	2	3	1
(*c*)	2	1	3
(*d*)	3	2	1

95. In a standard proctor test, 1.8 kg of moist soil was filling the mould (volume = 944 cc) after compaction. A soil sample weighing 23 g was taken from the mould and ovendried for 24 hours at a temperature of 110°C. Weight of the dry sample was found to be 20 g. Specific gravity of soil solids is G = 2.7. The theoretical maximum value of the dry unit weight of the soil at that water content is equal to

(*a*) 4.67 kN/m^3

(*b*) 11.5 kN/m^3

(*c*) 16.25 kN/m^3

(*d*) 18.85 kN/m^3

96. For the soil profile shown in figure below, the minimum number of precast concrete piles of 300 mm diameter required to safely carry the load for a given factor of safety of 2.5 (assuming 100% efficiency for the pile group) is equal to

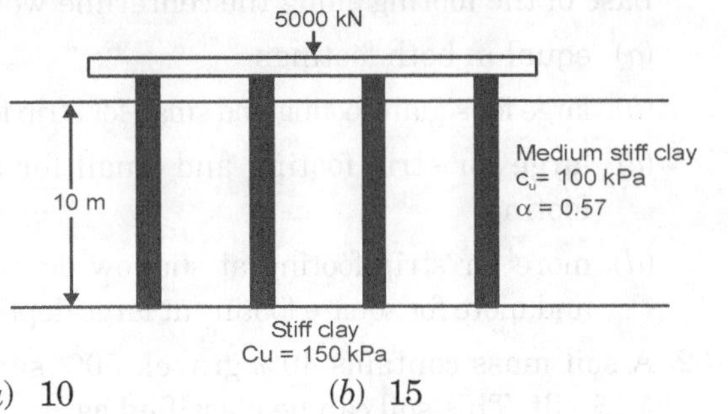

(*a*) 10 (*b*) 15

(*c*) 20 (*d*) 25

97. A sample of saturated cohesionless soil tested in a drained triaxial compression test showed an angle of internal friction of 30°. The deviatoric stress at failure for the sample at a confining pressure of 200 kPa is equal to

(*a*) 200 kPa (*b*) 400 kPa

(*c*) 600 kPa (*d*) 800 kPa

2005

98. A clayey soil has a maximum dry density of 16 kN/m^3 and optimum moisture content of 12%. A contractor during the construction of core of an earth dam obtained the dry density 15.2 kN/m^3 and water content 11%. This constructijn is acceptable because

(*a*) the density is less than the maximum dry density and water content is on dry side of optimum

(*b*) the compaction density is very low and water content is less than 12%

(*c*) the compaction is done on the dry side of the optimum

(*d*) both the dry density and water content of the compacted soil are within the desirable limits

99. Root time method is used to determine

(*a*) T, time factor

(*b*) c_v, coefficient of consolidation

(*c*) a_v, coefficient of compressibility

(*d*) m_v, coefficient of volume compressibility

100. Negative skin friction in a soil is considered when the pile is constructed through a

(*a*) fill material

(*b*) dense coarse sand

(*c*) over consolidated stiff clay

(*d*) dense fine sand

101. There are two footings resting on the ground surface. One footing is square of dimension 'B'. The other is strip footing of width 'B'. Both of them are subjected to a loading intensity of q. The pressure intensity at any depth below the base of the footing along the centerline would be

(*a*) equal in both footings

(*b*) large for square footing and small for strip footing

(*c*) large for strip footing and small for square footing

(*d*) more for strip footing at shallow depth (≤ B) and more for square footing at large depth (> B)

102. A soil mass contains 40% gravel, 50% sand and 10% silt. This soil can be classified as

(*a*) silty sandy gravel having coefficient of uniformity less than 60

(*b*) silty gravelly sand having coefficient of uniformity equal to 10

(*c*) gravelly silty sand having coefficient of uniformity greater than 60

(*d*) gravelly silty sand and its coefficient of uniformity cannot be determined.

103. A saturated soil mass has a total density 22kN/m^3 and a water content of 10%. The bulk density and dry density of this soil are

(*a*) 12 kN/m^3 and 20 kN/m^3 respectively

(*b*) 22 kN/m^3 and 20 kN/m^3 respectively

(*c*) 19.8 kN/m^3 and 19.8 kN/m^3 respectively

(*d*) 23.2 kN/m^3 and 19.8 kN/m^3 respectively.

104. In a constant head permeameter with cross section area of 10 cm^2, when the flow was taking place under a hydraulic gradient of 0.5, the amount of water collected in 60 seconds is 600 cc. The permeability of the soil is

(*a*) 0.002 cm/s (*b*) 0.02 cm/s

(*c*) 0.2 cm/s (*d*) 2.0 cm/s

105. For a triaxial shear test conducted on a sand specimen at a confining pressure of 100 kN/m^2 under drained conditions, resulted in a deviator stress ($\delta_1 - \delta_3$) at failure of 100 kN/m^2. The angle of shearing resistance of the soil would be

(*a*) 18.43° (*b*) 19.47°

(*c*) 26.56° (*d*) 30°

106. During the subsurface investigations for design of foundations, a standard penetration test was conducted at 4.5 m below the ground surface. The record of number of blows is given below.

Penetration depth (cm)	No. of blows
0 - 7.5	3
7.5 - 15	3
15 - 22.5	6
22.5 - 30	6
30 - 37.5	8
37.5 45	7

Assuming the water table at ground level, soil as fine sand and correction factor for overburden as 1.0, the corrected 'N' value for the soil would be

(*a*) 18 (*b*) 19

(*c*) 21 (*d*) 33

107. For two infinite slopes (one in dry condition and other in submerged condition) in a sand deposit having the angle of shearing resistance 30°, factor of safety was determined as 1.5 (for both slopes). The slope angles would have been

(*a*) 21.05° for dry slope and 21.05° for submerged slope

(*b*) 19.47° for dry slope and 18.40° for submerged slope

(*c*) 18.4° for dry slope and 21.05° for submerged slope

(*d*) 22.6° for dry slope and 19.47° for submerged slope

108. A strip footing (8m wide) is designed for a total settlement of 40mm. The safe bearing capacity (shear) was 150 kN/m^2 and safe allowable soil pressure was 100 kN/m^2. Due to importance of the structure, now the footing is to be redesigned for total settlement of 25 mm. The new width of the footing will be

(*a*) 5 m

(*b*) 8 m

(*c*) 12 m

(*d*) 12.8 m

109. A 3 m high retaining wall is supporting a saturated sand (saturated due to capillary action) of bulk density 18 kN/m^3 and angle of shearing resistance 30°. The change in magnitude of active earth pressure at the base due to rise in ground water table from the base of the footing to the ground surface shall (γ_w = 10 kN/m^3)

(*a*) increase by 20 kN/m^2
(*b*) decrease by 20 kN/m^2
(*c*) increase by 30 kN/m^2
(*d*) decrease by 30 kN/m^2

2004

110. The ratio of saturated unit weight to dry unit weight of a soil is 1.25. If specific gravity of solids (G_s) is 2.56, then void ratio of the soil is

(*a*) 0.625 (*b*) 0.663
(*c*) 0.944 (*d*) 1.325

111. A 10 m thick clay layer is underlain by a sand layer of 20 m depth (see figure below). The water table is 5 m below the surface of clay layer. The soil above the water table is capillary saturated. The value of g_{sat} is 19 kN/m^3. The unit weight of water is g_w. If now the water table rises to the surface, the effective stress at a point P on the interface will

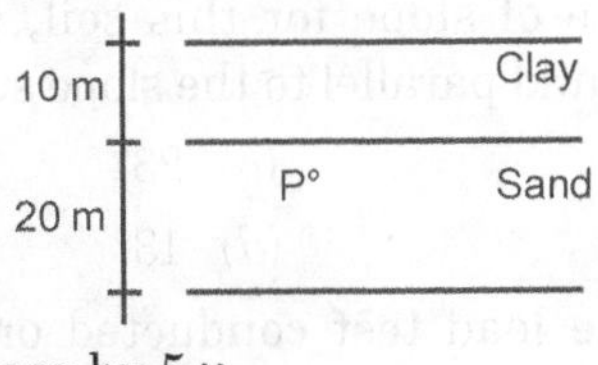

(*a*) increase by 5 γ_w
(*b*) remain unchanged
(*c*) decrease by 5 γ_w
(*d*) decrease by 10 γ_w

112. In an undrained triaxial test on a saturated clay, Poisson's ratio is

(*a*) $\frac{\sigma_3}{(\sigma_1+\sigma_3)}$ (*b*) $\frac{\sigma_3}{(\sigma_1-\sigma_3)}$

(*c*) $\frac{(\sigma_1-\sigma_3)}{\sigma_3}$ (*d*) $\frac{(\sigma_1+\sigma_3)}{\sigma_3}$

113. Two circular footings of diameters D_1 and D_2 are resting on the surface of the same purely cohesive soil. The ratio of their gross ultimate bearing capacities is

(*a*) $\frac{D_1}{D_2}$ (*b*) 1.0

(*c*) $\frac{D_1^1}{D_2^2}$ (*d*) $\frac{D_2}{D_1}$

114. An unit volume of a mass of saturated soil is subjected to horizontal seepage. The saturated unit weight is 22 kN/m^3 and the hydraulic gradient is 0.3. The resultant body force on the soil mass is

(*a*) 1.98 kN (*b*) 6.6 kN
(*c*) 11.49 kN (*d*) 22.97 kN

115. The undrained cohesion of a remoulded clay soil is 10 kN/m^2. If sensitivity of the clay is 20, then corresponding remoulded compressive strength is

(*a*) 5 kN/m^2 (*b*) 10 kN/m^2
(*c*) 20 kN/m^2 (*d*) 200 kN/m^2

116. The figure given below represents the contact pressure distribution underneath a

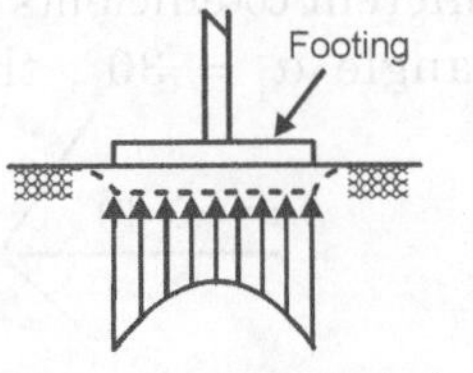

(*a*) rigid footing on saturated clay
(*b*) rigid footing on sand
(*c*) flexible footing on saturated clay
(*d*) flexible footing on sand

117. An infinite soil slope with an inclination of 35° is subjected to seepage parallel to its surface. The soil has c′ = 100 kN/m^2 and ϕ' = 30°. Using the concept of mobilized cohesion and friction, at a factor of safety of 1.5 with respect to shear strength, the mobilized friction angle is

(*a*) 20.02° (*b*) 21.05°
(*c*) 23.33° (*d*) 30.00°

118. A 6m thick clay layer undergoes 90% consolidation four times faster under two-way drainage as compared to one-drainage. In an identical clay layer of 15 m thickness, two-way drainage will be faster as compared to one-way drainage by

(*a*) 8 times (*b*) 4 times
(*c*) 2.5 times (*d*) 2 times

119. Using ϕ_u = 0 analysis and assuming planar failure as shown , the minimum factor of safety against shear failure of a vertical cut of height 4 m in a pure clay having c_u = 120 kN/m^2 and γ_{sat} = 20 kN/m^3 is

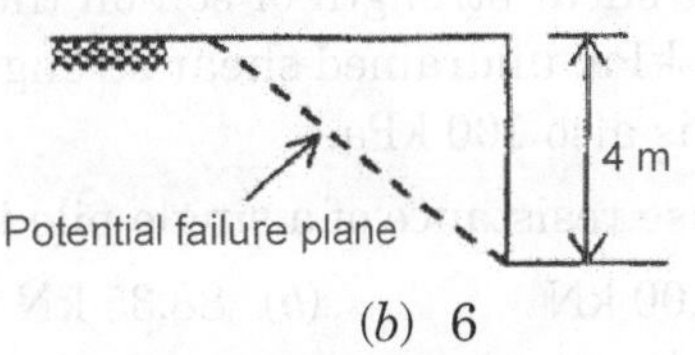

(*a*) 1 (*b*) 6
(*c*) 10 (*d*) 20

120. In the context of collecting undisturbed soil samples of high quality using a spoon sampler, following statements are made.

I. Area ratio should be less than 10%.

II. Clearance ratio should be less than 1 %.

With reference to above statements, which of the following applies ?

(*a*) Both the statements are true

(*b*) Statement II is true but I is false

(*c*) Statement I is true but II is false

(*d*) Both the statements are false

121. The figure below shows two flow lines for seepage across an interface between two soil media of different co-efficients of permeability. If entrance angle α_1 = 30°, the exit angle α_2 will be

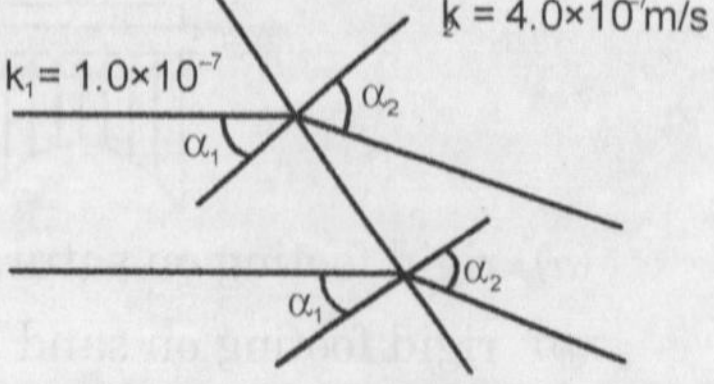

(*a*) 7.50°

(*b*) 14.03°

(*c*) 66.59°

(*d*) 75.96°

122. An unsupported excavation is made to the maximum possible depth in a clay soil having γ_t = 18 kN/m³, c = 100 kN/m³, φ = 30°. The active earth pressure, according to Rankine's theory, at the base level of the excavation is

(*a*) 115.47 kN/m² (*b*) 54.36 kN/m²

(*c*) 27.18 kN/m² (*d*) 13. kN/m²

123. A retaining wall of height 8 m retains dry sand. In the initial state, the soil is loose and has a void ratio of 0.5, γ_d = 17.8 kN/m³ and φ = 30°. Subsequently, the backfill is compacted to a state where void ratio is 0.4, γ_d = 18.8 kN/m³ and φ = 35°. The ratio of initial passive thrust to the final passive thrust, according to Rankine's earth pressure theory, is

(*a*) 0.38 (*b*) 0.64

(*c*) 0.77 (*d*) 1.55

Common Data Q. 124-125

A group of 16 piles of 10 m length and 0.5 m diameter is installed in a 10 m thick stiff clay layer underlain by rock. The pile-soil adhesion factor is 0.4; average shear strength of soil on the side on the sides is 100 kPa; undrained shear strength of the soil at the base is also 100 kPa.

124. The base resistance of a single pile is

(*a*) 40.00 kN (*b*) 88.35 kN

(*c*) 100.00 kN (*d*) 176.71 kN

125. Assuming 100% efficiency , the group side resistance is

(*a*) 5026.5 kN (*b*) 10000.0 kN

(*c*) 10053.1 kN (*d*) 20106.0 kN

2003

126. A masonry dam is founded on previous sand having porosity equal to 45% and specific gravity of sand particles is 2.65. For a desired factor of safety of 3 against sand boiling, the maximum permissible upward gradient will be

(*a*) 0.225 (*b*) 0.302

(*c*) 1.0 (*d*) none of these

127. A double draining clay layer, 6m thick, settles by 30mm in three years under the influence of a certain loads. Its final consolidation settlement has been estimated to be 120 mm. If a thin layer of sand having negligible thickness is introduced at a depth of 1.5m below the top surface, then final consolidation settlement of clay layer will be

(*a*) 60 mm (*b*) 120 mm

(*c*) 240 mm (*d*) None of these

128. A granular soil possesses saturated density of 20 kN/m³. Its effective angle of internal friction is 35 degrees. If desired factor of safety is 1.5, the safe angle of slope for this soil, when seepage occurs at and parallel to the slope surface, will be

(*a*) 25° (*b*) 23°

(*c*) 20° (*d*) 13°

129. In a plate load test conducted on cohesionless soil, a 600 mm square test plate settles by 15mm under a load intensity of 0.2 N/mm². All conditions remaining the same, settlement of a 1m square footing will be

(*a*) less than 15 mm (*b*) greater than 25 mm

(*c*) 15.60 mm (*d*) 20.50 mm

130. A 25 kN point load acts on the surface of an infinite elastic medium. The vertical pressure intensity in kN/m² at a point 6.0 m below and 4.0 m away from the load will be

(*a*) 132 (*b*) 13.2

(*c*) 1.32 (*d*) 0.132

131. Water is pumped from a well tapping an unconfined aquifer at a certain discharge rate and the steady state drawdown (X) in an observation well is monitored. Subsequently, pumping discharge is doubled and steady state drawdown in the same observation well is found to be more than double (i.e., more than 2X). This dispropor-tionate drawdown is caused by (*a*) well losses

(*b*) decrease in the saturated thickness of the aquifer

(*c*) nonlinear flow

(*d*) delayed gravity yield

132. Compaction of an embankment is carried out in 500 mm thick layers. The rammer used for compaction has a foot area of 0.05 sq. m and the energy imparted in every drop of rammer is 400 Nm. Assuming 50% more energy in each pass over the compacted area due to overlap, the number of passes required to develop compactive energy equivalent to Indian Standard light compaction for each layer would be

(*a*) 10 (*b*) 16

(*c*) 20 (*d*) 26

133. A braced cut, 5m wide and 7.5m deep is proposed in a cohesionless soil deposit having effective cohesion $c' = 0$ and effective friction angle, $\phi' = 36°$. The first row of struts is to be installed at a depth of 0.5 m below ground surface and spacing between the struts should be 1.5m. If horizontal spacing of struts is 3m and unit weight of the deposit is 20 kN/m^3, then maximum strut load will be

(*a*) 70.87 kN (*b*) 98.72 kN

(*c*) 113.90 kN (*d*) 151.86 kN

134. For the soil strata shown in figure, the water table is lowered by drainage by 2m and if the top 2m thick silty sand stratum remains saturated by capillary action even after lowering of water table, the increase in effective vertical pressure in kPa at mid-height of clay layer will be

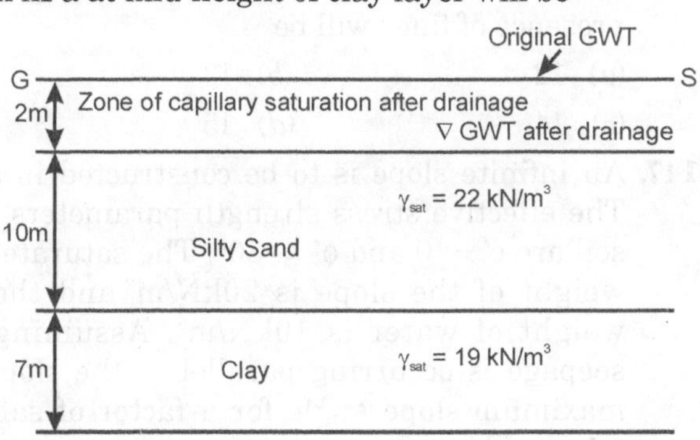

(*a*) 0.2 (*b*) 2

(*c*) 20 (*d*) 200

135. At a reclamation site for which the soil strata is shown in figure, a 3m thick layer of a fill material is to be laid instantaneously on the top surface. If the coefficient of volume compressibility, m_v for clay is 2.2×10^{-4} m^2/kN, the consolidation settlement of the clay layer due to placing of fill material will be

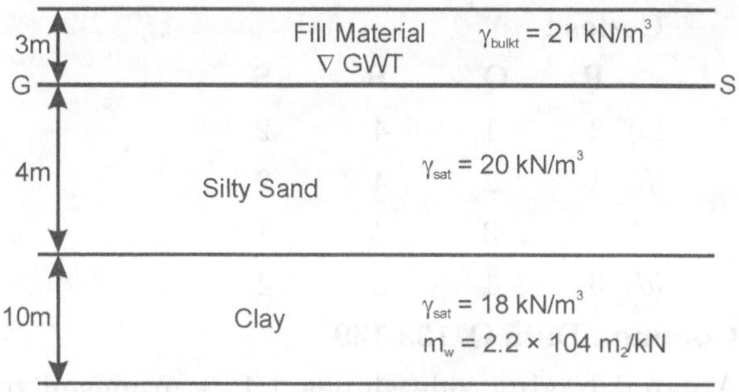

(*a*) 69.5 mm (*b*) 139 mm

(*c*) 228 mm (*d*) 278 mm

136. For the (3 × 3) pile group shown in the figure, the settlement of pile group, in a normally consolidated clay stratum having properties as shown in the figure, will be

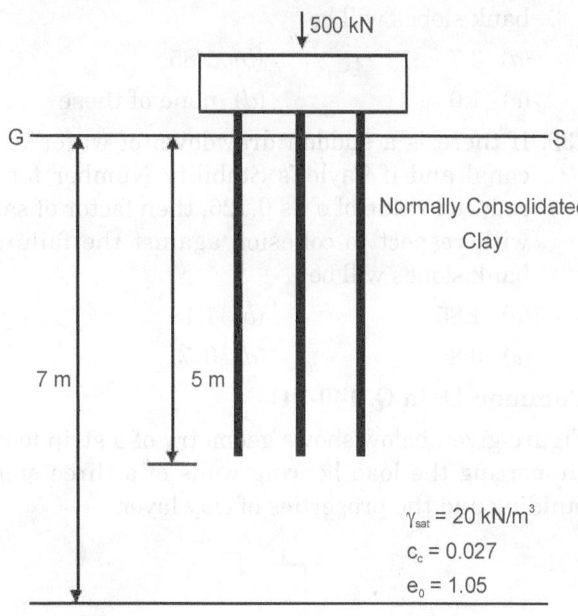

(*a*) 13.2 mm (*b*) 12.775 mm

(*c*) 7.345 mm (*d*) none of these

137. Match the items of the two lists and select the correct answer.

List I *(Boring Methods)*

P. Auger Boring

Q. Wash Boring

R. Percussion Drilling

S. Rotary Drilling

List II *(Field Conditions)*

1. Below water table in all soil types except hard soils and rocks

2. Large diameter boreholes over 150 mm in size

3. Explorations for shallow foundations and highways

4. Bouldery and gravelly strata

Codes :

	P	Q	R	S
(*a*)	3	1	4	2
(*b*)	1	2	4	3
(*c*)	2	3	4	1
(*d*)	3	1	2	4

Common Data Q. 138-139

A canal having side slopes 1:1 is proposed to be constructed in a cohesive soil to a depth of 10 m below the ground surface. The soil properties are:

$\phi_u = 15°$, $C_n = 12$ kPa, e = 1.0, Gs = 2.65.

138. If Taylor's Stability Number, Sn is 0.08 and if the canal flows full, the factor of safety with respect to cohesion against failure of the canal bank slopes will be

(*a*) 3.7 (*b*) 1.85

(*c*) 1.0 (*d*) none of these

139. If there is a sudden drawdown of water in the canal and if Taylor's Stability Number for the reduced value of ϕ_v is 0.126, then factor of safety with respect to cohesion against the failure of bank slopes will be

(*a*) 1.85 (*b*) 1.18

(*c*) 0.84 (*d*) 0.53

Common Data Q. 140-141

Figure given below shows geometry of a strip footing supporting the load bearing walls of a three storied building and the properties of clay layer.

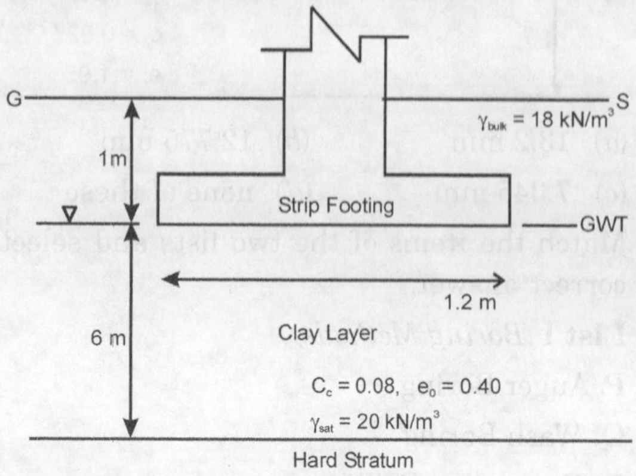

140. If pressure acting on the footing is 40 kPa, then consolidation settlement of the footing will be

(*a*) 0.89 mm (*b*) 8.9 mm

(*c*) 89.0 mm (*d*) none of these

141. If elastic modulus and Poisson's ratio of the clay layer are respectively 50×10^3 kPa and 0.4 and if influence factor for the strip footing is 1.75, then elastic settlement of the footing will be

(*a*) 0.41 mm (*b*) 1.41 mm

(*c*) 14.1 mm (*d*) None of these

2002

142. Data from a sieve analysis conducted on a given sample of soil showed that 67% of the particles passed through 75 micron IS sieve. The liquid limit and plastic limit of the finer fraction was found to be 45 and 33 percents respectively. The group symbol of the given soil as per IS:1498-1970 is

(*a*) SC (*b*) MI

(*c*) CH (*d*) MH

143. The avoid ratios at the densest, loosest and the natural states of a sand deposit are 0.2, 0.6 and 0.4, respectively. The relative density of the deposit is

(*a*) 100% (*b*) 75%

(*c*) 50% (*d*) 25%

144. The following data were obtained from a liquid limit test conducted on a soil sample.

Number of blows	17	22	25	28	34
Water content (%)	63.8	63.1	61.9	60.6	60.5

The liquid limit of the soil is

(*a*) 63.1% (*b*) 62.8%

(*c*) 61.9% (*d*) 60.6%

145. The specific gravity and insitu void ratio of a soil deposit are 2.71 and 0.85 respectively. The value of the critical hydraulic gradient is

(*a*) 0.82 (*b*) 0.85

(*c*) 0.92 (*d*) 0.95

146. The observed value of the standard penetration number (N) at 10m depth of a silty sand deposit is 13. The unit weight of the soil is 16 kN/m³. The N value after correcting for the presence of fines will be

(*a*) 12 (*b*) 13

(*c*) 14 (*d*) 15

147. An infinite slope is to be constructed in a soil. The effective stress strength parameters of the soil are c′ = 0 and ϕ' = 30°. The saturated unit weight of the slope is 20kN/m³ and the unit weight of water is 10kN/m³. Assuming that seepage is occurring parallel to the slope, the maximum slope angle for a factor of safety of 1.5 would be

(*a*) 10.89° (*b*) 11.30°

(*c*) 12.48° (*d*) 14.73°

148. If effective stress strength parameters of a soil are c′= 10kPa and ϕ = 30°, then shear strength on a plane within the saturated soil mass at a point where total normal stress is 300kPa and pore water pressure is 150 kPa, will be

(*a*) 90.5 kPa. (*b*) 96.6kPa.

(*c*) 101.5 kPa (*d*) 105.5 kPa

149. The time for a clay layer to achieve 85% consolidation is 10 years. If layer was half as thick, 10 times more permeable and 4 times more compressible then time that would be required to achieve the same degree of consolidation is

(*a*) 1 year (*b*) 5 years

(*c*) 12 years (*d*) 16 years

150. In a triaxial test carried out on a cohesionless soil sample with a cell pressure of 20 kPa, the observed value of applied stress at the point of failure was 40 kPa. The angle of internal friction of the soil is

(*a*) 10° (*b*) 15°

(*c*) 25° (*d*) 30°

151. In a falling head permeability test, the initial head of 1.0m dropped to 0.35 m in 3 hours, the diameter of the stand pipe being 5 mm. The soil specimen was 200 mm long and of 100 mm diameter. The coefficient of permeability of the soil is

(*a*) 4.86×10^{-5} cm/s (*b*) 4.86×10^{-6} cm/s

(*c*) 4.86×10^{-7}cm/s (*d*) 4.86×10^{-8}cm/s

2001

152. The following two statements are made with respect to different sand samples having the same relative density. Identify if they are TRUE or FALSE.

I. Poorly graded sands will have lower friction angle than the well graded sands.

II. The particle size has no influence on the friction angle of sand.

(*a*) II is TRUE but I is FALSE.

(*b*) Both are FALSE statements

(*c*) Both are TRUE statements.

(*d*) I is TRUE but II is FALSE.

153. The following two statements are made with reference to the calculation of net bearing capacity of a footing in pure clay soil ($\phi = 0$) using Terzaghi's bearing capacity theory. Identify if they are TRUE or FALSE.

I. Increase in footing width will result in increase in bearing capacity.

II. Increase in depth of foundation will result in higher bearing capacity.

(*a*) Both statements are TRUE

(*b*) Both statements are FALSE

(*c*) I is TRUE but II is FALSE

(*d*) I is FALSE but II is TRUE

154. The width and depth of a footing are 2 and 1.5 m respectively. The water table at the site is at a depth of 3 m below the ground level . The water table correction factor for the calculation of the bearing capacity of soil is

(*a*) 0.875 (*b*) 1.000

(*c*) 0.925 (*d*) 0.500

155. The void ratio and specific gravity of a soil are 0.65 and 2.72 respectively. The degree of saturation (in percent) corresponding to water content of 20% is

(*a*) 65.3 (*b*) 20.9

(*c*) 83.7 (*d*) 54.4

156. With respect to a c-ϕ soil in an infinite slope, identify if the following two statements are TRUE or FALSE.

I. The stable slope angle can be greater than ϕ

II. The factor of safety of the slope does not depend on the height of soil in the slope.

(*a*) Both statements are FALSE

(*b*) I is TRUE but II is FALSE

(*c*) I is FALSE but II is TRUE

(*d*) Both statements are TRUE

157. Identify the two FALSE statements from the following four statements.

I. The consolidation of soil happens due to the change in total stress.

II. When Standard Penetration Tests are performed in fine sands below the water table, the dilation correction is applied after the overburden correction is applied.

III. Over consolidated clays will have predominantly cohesive strength as compared to the frictional strength.

IV. Compaction of soils is due to expulsion of water.

(*a*) II and III

(*b*) I and IV

(*c*) I and III

(*d*) II and IV

158. The critical slip circle for a slope is shown below along with the soil properties.

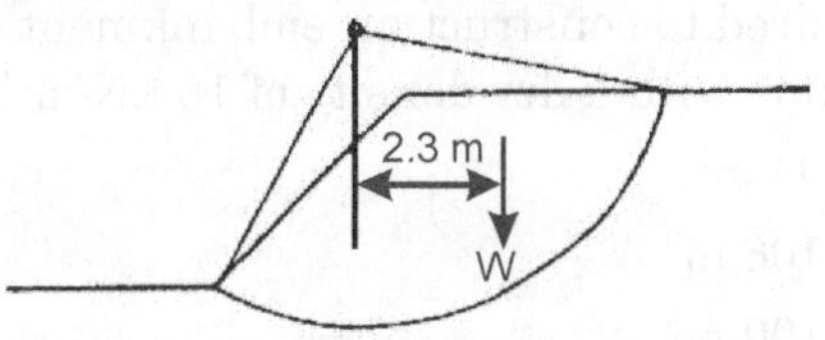

The length of the arc of the slip circle is 15.6 m and area of soil within the slip circle is 82 m².

The radius of the slip circles is 10.3 m. The factor of safety against the slip circle failure is nearly equal to

(a) 1.05 (b) 1.22

(c) 0.78 (d) 1.28

159. The coefficients of permeability of a soil in horizontal and vertical directions are 3.46 and 1.5m/day respectively. The base length of a concrete dam resting in this soil is 100 m. When the flow net is developed for this soil with 1:25 scale factor in the vertical direction, then reduced base length of the dam will be

(a) 2.63 m (b) 4.00 m

(c) 6.08 m (d) 5.43 m

160. A plate load test was conducted in sand on a 300 mm diameter plate. If the plate settlement was 5 mm at a pressure of 100 kPa, then settlement (in mm) of a 5m×8m rectangular footing at the same pressure will be

(a) 9.4 (b) 18.6

(c) 12.7 (d) 17.8

161. Identify the two TRUE statements from the following four statements.

I. Negative skin friction is higher on floating piles than on end bearing piles.

II. All other things being the same in footings on sand, the footing with smaller width will have lower settlement at the same net pressure.

III. The void ratio of soils is always less than 1.0.

IV. For determining depth of embedment of anchored sheet piles, net moment at the anchor elevation is set to zero.

(a) I and IV (b) I and III

(c) II and IV (d) II and III

2000

162. If a soil sample has a void ratio of 0.5, then and its porosity will be close to

(a) 50% (b) 66%

(c) 100% (d) 33%

163. A borrow pit soil has a dry density of 17 kN/m^3. How many cubic meters of this soil will be required to construct an embankment of 100 m^3 volume with a dry density of 16 kN/m^3?

(a) 94 m^3

(b) 106 m^3

(c) 100 m^3

(d) 90 m^3

164. The group efficiency of pile group

(a) will be always less than 100%

(b) will be always greater than 100%

(c) may be less than 100% or more than 100%

(d) will be more than 100% for pile groups in cohesionless soils and less than 100% for those in cohesive soils.

165. The two criteria for the determination of allowable bearing capacity of a foundation are

(a) tensile failure and compression failure.

(b) tensile failure and settlement.

(c) bond failure and shear failure.

(d) shear failure and settlement.

166. The ultimate bearing capacity of a soil is 300 kN/m^2. The depth of foundation is 1m and unit weight of soil is 20 kN/m^3. Choosing a factor of safety of 2.5, the net safe bearing capacity is

(a) 100 kN/m^2 (b) 112 kN/m^2

(c) 80 kN/m^2 (d) 100.5 kN/m^2

167. A deep cut of 7m has to be made in a clay with unit weight 16 kN/m^3 and a cohesion of 25 kN/m^2. What will be the factor of safety if one has to have a slope angle of 30° ? Stability number is given to be 0.178 (from Taylor's chart) for a depth factor of 3.

(a) 0.80 (b) 1.1

(c) 1.25 (d) 1.0

168. In a drained triaxial compression test, a saturated specimen of a cohesionless sand fails under a deviatoric stress of 3 kgf/cm^2 when the cell pressure is 1 kgf/cm^2. The effective angle of shearing resistance of sand is about

(a) 37° (b) 45°

(c) 53° (d) 20°

169. Two footings, one circular and the other square, and founded on the surface of a purely cohesionless soil. The diameter of the circular footing is same as that of the side of the square footing. The ratio of their ultimate bearing capacities is

(a) $\frac{3}{4}$ (b) $\frac{4}{3}$

(c) 1.0 (d) 1.3

170. To have zero active pressure intensity at the tip of a wall in cohesive soil, one should apply a uniform surcharge intensity of

(a) 2 c tan α (b) 2 c cot α

(c) – 2 c tan α (d) – 2 c tan α

1999

171. The values of liquid limit and plasticity index for soils having common geological origin in a restricted locality usually define

(*a*) a zone above A - line

(*b*) a staright line parallel to A - line

(*c*) a straight line perpendicular to A - line

(*d*) points may be anywhere in the plasticity chart

172. The toughness index of clayey soils is given by

(*a*) Plasticity index/Flow index

(*b*) Liquid limit/Plastic limit

(*c*) Liquidity index/Plastic limit

(*d*) Plastic limit/Liquidity index

173. Principle involved in the relationship between submerged unit weight and saturated weight of a soil is based on

(*a*) Equilibrium of floating bodies

(*b*) Archimedes' principle

(*c*) Stokes' law

(*d*) Darcy's law

174. For an anisotropic soil, permeabilities in x and y directions are k_x and k_y respectively in a two dimensional flow. The effective permeability k_{eff} for the soil is given by

(*a*) $k_x + k_y$ (*b*) $\frac{k_x}{k_y}$

(*c*) $(k_x^2 + k_y^2)^{½}$ (*d*) $(k_x k_y)^{½}$

175. Cohesion in soil

(*a*) decreases active pressure and increases passive resistance

(*b*) decreases both active pressure and passive resistance

(*c*) increases the active pressure and decreases the passive resistance

(*d*) increases both active pressure and passive resistance

176. Consolidation in soils

(*a*) is a function fo the effective stress

(*b*) does not depend on the present stress

(*c*) is a function of the pore water pressure

(*d*) is a function of the total stress

177. The two tubes shown in figure may be considered to be permeameters. Dimensions of the smaple in Fig. (*i*) and (*ii*) are alike, and the elevations of head water and tail water are the same for both the figures. A, B, ... etc. indicate points and AB, AE, ...etc. indicated heads. Head loss through these samples are

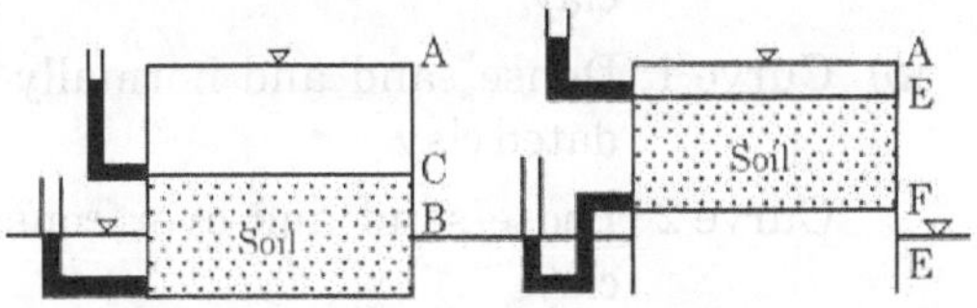

(i) (ii)

(*a*) (*i*) BD, (*ii*) FB

(*b*) (*i*) AC, (*ii*) AE

(*c*) (*i*) AD, (*ii*) AF

(*d*) (*i*) AB, (*ii*) AB

178. A river 5 m deep consists of a sand bed with saturatd unit weight of 20 kN/m³. v_w = 9.81 kN/m³. The effective vertical stress at 5m from the top of sand bed is

(*a*) 41 kN/m² (*b*) 51 kN/m²

(*c*) 55 kN/m² (*d*) 53 kN/m²

179. A soil sample in its natural state has mass of 2.290 kg and a volume of 1.15×10^{-3} m³. After being oven dried, the mass of the sample is 2.035 kg. G_s for soil is 2.68. The void ratio of the natural soil is

(*a*) 0.40 (*b*) 0.45

(*c*) 0.55 (*d*) 0.53

1998

180. Some of the structural strength of a clayey material that is lost by remoulding is slowly recovered with time. This property of soils to undergo an isothermal gel-to sol-to -gel transformation upon agitation and subsequent rest is termed

(*a*) Isotropy (*b*) Anisotropy

(*c*) Thixotropy (*d*) Allotropy

181. If soil is dried beyond its shrinkage limit, it will show

(*a*) Large volume change

(*b*) Moderate volume change

(*c*) Low volume change

(*d*) No volume change

182. The stress-strain behaviour of soils as shown in the Figure corresponds to:

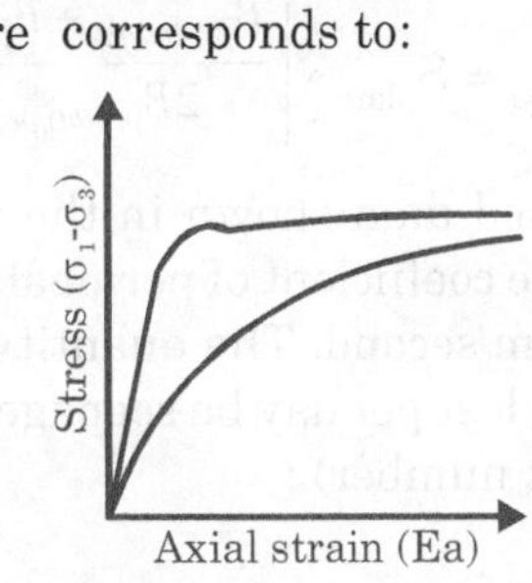

(a) Curve 1: Loose sand and normally consolidated clay
Curve 2: Loose sand and over consolidated clay

(b) Curve 1: Dense sand and normally consolidated clay
Curve 2: Loose sand and over consolidated clay

(c) Curve 1: Dense sand and over consolidated clay
Curve 2: Loose sand and normally consolidated clay

(d) Curve 1: Loose sand and over consolidated clay
Curve 2: Dense sand normally consolidated clay

183. In cohesive soils the depth of tension crack (Z_{cr}) is likely to be

(a) $Z_{cr} \geq \frac{2c}{\gamma}\sqrt{\tan\left(45° - \frac{\phi}{2}\right)}$

(b) $X_{cr} \geq \frac{2c}{\gamma}\sqrt{\tan\left(45° + \frac{\phi}{2}\right)}$

(c) $Z_{cr} \geq \frac{4c}{\gamma}\sqrt{\tan\left(45° - \frac{\phi}{2}\right)}$

(d) $Z_{cr} \geq \frac{4c}{\gamma}\sqrt{\tan\left(45° + \frac{\phi}{2}\right)}$

184. The settlement of prototype in granular material may be estimated using plate load test data from the following expression:

(a) $S_{prototype} = S_{plate} \times \left(\frac{B_{prototype}}{B_{plate}}\right)$

(b) $S_{prototype} = S_{plate} \times \left(\frac{B_{plate}}{B_{prototype}}\right)$

(c) $S_{prototype} = S_{plate} \times \left[\frac{2B_{prototype}}{B_{prototype} + B_{plate}}\right]^2$

(d) $S_{prototype} = S_{plate} \times \left[\frac{B_{prototype} + B_{plate}}{2B_{prototype}}\right]^2$

185. The proposed dam shown in the figure is 90 m long and the coefficient of permeability of the soil is 0.0013mm/second. The quantity of water (m^3) that will be lost per day be seepage is (rounded to the nearest number) :

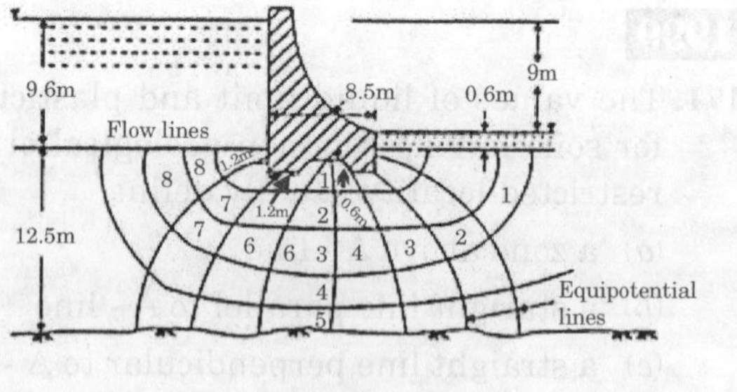

(a) 55 (b) 57
(c) 59 (d) 61

186. The time for a clay layer to achieve 90% consolidation is 15 years. The time required to achieve 90% consolidation, if the layer were twice as thick, 3 times more permeable and 4 times more compressible would be :

(a) 70 years (b) 75 years
(c) 80 years (d) 85 years

187. The total active thrust on a vertical wall 3m high retaining a horizontal sand backfill (unit weight γ_t = 20 kN/m^3, angle of shearing resistance ϕ' = 30°) when the water table is at the bottom of the wall, will be:

(a) 30 kN/m (b) 35 kN/m
(c) 40 kN/m (d) 45 kN/m

188. A 40° slope is excavated to a depth of 10 (d) depth of 10 m is a deep layer of saturated clay of unit weight 20 kN/m^3; the relevant shear strength parameters are c_u = 72 kN/m^2 and ϕ_u = 0. The rock ledge is at a great depth. The Taylor's stability coefficient for ϕ_u = 0 and 40° slope angle is 0.18. The factor of safety of the load is:

(a) 2.0 (b) 2.1
(c) 2.2 (d) 2.3

189. A point load of 700 kN is applied on the surface of thick layer of saturated clay. Using Boussinesq's elastic analysis, the estimated vertical stress (σ_v) at a depth of 2 m and a radial distance of 1.0 m from the point of application of the load is:

(a) 47.5 kPa (b) 47.6 kPa
(c) 47.7 kPa (d) 47.8 kPa

1997

190. If the porosity of a soil sample is 20%, the void ratio is

(a) 0.20 (b) 0.80
(c) 1.00 (d) 0.25

191. The shape of clay particle is usually

(a) angular (b) flaky
(c) tubular (d) rounded

192. Consistency Index for a clayer soil is [LL = Liquid Limit, PL = Plastic Limit, PI = Plasticity Index, W = natural moisture content]

(a) $\frac{LL - w}{PI}$ (b) $\frac{w - PL}{PI}$

(c) LL – PL (d) 0.5 w

193. According to Darcy's law for flow through porous media, the velocity is proportional to

(a) effective stress (b) hydraulic gradient

(c) cohesion (d) stability number

194. A soil mass has coefficients of horizontal and vertical permeability as 9×10^{-7}cm/s and 4×10^{-7}cm/s, respectively. The transformed coefficient of permeability of an equivalent isotropic soil mass is

(a) 9×10^{-7}cm/s (b) 4×10^{-7}cm/s

(c) 13×10^{-7}cm/s (d) 6×10^{-7}cm/s

195. In a compaction test, as the compaction effort is increased, the optimum moisture content

(a) decreases

(b) remains same

(c) increases

(d) increases first there after decreases

196. The vertical stress at depth. z directly below the point load p is (k is a constant)

(a) $k\frac{p}{z}$ (b) $k\frac{p}{z^3}$

(c) $k\frac{p}{z^2}$ (d) $k\frac{p}{\sqrt{z}}$

197. The slope of the e - log p curve for a soil mass gives

(a) coefficient of permeability, k

(b) coefficient of consolidation C_v

(c) compression index, C_c

(d) coefficient of volume compressibility, m_v

198. Sand and drains are used to

(a) reduce the settlement

(b) accelerate the consolidation

(c) increase the permeability

(d) transfer the load

199. Coulomb's theory of earth pressure is based on

(a) the theory of elasticity

(b) the therory of plasticity

(c) empirical rules

(d) wedge theory

200. The depth of tension crack in a soft clay ($\phi_u = 0$) is

(a) $\frac{4C_u}{\gamma}$ (b) $\frac{2C_u}{\gamma}$

(c) $\frac{C_u}{\gamma}$ (d) $\frac{C_u}{2\gamma}$

201. Vane tester is normally used for determining in situ shear strength of

(a) soflt clays (b) sand

(c) stiff clays (d) gravel

202. The unit for coefficient of subgrade modulus is

(a) kN/m³ (b) ku/m²

(c) ku/m (d) ku/m

203. The ratio of unconfined compressive strength of an undisturbed sample of soil to that of a remoulded sample, at the same water content, is known as

(a) activity (b) damping

(c) plasticity (d) sensitivity

204. Well foundation are commonly used as foundation for the following structures :

(a) Water tanks

(b) Bridges

(c) Buildings

(d) Reciprocating machines

1996

205. The particle size distribution curves are extremely useful for the classification of

(a) fine grained soils

(b) coarse grained soils

(c) both coarse grained and fine grained soils

(d) silts and clays

206. Which one of the following relations is not correct ?

(a) $e = \frac{n}{1-n}$ (b) $\gamma_{sat} = \frac{G+e}{1+e}\gamma_w$

(c) $n = \frac{e}{1-e}$ (d) $e = \frac{wG}{S}$

where e = void ration n-porosity w = water content, S = degree of saturation, γ_{sat} = saturated unit weight of soil, γ_w = unit weight of water

207. Seepage force per unit volume (j) can be expressed as

(a) $i\gamma_w L$ (b) iL

(c) $\gamma_w h$ (d) $i\gamma_w$

where i = hydraulic gradient, L = length of soil sample, h-hydraulic head, γ_w = unit weight of water

208. Skin frictional capacities of a 40 cm diameter driven concrete pile for the portions A, B and C are 17 kN, 63 kN and 503 kN.respectively, and point load capacity is 11000 kN/m^2 Total pile load capacity will be

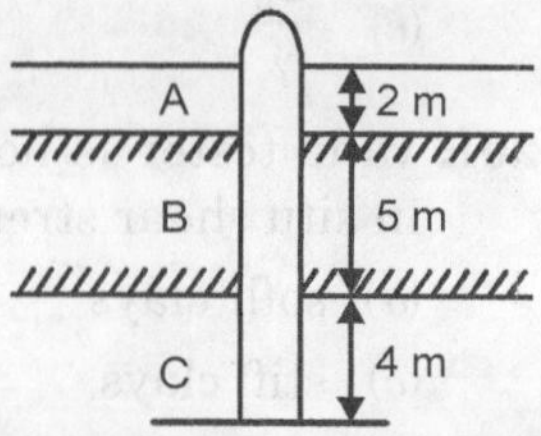

(*a*) 3743 kN
(*b*) 2864 kN
(*c*) 1965kN
(*d*) 1529 kN

209. Mechanical stabilization requires

(*a*) mixing of two or more types of natural soils
(*b*) addition of chemicals to soils
(*c*) addition of lime to soils
(*d*) addition of cementing material to soils

210. Piping in soil occurs when

(*a*) the soil is highly porous
(*b*) sudden change in permeability occurs
(*c*) effective pressure becomes zero
(*d*) the soil is highly stratified

211. The soils most susceptible to liquefaction are

(*a*) saturated dense sands
(*b*) saturated fine and medium sands of uniform particle size
(*c*) saturated clays of uniform size
(*d*) saturated gravels and cobbles

212. The most commonly used sample for obtaining a disturbed sample of the soil is

(*a*) split spoon sampler
(*b*) open drive sampler
(*c*) piston sampler
(*d*) thin wall shelby tube sampler

213. For the determination of earth pressure, Coulomb's wedge theory assumes that

(*a*) the back of wall is smooth and vertical
(*b*) the soil is non homogeneous and anisotropic
(*c*) the slip surface is circular
(*d*) the wall surface is rough

214. Two soil samples A and B have porosities n_A = 40% and n_B = 60 % respectively. What is the ratio of void ratios e_A:e_B?

(*a*) 2 : 3
(*b*) 3 : 2
(*c*) 4 : 9
(*d*) 9 : 4

215. A foundation is considered as shallow, if its depth is

(*a*) less than 1 meter
(*b*) greater than its width
(*c*) equal to or less than its width
(*d*) greater than 1 meter

216. What are the respective values of void ratio, porosity ratio and saturated density (in kN/m^3) for a soil sample which has saturation moisture content of 20% and specific gravity of grains as 2.6 ?

(Take density of water 10 kN/m^3)

(*a*) 0.52, 1.08, 18.07 (*b*) 0.52, 0.34, 18.07
(*c*) 0.77, 1.08, 16.64 (*d*) 0.52, 0.34, 20.14

217. The value of bearing capacity factor for cohesion, N'_c, for piles as per Meyerhof, is taken as

(*a*) 6.2 (*b*) 12.0
(*c*) 9.0 (*d*) 5.14

218. Negative skin friction occurs when

(*a*) an upward drg exists in the pile
(*b*) the surrounding soil settles more than the pile
(*c*) the pile passes continuously through a firm soil
(*d*) the driving operation begins

NUMERICAL TYPE QUESTIONS

2016

1. The porosity (n) and the degree of saturation (S) of a soil sample are 0.7 and 40%, respectively. In a 100 m^3 volume of the soil, the volume (expressed in m^3) of air is ______.

2. An undisturbed soil sample was taken from the middle of a clay layer (i.e., 1.5 m below GL), as shown in figure. The water table was at the top of clay layer. Laboratory test results are as follows:

Natural water content of clay	:	25%
Preconsolidation pressure of clay	:	60 kPa
Compression index of clay	:	0.50
Recompression index of clay	:	0.05
Specific gravity of clay	:	2.70
Bulk unit weight of sand	:	17 kN/m^3

A compacted fill of 2.5 m height with unit weight of 20 kN/m^3 is placed at the ground level.

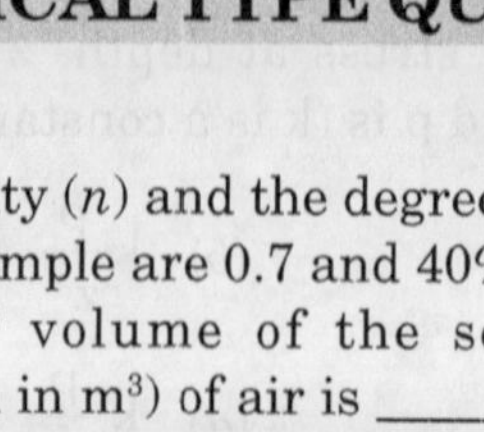

Assuming unit weight of water as 10 kN/m³, the ultimate consolidation settlement (expressed in mm) of the clay layer is ________.

3. A seepage flow condition is shown in the figure. The saturated unit weight of the soil γ_{sat} = 18 kN/m³. Using unit weight of water, γ_w = 9.81 kN/m³, the effective vertical stress (expressed in kN/m²) on plane X-X is ________.

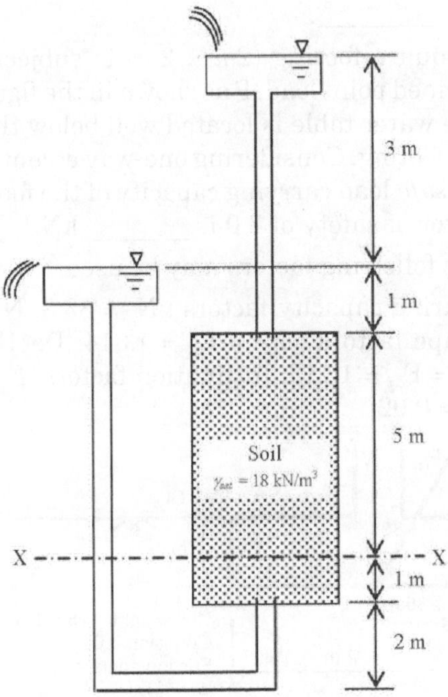

4. A drained triaxial compression test on a saturated clay yielded the effective shear strength parameters as c' = 15 kPa and ϕ' = 22°. Consolidated Undrained triaxial test on an identical sample of this clay at a cell pressure of 200 kPa developed a pore water pressure of 150 kPa at failure. The deviator stress (expressed in kPa) at failure is ________.

5. The seepage occurring through an earthen dam is represented by a flownet comprising of 10 equipotential drops and 20 flow channels. The coefficient of permeability of the soil is 3 mm/min and the head loss is 5 m. The rate of seepage (expressed in cm³/s per m length of the dam) through the earthen dam is ________

6. The soil profile at a site consists of a 5 m thick sand layer underlain by a c-ϕ soil as shown in figure. The water table is found 1 m below the ground level. The entire soil mass is retained by a concrete retaining wall and is in the active state. The back of the wall is smooth and vertical. The total active earth pressure (expressed in kN/m²) at point A as per Rankine's theory is ________.

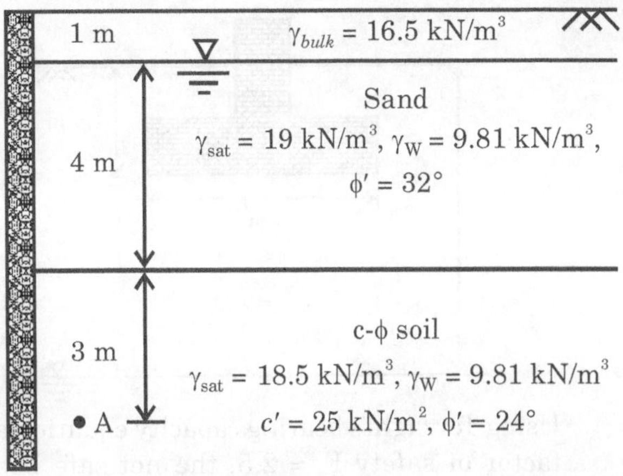

7. Water flows from P to Q through two soil samples, Soil 1 and Soil 2, having cross sectional area of 80 cm² as shown in the figure. Over a period of 15 minutes, 200 ml of water was observed to pass through any cross section. The flow conditions can be assumed to be steady state. If the coefficient of permeability of Soil 1 is 0.02 mm/s, the coefficient of permeability of Soil 2 (expressed in mm/s) would be

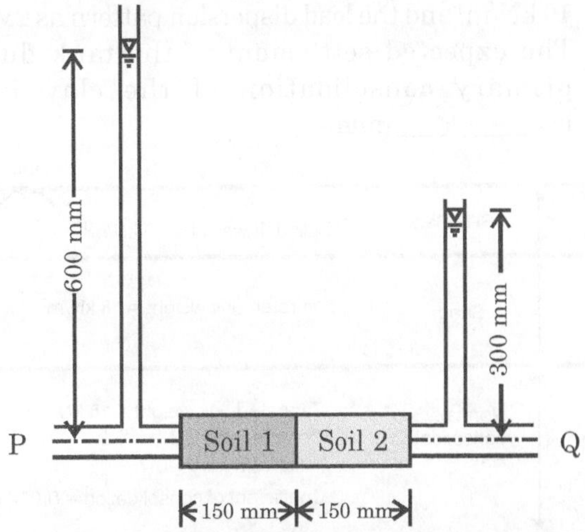

8. A 4 m wide strip footing is founded at a depth of 1.5 m below the ground surface in a c-ϕ soil as shown in the figure. The water table is at a depth of 5.5 m below ground surface. The soil properties are: c' = 35 kN/m², ϕ' = 28.63°, γ_{sat} = 19 kN/m³, γ_{bulk} = 17 kN/m³ and γ_w = 9.81 kN/m³. The values of bearing capacity factors for different ϕ' are given below.

ϕ'	N_c	N_q	N_γ
15°	12.9	4.4	2.5
20°	17.7	7.4	5.0
25°	25.1	12.7	9.7
30°	37.2	22.5	19.7

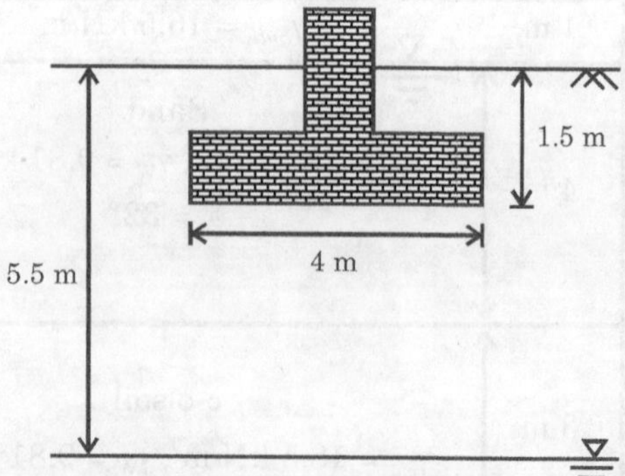

Using Terzaghi's bearing capacity equation and a factor of safety $F_s = 2.5$, the net safe bearing capacity (expressed in kN/m^2) for local shear failure of the soil is ________

2015

9. A water tank is to be constructed on the soil deposit shown in the figure below. A circular footing of diameter 3 m and depth of embedment 1 m has been designed to support the tank. The total vertical load to be taken by the footing is 1500 kN. Assume the unit weight of water as 10 kN/m^3 and the load dispersion pattern as 2V:1H. The expected settlement of the tank due to primary consolidation of the clay layer is ________ mm.

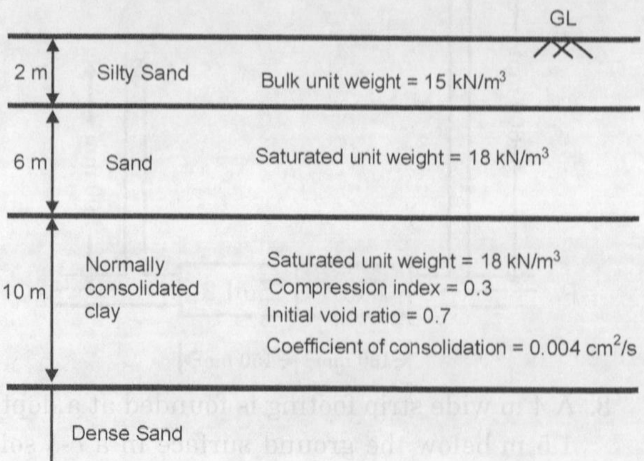

10. A 20 m thick clay layer is sandwiched between a silty sand layer and a gravelly sand layer. The layer experiences 30 mm settlement in 2 years.

Given

$$T_V = \begin{cases} \dfrac{\pi}{4}\left(\dfrac{U}{100}\right)^2 & \text{for } U \le 60\% \\ 1.781 - 0.933\log_{10}(100 - U) & \text{for } U > 60\% \end{cases}$$

Where T_V is the time factor and U is the degree of consolidation in %.

If the coefficient of consolidation of the layer is 0.003 cm^2/s, the deposit will experience a total of 50 mm settlement in the next ______ years.

11. A non-homogeneous soil deposit consists of a silt layer sandwiched between a fine-sand layer at top and clay layer below. Permeability of the silt layer is 10 times the permeability of the clay layer and one-tenth of the permeability of the sand layer. Thickness of the silt layer is 2 times the thickness of the sand layer and two-third of the thickness of the clay layer. The ratio of equivalent horizontal and equivalent vertical permeability of the deposit is ________.

12. A square footing ($2m \times 2m$) is subjected to an inclined point load, P as shown in the figure below. The water table is located well below the base of the footing. Considering one-way eccentricity, the *net safe* load carrying capacity of the footing for a factor of safety of 3.0 is________ kN.

The following factors may be used.

Bearing capacity factors : $N_q = 33.3$, $N_\gamma = 37.16$; Shape factors : $F_{qs} = F_{\gamma s} = 1.314$; Depth factors: $F_{qd} = F_{\gamma d} = 1.113$; Inclination factors: $F_{qi} = 0.444$, $F_{\gamma i} = 0.02$

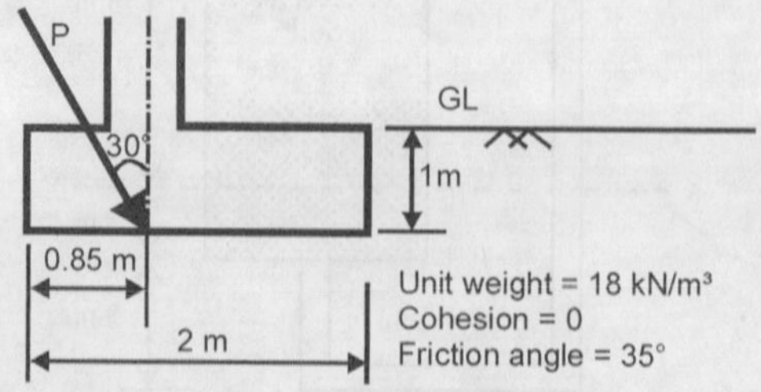

13. A 588 cm^3 volume of moist sand weighs 1010 gm. Its dry weight is 918 gm and specific gravity of solids, G is 2.67. Assuming density of water as 1 gm/cm^3, the void ratio is ________.

14. A 4 m thick layer of normally consolidated clay has an average void ratio of 1.30. Its compression index is 0.6 and coefficient of consolidation is 1 m^2/yr. If the increase in vertical pressure due to foundation load on the clay layer is equal to the existing effective overburden pressure, the change in the thickness of the clay layer is ______ mm.

15. A pile of diameter 0.4 m is fully embedded in a clay stratum having 5 layers, each 5 m thick as shown in the figure below. Assume a constant unit weight of soil as 18 kN/m^3 for all the layers.

Using λ -method ($\lambda = 0.15$ for 25 m embedment length) and neglecting the end bearing component, the ultimate pile capacity (in kN) is ________.

2014

16. The following data are given for the laboratory sample.

$\sigma_o' = 175$ kPa; $e_o = 1.1$; $\sigma_o' + \Delta\sigma_o' = 300$kPa; $e = 0.9$

If thickness of the clay specimen is 25mm, the value of coefficient of volume compressibility is ________ $\times 10^{-4}$ m^2 / kN

17. The flow net constructed for the dam is shown in the figure below. Taking the coefficient of permeability as 3.8×10^{-6} m/s, the quantity of flow (in cm^3/s) under the dam per meter of dam is ________.

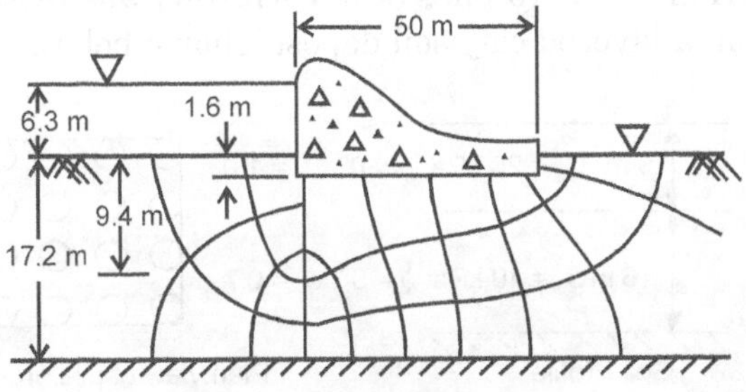

Impermeable stratum

18. A certain soil has the following properties: $G_s = 2.71$, $n = 40\%$ and $w = 20\%$. The degree of saturation of the soil (rounded off to the nearest percent) is ________

19. Water is flowing at a steady rate through a homogeneous and saturated horizontal soil strip of 10 m length. The strip is being subjected to a constant water head (H) of 5m at the beginning and 1 m at the end. If the governing equation of flow in the soil strip is $\frac{d^2H}{dx^2} = 0$ (where x is the distance along the soil strip), the value of H (in m) at the middle of the strip is ________

20. An infinitely long slope is made up of a c-φ soil having the properties: cohesion (c) = 20 kPa, and dry unit weight (γ_d) = 16 kN/m³. The angle of inclination and critical height of the slope are 40° and 5 m, respectively. To maintain the limiting equilibrium, the angle of internal friction of the soil (in degrees) is ________

21. A single vertical friction pile of diameter 500 mm and length 20 m is subjected to a vertical compressive load. The pile is embedded in a homogeneous sandy stratum where: angle of internal friction (φ) = 30°, dry unit weight (γ_d) = 20 kN/m³ and angle of wall friction (δ) = $2\varphi/3$. Considering the coefficient of lateral earth pressure (K) = 2.7 and the bearing capacity factor (N_q) = 25, the ultimate bearing capacity of the pile (in kN) is ________

2002

22. Soil has been compacted in an embankment at a bulk density of 2.15 Mg/m³ and a water content of 12%. The value of specific gravity of soil solids is 2.65. The water table is well below the foundation level. Estimate the dry density, void ratio, degree of saturation and air content of the compacted soil.

23. A retaining wall with a stratified backfill and a surcharge load is shown in the following figure. Draw earth pressure diagram detailing the values at the critical points. Also estimate the resultant thrust on the wall and its position.

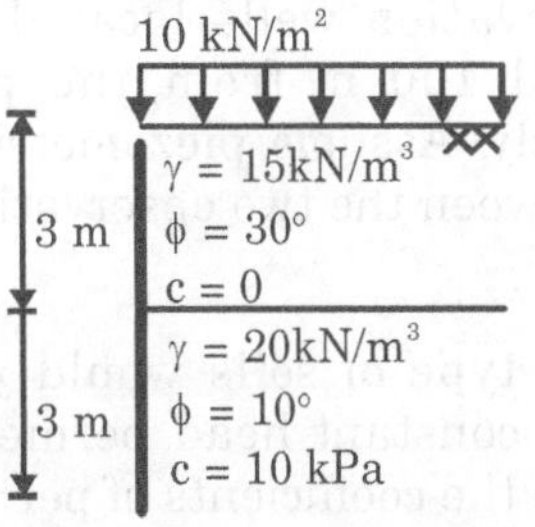

24. A group of 16 piles (diameter = 50 cm, length = 14m, centre to centre spacing = 1m) arrange in a square pattern passes through a recent fill (thickness = 3 m) overlying a soft clay deposit (thickness = 5m) which is consolidating under the fill load and rests in a stiff clay strata. All the strata are saturated. The soil properties of different strata are given below

Type of Sole	Unit Weight (γ) (kN/m³)	Strength Parameters		Adhesion Parameter (α)
		C_u(kPa)	ϕ_u	
Fill	16	50	0	0.60
Soft Clay	17	20	0	0.40
Stiff Clay	21	70	0	0.45

Estimate ultimate load carrying capacity (Q_g) of the pile group.

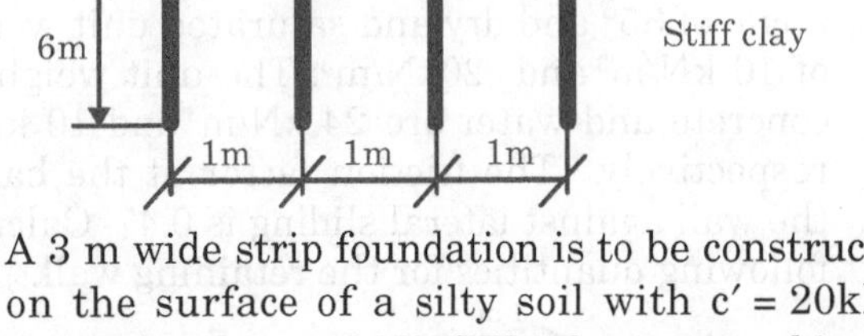

25. A 3 m wide strip foundation is to be constructed on the surface of a silty soil with c′ = 20kPa, ϕ' = 30° and γ = 18kN/m³. The footing is subjected to a vertical load of 200 kN/m run of wall at an eccentricity of 0.25 m together with a horizontal force of 50 kN/m run of wall. Assuming that the water table is well below the foundation level, estimate ultimate bearing capacity of the foundation.

26. A 30-cm diameter pumping well starts to pump water at 6:00 AM on a day at a rate of 2000 L/min from a confined aquifer (thickness = 30m, permeability = 30 m/day, and storage coefficient = 0.005). Find out slope of the hydraulic gradient at 9:00 PM on the same day between two observation wells located at distance of 50 m and 100 m from the pumping well, respectively. Assume piezometric surface to be linear between the two observation wells.

2001

27. For what type of soils would you use falling head and constant head permeability tests to determine the coefficients of permeability

28. Derive the formula to estimate the coefficient of permeability from falling head permeability test.

29. Estimate the flow quantity (in litres per second) through the soil in the pipe shown below. The pressure heads at two locations are shown in the figure. The internal diameter of the pipe is 1 m and coefficient of permeability of soil is 1×10^{-5}m/sec.

30. Referring to consolidated-undrained (C-U) triaxial compression tests, sketch the total stress and the corresponding effective stress Mohr circles along with the failure envelopes for the following soils.

(*i*) Normally consolidated clays

(*ii*) Over consolidated clays

Clearly mark total stress and the effective stress shear strength parameters on the figures. With respect to the slope stability analysis of embankments, when are the total and effective stress parameters used.

31. A C-U triaxial compression test was performed on saturated sand at a cell pressure of 100 kPA. The utlimate deviator stress was 350 kPa and the pore pressure at the peak stress was — 40 kPs (suction). Estimate total and effective stress shear strength parameters.

32. A concrete gravity type retaining wall, shown below, retains granular soil having a friction angle of 35° and dry and saturated unit weights of 16 kN/m³ and 20kN/m³. The unit weights of concrete and water are 24 kN/m³ and 10 kN/m³ respectively. The friction factor at the base of the wall against lateral sliding is 0.47. Calculate following quantities for the retaining wall.

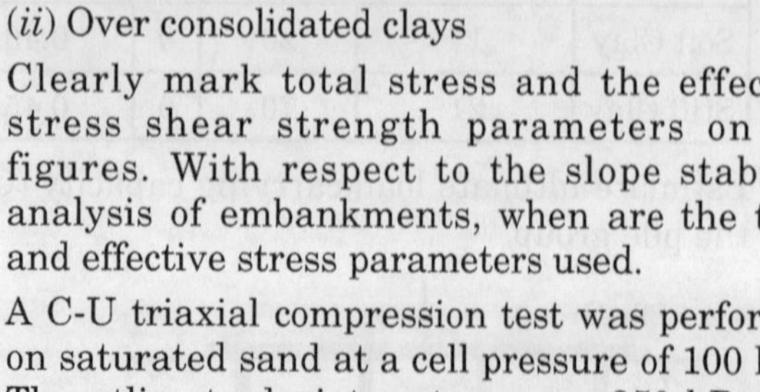

(*i*) Factor of safety against lateral sliding,

(*ii*) Factor of safety against overturning, and

(*iii*) Bearing pressure on foundation soil using Meyerhoffs method.

33. A group of 16 piles (4 in each row) was installed in a layered clay soil deposit shown below.

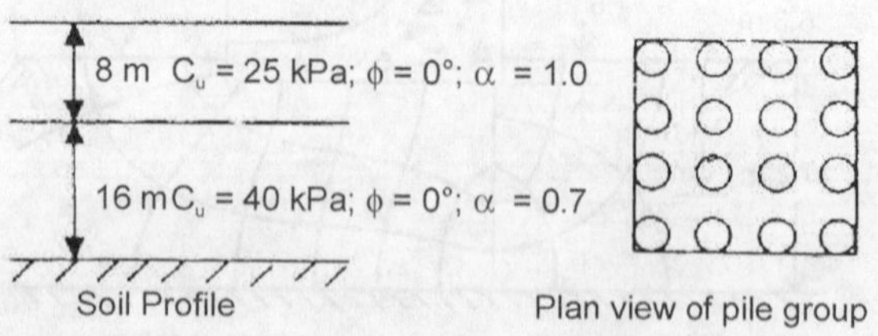

The diameter of each pile is 500 mm and their c/c distance is 1m. The length of the pile group is 18m. Estimate safe load capacity of the group with a factor of safety of 2.50. The adhesion factors (α) between pile and the soil in each soil layer are shown in the figure.

34. A spillway has its crest at elevation of + 144.0m and a horizontal apron at an elevation of + 105.0 m on the downstream side. Find tail water elevation required to form a hydraulic jump when elevation of energy line above the crest is +146.5 m. The C_d for the flow can be assumed as 0.73. The energy loss over the spillway face may be neglected.

35. A well of 300 mm diameter is located in a confined aquifer of 40m thick. The aquifer has a hydraulic conductivity of 25 m/day and the radius of influence is 300m. Determine discharge in m³/hour from the well if the draw down is 3m. Also determine percentage increase in the discharge if diameter is made to 450 mm, with other conditions remain same.

2000

36. A 5m thich clay layer lies between two layers of sand each 4m thick, the top of the upper layer of sand being at ground level. The water table is 2m below the piezometric surface being 4m above ground level. The saturated unit weight of clay is 20 kN/m³ and that of sand is 19 kN/m³. Above the water table unit weight of sand is 16.5 kN/m³. Calculate effective vertical stresses at the top and bottom of the layer. Also draw the total vertical stress diagram in the given soil layers.

37. In an oedometer test, a specimen of saturated clay 19 mm thick reaches 50% consolidation in 20 minutes. How long it would take a layer of this clay 5m thick to reach the same degree of consolidation under the same stress and drainage conditions ? How lond would it take for the clay layer to reach 30% consolidation ?

38. A footing 3m square carries a gross pressure of 350 kN/m^2 at a depth of 1.2 m in sand. The saturated unit weight of sand is 20 kN/m^3 and the unit weight above the water table is 17kN/m^3. The shear strength parameters are $c' = 0$ and $\phi' = 30°$. (For $\phi' = 30°$, $N_q = 22$ and $N_g = 20$). Determine the factor of safety with respect of shear failure for the following cases:

I. Water table is 5m below ground level.

II. Water table is at 1.2 m below ground level.

39. For the retaining wall shown in the figure below assume that the wall can yield sufficiently to develop active stage. Use Rankine's active earth pressure theory and determine

(*a*) active force per meter of the wall, and

(*b*) location of the resultant line of action.

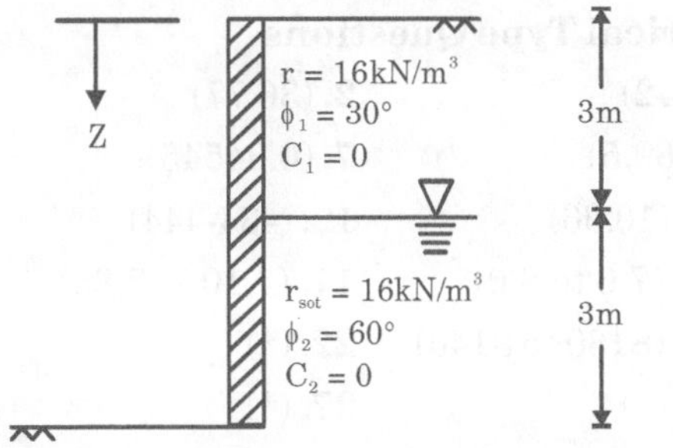

40. A pump is to deliver water at a rate of 1.4 m^3/s from a sump to a reservoir which is 10 km away through a steel pipe 1200 mm in diameter for a length of 5 km and the remaining through a PSC pipe of diameter 1000 mm. The elevation difference between full tank level of the delivery reservoir and the water level of the sump is 50 m. What is the Break Horse Power (in kW) requirement of the motor ? Assume a pump house loss of 3.75 m and minor losses to be 10% of the friction losses. Use Darcy-Weisbach friction factor (f) of 0.018 for steel pipe and 0.02 for PSC pipe with a pump efficiency of 80%.

ANSWERS

MCQ Type Questions

1. (*b*)	**2.** (*d*)	**3.** (*d*)	**4.** (*a*)	**5.** (*b*)	**6.** (*d*)	**7.** (*c*)	**8.** (*b*)	**9.** (*b*)	**10.** (*b*)
11. (*d*)	**12.** (*c*)	**13.** (*c*)	**14.** (*a*)	**15.** (*a*)	**16.** (*a*)	**17.** (*d*)	**18.** (*b*)	**19.** (*a*)	**20.** (*d*)
21. (*a*)	**22.** (*b*)	**23.** (*a*)	**24.** (*d*)	**25.** (*b*)	**26.** (*d*)	**27.** (*a*)	**28.** (*d*)	**29.** (*a*)	**30.** (*d*)
31. (*d*)	**32.** (*a*)	**33.** (*b*)	**34.** (*c*)	**35.** (*c*)	**36.** (*c*)	**37.** (*d*)	**38.** (*a*)	**39.** (*a*)	**40.** (*d*)
41. (*a*)	**42.** (*b*)	**43.** (*b*)	**44.** (*b*)	**45.** (*a*)	**46.** (*a*)	**47.** (*d*)	**48.** (*a*)	**49.** (*b*)	**50.** (*b*)
51. (*d*)	**52.** (*a*)	**53.** (*a*)	**54.** (*a*)	**55.** (*c*)	**56.** (*b*)	**57.** (*b*)	**58.** (*c*)	**59.** (*d*)	**60.** (*b*)
61. (*b*)	**62.** (*d*)	**63.** (*b*)	**64.** (*b*)	**65.** (*b*)	**66.** (*a*)	**67.** (*c*)	**68.** (*b*)	**69.** (*b*)	**70.** (*c*)
71. (*c*)	**72.** (*a*)	**73.** (*b*)	**74.** (*b*)	**75.** (*a*)	**76.** (*a*)	**77.** (*b*)	**78.** (*b*)	**79.** (*b*)	**80.** (*a*)
81. (*a*)	**82.** (*c*)	**83.** (*d*)	**84.** (*a*)	**85.** (*a*)	**86.** (*a*)	**87.** (*a*)	**88.** (*a*)	**89.** (*d*)	**90.** (*b*)
91. (*a*)	**92.** (*d*)	**93.** (*c*)	**94.** (*d*)	**95.** (*c*)	**96.** (*c*)	**97.** (*b*)	**98.** (*d*)	**99.** (*b*)	**100.** (*a*)
101. (*c*)	**102.** (*b*)	**103.** (*b*)	**104.** (*d*)	**105.** (*b*)	**106.** (*c*)	**107.** (*a*)	**108.** (*d*)	**109.** (*b*)	**110.** (*b*)
111. (*c*)	**112.** (*b*)	**113.** (*b*)	**114.** (*b*)	**115.** (*b*)	**116.** (*a*)	**117.** (*b*)	**118.** (*b*)	**119.** (*b*)	**120.** (*c*)
121. (*c*)	**122.** (*a*)	**123.** (*c*)	**124.** (*d*)	**125.** (*c*)	**126.** (*b*)	**127.** (*b*)	**128.** (*a*)	**129.** (*b*)	**130.** (*d*)
131. (*c*)	**132.** (*b*)	**133.** (*c*)	**134.** (*c*)	**135.** (*b*)	**136.** (*a*)	**137.** (*a*)	**138.** (*b*)	**139.** (*d*)	**140.** (*b*)
141. (*b*)	**142.** (*c*)	**143.** (*c*)	**144.** (*c*)	**145.** (*c*)	**146.** (*b*)	**147.** (*a*)	**148.** (*b*)	**149.** (*a*)	**150.** (*d*)
151. (*b*)	**152.** (*d*)	**153.** (*b*)	**154.** (*a*)	**155.** (*c*)	**156.** (*b*)	**157.** (*b*)	**158.** (*d*)	**159.** (*a*)	**160.** (*d*)
161. (*c*)	**162.** (*d*)	**163.** (*b*)	**164.** (*a*)	**165.** (*d*)	**166.** (*b*)	**167.** (*c*)	**168.** (*d*)	**169.** (*a*)	**170.** (*a*)
171. (*b*)	**172.** (*a*)	**173.** (*b*)	**174.** (*d*)	**175.** (*d*)	**176.** (*c*)	**177.** (*d*)	**178.** (*b*)	**179.** (*d*)	**180.** (*c*)
181. (*d*)	**182.** (*c*)	**183.** (*b*)	**184.** (*a*)	**185.** (*d*)	**186.** (*c*)	**187.** (*a*)	**188.** (*a*)	**189.** (*d*)	**190.** (*d*)
191. (*b*)	**192.** (*a*)	**193.** (*b*)	**194.** (*d*)	**195.** (*a*)	**196.** (*c*)	**197.** (*b*)	**198.** (*a*)	**199.** (*d*)	**200.** (*b*)
201. (*a*)	**202.** (*a*)	**203.** (*d*)	**204.** (*b*)	**205.** (*c*)	**206.** (*c*)	**207.** (*d*)	**208.** (*c*)	**209.** (*a*)	**210.** (*c*)
211. (*b*)	**212.** (*a*)	**213.** (*d*)	**214.** (*c*)	**215.** (*c*)	**216.** (*d*)	**217.** (*c*)	**218.** (*b*)		

Numerical Type Questions

1. (42)	**2.** (36.77)	**3.** (65.475)	**4.** (104.37)	**5.** (500)
6. (69.5)	**7.** (0.04545)	**8.** (298.48)	**9.** (53.236)	**10.** (4 to 5)
11. (10.96)	**12.** (434-444)	**13.** (0.71)	**14.** (314)	**15.** (1625.774)
16. (7.6 to 8.0)	**17.** (7.10 to 7.85)	**18.** (81.0 to 81.5)	**19.** (3.0 to 3.0)	**20.** (21.0 to 23.0)
21. (8130 to 8140)	**22.** (*)	**23.** (*)	**24.** (*)	**25.** (*)
26. (*)	**27.** (*)	**28.** (*)	**29.** (*)	**30.** (*)
31. (*)	**32.** (*)	**33.** (*)	**34.** (*)	**35.** (*)
36. (*)	**37.** (*)	**38.** (*)	**39.** (*)	**40.** (*)

"*" Refers to Subjective Question.

EXPLANATIONS

MCQ TYPE QUESTIONS

2. $w_P = 26\%$, $w_L = 48\%$

Since w_L lies in the range of 35 – 50%, it is intermediate compressible. Hence soils is CI.

4. Total active force, $F_a = \frac{1}{2} \times 40 \times 6 = 120$ kN

Taking moments about P

$F_a \times 2 = W \times 2$

Therefore, $W = F_a = 120$ kN

5. Equivalent hydraulic conductivity

$$k = \frac{Z_1 + Z_2 + Z_3}{\frac{Z_1}{K_1} + \frac{Z_2}{K_2} + \frac{Z_3}{K_3}}$$

$$= \frac{20 + 30 + 10}{\frac{20}{2} + \frac{30}{3} + \frac{10}{1}} = 2 \text{ m/day}$$

$$Q = k.i.A = K.\frac{h}{L}.A$$

$$= \frac{2 \times (15 - 10)}{60} \times 3 \times 1 = \frac{1}{2} = 0.5$$

6. Bearing capacity of clayey soild does not depend upon the size of footing.

Therefore, the bearing capacity remains same.

7.
$$i = \frac{G - 1}{1 + e}$$

$$n = 30\%$$

$$e = \frac{n}{1 - n}$$

$$= \frac{0.3}{0.7} = \frac{3}{7} = 0.428$$

Therefore, $G = i(1 + e) + 1$

$= i(1 + 0.428) + 1$

$= 1.43\,i + 1$

8.

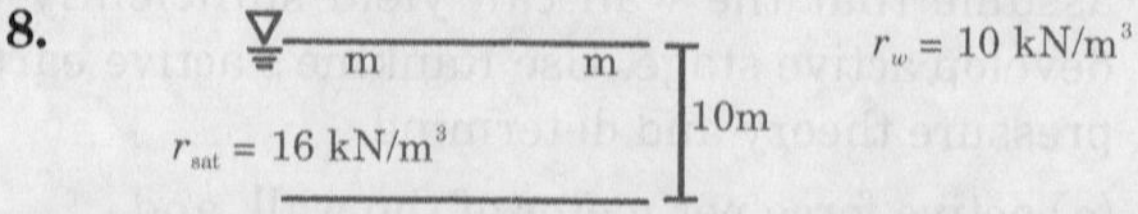

Pre-consolidation pressure = 90 kPa

Effective stress $= r_{sat} \,.\, 10 - r_w \,.\, 10$

$= 16 \times 10 - 10 \times 10$

$= 60$ kN/m²

Over-consolidation ratio

$$= \frac{\text{Pre-consolidation pressure}}{\text{Current effective stress}}$$

$$= \frac{90}{30} = \frac{3}{2}$$

$= 1.5$

9. Because energy imparted to the mould in case of standard proctor test is less as compared to modified proctor test. Therefore, more compaction is obtained for modified proctor test and hence high maximum dry density.

Similarily the moisture content at which maximum compaction is obtained will also be less for modified proctor test.

10. Discharge velocity (v) is quantity of water flowing in unit time through a unit gross cross-sectional area of soil at right angle to the direction of flow.

Seepage velocity is an average velocity through pore voids.

$$q = vA = v_S A_V$$

where $q \rightarrow$ discharge

$v \rightarrow$ discharge velocity

$v_S \rightarrow$ seepage velocity

$A \rightarrow$ area of soil

$A_V \rightarrow$ area of voids

Significantly, seepage velocity is always greater than discharge velocity.

11. Liquefaction is the phenomenon where saturated or partially saturated soil foses strength and stiffness in response to an applied stress, usually earthquake shaking or other sudden change in stress condition, causing it to behave like a liquid.

12. $I_P = W_L - W_P = (40 - 28)\% = 12\%$

$$\text{Now Activity} = \frac{I_P}{C} = \frac{12}{100-60} = 0.3$$

13. $$B = \frac{\text{Change in pore pressure}}{\text{Change in cell pressure}}$$

$$= \frac{\Delta U_C}{\Delta \sigma_3} = \frac{80}{250-150} = 0.8$$

15. Volume of embankment = 5000 m^3

Target dry unit weight = 16.2 kN/m^3

Specific gravity of soil = 2.7

Excavation volume of soil required from site 'X'

$$= \frac{V}{1+e_1} = \frac{5000}{1.6}$$

Excavation volume of soil required from site 'Y'

$$= \frac{5000}{1.7}$$

Excavation volume of soil required from site 'Z'

$$= \frac{5000}{1.64}$$

$$\text{Dry density of soil} = \frac{(Gyw)}{1+e} = \frac{2.7\times 10}{1+e}$$

$$= \frac{27}{1+e}\ kN/m^3$$

∵ Minimum volume will be excavated from X site and hence less transportation cost will be needed.

∴ X soil is chosen.

16. Friction Circle method assumes circular slip surface of radius (r). The resultant reaction between the two portions of the soil mass is tangential to a concentric smaller circle of radius of $r \sin\phi$ which is called friction circle.

17. For clay,

$$q_{nu} = {}_C N_C$$

No depth or size factor is present.

18.
$$Se = wG$$
$$S = 100\%$$
$$w = 100$$
$$\therefore \quad e = G$$

19.

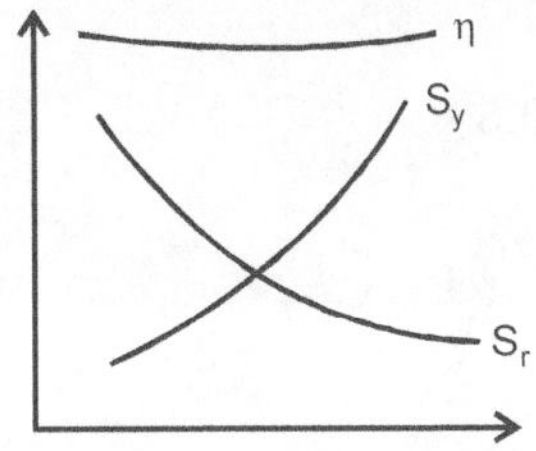

$$S_y + S_r = \eta$$

20. To eliminate tensile crack the surcharge loading

$$= \frac{2C}{\sqrt{K_a}}$$

21.

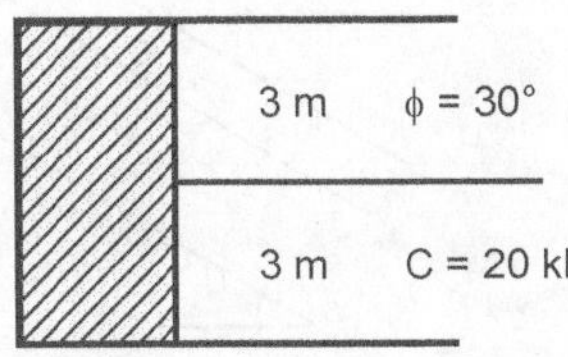

$\gamma = 18\ kN/m^3$

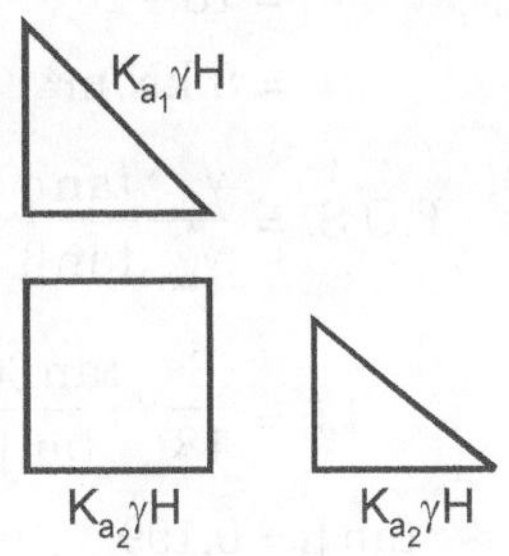

$$K_{a_1} = \frac{1}{3}$$

$$K_{a_2} = 1$$

$$P_a = \frac{K_{a_1}\gamma H^2}{2} + K_{a_2}\gamma H^2 + \frac{K_{a_2}\gamma H^2}{2} - 2CH\sqrt{K_{a_2}}$$

$$= \frac{1}{3}\times\frac{18}{2}\times 3 + (18\times 1\times 9) + \left(1\times\frac{18}{2}\times 9\right) - 2\times 20\times 3\sqrt{1}$$

$$= 150$$

22. $$q = 10\sqrt{3} + 0.5\,p$$

Stress path equation is

$$\frac{\sigma_1 - \sigma_3}{2} = C\cos\phi + \frac{\sigma_1+\sigma_3}{2}\sin\phi$$

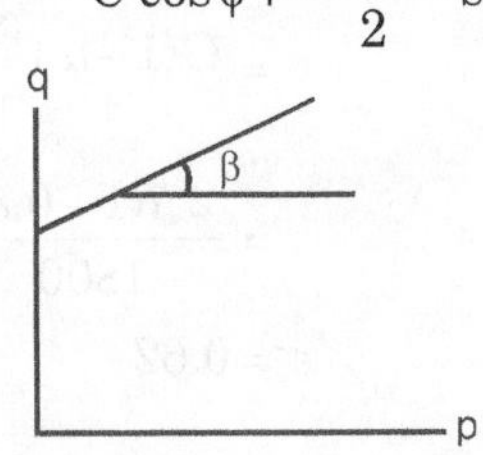

$\therefore \quad C\cos\phi = 10\sqrt{3}$

$\tan\beta = \sin\phi = 0.5$

$\phi = 30°$

$\therefore \quad C\cos\phi = 10\sqrt{3}$

$\Rightarrow \quad C\cos 30 = 10\sqrt{3}$

$C = 20$

24. Unconsolidated undrained test is used for completely saturated cohesive soil. In this test, no drainage is permitted during the first state as well as in the second stage.

26. $c' = 0, \phi' = 34°$

$\gamma_{sat} = 18$ kN/m³

$\gamma_w = 10$ kN/m³

F.O.S. = 1.5

$\gamma_{sub} = \gamma_{sat} - \gamma_w$

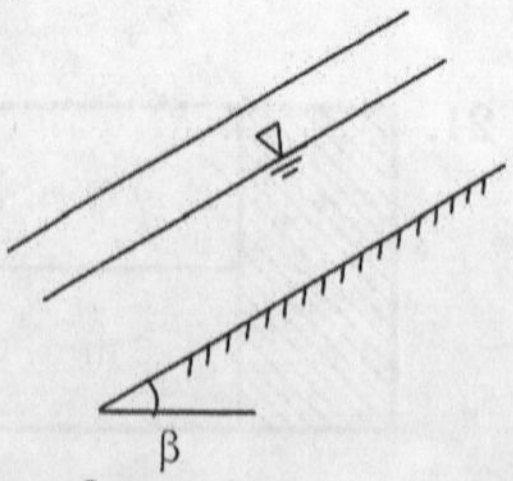

When water table rises to surface

$= 18 - 10$

$= 8$ kN/m³

$$\text{F.O.S.} = \frac{\gamma_{sub}}{\gamma_{sat}}\frac{\tan\phi}{\tan\beta}$$

$$\Rightarrow \quad 1.5 = \frac{8}{18}\times\frac{\tan 30°}{\tan\beta}$$

$$\Rightarrow \quad \tan\beta = 0.199$$

$$\Rightarrow \quad \beta = 11.30°$$

27. K → General shear failure

L → Local shear failure

J → Punching shear failure

General shear failure (Q) : Well define failure pattern(3)

Local shear failure (L) : Rankine passive zone developes(2)

Punching shear failure : No heaving of soil around footing

28. $e_{max} = 0.85, e_{min} = 0.5\ \rho_{field} = 1800$ kg/m³

$w = 8\%, \rho_w = 1000$ kg/m³, $G_s = 2.7$

$$\rho = \frac{G(1+w)}{1+e}\rho_w$$

$$\Rightarrow \quad 1 + e = G(1+w)\frac{\rho_w}{\rho}$$

$$= \frac{2.7(1+0.8)}{1800}\times 1000$$

$$\Rightarrow \quad e = 0.62$$

$$\text{Relative Density} = \frac{e_{max} - e}{e_{max} - e_{min}}\times 100$$

$$= \frac{0.85 - 0.62}{0.85 - 0.5}\times 100$$

$$= 65.71\%$$

29. As per Indian soil classification system (IS : 1498 – 1970)

An expression for A-line is ⇒ $I_P = 0.73(\omega_L - 20)$

I_P ⇒ Plasticity Index.

ω_L ⇒ Liquid limit

A-line dividing Inorganic clay from silt and organic soil from this equation.

Correct choise is (*a*).

30. Montmorillonite clay mineral primarily governing the swelling behavior of Black cotton soil. (It exhibits high shrinkage & swelling)

Kaolinite ⇒ shows relatively little swell on wetting

Illite ⇒ structure does not swell because of movement of water between the sheets.

So, correct choice is (*d*).

31. Minimum at centre & maximum at Edge.

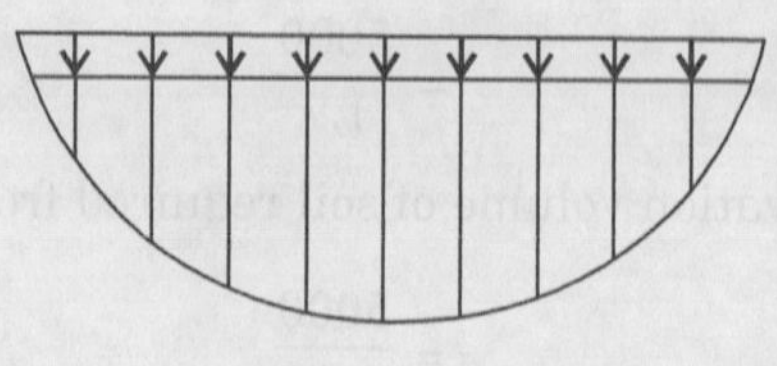

(b) Cohession less Sand

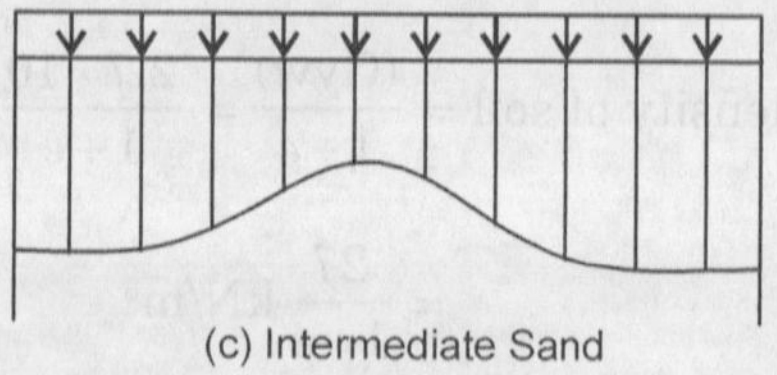

(c) Intermediate Sand

Fig. Contact pressure distribution under rigid footing.

32. Pressuremeter test – Menard's Modulus (E_m)

Static cone Penetration test – Skin Resistance (f_c)

Standard Penetration test – Number of blows (N)

Vane shear test – It is undrained cohesion (c_k)

the correct choice is (*a*)

34. Find the degree of saturation 'S_r' of a soil in natural condition.

$$\gamma = \frac{W}{V} = \frac{1.980\times 9.81}{0.001} = 19.42 \text{ kN/m}^3$$

$$\gamma_d = \frac{W_d}{V} = \frac{1.800\times 9.81}{0.001} = 17.658 \text{ kN/m}^3$$

$$\gamma_d = \frac{G.\gamma_w}{1+l}, \text{ where G = 2.7 given}$$

$$\gamma_w = 9.81 \text{ kN}/m^3$$

$$\therefore \quad 17.658 = \frac{2.7 \times 9.81}{1+l}$$

Voide ratio, $$l = \frac{2.7 \times 9.81}{17.658} - 1$$

$$l = 1.5 - 1$$

$$= 0.5$$

$$l = 0.5$$

In natural condition, $\gamma = \dfrac{\gamma_w (G + l \cdot S_r)}{1+l}$

Here, only S_r is known

$$19.42 = \frac{9.81(2.7 + 0.5 \times S_r)}{1+0.5}$$

$$29.13 = 9.81(2.7 + 0.5 \times S_r)$$

$$2.969 - 2.7 = 0.5 \times S_r$$

$$S_r = 0.5388$$

$\because$ Degree of saturation $S_r = 0.54$

35. The shape factor depends upon the dimenssion of flow space.

36. Calculate the quantity of water flowing into the lake from the sand layer through the silt layer per unit area of the lake bed.

By Darcy's law

$$q = \text{K i A}$$

We want q per unit area

$$\therefore \quad q = \frac{\text{K i A}}{\text{A}} = \text{K.i}$$

K = Co-efficient of permiability

$= 10^{-6}$ m/s

i = hydraulic gradient $= \dfrac{L}{h}$

L = Length of silt layer in direction of flow

L = 20 m. (from figure)

h = difference in water levels of the overhead and bottom of lake.

= 20 m. (from figure)

$$\therefore \quad q = \text{K. i} = \text{K} \times \frac{L}{h}$$

$$q = 10^{-6} \times \frac{20}{20}$$

$$q = 1 \times 10^{-6} \text{ m}^3/\text{sec}$$

39.

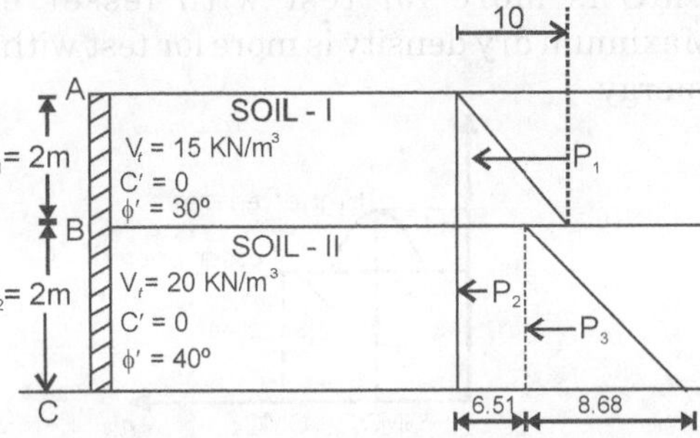

Calculate the co-efficient of active earth pressure for SOIL-I and SOIL-II, i.e. K_{a_1} and K_{a_2} respectively

$$K_{a_1} = \frac{1-\sin\phi}{1+\sin\phi} = \frac{1-\sin 30^\circ}{1+\sin 30^\circ} = \frac{0.5}{1.5} = \frac{1}{3} = 0.333$$

$$K_{a_2} = \frac{1-\sin 40^\circ}{1+\sin 40^\circ} = \frac{0.357}{1.643} = 0.217$$

Calculate the intensities and make the pressure diagram.

SOIL-I

Intensity of pr. at A, $p_A = 0$

Intensity of pr. at B, $p_B = K_a \cdot V_t \cdot H_1$

$$p_B = 0.3333 \times 15 \times 2$$

$$= 9.99 = 10 \text{ kN/m}$$

SOIL-II

This soil layer is fully under the SOIL-I, so SOIL-I is to be treated as a uniform surcharge of intensity q_1, where

$$q_1 = V_t \times H_1 = 15 \times 2.0 = 30 \text{ kN/m}^2$$

$\therefore$ pr. intensity at B due to surcharge load coming from soil - I.

$$p_B = K_{a_2}\, q_1 = 0\ 217 \times 30 = 6.51 \text{ kN/m}^2$$

pr. Intensity at C, $p_C = K_{a_2} \cdot V_t \cdot H_2$

$$= 0.217 \times 20 \times 2 = 8.68$$

Computation of active earth pr. per m length

$$P_1 = \frac{1}{2} \times 10 \times 2 = 10 \text{ kN/m}$$

$$P_2 = 6.51 \times 2 = 13.02 \text{ kN/m}$$

$$P_3 = \frac{1}{2} \times 8.68 \times 2 = 8.68 \text{ kN/m}$$

Total resultant active earth force/pr per unit length P_a

$$P_a = P_1 + P_2 + P_3 = 10 + 13.02 + 8.68 = 31.7 \text{ kN/m}$$

41. $\tau = c + \sigma \tan\phi$

For cohesionless soils, $c = 0$

$\tan\phi = \tan 38^\circ = 0.781$

$$\therefore \quad \tau = \sigma \times 0.781$$

$$\Rightarrow \quad \frac{\tau}{\sigma} = 0.781$$

42. OMC is more for test with lesser energy. Maximum dry density is more for test with higher energy

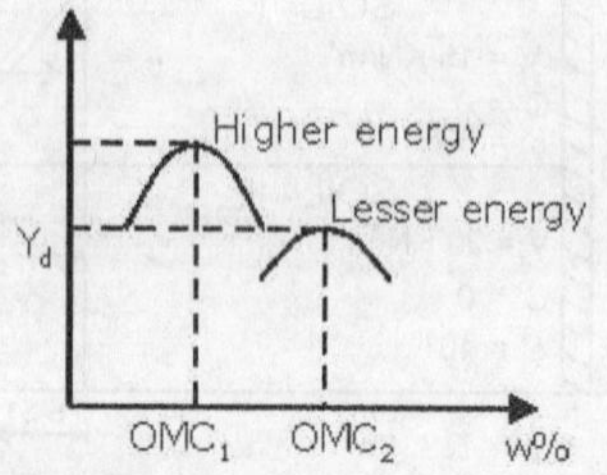

43. LL = 40% and PI = 28%

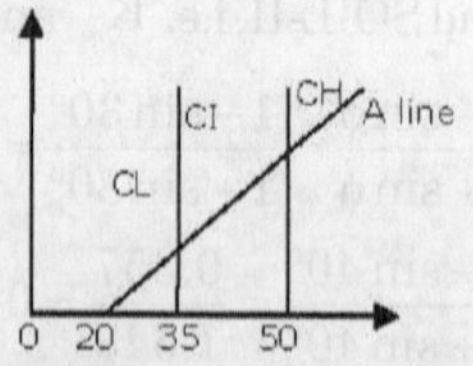

As per IS CODE,

A line, $\quad PI = 0.73\,(LL - 20)$

$= 14.6 < (28.1 \text{ given})$

So, above A line and 20% < LL = 40% < 50%

So, 'CI'.

46. For falling head test,

$$\ln\left(\frac{h_1}{h_2}\right) = \frac{kAt}{\partial L}$$

$$\frac{h_1}{h_2} = e^{kd} \quad \left[\alpha = \frac{At}{\partial L} = \text{constant}\right]$$

First specimen,

$$R_1 = \frac{h_1}{h_2} = 1.25 = e^{kd}$$

$$R_2 = \left(\frac{h_1}{h_2}\right) = e^{5kd} = (e^{kd})$$

$$= (R_1)^5 = (1.25)^5 = 3.05$$

47. At $\quad P, h = 1.5$

At $\quad R, h = 2.6$

So at $\quad Q, h = 2.05$

Hence effective stress

$$= \sigma - \upsilon = q_{sat} \times H - q_{\omega} \times h$$

$$= 18 \times 2 - 9.81 \times 2.05 = 15.89 \text{ kN/m}^2$$

49. Critical hydraulic gradient,

$$i_{cr} = \frac{G-1}{1+R} = \frac{2.67-1}{1+0.7} = 0.98$$

50.

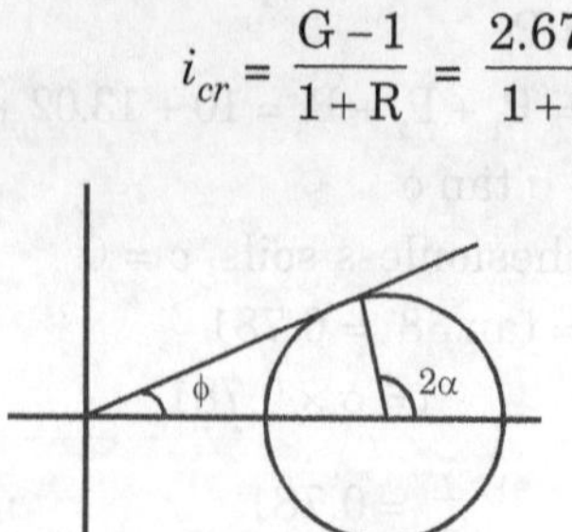

We know, $\quad 2\alpha = 90° + \phi$

$$\alpha = 45° + \frac{\phi}{2}$$

53. By Stoke's law

$$\text{terminal velocity } V = \frac{D^2}{18\mu}\,(\omega_s - \omega)$$

$$\therefore \quad V \propto D^2$$

$$\Rightarrow \quad \frac{V_1}{V_2} = \left(\frac{D_1}{D_2}\right)^2$$

$$\therefore \quad \frac{0.5}{V_2} = \left(\frac{0.075}{0.0075}\right)^2 = 100$$

$$\Rightarrow \quad V_2 = 0.005 \text{ mm/sec}$$

54. From IS 800 : 1934

(*i*) Vertical stiffness prevent the local buckling of the web.

(*ii*) Horizontal stiffness should be placed on the web at a distance from the compression flange equal to 2/5 of the distance from the compression flange to neutral axis.

So Correct choice is (*b*)

55.

$$T_v = \frac{C_v.t}{H^2}$$

$$\Rightarrow \quad t \propto H^2$$

$$\therefore \quad \frac{t_1}{t_2} = \left(\frac{H_1}{H_2}\right)^2$$

$$\Rightarrow \quad \frac{36}{t_2} = \left(\frac{10}{5}\right)^2 = 4$$

$$\Rightarrow \quad t_2 = 9 \text{ months.}$$

Time rate of considration settlement will be appreciafic altered because of reduced drainage path which becomes half in last *y* with respect to XSIH.

56.

$$T = c_u \times \pi\left[\frac{d^2R}{2} + \frac{d^3}{12}\right]$$

$$4.6 = c_u \times \pi\left[\frac{0.05^2 \times 0.1}{2} + \frac{0.05^3}{12}\right]$$

$$\Rightarrow \quad c_u = 10812.7 \text{ N/m}^2$$

$\therefore$ Undrained shear strength, $R = c_u$

$$= 10.81 \text{ kPa}$$

57. We know $\quad Se = wG$

$$\Rightarrow \quad e = \frac{wG}{S}$$

For zero air voids, soil must be fully saturated and hence S = 1

$$\therefore \quad e = wG$$

Hence $\quad \gamma_d = \dfrac{G\gamma_w}{1+e} = \dfrac{G\gamma_w}{1+wG}$

58.

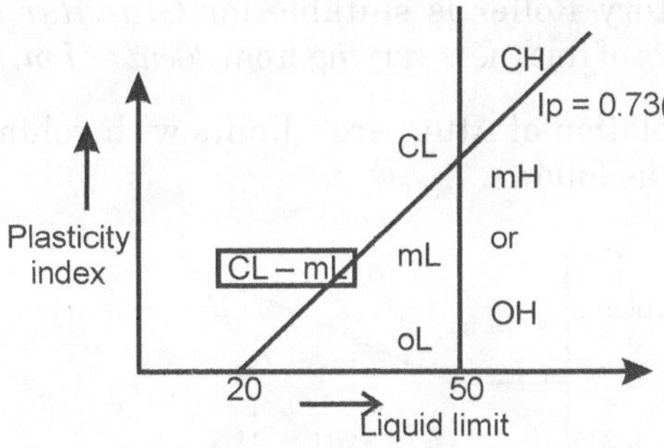

Given : Liquit limit, $w_L = 60$

Plasticity index

From chart,

$$I_{Pc} = 0.73\,(w_L - 20) = 0.73 \times (40) = 29.2$$

and $\quad I_P = 60 - 20 = 40$

$\therefore \quad I_P > I_{Pc}$

Hence it is CH.

59. Quick sand occurs when net effective pressure is zero

i.e. $\quad \bar{\sigma} = 0$

$\therefore \quad \gamma_{sub}.L - \gamma_w h = 0$

But $\quad P_s = q_w.\,h$

$\therefore \quad \gamma_{sub}.L = P_s$

Soquick sand condition occurs when upward seepage pressure in soil becomes equal to the submerged unit weight of the soil.

60. This condition represents over consolidated clay.

61. Coefficient of earth pressure at rest is equal to the ratio of effective horizontal stress to the effective vertical stress.

$$\sigma_h' = k_o \gamma z$$

$$\Rightarrow \quad \sigma_h' = k_o \sigma_v'$$

$$\Rightarrow \quad k_o = \frac{\sigma_h'}{\sigma_v'}$$

62.

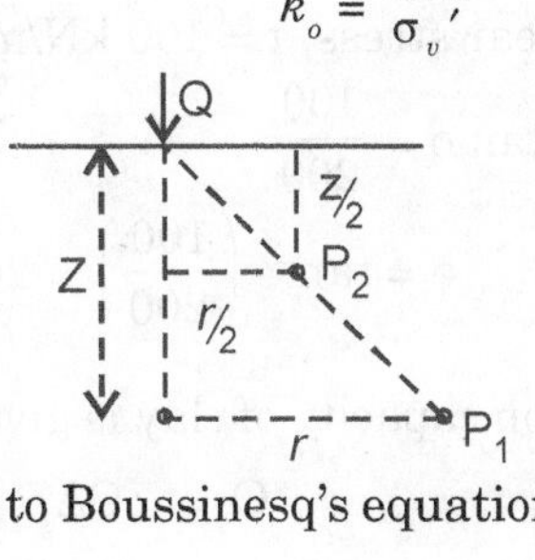

According to Boussinesq's equation

$$\sigma_2 = I_B \frac{Q}{z^2}$$

where, $I_B = \dfrac{3}{2\pi}\left[\dfrac{1}{1+\left(\dfrac{r}{z}\right)^2}\right]^{5/2}$

Given : $(\sigma_z)_{P1} = \sigma_z$

Since value of (r/z) for both the points P_1 and P_2 is same, therefore

$$(I_B)_{P1} = (I_B)_{P2}$$

$$\Rightarrow \quad (\sigma_Z)_{P1} = I_B \cdot \frac{Q}{z^2} = \sigma z$$

$$\Rightarrow \quad (\sigma_Z)_{P2} = I_B \cdot \frac{Q}{(z/2)^2} = 4I_B \cdot \frac{Q}{z^2} = 4\sigma_Z$$

63. Given : $k = 10^{-2}$ m/s

Distance travel = 0.75 m

$$\therefore \quad \text{Time required} = \frac{0.75}{10^{-2}} = 75\,\text{sec} = 75 \text{ sec}$$

64. Ultimate load bearing capacity is given by

$$Q_u = q_b A_b + q_s A_s = 9cA_b + \alpha\bar{c}A_s$$

$$\Rightarrow \quad Q_u = 9c \times B^2 + \alpha\bar{c}\,(uBL) \qquad ...(i)$$

Case I

L = 10m,

B = 500 mm

C = 40 kN/m²,

Q_u = 700 kN

Now $\quad Q_u = 9 \times 40 \times 0.5^2 + \alpha\bar{c}(4 \times 0.5 \times 10)$

$\therefore \quad 700 = 90 + \alpha\bar{c} \times 20$

$\Rightarrow \quad \alpha\bar{c} = 30.5$ kN/m²

Case II

L = 20 m,

B = 250 mm

$\therefore \quad Q_u = 9 \times 40 \times 0.25^2 + 30.5 \times 4 \times 0.25 \times 20 = 22.5 + 610 = 632.5$ kN

65. We know, $\quad Q = \dfrac{2\pi\, bk\,(s_1 - s_2)}{\log_e (r_2/r_1)}$

But $\quad T = bk$

$$\therefore \quad Q = \frac{2\pi\, T\,(s_1 - s_2)}{\log_e (r_2/r_1)}$$

$$\therefore \quad T = \frac{Q \log_e (r_2/r_1)}{2\pi(s_1 - s_2)}$$

$= 2720$ lit/minute

$= 2720 \times 10^{-3} \times 60 \times 24$ m³/day

$= 3916.8$ m³/day

where, t = transmissivity

Q = pumping rate

s_1 & s_2 = hydraulic heads at distances r_1 and r_2 from pumping well.

$$\therefore \quad T = \frac{3916.8 \ln(100/10)}{2\pi(3-0.5)}$$

$= 576$ m²/day

68. Given : W = 50×10^3 N

$h = 1$m

$\eta = 0.6$

S = 4mm

$c = 6$

From Modified Hiley formula

$$Q_v = \frac{Wh\eta}{\left(S + \frac{c}{2}\right)}$$

$$= \frac{50 \times 1 \times 0.6}{\left(4 + \frac{6}{2}\right) \times 10^{-3}}$$

$= 4285.71$ kN

69. Since more than 50% pass through 4.75 mm, so it will be *sand*.
As it contain fine more than 12%, so it will be *silt*
Hence this soil sample is *silty sand* (SM).

71.

Soil	Symbol
Sand	S
Silt	M
Clay	C

Hence for *Silty sand* means sand containing silt, *symbol* is SM *and for clayey sand* means sand containing a considerable percentage of clay, *symbol is SC.*

72.

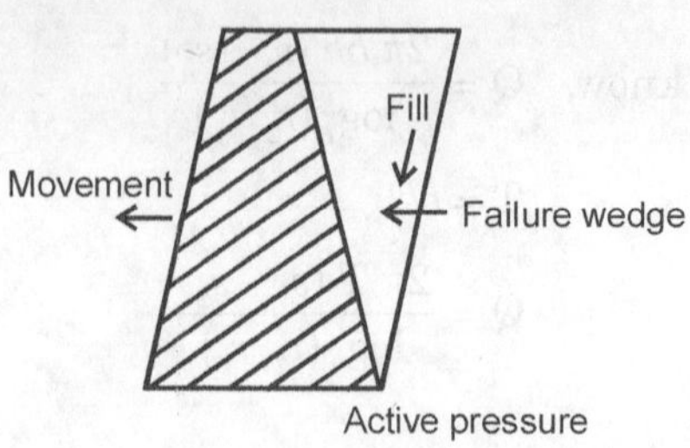

73. Vibratory Roller is suitable for *Granular soils deposits* of thickness varying from *30cm to 1 m.*

74. The relation of Atterberg's limits with volume of soil is as follows.

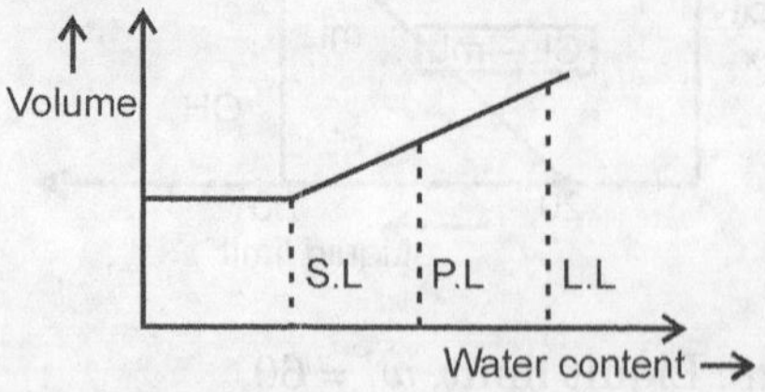

Hence from the graph,

LL > PL > SL

75.

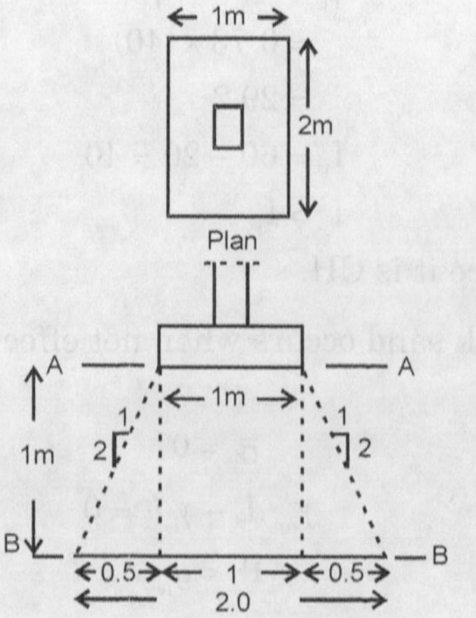

At level A – A,

Plan area = 2 × 1 = 2m²

At level B – B,

Plan area = {(2 + 2 × 0.5) (1 + 2 × 0.5)}

= 6m²

So, pressure intensitiy at level B – B (1m below the A – A)

$= 150 \times \frac{2}{6}$

$= 50$ kN/m²

76. *Given* : C = 0

Normal stress, $\sigma = 200$ kN/m²

Shear stress, $\tau = 100$ kN/m²

$$\therefore \quad \tan\phi = \frac{100}{200}$$

$$\Rightarrow \quad \phi = \tan^{-1}\left(\frac{100}{200}\right) = 26.6°$$

77. Skin friction capacity of clay is given by

$$Q_s = \alpha C A_s$$

where, A_s = effective area in developing friction

$= \pi \times d \times l$

d = diameter of pile

l = length of pile

α = adhesion factor

C = undrained cohesion

$$\therefore \quad Q_s = 0.6 \times 60 \times \pi \times 0.5 \times 10$$
$$= 565.4 \text{ kN}$$
$$\approx 565 \text{ kN}$$

78.

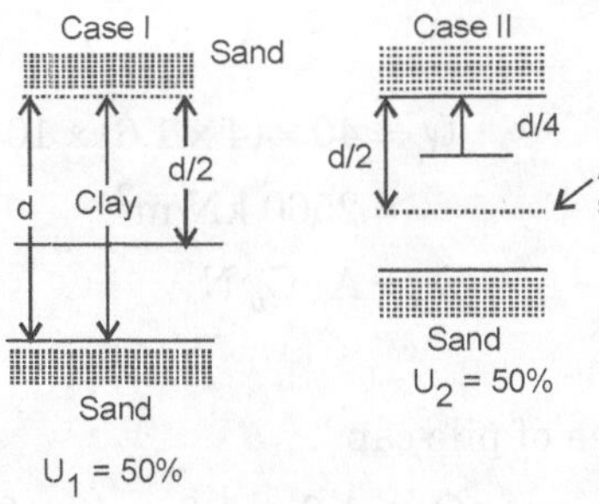

We know, $T_{V_1} = \dfrac{C_V . T_1}{(d/2)^2}$

and $T_{V_2} = \dfrac{C_V . T_2}{(d/4)^2}$

But $U_1 = U_2$, and C_V = constant

$\therefore \quad T_{V_1} = T_{V_2}$

$$\therefore \quad \frac{t_1}{(d/2)^2} = \frac{t_2}{(d/4)^2}$$

$$\Rightarrow \quad t_2 = \frac{t_1}{4}$$

$$= \frac{16}{4} = 4 \text{ years} \quad \text{since } t_1 = 16 \text{ years}$$

79. By plate load test, settlement of foundation in sand given by

$$S_f = S_p\left[\frac{B_f(B_p + 0.3)}{B_p(B_f + 0.3)}\right]^2$$

where, B_p = size of plate
= 30 cm
= 0.3m

B_f = width of footing
= 150 cm
= 1.5 m

S_p = selltement of plate
= 10 mm
= 1 cm
= 0.01m

$$\therefore \quad S_f = 0.01 \times \left[\frac{1.50(.30 + 0.3)}{0.3(1.5 + 0.3)}\right]^2$$
$$= 0.0278 \text{ m} = 27.8 \text{ mm}$$

80.

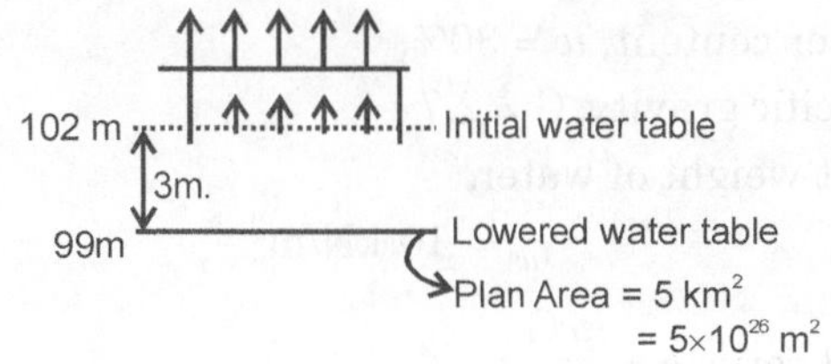

Volume of pumped water = 3×10^6 m^3

So, specific yield

$$= \frac{\text{Pumped out volume}}{\text{Total volume of saturated aquifier}}$$

$$= \frac{3 \times 10^6 m^3}{3 \times 5 \times 10^6 m^3} = 0.20$$

81. In a triaxial test under consolidated drained condition, the pore water pressure is always zero.

82. The first penetration of sampler is for seating of sampler. So it is not considered.
Hence, for SPT test, number of blows for total penetration

$$N = 8 + 10 = 18$$

83.

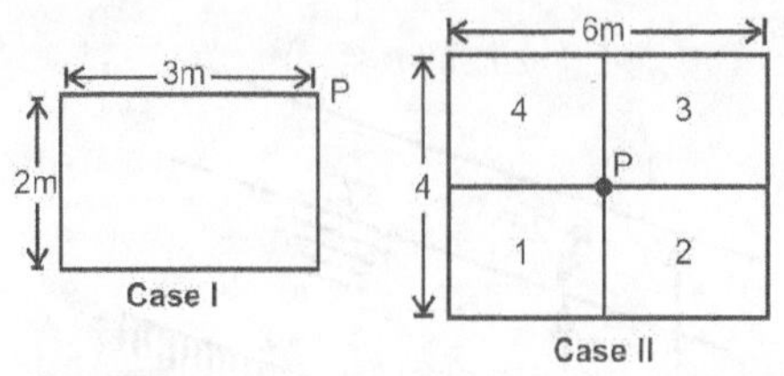

Case I:

Vertical stress (σ_2) at some depth below the corner (P) of 2m × 3m footing due to certain loading

$$(\sigma_2)_1 = 100 \text{ kN/m}^2$$

Case II:

Point P is at the corner of four equal areas of (3m × 2m) 1, 2, 3,4.

Hence vertical stress below point P due to same loading

$$(\sigma_2)_2 = 4 \times (\sigma_2)_2$$
$$= 4 \times 100$$
$$= 400 \text{ kN/m}^2$$

84. According to unified soil classification system, if more than 50% of soil is retained on 0.075 mm sieve, it is *coarse grained soil*.

Again if 50% or more of coarse grained soil is retained on 4.75 mm size, it is *gravel* otherwise it is termed as *sand* (S).

Hence, given sample is *sand*.

According to plasticity chart (USC) if liquid limit is less than 50% and plastic limit more than 7%, the soil is classified as *clay*.

Hense sample is *clayey sand* (SC)

85. Water content, $w = 30\%$,

Specific gravity, G = 2.7

Unit weight of water,

$$\gamma_w = 10 \text{ kN/m}^3$$

Void ratio, $e = \dfrac{wG}{Sr}$

where, $S_r = 1$

$$\therefore \quad e = 0.3 \times 2.7 = 0.81$$

Saturated unit weight,

$$g_{sat} = \left(\frac{G+e}{l+e}\right)\gamma_w$$

$$= \left(\frac{2.7+0.81}{1+0.81}\right) \times 10 = 19.4 \text{ kN/m}^3$$

86. *Given*: c = 0

$$\phi = 35°$$

$$\gamma_{dry} = 10 \text{ kN/m}^3$$

$$\gamma_{sat} = 29 \text{ kN/m}^3$$

Factor of safety of infinite slope,

$$F_S = \frac{S}{T}$$

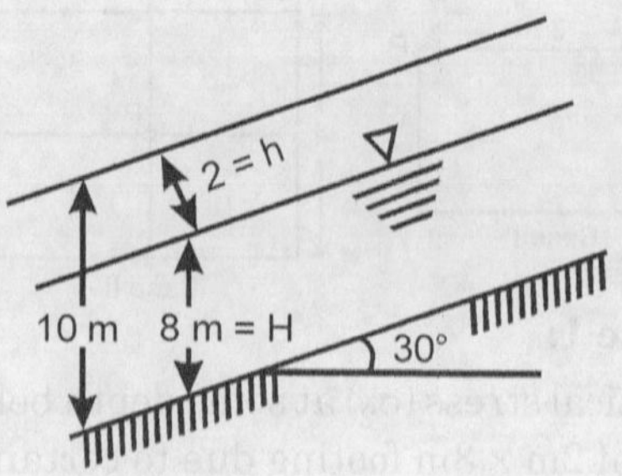

where, $S = (\gamma' H + h\,\gamma_{dry}) \cos^2 i \tan \phi'$

and $T = (\gamma_{sat} \times H + \gamma'_{dry} \times \lambda) \sin i \cos i$

$$\therefore \quad F_S = \frac{(\gamma' H + h\, g_{dry})\cos^2 i \tan\phi}{(\gamma_{sat} \times H + \gamma_{dry} \times h)\sin i \cos i}$$

$$= \frac{[(20-10)\times 8 + 2\times 16]\cos^2 30 \times \tan 35°}{[20\times 8 + 16\times 2]\sin 30\ \cos 30°}$$

$$= \mathbf{0.707}$$

88. Size of footing = 1.5m × 3m (B × L)

If water table is at the depth (b) equal or more than B from the footing, it has no effect on bearing capacity of footing

So when water table is at 3m below > (B = 1.5m),

bearing capacity = 600 kN/m² (same)

When water table at 7.5 m below

bearing capacity = 600 kN/m²

When water table rise upto B depth below base of the footing, bearing capacity starts decreasing.

89. *Given :* $C_u = 40 \text{ kN/m}^2$

$$Q_u = 0°$$

Total bearing capacity,

$$Q_u = Q_p + Q_S$$

But $\quad Q_S = C_u + p \times 1$

where, p = perimeter of plane

L = length of pile

= 10

$$\therefore \quad Q_s = 40 \times (4 \times 1.6) \times 10$$

$$= 2560 \text{ kN/m}^2$$

Again $\quad Q_p = A_p\, C_u\, N_c$

where, $N_c = 9$

and $\quad A_p$ = area of pile cap

$$\therefore \quad Q_p = 1.6 \times 1.6 \times 40 \times 9$$

$$= 921.6 \text{ kN/m}^2$$

Hence $\quad Q_u = 2560 + 921.6$

$$= 3481.6 \text{ kN/m}^2$$

90. *Given :* G = 2.67 ,

$$e = 0.67$$

$$\therefore \quad i_c = \frac{G-1}{1+e}$$

$$= \frac{2.67-1}{1+.67}$$

$$= 1$$

Hence void ratio range is 0.6 to 0.7

91. From given Mohr's diagram, we can see that length OP represent vertical stress at the base.

92. Capillary pressure is more in coarse aggregate.

93. *Given :* n = 0.35

G = 2.5

$$e = \frac{0.35}{1-0.35} = 0.538$$

$$\therefore \quad i_c = \frac{G-1}{1+e}$$

$$= \frac{2.5-1}{1+0.538}$$

$$= 0.975$$

Exit gradient with factor of safety

$$= \frac{0.975}{5}$$

$$= 0.195$$

94. $\gamma_w = \dfrac{1.8 \times 9.8}{944}$

$= 18.686 \text{ kN/m}^3$

$w = \dfrac{23-20}{20}$

$= \dfrac{3}{20} = .15$

$\gamma_d = \dfrac{G}{1+e}\gamma_\omega$

$= \dfrac{\gamma}{1+w}$

$= \dfrac{18.686}{1.15}$

$= 16.25 \text{ kN/m}^3$

97. $\sigma_1 = \sigma_3 \tan^2\alpha$

where, $\alpha = 45° + \dfrac{\phi}{2}$

$= 45° + \dfrac{30°}{2}$

$= 60°$

$\sigma_1 = \sigma_3 \tan^2 60°$

$= 3\sigma_3$

$= 3 \times 200$

$= 600 \text{ kPa}$

∴ Deviatoric stress $= \sigma_1 - \sigma_3$

$= 600 - 200$

$= 400 \text{ kPa}$

102. Since the soil contain :

40% gravel, 50% sand, and 10% silt

so, soil will be silty, gravelly sand.

103. *Given :* Bulk density = 22 kN/m^3

∴ Dry density, $\gamma_d = \dfrac{\gamma}{1+w}$

$= \dfrac{22}{1+0.1}$

$= 20 \text{ kN/m}^3$

104. In a constant head permeameter,

$Q = kiA$

where, Q = discharge

$= \dfrac{600}{60}$

$= 10 \text{ cc/cm}^2$

i = hydraulic gradient

= 0.5

A = area of cross-section

$= 10 \text{ cm}^2$

k = permeability of soil

∴ $k = \dfrac{Q}{iA}$

$= \dfrac{10}{0.5 \times 10} = 2 \text{ cm/sec}$

105. Confining pressure, $\sigma_3 = 100 \text{ kN/m}^2$

Deviator stress $= \sigma_1 - \sigma_3$

$= 100 \text{ kN/m}^2$

$\sigma_1 = 100 + s_3$

$= 100 + 100$

$= 200 \text{ kN/m}^2$

Also, $\sigma_1 = \sigma_3 N_\phi + 2c\sqrt{Nd}$

But given soil sample is sandy, hence c = 0

∴ $\sigma_1 = \sigma_3 N_\phi$

⇒ $200 = 100 N_\phi$

∴ $N_\phi = \dfrac{200}{100} = 2$

∴ $\tan^2\left(45° + \dfrac{\phi}{2}\right) = 2$

⇒ $\tan\left(45° + \dfrac{\phi}{2}\right) = \sqrt{2}$

⇒ $\left(45° + \dfrac{\phi}{2}\right) = 54.73$

∴ $\phi = (54.73 - 45) \times 2$

$= 19.47°$

107. In infinite slope, for dry soil,

$\text{F.S.} = \dfrac{\tan\phi}{\tan i}$

$i = \tan^{-1}\left(\dfrac{\tan\phi}{\text{F.S}}\right)$

$= \tan^{-1}\left(\dfrac{\tan 30}{1.5}\right)$

$= 21.05°$

For submersed condition,

$\text{F.S} = \dfrac{\tan\phi'}{\tan i}$

∴ $i = 21.05°$

110. *Given :* $\dfrac{\gamma_{sat}}{\gamma_{dry}} = 1.25$

$G_s = 2.65$

Now $\gamma_{sat} = \dfrac{G_s + e}{1+e}.\gamma_w,$

and $\gamma_{dry} = \dfrac{m.\gamma_w}{1+e}$

$$\therefore \quad \frac{\gamma_{sat}}{\gamma_{dry}} = \frac{G_s+e}{1+e}.\gamma_w, \times \frac{1+e}{G_s}.\frac{1}{\gamma_w}$$

$$= \frac{G_s+e}{G_s} = 1.25$$

$$\Rightarrow \quad 1.25 = \frac{2.65+e}{2.65}$$

$$\therefore \quad e = 1.25 \times 2.65 - 2.65$$

$$= 2.65\,(1.25-1)$$

$$= 2.65 \times 0.25 = 0.663$$

111. Case I

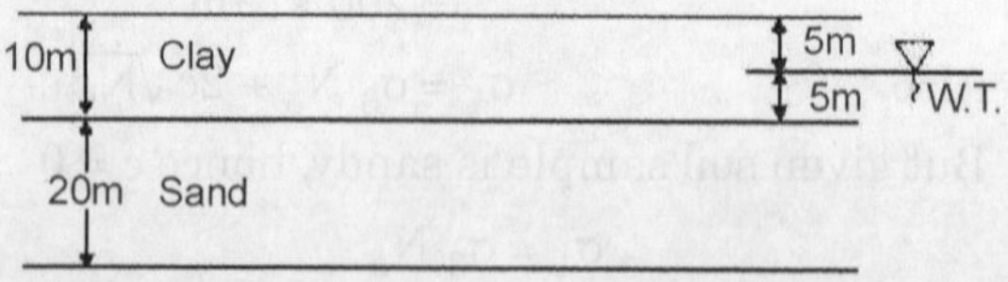

Effective stress, $\sigma' = 5\gamma_{sat} + 5\gamma'$

$$= 5\gamma_{sat} + 5\,(\gamma_{sat} + \gamma_w)$$

$$= 10\gamma_{sat} - 5\gamma_w$$

Case II

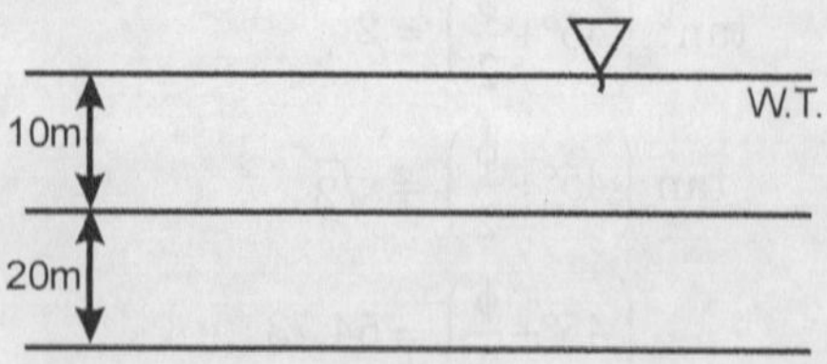

Effective stress, $\sigma' = 10\gamma_{sat} - 10\gamma_w$

$$= 10\gamma_{sat} - 5\,\gamma_w - 5\gamma_w.$$

Hence effective stress at a point P on the interface will decrease by $5\,\gamma_w$.

113. $q_{ult} = 1.3\,CN_c + \gamma DN_q + 0.5\,\gamma BN_q$

For clay, $N_C = 5.7;$

$N_q = 1$

$N_q = 0$

For D = 0,

$$q_{ult} = 1.3\,CN_C$$

For square footing,

$$q_{ult} = 1.3\,CN_C$$

$\therefore$ Required ratio = 1.0

114. Seepage force/ unit volume

$$= i\gamma_w$$

$$= 0.3 \times 22 = 6.6 \text{ kN}$$

115. Undrained cohesion of remoulded clay

$$= 10 \text{ kPa}$$

$$q_u = 2 \times 10 = 20$$

Now, Sensitivity = $\dfrac{\text{Undisturbed shear strength}}{\text{Disturbed shear strength}}$

Remoulded compressive strength = 20 kN/m²
(Since compressive strength itself is q_u)

116. External work done = $W_u \times \dfrac{L\theta}{2}$

Internal work done by positive moment

$$= M_p \times 3\theta$$

Equating external and internal work done,

$$W_u \times \frac{L\theta}{2} = M_P \times 3\theta$$

$$\therefore \quad W_u = \frac{6M_P}{L}$$

117. $$\tan\phi_m = \frac{\tan\phi}{\text{factor of safety}}$$

$$= \frac{\tan 30}{1.5}$$

$$\therefore \quad \phi_m = 21.05°$$

118. $$\frac{t_2}{t_1} = \frac{d_2^2}{d_1^2} = \frac{H_1^2}{4H_1^2}$$

$$\therefore \quad t_2 = \frac{t_2}{4}$$

Hence four times faster.

120. For spoon sampler

(*i*) Area ratio should preferably not exeed 10%

(*ii*) Inside clearance should be between 1 to 3%

121. We know, $\dfrac{k_1}{k_2} = \dfrac{\tan\alpha_1}{\tan\alpha_2}$

$$\therefore \quad \tan\alpha_2 = \frac{k_2}{k_1}\tan\alpha_1$$

$$= \frac{4\times10^{-7}}{1\times10^{-7}}\times\tan 30°$$

$$= 4 \times \tan 30°$$

$$= 2.3094$$

$$\therefore \quad \alpha = \tan^{-1} 2.3094$$

$$= 66.586$$

$$\approx 66.59°$$

122. Critical height of an unsupported vertical cut in cohesive soil,

$$H_C = \frac{4c}{\gamma}\tan\alpha = \frac{4\times100}{18}\tan\left(45° + \frac{\phi}{2}\right)$$

$$= \frac{400}{18}\tan\,(45° + 15)$$

$$= 38.49 \text{ m}$$

According to rankines theory, active earth pressure,

$$P_a = k_a\gamma H - 2c\,\sqrt{k_a}$$

where, $k_a = \dfrac{1-\sin 30°}{1+\sin 30°} = \dfrac{1}{3}$

$$\therefore \quad P_a = \frac{0.5}{1.5}\times 18 \times 38.49 - 2 \times 100\sqrt{\frac{0.5}{1.5}}$$

$$= 115.47 \text{ kN/m}^2.$$

123. *Before compaction*

void ratio, e = 0.5

H = 8m

$\gamma_{d_1} = 17.8$ kN/m³

$\phi = 30°$

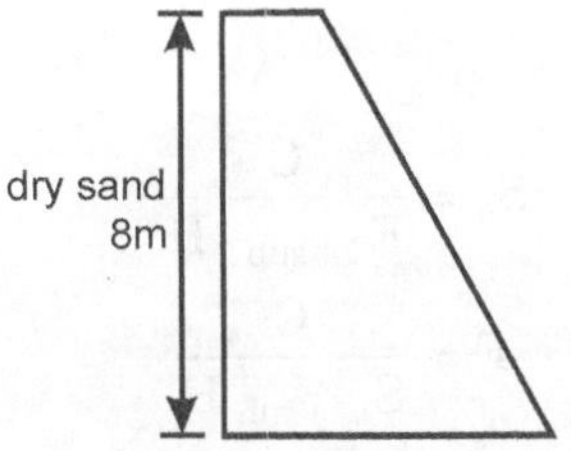

After compaction

e = 0.4

H = 8m

$\gamma_{d_2} = 18.8$ kN/m³

$\phi = 35$

$$\therefore \quad k_{P_1} = \frac{1+\sin\phi}{1-\sin\phi}$$

$$= \frac{1+\sin 30°}{1-\sin 30°} = 3$$

and $$k_{P_2} = \frac{1+\sin 35°}{1-\sin 35°} = 3.69$$

$$\therefore \quad \frac{P_{p1}}{P_{p2}} = \frac{\frac{1}{2}k_{p1}\cdot\gamma_{d1}\cdot H^2}{\frac{1}{2}k_{p2}\cdot\gamma_{d2}\cdot H^2}$$

$$= \frac{k_{p1}}{k_{p2}}\cdot\frac{\gamma_{d1}}{\gamma_{d2}}$$

$$= \frac{3.0}{3.69}\times\frac{17.8}{18.8}$$

$$= 0.7697 \approx 0.77$$

124. *Given* : k = 0.4

C = 100

$$Q_u = A_b f_b + A_s f_s$$

$$\therefore \quad A_s f_s = \pi dl\,\alpha C$$

$$= \pi \times 0.5 \times 10 \times 0.4 \times 100$$

$$= 628.32$$

and $$A_b f_b = \frac{\pi}{4}(0.5)^2 (CN_C)$$

$$= \frac{\pi}{4}(0.5)^2 (100 \times 9)$$

$$= 176.714 \text{ kN}$$

125. $$\eta = \frac{Q_{ug}}{nQ_u}$$

$$\therefore \quad Q_{ug} = nQ_u$$

$$= 16\,(A_s f_s)$$

$$= 16(628.32)$$

$$= 10053.12 \text{ kN}$$

126. $$\text{F.O.S.} = \frac{h_c}{h}$$

$$= \frac{\text{critical hydraulic gradient}}{\text{hydraulic gradient}}$$

$$\therefore \quad h = \frac{h_c}{\text{F.O.S.}}$$

$$= \frac{(1-\eta)(G-1)}{3}$$

$$= \frac{(1-0.45)(2.65-1)}{3}$$

$$= \mathbf{0.3025}$$

127. Final settlement will remain same, but rate of settlement will increase.

128. *Given* : $\gamma_{sat} = 20$ kN/m²,

$\phi = 35°$

F.O.S. = 1.5

But $$\text{F.O.S.} = \frac{\tan\phi}{\tan i} = \frac{\tan 35°}{\tan i}$$

$$\therefore \quad \tan i = 0.4668$$

or $$i = \mathbf{25°}$$

129. We know, $\rho_F = \rho_p\left[\frac{B_f(0.3+1.0)}{B_\rho(0.3+0.0.06)}\right]^2 > 25\text{mm}$

where, $B_f = 100$ mm

$B_P = 600$ mm

130. From Boussinesq's formula for point load

$$\sigma_z = \frac{3Q}{2\pi z^2}\left[\frac{1}{1+(r/z)^2}\right]^{5/2}$$

$$= \frac{3\times 25}{2\pi\times 6^2}\left[\frac{1}{1+(4/6)^2}\right]^{5/2}$$

$$= \mathbf{0.132\ kN/m^2}$$

131. The disproportionate drawdown is caused by nonlinear flow.

132. *In standard proctor test*

Energy imparted = 592 kJ/m³

$$\therefore \quad \text{Required energy} = \frac{400}{0.05\times 0.5\times 1000}$$

$$= 16 \text{ kJ/m}^3$$

$$\therefore \quad \text{Number of passes} = \frac{592}{16} = 32$$

$$\therefore \text{ Required number of passes} = \frac{50}{100}\times 32$$

$$= \mathbf{16.}$$

133. Strut load $= \frac{1}{2} k_a r z^2$

$$= \frac{1}{2} \times \frac{1-\sin 36°}{1+\sin 36°} \times 20 \times 5^2 \times 9$$

$$= \mathbf{113.9\ kN}$$

134. Increase in effective stress at mid-height of clay layer is

$$\sigma_e = -(\gamma_{ss}' \times 12 + \gamma'_{clay} \times 3.5) + \gamma_{sat\ ss} \times 2 + \gamma' \times 10 + \gamma'_{clay} \times 3.5$$

$$= 2 \times 22 + 10 \times (22-10) + 3.5 \times 9 - 12(12) - 3.5 \times 9$$

$$= 44 + 10 \times 22 - 12 \times 12$$

$$= \mathbf{20\ kPa}$$

Alternately

According to question, other conditions remaining the same, but the condition of top 2m layer is changed.

$\therefore$ Change in effective pressure $= 9.81 \times 2$

$$= 19.62\ kPa$$

$$\approx \mathbf{20\ kPa}$$

135. The consolidation settlement of the clay layer due to placing of fill material

$$\Delta H = m_v H \Delta\sigma$$

Here, $m_v = 2.2 \times 10^{-4}\ n^2/kN$

$H = 10\ m$

$\Delta_r = 21 \times 3 = 66\ kN/m^2$

$\therefore$ $\Delta H = 2.2 \times 10^{-4} \times 10 \times 66 \times 10^3$

$$= \mathbf{139\ mm.}$$

136. Thickness of compressible structure

$$= -\frac{2}{3} \times 5 = 3.67$$

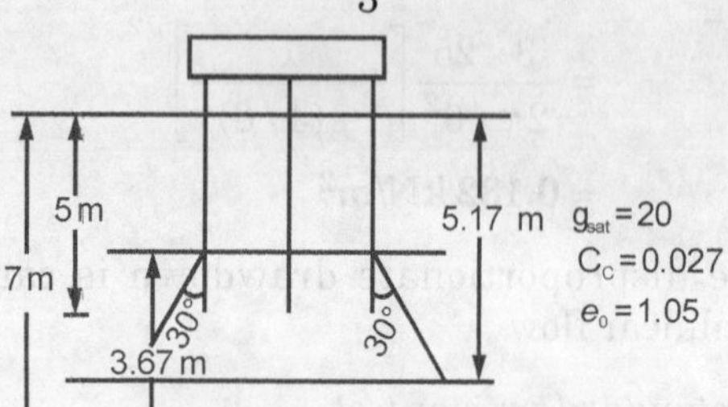

$$S = \frac{C_C H}{1+e_0} \log \frac{\sigma_0 + \Delta\sigma}{\overline{\sigma}_0}$$

Here, $\overline{\sigma}_0 = \gamma' D = (20-10)\ 5.17 = 51.7\ kN/m^2$

(Since equivalent raft at a depth 5.17)

Increased width $= 1.2 + \left(5.17 - \frac{2}{3} \times 5\right) \tan 30° \times 2$

$$= 3.32$$

$\Rightarrow$ $\Delta\sigma = \frac{500}{3.32^2} = 43.36\ kN/m^2$

$\therefore$ $S = \frac{0.027 \times 3.667}{1+1.05} \log \frac{51.7+43.36}{51.7}$ mm

$$= \mathbf{13.2\ mm}$$

138. Stability number

$$S_n = \frac{C}{F \cdot \gamma_{sub} \cdot H}$$

$\Rightarrow$ $F = \frac{C}{S_n\ \gamma_{sub}\ H}$

where, C = undrained cohesion

$$\gamma_{sub} = \gamma_{sat} - \gamma_\omega$$

$$= \left(\frac{G+e}{1+e}\right)\gamma_\omega - \gamma_\omega$$

$$= \gamma_\omega\left(\frac{G+e}{1+e} - 1\right)$$

$$= \left(\frac{2.65+1}{1+1} - 1\right) \times 10$$

$$= 8.25\ kN/m^2$$

H = depth of soil = 10 m

S_n = stability number = 0.08

$\therefore$ $F = \frac{12}{0.08 \times 8.25 \times 10}$

$$= \mathbf{1.822}$$

139. Factory of safety, $F = \frac{C}{S_n\ \gamma_{sat} \times H}$

Here, γ_{sat} because it is a case of sudden draw down.

$\therefore$ $F = \frac{12}{0.126 \times 18.25 \times 10}$

$$= \mathbf{0.53}$$

140. Here initial effective overburden pressure of middle of clay layer,

$$\sigma_0 = \gamma_{bulk} \times 1 + \gamma_{sub} \times 3$$

$$= 1 \times 18 + 3 \times (20 - 9.81)$$

$$= 48.57\ kN/m^2$$

Here increase in pressure at center of clay layer due to construction of footing

$$\Delta\sigma_0 = 40\ kN/m^2$$

Now consolidation settlement of the footing is calculated by

$$\Delta h = H \times \frac{C_c}{1+e_0} \log_{10} \frac{\sigma_0 + \Delta\sigma}{\sigma_0}$$

$$= 7 \times \frac{0.08}{1+0.4} \log_{10} \frac{48.57+40}{48.57}$$

$$= \frac{0.56}{1.4} \log_{10} \frac{88.57}{48.57}$$

$$= \frac{0.56}{1.4} \times \log_{10} 1.8235$$

$$= \frac{0.56}{1.4} \times 0.2609$$

$$= 0.089 \text{ m}$$

$$= \mathbf{8.9\ mm}$$

141. Elastic settlement

$$S_i = \frac{1-\mu^2}{E} qBI_f$$

$$= \frac{1-(0.4)^2}{50 \times 10^3} \times 40 \times 1.2 \times 1.75$$

$$= \frac{1.411}{10^3} = \mathbf{1.411\ m}$$

148. Shear strength, $S = c + \sigma' \tan\phi$

$$= 10 + (150) \tan 30$$

$$= 96.6 \text{ kPa}$$

149.

$$k = C_v m_v \gamma_w$$

$$T_v = \frac{\pi}{4}(v)^2 = \frac{C_v t}{h^2}$$

$$t_1 = \frac{h_1^2 v_1^2}{k_1} m_v \gamma_w$$

and

$$t_2 = \frac{h_2^2 v_2^2}{k_2} m_v \gamma_w$$

$$\therefore \quad \frac{t_1}{t_2} = \left(\frac{n^2}{n}\right)^2 \times \left(\frac{v_2}{v_1}\right)^2 \left(\frac{k_1}{k_2}\right) \left(\frac{m_{v_2}}{m_{v_1}}\right)^2$$

$$= \frac{1}{4} \times 1 \times \frac{1}{10} \times 4$$

$$= \frac{1}{10}$$

$$\therefore \quad t_2 = \frac{t_1}{10} = \frac{10}{10} = 1 \text{ year}$$

150. *Given :* $\sigma_1 = 60$

$\sigma_3 = 20$

Now $\quad \sigma_1 = \sigma_3 N_\phi + 2c\sqrt{N_\phi}$

where, $c = 0$

$$\therefore \quad \frac{\sigma_1}{\sigma_3} = N_\phi = \frac{60}{20} = 3$$

$$\therefore \quad \tan\left(45 + \frac{\phi}{2}\right) = \sqrt{3} = 1.732$$

$$\Rightarrow \quad 45 + \frac{\phi}{2} = 60$$

$$\Rightarrow \quad \frac{\phi}{2} = 60° - 45°$$

$$\therefore \quad \phi = 30°$$

151.

$$k = \frac{aL}{At} \log_e\left(\frac{n_1}{n_2}\right)$$

where, $a = \frac{\pi}{4}(.5)^2 = 0.196$

$$A = \frac{\pi}{4}(20)^2 = 78.54$$

$$L = 20 \text{ cm}$$

$$t = 3 \times 60 \times 60$$

$$\therefore \quad k = \frac{196 \times 20}{7.054 \times 60 \times 3 \times 60} \log\left(\frac{1}{35}\right)$$

$$= 4.86 \times 10^{-6} \text{ cm/sec}$$

153. $q_u = cN_C + \gamma D_f N_q + 0.5\gamma BN_q$

For pure along evil,

$\phi = 0,$

$N_C = 5.7,$

$N_q = 1$

$N_\gamma = 0$

$$\therefore \quad q_u = cN_C + \gamma D_f$$

$$q_{nu} = cN_C + \gamma D_f - \gamma D_f$$

$$= cN_C$$

154.

$$W\gamma = 0.5 + 0.5\frac{b}{B}$$

$$= 0.5 + 0.5 \times \frac{1.5}{2}$$

$$= 0.875$$

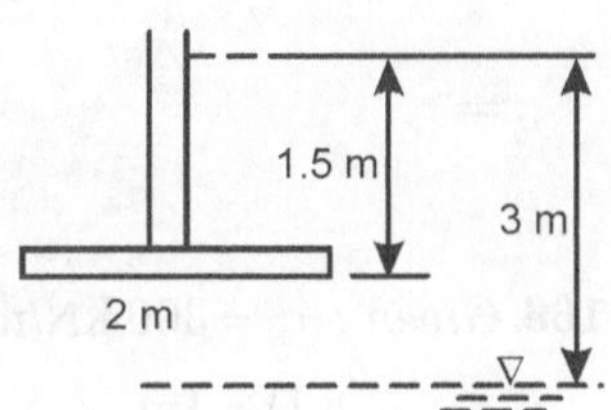

155. *Given :* $e = 0.65$

$G = 2.72$

$w = 20\%$

From $\quad e = \frac{wG}{S}$

$$\therefore \quad S = \frac{wG}{e}$$

$$= \frac{0.2 \times 2.72}{0.65}$$

$$= 0.84 \text{ or } 84\%$$

158. We know, $w \times d = \frac{Cu}{F} \times l_\alpha \times \gamma$

$\therefore \quad F = \frac{C_u \times l_a \times \gamma}{w \times d}$

$= \frac{30 \times 15.6 \times 10.3}{82 \times 1 \times 20 \times 2.3}$

$= 1.28$

160. For cohesionless soil,

$$S_f = S_p \left[\frac{B_f(B_p+.3)}{B_p(B_f+.3)}\right]^2$$

$$= 5\left[\frac{5(0.3+.3)}{.3(5+0.3)}\right]^2$$

$= 17.8$ mm

162. *Given* : $e = 0.5$

$\therefore \quad n = \frac{e}{1+e} = \frac{0.5}{1.5} = 33\%$

163. *Given* : $\gamma_{d1} = 17$ kN/m³

$\gamma_{d2} = 16$ kN/m³

$V_2 = 100$m³

$\therefore \quad \gamma_{s1} = \frac{W_s}{V_s} = \frac{W_s}{V_1} \times \gamma_{d1}$

and $\quad \gamma_{s2} = \frac{W_s}{V_2} \times \gamma_{d2}$

Since γ_s is constant for a given soil, therefore

$$\frac{\gamma_{d1}}{\gamma_{d2}} = \frac{V_1}{V_2}$$

$\Rightarrow \quad V_1 = \frac{17}{16} \times 100$

$= 106$ m³

166. *Given* : $q_f = 300$ kN/m²

$D = 1$m

$r = 20$ kN/m³

$F = 2.5$

$\therefore \quad q_{nf} = q_f - \gamma_D$

$= 300 - 20 \times 1$

$= 280$kN/m²

and $\quad NSBC = \frac{q_{nf}}{F}$

$= \frac{280}{2.5} = 112$kN/m²

167. *Given* : H = 7m

r = 16 kN/m³

C = 25 kN/m²

$s = 0.178$

For (D.F. = 3 and i = 30°). F_c = ?

From $\quad s = \frac{C}{F_c rH}$

$F_c = \frac{25}{0.178 \times 16 \times 7}$

$= 1.25$

168. Here, $\quad \sigma_1 - \sigma_3 = 3$ k$_g$f/cm²

Then $\quad \sigma_1 = (\sigma_1 - \sigma_3) -$ cell pressure

$= 3 - 1 = 2$

Also $\quad \sigma_1 = \sigma_3 \tan^2\alpha$

$\therefore \quad 1 \times \tan^2 \alpha = 2$

$\Rightarrow \quad \tan \alpha = \sqrt{2}$

$\Rightarrow \quad \alpha = 54$

Now $\quad \alpha = 45° + \frac{\phi}{2} = 54°$

$\therefore \quad \phi' = 20°$

169. Since $\quad C = 0; \; D = 0$

$\therefore \quad \frac{(q_f)_{cir}}{(q_f)_{sq}} = \frac{0.3\gamma\beta N_q}{0.4\gamma\beta N_q} = \frac{3}{4}$

170. $\quad p_a = r_z \cot^2\alpha - 2c \cot \alpha + q \cot^2 \alpha$

For z = 0, and $p_a = 0$

$2c \cot \alpha = q \cot^2 \alpha$

$\therefore \quad q = 2 c \tan \alpha$

175. $\quad p_a = K_a . \gamma z - 2c' \sqrt{K_a}$

$p_p = K_p \gamma z + 2c' \sqrt{K_p}$

179. We know $\quad G = \frac{\gamma_s}{\gamma_w}$

$\therefore \quad \gamma_s = 2.68 \times 9.81 = 26.291$ kN/m³

Again $\quad V_s = \frac{W}{\gamma_s}$

$= \frac{2.035 \times 9.81}{26.291 \times 10^{-3}} = 0.75 \times 10^{-3}$ m³

$V_v = V - V_s$

$= 1.15 \times 10^{-3} - 0.75 \times 10^{-3}$

$= 0.4 \times 10^{-3}$ m³

$\therefore \quad e = \frac{V_v}{V_s} = \frac{0.4 \times 10^{-3}}{0.75 \times 10^{-3}} = 0.53.$

182. The stress strain curves for dense and loose sands are similar in shape to the stress-strain curve for overconsolidated and normally consolidated clays.

184. $$\frac{S_f}{S_p} = \frac{B_f}{B_p}$$

where, S_f = settlement of foundation

S_p = settlement of plate

B_f = width of foundation

B_p = width of place

$$\therefore \quad S_{(prototype)} = S_{(plate)} \times \left(\frac{B_{prototype}}{B_{plate}}\right)$$

185. $$q = \frac{K.H_L.N_f}{N_d}$$

where, $K = 0.0013$ mm/sec

$= 0.0013 \times 10^{-3}$ m/sec

$= 1.3 \times 10^{-6}$ m/sec

$H_L = 9.6$ m

N_f = Number of flow channels = 5

N_d = Equipotential lines = 8

$$\therefore \quad q = \frac{1.3 \times 10^{-6} \times 9.6 \times 5}{8}$$

$= 7.8 \times 10^{-6}$ m³/sec/m

Given $L = 90$ m

$Q = q.L$

$= 7.8 \times 10^{-6} \times 90$ m³/sec

$= 7.8 \times 10^{-6} \times 90 \times 60 \times 60 \times 24$

$= 60.65 \approx 61$ m³/day

186. For 90% consolidation of the clay

$$\frac{t_1}{t_2} = \frac{H_1^2}{H_2^2} \quad [\text{since } t \propto d^2]$$

where $t_1 = 15$ years

$H_1 = H$(say)

and $H_2 = 2H$

$$\therefore \quad \frac{15}{t_2} = \left(\frac{H}{2H}\right)^2 = \frac{1}{4}$$

$\Rightarrow \quad t_2 = 15 \times 4 = 60$ years

As the clay layer is 3 times more permeable and 4 times more compressible, therefore time required for 90% consolidation

$$= 60 \times \frac{4}{3} = 80 \text{ years}$$

187. Total active thrust on a vertical wall is given by,

$$P = \frac{K_a.\gamma_t.H^2}{2}$$

where $$K_a = \frac{1-\sin\phi}{1+\sin\phi}$$

$$= \frac{1-\sin 30°}{1+\sin 30°}$$

$$= \frac{1}{3} \quad \text{... since } \phi = 30°$$

$\gamma_t = 20$ kN/m³(given)

and $H = 3$m

$$\therefore \quad p = \frac{1}{3} \times \frac{20 \times (3)^2}{2} = 30 \text{ kN/m}$$

188. Taylor's stability coefficient,

$$N_s = \frac{c}{F_c.\gamma.H}$$

where F_c = factor of safety

$H = 10$m (depth)

q = unit weight = 20 kN/m³

$c = c_u = 72$ kN/m²

$\therefore$ Factor of safety,

$$F_c = \frac{c}{N_s.\gamma.H} = \frac{72}{0.18 \times 20 \times 10} = 2$$

189. Given , $Q = 700$ kN

$z = 2$m

$r = 1.0$ m

$$\sigma_z = \frac{Q}{(z)^2} \times \frac{3}{2\pi}\left[\frac{1}{1+(r/z)^2}\right]^{5/2}$$

$$= \frac{700}{(2)^2} \times \frac{3}{2\pi}\left[\frac{1}{1+(1/2)^2}\right]^{5/2}$$

$= 47.8$ kPa (kN/m²)

190. Porosity, $\frac{e}{1+e}$, where e is void ratio

$$\therefore \quad 0.20 = \frac{e}{1+e}$$

$\Rightarrow \quad 0.20(1 + e) = e$

$\Rightarrow \quad 0.20 = (e - 0.20e)$

$\therefore \quad e = 0.25$

194. Transformed coefficient of permeability for an equivalent isotropic soil mass.

$$K = \sqrt{K_x.K_y}$$

$$= \sqrt{9 \times 10^{-7} \times 10^{-7} \times 4}$$

$= 6 \times 10^{-7}$ cm/sec

196. As per Boussinesq equation, for a point load the vertical stress at depth z, directly below the point load p is give by

$$\sigma_z = \frac{3P}{2\pi z^2}\left[\frac{1}{1+\left(\frac{r}{z}\right)^2}\right]^{5/2}$$

$$= \frac{P}{z^2}K$$

where, $K = \frac{2}{2\pi}\left[1+\frac{1}{(r/z)^2}\right]^{5/2}$

= constant known as Boussinesq influence factor.

208. Load carried by the pile by skin friction

$= m \times c \times A_p$

$= m \times c \times 4B \times L$

$= 0.8 \times 5 \times 4 \times 0.3 \times 10 = 48$ kN

214. Void ratio, $e_A = \frac{\eta_A}{1-\eta_A} = \frac{0.4}{1-0.4} = \frac{0.4}{0.6}$

Void ratio, $e_B = \frac{\eta_B}{1-\eta_B} = \frac{0.6}{1-0.6} = \frac{0.6}{0.4}$

$\therefore \quad \frac{e_A}{e_B} = \frac{0.4/0.6}{0.4/0.6} = \frac{16}{36} = \frac{4}{9}$

216. Since soil is saturated, therefore

degree of saturation, S = 1

Moisture content, $w = 0.2$

Specifie gravity, G = 2.6

Now Se = wG

∴ Void ratio, e = wG = 0.2 × 2.6 = 0.52

Porosity, $n = \frac{e}{1+e} = \frac{0.52}{1+0.52} = 0.34$

Saturated density, γ_{sat}

$= \frac{G+e}{1+e}\cdot\gamma_w$

$= \frac{2.6+0.52}{1+0.52}\times 9.81 = 20.14 \text{ kN/m}^3$

NUMERICAL TYPE QUESTIONS

1. n = 0.7, S = 40%, V = 100 m³

$n = \frac{V_v}{V} \Rightarrow V_v = n.V = 0.7 \times 100 = 70 \text{ m}^3$

$S = \frac{V_w}{V} \Rightarrow V_w = S.V_v = 0.4 \times 70 = 28 \text{ m}^3$

Therefore, Volume of air,

$V_a = V_V - V_W = 70 - 28 = 42 \text{ m}^3$

2. For clay: $e = \frac{wG}{S} = 0.25 \times 2.70 = 0.675$

$\gamma_{sat} = \gamma_w\left[\frac{G+e}{1+e}\right] = 10\left[\frac{2.70+0.675}{1+0.675}\right] = 20.15 \text{kN/m}^3$

At the centre of clay,

$= 1 \times 17 + 0.5[20.15 - 10]$

$= 22.075 \text{ kN/m}^2$

Applied load intensity,

$q = 2.5 \times 20 = 50 \text{ kN/m}^2$

Therefore, final effective stress at the end of consolidation at the centre of day,

$\sigma_f' = \sigma' + \Delta\sigma' = 22.075 + 50 = 72.075 \text{ kN/m}^2$

The ultimate consolidation settlement,

$S_f = H.\frac{C_R}{1+e_0}\log_{10}\left[\frac{\sigma_C'}{\sigma_O'}\right] + H\frac{C_C}{1+e_0}\log_{10}\left[\frac{\sigma_f'}{\sigma_C'}\right]$

$S_f = 1\times\frac{0.05}{1+0.0675}\log_{10}\left[\frac{60}{22.075}\right]$

$+1\times\frac{0.5}{1+0.675}\log_{10}\left[\frac{72.075}{60}\right] = 0.3677\text{m}$

$= 36.77$ m

3. $\sigma' = \gamma' z + \gamma_w.h$

$\sigma' = (18-9.81)\times 5 + 9.81\left(3-\frac{3}{6}\times 1\right) = 65.475 \text{ kPA}$

4. C = 15kPa, ∅′ = 22°

In CU test,

$\sigma_3 = 200$ kPa

$u = 150$ kPa

$\sigma'_3 = \sigma_3 - u = 200 - 150 = 50\text{k Pa}$

$\sigma'_1 = \sigma'_3 \tan^2\left[45+\frac{\varnothing'}{2}\right] + 2C'\tan\left[45+\frac{\varnothing'}{2}\right]$

$\sigma'_1 = 50.\tan^2\left[45+\frac{22}{2}\right] + 2\times 15\tan\left[45+\frac{22}{2}\right]$

$= 154.377$ kPa

$\sigma'_1 = \sigma_1 - u = (\sigma_3 + \sigma_d - u)$

$154.377 = (200 + \sigma_d - 150)$

Therefore,

$\sigma_d = 104.37$ kPa

5. $N_d = 10$,

$N_f = 20$

K = 3 mm/min,

$\Delta h = 5$ m

$Q = K.h.\frac{N_f}{N_d}$

$= \frac{\frac{3}{1000}\times 5}{60}\times\frac{20}{10}$

$= 500 \text{ cm}^3\text{/s}$

6. In c-ϕ soil

Earth pressure $Pa = k_a\sigma_v - 2c\sqrt{k_a}$

$K_a = \frac{1-\sin 24}{1+\sin 24} = 0.4217$

Now at point A

$\sigma_v = 1\times\gamma_b + 4\gamma_{sat} + 3\gamma_{sat}$

Earth pressure $Pa = k_a\sigma_v - 2c\sqrt{k_a}$

$= 0.4217\,(16.5 + 4(19 - 9.81) + 3(18.5 - 9.81))$

$+ 7 \times 9.81 - 2 \times 25 \times \sqrt{0.4217}$

$= 69.65 \text{ kN/m}^2$

7. $\Delta h = 600 - 300$ mm

$Q = 200$ ml/15 minutes

$$= \left(\frac{200}{1000}\right) \times \left(\frac{1000}{15 \times 60}\right)$$

$$= \frac{1000}{5 \times 900}$$

$= 0.222$ cm³/sec

$A = 80$ cm²

$$V = \frac{Q}{A} = \frac{1000}{(5 \times 900 \times 80)}$$

$= 0.0277$ mm/sec

$V = Ki$

$$V = k \times \left(\frac{300}{300}\right)$$

Therefore, $$K = \frac{H_1 + H_2}{\frac{H_1}{K_1} + \frac{H_2}{K_2}}$$

$$\frac{1}{K_2} = 72 - 50$$

$K_2 = 0.04545$ mm/sec

8. $$q_u = cN_C + 8D_f N_q + 0.5\,B\,\gamma N_\gamma$$

$$q_u = \frac{2}{3} \times 35 \times 17.7 + 17 + 1.5 \times 7.4 + 0.5 \times 4 \times 17 \times 5$$

$q_u = 771.7$ kN/m²

$q_{nu} = q_u - \gamma D_f$

$= 771.7 - 17 \times 1.5$

$q_{nu} = 746.2$ kN/m²

$$q_{net} = \frac{q_{nu}}{F}$$

$$= \frac{746.2}{2.5}$$

$= 298.48$ kN/m²

9. Depth at which settlement is required

$= (2 + 6 + 5)$

$= 13$ m below G.L.

$\because$ Distribution of load is 2V:1H

$\therefore$ Load at 13 m below G.L. $= (s\bar{\sigma}_0)$

$$= \frac{1500 \text{ kN}}{\frac{\pi}{4}(D+12)^2} = \frac{1500}{\frac{\pi}{4} \times (15)^2}$$

$= 8.488$ kN/m³

Effective principle stress at 13m below ground level (for a long period of time)

$$\bar{\sigma}_0 = (2 \times 15) + (6 \times 18) + (5 \times 18) - (11 \times 10)$$

$= 30 + 88 = 118$ kN/m³

$$\therefore \text{ Settlement} = \Delta H = \frac{C_C H_o}{1 + C_o} \log\left(\frac{\Delta\bar{\sigma}_o + \bar{\sigma}_o}{\bar{\sigma}_o}\right)$$

$$= \frac{0.3 \times 10}{(1 + 0.7)} \log\left(\frac{118 + 8.488}{118}\right)$$

$= 0.053$ m

$= 53.23$ mm

10. Degree of consolidation for 50 mm settlement

$$U = \frac{50}{61.22} = 0.817 = 81.7\%$$

$$Tv = 1.784 - 0.933 \log_{10}(100 - U)$$

$= 0.608$

$$0.608 = \frac{c_v \times t}{d^2}$$

$$\Rightarrow \quad t = \frac{0.608 \times (10^2)}{0.003 \times 10^{-4}}$$

$= 202666667$ sec $= 6.43$ year

Additional number of years

$= 6.43 - 2 = 4.43$ years

11. $$k_2 = 10k_3 = \frac{k_1}{10}$$

$k_1 H_1$	Fine Sand
$k_2 H_2$	Silt
$k_3 H_3$	Clay

$$\Rightarrow \quad k_1 = 100\,k_3$$

$$H_2 = 2H_1 = \frac{2}{3} H_3$$

$$\Rightarrow \quad H_1 = \frac{H_3}{3}$$

$$k_h = \frac{k_1 H_1 + k_2 H_2 + k_3 H_3}{H_1 + H_2 + H_3}$$

$$= \frac{100k_3 . \frac{H_3}{3} + 10k_3 . \frac{2}{3} H_3 + k_3 H_3}{\frac{H_3}{3} + \frac{2}{3} H_3 + H_3}$$

$$= \frac{100k_3 + 20k_3 + 3k_3}{H_3 + 2H_3 + 3H_3} = \frac{123}{6} k_3$$

$$k_z = \frac{\frac{H_3}{3} + \frac{2}{3} H_3 + H_3}{\frac{H_3}{3 \times 100k_3} + \frac{2H_3}{3 \times 10k_3} + \frac{H_3}{k_3}}$$

$$= \frac{6}{\left[\left(\frac{1}{100}\right) + \frac{2}{10} + 3\right] k_3} = \frac{600}{321} k_3$$

$$\frac{k_n}{k_z} = \frac{123}{6} \times \frac{321}{600} = 10.96$$

12. $q_{safe} = \dfrac{q_{nu}}{3}$

$q_{nu} = cN_c + qN_q + 0.5\gamma BN_\gamma - 8\Delta$

$C = 0$

$q_{nu} = q(N_q - 1) + 0.5\gamma BN_\gamma$

$q_{ns} = \dfrac{1}{3}(_q((N_q - 1)F_{qs} \times F_{qd} + F_{qp} + 0.5\gamma BN_\gamma \times F_{\gamma s} \times G_{\gamma o} \times F_{\gamma p})$

$q_{ns} = \dfrac{1}{3}(18 \times 1(33.3 - 1)\,1.314 \times 1.113 \times 0.444 + \dfrac{1}{2} \times 2 \times 18 \times 37.16 \times 1.314 \times 1.113 \times 0.02)$

$= \dfrac{397.03}{3} = 132.264 \text{ kN/m}^2$

For one way shear eccentricity area to be reduced.

Reduced area of footing = 2 × 1.7 = 3.4 m²

Load carrying capacity = 1.32 × 3.4

= 440 kN Appro.

13. Volume of moist sand = 588 m³

Weight of moist sand = 1010 gm

Dry weight of sand = 918 gm

Water weight = 1010 – 918 = 92 gm

Specific gravity (G) = 2.67

Dry density, $\rho_d = \dfrac{918}{588}$ gm/cm³

$= 1.56$ gm/cm³ ...(*i*)

Also, $\rho_d = \dfrac{G\gamma_\omega}{1+e}$

$= \dfrac{2.67 \times 1}{1+e}$ ...(*ii*)

By (*i*) and (*ii*)

$1.56 = \dfrac{2.67}{1+e} \Rightarrow e = 0.71$

14. $\Delta H = \dfrac{C_c}{1+e_0} H_0 \log_{10}\left(\dfrac{\sigma_0 + \Delta\sigma}{\sigma_0}\right)$

$= \dfrac{0.6}{1+1.3} \times 4 \times \log_{10}\left(\dfrac{2\sigma_0}{\sigma_0}\right)$

$[\because \Delta\sigma = \sigma_0]$

$= 0.314$ m

$\Delta H = 314$ mm

15. Diameter = 0.4 m

$\gamma = 18$ kN/m³

$\lambda = 0.15$

Neglected → End Bearing Capacity

The ultimate pile capacity (kN)

$= \lambda(\sigma_{avg} + 2\bar{c})(\text{Area})$

$= 0.15\left\{\dfrac{18 \times 25}{2} + 2 \times \dfrac{(40+50+60+70+80)}{5}\right\} \times (\pi \times 0.4 \times 25)$

= 1625.774 kN

16. Volume compressibility,

$m_v = \dfrac{a_v}{\Delta\sigma} = \dfrac{\dfrac{\Delta e}{1+e_0}}{\Delta\sigma}$

$= \dfrac{0.2}{1+1.1} \times \dfrac{1}{125}$

$= 7.62 \times 10^{-4}$ m²/kN

17. $Q = K.H.\dfrac{N_f}{N_d}$

N_f = No. of flow channels = 3

N_d = No. of equipotential drops = 10

$K = 3.8 \times 10^{-4}$ s/m

$H = 6.3$ m

$Q = 3.8 \times 10^{-6} \times 6.3 \times \dfrac{3}{10}$

$= 7.182 \times 10^{-6}$ m³/s/m

$= 7.182$ cm³/s/m

18. We know the relation

$\boxed{\omega \cdot G = e \cdot Sr.}$

$n = \dfrac{e}{e+1};\ e = \dfrac{n}{1-n}$

Given $n = 40\%$

$n = 0.4$

$e = \dfrac{0.4}{1-0.4} = 0.67$

$\boxed{e = 0.67}$ – Substitute in equation ...(*i*)

$\boxed{wG = eSr}$

$0.2 \times 2.71 \Rightarrow 0.67 \times Sr$

$S_r = 0.8089 = 81\%$

$\boxed{S_r = \mathbf{81\%\ to\ 81.5\%}}$

19. Length of step = 10 m

The governing equation of flow

$\Rightarrow \dfrac{d^2x}{dx^2} = 0$ ...(*i*)

Integrate the above equation (*i*) $\dfrac{dx}{dn} = c_1$ ...(*ii*) again integrate the above equation

$$H = G.x + C_2 \quad ...(iii)$$

if the $x = 0$, H = 5m (i.e. head of water of begining)

$\therefore \quad 5 = c_1 \times 0 + c_2$

$c_2 = 5$

if $x = 10$, H = 1m

$1 = c_1 \times 10 + c_2$

$c_2 = 5$ putting in above equation

$1 = c_1 \times 10 + 5$

$10\, c_1 = 1 - 5 = -4$

$$c_1 = -\frac{2}{5}$$

So the value of c_1 & c_2 putting in Eq. (*iii*)

$H = c_1 x + c_2$

$$H = -\frac{2}{5}x + 5$$

We required the value of H at middle of soil strip i.e., at x_2 5

$$\because \quad H = -\frac{2}{5}\times 5 + 5 = -2 + 5 = 3.0 \text{ m.}$$

20. For infinitly long slope applying the Equation

$$\frac{c}{\gamma H_c} = (\tan i - \tan \phi) \times \cos^2$$

$c = 20$ KN/m², $\gamma_d = \gamma = 16$ KN/m², $H_C = 5$m

$i = 40°$, $\phi = ?$

$$\frac{20}{16\times 5} = (\tan 40° - \tan \phi) \times \cos^2 40$$

$0.25 = (\tan 40° - \tan \phi) \times 0.587$

$\tan \phi = \tan 40° - 0.426 = 0.413$

$\phi = 22.4°$

22. *Given:* $\rho = 2.15$ Mg/m³, w = 12%, G = 2.65

(*i*)
$$\rho_d = \frac{\rho}{1+w} = \frac{2.15}{1+.12} = 1.92 \text{Mg/m}^3$$

(*ii*)
$$\rho_d = \frac{G\rho_w}{1+e}$$

$$\Rightarrow \quad 1 + e = \frac{G\rho_w}{\rho_d} = \frac{2.65\times 1}{1.92} = 1.38$$

$\Rightarrow \quad e = 1.38 - 1 = 0.38$

(*iii*) $Se = wG$

$$\Rightarrow \quad S = \frac{wG}{e} = \frac{0.12\times 2.65}{0.38} = 0.8368 = 83.68\%$$

(*iv*) Air content, $a_c = 1 - S$

$= 1 - 0.8368 = 0.1632$

23.

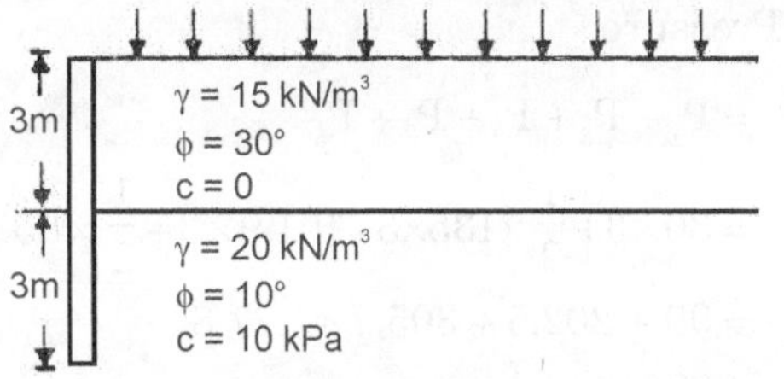

$$K_{p1} = \frac{1+\sin\phi}{1-\sin\phi} = \frac{1+\sin 30}{1-\sin 30} = \frac{1.5}{0.5} = 3$$

$$K_{p2} = \frac{1+\sin 10}{1-\sin 10} = \frac{1.1736}{0.3264} = 1.42$$

$$\therefore \quad \sqrt{K_{p2}} = 1.19$$

Passive pressure due to surcharge

At top layer

$p_1 = K_{p1} q$

$= 3 \times 10 = 30$kN/m²

At bottom

$p_2 = K_{p2} q$

$= 1.42 \times 10 = 14.2$kN/m²

$p_3 = K_{p1}\, \gamma H + 2C\sqrt{K_{p_1}}$

where, H = 3, C = 0

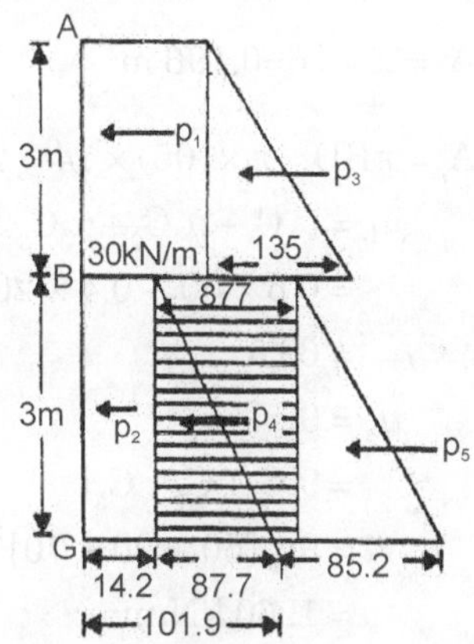

At top

$p_3 = 3 \times 15 \times 3 = 135$kN/m²

Bottom layer

$P_4 = K_{p2}\, \gamma H + 2C\sqrt{K_{p_2}}$

$= 1.42 \times 15 \times 3 + 2 \times 10 \times 1.19$

$= 87.7$kN/m²

At Point C

$P_5 = 10 + 15 \times 3 + 20 \times 3$

$= 115$kM/m³

$p_5 = 115 \times 1.42 + 2 \times 10\sqrt{1.42}$

$= 187.1$ KN/m³

$P_5 = 187.1 - (87.7 + 14.2)$
$= 85.2$

Total Pressure,

$$\bar{P} = P_1 + P_3 + P_2 + P_4 + P_5$$
$$= 30 \times 3 + \frac{1}{2} \times 135 \times 3 + 101.9 \times 3 + \frac{1}{2} \times 85.2 \times 3$$
$$= 90 + 202.5 + 305.7 + 127.8$$
$$= 726$$

Horizontal thrust is acting at high H from bottom

$$H = \frac{P_1 y_1 + P_3 y_2 + P_4 y_3 + P_5 y_4}{\bar{P}}$$

where, $P_4' = P_2 + P_4$

$$= \frac{90 \times 4.5 + 202.5 \times 4 + 305.7 \times 1.5 + 107.8 \times 1}{726}$$

$= 2.45$m from bottom

24. *Given:* D = depth of foundation = 14m

d = diameter of the pile = 0.5m

Spacing, S = 1m

Now, $Q_u = nQ_u n_g$

Also $Q_u = q_f A_f + q_b A_b$

where, $A_b = \frac{\pi}{4}(d)^2 = 0.196\ m^2$

$A_f = \pi BD = \pi \times (05) \times 14 = 21.99 \simeq 22m^2$

$$q_f = \alpha_1 C_1 + \alpha_2 C_2 + \alpha_3 C_3$$
$$= 0.6 \times 50 + 0.4 \times 20 + 0.45 \times 70$$
$$= 69.5$$
$$q_b = 9C$$
$$= 9[C_1 + C_2 + C_3]$$
$$= 9 \times [50 + 20 + 70]$$
$$\simeq 1260\ kN/m^2$$
$$Q_u = 69.5 \times 22 + 1260 \times 0.196$$
$$\simeq 1775.96 kN$$

Now, $$n_g = 1 - \frac{\theta}{90}\left[\frac{m(n-1) + n(m-1)}{mn}\right]$$

where, $\theta = \tan^{-1}\left(\frac{d}{s}\right) = \tan{-1}\left(\frac{5}{1}\right) = 26.56$

$\therefore$ $$n_g = 1 - \frac{26.56}{90}\left[\frac{4 \times 3 + 3 \times 4}{4 \times 4}\right]$$
$$= 1 - 0.44 = 0.56$$

Hence $Q_u = n\ Q_u n_g$
$$= 16 \times 0.56 \times 1775.96$$
$$= 15912.6\ kN$$

25. *Given* : Width = B = 3m,

$C' = 20$kPa,

$\phi = 30°$

$\gamma = 18kN/m^3$

$R_v = Q = 20kN/m$

$e = 0.25$m,

F = 50 Kk/m

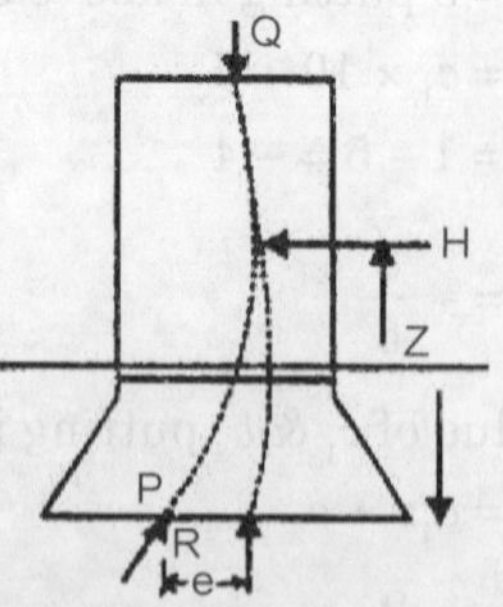

Depth of foundation Z

Taking moment about toe

$$R_v \times e = Z \times F$$
$$\Rightarrow \quad 20 \times 0.25 = Z \times 50$$
$$\Rightarrow \quad Z = \frac{5}{50} = 0.1m$$

Depth of foundation = 0.1m

According to Terzaghi and Hocrentolear equation

$$q_u = \gamma D_f \tan^4\alpha + \frac{\gamma B}{4}(\tan^5\alpha - \tan\alpha') + 2\,C(\tan^3\alpha' + \tan\alpha')$$

where, $\alpha' = 45 + \frac{\phi}{2}$

$$= 45 + \frac{30}{2} = 60°$$

$\tan^4 60 = 9$

$\tan^5 60 = 15.6$

$\tan 60 = 1.732$

$\tan^3 60 = 5.2$

$\therefore$ $$q_u = 18 \times 0.1 \times 9 + \frac{18 \times 3}{4}(15.6 - 1.732) + 2 \times 20 \times (5.2 + 1.732)$$
$$= 16.2 + 187.22 + 277.28$$
$$= 480.7kN/m^2$$

26. (*a*) Storage coefficient, S = 0.005

$Q = 2m^3$/mint

H = 30 m

$$K = \frac{30}{24 \times 60} \text{ m/mint}$$

Slope = ?

$t = 14 \times 60$ mint

$R = 50\text{m}, R_2 = 100\text{m}$

$T = KB$

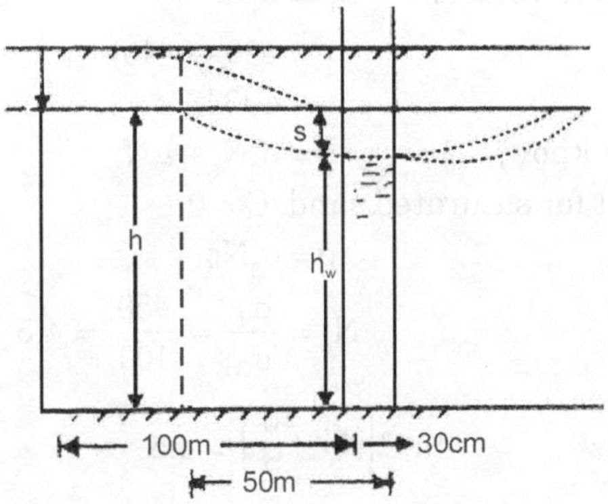

$$= \frac{30}{24 \times 60 \times 60} \times 30$$

$$= 1.04 \times 10^{-2} \text{m}^2/\text{sec}$$

$$u = \frac{r^2 s}{4Tt}, \text{ where } r = 100\text{m}$$

$$= \frac{(100)^2 \times 0.005}{4 \times 1.04 \times 10^{-2} \times 14 \times 60 \times 60}$$

$$= 2.385 \times 10^{-2}$$

$$\approx 0.024$$

$$W_u = -0.577216 - \text{In}u + u - \frac{u^2}{2 \times 2!} + \frac{u^3}{3 \times 3!}$$

$$= -0.577216 - \text{In}(0.024) + 0.024 - \frac{0.024^2}{2.21} + \frac{0.024^3}{3.31}$$

$$= -0.577216 + 3.73 + 0.024 - 144 \times 10^{-4} + 7.68 \times 10^{-7}$$

$$= -3.177$$

$$S_{100} = \frac{Q}{4\pi T} W_u$$

$$= \left(\frac{2}{60}\right) \times \frac{1}{4 \times \pi \times 1.04 \times 10^{-2}} \times 3.177$$

$$= 0.81 \text{ m}$$

$r = 50\text{m}$

(b) $$u = \frac{(50)^2 \times (0.005)}{4 \times 1.04 \times 10^{-2} \times 14 \times 60 \times 60}$$

$$= 0.00596$$

$$\simeq 0.006$$

$$W_u = -0.5772 - \text{In}(0.006) + 0.006 - \frac{0.006^2}{2 \times 2!} + \frac{0.006^3}{3 \times 3!}$$

$$= -0.5772 + 5.12 + 0.006$$

$$= 4.55$$

$$S_{50} = \frac{Q}{4pT}(W_u)$$

$$= \left(\frac{2}{60}\right) \times \frac{1}{4 \times \pi \times 1.04 \times 10^{-2}} \times 4.55$$

$$= 1.16\text{m}$$

Slope of hydraulic gradient,

$$\frac{\Delta H}{L} = \frac{S_{50} - S_{100}}{100 - 50}$$

$$= \frac{1.16 - 0.81}{50}$$

$$= 7 \times 10^{-3}$$

$$= \frac{1}{143}$$

$$= 1:143$$

27. Falling head permeability test is used for less pervious soil (i.e. finer soil) but constant head permeability test is used for more pervious soil (*i.e.* coarser soil).

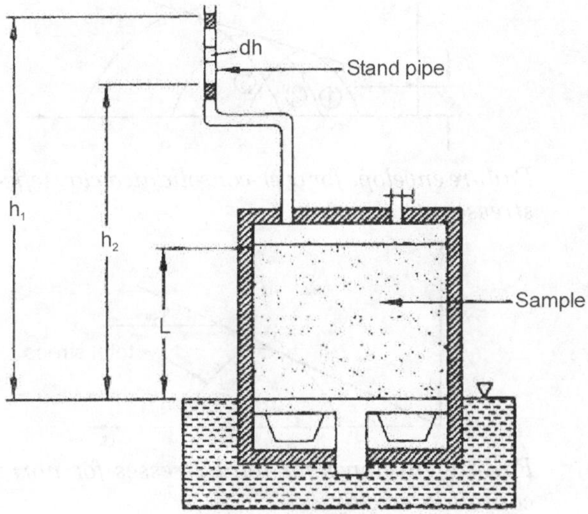

28. Consider instant when the head is h.

For infinitesimal small time dt, the head falls by dh.

Let discharge through the sample be q.

From continuity equation

$$adh = -qdt$$

where, q = cross-sectional area of the stand pipe.

$$\therefore \quad adh = -(A \times k \times i) \times dt$$

$$\Rightarrow \quad adh = -Ak\frac{h}{L} \times dt$$

$$\Rightarrow \quad \frac{Akdt}{aL} = -\frac{dh}{h}$$

$$\Rightarrow \quad \frac{Ak}{aL}\int_{t_1}^{t_2} dt = -\int_{h_1}^{t_2}\frac{dh}{h}$$

$$\Rightarrow \quad \frac{Ak}{aL}(t_2 - t) = \log_e\left(\frac{h_1}{h_2}\right)$$

$$\therefore \quad k = \frac{aL}{At}\log_e\left(\frac{h_1}{h_2}\right)$$

29. $Q = A \times v = A \times ki$

$$= A \times k \times \frac{h}{L} = \frac{\pi}{4}(1)^2 \times 1 \times 10^{-5} \times \frac{0.5}{10 \times 10}$$

$$= 3.93 \times 10^{-7} \text{m}^3/\text{s}$$

30.

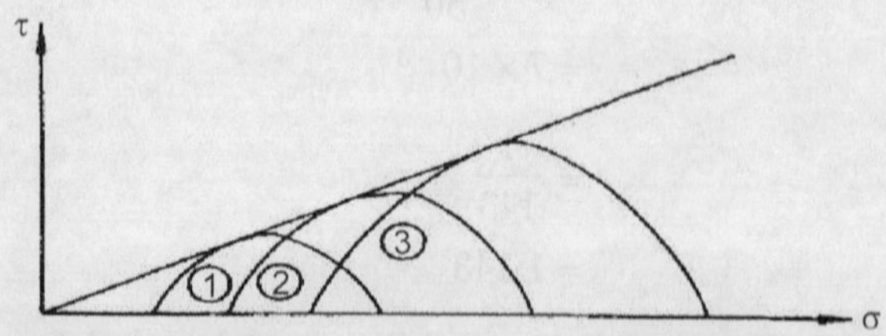

Failure envelope for normally consolidated clay (effective stresses)

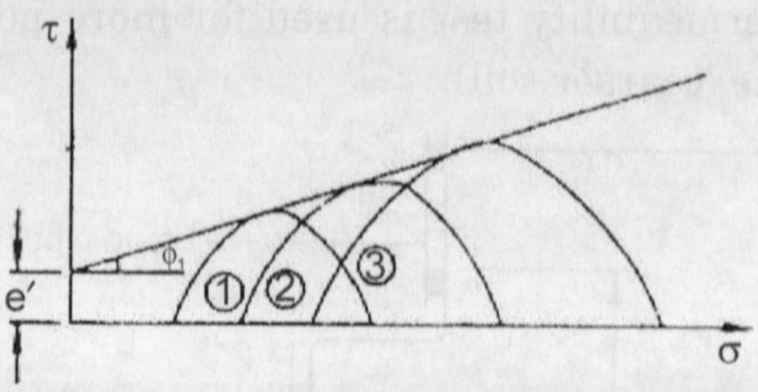

Failure envelope for over-consolidated clay (effective stresses)

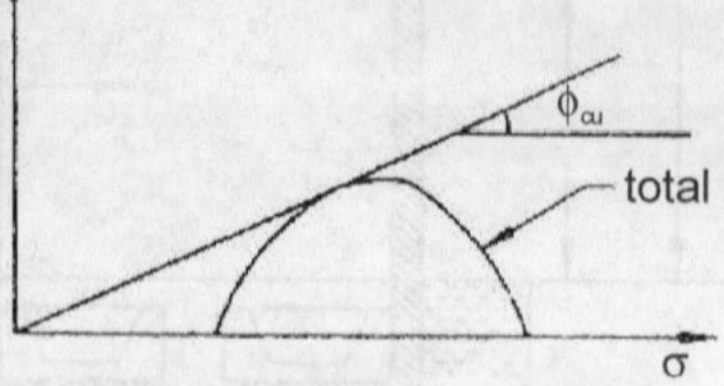

Failure envelopes for total stresses for normally consolidated clay

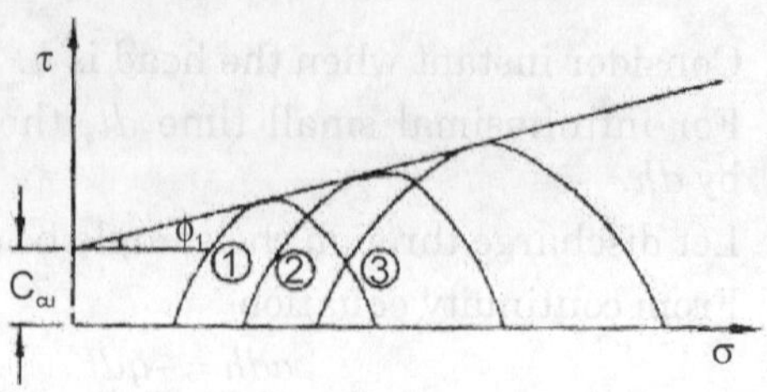

Failure envelope for over-consolidated clay (total stresses)

31. *Given :* $\sigma_3 = 100\text{kPa}$

$\sigma_d = 350 \text{ kPa}$

$$\therefore \quad \sigma_1 = \sigma_d + \sigma_3 = 450 \text{ kPa}$$

$$u = -40 \text{ kPa}$$

Total stress, $\sigma_1 = 450$ kPa

Effective stress $\bar{\sigma}_1 = \sigma_1 - u$

$$= 450 - (-40)$$

$$= 490 \text{ kPa}$$

We know, $\sigma = \sigma_3 N_\phi + uN_\phi$

But for saturated sand, C = 0

$$\therefore \quad \sigma = \sigma_3 N\phi$$

$$\therefore \quad N_\phi = \frac{\sigma_1}{\sigma_3} = \frac{450}{100} = 4.5$$

Now $\tan^2\left(45° + \frac{\phi}{2}\right) = 4.5$

$$\Rightarrow \quad \tan\left(45° + \frac{\phi}{2}\right) = 2.12$$

$$\Rightarrow \quad 45 + \frac{\phi}{2} = \tan^{-1}(2.12) = 64.75°$$

$$\Rightarrow \quad \frac{\phi}{2} = 19.74$$

$$\therefore \quad \phi = 39.5°$$

Hence $\sigma = 450°\text{kPa}$

$\sigma = 49°\text{kPa}$

$e = 0$

32.

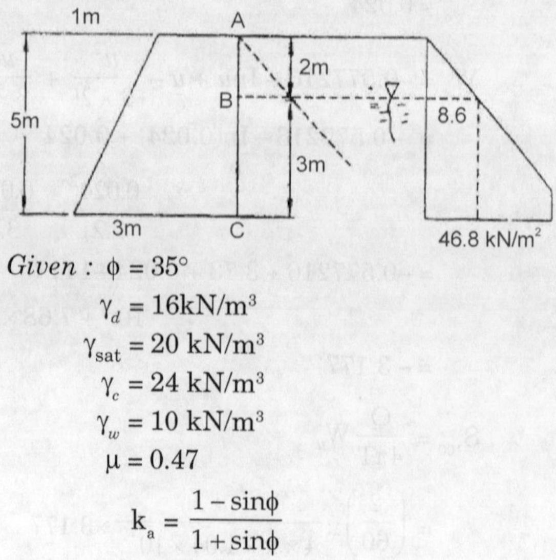

Given : $\phi = 35°$

$\gamma_d = 16\text{kN/m}^3$

$\gamma_{sat} = 20 \text{ kN/m}^3$

$\gamma_c = 24 \text{ kN/m}^3$

$\gamma_w = 10 \text{ kN/m}^3$

$\mu = 0.47$

$$k_a = \frac{1 - \sin\phi}{1 + \sin\phi}$$

$$= \frac{1 - \sin 35°}{1 + \sin 35°}$$

$$P_A = 0$$

$$P_B = k_a \gamma_d z$$

$$= 0.27 \times 16 \times 2$$

$$= 8.67 \, kN/\text{m}^2$$

$$P_C = k_a\gamma_d z_1 + k_a\gamma z_2 + \gamma_w z_2$$
$$= 0.27[16\times2+(20-10)\times3]+10\times3$$
$$= 0.27[32+30]+30 = 46.8\text{kN/m}^2$$

$$\text{Weight of the dam} = \frac{1}{2}(1+3)\times5\times1\times24$$
$$= 240\text{kN}$$

Net force in the horizontal direction

$$= \frac{1}{2}\times8.6\times2+8.6\times3+\frac{1}{2}\times38.2\times3$$
$$= 8.6 + 25.8 + 57.3$$
$$= 91.7 \text{ kN}$$

(*i*) Factor of safety against lateral sliding

$$= \frac{\mu W}{P_H}$$
$$= \frac{0.47\times240}{91.7} = 1.23$$

(*ii*) Factor of safety against overturning

$$= \frac{\text{Restoring moment}}{\text{Disturbing moment}}$$
$$= \frac{\left(\frac{1}{2}\times2\times5\times1\times24\right)\times\frac{2}{3}\times2+(1\times5\times1\times24)\times(2+0.5)}{\frac{1}{2}\times8.6\times2\left(3+\frac{1}{3}\times2\right)+8.6\times1.5+\frac{1}{2}\times38.2\times3\left(\frac{1}{3}\times3\right)}$$
$$= \frac{160+300}{31.53+38.7+57.3}$$
$$= 3.6.$$

(*iii*) $q_u = c'N_c + q_0N_q + 0.5\gamma BN_\gamma$

For cohenonless soil, c = 0

$$\therefore \quad q_u = q_0N_q + 0.5\gamma NN\gamma$$

Again, $\phi = 35°$

$N_q = 38$

$N_\gamma = 45$

$$\therefore \quad q_4 = (16\times2+20\times3)\,38+0.5\times20\times3\times45$$
$$= 3496 + 1350 = 4846\text{kN/m}^2.$$

33. $Q_g(u) = q_pA_g + \alpha e(P_gD)$

$$= 9\times25(3.5\times3.5)+1.0\times25\times(4\times3.5)\times8$$
$$+9\times40(3.5\times3.5)+0.7\times40(4\times3.5)\times10$$
$$= 13885.25\text{kN} \qquad (1)$$

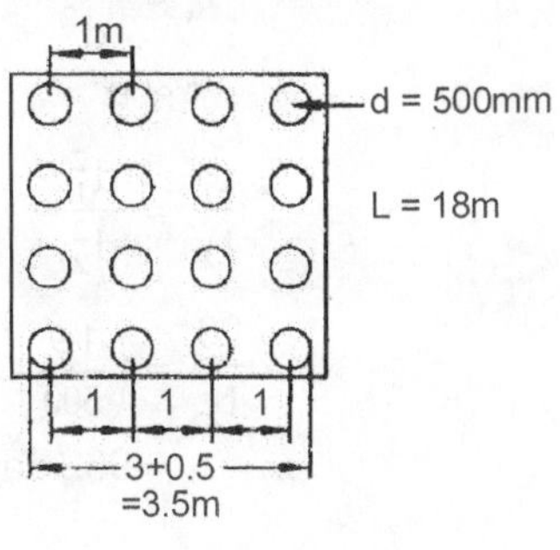

$$Q_u = q_pA_P + \alpha e(P\times D)$$
$$= (9\times25)\times\frac{\pi}{4}(0.5)^2+1.0\times25(\pi\times0.5)\times8$$
$$+9\times40\times\frac{\pi}{4}(0.5)^2+0.7\times40(\pi\times0.5)\times10$$
$$= 868.85\text{kN}$$

$$\therefore \quad Q_g(u) = NQ_u$$
$$= 16\times868.85 = 13901.5\text{kN} \qquad ...(2)$$

$$\text{Hence safe load} = \frac{\text{lesser of (1) and (2)}}{\text{F.S.}}$$
$$= \frac{13885.25}{2.5} = 5554.5\text{kN}$$

34.

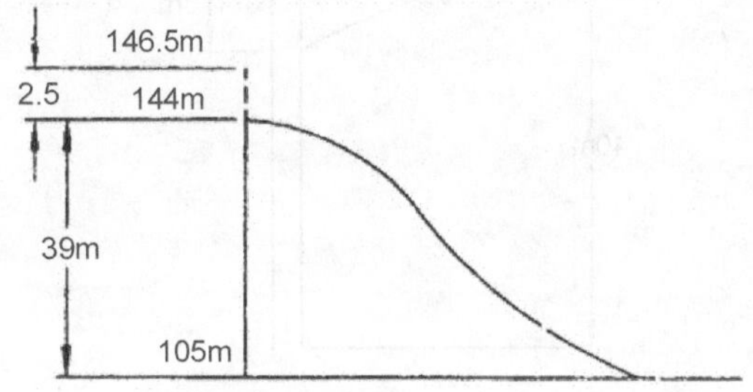

For unit length

$$Q = \frac{2}{3}C_dL\sqrt{2g}\,.\,h^{\frac{3}{2}}$$
$$= \frac{2}{3}\times0.73\times1\times\sqrt{2\times9.81}\times2.5^{\frac{3}{2}}$$
$$= 8.52$$

Average fall of water,

$$h_{av} = 39+\frac{2.5}{2} = 40.25$$

$$\text{Theoretical velocity} = \sqrt{2\times g\times h_{av}}$$
$$= \sqrt{2\times9.81\times40.25}$$
$$= 28.10\text{m/sec}$$

$$\text{Actual velocity} = 0.9\times28\times10$$
$$= 25.30 \text{ m/sec}$$

Depth of flow at foot of spillway,

$$y = \frac{Q}{V}$$
$$= \frac{8.52}{25.30} = 0.3370 \text{ m}$$

$$F_1 = \frac{V_1}{\sqrt{gy_1}}$$
$$= \frac{25.30}{\sqrt{9.81\times0.3370}}$$
$$= 13.91$$

$$y_2 = \frac{y_1}{2}\left[\sqrt{1+8F_1^2}-1\right]$$

$$= \frac{0.3370}{2}\left[\sqrt{1+8\times 13.9^2}-1\right]$$

$$= 6.46\text{m}$$

35. *Given :* diameter of well = 300mm

Thickness of confined aquifer = 40 m

k = 25m/day

R = 300m

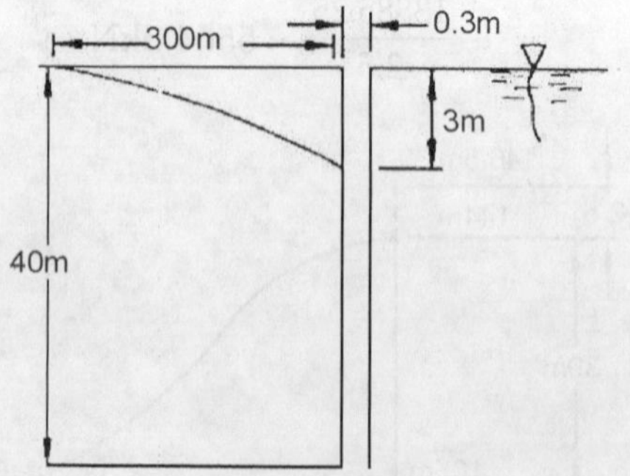

Now, $$k = \frac{q}{2\pi b(\tau_2-\tau_1)}\log_e\left(\frac{r_2}{r_1}\right)$$

$$\therefore \quad 25 = \frac{Q_1}{2\pi\times 40(40-37)}\log_e\left(\frac{300}{0.150}\right)$$

$$\Rightarrow \quad Q_1 = \frac{25\times 2\times\pi\times 40\times 3}{\log_e\left(\frac{300}{0.150}\right)}$$

$$= 2497.9 \text{ m}^3\text{/day}$$

$$= 103.3 \text{ m}^3\text{/hours.}$$

When diameter is made to 450 mm, then

$$25 = \frac{Q_2}{2\pi\times 40(40-37)}\log_e\left(\frac{300}{0.225}\right)$$

$$\Rightarrow \quad Q_2 = \frac{25\times 2\times\pi\times 40\times 3}{\log_e\left(\frac{300}{0.225}\right)}$$

$$= 2619.65 \text{ m}^3\text{/day}$$

$$= 1092 \text{ m}^3\text{/hour.}$$

36.

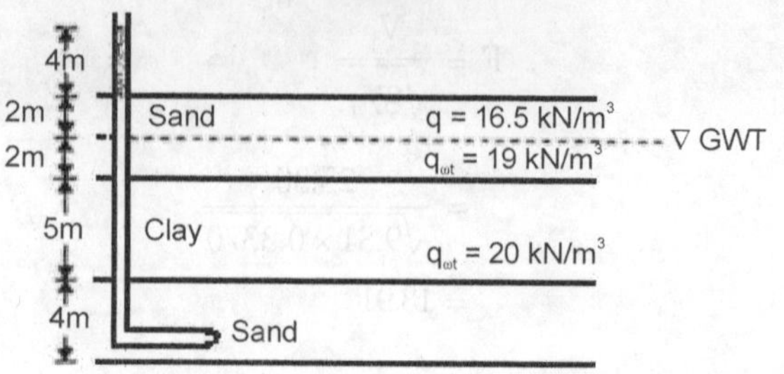

(a) At top surface of clay

Total stress, $\sigma = \sum qH$

$$= 2\times 16.5\times 2\times 19$$

$$= 71 \text{ kN/m}^2$$

$$\therefore \quad U = H_w\, q_w$$

$$= 2\times 10$$

$$= 20 \text{ kN/m}^2$$

Hence effective stress,

$$\sigma' = \sigma - U$$

$$= 71 - 20$$

$$= 51 \text{ kN/m}^2$$

(b) At bottom of clay layer

$$\sigma = 2\times 16.5 + 2\times 19 + 5\times 20$$

$$= 171 \text{ kN/m}^2$$

$$U = (4+4+5)\times 10$$

$$= 130 \text{ kN/mm}^2$$

$$\therefore \quad \sigma' = 171 - 130 = 41 \text{ kN/m}^2$$

At ground water table level

$$\sigma = 2\times 16.5 = 33 \text{ kN/m}^2$$

At bottom of lower sand

$$\sigma = 171 + 4\times 19$$

$$= 247 \text{ kN/m}^2$$

Total vertical stress diagram

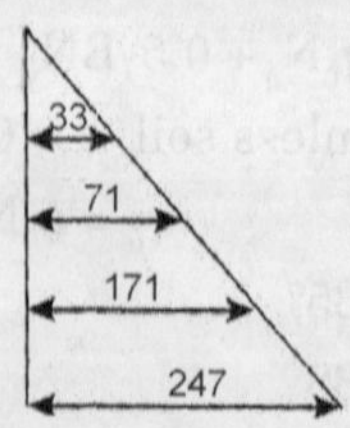

37. *Given :* d_i = 19mm

U_1 = 50%

t_i = 20min

d_2 = 5 m

Since for same degree of consolidation T_v and C_v are constants, therefore

$$T_v = \frac{C_v.t}{d^2}$$

$$t \propto d^2$$

$$\Rightarrow \quad \frac{t_1}{t_2} = \frac{d_1^2}{d_2^2}$$

$$\therefore \quad \frac{20}{t_2} = \left(\frac{1.9}{500}\right)^2$$

$$\Rightarrow \quad t_2 = 962 \text{ days}$$

Now $$(T_v)_{50\%} = \frac{\pi}{4}\left(\frac{U}{100}\right)^2$$

$$= \frac{\pi}{4}\left(\frac{50}{100}\right)^2$$

$$= 0.196$$

and $$(T_v)_{30\%} = \frac{\pi}{4}\left(\frac{30}{1000}\right)^2 = 0.0707$$

Then $$\frac{T_{u1}}{T_{u2}} = \frac{t_1}{t_3} \times \left(\frac{500}{1.9}\right)^2$$

$$\therefore \quad \frac{0.196}{0.0707} = \frac{20}{t_3}(263.15)^2$$

$$\Rightarrow \quad t_3 = 496.6 \times 10^3 \text{ min}$$

$$= 347 \text{ days.}$$

38. *Given :* $B = 3m$

$D = 1.2$ m

$C' = 0$

$q_{sat} = 20kN/m^3$

$N_q = 22$

$N_q = 20$

$\phi' = 30°$

$q = 17$ kN/m³

Net Safe bearing capacity = $q_{ns} = 350$ kN/m²

Now, $$q_{ns} = \frac{1}{F}[1.3\,CN_c + qD\,(N_q - 1)\,R_{w1} + 0.3\,qB\,N_q R_{w2}] \quad ...(i)$$

Case I: When water table in 5 m below ground level

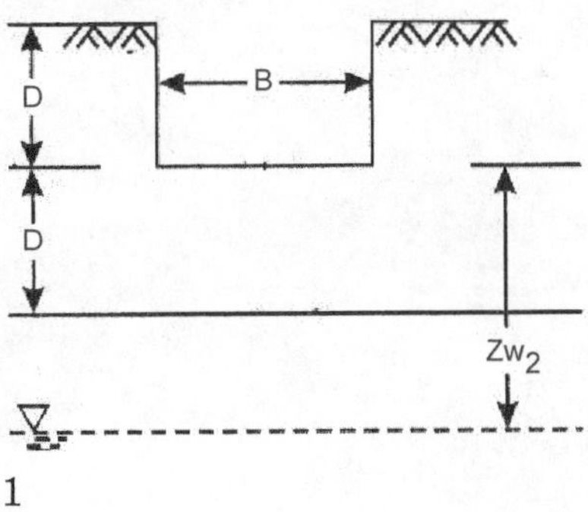

$R_{w1} = 1$

$R_{w2} = 1$

and $q = 17$

Hence from equation (i), we have

$$350 = \frac{1}{F}[17 \times 1.2\,(22 - 1) + 0.4 \times 17 \times 3 \times 20 \times 1]$$

$$\Rightarrow \quad F = \frac{1}{350}[428.4 + 408]$$

$$= 2.39$$

Case II: At D varies 0 to 1.2 m

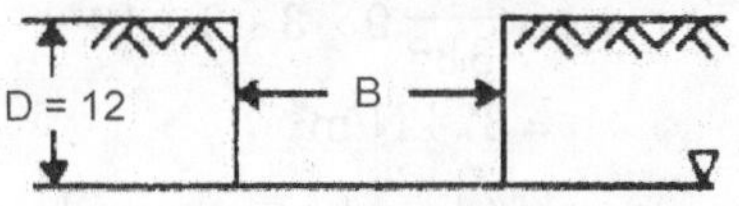

$R_{wl} = 1$ & $q = 17$ kN/m³

At D > 1.2 m; $R_{w2} = 0.5$ and $q = q_{sat} = 20$

From equation (i), we have

$$350 = \frac{1}{F}\left[qB(N_q - 1) + 0.4\,q_{sat}\,BN_q \times 0.5\right]$$

$$\text{or} \quad F = \frac{1}{350}[17 \times 1.2 \times 21 + 0.4 \times 20 \times 3 \times 20 \times 0.5]$$

$$= 1.91$$

39. (*a*) At 3 m below ground

$$k_{a1} = \frac{1 - \sin 30°}{1 + \sin 30°} = \frac{1}{3}$$

$$C_1 = 0$$

$$\Rightarrow \quad P_{a1} = ka_1\,q_1\,H_1$$

$$= \frac{1}{3} \times 16 \times 3$$

$$= 16 \text{ kN/m}^2$$

and $$P_a = \frac{1}{2}k_{a\,q}H^2$$

$$= \frac{1}{2} \times \frac{1}{3} \times 16 \times 9$$

$$= 24 \text{ kN/m}$$

It will act at 2 m below ground.

(b) At 6 m below ground

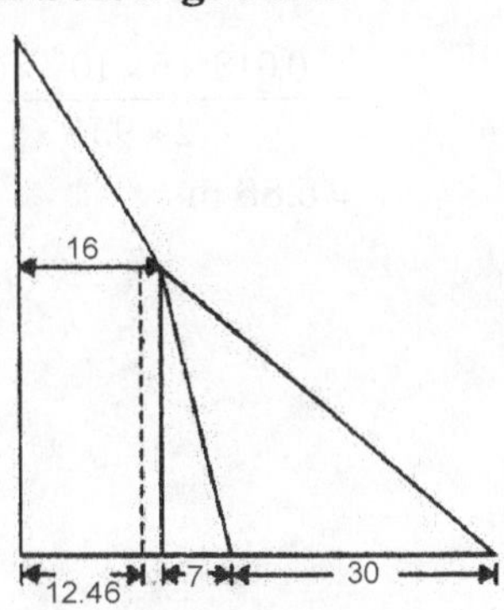

$$k_{a2} = \frac{1 - \sin 36°}{1 + \sin 36°} = \frac{1}{3.85}$$

$$k_{a2} < k_{a1}$$

$$P_{a2} = k_{a2}\,q_2\,H_2 + q_w H$$

$$\therefore \quad q_{sat} = q' - q_w$$

$$\Rightarrow \quad q' = 19 - 10$$

$$= 9 \text{ kN/m}^2$$

$$\therefore \quad p_{a2} = \frac{1}{3.85} 9 \times 3 + 3 \times 10$$
$$= 37 \text{ kN/m}^2$$

and $$p_{a2} = \left(\frac{1}{2} \times \frac{1}{3.85} \times 9 \times 9\right) + \frac{1}{2} \times 10 \times 9$$
$$= 55.5 \text{ kN/m}$$

At G.W.T., force diagram will be shift by

$$\frac{k_{a2}}{k_{a1}} = \frac{3}{3.85} = 0.78$$

$\therefore$ Ordinate of force = 16 × 0.78 = 1246

Resultant of force acts at $\bar{x}$ m from base

$$\bar{x} = \frac{24\,(3+1)+16\times1.5+37\times1}{24+16+37}$$
$$= 2.04 \text{ m from ground}$$

40. *Given :* Q = 1.4 m³/sec

L = 10 km

Using suffix S for steel and P for PSC pipes.

(*a*) In steel pipe :

$d_s = 1.2$ m

$L_s = 5$ m,

$f_s = 0.018$

$$\therefore \quad A_s = \frac{\pi}{4}(1.2)^2 = 1.131 \text{ m}^2$$

$$\Rightarrow \quad V_s = \frac{Q}{A_s}$$
$$= \frac{1.14}{1.13} = 1.238 \text{ m/sec.}$$

$$\therefore \text{ Head loss, } h_L = \frac{f_s L_s V_s^2}{2gd_s^2}$$
$$= \frac{0.018 \times 5 \times 10^3 \times (1.238)^2}{2 \times 9.18 \times 1.2}$$
$$= 5.86 \text{ m}$$

$\therefore$ Minor loss = 0.1 hL = 0.586 m

Hence total head loss in steel pipe,

$$H_{Ls} = 6.444 \text{ m}$$

(b) In PSC pipe

$d_p = 1$ m

$L_p = 5$ km

$f_p = 0.02$

$$\therefore \quad A_p = \frac{\pi}{4}(1)^2 = 0.7854$$

and $$V_p = \frac{1.4}{0.7854} = 1.78 \text{ m/sec}$$

$$\Rightarrow \quad h_{Lp} = \frac{0.02 \times 5 \times 10^3 \times 1.78}{2 \times 9.81 \times 1}$$
$$= 9.07 \text{ m}$$

$\therefore$ Minor losses = 0.907 m

$$\Rightarrow \quad H_{Lp} = 9.98 \text{ m}$$

Now level difference in sump & reservior = 50 m

$\therefore$ Total head lift required,

$$H' = 50 + 6.444 + 9.98$$
$$= 66.42$$

B.H.P. of pump required for η = 0.80

$$\text{H.P.} = \frac{WQH'}{75\eta}$$
$$= \frac{1000 \times 1.4 \times 66.42}{75 \times 0.8}$$
$$= 1550 \text{ rpm}$$

■■

8 CHAPTER Fluid Mechanics and Hydraulics

MCQ TYPE QUESTIONS

2016

1. A 4 m wide rectangular channel, having bed slope of 0.001 carries a discharge of 16 m^3/s. Considering Manning's roughness coefficient = 0.012 and g = 10 m/s^2, the category of the channel slope is

(*a*) horizontal (*b*) mild
(*c*) critical (*d*) steep

2. Group I contains the types of fluids while Group II contains the shear stress - rate of shear relationship of different types of fluids, as shown in the figure.

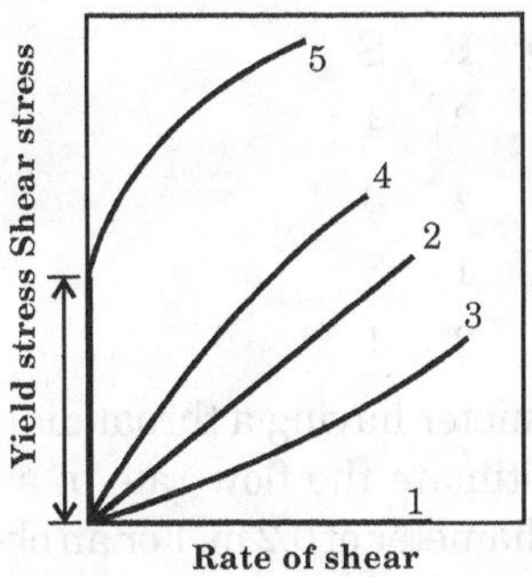

Group I	Group II
P. Newtonian fluid	1. Curve 1
Q. Pseudo plastic fluid	2. Curve 2
R. Plastic fluid	3. Curve 3
S. Dilatant fluid	4. Curve 4
	5. Curve 5

The correct match between Group I and Group II is

(*a*) P-2, Q-4, R-1, S-5
(*b*) P-2, Q-5, R-4, S-1
(*c*) P-2, Q-4, R-5, S-3
(*d*) P-2, Q-1, R-3, S-4

2015

3. For steady incompressible flow through a closed-conduit of uniform cross-section, the direction of flow will always be:

(*a*) from higher to lower elevation
(*b*) from higher to lower pressure
(*c*) from higher to lower velocity
(*d*) from higher to lower piezometric head

4. In a two-dimensional steady flow field, in a certain region of the *x–y* plane, the velocity component in the *x*-direction is given by $V_x = x^2$ and the density varies as $\rho = \frac{1}{x}$. Which of the following is a valid expression for the velocity component in the y-direction, v_y ?

(*a*) $v_y = -\frac{x}{y}$ (*b*) $v_y = \frac{x}{y}$
(*c*) $v_y = -xy$ (*d*) $v_y = xy$

5. The drag force, F_D, on a sphere kept in a uniform flow field depends on the diameter of the sphere, D, flow velocity, V; fluid density, ρ; and dynamic viscosity, μ. Which of the following options represents the non-dimensional parameters which could be used to analyze this problem?

(*a*) $\frac{F_D}{VD}$ and $\frac{\mu}{\rho VD}$ (*b*) $\frac{F_D}{\rho VD^2}$ and $\frac{\rho VD}{\mu}$
(*c*) $\frac{F_D}{\rho V^2D^2}$ and $\frac{\rho VD}{\mu}$ (*d*) $\frac{F_D}{\rho V^3D^3}$ and $\frac{\mu}{\rho VD}$

6. The relationship between the length scale ratio (L_r) and the velocity scale ratio (V_r) in hydraulic models, in which Froude dynamic similarity is maintained, is

(*a*) $V_r = L_r$ (*b*) $L_r = \sqrt{V_r}$
(*c*) $V_r = L_r^{1.5}$ (*d*) $V_r = \sqrt{L_r}$

7. A hydraulic jump takes place in a frictionless rectangular channel. The pre-jump depth is y_p. The alternate and sequent depths corresponding to y_p are y_a and y_s respectively. The correct relationship among y_p, y_a and y_s is

(*a*) $y_a < y_s < y_p$ (*b*) $y_p < y_s < y_a$
(*c*) $y_p < y_s = y_a$ (*d*) $y_a = y_s = y_p$

8. The velocity components of a two dimensional plane motion of a fluid are

$$u = \frac{y^3}{3} + 2x - x^2y \text{ and } v = xy^2 - 2y - \frac{x^3}{3}$$

The correct statement is:

(*a*) Fluid is incompressible and flow is irrotational
(*b*) Fluid is incompressible and flow is rotational
(*c*) Fluid is compressible and flow is irrotational
(*d*) Fluid is compressible and flow is rotational

2014

9. The dimension for kinematic viscosity is

(a) $\frac{L}{MT}$ (b) $\frac{L}{T^2}$

(c) $\frac{L^2}{T}$ (d) $\frac{ML}{T}$

10. A horizontal jet of water with its cross-sectional area of 0.0028 m² hits a fixed vertical plate with a velocity of 5 m/s. After impact the jet splits symmetrically in a plane parallel to the plane of the plate. The force of impact (in N) of the jet on the plate is

(a) 90 (b) 80

(c) 70 (d) 60

11. A rectangular channel having a bed slope of 0.0001, width 3.0 m and Manning's coefficient 'n' 0.015, carries a discharge of 1.0 m³/s. Given that the normal depth of flow ranges between 0.76 m and 0.8 m. The minimum width of a throat (in m) that is possible at a given section, while ensuring that the prevailing normal depth is not exceeded along the reach upstream of the contraction, is approximately equal to (assume negligible losses)

(a) 0.64 (b) 0.84

(c) 1.04 (d) 1.24

12. Three rigid buckets, shown as in the figures (1), (2) and (3), are of identical heights and base areas. Further, assume that each of these buckets have negligible mass and are full of water. The weights of water in these buckets are denoted as W_1, W_2, and W_3 respectively. Also, let the force of water on the base of the bucket be denoted as F_1, F_2, and F_3 respectively. The option giving an accurate description of the system physics is

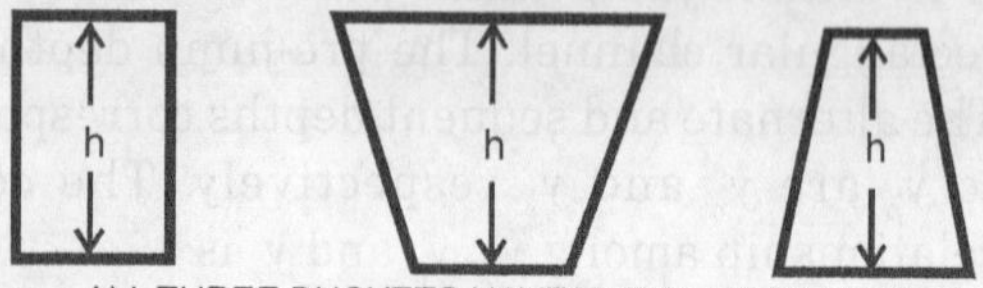

ALL THREE BUCKETS HAVE THE SAME BASE AREA

(1) (2) (3)

(a) $W_2 = W_1 = W_3$ and $F_2 > F_1 > F_3$

(b) $W_2 > W_1 > W_3$ and $F_2 > F_1 > F_3$

(c) $W_2 = W_1 = W_3$ and $F_1 = F_2 = F_3$

(d) $W_2 > W_1 > W_3$ and $F_1 = F_2 = F_3$

13. A plane flow has velocity components $u = \frac{x}{T_1}, v = -\frac{y}{T_2}$ and $w = 0$ along x, y and z directions respectively, where $T_1 (\neq 0)$ and $T_2 (\neq 0)$ are constants having the dimension of time. The given flow is incompressible if

(a) $T_1 = -T_2$ (b) $T_1 = -\frac{T_2}{2}$

(c) $T_1 = \frac{T_2}{2}$ (d) $T_1 = T_2$

14. **Group I** lists a few devices while **Group II** provides information about their uses. Match the devices with their corresponding use.

Group I	Group II
P. Anemometer	1. Capillary potential of soil water
Q. Hygrometer	2. Fluid velocity at a specific point in the flow stream
R. Pitot Tube	3. Water vapour content of air
S. Tensiometer	4. Wind speed

Codes:

	P	Q	R	S
(a)	1	2	3	4
(b)	2	1	4	3
(c)	4	2	1	3
(d)	4	3	2	1

15. A venturimeter having a throat diameter of 0.1 m is used to estimate the flow rate of a horizontal pipe having a diameter of 0.2 m. For an observed pressure difference of 2 m of water head and coefficient of discharge equal to unity, assuming that the energy losses are negligible, the flow rate (in m³/s) through the pipe is approximately equal to

(a) 0.500

(b) 0.150

(c) 0.050

(d) 0.015

16. With reference to a standard Cartesian (x, y) plane, the parabolic velocity distribution profile of fully developed laminar flow in x-direction between two parallel, stationary and identical plates that are separated by distance, h, is given by the expression

$$u = -\frac{h^2}{8\mu}\frac{dp}{dx}\left[1 - 4\left(\frac{y}{h}\right)^2\right]$$

In this equation, the y = 0 axis lies equidistant between the plates at a distance h/2 from the two plates, p is the pressure variable and μ is the dynamic viscosity term. The maximum and average velocities are, respectively

(a) $u_{max} = -\frac{h^2}{8\mu}\frac{dp}{dx}$ and $u_{average} = \frac{2}{3}u_{max}$

(b) $u_{max} = \frac{h^2}{8\mu}\frac{dp}{dx}$ and $u_{average} = \frac{2}{3}u_{max}$

(c) $u_{max} = -\frac{h^2}{8\mu}\frac{dp}{dx}$ and $u_{average} = \frac{3}{8}u_{max}$

(d) $u_{max} = \frac{h^2}{8\mu}\frac{dp}{dx}$ and $u_{average} = \frac{3}{8}u_{max}$

2013

17. For subcritical flow in an open channel, the control section for gradually varied flow profiles is

(a) at the downstream end

(b) at the upstream end

(c) at both upstream and downstream ends

(d) at any intermediate section

18. Group-I contains dimensionless parameters and Group- II contains the ratios.

Group-I	Group -II
P. Mach Number	1. Ratio of inertial force and gravitational force
Q. Reynolds Number	2. Ratio of fluid velocity and velocity of sound
R. Weber Number	3. Ratio of inertial force and viscous force
S. Froude Number	4. Ratio of inertial force and surface tension force

The correct match of dimensionless parameters in Group- I with ratios in Group-II is

(a) P-3, Q-2, R-4, S-1 (b) P-3, Q-4, R-2, S-1

(c) P-2, Q-3, R-4, S-1 (d) P-1, Q-3, R-2, S-4

19. For a two dimensional flow field, the stream function ψ is given n as $\psi = \frac{3}{2}(y^2 - x^2)$. The magnitude of discharge occurring between the stream lines passing through points (0,3) and (3,4) is:

(a) 6 (b) 3

(c) 1.5 (d) 2

2012

20. A trapezoidal channel is 10.0 m wide at the base and has a side slope of 4 horizontal to 3 vertical. The bed slope is 0.002. The channel is lined with smooth concrete (Manning's n = 0.012). The hydraulic radius (in m) for a depth of flow of 3.0 m is

(a) 20.0 (b) 3.5

(c) 3.0 (d) 2.1

21. A rectangular open channel of width 5.0 m is carrying a discharge of 100 m³/s. The Froude number of the flow is 0.8. The depth of flow (in m) in the channel is

(a) 4 (b) 5

(c) 16 (d) 20

22. The circular water pipes shown in the sketch are flowing full. The velocity of flow (in m/s) in the branch pipe "R" is

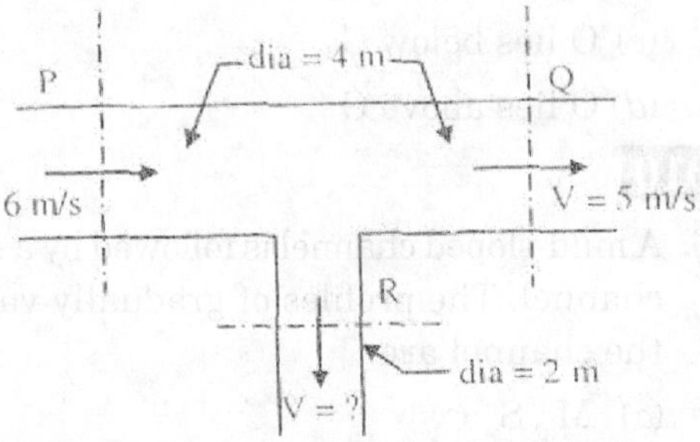

(a) 3 (b) 4

(c) 5 (d) 6

23. The top width and the depth of flow in a triangular channel were measured as 4 m and 1 m, respectively. The measured velocities on the centre line at the water surface, 0.2 m and 0.8 m below the surface are 0.7 m/s, 0.6 m/s and 0.4 m/s, respectively. Using two-point method of velocity measurement, the discharge (in m³/s) in the channel is

(a) 1.4

(b) 1.2

(c) 1.0

(d) 0.8

24. Group I contains parameters and **Group II** lists methods/instruments.

Group I	Group II
P. Stream flow velocity	1. Anemometer
Q. Evapo-transpiration rate	2. Penman's method
R. Infiltration rate	3. Horton's method
S. Wind velocity	4. Current meter

The **CORRECT** match of **Group I** with **Group II** is

(a) P – 1, Q – 2, R – 3, S – 4

(b) P – 4, Q – 3, R – 2, S – 1

(c) P – 4, Q – 2, R – 3, S – 1

(d) P – 1, Q – 3, R – 2, S – 4

2011

25. For a body completely submerged in a fluid, the centre of gravity (G) and centre of Buoyancy (O) are known. The body is considered to be in stable equilibrium if

(*a*) O does not coincide with the centre of mass of the displaced fluid

(*b*) G coincides with the centre of mass of the displaced fluid

(*c*) O lies below G

(*d*) O lies above G

2010

26. A mild-sloped channel is followed by a steep-sloped channel. The profiles of gradually varied flow in the channel are

(*a*) M_3, S_2

(*b*) M_3, S_3

(*c*) M_2, S_1

(*d*) M_2, S_2

27. The flow in a rectangular channel is subcritical. If width of the channel is reduced at a certain section, the water surface under no-choke condition will

(*a*) drop at a downstream section

(*b*) rise at a downstream section

(*c*) rise at an upstream section

(*d*) not undergo any change

28. Group -I gives a list of devices and Group -II gives the list of uses.

Group-I	Group-II
P. Pitot tube	1. measuring pressure in a pipe
Q. Manometer	2. measuring velocity of low in a pipe
R. Venturimeter	3. measuring air and gas velocity
S. Anemometer	4. measuring discharge in a pipe

The correct match of Group-I with Group-II is

Codes :

	P	Q	R	S
(*a*)	1	2	4	3
(*b*)	2	1	3	4
(*c*)	2	1	4	3
(*d*)	4	1	3	2

29. For a rectangular channel seetion, Group-I lists geometrical elements and Group-II gives proportions for hydraulically efficient section.

Group-I	Group-II
P. Top width	1. $\frac{y_e}{2}$
Q. Perimeter	2. y_e
R. Hydraulic Radius	3. $2y_e$
S. Dydraulic Depth	4. $4y_e$

y_e is flow depth corresponding to hydraulically efficient section.

The correct match of Group-I with Group-II is

Codes :

	P	Q	R	S
(*a*)	2	4	1	3
(*b*)	3	1	4	2
(*c*)	3	4	1	2
(*d*)	3	4	2	1

30. The Froude number of flow in a rectangular channel is 0.8. If the depth of flow is 1.5 m, then critical depth is

(*a*) 1.80 m (*b*) 1.56 m

(*c*) 1.36 m (*d*) 1.29 m

2009

31. Water flows through a 100 mm diameter pipe with a velocity of 0.015 m/sec. If kinematic viscosity of water is 1.13×10^{-6} m^2/sec, then friction factor of the pipe material is

(*a*) 0.0015 (*b*) 0.032

(*c*) 0.037 (*d*) 0.048

32. A rectangular open channel of width 4.5 m is carrying a discharge of 100 m^3/sec. The critical depth of the channel is

(*a*) 7.09 m (*b*) 3.69 m

(*c*) 2.16 m (*d*) 1.31 m

33. Water (γ_w = 9.879 kN/m^3) flows with flow rate of 0.3 m^3/sec through a pipe AB of 10 m length and of uniform cross section. The end 'B' is above end 'A' and the pipe makes an angle of 30° to the horizontal. For a pressure of 12 kN/m^2 at the end 'B', corresponding pressure at the end 'A' is

(*a*) 12.0 kN/m^2

(*b*) 17.0 kN/m^2

(*c*) 56.4 kN/m^2

(*d*) 61.4 kN/m^2

2008

34. The flow of water (mass density = 1000 kg/m³ and kinematic viscosity = 10^{-6} m²/s) in a commercial pipe, having equivalent roughness k_s as 0.12 mm, yields an average shear stress at the pipe boundary = 600 N/m². The value of k_s/δ'(δ' being thickness of laminar sub-layer) for this pipe is

(*a*) 0.25 (*b*) 0.50
(*c*) 6.0 (*d*) 8.0

2007

35. At two points 1 and 2 in a pipeline, the velocities are V and 2V, respectively. Both the points are at the same elevation. The fluid density is ρ. The flow can be assumed to be in compressible, inviscid, steady and irrotational. The difference in pressures P_1 and P_2 at points 1 and 2 is

(*a*) 0.5 ρV² (*b*) 1.5 ρV²
(*c*) 2 ρV² (*d*) 3 ρV²

36. A horizontal water jet with a velocity of 10 m/s and cross sectional area of 10 mm² strikes a flat plate held normal to the flow direction. The density of water is 1000 kg/m³. The total force on the plate due to the jet is

(*a*) 100 N (*b*) 10 N
(*c*) 1 N (*d*) 0.1 N

37. A 1 : 50 scale model of a spillway is to be tested in the laboratory. The discharge in the prototype is 1000 M³/s. The discharge to be maintained in the model test is

(*a*) 0.057 m³/s (*b*) 0.08 m²/s
(*c*) 0.57 m³/s (*d*) 5.7 m³/s

38. A triangular open channel has a vertex angle of 90° and carries flow at a critical depth of 0.30 m. The discharge in the channel is

(*a*) 0.08 m³/s (*b*) 0.11 m³/s
(*c*) 0.15 m³/s (*d*) 0.2 m³/s

39. Flow rate of a fluid (density = 1000 kg/m³) in a small diameter tube is 800 mm³/s. The length and the diameter of the tube are 2 m and 0.5 mm, respectively. The pressure drop in 2 m length is equal to 2.0 MPa. The viscosity of the fluid is

(*a*) 0.025 N.s/m² (*b*) 0.012 N.s/m²
(*c*) 0.00192 N.s/m² (*d*) 0.00102 N.s/m²

40. The flow rate in a wide rectangular open channel is 2.0 m³/s per metre width. The channel bed slope is 0.002. The Manning's roughness coefficient is 0.012. The slope of the channel is classified as

(*a*) Critical (*b*) Horizontal
(*c*) Mild (*d*) Steep

2006

41. Identify the FALSE statement from the following: The specific speed of the pump increases with

(*a*) increase in shaft speed
(*b*) increase in discharge
(*c*) decrease in gravitational acceleration
(*d*) increase in head

42. For steady flow to a fully penetrating well in a confined aquifer, the drawdowns at radial distances of r_1 and r_2 from the well have been measured as s_1 and s_2 respectively, for a pumping rate of Q. The transmissivity of the aquifer is equal to

(*a*) $\dfrac{Q}{2\pi}\dfrac{\ln\dfrac{r_2}{r_1}}{(s_1 - s_2)}$ (*b*) $\dfrac{Q}{2\pi}\dfrac{\ln(r_2 - r_1)}{(s_1 - s_2)}$

(*c*) $\dfrac{Q}{2\pi}\ln\left(\dfrac{r_2/r_1}{s_2/s_1}\right)$ (*d*) $2\pi Q\dfrac{r_2 - r_1}{\ln\left(\dfrac{s_2}{s_1}\right)}$

43. The thickness of the laminar boundary layer on a flat plate at a point A is 2 cm and at a point B, 1m downstream of A, is 3 cm. What is the distance of A from the leading edge of the plate?

(*a*) 0.50 m (*b*) 0.80 m
(*c*) 1.00 m (*d*) 1.25 m

44. The velocity field for flow is given by

$$\vec{V} = (5x + 6y + 7z)\hat{i} + (6x + 5y + 9z)\hat{j} + (3x + 2y + \lambda z)\hat{k}$$

and density varies as $\rho = \rho_0 \exp(-2t)$.

In order that the mass is conserved, the value of λ should be

(*a*) – 12 (*b*) – 10
(*c*) – 8 (*d*) 10

45. A hydraulic jump occurs in a rectangular, horizontal, frictionless channel. What would be the pre-jump depth if the discharge per unit width is 2m³/s/m and the energy loss is 1 m ?

(*a*) 0.2 m (*b*) 0.3 m
(*c*) 0.8 m (*d*) 0.9 m

46. A very wide rectangular channel is designed to carry a discharge of 5 m³/s per meter width. The design is based on the Manning's equation with the roughness coefficient obtained from the grain size using Strickler's equation and results in a normal depth of 1.0 m. By mistake, however, the engineer used the grain diameter in mm in the Strickler's equation instead of in meter. What should be the correct normal depth ?

(*a*) 0.32 m (*b*) 0.50 m
(*c*) 2.00 m (*d*) 3.20 m

47. The flow of glycerin (kinematic viscosity $v = 5 \times 10^{-4}$ m²/s) in an open channel is to be modeled in a laboratory flume using water ($v = 10^{-6}$ m²/s) as the flowing fluid. If both gravity and viscosity are important, what should be the length scale (*i.e.* ratio of prototype to model dimensions) for maintaining dynamic similarity?

(*a*) 1 (*b*) 22

(*c*) 63 (*d*) 500

2005

48. An inert tracer is injected continuously from a point in an unsteady flow field. The locus of locations of all the tracer particles at an instance of time represents

(*a*) Streamline (*b*) Pathline

(*c*) Steamtube (*d*) Streakline

49. A horizontal bed channel is followed by a steep bed channel as shown in the figure. The gradually-varied profiles over the horizontal and steep beds are

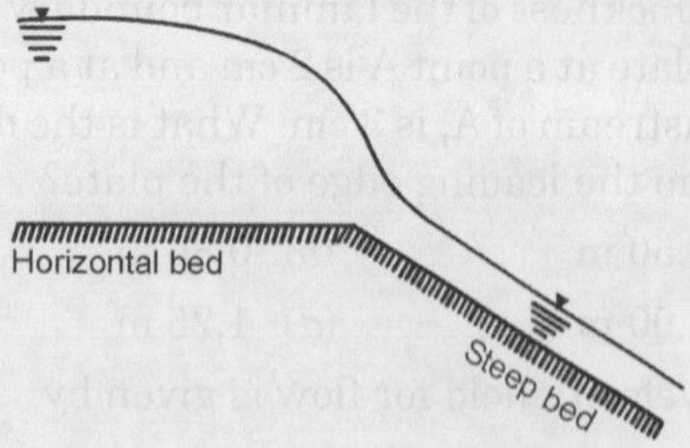

(*a*) H_2 and S_2 respectively

(*b*) H_2 and S_1 respectively

(*c*) H_3 and S_2 respectively

(*d*) H_3 and S_1 respectively

50. The reading of differential manometer of a Venturimeter, placed at 45° to the horizontal is 11 cm. If Ventruimeter is turned to horizontal position, the manometer reading will be

(*a*) zero (*b*) $\frac{11}{\sqrt{2}}$ cm

(*c*) 11 cm (*d*) $11\sqrt{2}$ cm

51. Critical depth at a section of a rectangular channel is 1.5 m. The specific energy at that section is

(*a*) 0.75 m (*b*) 1.0 m

(*c*) 1.5 m (*d*) 2.25 m

52. A partially open sluice gate discharges water into a rectangular channel. The tail water depth in the channel is 3 m and Froude number is $\frac{1}{2\sqrt{2}}$. If a free hydraulic jump is to be formed at a downstream of the sluice gate after the vena contracta of the jet coming out from the sluice gate, the sluice gate opening should be (coefficient of contraction, $C_c = 0.9$)

(*a*) 0.3 m (*b*) 0.4 m

(*c*) 0.69 m (*d*) 0.9 m

53. A stream function is given by

$$Y = 2x^2y + (x + 1)y^2$$

The flow rate across a line joining points A(3, 0) and B(0, 2) is

(*a*) 0.4 units

(*b*) 1.1 units

(*c*) 4 units

(*d*) 5 units

54. Cross-section of an object (having same section normal to the paper) submerged into a fluid consists of a square of sides 2 m and triangle as shown in the figure. The object is hinged at point P that is one meter below the fluid free surface. If the object is to be kept in the position as shown in the figure, the value of '*x*' should be

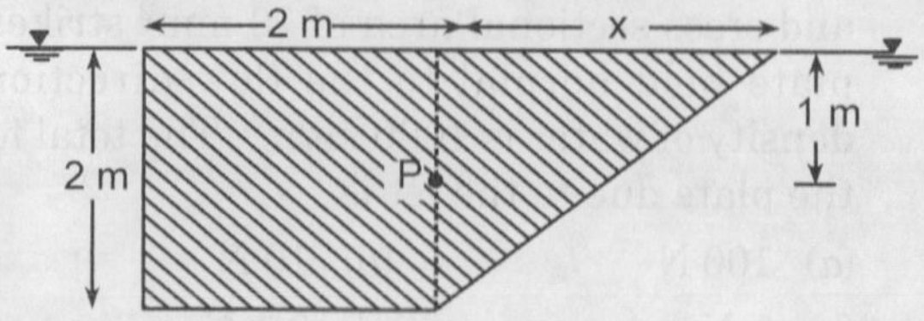

(*a*) $2\sqrt{3}$ (*b*) $4\sqrt{3}$

(*c*) 4 m (*d*) 8 m

55. The circulation 'G' around a circle of radius 2 units for the velocity field $u = 2x + 3y$ and $v = -2y$ is

(*a*) -6π units

(*b*) -12π units

(*c*) -18π units

(*d*) -24π units

56. A tank and a deflector are placed on a frictionless trolley. The tank issues water jet (mass density of water = 1000 kg/m³), which strikes the deflector and truns by 45°. If the velocity of jet leaving the deflector is 4 m/s and discharge is 0.1 m³/s, then force recorded by the spring will be

(*a*) 100 N (*b*) $100\sqrt{2}$ N

(*c*) 200 N (*d*) $200\sqrt{2}$ N

2004

57. In the inclined manometer shown in the figure below, the reservoir is large. Its surface may be assumed to remain at a fixed elevation. A is connected to a gas pipeline and the deflection noted on the inclined glass tube is 100 mm. Assuming θ = 30° and the manometric fluid as oil with specific gravity of 0.86, the pressure at A is

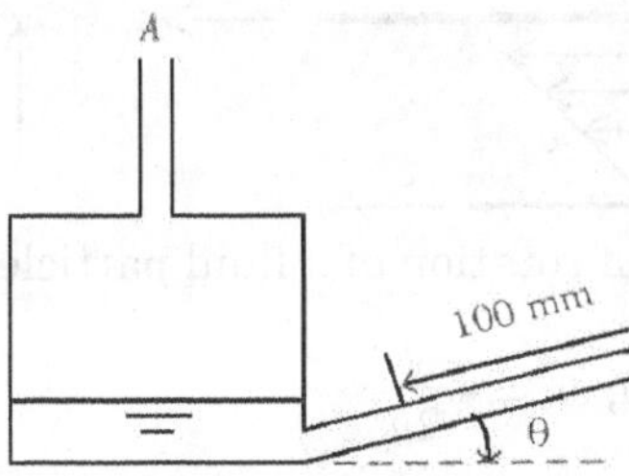

(*a*) 43 mm water (vacuum)

(*b*) 43 mm water

(*c*) 86 mm water

(*d*) 100 mm water

58. The x component of velocity in a two dimensional incompressible flow is given by

u = 1.5 x

At the point (*x*, *y*) = (1, 0), the y component of velocity v = 0. The equation for the y component of velocity is

(*a*) v = 0 (*b*) v = 1.5 y

(*c*) v = –1.5x (*d*) v = –1.5y

59. An aircraft is flying in level flight at a speed of 200 km/hr through air (*density*, ρ = 1.2 kg/m^2, and viscosity m = 1.6 × 10^{-5} N-s/m^2). The lift co-efficient at this speed is 0.4 and the drag co-efficient is 0.0065. The mass of the aircraft is 800 kg. The effective lift area of the aircraft is

(*a*) 21.2 m^2 (*b*) 10.6 m^2

(*c*) 2.2 m^2 (*d*) 1.1 m^2

60. A frictionless fluid of density r flow through a bent pipe as shown below. If A is the cross sectional area and V is the velocity of flow, the forces exerted on segment 1 – 2 of the pipe in the x and y directions are, respectively

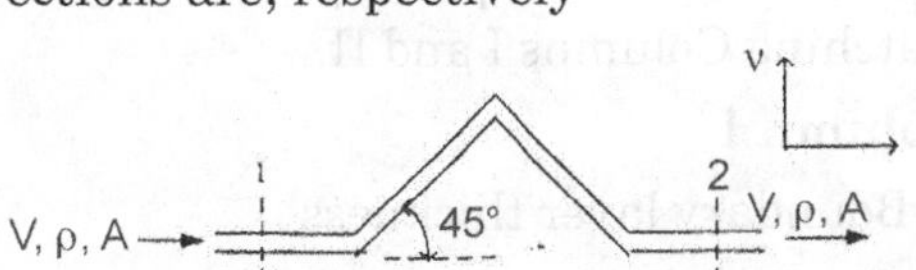

(*a*) ρAV^2; 0 (*b*) rAV^2; $\sqrt{2}$ ρAV^2

(*c*) 0 ; 0 (*d*) 0; $\frac{1}{\sqrt{2}}$ ρAV^2

61. For a pipe of radius, r, flowing half full under the action of gravity, the hydraulic depth is

(*a*) *r* (*b*) $\frac{\pi r}{4}$

(*c*) $\frac{r}{2}$ (*d*) 0379 *r*

62. A volocity field is given as

$$\vec{v} = 2y\hat{i} + 3x\hat{j}$$

where *x* and *y* are in metres.

The acceleration of the a fluid particle at (x, y) = (1, 1) in the x direction is

(*a*) 0 m/s^2

(*b*) 5.00 m/s^2

(*c*) 6.00 m/s^2

(*d*) 8.48 m/s^2

63. A thin flat plate 0.5 m × 0.7 m in size settles in a large tank of water with a terminal velocity of 0.12 m/s. The co-efficient of drag $C_D = \frac{1.328}{\sqrt{R_L}}$ for a laminar boundary layer and $C_D = \frac{0.072}{(R_L)^{1/5}}$ for a turbulent boundary layer, where R_L is the plate Reynolds number.

Assume μ = 10^{-3} N-s/m^2 and ρ = 1000 kg/m^3.

The submerged weight of the plate is

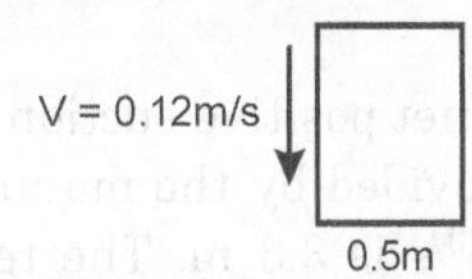

(*a*) 0.0115 N (*b*) 0.0118 N

(*c*) 0.0231 N (*d*) 0.0376 N

64. A fire protection system is supplied from a water tower with a bent pipe as shown in the figure. The pipe friction f is 0.03. Ignoring all minor losses, the maximum discharge, Q, in the pipe is

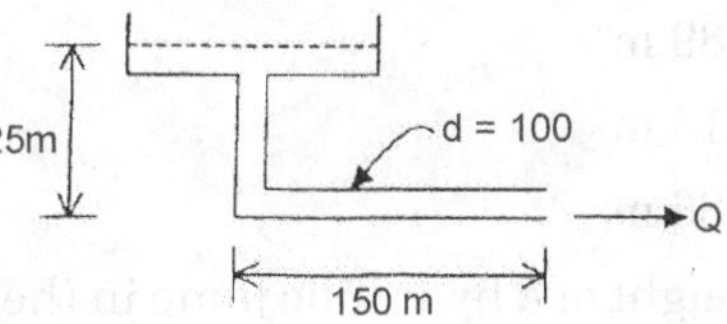

(*a*) 31.7 lit/sec (*b*) 24.0 lit/sec

(*c*) 15.9 lit/sec (*d*) 12.0 lit/sec

65. A steady flow occurs in an open channel with lateral inflow of q m^3/s per unit width as shown in the figure. The mass conservation equation is

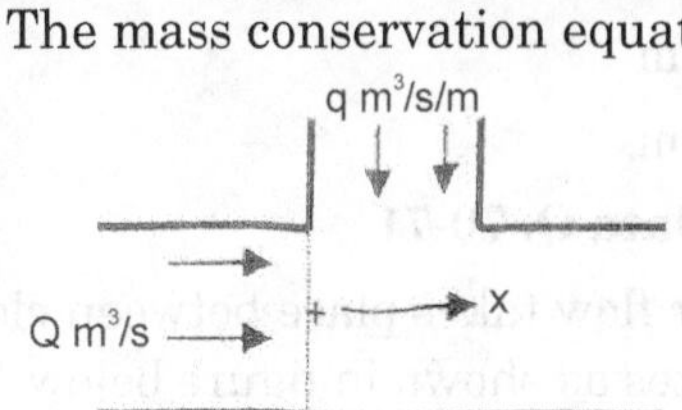

(*a*) $\frac{\partial q}{\partial x} = 0$

(*b*) $\frac{\partial Q}{\partial x} = 0$

(*c*) $\frac{\partial Q}{\partial x} - q = 0$

(*d*) $\frac{\partial Q}{\partial x} + q = 0$

66. A steep wide rectangular channel takes off from a reservoir having an elevation of 101.2 m. At the entrance, bottom elevation of the channel is 100 m. If slope of the channel is increased by 4%, then discharge per unit length in channel will be

(*a*) 2.24 m^2/s

(*b*) higher than 2.24 m^2/s by 4%

(*c*) higher than 2.24 m^2/s by 2%

(*d*) choked

67. The velocity in m/s at a point in a two-dimensional flow is given as $\vec{V} = 2\hat{i} + 3\hat{j}$. The equation of the stream line passing through the point is

(*a*) $3dx - 2dy = 0$

(*b*) $2x + 3y = 0$

(*c*) $3dx + 2dy = 0$

(*d*) $xy = 6$

68. The allowable net positive suction head (NPSH) for a pump provided by the manufacturer for a flow of 0.05 m^3/s is 3.3 m. The temperature of water is 30°C (vapour pressure head absolute = 0.44m), atmospheric pressure is 100 kPa absolute and the head loss from the reservoir to pump is 0.3 N-m/N. The maximum height of the pump above the suction reservoir is

(*a*) 10.19 m

(*b*) 6.89 m

(*c*) 6.15 m

(*d*) 2.86 m

69. The height of a hydraulic jump in the stilling pool of 1.25 scale model was observed to be 10 cm. The corresponding prototype height of the jump is

(*a*) not determinable from the data given

(*b*) 2.5 m

(*c*) 0.5 m

(*d*) 0.1 m

Common Data Q. 70-71

The laminar flow takes place between closely spaced parallel plates as shown in figure below. The velocity profile is given by

$$u = V\frac{y}{h}.$$

The gap height, h, is 5 mm and the space is filled with oil (specific gravity = 0.86, viscosity m = 2×10^{-4} N-s/m^2). The bottom plate is stationary and the top plate moves with a steady velocity of V = 5 cm/s. The area of the plate is 0.25 m^2.

70. The rate of rotation of a fluid particle is given by

(*a*) $\omega_y = 0; \omega_z = -\frac{y}{2h}$

(*b*) $\omega_y = 0; \omega_z = -\frac{y}{h}$

(*c*) $\omega_y = \frac{y}{h}; \omega_z = \frac{y}{h}$

(*d*) $\omega_y = \frac{y}{h}; \omega_z = 0$

71. The power required to keep the plate in steady motion is

(*a*) 5×10^{-4} watts (*b*) 10^{-5} watts

(*c*) 2.5×10^{-5} watts (*d*) 5×10^{-5} watts

2003

72. For a two-dimensional irrotational flow, the velocity potential is defined as

$$\phi = \ln(x^2 + y^2)$$

Which of the following is a possible stream function, ψ, for this flow ?

(*a*) $\frac{1}{2}\tan^{-1}\left(\frac{y}{x}\right)$ (*b*) $\tan^{-1}\left(\frac{y}{x}\right)$

(*c*) $2\tan^{-1}\left(\frac{y}{x}\right)$ (*d*) $2\tan^{-1}\left(\frac{y}{x}\right)$

73. A flat plate is kept in an infinite fluid medium. The fluid has a uniform free-stream velocity parallel to the plate. For the laminar boundary layer formed on the plate, pick the correct option matching Columns I and II.

Column I

P. Boundary layer thickness

Q. Shear stress at the plate

R. Pressure gradient along the plate

Column II

1. decreases in the flow direction

2. Increases in the flow direction

3. remains unchanged

Codes :

	P	Q	R
(*a*)	1	2	3
(*b*)	2	2	2
(*c*)	1	1	1
(*d*)	2	1	3

74. A laboratory model of a river is built to a geometric scale of 1:100. The fluid used in the model is oil of mass density 900 kg/m^3. The highest flood in the river is 10,000 m^3/s. The corresponding discharge in the model shall be

(*a*) 0.095 m^3/s.

(*b*) 0.100 m^3/s.

(*c*) 0.105 m^3/s.

(*d*) 10.5 m^3/s.

75. The vertical hydraulic conductivity of the top soil at certain is 0.2 cm/hr. A storm of intensity 0.5 cm/hr occurs over the soil for an indefinite period. Assuming the surface drainage to be adequate, the infiltration rate after the storm has lasted for a very long time, shall be

(*a*) smaller than 0.2 cm/hr

(*b*) 0.2 cm/hr

(*c*) between 0.2 and 0.5 cm/hr

(*d*) 0.5 cm/hr

76. A horizontal jet strikes a frictionless vertical plate (the plan view is shown in the figure). It is then divided into two parts, as shown in the figure. If the impact loss can be neglected, what is the value of θ ?

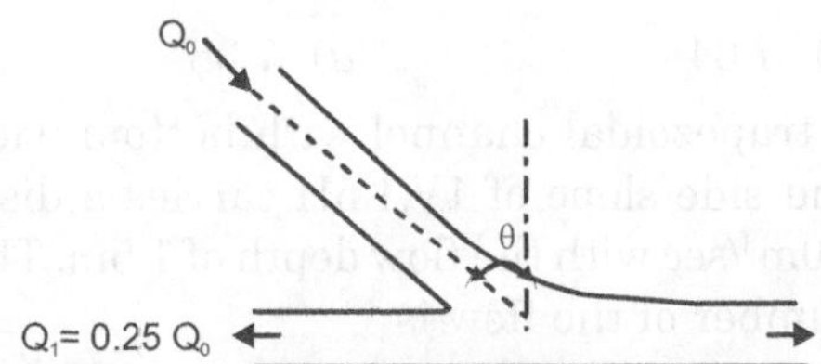

(*a*) 15° (*b*) 30°

(*c*) 45° (*d*) 60°

77. A hydraulic jump takes place in a triangular channel of vertex angle 90°, as shown in the figure. The discharge is 1m^3/s and the pre-jump depth is 0.5 m. What will be the post-jump depth ? (Take g = 9.81 m/s^2)

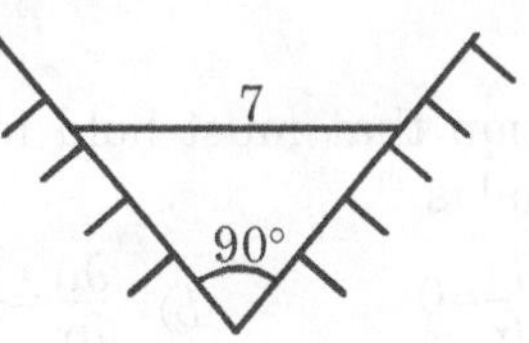

(*a*) 0.57 m (*b*) 0.91 m

(*c*) 1.02 m (*d*) 1.57 m

78. Two pipelines, one carrying oil (mass density 900 kg/m^3) and the other water, are connected to a manometer as shown in the figure. By what amount the pressure in the water pipe should be increased so that the mercury levels in both the limbs of the manometer become equal ? (Mass density of mercury = 13,550 kg/m^3 and g = 9.81 m/s^2)

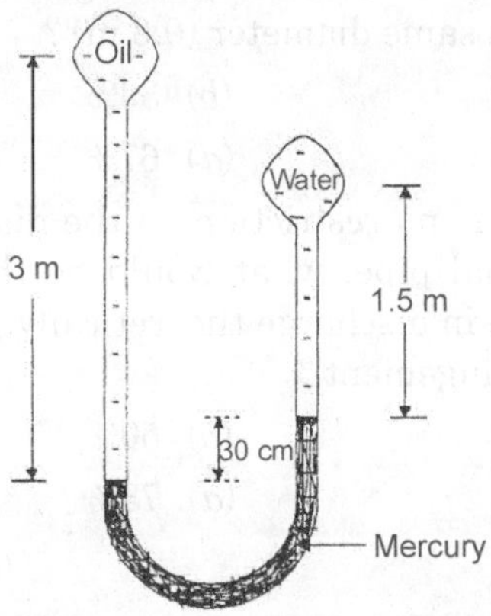

(*a*) 24.7 kPa (*b*) 26.5 kPa

(*c*) 26.7 kPa (*d*) 28.9 kPa

79. A solid sphere (diameter 6 mm) is rising through oil (mass density 900 kg/m^3, dynamic viscosity 0.7 kg/ms) at a constant velocity of 1 cm/s. What is the specific weight of the material from which the sphere is made ? (Take g = 9.81 m/s^2)

(*a*) 4.3 kN/m^3 (*b*) 5.3 kN/m^3

(*c*) 8.7 kN/m^3 (*d*) 12.3 kN/m^3

Common Data Q. 80-81

A very wide rectangular channel carries a discharge of 8 m^3/s per m width. The channel has a bed slope of 0.004 and Manning's roughness coefficient, n = 0.015. At a certain section of the channel, the flow depth is 1m.

80. What Gradually Varied Flow profile exists at this secton ?

(*a*) M_2 (*b*) M_3

(*c*) S_2 (*d*) S_3

81. At what distance from this section the flow depth will be 0.9 m? (Use the direct step method employing a single step)

(*a*) 65 m downstream

(*b*) 50 m downstream

(*c*) 50 m downstream

(*d*) 65 m downstream

Common Data Q. 82-83

A pipeline (diameter 0.3 m, length 3 km) carries water from point P to point R (see figure). The piezometric heads at P and R are to be maintained at 100 m and 80 m, respectively. To increase the discharge, a second pipe is added in parallel to the existing pipe from Q to R. The length of the additional pipe is also 2 km. Assume the friction factor, f = 0.04 for all pipes and ignore minor losses.

82. What is the increase in discharge if the additional pipe has same diameter (0.3 m) ?

(a) 0% (b) 33%

(c) 41% (d) 67%

83. If there is no restriction on the diameter of the additional pipe, what would be the maximum increase in discharge theoretically possible from this arrangement ?

(a) 0% (b) 50%

(c) 67% (d) 73%

2002

84. The force 'F' required at equilibrium on the semi-cylindrical gate shown below is

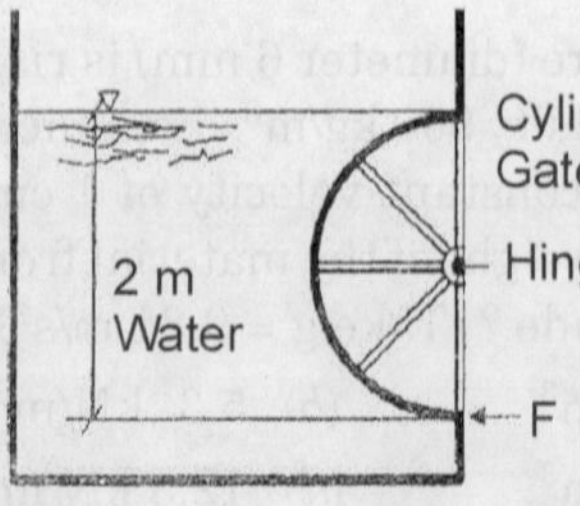

(a) 9.81kN (b) 0.0kN

(c) 19.62kN (d) none of these

85. Velocity distribution in a boundary layer flow over a plate is given by

$$u/u_m = 1.5\eta$$

where, $\eta = \frac{y}{\delta}$; y is the distance measured normal to the plate; δ is the boundary layer thickness; and u_m is the maximum velocity at $y = \delta$.

If shear stress τ, acting on the plate is given by

$$\tau = \frac{K(\mu_m)}{\delta}$$

where, μ is dynamic viscosity of the fluid, K takes the value of

(a) 0 (b) 1

(c) 1.5 (d) none of these

86. A pump can lift water at a discharge of 0.15m³/s to a head of 25 m. The critical cavitation number (σ_c) for the pump is found to be 0.144. The pump is to be installed at a location where the barometric pressure is 9.8 m of water and the vapour pressure of water is 0.30 of water. The intake pipe friction loss is 0.40 m. Using the minimum value of NPSH (Net Positive Suction Head), the maximum allowable elevation above the sump water surface at which the pump can be located is

(a) 9.80 m (b) 6.20 m

(c) 5.50 m (d) none of these

2001

87. In a Bernoulli equation, used in pipe flow, each term represents

(a) energy per unit weight

(b) energy per unit mass

(c) energy per unit volume

(d) energy per unit flow length

88. The stage-discharge relation in a river during the passage of flood is measured. If q_f is the discharge at the stage when water surface is falling and q_r is the discharge at the same stage when water surface is rising , then

(a) $q_f = q_r$

(b) $q_f < q_r$

(c) $q_f > q_r$

(d) $\frac{q_f}{q_r}$ = constant for all stages

89. A 15 cm length of steel rod with relative density of 7.4 is submerged in a two layer fluid. The bottom layer is mercury and the top layer is water. The height of top surface of the rod above the liquid interface in 'cm' is

(a) 8.24 (b) 7.82

(c) 7.64 (d) 7.38

90. A trapezoidal channel with bottom width of 3m and side slope of 1V:1.5H carries a discharge of 8.0m³/sec with the flow depth of 1.5m. The Froude number of the flow is

(a) 0.066 (b) 0.132

(c) 0.265 (d) 0.528

91. In a 1/50 model of a spillway, the discharge was measured to be 0.3m³/sec. The corresponding prototype discharge in m³/sec is

(a) 2.0 (b) 15.0

(c) 106.0 (d) 5303.0

2000

92. The relation that must hold for the flow to be irrotational is

(a) $\frac{\partial u}{\partial y} - \frac{\partial v}{\partial x} = 0$ (b) $\frac{\partial u}{\partial x} = \frac{\partial v}{\partial y}$

(c) $\frac{\partial^2 u}{\partial x^2} + \frac{\partial^2 v}{\partial y^2} = 0$ (d) $\frac{\partial u}{\partial y} = -\frac{\partial v}{\partial x}$

93. Cavitation is caused by

(a) high velocity

(b) low pressure

(c) high pressure

(d) high temperature

94. If pump head is 75m, discharge is 0.464 m^3/s and the motor speed is 1440 rpm at rated condition, then specific-speed of the pump is about

(*a*) 4 (*b*) 26
(*c*) 38 (*d*) 1440

95. Water flows at a depth of 0.1m with a velocity of 6m/s in a rectangular channel. The alternate depth is

(*a*) 0.30 m (*b*) 0.40 m
(*c*) 0.86 m (*d*) 0.81 m

96. Water flows at a rate of 10m^3/s in a rectangular channel 3 m wide. The critical depth of flow is

(*a*) 1.13 m (*b*) 2 m
(*c*) 1.45 m (*d*) 1.04 m

97. A 1 : 49 scale model of a proposed dam is used to predict prototype flow conditions. The design flood discharge near the spillway is 15000 m^3/s. Establish the relation between velocities in the model and prototype. What flow rate should be established in the model to simulate this flow ? If a velocity of 1.2 m/s is measured at a point in the model what is the velocity at a corresponding point in the prototype ?

1999

98. The sequent depth ratio of a hydraulic jump in a rectangular horizontal channel is 10.30. The Froude Number at the beginning of the jump is

(*a*) 5.64 (*b*) 7.63
(*c*) 8.05 (*d*) 13.61

99. In an iceberg, 15% of the volume projects above the sea surface. It the specific weight of sea water is 10.5 kN/m^3, the specific weight of iceberg in kN/m^3 is

(*a*) 12.52 (*b*) 9.81
(*c*) 8.93 (*d*) 7.83

100. A hydraulic turbine has a discharge of 5 m^3/s, when opertating under a head of 20 m with a speed of 500 rpm. If it is to operate under a head of 15 m, for the same discharge, the rotational speed in rpm will approximately be

(*a*) 433 (*b*) 403
(*c*) 627 (*d*) 388

1998

101. In which one of the following arrangement would the vertical force on the cylinder due to water be the maximum?

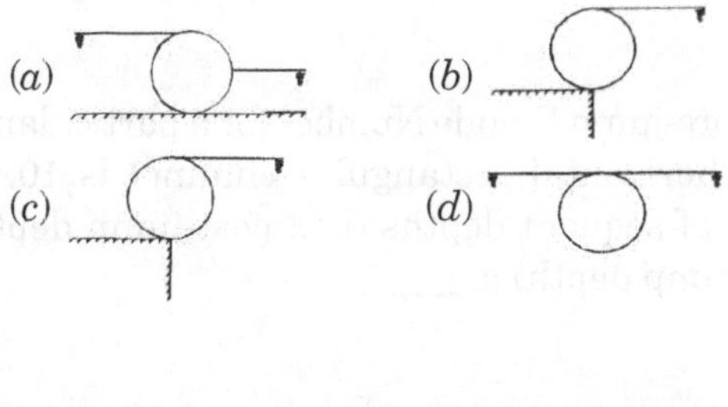

102. At the same mean velocity, the ratio of head loss per unit length for a sewer pipe flowing full to that for the same pipe flowing half full would be

(*a*) 2.0 (*b*) 1.63
(*c*) 1.00 (*d*) 0.61

103. Three reservoirs A, B and C are interconnected by pipes as shown in the Figure. Water surface elevations in the reservoirs and the Piezometric head at the junction J are indicated in the Figure.

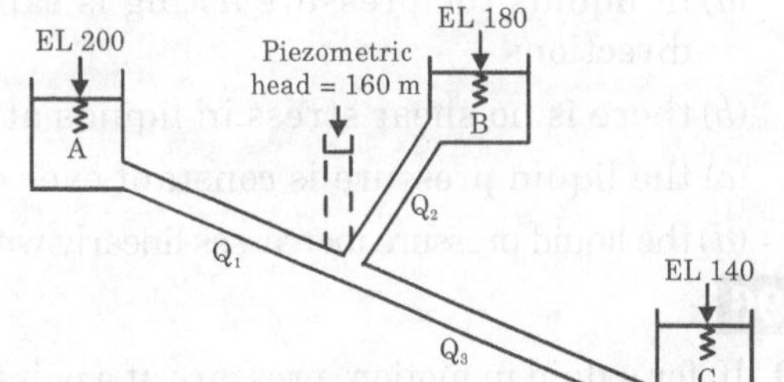

Discharge Q_1, Q_2 and Q_3 are related as

(*a*) $Q_1+Q_2 = Q_3$ (*b*) $Q_1 = Q_2+Q_3$
(*c*) $Q_2 = Q_1+Q_3$ (*d*) $Q_1+Q_2+Q_3 = 0$

104. The comparison between pumps operating in series and in parallel is

(*a*) Pumps operating in series boost the discharge, whereas pumps operating in parallel boost the head.

(*b*) Pumps operating in parallel boost the discharge, whereas pumps operating in series boost the head.

(*c*) In both cases there would be a boost in discharge only

(*d*) In both case there would be a boost in head only.

105. A nozzle discharging water under head H has an outlet area "*a*" and discharge coefficient $c_d = 1.0$. A vertical plate is acted upon by the fluid force F_j when held across the free jet and by the fluid force F_n when held against the nozzle to stop the flow. The ratio $\frac{F_j}{F_n}$ is

(*a*) 1/2 (*b*) 1
(*c*) $\sqrt{2}$ (*d*) 2

106. A body moving through still water at 6m/sec produces a water velocity of 4m/sec at a point 1m ahead. The difference in pressure between the nose and the point 1m ahead would be

(*a*) 2,000N/m^2
(*b*) 10,000N/m^2
(*c*) 19,620N/m^2
(*d*) 98,100 N/m^2

1997

107. The unit of dynamic viscosity of a fluid is

(*a*) m²/s (*b*) Ns/m²

(*c*) Pas/m² (*d*) kg s²/m²

108. The centre of pressure of a liquid on a plane surface immersed vertically in a static body of liquid, always lies below the centroid of the surface area, because

(*a*) in liquids the pressure acting is same in all directions

(*b*) there is no shear stress in liquids at rest

(*c*) the liquid pressure is constant over depth

(*d*) the liquid pressure increasses linearly with depth

1996

109. If, for a fluid in motion, pressure at a point is same in all directions, then the fluid is

(*a*) a real fluid (*b*) a Newtonian fluid

(*c*) an ideal fluid (*d*) a non-Newtonian fluid

110. A vertical triangular plane area, submerged in water, with one side in the free surface, vertex downward and latitude '*h*' has the pressure centre below the free surface by

(*a*) $\frac{h}{4}$ (*b*) $\frac{h}{3}$

(*c*) $\frac{2h}{3}$ (*d*) $\frac{h}{2}$

111. The equation $gz + \frac{v^2}{2} + \int \frac{dp}{p}$ = constant along a stream line holds true for

(*a*) steady, frictionless, compressible fluid

(*b*) steady, uniform, incompressible fluid

(*c*) steady, frictionless, incompressible fluid

(*d*) unsteady, incompressible fluid

112. The Prandtl mixing length for turbulent flow through pipes is

(*a*) independent of shear stress

(*b*) a universal constant

(*c*) zero at the pipe wall

(*d*) independent of radial distance from pipe axis

113. In network of pipes

(*a*) the algebraic sum of discharges around each circuit is zero

(*b*) the algebraic sum of discharges around each circuit is zero

(*c*) the elevation of hydraulic grade line is assumed for each junction point

(*d*) elementary circuits are replaced by equivalent pipes

114. Flow at critical depth takes place in an open channel when

(*a*) for a given specific energy, discharge is maximum

(*b*) for a given discharge, specific energy is maximum

(*c*) discharge is minimum for a given specific force

(*d*) discharge is maximum for a given specific force

115. On an immersed body in a flowing fluid, the lift force is

(*a*) due to buoyant force

(*b*) always in the opposite direction on gravity

(*c*) due to wake phenomenon

(*d*) the dynamic fluid-force component normal to approach velocity

116. The repeating variables in dimensional analysis should

(*a*) include the dependent variable

(*b*) have amongst themselves all the basic dimensions

(*c*) be derivable from one another

(*d*) exclude one of the basic dimensions

117. For the sluice gate shown in the figure, force required to hold the gate in place will be (assume gate width to be 2.0 m.)

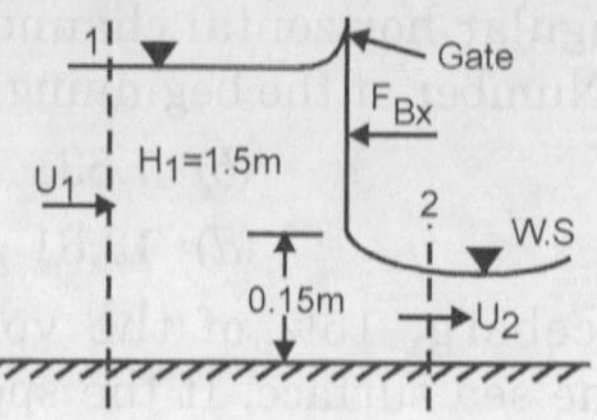

(*a*) 15.2 Nm (*b*) 16.2 Nm

(*c*) 17.2 Nm (*d*) 18.2 Nm

118. At a rated capacity of 44 cumecs, a centrifugal pump develops 36 m of head when operating at 1450 rpm. Its specific speed is

(*a*) 654 (*b*) 509

(*c*) 700 (*d*) 90

119. X–component of velocity in *a* 2–D incompressible flow is given by $u = y^2 + 4xy$. If Y – component of velocity '*v*' equals zero at $y = 0$, the expression for '*v*' is given by

(*a*) $4y$ (*b*) $2y^2$

(*c*) $-2y^2$ (*d*) $2xy$

NUMERICAL TYPE QUESTIONS

2016

1. The pre-jump Froude Number for a particular flow in a horizontal rectangular channel is 10. The ratio of sequent depths (i.e., post-jump depth to pre-jump depth) is _____.

2. A sector gate is provided on a spillway as shown in the figure. Assuming $g = 10$ m/s^2, the resultant force per meter length (expressed in kN/m) on the gate will be ______.

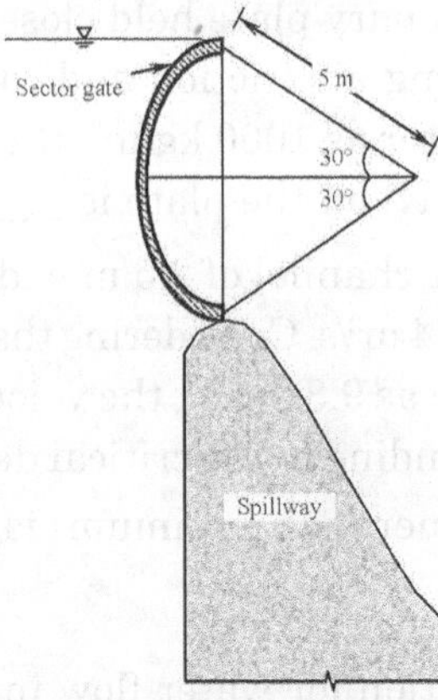

3. A square plate is suspended vertically from one of its edges using a hinge support as shown in figure. A water jet of 20 mm diameter having a velocity of 10 m/s strikes the plate at its mid-point, at an angle of 30° with the vertical. Consider g as 9.81 m/s^2 and neglect the self-weight of the plate. The force F (expressed in N) required to keep the plate in its vertical position is ______

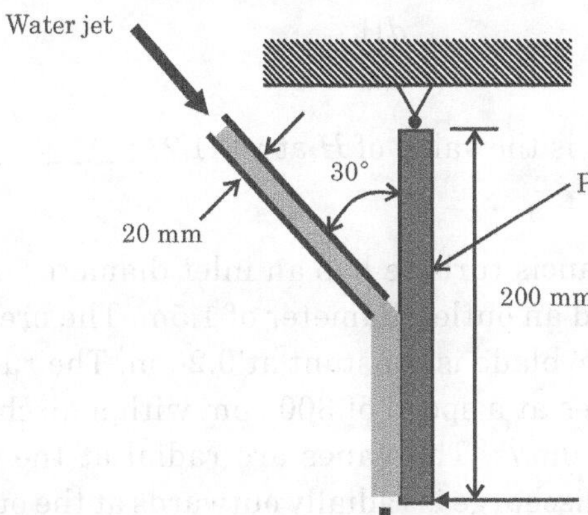

4. A 3 m wide rectangular channel carries a flow of 6 m^3/s. The depth of flow at a section P is 0.5 m. A flat-topped hump is to be placed at the downstream of the section P. Assume negligible energy loss between section P and hump, and consider g as 9.81 m/s^2. The maximum height of the hump (expressed in m) which will not change the depth of flow at section P is ______

5. A penstock of 1 m diameter and 5 km length is used to supply water from a reservoir to an impulse turbine. A nozzle of 15 cm diameter is fixed at the end of the penstock. The elevation difference between the turbine and water level in the reservoir is 500 m. Consider the head loss due to friction as 5% of the velocity head available at the jet. Assume unit weight of water = 10 kN/m^3 and acceleration due to gravity (g) = 10 m/s^2. If the overall efficiency is 80%, power generated (expressed in kW and rounded to nearest integer) is ______

2015

6. Two reservoirs are connected through a 930 m long, 0.3 m diameter pipe, which has a gate valve. The pipe entrance is sharp (loss coefficient = 0.5) and the valve is half-open (loss coefficient = 5.5). The head difference between the two reservoirs is 20 m. Assume the friction factor for the pipe as 0.03 and g = 10 m/s^2. The discharge in the pipe accounting for all minor and major losses is ______ m^3/s.

7. A short reach of a 2 m wide rectangular open channel has its bed level rising in the direction of flow at a slope of 1 m in 10000. It carries a discharge of 4 m^3/s and its Manning's roughness coefficient is 0.01. The flow in this reach is gradually varying. At a certain section in this reach, the depth of flow was measured as 0.5 m. The rate of change of the water depth with distance, dy/dx, at this section is ______ (use g = 10 m/s^2).

8. A hydraulic jump is formed in a 2 m wide rectangular channel which is horizontal and frictionless. The post-jump depth and velocity are 0.8 m and 1 m/s, respectively. The pre-jump velocity is ______ m/s. (use g = 10 m/s^2)

9. A nozzle is so shaped that the average flow velocity changes linearly from 1.5 m/s at the beginning to 15 m/s at its end in a distance of 0.375 m. The magnitude of the convective acceleration (in m/s^2) at the end of the nozzle is ______.

10. A pipe of 0.7 m diameter has a length of 6 km and connects two reservoirs A and B. The water level in reservoir. A is at an elevation 30 m above the water level in reservoir B. Halfway along the pipe line, there is a branch through which water can be supplied to a third reservoir C. The friction factor of the pipe is 0.024. The quantity of water discharged into reservoir C is 0.15 m^3/s. Considering the acceleration due to gravity as 9.81 m/s^2 and neglecting minor losses, the discharge (in m^3/s) into the reservoir B is ______.

11. A triangular gate with a base width of 2 m and a height of 1.5 m lies in a vertical plane. The top vertex of the gate is 1.5 m below the surface of a tank which contains oil of specific gravity 0.8. Considering the density of water and acceleration due to gravity to be 1000 kg/m^3 and 9.81 m/s^2, respectively, the hydrostatic force (in kN) exerted by the oil on the gate is ______.

2014

12. An incompressible homogeneous fluid is flowing steadily in a variable diameter pipe having the large and small diameters as 15 cm and 5 cm, respectively. If the velocity at a section at the 15 cm diameter portion of the pipe is 2.5 m/s, the velocity of the fluid (in m/s) at a section falling in 5 cm portion of the pipe is ______.

13. A particle moves along a curve whose parametric equations are: $x = t^3 + 2t$, $y = -3e^{-2t}$ and $z = 2\sin(5t)$, where x, y and z show variations of the distance covered by the particle (in cm) with time t (in s). The magnitude of the acceleration of the particle (in cm/s^2) at $t = 0$ is ______.

14. A venturimeter, having a diameter of 7.5 cm at the throat and 15 cm at the enlarged end, is installed in a horizontal pipeline of 15 cm diameter. The pipe carries an incompressible fluid at a steady rate of 30 litres per second. The difference of pressure head measured in terms of the moving fluid in between the enlarged and the throat of the venturimeter is observed to be 2.45 m. Taking the acceleration due to gravity as 9.81 m/s^2, the coefficient of discharge of the venturimeter (correct up to two places of decimal) is ______

15. An incompressible fluid is flowing at a steady rate in a horizontal pipe. From a section, the pipe divides into two horizontal parallel pipes of diameters d_1 and d_2 (where $d_1 = 4d_2$) that run for a distance of L each and then again join back to a pipe of the original size. For both the parallel pipes, assume the head loss due to friction only and the Darcy-Weisbach friction factor to be the same. The velocity ratio between the bigger and the smaller branched pipes is ______.

16. A horizontal nozzle of 30 mm diameter discharges a steady jet of water into the atmosphere at a rate of 15 litres per second. The diameter of inlet to the nozzle is 100 mm. The jet impinges normal to a flat stationary plate held close to the nozzle end. Neglecting air friction and considering the density of water as 1000 kg/m^3, the force exerted by the jet (in N) on the plate is ______

17. A rectangular channel of 2.5 m width is carrying a discharge of 4 m^3/s. Considering that acceleration due to gravity as 9.81 m/s^2, the velocity of flow (in m/s) corresponding to the critical depth (at which the specific energy is minimum) is ______

2013

18. Laplace equation for water flow in soils is given below.

$$\frac{\partial^2 H}{\partial x^2} + \frac{\partial^2 H}{\partial y^2} + \frac{\partial^2 H}{\partial z^2} = 0$$

Head H does not vary in y and z directions.

Boundary conditions are: at $x = 0$, H = 5;

and $$\frac{dH}{dx} = -1.$$

What is the value of H at $x = 1.2$? ______

2002

19. A Francis turbine has an inlet diameter of 2.5 m and an outlet diameter of 1.5m. The breadth of the blade is constant at 0.20 m. The runner rotates at a speed of 300 rpm with a discharge of 10.0m^3/s. The vanes are radial at the inlet and discharge is radially outwards at the outlet. Calculate the angle of guide vane to be set at the inlet and the blade angle at the outlet.

ANSWERS

MCQ Type Questions

1. (*b*)	**2.** (*c*)	**3.** (*d*)	**4.** (*c*)	**5.** (*c*)	**6.** (*d*)	**7.** (*b*)	**8.** (*a*)	**9.** (*c*)	**10.** (*c*)
11. (*b*)	**12.** (*d*)	**13.** (*d*)	**14.** (*d*)	**15.** (*c*)	**16.** (*a*)	**17.** (*a*)	**18.** (*c*)	**19.** (*b*)	**20.** (*d*)
21. (*a*)	**22.** (*b*)	**23.** (*c*)	**24.** (*c*)	**25.** (*d*)	**26.** (*d*)	**27.** (*a*)	**28.** (*c*)	**29.** (*c*)	**30.** (*d*)
31. (*d*)	**32.** (*b*)	**33.** (*d*)	**34.** (*d*)	**35.** (*b*)	**36.** (*c*)	**37.** (*a*)	**38.** (*b*)	**39.** (*c*)	**40.** (*c*)
41. (*d*)	**42.** (*a*)	**43.** (*b*)	**44.** (*b*)	**45.** (*c*)	**46.** (*b*)	**47.** (*c*)	**48.** (*d*)	**49.** (*a*)	**50.** (*c*)
51. (*d*)	**52.** (*c*)	**53.** (*c*)	**54.** (*a*)	**55.** (*b*)	**56.** (*d*)	**57.** (*b*)	**58.** (*d*)	**59.** (*d*)	**60.** (*c*)
61. (*b*)	**62.** (*d*)	**63.** (*a*)	**64.** (*b*)	**65.** (*b*)	**66.** (*c*)	**67.** (*a*)	**68.** (*c*)	**69.** (*b*)	**70.** (*a*)
71. (*c*)	**72.** (*c*)	**73.** (*b*)	**74.** (*b*)	**75.** (*b*)	**76.** (*b*)	**77.** (*d*)	**78.** (*b*)	**79.** (*d*)	**80.** (*d*)
81. (*b*)	**82.** (*c*)	**83.** (*c*)	**84.** (*a*)	**85.** (*b*)	**86.** (*a*)	**87.** (*a*)	**88.** (*b*)	**89.** (*d*)	**90.** (*)
91. (*d*)	**92.** (*a*)	**93.** (*c*)	**94.** (*c*)	**95.** (*d*)	**96.** (*d*)	**97.** (*)	**98.** (*b*)	**99.** (*c*)	**100.** (*a*)
101. (*b*)	**102.** (*b*)	**103.** (*b*)	**104.** (*b*)	**105.** (*b*)	**106.** (*b*)	**107.** (*b*)	**108.** (*d*)	**109.** (*a*)	**110.** (*c*)
111. (*a*)	**112.** (*d*)	**113.** (*a*)	**114.** (*a*)	**115.** (*a*)	**116.** (*d*)	**117.** (*c*)	**118.** (*a*)	**119.** (*a*)	

Numerical Type Questions

1. (13.651)	**2.** (127.04)	**3.** (7.85)	**4.** (0.20)	**5.** (6570)
6. (0.1413)	**7.** (0.0032)	**8.** (4.828)	**9.** (540)	**10.** (0.5716)
11. (29.430)	**12.** (22 to 23)	**13.** (12 to 12)	**14.** (0.93 to 0.96)	**15.** (2 to 2)
16. (318 to 319)	**17.** (2.45 to 2.55)	**18.** (3.8)	**19.** (*)	

"*" Refers to subjective question.

EXPLANATIONS

MCQ TYPE QUESTIONS

1.

$b = 4\text{m},$

$S_o = 0.001$

$Q = 16 \text{ m}^3/\text{sec}$

$R = \frac{A}{P} = \frac{4y_n}{4+2y_n}$

$n = 0.012$

$g = 10 \text{ m/s}^2$

$$y_c = \left(\frac{Q^2}{g.b^2}\right)^{\frac{1}{3}}$$

$$= \left(\frac{16^2}{10 \times 4^2}\right)^{\frac{1}{3}}$$

$= 1.16$ m

$Q = A.V$

$$16 = 4.y_n\left(\frac{1}{0.012}\right)\left(\frac{4.y_n}{4+2y_n}\right)^{2/3} 0.001^{1/2}$$

$$y_n = \left(\frac{4.y_n}{4+2y_n}\right)^{\frac{2}{3}} = \frac{4 \times 0.012}{\sqrt{0.001}} = 1.55$$

Assume $y_n = 1.16$ m

LHS = 1.79

$Y_n > y_c$

$S_o < S_c$

Therefore, Mild slope.

2. P : Newtonian fluid – curve 2

Q – Pseudoplastic fluid – curve 4

R – Plastic fluid – curve 5

S = Dilatent fluid – curve 3

4. Continuity equation in differential form says :

$$\left[\frac{\partial}{\partial x}(\text{P}u)+\frac{\partial}{\partial y}(\text{P}v)+\frac{\partial}{\partial z}(\text{P}w)+\frac{\partial \text{P}}{\partial t}=0\right]$$

Steady and 2 D flow → 3rd and 4th term neglected

$$\frac{\partial}{\partial x}[\text{P}.u]+\frac{\partial}{\partial y}[\text{P}.v] = 0$$

Where, u = velocity in x direction

v = velocity in y direction

$$\frac{\partial}{\partial x}\left[\frac{1}{x}.x^2\right]+\frac{\partial}{\partial y}\left[\frac{\text{V}}{x}\right] = 0$$

$$\frac{\partial x}{\partial x}+\frac{\partial}{\partial y}\left[\frac{\text{V}}{x}\right] = 0$$

$$\frac{\partial \text{V}}{\partial y} = -x$$

$$\Rightarrow \quad \text{V} = -xy$$

5. $$\frac{\rho \text{VD}}{\mu} = \frac{\text{kg}}{\text{m}^3}\times\frac{\text{m}}{\text{S}}\times\frac{\text{m}}{\text{N.S}}\text{m}^2$$

$$= \frac{\text{kg m}}{\text{S}^2}\times\frac{1}{\text{N}}$$

$$= \text{N}\times\frac{1}{\text{N}} = 1$$

$$\frac{\text{F}_\text{D}}{\rho \text{V}^2\text{D}^2} = \frac{\text{N}}{\frac{\text{kg}}{\text{m}^3}\times\frac{\text{m}^2}{\text{S}^2}\times\text{m}^2}$$

$$= \frac{\text{N}}{\frac{\text{km}}{\text{S}^2}} = \frac{\text{N}}{\text{N}} = 1$$

6. $$\text{F} = \frac{\text{V}}{\sqrt{g\text{L}}}$$

Froude dynamic similarity,

$$\frac{\text{V}_r}{\sqrt{\text{L}_r g_r}} = 1$$

$$\text{V}_r = \sqrt{\text{L}_r g_r}$$

Acceleration ratio,

$$a_r = g_r = 1$$

$$\therefore \quad \text{V}_r = \sqrt{\text{L}_r}$$

8. For incompressible flow, $\frac{\partial \text{U}}{\partial x}+\frac{\partial \text{V}}{\partial y} = 0$

$$\frac{\partial}{\partial x}\left(\frac{y^3}{3}+2x-x^2y\right)+\frac{\partial}{\partial y}\left(xy^2-2y-\frac{x^3}{3}\right) = 0$$

$$2-2xy+2xy-2 = 0$$

Hence flow is incompressible

For irrotational flow, $\left(\frac{\partial \text{V}}{\partial x}-\frac{\partial \text{U}}{\partial y}\right) = 0$

$$\frac{\partial}{\partial x}\left(xy^2-2y-\frac{x^3}{3}\right)-\frac{\partial}{\partial y}\left(\frac{y^3}{3}+2x-x^2y\right)$$

$$\left[y^2-\frac{3x^2}{3}-\frac{3y^2}{3}+x^2\right] = 0$$

10.

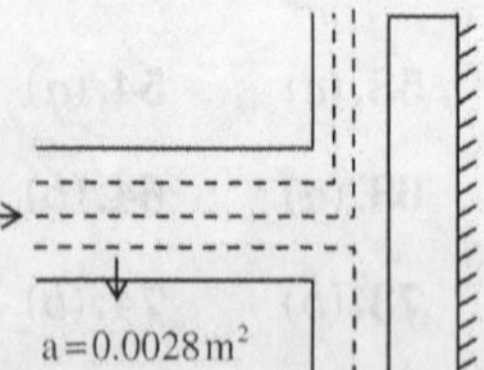

Force on plate, $\text{F} = \rho.\text{a.v}^2$

$= 1000 \times 0.0028 \times 5^2$

$= 70$ N

11. n =0.015, Q = 1m³/s, B = 3.0m

Normal depth of flow ranges between 0.76m to 0.8 m

If prevailing normal depth of flow is not exceeded, there must not be choking of the section or it should be at boundary condition of choking.

So, width of section should be such that there should be critical flow corresponding to the prevailing specific energy.

i.e. $\frac{3}{2}\left(\frac{q^2}{g}\right)^{1/3} = \text{E}_c = \text{E}_{\text{initial}}$

$$q = \frac{\text{Q}}{\text{B}_{\min}},\ \text{So}\ \frac{3}{2}\left[\frac{\left(\frac{\text{Q}}{\text{B}_{\min}}\right)^2}{g}\right]^{1/3} = \text{E}_{\text{initial}}$$

$$\text{E}_{\text{initial}} = y+\frac{q^2}{2gy^2}$$

$$\text{Q} = \frac{1}{n}.\text{A.R}^{2/3}\text{S}^{1/2}$$

$$\Rightarrow 1 = \frac{1}{0.015}.(\text{B}.y.).\left(\frac{\text{B}.y}{\text{B}+2y}\right).\text{S}^{1/2}$$

$$\Rightarrow 1 = \frac{1}{0.015}.(3y).\left(\frac{3y}{3+2y}\right).(0.0001)^{1/2}$$

$$\Rightarrow y = 0.78 \text{ m}$$

$$\text{So, } E_{initial} = 0.78 + \frac{\left(\frac{1}{3}\right)^2}{2 \times 9.81 \times (0.78)^2} = 0.789 \text{ m}$$

$$\text{So, } \frac{3}{2}\left(\frac{Q}{g.(B_{min})^2}\right)^{1/3} = 0.789$$

$\Rightarrow B_{min} = 0.84$ m

12.

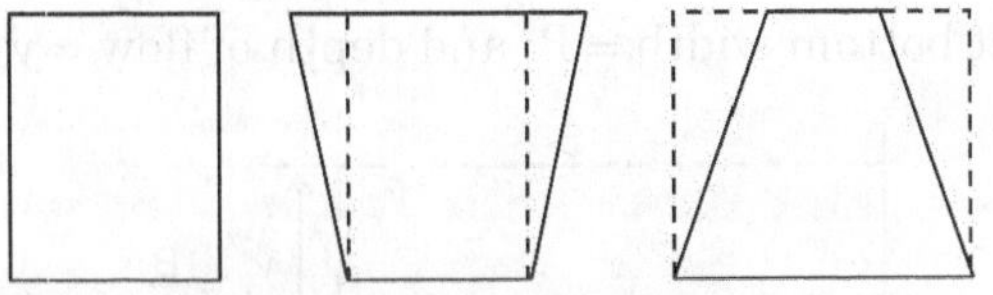

Force on base of Bucket,

$$F = \gamma_w . h . A_{base}$$

$\because$ Base area of all buckets is same.

So, $F_1 = F_2 = F_3$

Weight of water, $W = \gamma_w . V$

Since $V_2 > V_1 > V_3$

So, $W_2 > W_1 > W_3$

13. Hence, $u = \frac{x}{T_1}, v = -\frac{y}{T_2}$

differentiate both

$$\frac{\partial u}{\partial x} = \frac{1}{T_1} \quad \& \quad \frac{\partial v}{\partial y} = -\frac{1}{T_2}$$

For a flow to exist.

$$\frac{\partial u}{\partial x} + \frac{\partial v}{\partial y} = 0$$

$$\frac{1}{T_1} - \frac{1}{T_2} = 0$$

$$\therefore \quad \frac{1}{T_1} = \frac{1}{T_2}$$

$\therefore$ $\boxed{T_1 = T_2}$ right choise is (d)

14. P – 4; Q – 3; R – 2; S – 1

(P) Anemometer – (4) wind & speed.

(Q) Hygrometer – (3) water vapours content of air.

(R) Pitot tube – (2) Fluid velocity at a specific point in the flow stream

(S) Tension meter – (1) Capillary potential of soil water.

15.

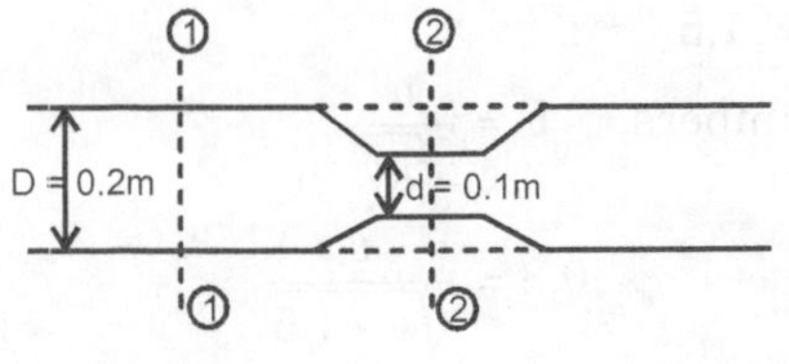

head of water = 2.0 m

coefficient of discharge $C_D = 1$

Energy losses are negligible, so

$$\text{Discharge } Q = \frac{C_D A_1 A_2 \sqrt{2gh}}{\sqrt{A_1^2 - A_2^2}}$$

$$A_1 = \frac{\pi}{4} \times (0.2)^2 = 0.0314$$

$$A_2 = \frac{\pi}{4} \times (0.1)^2 = 0.0078$$

$$Q = \frac{1 \times 0.0314 \times 0.0078 \sqrt{2 \times 9.81 \times 2}}{\sqrt{0.0314^2 - 0.0078^2}}$$

$$= \frac{1.53 \times 10^{-3}}{0.0304}$$

$$= 0.0504$$

$$\approx 0.050$$

Correct choice is (*c*).

17. For subcritical flow in an open channel, the control section for gradually varied flow profile is at the down stream end.

18. – Mach number is ratio of fluid velocity and velocity of sound.

– Reynolds number is ratio of internal force and viscous force.

– Weber number is ratio of inertial force and surface tension.

– Froude number is ratio of inertial force and gravitational force.

19. $Q = (\psi)_{0,3} - (\psi)_{3,4}$,

$$\psi = \frac{3}{2}(y^2 - x^2)$$

$$\psi_{0,3} = \frac{3}{2}(3^2 - 0^2) = 13.5$$

$$\psi_{3,4} = \frac{3}{2}(4^2 - 3^2) = 10.5$$

$$\therefore \quad Q = 13.5 - 10.5 = 3$$

20.

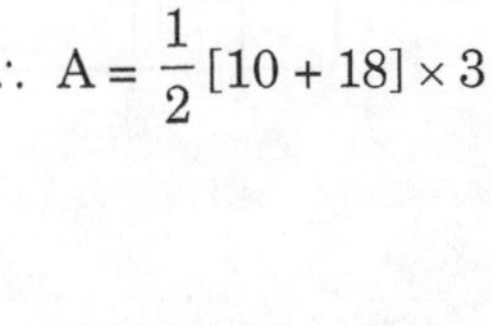

Top width = 4 + 10 + 4 = 18m

$$\therefore A = \frac{1}{2}[10 + 18] \times 3 = 42 \text{ m}^2$$

Side length = $\sqrt{3^2+4^2} = 5$

Perimeters = $2 \times 5 + 10 = 20$

Hydraulic radius = $\dfrac{A}{P} = \dfrac{42}{20} = 2.1$

21. Given : $Q = 100\ m^3/s$; $B = 5m$

Now, $F_r = \dfrac{V}{\sqrt{gd}} = 0.8$

$\Rightarrow \dfrac{Q}{Bd\sqrt{gd}} = 0.8$

Solving, we get

$$d = 4\ m$$

22. Using continuity equtaion

$$(Av)_P = (Av)_R + (Av)_Q$$

$\therefore \dfrac{\pi}{4} \times 4^2 \times 6 = \dfrac{\pi}{4} \times 2^2 \times v + \dfrac{\pi}{4} \times 4^2 \times 5$

Solving we get $v = 4$m/s

23. As per two point method of velocity measurement

$$V_{mean} = \frac{V_{0.2} + V_{0.8}}{2}$$

$$= \frac{0.6+0.4}{2} = 0.5\ m/s$$

$\therefore Q = V \times A$

$$= 0.5 \times \left[\frac{1}{2} \times 4 \times 1\right] = 1\ m^3/s$$

26.

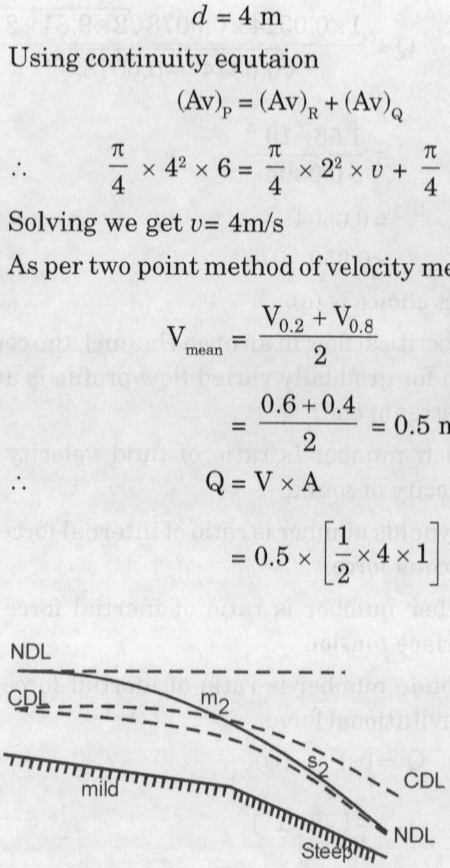

27. In the subcritical region of flow for a given specific energy, as the discharge per unit width q is increased due to reducing width at the section, the depth of flow will decrease.

Therefore water surface drop at a downstream section.

Alternately

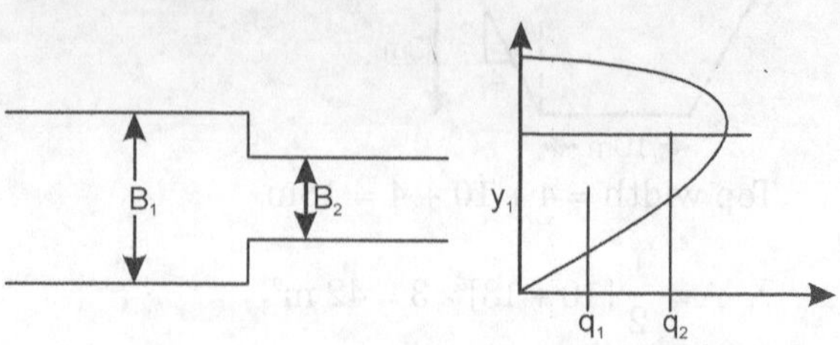

Here $q_1 = \dfrac{\theta}{B_1}$; $q_2 = \dfrac{\theta}{B_2}$

$\therefore q_2 > q_1$

From discharge diagram, it can be seen that, for subcritical flow as q increases, depth decreases.

Therefore at the reduced section, depth decreases.

29. Let bottom width = B, and depth of flow = y

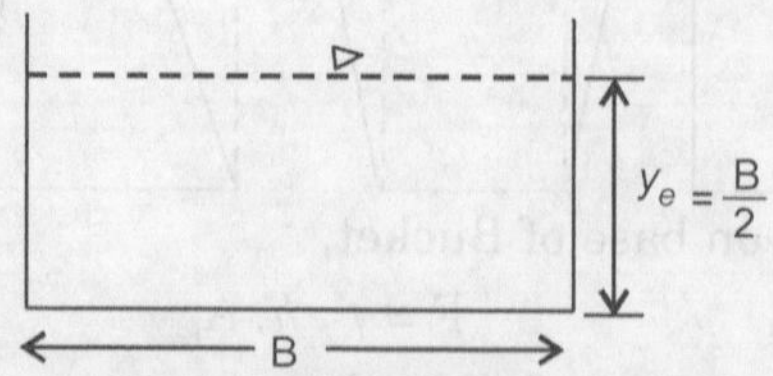

Area of flow, $A = By_e$

Wetted perimeter, $P = B + 2y_e$

$$= \frac{A}{y_e} + 2y_e$$

If P is minimum with A constant, then

$$\frac{dP}{dy_e} = -\frac{A}{y_e^2} + 2$$

For hydraulically efficient section

$$\frac{dP}{dy_e} = 0$$

$\Rightarrow A = 2y_e^{\,2}$

$\Rightarrow B = 2y_e$

Hydraulic radius, $R = \dfrac{A}{P} = \dfrac{By}{4y_e} = \dfrac{2y_e^2}{4y_e} = \dfrac{y_e}{2}$

Top width = $2y_e = B$

Perimeter, $P = B + 2y_e = 4y_e$

Hydraulic depth = $\dfrac{A}{T}$

$$= \frac{2y_e^{\,2}}{2y_e}$$

$$= y_e$$

$\therefore$ Top width = $2y_e$

Perimeter = $4y_e$

Hydraulic radius = $\dfrac{y_e}{2}$

and Hydraulic depth = y_e

$\therefore$ P–3, Q–4, R–1, S–2

30. Given : $F = 0.8$,

$$y = 1.5$$

Froude numbers, $F = \dfrac{v}{\sqrt{gy}}$

$\therefore 0.8 = \dfrac{v}{\sqrt{g \times 1.5}}$

$\Rightarrow \quad v = 3.0688$ m/s

Now $\quad E = y + \frac{v^2}{2g}$

$= 1.5 + \frac{3.0688^2}{2 \times g}$

$= 1.979$ m

But $\quad E_1 = E_C$

$\therefore \quad y_c = \frac{2}{3}E_c = 1.3$ m

Alternately

Froude number, $\quad F = \frac{v}{\sqrt{gA/T}} = \frac{Q}{\sqrt{gA^3/T}}$

where, T = top width

A = area of rectangular section

= T.y.

$\therefore \quad F = \frac{Q}{\sqrt{gTy^3/T}} = \frac{Q}{\sqrt{gy^3}}$

$\Rightarrow \quad 0.8 = \frac{Q}{\sqrt{g(1.5)^3}}$

$\Rightarrow \quad \frac{Q}{\sqrt{g}} = 1.47$

For critical flow, $\quad F = 1.0$

$\therefore \quad \frac{Q}{\sqrt{gy_c^3}} = 1$

$\Rightarrow \quad \frac{1}{(y_c)^{3/2}} \times 1.47 = 1$

$\Rightarrow \quad y_c = (1.47)^{2/3} = 1.29$ m

31. Friction factor, $f = \frac{64}{Re}$

Reynold number, $Re = \frac{\rho VD}{\nu}$

$= \frac{VD}{v}$

$= \frac{0.015 \times 0.1}{1.13 \times 10^{-6}}$

$= 1327.4.3$

$\therefore \quad f = \frac{64}{1327.43} = 0.048$

32. For rectangular channel,

$$y_c^3 = \frac{q_c^2}{g}$$

Here, $q_c = \frac{Q}{B}$

$= \frac{100}{4.5} = 22.22$

$\therefore \quad y_c^3 = \frac{(22.22)^2}{9.81}$

$\Rightarrow \quad y = 3.69$ m

33. BD = sin 30° × 10 = 5

Apply Bernoulli's equation between A and B (neglecting losses)

$$\frac{p_A}{w} + \frac{v_1^2}{2g} = \frac{p_B}{w} + \frac{v_2^2}{2g} + h$$

But $\quad v_1 = v_2$

$\therefore \quad p_A - p_B = hw$

$= 5 \times 9.879$

$= 49.395$

Given : $p_A = 12$ kN/m²

$\therefore \quad p_B = 12 + 49.395$

$= 61.395$

≈ 61.4 kN/m²

34. $\quad \delta = \frac{11.6\nu}{v}$

where, ν = kinematic viscosity

$= 10^{-6}$ m²/sec

$\therefore \quad \delta = \frac{11.6 \times 10^{-6}}{0.744}$

$= 1.4987 \times 10^{-5}$ m

Hence $\quad \frac{k_s}{\delta} = \frac{0.12 \times 10^{-3}\,m}{1.4987 \times 10^{-5}\,m}$

35. Applying Bernoulli's theorem

$$\frac{\rho V_1^2}{2} + P_1 = \frac{\rho V_2^2}{2} + P_2$$

$$\Rightarrow \quad \frac{\rho V^2}{2} + P_1 = \frac{\rho(2V)^2}{2} + P_2$$

$$\Rightarrow \quad P_1 - P_2 = \rho\left(\frac{4-1}{2}\right)V^2$$

$$\Rightarrow \quad \Delta P = 1.5\,\rho V^2$$

36. Density of water, ρ = 1000 kg/m³

Velocity of water, v = 10 m/s

Cross section of jet = 10 mm²

Force on plate due to jet of water

$F = \rho v^2 A$

$= 1000 \times (10)^2 \times 10 \times 10^{-6}$

$= 1$ kN

37. Length scale ratio, $\quad L_r = \frac{1}{50}$

Depth scale ratio, $\quad D_r = L_r = \frac{1}{50}$

Discharge scale ratio, $Q_r = \frac{Q_m}{Q_p}$

$$Q_r = A_r V_r$$
$$= L_r D_r \times (D_r)^{1/2}$$

But $L_r = D_r$

$\therefore$ $Q_r = (L_r)^{5/2}$

Discharge in prototype,

$$Q_P = 1000 \text{ m}^3\text{s}$$

$\therefore$ Discharge in model,

$$Q_m = Q_p \times Q_r$$
$$= 1000 \times (L_r)^{3/2}$$
$$= 1000 \times \left(\frac{1}{50}\right)^{3/2} = 0.0565 \text{ m}^3/\text{s}$$

38. Critical depth, $y_c = 0.3$m,

$$z = \tan 45° = 1$$

$\therefore$ $$y_c = \left(\frac{2Q_c^2}{gz^2}\right)^{1/5}$$

$\therefore$ Discharge, $Q_c = y_c^{5/2}\sqrt{\frac{g}{2}} \times z$

$$= (0.3)^{5/2}\sqrt{\frac{9.81}{2}} \times 1$$
$$= \mathbf{0.11\ m^3/sec}$$

39. By Hagen Poiseuille equation,

$$\Delta P = P_1 - P_2 = \frac{32\mu VL}{D^2}$$

$\Rightarrow$ $$\mu = \frac{\Delta P \times D^2}{32 \times V \times L}$$

where, ΔP = pressure drop

= 20 MPa

$\therefore$ Velocity, $V = \frac{Q}{A}$

$$= \frac{800}{\frac{\pi}{4}D^2}$$
$$= \frac{800}{\frac{\pi}{4} \times (0.5)^2}$$
$$= 4074.36 \text{ mm/sec.}$$

Diameter of pipe, D = 0.5 mm

Length of pipe, L = 2 m

$\therefore$ $$\mu = \frac{2 \times (0.5)^2}{32 \times 4074.36 \times 2 \times 1000 \times 10^{-6}}$$
$$= 0.00192 \text{ N s/m}^2$$

40. Flow rate in wide rectangular open channel $(q) = 2\text{m}^3$/sec per metre with

Bed slope, S = 0.002

Manning coefficient, $n = 0.012$

$$y_c = \left(\frac{q^2}{g}\right)^{1/3}$$
$$= \left[\frac{(2)^2}{9.81}\right]^{1/3}$$
$$= 0.742 \text{ m}$$
$$R = \frac{y}{2}$$

Now $$V = \frac{1}{n}R^{2/3}S_o^{1/2}$$

$\therefore$ $$\frac{Q}{A} = \frac{1}{n}R^{2/3}S_o^{1/2}$$

$\Rightarrow$ $$\frac{Q}{b \times y_n} = \frac{1}{n} \times \left(\frac{y_n}{2}\right)^{2/3} S_0^{1/2} \quad \left[\frac{Q}{b} = 9\right]$$

$\therefore$ $$\frac{q}{y_n} = \frac{1}{0.012} \times \frac{(y_n)^{2/3}}{2^{2/3}} \times (0.002)^{1/2}$$

$\Rightarrow$ $y_n = 0.908$

Since $y_n > y_c$, hence slop in mild.

41. $N_S = \frac{N\sqrt{Q}}{H_m^{3/4}}$

Specific speed of pump decrease with increase in head.

42. Discharge through combiner aquified is given by

$$Q = \frac{2\pi kb\,(H - h)}{\log_e \frac{R}{r}}$$

$\therefore$ Transmissivity, T = kb

$\therefore$ $$Q = \frac{2\pi T\,(H - h)}{\log_e \frac{R}{r}}$$

or $$T = \frac{Q \log_e \frac{R}{r}}{2\pi\,(H - h)}$$

If drawdowns at radial discharge of r_1 and r_2 from well, S measured as S_1 and S_2, then

$$T = \frac{Q}{2\pi} \frac{\log\left(\frac{r_2}{r_1}\right)}{(S_1 - S_2)}$$

43. Since $x \propto t^2$

$\therefore$ $$\frac{x}{x+1} = \frac{k(2)^2}{k(3)^2}$$

$\Rightarrow$ $x = 0.8$m

44. From continuity equation,

Div V = 0

$$\Rightarrow \quad \frac{\partial V}{\partial x} + \frac{\partial V}{\partial y} + \frac{\partial V}{\partial z} = 0$$

$$\Rightarrow \quad \frac{\partial}{\partial x}(5x + 6y + 7z) + \frac{\partial}{\partial y}(6x + 5y + 9z) + \frac{\partial}{\partial z}(3x + 2y + lz) = 0$$

$$\Rightarrow \quad 5 + 5 + \lambda = 0$$

$$\Rightarrow \quad \lambda = -10$$

45. $$\Delta E = \frac{(y_1 - y_2)^3}{4y_1 y_2}$$

where, $y_c = 0.74$

$y_1 = 0.8$ m.

47. *Given :* $$\frac{V_r L_r}{\upsilon_r} = 1$$

$$\frac{V_r}{\sqrt{L_r}} = 1$$

$$\therefore \quad L_r\sqrt{L_r} = \upsilon_r$$

$$\Rightarrow \quad L_r^{3/2} = \upsilon_r$$

$$\Rightarrow \quad L_r^{3/2} = \frac{5 \times 10^{-4}}{10^{-6}}$$

$$\therefore \quad L_r = 63$$

$$= 8$$

50. Reading of differential manometer of a venturimeter is independent of angle of inclination of venturimeter.

51. For a rectangular channel, specific energy at critical depth of 1.5 m

$$= \frac{3}{2} y_c$$

$$= \frac{3}{2} \times 1.5$$

$$= 2.25\text{m}$$

Here, y_c = critical depth of rectangular channel

53. Flow rate across the line joining A (3,0) and B (0, 2)

$$Q = (\psi_2 - \psi_1) \times 1$$

where, ψ_2 = value of the stream function for the curve boundary through point B

$$= \{2x^2y + (x + 1)\,y^2\}_{(0,\,2)}$$

$$= \{2 \times 0^2 \times 2 + (0 + 1)2^2\}$$

$$= 4$$

ψ_1 = value of stream function for straight horizontal boundary through point A.

$$= \{(x + 1)\,y^2\}_{(3,\,0)}$$

$$= (3 + 1) \times 0^2$$

$$= 0$$

$$\therefore \quad Q = (4 - 0) = 4 \text{ unit.}$$

54. Net down force in I = v (d – r) g = weigh buoyancy

where, v = velocity of I

d = density of object

ρ = density of fluid

Moment of I = Moment of II about P.

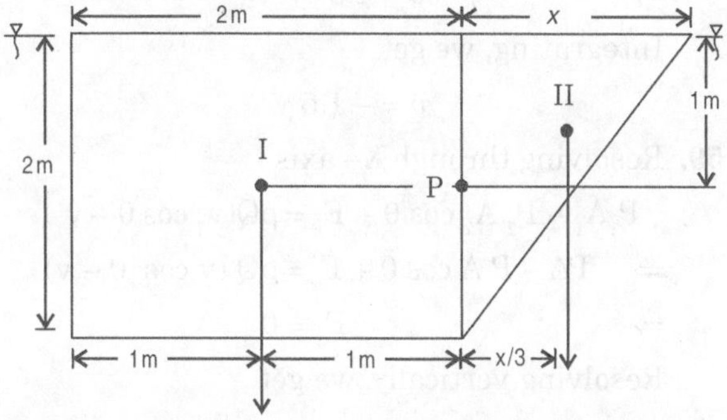

Taking unit length of object, let v_1 be the velocity of II.

$$\therefore \; v\,(d - r)\,g \times \frac{2}{2} = v_1\,(d - r)\,g \times \frac{x}{3}$$

$$\Rightarrow \quad x^2 = 12$$

$$\Rightarrow \quad x = 2\sqrt{3}$$

56. Force exerted on fluid in x-direction

$$= rQ\,(V_{2x} - V_{1x})$$

$$= rQ\,(v\cos 45° - 0)$$

$$= rg\,Q\,v\cos 45°$$

$$= 1000 \times 0.1 \times 4 \times \frac{1}{\sqrt{2}}$$

$$= \frac{400}{\sqrt{2}} = 200\sqrt{2} \text{ N.}$$

$\therefore$ Force recorded by spring

= Force exerted on fluid in x direction

$$= 200\sqrt{2} \text{ N}$$

57. We know, $$\frac{P}{w} = 100 \sin 30°$$

$$\Rightarrow \quad P = 100 \sin 30° \times w$$

$$= 100 \sin° \times v_1 \times \gamma_\omega$$

where, q_w = unit weight of water

$$\therefore \quad P = 100 \sin 30° \times 0.86 \times 1$$

$$= 43 \text{ mm water.}$$

58. For two - dimensional incompressible flow,

$$\frac{\partial u}{\partial x}+\frac{\partial v}{\partial y}=0,$$

and $u = 1.5\,x$

$$\Rightarrow \quad \frac{\partial u}{\partial u} = 1.5$$

$$\therefore \quad 1.5 + \frac{\partial v}{\partial y} = 0$$

$$\Rightarrow \quad \frac{\partial v}{\partial y} = -1.5$$

$$\Rightarrow \quad \partial v = 1.5\,\partial y$$

Integrating, we get

$$v = -1.5\,y.$$

59. Resolving through X– axis

$$P_1A_1 - P_2A_2\cos\theta + F_x = \rho Q(v_2\cos\theta - v_1)$$

$$\Rightarrow \quad PA - PA\cos 0 + F_x = \rho Q\,(v\cos 0 - v)$$

$$\Rightarrow \quad F_x = 0$$

Resolving vertically, we get

$$-P_2A_2\sin\theta + F_y = \rho Q(v_2\sin\theta)$$

$$\Rightarrow \quad -PA\sin 0 + F_y = \rho Q\,(v\sin\theta)$$

$$F_y = 0$$

$$A = \frac{2F_D}{C_L\rho v^2}$$

$$= \frac{2\times 8000}{0.4\times1.2\times(200\times0.278)^2}$$

$$= 1.08$$

$$\approx 1.1\ \text{m}^2$$

61.

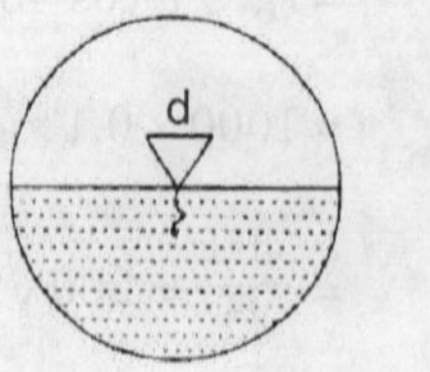

$$\text{Hydraulic depth} = \frac{\text{Area}}{\text{Top width}}$$

$$\text{Area} = \frac{\pi r^2}{2}$$

$$\text{Top width} = 2r$$

$$\therefore \quad \text{Hydraulic depth} = \frac{\pi r^2}{2\times 2r} = \frac{\pi r}{4}$$

62. *Given :* $\vec{v} = 2y\,\vec{i} + 3x\,\vec{j}$

At (1, 1), u = 2y, v = 3x

$$\therefore \quad a_x = 4.\frac{\partial y}{\partial x} + v.\frac{\partial y}{\partial y}$$

$$= 2y\times 0 + 3x\times 2$$

$$= 6x = 6\times 1 = 6 \text{ units}$$

and

$$a_y = 4.\frac{\partial v}{\partial x} + v.\frac{\partial v}{\partial y}$$

$$= 2y\times 3 + 3x\times 0$$

$$= 6y = m6\times 1 = 6 \text{ units}$$

$$\therefore \quad \text{Acceleration, } A = a_x\,\vec{i} + a_y\,\vec{j}$$

$$= 6\,\vec{i} + 6\,\vec{j}$$

$$\therefore \quad \text{Resultant } A = \sqrt{6^2+6^2}$$

$$= 8.48\ \text{m/sec}^2$$

63. Kinematic viscosity, $\nu = \dfrac{M}{\text{density}}$

$$R_L = \frac{vL}{u}$$

$$= \frac{0.12\times0.7\times1000}{10^{-3}}$$

$$= 84000$$

This is a Laminar flow.

$$\text{Hence } C_d = \frac{1.328}{\sqrt{R_L}} = 0.00458$$

$$\therefore \quad F = C_d\,\frac{\rho A v^2}{2}$$

$$= 0.00458\times1000\times(0.5\times0.7)\,\frac{(0.12)^2}{2}$$

$$= 0.0115\ \text{N}$$

66. Let discharge in the channel be Q_0 when the slope is S_0.

Then discharge is given by

$$Q_0 = \frac{1}{n}.\,R^{2/3}.\,S_0^{1/2}$$

$$= k\sqrt{S_0} \qquad \text{...}(i)$$

From the question, when the slope increases by 4%, then new discharge

$$\therefore \quad Q_1 = k\sqrt{1.04S_0} \qquad \text{...}(ii)$$

Dividing equation (*ii*) by (*i*), we have

$$\frac{Q_1}{Q_0} = \frac{k\sqrt{1.04S_0}}{k\sqrt{S_0}}$$

$$\therefore \quad Q_1 = 1.019\,Q_0$$
$$\approx 1.02\,Q_0$$

Hence discharge increase by 2%.

67. $\text{Slope} = \dfrac{dy}{dx} = \dfrac{3}{2}$

$$\Rightarrow \quad 3dx - 2dy = 0$$

68. $\text{NPSH} = \dfrac{P_a}{\gamma} - \dfrac{P_v}{\gamma} - h_s - h_{fs}$

$$\therefore \quad 3.3 = \frac{100}{9.81} - 0.44 - h_s - 0.3$$
$$\Rightarrow \quad h_s = 6.1536\text{m}$$

69. *Given :* Linear scale ratio, $L_r = 25$

Height of the hydraulic jump in model,

$$H_m = 0.10 \text{ m}$$

Let height of hydraulic jump in the prototype be H_p.

Then $\dfrac{H_p}{H_m} = L_r = 25$

$$H_p = H_m \times 25 \text{ m.}$$
$$= 0.1 \times 25$$
$$= 2.5 \text{ m.}$$

70. *Given :* $u = \dfrac{y}{h}$

$$\Rightarrow \quad \frac{\partial u}{\partial y} = \frac{v}{h}$$

Rotational component are,

$$w_z = q\frac{1}{2}\left(\frac{\partial v}{\partial x} - \frac{\partial u}{\partial y}\right)$$
$$= \frac{1}{2}\left(0 - \frac{v}{h}\right)$$
$$= \frac{v}{2h}$$
$$w_y = \frac{1}{2}\left(\frac{\partial u}{\partial z} - \frac{\partial w}{\partial x}\right)$$
$$= \frac{1}{2}[0-0] = 0$$

72. *Velocity potential :*

$$u = -\frac{\partial \phi}{\partial x} \qquad ...(i)$$

Stream function :

$$\frac{\partial \Psi}{\partial y} = -u \qquad ...(ii)$$

From equations (*i*) and (*ii*)

$$\therefore \quad \frac{\partial \phi}{\partial x} = \frac{\partial \Psi}{\partial y}$$

$$\Rightarrow \quad \frac{\partial\{\ln(x^2+y^2)\}}{\partial n} = \frac{\partial \Psi}{\partial y}$$

Integrating, we get

$$\Psi = 2\tan^{-1}\left(\frac{y}{x}\right)$$

73. (*i*)

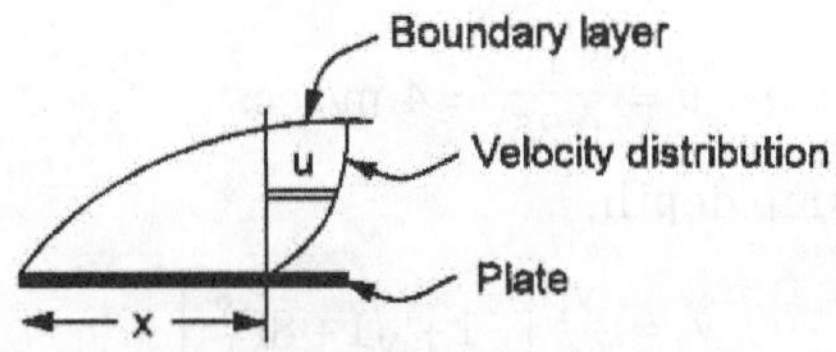

(*ii*)

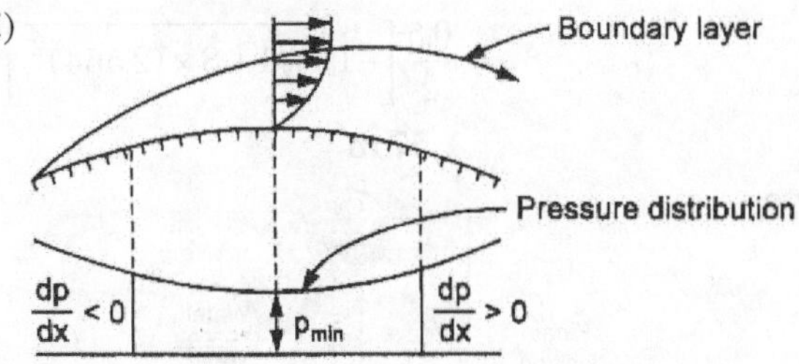

74. Corresponding discharge in model will be

$$Q_m = Q_p \times (L_r)^{5/2}$$
$$= 10000 \times \left(\frac{1}{100}\right)^{5/2}$$
$$= 10{,}000 \times 10^{-5}$$
$$= \mathbf{0.1}\ \boldsymbol{m^3/\text{sec.}}$$

75. If storm has lasted for a very long times, then infilteration rate will be the vertical hydraulic conductivity, *i.e.* 0.2 cm/hr.

76. From the figure,

$$Q_0 \sin\theta + 0.25\,Q_0 = 0.75\,Q_0$$
$$\Rightarrow \quad Q_0 \sin\theta = 0.50\,Q_0$$
$$\Rightarrow \quad \sin\theta = \frac{1}{2}$$
$$\Rightarrow \quad \theta = \mathbf{30°}$$

77. For triangle channel,

$$\text{Froude Number, } F = \frac{V}{\sqrt{g\dfrac{A}{T}}}$$
$$= \frac{L}{\sqrt{9.81 \times \dfrac{0.25}{1}}}$$
$$= 2.554$$

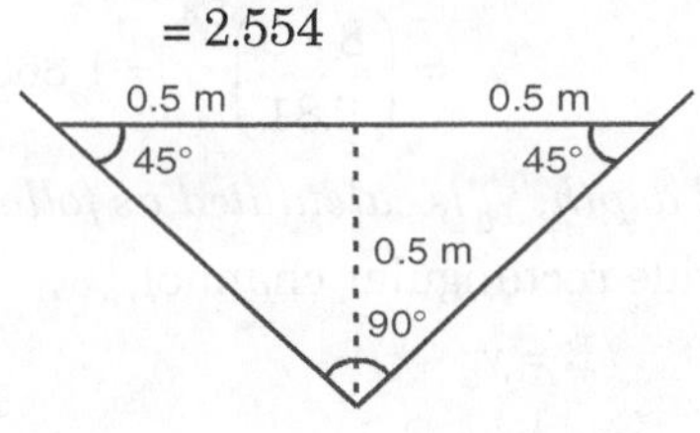

∵ Area of triangular channel,

$$A = \frac{1}{2} \times (0.5 + 0.5) \times 0.5$$

$$= 0.25 \text{ m}^2$$

Top width, T = (0.5 + 0.5) = 1m

Now, $Q = Av$

$$\Rightarrow \quad v = \frac{1}{0.25} = 4 \text{ m/sec.}$$

Last jump depth,

$$y_2 = \frac{y_1}{2}\left[-1 + \sqrt{1 + 8F_1^2}\right]$$

$$= \frac{0.5}{2}\left[-1 + \sqrt{1 + 8 \times (2.554)^2}\right]$$

$$= \mathbf{1.57\ m}$$

78.

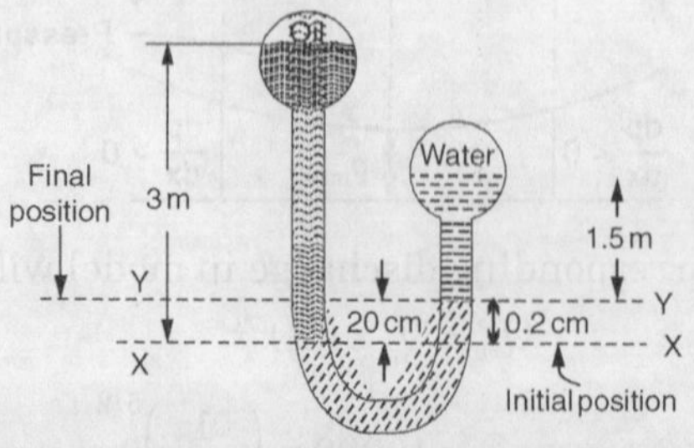

Equating pressure of section Y – Y, we have

$$1.7 + \frac{p_w}{\gamma_\omega} = 0.2 \times 13.57 + (3 - 0.2) \times 0.9$$

Solving, we get $p_w = 26.5\ kPa$

79. From Stoke's formula,

$$v_s = \frac{g}{18}[\rho_s - 1]\frac{d^2}{\nu}$$

$$\Rightarrow \quad 1 \times 5^2 = \frac{9.81}{18}[\rho_s - 1] \times \frac{(6 \times 10^{-3})^2}{\mu / \rho}$$

Solving this equation, we get

specific weight of material = **12.3 kN/m³**

80. *Given :* $q = 8$ m³/sec/m

Manning's roughness coefficient,

$$n = 0.015$$

Bed slope = 0.004

Depth = y_0 = 1m

$$\text{Critical depth, } y_c = \left(\frac{q^2}{g}\right)^{1/3}$$

$$= \left(\frac{8 \times 8}{9.81}\right)^{1/3} = 1.868 \text{ m}$$

Normal depth, y_0 is calculated as follows.

For *a* wide rectangular channel,

$$R = y_0$$

$$\therefore \quad q = \frac{1}{n} y_0 (y_0)^{2/3} S_0^{1/2}$$

$$\Rightarrow \quad y_0 = \left(\frac{nq}{\sqrt{S_0}}\right)^{3/5}$$

$$= \left(\frac{0.015 \times 8}{\sqrt{0.004}}\right)^{3/5} = 1.468 \text{ m}$$

∴ Critical depth, y_c = 1.868 m

and actual depth, y = 1 m

Since $y_c > y_0 > y$

hence Gradually profile will be of S_3 type.

81.

$$\Delta x = \frac{E_2 - E_1}{S_0 - \overline{S}_f}$$

$$\text{where, } \overline{S}_f = \frac{S_{f_1} + S_{f_2}}{2}$$

= Average friction slope of reach

$$S_{f_1} = \frac{n^2 v^2}{R^{4/3}}$$

$$= \frac{n^2 q^2}{y^{10/3}}$$

$$= \frac{(0.015)^2 \times 64}{(1)^{10/3}}$$

$$= 64\,(0.015)^2 = 0.144$$

$$S_{f_2} = \frac{n^2 q^2}{y^{10/3}}$$

$$= \frac{(0.015)^2 \times 64}{(0.9)^{10/3}} = 0.02 \text{ m}$$

Specific energy at section I,

$$E_1 = y_1 + \frac{v_1^2}{2g} = y_1 + \frac{q^2}{2g \times y_1^2}$$

$$= 1 + \frac{(8)^2}{2 \times 9.81 \times (1)^2}$$

$$= 1 + \frac{64}{19.62} = 4.26 \text{ m}$$

Similarly, specific energy of section II,

$$E_2 = y_2 + \frac{v_1^2}{2g}$$

$$= y_2 + \frac{q^2}{2g\, y_2^2}$$

$$= 0.9 + \frac{64}{19.62 \times (0.9)^2}$$

$$= 4.927 \text{ m}$$

Now, $\Delta x = \dfrac{E_2 - E_1}{S_0 - \bar{S}_f}$

where, $S_0 = 0.004$

$$\bar{S}_f = \frac{S_{f1} + S_{g2}}{2}$$

$$= \frac{0.0144 + 0.02}{2} = 0.017$$

$$\therefore \quad \Delta x = \frac{4.927 - 4.26}{0.004 - 0.017} = -50.3 \text{ m}$$

Hence down stream.

82. Before introduction of parallel pipe

$$h_f = 100 - 80 = 20 \text{ m}$$

Also, $h_f = \dfrac{f L v^2}{v \times 2g}$

$$\therefore \quad 20 = \frac{0.04 \times 3 \times 1000 \times v^2}{0.3 \times 19.62}$$

$$\Rightarrow \quad v = \sqrt{\frac{6 \times 19.62}{0.04 \times 3 \times 1000}}$$

$$= 0.99 \text{ m/sec}$$

$\therefore$ Discharge, $Q_0 = \dfrac{\pi}{4}(0.3)^2 \times 0.99$

$$= 0.07 \text{ m}^3\text{/sec}$$

After introduction of parallel pipe of 2 km length,

$f = 0.04$ for all pipe

$d = 0.3$ m for all pipe

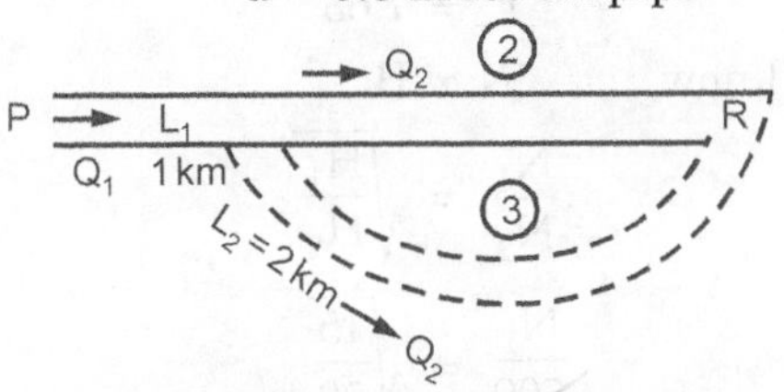

Since pipes 2 and 3 are connected in arallel,

$$h_{f1} = h_{f2}$$

Here, $d_2 = d_3$

$L_2 = L_3$

$f_2 = f_3$

$$\therefore \quad \frac{f L_2 v_2^2}{d_2 \times 2g} = \frac{f L_3 v_3^3}{d_3 \times 2g}$$

$$\Rightarrow \quad v_2 = v_3$$

$$\therefore \quad Q_2 = Q_3$$

Now applying continuity equation at joint Q,

$$Q_1 = Q_2 + Q_3 = 2Q_2 \qquad ...(i)$$

Now $\quad 100 - 80 = \dfrac{f L_1 v_1^2}{d_1 \times 2g} + \dfrac{f L_2 v_2^2}{d_2 \times 2g}$

$$\therefore \quad 20 = \frac{0.04}{2 \times 9.81 \times 0.3}[1000 \times v_1^2 + 2000 \times v_2^2]$$

$$\Rightarrow \quad v_1^2 + 2v_2^2 = 2.943 \qquad ...(ii)$$

From equation *(i)*,

$$Q_1 = 2Q_2$$

$$\Rightarrow \quad v_1 = 2v_2 \qquad ...(iii)$$

Solving equations *(ii)* and *(iii)*, we get

$$4v_2^2 + 2v_2^2 = 2.943$$

$$\Rightarrow \quad v_2 = \sqrt{\frac{2.943}{6}}$$

$$= \sqrt{0.4905}$$

$$= 0.7 \text{ m/sec}$$

$$\therefore \quad Q_2 = \frac{\pi}{4}(0.3)^2 \times 0.7$$

$$= 0.05 \text{ m}^3\text{/sec}$$

and $\quad Q_1 = 2 \times Q_2$

$$= 2 \times 0.05$$

$$= 0.10 \text{ m/sec}$$

Hence % increase in discharge

$$= \frac{Q_1 - Q_0}{Q_0} \times 100$$

$$= \frac{0.01 - 0.07}{0.07} \times 100 = \mathbf{41\%}$$

83. If there is no restriction on diamter of pipe, then add parallel pipe at mid point of the length of pipe. In that case

$$20 = \frac{f L_1 v_1^2}{d_1 \times 2g} + \frac{f L_2 v_2^2}{d_2 \times 2g}$$

Here, $L_1 = L_2 = 1.5$ km.

$$v_1 = v_2 = 0.496 \text{ m/sec.}$$

$$Q = Q_1 + Q_2$$

$$= \frac{\pi}{4}(0.03)^2 [0.496 + 0.496] \times 2$$

$$= 2 \times 0.058$$

$$= 0.1169 \text{ m}^3\text{/sec}$$

$\therefore$ % increase in discharge

$$= \frac{Q - Q_0}{Q_0} \times 100$$

$$= \frac{0.1169 - 0.07}{0.07} \times 100$$

$$= \mathbf{67\%}$$

89. $15 \times A \times 7.4 = x \times 1 \times A + (15 - x)\, 4 \times 13.6$

$$\Rightarrow \quad x = \frac{15(13.6 - 7.4)}{12.6} = 7.38$$

90. *Given* : $Q = 8\ m^3/s$

$$A = \frac{1}{2}(3 + 7.5) \times 1.5 = 7.875 m^2$$

$$\therefore \quad V = \frac{Q}{A} = \frac{8}{7.875} = 1.015 m/s$$

Now, $\frac{1}{1.5} = \frac{1.5}{x}$

$\Rightarrow$ $x = 2.25$

$$T = 3 + 2 \times 2.25 = 7.5 m$$

$$D = \frac{A}{T} = \frac{7.875}{7.5} = 1.05$$

$$\therefore \quad F = \frac{V}{\sqrt{gD}}$$

$$= \frac{1.05}{\sqrt{9.81 \times 1.05}} = 0.316$$

Hence no any option is correct.

91. $Q_r = L_r^{2.5}$

$$\frac{Q_p}{Q_m} = \left(\frac{L_p}{L_m}\right)^{2.5} = (50)^{2.5}$$

$$\therefore \quad Q_p = 0.3 \times (50)^{2.5} = 5303.3\ m^3/s$$

92. For irrotational flow,

$$\text{vorticity, } \xi = \frac{\partial u}{\partial y} - \frac{\partial v}{\partial x} = 0$$

94. Here, $H = 75m$

$Q = 0.464\ m^3/sec$

$N = 1440$ rpm

$$\therefore \quad N_s = \frac{N\sqrt{Q}}{H^{3/4}} = \frac{1440\sqrt{0.464}}{(75)^{3/4}} = 38$$

95. $q = a.v$

$= 0.1 \times 1 \times 6$

$= 0.6\ m^3/sec.$

$\therefore$ Alternate depth,

$$y_2 = -\frac{y_1}{2} + \sqrt{\left(\frac{y_1}{2}\right)^2 + \frac{2q^2}{gy_1}}$$

$$= -\frac{0.1}{2} + \sqrt{\left(\frac{0.1}{2}\right)^2 + 2\frac{(0.6)^2}{9.81 \times 0.1}}$$

$$= -0.05 + 8.58 = 0.81\ m.$$

96. $q = \frac{Q}{B} = \frac{10}{3} = 3.333$

$$\therefore \quad y_c = \left[\frac{q^2}{g}\right]^{1/3} = \left[\frac{(3.333)^2}{9.81}\right]^{1/3} = 1.04\ m$$

97. *Given*: Scale ratio, $L_r = \frac{L_p}{L_m} = 49:1$

Discharge, $Q_p = 15{,}000$

$\therefore$ Scale ratio for discharge,

$$Q_r = (L_r)^{2.5}$$

$$\therefore \quad \frac{Q_p}{Q_m} = (L_r)^{2.5}$$

$$\Rightarrow \quad Q_m = \frac{15000}{(49)^{2.5}} = 0.89\ m^3/sec$$

Velocity ratio, $\frac{V_p}{V_m} = \sqrt{L_r} = 7$

Given : $V_m = 1.2$ m/sec

$\therefore$ $V_p = 7 \times 1.2 = 8.4$ m/sec.

Fraud's Number, $F_r = \frac{V}{\sqrt{Lg}}$

But F_r = constant

$$\therefore \quad \frac{V_p}{V_m} = \sqrt{\frac{L_p}{L_m}}$$

$$\Rightarrow \quad V_r = \sqrt{L_r}$$

98. $$\frac{y_2}{y_1} = \frac{1}{2}\left[-1 + \sqrt{1 + 8Fr_1^2}\right]$$

or $$10.3 = \frac{1}{2}\left[-1 + \sqrt{1 + 8Fr_1^2}\right]$$

$\therefore$ $F_r = 7.63$

100. We know, $N \propto H^{1/2}$

$$\therefore \quad \frac{N_1}{N_2} = \sqrt{\frac{H_1}{H_2}}$$

or $$\frac{N_1}{500} = \sqrt{\frac{15}{20}}$$

$\therefore$ $N_1 = 433$ rpm

101. Head loss per unit length is independent of sewer running full or half full condition.

106. Velocity at nose, $v_1 = 6 m/sec$

Velocity at 1m. head, $v_2 = 4$ m/sec

Let pressure at nose and 1 m head be p_1 and p_2 respectively, then

$$\frac{p_1}{w} + z + \frac{v_1^2}{2g} = \frac{p_2}{w} + z + \frac{v_2^2}{2g}$$

or $$\frac{(P_1 - P_2)}{w} = \frac{v_2^2 - v_1^2}{2g} = \frac{(6)^2 - (4)^2}{2 \times 9.81} \times 9810$$

$$= 10{,}000\ N/m^2$$

107. Dynamic viscosity,

$$\mu = \frac{\text{shear stress}}{\frac{\text{change of velocity}}{\text{change of distance}}}$$

$$= \frac{\text{Force / (length)}^2}{1/\text{time}}$$

$$= \frac{\text{Force} \times \text{Time}}{(\text{length})^2}$$

$$= \frac{\text{N.s}}{\text{m}^2}$$

117. Assume, $C_c = 0.60$

$$C_d = \frac{0.60}{\sqrt{1+0.6\frac{0.15}{1.50}}} = 0.583$$

$$q = C_d\, a\, \sqrt{2gh}$$

$$= 0.583 \times 0.15 \times \sqrt{2 \times 0.806 \times 1.5}$$

$$= 0.474 \text{ m}^3/sm$$

$$U_1 = \frac{0.474}{1.50} = 0.316 \text{ m/s}$$

$$H_2 = 0.150 \times 6.0 = 0.090 \text{ m}$$

and $$U_2 = \frac{0.474}{0.09} = 5.267 \text{ m/s}$$

Apply moment equation between sections 1 and 2. Boundary frictional force F_t is assumed to be very small and hence neglected.

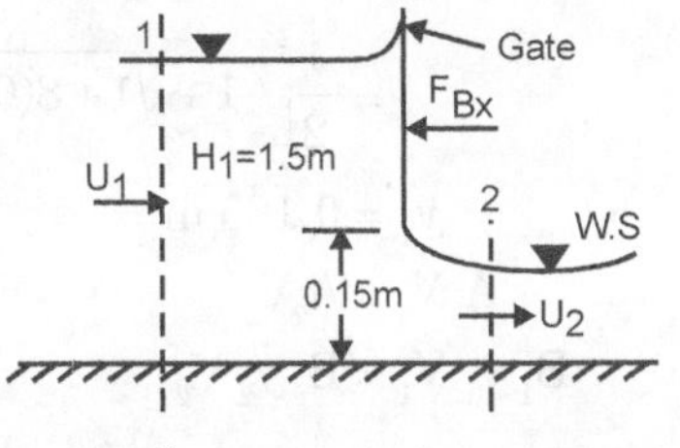

$$\therefore\ \frac{1}{2}\gamma H_1^2 B - \frac{1}{2}\gamma H_2^2 B - F_{BX} = Q\rho(U_2 - U_1)$$

$$\Rightarrow \frac{1}{2} \times 2 \times 9787(1.5^2 - 0.09^2) - F_{Bx} = 0.474 \times 2 \times 998(5.267 - 0.316)$$

$$\Rightarrow F_{Bx} = 17255.94 \text{ N or } 17.256 \text{ kN}$$

NUMERICAL TYPE QUESTIONS

1. Given, $F_r = 10$

$$\frac{y_2}{y_1} = \frac{1}{2}\left[-1+\sqrt{1+8F_r^2}\right]$$

$$= \frac{1}{2}\left[-1+\sqrt{1+8\times 10^2}\right] = 13.651$$

2. Vertical Height of sector gate (AB)

AB = 5 m (equilateral triangle)

Area of segment = area of sector – area of triangle

$$A = \frac{1}{6}(\pi \times 5^2) - \left(\frac{1}{6} \times AB \times OD\right)$$

$$= \frac{1}{6}(\pi \times 5^2) - \left(\frac{1}{6} \times 5 \times 5\cos 30°\right)$$

$$A = 13.09 - 10.82$$

$$= 2.27 \text{ m}^2$$

Horizontal component of hydrostatic force on sector gate

$$F_H = \rho.g.\bar{h}.A_{\textit{projected vertical plane}}$$

$$F_H = 1000 \times 10 \times \frac{5}{2} \times (5 \times 1) \text{ N/m}$$

$$F_H = 125 \text{ kN/meter length}$$

Vertical component of hydrostatic force on the sector gate,

$$F_V = \rho.g. \text{ volume displaced}$$

$$F_V = \rho.g.A.L$$

$$F_V = 1000 \times 10 \times 2.27 \times 1$$

$$FV = 22.7 \text{ kN/meter length}$$

Resultant hydrostatic force,

$$= \sqrt{F_H^2 + F_V^2}$$

$$= \sqrt{125^2 + 22.7^2}$$

$$= 127.04 \text{ kN/m}$$

3.

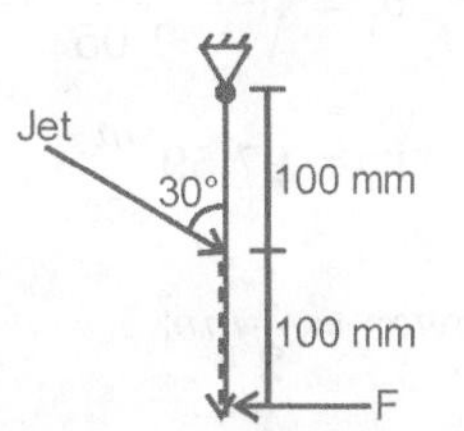

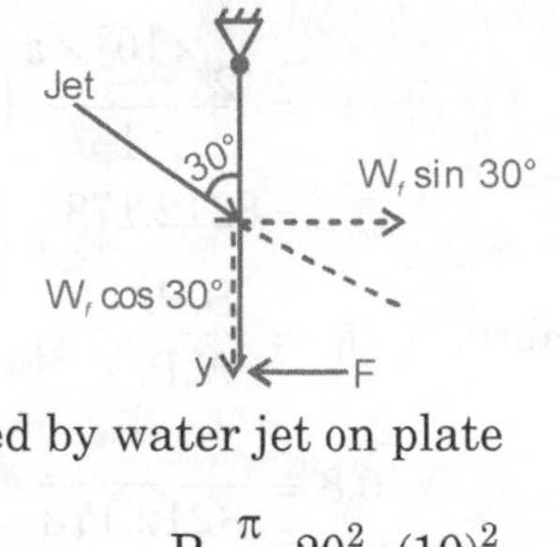

Force exerted by water jet on plate

$$= P \cdot \frac{\pi}{4} \cdot 20^2 \cdot (10)^2$$

$$= P \cdot \frac{\pi}{4} \cdot \left(\frac{20}{100}\right)^2 \cdot (100)$$

$$= 10\pi = 31.4 \text{ N}$$

$$W_f \sin 30° \times 100 = F \times 200$$

Therefore, $F = \frac{1}{2} W_f \sin 30°$

$= \frac{W_f}{4} = \frac{31.4}{4}$

$= 7.85$ N

4. $E_1 = E_{2min} + \Delta Z_{max}$

$$E_{2min} = \left(\frac{q^2}{g}\right)^{1/3} \times \frac{3}{2}$$

$$= \left(\frac{(6/3)^2}{9.81}\right)^{1/3} \times \frac{3}{2} = 1.1123 \text{ m}$$

$$E_1 = y + \frac{q^2}{2gy^2}$$

$$= 0.5 + \frac{2^2}{2 \times 9.81 \times 0.5^2}$$

$= 1.3155$ m

$\therefore \quad \Delta Z_{max} = 1.3155 - 1.1123$

$= 0.20$ m

5. Apply energy equation of free surface of reserveoir,

$$500 = \text{head loss} + \frac{v_1^2}{2g}$$

$$= 0.05\frac{v_1^2}{2g} + \frac{v_1^2}{2g}$$

$$v_1 = \sqrt{\frac{2 \times 10 \times 500}{1.05}}$$

$$v_1 = 97.59 \frac{m}{\text{sec}}$$

$$\text{Water power} = \frac{1}{2} m v_1^2$$

$$= \frac{\frac{1}{2} \times 10^3 \times \pi}{4}(0.15^2) \times 97.59$$

$= 8212.178$

Therefore, $\eta_0 = \frac{\text{S.P.}}{\text{W.P.}}$

$\Rightarrow \quad 0.8 = \frac{\text{S.P.}}{8212.178}$

$\Rightarrow \quad$ S.P. $= 6570$ kW

6. Total Loss = 20 mm

$$\Rightarrow 20 = \frac{0.5v^2}{29} + \frac{f \times L}{d} \times \frac{v^2}{2g} + \frac{5.5v^2}{2g} + \frac{v^2}{2g}$$

$$\Rightarrow 20 \times 2 \times 10 = 0.5v^2 + \frac{0.03 \times 930 \times v^2}{0.3} = 5.5v^2 + v^2$$

$$\Rightarrow v^2 = \frac{400}{100} = 4$$

$\Rightarrow v = 2$ m/s

$$\Rightarrow \theta = \frac{\pi}{4} d^2 \times v = \frac{\pi}{4} \times (0.3)^2 \times 2 = 0.1413 \text{ m}^3/\text{s}$$

7. Adverse slope $= -\frac{1}{10000}$

$\theta = 4$ m³/s, $n = 0.01$, $y = 0.5$ m

$$\frac{dy}{dx} = \frac{S_0 - S_f}{1 - F_r^2}$$

$$F_r = \frac{V}{\sqrt{gy}} = \frac{\frac{Q}{By}}{\sqrt{gy}}$$

$$= \frac{4}{2 \times 0.5 \times \left(\frac{4 \times 0.01}{2+1}\right)^{\frac{2}{3}}} = 1.79$$

Therefore, by using above formula,

$$\frac{dy}{dx} = 0.0032$$

8. $y_2 = 0.8$ m

$V_2 = 1$ m/s, $V_1 = ?$

$$F_2 = \frac{V_2}{\sqrt{gy_2}} = \frac{1}{\sqrt{10 \times 0.8}} = 0.353$$

and $$\frac{y_1}{y_2} = \frac{1}{2}\left[-1 + \sqrt{1 + 8F_2^2}\right]$$

$$= \frac{1}{2}\left[-1 + \sqrt{1 + 8(0.353)^2}\right]$$

$y_1 = 0.165$ m

Now, $A_1 V_1 = A_2 V_2$

$\Rightarrow \quad B_1 y_1 V_1 = B_2 y_2 V_2$

$$\therefore \quad V_1 = \frac{0.8 \times 1}{0.165} = 4.828 \text{ m/s}$$

9. Convective Acceleration,

$$a_c = u\frac{du}{dx} + v\frac{du}{dy} + w\frac{du}{dz}$$

$$\frac{du}{dy} = \frac{du}{dz} = 0$$

$$\therefore \quad a_c = u\frac{du}{dx}$$

$$= \frac{15(15 - 1.5)}{0.375}$$

$= 540$ m/s² at the end of nozzle.

10. $$h_f = \frac{flQ^2}{12.1d^5}$$

Let discharge in reservoir A be Q.

In C = 0.15 m³/s

$\therefore$ discharge in reservoir B = (Q – 0.15)

Now In reservoir A water level is 30 m above water level in B.

$$\therefore\ h_{fa} + h_{fb} = 30$$

$$\frac{0.024\times 3000\times Q^2}{12.1\times 0.7^5} + \frac{0.024\times 3000\times (Q-0.15)^2}{12.1\times 0.7^5} = 30$$

$$\therefore\ Q = 0.7216$$

$$Q_B = 0.7216 - 0.15 = 0.5716$$

11.

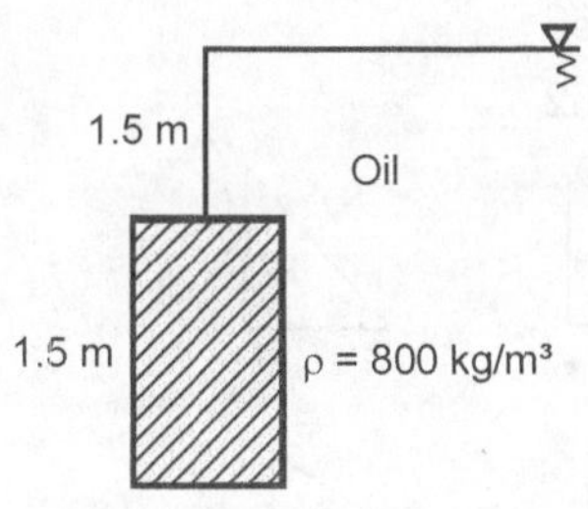

Hydrostatic Force = Area × Pressure

$$= \frac{1.5\times 2}{2}\times \rho g \times h$$

$$= 1.5\times 800\times 9810\times\left[1.5+\frac{2}{3}\times 1.5\right]$$

$$= 1200 \times 9810 \times 2.5$$

$$= 29.430 \text{ kN}$$

12. By continuity equation between two sections

$$A_1V_1 = A_2V_2$$

$$\Rightarrow\ \frac{\pi}{4}\times (d_1)^2.V_1 = \frac{\pi}{4}\times (d_2)^2.V_2$$

$$\Rightarrow\ (15)^2 \times 2.5 = (5)^2 \times V_2$$

$$\Rightarrow\ V_2 = \frac{225\times 2.5}{25} = 22.5 \text{ m/s}$$

13. $x = t^3 + 2t,\ y = -3^{c-2t},\ z = 2\sin(5t)$

$$a_x = \frac{\partial^2 x}{\partial t^2} = 6t$$

$$a_y = \frac{\partial^2 y}{\partial t^2} = -12e^{-2t}$$

$$a_z = \frac{\partial^2 z}{\partial t^2} = -50\sin 5t$$

$$a = a_x\hat{i} + a_y\hat{j} + a_z\hat{k}$$

$$= (6t)\hat{i} - (12e^{-2t})\hat{j} - (50\sin 5t)\hat{k}$$

$$a(t=0) = -12$$

So, $|a|_{t=0} = 12 \text{ cm/s}^2$

14. $$Q = \frac{c_d.a_1a_2}{\sqrt{a_1^2 - a_2^2}}.\sqrt{2gh}$$

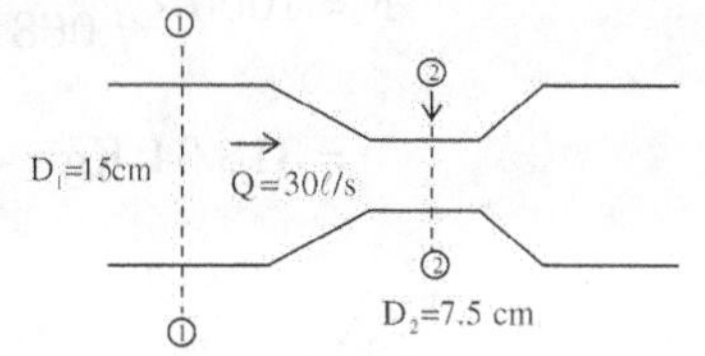

$$a_1 = \frac{\pi}{4}\times(.15)^2$$

$$= 0.0176 \text{ m}^2$$

$$a_2 = \frac{\pi}{4}\times(.075)^2$$

$$= 0.0044 \text{ m}^2$$

$$\Rightarrow\ 30\times 10^{-3} = C_d.\frac{0.0176\times 0.0044}{\sqrt{(0.0176)^2-(0.0044)^2}}\times\sqrt{2\times 9.81\times 2.45}$$

$$\Rightarrow\ C_d = 0.95$$

15. Since pipes are in parallel, so, head loss will be same

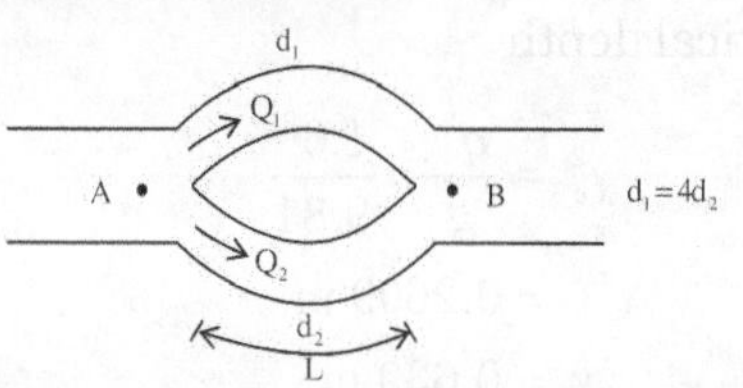

$$\frac{fL}{d_1}.\frac{V_1^2}{2g} = \frac{fL}{d_2}.\frac{V_2^2}{2g}$$

$$\Rightarrow\ \frac{V_1^2}{d_1} = \frac{V_2^2}{d_2}$$

$$\Rightarrow\ \frac{V_1^2}{V_2^2} = \frac{d_1}{d_2} = 4$$

$$\Rightarrow\ \frac{V_1}{V_2} = 2$$

16. Nozzle diameter = 30 mm = 0.03m

Q = 15 *l*/sec

$$[1000l = 1\text{m}^3 \Rightarrow 1l = \frac{1}{1000}\text{m}^3]$$

Diameter of inlet = 100 mm

Density of water = 1000 kg/m³

Force exerted by the jet on the plate 'F' = ?

$$F = e.a.v^2$$

$$v = \frac{Q}{a}$$

$$\because\ F = e.a.\left(\frac{Q^2}{a^2}\right) = e.\frac{Q^2}{a}$$

$$a = \frac{\pi}{4}(0.03)^2 = 7.068\times 10^{-4} \text{ m}^2$$

$$Q = \frac{15}{1000}\text{ m}^3/\text{sec} = 0.015 \text{ m}^3/\text{sec}$$

$$F = 1000 \times \frac{0.015^2}{7.068\times10^{-4}}\ \frac{Kg}{m^3} \times \frac{m^6}{m^2\,sec^2}$$

$$= 318.34\ Kg \times \frac{m}{sec^2} = 318.34N$$

$$\left[\because\ 1Kg \times \frac{m}{sec^2} = 1N\right]$$

17. At critical depth,

$$\text{Velocity head} = \frac{\gamma_c}{2}$$

$$\frac{V^2}{2g} = \frac{\gamma_c}{2} \qquad ...(i)$$

We have, Q = 4 m³/sec, B = 2.5 m

$$\because \quad q = \frac{Q}{B} = \frac{4}{2.5} = 1.6\ m^3/sec/m$$

critical depth

$$\gamma_c^2 = \frac{q^2}{g} = \frac{1.6^2}{9.81}$$

$$= 0.2609\ m$$

$$\therefore \quad \gamma_c = 0.639\ m$$

From equation (*i*)

$$\frac{V^2}{2g} = \frac{\gamma_c}{2}$$

$$V^2 = \frac{0.639\times2\times9.81}{2}$$

$$V = 2.503 \cong 2.5\ m/sec$$

19. *Given :* $\theta_1 = 2.5m$

$\beta_1 = \beta_2 = 0.2m$

$\theta_1 = 0.5\ m$

$N = 300rpm$

$$\therefore \quad u_1 = \frac{\pi\theta_1 N}{60}$$

$$= \frac{\pi\times2.5\times300}{60} = 39.27$$

$$u_2 = \frac{\pi\theta_2 N}{60}$$

$$= \frac{\pi\times1.5\times300}{60} = 23.56$$

$V_{f1} = 6.366$

$V_{f2} = 10.62$

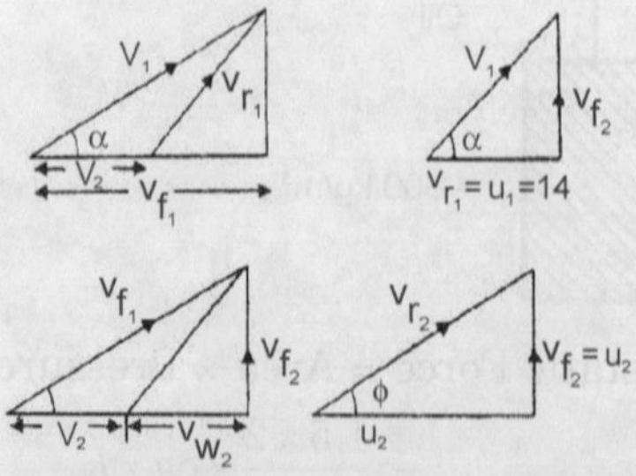

$$\therefore \quad \tan\alpha = \frac{V_{f_1}}{u_1} = \frac{6.366}{39.26}$$

$$\Rightarrow \quad \alpha = 9.21°$$

$$\tan\phi = \frac{V_{f_2}}{u_2} = \frac{10.62}{23.56} = 24.26°$$

9 CHAPTER Hydrology

MCQ TYPE QUESTIONS

2016

1. The type of flood routing (Group I) and the equation(s) used for the purpose (Group II) are given below.

	Group I	**Group II**
P.	Hydrologic flood routing	1. Continuity equation
Q.	Hydraulic flood routing	2. Momentum equation
		3. Energy equation

The correct match is

(*a*) P - 1; Q - 1, 2 & 3

(*b*) P - 1; Q - 1 & 2 only

(*c*) P -1 & 2; Q - 1 only

(*d*) P - 1 & 2; Q - 1 & 2

2015

2. The average surface area of a reservoir in the month of June is 20 km^2. In the same month, the average rate of inflow is 10 m^3/s, outflow rate is 15 m^3/s, monthly rainfall is 10 cm, monthly seepage loss is 1.8 cm and the storage change is 16 million m^3. The evaporation (in cm) in that month is

(*a*) 46.8 (*b*) 136.6

(*c*) 13.6 (*d*) 23.4

2014

3. A conventional flow duration curve is a plot between

(*a*) Flow and percentage time flow is exceeded

(*b*) Duration of flooding and ground level elevation

(*c*) Duration of water supply in a city and proportion of area receiving supply exceeding this duration

(*d*) Flow rate and duration of time taken to empty a reservoir at that flow rate

4. In reservoirs with an uncontrolled spillway, the peak of the plotted outflow hydrograph

(*a*) lies outside the plotted inflow hydrograph

(*b*) lies on the recession limb of the plotted inflow hydrograph

(*c*) lies on the peak of the inflow hydrograph

(*d*) is higher than the peak of the plotted inflow hydrograph

5. An isolated 3-h rainfall event on a small catchment produces a hydrograph peak and point of inflection on the falling limb of the hydrograph at 7 hours and 8.5 hours respectively, after the start of the rainfall. Assuming, no losses and no base flow contribution, the time of concentration (in hours) for this catchment is approximately

(*a*) 8.5 (*b*) 7.0

(*c*) 6.5 (*d*) 5.5

6. The Muskingum model of routing a flood through a stream reach is expressed as $O_2 = K_0I_2 + K_1I_1 + K_2O_1$, where K_0, K_1 and K_2 are the routing coefficients for the concerned reach, I_1 and I_2 are the inflows to the reach, and O_1 and O_2 are the outflows from the reach corresponding to time steps 1 and 2 respectively. The sum of K_0, K_1 and K_2 of the model is

(*a*) – 1 (*b*) – 0.5

(*c*) 0.5 (*d*) 1

2013

7. A 1-h rainfall of 10 cm magnitude at a station has a return period of 50 years. The probability that a 1-h rainfall of magnitude 10 cm or more will occur in each of two successive years is

(*a*) 0.04 (*b*) 0.2

(*c*) 0.02 (*d*) 0.0004

8. An isohyet is a line joining points of

(*a*) equal temperature

(*b*) equal humidity

(*c*) equal rainfall depth

(*d*) equal evaporation

9. The normal depth in a wide rectangular channel is increased by 10%. The percentage increase in the discharge in the channel is

(*a*) 20.1 (*b*) 15.4

(*c*) 10.5 (*d*) 17.2

2012

10. The ratio of actual evapo-transpiration to potential evapo-transpiration is in the range

(*a*) 0.0 to 0.4 (*b*) 0.6 to 0.9

(*c*) 0.0 to 1.0 (*d*) 1.0 to 2.0

2011

11. A watershed got transformed from rural to urban over a period of time. The effect of urbanization on storm runoff hydrograph from the watershed is to
(a) decrease the volume of runoff
(b) increase the time to peak discharge
(c) decrease the time base
(d) decrease the peak discharge

12. For a given discharge, the critical flow depth in an open channel depends on
(a) channel geometry only
(b) channel geometry and bed slope
(c) channel geometry, bed slope and roughness
(d) channel geometry, bed slope, roughness and Reynolds number

13. The flow in a horizontal, frictionless rectangular open channel is supercritical. A smooth hump is built on the channel floor. As the height of hump is increased, choked condition is attained. With further increase in the height of the hump, the water surface will
(a) rise at a section upstream of the hump
(b) drop at a section upstream of the hump
(c) drop at the hump
(d) rise at the hump

14. A single pipe of length 1500 m and diameter 60 cm connects two reservoirs having a difference of 20 m in their water levels. The pipe is to be replaced by two pipes of the same length and equal diameter d to convey 25% more discharge under the same head loss. If the friction factor is assumed to be the same for all the pipes, the value of d is approximately equal to which of the following options?
(a) 37.5 cm (b) 40.0 cm
(c) 45.0 cm (d) 50.0 cm

15. A spillway discharges flood at a rate of 9 m^3/s per metre width. If the depth of flow on the horizontal apron at the toe of the spillway is 46 cm, the tail water depth needed to form a hydraulic jump is approximately given by which of the following options?
(a) 2.54 m (b) 4.90 m
(c) 5.77 m (d) 6.23 m

16. In an aquifer extending over 150 hectare, the water table was 20 m below ground level. Over a period of time the water table dropped to 23 m below the ground level. If the porosity of aquifer is 0.40 and the specific retention is 0.15, what is the change in ground water storage of the aquifer?
(a) 67.5 ha-m (b) 112.5 ha-m
(c) 180.0 ha-m (d) 450.0 ha-m

2010

17. The correct match of Group -I Group-II is

Group-I	Group-II
P. Evapotranspiration	1. Penman method
Q. Infiltration	2. Snyder's method
R. Synthetic unit hydrography	3. Muskingum method
S. Channel Routing	4. Horton's method

Codes :

	P	Q	R	S
(a)	1	3	4	2
(b)	1	4	2	3
(c)	3	4	1	2
(d)	4	2	1	3

2009

18. Direct step method of computation for gradually varied flow is
(a) applicable to non-prismatic channels
(b) applicable to prismatic channels
(c) applicable to both prismatic and non-prismatic channels
(d) not applicable to both prismatic and non-prismatic channels

19. The relationship among specific yield (S_y), specific ratention (S_r) and porosity (η) of an aquifer is
(a) $S_y = S_r + \eta$ (b) $S_y = S_r - \eta$
(c) $S_y = \eta - S_r$ (d) $S_y = S_r + 2\eta$

2008

20. A flood wave with a known inflow hydrograph is routed through a large reservoir. The outflow hydrograph will have
(a) attenuated peak with reduced time-base
(b) attenuated peak with increased time-base
(c) increased peak with increased time-base
(d) increased peak with reduced time-base

21. A stable channel is to be designed for a discharge of Q m³/s with silt factor f as per Lacey's method.The mean flow velocity (m/s) in the channel is obtained by

(*a*) $(Q f^2/140)^{1/6}$

(*b*) $(Qf/140)^{1/3}$

(*c*) $(Q^2 f^2/140)^{1/6}$

(*d*) $0.48\,(Q/f)^{1/3}$

22. Water emerges from an ogee spillway with velocity = 13.72 m/s and depth = 0.3 m at its toe. The tail water depth required to form a hydraulic jump at the toe is

(*a*) 6.48 m

(*b*) 5.24 m

(*c*) 3.24 m

(*d*) 2.24 m

23. A river reach of 2.0 km long with maximum flood discharge of 10000 m³/s is to be physically modeled in the laboratory where maximum available discharge is 0.20 m³/s. For a geometrically similar model based on equality of Froude number, length of the river reach (m) in the model is

(*a*) 26.4 (*b*) 25.0

(*c*) 20.5 (*d*) 18.0

2007

24. There is a free overfall at the end of a long open channel. For a given flow rate, the critical depth is less than the normal depth. What gradually varied flow profile will occur in the channel for this flow rate ?

(*a*) M_1 (*b*) M_2

(*c*) M_3 (*d*) S_1

25. As per the Lacey's method for design of alluvail channels, identify the true statement from the following.

(*a*) Wetted perimeter increases with an increase in design discharge

(*b*) Hydraulic radius increases with an increase in silt factor.

(*c*) Wetted perimeter decreases with an increase in design discharge

(*d*) Wetted perimeter increases with an increase in silt factor

26. The culturable command area for a distributary channel is 20,000 hectares. Wheat is grown in the entire area and the intensity of irrigation is 50%. The *kor* period for wheat is 30 days and the *kor* water depth is 120 mm. The outlet discharge for the distributary should be

(*a*) 2.85 m³/s (*b*) 3.21 m³/s

(*c*) 4.63 m³/s (*d*) 5.23 m³/s

27. An isolated 4-hour storm occurred over a catchment as follows

Time	1st hour	2nd hour	3rd hour	4th hour
Rainfall (mm)	9	28	12	7

The ϕ index for the catchment is 10 mm/h. The estimated runoff depth from the catchment due to the above storm is

(*a*) 10 mm (*b*) 16 mm

(*c*) 20 mm (*d*) 23 mm

2006

28. During a 3 hour storm event, it was observed that all abstractions other than infiltration are negligible. The rainfall was idealized as 3 one hour storms of intensity 10 mm/hr, 20 mm/hr and 10 mm/hr respectively and the infiltration was idealized as a Horton curve, $f = 6.8 + 8.7 \exp(-t)$ (f in mm/hr and t in hr).

What is the effective rainfall?

(*a*) 10.00 mm

(*b*) 11.33 mm

(*c*) 12.43 mm

(*d*) 13.63 mm

2005

29. The intensity of rainfall and time interval of a typical storm are

Time interval (minutes)	Intensity of rainfall (mm/minute)
0 - 10	0.7
10 - 20	1.1
20 - 30	2.2
30 - 40	1.5
40 - 50	1.2
50 - 60	1.3
60 - 70	0.9
70 - 80	0.4

The maximum intensity of rainfall for 20 minutes duration of the storm is

(*a*) 1.5 mm/minute (*b*) 1.85 mm/minute

(*c*) 2.2 mm/minute (*d*) 3.7 mm/minute

30. Which one of the following equations reprsents the downstream profile of Ogee spillway with vertical upstream face?

{(x, y) are co-ordinates of the point on the downstream profile with origin at the crest of the spillway and H_d is the design head}

(a) $\frac{y}{H_d} = -0.5\left(\frac{x}{H_x}\right)^{1.85}$

(b) $\frac{y}{H_d} = -0.5\left(\frac{x}{H_d}\right)^{1/1.85}$

(c) $\frac{y}{H_d} = -2.0\left(\frac{x}{H_d}\right)^{1.85}$

(d) $\frac{y}{H_d} = -2.0\left(\frac{x}{H_d}\right)^{1/1.85}$

31. Assuming that a river bed level does not change and the depth of water in river was 10 m, 15 m and 8 m during the months of February, July and December respectively of a particular year. The average bulk density of the soil is 20 kN/m^3. The density of water is 10 kN/m^3. The effective stress at a depth of 10 m below the river bed during these months would be

(a) 300 kN/m^2 in February, 350 kN/m^2 July and 320 kN/m^2 in December

(b) 100 kN/m^2 in February, 100 kN/m^2 July and 100 kN/m^2 in December

(c) 200 kN/m^2 in February, 250 kN/m^2 July and 180 kN/m^2 in December

(d) 300 kN/m^2 in February, 350 kN/m^2 July and 280 kN/m^2 in December

32. Two observation wells penetrated into a confined aquifer and located 1.5 km apart in the direction of flow, indicate head of 45 m and 20 m. If the coefficient of permeability of the aquifer is 30 m/day and porosity is 0.25, then time of travel of an inert tracer from one well to another is

(a) 416.7 days (b) 500 days

(c) 750 days (d) 3000 days

Linked Answer Q. 33-34

A four hour unit hydrograph of a catchment is triangular in shape with base of 80 hours. The area of the catchment is 720 km^2. The base flow and f-index are 30 m^3/s and 1 mm/h, respectively. A storm of 4 cm occurs uniformly in 4 hours over the catchment.

33. The peak discharge of four hour unit hydrograph is

(a) 40 m^3/s (b) 50 m^3/s

(c) 60 m^3/s (d) 70 m^3/s

34. The peak flood discharge due to the storm is

(a) 210 m^3/s (b) 230 m^3/s

(c) 260 m^3/s (d) 720 m^3/s

2004

35. The rainfall during three successive 2 hour periods are 0.5, 2.8 and 1.6 cm. The surface runoff resulting from this storm is 3.2 cm. The ϕ index value of this storm is

(a) 0.20 cm/hr (b) 0.28 cm/hr

(c) 0.30 cm/hr (d) 0.80 cm/hr

36. The average rainfall for a 3 hour duration storm is 2.7 cm and the loss rate is 0.3 cm/hr. The flood hydrograph has a base flow of 20 m^3/s and produces a peak flow of 210 m^3/s. The peak of a 3-h unit hydrograph is

(a) 125.50 m^3/s (b) 105.50 m^3/s

(c) 77.77 m^3/s (d) 70.37 m^3/s

2003

37. While applying the Rational formula for computing the design discharge, the rainfall duration is stipulated as the time of concentration because

(a) this leads to the largest possible rainfall intensity

(b) this leads to the smallest possible rainfall intensity

(c) the time of concentration is the smallest rainfall duration for which the Rational formula is applicable

(d) the time of concentration is the largest rainfall duration for which the Rational formula is applicable

Common Data Q. 38-39

An average rainfall of 16 cm occurs over a catchment during a period of 12 hours with a uniform intensity. The unit hydrograph (unit depth = 1 cm, duration = 6 hours) of the catchment rises linearly from 0 to 30 cumecs in six hours and then falls linearly from 30 to 0 cumecs in the next 12 hours. ϕ index of the catchment is known to be 0.5 cm/hr. Base flow in the river is known to be 5 cumecs.

38. Peak discharge of the resulting direct runoff hydrograph shall be

(a) 150 cumecs (b) 225 cumecs

(c) 230 cumecs (d) 360 cumecs

39. Area of the catchment in hectares is

(a) 97.20 (b) 270

(c) 9720 (d) 27000

2002

40. A 6-hour Unit Hydrograph (UH) of a catchment is triangular in shape with a total time base of 36 hours and a peak discharge of 18 m^3/s. The area of the catchment(in sq. km)is

(a) 233

(b) 117

(c) 1.2

(d) sufficient information not available

41. When there is an increase in the atmospheric pressure, the water level in a well penetrating in a confined aquifer

(a) increases

(b) decreases

(c) may increase or decrease depending on the nature of the aquifer

(d) does not undergo any change

42. The rainfall on five successive days in a catchment were measured as 3, 8, 12, 6, and 2 cms. If the total runoff at the outlet from the catchment was 15 cm, the value of then ϕ- index (in mm/hour) is

(a) 0.0

(b) 1.04

(c) 1.53

(d) sufficient information not available

2001

43. Isopleths are lines on a map through points having equal depth of

(a) rainfall

(b) infiltration

(c) evapotranspiration

(d) total runoff

44. A linear reservoir is one in which Storage varies linearly with

(a) time

(b) outflow rate

(c) inflow rate

(d) elevation

45. The direct runoff hydrograph of a storm obtained from a catchment is triangular in shape and has a base period of 80 hours. The peak flow rate is 30m^3/sec and catchment area is 86.4 km^2. The rainfall excess that has resulted the above hydrograph is

(a) 5 cm (b) 8 cm

(c) 10 cm (d) 16 cm

46. The 4-hour unit hydrograph (UH) for a catchment having an area of 536 km^2 is shown in the Table below. What will be the peak discharge when a 3-hour period of rainfall excess with intensity of 6 mm/hr was realised in the catchment ? Assume that there is no base flow.

Time (hours)	UH ordinate (m^3/sec)	Time (hours)	UH ordinate (m^3/sec)
0	0	11	73
1	10	12	59
2	60	13	48
3	120	14	36
4	170	15	28
5	200	16	20
6	180	17	13
7	150	18	8
8	124	19	3
9	104	20	0
10	88	21	0

(a) 184 m^3/sec (b) 284 m^3/sec

(c) 384 m^3/sec (d) 584 m^3/sec

2000

47. In an area of 200 hectare, water table drops by 4 m. If porosity is 0.35 and specific retention is 0.15, then change in volume of storage in the aquifer is

(a) 160 m^3 (b) 1.6 × 10^6 m

(c) 8 × 10^6 m^3 (d) 1.6 × 10^3 m^3

48. A tube well having a capacity of 4m^3/hour operates for 20 hours each day during the irrigation season. How much area can be commanded if the irrigation interval is 20 days and depth of irrigation is 7 cm ?

(a) 1.71 × 10^4 m^2

(b) 1.14 × 10^4 m^2

(c) 22.9 × 10^4 m^2

(d) 2.29 × 10^4 m^2

49. A 6 hour unit hydrograph of a watershed is given below. Calculate 18 hour unit hydrograph using S- curve method and tabulate the results.

Time (hour)	6hr unit hydrograph [m^3/(cm)]
0	0
6	1.8
12	30.9
18	85.6
24	41.8
30	14.6
36	5.5
42	1.8

1999

50. The ordinate of the Instantaneous Unit Hydrograph (IUH) of a catchment at any time t, is

(*a*) the slope of the 1-hour unit hydrograph at that time

(*b*) the slope of the direct runoff unit hydrograph at that time

(*c*) difference in the slope of the S - curve and 1 - hour unit hydrograph

(*d*) the slope of the S - curve with effective rainfall intensity of 1cm/hr

51. An isochrone is a line on the basin map

(*a*) joining raingauge stations having equal rainfall duration

(*b*) joining points having equal rainfall depth in a given time interval

(*c*) joining points having equal time of travel of surface runoff to the catchment outlet

(*d*) joining ponts which are at equal distance from the catchment outlet.

52. A soil formation through which only seepage is possible, being partly permeable and capable of giving insignificant yield, is classified as

(*a*) Aquifer (*b*) Aquitard

(*c*) Aquifuge (*d*) Aquiclude

53. A direct runoff hydrograph due to an isolated storm with an effective rainfall of 2 cm was trapezoidal in shape as shown in the figure. The hydrograph corresponds to a catchment area (in sq. km.) of

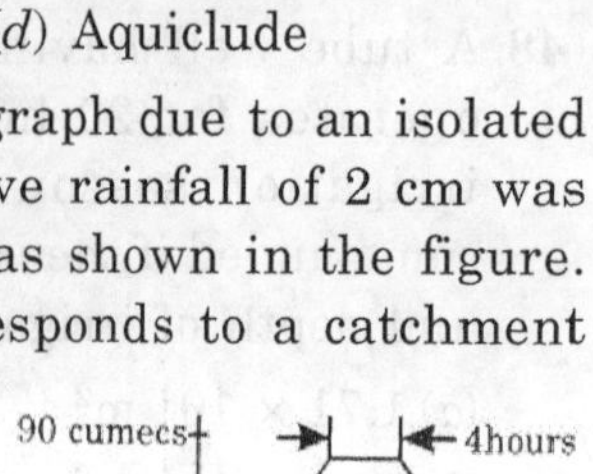

(*a*) 790.2

(*b*) 599.4

(*c*) 689.5

(*d*) 435.3

54. In a river, discharge is 173 mp³/s; water surface slope is 1 in 6000; and stage at the gauge station is 10.0 m. If during a flood, the stage at the gauge station is same and the water surfae slope is 1 in 2000, the flood discharge in m^3/s, is approximately

(*a*) 371 (*b*) 100

(*c*) 519 (*d*) 300

1998

55. The Bowen ratio is defined as

(*a*) Ratio of heat and vapour diffusivities

(*b*) Proportionality constant between vapour heat flux and sensible heat flux.

(*c*) Ratio of actual evapotranspiration and potential evapotranspiration.

(*d*) Proportionality constant between heat energy used up in evaporation and the bulk radiation from a water body.

56. The return period for the annual maximum flood of a given magnitude is 8 years. The probability that this flood magnitude will be exceeded once during the next 5 years is

(*a*) 0.625 (*b*) 0.966

(*c*) 0.487 (*d*) 0.529

57. Two completely penetrating wells are located L (in meters) apart, in a homogeneous confined aquifer. The drawdown measured at the mid point between the two wells (at a distance of 0.5L from both the wells) is 2.0 m when only the first well is being pumped at the steady rate of $Q_1 m^3$/sec. When both the wells are being pumped at identical steady rate of $Q_2 m^3$/sec, the drawdown measured at the same location is 8.0m. It may be assumed that the drawdown at the wells always remains at 10.0 m when being pumped and the radius of influence is larger than 0.5L $\frac{Q_1}{Q_2}$ is equal to

(*a*) $\frac{8}{5}$ (*b*) $\frac{4}{3}$

(*c*) $\frac{5}{8}$ (*d*) $l_n\left(\frac{L}{2}\right)$

1997

58. A hyetograph is a graph representing

(*a*) rainfall volume with time

(*b*) rainfall intensity with time

(*c*) rainfall intensity with duration

(*d*) rainfall intensity over an area

59. Mukingham method for routing of flood

(*a*) is used for routing floods through reservoirs

(*b*) is a method of routing that uses continuity and momentun equations

(*c*) is a hydrologic method of routing floods through streams

(*d*) is one is which only energy equation is used

60. Storage coefficient of a compressible confined aquifer is a function of

(*a*) specific weight of water, thickness of the aquifer, compressibility of the aquifer and that of water.

(*b*) permeability, thickness and compressibility of aquifer and compressibility of water

(*c*) transmissibility of the auifer and compressibility of water

(*d*) transmissibility of aquifer and specific yield of aquifer

NUMERICAL TYPE QUESTIONS

2016

1. The direct runoff hydrograph in response to 5 cm rainfall excess in a catchment is shown in the figure. The area of the catchment (expressed in hectares) is_____

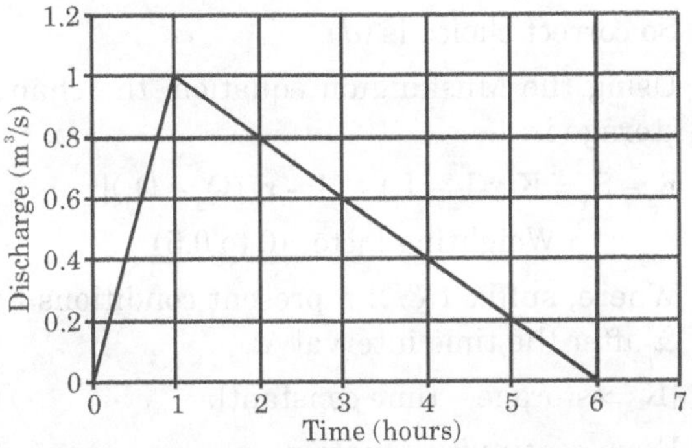

2. Water table of an aquifer drops by 100 cm over an area of 1000 km^2. The porosity and specific retention of the aquifer material are 25% and 5%, respectively. The amount of water (expressed in km^3) drained out from the area is ________

3. The ordinates of a one-hour unit hydrograph at sixty minute interval are 0, 3, 12, 8, 6, 3 and 0 m^3/s. A two-hour storm of 4 cm excess rainfall occurred in the basin from 10 AM. Considering constant base flow of 20 m^3/s, the flow of the river (expressed in m^3/s) at 1 PM is ________

2015

4. The 4-hr unit hydrograph for a catchment is given in the table below. What would be the maximum ordinate of the S-curve (in m^3/s) derived from this hydrograph?

Time (hr)	0	2	4	6	8	10	12	14	16	18	20	22	24
Unit hydrograph ordinate (m^3/s)	0	0.6	3.1	10	13	9	5	2	0.7	0.3	0.2	0.1	0

5. In a catchment, there are four rain-gauge stations, P, Q, R and S. Normal annual precipitation values at these stations are 780 mm, 850 mm, 920 mm and 980 mm respectively. In the year 2013, stations Q, R and S, were operative but P was not. Using the normal ratio method, the precipitation at station P for the year 2013 has been estimated as 860 mm. If the observed precipitation at stations Q and R for the year 2013 were 930 mm and 1010 mm, respectively, what was the observed precipitation (in mm) at station S for that year?

6. A field channel has cultivable commanded area of 2000 hectares. The intensities of irrigation for gram and wheat are 30% and 50% respectively. Gram has a kor period of 18 days, kor depth of 12 cm, while wheat has a kor period of 18 days and a kor depth of 15 cm. The discharge (in m^3/s) required in the field channel to supply water to the commanded area during the kor period is _____.

2002

7. A and B are two concentration points along a river in a catchment of area3.5 km^2. The concentration point A is 1 km down stream of concentration point B. The base flow in the river reach AB is 10m^3/s. A 2-hour rainfall event occurs on this catchment having rainfall intensities of 4 cm/hr during the first hour and 2 cm/hr during the second hour. A 2-hr UH at concentration point B is given below.

Time (hours)	0	3	6	9	12	15	18	21
UH Ordinate (m^3/s)	0	25	50	160	110	45	8	0

Compute the ordinates of the flood hydrograph at concentration point A resulting from the 2-hr rainfall event.

(Assume ϕ-index = 2cm/hour; Muskingum routing constants K = 6 hours and x = 0.2).

ANSWERS

MCQ Type Questions

1. (*b*) 2. (*d*) 3. (*a*) 4. (*b*) 5. (*d*) 6. (*d*) 7. (*d*) 8. (*c*) 9. (*d*) 10. (*c*)

11. (*c*) 12. (*a*) 13. (*b*) 14. (*d*) 15. (*c*) 16. (*b*) 17. (*b*) 18. (*b*) 19. (*c*) 20. (*)

21. (*a*) 22. (*c*) 23. (*) 24. (*b*) 25. (*a*) 26. (*c*) 27. (*c*) 28. (*d*) 29. (*b*) 30. (*a*)

31. (*b*) 32. (*c*) 33. (*b*) 34. (*a*) 35. (*c*) 36. (*b*) 37. (*a*) 38. (*d*) 39. (*c*) 40. (*a*)

41. (*d*) 42. (*c*) 43. (*c*) 44. (*b*) 45. (*a*) 46. (*c*) 47. (*d*) 48. (*d*) 49. (*) 50. (*d*)

51. (*c*) 52. (*b*) 53. (*b*) 54. (*d*) 55. (*d*) 56. (*c*) 57. (*d*) 58. (*b*) 59. (*b*) 60. (*d*)

Numerical Type Questions

1. (21.6) 2. (0.2) 3. (60) 4. (22) 5. (1094) 6. (1.4275) 7. (*)

"*" Refers to subjective question

EXPLANATIONS

MCQ TYPE QUESTIONS

3.

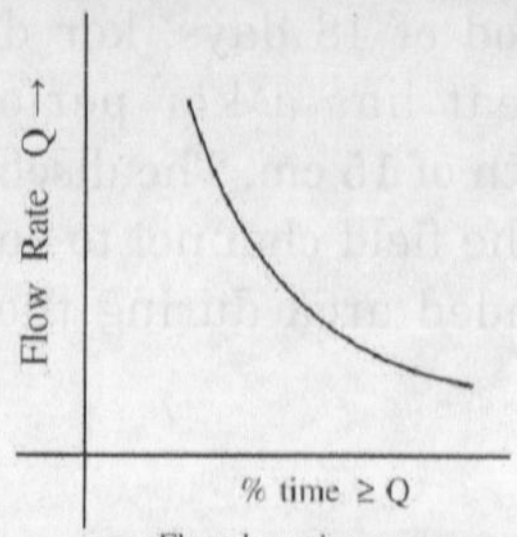

Flow duration curve

4.

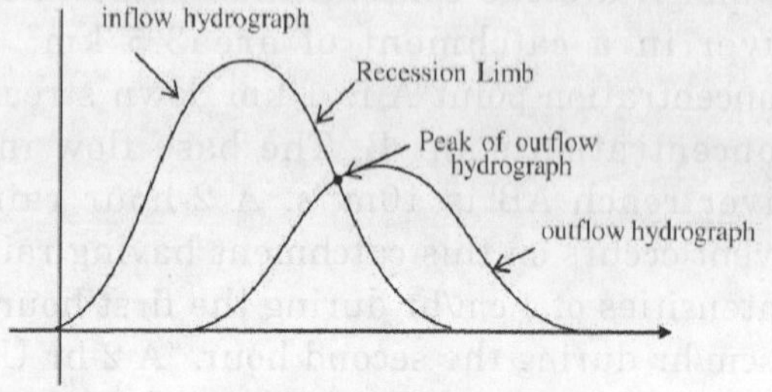

5.

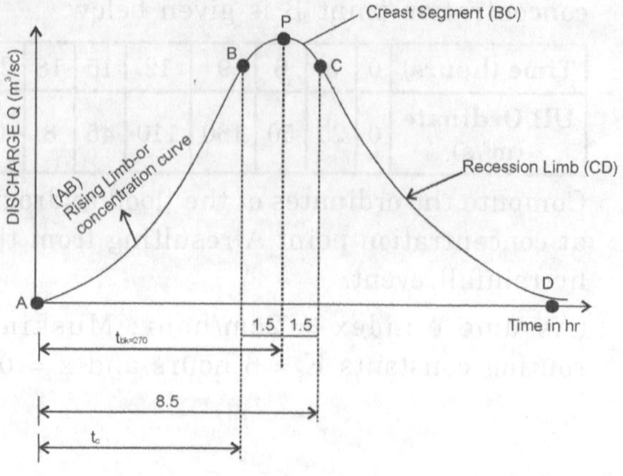

Peak make in between B & C. So distance between P to C is 1.5 hr.

So the distance between P to B is also 1.5 hr.

$\because \; t_c = 7.0 - 1.5 = 5.5$ hr

So correct choice is (*d*)

6. Using the Muskingum equation, the change in storage is:

$$S_2 - S_1 = K[x(I_2 - I_1) + (1 - x)(Q_2 - Q_1)]$$

$x \Rightarrow$ Weighting factor (0 to 0.5)

Where, suffix 1 & 2 represent conditions before & after the time interval Δt.

[$K \Rightarrow$ storage – time constant].

From continuity equation.

$$S_2 - S_1 = \left[\frac{I_2 + I_1}{2}\right]\Delta t - \left[\frac{Q_2 + Q_1}{2}\right]\Delta t$$

Q_2 is evaluated as:

$$\boxed{Q_2 = C_o I_2 + C_1 I_1 + C_2 Q_1}$$

Note that the sum of $\boxed{C_0 + C_1 + C_2 = 1.0}$

So that in given question the sum of

$$K_0 + K_1 + K_2 = 1.0$$

7. Given

1 – *h* rainfall of 10 cm magnitude

Return period T = 50 years

Successive year $n = 2$

find the probability of 1 – h rainfall of magnitude 10 cm or more will occur in each of two successive year, i.e.

$\because \quad r = 2$

$$P_{2,n} = \frac{\lfloor n}{\lfloor (n-2) \lfloor 2} p^2 \cdot q^{n-2}$$

$$q = 1 - P$$

$$P = \frac{1}{T} = \frac{1}{50} = 0.02$$

$$q = 1 - 0.02 = 0.98$$

$$P_{2,n} = \frac{2!}{(2-2)!2!}(0.02)^2 \times (0.98)^{2-2}$$

$$P_{2,n} = 1 \times (0.02)^2 \times 10 = 0.0004$$

$\because$ Force in member RS = 7.5 (comp.)

8. An isohyet is a line joining point of equal rainfall magnitude.

$\because$ (c) is correct Answer. i.e. equal rainfall depth.

9. Assume the depth of rectangular channel = Y the width of rectangular channel = B

$\because$ Hydraullic mean depth of rectangular channel

$$R = \frac{Y}{2}$$

Discharge Q before increasing the depth

$$Q = A \times V = Y \times B \times \frac{1}{n} (R)^{\frac{2}{3}} (S)^{\frac{1}{2}}$$

$$Q = Y \times B \times \frac{1}{n} \left(\frac{Y}{2}\right)^{\frac{2}{3}} (S)^{\frac{1}{2}}$$

Discharge Q after increasing the depth = 10%

$$\because \quad Q_1 = 1.1Y \times B \times \frac{1}{n} \left(\frac{1.1Y}{2}\right)^{\frac{2}{3}} \cdot (S)^{\frac{1}{2}}$$

Ratio of Q and Q_1

$$\frac{Q}{Q_1} = \frac{Y \times B \times \frac{1}{n}\left(\frac{Y}{2}\right)^{\frac{2}{3}} (S)^{\frac{1}{2}}}{1.1Y \times B \times \frac{1}{n}\left(\frac{1.1Y}{2}\right)^{\frac{2}{3}} (S)^{\frac{1}{2}}}$$

$$\frac{Q}{Q_1} = \frac{1}{(1.1)^{\frac{5}{3}}}$$

$$Q = 1.172\ Q_1$$

$\because$ Discharge will increase 17.2%

11. Vegetation and forests increase infiltration and storage capacity of the soils. Further, they cause considerable retardance to the overland flow. Thus, vegetal cover reduces peak flow and increases the time loss with the change of land use from rural to urban, there will be increase of discharge and reduction of time loss.

14. Let D be the diameter of the single large pipe which is to be replaced by 2 parallel pipes of equal diameter d to increase discharge by 25%. Let all the pipes be of equal length L.

$$\therefore \quad \text{Heat loss, } h_f = \frac{fL}{D} \cdot \frac{V^2}{2g}$$

$\therefore$ Velocity in large pipe,

$$V = \sqrt{\frac{h_f \times 2g}{f}} \sqrt{\frac{D}{L}}$$

Discharge in large pipe,

$$Q = \frac{\pi D^2}{4} \cdot V$$

$$= \frac{\pi D^2}{4} \times \sqrt{\frac{h_f \times 2g}{f}} \times \sqrt{\frac{D}{L}}$$

$$= \frac{\pi}{4} \sqrt{\frac{2gh_f}{f}} \left(\frac{D^{5/2}}{L^{1/2}}\right)$$

Frictional head loss in each of the parallel pipes = Frictional head less in large pipe = h_f

Similarly, discharge in each of parallel pipes

$$q = \frac{\pi}{4} \sqrt{\frac{2gh_f}{f}} \cdot \left(\frac{d^{5/2}}{L^{1/2}}\right)$$

But $2q = 1.25\ \alpha$

$$\therefore \quad 2 \times \frac{\pi}{4} \sqrt{\frac{2gh_f}{f}} \left(\frac{d^{5/2}}{L^{1/2}}\right) = 1.25 \times \frac{\pi}{4} \sqrt{\frac{2gh_f}{f}} \left(\frac{D^{5/2}}{L^{1/2}}\right)$$

$$\Rightarrow \quad 2(d^{5/2}) = 1.25\ D^{5/2}$$

$$\Rightarrow \quad d^{5/2} = 0.625\ (60)^{5/2}$$

$$\Rightarrow \quad d = 49.7 \approx 50 \text{ cm.}$$

15. In hydraulic jump formation

$$y_2 = \frac{y_1}{2} \left[-1 + \sqrt{\frac{1 + 8q^2}{gy_1^3}}\right]$$

$$= \frac{0.46}{2} \left[-1 + \sqrt{\frac{1 + 8 \times 9^2}{9.81 \times 0.46^3}}\right]$$

$$= 5.677$$

$$\approx 5.77$$

16. Porosity, n = specific yield (S_y) + specific retention S_r

$$0.40 = S_y + 0.15$$

$$\Rightarrow \quad S_y = 0.25$$

Volume pumped out

= Area × drop in water table × specific yield (S_y)

= 150 × 3 × 0.25 = 112.5 ha. m

Specific yield.

The actual volume of water that can be extracted by the force of gravity from a unit volume of aquifer is called *specific yield*.

19. We know, $S_y = \dfrac{V_w}{V}$

$$S_r = \frac{V_r}{V}$$

and $\quad V_w + V_r = V_{void}$

$$\therefore \quad S_y + S_r = \frac{V_w + V_r}{V} = \frac{V_v}{V}$$

$$S_y + S_r = n$$

where, n = porosity

$$S_y = n - S_r$$

21. According to Lacey, the relation between cross sectional area A, mean velocity V and silt factor f is given by

$$A \, . \, f^2 = 140\, V^5$$

$$\Rightarrow \quad V.\, A.\, f^2 = 140\, V^5$$

$$\Rightarrow \quad Q\, f^2 = 140\, V^6$$

$$\Rightarrow \quad \left(\frac{Qf^2}{140}\right)^{1/6} = V$$

which is the required expression.

22. Here, V_1 = 13.72 m/s

y_1 = pre jump depth

y_2 = post jump depth

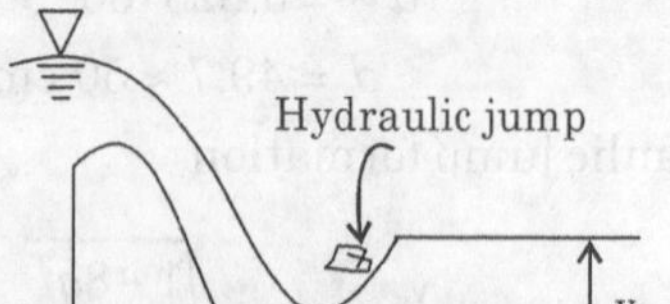

Froude number, $F_1 = \dfrac{V_1}{\sqrt{gy_1}}$

$$= \frac{13.72}{\sqrt{9.8 \times 0.3}} = 8$$

We know, $\dfrac{y_2}{y_1} = \dfrac{1}{2}\left(-1 + \sqrt{1 + 8F_1^2}\right)$

$$\Rightarrow \quad y_2 = 0.3 \times \frac{1}{2}\left(-1 + \sqrt{1 + 8 \times 8^2}\right)$$

$$= 3.24 \text{ m}$$

24. For a given flow rate, if crifical depth (y_c) is less than normal depth (y_n)

i.e. if $\quad y_c < y_n$

then flow profile will be M_2.

25. According to Lacey's method for design of alluvial channels, watted perimeter increases with an increase in design discharge

26. Total command area, A = 20,000 heat

Intensity of irrigation, i = 50% = 0.5

Kor period, t = 30 days

Kor water depth, d = 120 mm

Outlet discharge factor,

$$F = \frac{8.64 \times t}{d}$$

$$= \frac{8.64 \times 30}{120 \times 10^{-3}}$$

= 2160 hect/cumec

$$\therefore \text{ Outlet discharge, } Q = \frac{Ai}{F}$$

$$= \frac{0.5 \times 20{,}000}{2160}$$

$$= \mathbf{4.63\ m^3/sec}$$

27.

Time	1	2	3	4
Rainfall	9	28	12	7
ϕ– index	10	10	10	10
Rainfall excess	–1	18	2	–3

∴ Runoff depth = 18 + 2 = 20 mm

29. It is clear from figure that intensity for 20 minutes duration will be maximum for 20 – 30 – 40 minutes.

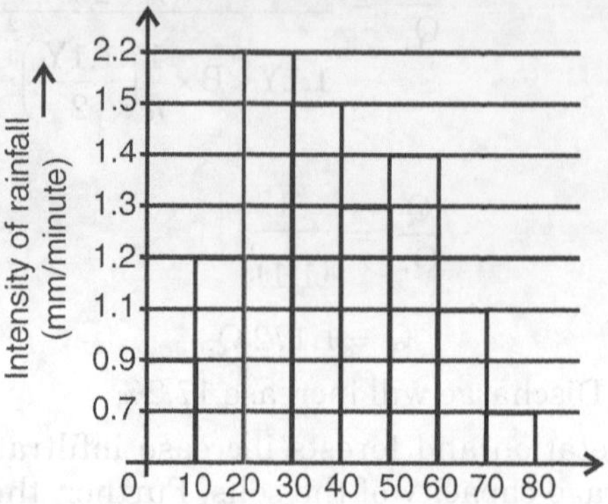

$$\text{Intensity} = \frac{\text{Rain fall}}{\text{Time}}$$

$$= \frac{2.2 \times 10 + 1.5 \times 10}{20}$$

$$= 1.85 \text{ mm/minute}$$

30. If down-stream profile of Ogee spillway whose up-stream face is vertical, then centre of crest

$$x^{1.85} = -2H_d^{0.85} . y$$

$$\Rightarrow \left(\frac{x}{H_d}\right)^{1.85} = -2\frac{H_d^{0.85}}{H_d^{1.85}}.y$$

$$= -2\frac{y}{Hd}$$

$$\Rightarrow \left(\frac{y}{H_d}\right) = -0.5\left(\frac{x}{H_d}\right)^{1.85}$$

31. Since effective stress at any depth does not depends upon variation of water level above the bed of river

$$\therefore \quad \bar{\sigma} = \sigma - u$$

$$= \gamma h - \gamma_w h$$

$$= 20 \times 10 - 10 \times 10 = 100 \text{ kN/m}^2$$

Hence, 100 kN/m^2 in February 100 kN/m^2 in July and 100 kN/m^2 in December.

32. $\text{Velocity of travel} = \frac{\text{discharge velocity}}{\text{Porosity}}$

$$= \frac{ki}{n}$$

$$= \frac{30 \times (45 - 20)}{1.5 \times 1000 \times 0.25}$$

$$= 2 \text{ m/day}$$

$$\therefore \quad \text{Time taken} = \frac{1.5 \times 1000}{2}$$

$$= 750 \text{ days}$$

33. Let peak discharge = x m^3/sec

Volume represented by area of UH

= Volume of 1 cm depth over catchment

$$\therefore \quad \frac{1}{2} \times 80 \times x \times 60 \times 60 = 720 \times 10^6 \times \frac{1}{100}$$

$$\Rightarrow \quad x = \frac{720 \times 10^6 \times 2}{80 \times 60 \times 60 \times 100}$$

$$= \mathbf{50 \text{ m}^3\text{/sec}}$$

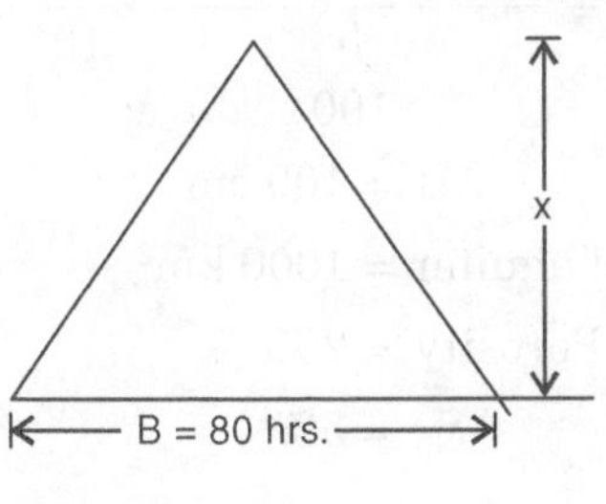

34. Total depth of rainfall = 4 cm

Loss ≅ 1 mm/hr for 4 hr = 4 mm = 0.4 cm

Rainfall excess = 4 – 0.4 = 3.6 cm

$$\text{Peak of 3 hr. unit hydrograph} = \frac{\text{peak of DRH}}{\text{Rainfall excess}}$$

Here, Peak of DRH = 3.6 × 50 = 180 cm

∴ Peak discharge due to storm

$$= 180 + 30$$

$$= \mathbf{210 \text{ m}^3\text{/sec}}$$

($\because$ base flow = 30 m^3/sec)

35.

$$\phi_1 = \frac{P - R}{\epsilon_e}$$

$$= \frac{(0.5 + 2.8 + 1.6) - 32}{3}$$

$$= 0.5666$$

$$\therefore \quad \phi_2 = \frac{(2.8 + 1.6) - 3.2}{2}$$

$$= 0.6 \text{ cm/2 hr}$$

[because 0.5 < 0.5666, so no infiltration]

and $\phi_{index} = 0.3$ cm/hr

36. Given: Total depth of rainfall = 2.7 cm

Loss @ 0.3 cm/hr for 3 hour = 0.9 cm

Rainfall excess = 2.7 – 0.9 cm

= 1.8 cm

Peak of flood hydrograph = 210 m^3/sec

Base flow = 20 m^3/sec

Peak of DRH = 210 – 20

= 190 m^3/sec

∴ Peak of 3-hour unit hydrograph

$$= \frac{\text{Peak of DRH}}{\text{rainfall excess}}$$

$$= \frac{190}{1.8} = 105.50 \text{ m}^3\text{/sec.}$$

37. The rainfall duration is stipulated as the time of concentration in rational formula for computing the design discharge because this lead to the largest possible rainfall intensity.

By rational formula,

$$Q_P = \frac{1}{3.6} C\,(i_{t_c, p})A$$

where t_c = time of concentration

38. Here, total rainfall = 16 cm

and total infiltration = 0.5 × 12 = 6 cm

∴ Net effective rainfall = (16 – 6) = 10 cm

∴ Peak of direct run off hydrograph

= 30 × 10 + 5 × 12

= 360 cumecs

[assuming base flow is varied 5 cumecs/hr, because it is not given that base flow is constant or not.]

39.

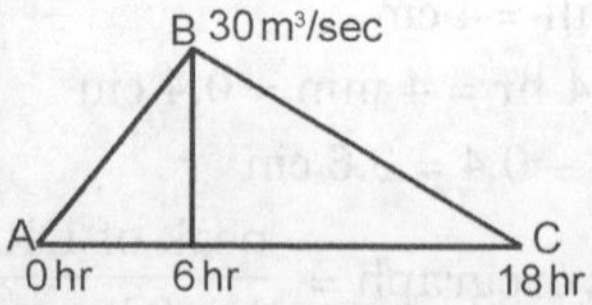

$$\text{Area of catchment} \times \frac{1}{100} m$$

$$= \text{Area under graph ABC}$$

$$\therefore \quad A \times \frac{1}{100} = \frac{1}{2} \times 18 \times 3600 \times 30$$

$$\Rightarrow \quad A = (50 \times 18 \times 3600 \times 30)\ m^2$$

$$\therefore \quad \text{Area in hectare} = \frac{50 \times 18 \times 3600 \times 30}{10^4}$$

$$= \mathbf{9720\ hectares}$$

45.

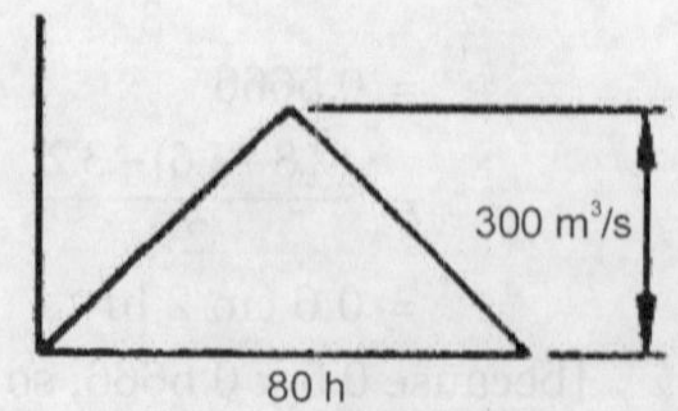

$$A = 56.4\ km^2$$

$$\text{Rainfall excess} = \frac{1}{2} \times \frac{30 \times 80 \times 60 \times 60 m^3}{86.4 \times 10^6 m^2}$$

$$= \frac{1}{10}\ m = \frac{100}{20}\ cm = 5\ cm$$

46.

1	2	3	4	5	3 hr U.H
Time	**U.H. $1m^3$/sec**	**S-curve addition**	**S-curve ordinate**	**Lagged by 3 hr.**	**$\frac{\text{Col.4 – Col.5}}{(3/4)}$ (m^3/sec)**
0	0	–	0	0	0
1	10	–	10	0	13.3
2	60	–	60	0	80
3	120	–	120	0	16
4	170	0	170	10	213.33
5	200	10	210	60	200
6	180	60	240	120	160
7	150	120	270	170	133.33
8	124	170	294	210	112
9	104	210	314	240	98.67
10	88	240	328	270	77.33
11	73	270	343	294	65.33
12	59	294	353	314	52
13	48	314	362	328	45.33
14	36	328	364	343	28
15	28	343	371	353	24
16	20	353	373	362	19.67
17	10	362	375	364	17.33
18	8	364	375	371	5.33
19	3	371	375	373	2.67
20	0	373	375	375	0
21	0	375	–	–	0
22	–	375	–	–	0
23	–	375	–	–	–
24	–	375	–	–	–

In 3 hr, $\quad$ rainfall = 3 . 6 = 18mm = 1.8 cm

$\therefore$ Peak discharge = 213.33 × 1.8 = 384m^3/sec

47. Here, $V = 200 \times 10^4 \times 4 = 800 \times 10^4\ m^3$

$$S_r = 0.15$$

$$n = 0.35$$

$\therefore$ Specific yield, $S_y = n - s_r = 0.35 - 0.15 = 0.20$

From $\quad S_y = \frac{\Delta V}{V}$

$$\therefore \quad \Delta V = 0.2 \times 800 \times 10^4$$

$$= 1.6 \times 10^6 m^3$$

48. $Q = 4 \frac{m^3}{h_r} \times 20 \frac{hr}{day} = 80\ m^3/day$

$$\Rightarrow \quad Q = 9.26 \times 10^{-4}\ m^3/sec$$

Given : B = 20 days

$$\Delta = 7\ cm = 0.07m$$

$\therefore$ Duty at outlet

$$D = \frac{8.64B}{\Delta} = \frac{8.64 \times 20}{0.07}$$

$$= 2468\ \text{hect/cumec}$$

Area, $A = D \times Q$

$$= 2468 \times 9.26 \times 10^{-4}$$

$$= 2.29\ \text{hect.}$$

$$= 2.29 \times 10^4\ m^2$$

52. Aquitard is that geological formation which does not yield water freely to wells due to its lesser permeability, although seepage is possible through it. The yield from such formation is thus insignificant.

56. Give, Return period, T_r = 8 years

$$\therefore \quad P = \frac{1}{T_r} = \frac{1}{8} = 0.125$$

The probability of this flood magnitude will be exceeded once in the next 5 years is given by

$$R = 1-(1-P)^N, \quad \text{where} \quad N = 5\text{years}$$

$$\therefore \quad R = 1-(1-0.125)^5 = 0.487$$

NUMERICAL TYPE QUESTIONS

1. Area of catchment

$$= \frac{\frac{1}{2} \times 1 \times 6 \times 60 \times 60}{\frac{5}{100}} \times \frac{1}{10^4} hec = 21.6\ hec$$

2. $\quad \Delta H = 100$ cm

Area of aquifer = 1000 km^2

Porosity = 25%

S_R = 5%

Total volume of water reduced

$$= \frac{10}{100} \times 1000 \text{ km}^3$$
$$= 1 \text{ km}^3$$
$$= 10^9 \text{ m}^3$$

Also, $S_R = \dfrac{\text{volume of water rate } (V_R)}{\text{total volume of water}}$

$$\frac{5}{100} \times 10^9 = \text{Volume of water rate}$$

Therefore, $V_R = 5 \times 10^7 \text{ m}^3$

$$S_R + S_Y = n$$
$$5\% + S_Y = 25\%$$
$$S_Y = 20\%$$

Volume of water drained = 20% of total water

$$= \frac{20}{100} \times 10^9$$
$$= 2 \times 10^8 \text{ m}^3$$
$$= 0.2 \text{ km}^3$$

3.

Time	Ordinate of 1 hr UH (m^3/sec)	Offset Ordinate (m^3/sec)	Ordinate of 2 hr UH (m^3/sec)	Ordinate of DRH 4cm rainfall excess(m^3/sec)	Ordinate of flood hydrograph (m^3/sec)
10 : 00	0	-	0	0	20
11 : 00	3	0	3	6	26
12 : 00	12	3	15	30	50
1 : 00	8	12	20	40	60
2 : 00	6	8	14	28	48
3 : 00	3	6	9	18	38
4 : 00	0	3	3	6	26
5 : 00		0	0	0	20

Ordinate of 1 : 00 pm = 60 m^3/sec

4.

Time	Ordinate	Lagging by 4-hrs & adding	S-curve
0	0	—	0
2	0.6	—	0.6
4	3.1	0	3.1
6	10	0.6	10.6
8	13	3.1	16.1
10	9	10.6	19.6
12	5	16.1	21.1
14	2	19.6	21.6
16	0.7	21.1	21.8
18	0.3	21.6	21.9
20	0.2	21.8	22
22	0.1	21.9	22
24	0	22	22

5.
$$\frac{P_S}{N_S} = \frac{1}{3}\left[\frac{P_P}{N_P} + \frac{P_Q}{N_Q} + \frac{P_R}{N_R}\right]$$

$$\Rightarrow \frac{P_S}{980} = \frac{1}{3}\left[\frac{860}{780} + \frac{930}{850} + \frac{1010}{920}\right]$$

$$\Rightarrow P_S = 1094 \text{ mm}$$

6. Cultivable command area = 2000 hectares

Irrigation intensities ⇒ 30% for Gram
50% for Wheat

Kor period ⇒ 18 days for both

Kor depth ⇒ 12 cm for Gram
15 cm for Wheat

$$\text{Discharge} = \frac{\left[0.3 \times (2000 \times 10^4) \times 0.12\right] + \left[0.5 \times (2000 \times 10^4) \times 0.15\right]}{18 \times 24 \times 60 \times 60}$$

$$= 1.4275$$

7. Here, A = 3.5 km^2

Base flow = 10m^3/sec

Rain fall intensity = 4 cm/hr in first hour
= 2 cm/h second hour

$$C_0 = -\frac{-k_x + 0.5\,\Delta t}{k - k_c + 0.5\,\Delta t}$$

$$= \frac{-6 \times 0.2 + 0.5 \times 3}{6 - 6 \times 0.2 + 0.5 \times 3}$$

$$= 0.0476$$

$$C_1 = \frac{k_x + 0.5\,\Delta t}{k - k_x + 0.5\,\Delta t}$$

$$= \frac{6 \times 0.2 + 0.5 \times 3}{6.3} = 0.428$$

$$C_2 = \frac{k - k_x - 0.5\Delta t}{k - k_x + 0.5\,\Delta t}$$

$$= \frac{6 - 6 \times .2 - 0.5 \times 3}{6.3} = 0.524$$

Excess rainfall = 4 + 4 − 4 = 4 cm

Time (i)	U.H. Ordinate (ii)	ordinate × 4cm (iii)		C_0I_2 .048I	C_1I_1 $0.43I_1$	C_2Q_1 0.524	Q	Remark
0	0	0	0	4.8	0	5.24	10	
3	25	100	100	9.6	43.0	5.26	10.04	
6	50	200	200	30.72	86.0	27.56	52.6	
9	160	640	640	21.12	275.2	61.16	116.72	
12	110	440	440	8.64	189.2	155.27	296.32	
15	45	180	180	1.54	77.4	103.67	197.84	
18	8	32	32	0	13.76	41.36	78.94	
21	0	0	0	—	—	7.21	13.76	

■■

10 CHAPTER

Irrigation

MCQ TYPE QUESTIONS

2016

1. Profile of a weir on permeable foundation is shown in figure I and an elementary profile of 'upstream pile only case' according to Khosla's theory is shown in figure II. The uplift pressure heads at key points Q, R and S are 3.14 m, 2.75 m and 0 m, respectively (refer figure II).

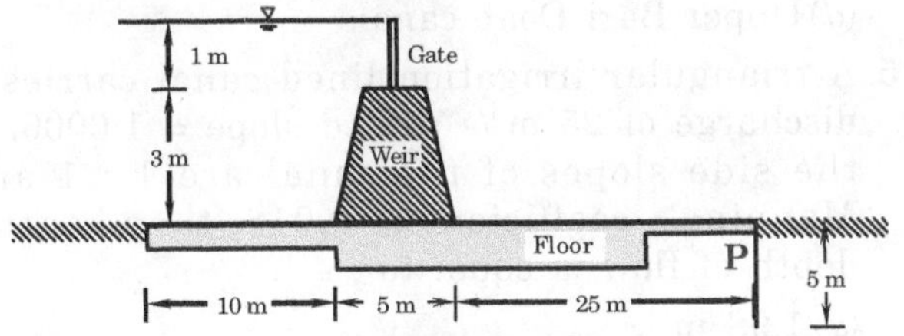

Figure I

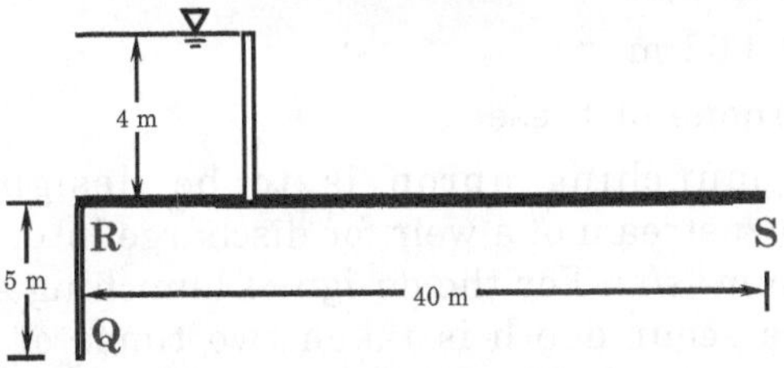

Figure II

What is the uplift pressure head at point P downstream of the weir (junction of floor and pile as shown in the figure I)?

(a) 2.75 m

(b) 1.25 m

(c) 0.8 m

(d) Data not sufficient

2015

2. The two columns below show some parameters and their possible values

Parameter	Value
P. Gross Command Area	I. 100 hectares/cumec
Q. Permanent Wilting Point	II. 6°C
R. Duty of canal water	III. 1000 hectares
S. Delta of wheat	IV. 1000 cm
	V. 40 cm
	VI. 0.12

Which of the following options matches the parameters and the values correctly?

Codes :

	P	Q	R	S
(a)	I	II	III	IV
(b)	III	VI	I	V
(c)	I	V	VI	II
(d)	III	II	V	IV

2013

3. The transplantation of rice requires 10 days and total depth of water required during transplantation is 48 cm. During transplantation, there is an effective rainfall (useful for irrigation) of 8 cm. The duty of irrigation water (in hectares/cumec) is

(a) 612 (b) 216

(c) 300 (d) 108

2012

4. Wheat crop requires 55 cm of water during 120 days of base period. The total rainfall during this period is 100 mm. Assume the irrigation efficiency to bo 60%. The area (in ha) of the land which can be irrigated with a canal flow of 0.01 m^3/s is

(a) 13.82 (b) 18.85

(c) 23.04 (d) 230.40

2009

5. The depth of flow in an alluvial channel is 1.5 m. If critical velocity ratio is 1.1 and Manning's n is 0.018, the critical velocity of the channel as per Kennedy's method is

(a) 0.713 m/s (b) 0.784 m/s

(c) 0.879 m/s (d) 1.108 m/s

6. An agricultural land of 437 ha is to be irrigated for a particular crop. The base period of the crop is 90 days and the total depth of water required by the crop is 105 cm. If a rainfall of 15 cm occurs during the base period, the duty of irrigation water is

(a) 437 ha/cumec (b) 486 ha/cumec

(c) 741 ha/cumec (d) 864 ha/cumec

2008

7. A person standing on the bank of a canal drops a stone on the water surface. He notices that the disturbance on the water surface in not traveling upstream. This is because the flow in the canal is

(a) sub-critical (b) super-critical

(c) steady (d) uniform

8. The base width of an elementary profile of gravity dam of height H is b. The specific gravity of the material of the dam is G and uplift pressure coefficient is K. The correct relationship for no tension at the heel is given by

(a) $\frac{b}{H} = \frac{1}{\sqrt{G-K}}$ (b) $\frac{b}{H} = \sqrt{G-K}$

(c) $\frac{b}{H} = \frac{1}{G-K}$ (d) $\frac{b}{H} = \frac{1}{K\sqrt{G-K}}$

9. A weir on a permeable foundation with downstream sheet pile is shown in the figure below. The exit gradient as per Khosla's method is

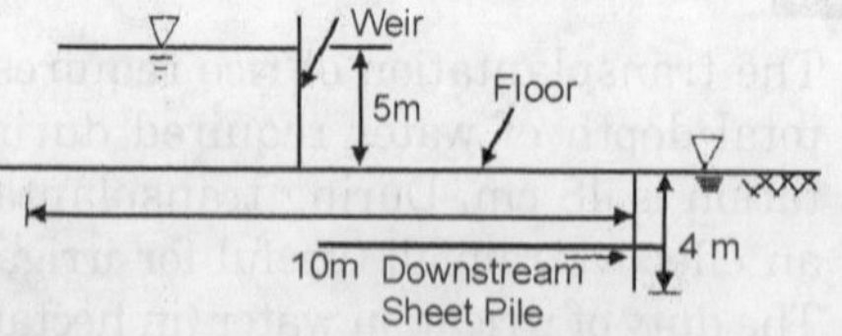

(a) 1 in 6.0 (b) 1 in 5.0

(c) 1 in 3.4 (d) 1 in 2.5

10. An outlet irrigates an area of 20 ha. The discharge (l/s) required at this outlet to meet the evapotranspiration requirement of 20 mm occurring uniformly in 20 days neglecting other field losses is

(a) 2.52 (b) 2.31

(c) 2.01 (d) 1.52

2007

11. The consumptive use of water for a crop during a particular stage of growth is 2.0 mm/day. The maximum depth of available water in the root zone is 60 mm. Irrigation is required when the amount of available water is 50% of the maximum available water in the root zone. Frequency of irrigation should be

(a) 10 days (b) 15 days

(c) 20 days (d) 25 days

2006

12. In a cultivated area, the soil has porosity of 45% and field capacity of 38%. For a particular crop, the root zone depth is 1.0 m, the permanent wilting point is 10% and the consumptive use is 15 mm/d. If the irrigation efficiency is 60%, what should be the frequency of irrigation such that the moisture content does not fall below 50% of the maximum available moisture ?

(a) 5 d (b) 6 d

(c) 9 d (d) 15 d

2005

13. When the outflow from a storage reservoir is uncontrolled as in a freely operating spillway, then peak of outflow hydrograph occurs at

(a) the point of inter-section of the inflow and outflow hydrographs

(b) a point, after the inter-section of the inflow and outflow hydrographs

(c) the tail of inflow hydrographs

(d) a point, before the inter-section of the inflow and outflow hydrographs

14. On which of the canal systems, R.G. Kennedy, executive engineer in the Punjab Irrigation Department made his observations for proposing his theory on stable channels?

(a) Krishna Western Delta canals

(b) Lower Bari Doab canals

(c) Lower Chenab canals

(d) Upper Bari Doab canals

15. A triangular irrigation lined canal carries a discharge of 25 m^3/s at bed slope = 1/6000. If the side slopes of the canal are 1 : 1 and Manning's coefficient is 0.018, then central depth of flow is equal to

(a) 1.98 m

(b) 2.98 m

(c) 3.62 m

(d) none of these

16. A launching apron is to be designed at downstream of a weir for discharge intensity of 6.5 $m^3/s/m$. For the design of launching aprons the scour depth is taken two times of Lacey scour depth. The silt factor of the bed material is unity. If the tailwater depth is 4.4 m, the length of launching apron in the launched position is

(a) $\sqrt{5}$ m (b) 4.7 m

(c) 5 m (d) $5\sqrt{5}$ m

17. Uplift pressures at points E and D (Figure A) of a straight horizontal floor of negligible thickness with a sheet pile at downstream end are 28% and 20%, respectively. If the sheet pile is at upstream end of the floor (Figure B), the uplift pressures at points D_1 and C_1 are

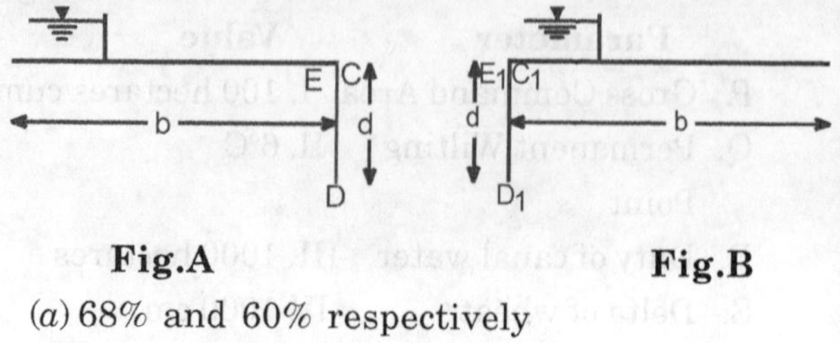

Fig.A Fig.B

(a) 68% and 60% respectively

(b) 80% and 72% respectively

(c) 88% and 70% respectively

(d) 100% and zero respectively

18. The culturable commanded area for a distributary is 2×10^8 m^2. The intensity of irrigation for a crop is 40%. If kor water depth and kor period for the crop are 14 cm and 4 weeks, respectively, the peak demand discharge is

(*a*) 2.63 m^3/s (*b*) 4.63 m^3/s

(*c*) 8.58 m^3/s (*d*) 11.58 m^3/s

2004

19. A sprinkler irrigation system is suitable when

(*a*) the land gradient is steep and the soil is easily erodible

(*b*) the soil is having low permeability

(*c*) the water table is low

(*d*) the crops to be grown have deep roots

20. A canal irrigates a portion of a culturable command area to grow sugarcane and wheat. The average discharges required to grow sugarcane and wheat are, respectively, 0.36 and 0.27 cumecs. The time factor is 0.9. The required design capacity of the canal is

(*a*) 0.36 cumecs (*b*) 0.40 cumecs

(*c*) 0.63 cumecs (*d*) 0.70 cumecs

2003

21. The moisture content of soil in the root zone of an agricultural crop at certain stage is found to be 0.05. The field capacity of the soil is 0.15. The root zone depth is 1.1m. The consumptive use of crop at this stage is 2.5 mm/day and there is no precipitation during this period. Irrigation efficiency is 65%. It is intended to raise the moisture content to the field capacity in 8 days through irrigation. The necessary depth of irrigation is

(*a*) 115 mm (*b*) 169 mm

(*c*) 200 mm (*d*) 285 mm

22. While designing a hydraulic structure, the piezometric head at bottom of the floor is computed as 10m. The datum is 3m below floor bottom. The assured standing water depth above the floor is 2m. The specific gravity of the floor material is 2.5. The floor thickness should be

(*a*) 2.00 m (*b*) 3.33 m

(*c*) 4.40 m (*d*) 6.00 m

23. The plan area of a reservoir is 1 km^2. The water level in the reservoir is observed to decline by 20 cm in a certain period. During this period the reservoir receives a surface inflow of 10 hectare-meters, and 20 hectare-meters are abstracted from the reservoir for irrigation and power. The pan evaporation and rainfall recorded during the same period at a near by meteorological station are 12 cm and 3 cm respectively. The calibrated pan factor is 0.7. The seepage loss from the reservoir during this period in hectare-meters is

(*a*) 0.0 (*b*) 1.0

(*c*) 2.4 (*d*) 4.6

2002

24. During a 6-hour storm, the rainfall intensity was 0.8 cm/hour on a catchment of area 8.6 km^2. The measured runoff volume during this period was 2,56.000 m^3. The total rainfall was lost due to infiltration, evaporation, and transpiration (in cm/hour) is

(*a*) 0.80

(*b*) 0.304

(*c*) 0.496

(*d*) sufficient information not available

25. A canal was designed to supply the irrigation needs of 1200 hectares to land growing rice of 140 days base period having a Delta of 134 cms. If this canal water is used to irriagate wheat of base period 120 days having a Delta of 52 cm, the area (in Hectares) that can be irrigated is

(*a*) 2650 (*b*) 3608

(*c*) 543 (*d*) none of these

2001

26. A field was supplied water from an irrigation tank at a rate of 120 lit/sec to irrigate an area of 2.5 hectares. The duration of irrigation is 8 hours. It was found that the actual delivery at the field, which is about 4 km from the tank, was 100 lit/sec. The runoff loss in the field was estimated as 800 m^3. The application efficiency in this situation is

(*a*) 62% (*b*) 72%

(*c*) 76% (*d*) 80%

2000

27. If duty (D) is 1428 hectares/cumec and base period (B) is 120 days for an irrigated crop, then delta (Δ) in meters is given by

(*a*) 102.8 (*b*) 0.73

(*c*) 1.38 (*d*) 0.01

28. The parameters in Horton's infiltration equation

$$f(t) = f_c + (f_o - f_c)^{e-kt}$$

are given as,

f_o = 7.62 cm/hour

f_c = 1.34 cm/hour

and k = 4.182/hour.

For assumed continuous ponding, the cumulative infiltration at the end of 2 hours is

(*a*) 2.68 cm (*b*) 1.50 cm

(*c*) 1.34 cm (*d*) 4.18 cm

1999

29. In a steady radial flow into an intake, the velocity is found to vary as $(1/r^2)$, where r is the radial distance. The acceleration of the flow is proportional to

(*a*) $\frac{1}{r^5}$ (*b*) $\frac{1}{r^3}$

(*c*) $\frac{1}{r^4}$ (*d*) $\frac{1}{r}$

1998

30. In connection with the design of a barrage, identify the correct matching of the criteria of design with the items of design

Item of design

(*i*) Width of waterway

(*ii*) Level and length of downstream floor

(*iii*) Depth of sheet piles and total length of arrage floor

(*iv*) Barrage floor thickness

Criteria of design

(A) Scour depth and exit gradient

(B) Lacey's formula for wetted perimeter and discharge capacity of the barrage as computed by weir equations

(C) Uplift pressure variation

(D) Hydraulic jump considerations

Codes:

	(*i*)	(*ii*)	(*iii*)	(*iv*)
(*a*)	A	B	C	D
(*b*)	D	C	B	A
(*c*)	B	A	D	C
(*d*)	B	D	A	C

1997

31. For a 'best' symmetrical trapezoidal section of an open channel with a given area of section and side slopes, one of the following statements holds true:

(*a*) Half the top width is equal to one of the side slope

(*b*) Half the top width plus the bottom width is equal to both the side slopes put together

(*c*) Water depth is equal to half bottom width

(*d*) Hydraulic mean depth is equal to half the top width

32. Lysimeter and Tensiometer are used to measure respectively, one of the following groups of quantities :

(*a*) Capillary potential and permeability

(*b*) Evapotranspiration and capillary potential

(*c*) Velocity in channels and vapour pressure

(*d*) Velocity in pipes and pressure head

1996

33. Delta (Δ) in cm, Duty (Δ) in hectare/cumec and Base period(B) in days are related as

(*a*) D = 864 B/Δ (*b*) B = 864 D/Δ

(*c*) B = 864 Δ/D (*d*) D = 8.64 B/Δ

34. Water table drops by 3 m in an irrigable land of 50 hectare. If porosity and specific retention are 0.30 and 0.10 respectively, then change in storage in hectare-meter is

(*a*) 15 (*b*) 30

(*c*) 45 (*d*) 60

NUMERICAL TYPE QUESTIONS

2016

1. A concrete gravity dam section is shown in the figure. Assuming unit weight of water as 10 kN/m^3 and unit weight of concrete as 24 kN/m^3, the uplift force per unit length of the dam (expressed in kN/m) at PQ is ______

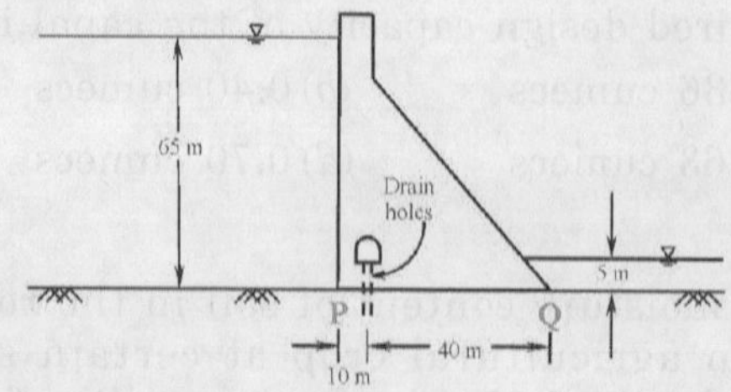

2014

2. Irrigation water is to be provided to a crop in a field to bring the moisture content of the soil from the existing 18% to the field capacity of the soil at 28%. The effective root zone of the crop is 70 cm. If the densities of the soil and water are 1.3 g/cm^3 and 1.0g/cm^3 respectively, the depth of irrigation water (in mm) required for irrigating the crop is ______

2002

3. A culturable command area for a distributory is 12,000 hectares. The intensity of irrigation is 90-percent for Rabi and 80-percent for Kharif crop. The outlet factors for Rabi and Kharif crops are 720 ha/cumec and 775 ha/cumec, respectively. Design a lined canal having hydraulically efficient trapezoidal shape with 2 H: IV side slopes and a bed slope of 1/5000. Assume Manning 's to be 0.014.

2000

4. An area of 1 hectare is irrigated through a stream of 0.03 m^3/s. Depth of the root zone is 1 m and available moisture holding capacity is 16 cm/m. Irrigation water is supplied when 50% of the available moisture is depleted. Water application efficiency is 60% Determine storage efficiency.

ANSWERS

MCQ Type Questions

1. (*b*)	**2.** (*b*)	**3.** (*b*)	**4.** (*a*)	**5.** (*b*)	**6.** (*d*)	**7.** (*b*)	**8.** (*a*)	**9.** (*c*)	**10.** (*b*)
11. (*b*)	**12.** (*c*)	**13.** (*a*)	**14.** (*d*)	**15.** (*a*)	**16.** (*c*)	**17.** (*b*)	**18.** (*b*)	**19.** (*a*)	**20.** (*d*)
21. (*c*)	**22.** (*c*)	**23.** (*d*)	**24.** (*b*)	**25.** (*a*)	**26.** (*b*)	**27.** (*b*)	**28.** (*a*)	**29.** (*a*)	**30.** (*d*)
31. (*a*)	**32.** (*b*)	**33.** (*d*)	**34.** (*a*)						

Numerical Type Questions

1. (10500 kN/m) **2.** (91 to 91) **3.** (*) **4.** (*)

"*" Refers to subjective question.

EXPLANATIONS

MCQ TYPE QUESTIONS

1.
$$\phi_R = \frac{2.75}{4}\times 100 = 68.75\%$$
$$\phi_P = 100 - \phi_R = 31.25\%$$
$$\phi_P = \frac{\text{Pressure head to point P}}{\text{total head}}\times 100$$
$$\Rightarrow \quad 31.25 = \frac{h}{4}\times 100$$
$$h = 1.25 \text{ m}$$

2. The unit of Gross Command Area is hectares Permanent Wilting Point is the moisture content and is given in %

∴ 12% or 0.12.

The unit of duty is hectares/cumec

The unit of delta is cm.

3. Duty $\quad D = \frac{864B}{\Delta}$

where, Δ is in cm, B is in days, then get duty is in hectare/cumecs

$B = 10$ days

Total depth of water required during transplantation of rise = 48 cm

effective rainfall (usefull for irrigation) = 8 cm

So, actual depth of water required = 48 – 8 = 40 cm

$$\therefore \; D = \frac{864\times 10}{40} = 216 \text{ hectare/cumecs}$$

5. Critical velocity ratio, m = 1.1

D = 1.5 *m* $\quad$ *n* = 0.018

From Kennedy method

$V_0 = 0.55\, mD^{0.64} = 0.55\times 1.1\times (1.5)^{0.64} = 0.784$ m/sec

6. Effective rainfall = Total required rainfall

Required depth = 105 – 15 = 90 cm

Volume of water required

= Area × depth required = $437\times 10^4\times 0.90$

$$\therefore \quad \text{Discharge, } Q = \frac{v}{t} = \frac{437\times 10^4\times 0.90}{90\times 24\times 60\times 60} = 437\text{ha}$$

∴ Duty = Area irrigation by unit discharge

$$= \frac{437}{Q} = 864 \text{ ha.cm/sec}$$

7. Super critical because in supercrifical state of flow, the velocity is much high, that will prevent the propogation of water surface wave.

8.

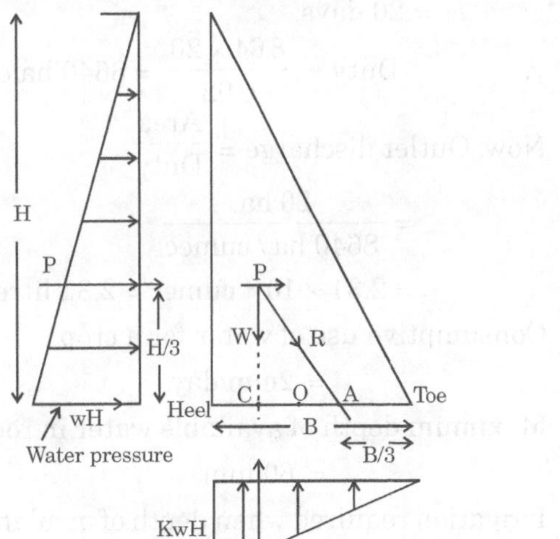

For no tension at heal, the resultant R must pass through middle third.

$$\therefore \; OA = \frac{b}{6} \quad \text{and } AC = \frac{b}{3}$$

Also $\quad OC = \frac{b}{6}$

Let W = Self weight of dam = wG

Now taking moments of all forces about A

$$W\left(\frac{b}{3}\right) - U\left(\frac{b}{3}\right) - P\left(\frac{H}{3}\right) = 0$$

$$\Rightarrow \frac{1}{2}\text{ b.H.Gw. }\frac{b}{3} - \frac{1}{2}\text{ kwH. b. }\frac{b}{3} - \frac{1}{2}\text{ wH.H. }\frac{H}{3} = 0$$

Solving, we get

$$b^2 (G - k) = H^2$$

$$\frac{b}{H} = \frac{1}{\sqrt{G-k}}$$

9. Here, H = 5m,

b = 10m,

d = 4m.

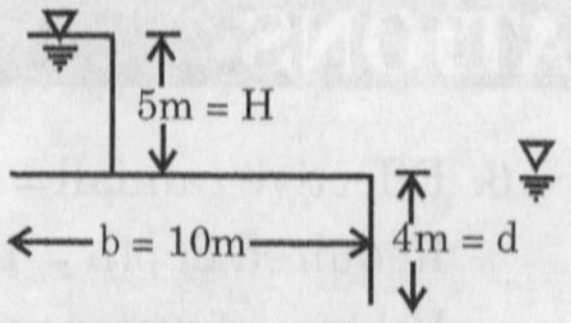

$$\therefore \quad \alpha = \frac{b}{d} = \frac{10}{4} = 2.5$$

$$\text{and} \quad \lambda = \frac{1+\sqrt{1+\alpha^2}}{2} = \frac{1+\sqrt{1+2.5^2}}{2} = 1.846$$

According to Khosla's method, exit gradient is given by

$$G_E = \frac{H}{d}.\frac{1}{\pi\sqrt{\lambda}} = \frac{5}{4}.\frac{1}{\pi\sqrt{1.846}}$$

$$= \frac{1}{3.41}, \text{ i.e. 1 in 3.4}$$

10. Duty, $D = \frac{8.64\ B}{\Delta}$ ha./cumec

Here, $\Delta = 0.02\text{m } (20 \times 10^{-3} \text{ m.})$

B = base period

= 20 days.

$$\therefore \quad \text{Duty} = \frac{8.64 \times 20}{0.02} = 8640 \text{ ha/cumec}$$

$$\text{Now, Outlet discharge} = \frac{\text{Area}}{\text{Duty}}$$

$$= \frac{20 \text{ ha.}}{8640 \text{ ha / cumec}}$$

$$= 2.31 \times 10^{-3} \text{ cumec} = 2.31 \text{ litre/sec}$$

11. Consumptive use of water for a crop

= 2mm/day.

Maximum depth of available water in root zone

= 60 mm.

Irrigation required when depth of available water in root zone $= 0.5 \times 60 = 30$ mm

Water aviable for crop = 60 – 30 = 30 mm.

Frequency of irrigation (f)

$$= \frac{\text{Water available for crop}}{\text{consumptive use of water}}$$

$$= \frac{30}{2} = 15 \text{ days}$$

15. $V = \frac{1}{n} R^{2/3} S^{1/2}$

$$Q = \frac{1}{n} R^{2/3} A^{1/2}.A,$$

$$\therefore \quad 25 = \frac{1}{0.018} \times \left(\frac{A}{P}\right)^{2/3} \times \left(\frac{1}{6000}\right)^{1/2} .A$$

$$\therefore \quad 25 \times 0.018 \times \sqrt{6000} = \frac{A^{5/3}}{P^{2/3}}$$

$$= \frac{y^{2\times 5/3}}{(2\sqrt{2})^{2/3} y^{2/3}}$$

But $\theta = 45^o$

$$\therefore \quad 25 \times 0.018 \times \sqrt{6000} \times \left(2\sqrt{2}\right)^{2/3} = y^{\frac{10}{3}-\frac{2}{3}} = y^{8/3}$$

$$\therefore \quad A = \frac{1}{2} \times 2\,y \times y = y^2$$

$$\therefore \quad y = 4.91$$

$$P = 2 \times \sqrt{y^2 + y^2}$$

$$= 2\sqrt{2}y$$

Remarks : Given answer is wrong.

17.

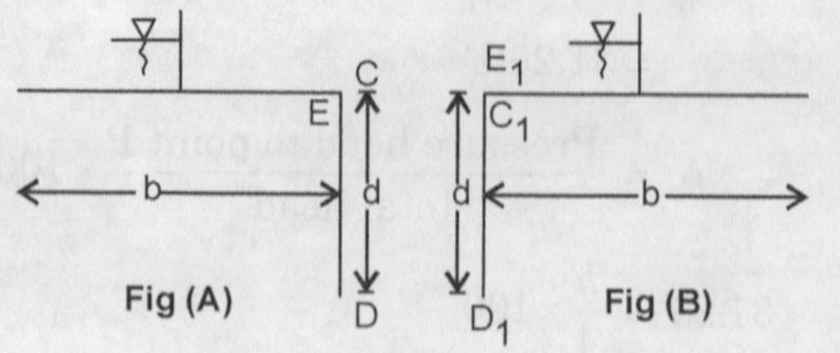

$$\phi\,(C_1) = 100 - \phi_{(\varepsilon)} = 100 - 28 = 72\%$$

$$\phi\,(D_1) = 100 - \phi_B = 100 - 20 = 80\%$$

18. Kor water depth = 14 cm

Kor period for crop = 4 weeks

= 28 days

Kor demand (duty for crop)

$$= \frac{864 \times \text{kor period for crop}}{\text{kor water depth for crop}}$$

$$= \frac{864 \times 28}{14} = 1728 \text{ ha/cumec}$$

Intensity of irrigation for crop = 40%

Crop area $= 0.4 \times 2 \times 10^8$ m^2

$= 0.8 \times 10^4$ ha

$$\therefore \quad \text{Peak demand} = \frac{0.8 \times 10^4}{1728}$$

$$= 4.63 \text{ m}^3\text{/sec}$$

20. Design capacity $= \frac{\text{Capacity of canal}}{\text{Time factor}}$

$$= \frac{0.36 + 0.27}{0.9} = 0.70 \text{ cumecs}$$

21. Moisture content = 0.05

Field capacity = 0.15

Root zone depth = 1.1 m

Consumptive use = 2.5 mm/day

Irrigation efficiency = 65%

Since 2.5 mm of water is required per day

∴ Requirement in 8 days = 2.5 × 8 = 20 mm

Depth of water stored in root zone between these two limits

$$= \frac{\gamma_d d}{\gamma_w}$$

$$= \mathbf{200\ mm}$$

[F.C. – optimum moisture content].

22. Thickness of floor

$$= 1.33 \times \frac{h}{G-1}$$

$$= 1.33 \times \frac{10-2-3}{2.5-1} = \mathbf{4.40\ mm}$$

23. Evapouration loss from reservoir

= 0.7 × 12 = 8.4 cm

Water level in reservoir

= 20 cm

Surface inflow = 10 cm

Abstraction = 20 cm

Rainfall = 3 cm

Hence depth of water in the reservoir, if there is no loss

= 20 + 10 – 20 + 3 = 13 cm

But depth is 8.4 cm, therefore

seepage loss = 13 – 8.4

= **4.6 cm**

26. Water application efficiency

$$\eta_\alpha = \frac{\text{Water stored in the roof zone during irrigation}}{\text{Water delivered to the field}} \times 100$$

Water supplied to field during 8 hours @ 100 lit/s

= 100 × 8 × 60 × 60 = 2880 m³

Runoff loss in the field = 800m³

∴ Water stored in the root zone

= 2880 – 800 = 2080 m³

$$\therefore \quad \eta_\alpha = \frac{2080}{2880} \times 100 = 72.72\%$$

27. Here, D = 1428 hect/cum

B = 120 days

$$\therefore \quad \Delta = \frac{8.64B}{D} = \frac{8.64 \times 120}{1428} = 0.73\ m$$

28.

$$f(t) = f_c + (f_o - f_c)\,e^{-kt}$$

$$\Rightarrow \quad f(t) = 1.34 + (7.62 - 1.34)e^{-4.182 \times 2}$$

$$= 1.34 + 6.28 \times 2.33 \times 10^{-4}$$

= 1.34 cm for one hour

Hence cumulative infiltration at end of 2 hours

= 2 × 1.34 = 2.68 cm.

NUMERICAL TYPE QUESTIONS

1. Area of uplift pressure distribution diagram gives uplift force

Total area,

$$= (10 \times 250) + \frac{1}{2} \times 10 \times 400 + (40 \times 50) + \frac{1}{2} \times 40 \times 200$$

= 10500 kN/m

2. The depth of irrigation water required $= \frac{\gamma \cdot d}{\gamma_\omega}$

[Field capacity available – Moisture content]

d = effective depth of root zone

= 70 cm = 700 mm

Field capacity = 28%

Available moisture constent = 18%

density of soil = 1.3 gm/cm³ = γ

density of water = γ_ω = 1.0gm/cm³

∵ Depth of irrigation water required

$$= \frac{1.3 \times 700[0.28 - 0.18]}{1.0} = 91\ mm$$

3. Total area = 12000ha

Rabi = 12000 × 0.9

= 10800ha

Kharif = 12000 × 0.8

= 9600ha

Intensity of Rabi = 720 ha/cm

Intensity of Kharif = 775ha/cm

$$\text{Discharge for Rabi} = \frac{10800}{720}$$

= 15m³/sec

$$\text{Discharge for Kharif} = \frac{9600}{775}$$

= 12.39m³/sec

∴ Discharge of Rabi > Discharge of Kharif

Hence design should be taken as Rabi discharge

∴ Q = 15m³.sec

$$\text{Slope} = \frac{1}{5000}$$

n = 0.014

From Manning's equation

$$Q = \frac{A}{n}(R)^{\frac{2}{3}}(S)^{\frac{1}{2}}$$

Hydraulically efficient section is trapezoidal.

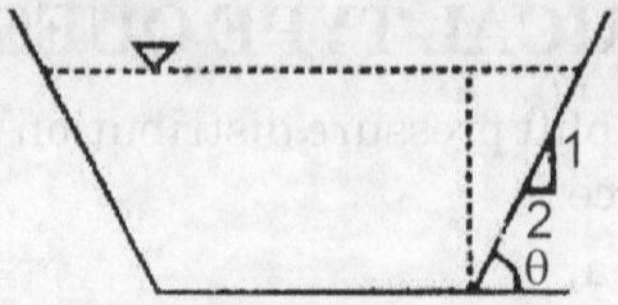

$$A_e = \left(2\sqrt{1+m^2} - m\right)y_c^2$$

$$= \left(2\sqrt{1+4} - 2\right)y_c^2$$

$$= 2.47\, y_e^2$$

$$Re = \frac{y_c}{2}$$

$$\therefore \quad Q = \frac{2.47 y_c^2}{0.014}\left(\frac{y_c}{2}\right)^{\frac{2}{3}}\left(\frac{1}{5000}\right)^{\frac{1}{2}}$$

$$\Rightarrow \quad 15 = \frac{176.43\, y_c^{8/3}}{2^{\frac{2}{3}}} \times 0.014$$

$$\Rightarrow \quad \frac{15 \times 1.587}{0.014 \times 176.43} = y_c^{8/3}$$

$$\Rightarrow \quad 9.64 = y_c^{8/3}$$

$$\Rightarrow \quad y_c = (9.64)^{\frac{3}{8}}$$

$$= 2.34 \text{ m}$$

$$\therefore \quad A = 2.47\, y_c^2$$

$$= 2.47 \times (2.34)^2 = 13.5 \text{m}^2$$

$$B = 2y_c\left(\sqrt{1+m^2} - m\right)$$

$$= 4.68\left(\sqrt{5} - 2\right) = 1.1 \text{m}$$

$\therefore$ Width of channel = 1.1

Area of channel = 13.5m²

Depth of channel = 2.34

Site slope = 2:1

4. Resu lts are calculated in tabular from

Time	Ordinate of 6 hours hydrograph	S-Curve addition	S-Curve ordinate	S-Curve lagged by 18 hours	Column (4)-(5)	$\left[\frac{\text{Column}(6)}{18/6}\right]$
(1)	(2)	(3)	(4)	(5)	(6)	(7)
0	0		0	-	0	0
6	1.8	0	1.8	-	1.8	0.6
12	30.9	1.8	32.7	-	32.7	10.9
18	85.6	32.7	118.3	0	118.3	39.43
24	41.8	118.3	160.1	1.8	158.3	52.77
30	14.6	160.1	174.7	30.9	143.6	47.93
36	5.5	174.7	180.2	85.6	89.1	29.7
42	1.8	180.2	182	41.8	138.4	46.13
	0	182	182	14.6	167.4	55.6
		182	182	5.5	176.5	58.83
		182	182	1.8	180.2	60.06
				0		

■■

11 CHAPTER Water Supply

MCQ TYPE QUESTIONS

2015

1. A circular pipe has a diameter of 1m, bed slope of 1 in 1000, and Manning's roughness coefficient equal to 0.01. It may be treated as an open channel flow when it is flowing just full, i.e., the water level just touches the crest. The discharge in this condition is denoted by Q_{full}. Similarly, the discharge when the pipe is flowing half-full, *i.e.,* with a flow depth of 0.5 m, is denoted by Q_{half}. The ratio Q_{full}/Q_{half} is :

(*a*) 1 (*b*) $\sqrt{2}$

(*c*) 2 (*d*) 4

2014

2. Some of the nontoxic metals normally found in natural water are

(*a*) arsenic, lead and mercury

(*b*) calcium, sodium and silver

(*c*) cadmium, chromium and copper

(*d*) iron, manganese and magnesium

3. 16 MLD of water is flowing through a 2.5 km long pipe of diameter 45 cm. The chlorine at the rate of 32 kg/d is applied at the entry of this pipe so that disinfected water is obtained at the exit. There is a proposal to increase the flow through this pipe to 22 MLD from 16 MLD. Assume the dilution coefficient, $n = 1$. The minimum amount of chlorine (in kg per day) to be applied to achieve the same degree of disinfection for the enhanced flow is

(*a*) 60.50 (*b*) 44.00

(*c*) 38.00 (*d*) 23.27

4. For a sample of water with the ionic composition shown in the figure below, the carbonate and non-carbonate hardness concentrations (in mg/l as $CaCO_3$), respectively are:

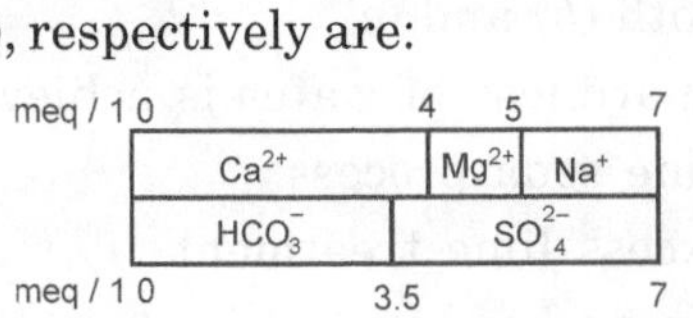

(*a*) 200 and 50 (*b*) 175 and 75

(*c*) 75 and 175 (*d*) 50 and 200

2013

5. Some of the water quality parameters are measured by titrating a water sample with a titrant. Group-I gives a list of parameters and Group-II gives the list of titrants.

Group-I	Group-II
P. Alkalinity	1. N/35.5 $AgNO_3$
Q. Hardness	2. N/40 $Na_2S_2O_3$
R. Chloride	3. N/50 H_2SO_4
S. Dissolved oxygen	4. N/50 EDTA

The correct match of water quality parameters in Group-I with titrants in Group-II is

(*a*) P-1, Q-2, R-3, S-4

(*b*) P-3, Q-4, R-1, S-2

(*c*) P-2, Q-1, R-4, S-3

(*d*) P-4, Q-3, R-2, S-1

6. A 2 km long pipe of 0.2 m diameter connects two reservoirs. The difference between water levels in the reservoirs is 8 m. The Darcy-Weisbachfriction factor of the pipe is 0.04. Accounting for frictional, entry and exit losses, the velocity in the pipe (in m/s) is

(*a*) 0.63 (*b*) 0.35

(*c*) 2.52 (*d*) 1.25

2012

7. A water sample has a pH of 9.25. The concentration of hydroxyl ions in the water sample is

(*a*) $10^{-9.25}$ moles/L

(*b*) $10^{-4.75}$ mmoles/L

(*c*) 0.302 mg/L

(*d*) 3.020 mg/L

2007

8. The presence of hardness in excess of permissible limit causes

(*a*) cardio vascular problems

(*b*) skin discolouration

(*c*) calcium deficiency

(*d*) increased laundry expenses

9. The alkalinity and the hardness of a water sample are 250 mg/L and 350 mg/L as $CaCO_3$, respectively. The water has

(*a*) 350 mg/L carbonate hardness and zero non-carbonate hardness

(*b*) 250 mg/L carbonate hardness and zero non-carbonate hardness

(*c*) 250 mg/L carbonate hardness and 350 mg/L non-carbonate hardness

(*d*) 250 mg/L carbonate hardness and 100 mg/L non-carbonate hardness

2005

10. In aerobic environment, nitrosomonas convert

(*a*) NH_3 to NO_2 (*b*) NO_2^- to NO_3^-

(*c*) NH_3 to N_2O (*d*) NO_2^- to HNO_3

11. Match the following

Group 1	**Group 2**
P. Release valve	**1.** Reduce high inlet pressure to lower outlet pressure
Q. Check valve	**2.** Limit the flow of water to single direction
R. Gate valve	**3.** Remove air from the pipeline
S. Pilot valve	**4.** Stopping the flow of water in the pipeline

Codes :

	P	**Q**	**R**	**S**
(*a*)	3	2	4	1
(*b*)	4	2	1	3
(*c*)	3	4	2	1
(*d*)	1	2	4	3

Linked Answer Q. 12-13

A city is going to install the rapid sand filter after the sedimentation tanks. Use the following data.

Design loading rate to the filter = 200 m^3/m^2d

Design flow rate = 0.5 m^3/s

Surface area per filter box = 50 m^2

12. The surface area required for the rapid sand filter will be

(*a*) 210 m^2 (*b*) 215 m^2

(*c*) 216 m^2 (*d*) 218 m^2

13. The number of filters required shall be

(*a*) 3 (*b*) 4

(*c*) 6 (*d*) 8

2004

14. Most of the turbidity meters work on the scattering principle. The turbidity value so obtained is expressed in

(*a*) CFU (*b*) FTU

(*c*) JTU (*d*) NTU

15. Hardness of water is directly measured by titration with ethylene-di-amine-tetracetic acid(EDTA) using

(*a*) eriochrome black T indicator

(*b*) ferroin indicator

(*c*) methyl orange indicator

(*d*) phenolphthalein indicator

16. The organism, which exhibits very nearly the characteristics of an ideal pathogenic indicator is

(*a*) Entamoeba histolytica

(*b*) Escherichia coli

(*c*) Salmonella typhi

(*d*) Vibrio comma

17. Chlorine gas used for disinfection combines with water to form hypochlorous acid (HOCl). The HOCl ionizes to form hypochlorite (OCl^-) in a reversible reaction:

$HOCl \Leftrightarrow H^+ + OCl^-$ ($k = 2.7 \times 10^{-8}$ at 20° C)

the equilibrium of which is governed by pH. The sum of HOCl and OCl^- is known as free chlorine residual and HOCl is the more effective disinfactant. The 90% fraction of HOCl in the free chlorine residual is available at a pH value

(*a*) 4.8

(*b*) 6.6

(*c*) 7.5

(*d*) 9.4

2003

18. The results of analysis of a raw water sample are given below

Turbidity	:	5 mg/l
pH	:	7.4
Fluorides	:	2.5 mg/l
Total Hardness	:	300 mg/l
Iron	:	3.0 mg/l
MPN	:	50 per 100 ml

From the data given above, it can be inferred that water needs removal of

(*a*) Turbidity followed by disinfection

(*b*) Fluorides and Hardness

(*c*) Iron followed by disinfection

(*d*) Both (*b*) and (*c*)

19. Zero hardness of water is achieved by

(*a*) lime soda process

(*b*) excess lime treatment

(*c*) ion exchange treatment

(*d*) excess alum and lime treatment

20. Results of a water sample analysis are as follows :

Cation	Concentration (mg/l)	Equivalent Weight
Na^+	40	23
Mg^{+2}	10	12.2
Ca^{+2}	55	20
K^+	2	39

(milliequivalent weight of $CaCO_3$ = 50 mg/meq).
Hardness of the water sample in mg/l as $CaCO_3$ is

(*a*) 44.8 (*b*) 89.5
(*c*) 179 (*d*) 358

21. An ideal horizontal flow settling basin is 3m deep having surface area 900 m^2. Water flows at the rate of 8000 m^3/d, at water temperature 20°C (m = 10^{-3} kg/m.s) and ρ = 1000 kg/m^3). Assuming Stokes law to be valid, the proportion (percentage) of spherical sand particles (0.01 mm in diameter with specific gravity 2.65), that will be removed, is

(*a*) 32.5 (*b*) 67
(*c*) 87.5 (*d*) 95.5

2002

22. In disinfection, which of the following forms of chlorine is most effective in killing the pathogenic bacteria?

(*a*) Cl (*b*) OCl
(*c*) NH_2Cl (*d*) HOCl

23. In natural water, hardness is mainly caused by

(*a*) Ca^{++} and Mn^{++} (*b*) Ca^{++} and Fe^{++}
(*c*) Na^+ and K^+ (*d*) Ca^{++} and Mg^{++}

24. The theoretical oxygen demand of a 0.001 mol/L glucose solution is

(*a*) 180 mg/L (*b*) 192 mg/L
(*c*) 90mg/L (*d*) 96 mg/L

2001

25. Aeration of water is done to remove

(*a*) suspended impurities
(*b*) colour
(*c*) dissolved salts
(*d*) dissolve gases

26. Which of following chemical is used for coagulation ?

(*a*) Ammonium Chloride
(*b*) Aluminium Chloride
(*c*) Aluminium Sulphate
(*d*) Copper Sulphate

27. The Ca^{2+} concentration and Mg^{2+} concentration of a water sample are 160mg/lit and 40 mg/lit as their ions respectively. The total hardness of this water sample in terms of $CaCO_3$ in mg/lit is approximately equal to

(*a*) 120 (*b*) 200
(*c*) 267 (*d*) 567

2000

28. The following characteristics pertain to the sand filters used in water industry.

I. Filtration rate is 1 to 4 m^3/(m^2 day).
II. Typical duration of operation in one run is 24 to 72 hours.
III. Operating cost is low.

Which of the above characteristics pertain to slow sand filters ?

(*a*) I, II and III (*b*) I and II
(*c*) II and III (*d*) I and III

1999

29. Temporary hardness in water is caused by the presence of

(*a*) Bicarbonates of Ca and Mg
(*b*) Sulphates of Ca and Mg
(*c*) Chlorides of Ca and Mg
(*d*) Nitrates of Ca and Mg

30. Blue baby disease (methaemoglobinemia) in children is caused by the presence of excess

(*a*) Chlorides (*b*) Nitrates
(*c*) Fluoride (*d*) Lead

31. The minimum dissolved oxygen content (ppm) in a river necessary for the survival of aquatic life is

(*a*) 0
(*b*) 2
(*c*) 4
(*d*) 8

32. Chlorine is sometimes used in sewage treatment

(*a*) to avoid flocculation
(*b*) to increase biological activity of bacteria
(*c*) to avoid bulking of activated sludge
(*d*) to help in grease separation

33. Two samples of water A and B have pH values of 4.4 and 6.4 respectively. How many times more acidic sample A is than sample B?

(*a*) 0 (*b*) 15
(*c*) 100 (*d*) 200

1998

34. An inverted siphon is a
    (a) device for distributing septic tank effluent to a soil absorption system
    (b) device for preventing overflow from elevated water storage tank
    (c) device for preventing crown corrosion of sewer
    (d) section of sewer which is dropped below the hydraulic grade line in order to avoid an obstacle.

35. Water distribution systems are sized to meet the
    (a) maximum hourly demand
    (b) Average hourly demand
    (c) maximum daily demand and fire demand
    (d) average daily demand and fire demand.

36. For a flow of 5.7 MLD (million litres per day) and a detention time of 2 hours, the surface area of a rectangular sedimentation tank to remove all particles have settling velocity of 0.33 mm/s is
    (a) $20m^2$
    (b) $100m^2$
    (c) $200m^2$
    (d) $400m^2$

1997

37. Flocculation is a process
    (a) that removes algae from stabilization pond effluent
    (b) that promotes the aggregation of small particles into larger particles to enchance their removal by gravity

38. Pathogens are usually removed by
    (a) chemical precipitation
    (b) sedimentation
    (c) activated sludge process
    (d) chlorination

39. The 'sag' in the dissolved oxygen curve results because
    (a) it is a function of the rate of addition of oxygen to the stream
    (b) it is a function of the rate of depletion of oxygen from the stream
    (c) it is a function of both addition and depletion of oxygen from the stream
    (d) the rate of addition is linear but the rate of depletion is non-linear.

40. Alkalinity of water can be defined correctly in one of the following ways
    (a) it is the measure of ability of water to neutralize oxygen
    (b) it is the measure of ability of water to neutralize carbonates
    (c) it is the presence of ions in water that will neutralize hydrogen ions
    (d) it is the measure of ability of water to neutralize hydroxides.

1996

41. The most important water quality parameter for domestic use of water is
    (a) carbonate hardness
    (b) non-carbonate hardness
    (c) coliform group of organisms
    (d) chlorides

42. Presence of fluoride in water greater than permissible level of 1.5 mg/L causes
    (a) cardiovascular disease
    (b) methemoglobinemia
    (c) hepatitis
    (d) dental fluorosis

43. A single rapid test to determine the pollutional status of river water is
    (a) biochemical oxygen demand
    (b) chemical oxygen demand
    (c) total organic solids
    (d) dissolved oxygen

44. Presence of excess nitrates in river water indicates
    (a) recent pollution of water with sewage
    (b) past pollution of water with sewage
    (c) immediate pollution of water with sewage
    (d) no pollution of water with sewage

NUMERICAL TYPE QUESTIONS

2016

1. The hardness of a ground water sample was found to be 420 mg/L as $CaCO_3$. A softener containing ion exchange resins was installed to reduce the total hardness to 75 mg/L as $CaCO_3$ before supplying to 4 households. Each household gets treated water at a rate of 540 L/day. If the efficiency of the softener is 100%, the bypass flow rate (expressed in L/day) is ________

2014

2. A straight 100 m long raw water gravity main is to carry water from an intake structure to the jack well of a water treatment plant. The required flow through this water main is 0.21 m^3/s. Allowable velocity through the main is 0.75 m/s. Assume $f = 0.01$, $g = 9.81\ m/s^2$. The minimum gradient (in cm/100 m length) to be given to this gravity main so that the required amount of water flows without any difficulty is _______.

2000

3. Consider a glucose solution ($C_6H_{12}O_6$) of molarity 1.75×10^{-3} that is completely oxidized to CO_2 and H_2O. Find the amount of oxygen required for this reaction. The chemical mass balance equation for the above reaction is given as

$$C_6H_{12}O_6\ 6O_2 \rightarrow 6CO_2 + 6H_2O$$

(Atomic weights are : C = 12; H = 1; O = 16).

ANSWERS

MCQ Type Questions

1. (*c*)	**2.** (*d*)	**3.** (*a*)	**4.** (*b*)	**5.** (*b*)	**6.** (*a*)	**7.** (*c*)	**8.** (*d*)	**9.** (*d*)	**10.** (*a*)
11. (*a*)	**12.** (*c*)	**13.** (*c*)	**14.** (*c*)	**15.** (*a*)	**16.** (*b*)	**17.** (*)	**18.** (*d*)	**19.** (*c*)	**20.** (*c*)
21. (*c*)	**22.** (*d*)	**23.** (*d*)	**24.** (*d*)	**25.** (*d*)	**26.** (*c*)	**27.** (*d*)	**28.** (*d*)	**29.** (*a*)	**30.** (*b*)
31. (*c*)	**32.** (*c*)	**33.** (*c*)	**34.** (*d*)	**35.** (*c*)	**36.** (*c*)	**37.** (*b*)	**38.** (*d*)	**39.** (*c*)	**40.** (*c*)
41. (*c*)	**42.** (*d*)	**43.** (*d*)	**44.** (*d*)						

Numerical Type Questions

1. (1774.29) **2.** (4.7 to 4.9) **3.** (*)

"*" Refers to subjective question.

EXPLANATIONS

MCQ TYPE QUESTIONS

1.
$$Q_{full} = V_{full} \times A$$
$$Q_{half} = V_{half} \times \frac{A}{2}$$
But $V_{full} = V_{half} = V$

So, $$\frac{Q_{full}}{Q_{half}} = \frac{V \times A}{V \times A/2} = 2$$

3. For disinfection, we have $tc^n = K$

Where t = time required to kill all organisms

c = concentration of disinfectant

n = dilution coefficient

k = constant

So, $t_1c_1^n = t_2c_2^n$

here, $n = 1$

$$t_1c_1 = t_2c_2$$

$t_1 = \frac{L}{V_1}$; L = length of pipe
V_1 = velocity of flow

$= \frac{L}{Q_1/A} = \frac{L.A}{Q_1}$;

Q_1 = discharge per day

$C_1 = \frac{W_1}{Q_1}$,

W_1 = weight of disinfectant per day

So, $\frac{LA}{Q_1}\frac{W_1}{Q_1} = \frac{LA}{Q_2}\cdot\frac{W_2}{Q_2}$

$$W_2 = \left(\frac{Q_2}{Q_1}\right)^2 .W_1 = \left(\frac{22}{16}\right)^2 \times 32$$

$= 60.5$ kg/d

4. Carbonate hardness CH = 3.5 × 50 mg/ℓ as $CaCO_3$ = 175 mg/ℓ

Non Carbonate hardness,

NCH = Total Hardness – Carbonate hardness

= 50 × 50 – 175

= 75 mg/ℓ as $CaCO_3$

5. Alkalinity measured by titrating a water sample with a titrant $\frac{N}{50} H_2SO_4$

Hardness measured by titrating a water sample with a titrant $\frac{N}{50}$ EDTA

Chloride measured by titrating a water sample with a titrant $\frac{N}{35-5} AgNO_3$

Dissolved oxygen measured by titrating a water sample with a titrant $\frac{N}{40} Na_2S_2O_3$

6. Length of pipe (L) = 2 km = 2000 m

Diameter of pipe (D) = 0.2 m

Head loss (H_L) = 8 m

Darcy-Weisbach friction factor

$f = 0.04$

Valocity in pipe v = ?

$$H_L = \frac{f \cdot L \cdot v^2}{2gd}$$

$$\delta = \frac{0.04 \times 2000 \times v^2}{2 \times 9.81 \times 0.2}$$

$$v = \sqrt{\frac{8 \times 2 \times 9.81 \times 0.2}{0.04 \times 2000}}$$

$= 0.6264$ m/sec

$v = 0.63$ m/sec

8. Due to hardness, soap consumption increases hence, presence of hardness in excess of permissible limit causes increased laundry expenses.

9. *Given :*

Alkalinity of water sample = 250 mg/L as $CaCO_3$

Hardness of water sample = 350 mg/L as $CaCO_3$

Then, carbonate hardness = 250 mg/L

and Noncarbonate hardness = 350 – 250

= 100 mg/L

10. In an aerobic environment, oxidation occurs. Hence, NH_3 convert into NO_2

12. Design loading rate to filter = 200 m^3/m^2 day

Design flow rate = 0.5 m^3/sec

= $0.5m^3 \times 3600 \times 24$ m^2/day

Surface area per filter box = 50m^2

Surface area required for Rapid sand filter

$$= \frac{\text{Design flow rate}}{\text{design loading rate to filter}}$$

$$= \frac{0.5 \times 3600 \times 24}{200}$$

= 216 m^2

13. Number of filters required = $\frac{216}{50}$ = 4.32.

Accept 6 from given alternatives.

18. According to Indian Standards : Drinking Water Specifications are

		Standard
Turbidity	5 to 10 mg/l	5 mg/l
pH	6.5 to 8	7.4
Flourides	1.0 mg/l	2.5 mg/l
Total Hardness	75 to 115 mg/l	300 mg/l
Iron	upto 0.3 mg/l	3 mg/l
MPN	1 per 100 ml	50 per 100 ml

19. Zero hardness of water is achieved by ion exchange treatment. It is also called *Zeolite process or Base exchange treatment*. Zeolite are natural or the synthetic cation or base exchange hydrated silicates of sodium and alumnium having the general formula as

$$Na_2O\ Al_2O_3.\ xSiO_2.\ yH_2O.$$

Other process such as lime soda process can produce water of hardness not less than about 50 mg/l.

20. Total Hardness as $CaCO_3$

$$= Ca^{2+} \times \frac{50}{20} + Mg^{2+} \times \frac{50}{12}$$

$$= \frac{55 \times 50}{20} + \frac{50}{12} \times 10$$

$$= \mathbf{179\ mg/l}$$

21. According to Stoke's law

$$V_S = 418\ (S_S - 1)\ d^2 \left[\frac{3T + 70}{100}\right] \text{ (for } d < 0.1 \text{ mm)}$$

$$= \mathbf{0.089\ mm/sec.}$$

Alternatively

$$V_S = \frac{9}{18}\ (S_S - 1)\ \frac{d^2}{v}$$

$$\text{where, } v = \frac{\mu}{P} = \frac{10^{-3}}{10^3}$$

$$= 10^{-6}\ m/s$$

$$= \mathbf{0.089\ mm/sec}$$

$$\text{Now,} \quad V_0 = \frac{a}{A}$$

$$= \frac{8000}{24 \times 3600 \times 900}$$

$$= 0.1028\ mm/sec.$$

∴ % of particle removed

$$= \frac{V_S}{V_0} \times 100$$

$$= \frac{0.089}{0.1028} \times 100$$

$$= \mathbf{87.5\%}$$

27. Total hardness

$$= Ca^{++}\ mg/I \left(\frac{50}{20}\right) + Mg^{++} mg/I \times \left(\frac{50}{12}\right)$$

$$= 160 \times \frac{50}{20} + 40 \times \frac{50}{12} = 566.67\ mg/I$$

36. Flow, *i.e.* Discharge,

$$Q = 5.7\ MLD$$

$$= 5.7 \times 10^6\ litres/day$$

$$= \frac{5.7 \times 10^6}{24 \times 10^3}$$

$$= 237.7\ m^3/hour$$

Given, setting velocity,

$$V = 0.33 m/sec = 1.188\ m/hour$$

Surface area of rectangular sedimentation tank

$$= \frac{Q}{V} = \frac{237.5}{1.188}$$

$$= 199.9\ m^2 \approx 200 m^2$$

NUMERICAL TYPE QUESTIONS

1. Hardness = 420 mg/l as $CaCO_3$

Reduced hardness = 75 mg/l as $CaCO_3$

Total hardness removed = 345 mg/l as $CaCO_3$

Total quantity of treated water = 540 × 4

= 2160 1/day with 75 mg/l hardness

Total hardness for supply water

= 2160 × 75 mg/day

= 162000 mg/day

Volume of water required not to be treated

$$= \frac{162000}{420}\ \frac{mg}{day}$$

$$= 385.71 \frac{ltr}{day}$$

Therefore, volume of water required to be treated

= 2160 – 385.71

= 1774.29 ltrs/day

2. $Q = 0.21\ m^3/s$

$V_a = 0.75$ m/s,

$f = 0.01$, $g = 9.81\ m/s^2$

$$A = \frac{Q}{V}$$

$$\Rightarrow \quad \frac{\pi}{4}.d^2 = \frac{0.21}{0.75}$$

$$\Rightarrow \quad d = 0.60\ m$$

$$h_f = \frac{f \times L}{d} \times \frac{V^2}{2g}$$

$$= \frac{0.01 \times 100 \times (0.75)^2}{0.60 \times 2 \times 9.81}$$

$$= 0.047\ m = 4.7\ cm$$

$$\text{Minimum gradient} = \frac{h_f}{L} = \frac{4.7\,cm}{100\,m}$$

3. Molecular weight of $C_6H_{12}O_6$

$$= 6 \times 12 + 12 \times 1 + 6 \times 16 = 180$$

$$\text{Molarity} = \frac{\text{Wt. of } C_6H_{12}O_6 \text{ in gm}}{\text{Molecular wt. of } C_6H_{12}O_6} \times 100$$

Molecular weight of O_2

$$= 2 \times 16 = 32$$

$$\therefore\ 1.75 \times 10^{-3} = \frac{W_1}{180} \times 1000$$

or $\quad W_1 = 3.15 \times 10^{-4}$ gm

Now 1 mole = $C_6H_{12}O_6$ require O_2 = 6 mole

$\Rightarrow\quad$ 180 gm $C_6H_{12}O_6$ require $O_2 = 6 \times 32$ gm

$\therefore\ $ 1 gm $C_6H_{12}O_6$ require $O_2 = \frac{6 \times 32}{180}$ gm

$\Rightarrow$ 3.15×10^{-4} gm $C_6H_{12}O_6$ require

$$O_2 = \frac{6 \times 32}{180} \times 3.15 \times 10^{-4}$$

$$= 3.36 \times 10^{-4} \text{ gm}$$

Hence amount of O_2 required to oxidize 1 lit of glucose is 3.36×10^{-4} gm.

■■

12 CHAPTER Solid Waste and Waste Water Treatment

MCQ TYPE QUESTIONS

2016

1. Crown corrosion in a reinforced concrete sewer is caused by:

(a) H_2S (b) CO_2

(c) CH_4 (d) NH_3

2. Match the items in Group – I with those in Group – II and choose the right combination.

Group - I

P. Activated sludge process

Q. Rising of sludge

R. Conventional nitrification

S. Biological nitrogen removal

Group - II

1. Nitrifiers and denitrifiers
2. Autotrophic bacteria
3. Heterotrophic bacteria
4. Denitrifiers

(a) P-3, Q-4, R-2, S-1 (b) P-2, Q-3, R-4, S-1

(c) P-3, Q-2, R-4, S-1 (d) P-1, Q-4, R-2, S-3

2015

3. Solid waste generated from an industry contains only two components, X and Y as shown in the table below

Component	Composition (% weight)	Density (kg/m³)
X	c_1	ρ_1
Y	c_2	ρ_2

Assuming $(c_1 + c_2) = 100$, the composite density of the solid waste (ρ) is given by :

(a) $\dfrac{100}{\left(\dfrac{c_1}{\rho_1}+\dfrac{c_2}{\rho_2}\right)}$ (b) $100\left(\dfrac{\rho_1}{c_1}+\dfrac{\rho_2}{c_2}\right)$

(c) $100(c_1\rho_1 + c_2\rho_2)$ (d) $100\left(\dfrac{\rho_1\rho_2}{c_1\rho_1 + c_2\rho_2}\right)$

4. Total Kjeldahl Nitrogen (TKN) concentration (mg/L as N) in domestic sewage is the sum of the concentrations of

(a) organic and inorganic nitrogen in sewage

(b) organic nitrogen and nitrate in sewage

(c) organic nitrogen and ammonia in sewage

(d) ammonia and nitrate in sewage

5. Ultimate BOD of a river water sample is 20 mg/L. BOD rate constant (natural log) is 0.15 day^{-1}. The respective values of BOD (in %) exerted and remaining after 7 days are:

(a) 45 and 55

(b) 55 and 45

(c) 65 and 35

(d) 75 and 25

2014

6. The potable water is prepared from turbid surface water by adopting the following treatment sequence.

(a) Turbid surface water → Coagulation → Flocculation → Sedimentation → Filtration → Disinfection → Storage & Supply

(b) Turbid surface water → Disinfection → Flocculation → Sedimentation → Filtration → Coagulation → Storage & Supply

(c) Turbid surface water → Filtration → Sedimentation → Disinfection → Flocculation → Coagulation → Storage & Supply

(d) Turbid surface water → Sedimentation → Flocculation → Coagulation → Disinfection → Filtration → Storage & Supply

7. The dominating microorganisms in an activated sludge process reactor are

(a) aerobic heterotrophs

(b) anaerobic heterotrophs

(c) autotrophs

(d) phototrophs

8. A waste water stream (flow = 2 m³/s, ultimate BOD = 90 mg/l) is joining a small river (flow = 12 m³/s, ultimate BOD = 5 mg/l). Both water streams get mixed up instantaneously. Cross-sectional area of the river is 50 m². Assuming the de-oxygenation rate constant, k = 0.25/day, the BOD (in mg/l) of the river water, 10 km downstream of the mixing point is

(a) 1.68 (b) 12.63

(c) 15.46 (d) 1.37

2013

9. A settling tank in a water treatment plant is designed for a surface overflow rate of $30 \frac{m^3}{day.m^2}$.

Assume specific gravity of sediment particles = 2.65, density of water (ρ) = 1000 kg/m^3, dynamic viscosity of water (μ) = 0.001 N.s/m^2, and *Stokes' law* is valid. The approximate minimum size of particles that would be completely removed is

(*a*) 0.01mm (*b*) 0.02 mm

(*c*) 0.03 mm (*d*) 0.04 mm

10. Elevation and temperature data for a place are tabulated below:

Elevation, m	Temperature, °C
4	21.25
444	15.70

Based on the above data, lapse rate can be referred as

(*a*) Super-adiabatic

(*b*) Neutral

(*c*) Sub-adiabatic

(*d*) Inversions

2012

11. A sample of domestic sewage is digested with silver sulphate, sulphuric acid, potassium dichromate and mercuric sulphate in chemical oxygen demand (COD) test. The digested sample is then titrated with standard ferrous ammonium sulphate (FAS) to determine the un-reacted amount of

(*a*) mercuric sulphate

(*b*) potassium dichromate

(*c*) silver sulphate

(*d*) sulphuric acid

12. Assertion [A]: At a manhole, the crown of the outgoing sewer should not be higher than the crown of the incoming sewer.

Reason [R]: Transition from a larger diameter incoming sewer to a smaller diameter outgoing sewer at a manhole should not be made.

The **CORRECT** option evaluating the above statements is :

(*a*) Both [A] and [R] are true and [R] is the correct reason for [A]

(*b*) Both [a] and [R] are true but [R] is not the correct reason for [A]

(*c*) Both [A] and [R] are false

(*d*) [A] is true but [R] is false

13. A town is required to treat 4.2 m^2/min of raw water for daily domestic supply. Flocculating particles are to be produced by chemical coagulation. A column analysis indicated that an overflow rate of 0.2 mm/s will produce satisfactory particle removal in a settling basin at a depth of 3.5 m. The required surface area (in m^2) for settling is

(*a*) 210

(*b*) 350

(*c*) 1728

(*d*) 21000

2011

14. Consider the following unit processes commonly used in water treatment: rapid mixing (RM), flocculation (F), primary sedimentation (PS), secondary sedimentation (SS), chlorination (C) and rapid sand filtration (RSF). The order of these unit processes (first to last) in a conventional water treatment plant is

(*a*) PS → RSF → F → RM → SS → C

(*b*) PS → F → RM → RSF → SS → C

(*c*) PS → F → SS → RSF → RM → C

(*d*) PS → RM → F → SS → RSF → C

2010

15. A coastal city produces municipal solid waste (MSW) with high moisture content, high organic materials, low calorific value and low inorganic materials. The most effective and sustainable option for MSW management in that city is

(*a*) Composting (*b*) Dumping is sea

(*c*) Incineration (*d*) Landfill

16. If BOD_3 of a wastewater sample is 75 mg/L and reaction rate constant k (base e) is 0.345 per day, then amount of BOD remaining in the given sample after 10 days is

(*a*) 3.21 mg/L (*b*) 3.45 mg/L

(*c*) 3.69 mg/L (*d*) 3.92 mg/L

2009

17. A horizontal flow primary clarifier treats wastewater in which 10%, 60% and 30% of particles have settling velocities of 0.1 mm/s, 0.2 mm/s, and 1.0 mm/s respectively. What would be the total percentage of particles removed if clarifier operates at a Surface Overflow Rate (SOR) of 43.2 m^3/m^2.d?

(*a*) 43% (*b*) 56%

(*c*) 86% (*d*) 100%

18. An aerobic reactor receives wastewater at a flow rate of 500 m^3/d having a COD of 2000 mg/L. The effluent COD is 400 mg/L. Assuming that wastewater contains 80% biodegradable waste, the daily volume of methane produced by the reactor is

(*a*) 0.224 m^3

(*b*) 0.280 m^3

(*c*) 224 m^3

(*d*) 280 m^3

19. Column I

P. Grit chamber

Q. Secondary settling tank

R. Activated sludge process

S. Trickling filter

Column II

1. Zone settling

2. Stoke's law

3. Aerobic

4. Contact stabilisation

The correct match of **Column I** with **Column II** is

Codes :

	P	Q	R	S
(*a*)	1	2	3	4
(*b*)	2	1	3	4
(*c*)	1	2	4	3
(*d*)	2	1	4	3

2008

20. Two biodegradable components of municipal solid waste are

(*a*) plastics and wood

(*b*) cardboard and glass

(*c*) leather and tin cans

(*d*) food wastes and garden trimmings

21. A wastewater sample contains $10^{-5.6}$ mmol/ I of OH^- ions at 25° C. The pH of this sample is

(*a*) 8.6

(*b*) 8.4

(*c*) 5.6

(*d*) 5.4

22. Group I lists estimation methods of some of the water and wastewater quality parameters. Group II lists the indicators used in the estimation methods. Match the estimation method (Group I) with the corresponding indicator (Group II).

Group I	Group II
P. Azide modified Winkler method for dissolved oxygen	**1.** Eriochrome Black T
Q. Dichromate method for chemical oxygen demand	**2.** Ferrion
R. EDTA titrimetric method for hardness	**3.** Potassium chromate
S. Mohr or Argentometric method for chlorides	**4.** Starch

Codes :

	P	Q	R	S
(*a*)	3	2	1	4
(*b*)	4	2	1	3
(*c*)	4	1	2	3
(*d*)	4	2	3	1

23. Determine the correctness or otherwise of the following **Assertion [a]** and the **Reason [r]**

Assertion : The crown of the outgoing larger diameter sewer is always matched with the crown of incoming smaller diameter sewer.

Reason : It eliminates backing up of sewage in the incoming smaller diameter sewer.

(*a*) Both [a] and [r] are true and [r] is the correct reason for [a]

(*b*) Both [a] and [r] are true but [r] is not the correct reason for [a]

(*c*) Both [a] and [r] are false

(*d*) [a] is true but [r] is false

24. The 5-day BOD of a wastewater sample is obtained as 190 mg/I (with k= $0.01h^{-1}$). The ultimate oxygen demand (mg/I) of the sample will be

(*a*) 3800 (*b*) 475

(*c*) 271 (*d*) 190

25. A water treatment plant is required to process 28800 m^3/d of raw water (density = 1000 kg/m^3, kinematic viscosity = 10^{-6} m^2/s). The rapid mixing tank imparts a velocity gradient of $900s^{-1}$ to blend 35 mg/I of alum with the flow for a detention time of 2 minutes. The power input (W) required for rapid mixing is

(*a*) 32.4 (*b*) 36

(*c*) 324 (*d*) 32400

26. Match **Group I** *(Terminology)* with **Group II** *(Definition / Brief Description)* for wastewater treatment systems

Group I

P. Primary treatment

Q. Secondary treatment

R. Unit operation

S. Unit process

Group II

1. Contaminant removal by physical forces
2. Involving biological and /or chemical reaction
3. Conversion of soluble organic matter to biomass
4. Removal of solid materials from incoming wastewater

Codes :

	P	Q	R	S
(a)	4	3	1	2
(b)	4	3	2	1
(c)	3	4	2	1
(d)	1	2	3	4

2006

27. To determine the BOD_5 of a wastewater sample, 5, 10 and 50 mL of the waste water were diluted to 300 mL and incubated at 20°C in BOD bottles for 5 days. The results were as follows.

S.No.	Waste water Volume mL	Initial DO $\frac{mg}{L}$	DO After 5 days $\frac{mg}{L}$
1.	5	9.2	6.9
2.	10	9.1	4.4
3.	50	8.4	0.0

Based on the given data, average BOD_5 of the wastewater is equal to

(*a*) 139.5 mg/L (*b*) 126.5 mg/L

(*c*) 109.8 mg/L (*d*) 72.2 mg/L

28. A synthetic sample of water is prepared by adding 100 mg Kaolinite (a clay mineral), 200 mg glucose, 168 mg NaCl, 120 mg $MgSO_4$, and 111 mg $CaCl_2$ to 1 liter of pure water. The concentrations of total solids (TS) and fixed dissolved solids (FDS)respectively in the solution in mg/L are equal to

(*a*) 699 and 599 (*b*) 599 and 399

(*c*) 699 and 199 (*d*) 699 and 399

29. The composition of a certain MSW sample and specific weights of its various components are given below.

Component	Percent by Weight	Specific Weight $\left(\frac{kg}{m^3}\right)$
Food waste	50	300
Dirt and Ash	30	500
Plastics	10	65
Wood and Yard waste	10	125

Specific weight (kg/m^3) of the MSW sample is

(*a*) 319 (*b*) 217

(*c*) 209 (*d*) 199

2005

30. Bulking sludge refers to having

(*a*) $F/M < \frac{0.3}{d}$ (*b*) $\frac{0.3}{d} < F/M < \frac{0.6}{d}$

(*c*) F/M = zero (*d*) $F/M > \frac{0.6}{d}$

31. Group 1 contains some properties of water/ waste water and group 2 contains list of some tests on water/waste water. Match the property with corresponding test.

Group 1	Group 2
P. Suspended solids concentration	1. BOD
Q. Metabolism of biodegradable organics	2. MPN
R. Bacterial concentration	3. Jar test
S. Coagulant dose	4. Turbidity

Codes :

	P	Q	R	S
(a)	2	1	4	3
(b)	4	1	2	3
(c)	2	4	1	3
(d)	4	2	1	3

32. Match the following

Group 1	Group 2
P. Thickening of sludge	1. Decrease in volume of sludge by chemical oxidation
Q. Stabilization of sludge	2. Separation of water by heat or chemical treatment
R. Conditioning of sludge	3. Digestion of sludge

	S. Reduction of sludge	4. Separation of water by flotation or gravity

Codes :

	P	Q	R	S
(a)	4	3	1	2
(b)	3	2	4	1
(c)	4	3	2	1
(d)	2	1	3	4

33. In a certain situation, waste water discharged into a river, mixes with the river water instantaneously and completely. Following is the data available:

Wastewater :

DO = 2.00 mg/L

Discharge rate = 1.10 m^3/s

River water:

DO = 8.3 mg/L

Flow rate = 8.70 m^3/s

Temperature = 20°C

Initial amount of DO in the mixture of waste and river shall be

(a) 5.3 mg/L (b) 6.5 mg/L

(c) 7.6 mg/L (d) 8.4 mg/L

34. A circular primary clarifier processes an average flow of 5005 m^3/d of municipal wastewater. The overflow rate is 35 m^3/m^2d. The diameter of clarifier shall be

(a) 10.5 m (b) 11.5 m

(c) 12.5 m (d) 13.5 m

2004

35. An existing 300 mm diameter circular sewer is laid at a slope of 1:28 and carries a peak discharge of 1728 m^3/d. Use the partial flow diagram shown in the figure below and assume Manning's n = 0.015.

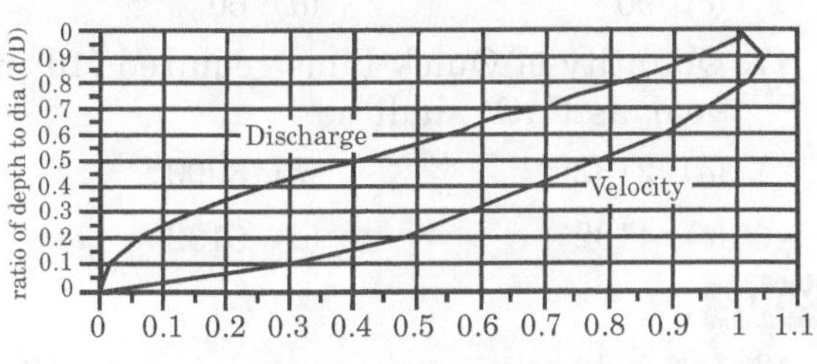

Partial flow diagram for a circular sewer

At the peak discharge, the depth of flow and the velocity are, respectively

(a) 45 mm and 0.28 m/s

(b) 120 mm and 0.50 m/s

(c) 150 mm and 0.57 m/s

(d) 300 mm and 0.71 m/s

36. An analysis for determination of solids in the return sludge of activated sludge process was done as follows: (1) A crucible was dried to a constant mass of 62.485 g. (2) 75 ml of a well-mixed sample was taken in the crucible. (3) The crucible with the sample was dried to a constant mass of 65.020 g in a drying oven at 104° C. (4) The crucible with the dried sample was placed in a muffle furnace at 600° C. for an hour. After cooling, the mass of the crucible with residues was 63.145 g.

The concentration of organic fraction of solids present in the return sludge sample is

(a) 8800 mg/l (b) 25000 mg/l

(c) 33800 mg/l (d) 42600 mg/l

37. Water samples (X and Y) from two different sources were brought to the laboratory for the measurement of dissolved oxygen (DO) using modified Winkler method. Samples were transferred to 300 ml BOD bottles. 2 ml of $MnSO_4$ solution and 2 ml of alkaliodide-azide reagent were added to the bottles and mixed. Sample X developed a brown precipitate, whereas sample Y developed a white precipitate.

In reference to these observations, the correct statement is

(a) Both the samples were devoid of DO

(b) Sample X was devoid of DO while sample Y contained DO

(c) Sample X contained DO whiloe sample Y was devoid of DO

(d) Both the samples contained DO

38. A portion of wastewater sample was subjected to standard BOD test (5 days, 20° C), yielding a value of 180 mg/l. The reaction rate constant (to the base 'e') at 20° C was taken as 0.18 per day. The reaction rate constant at other temperature may be estimated by

$$k_T = k_{20} (1.047)^{T-20}.$$

The temperature at which the other portion of the sample should be tested, to exert the same BOD in 2.5 days, is

(a) 4.9°C (b) 24.9°C

(c) 31.7°C (d) 35.0°C

2003

39. Which of the following sewage treatment methods has inherent problems of odour, ponding, and fly nuisance ?

(a) UASB system

(b) Activated sludge process

(c) Trickling filters

(d) Stabilization ponds

40. From amongst the following sewage treatment options, largest land requirements for a given discharge will be needed for
(*a*) trickling filter (*b*) anaerobic pond
(*c*) oxidation ditch (*d*) oxidation pond

41. Settling test on a sample drawn from Aeration Tank liquor of ASP (MLSS = 2800 mg/l) was carried out with 1 litre sample. The test yielded a settled volume of 200 ml. The value of Sludge Volume Index shall be
(*a*) 14.0 (*b*) 34.2
(*c*) 71.4 (*d*) 271

42. Match the following :

Group I *(Characteristics of sewage discharged into inland waters)*	Group II *(Allowable limit, mg/l)*
P. BOD_5	**1.** 250
Q. COD	**2.** 30
R. Oil and Grease	**3.** 20
S. Total Suspended Solids	**4.** 10
	5. 5
	6. 3

Codes :

	P	Q	R	S
(*a*)	2	5	4	2
(*b*)	4	1	6	4
(*c*)	3	1	4	2
(*d*)	2	1	6	3

43. Match the following :

Group I *(Type of water impurity)*	Group II *(Method of treatment)*
P. Hardness	**1.** Reverse Osmosis
Q. Brackish water from sea	**2.** Chlorination
R. Residual MPN from filters	**3.** Zeolite Treatment
S. Turbidity	**4.** Coagulation and Flocculation
	5. Coagulation, Flocculation, and Filtration

Codes :

	P	Q	R	S
(*a*)	1	2	4	5
(*b*)	3	2	2	4
(*c*)	2	1	3	5
(*d*)	3	1	2	5

Common Data Q. 44-45

A conventional Activated Sludge Plant treating 1000 m^3/d of municipal waste water disposes of its anaerobically digested sludge on relatively impervious farmland. Use the following data.

1. **Raw Sewage :**
SS = 225 mg/l (70% volatile); BOD = 190 mg/l (Excess activated sludge returned to primary)
2. **Primary Setting :**
SS – 50% removal; BOD – 30% removal
3. **Excess Activated Sludge :**
0.4 g VSS produced per g BOD applied (80% Volatile of total)
4. **Anaerobic Digester :**
VSS reduced 50% Digested Sludge Concentration–60%; Sludge Specific Gravity—1
5. **Application on farmland :**
2 m^3/ha.d

44. Total volatile suspended solids to be anaerobically digested (kg/d VSS) shall be
(*a*) 133 (*b*) 168
(*c*) 233 (*d*) 245

45. Area requirements (ha) for disposal of the sludge on farmland shall be
(*a*) 2.95 (*b*) 1.95
(*c*) 0.95 (*d*) 0.55

Common Data Q. 46-47

A water treatment plant treating 10 mld of water requires 20 mg/l of filter Alum, $Al_2(SO_4)_3 \cdot 18\,H_2O$. The water has 6 mg/l of alkalinity as $CaCO_3$ (Al = 26.97, S = 32, O = 16, H =1, Ca = 40, and C = 12).

46. Total alkalinity requirement (10^6 mg per day as $CaCO_3$) matching filter Alum, shall be
(*a*) 180 (*b*) 120
(*c*) 90 (*d*) 60

47. Quantity of Quick Lime required (10^6 mg per year as CaO) shall be
(*a*) 2132 (*b*) 3000
(*c*) 4132 (*d*) 6132

2002

48. In a domestic wastewater sample, COD and BOD were measured. Generally which of the following statement is true for their relative magnitude?
(*a*) COD = BOD
(*b*) COD > BOD
(*c*) COD < BOD
(*d*) Nothing can be said

49. A Trickling filter is designed to remove

(*a*) Settleable solids

(*b*) Colloidal solids

(*c*) Dissolved organic matter

(*d*) None of these

50. For a water treatment plant having a flow rate of 432m³/hr, what is the required plan area of a Type I setting tank to remove 90% of the particles having a settling velocity of 0.12 cm/sec is

(*a*) 120m² (*b*) 111m²

(*c*) 90m² (*d*) 100m²

51. A wastewater sample has an initial BOD of 222mg/L. The first order BOD decay coefficient is 0.4/day. The BOD consumed (in mg/L) in 5 days is

(*a*) 150 (*b*) 192

(*c*) 30 (*d*) 50

2001

52. The unit in which both sedimentation and digestion processes of sludge take place simultaneously is

(*a*) Skimming tank (*b*) Imhoff tank

(*c*) Detritus tank (*d*) Digestion tank

53. If $BOD_{5,20}$ of waste is 150mg/L and reaction rate constant (to the base 'e') at 20°C is 0.35/day, then ultimate BOD in mg/L is

(*a*) 97.5 (*b*) 181.5

(*c*) 212.9 (*d*) 230.5

54. A wastewater treatment plant discharges 1.5m³/sec of effluent having an ultimate BOD of 40.0mg/lit into a stream flowing at 10m³/sec. Just upstream of the discharge point, the stream has an ultimate BOD of 3.0mg/lit. The deoxygenation constant to the base 'e' is estimated as 0.32/day.

(*a*) Assuming complete mixing, find the ultimate BOD of the mixture of waste and stream just downstream of the outfall.

(*b*) Assuming a constant cross section area for the stream equal of 50m², estimate the BOD of the stream at a point 2.5 km downstream from the outfall.

2000

55. A circular sewer 2m diameter has to carry a discharge of 2m³/s when flowing nearly full. What is the minimum required slope to initiate the flow? Assume Manning's n = 0.015.

(*a*) 0.00023

(*b*) 0.000036

(*c*) 0.000091

(*d*) 0.000014

56. The dissolved oxygen (DO) in an unseeded sample of diluted waste having an initial DO of 9.5 mg/l is measured to be 3.5 mg/l after 5 days. The dilution factor is 0.03 and the reaction rate constant, k = 0.22/day (to the base 'e' in the decay curve at 20°C.) Estimate

(*a*) 5 day BOD of the waste at 20° C.

(*b*) Ultimate carbonaceous BOD.

(*c*) Remaining oxygen demand after 5 days.

(*d*) 5 day BOD of this waste at 25° C assuming a temperature coefficient of 1.047.

1999

57. Standard 5-day BOD of a waster water sample is nearly x% of the ultimate BOD, where x is

(*a*) 48 (*b*) 58

(*c*) 68 (*d*) 78

1998

58. Sewage treatment in an oxidation pond is accomplished primarily by

(*a*) alga-bacterial symbiosis

(*b*) algal photosynthesis only

(*c*) bacterial oxidation only

(*d*) chemical oxidation only

59. In a BOD test using 5% dilution of the sample (15 ML of sample and 285 mL of dilution water), dissolved oxygen values for the sample and dilution water blank bottles after five days incubation at 20°C were 3.80 and 8.80 mg/L, respectively. Dissolved oxygen originally present in the undiluted sample was 0.80 mg/L. The 5-day 20°C BOD of the sample is

(*a*) 116 mg/L

(*b*) 108 mg/L

(*c*) 100 mg/L

(*d*) 92 mg/L

1997

60. The absorbent most commonly used in water and waste treatment is

(*a*) Sand of grain size from 0.1 to 2 mm

(*b*) Activated carbon granules of size 0.1 to 2 mm

(*c*) Ordinary wood shavings of fine size

(*d*) Coal-tar.

61. Among the following disinfectants of waste water, the one that is most commonly used is

(*a*) Chlorine dioxide

(*b*) Chlorine

(*c*) Ozone

(*d*) UV-radiation.

62. A typical biological process in treating waste water using aerated lagoon can be described by one of the following schemtic diagrams

(*a*) 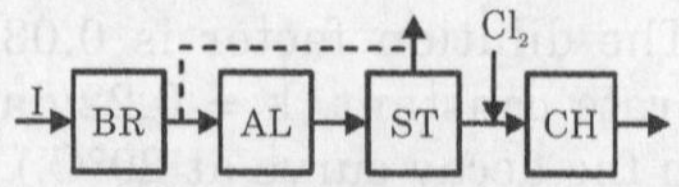

(*b*) 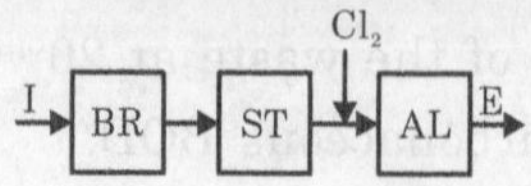

(*c*) 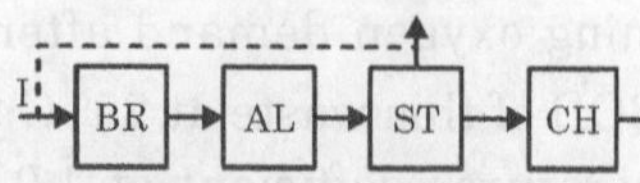

(*d*) 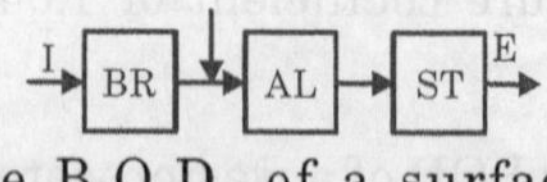

63. The B.O.D_5 of a surface water sample is 200 mg/litre at 20°C. The value of the reaction constant is K = 0.2 day^{-1} with base 'e'. The ultimate B.O.D. of the sample is

(*a*) 126 mg/litre

(*b*) 544 mg/litre

(*c*) 146 mg/litre

(*d*) 316 mg/litre

64. MPN index is measure of one of the following :

(*a*) Coliform bacteria

(*b*) BOD_5

(*c*) Dissolved Oxygen content

(*d*) Hardness

1996

65. Chemical oxygen Demand (COD) of a sample is always greater than Biochemical Oxygen Demand (BOD) since it represents

(*a*) biodegradable organic matter only

(*b*) biodegradable and non-biodegradable organic matter

(*c*) non-biodegradable organic matter

(*d*) inorganic matter

66. A waste water sample diluted to 100 times with aeration water had an initial dissolved oxygen (DO) of 7.0 mg/L and after 5 days of incubation at 20°C, then DO was zero. The BOD of waste water is

(*a*) 700 mg/L

(*b*) 100 mg/L

(*c*) cannot be determined

(*d*) 7 mg/L

67. The treatment that should be given to the water from a deep tube well is

(*a*) pre-settling only

(*b*) coagulation and flocculation only

(*c*) filtration only

(*d*) disinfection only

68. The drop manholes are provided in a sewerage system when there is

(*a*) change in alignment of sewer line

(*b*) change in size of sewers

(*c*) change in the elevation of ground level

(*d*) change from gravity system to pressure system

69. The removal of dissolved organic matter occurs in

(*a*) slow sand filters

(*b*) trickling filters

(*c*) rapid sand filters

(*d*) dual media filters

70. The main constituents of gas generated during the anaerobic digestion of sewage sludge are

(*a*) carbon dioxide and methane

(*b*) methane and ethane

(*c*) carbon dioxide and carbon monoxide

(*d*) carbon monoxide and nitrogen

NUMERICAL TYPE QUESTIONS

2016

1. A water supply board is responsible for treating 1500 m^3/day of water. A settling column analysis indicates that an overflow rate of 20 m/day will produce satisfactory removal for a depth of 3.1 m. It is decided to have two circular settling tanks in parallel. The required diameter (expressed in m) of the settling tanks is ________

2015

2. Consider a primary sedimentation tank (PST) in a water treatment plant with Surface Overflow Rate (SOR) of 40 $m^3/m^2/d$. The diameter of the spherical particle which will have 90 percent theoretical removal efficiency in this tank is ____ μm. Assume that settling velocity of the particles in water is described by Stokes's Law.

Given : Density of water = 1000 kg/m^3; Density of particle = 2650 kg/m^3; g = 9.81 m/s^2; Kinematic viscosity of water (ν) = 1.10×10^{-6} m^2/s.

3. A groundwater sample was found to contain 500 mg/L total dissolved solids (TDS). TDS (in %) present in the sample is ______.

4. A water treatment plant of capacity, 1 m^3/s has filter boxes of dimensions 6 m × 10 m. Loading rate to the filters is 120 $m^3/day/m^2$. When two of the filters are out of service for back washing, the loading rate (in $m^3/day/m^2$) is ______

5. In a wastewater treatment plant, primary sedimentation tank (PST) designed at an overflow rate of 32.5 $m^3/day/m^2$ is 32.5 m long, 80 m wide and liquid depth of 2.25 m. If the length of the weir is 75 m, the weir loading rate (in $m^3/day/m$) is ________.

6. A landfill is to be designed to serve a population of 200000 for a period of 25 years. The solid waste (SW) generation is 2 kg/person/day. The density of the un-compacted SW is 100 kg/m^3 and a compaction ratio of 4 is suggested. The ratio of compacted fill (i.e. SW + cover) to compacted SW is 1.5. The landfill volume (in million m^3) required is ______.

2014

7. A suspension of sand like particles in water with particles of diameter 0.10 mm and below is flowing into a settling tank at 0.10 m^3/s. Assume g = 9.81 m/s^2, specific gravity of particles = 2.65, and kinematic viscosity of water = 1.0105×10^{-2} cm^2/s. The minimum surface area (in m^2) required for this settling tank to remove particles of size 0.06 mm and above with 100% efficiency is ________

8. A surface water treatment plant operates round the clock witha flow rate of 35 m^3/min. The water temperature is 15 °C and jar testing indicated an alum dosage of 25 mg/*l* with flocculation at a G*t* value of 4×10^4 producing optimal results. The alum quantity required for 30 days (in kg) of operation of the plant is ________

9. An effluent at a flow rate of 2670 m^3/d from a sewage treatment plant is to be disinfected. The laboratory data of disinfection studies with a chlorine dosage of 15 mg/l yield the model $N_t = N_o e^{-0.145t}$ where N_t = number of micro-organisms surviving at time *t* (in min.) and N_o = number of micro-organisms present initially (at t = 0). The volume of disinfection unit (in m^3) required to achieve a 98% kill of micro-organisms is __________

2013

10. A water treatment plant is designed to treat 1 m^3/s of raw water. It has 14 sand filters. Surface area of each filter is 50 m^2. What is the loading rate $\left(\text{in}\ \frac{m^3}{day.m^2}\right)$ with two filters out of service for routine backwashing ________

11. A student began experiment for determination of 5-day, 20°C BOD on Monday. Since the 5th day fell on Saturday, the final DO readings were taken on next Monday. On calculation, BOD (i.e. 7 day, 20°C) was found to be 150 mg/L. What would be the 5-day, 20°C BOD (in mg/L)? Assume value of BOD rate constant (*k*) at standard temperature of 20°C as 0.23/day (base *e*) _________

2001

12. The data of an activated sludge process are as follows :

MLSS = 5000mg/lit

Flow = 0.15m^3/sec

SS of inflow = 400mg/lit

Solids settled after 30 min = 25%

Volume of Aeration tank = 3000m^3

Sludge Wastage Rate

= 120 m^3/day with VSS of 15000mg/lit.

Calculate Sludge Volume Index (SVI), Sludge Density Index (SDI), Return Sludge Ratio (Q_r/Q) and Mean cell residence time (θ_c).

ANSWERS

MCQ Type Questions

1. (a)	2. (a)	3. (a)	4. (c)	5. (c)	6. (a)	7. (a)	8. (c)	9. (b)	10. (a)
11. (b)	12. (a)	13. (b)	14. (d)	15. (a)	16. (c)	17. (b)	18. (c)	19. (b)	20. (b)
21. (d)	22. (b)	23. (a)	24. (c)	25. (d)	26. (d)	27. (c)	28. (d)	29. (b)	30. (a)
31. (b)	32. (a)	33. (c)	34. (d)	35. (c)	36. (c)	37. (c)	38. (d)	39. (c)	40. (d)
41. (a)	42. (d)	43. (d)	44. (a)	45. (d)	46. (c)	47. (d)	48. (b)	49. (c)	50. (c)
51. (b)	52. (b)	53. (b)	54. (*)	55. (a)	56. (*)	57. (c)	58. (b)	59. (d)	60. (b)
61. (b)	62. (a)	63. (d)	64. (a)	65. (b)	66. (a)	67. (d)	68. (a)	69. (b)	70. (a)

Numerical Type Questions

1. (6.909) 2. (22.58) 3. (0.05) 4. (144) 5. (112.667) 6. (13.6875)

7. (31.0 to 32.0) 8. (37800 to 37800) 9. (49.0 to 51.0) 10. (143 to 145) 11. (127 to 132)

12. (*)

"*" Refers to subjective question.

EXPLANATIONS

MCQ TYPE QUESTIONS

4. TKN is the sum of organic nitrogen, ammonia (NH_3), and ammonimum (NH_4^+) in the chemical analysis of wastewater.

5.

$$BOD_u = 20 \text{ mg/L}$$

$$BOD_7 = BOD_u\, e^{-k \times 7}$$

$$= 20 \times e^{-0.15 \times 7}$$

$$= 6.998 \approx 7$$

$$\% \text{ of BOD remaining after 7 days} = \frac{7}{20} = 35\%$$

$$\% \text{ exerted} = 100 - 35 = 65\%$$

7. Aerobic heterotrophs :

They are organisms that cannot live without free oxygen and do not produce their own food.

8. Flow = 2 m³/sec = Q_1, ultimate BOD = Y_{u_1} = 90 mg/*l*

Flow = 12 m³/sec = Q_2, ultimate BOD = Y_{u_2} = 5 mg/*l*

$$\text{BOD of mixture } (Y_0) = \frac{Q_1 \times Y_{u_1} + Q_2 \times Y_{u_2}}{Q_1 + Q_2}$$

$$= \frac{2 \times 90 + 12 \times 5}{2 + 12} = 17.14 \text{ gm/m}^3$$

the BOD of mixture at 10 km downstream of mixing point Y_f

$$Y_f = Y_0\left[1 - \left(10^{-k_D t}\right)\right]$$

$$K_D = 0.434\, k'$$

$$= 0.434 \times 0.25 = 0.1085$$

time taken to reach 10 km

$$t = \frac{\text{length}}{\text{velocity}}$$

$$Q_2 = Q_1 + Q_2 = 2 + 12 = 14$$

$$\text{velocity} = \frac{Q}{A} = \frac{14}{50} = 0.28 \text{ m/sec}$$

$$t = \frac{10 \times 1000}{0.28} \text{ sec}$$

$$= \frac{35714.28}{60 \times 60} = 9.921 \text{ days}$$

$$\therefore \quad Y_t = Y_0\left[1-(10)^{-k_D t}\right]$$

$$= 17.143\left[1-(10)^{-0.1085 \times 9.921}\right]$$

$$= 15.70 \text{ mg}/l$$

So, the correct choise in (*c*)

9. A settling tank in a water treatment plant

$$\text{Surface overflow rate} = 30\frac{m^3}{\text{day} - m^2} = v_s$$

Specific gravity $S_s = 2.65$

Density of water $\rho = 1000 \text{ kg/m}^3$

Dynamic viscosity of water $\mu = 0.001 \text{ N.S/m}^2$

By Soteks law

$$v_s = \frac{\gamma_w}{18}\left(\frac{S_s - 1}{\mu}\right) \cdot d^2$$

Where,

γ_w is in kg/m³

v_s is in $\frac{m^3}{\text{sec} - m^2}$

μ is in kg–s/m

then particle size get in m.

Therefore,

$$\gamma_w = 1000 \text{ kg/m}^3$$

$$v_s = \frac{30}{24 \times 3600} = 3.47 \times 10^{-4} \frac{s^3}{\text{sec} - m^3}$$

$$\rho = \gamma_w = 1000 \text{ kg/m}^3$$

$$\mu = \frac{0.001}{9.81} = 1.02 \times 10^{-4} \text{ kg – s/m}$$

$$\because \quad v_s = \frac{\gamma_w}{18}\left(\frac{S_s - 1}{\mu}\right) d^2$$

$$\Rightarrow 3.47 \times 10^{-4} = \frac{1000}{18}\left(\frac{2.65 - 1}{1.02 \times 10^{-4}}\right) d^2$$

$$d^2 = 3.85 \times 10^{-10}$$

$$d = 1.96 \times 10^{-5} \text{ M}$$

$$d = 0.0196 \text{ mm} \approx 0.02 \text{ mm}$$

10. As per every 165 m temperature reduces by 1°C

height = 444 – 4 = 440 m

So, the average temp. reduction

$$= \frac{440}{165} = 2.67°\text{C}$$

The temperature after reduction

= 21.25 – 2.67

= 18.58°C > 15.70

So lapse rate is more than natural

$\because$ lapse rate referred as super adiabatic.

15. Composting is the effective and sustainable option for MSW management because waste contains good amount of organic matter.

16. *Given :* $BOD_3 = 75$ mg/l

Reaction rate constant, $k = 0.345$ per day

$$\therefore \quad k_D = 0.434 \times k$$

$$= 0.434 \times 0.345 = 0.15$$

Using equation, $Y_3 = L_o\left[1 - 10^{-k_D \times t}\right]$

$$\therefore \quad 75 = L_o\left[1 - 10^{-.15 \times 3}\right]$$

$$\Rightarrow \quad L_o = 116.245 \text{ mg/lit}$$

Now, $Y_{10} = 116.245 \times \left[1 - 10^{-.15 \times 10}\right]$

$$= 112.569 \text{ mg/lit}$$

$\therefore$ Remaining BOD = 116.245 – 112.569 = 3.69 mg/l or ppm

17. *Given :* $V_0 = 43.2 \text{ m}^3/\text{m}^2\text{d}$

$$= \frac{43.2}{24 \times 60 \times 60} = 0.5 \text{ mm/sec}$$

For 10% particles :

$$v_1(= 0.1) < v_0$$

$$\frac{v_1}{v_0} = \frac{0.1}{0.5} = 20\%$$

$\therefore$ Net = 20% of 10% = 0.2 × 0.1 = 2%

For 60% particles :

$$v_2(= 0.2) < v_0$$

$$\frac{v_2}{v_0} = \frac{0.2}{0.5} = 40\%$$

$\therefore$ Net = 40% of 60% = 0.4 × 0.6 = 24%

For 30% particles :

$$v_3(= 1) > v_0 - 100\%$$

$\therefore$ Net = 100% of 30% = 30%

Total = 30 + 24 + 2 = 56%

21. $[OH^-] = 10^{-5.6}$ m.mol/litre

$= 10^{-5.6} \times 10^{-3}$ mol/litre

$= 10^{-8.6}$ mol/litre

But $[H^+][OH^-] = 10^{-14}$

$\Rightarrow$ $[H^+] = 10^{-14} \times 10^{8.6} = 10^{-5.4}$ mol/litre

$\therefore$ $p^H = -\log_{10}[H^+]$

$= -\log_{10} 10^{-5.4} = 5.4 \log_{10} 10 = 5.4$

24. $k_D = 0.434.k = 0.434 \times 0.01\ h^{-1}$

$\therefore$ $k_D.t = 0.434 \times 0.01 \times 5 \times 24 = 0.5208$

Given: $(BOD)_5 = 190$ mg/l

Now $(BOD)_5 = (BOD)_u \{1 - 10^{-kD.t}\}$

where $(BOD)_4$ = ultimate BOD

$\therefore$ 190 mg/l $= (BOD)_u \{1 - 10^{-0.5208}\}$

$\Rightarrow$ $(BOD)_u = \dfrac{190}{0.699} = 271$ mg/l

25. Velocity gradient, $G = \left\{\dfrac{P}{\mu.V}\right\}^{1/2}$

where, P = power

μ = dynamic viscosity

V = volume

$\therefore$ $P = G^2.\ \mu \times V$

$= (900)^2 \times (10^{-6} \times 1000) \times \dfrac{(28800 \times 2)}{24 \times 60}$

= 32400 watts

27.

S.No.	Waste Water Volume (mL)	Dilute factor	D.O. used = initial – final	BO D5 Dilute × DO used
1.	5	$\frac{300}{5}$ = 60	9.2 – 6.9 = 2.3	60 × 2.3 = 138
2.	10	$\frac{300}{10}$ = 30	4.7	141
3.	50	$\frac{300}{50}$ = 6	8.4	50.4

$\therefore$ Average, $BOD_5 = \dfrac{138 + 141 + 50.4}{3}$

= 109.8 mg/L

29. From given data,

$\left(\dfrac{0.5x}{300} + \dfrac{0.3x}{500} + \dfrac{0.1x}{65} + \dfrac{0.x}{125}\right) = 1$

$\Rightarrow$ $4.6 \times 10^{-3}x = 1$

$\Rightarrow$ $x = 217$

33. *Given :*

D.O. of waste water = 2.00 mg/L

D.O. of river water = 8.3 mg/L

$\therefore$ D.O. in mixture $= \dfrac{2 \times 1.1 + 8.3 \times 8.7}{(1.1 + 8.7)}$

$= 7.59\ m^3/sec$

34. Over flow rate = 35 m^3/m^2d

Average flow rate = 5005 m^3/d

Flow area $= \dfrac{\text{average flow rate}}{\text{over flow rate}}$

$= \dfrac{5005}{35} = 143\ m^2$

$\therefore$ $\dfrac{\pi}{4}d^2 = 143$

$\Rightarrow$ $d = \sqrt{\dfrac{143 \times 4}{\pi}}$

= 13.5 m

38. *Given :* $(B.O.D.)_5 = 180$ mg/lit

$k_T = k_{20}\ (1.047)^{T-20}$

$y_{2.5} = y_4\ [1 - e^{-kt})$

$= \dfrac{180[1 - e^{-k \times 2.5}]}{[1 - e^{-0.18 \times 5}]}$

$\Rightarrow$ $180 = \dfrac{180[1 - e^{-k \times 2.5}]}{[1 - e^{-0.18 \times 5}]}$

$\Rightarrow$ $k = 0.36$

$\therefore$ $0.36 = 0.18(1.047)^{T-20}$

$\Rightarrow$ $T = 35.0°C$

39. The filter fly, psychoda, which develops in the trickling filter particles, may prove to be nuisance. When fixed nozzles are used in it, H_2S and other odourous gases are frequently released from the sprays into the atmosphere. Sometimes, voids in the filter media gets clogged due to heavy growth of fungi and algae. This may result in ponding of the sewage over the filter bed.

41. *Given :* MLSS = 2800 mg/l

Settled volume = 200 ml

$\therefore$ Sludge volume index,

$SVI = \dfrac{V}{X}$

$= \dfrac{2800}{200} = \mathbf{14}$

44. SS (Raw sewage) = 225 mg/l

BOD (Raw sewage) = 190 mg/l

SS (Primary setting) = 50% removal

= 112.5 mg/l remain

BOD (Primary setting) = 30% removal

= 133 mg/l remain

0.4 g VSS produce per g BOD applied in excess activated sludge plant, and VSS reduced 50% in anaerobic digester.

Digested study concentration = 6%

S.S.C. = 1

Total volatile suspended solid to be anaerobically digested = 133 (kg/d VSS).

46. Chemical reaction that takes place are

$Al_2(SO_4)_3 . 18H_2O + 3Ca(HCO_3)_2$

$\rightarrow 2Al(OH)_3 \downarrow + 3CaSO_4 + 18H_2O + 6CO_2 \uparrow$

$CaCO_3 + H_2O + CO_2 \rightarrow Ca(HCO_3)_2$

$CaCO_3 \rightarrow CaO + CO_2$

The molecular weight are

Mol. weight of $Al_2(SO_4)_3 \cdot 18H_2O$

$= 2 \times 27 + 3(32 + 64) + 18(2 + 18)$

$= 666$

Mol. weight of $Ca(HCO_3)_2$

$= 40 + 2(1 + 12 + 48) = 162$

Mol. weight of $CaCO_3$

$= 40 + 12 + 48 = 100$

Mol. weight of CaO

$= 40 + 16 = 56$

Now from above equation, it is clear that alum requires [3 × 162] parts of natural alkalinity as $Ca(CHO_3)_2$.

For every 665 parts of alum (3 × 162) parts of alkalinity as $(Ca(HCO_3)_2$ is equivalent to 3 × 100 parts of alkalinity required as $CaCO_3$.

Hence alkalinity required as $CaCO_3$, for the water containing alum of 20 mg/l

$$= \frac{3 \times 100}{666} \times 20 = 9 \text{ mg/l}$$

Hence total alkalinity requirement [10^6 mg per day as $CaCO_3$] matching filter alum shall be

$= 9\text{mg/l} \times 10^6 \times 10$

$= 90 \times 10^6$ mg per day as $CaCO_3$

47. Natural alkalinity available of $CaCO_3$ = 6 mg/l

Hence additional alkalinity required to be added in the form of lime

$= (9 - 6)$

$= 3$ mg/l as $CaCO_3$.

Now 100 parts of $CaCO_3$ 5 produced by 56 parts of CaO.

∴ Quantity of quick lime [CaO] required

$$= \frac{3 \times 56}{100}$$

$= 1.68$ mg/l

Hence yearly consumption of quick lime

$= 1.88 \times 10 \times 365 \times 10^5$

$= \mathbf{6132 \times 10^6}$ **mg per yr. as CaO**

53. Here, $k_D = \dfrac{0.35}{2.3} = 0.15/\text{day}$

Now $Y_t = L[1 - 10^{-k_D t}]$

∴ $150 = L[1 - 10^{-0.15 \times 5}] = 0.822L$

∴ $L = \dfrac{150}{0.822} = 182.4$ mg/l

54.

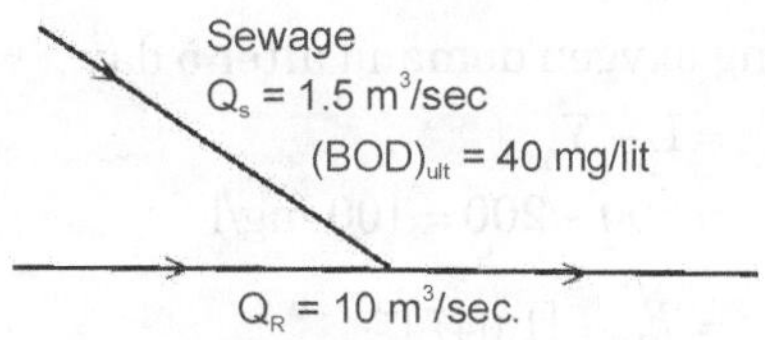

Given : $Q_R = 10\text{m}^3/\text{sec}$

$(BOD)_{ult} = 3$ mg/l.

∴ $k_D = \dfrac{0.32}{2.303} = 0.139/\text{day}$

(*a*) $(Y_5)_{sewage} = 40(1 - 10 - 0.139 \times 5)$

$= 40 \times 0.798 = 31.92$mg/l

$(Y_5)_{river} = 3 \times 0.798 = 2.394$mg/l

$$(Y_5)_{mixture} = \frac{31.92 \times 1.5 + 2.394 \times 10}{11.5}$$

$= 6.24$mg/l

Ultimate BOD of mixture

$$= \frac{6.24}{0.798} = 7.82\text{mg/l}$$

(b) Volume of flow $= \dfrac{11.5}{50} = 0.23$m/sec

Time taken $= \dfrac{2.5 \times 1000}{0.23}$

$= 10870$ sec $= 0.126$days

$Y_t = L\left[1 - 10^{k_D . t}\right]$

$= 7.82[1 - 10 - 0.139 \times 0.126]$

$= 0.3086$ mg/l.

55. S = 0.00023

56. Given : Initial D.O. = 9.5 mg/l

D.O. after 5 days = 3.5 mg/l

∴D.O. consumption in 5 days

= 9.5 – 3.5 = 6 mg/l

(*a*) 5 Day BOD,

$$Y_5 = \frac{\text{D.O. consumed}}{\text{dilution factor}} = \frac{6}{0.03} = 200 \text{ mg/l}$$

(*b*) *Given :* $k = 0.22$/day

$\Rightarrow \quad k_D = 0.434\,K = 0.095$/day

$Y_5 = 200$ mg/l

$t = 5$ days

From $Y_5 = L\left[1-(10)^{-k_D t}\right]$

$200 = L\left[1-(10)^{-0.095} \times 5\right]$

∴ $L = 300$ mg/l

(*c*) Remaining oxygen demand after 5 day

$= L - Y_5$

$= 300 - 200 = 100$ mg/l

(*d*) $k_{D_{25°}} = k_{D_{20°}}\,[1.047]^{20°-20}$

$= 0.095\,[1.047]^5 = 0.12$/day

∴ $Y_5 = 300\left[1-(10)^{-012\times 5}\right] = 225$ mg/l

59. Given, Dilution of sample = 5%

Dissolved oxygen of blank = 8.80 mg/L

Dissolved oxygen of diluted sample = 3.80 mg/L

Dissolved oxygen of original sample = 0.8 mg/L

Now initial dissolved oxygen present in the

diluted sample $= 8.00 \times \frac{95}{100} + 0.8 \times \frac{5}{100}$

= 8.4 mg/lit

Dissolve oxygen after 5 days of incubation

= 3.8 mg/lit

$$5 \text{ day B.O.D} = [8.4 - 3.8] \times \frac{100}{5} = 92 \text{ mg/L.}$$

63. BOD_5 of surface water sample

=200 mg/litre at 20 C

Reaction constant, $K = 0.2\ \text{day}^{-1}$ with base 'e'

Now, $y = L_a[1 - e^{-kt1}]$

where $y = 200$ mg/litre

and L_a = ultimate B.O.D.

∴ $200 = L_a[1 - e^{-0.2\times 5}]$

or $200 = L_a[1 - 0.368]$

or $L_a = \frac{200}{0.632} = 316$ mg/litre.

NUMERICAL TYPE QUESTIONS

1. $Q = 1500\ m^3$/day

Overflow rate = 20 m/day

Depth of tank = 3.1 m

Area of settling tank required (A)

$$= \frac{Q}{\text{O.F.R.}}$$

$$= \frac{1500}{20} = 75\ m^2$$

Since provision of two tanks is there

Therefore area of each tank

$$= \frac{75}{2} = 37.5\ m^2$$

Therefore, diameter of settling tank (D)

$$= \sqrt{\frac{\text{Area}}{\frac{\pi}{4}}} = \frac{37.5\times 4}{\pi}$$

$$= 6.909 \text{ m}$$

2. $$\%\text{removal} = \frac{V_s}{V}\times 100$$

$$V'_s = 0.9\,V_s$$

$$= \frac{0.9\times 40}{86400} \text{ m/s}$$

$$\Rightarrow \frac{1}{18}\times d^2 \times \frac{g}{\mu}(\rho_s - \rho_w) = \frac{0.9\times 40}{86400}$$

$$\Rightarrow \sqrt{\frac{0.9\times 40\times 18\times V.\rho_w}{86480(G_S - 1)\times \rho_w \times g}}$$

$$\Rightarrow d = 22.58\ \mu m$$

3. Total dissolved solids (TDS) = 500 *mg/l*

$$\text{TDS \%} = \frac{500}{1000\times 1000}\times 100 = 0.05$$

4. Capacity = $1\ m^3/s$

$= 3600 \times 24\ m^3$/day

Loading Rate per day

$= 120 \times 60\ m^3$/day

$= 7200\ m^3$/day

No. of filters $= \frac{3600\times 24}{7200} = 12$

Now, if 2 filters are out of service, no. of available filter = 10

So, loading rate becomes $= \frac{3600\times 24}{10\times(6\times 10)} = 144$

5. Overflow rate = 32.5 $m^3/day/m^2$

Length = 32.5 m

Width = 8.0 m

Depth = 2.25 m

Capacity of tank = Overflow rate × Area of tank

= 32.5 × 32.5 × 8

$$\text{Weir loading rate} = \frac{\text{Capacity}}{\text{Length of weir}}$$

$$= \frac{32.5 \times 32.5 \times 8}{75}$$

= 112.667

6. Population = 2,00,000

Years = 25; Compaction Ratio = 4

Waste Generation = 2 kg/person/day

Density of uncompacted solid waste = 100 kg/m^3

Total waste generated in 25 years

= 25 × 365 × 200000 × 2 kg

= 365 × 10^7 kg

Volume of uncompacted,

$$SW = \frac{365 \times 10^7}{100} m^3$$

= 365 × 10^5 m^3

Compacted volume = 365 × 10^5 m^3/4 = 9125000 m^3

$$\text{Also, } \frac{\text{Compacted fill (SW + Cover)}}{\text{Compacted SW}} = 1.5$$

∴ Volume of cover

= (1.5 – 1) (Vol. of compacted SW)

Total volume required

= Vol. of cover + Compacted Vol. of SW

$$= \frac{9125000}{2} + 9125000$$

= 13687500 m^3

= 13.6875 million m^3

7. Settling velocity $V_S = \dfrac{Q}{\text{Surface Area of tank}}$

Surface Area of tank = B × L

∵ First we calculating settling velocity

$$V_S = \frac{(a-1)gd^2}{18\gamma}$$

$$= \frac{(2.65-1) \times 9.81 \times (0.06)^2}{18 \times 1.0105 \times 10^{-2}}$$

= 0.3204 cm/sec

$$= \frac{0.3204}{100} \text{ m/sec}$$

= 3.204 × 10^{-3} m/sec

$$V_s = \frac{Q}{\text{Surface Area of tank}}$$

$$\text{Surface Area of tank} = \frac{Q}{V_s}$$

$$= \frac{0.1}{3.204 \times 10^{-3}} = 31.214 \text{ m}^2$$

8. Flow rate = 35 m^3/min

water tamperature = 15°C

alum dosage = 25 mg/*l*

∵ the alum quantity required for 30 days

$$= 35 \frac{m^3}{\text{min}} \times 25 \times 10^3 \frac{mg}{m^3} \times 30 \times 24 \times 60 \text{ min}$$

$$= 3.78 \times 10^{10} mg$$

$$= \frac{3.78 \times 10^{10}}{10^6} \text{Kg} = 37800 \text{ Kg}$$

10. Raw water discharge rate of plant = 1 m^3/sec

= 1 × 24 × 3600

= 86400 m^3/day

Two filter out of service for routine backwashing.

Number of active filter = 14 – 2

= 12 Nos

Surface area of each filter = 50 m^2

$$\text{So loading rate} = \frac{86400}{12 \times 50}$$

= 144 $m^3/day.m^2$

11. BOD_5 = ? at 20° C

BOD_7 = 150 mg/L at 20° C

K = 0.23/day (base e) at 20° C

∴ K_d = 0.434 K = 0.434 × 0.23 = 0.099

$$Y_T = L\left[1-(10)^{-KD \times T}\right]$$

t = time in days

For seven days BOD

$$Y_7 = L\left[1-(10)^{-0.099 \times 7}\right]$$

$$150 = L\left[1-(10)^{-0.099 \times 7}\right]$$

L = 188.15 mg/L

So, BOD_5 at 20°C temperature

$$Y_5 = L\left[1-(10)^{-KD \times 5}\right]$$

$$Y_5 = 188.15\left[1-(10)^{-KD \times 5}\right]$$

Y_5 = 127.96 Mg/L ≃ 128 mg/L

12. *Given :* MLSS = 5000mg/lit

FLOW = 0.15 m^3/secc

SS of inflow = 400 mg/lit.

Solids settled after 30 min. = 25%

Volume of Aeration tank = 3000m^3

Sludge wastage rate = 120m^3/day with VSS of 15000 mg/lit

Sludge volume index

$$= \frac{V(ml/l)}{X(mg/l)}$$

$$= \frac{V}{X} \text{ ml/mg}$$

$$= \frac{0.25 \times 400}{400} \times 1000$$

$$= 250 \text{ml/gm}$$

Return sludge ratio,

$$\frac{Q_R}{Q} = \frac{x_t}{\left(\frac{10^6}{SVI} - X_t\right)}$$

$$= \frac{5000}{\frac{10^6}{250} - 5000} = 0.14$$

$$Q_C = \frac{X_t V}{Q_w X_R - (Q - Q_w) X_E}$$

$$= \frac{5000 \times 10^{-6} \times 3000 \times 10^3}{120 \times 15000 + (0.15 \times 24 \times 60 \times 60 - 120)400}$$

$$= \frac{15000}{120 \times 15000 \times 10^3 \left(\frac{lit}{day}\right) \times 10^{\pm 6} \left(\frac{kg}{lit}\right) + 12800 \times 10^3 \left(\frac{lit}{day}\right) \times 400 \times 10^{\pm 6} \left(\frac{kg}{lit}\right)}$$

$$= \frac{15000}{120 \times 15 + 12840 \times 0.4} \text{ day}$$

$$= \frac{15000}{1800 + 5136} = 2.16 \text{ days.}$$

13

CHAPTER

Air Pollution

MCQ TYPE QUESTIONS

2016

1. Pre-cursors to photochemical oxidants are

(a) NO_X, VOCs and sunlight

(b) SO_2, CO_2 and sunlight

(c) H_2S, CO and sunlight

(d) SO_2, NH_3 and sunlight

2. It was decided to construct a fabric filter, using bags of 0.45 m diameter and 7.5 m long, for removing industrial stack gas containing particulates. The expected rate of airflow into the filter is 10 m^3/s. If the filtering velocity is 2.0 m/min, the minimum number of bags (rounded to nearest higher integer) required for continuous cleaning operation is

(a) 27 (b) 29

(c) 31 (d) 32

3. The atmospheric layer closest to the earth surface is

(a) the mesosphere (b) the stratosphere

(c) the thermosphere (d) the troposphere

2015

4. SO_2 and CO adversely affect

(a) oxygen carrying capacity of blood and functioning of lungs respectively

(b) functioning of the respiratory system and brain respectively

(c) functioning of the respiratory system and oxygen carrying capacity of blood respectively

(d) functioning of air passages and chest respectively.

2014

5. The two air pollution control devices that are usually used to remove very fine particles from the flue gas are

(a) Cyclone and Venturi Scrubber

(b) Cyclone and Packed Scrubber

(c) Electrostatic Precipitator and Fabric Filter

(d) Settling Chamber and Tray Scrubber

2011

6. Anaerobically treated effluent has MPN of total coliform as 10^6/100 mL. After chlorination, the MPN value declines to 10^2/100 mL. The percent removal (%R) and log removal (log R) of total coliform MPN is

(a) % R = 99.90; log R = 4 (b) %R = 99.90; log R = 2

(c) %R = 99.99; log R = 4 (d) %R = 99.99; log R = 2

7. Consider four common air pollutants found in urban environments, NO, SO_2, Soot and O_3. Among these which one is the secondary air pollutant?

(a) O_3 (b) NO

(c) SO_2 (d) Soot

8. Total suspended particulate matter (TSP) concentration in ambient air is to be measured using a high volume sampler. The filter used for this purpose had an initial dry weight of 9.787 g. The filter was mounted in the sampler and the initial air flow rate through the filter was set at 1.5 m^3/min. Sampling continued for 24 hours. The airflow after 24 hours was measured to be 1.4 m^3/min. The dry weight of the filter paper after 24 hour sampling was 10.283g. Assuming a linear decline in the air flow rate during sampling, what is the 24 hour sampling was 10.283g. Assuming a linear decline in the air flow rate during sampling, what is the 24 hour average TSP concentration in the ambient air?

(a) 59.2 $\mu g/m^3$ (b) 118.6 $\mu g/m^3$

(c) 237.5 $\mu g/m^3$ (d) 574.4 $\mu g/m^3$

9. Chlorine gas (8 mg/L as Cl_2) was added to a drinking water sample. If the free chlorine residual and pH was measured to be 2 mg/L (as Cl_2) and 7.5, respectively, what is the concentration of residual OCl^- ions in the water? Assume that the chlorine gas added to the water is completely converted to HOCl and OCl^-. Atomic Weight of Cl: 35.5

Given: $OCl^- + H^+ \xrightleftharpoons{K} HOCl$, $K = 10^{7.5}$

(a) 1.408×10^{-5} moles /L

(b) 2.817×10^{-3} moles/L

(c) 5.634×10^{-5} moles/L

(d) 1.127×10^{-4} moles/L

2010

10. An air parcel having 40°C temperature moves from ground level to 500 m elevation in dry air following the *adiabatic lapse rate*. The resulting temperature of air parcel at 500 m elevation will be

(*a*) 35°C (*b*) 38°C
(*c*) 41°C (*d*) 44°C

2009

11. Particulate matter (fly ash) carried in effluent gases from the furnaces burning fossil fuels are better removed by

(*a*) Cotton bag house filter
(*b*) Electrostatic precipitator (ESP)
(*c*) Cyclone
(*d*) Wet scrubber

2008

12. Two primary air pollutants are

(*a*) sulphur oxide and ozone
(*b*) nitrogen oxide and peroxyacetylnitrate
(*c*) sulphur oxide and hydrocarbon
(*d*) ozone and peroxyacetynitrate

2007

13. The dispersion of pollutants in atmosphere is maximum when

(*a*) environmental lapse rate is greater than adiabatic lapse rate
(*b*) environmental lapse rate is less than adiabatic lapse rate
(*c*) environmental lapse rate is equal to adiabatic lapse rate
(*d*) maximum mixing depth is equal to zero

14. Two electrostatic precipitators (ESPs) are in series. The fractional efficiencies of the upstream and downstream ESPs for size d_p are 80% and 65%, respectively. What is the overall efficiency of the system for the same d_p ?

(*a*) 100 % (*b*) 93 %
(*c*) 80 % (*d*) 65 %

15. 50 g of CO_2 and 25 g of CH_4 are produced from the decomposition of municipal solid waste (MSW) with a formula weight of 120 g. What is the average per capita green house gas production in a city of 1 million people with a MSW production rate of 500 ton/day?

(*a*) 104 g/day (*b*) 120 g/day
(*c*) 208 g/day (*d*) 313 g/day

2006

16. The mean indoor airborne Chloroform ($CHCl_3$) concentration in a room was determined to be 0.4 μg/m³.

Use the following data:

T = 293 K P = 1 atmosphere

$R = 82.05 \times 10^{-6}$ atm.m³/mol-K

Atomic weights: C = 12, H = 1, Cl = 35.5.

This concentration expressed in parts per billion (volume basis, ppbv) is equal to

(*a*) 1.00 ppbv (*b*) 0.20 ppbv
(*c*) 0.10 ppbv (*d*) 0.08 ppbv

NUMERICAL TYPE QUESTIONS

2015

1. The concentration of sulfur Dioxide (SO_2) in ambient atmosphere was measured as 30 μg/m³. Under the same conditions, the above SO_2 concentration expressed in ppm is __________.

Given : $\frac{P}{(RT)} = 41.6$ mol/m³,

where,

P = Pressure; T = Temperature;

R = Universal gas constant.

Molecular weight of SO_2 = 64.

2014

2. The amount of CO_2 generated (in kg) while completely oxidizing one kg of CH_4 to the end products is __________.

ANSWERS

MCQ Type Questions

1. (*a*) 2. (*b*) 3. (*d*) 4. (*c*) 5. (*c*) 6. (*d*) 7. (*a*) 8. (*c*) 9. (*b*) 10. (*a*)
11. (*b*) 12. (*c*) 13. (*a*) 14. (*b*) 15. (*d*) 16. (*d*)

Numerical Type Questions

1. (0.0113) 2. (2.7 to 2.8)

EXPLANATIONS

MCQ TYPE QUESTIONS

2. $v = \frac{2}{60}$ m/sec

Total surface area of bags required $= \frac{Q}{v}$

$$\frac{10}{\frac{2}{60}} = 300\text{m}^2$$

Surface area of each bag $= \pi dH = \pi \times 0.45 \times 7.5$

Number of bags required

$$= \frac{300}{\pi \times 0.45 \times 7.5} = 28.29 \simeq 29$$

3. Figure

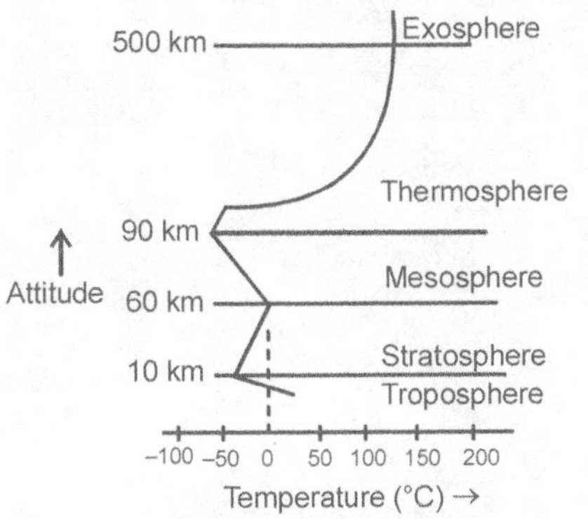

4. SO_2 and CO affects the respiratory system by causing inflammation in respiratory tract.

CO combines with blood hemoglobin and reduces its oxygen carrying capacity.

5. Electrostatic Precipitator and Fabric filter are used to remove very fine particles from flue gas.

10. *Given :* Adiabatic lapse rate = 0.98°C per 100 m

For 500 m, decrease in temperature

$$= 0.98 \times 5 = 4.9° \approx 5°\text{C}$$

Thus temperaturing air parcel at 500m elevation

$$= 40 - 5 = 35°\text{C}$$

12. Primary pollutants are :

SO_2, NO_x, Hydrocarbons.

Note : Ozone is secondary pollutant.

13. The dispersion of pollutants in atmosphere is maximum when environmental lapse rate is greater than adiabatic lapse rate.

14. *Given :*

Upstream efficiency, $\eta_u = 0.8$

Downstream efficiency, $\eta_d = 0.65$

$\therefore$ Overall efficiency ,

$$\eta = \eta_u + (1 - \eta_u) \times \eta_d$$
$$= 0.8 + (1 - 0.8)0.65$$
$$= 0.8 + 0.13 = 0.93 = 93\%$$

15. *Given :* Population = 1 million

Green house gas production is (50 + 25) 75 gm from 120 gm of MSW

MSW production rate = 500 ton/day

Per capita green house gas production

$$= \frac{75}{120} \times \frac{500 \times 1000 \times 10^3}{1,000,000} = 312.5 \text{ gm/day}$$

NUMERICAL TYPE QUESTIONS

1. $$\text{ppm } SO_2 = \mu g\ SO_2/m^3 \times 3.82 \times 10^{-4}$$
$$= 30 \times 3.82 \times 10^{-4} = 0.01146$$

2. $\underset{16g}{CH_4} + 2O_2 \rightarrow \underset{44g}{CO_2} + 2H_2O$

When completely oxidized, 16g CH_4 liberates = 44g CO_2

So, 1 kg CH_4 will liberate $= \frac{44}{16} \times 1 = 2.75$ kg

■■

14 CHAPTER

Noise Pollution

MCQ TYPE QUESTIONS

2016

1. The sound pressure (expressed in μPa) of the faintest sound that a normal healthy individual can hear is
   (*a*) 0.2 (*b*) 2
   (*c*) 20 (*d*) 55

2010

2. According to the Noise Pollution (Regulation and Control) Rules, 2000, of the Ministry of Environment and Forests, India, the day time and night time noise level limits in ambient air for residential areas expressed in *d*B(A) L_{eq} are
   (*a*) 50 and 40 (*b*) 55 and 45
   (*c*) 65 and 55 (*d*) 75 and 70

2009

3. The reference pressure used in the determination of sound pressure level is
   (*a*) 20 μPa (*b*) 20 db
   (*c*) 10 μPa (*d*) 10 db

2006

4. The cumulative noise power distribution curve at a certain location is given below.

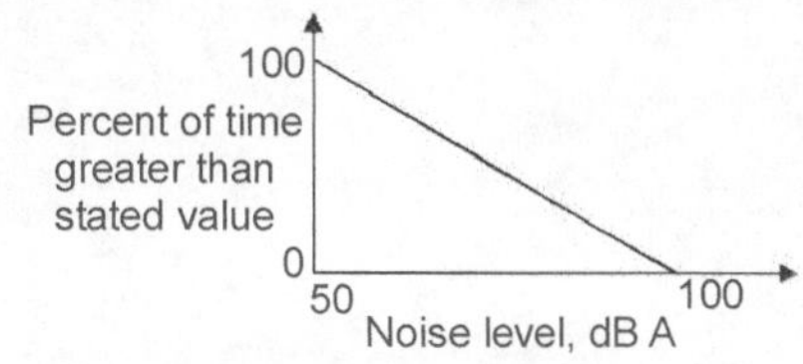

The value of L_{40} is equal to
   (*a*) 90 dBA (*b*) 80 dBA
   (*c*) 70 dBA (*d*) 60 dBA

ANSWERS

MCQ Type Questions

1. (*c*) **2.** (*b*) **3.** (*a*) **4.** (*b*)

EXPLANATIONS

MCQ TYPE QUESTIONS

1. The faintest sound we can hear is 0.00002 Pa and loudest round we can hear is 20 Pa

4.
$$\frac{100-50}{100-0} = \frac{x-50}{100-40}$$
$$\Rightarrow \quad 30 = x - 50$$
$$\Rightarrow \quad x = 80 \text{ dBA.}$$

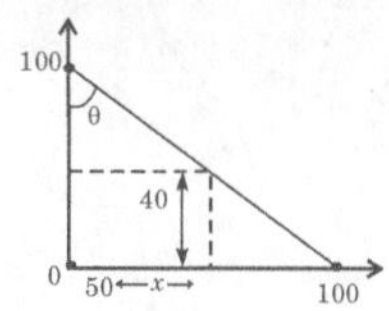

15 CHAPTER Transportation Engineering

MCQ TYPE QUESTIONS

2016

1. During a forensic investigation of pavement failure, an engineer reconstructed the graphs P, Q, R and S, using partial and damaged old reports.

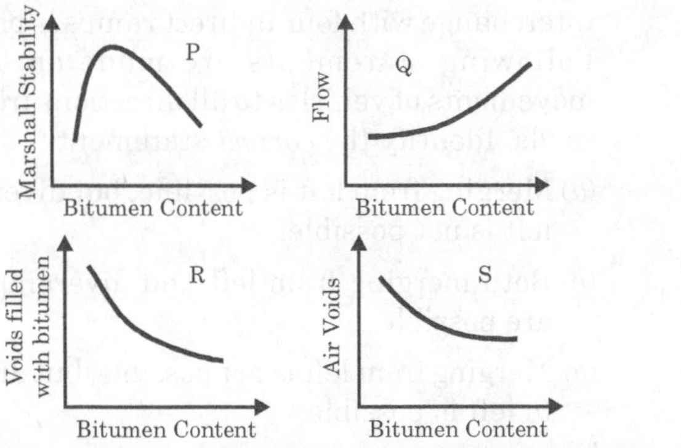

Theoretically plausible correct graphs according to the 'Marshall mixture design output' are

(*a*) P, Q, R
(*b*) P, Q, S
(*c*) Q, R, S
(*d*) R, S, P

2. In the context of the IRC 58-2011 guidelines for rigid pavement design, consider the following pair of statements.

I. Radius of relative stiffness is directly related to modulus of elasticity of concrete and inversely related to Poisson's ratio

II. Radius of relative stiffness is directly related to thickness of slab and modulus of subgrade reaction.

Which one of the following combinations is correct?

(*a*) I : True; II : True
(*b*) I : False; II : False
(*c*) I : True; II : False
(*d*) I : False; II : True

3. If the total number of commercial vehicles per day ranges from 3000 to 6000, the minimum percentage of commercial traffic to be surveyed for axle load is

(*a*) 15 (*b*) 20
(*c*) 25 (*d*) 30

2015

4. Which of the following statements CANNOT be used to describe free flow speed (u_f) of a traffic stream?

(*a*) u_f is the speed when flow is negligible
(*b*) u_f is the speed when density is negligible
(*c*) u_f is affected by geometry and surface conditions of the road
(*d*) u_f is the speed at which flow is maximum and density is optimum

5. A superspeedway in New Delhi has among the highest super-elevation rates of any track on the Indian Grand Prix circuit. The track requires drivers to negotiate turns with a radius of 335 m and 33° banking. Given this information, the coefficient of side friction required in order to allow a vehicle to travel at 320 km/h along the curve is

(*a*) 1.761
(*b*) 0.176
(*c*) 0.253
(*d*) 2.530

6. The following statements are made related to the lengths of turning lanes at signalised intersections

(*i*) 1.5 times the average number of vehicles (by vehicle type) that would store in turning lane per cycle during the peak hour.

(*ii*) 2 times the average number of vehicles (by vehicle type) that would store in turning lane per cycle during the peak hour.

(*iii*) Average number of vehicles (by vehicle type) that would store in the adjacent through lane per cycle during the peak hour.

(*iv*) Average number of vehicles (by vehicle type) that would store in all lanes per cycle during the peak hour.

As per the IRC recommendations, the correct choice for design length of storage lanes is

(*a*) Maximum of (*ii* and *iii*)
(*b*) Maximum of (*i* and *iii*)
(*c*) Average of (*i* and *iii*)
(*d*) Only (*iv*)

7. In Marshall method of mix design, the coarse aggregate, fine aggregate, fines and bitumen having respective values of specific gravity 2.60, 2.70, 2.65 and 1.01, are mixed in the relative proportions (% by weight) of 55.0, 35.8, 3.7 and 5.5 respectively. The theoretical specific gravity of the mix and the effective specific gravity of the aggregates in the mix respectively are

(*a*) 2.42 and 2.63 (*b*) 2.42 and 2.78

(*c*) 2.42 and 2.93 (*d*) 2.64 and 2.78

2014

8. The minimum value of 15 minute peak hour factor on a section of a road is

(*a*) 0.10 (*b*) 0.20

(*c*) 0.25 (*d*) 0.33

9. The following statements are related to temperature stresses developed in concrete pavement slabs with free edges (without any restraint):

P. The temperature stresses will be zero during both day and night times if the pavement slab is considered weightless

Q. The temperature stresses will be compressive at the bottom of the slab during night time if the self-weight of the pavement slab is considered

R. The temperature stresses will be compressive at the bottom of the slab during day time if the self-weight of the pavement slab is considered

The TRUE statement(s) is(are)

(*a*) P only

(*b*) Q only

(*c*) P and Q only

(*d*) P and R only

10. A road is being designed for a speed of 110 km/hr on a horizontal curve with a super elevation of 8%. If the coefficient of side friction is 0.10, the minimum radius of the curve (in m) required for safe vehicular movement is

(*a*) 115.0 (*b*) 152.3

(*c*) 264.3 (*d*) 528.5

11. A student riding a bicycle on a 5 km one-way street takes 40 minutes to reach home. The student stopped for 15 minutes during this ride. 60 vehicles overtook the student (assume the number of vehicles overtaken by the student is zero) during the ride and 45 vehicles while the student stopped. The speed of vehicle stream on that road (in km/hr) is

(*a*) 7.5 (*b*) 12

(*c*) 40 (*d*) 60

12. On a section of a highway the speed-density relationship is linear and is given by $v = [80 - \frac{2}{3}k]$; where v is in km/h and k is in veh/km. The capacity (in veh/h) of this section of the highway would be

(*a*) 1200 (*b*) 2400

(*c*) 4800 (*d*) 9600

2013

13. For two major roads with divided carriageway crossing at right angle, a full clover leaf interchange with four indirect ramps is provided. Following statements are made on turning movements of vehicles to all directions from both roads. Identify the *correct* statement:

(*a*) Merging from left is possible, but diverging to left is not possible.

(*b*) Both merging from left and diverging to left are possible.

(*c*) Merging from left is not possible, but diverging to left is possible.

(*d*) Neither merging from left nor diverging to left is possible.

14. The latitude and departure of a line AB are +78 m and – 45.1 m, respectively. The whole circle bearing of the line AB is

(*a*) 30° (*b*) 150°

(*c*) 210° (*d*) 330°

2012

15. Two major roads with two lanes each are crossing in an urban area to form an un-controlled intersection. The number of conflict points when both roads are one-way is "X" and when both roads are two-way is "Y". The ratio of X to Y is

(*a*) 0.25 (*b*) 0.33

(*c*) 0.50 (*d*) 0.75

16. Two bitumen samples "X" and "Y" have softening points 45°C and 60°C, respectively. Consider the following statements:

I. Viscosity of "X" will be higher than that of "Y" at the same temperature.

II. Penetration value of "X" will be lesser than that of "Y" under standard condition:

The **CORRECT** option evaluating the above statements is

(*a*) Both I and II are TRUE

(*b*) I is FALSE and II is TRUE

(*c*) Both I and II are FALSE

(*d*) I is TRUE and II is FALSE

17. Road roughness is measured using

(*a*) Benkelman beam

(*b*) Bump integrator

(*c*) Dynamic cone penetrometer

(*d*) Falling weight deflectometer

18. A pavement designer has arrived at a design traffic of 100 million standard axles for a newly developing national highway as per IRC:37 guidelines using the following data: design life = 15 years, commercial vehicle count before pavement construction = 4500 vehicles/day, annual traffic growth rate = 8%. The vehicle damage factor used in the calculation was

(*a*) 1.53

(*b*) 2.24

(*c*) 3.66

(*d*) 4.14

19. The following data are related to a horizontal curved portion of a two-lane highway: length of curve = 200 m, radius of curve = 300 m and width of pavement = 7.5 m. In order to provide a stopping sight distance (SSD) of 80 m, the set back distance (in m) required from the centre line of the inner lane of the pavement is

(*a*) 2.54 (*b*) 4.55

(*c*) 7.10 (*d*) 7.96

20. A two-lane urban road with one-way traffic has a maximum capacity of 1800 vehicles/hour. Under the jam condition, the average length occupied by the vehicles is 5.0 m. The speed versus density relationship is linear. For a traffic volume of 1000 vehicles/hour, the density (in vehicles/km) is

(*a*) 52

(*b*) 58

(*d*) 67

(*d*) 75

2011

21. The probability that k number of vehicles arrive (i.e. cross a predefined line) in time t is given as $(\lambda t)^k e^{-\lambda t}/k!$, where λ is the average vehicle arrival rate. What is the probability that the time headway is greater than or equal to time t_1?

(*a*) $\lambda e^{\lambda t_1}$ (*b*) λe^{-t_1}

(*c*) $e^{\lambda t_1}$ (*d*) $e^{-\lambda t_1}$

22. A vehicle negotiates a transition curve with uniform speed v. If the radius of the horizontal curve and the allowable jerk are R and J, respectively, the minimum length of the transition curve is

(*a*) $R^3/(vJ)$ (*b*) $J^3/(Rv)$

(*c*) v^2R/J (*d*) $v^3/(RJ)$

23. In Marshall testing of bituminous mixes, as the bitumen content increases the *flow* value

(*a*) remains constant

(*b*) decreases first and then increases

(*c*) increases monotonically

(*d*) increases first and then decreases

24. If the *jam density* is given as k_f and the *free flow speed* is given as μ_f, the maximum flow for a linear traffic speed-density model is given by which of the following options?

(*a*) $\frac{1}{4} k_f \times u_f$ (*b*) $\frac{1}{3} k_f \times u_f$

(*c*) $\frac{3}{5} k_f \times u_f$ (*d*) $\frac{2}{3} k_f \times u_f$

25. If v is the initial speed of a vehicle, g is the gravitational acceleration, G is the upward longitudinal slope of the road and f_r is the coefficient of rolling friction during braking, the braking distance (measured horizontally) for the vehicle to stop is

(*a*) $\frac{v^2}{g(G+f_r)}$ (*b*) $\frac{v^2}{2g(G+f_r)}$

(*c*) $\frac{vg}{(G+f_r)}$ (*d*) $\frac{vf_r}{(G+g)}$

26. The cumulative arrival and departure curve of one cycle of an approach lane of a signalized intersection is shown in the adjoining figure. The cycle time is 50s and the effective red time is 30s and the effective green time is 20s. What is the average delay?

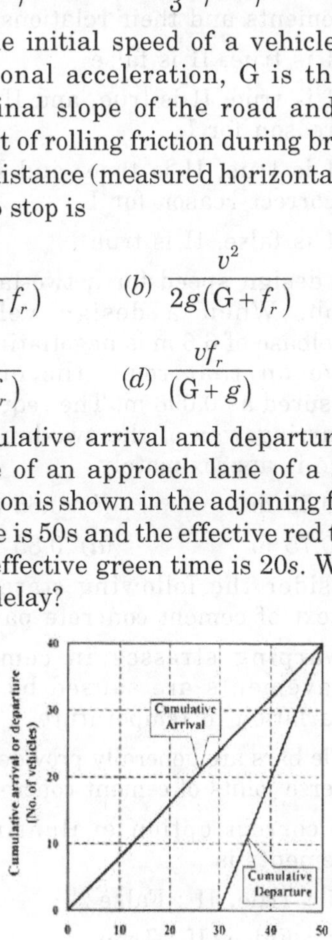

(*a*) 15 *s* (*b*) 25 *s*

(*c*) 35 *s* (*d*) 45 *s*

2010

27. Aggregate impact value indicates which of the following property of aggregates ?

(*a*) Durability (*b*) Toughness

(*c*) Hardness (*d*) Strength

28. As per IRC : 67 : 2001, a traffic sign indicating the Speed Limit on a road should be of

(*a*) Circular Shape with white Background and Red Border

(*b*) Triangular Shape with White Background and Red Border

(*c*) Triangular Shape with Red Background and White Border

(*d*) Circular Shape with Red Background and White Border

29. Consider the following statements in the context of geometric design of roads.

I. A simple parabolic curve is an acceptable shape for summit curves

II. Comfort to passengers is an important consideration in the design of summit curves

The correct option evaluating the above statements and their relationship is

(*a*) I is true, II is false

(*b*) I is true, II is true, and II is the correct reason for I

(*c*) I is true, II is true, and II is NOT the correct reason for I

(*d*) I is false, II is true

30. The design speed for a two-lane road is 80 kmph. When a design vehicle with a wheelbase of 6.6 m is negotiating a borizontal curve on that road, the off-tracking is measured as 0.096 m. The required widening of carriageway of the two-lane road on the curve is approximately

(*a*) 0.55 m (*b*) 0.65 m

(*c*) 0.75 m (*d*) 0.85 m

31. Consider the following statements in the context of cement concrete pavements

I. Warping stresses in cement concrete pavements are saused by the seasonal variation in temperature

II. Tie bars are generally provided across transverse joints of cement concrete pavements

The correct option evaluating the above statements is

(*a*) I : True, II : False

(*b*) I : False, II : True

(*c*) I : True, II : True

(*d*) I : False, II : False

2009

32. The value of lateral friction or side friction used in the design of horizontal curve as per Indian Roads Congress guidelines is

(*a*) 0.40 (*b*) 0.35

(*c*) 0.24 (*d*) 0.15

33. During a CBR test, the load sustained by a remolded soil specimen at 5.0 mm penetration is 50 kg. The CBR value of the soil will be

(*a*) 10.0 % (*b*) 5.0 %

(*c*) 3.6 % (*d*) 2.4 %

34. On a specific highway, the speed-density relationship follows the Greenberg's model

$$v = v_f \ln (k_j / k)$$

where, v_f and k_j are free flow speed and jam density respectively.

When the highway is operating at capacity, the density obtained as per this model is

(*a*) $e.k_j$ (*b*) k_j

(*c*) $k_j/2$ (*d*) k_j/e

35. A three-phase traffic signal at an intersection is designed for flows shown in the figure below. There are six groups of flows identified by the numbers 1 throug 6. Among these 1, 3, 4, and 6 are through flows and, 2 and 5 are right turning. Which phasing scheme is **not feasible** ?

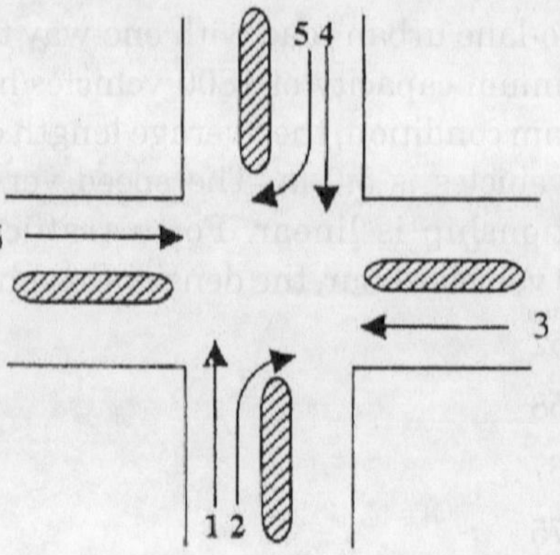

Combination choice	Phase I	Phase II	Phase III
P	1, 4	2, 5	3, 6
Q	1, 2	4, 5	3, 6
R	1, 5	1, 3	4, 6
S	1, 4	2, 6	3, 5

(*a*) P (*b*) Q

(*c*) R (*d*) S

2008

36. The specific gravity of paving bitumen as per IS:73-1992 lies between

(*a*) 1.10 and 1.06 (*b*) 1.06 and 1.02

(*c*) 1.02 and 0.97 (*d*) 0.97 and 0.92

37. A combined value of flakiness and elongation index is to be determined for a sample of aggregates. The sequence in which the two tests are conducted is

(*a*) elongation index test followed by flakiness index test on the whole sample

(*b*) flakiness index test followed by elongation index test on the whole sample

(*c*) flakiness index test followed by elongation index test on the non-flaky aggregates

(*d*) elongation index test followed by flakiness index test on non-elongated aggregates

38. The capacities of "One-way 1.5 m wide sidewalk(persons per hour)" and "One-way 2-lane urban road (PCU per hour, with no frontage access, no standing vehicles and very little cross traffic)" are respectively

(*a*) 1200 and 2400 (*b*) 1800 and 2000

(*c*) 1200 and 1500 (*d*) 2000 and 1200

39. A roundabout is provided with an average entry width of 8.4 m, width of weaving section as 14 m, and length of the weaving section between channelizing islands as 35 m. The crossing traffic and total traffic on the weaving section are 1000 and 2000 PCU per hour respectively. The nearest rounded capacity of the roundabout (in PCU per hour) is

(*a*) 3300 (*b*) 3700

(*c*) 4500 (*d*) 5200

40. The shape of the STOP sign according to IRC :67-2001 is

(*a*) circular (*b*) triangular

(*c*) octagonal (*d*) rectangular

41. Design parameters for a signalized intersection are shown in the figure below. The green time calculated for major and minor roads are 34 and 18s, respectively.

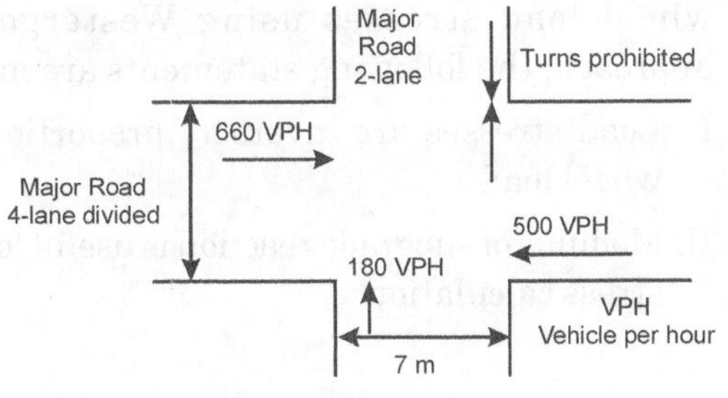

The critical lane volume on the major road changes to 440 vehicles per hour per lane and the critical lane volume on the minor road remains unchanged. The green time will

(*a*) increase for the major road and remain same for the minor road

(*b*) increase for the major road and decrease for the minor road

(*c*) decrease for both the roads

(*d*) remain unchanged for both the roads

42. It is proposed to widen and strengthen an existing 2-lane NH section as a divided highway. The existing traffic in one direction is 2500 commercial vehicles (CV) per day. The construction will take 1 year. The design CBR of soil subgrade is found to be 5 percent.

Given: traffic growth rate for CV = 8 percent,
vehicle damage factor
= 3.5 (standard axles per CV),
design life = 10 years and
traffic distribution factor = 0.75.

The cumulative standard axles (msa) computed are

(*a*) 35 (*b*) 37

(*c*) 65 (*d*) 70

43. A linear relationship is observed between speed and density on a certain section of a highway. The free flow speed is observed to be 80 km per hour and the jam density is estimated as 100 vehicles per km length. Based on the above relationship, maximum flow expected on this section and the speed at the maximum flow will respectively be

(*a*) 8000 vehicles per hour and 80 km per hour

(*b*) 8000 vehicles per hour and 25 km per hour

(*c*) 2000 vehicles per hour and 80 km per hour

(*d*) 2000 vehicles per hour and 40 km per hour

2007

44. The consistency and flow resistance of bitumen can be determined from which of the following ?

(*a*) Ductitility test (*b*) Penetration test

(*c*) Softening point test (*d*) Viscosity test

45. If a two-lane national highway and a two-lane state highway intersect at right angles, the number of potential conflict points at the intersection, assuming that both the roads are two-way is

(*a*) 11 (*b*) 17

(*c*) 24 (*d*) 32

46. In signal design as per Indian Roads Congress specifications, if sum of the ratios of normal flows to saturation flow of two directional traffic flow is 0.50 and total lost time per cycle is 10 seconds, then optimum cycle length in seconds is

(*a*) 100 (*b*) 80
(*c*) 60 (*d*) 40

47. If standard deviation of the spot speed of vehicles in a highway is 8.8 kmph and mean speed of the vehicles is 33 kmph, then coefficient of variation in speed is

(*a*) 0.1517 (*b*) 0.1867
(*c*) 0.2666 (*d*) 0.3646

48. The extra widening required for a two-lane national highway at a horizontal curve of 300 m radius, considering a wheel base of 8 m and a design speed of 100 kmph is

(*a*) 0.42 m (*b*) 0.62 m
(*c*) 0.82 m (*d*) 0.92 m

49. While designing a hill road with a ruling gradient of 6%, if a sharp horizontal curve of 50 m radius is encountered, the compensated gradient at the curve as per the Indian Roads Congress specifications should be

(*a*) 4.4 % (*b*) 4.75%
(*c*) 5.0 % (*d*) 5.25 %

50. The design speed on a road is 60 kmph. Assuming the driver reaction time of 2.5 seconds and coefficient of friction of pavement surface as 0.35, the required stopping distance for two-way traffic on a single lane road is

(*a*) 82.1 m (*b*) 102.4 m
(*c*) 164.2 m (*d*) 186.4 m

51. The width of the expansion joint is 20 mm in a cement concrete pavement. The laying temperature is 20°C and the maximum slab temperature in summer is 60°C. The coefficient of thermal expansion of concrete is 10×10^{-6} mm/mm/°C and the joint filler compresses up to 50% of the thickness. The spacing between expansion joints should be

(*a*) 20 m
(*b*) 25 m
(*c*) 30 m
(*d*) 40 m

52. The following data pertains to the number of commercial vehicles per day for the design of a flexible pavement for a national highway as per IRC:37-1984.

Type of commercialNumber of vehicles per day Vehicle Damage Factor
vehicleconsidering the number of lanes
Two axle trucks 2000 5
Tandem axle trucks 200 6

Assuming a traffic growth factor of 7.5 % per annum for both the types of vehicles, the cumulative number of standard axle load repetitions (in million) for a design life of ten years is

(*a*) 44.6 (*b*) 57.8
(*c*) 62.4 (*d*) 78.7

53. Match the following tests on aggregate and its properties

TEST	PROPERTY
P. Crushing test	1. Hardness
Q. Los Angeles abrasion test	2. Weathering
R. Soundness test	3. Shape
S. Angularity test	4. Strength

Codes :

	P	Q	R	S
(*a*)	2	1	4	3
(*b*)	4	2	3	1
(*c*)	3	2	1	4
(*d*)	4	1	2	2

2006

54. If aggregate size of 50-40 mm is to be tested for finding out the portion of elongated aggregates using length gauge, then slot length of the gauge should be

(*a*) 81 mm (*b*) 45 mm
(*c*) 53 mm (*d*) 90 mm

55. Name the traffic survey data which is plotted by means of "*Desire lines*".

(*a*) Accident
(*b*) Classified volume
(*c*) Origin and Destination
(*d*) Speed and Delay

56. In case of governing equations for calculating wheel load stresses using Westergaard's approach, the following statements are made.

I. Load stresses are inversely proportional to wheel load

II. Modulus of subgrade reaction is useful for load stress calculation

Of these statements

(a) Both statements are **True**

(b) I is **True** and II is **False**

(c) Both statements are **False**

(d) I is **False** and II is **True**

57. A subgrade soil sample was tested using standard CBR apparatus and the observations are given below.

Load, kg	Penetration, mm
60.5	2.5
80.5	5.0

Assuming that the load-penetration curve is convex throughout, CBR value (%) of the sample is

(a) 6.5 (b) 5.5

(c) 4.4 (d) 3.9

58. A vehicle moving at 60 kmph on an ascending gradient of a highway has to come to stop position to avoid collision with a stationary object. The ratio of lag to brake distance is 6 : 5. Considering total reaction time of the driver as 2.5 seconds and coefficient of longitudinal friction as 0.36, the value of ascending gradient (%) is

(a) 3.3 (b) 4.8

(c) 5.3 (d) 6.8

59. For designing a 2-phase fixed type signal at an intersection having North-South and East -West road where only straight ahead traffic is permitted, the following data is available.

Parameter	North	South	East	West
Design Hour Flow (PCU/hr)	1000	700	900	550
Saturation Flow (PCU/hr)	2500	2500	3000	3000

Total time lost per cycle is 12 seconds. The cycle length (seconds) as per Webster's approach is

(a) 67 (b) 77

(c) 87 (d) 91

60. On an urban road, the free mean speed was measured as 70 kmph and the average spacing between the vehicles under jam condition as 7.0 m. The speed-flow -density equation is given by

$$U = U_{sf}\left[1 - \frac{k}{k_j}\right] \quad \text{and} \quad q = Uk$$

where U = space-mean speed (kmph)

U_{sf} = free mean speed (kmph)

k = density (veh/km)

k_j = jam density (veh/km)

q = flow (veh/hr)

The maximum flow (veh/hr) per lane for this condition is equal to

(a) 2000

(b) 2500

(c) 3000

(d) None of these

61. At a horizontal curve portion of a 4 lane undivided carriageway, a transition curve is to be introduced to attain required superelevation. The design speed is 60 kmph and radius of the curve is 245 m. Assume length of wheel base of a longest vehicle as 6 m, superelevation rate as 5% and rate of introduction of this superelevation as 1 in 150. The length of the transition curve (m) required, if the pavement is rotated about inner edge is

(a) 81.4 (b) 85.0

(c) 91.5 (d) 11.25

62. Using IRC: 37 - 1984 "*Guidelines for the Design of Flexible Pavements*" and the following data, choose the total thickness of the pavement.

Number of commercial vehicles when construction is completed = 2723 veh/day

Annual growth rate of the traffic = 5.0%

Design life of the pavement = 10 years

Vehicle damage factor = 2.4

CBR value of the subgrade soil = 5%

Data for 5% CBR value

Number of Standard Axles, msa	Total Thickness, mm
20	620
25	640
30	670
40	700

(a) 620 mm

(b) 640 mm

(c) 670 mm

(d) 700 mm

2005

63. Pradhan Mantri Gram Sadak Yojna (PMGSY), launched in the year 2000, aims to provide rural connectivity with all-weather roads. It is proposed to connect the habitations in plain areas of population more than 500 persons by the year

(*a*) 2005 (*b*) 2007
(*c*) 2010 (*d*) 2012

64. Group I contains some properties of Bitumen. Group II gives a list of Laboratory Tests conducted on Bitumen to determine the properties. Match the property with the corresponding test

Group I	Group II
P. Resistance to flow	**1.** Ductility test
Q. Ability to deform under load	**2.** Penetration test
R. Safety	**3.** Flash and Fire point test

Codes :

	P	Q	R
(*a*)	2	1	3
(*b*)	2	3	1
(*c*)	1	2	3
(*d*)	3	1	2

65. The length of Summit Curve on a two lane two way highway depends upon

(*a*) allowable rate of change of centrifugal acceleration
(*b*) coefficient of lateral friction
(*c*) required Stopping Sight Distance
(*d*) required Overtaking Sight Distance

66. Bituminous concrete is a mix comprising of

(*a*) fine aggregate, filler and bitumen
(*b*) fine aggregate and bitumen
(*c*) coarse aggregate, fine aggregate, filler and bitumen
(*d*) coarse aggregate, filler and bitumen

67. For a 25 cm thick cement concrete pavement, analysis of stresses gives the following values

Wheel load stress due to corner loading
= 30 kg/cm^2

Wheel load stress due to edge loading
= 32 kg/cm^2

Warping stress at corner region during summer
= 9 kg/cm^2

Warping stress at corner region during winter
= 7 kg/cm^2

Warping stress at edge region during summer
= 8 kg/cm^2

Warping stress at edge region during winter
= 6 kg/cm^2

Frictional stress during summer = 5 kg/cm^2

Frictional stress during winter = 4 kg/cm^2

Most critical stress value for this pavement is

(*a*) 40 kg/cm^2
(*b*) 42 kg/cm^2
(*c*) 44 kg/cm^2
(*d*) 45 kg/cm^2

68. A transport company operates a scheduled daily truck service between city P and city Q. One-way journey time between these two cities is 85 hours. A minimum layover time of 5 hours is to be provided at each city. How many trucks are required to provide this service?

(*a*) 4 (*b*) 6
(*c*) 7 (*d*) 8

69. A single lane unidirectional highway has a design speed of 65 kmph. The perception-brake-reaction time of drivers is 2.5 seconds and the average length of vehicles is 5 m. The coefficient of longitudinal friction of the pavement is 0.4. The capacity of this road in terms of '*vehicles per hour per lane* is

(*a*) 1440 (*b*) 750
(*c*) 710 (*d*) 680

70. The following observations were made of an axle-load survey on a road

Axle Load (kN)	Repetitions per day
35 - 45	800
75 - 85	400

The standard axle-load is 80 kN. Equivalent daily number of repetitions for the standard axle-load are

(*a*) 450 (*b*) 480
(*c*) 800 (*d*) 1200

71. A road is having a horizontal curve of 400 m radius on which a super-elevation of 0.07 is provided. The coefficient of lateral friction mobilized on the curve when a vehicle is travelling at 100 kmph is

(*a*) 0.07 (*b*) 0.13
(*c*) 0.15 (*d*) 0.4

2004

72. The Star and Grid pattern of road network was adopted in

(a) Nagpur Road Plan (b) Lucknow Road Plan
(c) Bombay Road Plan (d) Delhi Road Plan

73. The road geometrics in India are designed for the

(a) 98th highest hourly traffic volume
(b) 85th highest hourly traffic volume
(c) 50th highest hourly traffic volume
(d) 30th highest hourly traffic volume

74. In the context of flexible pavement design, the ratio of contact pressure to tyre pressure is called Rigidity Factor. This factor is less than unity when the tyre pressure is

(a) less than 0.56 N/mm^2
(b) equal to 0.56 N/mm^2
(c) equal to 0.7 N/mm^2
(d) more than 0.7 N/mm^2

75. For a road with camber of 3% and design speed of 80 km/hr, the minimum radius of the curve, beyond which NO superelevation is needed, is

(a) 1680 m (b) 948 m
(c) 406 m (d) 280 m

76. Three new roads P, Q, and R are planned in a district. The data for these roads are given in the table below.

Road	Length (km)	Number of villages with population		
		Less than 200	2000-5000	More than 5000
P	20	8	6	1
Q	28	19	8	4
R	12	7	5	2

Based on the principle of maximum utility, the order of priority for these three roads should be

(a) P, Q, R
(b) Q, R, P
(c) R, P, Q
(d) R, Q, P

77. A Marshall specimen is prepared for bituminous concrete with a bitumen content of 5 percent by weigh of total mix. The theoretical and the measured unit weights of the mix are 2.442 g/cm^3 and 2.345 g/cm^3, respectively. The bitumen has a specific gravity of 1.02. The percent voids in mineral aggregate filled with bitumen (VFB) are

(a) 34.55 (b) 35.9
(c) 73.55 (d) 74.3

78. The data given below pertain to the design of a flexible pavement:

Initial traffic = 1213 cvpd
Traffic growth rate = 8 percent per annum
Design life = 12 years
Vehicle damage factor = 2.5
Distribution factor = 1.0

The design traffic in terms of million standard axles (msa) to be catered would be

(a) 0.06 msa (b) 8.40 msa
(c) 21.00 msa (d) 32.26 msa

79. The co-efficient of friction in the longitudinal direction of a highway is estimated as 0.396. The braking distance for a car moving at a speed of 65 km/hr is

(a) 87 m (b) 45 m
(c) 42 m (d) 40 m

2003

80. Temperature stresses in concrete pavements may cause the slab to crack. If a slab cools uniformly, then the crack will develop at the following locations of the slab

(a) at centre (b) near edges
(c) at corners (d) both (b) and (c)

81. The speed and delay studies on a defined section of highway are conducted by

(a) radar gun (b) traffic counters
(c) moving car method (d) enoscope

82. The design speed for a National Highway is 100 kmph. If the maximum permissible superelevation is 0.10 and the coefficient of lateral friction is 0.15, the ruling minimum radius of horizontal curve on the highway should be

(a) 260 m (b) 315 m
(c) 380 m (d) 410 m

83. A traffic stream in a particular direction of a two lane road is moving with a constant speed of 50 kmph, with an average headway of 2.52 seconds. The longitudinal distance between two consecutive vehicles is

(a) 30 m (b) 35 m
(c) 38 m (d) 42 m

84. In the Marshall method of mix design, the coarse aggregates, fine aggregates, filler and bitumen, having respective specific gravities of 2.62, 2.72, 2.70 and 1.02, are mixed in the ratio of 55, 34.6, 4.8 and 5.6 percent, respectively. The theoretical specific gravity of the mix would be

(a) 2.36 (b) 2.40
(c) 2.44 (d) 2.50

85. The plate load test conducted with a 75 cm diameter plate on soil subgrade yielded a deflection of 2.5 mm under a stress of 800 N/cm^2. The modulus of elasticity of the subgrade soil, in kN/cm^2, is

(*a*) 141.6

(*b*) 154.6

(*c*) 160.0

(*d*) 185.4

86. Column I below gives a list of physical properties of aggregates which should be determined to judge their suitability in road construction. Column II gives a list of laboratory tests which are conducted to determine these properties.

Column I	Column II
P. Hardness	**1.** Water adsorption
Q. Porosity	**2.** Impact test
R. Toughness	**3.** Soundness test
S. Durability	**4.** Abrasion test

Which of the following matches is correct ?

Codes :

	P	Q	R	S
(*a*)	1	2	3	4
(*b*)	4	1	2	3
(*c*)	3	4	1	2
(*d*)	2	3	4	1

2002

87. Bitumen is derived from

(*a*) destructive distillation of coal tar

(*b*) destructive distillation of petroleum

(*c*) fractional distillation of petroleum

(*d*) naturally occurring ores

88. Dowel bars in concrete pavement are placed

(*a*) along the direction of traffic

(*b*) perpendicular to the direction of traffic

(*c*) along 45° to the direction of traffic

(*d*) can be placed along any direction

89. Stopping sight distance and frictional co-efficients are

(*a*) directly proportional to each other

(*b*) inversely proportional to each other

(*c*) unrelated

(*d*) either directly or inversely proportional to each other depending on the nature of pavement

90. A car is moving at a speed of 72 km/hr on a road having 2% upward gradient. The driver applies brakes when he sees an obstruction. If his reaction time is 1.5 seconds, assuming that the co-efficient of friction between pavement and tyre as 0.15, what is the distance traversed before the car finally stops?

(*a*) 24 m (*b*) 150 m

(*c*) 1056 m (*d*) 324 m

91. The average daily traffic on a stretch of road is 300 commercial vehicles per lane per day. Design traffic repetitions for 10 years when vehicle damage factor is 2.5 and traffic growth rate is 7%, is

(*a*) 3.8 msa (*b*) 23.5 msa

(*c*) 45.4 msa (*d*) 16 msa

92. In a concrete pavement

A. temperature stress is tensile at bottom during day time

B. load stress is compressive at bottom

(*a*) Both the statements are correct

(*b*) Statement (A) is correct and (B) is wrong

(*c*) Statement (B) is wrong and (A) is correct

(*d*) Both statements are wrong

93. A vehicle is maneuvering a horizontal curve of radius R with superelevation Q°. Derive the expression for maximum speed beyond which it would overturn outward if coefficient of friction between the tyre and the pavement is'f'. Find the value of this speed if R = 250 m, θ = 5° f = 0.15, h = 1 m and b= 0.75 m (Refer to the following figure).

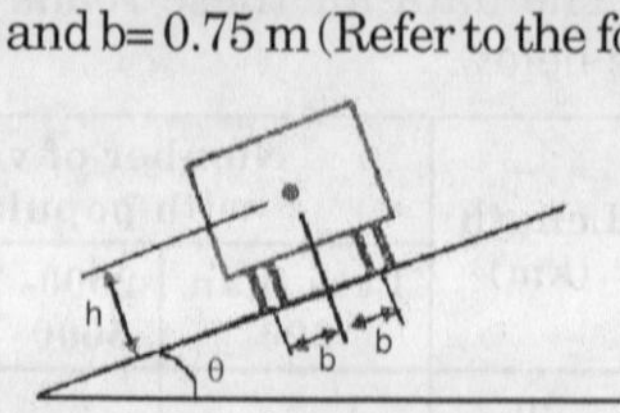

2001

94. The design value of lateral friction coefficient on highway is

(*a*) 1.5 (*b*) 0.50

(*c*) 0.35 (*d*) 0.15

95. Camber on highway pavement is provided to take care of

(*a*) Centrifugal force

(*b*) Drainage

(*c*) Sight distance

(*d*) Off-tracking

96. The minimum value of CBR(%) required for granular sub-base as per Ministry of Surface Transport (MOST) specification is

(*a*) 5 (*b*) 10

(*c*) 15 (*d*) 20

97. A town has an existing horizontal flow sedimentation tank with an overflow rate of 17m³/day/m², and it is desirable to remove particles that have settling velocity of 0.1mm/sec. Assuming the tank is an ideal sedimentation tank, the percentage of particles removal is approximately equal to

(*a*) 30% (*b*) 50%

(*c*) 70% (*d*) 90%

98. A valley curve has a descending gradient of 1 in 40 followed by an ascending gradient of 1 in 50. The length of the valley curve required for a design speed of 80 km/hour for comfort condition is

(*a*) 199 m (*b*) 116 m

(*c*) 58 m (*d*) 37 m

99. The radius of relatives stiffness for a 20 cm thick slab with E= 3×10^5 kg/cm² and Poisson's ratio = 0.15, resting on a subgrade having modulus of 5kg/cm³ is

(*a*) 10 cm

(*b*) 80 cm

(*c*) 120 cm

(*d*) 320 cm

2000

100. The ruling minimum radius of horizontal curve of a national highway in plain terrain for a fuling design speed of 100 km/hour with e = 0.07 and f = 0.15 is close to

(*a*) 250 m

(*b*) 360 m

(*c*) 36 m

(*d*) 300 m

101. Design rate of super eleZvation for horizontal highway curve of radius 450 m for a mixed traffic condition, having a speed of 125 km/hour is

(*a*) 1.0 (*b*) 0.05

(*c*) 0.07 (*d*) 0.154

1999

102. The total length (in km) of the existing National Highways in India is in the range of

(*a*) 15,000 to 25,000 (*b*) 25,000 to 35,000

(*c*) 35,000 to 45,000 (*d*) 45,000 to 55,000

103. The relationship between the length (l) and radius (r) of an ideal transition curve is given by

(*a*) $l \propto r$ (*b*) $l \propto r^2$

(*c*) $l \propto \frac{1}{r}$ (*d*) $l \propto \frac{1}{r^2}$

104. Rapid curing cutback bitumen is produced by lending bitumen with

(*a*) Kerosene

(*b*) Benzene

(*c*) Diesel

(*d*) Petrol

105. A parabolic curve is used to connect a 4% upgrade with 2% downgrade as shown in the figure. The highest point on the summit is at a distance of (measured horizontally from the first tangent point-FTP)

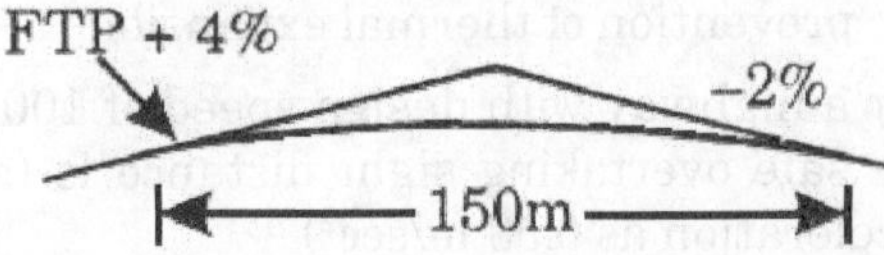

(*a*) 50 m (*b*) 60 m

(*c*) 75 m (*d*) 100 m

106. A two lane single carriage-way is to be designed for a design life period of 15 years. Total two - way traffic intensity in the year of completion of construction is expected to be 2000 commercial vehicles per day. Vehicle damage factor = 3.0, Lane distribution factor = 0.75. Assuming an annual rate of traffic growth as 7.5%, the design traffic expressed as cumulative number of standard axles, is

(*a*) 42.9×16^6

(*b*) 22.6×16^6

(*c*) 10.1×16^6

(*d*) 5.3×16^6

1998

107. At highway stretches where the required overtaking sight distance cannot be provided, it is necessary to incorporate in such sections the following

(*a*) at least twice the stopping sight distance

(*b*) half of the required overtaking sight distance

(*c*) one third of the required overtaking sight distance

(*d*) three times the stopping sight distance

108. The modulus of subgrade reaction is obtained from the plate bearing test in the form of load-deformation curve. The pressure corresponding to the following settlement value should be used for computing modulus of subgrade reaction

(*a*) 0.375 cm

(*b*) 0.175 cm

(*c*) 0.125 cm

(*d*) 0.250 cm

109. In the plate bearing test, if the load applied is in the form of an inflated type of wheel, then this mechanism corresponds to

(*a*) rigidplate
(*b*) flexible plate
(*c*) semi-rigid plate
(*d*) semi-elastic plate

110. Base course is used in rigid pavements for

(*a*) prevention of subgrade settlement
(*b*) prevention of slab cracking
(*c*) prevention of pumping
(*d*) prevention of thermal expansion

111. For a highway with design speed of 100 kmph, the safe overtaking sight distance is (assume acceleration as 0.53 m/sec^2)

(*a*) 300 m (*b*) 750 m
(*c*) 320 m (*d*) 470 m

112. What is the equivalent single wheel load of a dual wheel assembley carrying 20,440 N each for pavement thickness of 20 cm? Centre to centre spacing of tyres is 27 cm and the distance between the walls of tyres is 11 cm.

(*a*) 27600N
(*b*) 32300N
(*c*) 40880N
(*d*) 30190N

113. Plate bearing test with 20 cm diameter plate on soil subgrade yielded a pressure of 1.25×10^5N/m^2 at 0.5 cm deflection. What is the elastic modulus of subgrade?

(*a*) 56.18×10^5N/m^2 (*b*) 22.10×10^5N/m^2
(*c*) 44.25×10^5N/m^2 (*d*) 50.19×10^5N/m^2

1997

114. The value of the camber recommended for cement concrete roads in areas of heavy rainfall is

(*a*) 1 in 25 (*b*) 1 in 33
(*c*) 1 in 40 (*d*) 1 in 50

115. The reaction time for calculation of stopping distance may be assumed as

(*a*) 5 secs (*b*) 2.5 secs
(*c*) 0.5 secs (*d*) 10.0 secs

116. The superelevation needed for a vehicle travelling at a speed of 60 kmph on a curve of radius 128 m on a surface with a coefficient of friction 0.15 is

(*a*) 0.71 (*b*) 0.15
(*c*) 0.22 (*d*) 0

117. If an ascending gradient of 1 in 50 meets another ascending gradient of 1 in 30 then the deviation angle is

(*a*) $\frac{1}{50}$ (*b*) $\frac{1}{75}$
(*c*) $\frac{1}{30}$ (*d*) $\frac{8}{150}$

118. The penetration test for bitumen is conducted at a temperature of

(*a*) 60°C (*b*) 37°C
(*c*) 25°C (*d*) 50°C

119. The group index of a soil subgrade is 7. The subgrade soil is rated as

(*a*) poor (*b*) very poor
(*c*) good (*d*) fair

120. The total thickness of pavement by CBR method depends on the CBR value of

(*a*) base course (*b*) surface course
(*c*) subgrade (*d*) all layers

121. The width of expansion joint gap is 2.5 cm in a cement concrete pavement. The spacing between expansion joint for a maximum rise in temperature of 25°C is, (assuming coefficient of thermal expansion of concrete as 10×10^{-6} per degree C)

(*a*) 5 m (*b*) 50 m
(*c*) 100 m (*d*) 25 m

122. Mud pumping is commonly associated with

(*a*) bituminous penetration macadam construction
(*b*) cement concrete pavement on granular subgrage
(*c*) premixed bituminous construction
(*d*) cement concrete pavement on clay subgrage

123. Alligator or map cracking is the common type of failure in

(*a*) concrete pavements
(*b*) bituminous surfacing
(*c*) gravel roads
(*d*) WBM construction

1996

124. One of the probable causes of rutting on flexible pavements is

(*a*) excessive stripping of binder material from the wearing course
(*b*) use of flaky aggregates in the wearing course
(*c*) inadequate compaction of pavement layers
(*d*) high wind speeds

125. An important purpose of prime coats is to

(*a*) promote the bond between the base and the wearing courses

(*b*) promote the adhesion between an exisiting wearing surface and a subsequent wearing surface

(*c*) promote the bond between the sub-base course and the sub-grade

(*d*) increase the stability of the sub-grade

126. Flexible pavements derive stability primarily from

(*a*) aggregate interlock, particle friction and cohesion

(*b*) cohesion alone

(*c*) the binding power of bituminous materials

(*d*) the flexural strength of the surface course

127. Transition curves are provided on the approach to horizontal curves in order to

(*a*) increase jerk to allowable levels

(*b*) minimize the length of the horizontal curve

(*c*) simplify the laying-out and construction of the horizontal curve

(*d*) reduce jerk to allowable levels

128. The following general statement may be made about the penetration value and softening point of bitumen.

(*a*) Higher the penetration value, higher is the softening point

(*b*) Higher the penetration value, lower is the softening point

(*c*) For very high and very low penetration value, the softening point is very low

(*d*) Absolutely no correlation can be drawn between penetration value and softening point of bitumen

129. On sag (or valley) curves the available sight distance is determined based on

(*a*) design speed

(*b*) height of obstacle

(*c*) height of driver eye

(*d*) night-time driving conditions

130. Bituminous materials are used in highway construction primarily because of their

(*a*) cementing and water-proofing properties

(*b*) load bearing capacity

(*c*) high specific gravity

(*d*) black colour which facilitates road markings

131. Maximum allowable grades are lower for railways than for highways because

(*a*) construction costs become prohibitive for railways at high grades

(*b*) trains are longer than vehicles which use the highways

(*c*) high grades cause discomfort to passengers

(*d*) steel wheels on steel rails have lower frictional coefficient than rubber tyres on pavements

132. Since the moisture content at the bottom of a rigid pavement slab is generally more than at the top

(*a*) the bottom of the slab is generally in tension

(*b*) the top of the slab is generally in compression

(*c*) the bottom of the slab is generally in compression

(*d*) the bottom of the slab is neither in compression nor in tension

133. Sliding considerations for stopped vehicles on super elevated horizontal curves provide the following bound on the amount of super elevation, *e*,

(*a*) $e \geq$ coefficient of rolling friction

(*b*) $e \geq$ coefficient of side friction

(*c*) $e <$ coefficient of side friction

(*d*) $e \leq$ coefficient of side friction

NUMERICAL TYPE QUESTIONS

2016

1. In a one-lane one-way homogeneous traffic stream, the observed average headway is 3.0 s. The flow (expressed in vehicles/hr) in this traffic stream is______.

2015

2. The penetration value of a bitumen sample tested at 25°C is 80. When this sample is heated to 60°C and tested again, the needle of the penetration test apparatus penetrates the bitumen sample by d mm. The value of d CANNOT be less than ______ mm.

3. A sign is required to be put up asking drivers to slow down to 30 km/h before entering Zone Y (see figure). On this road, vehicles require 174 m to slow down to 30 km/h (the distance of 174 m includes the distance travelled during the perception-reaction time of drivers). The sign can be read by 6/6 vision drivers from a distance of 48 m. The sign is placed at a distance of *x* m from the start of Zone Y so that even a 6/9 vision driver can slow down to 30 km/h before entering the zone. The minimum value of x is ________ m.

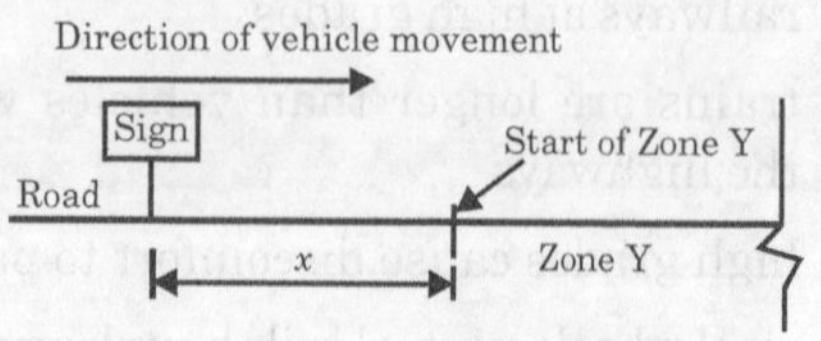

4. The acceleration-time relationship for a vehicle subjected to non-uniform acceleration is,

$$\frac{dv}{dt} = (\alpha - \beta v_0)e^{-\beta t}$$

Where, v is the speed in m/s, t is the time in s, α and β are parameters, and v_0 is the initial speed in m/s. If the accelerating behavior of a vehicle, whose drive intends to overtake a slow moving vehicle ahead, is described as,

$$\frac{dv}{dt} = (\alpha - \beta v)$$

Considering $\alpha = 2\ m/s^2$, $\beta = 0.05\ s^{-1}$ and $\frac{dv}{dt} = 1.3\ m/s^2$ at $t = 3\ s$, the distance (in m) travelled by the vehicle in 35 s is ________.

5. The relation between speed u (in km/h) and density k (number of vehicles/km) for a traffic stream on a road is $u = 70 - 0.7k$. The capacity on this road is ________ *vph* (vehicles/hour).

2014

6. A traffic office imposes on an average 5 number of penalties daily on traffic violators. Assume that the number of penalties on different days is independent and follow a Poisson distribution. The probability that there will be less than 4 penalties in a day is ________.

7. A traffic survey conducted on a road yields an average daily traffic count of 5000 vehicles. The axle load distribution on the same road is given in the following table:

Axle load (tonnes)	Frequency of traffic (%)
18	10
14	20
10	35
8	15
6	20

The design period of the road is 15 years, the yearly traffic growth rate is 7.5% and the load safety factor (LSF) is 1.3. If the vehicle damage factor (VDF) is calculated from the above data, the design traffic (in million standard axle load, MSA) is________.

8. The perception-reaction time for a vehicle travelling at 90 km/h, given the coefficient of longitudinal friction of 0.35 and the stopping sight distance of 170 m (assume g = 9.81 m/s²), is ________ seconds.

9. The speed-density (u-k) relationship on a single lane road with unidirectional flow is $u = 70 - 0.7k$, where u is in km/hr and k is in veh/km. The capacity of the road (in veh/hr) is ________.

10. An isolated three-phase traffic signal is designed by Webster's method. The critical flow ratios for three phases are 0.20, 0.30 and 0.25 respectively, and lost time per phase is 4 seconds. The optimum cycle length (in seconds) is ________.

11. The average spacing between vehicles in a traffic stream is 50 m, then the density (in veh/km) of the stream is ________

12. An observer counts 240 veh/h at a specific highway location. Assume that the vehicle arrival at the location is Poisson distributed, the probability of having one vehicle arriving over a 30-second time interval is ________

13. In a Marshall sample, the bulk specific gravity of mix and aggregates are 2.324 and 2.546 respectively. The sample includes 5% of bitumen (by total weight of mix) of specific gravity 1.10. The theoretical maximum specific gravity of mix is 2.441. The void filled with bitumen (VFB) in the Marshall sample (in %) is ________

14. A pre-timed four phase signal has critical lane flow rate for the first three phases as 200, 187 and 210 veh/hr with saturation flow rate of 1800 veh/hr/lane for all phases. The lost time is given as 4 seconds for each phase. If the cycle length is 60 seconds, the effective green time (in seconds) of the fourth phase is ________

2013

15. It was observed that 150 vehicles crossed a particular location of a highway in a duration of 30 minutes. Assuming that vehicle arrival follows a negative exponential distribution, find out the number of time headways greater than 5 seconds in the above observation? ________

2001

16. The speeds of overtaking and overtaken vehicles on a highway are 85kmph and 70kmph respectively. Calculate overtaking sight-distance needed for two way traffic. Assume acceleration of the overtaking vehicle as 2.5 kmph per second and speed of the vehicle in the oppsite direction as 85 kmph.

17. Calculate spacing between contraction joints for a two lane 250 mm thick concrete road having 3.5 m wide slab.

Unit weight of concrete = 24 kN/m^3.

Ultimate stress in tension = 0.16 N/mm^2

Coefficients of Friction at interface = 1.5.

Factor of Safety = 2.

Also calculate the spacing between expansion joints if the increase in temperature is 20°C, the expansion joint gap is 24 mm and thermal coefficient = 10×10^{-6} per°C.

2000

18. The load penetration date from a California Bearing Ratio (CBR) test is provided in the following table. Indicate whether any correction is required for the calculated CBR value. Find the CBR value of the soil from the data provided.

Table : Load penetration data

Penetration (in mm)	Load in kgf (kg force)
0	0
0.5	4
1.0	13
1.5	29
2.0	40
2.5	50
3.0	58
4.0	70
5.0	78
7.5	93
10	103
12.5	112

Area of plunger is given as 19.6 cm^2. Pressure for standard crushed stones at 2.5 mm and 5.0 mm are 70 kgf/cm^2 and 105 kgf/cm^2 respectively.

19. An ascending gradient of 1 in 45 meets a descending gradient of 1 in 60. A summit curve has to be designed for design speed of 80 km/h so as to provide a safe overtaking sight distance of 230 m. Estimate length of the summit curve.

ANSWERS

MCQ Type Questions

1. (b)	**2.** (b)	**3.** (a)	**4.** (d)	**5.** (a)	**6.** (b)	**7.** (a)	**8.** (c)	**9.** (c)	**10.** (d)
11. (d)	**12.** (b)	**13.** (b)	**14.** (d)	**15.** (a)	**16.** (c)	**17.** (b)	**18.** (b)	**19.** (b)	**20.** (c)
21. (d)	**22.** (d)	**23.** (c)	**24.** (a)	**25.** (b)	**26.** (a)	**27.** (b)	**28.** (a)	**29.** (a)	**30.** (c)
31. (d)	**32.** (d)	**33.** (d)	**34.** (d)	**35.** (c)	**36.** (c)	**37.** (*)	**38.** (a)	**39.** (b)	**40.** (c)
41. (a)	**42.** (b)	**43.** (d)	**44.** (d)	**45.** (c)	**46.** (d)	**47.** (c)	**48.** (c)	**49.** (a)	**50.** (c)
51. (b)	**52.** (b)	**53.** (d)	**54.** (a)	**55.** (c)	**56.** (d)	**57.** (c)	**58.** (b)	**59.** (b)	**60.** (b)
61. (d)	**62.** (c)	**63.** (b)	**64.** (a)	**65.** (c)	**66.** (c)	**67.** (b)	**68.** (d)	**69.** (c)	**70.** (a)
71. (b)	**72.** (a)	**73.** (d)	**74.** (d)	**75.** (d)	**76.** (d)	**77.** (c)	**78.** (c)	**79.** (c)	**80.** (a)
81. (c)	**82.** (b)	**83.** (b)	**84.** (c)	**85.** (a)	**86.** (b)	**87.** (b)	**88.** (a)	**89.** (b)	**90.** (b)
91. (a)	**92.** (a)	**93.** (*)	**94.** (d)	**95.** (b)	**96.** (d)	**97.** (b)	**98.** (c)	**99.** (b)	**100.** (b)
101. (c)	**102.** (b)	**103.** (c)	**104.** (d)	**105.** (d)	**106.** (a)	**107.** (a)	**108.** (c)	**109.** (c)	**110.** (c)
111. (b)	**112.** (d)	**113.** (c)	**114.** (a)	**115.** (b)	**116.** (a)	**117.** (b)	**118.** (c)	**119.** (a)	**120.** (d)
121. (a)	**122.** (a)	**123.** (b)	**124.** (c)	**125.** (a)	**126.** (a)	**127.** (c)	**128.** (d)	**129.** (c)	**130.** (a)
131. (c)	**132.** (c)	**133.** (c)							

Numerical Type Questions

1. (1200)	**2.** (8)	**3.** (142)	**4.** (900.83)	**5.** (1750)
6. (0.26 to 0.27)	**7.** (307 to 310)	**8.** (3.1 to 3.2)	**9.** (1750 to 1750)	**10.** (90 to 95)
11. (20.0 to 20.0)	**12.** (0.25 to 0.28)	**13.** (62 to 68)	**14.** (14.0 to 18.0)	**15.** (*) **16.** (*)
17. (*)	**18.** (*)	**19.** (*)		

"*" Refers to subjective question

EXPLANATIONS

MCQ TYPE QUESTIONS

2. $$l = \sqrt{\frac{Eh^3}{12(1-\mu^2)}}$$

Where l = radius of rel. stiffness in cm

E = modulud of elasticity of cement concrete in kg/cm^2

H = slab thickness

μ = Poission's ratio

K = Modulus of subgrade reaction

3. Total number of commercial vehicles per day = 3000 to 6000

Minimum % of commercial traffic to be surveyed for axle load = 15

4. Free flow speed is defined as the speed that occurs when density and flow are zero.

5. $R = 335$ m

$$e = \tan\theta = \tan 33° = 0.649$$

$$e + f = \frac{V^2}{127\,R}$$

$$0.649 + f = \frac{(320)^2}{127 \times 335}$$

$$f = 1.756 \approx 1.76$$

7. $$\text{Theoretical } G = \frac{w_1 + w_2 + w_3 + w_4}{\frac{w_1}{G_1} + \frac{w_2}{G_2} + \frac{w_3}{G_3} + \frac{w_4}{G_4}}$$

$$= \frac{100}{\frac{55}{2.0} + \frac{35.8}{2.7} + \frac{3.7}{2.65} + \frac{5.5}{1.01}} = 2.42$$

Effective G is of fine and coarse aggregates,

$$G_{eff} = \frac{55\times2.6+35.8\times2.7}{55+35.8} = 2.64$$

8. 15 minute peak hour factor is used to design traffic intersections

$$PHF = \frac{\left(V/4\right)}{V_{15}}$$

Where V = peak hour volume (veh/hr)

V_{15} = Maximum 15 minimum volume within the peak hour.

Minimum value is 0.25

9. Temperature stress = 0 if pavement is weightless

Temperature stress

= Tensile at bottom of slab in day time

= Compressive at bottom of slab in night time

10. Given data : V = 110 km/hr

e = 8% (Super elevation)

f = 0.10 (coefficient of side friction)

R = the minimum radius of curve in (m)

$$e + f = \frac{V^2}{127R}$$

$$R = \frac{V^2}{127(e+f)} = \frac{(110)^2}{127\left(\frac{8}{100} + 0.10\right)}$$

= 529.30 m

R = 528.5 m

11. Total riding minute = 40–15 = 25 min

$$\therefore \quad \text{speed of student} = \frac{\text{distance}}{\text{time}} = \frac{5\,\text{km}}{25\,\text{min}}$$

$= 0.2 \times 60$ km/hr

= 12 km/hr.

$$\left(\frac{\text{Vehicle/min}}{\text{Relative speed of vehicle w.r.t. student}}\right)_{\text{moving}}$$

$$= \left(\frac{\text{Vehicle/min}}{\text{Relative speed of vehicle w.r.t. student}}\right)_{\text{standing}}$$

Assume x be the speed of vehicle

$$\frac{60/25}{x-12} = \frac{45/15}{x-0}$$

$$(2.4)x = 3(x-12)$$

$$3x - 2.4x = 36$$

$$x = 60 \text{ km/hr}$$

correct choice is (*d*)

12. $$v = 80 - \frac{2}{3}K$$

capacity of the section $q = v \times K$

$$q = 80K - \frac{2}{3}K^2$$

For maximum capacity $\frac{dq}{dK} = 0$

$$\frac{dq}{dK} = 80 - \frac{4K}{3} = 0$$

$$K = \frac{80 \times 3}{4}$$

$$= 60 \text{ vehicle/km}$$

$\because$ maximum capacity $q = 80K - \frac{2}{3}K^2$

$$= 80 \times 60 - \frac{2}{3} \times (60)^2$$

$$= 2400$$

the correct choice is (*b*)

13. The correct statement is both merging from left and diverging to left are possible.

14. Draw to the scale, L = +78 m, D = – 45.1

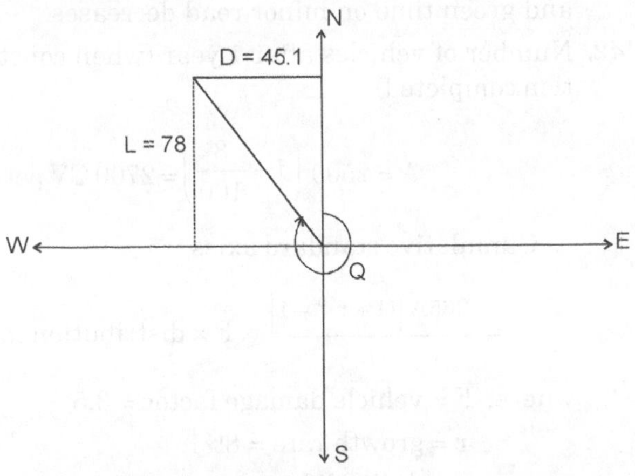

WCB = 330°

15. Number of conflict points for one-way roads = 6(x)
Number of conflict points for two-way roads = 24 (y)

So, $\frac{x}{y} = \frac{6}{24} = 0.25$

16. Lower the softening, lower the viscosity
Second higher the penetration value.

18. $$mSA = \frac{365A[(1+r)^n - 1]}{r} \times V_{DF}$$

$$\therefore\ 100 \times 10^6 = 365 \times 4500\ \frac{[(1+0.08)^{15} - 1]}{r} \times V_{DF}$$

$\Rightarrow V_{DF} = 2.24$

29. When a fast moving vehicle travels along a summit curve, the centrifugal force will act upwards, against gravity and hence a part of the pressure on the tyres and spring of the vehicle suspensions is relieved. So there is no problem of discomfort to passengers on summit curves, particalarly because the deviation angles on the roads are quite small and if summit curve is designed to have adequate sight distance, length of the curve would be long enough to ease the shock due to change in gradients.

Circular summit curves are ideal as the sight distance available throughout the length of the circular curve is constant. The deviation angle in vertical curves of the highways are very small and so between same tangent points, a simple parabola is nearly congruent with a circular arc; also a parabola is very easy for airthmetical manipulation for computing ordinates. The use of simple parabola as summit curve is found to give good riding comfort too. Because of these reasons in actual practice, a simple parabolic curve is used as summit curve instead of the circular arc

Hence answer I is true, II is false.

30. *Given :* V = 80 kmph

$l = 6.6$m

$W_m = 0.096$m

Total widening of carriage way on the curve,

$$W_e = W_m + W_{psy}$$

where, $$W_m = \frac{nl^2}{2R}$$

As W_m is measured only for one vehicle, so

$$n = 1$$

$$\therefore\quad 0.096 = \frac{6.6^2}{2 \times R}$$

$$\Rightarrow\quad R = 226.9 \text{ m}$$

$$W_{psy} = \frac{V}{9.5\sqrt{R}}$$

$$= \frac{80}{9.5\sqrt{226.9}} = 0.559 \text{ m}$$

$\therefore$ Total widening of two lanes

$$= 2 \times 0.096 + 0.559 = 0.751 \text{ m}$$

Alternately

$$R^2 = (R + W_m)^2 + l^2$$

$$W_m = \frac{l^2}{2R}$$

$$\text{Off tracking} = \frac{l^2}{2R}$$

$$\therefore\quad 0.096 = \frac{(6.6)^2}{2R}$$

$$\Rightarrow\quad R = 226.875 \text{ m}$$

$\therefore$ Extra widening $= \frac{nl^2}{2R} + \frac{V}{9.5\sqrt{R}}$

$= \frac{2\times 6.6^2}{2\times 226.875} + \frac{80}{9.5\sqrt{226.875}}$

$= 0.75$m

31. I. During mid-day, the top pavement slab has higher temperature than bottom of the slab. This causes top fibers of the slab to expand more than the bottom fibres, and the slab curls at the edges. This phenomenm is called *warping down of the slab.*

By the mid-night, temperature of the bottom slab is higher than temperature of the slab at the top. Hence slab warps up during this time.

II. Tie bars are provided in the longitudinal joints to hold the adjacent slabs together.

Thus, I and II are False.

33. For 5mm, Load = 50 kg

Standard = 2055 kg

$\therefore$ C.B. R value $= \frac{\text{Load}}{\text{Standard Load}} \times 100$

$= \frac{500}{2055} \times 100 = 2.43\%$

34. $v = v_f \ln\left(\frac{k_j}{k}\right)$

$\Rightarrow e^{\left(\frac{v}{v_f}\right)} = \frac{k_j}{k}$

$\Rightarrow k = \frac{k_j}{e^{\frac{v}{v_f}}}$

For operating capacity,

$v = v_f$

$\therefore k = \frac{k_j}{e}$

36. As per IS. 73 : 1992, the specific gravity of bitumen lies between 1.02 and 0.97.

38. As per IRC, it is 1200 and 2400.

39. Capacity of a Rotary, $C = \frac{280\,W\left(1+\frac{e}{W}\right)\left(1-\frac{p}{3}\right)}{\left(1+\frac{W}{L}\right)}$

where, W = width of weaving section = 14 m.

L = Length of weaving section = 35m

e_i = entry width = 8.4

p = weaving proportion

$= \frac{\text{crossing traffic}}{\text{total traffic}} = \frac{1000}{2000} = 0.5$

Putting all these values, we get

C = 3733.33 say 3700 PCU per hour

40. Octagonal (red colour, white border)

41. As the flow on major road increases, the optimum cycle length, C_o increases; green time increases and green time on minor road decreases.

42. Number of vehicles after 1 year (when construction completed)

$A = 2500\left(1+\frac{8}{100}\right) = 2700$ CV per day.

$\therefore$ Cumulative standard axles

$= \frac{365A\left[(1+r)^n - 1\right]}{r} \times F \times \text{distribution factor}$

where, F = vehicle damage factor = 3.5

r = growth rate = 8%

n = design life

$\therefore$ Cumulative standard axles

$= 365 \times 2700 \frac{\left[(1+0.08)^{10} - 1\right]}{0.08} \times 3.5 \times 0.75$

$= 37.34 \times 10^6 = 37.34$ msa, say 37 msa

43. Free flow speed, v = 80 kmph

Jam density ρ = 100 vehicles /km

$\therefore$ Maximum flow $= \frac{v}{2}\cdot\frac{\rho}{2}$

$= \frac{80}{2}\cdot\frac{100}{2} = 2000$ vehicles per hour.

$\therefore$ Speed at maximum flow $= \frac{80}{2} = 40$ kmph

44. The consistency and flow resistance of bitumen is determined by viscosity test.

45.

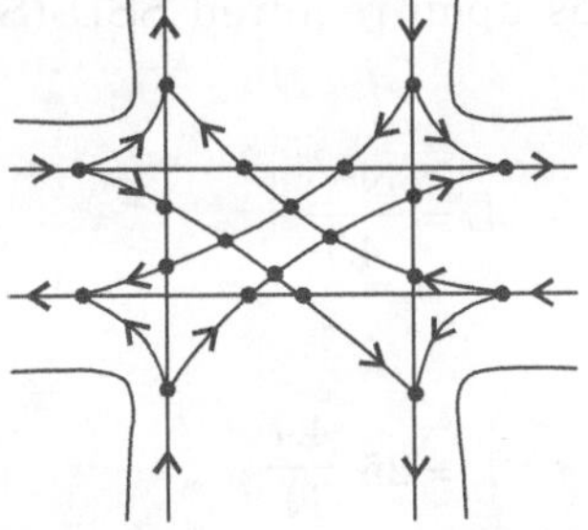

46. Optimum cycle length, $C_0 = \dfrac{1.5\,L + 5}{1 - Y}$

where, Y = sum of ratio of normal flow to saturation flow of two direction traffic flow

= 0.5

L = total lost time per cycle = 10 seconds

$\therefore \quad C_0 = \dfrac{1.5 \times 10 + 5}{1 - 0.5} = 40$ seconds.

47. Standard deviation of spot speed of vehicles = 8.8 kmph

Mean speed of vehicles = 33 kmph

Coefficient of variation of speed

$= \dfrac{\text{standard deviation of spot speed}}{\text{Mean speed of vehicles}}$

$= \dfrac{8.8}{33} = \mathbf{0.2666}$

48. Radius, R = 300 m

Wheel base, l = 8 m

Design speed, V = 100 kmph

Number of lanes, n = 2

Extra widening, $W_e = \dfrac{n\ell^2}{2R} + \dfrac{V}{9.5\sqrt{R}}$

$= \dfrac{2 \times (8)^2}{2 \times 300} + \dfrac{100}{9.5\sqrt{300}} = 0.82$ m.

49. Ruling gradient = 6%

Radius of horizontal curve, R = 500 m

Grade compensation $= \dfrac{30 + R}{R} = \dfrac{30 + 50}{50} = 1.6\%$

Compensated gradient = 6 – 1.6 = 4.4%

50. Design speed, V = 60 kmph

$= \dfrac{60}{3.6} = 16.66$ m/sec

Reaction time, t = 2.5 sec

Coefficient of friction, f = 0.35

Stopping distance

= lag distance + breaking distance

$= Vt + \dfrac{V^2}{2gf}$

$= 16.66 \times 2.5 + \dfrac{(16.66)^2}{2 \times 9.81 \times 0.35}$

= 82.06 m

Stopping sight distance for two way traffic with single lane

= 2 × stopping distance

= 2 × 82.06 = **164.2 m**

51. Coefficient of thermal expansion,

$\alpha = 10 \times 10^{-6}$ mm/mm/°C

Width of expansion joint = 20 mm

Compression of filler, Δ = 20 × 0.5 = 10 mm

Temperature difference, $\Delta T = T_2 - T_1$

= 60 – 20 = 40°

Now, $\Delta = L\,\alpha\,T$

$\Rightarrow \quad L = \dfrac{\Delta}{\alpha T}$

$= \dfrac{10}{10 \times 10^{-6} \times 40} = \mathbf{25\ m}$

52. Cumulative number of standard axle load repetitions

$N_s = 365\,A\left[(1+r)^n - 1\right]F$

where, A = number of vehicles per day considering number of lanes

r = growth factor = 7.5 %

Vehicle damage factor (F)

For two axle trucks :

n = design life = 10 years

$A_1 = 2000$

$F_1 = 5$

For tandem axle trucks :

$A_2 = 200$

$F_2 = 6$

$\therefore \quad N_s = \dfrac{365A_1\left[(1+r)^n - 1\right]F_1}{r} + \dfrac{365A_2\left[(1+r)^n - 1\right]F_2}{r}$

$= \dfrac{365}{7.5}\left[(1 + 0.75)^{10} - 1\right]\left[2000 \times 5 + 200 \times 6\right]$

= **57.8 millions**

54. Mean size = $\frac{40+50}{2} = 45$

Slot length = 45 × 1.8 = 81 mm.

55. O and D provides the basic data for determining the desired direction of flow or the **desire lines**.

56.
$$S_l = \frac{0.316P}{h^2}\left\{4\log_{10}\left(\frac{l}{b}\right)+1.069\right\} \quad ...(i)$$

$$l = \left[\frac{Eh^3}{12k(1-\mu^2)}\right]^{1/4} \quad ...(ii)$$

From equations *(i)*, load stress is directly propotinal to wheel load.

From equations *(ii)*, it also depends on modulus of subgrade reaction

58. *Given* : V = 60 kmph = 16.67 m

$g = 9.81$ m^2/s

$f = 0.36$

t = 2.5 sec

We know,
$$\frac{\frac{Vt}{V^2}}{2g\left(f \pm \frac{n}{100}\right)} = \frac{6}{5}$$

Solving, we get $n = 4.8$

60.
$$k_i = \frac{1000 \times V}{S}$$

$$= \frac{1000 \times 1}{7} = 142.857$$

$$\text{Flow} = k_i \times \frac{V_{st}}{4}$$

$$= \frac{142.857 \times 70}{4} = 2500$$

61.
$$L = eBN$$

$$= \frac{5}{100}(4 \times 3.75) \times 150 = 11.25 \text{ m.}$$

62. *Given:* A = 2723 ueh/day

$r = 5\%$

$n = 10$ years.

F = 2.4

$$\therefore \quad N_s = \frac{365A\left[(1+r)^n - 1\right]F}{r}$$

= 30 msA.

From table, $t = 670\ mm$

65. Length of Summit curve on a two lane two way highway depends upon required SSD (Stoping Sight Distance)

(i) If L > SSD $\quad L = \frac{NS^2}{4.4}$

(ii) If L < SSD

then, $\quad L = 25\ \frac{4.4}{N}$

where, N = deviation angle

S = SSD

L = length of summit curve

66. Bituminous concrete is mixture comprising of coarse aggregate, fine aggregate, filler and bitumen.

67. I. Critical stress combination (during summer)

= (load stress + warping stress – frictional stress) (at edge region)

= 32 + 8 – 5 = 35 kg/cm^2

II. Critical stress combination (during win)

= (load stress + warping stress + frictional stress) (at edge)

= 32 + 6 + 4 = 42 kg/cm^2

III. Critical stress combination in summer at midnight (at corner)

= load stress + warping stress

= 30 + 9 = 39 kg/cm^2

IV. Critical stress combination at mid-night (winter) at corner

= load stress + warping stress

= 30 + 9 = 39 kg/cm^2

Hence, most critical section for the pavement

= 42 kg/cm^2

69.
$$C = 1000\frac{V}{S}$$

and
$$S = 0.278Vt + \frac{V^2}{254f} + L$$

where, V = velocity of vehicle in km/h

L = length of vehicle (m)

$$\therefore \quad S = 0.278 \times 65 \times 2.5 + \frac{65^2}{254 \times 0.4} + 5$$

$$= 45.175 + 41.58 + 5 = 91.75 \text{ m}$$

$$\therefore \quad C = 1000\frac{V}{S} = \frac{1000 \times 65}{91.75}$$

= 708.32, i.e. 710 vehicles/hr.

71. Given: R = 400 m

$e = 0.7$

V = 100 km/hr

From formula,

$$e + f = \frac{V^2}{127R}$$

$$\therefore \quad f = \frac{V^2}{127R} - e = \frac{(100)^2}{127 \times 400} - 0.07 = 0.126 \approx \mathbf{0.13}$$

72. The Nagpur road plan formulae were prepared assuming star and grid pattern.

73. For the purpose of highway goemetric design, the 99th percentile speed is taken.

74. The ratio of compact pressure to tyre pressure is defined as rigidit factor. For rigidity factor less than units for tyre pressure higher than 7 kg/cm² or more than 0.7 N/mm²

75.
$$e + f = \frac{v^2}{127R}$$

where, e = camber = 0.03

f = 0.15

$$\therefore \quad R = \frac{80^2}{(0.18)^{127}} = 279.96 \text{ m}$$

76. Here, P = 8 × 0.25 + 6 × 0.5 + 1 × 1 = 6

Q = 19 × 0.25 + 8 × 0.5 + 4 × 1 = 12.75

R = 7 × 0.25 + 5 × 0.5 + 2 × 7 = 7.75

$$\therefore \quad P = \frac{6}{20} = \frac{6}{\text{length}} = 0.3$$

$$Q = \frac{12.75}{28} = 0.4553$$

$$R = \frac{7.75}{12} = 0.645$$

Hence, Order of priority = R, Q, P

77. We know, $VFB = \dfrac{100V_b}{VMA}$

where, $V_b = G_M \times \dfrac{W_u}{G_u} = 2.345 \times \dfrac{0.05}{1.02} = 0.1149$

$$V_v = \frac{G_T - G_M}{G_M} = \frac{2.442 - 2.345}{2.345} = 0.04136$$

$$VMA = V_v + V_b$$

$$\therefore \quad VFB = \frac{100 \times 0.1149}{(0.1149 + 0.04136)}$$

$$= 73.53\%$$

78. Cumulative number of standard axle load

$$N_s = \frac{365A[1+r)^n - 1]}{r} \times F$$

where, A = number of commercial vehicle or initial traffic

= 1213 cvpd

r = traffic growth rate = 8% per annum

n = design life = 2 years

f = vehicle damage factor = 2.5

$$\therefore \quad N_S = \frac{365 \times 1213[(1+0.08)^{12} - 1]}{0.08} \times 2.5$$

= 21.00 msa

79. Breaking distance $= \dfrac{v^2}{2gf} = \dfrac{(0.278 \times 65)^2}{19.62 \times 0.396} = 42$ m

80. Temperature stress are developed in cement concrete pavement due to variation in slab temperature.

Temperature produces two type of stresses in a concrete parement.

(*i*) Warping stress

(*ii*) Frictional stress

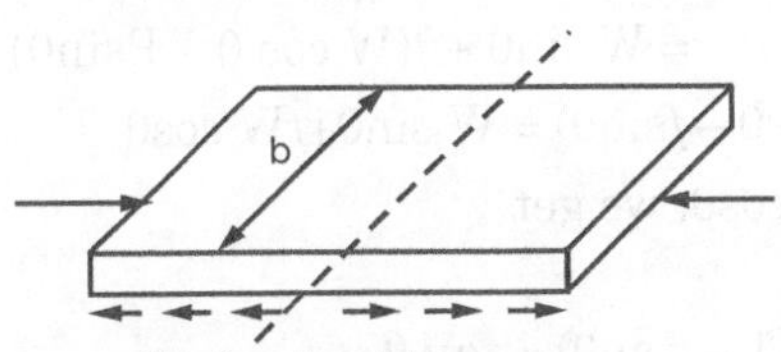

81. Speed and delay studies are useful in detecting the spots of congestion.

Methods of carrying out speed and delay study.

(*i*) Moving car or floating car method

(*ii*) License plate method

(*iii*) Interview method

(*iv*) Elevated observation

(*v*) Photographic technique

82. We know $e + f = \dfrac{v^2}{gR}$

Ruling minimum radius of horizontal curve,

$$R = \frac{v^2}{g(e+g]}$$

Here, $v = 100 \times \dfrac{5}{18} = 27.8$ m/sec

$$\therefore \quad R = \frac{(27.8)^2}{9.81 \times (0.10 + 0.15)}$$

= **315 m**

83. *Given* : v = 50 kmph = 13.88 m/s

Average headway = 2.52 seconds

∴ Longitudinal distance = vt

$$= 13.88 \times 2.52 = \mathbf{35\ m}$$

84. Theoretical specific gravity of mix

$$G_t = \frac{100}{\frac{55}{2.62}+\frac{34.6}{2.73}+\frac{4.8}{2.7}+\frac{5.6}{1.02}}$$

$$= \frac{100}{40.98} = \mathbf{2.44}$$

85.
$$E = 1.18\frac{Pa}{\Delta}$$

$$= \frac{1.18}{1000}\times\frac{800\times 37.5}{0.25}\ \text{kN/cm}^2$$

$$= \mathbf{141.6\ kN/cm^2}$$

93. Superelevation, $e = \frac{NL}{ML} = \tan\theta$

For equilibrium condition

$$P\cos\theta = W\sin\theta + F_A + F_B$$
$$= W\sin\theta + fR_A + fR_B$$
$$= W\sin\theta + f(W\cos\theta + P\sin\theta)$$

$\Rightarrow$ $P(\cos\theta - f\sin\theta) = W\sin\theta + fW\cos\theta$

Divided by cosθ, we get

$$\frac{P}{W}(1 - f\tan\theta) = \tan\theta + f$$

$$\Rightarrow \quad \frac{P}{W} = \frac{f+\tan\theta}{1-f\tan\theta}$$

Also $\quad \frac{P}{W} = \frac{V^2}{gR} = \frac{f+\tan\theta}{1-f\tan\theta}$

But $\quad 1 - f\tan\theta \simeq 1$

$$\therefore \quad \frac{V^2}{gR} = f + \tan\theta$$

where, $\tan\theta = e$

$$\therefore \quad \frac{V^2}{gR} = f + e$$

$$\Rightarrow \quad V = \sqrt{(P+e)gR}$$
$$= (0.15 + 0.087)\times 9.81\times 250$$
$$= 24.11\text{m/sec.}$$

97. Overflow rate = 17m³/day/m²

$$= 17\text{m/day}$$

$$= \frac{17\times 1000}{24\times 60\times 60}$$

$$= 0.196\text{mm/s}$$

Percent particle removal $= \frac{0.1}{0.196}\times 100$

$$= 50\%$$

98. $L = 2\left[\frac{NV^3}{C}\right]^{1/2}$

$= 0.38[NV^3]^{1/2}$ ...[since e = 0.6m/s³]

$$N = \frac{1}{40} - \left(-\frac{1}{50}\right)$$

$$= \frac{1}{40} + \frac{1}{50}$$

$$= 0.045$$

∴ $L = 0.38[0.045\times 80^3]^{1/2}$

$= 57.67$

99. *Given* : h = 20cm

E = 3×10⁵kg/cm²

V = 0.15

k = 5kg/cm³

$$\therefore \quad l = \left[\frac{Eh^3}{12k(1-\mu^2}\right]^{1/4}$$

$$= \left[\frac{3\times 10^5\times 20^3}{12\times 5[1-(0.15)^2}\right]^{1/4}$$

$= 79.98$ cm

$= 80$cm

100.
$$e + f = \frac{V^2}{127R}$$

$$\Rightarrow \quad R = \frac{(100)^2}{127\times(0.07+0.15)}$$

$= 358$ m

≈ 360 m.

101.
$$e = \frac{V^2}{225R} = \frac{(125)^2}{225\ 450} = 0.154$$

But $\quad 0.154 > e_{max}$ (= 0.07)

Hence e > 0.07 is not provided.

∴ Design rate of super devation,

e = 0.07

102. Highest point on the summit curve is a distance $\frac{Ln_1}{N}$ from the FTP on the first grade n_1

$$\therefore \quad \frac{Ln_1}{N} = 150 \times \frac{4}{4-(-2)} = 100m$$

111.
$$V = 100 \text{ Kmph}$$
$$V_b = (V-16) = (100-16) = 84 \text{ Kmph}$$
Take t = reactive time = 2 sec
$$A = 0.53 \text{ m/sec}^2 = 0.53 \times 36$$
$$= 1.92 \text{ kN/sec}^2$$
Now $d_1 = 0.28 \times V_b \times t = 0.28 \times 84 \times 2 = 47.04$ m
$$d_2 = 0.28\, V_b \times T$$
$$T = \sqrt{\frac{14.4s}{A}}$$
where, $s = 0.2\, V_b + 6 = 0.2 \times 84 + 6 = 22.8$ m
$$\therefore T = \sqrt{\frac{14.4 \times 22.8}{1.92}} = 13.11 \text{ sec.}$$
$$d_2 = 0.28 \times 84 \times 13.11 = 354.13 \text{ m}$$
$$d_3 = 0.28\, V \times T$$
$$= 0.28 \times 100 \times 13.11 = 367.08 \text{ m}$$
$\therefore$ safe overtaking sight distance
$$= d_1 + d_2 + d_3$$
$$= 470.4 + 354.13 + 367.08$$
$$= 768.25\text{m} \approx 750\text{m}$$

112. Given, $P = 20440$ N
$$2P = 40880 \text{ N}$$

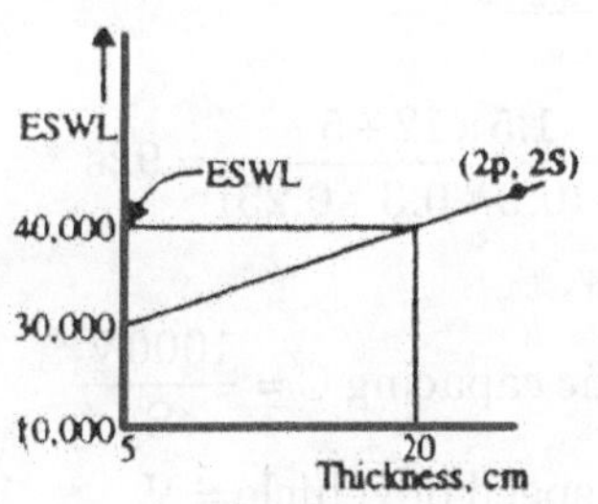

Pavement thickness = 20 cm.

Distance of wall of types, d = 11cm

c/c tyre spacing, s = 27 cm

To calculate ESWL, X and Y points are plotted on log-log graph between ESWL and pavement thickness.

X has coordinates $(p, d/2)$ = (20440, 5.5)

Y has coordinates $(2p, 2s)$ (40880, 54)

From the graph

ESWL = 30130 M

113. None

Elastic modulus of subgrade reaction,
$$K = \frac{\text{pressure, P}}{\text{deflection, }\Delta} = \frac{1.25 \times 10^5 \text{N/m}^2}{(0.5/100)}$$
$$= 2.5 \times 10^7 \text{ N/m}$$

116. Speed, $V = 60$ kmph,

Radius $R = 128$ m,

Coefficient of friction, $f = 0.15$

Now, $e = \frac{V^2}{127R} - f = \frac{(60)^2}{127 \times 128} - 0.15 = 0.71$

117. Deviation angle, $\delta = \frac{1}{30} - \frac{1}{50} = \frac{1}{75}$

121.
$$\delta_1 = \frac{25}{2} = 1.25 \text{ cm}$$
$$t_2 - t_1 = 25 \text{ C}$$
Coefficient of thermal expansion,
$$C = 10 \times 10^{-6} \times \text{per C}$$
$\therefore$ Spacing of expansion joint,
$$L = \frac{\delta_1}{100C(t_2 - t_1)}$$
$$= \frac{1.25}{100 \times 10 \times 10^{-6} \times 25} = 5 \text{ meters}$$

NUMERICAL TYPE QUESTIONS

1. Average time head way, H_t = 3s

   The flow of traffic stream,
$$q = \frac{3600}{H_t} = \frac{3600}{3} = 1200 \text{ vehicles/hr}$$

2. The correct answer is 80. Because as the temperature rises form 25 to 60 degree celsius, the viscosity of bitumen decreases and thus the penetration value should be more than 80 mm in any case.

3.

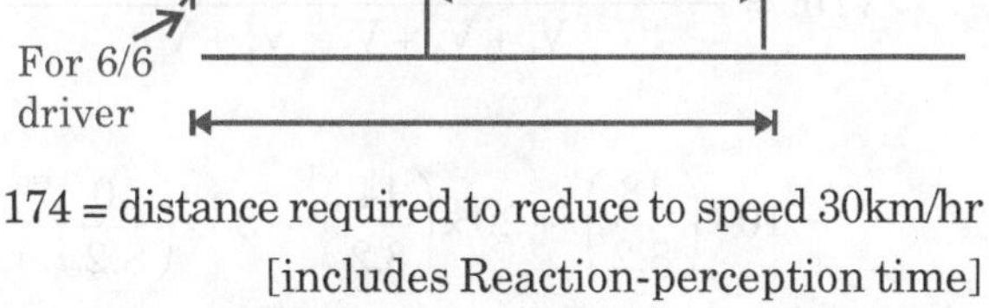

174 = distance required to reduce to speed 30km/hr [includes Reaction-perception time]

$\therefore$ x for 6/9 driver = 174 m

Minimum value of x = 174 – 32 = 142 m

4. $\frac{dv}{dt} = (\alpha - \beta v_0)e^{-\beta t}$

$\therefore\ dv = (\alpha - \beta v_0)e^{-\beta t}\,dt$

Now integrating on both side we get

$$\int dv = \int (\alpha - \beta v_0)e^{-\beta t}dt$$

$$V = \frac{(\alpha - \beta v_0)e^{-\beta t}}{-\beta}$$

Now at $t = 0$, $v = v_o$

$$\therefore\ v_0 = \left(\frac{\alpha - \beta v_o}{-\beta}\right) + c$$

$$\therefore\ c = v_0 + \left(\frac{\alpha - \beta V_0}{-\beta}\right)$$

5. $U = 70 - 0.7\,k$

Capacity, $q = U \times k = Uk$

$= (70 - 0.7\,k)k$

$= 70\,k - 0.7\,k^2$

$\frac{dq}{dk} = 0 \Rightarrow 70 - 0.7 \times 2 \times k = 0$

$k = 50$ V/km

$q = 70 \times 50 - 0.7 \times (50)^2$

$= 1750$ vph

6. $P(x) = \frac{e^{-\lambda}\lambda^x}{x!}, \lambda = 5$

$P(x < 4) = P(x = 0) + P(x = 1) + P(x = 2) + P(x = 3)$

$$= \frac{e^{-5}5^0}{\underline{|0}} + \frac{e^{-5}5^1}{\underline{|1}} + \frac{e^{-5}5^2}{\underline{|2}} + \frac{e^{-5}5^3}{\underline{|3}}$$

$$= e^{-5}\left[1 + 5 + \frac{25}{2} + \frac{125}{6}\right]$$

$= 0.265$

7. Vehicle damage factor

$$VDF = \frac{V_1 \times \left(\frac{W_1}{W_s}\right)^4 + V_2.\left(\frac{W_2}{W_s}\right)^4 + \ldots V_s.\left(\frac{W_s}{W_s}\right)^4}{V_1 + V_2 + V_3 + V_4 + V_5}$$

$$= \frac{10 \times \left(\frac{18}{8.2}\right)^4 + 20 \times \left(\frac{14}{8.2}\right)^4 + 35 \times \left(\frac{10}{8.2}\right)^4 + 15 \times \left(\frac{8}{8.2}\right)^4 + 20 \times \left(\frac{16}{8.2}\right)^4}{10 + 20 + 35 + 15 + 20} = 4.99$$

$$N = \frac{365 \times A[(1 + r)^n - 1]}{r} \times VDF$$

$$= \frac{365 \times 5000 \times [(1.075)^n - 1]}{0.075} \times 4.99$$

$= 237.785$ MSA

So, Design traffic $= 237.785 \times 1.3$

$= 309.21$ MSA

8. $SSD = Vt + \frac{V^2}{2gf}$

$$170 = \left(90 \times \frac{5}{18}\right) + \frac{\left(90 \times \frac{5}{18}\right)^2}{2 \times 9.81 \times 0.35}$$

$\Rightarrow t = 3.15$ s

9. $U = 70 - 0.7\,K$

$\text{capacity} = \frac{U_f \times K_j}{4}$ U_f = free velocity, K_j = jam density

At K_j, $U = 0$

So $K_j = \frac{70}{7} = 100$ veh/km

At $K = 0$, $U = U_f = 70$ km/hr

So, capacity $= \frac{70 \times 100}{4} = 1750$ veh/hr

10. Total time lost in a cycle,

$L = 4 \times 3 = 12$ sec

$$C = \frac{1.5L + 5}{1 - y}$$

$$= \frac{1.5 \times 12 + 5}{1 - (0.2 + 0.3 + 0.25)} = 92s$$

11. We know,

Traffic capacing $C = \frac{1000V}{S}$

Assume speed of vehicle = V

S = 50 m.

$$C = \frac{1000 \times V}{50}$$

$\boxed{C = 20V}$ (Capacity = volume)

Traffic volume = Traffic density × traffic speed.

20V = Traffic density × V

Traffic density $= \frac{20V}{V}$

Traffic density = 20 veh/km.

13. Volume of air voids in percent

$$V_v = \frac{G_f - G_m}{G_m} \times 100$$

G_t = theoretical specific gravity of mixture = 2.441
G_m = bulk dencity/mass density of specimen = 2.324

$$\therefore \; V_v = \frac{2.441 - 2.324}{2.324} \times 100 = 5.03\%$$

Volume of bitumen in percent $V_b = G_m \cdot \frac{W_b}{G_b}$

W_b = percent by weight of bitumen in total mix = 5
G_b = specific gravity of bitumen = 1.1

$$\therefore \; V_b = 2.324 \times \frac{5}{1.1} = 10.56$$

percent voids in mineral Aggregate (VMA) = $V_v + V_b$
VMA= 5.03 + 10.56 = 15.59
Percent voids filled with Bitumen (VFB)

$$VFB = \frac{100 \times V_b}{VMA} = \frac{100 \times 10.56}{15.59}$$

$$= 67.71$$

16.

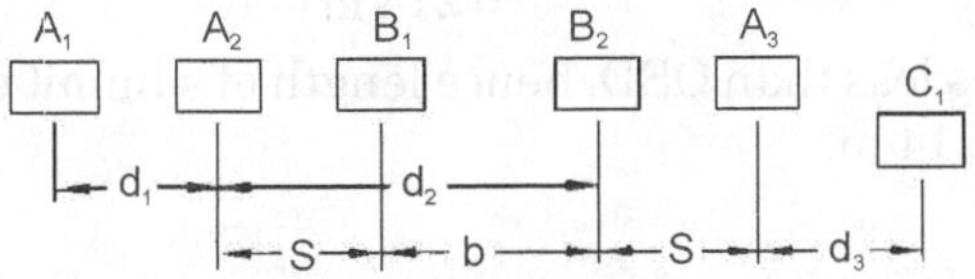

Here, t = reaction time = 2sec
V_b = velocity of overtaken vehicle
= 70 kmph
= 19.44 m/sec

Now $S = 0.7\,V_b + 6$
$= 0.7 \times 19.44 + 6 = 19.60$

and $T = \sqrt{\frac{4S}{a}}$

where, $a = 2.5 \times \frac{5}{18}$ m/sec²

= 0.694 m/sec

$$\therefore \quad T = \sqrt{\frac{4 \times 19.60}{0.694}} = 10.62 \text{ sec}$$

Now $d_1 = V_b \times t$
$= 19.44 \times 2 = 38.88$m
$d_2 = b + 2S$
$= V_b T + 2S$
$= 19.44 \times 10.62 + 2 \times 19.60$
= 245.75m

$d_3 = VT$

$$= 85 \times \frac{5}{18} \times 10.62 = 250.75 \text{ m}$$

∴ Overtaking side distance,
$OSD = d_1 + d_2 + d_3$
$= 38.88 + 245.75 + 250.75$
= 535.38m

17. Spacing of expansion joint,

$$L_e = \frac{\delta'}{1000C(T_2 - T_1)}$$

where, $\delta = \frac{24}{2} = 12 \text{ mm} = 1.2 \text{ cm}$

$$\therefore \quad L_e = \frac{1.2}{100 \times 10 \times 10^{-6} \times 20} = 60\text{m}$$

For contraction joint,

$$\text{frictional resistance} = w \times b \times \frac{h}{1000} \times \frac{L_e}{2} \times f \quad .....(i)$$

$$\text{allowable tension} = \frac{S_e}{F} \times h \times b \times 1000 \quad .....(ii)$$

Equating equations (*i*) and (*ii*), we get

$$wb\frac{h}{1000} \times \frac{L_e}{2} \times f = \frac{S_e}{F} \times h \times b \times 1000$$

$$\Rightarrow \quad L_e = \frac{2 \times 1000^2 Se}{Wf}$$

$$= \frac{2 \times 1000^2 \times 1.6}{24000 \times 1.5} = 8.8\text{m}$$

Taking length of contration joint = 8m
and length of expansion joint = 8 × 8 = 64m.

18. The graphical plot is given here. From the graph, we find correction needed to calculate CBR value. That are made in graph.
After correction, load at 2.5 mm penetration
= 51 kg

5 mm penetration = 83 kg
Area of plunger = 19.6 cm²
∴Pressure for 2.5 mm penetration

$$= \frac{51}{19.6} \text{ kg/cm}^2$$

Pressure for 5.0 mm penetration

$$= \frac{83}{19.6} \text{ kg/cm}^2$$

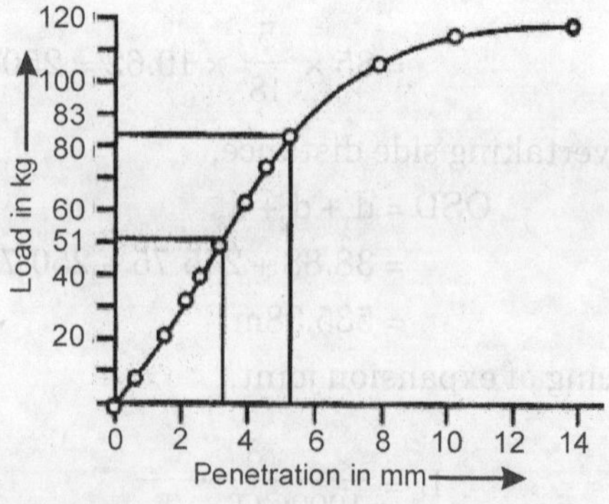

(*i*) CRB value of soil at 2.5 mm

$$= \frac{\text{Pressure of plunger at 2.5 mm pentration of soil}}{\text{Pressure as for standard crushed store at 2.5 mm}} \times 100$$

$$= \frac{51}{19.6} \times \frac{100}{70}$$

$$= 3.71\%$$

(*ii*) CRB of soil at 5.0 mm

$$= \frac{83}{19.6} \times \frac{100}{105} = 4\%$$

Hence adopt CBR value = 4%

19. *Given :* $n_1 = \frac{1}{45}$

$$n_2 = -\frac{1}{60}$$

$$\therefore \quad N = \frac{1}{45} - \left(\frac{1}{60}\right)$$

$$= \frac{4+3}{180} = \frac{7}{180}$$

If L > OSD, then length of summit curve,

$$L = \frac{NS^2}{9.6}$$

$$= \frac{7}{180} \times \frac{(230)^2}{9.6}$$

$$= 214.3$$

This value is less than OSD = 230 m

Assuming L < OSD

$$L = 2S - \frac{9.6}{N}$$

$$= 2 \times 230 - 9.6 \times \frac{180}{7}$$

$$= 213.14 \text{ m}$$

$$= 214 \text{ m (say)}$$

It is less than OSD, hence length of summit curve is 214 m.

16
CHAPTER

Surveying

MCQ TYPE QUESTIONS

2016

1. The minimum number of satellites needed for a GPS to determine its position precisely is

(a) 2 (b) 3
(c) 4 (d) 24

2. The system that uses the Sun as a source of electromagnetic energy and records the naturally radiated and reflected energy from the object is called

(a) Geographical Information System
(b) Global Positioning System
(c) Passive Remote Sensing
(d) Active Remote Sensing

3. The staff reading taken on a workshop floor using a level is 0.645 m. The inverted staff reading taken to the bottom of a beam is 2.960 m. The reduced level of the floor is 40.500 m. The reduced level (expressed in m) of the bottom of the beam is

(a) 44.105 (b) 43.460
(c) 42.815 (d) 41.145

4. Optimal flight planning for a photogrammetric survey should be carried out considering

(a) only side-lap
(b) only end-lap
(c) either side-lap or end-lap
(d) both side-lap as well as end-lap

5. The reduced bearing of a 10 m long line is N30°E. The departure of the line is

(a) 10.00 m (b) 8.66 m
(c) 7.52 m (d) 5.00 m

6. A circular curve of radius R connects two straights with a deflection angle of 60°. The tangent length is

(a) 0.577 R (b) 1.155 R
(c) 1.732 R (d) 3.464 R

2015

7. In a closed loop traverse of 1 km total length, the closing errors in departure and latitude are 0.3 m and 0.4 m, respectively. The relative precision of this traverse will be

(a) 1 : 5000 (b) 1 : 4000
(c) 1 : 3000 (d) 1 : 2000

8. Which of the following statements is FALSE ?

(a) Plumb line is along the direction of gravity
(b) Mean Sea Level (MSL) is used as a reference surface for establishing the horizontal control
(c) Mean Sea Level (MSL) is a simplification of the Geoid
(d) Geoid is an equi-potential surface of gravity

9. In a survey work, three independent angles X, Y and Z were observed with weight W_X, W_Y, W_Z respectively. The weight of the sum of angles X, Y and Z is given by:

(a) $\dfrac{1}{\left(\dfrac{1}{W_X}+\dfrac{1}{W_Y}+\dfrac{1}{W_Z}\right)}$ (b) $\left(\dfrac{1}{W_X}+\dfrac{1}{W_Y}+\dfrac{1}{W_Z}\right)$
(c) $W_X + W_Y + W_Z$ (d) $W_X^2 + W_Y^2 + W_Z^2$

10. In a region with magnetic declination of 2°E, the magnetic Force bearing (FB) of a line AB was measured as N79°50′E. There was a local attraction at A. To determine the correct magnetic bearing of the line, a point O was selected at which there was no local attraction. The magnetic FB of line AO and OA were observed to be S52°40′E and N50°20′W, respectively. What is the true FB of line AB?

(a) N81°50′E (b) N82°10′E
(c) N84°10′E (d) N77°50′E

11. On a circular curve, the rate of super elevation is *e*. While negotiating the curve a vehicle comes to a stop. It is seen that the stopped vehicle does not slide inwards (in the radial direction). The coefficient of side friction is *f*. Which of the following is true?

(a) $e \le f$ (b) $f < e < 2f$
(c) $e \ge 2f$ (d) None of these

12. The combined correction due to curvature and refraction (in m) for distance of 1 km on the surface of Earth is

(a) 0.0673 (b) 0.673
(c) 7.63 (d) 0.763

13. The bearings of two inaccessible stations, S_1 (Easting 500 m, Northing 500 m) and S_2 (Easting 600 m, Northing 450 m) from a station S_3 were observed as 225° and 153° 26′ respectively. The independent Easting (in m) of station S_3 is:

(a) 450.000 (b) 570.710
(c) 550.000 (d) 650.000

2014

14. A levelling is carried out to establish the Reduced Levels (RL) of point R with respect to the Bench Mark (BM) at P. The staff readings taken are given below.

Staff Station	BS	IS	FS	RL
P	1.655 m			100.000m
Q	−0.950 m		−1.500 m	
R			0.750 m	?

If RL of P is +100.000 m, them RL (in m) of R is

(*a*) 103.355 (*b*) 103.155

(*c*) 101.455 (*d*) 100.355

15. Group I lists tool/instrument while Group II lists the method of surveying. Match the tool/instrument with the corresponding method of surveying.

Group I	Group II
P. Alidade	1. Chain surveying
Q. Arrow	2. Levelling
R. Bubble tube	3. Plain table surveying
S. Stadia hair	4. Theodolite surveying

	P	Q	R	S
(*a*)	3	2	1	4
(*b*)	2	4	3	1
(*c*)	1	2	4	3
(*d*)	3	1	2	4

16. The survey carried out to delineate natural features, such as hills, rivers, forests and man-made features, such as towns, villages, buildings, roads, transmission lines and canals is classified as

(*a*) engineering survey (*b*) geological survey

(*c*) land survey (*d*) topographic survey

17. The chainage of the intersection point of two straights is 1585.60 m and the angle of intersection is 140°. If the radius of a circular curve is 600.00 m, the tangent distance (in m) and length of the curve (in m), respectively are

(*a*) 418.88 and 1466.08 (*b*) 218.38 and 1648.49

(*c*) 218.38 and 418.88 (*d*) 418.88 and 218.38

2012

18. The horizontal distance between two stations P and Q is 100 m. The vertical angles from P and Q to the top of a vertical tower at T are 3° and 5° above horizontal, respectively. The vertical angles from P and Q to the base of the tower are 0.1° and 0.5° below horizontal, respectively. Stations P, Q and the tower are in the same vertical plane with P and Q being on the same side of T. Neglecting earth's curvature and atmospheric refraction, the height (in m) of the tower is

(*a*) 6.972 (*b*) 12.387

(*c*) 12.540 (*d*) 128.745

2011

19. Curvature correction to a staff reading in a differential leveling survey is

(*a*) always subtractive

(*b*) always zero

(*c*) always additive

(*d*) dependent on latitude

20. The observations from a closed loop traverse around an obstacle are

Segment	Observation from station	Length (m)	Azimuth (clockwise from magnetic north)
PQ	P	Missing	33.7500°
QR	Q	300.000	86.3847°
RS	R	354.524	169.3819°
ST	S	450.000	243.9003°
TP	T	268.000	317.5000°

What is the value of the missing measurement (rounded off to the nearest 10 mm)?

(*a*) 396.86 m (*b*) 396.79 m

(*c*) 396.05 m (*d*) 396.94 m

2010

21. The local mean time at a place located in longitude 90° 40′ E when the standard time is 6 hours and 30 minutes and the standard meridian is 82° 30′ E is

(*a*) 5 hours, 2 minutes and 40 seconds

(*b*) 5 hours, 57 minutes and 20 seconds

(*c*) 6 hours and 30 minutes

(*d*) 7 hours, 02 minutes and 40 seconds

22. A bench mark has been established at the soffit of an ornamental arch at the known elevation of 100.0 m above mean sea level. The back sight used to establish height of instrument is a inverted staff reading of 2.105 m. A forward sight reading with normally held staff of 1.105 m is taken on a recently constructed plinth. The elevation of the plinth is

(*a*) 103.210 m (*b*) 101.000 m

(*c*) 99.000 m (*d*) 96.790 m

2009

23. In quadrantal bearing system, bearing of a line varies from

(*a*) 0° to 360° (*b*) 0° to 180°
(*c*) 0° to 90° (*d*) 0° N to 90° S

24. The magnetic bearing of a line AB was N 59° 30′ W in the year 1967, when the declination was 4° 10′ E. If present declination is 3° W, the whole circle bearing of the line is

(*a*) 299° 20′ (*b*) 307° 40′
(*c*) 293° 20′ (*d*) 301° 40′

25. Determine the correctness or otherwise of the following **Assertion [a]** and the **Reason [r]:**

Assertion [*a*] :

Curvature correction must be applied when the sights are long.

Reason [*r*] :

Line of collimation is not a level line but is tangential to the level line.

(*a*) Both [*a*] and [*r*] are true and [*r*] is the correct reason for [*a*].
(*b*) Both [*a*] and [*r*] are true but [*r*] is **not** the correct reason for [*a*].
(*c*) Both [*a*] and [*r*] are false.
(*d*) [*a*] is false but [*r*] is true.

2008

26. The type of surveying in which curvature of the earth is taken into account is called

(*a*) Geodetic surveying
(*b*) Plane surveying
(*c*) Preliminary surveying
(*d*) Topographical surveying

27. The plan of a survey plotted to a scale of 10 m to 1 cm is reduced in such a way that a line originally 10 cm long now measures 9 cm. The area of the reduced plan is measured as 81 cm^2. The actual (m^2) of the survey is

(*a*) 10000 (*b*) 6561
(*c*) 1000 (*d*) 656

28. The length and bearings of a closed traverse PQRSP are given below.

Line	Length (m)	Bearing (WCB)
PQ	200	0°
QR	1000	45°
RS	907	180°
SP	?	?

The missing length and bearing , respectively of the line SP are

(*a*) 207 m and 270° (*b*) 707 m and 270°
(*c*) 707 m and 180° (*d*) 907 m and 270°

29. The focal length of the object glass of a tacheometer is 200 mm, the distance between the vertical axis of the tacheometer and the optical centre of the object glass is 100 mm and the spacing between the upper and lower line of the diaphragm axis is 4 mm. With the line of collimation perfectly horizontal, the staff intercepts are 1 m (top), 2 m (middle), and 3 m (bottom). The horizontal distance (m) between the staff and the instrument station is

(*a*) 100.3 (*b*) 103.0
(*c*) 150.0 (*d*) 153.0

30. A light house of 120 m height is just visible above the horizon from a ship. The correct distance (m) between the ship and the light house considering combined correction for curvature and refraction, is

(*a*) 39.098 (*b*) 42.226
(*c*) 39098 (*d*) 42226

2007

31. The plan of a map was photo copied to a reduced size such that a line originally 100 mm, measures 90 mm. The original scale of the plan was 1 : 1000. The revised scale is

(*a*) 1 : 900 (*b*) 1 : 1111
(*c*) 1 : 1121 (*d*) 1 : 1221

32. The following table gives data of consecutive coordinates in respect of a closed theodolite traverse PQRSP.

Station	Northing, M	Southing, M	Easting, M	Westing, M
P	400.75			300.5
Q	100.25		199.25	
R		199.0	399.75	
S		300.0		200.5

The magnitude and direction of error of closure in whole circle bearing are

(*a*) 2.0 m and 45° (*b*) 2.0 m and 315°
(*c*) 2.82 m and 315° (*d*) 3.42 m and 45°

33. The following measurements were made during testing a levelling instrument.

Instrument at	Staff Reading at	
	P_1	Q_1
P	2.800 m	1.700 m
Q	2.700 m	1.800 m

P_1 is close to P and Q_1 is close to Q. If the reduced level of station P is 100.000 m, then reduced level of station Q is

(a) 99.000 m (b) 100.000 m

(c) 101.000 m (d) 102.000 m

34. Two straight lines intersect at an angle of 60°. The radius of a curve joining the two straight lines is 600 m. The length of long chord and mid-ordinates in metres of the curve are

(a) 80.4, 600.0 (b) 600.0, 80.4

(c) 600.0, 39.89 (d) 40,89, 300

35. The magnetic bearing of a line AB is S 45° E and the declination is 5° West. The true bearing of the line AB is

(a) S 45° E (b) S 40° E

(c) S 50° E (d) S 50° W

2006

36. The observed magnetic bearing of a line OE was found to be 185°. It was later discovered that station O had a local attraction of + 1.5°. The true bearing of the line OE, considering a magnetic a magnetic declination of 3.5° E shall be

(a) 180° (b) 187°

(c) 190° (d) 193°

37. A Bench Mark (BM) with Reduced Level (RL) = 155.305 m has been established at the floor of a room. It is required to find out the RL of the underside of the roof (R) of the room using Spirit Leveling. The Back Sight (BS) to the BM has been observed as 1.500 m whereas the Fore Sight (FS) to R has been observed as 0.575 m (Staff held inverted). The RL (m) of R will be

(a) 155.880 (b) 156.230

(c) 157.380 (d) 157.860

38. Consider the following figure, which is an extract from a contour map (scale = 1:20,000) of an area. An alignment of a road at a ruling gradient of 4% is to be fixed from the point O and beyond. What should be the radius of the arc with O as the center to get the point of alignment on the next contour on the map ?

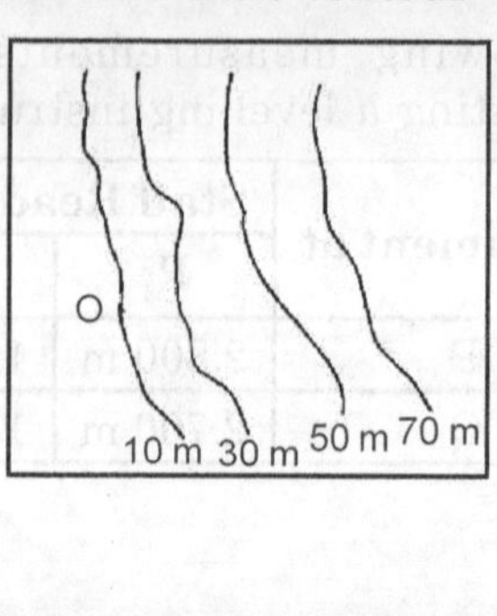

(a) 0.025 cm (b) 0.25 cm

(c) 2.5 cm (d) 5.0 cm

39. In the figure given below, lengths PQ (WCB : 30°) and QR (WCB : 45°) respectively up to three places of decimal are

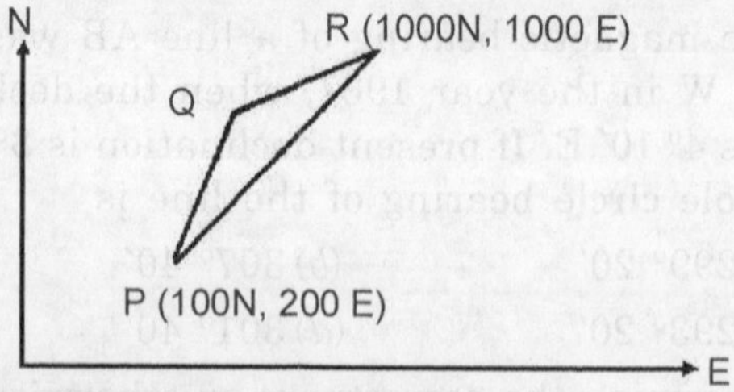

(a) 273.205, 938.186

(b) 273.205, 551.815

(c) 551.815, 551.815

(d) 551.815, 938.186

40. During a leveling work along a falling gradient using a Dumpy Level and a Staff of 3m length, following successive readings were taken : 1.785, 2.935, 0.360, 1.320.

What will be the correct order of booking these four readings in a level book ?

(BS : Back Sight, IS : Intermediate Sight, FS : Fore Sight)

(a) BS, FS, BS, FS.

(b) BS, IS, FS, FS

(c) BS, IS, IS, FS

(d) BS, IS, BS, FS

NUMERICAL TYPE QUESTIONS

2015

1. In a leveling work, sum of the Back Sight (B.S.) and Fore Sight (F.S.) have been found to be 3.085 m and 5.645 m respectively. If the Reduced Level (R.L.) of the starting station is 100.000 m, the R.L. (in m) of the last station is ——

2. Two pegs A and B were fixed on opposite banks of a 50 m wide river. The level was set up at A and the staff readings on Pegs A and B were observed as 1.350 m and 1.550 m, respectively.

Thereafter the instrument was shifted and set up at B. The staff readings on Pegs B and A were observed as 0.750 m and 0.550 m, respectively. If the R.L. of Peg A is 100.200 m, the R.L. (in m) of Peg B is ——

2014

3. A tacheometer was placed at point P to estimate the horizontal distances PQ and PR. The corresponding stadia intercepts with the telescope kept horizontal, are 0.320 m and 0.210 m, respectively. The $\angle$QPR is measured to be 61° 31′ 30″. If the stadia multiplication constant = 100 and stadia addition constant = 0.10 m, the horizontal distance (in m) between the points Q and R is ________

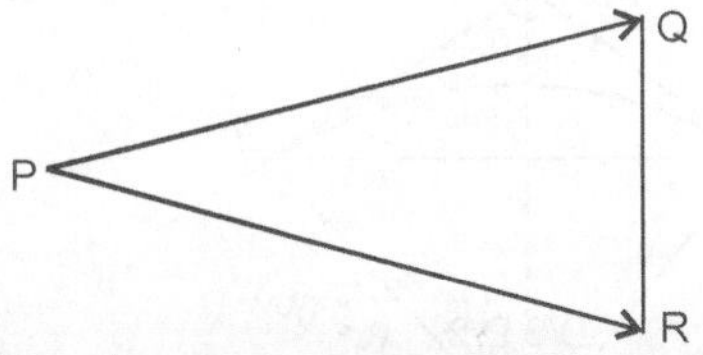

2013

4. A theodolite is set up at station A and a 3 m long staff is held vertically at station B. The depression angle reading at 2.5 m marking on the staff is 6°10′. The horizontal distance between A and B is 2200 m. Height of instrument at station A is 1.1 m and R.L. of A is 880.88 m. Apply the curvature and refraction correction, and determine the R.L. of B (in m). ________

ANSWERS

MCQ Type Questions

1. (*c*)	**2.** (*c*)	**3.** (*d*)	**4.** (*d*)	**5.** (*d*)	**6.** ($R\tan\left(\frac{60}{2}\right)$)	**7.** (*d*)	**8.** (*b*)	**9.** (*a*)	
10. (*c*)	**11.** (*a*)	**12.** (*a*)	**13.** (*c*)	**14.** (*c*)	**15.** (*d*)	**16.** (*d*)	**17.** (*c*)	**18.** (*b*)	**19.** (*a*)
20. (*b*)	**21.** (*d*)	**22.** (*d*)	**23.** (*c*)	**24.** (*b*)	**25.** (*a*)	**26.** (*a*)	**27.** (*a*)	**28.** (*b*)	**29.** (*a*)
30. (*d*)	**31.** (*b*)	**32.** (*c*)	**33.** (*c*)	**34.** (*a*)	**35.** (*b*)	**36.** (*b*)	**37.** (*c*)	**38.** (*c*)	**39.** (*a*)
40. (*a*)									

Numerical Type Questions

1. (97.44) **2.** (100) **3.** (28.0 to 29.0) **4.** (641.9 to 642.3)

EXPLANATIONS

MCQ TYPE QUESTIONS

3. Floor reduced level = 40.5 m

RL of bottom of beam = 40.5 + 0.645 + 2.96

= 44.105 m

4. For optimal flight planning for a photogrammateric survey, both side-lap as well as end lap are considered.

5.

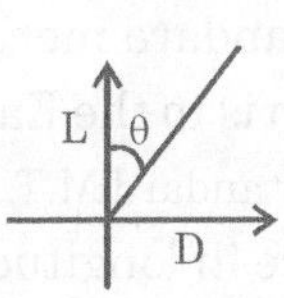

$L = 10$ m

θ = N. 30° E

Departure $= L \sin\theta$

$= 10 \sin 30°$

$= 10 \times (1/2) = 5$ m

6. Radius (R)

Deflection, L = 60°

$$\text{Tangent length} = R\tan\frac{\Delta}{2}$$

$$= R\tan\left(\frac{60}{2}\right)$$

7. Given $\Sigma L = 0.4$

& $\Sigma D = 0.3$ m

Closing error, $e = \sqrt{\Sigma L^2 + \Sigma D^2}$

$= \sqrt{(0.4)^2 + (0.3)^2} = 0.5$

The relative precision $= \frac{e}{p}$

$= \frac{1}{p/e} = \frac{1}{1000/0.5} = \frac{1}{2000}$

where $p = 1$ km $= 1000$ m

8. MSL is used as a usual datum for the purpose of measuring the heights of objectes on the land. This is a vertical datum.

9. In surveying by Law of weights, the weight of sum of two measurements is given as,

$$A + B = \frac{1}{\frac{1}{W_1} + \frac{1}{W_2}}$$

Here, $$X + Y + Z = \frac{1}{\frac{1}{W_X} + \frac{1}{W_Y} + \frac{1}{W_Z}}$$

10. d = 2°E

Magnetic F.B. of AB = N79°50′E = 79°50

Correct F.B. of OA = N50°20°W = 309°40′

$\therefore$ Correct B.B. of OA = 129°40′

Since, observed F.B. of AO = observed B.B. of OA

= 552°40′ E = 127°20′

Error = M.V. – T.V. = 2°20′

Therefore, Correction is + 2°20′

T.B. of F.B. of AB = N79°50′E + 2 + 2°20′

= N84°10′E

11. For no skidding,

Centrifugal force $\leq$ frictional force

and super elevation conteracts the effect of centrifuagal force.

$\therefore$ For no skidding $e \leq f$.

12. Combined correction due to curvature and refraction is given as,

$$C = 0.0673\,d^2 = 0.0673 \times (1)^2 = 0.0673$$

13.

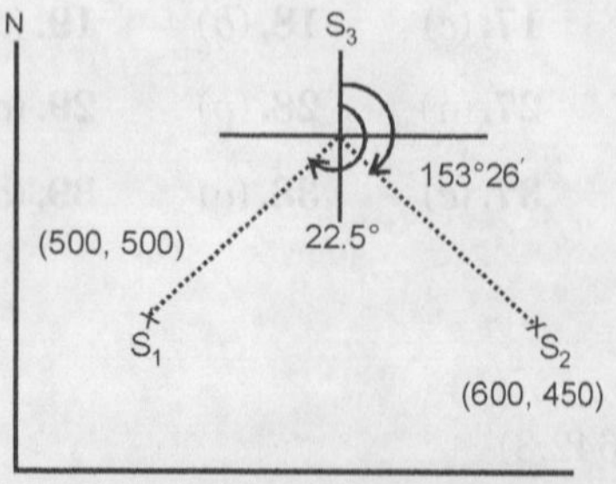

Now, considering the northings of point S_3 as below

$\Rightarrow$ Northing of S_3 = Northing of S_1 + S_1S_3 cos 45°

Also it can be written as (in terms of S_2)

= Northing of S_2 + S_2S_3 cos (26°34′)

Equating the above northings from S_1 and S_2, we get

Northing of S_1 + S_1S_3 cos 45°

= Northing of S_2 + S_2S_3 cos 26°34′

500 + S_1S_3 cos 45° = 450 + S_2S_3 cos 26° 34′

S_1S_3 cos 45° – S_2S_3 cos 26° 34′ = –50 ...(*i*)

Similarly easting equation is

500 + S_1S_3 sin 45° = 600 – S_2S_3 sin 26° 34′

S_1S_3 sin 45° + S_2S_3 sin 26° 34′ = 100 ...(*ii*)

By (*i*) and (*ii*) $\Rightarrow$ S_2S_3 = 111.80

Easting = 600 – 50 = 550

14. RL of P = 100.00 m

RL of Q = 100 + 1.655 + 1.5 = 103.155 m

RL of R = 103.155 – (0.95 + 0.75)

= 101.455 m

15. Alidade → used in plane table surveying

Arrow → chain surveying

Bubble tube → leveling

Stadia hair → Theodolite surveying

16. **Topographic Survey:** Natural features, such as hills, river, forests and man-made features, such as towns, villages, building roads, transmission lines and canals.

The correct choise is (*d*).

17. Length of curve $= \dfrac{\pi R}{180} D = \dfrac{\pi \times 600}{180} \times 40°$

= 418.88m

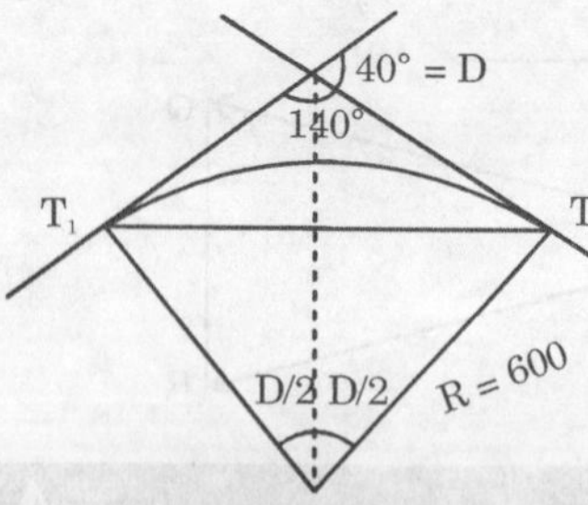

Tangent distance= R.tan D/2 = 600 tan 20° = 218.38

the correct choice is (*c*).

18. $x(\tan 3 + \tan 0.1) = (x - 100)(\tan 5° + \tan 0.5°)$

$\Rightarrow$ $x = 228.758$

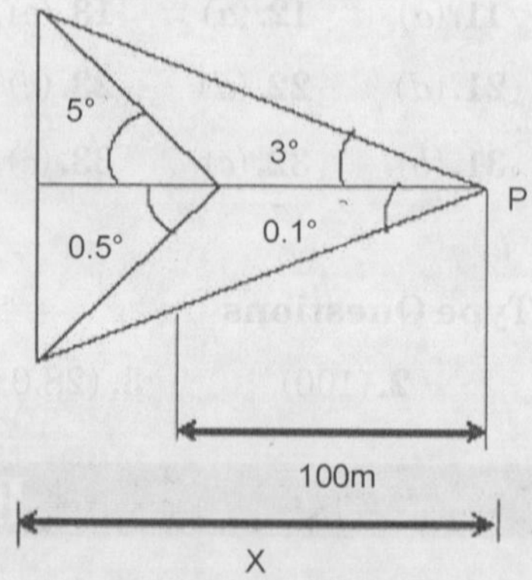

Height of tower = 228.758 (tan3° + tan0.1°)

= 12.387 m

21. Local Mean Time (L.M.T)

= Standard M.T. ± Difference in longitudes $\left(\dfrac{E}{W}\right)$

Use (+) sign if meridian of place is to the East of the standard meridian, and (–) sign if it is to the West of the standard meridian.

Here meridian is to the East of standard meridian

$\therefore$ L.M.T. = Standard M.T. + Difference in longitudes

$\therefore$ Difference in longitudes = 90° + 40° – 82° 30′

= 8° 10′

Now 24 hours = 360°

$\Rightarrow$ 1° = 4° minutes

$\Rightarrow$ 1′ = 4° seconds

$\therefore$ 8° 10′ = 32 minutes 40 seconds

So, L.M.T = 6 hrs 30 min + 32 min 40 sec

= 7 hrs 2 min 40 sec.

22.

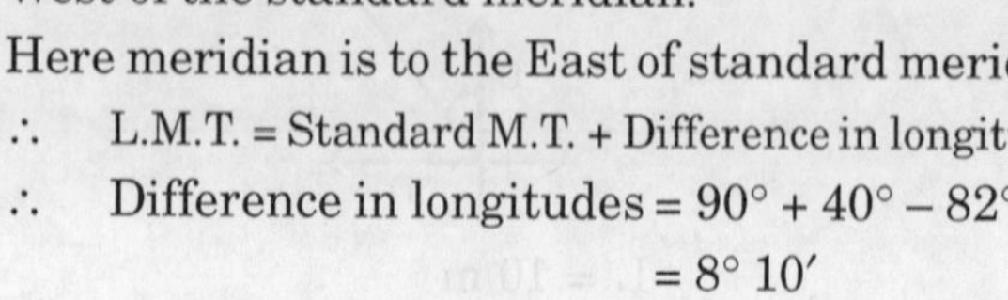

Elevation of line of sight

$= 100 - \text{Staff reading}$

$= 100 - 2.105 = 97.895 \text{ m}$

Elevation of plinth $= 97.895 - 1.105 = 96.790$ m

24. True bearing = Magnetic Bearing + Declination

+ East
– West

Initially, magnetic bearing in WCB

$= 360° - 59\ 30' = 300°\ 30'$

Initially, true bearing

$= 300°\ 30° + 4°\ 10'$

$= 304°\ 40'$

Now, $304°\ 40' = \text{Magnetic bearing} - 3°$

$\therefore$ Magnetic bearing $= 304°\ 40' + 3° = 307°\ 40'$

26. In the Geodatic Surveying, shape of the earth is taken into account, i.e. curvature of surface.

27. Modified length $= \dfrac{\text{Modified R.F.}}{\text{Actual R.F.}} \times \text{Actual Lenght.}$

$$\therefore \quad 9 \text{ cm} = \frac{(R.F.)_1}{\frac{1}{1000}} \times 10$$

$$\text{or} \quad (\text{R. F}_1) = \frac{9}{10000} = \frac{1}{1111.11}$$

$$\text{Modified area} = \left(\frac{\text{Modified.R.F}}{\text{Actual R.F}}\right)^2 \times \text{Actual area}$$

$$\therefore \quad \text{Actual area} = \frac{81}{\left(\frac{1/1111.11}{1/1000}\right)^2}$$

$$= 81 \times \left(\frac{1111.11}{1000}\right)^2 = 100 \text{ cm}^2$$

But actual scale is 1 cm to 10 m, therefore

Area of field $= 100 \times (10)^2 = 10000 \text{ m}^2$

28. Let reduced bearing of line SP is Sθ° W.

Let length of line SP is l.

Line	Length	WCB	Reduced Bearing	Latitude	Departure
PQ	200	0°	N0°	+200	0
QR	1000	45°	N 45°E	+707	+707
RS	907	180°	S0°	–907	0
SP	l		Sθ°W	$-l\cos\theta$	$-l\sin\theta$

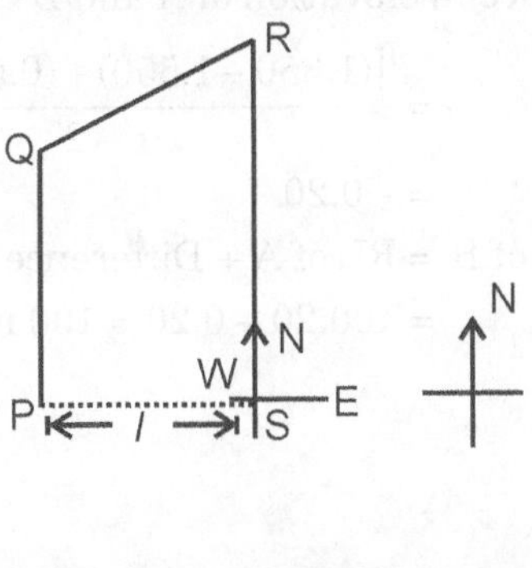

But $\quad \Sigma \text{ Latitude} = 0$

$\Rightarrow \quad l\cos\theta = 0$

Now $\quad \Sigma \text{ Departure} = 0$

$\Rightarrow \quad \theta = 90°$

$\Rightarrow \quad 707 - l\sin 90° = 0$

$\Rightarrow \quad l = 707 \text{ m}$

$\therefore$ Bearing $= 180 + 90° = 270°$ (WCB)

29. Focal length, $\quad f = 200$ mm

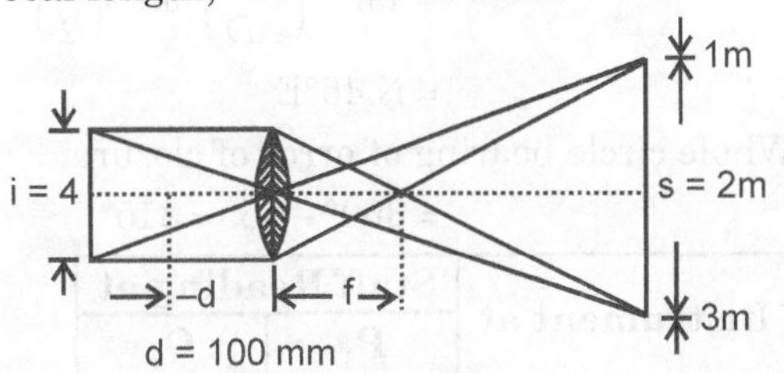

Distance between instrument station and staff station

$$= \frac{f}{i} \times s + (f + d)$$

$$= \frac{200}{4} \times 2 + \frac{(200 + 100)}{1000}$$

$$= 50 \times 2 + 0.3 = 100.3 \text{ m}$$

30. We know, $\quad h = \dfrac{D^2}{2R}(1 - 2m)$

Here, R = radius of earth

m = mean refraction coefficient

= 0.08 (for sight over sea)

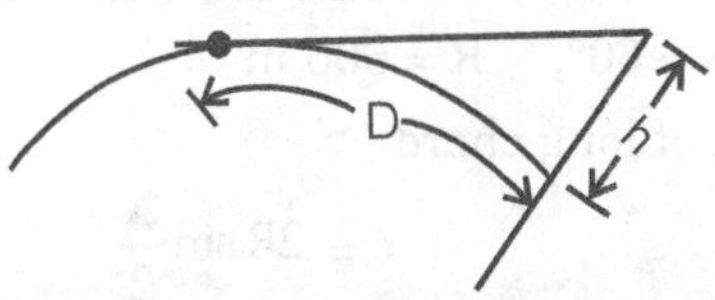

$\therefore \quad h = 0.066142\ D^2$

where, h in meter and D in kilometer

Given: $\quad h = 120$ m.

then $\quad D = \sqrt{\dfrac{120}{0.066142}} = 42.226 \text{ km} = 42226 \text{ m}$

31. $\quad 100 : x :: 1 : 1000$

$\Rightarrow \quad x = 1000 \times 100$

Again, $\quad 90 : x :: 1 : y$

$\Rightarrow \quad x = 90xy$

$\Rightarrow \quad y = \dfrac{x}{90} = \dfrac{1000 \times 100}{90} = 1111.11$

Hence, revised scale

$1 : y = 1 : 1111$

32.

Station	Latitude (L)	Departure (D)
P	+ 400.75	– 300.5
Q	+ 100.25	+199.25
R	–199.0	+299.75
S	– 300.0	–200.5

$\Delta L = 2,\ \Delta D = 2$

$$\text{Error of closure} = \sqrt{(\Delta L^2)^2 + (\Delta D)^2}$$
$$= \sqrt{(2)^2 + (2)^2}$$
$$= \sqrt{8} = 2.82 \text{ m}$$

Bearing of error of closure

$$= \tan^{-1}\left(\frac{\Delta L}{\Delta D}\right) \tan^{-1}\left(\frac{2}{2}\right)$$
$$= \text{N } 45°\text{E}$$

Whole circle bearing of error of closure

$$= 360° - 45° = 315°$$

33.

Instrument at	Staff Reading at	
	P_1	Q_1
P	2.800 m	1.700 m
Q	2.700 m	1.800 m

Since point P_1, Q_1 are very near to point P & Q respectively, therefore

dip correction of level line

$$= 2.8 - 2.7 = 0.1 \text{ m}$$

R.L. of point Q = R.L. of P
+ staff reading at P_1
– dip correction
– staff reading at Q_1
= 100 + 2.8 – 0.1 – 1.7 = **101 m**

34. Here, $\Delta = 60°$, R = 600 m

Length of long chord

$$\ell = 2R\sin\frac{\Delta}{2}$$
$$= 2 \times 600 \times \sin 30° = 600 \text{ m}$$

Length of mid ordinate

$$= R\left(1 - \cos\frac{\Delta}{2}\right)$$
$$= 600\,(1 - \cos 30°) = \mathbf{80.38\ m}$$

35. Magnetic bearing = S 45° E

Declination = 5°W

True bearing = Magnetic bearing ± Declination (E/W)
= S 45E – 5° = **S 40° E**

36. True bearing of line = 185° + 3.5° – 1.5° = 187°

37.

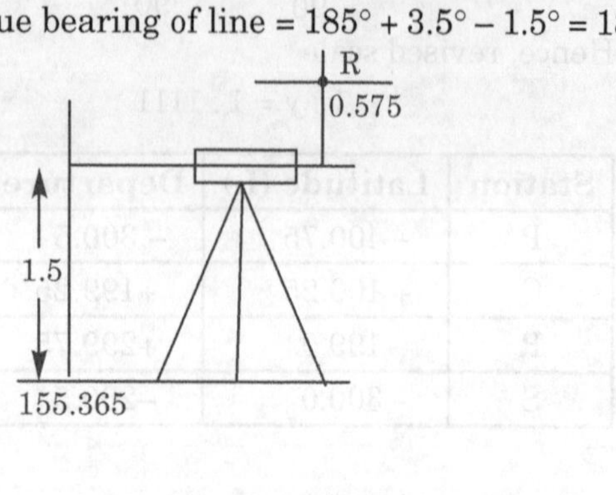

Reduced level (RL) of R

$$= 155.305 + 0.575 + 1.5 = 157.38 \text{ m}$$

39. From figure,

$$PR = \sqrt{800^2 + 900^2} = 1204.159$$
$$\angle P = 90° - 30° - \tan^{-1}\left(\frac{900}{800}\right)$$
$$= 90° - 30° - 48°\,21'\,59.26''$$
$$= 11°\,38'\,0.74'',$$
$$\angle Q = 180° + 30° - 45° = 165°$$

and $\angle R = 3°21'\,59.26''$

By Sine rule,

$$\frac{PR}{\sin Q} = \frac{QR}{\sin P}$$
$$\therefore \quad \frac{1204.159}{\sin 165°} = \frac{QR}{\sin(11°38'0.74'')}$$
$$\Rightarrow \quad QR = 938.186$$

Similarly,

$$\frac{PQ}{\sin 3°21'59.26''} = \frac{1204.159}{\sin 165°}$$
$$\Rightarrow \quad PQ = 273.205$$

40.

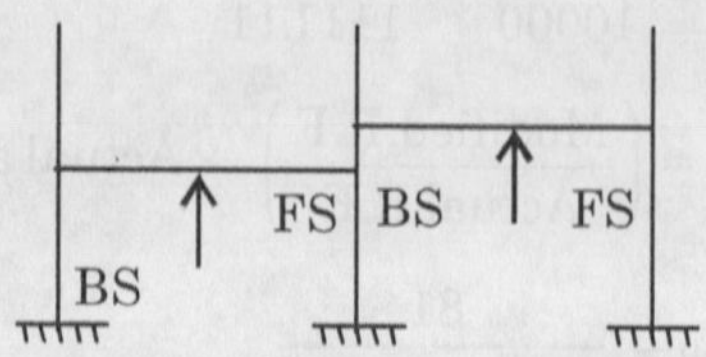

NUMERICAL TYPE QUESTIONS

1. B.S. = 3.085 m
FS = 5.645 m
RL (first) = 100 m
RL (last) = 100 + 3.085 – 5.645 = 97.44 m

2.

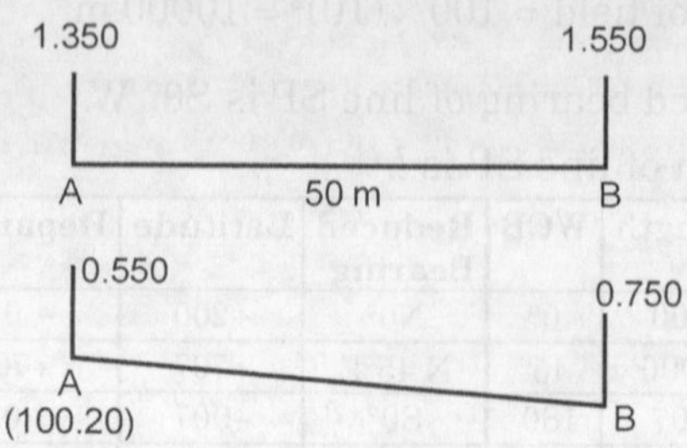

Difference in elevation of A and B

$$= \frac{[(1.350 - 1.550) + (0.550 - 0.750)]}{2}$$
$$= -0.20$$

$\therefore$ RL of B = RL of A + Difference
= 100.20 – 0.20 = 100 m

SOLVED PAPER - 2017

INSTRUCTIONS

1. Total of 65 questions carrying 100 marks, out of which 10 questions carrying a total of 15 marks are in General Aptitude (GA)
2. The Engineering Mathematics will carry around **15% of the total marks,** the General Aptitude section will carry **15% of the total marks** and the **remaining 70% of the total marks.**
3. **Types of Questions**
   (*a*) **Multiple Choice Questions (MCQ)** carrying 1 or 2 marks each in all papers and sections. These questions are objective in nature, and each will have a choice of four options, out of which the candidate has to mark the correct answer(s).
   (*b*) **Numerical Answer Questions** of 1 or 2 marks each in all papers and sections. For these questions the answer is a real number, to be entered by the candidate using the virtual keypad. No choices will be shown for these type of questions.
4. For **1-mark** multiple-choice questions, **1/3 marks** will be deducted for a wrong answer. Likewise, for **2-marks** multiple-choice questions, **2/3** marks will be deducted for a wrong answer. There is no negative marking for numerical answer type questions.

Chapter-Wise Analysis

#	Chapters	Marks		
		1	2	Total
1	General Aptitude	5	5	15
2	Engineering Mathematics	4	4	12
3	Engineering Mechanics	-	1	2
4	Solid Mechanics	1	5	11
5	Structural Analysis	1	1	3
6	Construction Materials and Management	2	-	2
7	Concrete Structures	1	2	5
8	Steel Structures	-	1	2
9	Soil Mechanics and Foundation Engineering	5	4	13

#	Chapters	Marks		
		1	2	Total
10	Fluid Mechanics and Hydraulics	1	2	5
11	Hydrology	2	1	4
12	Irrigation	-	1	2
13	Water Supply	-	-	-
14	Solid Waste and Waste Water Treatment	-	2	4
15	Air Pollution	1	1	3
16	Noise Pollution	1	1	3
17	Transportation Engineering	3	2	7
18	Surveying	3	2	7

TECHNICAL SECTION

(Q.1 – 25) : Carry One Mark Each

1. The VPI (vertical point of intersection) is 100 m away (when measured along the horizontal) from the VPC (vertical point of curvature). If the vertical curve is parabolic the length of the curve (in meters and measured along the horizontal) is ______

2. Consider a rigid retaining wall with partially submerged cohesionless backfill with a surcharge. Which one of the following diagrams closely represents the Rankine's active earth pressure distribution against this wall?

(*a*)  (*b*) 

(*c*)  (*d*) 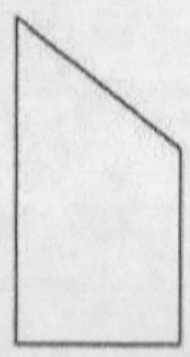

3. The most important type of species involved in the degradation of organic matter in the case of activated sludge process is

(*a*) autotrophs (*b*) heterotrophs
(*c*) prototrophs (*d*) photo-autotrophs

4. The infiltration capacity of a soil follows the Horton's exponential model $f = c_1 + c_2 e^{-kt}$. During an experiment, the initial infiltration capacity was observed to be 200 mm/hr. After a long time the infiltration capacity was reduced to 25 mm/h, if the infiltration capacity after 1 hour was 90 mm/h the value of the decay rate constant k (in h^{-1} up to two decimal places) is______

5. Let G be the specific gravity of soil solids, w the water content in the soil sample, γ_w the unit weight of water, and γ_d the dry unit weight of the soil. The equation for the zero air voids line in a compaction test plot is

(*a*) $\gamma_d = \dfrac{G\gamma_w}{1+Gw}$

(*b*) $\gamma_d = \dfrac{G\gamma_w}{Gw}$

(*c*) $\gamma_d = \dfrac{G_w}{1+\gamma_w}$

(*d*) $\gamma_d = \dfrac{G_w}{1-\gamma_w}$

6. A two-faced fair coin has its faces designated as head (H) and tail (T). This coin is tossed three times in succession to record the following outcomes: H.H.H. If the coin is tossed one more time, the probability (up to one decimal place) of obtaining H again given the previous realizations of H, H and H. would be ______

7. Consider the following statements related to the pore pressure parameters. A and B:

P. A always lies between 0 and 1.0

Q. A can be less than 0 or greater than 1.0

R. B always lies between 0 and 1.0

S. B can be less than 0 or greater than 1.0

For these statements, which one of the following options is correct ?

(*a*) P and R (*b*) P and S
(*c*) Q and R (*d*) Q and S

8. During a storm event in a certain period, the rainfall intensity is 3.5 cm/hour and the ϕ – index is 1.5 cm/hour. The intensity of effective rainfall (in cm/hour up to one decimal place) for this period is ________

9. While aligning a hill road with a ruling gradient of 6%, a horizontal curve of radius 50 m is encountered. The grade compensation (in percentage up two equal places) to be provided for this case would be ________

10. Let the characteristic strength be defined as that value, below which not more than 50% of the results are expected to fall. Assuming a standard deviation of 4 MPa, the target mean strength (in MPa) to be considered in the mix design of a M25 concrete would be

(*a*) 18.42 (*b*) 21.00
(*c*) 25.00 (*d*) 31.58

11. Following observations have been made for the elevation and temperature to ascertain the stability of the atmosphere.

Elevation (in m)	Temperature (in °C)
10	15.5
60	15.0
130	14.3

The atmosphere is classified as

(*a*) Stable

(*b*) Unstable

(*c*) Neutral

(*d*) Inverse

12. A sheet pile has an embedment depth of 12 m in a homogeneous soil stratum. The coefficient of permeability of soils is 10^{-6} m/s. Difference in the water levels between the two sides of the sheet pile is 4m. The flow net is constructed with five number of flow lines and eleven number of equipotential lines. The quantity of seepage (in cm^3/s per m. up to one decimal place) under the sheet pile is_____

13. In a material under a state of plane strain, a 10 × 10 mm square centred at a point gets deformed as shown in the figure.

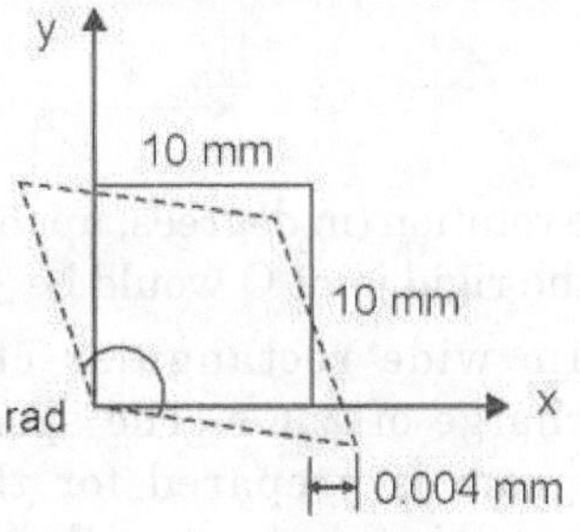

If the shear strain γ_{xy} at this point is expressed as 0.001 k (in rad.) the value of k is

(*a*) 0.50 (*b*) 0.25

(*c*) –0.25 (*d*) –0.50

14. The divergence of the vector field $V = x^2\,i + 2y^3\,j + z^4\,k$ at x = 1, y = 2, z = 3 is _________

15. The safety within a roundabout and the efficiency of a roundabout can be increased respectively by

(*a*) increasing the entry radius and increasing the exit radius.

(*b*) increasing the entry radius and decreasing the exit radius.

(*c*) decreasing the entry radius and increasing the exit radius.

(*d*) decreasing the entry radius and decreasing the exit radius.

16. Given that the scope of the construction work is well-defined with all its drawings, specifications, quantities and estimates. Which one of the following types of contract would be most preferred?

(*a*) EPC contract

(*b*) Percentage rate contract

(*c*) Item rate contract

(*d*) Lump sum contract

17. As per Noise pollution (Regulation and Control) Rules 2000 of India, the day time noise limit for a residential zone expressed in dB (A) L_{eq} is

(*a*) 55 (*b*) 65

(*c*) 75 (*d*) 85

18. The method of orientation used, when the plane table occupies a position not yet located on the map is called as

(*a*) traversing (*b*) radiation

(*c*) levelling (*d*) resection

19. Consider the following simultaneous equation (with c_1 and c_2 beings constants):

$$3x_1 + 2x_2 = c_1$$
$$4x_1 + x_2 = c_2$$

The characteristic equation for these simultaneous equations is

(*a*) $\lambda^2 - 4\lambda - 5 = 0$ (*b*) $\lambda^2 - 4\lambda + 5 = 0$

(*c*) $\lambda^2 + 4\lambda - 5 = 0$ (*d*) $\lambda^2 + 4\lambda + 5 = 0$

20. If a centrifugal pump has an impeller speed of N (in rpm) discharge Q (in m^3/s) and the total head H (in m), the expression for the specific speed N_s of the pump is given by _____

(*a*) $N_s = \frac{NQ^{0.5}}{H^{0.5}}$ (*b*) $N_s = \frac{NQ^{0.5}}{H}$

(*c*) $N_s = \frac{NQ^{0.5}}{H^{0.75}}$ (*d*) $N_s = \frac{NQ}{H^{0.75}}$

21. For a broad gauge railway track on a horizontal curve of radius R (in m) the equilibrium cant *e* required for a train moving at a speed of V (in km per hour) is _____

(*a*) $e = 1.676\frac{V^2}{R}$

(*b*) $e = 1.315\frac{V^2}{R}$

(*c*) $e = 0.80\frac{V^2}{R}$

(*d*) $e = 0.60\frac{V^2}{R}$

22. Consider the frame shown in figure.

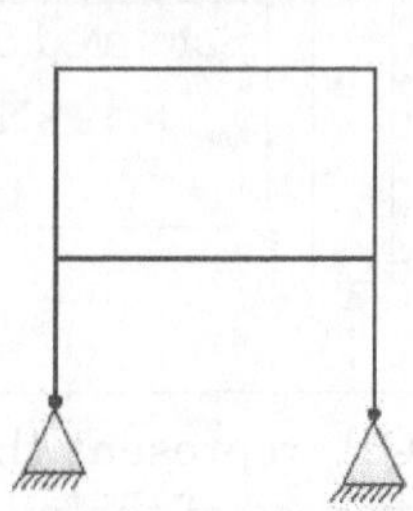

If the axial and shear deformations is different members of the frame are assumed to be negligible the reduction in the degree of kinematical indeterminacy would be equal to

(*a*) 5 (*b*) 6

(*c*) 7 (*d*) 8

23. Let w = (f(x, y), where x and y are functions of t. Then according to the chain rule $\frac{dw}{dt}$ is equal to

(*a*) $\frac{dw}{dx}\frac{dx}{dt}+\frac{dw}{dy}\frac{dt}{dt}$ (*b*) $\frac{\partial w}{\partial x}\frac{\partial x}{\partial t}+\frac{\partial w}{\partial y}\frac{\partial y}{\partial t}$

(*c*) $\frac{\partial w}{\partial x}\frac{dx}{dt}+\frac{\partial w}{\partial y}\frac{dy}{dt}$ (*d*) $\frac{dw}{dx}\frac{\partial x}{\partial t}+\frac{dw}{dy}\frac{\partial y}{\partial t}$

24. For a construction project the mean and standard deviation of the completion time are 200 days and 6.1 days respectively. Assume normal distribution and use the value of standard normal deviate z = 1.64 for the 95% confidence level. The maximum time required (in days) for the completion of the project would be ______

25. The plate load test was conducted on a clayey strate by using a plate of 0.3m × 0.3 m dimension, and the ultimate load per unit area for the plate was found to be 180 kPa. The ultimate bearing capacity (in kPa) of a 2m wide square footing would be

(*a*) 27 (*b*) 180

(*c*) 1200 (*d*) 2000

(Q.26 – 55) : Carry Two Marks Each

26. Two identical concrete piles having the plan dimensions 50 cm × 50 cm are driven into a homogeneous sandy layer as shown in the figures. Consider the bearing capacity factor Nq for ϕ = 30° as 24.

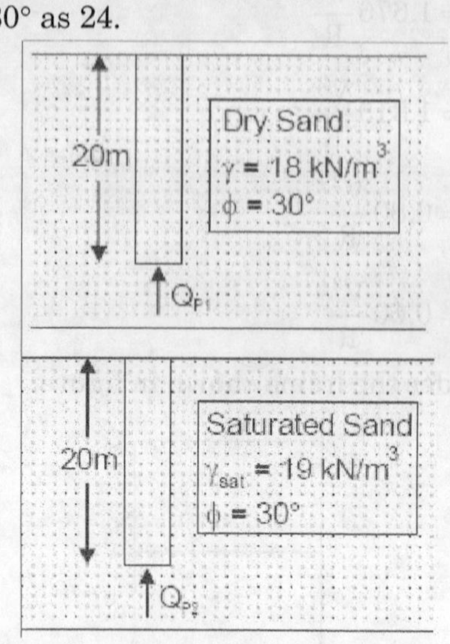

If QP_1 and QP_2 represent the ultimate point bearing resistances of the piles under dry and submerged conditions, respectively, which one of the following statements is correct?

(*a*) $QP_1 > QP_2$ by about 100%

(*b*) $QP_1 < QP_2$ by about 100%

(*c*) $QP_1 > QP_2$ by about 5%

(*d*) $QP_1 < QP_2$ by about 5%

27. Consider the portal frame shown in the figure and assume the modulus of elasticity, E = 2.5 × 10^4 MPa and the moment of mertia, I = 8 × 10^8 mm^4 for all the members of the frame.

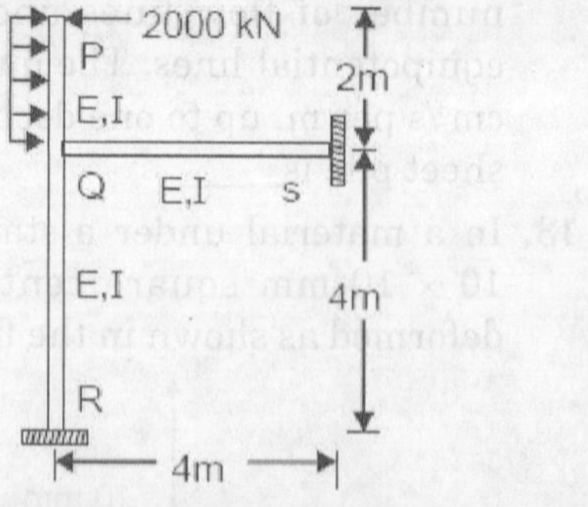

The rotation (in degrees, up to one decimal place) at the rigid joint Q would be ______

28. A 1m wide rectangular channel carries a discharge of $2m^3/s$. The specific energy-depth diagram is prepared for the channel. It is observed in this diagram that corresponding to a particular specific energy. The subcritical depth is twice the supercritical depth. The subcritical depth (in meters, up to two decimal places) is equal to ____.

29. For a given water sample, the ratio between $BOD_{5\text{-day, }20°C}$ and the ultimate BOD is 0.68. The value of the reaction rate constant *k* (on base *e*) in day^{-1}, up to two decimal places) is ______.

30. The tangent to the curve represented by $y = x \ln x$ is required to have 45° inclination with the *x*-axis. The coordinates of the tangent point would be

(*a*) (1, 0) (*b*) (0, 1)

(*c*) (1, 1) (*d*) $(\sqrt{2}, \sqrt{2})$

31. Consider the following statements:

P. Walls of one brick thick are measured in square meters.

Q. Walls of one brick thick are measured in cubic meters.

R. No deduction in the brickwork quantity is made for openings in walls up to $0.1m^2$ area.

S. For the measurement of excavation from the borrow pit in a fairly uniform ground, deadmen are left at suitable intervals.

For the above statements, the correct option is

(*a*) P – False; Q – True: R – False: S – True

(*b*) P – False; Q – True: R – False: S – False

(*c*) P – True; Q – False: R – True: S – False

(*d*) P – True; Q – False: R – True: S – True

32. An observer standing on the deck of a ship just sees the top of a lighthouse. The top of the lighthouse is 40m above the sea level and the height of the observer's eye is 5m above the sea level. The distance (in km. up to one decimal place) of the observer from the lighthouse is______

33. Two prismatic beams having the same flexual rigidity of 1000 kN-m² are shown in the figures.

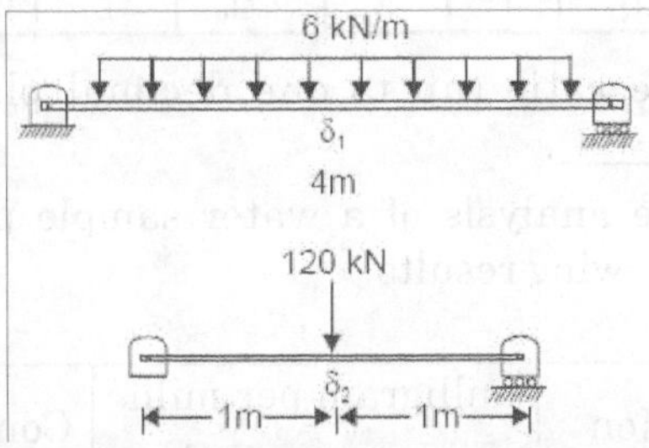

If the mid-span diflections of these beams are denoted by δ_1 and δ_2 (as indicated in the figures), the correct option is

(*a*) $\delta_1 = \delta_2$ (*b*) $\delta_1 < \delta_2$

(*c*) $\delta_1 > \delta_2$ (*d*) $\delta_1 >> \delta_2$

34. The composition of a municipal solid waste sample is given below :

Component	Percent by Mass	Moisture Content (%)	Energy Content (kJ/kg. on as – discarded basis)
Food Waste	20	70	2500
Paper	10	4	10000
Cardboard	10	4	8000
Plastics	10	1	14000
Garden Trimmings	40	60	3500
Wood	5	20	14000
Tin Cans	5	2	100

The difference between the energy content of the waste sample calculated on dry basis and as discarded basis (in kJ/kg) would be________.

35. Two plates of 8 mm thickness each are connected by a fillet weld of 6 mm thickness as shown in the figure.

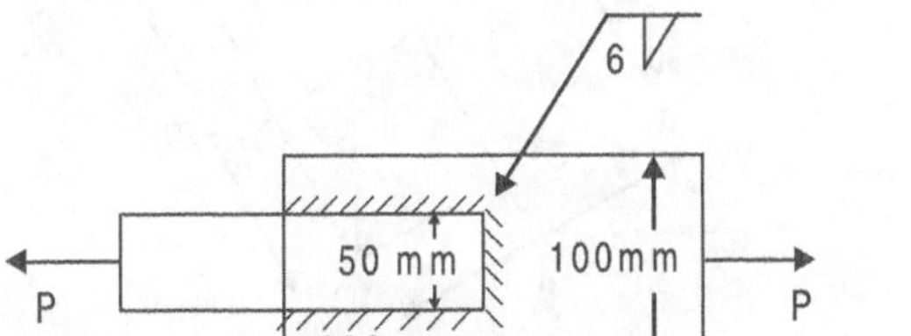

The permissible stresses in the plate and the weld are 150 MPa and 110 MPa. respectively. Assuming the length of the weld shown in the figure to be the effective length the permissible load P(in kN) is ______.

36. For the construction of a highway a cut is to be made as shown in the figure.

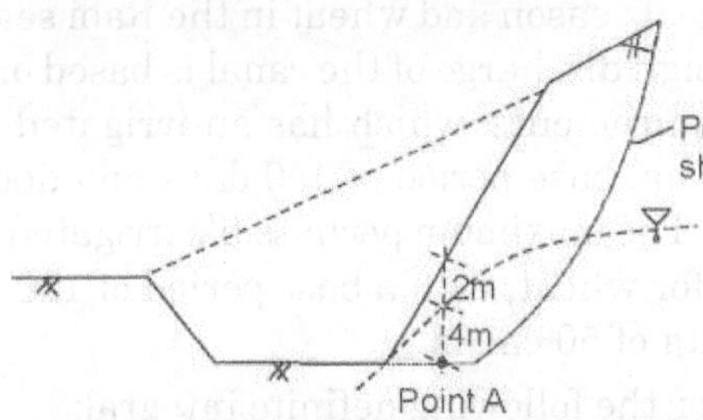

The soil exhibits $c' = 20$ kPa, $\phi' = 18°$, and the undrained shear strength = 80 kPa. The unit weight of water is 9.81 kN/m³. The unit weights of the soil above and below the ground water table are 18 and 20 kN/m³, respectively. If the shear stress at Point A is 50 kPa, the factors of safety against the shear failure at this point, considering the undrained and drained conditions respectively, would be

(*a*) 1.6 and 0.9 (*b*) 0.9 and 1.6

(*c*) 0.6 and 1.2 (*d*) 1.2 and 0.6

37. A municipal corporation is required to treat 1000 m³/day of water. It is found that an overflow rate of 20 m/day will produce a satisfactory removal of the discrete suspended particles at a depth of 3m. The diameter (in meters, rounded to the nearest integer) of a circular settling tank designed for the removal of these particles would be _____.

38. Consider the following second-order differential equation:

$$y'' - 4y' + 3y = 2t - 3t^2$$

The particular solution of the differential equation is

(*a*) $-2 - 2t - t^2$ (*b*) $-2t - t^2$

(*c*) $2t - 3t^2$ (*d*) $-2 - 2t - 3t^2$

39. Two cars P and Q are moving in a racing track continuously for two hours. Assume that no other vehicles are using the track during this time. The expressions relating the distance travelled d (in km) and time t (in hour) for both the vehicles are given as

P : $d = 60t$

Q : $d = 60t^2$

Within the first one hour, the maximum space headway would be

(*a*) 15 km at 30 minutes

(*b*) 15 km at 15 minutes

(*c*) 30 km at 30 minutes

(*d*) 30 km at 15 minutes

40. The culturable command area of a canal is 10,000 ha. The area grows only two crops-rice in the Kharif season and wheat in the Rabi season. The design discharge of the canal is based on the rice requirements, which has an irrigated area of 2500 ha, base period of 150 days and delta of 130 cm. The maximum permissible irrigated area (in ha) for wheat, with a base period of 120 days and delta of 50 cm. is ________

41. Consider the following definite integral:

$$I = \int_0^1 \frac{(\sin^{-1} x)^2}{\sqrt{1-x^2}}\,dx$$

The value of the integral is

(*a*) $\frac{\pi^3}{24}$ (*b*) $\frac{\pi^3}{12}$

(*c*) $\frac{\pi^3}{48}$ (*d*) $\frac{\pi^3}{64}$

42. If $A = \begin{bmatrix} 1 & 5 \\ 6 & 2 \end{bmatrix}$ and $B = \begin{bmatrix} 3 & 7 \\ 8 & 4 \end{bmatrix}$. AB^T is equal to

(*a*) $\begin{bmatrix} 38 & 28 \\ 32 & 56 \end{bmatrix}$ (*b*) $\begin{bmatrix} 3 & 40 \\ 42 & 8 \end{bmatrix}$

(*c*) $\begin{bmatrix} 43 & 27 \\ 34 & 50 \end{bmatrix}$ (*d*) $\begin{bmatrix} 38 & 32 \\ 28 & 56 \end{bmatrix}$

43. Consider the three prismatic beams with the clamped supports, P, Q and R as shown in the figures.

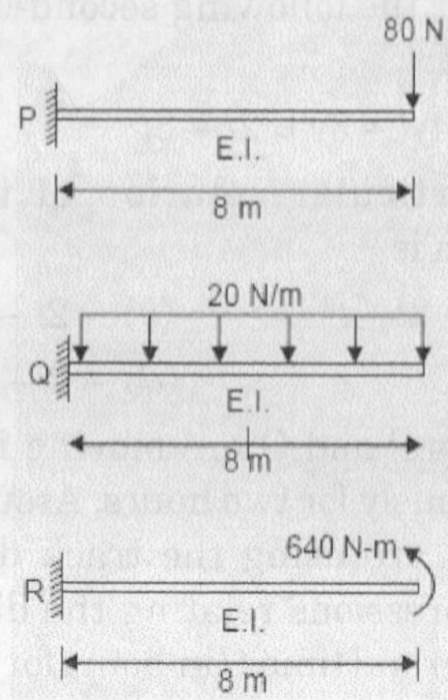

Given that the modulus of elasticity. E is 2.5×10^4 MPa. and the moment of inertia. I is 8×10^8 mm^4, the correct comparison of the magnitudes of the shear force S and the bending moment M developed at the supports is

(*a*) $S_P < S_Q < S_R$, $M_P = M_Q = M_R$

(*b*) $S_P = S_Q > S_R$, $M_P = M_Q > M_R$

(*c*) $S_P < S_Q > S_R$, $M_P = M_Q = M_R$

(*d*) $S_P < S_Q < S_R$, $M_P < M_Q < M_R$

44. The radii of relative stiffness of the rigid pavement P and Q are denoted by I_P and I_Q respectively. The geometric and material properties of the concrete slab and underlying soil are given below.

Pavement	Concrete					Soil
	Length of Slab	Breadth of Slab	Thickness of Slab	Modulus of Elasticity	Poisson's Ratio	Subgrade Reaction Modulus
P	L	B	h	E	μ	K
Q	L	B	0.5h	E	μ	2K

The ratio (up to one decimal place) of I_P/I_Q is ______.

45. The analysis of a water sample produces the following results

Ion	milligram per milli – equivalent for the ion	Concentration (mg L)
Ca^{2+}	20.0	60
Mg^{2+}	12.2	36.6
Na^+	23.0	92
K^+	39.1	78.2
Cl^-	35.5	71
SO_4^{2-}	48.0	72
HCO_3^-	61.0	122

The total hardness (in mg/L as CaO_3) of the water sample is ______.

46. Consider a square-shaped area ABCD on the ground with its centre at M as shown in the figure. Four concentrated vertical loads of P = 5000 kN are applied on this area, one at each corner.

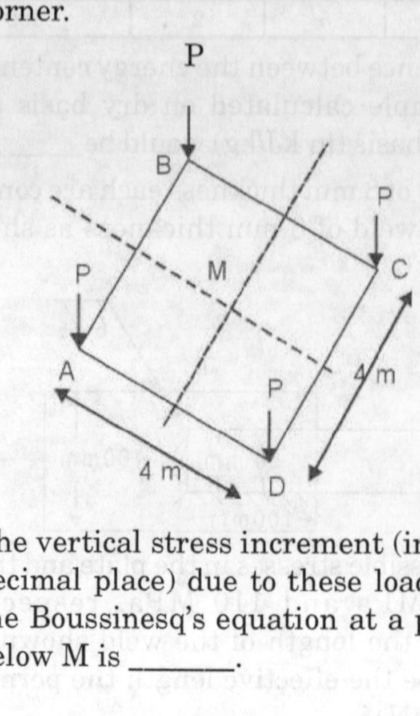

The vertical stress increment (in kPa up to one decimal place) due to these loads according to the Boussinesq's equation at a point 5 m right below M is ______.

47. Two towers, A and B standing vertically on a horizontal ground appear in a vertical aerial photograph as shown in the figure.

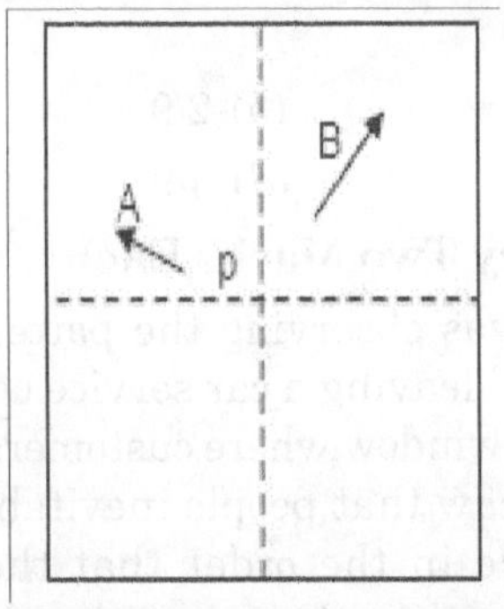

The length of the image of the tower A on the photograph is 1.5 cm and of the tower B is 2.0 cm. The distance of the top of the tower A (as shown in the arrowhead is 4.0 cm and the distance of the top of the tower B is 6.0 cm. as measured from the principal point p of the photograph. If the height of the tower B is 80 m the height (in meters) of the tower A is______

48. Water is pupped at steady uniform flow rate of 0.01 m^3/s though a horizontal smooth circular pipe of 100 mm diameter. Given that the Reynold number is 800 and g is 9.81 m/s^2, the head loss (in meters upto one decimal place) per km length due to friction would be ________

49. A 2 m long axially loaded mild steel rod of 8 mm diameter exhibits the load-displacement (P – δ) behavior as shown in the figure.

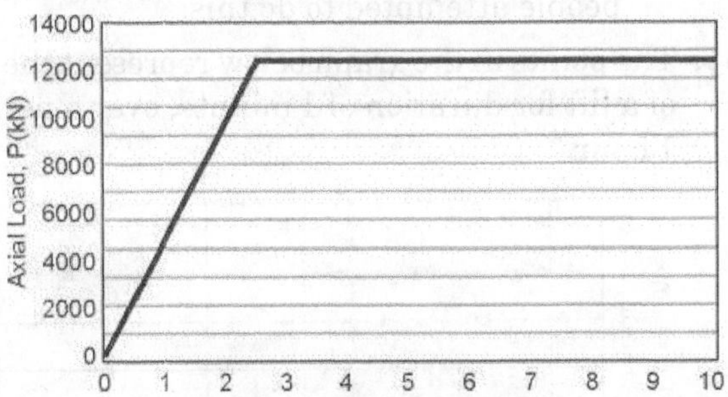

Assume the yield stress of steel as 250 MPa. The complementary strain energy in N-mm) stored in the bar up to its linear elastic behavior will be _____

50. A simply supported rectangular concrete beam of span 8 m has to be prestressed with a force of 1600 kN. The tendon is of parabolic profile having zero eccentricity at the supports. The beam has to carry an external uniformly distributed load of intensity 30 kN/m. Neglecting the self-weight of the beam, the maximum dip (in meters upto two decimal place) of the tendon at the mid-span to balance the external load should be______.

51. A catchment is idealized as a 25 km × 25 km square. It has five rain gauges, one at each corner and one at the centre, as shown in the figure.

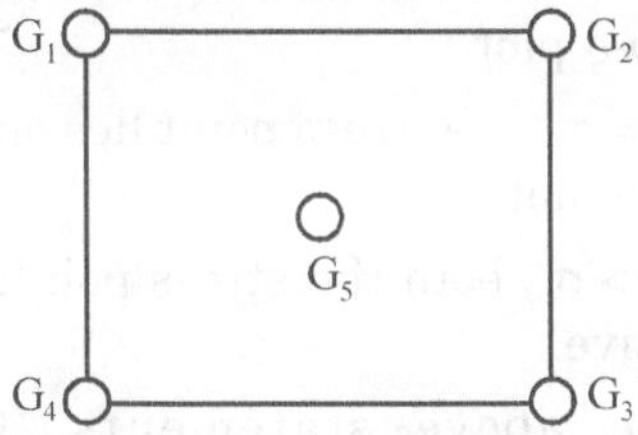

During a month the precipitation at these gauges is measured as G_1 = 300 mm, G_2 = 285 mm, G_3 = 272 m, G_4 = 290 mm and G_5 = 288 mm. The average precipitation (in mm up to one decimal place) over the catchment during this month by using the Thiessen polygon method is____.

52. The figure shows a U-tube having a 5 mm × 5 mm square cross-section filled with mercury (specific gravity = 13.6) upto a height of 20 cm in each limb (open to the atmosphere).

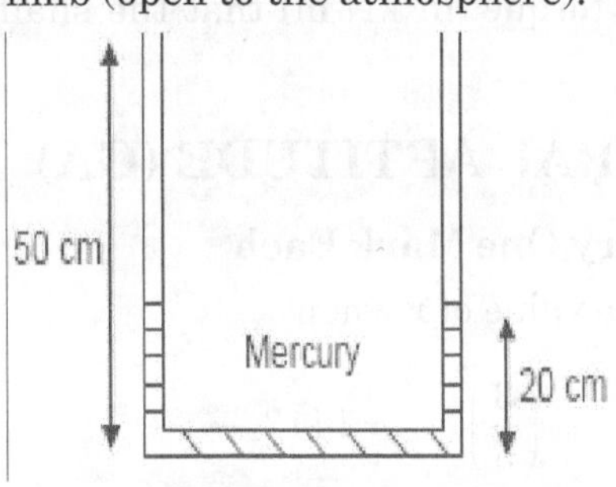

If 5 cm^3 of water is added to the right limb, the new height (in cm up to two decimal places) of mercury in the LEFT limb will be______.

53. Group I gives a list of test methods and test apparatus for evaluating some of the properties of Ordinary Portland Cement (OPC) and concrete Group II gives the list of these properties.

Group I

P. Le Chatelier test

Q. Vee-Bee test

R. Blaine air permeability test

S. The Vicat apparatus

Group II

1. Soundness of OPC
2. Consistency and setting time of OPC
3. Consistency or workability of concrete
4. Fineness of OPC

The correct match of the items in Group I with items in Group II is

(*a*) P-1, Q-3, R-4, S-2

(*b*) P-2, Q-3, R-1, S-4

(*c*) P-4, Q-2, R-4, S-1

(*d*) P-1, Q-4, R-2, S-3

54. Following are the statements related to the stress paths in a triaxial testing of soils.

P. If $\sigma_1 = \sigma_3$, the stress point lies at the origin of the p-q plot

Q. If $\sigma_1 = \sigma_3$, the stress point lies on the p-axis of to p-q plot

R. If $\sigma_1 > \sigma_3$, both the stress points p and q are positive.

For the above statements, the correct combination is

(*a*) P-False; Q-True; R-True

(*b*) P-True; Q-False; R-True

(*c*) P-False; Q-True; R-False

(*d*) P-True; Q-True; R-False

55. A hollow circular shaft has an outer diameter of 100 mm and inner diameter of 50 mm. If the allowable shear stress is 125 MPa, the maximum torque (in kN-m) that the shaft can resist is ________

GENERAL APTITUDE (GA)

(Q.1 – 5) : Carry One Mark Each

1. What is the value of x when

$$81 \times \left(\frac{16}{25}\right)^{x+2} \div \left(\frac{3}{5}\right)^{2x+4} = 144?$$

(*a*) 1 (*b*) – 1

(*c*) – 2 (*d*) Cannot be determined

2. There was no doubt that their work was thorough Which of the words is closest in meaning to the underlined word above?

(*a*) pretty (*b*) complete

(*c*) sloppy (*d*) haphazard

3. Four cards lie on a table. Each card has a number printed on the one side and a color on the other. The faces visible on the cards are 2, 3 red and blue.

Proposition. If a card has an even value on one side then its opposite face is red.

The cards which MUST be turned over to verify the above porposition are

(*a*) 2, red (*b*) 2, 3, red

(*c*) 2, blue (*d*) 2, red, blue

4. The event would have been successful if you _____ able to come.

(*a*) are (*b*) had been

(*c*) have been (*d*) would have been

5. Two dice are thrown simultaneously. The probability that the product of the numbers appearing on the top faces of the dice is a perfect square is

(*a*) 1/9 (*b*) 2/9

(*c*) 1/3 (*d*) 4/9

(Q.6 – 10) : Carry Two Marks Each

6. Bhaiclung was observing the pattern of people entering and leaving a car service centre. There was a single window where customers were being served. He saw that people inevitably came out of the centre in the order that they went in. However, the time they spent inside seemed to vary a lot some people came out in a matter of minutes while for others it took much longer.

From this what can one conclude?

(*a*) The centre operates on a first-come-first-served basis, but with variable service times depending on specific customer needs.

(*b*) Customers were served in an arbitrary order since they took varying amounts of time for service completion in the centre.

(*c*) Since some people came out within a few minutes of entering the centre, the system is likely to operate on a last-come-first-served basis.

(*d*) Entering the centre early ensured that one would have shorter service times and most-people attempted to do this.

7. The points in the graph below represent the halts of a lift for duration of 1 minute, over a period of 1 hour.

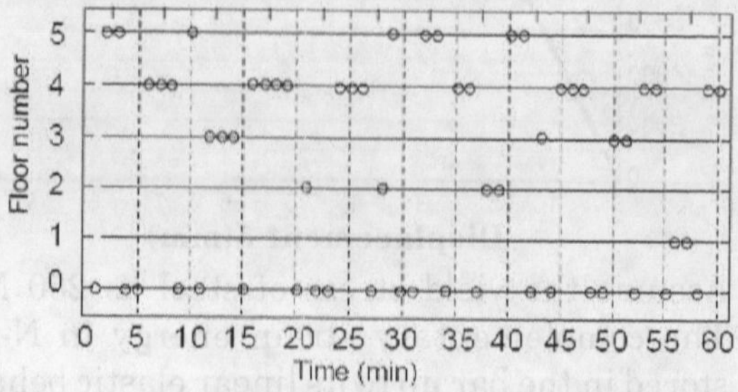

Which of the following statements are correct?

(*i*) The elevator never moves directly from any non-ground floor to another non-ground floor over the one hour period.

(*ii*) The elevator stays on the fourth floor for the longest duration over the one hour period.

(*a*) only (*i*)

(*b*) only (*ii*)

(*c*) Both (*i*) and (*ii*)

(*d*) Neither (*i*) nor (*ii*)

8. A map shows the elevations of Darjeeling, Gangtok, Kalimpong, Pelling and Siliguri. Kalimpong is at a lower elevation than Gangtok. Pelling is at a lower elevation than Gangtok Pelling is at a higher elevation than Siliguri. Darjeeling is at a higher elevation than Gangtok.

Which of the following statements can be inferred from the paragraph above?

(*i*) Pelling is at a higher elevation than Kalimpong

(*ii*) Kalimpong is at a lower elevation than Darjeeling

(*iii*) Kalimpong is at a higher elevation than Siliguri

(*iv*) Siliguri is at a lower elevation than Gangtok.

(*a*) Only (*ii*)

(*b*) Only (*ii*) and (*iii*)

(*c*) Only (*ii*) and (*iv*)

(*d*) Only (*iii*) and (*iv*)

9. P, Q, R, S, T and U are sealed around a circular table. R is seated two places to the right of Q. P is seated three places to the left of R. S is seated opposite U. If P and U now switch seats which of the following must necessarily be true.

(*a*) P is immediately to the right of R

(*b*) T is immediately to the left of P

(*c*) T is immediately to the left of P or P is immediately to other right of Q.

(*d*) U is immediately to the right of R or P is immediately to the left of T.

10. Budhan covers a distance of 19 km in 2 hours by cycling one fourth of the time and walking the rest. The next day he cycles (at the same speed as before) for half the time and walks the rest (at the same speed as before) and covers 26 km in 2 hours. The speed in km/h at which Budhan walks is

(*a*) 1 (*b*) 4

(*c*) 5 (*d*) 6

ANSWERS

Technical Section

1. (200)	**2.** (*b*)	**3.** (*b*)	**4.** (0.99)	**5.** (*a*)	**6.** (0.5)	**7.** (*c*)	**8.** (2.0)
9. (1.5)	**10.** (*c*)	**11.** (*b*)	**12.** (1.6)	**13.** (*d*)	**14.** (134)	**15.** (*c*)	**16.** (*d*)
17. (*a*)	**18.** (*d*)	**19.** (*a*)	**20.** (*c*)	**21.** (*b*)	**22.** (*b*)	**23.** (*c*)	**24.** (210)
25. (*b*)	**26.** (*a*)	**27.** (1.0)	**28.** (1.07)	**29.** (0.23)	**30.** (*a*)	**31.** (*d*)	**32.** (33 .0)
33. (*a*)	**34.** (3870)	**35.** (60)	**36.** (*a*)	**37.** (8)	**38.** (*a*)	**39.** (*a*)	**40.** (5199.97)
41. (*a*)	**42.** (*a*)	**43.** (*c*)	**44.** (2)	**45.** (300)	**46.** (190.84)	**47.** (90)	**48.** (66.1)
49. (15707.96)		**50.** (0.15)	**51.** (287.375)	**52.** (20.74)	**53.** (*a*)	**54.** (*a*)	**55.** (23)

General Aptitude (GA)

1. (*b*)	**2.** (*b*)	**3.** (*a*)	**4.** (*b*)	**5.** (*b*)	**6.** (*a*)	**7.** (*d*)	**8.** (*c*)	**9.** (*c*)	**10.** (*d*)

EXPLANATIONS

TECHNICAL SECTION

1.

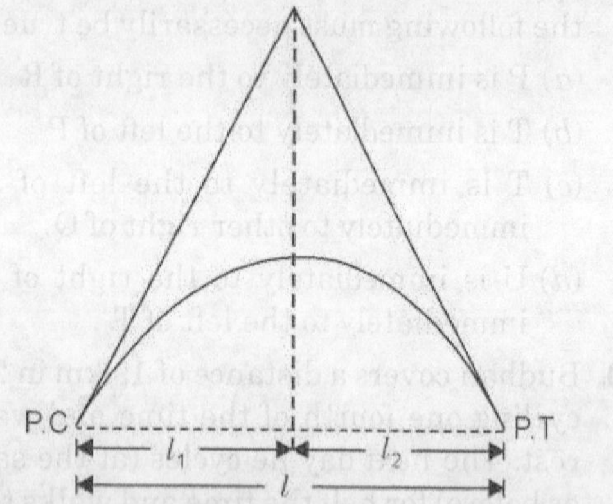

For flat vertical curve we assume that $l_1 = l_2$

$$l = l_1 + l_2$$
$$= (100 + 100)\text{ m}$$
$$= 200\text{ m}$$

2. Pressure diagram for partially submerged soil with surcharge

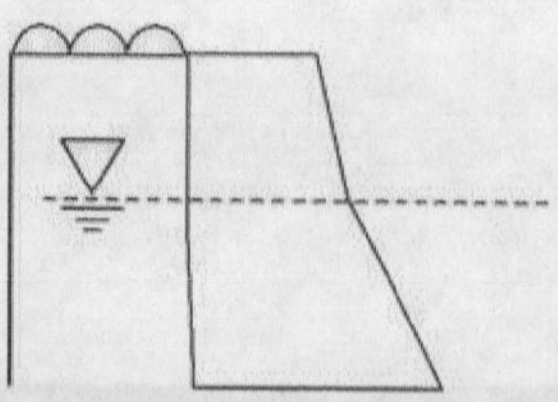

But no option is matching.

If the water table is at ground table then

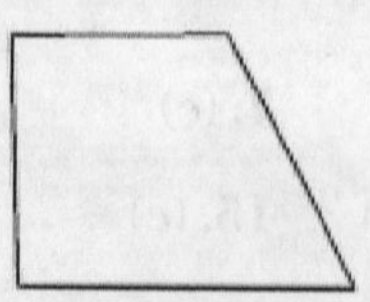

Option (*b*) can be the answer.

3. In activated sludge process, in the presence of food mass and oxygen heterotrophic bacteria converts food mass into biomass.

4.

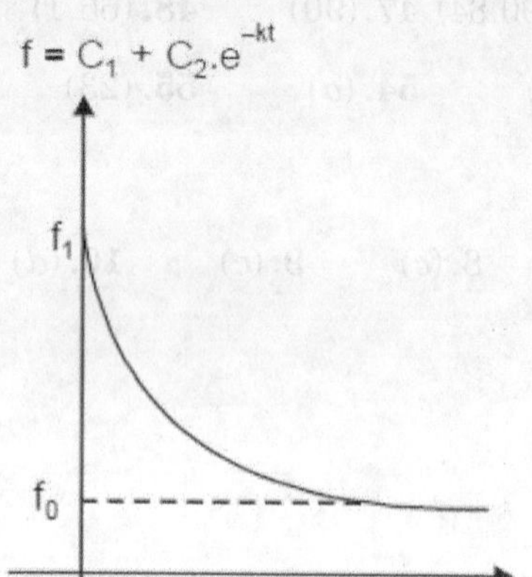

$$C_1 = f_0 \qquad C_2 = f_1 - f_0$$
$$= 25\text{ mm/hr} \qquad = 200 - 25$$
$$= 175\text{ mm/hr}$$

$f = 90$ mm/hr after 1 hr

Hence, $90 = 25 + 175 \times e^{-k \times 1}$

$$\Rightarrow \quad e^{-k} = \frac{90-25}{175} = 0.3715$$

Taking $\log_e$ on both side

$$-x \log_e e = \log_e (0.3714)$$
$$\Rightarrow \quad -x = -0.990$$
$$\Rightarrow \quad x = 0.99$$

5. $\because \quad \gamma_d = \dfrac{(1-\eta_a)G_s\gamma_w}{1+e}$

For zero air voids $\Rightarrow \eta_a = 0$ ... (i)

and $\quad S + a_c = 1$

where $\quad a_c$ = air content

For zero air voids $\Rightarrow a_c = 0$

Hence, $\quad S = 1$ ... (ii)

From (i) and (ii)

$$\therefore \quad \gamma_d = \frac{(1-0)G_s\gamma_w}{1+\dfrac{wG_s}{S}}$$

$$\gamma_d = \frac{G_s\gamma_w}{1+wG_s}$$

6. Probability of getting (H.H.H)

$$= \frac{1}{2}\times\frac{1}{2}\times\frac{1}{2}$$
$$= \frac{1}{8}$$

Probability of getting (H.H.H.H)

$$= \frac{1}{2}\times\frac{1}{2}\times\frac{1}{2}\times\frac{1}{2}$$
$$= \frac{1}{16}$$

$\because$ Now, Conditional probability of getting next H after (H.H.H)

$$= \frac{\frac{1}{16}}{\frac{1}{8}}$$

$$= \frac{1}{16}\times\frac{8}{1} = \frac{1}{2} = 0.5$$

7. Value of B lies between 0 and 1.0

For completely saturated soil, B = 1

For completely dry soil, B = 0

Value of A may be as large as 2 to 3 for very loose saturated fine sand and it can be less than zero for over consolidated clay.

8.

$$\text{Rainfall intensity} = 3.5 \text{ cm/hr}$$

$$\phi - \text{index} = 1.5 \text{ cm/hr}$$

$$\text{Intensity of effective rainfall} = (3.5 - 1.5) \text{ cm/hr} = 2 \text{ cm/hr}$$

9.

$$R = 50$$

$$\text{Grade compensation} = \frac{30+R}{R} = \frac{30+50}{50} = \frac{80}{50} = 1.6$$

$$(\text{Grade compensation})_{max} = \frac{75}{R} = \frac{75}{50} = 1.5$$

The compensated gradient should be greater than 4%.

Compensated gradient = 6 – 1.5 = 4.5

Hence, $(\text{grade compensation})_{max} = 1.5$

10. Since characteristic strength has been defined as the value below which not more than 50% of results are likely to fall.

Hence, Target mean strength

= Characteristic strength

$= f_{ck}$

$= 25 \text{ N/mm}^2$

11.

Elevation (in m)	Temperature (in °C)
10	15.5
60	15.0
130	14.3

$$ELR_1 = \frac{\Delta T}{\Delta h} = \frac{15.5 - 15.0}{60 - 10} = 0.01 \text{ °c/m}$$

$$ELR_2 = \frac{15.0 - 14.3}{130 - 60} = 0.01 \text{ °c/m}$$

∵ Adiabatic lapse rate (ALR) is 9.8°C/km for dry adiabatic and 6°C/km for wet adiabatic condition.

$$ELR = 0.01 \times 1000 = 10\text{°C/km}$$

$$\because \quad ELR > ALR$$

Hence it is said to be super adiabatic and the atmosphere is said to be highly unstable.

12.

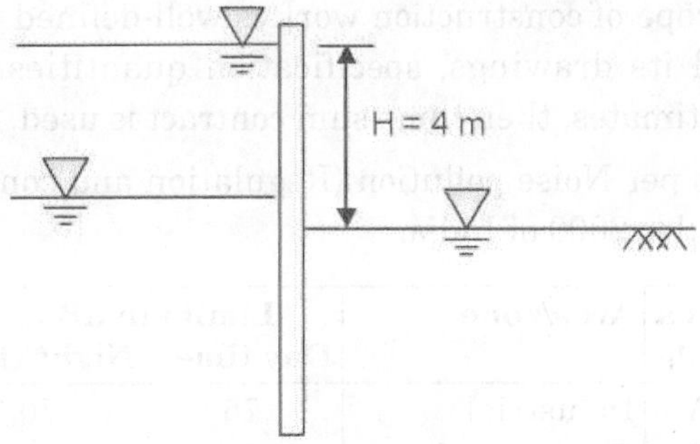

$$N_f = 5 - 1 = 4$$

$$N_d = 11 - 1 = 10$$

$$H = 4 \text{ m}$$

$$\therefore \quad Q = \frac{kHN_f}{N_D}$$

$$= 10^{-6} \text{ m/s} \times 4\text{m} \times \frac{4}{10} \text{ per meter width}$$

$$= 1.6 \times 10^{-6} \text{ m}^3\text{/s/m width}$$

$$= 1.6 \text{ cm}^3\text{/s/meter width}$$

13. Shear strain in an element is positive when the angle between two positive faces (or two negative faces) is reduced.

The strain is negative when the angle between two positive (or two negative) faces increase.

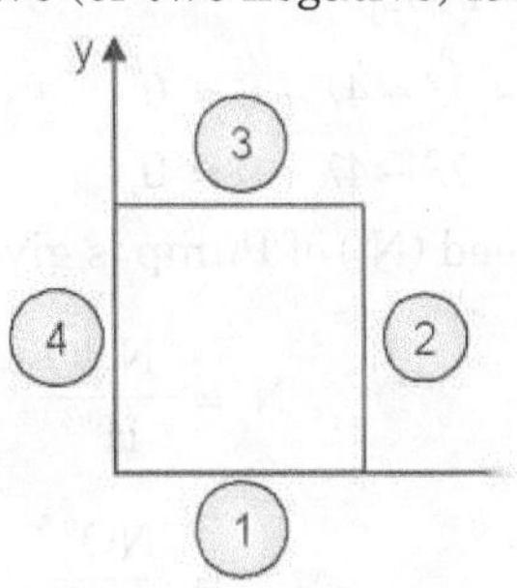

Face ② & ③ are +ve face

Face ① & ④ are –ve face.

Angle between ① & ④ is increased by 0.0005 rad.

$$\therefore \quad \gamma_{xy} = -0.0005 = 0.001 \text{ K}$$

$$\therefore \quad K = -0.50$$

14.

$$V = x^2\hat{i} + 2y^3\hat{j} + z^4\hat{k}$$

Divergence (∀)

$$= \frac{\partial}{\partial x}(x^2)\hat{i} + \frac{\partial}{\partial y}(2y^3)\hat{j} + \frac{\partial}{\partial x}(z^4)\hat{k}$$

$$2x + 6y^2 + 4z^3$$

$$[\text{Divergence (V)}]_{(x = 1, y = 2, z = 3)}$$

$$= 2 \times 1 + 6 \times 2^2 + 4 \times 3^3$$

$$= 2 + 24 + 108$$

$$= 134$$

16. Scope of construction work is well-defined with all its drawings, specification quantities and estimates, then lump sum contract is used.

17. As per Noise pollution (Regulation and control) Rules 2000 of India.

Area code	Area/zone	Limits in dB L_{eq} Day time	Night time
A	Industrial	75	70
B	Commerical area	65	55
C	Residential area	55	45
D	Silence Zone	50	40

18. Resection method of orientation is employed when the plane table occupies a position not yet plotted on the drawing sheet.

19.
$$3x_1 + 2x_2 = c_1$$
$$4x_1 + x_2 = c_2$$
$$\begin{bmatrix} 3 & 2 \\ 4 & 1 \end{bmatrix} = [A]$$
$$|[A] - \lambda[1]| = 0$$
$$\begin{bmatrix} 3-\lambda & 4 \\ 2 & 1-\lambda \end{bmatrix} = 0$$
$$3 + \lambda^2 - 4\lambda - 8 = 0$$
$$\lambda^2 - 4\lambda - 5 = 0$$

20. Specific speed (N_s) of Pump is given by
$$N_s = \frac{N\sqrt{Q}}{H^{3/4}}$$
$$= \frac{NQ^{0.5}}{H^{0.75}}$$

21. We know that,

$$\text{Equilibrium cant} = \frac{GV^2}{127R}$$

V in kmph and R in m.

For B.G track G = 1.676

$$\text{Cant} = \frac{1.676V^2}{127R}\text{ m}$$
$$= 0.01319\frac{V^2}{R}\text{ m}$$
$$e = \frac{1.319V^2}{R}\text{ cm}$$

The unit of *e* is not specified in question but option *b* is matching in cm units.

22.

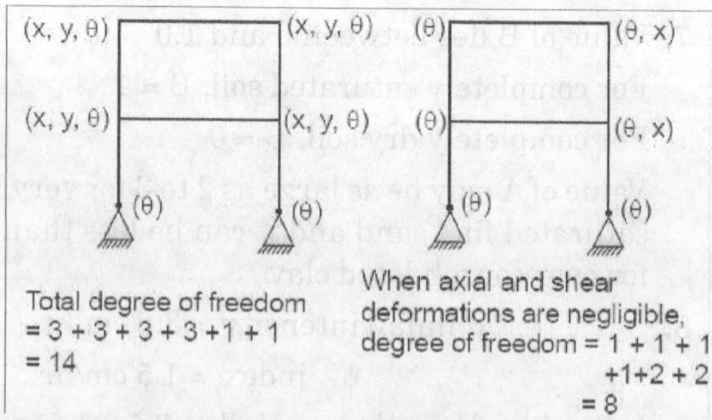

Hence, reduction in degree of kinematical indetermiancy = 14 − 8 = 6

23.
$$w = f(x, y)$$
$$\frac{dw}{dt} = \frac{\partial w}{\partial x}\frac{dx}{dt} + \frac{\partial w}{\partial y}\frac{dy}{dt}$$

24.
$$Z = \frac{T_s - T_E}{\sigma}$$
$$1.64 = \frac{T_s - 200}{6.1}$$
$$\Rightarrow \quad T_s = (1.64 \times 6.1 + 200)$$
$$T_s = 210 \text{ days}$$

25. In plate load test bearing capacity of clay does not depend upon size of footing.
$$q_{up} = q_{uf} \text{ (for clay)}$$

∴ Ultimate bearing capacity of a 2m wide square footing = 180 kPa

26.
$$Q_{P1} = qN_q + 0.4B\gamma N_\gamma$$
$$= 18 \times 20 \times N_q + 0.4 \times 0.5 \times 18 \times N_\gamma$$
$$= 18(20\,N_q + 0.2\,N_\gamma)$$
$$Q_{P2} = qN_q + 0.4B\gamma N_\gamma$$
$$= (19-10) \times 20 \times N_q + 0.4 \times 0.5 \times (19-10)N_\gamma$$
$$= 9(20\,N_q + 0.2\,N_\gamma)$$

(Assuming γ_w = 10 kN/m³)

$$\frac{Q_{P1}}{Q_{P2}} = 2$$

27. Moment at Q = M_Q
$$= 2000 \times 2 - 1650 \times \frac{2^2}{2}$$
$$= 700 \text{ kNm anticlockwise}$$

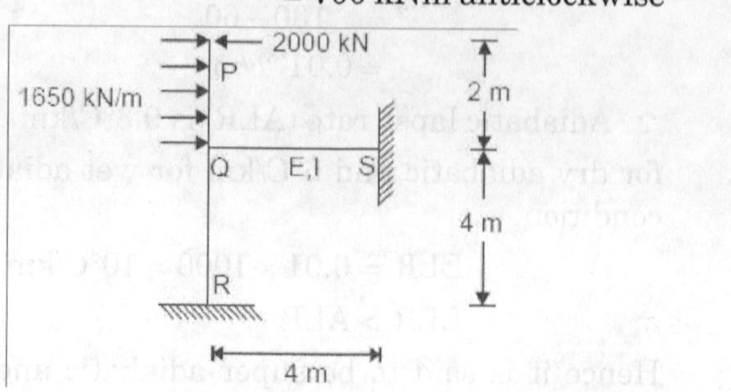

$$K_{QR} = \frac{4EI}{4} = \frac{EI}{1m}$$

$$K_{QS} = \frac{4EI}{4} = \frac{EI}{1m}$$

$$\therefore \quad K_Q = K_{QR} + K_{QS} = \frac{2EI}{1m}$$

$$M_Q = K_Q\theta_Q$$

$$\therefore \quad \theta_Q = \frac{M_Q}{K_Q}$$

$$= \frac{700\times10^6\times1000}{2\times2.5\times10^4\times8\times10^8}$$

$$= 0.0175 \text{ rad}$$

$$= 0.0175\times\frac{180}{\pi}$$

$$= 1.003°$$

$$= 1.0°$$

28. Let, Subcritical depth = y

then, super critical depth = $\frac{y}{2}$

$\because$ Specific energy subcritical depth

= Specific energy at super-critical depth

$$\Rightarrow \quad y + \frac{Q^2}{2gB^2y^2} = \frac{y}{2} + \frac{Q^2}{2gB^2\left(\frac{y}{2}\right)^2}$$

$$\Rightarrow \quad \left(y - \frac{y}{2}\right) = \frac{Q^2}{2gB^2}\left(\frac{-1}{y^2} + \frac{4}{y^2}\right)$$

$$\Rightarrow \quad \frac{y}{2} = \frac{2^2}{2\times9.81\times1^2}\times\frac{3}{y^2}$$

$$\Rightarrow \quad \frac{y}{2} = \frac{12}{2\times9.81\times y^2}$$

$$\therefore \quad y = 1.069 = 1.07$$

29. Let$BOD_{5day} = L_0(1 - e^{-5k})$

Ultimate BOD = L_0

$$\therefore \quad \frac{L_0(1-e^{-5k})}{L_0} = 0.68$$

$$\Rightarrow \quad 1 - e^{-5k} = 0.68$$

$$\Rightarrow \quad e^{-5k} = 0.32$$

$$\therefore \quad K = \frac{1}{5}\ln\left(\frac{1}{0.32}\right)$$

$$= 0.23$$

30. Target is having inclination of 45° with x-axis

$$\Rightarrow \quad \frac{dy}{dx} = \tan 45°$$

$$\Rightarrow \quad \frac{d(x\ ln\ x)}{dx} = 1$$

$$\Rightarrow \quad ln\ x + \frac{x}{x} = 1$$

$\therefore$ At $x = 1, y = 1 \times ln'\ 1 = 0$

(1, 0)

32. $0.0673\,(d_1)^2 = 5$ and, $0.0673\,(d_2)^2 = 40$

By solving, we get

$$d_1 = 8.62 \text{ k},$$

$$d_2 = 24.38 \text{ km}$$

$\therefore$ Distance of observer from the lighthouse

$$= d = d_1 + d_2$$

$$\Rightarrow \quad d = (8.62 + 24.38) \text{ km}$$

$$\Rightarrow \quad d = 33 \text{ km}$$

33.

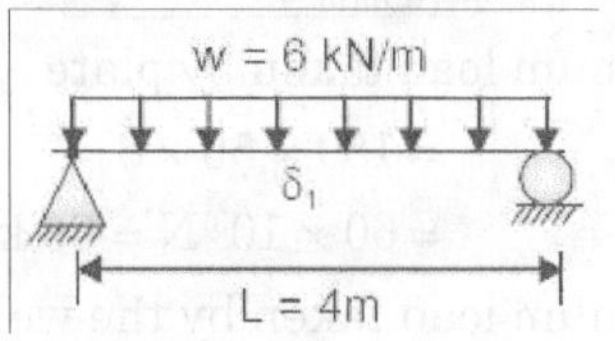

$$\delta_1 = \frac{5}{384}\frac{wL^4}{EI}$$

$$= \frac{5}{384}\times\frac{6\times4^4}{10000\text{ m}} = 20$$

$$P = 120 \text{ kN}$$

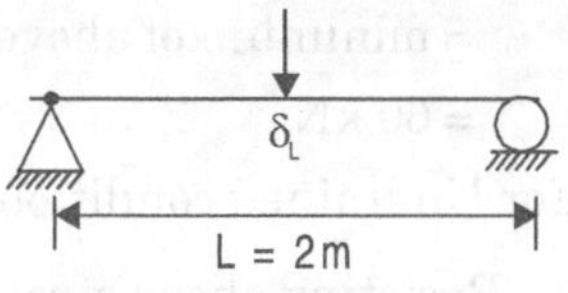

$$\delta_L = \frac{PL^3}{48EI}$$

$$= \frac{120\times2^3}{48\times10000\text{ m}} = 20 \text{ mm}$$

$$\therefore \quad \delta_1 = \delta_2$$

34.

Component (1)	% By mass (2)	Moisture Content % (3)	Dry content% (4) = 100 – (3)	Energy as discarded (5) (kJ/kg)	Total dry mass (6) (2) × (4)	Total energy as discarded (7) (2)× (5)
					100	100
Food waste	20	70	30	2500	6	500
Paper	10	4	96	10000	9.6	1000
Card board	10	4	96	8000	9.6	800
Plastics	10	1	99	14000	9.9	1400
Garden trimming	40	60	40	3500	16	1400
Wood	5	20	80	1400	4	700
Tin Cars	5	2	98	10	4.9	5
					= 60	= 5805 kJ/kg

By question,

Energy on dry basis: It will be the total energy when whole mass is dry.

By question,

In current situation only 60% mass is dry energy corresponding to 60% dry mass = 5805 kJ/kg

Energy corresponding to 100% dry mass

$$= \frac{5805}{60} \times 100$$

$$= 9675 \text{ kJ/kg}$$

So, energy on dry basis = 9675 kJ/kg

Energy based on discarded basis = 5805 kJ/kg

So, difference = (9675 – 5805) kJ/kg

= 3870 kJ/kg

35.

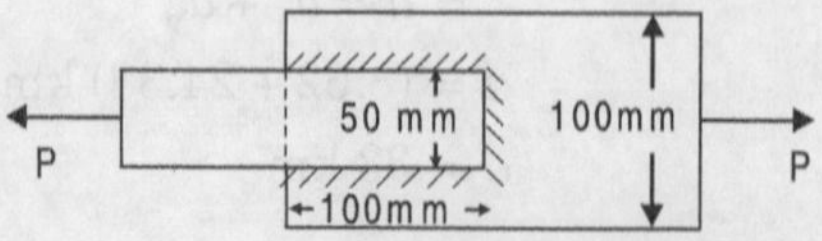

Maximum load taken by plate

$= 150 \times 50 \times 8$

$= 60 \times 10^3 \text{ N} = 60 \text{ kN}$

Maximum load taken by the weld

$= \text{KS } l_{eff}\, \tau$

$= 0.7 \times 6 \times (100 + 50 + 100) \times 110$

$= 4.2 \times (250) \times 110 = 115.5 \times 10^3 \text{ N}$

$= 115.5 \text{ kN}$

∴ Permissible load

= minimum of above two

= 60 kN

36. Case-I : For Undrained condition

$$\text{F.o.S} = \frac{\text{Resisting shear strength}}{\text{Acting shear stress}}$$

$$= \frac{80}{50} = 1.6$$

Case-II : For Drained condition

$$\text{F.o.S} = \frac{\bar{\sigma} \tan \phi' + c'}{\text{Acting shear stress}}$$

$$= \frac{[2 \times 18 + 4(20 - 9.81)] \times \tan 18° + 20}{50}$$

$$= \frac{[36 + 4 \times 10.19] \times \tan 18° + 20}{50}$$

$$= 0.9$$

37. $Q = 1000 \dfrac{m^3}{\text{day}}$

Overflow Rate = 20 m/day

H = 3 m

Surface Area of circular Tank = $\dfrac{1000}{20} = 50\ m^2$

Assuming diameter to be 'd'

Hence, $\dfrac{\pi}{4} \times d^2 = 50$

$$d = \sqrt{\frac{50 \times 4}{\pi}}$$

$= 7.978$

≈ 8 m

38. Given second order differential equation is

$$y'' - 4y' + 3y = 2t - 3t^2$$

$$f(\text{D}) = \text{D}^2 - 4\text{D} + 3$$

$$\text{P.I.} = \frac{\text{I}}{f(\text{D})}(2t - 3t^2)$$

$$= \frac{1}{\text{D}^2 - 4\text{D} + 3}(2t - 3t^2)$$

$$= \left[\frac{1}{1-D} - \frac{1}{3\left(1 - \frac{\text{D}}{3}\right)}\right]\left(t - \frac{3t^2}{2}\right)$$

$$(1 + \text{D} + \text{D}^2 + ...)\left(t - \frac{3t^2}{2}\right) - \frac{1}{3}\left(t - \frac{\text{D}}{3} + \frac{\text{D}^2}{9} + ...\right) \times \left(t - \frac{32t^2}{2}\right)$$

$$= \left(t - \frac{3t^2}{2} + 1 - 3t - 3\right) - \frac{1}{3}\left(1 - \frac{3t^2}{2} + \frac{1}{3} - 8 - \frac{1}{3}\right)$$

$$= -2 - 2t - t^2$$

39. P : $d = 60t$

Q : $d = 60t^2$

Space Headway (S)

$= 60t^2 - 60t$

For space headway to be max.,

$$\frac{ds}{dt} = 0$$

$$120t - 60 = 0$$

$\Rightarrow$ $60(2t - 1) = 0$

$\Rightarrow$ $t = \dfrac{1}{2} = 30$ minutes

Maximum space Headway

$$= 60 \times \frac{1}{2} - 60 \times \left(\frac{1}{2}\right)^2$$

$$= 60 \times \frac{1}{2} - 60 \times \frac{1}{4}$$

$= 30 - 15 = 15$ km

40. For Rice, culturable command area,

$$CCA = 10000 \text{ ha}$$
$$B.P. = 150 \text{ days}$$
$$\Delta = 130 \text{ cm} = 1.3 \text{ m}$$
$$D = \frac{B.P. \times 8.64}{\Delta}$$
$$= \frac{150 \times 8.64}{1.3}$$
$$= 996.923 \text{ ha/m}^3\text{/s}$$
$$\text{Area irrigated} = 2500 \text{ ha}$$

$\therefore$ Design Discharge

$$= \frac{2500}{996.923}$$
$$= 2.5077 \text{ m}^3\text{/s}$$

For, wheat, $B.P. = 120$ days

$$\Delta = 50 \text{ cm}$$
$$D = \frac{8.64 \times 120}{0.5}$$
$$= 2073.6 \text{ hect./m}^3\text{/s}$$

Area irrigated = Duty × Design Discharge (Q)

$$= 2073.6 \times 2.5077$$
$$= 5199.97 \text{ ha.}$$

41. Given, $$I = \int_0^1 \frac{(\sin^{-1} x)^2}{\sqrt{1-x^2}} dx$$

Put $\sin^{-1} x = t$

Now, $$\frac{dx}{\sqrt{1-x^2}} = dt$$

$$= \int_0^{\pi/2} t^2 dt$$
$$= \left[\frac{t^3}{3}\right]_0^{\pi/2}$$
$$= \frac{\pi^3}{24}$$

42. $$A = \begin{bmatrix} 1 & 5 \\ 6 & 2 \end{bmatrix} [B] = \begin{bmatrix} 3 & 7 \\ 8 & 4 \end{bmatrix}$$

$$AB^T = \begin{bmatrix} 1 & 5 \\ 6 & 2 \end{bmatrix} \begin{bmatrix} 3 & 8 \\ 7 & 4 \end{bmatrix}$$
$$= \begin{bmatrix} 3+35 & 8+20 \\ 18+14 & 48+8 \end{bmatrix}$$
$$= \begin{bmatrix} 38 & 28 \\ 32 & 56 \end{bmatrix}$$

43.

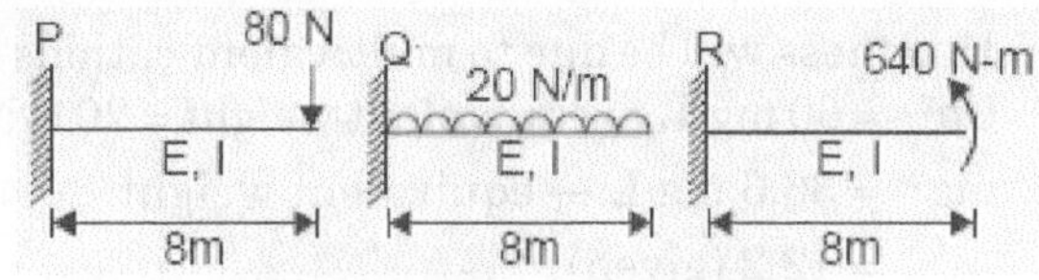

$$E = 2.5 \times 10^4 \text{ MPa}$$
$$I = 8 \times 10^8 \text{ mm}^4$$

For P,

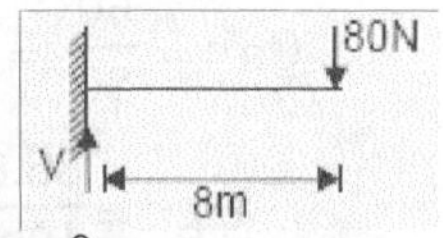

$$\Sigma f_V = 0$$
$$\Rightarrow \boxed{V = 80 \text{ N}}$$

$\therefore$ Shear force at support = Sp = 80N

$\Rightarrow$ Moment at support = 80 × 8 = 640 Nm.

$\Rightarrow$ Mp = 640 Nm

for Q

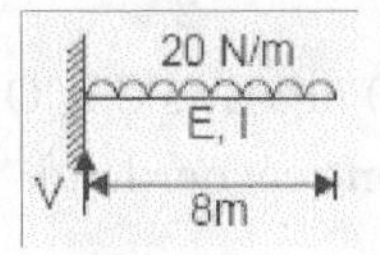

$$\Sigma f_V = 0$$
$$\Rightarrow \boxed{V = 20 \times 8 = 160 \text{ N}}$$

$\therefore$ Shear force at support = S_Q = 160N

$$\Rightarrow \text{Moment at support} = M_Q = 20 \times 8 \times \frac{8}{2} \text{ Nm}$$
$$= 640 \text{ Nm}$$

For R,

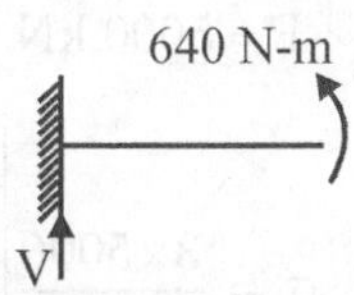

$$\Sigma f_V = 0$$
$$V = 0 = S_R$$

Moment at support = M_R = 640 Nm

Hence, by above values of S_P, S_Q, S_R and M_P, M_Q and M_R

We conclude that

$$\boxed{S_P < S_Q > S_R} \ \& \ \boxed{M_P = M_Q = M_R}$$

44. $$l = \left[\frac{Eh^3}{12(1-\mu^2)k}\right]^{1/4}$$

$$\frac{l_P}{l_Q} = \left[\frac{E_P}{E_Q} \times \frac{h_P^3}{h_Q^3} \times \frac{k_Q}{k_P}\right]^{1/4}$$
$$= \left[\frac{E}{E} \times \frac{h^3}{(0.5h)^3} \times \frac{2k}{K}\right]^{1/4}$$
$$= [16]^{1/4} = 2$$
$$\frac{l_P}{l_Q} = 2$$

45. Hardness will be due to multivalent cations only

Ca^{2+} = 60 mg/L → equivalent weight = 20 (given)

mg^{2+} = 36.6 mg/L → equivalent weight = 12.2 (given)

miliequivalent of

$$Ca^{2+} = \frac{60}{20} = 3$$

$$mg^{2+} = \frac{36.6}{12.2} = 3$$

Total hardness (mg/L as $CaCO_3$)

= (miliequivalent of Ca^{2+} + miliequivalent of mg^{2+}) × equivalent weight of $CaCO_3$

$$= (3 + 3) \times \frac{100}{2}$$

= 300 mg/L as $CaCO_3$

46. Due to one corner load

$$\sigma_2 = \frac{3P}{3\pi z^2}\left(\frac{1}{1+\left(\frac{r}{z}\right)^2}\right)^{5/2}$$

Given, $r = \sqrt{2^2 + 2^2} = 2\sqrt{2}\ m$

$z = 5\ m$

P = 5000 kN

$$\sigma_2 = \frac{3 \times 5000}{2\pi \times 5^5}\left[\frac{1}{1+\left(\frac{2\sqrt{2}}{5}\right)^2}\right]^{5/2}$$

= 47.709 kN/m²

So due to four load vertical stress = 4 σ_2

= 4 × 47.709

= 190.84 kN/m²

47.

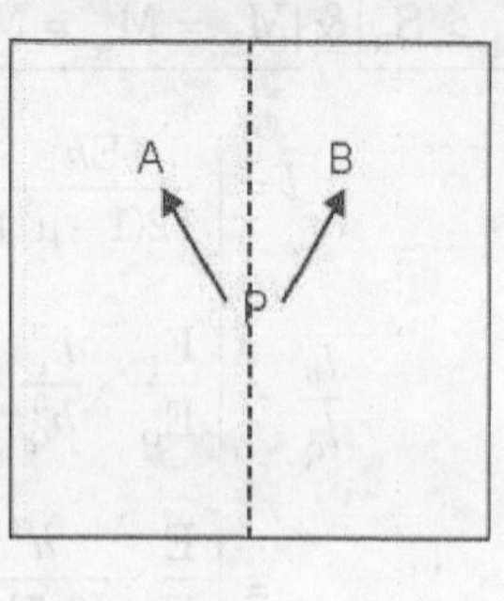

for Tower B,

Radial distance of top of tower = r = 6cm

Length of image = d = 2cm

Height of tower = 80 m = h_2

∴ Flying height above datum = H – h_1

$$\because \quad d = \frac{rh_2}{H - h_1}$$

$$\Rightarrow \quad H - h_1 = \frac{rh_2}{d}$$

$$= \frac{6 \times 80}{2}$$

$$= \left(\frac{480}{2}\right) m = 240\ m$$

for Tower A,

∵ Both on same plane

$$d = \frac{rh_2}{H - h_1}$$

r = 4 cm

d = 1.5 cm

H – h_1 = 240 m

$$h_2 = \frac{d \times H - h_1}{r}$$

$$= \frac{1.5 \times 240}{4}$$

h_2 = 90 m

48. Given that, Q = 0.01 m³/s

d = 100 mm

R_e = 800

g = 9.81 m/s²

Since R_e is less than 2000 the flow will be laminar flow

Hence, $$f = \frac{64}{R_e} = \frac{64}{800} = 0.08$$

Also, $$h_f = \frac{flQ^2}{12.1d^5}$$

$$= \frac{0.08 \times l \times (0.01)^2}{12.1 \times (0.1)^5}$$

$$= 0.0661l$$

Hence head loss per km length

= 0.0661 × 1000 = 66.1 m

49.

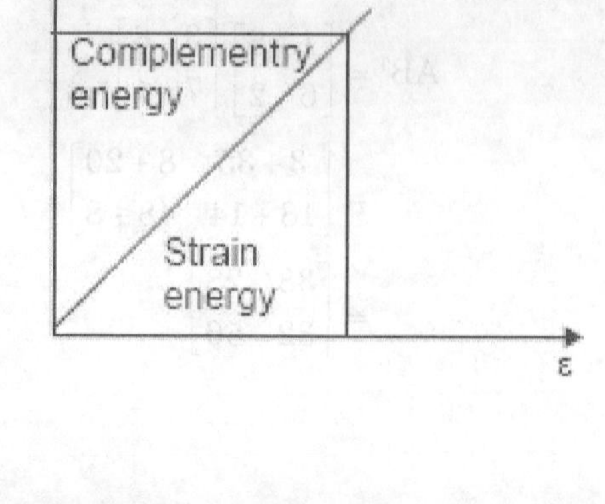

The area enclosed by the inclined line and the vertical axis is called complementary strain energy

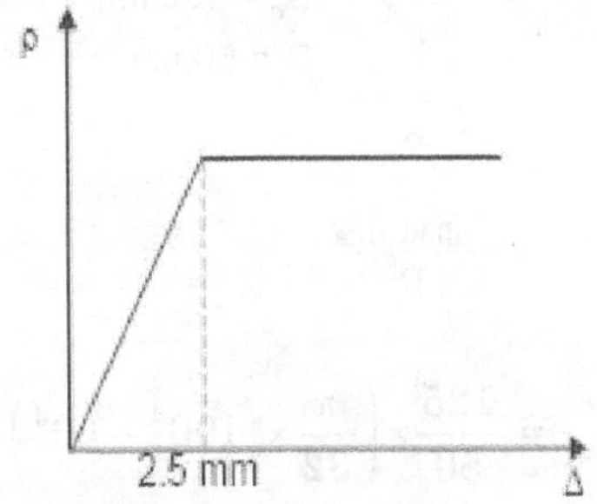

$$\in = \frac{\delta}{l}$$

$$= \frac{2.5}{2000}$$

$$= \frac{1}{800}$$

$$\sigma_y = 250 \text{ MPa}$$

Complementry strain energy

$$= \text{strain energy}$$

$$= \frac{1}{2} \times f_y \times \varepsilon \times \text{volume of bar}$$

$$= \frac{1}{2} \times 250 \times \frac{1}{800} \times \frac{\pi}{4} \times 8^2 \times 2000$$

$$= 15707.96 \text{ Nm}$$

50.

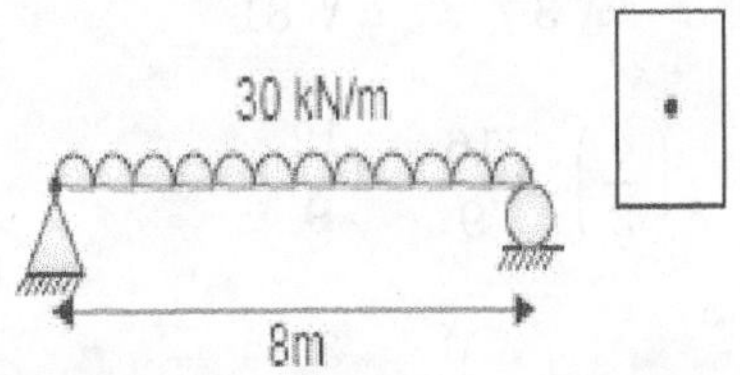

Prestressing force = 1600 kN

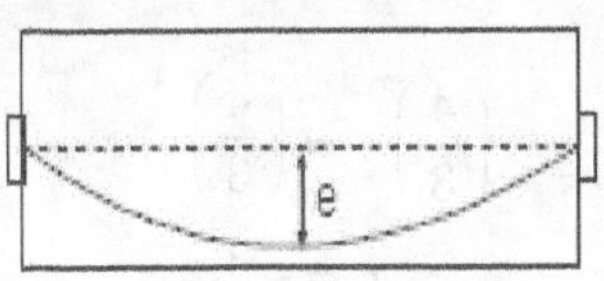

We know, $M_{max} = Pe$

$$M_{max} = \frac{wl^2}{8}$$

$$= \frac{30 \times 8^2}{8}$$

$$= 30 \times 8$$

$$= 240 \text{ kN–m}$$

Shape of tendon profile will follow the shape of bending moment diagram and for equillibrium of the section having maximum Bending Moment.

$$M_{max} = Pe$$

$$\Rightarrow \quad 240 \text{ kN-m} = 1600 \text{ kN} \times e$$

$$\Rightarrow \quad \frac{240 kN \times 10^3 \text{ mm}}{1600 kN} = e$$

$$e = 150 \text{ mm}$$

$$= 0.15 \text{ m}$$

51.

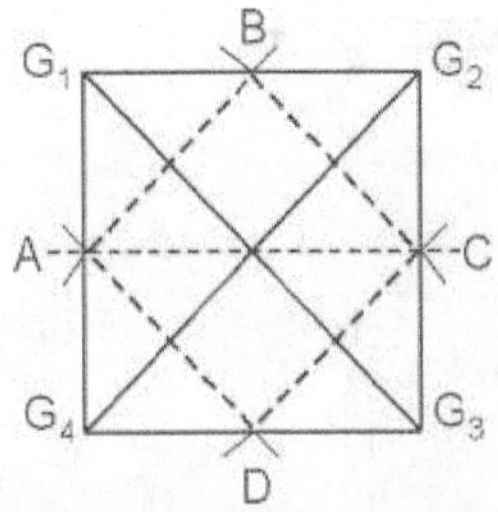

For G_1 area is AG_1B
For G_2 area is BG_2C
For G_3 area is CG_3D
For G_4 area is DG_4A
For G_5 area is ABCD

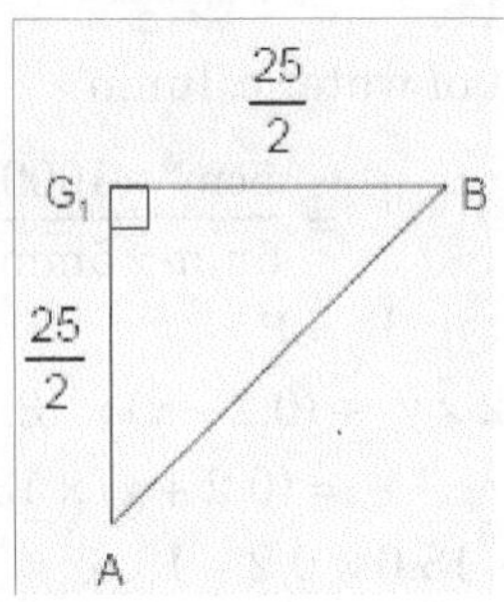

Take ΔAG_1B,

$$AB = \frac{25}{2}\sqrt{2}$$

$$= 17.67 \text{ km}$$

$$\text{Area of ABCD} = \left(\frac{25}{\sqrt{2}}\right)^2$$

$$= 312.5 \text{ km}^2$$

$$\text{Area of } AG_1B = \text{area of } AG_4D$$

$$= \text{Area of } CG_3D$$

$$= \text{area of } BG_2C$$

$$= \frac{625 - 312.5}{4}$$

$$= 78.125 \text{ km}^2$$

Average presipitation

$$= \frac{G_1A_1 + G_2A_2 + G_3A_3 + G_4A_4 + G_5A_5}{A_1 + A_2 + A_3 + A_4 + A_5}$$

$$= \frac{(300 + 285 + 272 + 290) \times 78.125 + 288 \times 312.5}{625}$$

$$= 287.375 \text{ mm}$$

52.

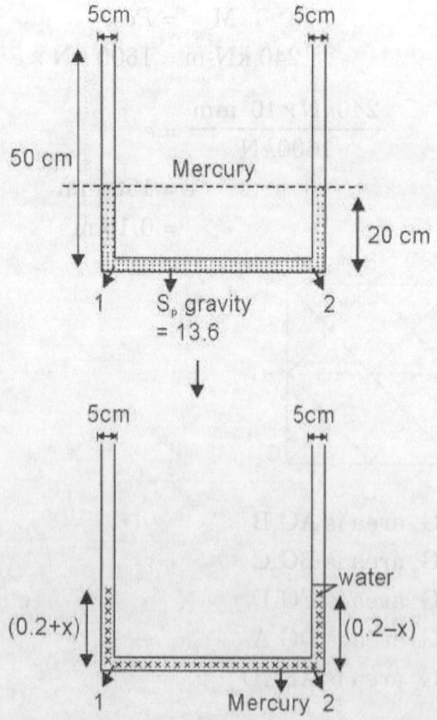

$\therefore$ Height of water column

$$= \frac{5\text{cm}^3 \times 1000}{5\text{mm} \times 5\text{mm}} = 20\text{cm} = 0.2\text{m}$$

Since, $P_2 = P_1$

Hence, $0.2 \times \gamma_w + (0.2 - x) \times 13.6 \times \gamma_w$

$$= (0.2 + x) \times 13.6 \times \gamma_w$$

$\Rightarrow$ $0.2 + 13.6 \times 0.2 - 13.6x$

$$= 13.6 \times 0.2 + 13.6x$$

$\Rightarrow$ $0.2 = 27.2x$

$\Rightarrow$ $x = 7.35 \times 10^{-3}\text{m} = 0.735 \text{ cm}$

New height = 20.735 cm

54.

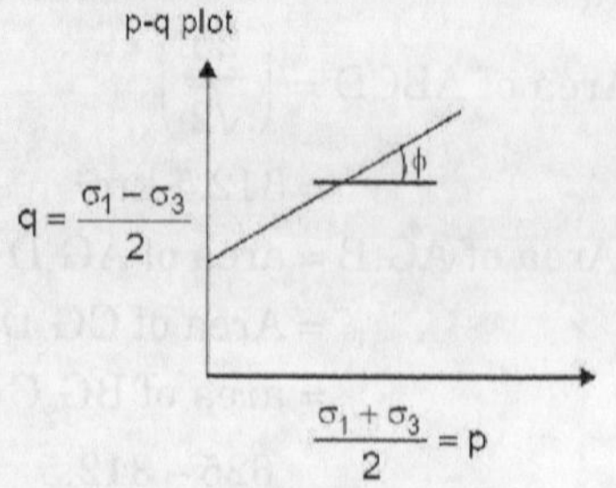

- When $\sigma_1 = \sigma_3$
- $q = 0, p = \frac{2\sigma_1}{2} = \sigma_1$
- So stress point lies on p-axis of p-q plot
- If $\sigma_1 > \sigma_3$

$$q = \frac{\sigma_1 - \sigma_3}{2} \text{ so } q > 0$$

$$p = \frac{\sigma_1 + \sigma_3}{2} \text{ so } p > 0$$

So, stress point p and q are positive.

55. Given that,

$$\tau_{\text{allowable}} = 125 \frac{\text{N}}{\text{mm}^2}$$

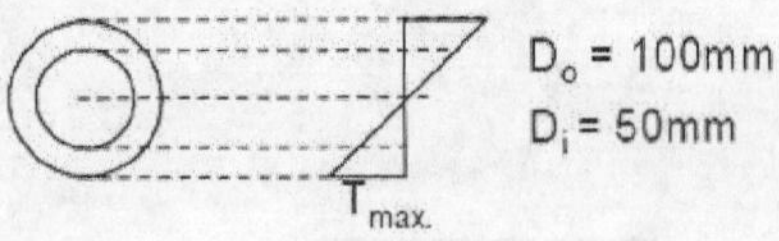

$$\frac{T_{max.}}{J} = \frac{\tau_{\text{allowable}}}{R}$$

$$\Rightarrow \quad T_{max.} = \frac{125}{50} \times \left(\frac{\pi}{32} \times \left(100^4 - 50^4\right)\right)$$

$$= 23009711.82 \text{ N-mm}$$

$$= 23.00 \times 10^{+6} \text{ N-mm}$$

$$= 23 \text{ kN-m}$$

GENERAL APTITUDE (GA)

1. $81 \times \left(\frac{16}{25}\right)^{x+2} \div \left(\frac{3}{5}\right)^{2x+4} = 144$

$$\Rightarrow \quad \left(\frac{4}{5}\right)^{2x+4} \times \left(\frac{5}{3}\right)^{2x+4} = \frac{144}{81}$$

$$\Rightarrow \quad \left(\frac{4}{3}\right)^{2x+4} = \frac{144}{81}$$

$$\Rightarrow \quad \left(\frac{4}{3}\right)^{x+2} = \sqrt{\frac{144}{81}}$$

$$\Rightarrow \quad \left(\frac{4}{3}\right)^{x} \times \frac{16}{9} = \frac{12}{9}$$

$$\Rightarrow \quad \left(\frac{4}{3}\right)^{x} = \frac{3}{4}$$

$$\Rightarrow \quad \left(\frac{4}{3}\right)^{x} = \left(\frac{4}{3}\right)^{-1}$$

$$x = -1$$

3. Card with visible face 2 should have red on the other side.

Card with visible face 3 may have any color on other side.

Card with visible face red should have even number on other side.

Card with visible face blue may have any number on other side.

Hence, to verify the above proposition, cards with face 2 and Red must be turned.

5. Favourable outcomes = {(1, 1), (1, 4), (2, 2), (3, 3), (4, 1), (4, 4), (5, 5), (6, 6)}

No. of favourable outcomes = 8

Total outcomes = 36

$$\therefore \quad \text{Probability} = \frac{8}{36} = \frac{2}{9}$$

7. The elevators has moved from 2nd to 5th floor between time 25 min to 30 min.

Elevation stayed at 4th floor for

$= 3 + 4 + 3 + 2 + 3 + 2 + 2 = 19$ min

Elevation stayed at ground floor for

$= 1 + 2 + 1 + 1 + 1 + 1 + 2 + 1 + 2 + 1 + 1 + 1 + 1 + 2 + 1 + 1 + 1$

$= 21$ min.

Hence, (*i*) and (*ii*) both are false.

8. $K < G, P < G, P > S, G < D$

if $K < G$ and $G < D$

Hence, $K < D$

If $S < P$ and $P < G$

Hence, $S < G$

Hence, (*ii*) and (*iv*) are correct.

9.

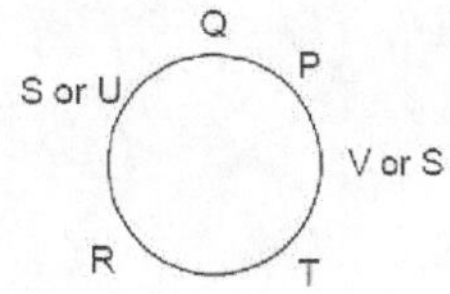

Before switching of seats of P and U

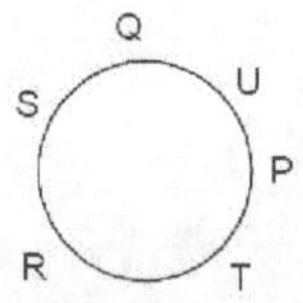

Ist possibility

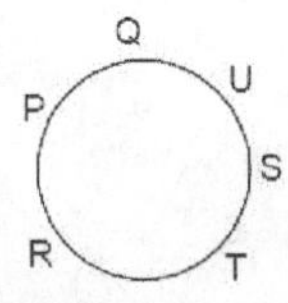

2nd possibility

10. Let C = Speed by cycling (km/h)

W = Speed of walking (km/h)

Now, by according to questions,

$$C \times \frac{2}{4} + W \times 2 \times \frac{3}{4} = 19$$

$$\Rightarrow \quad C + 3W = 38 \qquad ...(i)$$

and, $$C \times \frac{2}{2} + W \times \frac{2}{2} = 26$$

$$\Rightarrow \quad C + W = 26 \qquad ...(ii)$$

By solving equations (*i*) and (*ii*) we get,

$W = 6$ km/h

SOLVED PAPER 2018

INSTRUCTIONS

1. Total of 65 questions carrying 100 marks, out of which 10 questions carrying a total of 15 marks are in General Aptitude (GA)
2. The Engineering Mathematics will carry around **15% of the total marks,** the General Aptitude section will carry **15% of the total marks** and the **remaining 70% of the total marks.**
3. **Types of Questions**
   (*a*) **Multiple Choice Questions (MCQ)** carrying 1 or 2 marks each in all papers and sections. These questions are objective in nature, and each will have a choice of four options, out of which the candidate has to mark the correct answer(s).
   (*b*) **Numerical Answer Questions** of 1 or 2 marks each in all papers and sections. For these questions the answer is a real number, to be entered by the candidate using the virtual keypad. No choices will be shown for these type of questions.
4. For **1-mark** multiple-choice questions, **1/3 marks** will be deducted for a wrong answer. Likewise, for **2-marks** multiple-choice questions, **2/3** marks will be deducted for a wrong answer. There is no negative marking for numerical answer type questions.

Chapter-Wise Analysis

#	Chapters	Marks		
		1	2	Total
1	General Aptitude	5	5	15
2	Engineering Mathematics	3	4	11
3	Engineering Mechanics	0	2	4
4	Solid Mechanics	2	1	4
5	Structural Analysis	1	3	7
6	Constructional Material & Management	0	0	0
7	Concrete Structure	4	3	7

#	Chapters	Marks		
		1	2	Total
8	Steel Structure	1	1	3
9	Soil Mechanics and Foundation Engineering	5	5	15
10	Fluid Mechanics and Hydraulics	4	2	6
11	Hydrology & Irrigation	1	3	7
12	Environmental Engineering	2	2	6
13	Transportation Engineering	1	2	5
14	Surveying	1	2	5

GENERAL APTITUDE

1. "The driver applied the ________ as soon as she approached the hotel where she wanted to take a ________."

The words that best fill the blanks in the above sentence are

(*a*) brake, break (*b*) break, break

(*c*) brake, brake (*d*) break, brake

2. "It is no surprise that every society has had codes of behavior; however, the nature of these codes is often ________."

The word that best fills the blank in the above sentence is

(*a*) unpredictable (*b*) simple

(*c*) expected (*d*) strict

3. Hema's age is 5 years more than twice Hari's age. Suresh's age is 13 years less than 10 times Hari's age. If Suresh is 3 times as old as Hema. How old is Hema?

(*a*) 14 (*b*) 17

(*c*) 18 (*d*) 19

4. Tower A is 90 m tall and tower B is 140 m tall. They are 100 m apart. A horizontal skywalk connects the floors at 70 m in both the towers. If a taut rope connects the top of tower A to the bottom of tower B, at what distance (in meters) from tower A will the rope intersect the skywalk?

5. The temperature T in a room varies as a function of the outside temperature T_0 and the number of persons in the room p, according to the relation $T = K(\theta p + T_0)$, where θ is K are constants. What would be the value of θ given the following data?

T_0	p	T
25	2	32.4
30	5	42.0

(*a*) 0.8 (*b*) 1.0

(*c*) 2.0 (*d*) 10.0

6. A fruit seller sold a basket of fruits at 12.5% loss. Had he sold it for Rs. 108 more, he would have made a 10% gain. What is the loss in Rupees incurred by the fruit seller?

(*a*) 48

(*b*) 52

(*c*) 60

(*d*) 108

7. The price of a wire made of a superalloy material is proportional to the square of its length. The price of 10 m length of the wire is Rs. 1600. What would be the total price (in Rs.) of two wires of lengths 4 m and 6 m?

(*a*) 768 (*b*) 832

(*c*) 1440 (*d*) 1600

8. What of the following function(s) in an accurate description of the graph for the range(s) indicated?

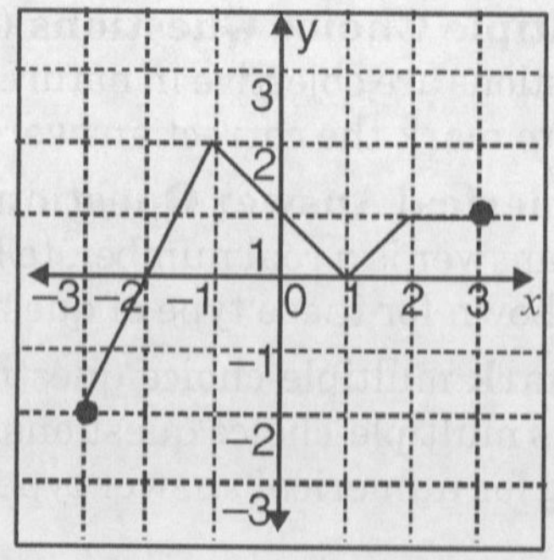

(*i*) $y = 2x + 4$ for $-3 \le x \le -1$

(*ii*) $y = |x - 1|$ for $-1 \le x \le 2$

(*iii*) $y = ||x| - 1|$ for $-1 \le x \le 2$

(*iv*) $y = 1$ for $2 \le x \le 3$

(*a*) (*i*), (*ii*) and (*iii*) only

(*b*) (*i*), (*ii*) and (*iv*) only

(*c*) (*i*) and (*iv*) only

(*d*) (*ii*) and (*iv*) only

9. Consider a sequence of number $a_1, a_2, a_3, \ldots, a_n$ where $a_n = \frac{1}{n} - \frac{1}{n+2}$, for each integer $n > 0$.

What is the sum of the first 50 terms?

(*a*) $\left(1+\frac{1}{2}\right) - \frac{1}{50}$

(*b*) $\left(1+\frac{1}{2}\right) + \frac{1}{50}$

(*c*) $\left(1+\frac{1}{2}\right) - \left(\frac{1}{51} + \frac{1}{52}\right)$

(*d*) $1 - \left(\frac{1}{51} + \frac{1}{52}\right)$

10. Each of the letters arranged as below represents a unique from 1 to 9. The letters are positioned in the figure such that (A × B × C), (B × G × E) and (D × E × F) are equal. Which integer among the following choices cannot be represented by the letters A, B, C, D, E, F or G?

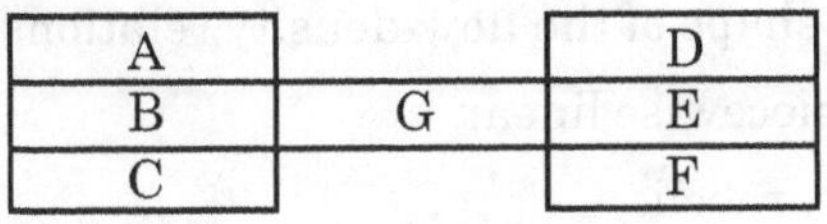

(*a*) 4 (*b*) 5

(*c*) 6 (*d*) 9

CIVIL ENGINEERING

1. Which one of the following matrices is singular?

(*a*) $\begin{bmatrix} 2 & 5 \\ 1 & 3 \end{bmatrix}$ (*b*) $\begin{bmatrix} 3 & 2 \\ 2 & 3 \end{bmatrix}$

(*c*) $\begin{bmatrix} 2 & 4 \\ 3 & 6 \end{bmatrix}$ (*d*) $\begin{bmatrix} 4 & 3 \\ 6 & 2 \end{bmatrix}$

2. For the given orthogonal matrix Q,

$$Q = \begin{bmatrix} \frac{3}{7} & \frac{2}{7} & \frac{6}{7} \\ -\frac{6}{7} & \frac{3}{7} & \frac{2}{7} \\ \frac{2}{7} & \frac{6}{7} & -\frac{3}{7} \end{bmatrix}$$

The inverse is

(*a*) $\begin{bmatrix} \frac{3}{7} & \frac{2}{7} & \frac{6}{7} \\ -\frac{6}{7} & \frac{3}{7} & \frac{2}{7} \\ \frac{2}{7} & \frac{6}{7} & -\frac{3}{7} \end{bmatrix}$ (*b*) $\begin{bmatrix} -\frac{3}{7} & -\frac{2}{7} & -\frac{6}{7} \\ \frac{6}{7} & -\frac{3}{7} & -\frac{2}{7} \\ -\frac{2}{7} & -\frac{6}{7} & \frac{3}{7} \end{bmatrix}$

(*c*) $\begin{bmatrix} \frac{3}{7} & \frac{6}{7} & \frac{2}{7} \\ \frac{2}{7} & \frac{3}{7} & \frac{6}{7} \\ \frac{6}{7} & \frac{2}{7} & -\frac{3}{7} \end{bmatrix}$ (*d*) $\begin{bmatrix} \frac{3}{7} & \frac{6}{7} & -\frac{2}{7} \\ -\frac{2}{7} & -\frac{3}{7} & \frac{6}{7} \\ \frac{6}{7} & -\frac{2}{7} & \frac{3}{7} \end{bmatrix}$

3. At the point $x = 0$, the function $f(x) = x^3$ has

(*a*) local maximum

(*b*) local minimum

(*c*) both local maximum and minimum

(*d*) neither local maximum nor local minimum

4. A column of height h with a rectangular cross-section of size a × 2a has a buckling load of P. If the cross-section is changed to 0.5a × 3a and its height changed to 1.5h, the buckling load of the redesigned column will be

(*a*) $\frac{P}{12}$ (*b*) $\frac{P}{4}$

(*c*) $\frac{P}{2}$ (*d*) $\frac{3P}{4}$

5. A steel column of ISHB 350 @ 72.4 kg/m is subjected to a factored axial compressive load of 2000 kN. The load is transferred to a concrete pedestal of grade M20 through a square base plate. Consider bearing of concrete as 0.45 f_{ck}, where f_{ck} is the characteristic strength of concrete. Using limit state method and neglecting the self weight of base plate and steel column, the length of a side of the base plate to be provided is

(*a*) 39 cm (*b*) 42 cm

(*c*) 45 cm (*d*) 48 cm

6. The Le Chatelier apparatus is used to determine

(*a*) compressive strength of cement

(*b*) fineness of cement

(*c*) setting time of cement

(*d*) soundness of cement

7. The deformation in concrete due to sustained loading is

(*a*) creep (*b*) hydration

(*c*) segregation (*d*) shrinkage

8. A solid circular beam with radius of 0.25 m and length of 2 m is subjected to a twisting moment of 20 kNm about the z-axis at the free end, which is the only load acting as shown in the figure. The stress component τ_{xy} at Point 'M' in the cross-section of the beam at a distance of 1 m from the fixed end is

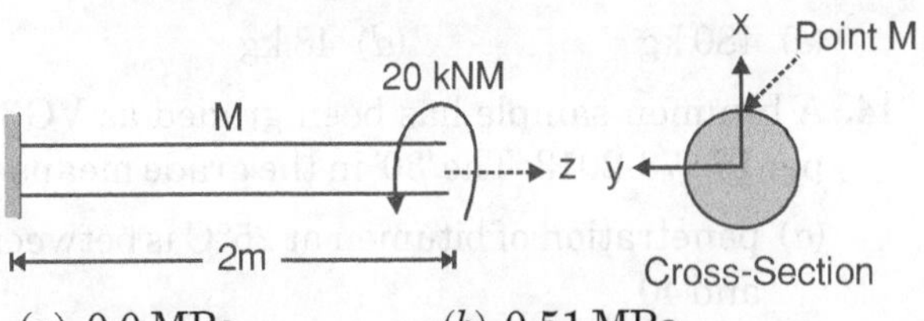

(*a*) 0.0 MPa (*b*) 0.51 MPa

(*c*) 0.815 MPa (*d*) 2.0 MPa

9. Two rectangular under-reinforced concrete beam sections X and Y are similar in all aspects except that the longitudinal compression reinforcement in section Y is 10% more. Which one of the following is the correct statement?

(*a*) Section X has less flexural strength and is less ductile than section Y

(*b*) Section X has less flexural strength but is more ductile than section Y

(*c*) Section X and Y have equal flexural strength but different ductility

(*d*) Sections X and Y have equal flexural strength and ductility.

10. The percent reduction in the bearing capacity of a strip footing resting on sand under flooding condition (water level at the base of the footing) when compared to the situation where the water level is at a depth much greater than the width of footing, is approximately

(*a*) 0 (*b*) 25

(*c*) 50 (*d*) 100

11. The width of a square footing and the diameter of a circular footing are equal. If both the footings are placed on the surface of sandy soil, the ratio of the ultimate bearing capacity of circular footing to that of square footing will be

(*a*) $\frac{4}{3}$ (*b*) 1

(*c*) $\frac{3}{4}$ (*d*) $\frac{2}{3}$

12. Bernoulli's equation is applicable for

(*a*) viscous and compressible fluid flow

(*b*) inviscid and compressible fluid flow

(*c*) inviscid and incompressible fluid flow

(*d*) viscous and incompressible fluid flow

13. There are 20,000 vehicles operating in a city with an average annual travel of 12,000 km per vehicle. The NO_x emission rate is 2.0 g/km per vehicle. The total annual release of NO_x will be

(*a*) 4,80,000 kg (*b*) 4,800 kg

(*c*) 480 kg (*d*) 48 kg

14. A bitumen sample has been graded as VG30 as per IS : 73-2013. The '30' in the grade means that

(*a*) penetration of bitumen at 25°C is between 20 and 40

(*b*) viscosity of bitumen at 60°C is between 2400 and 3600 Poise

(*c*) ductility of bitumen at 27°C is more than 30 cm

(*d*) elastic recovery to bitumen at 15°C is more than 30%

15. The speed-density relationship for a road section is shown in the figure.

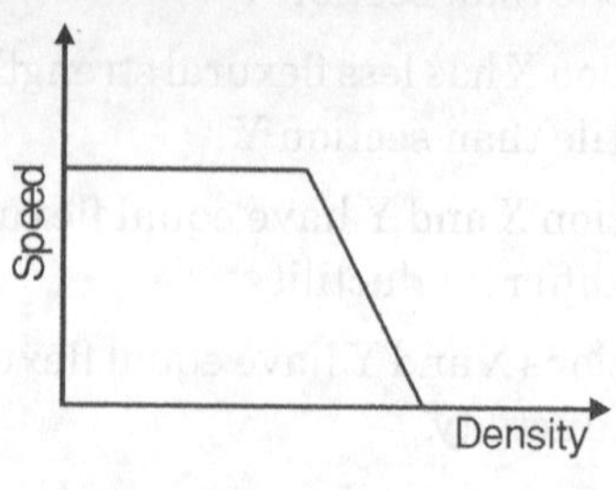

The shape of the flow-density relationship is

(*a*) piecewise linear

(*b*) parabolic

(*c*) initially linear then parabolic

(*d*) initially parabolic then linear

16. A well-designed signalized intersection is one in which the

(*a*) crossing conflicts are increased

(*b*) total delay is minimized

(*c*) cycle time is equal to the sum of red and green times in all phases

(*d*) cycle time is equal to the sum of red and yellow times in all phases

17. A flow field is given by $u = y^2, v = -xy, w = 0$. Value of the z-component of the angular velocity (in radians per unit time, up to two decimal places) at the point (0, –1, 1) is ________.

18. The frequency distribution of the compressive strength of 20 concrete cube specimen is given in the table.

f (MPa)	Number of specimens with compressive strength equal to *f*
23	4
28	2
22.5	5
31	5
29	4

If μ is the mean strength of the specimens and σ is the standard deviation, the number of specimens (out of 20) with compressive strength less than μ – 3σ is ________.

19. In a fillet weld, the direct shear stress and bending tensile stress are 50 MPa and 150 MPa, respectively. As per IS 800 : 2007, the equivalent stress (in MPa, up to two decimal places) will be ________.

20. In a shrinkage limit test, the volume and mass of a dry soil pat are found to be 50 cm³ and 88 g, respectively. The specific gravity of the soil solids is 2.71 and the density of water is 1 g/cc. The shrinkage limit (in % up to two decimal places) is ________.

21. A core cutter of 130 mm height has inner and outer diameters of 100 mm and 106 mm. respectively. The area ratio of the core cutter (in %, up to two decimal places) is ________.

22. A 1 : 50 model of a spillway is to be tested in the laboratory. The discharge in the prototype spillway is 1000 m^3/s. The corresponding discharge (in m^3/s up to two decimal places) to be maintained in the model, neglecting variation in acceleration due to gravity, is ________.

23. A 10 m wide rectangular channel carries a discharge of 20 m^3/s under critical condition. Using $g = 9.81\ m/s^2$, the specific energy (in m, up to two decimal places) is ________.

24. For routing of flood in a given channel using the Muskingum method, two of the routing coefficients are estimated as $C_0 = -0.25$ and $C_1 = 0.55$. The value of the third coefficient C_2 would be ________.

25. A city generates 40×10^6 kg of municipal solid waste (MSW) per year, out of which only 10% is recovered/recycled and the rest goes to landfill. The landfill has a single lift of 3 m height and is compacted to a density of 550 kg/m^3. If 80% of the landfill is assumed to be MSW, the landfill area (in m^2, up to one decimal place) required would be ________.

26. The value of the integral $\int_0^{x} x\cos^2 x dx$ is.

(*a*) $\frac{\pi^2}{8}$ (*b*) $\frac{\pi^2}{4}$

(*c*) $\frac{\pi^2}{2}$ (*d*) π^2

27. A cantilever beam of length 2 m with a square section of side length 0.1 m is loaded vertically at the free end. The vertical displacement at the free end is 5 mm. The beam is made of steel with Young's modulus of $2.0 \times 10^{11}\ N/m^2$. The maximum bending stress at the fixed end of the cantilever is

(*a*) 20.0 MPa (*b*) 37.5 MPa

(*c*) 60.0 MPa (*d*) 75.0 MPa

28. A cylinder of radius 250 mm and weight, W = 10 kN is rolled up an obstacle of height 50 mm by applying a horizontal force P at its centre as shown in the figure.

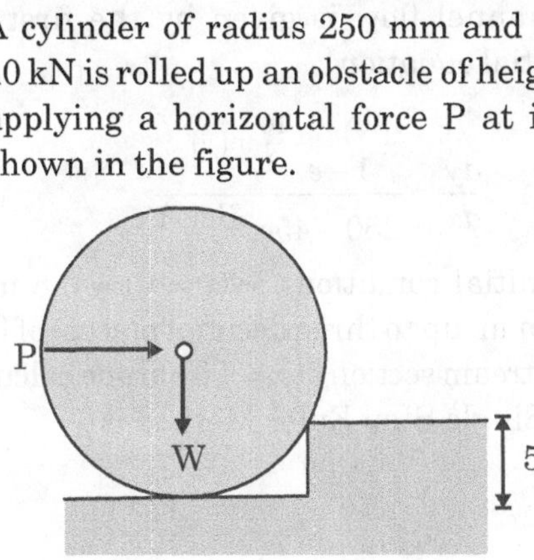

All interfaces are assumed frictionless. The minimum value of P is

(*a*) 4.5 kN (*b*) 5.0 kN

(*c*) 6.0 kN (*d*) 7.5 kN

29. A plate in equilibrium is subjected to uniform stresses along its edges with magnitude $\sigma_{xx} = 30$ MPa and $\sigma_{yy} = 50$ MPa as shown in the figure

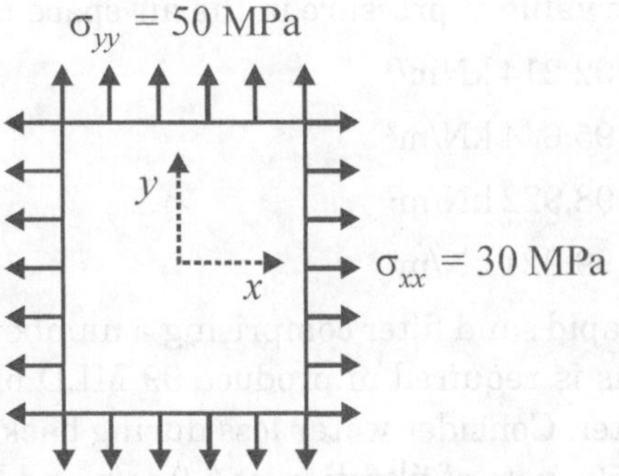

The Young's modulus of the material is $2 \times 10^{11}\ N/m^2$ and the Poisson's ratio is 0.3. If σ_{zz} is negligibly small and assumed to be zero, then the strain ε_{zz} is

(*a*) -120×10^{-6} (*b*) -60×10^{-6}

(*c*) 0.0 (*d*) 120×10^{-6}

30. The figure shows a simply supported beam PQ of uniform flexural rigidity EI carrying two moments M and 2M

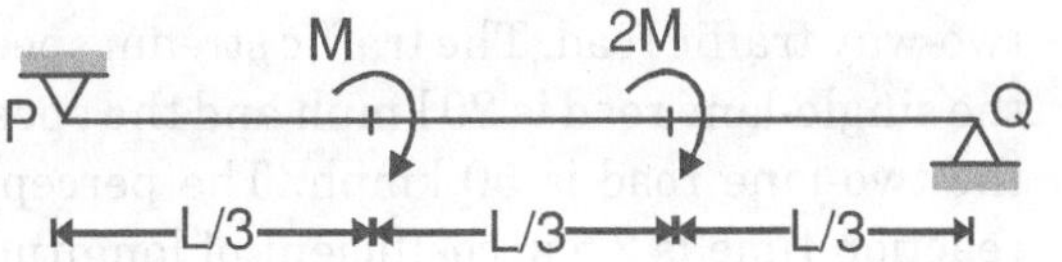

The slope at P will be

(*a*) 0

(*b*) ML/(9EI)

(*c*) ML/(6EI)

(*d*) ML/(3EI)

31. A 0.5 m × 0.5 m square concrete pile is to be driven in a homogeneous clayey soil having undrained shear strength, $c_u = 50$ kPa and unit weight, $\gamma = 18.0\ kN/m^3$. The design capacity of the pile is 500 kN. The adhesion factor α is given 0.75. The length of the pile required for the above design load with a factor of safety of 2.0 is

(*a*) 5.2 m

(*b*) 5.8 m

(*c*) 11.8 m

(*d*) 12.5 m

32. A closed tank contains 0.5 m thick layer of mercury (specific gravity = 13.6) at the bottom. A 2.0 m thick layer of water lies above the mercury layer. A 3.0 m thick layer of oil (specific gravity = 0.6) lies above the water layer. The space above the oil layer contains air under pressure. The gauge pressure at the bottom of the tank is 196.2 kN/m^2. The density of water is 1000 kg/m^3 and the acceleration due to gravity is 9.81 m/s^2. The value of pressure in the air space is

(*a*) 92.214 kNm/^{2}

(*b*) 95.644 kN/m^2

(*c*) 98.922 kN/m^2

(*d*) 99.321 kN/m^2

33. A rapid sand filter comprising a number of filter beds is required to produce 99 MLD of potable water. Consider water loss during backwashing as 5%, rate of filtration as 6.0 m/h and length to width ratio of filter bed as 1.35. The width of each filter bed is to be kept equal to 5.2 m. One additional filter bed is to be provided to take care of break-down repair and maintenance. The total number of filter beds required will be

(*a*) 19 (*b*) 20

(*c*) 21 (*d*) 22

34. A priority intersection has a single-lane one-way traffic road crossing an undivided two-lane two-way traffic road. The traffic stream speed on the single-lane road is 20 kmph and the speed on the two-lane road is 50 kmph. The perception-reaction time is 2.5 s, coefficient of longitudinal friction is 0.38 and acceleration due to gravity is 9.81 m/s^2. A clear sight triangle has to be ensured at this intersection. The minimum lengths of the sides of the sight triangle along the two-lane road and the single-lane road, respectively will be

(*a*) 50 m and 20 m (*b*) 61 m and 18 m

(*c*) 111 m and 15 m (*d*) 122 m and 36 m

35. The following details refer to a closed traverse:

Line	Consecutive coordinate			
	Northing (m)	Southing (m)	Easting (m)	Westing (m)
PQ	—	437	173	—
QR	101	—	558	—
RS	419	—	—	96
SP	—	83	—	634

The length and direction (whole circle bearing) of closure, respectively are

(*a*) 1 m and 90°

(*b*) 2 m and 90°

(*c*) 1 m and 270°

(*d*) 2 m and 270°

36. A square area (on the surface of the earth) with side 100 m and uniform height, appears as 1 cm^2 on a vertical aerial photograph. The topographic map shows that a contour of 650 m passes through the area. If focal length of the camera lens is 150 mm, the height from which the aerial photograph was taken, is

(*a*) 800 m

(*b*) 1500 m

(*c*) 2150 m

(*d*) 3150 m

37. The solution at $x = 1$, $t = 1$ of the partial differential equation $\frac{\partial^2 u}{\partial x^2} = 25\frac{\partial^2 u}{\partial t^2}$ subject to initial conditions of $u(0) = 3x$ and $\frac{\partial u}{\partial x}(0) = 3$ is ________.

(*a*) 1

(*b*) 2

(*c*) 4

(*d*) 6

38. The solution (up to three decimal places) at $x = 1$ of the differential equation $\frac{d^2y}{dx^2} + 2\frac{dy}{dx} + y = 0$ subject to boundary conditions $y(0) = 1$ and $\frac{dy}{dx} = (0) = -1$ is ________.

39. Variation of water depth (y) in a gradually varied open channel flow is given by the first order differential equation

$$\frac{dy}{dx} = \frac{1 - e^{-\frac{10}{3}\text{In}(y)}}{250 - 45e^{-3\text{In}(y)}}$$

Given initial conditions : $y(x = 0) = 0.8$ m. The depth (in m, up to three decimal places) of flow at a downstream section at $x = 1$ from one calculation step of Single Step Euler Method is ________.

40. An RCC short column (with lateral ties) of rectangular cross-section of 250 mm × 300 mm in reinforced with four member of 16 mm diameter longitudinal bars. The grades of steel and concrete are Fe415 and M20, respectively. Neglect eccentricity effect. Considering limit state of collapse in compression (IS 456 : 2000), the axial load carrying capacity of the column (in kN, up to one decimal place) is ________.

41. An RCC beam of rectangular cross-section has factored shear of 200 kN at its critical section. Its width b is 250 mm and effective depth d is 350 mm. Assume design shear strength τ_C of concrete as 0.62 N/mm² and maximum allowable shear stress $\tau_{c\ max}$ in concrete as 2.8 N/mm². If two legged 10 mm diameter vertical stirrups of Fe250 grade steel are used, then the required spacing (in cm, up to one decimal place) as per limit state method will be ________.

42. The dimensionless of a symmetrical welded I-section are shown in the figure.

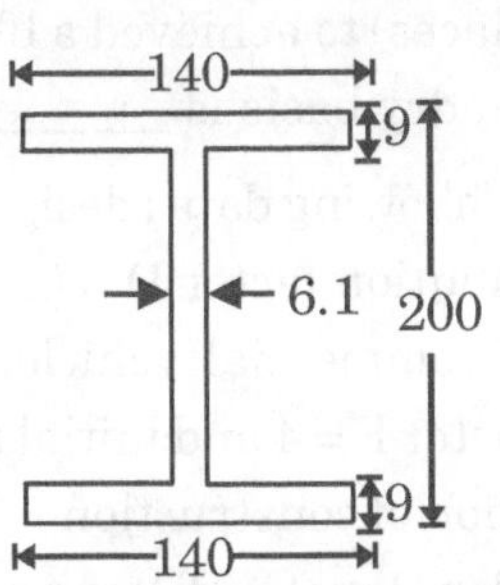

(All dimension are in mm)

The plastic section modulus about the weaker axis (in cm³, up to one decimal place) is ________.

43. Consider the deformable pin-jointed truss with loading, geometry and section properties as shown in figure.

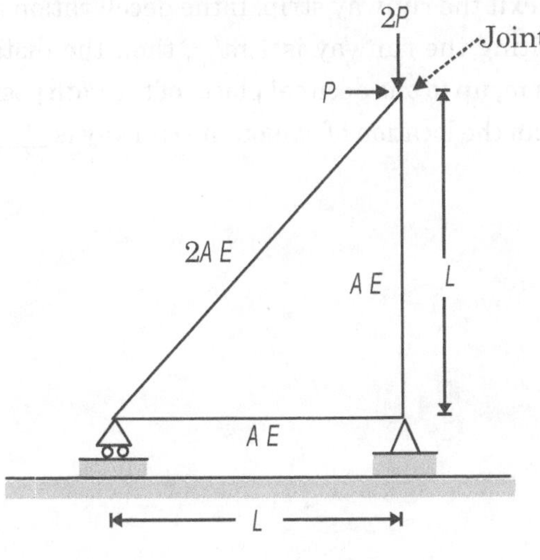

Given that $E = 2 \times 10^{11}$ N/m², A = 10 mm², L = 1 m and P = 1 kN, the horizontal displacement of Joint C (in mm, up to one decimal place) is ________.

44. At a construction site, a contractor plans to make an excavation as shown in the figure.

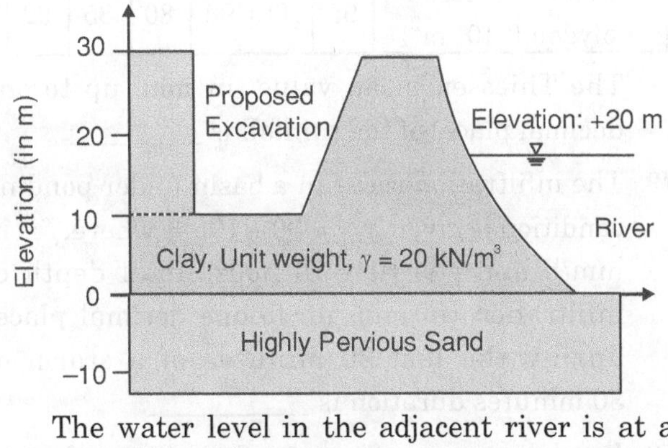

The water level in the adjacent river is at an elevation of +20.0 m. Unit weight of water is 10 kN/m³. The factor of safety (up to two decimal places) against sand boiling for the proposed excavation is ________.

45. A conventional drained triaxial compression test was conducted in a normally consolidated clay sample under an effective confining pressure of 200 kPa. The deviator stress at failure was found to be 400 kPa. An identical specimen of the same clay sample is isotropically consolidated to a confining pressure of 200 kPa and subjected to standard undrained triaxial compression test. If the deviator stress at failure is 150 kPa, the pore pressure developed (in kPa, up to one decimal place) is ________.

46. The void ratio of a soil is 0.55 at an effective normal stress of 140 kPa. The compression index of the soil is 0.25. In order to reduce the void ratio to 0.4, an increase in the magnitude of effective normal stress (in kPa, up to one decimal place) should be ________.

47. A rigid smooth retaining wall of height 7 m with vertical backface retains saturated clay as backfill. The saturated unit weight and undrained cohesion of the backfill are 17.2 kN/m³ and 20 kPa, respectively. The difference in the active lateral forces on the wall (in kN per meter length of wall, up to two decimal places), before and after the occurrence of tension cracks is ________.

48. Rainfall depth over a watershed is monitored through six number of well distributed rain gauges. Gauged data are given below:

Rain Gauge Number	1	2	3	4	5	6
Rainfall Depth (mm)	470	465	435	525	480	510
Area of Thiessen Polygon ($\times 10^4$ m^2)	95	100	98	80	85	92

The Thiessen mean value (in mm, up to one decimal place) of the rainfall is ________.

49. The infiltration rate f in a basin under ponding condition is given by $f = 30 + 10e^{-2t}$, where, f is in mm/h and t is time in hour. Total depth of infiltration (in mm, up to one decimal place) during the last 20 minutes of a storm of 30 minutes duration is ________.

50. In a laboratory, a flow experiment is performed over a hydraulic structure. The measured values of discharge and velocity are 0.05 m^3/s and 0.25 m/s, respectively. If the full scale structure (30 times bigger) is subjected to a discharge of 270 m^3/s, then the time scale (model to full scale) value (up to two decimal places) is ________.

51. A water sample analysis data is given below:

Ion	Concentration, mg/L	Atomic Weight
Ca^{2+}	60	40
Mg^{2+}	30	24.31
HCO_3	400	61

The carbonate hardness (expressed as mg/L of $CaCO_3$, up to one decimal place) for the water sample is ________.

52. The ultimate BOD (L_0) a wastewater sample is estimated as 87% of COD. The COD of this wastewater is 300 mg/L. Considering first order BOD reaction rate constant k (use natural log) = 0.23 per day and temperature coefficient $\theta = 1.047$, the BOD value (in mg/L, up to one decimal place) after three days of incubation at 27°C for the wastewater will be ________.

53. A waste activated sludge (WAS) is to be blended with green waste (GW). The carbon (C) and nitrogen (N) contents, per kg of WAS and GW on dry basis are given in the table.

Parameter	**WAS**	**GW**
Carbon (g)	54	360
Nitrogen (g)	10	6

The ratio of WAS to GW required (up to two decimal places) to achieved a blended C : N ratio of 20 : 1 on dry basis is ________.

54. Given the following data : design life n = 15 years, lane distribution factor D = 0.75, annual rate of growth of commercial vehicles r = 6%, vehicle damage factor F = 4 and initial traffic in the year of completion of construction = 3000 Commercial Vehicles Per Day (CVPD). As per IRC : 37-2012, the design traffic in terms of cumulative number of standard axles (in million standard axles, up to two decimal places) is ________.

55. An aircraft approaches the threshold of a runway strip at a speed of 200 km/h. The pilot decelerates the aircraft at a rate of 1.697 m/s^2 and takes 18 s to exit the runway strip. If the deceleration after exiting the runway is 1 m/s^2, then the distance (in m, up to one decimal place) of the gate position from the location of exit on the runway is ________.

ANSWERS

General Aptitude (GA)

1. (*a*)	**2.** (*a*)	**3.** (*d*)	**4.** 22.22	**5.** (*b*)	**6.** (*c*)	**7.** (*b*)	**8.** (*b*)	**9.** (*c*)	**10.** (*b*)

Civil Engineering

1. (*c*)	**2.** (*c*)	**3.** (*d*)	**4.** (*a*)	**5.** (*d*)	**6.** (*d*)	**7.** (*a*)
8. (*a*)	**9.** (*a*)	**10.** (*c*)	**11.** (*c*)	**12.** (*c*)	**13.** (*a*)	**14.** (*b*)
15. (*c*)	**16.** (*b*)	**17.** 1.5	**18.** 0	**19.** 173.21	**20.** 19.90	**21.** 12.36
22. 0.06	**23.** 1.11	**24.** 0.7	**25.** 27272.7	**26.** (*b*)	**27.** (*b*)	**28.** (*d*)
29. (*a*)	**30.** (*c*)	**31.** (*c*)	**32.** (*a*)	**33.** (*c*)	**34.** (*b*)	**35.** (*a, c*)
36. (*c*)	**37.** (*d*)	**38.** 0.36	**39.** 0.793	**40.** 918.1	**41.** 8.2	**42.** 89.9
43. 3.4	**44.** 1	**45.** 125	**46.** 417.3	**47.** 46.72	**48.** 479.1	**49.** 11.7
50. 0.18	**51.** 273.4	**52.** 160.2	**53.** 1.64	**54.** 76.45	**55.** 312.8	

EXPLANATIONS

GENERAL APTITUDE

2. Unpredictable - Contrary word required

3. Let the age of Hari be 'x'

$\therefore$ Hema age $= 5 + 2x$

and Suresh age $= 10x - 13$

Now Suresh is 3 time old as Hema

$\therefore \quad (10x - 13) = 3(5 + 2x)$

$\therefore \quad 10x - 13 = 15 + 6x$

$\therefore \quad 4x = 28$

$\therefore \quad x = 7$

$\therefore$ Hema's age $= 5 + 2x$

$= 5 + 14$

$= 19$

4.

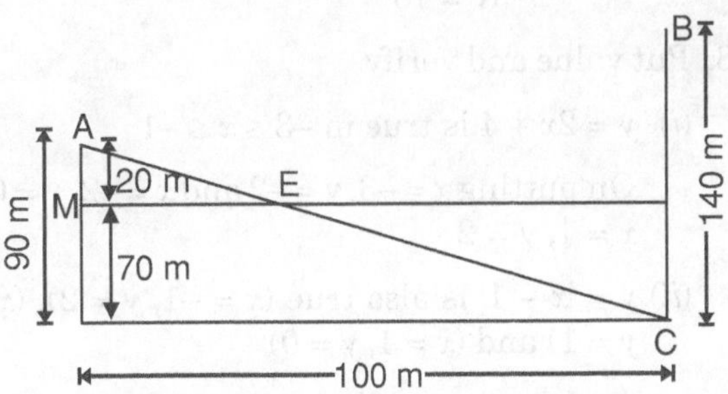

$$\Delta AME \approx \Delta APC$$

$$\frac{AM}{AP} = \frac{ME}{PC}$$

$$\therefore \quad \frac{20}{90} = \frac{ME}{100}$$

$$\therefore \quad ME = \frac{20 \times 100}{90} = 22.22$$

5. For $T = 32.4,$

$p = 2$

and $T_0 = 25$

$$T = K(\theta p + T_0)$$

$$\therefore \quad 32.4 = K(2\theta + 25) \quad ...(i)$$

Now for $T = 42, p = 5$ and $T_0 = 30$

$$T = K(\theta p + T_0)$$

$$\therefore \quad 42 = K(5\theta + 30) \quad ...(ii)$$

Solving equation (*i*) and (*ii*) we get

$$\theta = 1$$

and $K = \dfrac{78}{65}$

6. $12.5\%x + 10\%x = 108$

$$x = \frac{108}{22.5}$$

So loss $108 \times \dfrac{12.5}{22.5} = 60$

7. Total price $(p) = p_1 + p_2$

$$= 16(4)^2 + 16(6)^2$$

$$= 16[16 + 36]$$

$$= 16[52] = 832$$

Price $\propto L^2$

Price $= KL^2$ (where K is constant)

Now according to question

$$1600 = K(10)^2$$

$$\therefore \quad K = 16$$

8. Put value and verify

(*i*) $y = 2x + 4$ is true in $-3 \le x \le -1$

On putting $x = -3, y = -2$ and $x = -2, y = 0$ and $x = 1, y = 2$

(*ii*) $y = |x - 1|$ is also true $(x = -1, y = 2)$, $(x = 0, y = 1)$ and $(x = 1, y = 0)$

(*iv*) $y = 1$ in $(2 \le x \le 3)$ always true

(*i*), (*ii*) and (*iv*) are true.

9. Sum of series will be

$$= \left(1 - \frac{1}{3}\right) + \left(\frac{1}{2} - \frac{1}{4}\right) + \left(\frac{1}{3} - \frac{1}{5}\right) .... \left(\frac{1}{48} - \frac{1}{50}\right)$$

$$+ \left(\frac{1}{49} - \frac{1}{51}\right) + \left(\frac{1}{50} - \frac{1}{52}\right)$$

All like terms will cancel out and we will be left with

$$= \left(1 + \frac{1}{2}\right) - \left(\frac{1}{51} + \frac{1}{52}\right)$$

10. $A \times B \times C = B \times G \times E$

$$= D \times E \times F$$

$$= 72$$

$$8 \times 9 \times 1 = 9 \times 2 \times 4$$

$$= 3 \times 4 \times 6$$

$$= 72$$

Any of A, B, C, D, E, F, G cannot be 5.

CIVIL ENGINEERING

1. If $|A| = 0$, then the given matrix is singular.

Now from options.

From Option (a) : $|A| = 6 - 5 = 1$

From Option (b) : $|A| = 9 - 4 = 5$

From Option (c) : $|A| = 12 - 12 = 0$

From Option (d) : $|A| = 8 - 18 = -10$

Hence matrix (c) is singular.

2.

$$|Q| = \frac{3}{7}\left(-\frac{9}{49} - \frac{12}{49}\right) - \frac{2}{7}\left(\frac{18}{49} - \frac{4}{49}\right) + \frac{6}{7}\left(\frac{-36}{49} - \frac{6}{49}\right) = -1$$

$$(\text{Adj. Q}) = \begin{bmatrix} -\frac{21}{49} & \frac{42}{49} & -\frac{14}{49} \\ -\frac{14}{49} & -\frac{21}{49} & -\frac{42}{49} \\ -\frac{42}{49} & -\frac{14}{49} & \frac{21}{42} \end{bmatrix}$$

$$= \begin{bmatrix} -\frac{3}{7} & \frac{6}{7} & -\frac{2}{7} \\ -\frac{2}{7} & -\frac{3}{7} & -\frac{6}{7} \\ -\frac{6}{7} & -\frac{2}{7} & \frac{3}{7} \end{bmatrix}$$

Now, $$Q^{-1} = \frac{\text{AdjQ}}{|Q|} = \begin{bmatrix} \frac{3}{7} & -\frac{6}{7} & \frac{2}{7} \\ \frac{2}{7} & \frac{3}{7} & \frac{6}{7} \\ \frac{6}{7} & \frac{2}{7} & -\frac{3}{7} \end{bmatrix}$$

3. $f(x) = x^3$ at $x = 0$

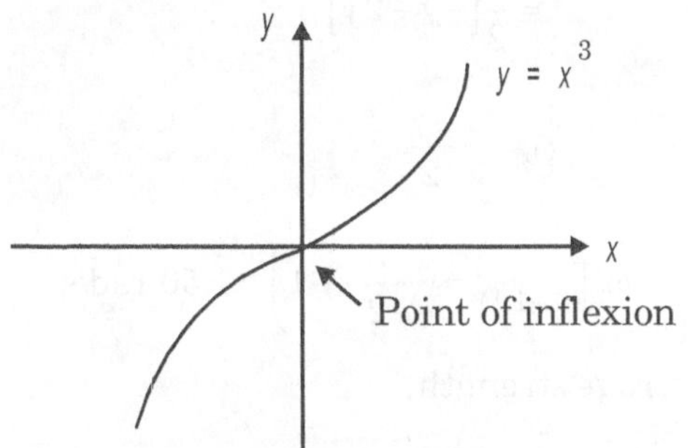

$$F(x) = x^3$$

Differentiating $f(x)$ w.r.t. to 'x' we and equating it to zero we get

$$f'(x) = 3x^2 = 0$$

and $$f''(x) = 6x$$

Now $$f''(x)\big|_{x=0} = 0$$

4. For column, $$P = \frac{\pi^2 EI_{min}}{L^2}$$

$$= \frac{\pi^2 E\left(\frac{2a \times a^3}{12}\right)}{h^2}$$

$$= \frac{\pi^2 E(2a^4)}{12h^2}$$

For new column, $$P' = \frac{\pi^2 E\left[\frac{3a \times (0.5a)^3}{12}\right]}{(1.5h)^2}$$

$$= \frac{1}{12} \times \frac{\pi^2 Ea^4}{6h^2} = \frac{P}{12}$$

5. Given data,

Factor Load = 2000 kN

Area required for base plate

$$= \frac{\text{Factored load}}{\text{Bearing capacity of concrete}}$$

$$= \frac{2000 \times 10^3}{0.45 \times 20} = 222222.222 \text{ mm}^2$$

Now, Side of base plate = $\sqrt{\text{Area of base plate}}$

$$= \sqrt{222222.22}$$

$$= 471.4 \text{ mm}$$

$$= 47.14 \text{ cm}$$

Since, given area must be more than required area

$\therefore$ Correct answer is 48 cm.

6. **Le Chatelier Apparatus** is used to determine the soundness of cement as per IS code 4031 (part 3); this cement testing procedure is called Le Chatelier test.

7. Creep is deformation due to sustained loading.

8.

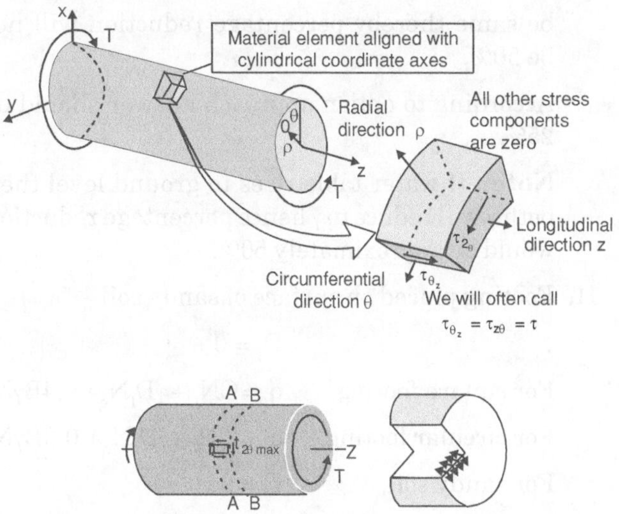

The only non-zero stresses are $\tau_{\theta z} = \tau_{\theta q} = \tau$, if θ is 90° then $\theta = y$

Hence $\tau_{zy} = \tau_{yz} = \tau_{max} = 16T/\pi d^3$

$= 0.815$ MPa

But in rest of the planes shear stresses are zero,

$\Rightarrow \quad \tau_{xy} = \tau_{yx} = 0$

9.

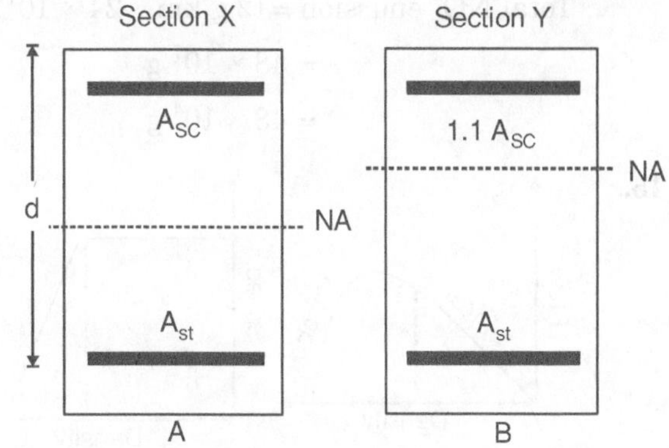

Due to presence of more compression steel in section Y, NA of section of Y is more than as of X. It means Y is more under-reinforced than X so ductility of Y is more. Now as compression steel of Y is more

$\Rightarrow$ flexure resistance of X is less than that of Y.

10. For strip footing on sand (c = 0)

$$q_u = \gamma D_f N_q + 0.5B\gamma N_\gamma$$

In flooding condition water level rises to base of footing hence IIIrd term unit weight of soil will change and IInd term unit weight will be unaffected.

$$\therefore \quad q_u = \gamma D_f N_q + 0.5B\gamma N_\gamma$$

$$\therefore \quad \gamma = \frac{1}{2}\gamma_{sat}$$

Hence third term reduced and second term will be same thereby percentage reduction will not be 50%.

According to option approach answer should be 25%.

Note : If water table rises to ground level then both γ will reduce to γ hence percentage reduction would be approximately 50%.

11. Footing placed on surface of sandy soil

$$\therefore \quad D_f = 0$$

For square footing, $q_u = CN_C + \gamma D_f N_q + 0.4B\gamma N_\gamma$

For circular footing, $q_u = CN_C + \gamma D_f N_q + 0.3B\gamma N_\gamma$

For sandy soil, $C = 0$

$$\therefore \quad \text{Ratio}\frac{(q_u)_{circular}}{(q_u)_{square}} = \frac{3}{4}$$

13. Total number of kms. travelled by all the vehicles

$$= (20000 \times 12000) \text{ km}$$
$$= 24 \times 10^7 \text{ km}$$

$\therefore$ Total NO_x emission = (2 g/km × 24 × 10^7) km

$$= 48 \times 10^7 \text{ g}$$
$$= 48 \times 10^4 \text{ g}$$

15.

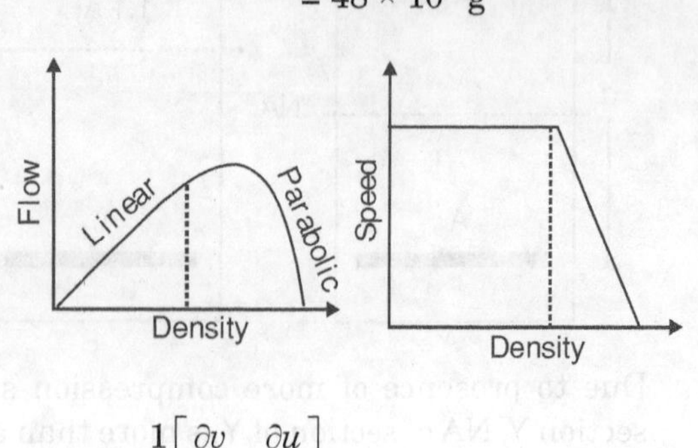

17.
$$\omega_z = \frac{1}{2}\left[\frac{\partial v}{\partial x} - \frac{\partial u}{\partial y}\right]$$

$$= \frac{1}{2}\left[\frac{\partial}{\partial x}(-xy) - \frac{\partial}{\partial y}(y)^2\right]$$

$$= \frac{1}{2}[-y - 2y]$$

$$\omega_z = -\frac{3y}{2}$$

$$\omega_z\big|_{(0,-1,1)} = \left(-\frac{3}{2} \times -1\right) = 1.50 \text{ rad/s}$$

18. Average strength,

$$\mu = \frac{(4 \times 23) + (2 \times 28) + (5 \times 22.5) + (5 \times 31) + (4 \times 29)}{20} = 26.575 \text{ MPa}$$

Now $\quad \sigma = \sqrt{\dfrac{\Sigma(\mu - f)^2}{n-1}}$

$$= \sqrt{\frac{(26.575-23)^2 \times 4 + (26.575-28)^2 \times 2 + (26.575-22.5)^2 \times 5 + (26.575-31)^2 \times 5 + (26.575-29)^2 \times 4}{(20-1)}} = 3.7$$

Now, $\quad \mu - 3\sigma = 26.575 - (3 \times 3.7)$

$$= 26.575 - 11.1$$
$$= 15.48$$

Thus, no specimen is having compressive strength less than $\mu - 3\sigma$.

19. Direct bending tensile stress, $(f_a) = 150$ MPa

Direct shear stress, $(q) = 50$ MPa

Now according to IS 800 : 2007, clause 10.5.10.1.1

The equivalent stress, $(f_e) = \sqrt{f_a^2 + 3q^2} \le \dfrac{f_u}{\sqrt{3}\gamma_{mw}}$

$$\therefore \quad f_e = \sqrt{150^2 + 3 \times 50^2}$$
$$= \sqrt{22500 + 7500}$$
$$= 173.21 \text{ MPa}$$

$$\Rightarrow \quad f_e \le \frac{f_u}{\sqrt{3}\gamma_{mw}}\left(= \frac{400}{\sqrt{3} \times 1.25} = 184.75 \text{ MPa}\right)$$

(Hence OK)

Note : The above check for combination of stresses need not be done for fillet welds where the sum of normal and shear stresses does not exceed f_{wd} [Clause 10.5.10.1.2 (b)].

Thus, $f_a + q = 150 + 50 = 200$ MPa

$f_a + q > f_{wd}$ (= 184.75 MPa)

$\Rightarrow$ $f_e = 173.21$ MPa

Thus, the weld is designed for an equivalent stress of 173.21 MPa.

20. Dry soil mass = 88 gm

Volume of dry soil = 50 cc

Dry density of soil mass is given by

$$(\rho_d) = \frac{M_d}{V_d} = \frac{88}{50} \text{gm/cc}$$

$\because$ Shrinkage limit,

$$(W_s) = \frac{1}{R} - \frac{1}{G} = \frac{1}{(\rho_d/\rho_w)} - \frac{1}{G} \quad \left(\because R = \rho_d/\rho_w\right)$$

$$\therefore \quad W_s = \left(\frac{\rho_w}{\rho_d} - \frac{1}{G}\right) = \left(\frac{1}{1.76} - \frac{1}{2.71}\right)$$

$$= 0.1990 = 19.90\%$$

21. Area ratio

$$= \left[\frac{A_{outer} - A_{inner}}{A_{inner}} \times 100\right]\%$$

$$= \left[\frac{\frac{\pi}{4}D_0^2 - \frac{\pi}{4}D_i^2}{\frac{\pi}{4}D_i^2} \times 100\right]\%$$

$$= \left[\frac{\frac{\pi}{4}(D_0 - D_i)(D_0 + D_i)}{\frac{\pi}{4}D_i^2} \times 100\right]\%$$

$$= \left[\frac{(106-100)(106+100)}{(100)^2} \times 100\right]\%$$

$$= \left[\frac{6 \times 206 \times 100}{10000}\right]\% = 12.36\%$$

22. Froude law is valid

$$Q_r = L_r^{2.5}$$

$$\frac{Q_m}{Q_p} = \left(\frac{1}{50}\right)^{2.5}$$

$$\therefore \quad \frac{Q_m}{1000} = \left(\frac{1}{50}\right)^{2.5}$$

$$\therefore \quad Q_m = 0.0566 \text{ m}^3/\text{s}$$

So, $Q_m \simeq 0.06 \text{m}^3/\text{s}$

23. $$E_C = \frac{3}{2}y_C$$

$$y_C = \left(\frac{q^2}{g}\right)^{1/3}$$

Here, $q = \frac{20}{10} = 2$

$$\Rightarrow \quad E_C = \frac{3}{2}\left(\frac{2^2}{9.81}\right)^{1/3} = 1.11$$

24. In Muskingum flood routing method

$$C_0 + C_1 + C_2 = 1$$

$$\therefore \quad C_2 = 1 - C_0 - C_1$$

$$\therefore \quad C_2 = 1 - (-0.25) - 0.55$$

$$\therefore \quad = 1 - 0.3 = 0.7$$

25. Total weight generated by city = 40×10^6 kg/year

Weight of MSW going into landfill

$= (0.9 \times 40 \times 10^6)$ kg/year

$= (36 \times 10^6)$ kg/year

Compacted density $= 550 \text{ kg/m}^3$

$$\text{Compacted volume of MSW} = \frac{\text{weight of MSW}}{\text{compacted density}}$$

$$= \frac{36 \times 10^6 \text{ kg/year}}{550 \text{ kg/m}^3} = 65454.5454 \text{ m}^3/\text{year}$$

Given, Volume of MSW = $0.8 \times$ Total landfill volume

$\therefore$ Volume of cover = $0.2 \times$ Total landfill volume

Total landfill volume = Volume of MSW + Volume of cover

$\therefore$ Total landfill volume

$$= \frac{65454.5454}{0.8} \text{m}^3/\text{year} = 81818.18175 \text{m}^3/\text{year}$$

Height of landfill = 3 m

$$\therefore \quad \text{Area of landfill} = \frac{81818.18175}{3}$$

$$= 27272.7 \text{m}^3/\text{year}$$

26. The value of $\int_0^{\pi} x\cos^2 x dx$

$$= \int_0^{\pi} x\left(\frac{1+\cos 2x}{2}\right)dx \quad \left[\because \cos 2\theta = 2\cos^2\theta - 1\right]$$

$$= \int_0^{\pi}\left(\frac{x}{2}+\frac{x\cos 2x}{2}\right)dx$$

$$= \left[\frac{x^2}{4}\right]_0^{\pi} + \frac{1}{2}\left[\frac{x\sin 2x}{2}+\frac{\cos 2x}{4}\right]_0^{\pi}$$

$$= \frac{\pi^2}{4}+\frac{1}{2}\left[\frac{\pi}{2}\times\sin(2\pi)+\frac{\cos 2\pi}{4}-\frac{1}{4}\right]$$

$$= \frac{\pi^2}{4}+\frac{1}{2}\left[\frac{1}{4}-\frac{1}{4}\right] = \frac{\pi^2}{4}$$

27.

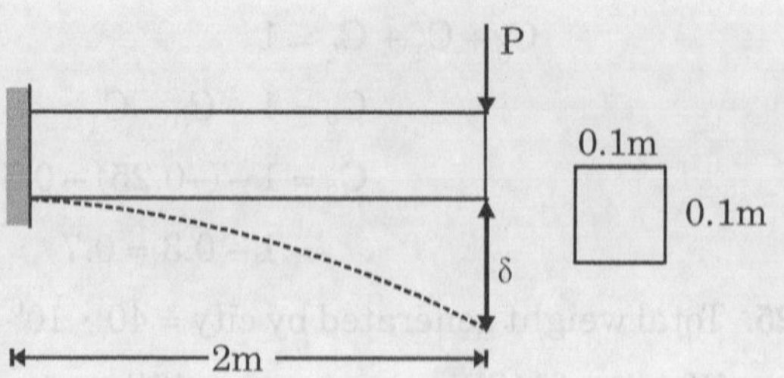

$$I = \frac{(0.1)^4}{12}$$

Deflection $\quad \delta = \frac{Pl^3}{3EI}$

$$\therefore \quad 5\times10^{-3}\text{ m} = \frac{P(2)^3}{3\times2\times10^{11}\times\frac{(0.1)^4}{12}} \quad (\because l = 2)$$

$\Rightarrow \quad P = 3125$ N

Now, $\quad M = Pl = 3125 \times 2 = 6250$ Nm

As, $\quad \frac{M}{I} = \frac{\sigma}{y} = \frac{E}{R}$

$$\sigma_{max} = \frac{M}{Z} = \frac{6250}{\frac{(0.1)^3}{6}}$$

$$\Rightarrow \quad = 37.5\times10^6 \text{ N/m}^2 = 37.5 \text{ MPa}$$

28.

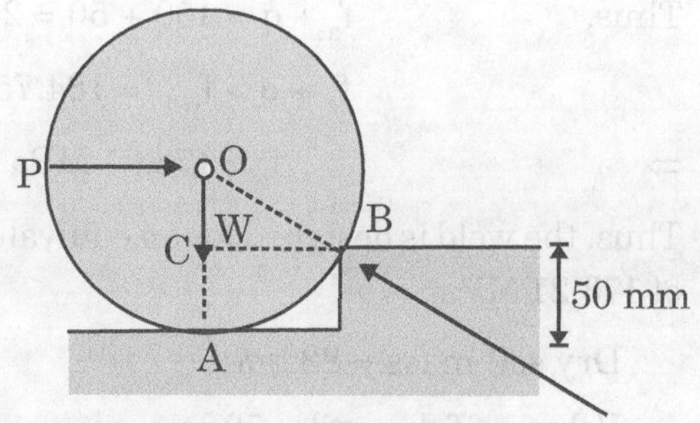

Given: r = 250 mm, W = 10 kN

$$\Sigma\vec{M}_B = 0$$

$$P\times OC - W\times BC = 0$$

In ΔOCB,

$$OC^2 + CB^2 = OB^2$$

$$CB = \sqrt{250^2 - 200^2} = 150$$

$$\therefore \quad P = \frac{10\times10^3\times150}{200} = 2.5\times3\text{kN} = 7.5\text{kN}$$

29. $\sigma_{xx} = 30$ MPa, $\sigma_{yy} = 50$ MPa, $\sigma_{zz} = 0$

$$\varepsilon_{zz} = \frac{\sigma_{zz}}{E} - \mu\frac{\sigma_{xx}}{E} - \mu\frac{\sigma_{yy}}{E} = -\frac{\mu}{E}(\sigma_{xx}+\sigma_{yy})$$

$$= -\frac{0.3}{2\times10^5}(30+50) = \frac{-0.3\times80}{2\times10^5}$$

$$= -120\times10^{-6}$$

30.

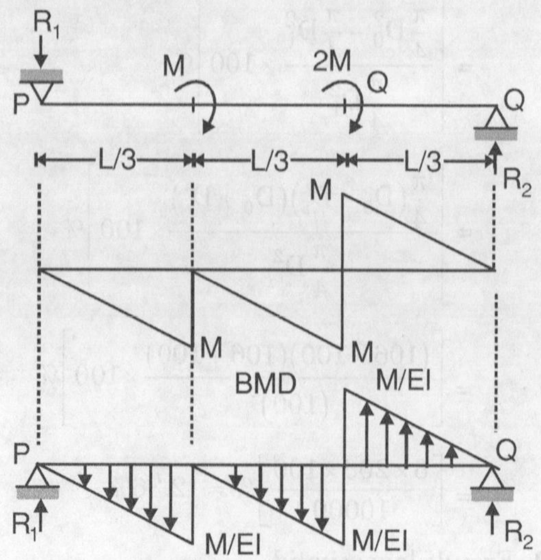

Conjugate Beam

Now $\quad R_1 + R_2 = \frac{1}{2}\times\frac{L}{3}\times\frac{M}{EI} = \frac{ML}{6EI}$

$$\Sigma M_Q = 0$$

$$\therefore \quad R_1 L - \frac{1}{2}\times\frac{L}{3}\times\frac{M}{EI}\left(\frac{2L}{3}+\frac{L}{9}\right) - \frac{1}{2}\times\frac{L}{3}\times\frac{M}{EI}$$

$$\left(\frac{L}{3}+\frac{L}{9}\right)+\frac{1}{2}\times\frac{L}{3}\times\frac{M}{EI}\times\frac{2L}{9}=0$$

$$\therefore\quad R_1=\frac{7ML}{54EI}+\frac{4ML}{54EI}-\frac{ML}{27EI}$$

$$=\frac{ML}{6EI}$$

$$\therefore\quad \text{Slope at P, } \theta_p = R_1=\frac{ML}{6EI}$$

31. Given data,

$C_u = 50$ kPa,

$\gamma = 18$ kN/m^3,

FOS = 2

Now, $Q_{up} = 9C\times B^2+\alpha\bar{C}(4BL)$

$\therefore\quad 1000 = 9\times 50\times 0.5\times 0.5+0.75\times 50\,(4\times 0.5\,L)$

$\therefore\quad L = 11.83$ m

33. Total water to be filtered = (99 × 1.05) MLD

= 103.95 MLD

(Addition of 5% to be used for backwashing)

$$\frac{L}{B}=1.35$$

$\therefore\quad L = 1.35\times B$

$= 1.35\times 5.2$ where B = 5.2 m

$= 7.02$ m

$\therefore$ Suface area of each filter = 36.504 m^2

Total surface area required =

$$\frac{\text{Discharge through filter}}{\text{Rate of filtration}}=\frac{103.95\times10^3}{6\times24}$$

$= 721.875$ m^2

Total number of working units required

$$=\frac{721.875}{36.504}=19.77\ \text{filters}$$

= 19.77 filters

≈ 20 filters

1 unit is to added as standby, thus total number of units required = 21

34.

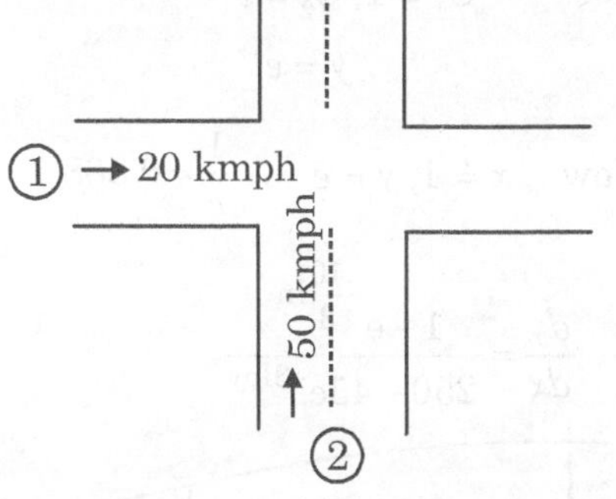

$$SSD_2 = 0.278\times Vt_R+\frac{V^2}{254f}$$

$$=\left(0.278\times50\times2.5+\frac{50^2}{254\times0.38}\right)\simeq 61\text{ m}$$

$$SSD_1=\left(0.278\times20\times2.5+\frac{20^2}{254\times0.38}\right)\simeq 18\text{ m}$$

35. $\Sigma L = (101+419)-437-83$

$= 520-520=0$

$\Sigma D = (173+558)-96-634$

$= 731-730=1$

Therefore departure of closure is 1 m and WCB is 90°.

36. $A = 100\times100$ m^2

Area on photo, $a = 1$ cm^2

Scale 1 cm = 100 m

f = 150 mm

h = 650 m

$$\text{Scale}=\frac{1}{100}=\frac{1}{100\times10^2}=\frac{1}{10000}=\frac{f}{H-h}$$

$$\therefore\quad \frac{1}{10000}=\frac{150\times10^{-3}}{H-650}$$

$\therefore\quad H = 2150$ m

38. $(D^2+2D+1)y=0$

$CF = (C_1+C_2x)e^{-x}$

$\therefore\quad y = C_1e^{-x}+C_2xe^{-x}$ ...(*i*)

Now $y(0)=1$ $\quad 1 = C_1$ ...(*ii*)

Now differenting equation (*i*) w.r.t to '*x*' we get

$y' = C_1^{e-x}+C_2\,(e^{-x}-xe^{-x})$

Now $y'(0)=-1$, $\quad -1=-C_1+C_2$ ...(*iii*)

From equation (*ii*) and (*iii*) we get

$$C_1 = 1, C_2 = 0$$

$$\therefore \quad y = e^{-x}$$

Now $x = 1, y = e^{-1} = \frac{1}{e} = 0.368$

39. $$\frac{dy}{dx} = \frac{1 - e^{-\frac{10}{3}\text{In}y}}{250 - 45e^{-3\text{In}y}}$$

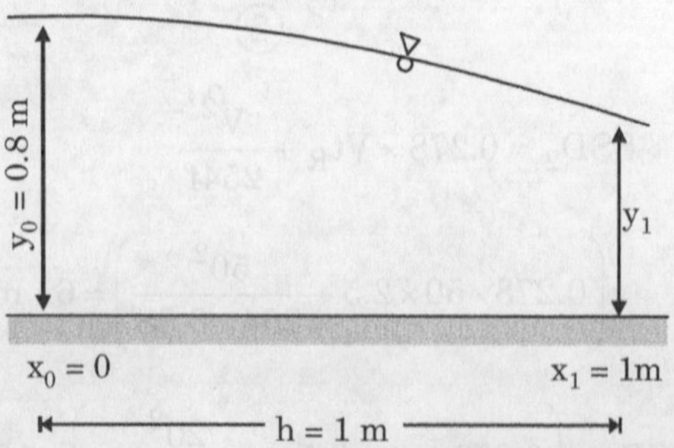

$$y_1 = y_0 + hf(x_0, y_0)$$

$$\Rightarrow \quad y_1 = 0.8 + 1\left[\frac{1 - e^{-\frac{10}{3}\text{In}0.8}}{250 - 45e^{-3\text{In}0.8}}\right]$$

$$= 0.8 + 1\left(\frac{-1.1039}{162.109}\right)$$

$$= 0.793 \text{ m}$$

40. Since eccentricity effect is being neglected so column can be considered as concentrically loaded.

Ultimate axial load carrying capacity of column.

$$(P_u) = 0.45\, f_{ck}A_g + \left(0.75\, f_y - 0.45\, f_{ck}\right)A_{sc}$$

$$= (0.45 \times 20 \times 250 \times 300) + (0.75 \times 415 - 0.45 \times 20)\; 4 \times \frac{\pi}{4} \times 16^2$$

$$= 918.1 \text{ kN}$$

41. Nominal shear stress

$$(\tau_v) = \frac{V_u}{bd} = \frac{200 \times 10^3}{250 \times 350} = 2.286 \text{ N/mm}^2 < \tau_{c,max}$$

(OK)

$$\text{SF taken by stirrups} = (\tau_v - \tau_c)\,bd$$

$$= (2.286 - 0.62) \times 250 \times 350$$

$$= 1.66 \times 250 \times 350$$

$$= 145.75 \text{ kN}$$

Now, $V_{us} = \frac{0.87 f_y A_{sv} d}{S_v}$

$$\therefore \quad S_v = \frac{0.87 f_y A_{SV} d}{V_{us}}$$

$$\therefore \quad S_v = \frac{0.87 \times 250 \times 350 \times 2 \times \frac{\pi}{4} \times 10^2}{145.75 \times 10^3}$$

$$\therefore \quad S_v = 82 \text{ mm}$$

$$= 8.2 \text{ cm}$$

42.

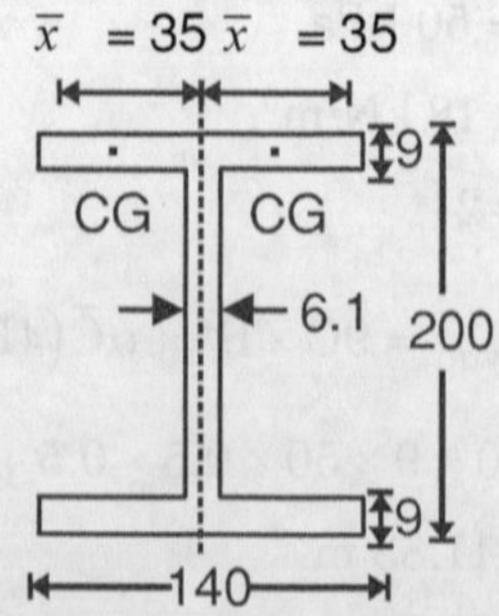

(All dimension are in mm)

$$Z_p = \frac{A}{2}(\bar{x} + \bar{x})$$

$$= \left[\frac{140 \times 9}{2}(35 + 35)\right] \times 2 + \left[\frac{(200 - 18) \times 6.1}{2}\right] \times \left[\frac{3.05}{2} + \frac{3.05}{2}\right]$$

$$= (88200 + 1693.055) \text{ mm}^3$$

$$= 89893.055 \text{ mm}^3$$

$$\simeq 89.9 \text{ cm}^3$$

43. Force is each member due to applied loading.

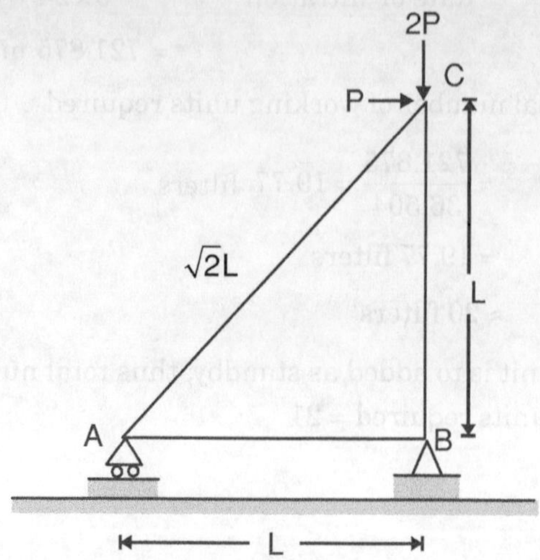

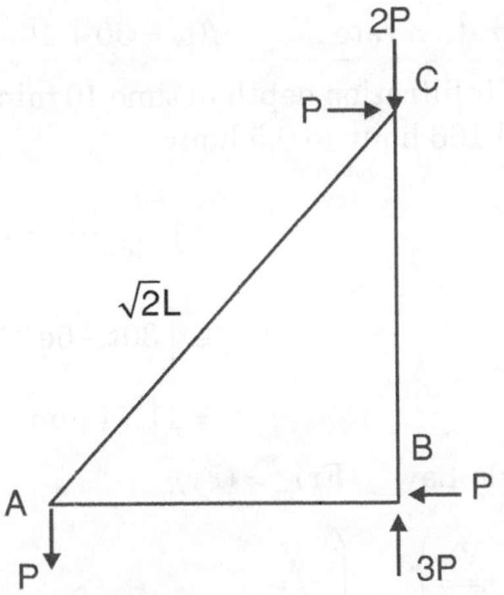

$F_{AB} = P$ (Comp.)

$F_{BC} = 3P$ (Comp.)

$F_{AC} = \sqrt{2}P$ (Tension)

Force in each member due to unit load.

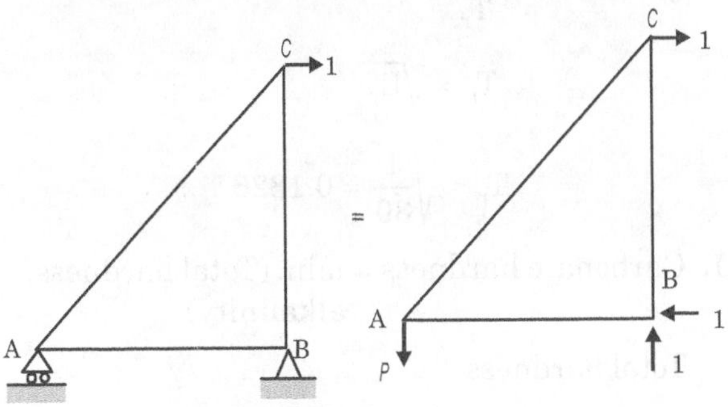

$F_{AB} = 1$ (Comp.)

$F_{BC} = 1$ (Comp.)

$F_{AC} = \sqrt{2}P$ (Tension)

Member	P	k	L	AE	$\frac{PkL}{AE}$
AB	–P	-1	L	AE	$\frac{PL}{AE}$
BC	-3P	-1	L	AE	$\frac{3PL}{AE}$
CA	$\sqrt{2}P$	$\sqrt{2}$	$\sqrt{2}L$	2AE	$\frac{2\sqrt{2}PL}{2AE}$
					$\Sigma = \frac{5.414PL}{AE}$

$\therefore$ Total deflection $= \delta_{H_C} = \Sigma\frac{PkL}{AE}$

$$= \frac{5.414 \times (1000N) \times (1000mm)}{(10mm^2) \times (2 \times 10^5 N/mm^2)}$$

$$= \frac{5.414}{2} = 2.7 \text{ mm}$$

44. $$FOS = \frac{10 \times \gamma_{sat}}{20 \times \gamma_w} = \frac{10 \times 20}{20 \times 10} = 1$$

$$\left(\because \frac{\gamma_{sat}}{\gamma_w} = \frac{20}{10}\right)$$

45. I[st] Specimen : Drained condition

$\bar{\sigma}_3 = 200$ kPa : $\sigma_d = 400$ kPa : $\bar{\sigma}_1 = 600$ kPa

IInd Specimen : Undrained condition

$\sigma_3 = 200$ kPa : $\sigma_d = 150$ kPa :

$$\sigma_1 = \sigma_3 + \sigma_d = 350 \text{ kPa}$$

Let pore pressure developed is u

$$\therefore \quad \bar{\sigma}_3 = (200 - u)$$

$$\bar{\sigma}_1 = (350 - u)$$

From stress relationship

$$\bar{\sigma}_1 = \bar{\sigma}_3 \tan^2\left(45° + \frac{\phi}{2}\right) + 2c \tan\left(45° + \frac{\phi}{2}\right)$$

For clay under drained condition c = 0

$$\therefore \quad \bar{\sigma}_1 = \bar{\sigma}_3 \tan^2\left(45° + \frac{\phi}{2}\right)$$

$$\therefore \quad 600 = 200\tan^2\left(45° + \frac{\phi}{2}\right)$$

$$\phi = 30°$$

For second specimen,

$$\bar{\sigma}_1 = \bar{\sigma}_3 \tan^2\left(45° + \frac{\phi}{2}\right)$$

$$(350 - u) = (200 - u)\tan^2\left(45° + \frac{30°}{2}\right)$$

$$u = 125 \text{ kPa}$$

47.

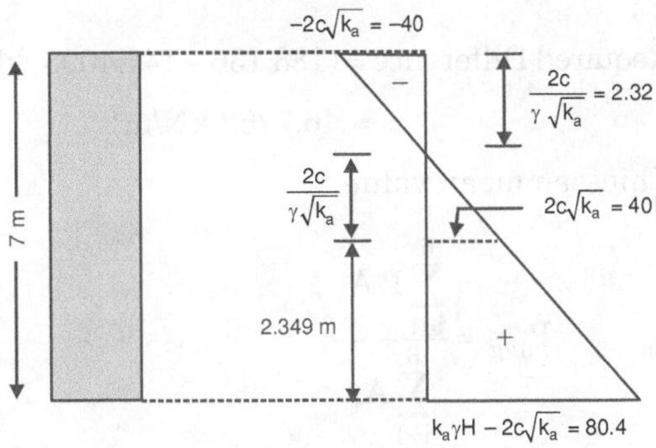

For clay $\phi = 0$

$$\therefore \quad k_a = \frac{1 - \sin\phi}{1 + \sin\phi} = \frac{1 - 0}{1 + 0} = 1$$

$$= \frac{1 - \sin 0}{1 + \sin 0} = 1$$

Earth pressure when tension cracks are not developed.

$$P_a = \frac{1}{2}(40 + 80.4) \times 2.349$$

$$= \frac{1}{2} \times (120.4) \times 2.349$$

$$= 141.4098$$

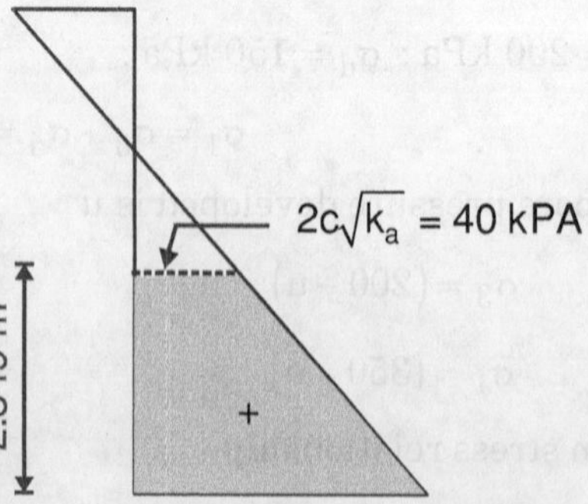

$k_a\gamma H - 2c\sqrt{k_a} = 80.4$ kPa

Earth pressure when tension crackes are developed

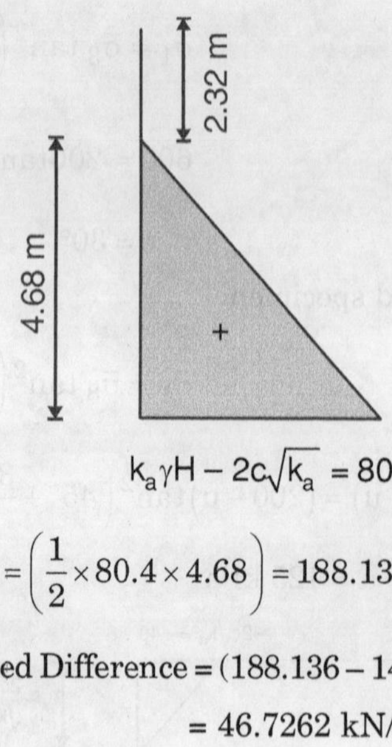

$k_a\gamma H - 2c\sqrt{k_a} = 80.4$

$$P_a = \left(\frac{1}{2} \times 80.4 \times 4.68\right) = 188.136$$

Required Difference = (188.136 – 141.4098) kN/m^2

$$= 46.7262 \text{ kN/m}^2$$

48. Thiessen mean value;

$$P_{avg} = \frac{\sum_{i=1}^{6} P_i A_i}{\sum_{i=1}^{6} A_i}$$

$$\Rightarrow P_{avg} = \frac{(470 \times 95) + (465 \times 100) + (435 \times 98) + (525 \times 80) + (480 \times 85) + (510 \times 92)}{95 + 100 + 98 + 80 + 85 + 92}$$

$$= 479.09 \text{ mm}$$

49. Infiltration rate $f(t) = 30 + 10e^{-2t}$

Total infiltration depth in time 10 min. to 30 min. i.e., 0.166 hour to 0.5 hour

$$= \int_{0.166}^{0.5} \left(30 + 10e^{-2t}\right) dt$$

$$= \left[30t - 5e^{-2t}\right]_{0.166}^{0.5}$$

$$= 11.74 \text{ mm}$$

50. Froude Law $(Fr)_m = (Fr)_p$

$$\left(\frac{V}{\sqrt{Lg}}\right)_m = \left(\frac{V}{\sqrt{Lg}}\right)_p \qquad (g_m = g_p)$$

$$V_r = \sqrt{L_r}$$

or $$\frac{L_r}{T_r} = \sqrt{L_r}$$

$$T_r = \sqrt{L_r}$$

$$T_r = \sqrt{\frac{1}{30}} = 0.1826$$

51. Carbonate hardness = min. (Total hardness, alkalinity)

Total hardness

$$= \left(\frac{60}{20} \times 50 + \frac{30}{12.155} \times 50\right) \text{ mg/l as } CaCO_3$$

$$= (150 + 123.406)$$

$$= 273.406 \text{ mg/l as } CaCO_3$$

$$\text{Alkalinity} = \left(\frac{400}{61} \times 50\right) \text{ mg/l as } CaCO_3$$

$$= 327.868 \text{ mg/l as } CaCO_3$$

$\therefore$ CH = 273.4 mg/l as $CaCO_3$

52. Ultimate BOD= 0.87 COD = 0.87 × 300=261 mg/l

$$k_{27} = k_{20}\,(1.047)^{T-20}$$

$$= 0.23\,(1.047)^{27-20} = 0.317 \text{ days}^{-1}$$

$$BOD_3 = L_0\left(1 - e^{-k_{27} \times 3}\right)$$

$$BOD_3 = 261\,(1 - e^{-0.317 \times 3}) = 160.226 \text{ mg/l}$$

53. Let 20 kg of C and 1 kg of N is required

Let x kg of WAS is taken

$\therefore$ Carbon in x kg = 0.054 x kg

Nitrogen in x kg = 0.010 x kg

Let y kg of GW is taken

∴ Carbon in y kg = 0.360 y kg

Nitrogen in y kg = 0.006 y kg

Total Carbon,

$$0.054x + 0.36\,y = 20 \text{ kg} \quad ....(i)$$

Total Nitrogen,

$$0.01x + 0.006\,y = 1 \text{ kg} \quad ....(ii)$$

Solving equation (i) and (ii) we get

$$x = 73.26 \text{ kg}$$

$$y = 44.566 \text{ kg}$$

∴ Required ratio $\left(\frac{x}{y}\right) = \frac{\text{WAS}}{\text{GW}} = \frac{73.26}{44.56}$

≈ 1.64

54. $$N_s = \frac{365A\left[\left(1+\frac{r}{100}\right)^n - 1\right] \times VDF \times LDF}{\frac{r}{100}}$$

$$= \frac{365 \times 3000\left[1.06^{15} - 1\right] \times 4 \times 0.75}{0.06}$$

= 76.45 msa

55. Speed of aircraft, (u_i) = 200 km/hr

$$= \left(200 \times \frac{5}{18}\right) = 55.56 \text{m/s}$$

$$\left[\because 1\text{km/hr} = \frac{5}{18}\text{m/s}\right]$$

Deceleration of aircraft on the runway

$$= 1.697 \text{m/s}^2$$

Now, Aircraft takes 18 sec to exit the runway strip.

Now Speed of the aircraft at the exit of the runway

$$u_t = u_i + at$$

$$= 55.56 - (1.697 \times 18)$$

$$= 55.56 - 30.546$$

$$= 25.014 \text{ m/s}$$

After runway aircraft decelerate with 1 m/s^2

Total distance travelled by the aircraft from the location of exit on the runway

$$U^2 = U^2 + 2aS$$

$$0 = 25.014^2 - 2 \times 1 \times S$$

$$S = 312.8 \text{ m}$$

SOLVED PAPER - 2019

INSTRUCTIONS

1. Total of 65 questions carrying 100 marks, out of which 10 questions carrying a total of 15 marks are in General Aptitude (GA)
2. The Engineering Mathematics will carry around **15% of the total marks,** the General Aptitude section will carry **15% of the total marks** and the **remaining 70% of the total marks.**
3. **Types of Questions**
   (*a*) Multiple Choice Questions (MCQ) carrying 1 or 2 marks each in all papers and sections. These questions are objective in nature, and each will have a choice of four options, out of which the candidate has to mark the correct answer(s).
   (*b*) Numerical Answer Questions of 1 or 2 marks each in all papers and sections. For these questions the answer is a real number, to be entered by the candidate using the virtual keypad. No choices will be shown for these type of questions.
4. For **1-mark** multiple-choice questions, **1/3 marks** will be deducted for a wrong answer. Likewise, for **2-marks** multiple-choice questions, **2/3** marks will be deducted for a wrong answer. There is no negative marking for numerical answer type questions.

Chapter-Wise Analysis

#	Subject	Marks		
		1	2	Total
1	General Aptitude	5	5	15
2	Engineering Mathematics	3	4	11
3	Engineering Mechanics	2	0	2
4	Solid Mechanics	2	1	4
5	Structural Analysis	0	3	6
6	Construction Materials and Management	0	1	2
7	Concrete Structures	3	4	11

#	Subject	Marks		
		1	2	Total
8	Steel Structures	1	1	3
9	Soil Mechanics and Foundation Engineering	4	5	14
10	Fluid Mechanics and Hydraulics	2	3	8
11	Hydrology & Irrigation	1	3	7
12	Environmental Engineering	2	3	8
13	Transportation Engineering	2	1	4
14	Surveying	3	1	5

GENERAL APTITUDE

(Q.1 – 5) : Carry One Mark Each

1. The lecture was attended by quite _______ students, so the hall was not very _______.

(a) few, quite (b) few, quiet

(c) a few, quiet (d) a few, quite

2. They have come a long way in _________ trust among the users.

(a) creating (b) create

(c) created (d) creations

3. If E = 10; J = 20; O = 30; and T = 40, what will be P + E + S + T?

(a) 51 (b) 120

(c) 82 (d) 164

4. The CEO's decision to quit was as shocking to the Board at it was to _______.

(a) myself (b) I

(c) my (d) me

5. On a horizontal ground, the base of a straight ladder is 6m away from the base of a vertical pole. The ladder makes an angle of 45° to the horizontal. If the ladder is resting at a point located at one-fifth of the height of the pole from the bottom, the height of the pole is _______ meters.

(a) 35 (b) 30

(c) 15 (d) 25

(Q.6 – 10) : Carry Two Marks Each

6. "The increasing interest in tribal characters might be a mere coincidence, but the timing is of interest. None of this, though, is to say that the tribal here has arrived in Hindi cinema, or that the new crop of characters represents the acceptance of the tribal character in the industry. The films and characters are too few to be described as a pattern."

What does the word „arrived. mean in the paragraph above?

(a) went to a place

(b) came to a conclusion

(c) attained a status

(d) reached a terminus

7. In a sports academy of 300 people, 105 play only cricket, 70 play only hockey, 50 play only football, 25 play both cricket and hockey, 15 play both hockey and football and 30 play both cricket and football. The rest of them play all three sports. What is the percentage of people who play at least two sports?

(a) 28.00 (b) 23.30

(c) 50.00 (d) 25.00

8. P, Q, R, S and T are related and belong to the same family. P is the brother of S. Q is the wife of P. R and T are the children of the siblings P and S respectively. Which one of the following statements is necessarily FALSE?

(a) S is the aunt of T

(b) S is the aunt of R

(c) S is the sister-in-law of Q

(d) S is the brother of P

9. The new cotton technology, Bollgard-II, with herbicide tolerant traits has developed into a thriving business in India. However, the commercial use of this technology is not legal in India. Now with standing that, reports indicate that the herbicide tolerant Bt cotton had been purchased by farmers at an average of Rs 200 more than the control price of ordinary cotton, and planted in 15% of the cotton growing area in the 2017 Kharif season.

Which of the following statements can be inferred from the given passage?

(a) Farmers want to access the new technology by paying high price

(b) Farmers want to access the new technology if India benefits from it

(c) Farmers want to access the new technology for experimental purposes

(d) Farmers want to access the new technology even if it is not legal

10. A square has sides 5 cm smaller than the sides of a second square. The area of the larger square is four times the area of the smaller square. The sided of the larger square is _________ cm.

(a) 18.50 (b) 10.00

(c) 15.10 (d) 8.50

CIVIL ENGINEERING

(Q.1 – 25) : Carry One Mark Each

1. Consider a two-dimensional flow through isotropic soil along x direction and z direction. If h is the hydraulic head, the Laplace's equation of continuity is expressed as

(a) $\frac{\partial h}{\partial x}+\frac{\partial h}{\partial z}=0$

(b) $\frac{\partial^2 h}{\partial x^2}+\frac{\partial^2 h}{\partial x\partial z}+\frac{\partial^2 h}{\partial z^2}=0$

(c) $\frac{\partial^2 h}{\partial x^2}+\frac{\partial^2 h}{\partial z^2}=0$

(d) $\frac{\partial h}{\partial x}+\frac{\partial h\partial h}{\partial x\partial z}+\frac{\partial h}{\partial z}=0$

2. The coefficient of average rolling friction of a road is f_r and its grade is +G%. If the grade of this road is doubled, what will be the percentage change in the braking distance (for the design vehicle to come to a stop) measured along the horizontal (assume all other parameters are kept unchanged?

(*a*) $\frac{0.02G}{f_r + 0.01G} \times 100$

(*b*) $\frac{f_r}{f_r + 0.02G} \times 100$

(*c*) $\frac{2f_r}{f_r + 0.01G} \times 100$

(*d*) $\frac{0.01G}{f_r + 0.02G} \times 100$

3. In the reinforced beam section shown in the figure (not drawn to scale), the nominal cover provided at the bottom of the beam as per IS 456-2000, is

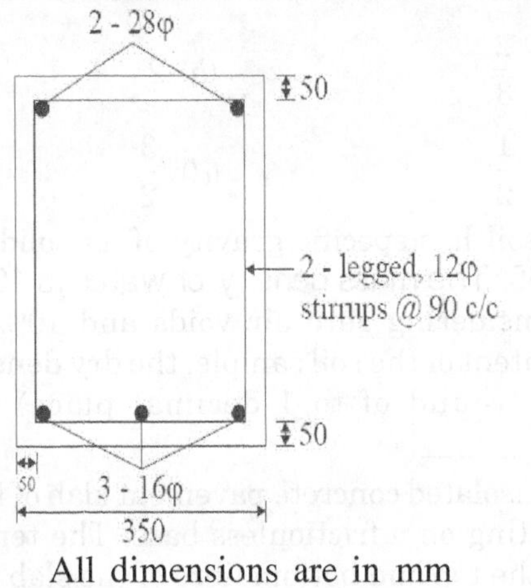

All dimensions are in mm

(*a*) 36 mm (*b*) 50 mm

(*c*) 30 mm (*d*) 42 mm

4. A plane truss is shown in the figure (not drawn to scale).

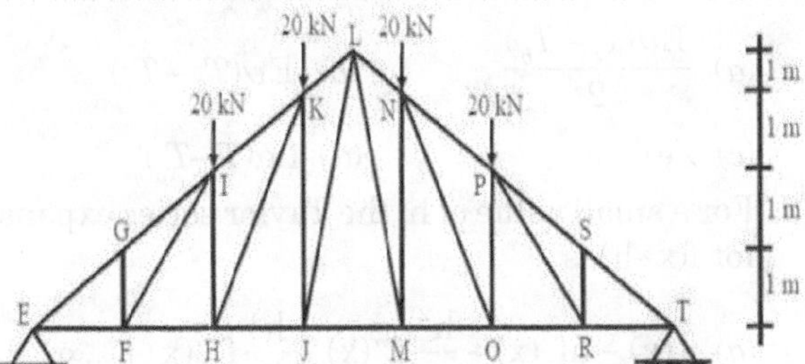

Which one of the options contains ONLY zero force members in the truss?

(*a*) FG, FI, HI, RS (*b*) FI, FG, RS, PR

(*c*) FI, HI, PR, RS (*d*) FG, FH, HI, RS

5. Which one of the following is a secondary pollutant?

(*a*) Ozone

(*b*) Carbon Monoxide

(*c*) Volatile Organic Carbon (VOC)

(*d*) Hydrocarbon

6. If the path of an irrigation canal is below the bed level of a natural stream, the type of cross-drainage structure provided is

(*a*) Aqueduct (*b*) Sluice gate

(*c*) Super passage (*d*) Level crossing

7. A circular duct carrying water gradually contracts from diameter of 30 cm to 15cm. The figure (not drawn to scale) shows the arrangement of differential manometer attached to the duct.

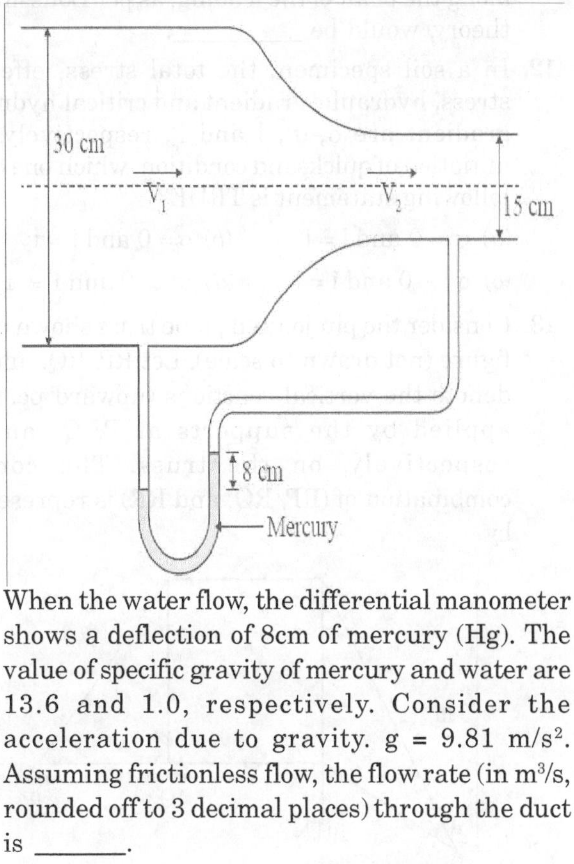

When the water flow, the differential manometer shows a deflection of 8cm of mercury (Hg). The value of specific gravity of mercury and water are 13.6 and 1.0, respectively. Consider the acceleration due to gravity. g = 9.81 m/s^2. Assuming frictionless flow, the flow rate (in m^3/s, rounded off to 3 decimal places) through the duct is ______.

8. For a given loading on a rectangular plain concrete beam with an overall depth of 500 mm, the compressive strain and tensile developed at the extreme fibers are of the same magnitude of 2.5×10^{-4}. The curvature in the beam cross-section (in m^{-1}, round off to 3 decimal places), is ______.

9. The maximum number of vehicles observed in any five minute period during the break hour is 160. If the total flow in the peak hour is 1000 vehicles, the five minute peak hour factor (round off to 2 decimal places) is ______.

10. The interior angles of four triangles are given below:

Triangle	Interior Angles
P	85°,50°,45°
Q	100°,55°,25°
R	100°,45°,35°
S	130°,30°,20°

Which of the triangles are ill-conditioned and should be avoided in Triangulation surveys?

(*a*) Both Q and S (*b*) Both Q and R

(*c*) Both P and R (*d*) Both P and S

11. A concentrated load of 500 kN is applied on an elastic half space. The ratio of the increase in vertical normal stress at depths of 2m and 4m along the point of the leading, as per Boussinesq's theory, would be ________.

12. In a soil specimen, the total stress, effective stress, hydraulic gradient and critical hydraulic gradient are σ, σ', i and i_c, respectively. For initiation of quicksand condition, which one of the following statement is TRUE?

(*a*) $\sigma' = 0$ and $i = i_c$ (*b*) $\sigma = 0$ and $i = i_c$

(*c*) $\sigma' \neq 0$ and $i = i_c$ (*d*) $\sigma' \neq 0$ and $i \neq i_c$

13. Consider the pin-jointed plane truss shown in the figure (not drawn to scale). Let RP, RQ, and RR denote the vertical reactions (upward positive) applied by the supports at P, Q, and R, respectively, on the truss. The correct combination of (RP, RQ, and RR) is represented by

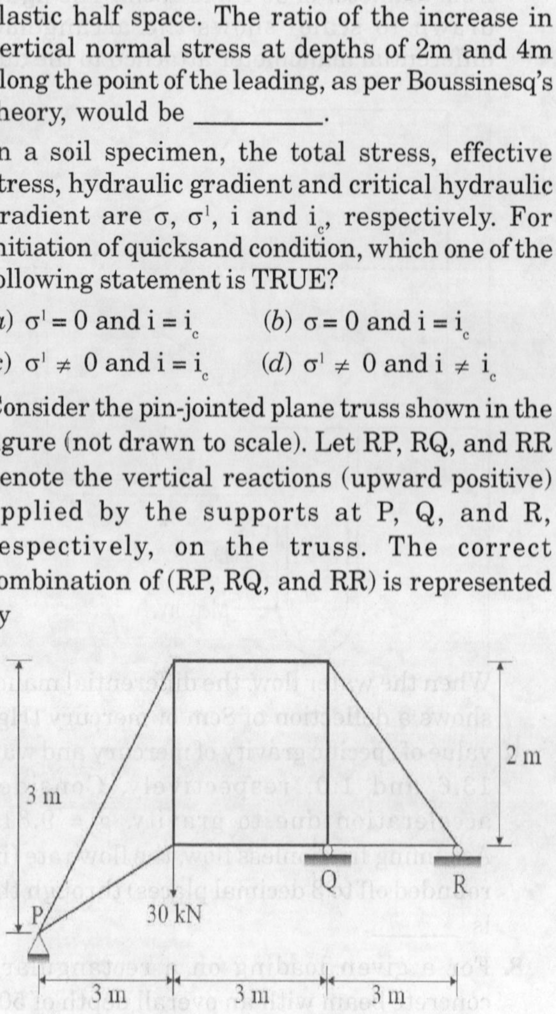

(*a*) (30,–30,30)kN (*b*) (10,30,–10)kN

(*c*) (20,0,10)kN (*d*) (0,60,–30)kN

14. A retaining wall of height H with smooth vertical backface supports a backfill inclined at an angle β with the horizontal. The backfill consists of cohesionless soil having angle of internal friction ϕ. If the active lateral thrust acting on the wall is P_a which one of the following statements is TRUE?

(*a*) P_a acts at a height H/3 from the base of the wall and at an angle β with the horizontal

(*b*) P_a acts at a height H/3 from the base of the wall and at an angle ϕ with the horizontal

(*c*) P_a acts at a height H/2 from the base of the wall and at an angle β with the horizontal

(*d*) P_a acts at a height H/2 from the base of the wall and at an angle ϕ with the horizontal

15. Assuming that there is no possibility of shear buckling in the web, the maximum reduction permitted by IS 800-2007 in the (low-shear) design bending strength of a semi-compact steel section due to high shear is

(*a*) 50%

(*b*) governed by the area of the flange

(*c*) zero

(*d*) 25%

16. In a rectangular channel, the ratio of the velocity head to the flow depth for critical flow condition, is

(*a*) $\frac{2}{3}$ (*b*) 2

(*c*) $\frac{1}{2}$ (*d*) $\frac{3}{2}$

17. A soil has specific gravity of its solids equal to 2.65. The mass density of water is 1000 kg/m³. Considering zero air voids and 10% moisture content of the soil sample, the dry density (in kg/m³, round of to 1 decimal place) would be ________.

18. An isolated concrete pavement slab of length L is resting on a frictionless base. The temperature of the top and bottom fibre of the slab are T_t and T_b, respectively. Given: the coefficient of thermal expansion = α and the elastic modulus = E. Assuming $T_t > T_b$ and the unit weight of concrete as zero, the maximum thermal stress is calculated as

(*a*) $\frac{E\alpha(T_t - T_b)}{2}$ (*b*) $E\alpha(T_t - T_b)$

(*c*) Zero (*d*) $L\alpha(T_t - T_b)$

19. For a small value of h, the Taylor series expansion for f(x+h) is

(*a*) $f(x) - hf'(x) + \frac{h^2}{2}f''(x) - \frac{h^3}{3}f'''(x) + ...\infty$

(*b*) $f(x) - hf'(x) + \frac{h^2}{2!}f''(x) - \frac{h^3}{3!}f'''(x) + ...\infty$

(*c*) $f(x) + hf'(x) + \frac{h^2}{2}f''(x) + \frac{h^3}{3}f'''(x) + ...\infty$

(*d*) $f(x) + hf'(x) + \frac{h^2}{2!}f''(x) + \frac{h^3}{3!}f'''(x) + ...\infty$

20. Which of one of the following is correct?

(*a*) $\lim_{x\to 0}\left(\frac{\sin 4x}{\sin 2x}\right) = 1$ and $\lim_{x\to 0}\left(\frac{\tan x}{x}\right) = 1$

(*b*) $\lim_{x\to 0}\left(\frac{\sin 4x}{\sin 2x}\right) = \infty$ and $\lim_{x\to 0}\left(\frac{\tan x}{x}\right) = 1$

(*c*) $\lim_{x\to 0}\left(\frac{\sin 4x}{\sin 2x}\right) = 2$ and $\lim_{x\to 0}\left(\frac{\tan x}{x}\right) = \infty$

(*d*) $\lim_{x\to 0}\left(\frac{\sin 4x}{\sin 2x}\right) = 2$ and $\lim_{x\to 0}\left(\frac{\tan x}{x}\right) = 1$

21. An element is subjected to biaxial normal tensile strains of 0.0030 and 0.0020. The normal strain in the plane of maximum shear strain is

(*a*) 0.0010 (*b*) Zero

(*c*) 0.0050 (*d*) 0.0025

22. The probability that the annual maximum flood discharge will exceed 25000 m^3/s, at least once in next 5 years is found to be 0.25. The return period of this flood event (in years, round off to 1 decimal place) is ______.

23. A completely mixed dilute suspension of sand particles having diameters 0.25, 0.35, 0.40, 0.45 and 0.50mm are filled in a transparent glass column of diameter 10 cm and height 2.50 m. The suspension is allowed to settle without any disturbance. It is observed that all particles of diameter 0.35 mm settle to the bottom of the column in 30 s. For the same period of 30s, the percentage removal (round off to integer value) of particles of diameters 0.45 and 0.50 mm from the suspension is ______.

24. A simple mass-spring oscillatory system consists of a mass m, suspended from a spring of stiffness k. Considering z as the displacement of the system at any time t, the equation of motion for the free vibration of the system is $m\ddot{z} + kz = 0$. The natural frequency of the system is

(*a*) $\frac{k}{m}$ (*b*) $\sqrt{\frac{m}{k}}$

(*c*) $\sqrt{\frac{k}{m}}$ (*d*) $\frac{m}{k}$

25. A catchment may be idealised as a rectangle There are three rain gauges located inside the catchment at arbitrary locations. The average precipitation over the catchment is estimated by two methods: (i) Arithmetic mean (P_A) and (ii) Thiessen polygon (P_T). Which of the following statements is correct?

(*a*) P_A is always equal to P_T

(*b*) There is no definite relationship between P_A and P_T

(*c*) P_A is always greater than P_T

(*d*) P_A is always smaller than P_T

(Q.26 – 55) : Carry Two Marks Each

26. The network of a small construction project awarded to a contractor is shown in the following figure. The normal duration, crash duration, normal cost, and crash cost of all the activities are shown in the table. The indirect cost incurred by the contractor is INR 5000 per day.

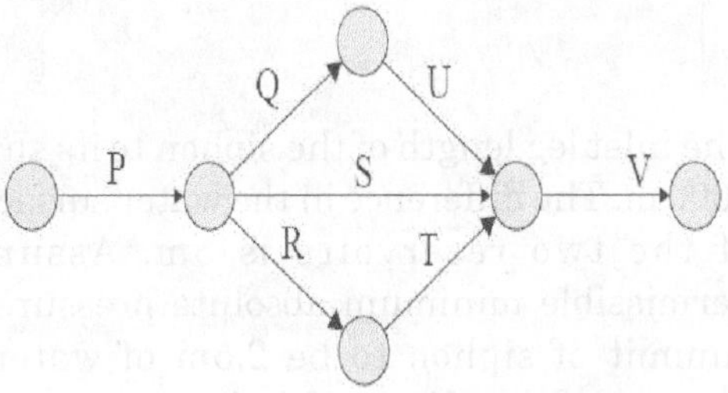

Activity	Normal Duration (days)	Crash Duration (days)	Normal cost (INR)	Crash Cost (INR)
P	6	4	15000	25000
Q	5	2	6000	12000
R	5	3	8000	9500
S	6	3	7000	10000
T	3	2	6000	9000
U	2	1	4000	6000
V	4	2	20000	28000

If the project is targeted for completion in 16 days, the total cost (in INR) to be incurred by the contractor would be ________.

27. A square footing of 4m side is placed at 1 m depth in a sand deposit. The dry unit weight (γ) of sand is 15 kN/m^3. This footing has an ultimate bearing capacity of 600 kPa. Consider the depth factors; $d_q = d_\gamma = 1.0$ and the bearing capacity factor: $N_\gamma = 18.75$. This footing is placed at a depth of 2m in the same soil deposit. For a factor of safety of 3.0 per Terzaghi's theory, the safe bearing capacity (in kPa) of this footing would be ______.

28. Which one of the following is NOT a correct statement?

(*a*) The function $|x|$ has the global minima at $x = 0$

(*b*) The function $\sqrt[x]{x}$, $(x > 0)$, has the global minima at $x = e$

(*c*) The function $\sqrt[x]{x}$, $(x > 0)$, has the global maximum at $x = e$

(*d*) The function x^3 has neither global minima nor global maxima

29. Two water reservoirs are connected by a siphon (running full) of total length 5000 m and diameter of 0.10 m, as shown below (figure not drawn to scale).

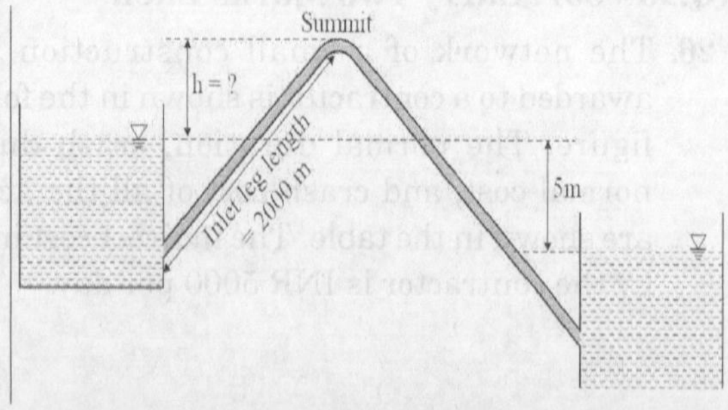

The inlet leg length of the siphon to its summit is 2000 m. The difference in the water surface levels of the two reservoirs is 5m. Assume the permissible minimum absolute pressure at the summit of siphon to be 2.5m of water when running full. Given: friction factor f = 0.02 throughout, atmospheric pressure = 10.3 m of water, and acceleration due to gravity g = 9.81 m/s². Considering only major loss using Darcy Weisbach equation the maximum height of the summit of siphon from the water level of upper reservoir, h (in m round off to 1 decimal place) is ________.

30. The rigid-jointed plane frame QRS shown in the figure is subjected to a load P at the joint R. Let the axial deformations in the frame be neglected. If the support S undergoes a settlement of $\Delta = \frac{PL^3}{\beta EI}$, the vertical reaction at the support S will become zero when β is equal to

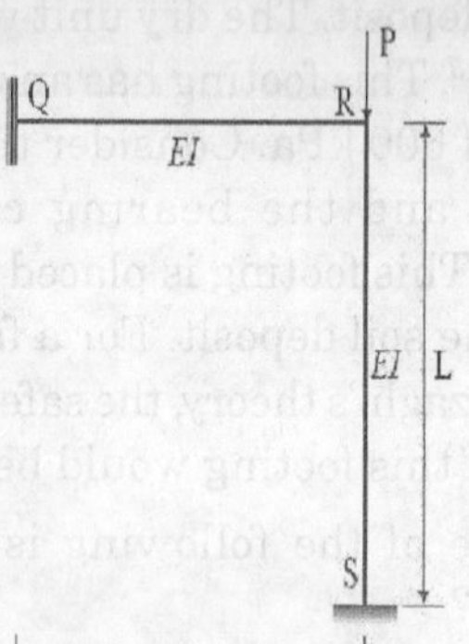

(*a*) 0.1 (*b*) 7.5

(*c*) 3.0 (*d*) 48.0

31. If the section shown in the figure turns from fully-elastic to fully-plastic, the depth of neutral axis (N.A), $\bar{y}$, decreases by

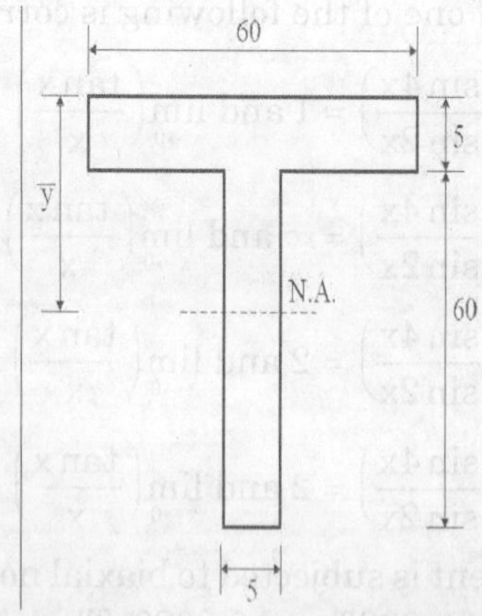

(*a*) 13.75 mm (*b*) 15.25 mm

(*c*) 10.75 mm (*d*) 12.25 mm

32. Traffic on a highway is moving at a rate 360 vehicles per hour at a location. If the number of vehicles arriving on this highway follows Poisson distribution, the probability (round off to 2 decimal places) that the headway between successive vehicles lies between 6 and 10 seconds is _____

33. For the following statements:

P – The lateral stress in the soil while being tested in an oedometer is always at-rest.

Q – For a perfectly rigid strip footing at deeper depths in a sand deposit, the vertical normal contact

R – The corrections for overburden pressure and dilatancy are not applied to measured SPT-N values in case of clay deposits.

The correct combination of the statements is

(*a*) P – TRUE; Q – TRUE; R– TRUE

(*b*) P – FALSE; Q – FALSE; R – TRUE

(*c*) P – TRUE; Q– TRUE; R– FALSE

(*d*) P – FALSE; Q– FALSE; R – FALSE

34. Average free flow speed and the jam density observed on a road stretch are 60 km/h and 120 vehicles/km, respectively. For a linear speed-density relationship, the maximum flow on the road stretch (in vehicles/h) is ____.

35. Consider the ordinary differential equations

$x^2\frac{d^2y}{dx^2} - 2x\frac{dy}{dx} + 2y = 0$ Given the values of y(1) = 0 and y(2) = 2, the value of y(3) (round off to 1 decimal place), is ______.

36. Sedimentation basin in a water treatment plant is designed for a flow rate of 0.2 m³/s. The basin is rectangular with a length of 32m, width of 8m, and depth of 4m. Assume that the settling velocity of these particles is governed by the Stokes. law.

Given: density of the particles = 2.5 g/cm^3; density of water = 1 g/cm^3; dynamic viscosity of water = 0.01 g/(cm.s); gravitational acceleration = 980 cm/s^2. If incoming water contains particles of diameter 25μm. (spherical and uniform), the removal efficiency of these particles is

(a) 78% (b) 51%

(c) 100% (d) 65%

37. A wastewater is to be disinfected with 35mg/L of chlorine to obtain 99% kill of microorganisms. The number of micro-organisms remaining alive (N_t) at time t, is modelled by $N_t = N_0 e^{-kt}$, where N_0 is number of micro-organisms at t = 0, and k is the rate of kill. The wastewater flow rate is 36 m^3/h, and k = 0.23 min^{-1}. If the depth and width of the chlorination tank are 1.5 m and 1.0m, respectively, the length of the tank (in m, round off to 2 decimal places) is ________.

38. A reinforced concrete circular pile of 12m length and 0.6 m diameter is embedded in stiff clay which has an undrained unit cohesion of 110 kN/m^2. The adhesion factor is 0.5. The Net Ultimate Pullout (Uplift) Load for the pile (in kN, round off to 1 decimal place is) is ________.

39. A granular soil has a saturated unit weight of 20 kN/m^3 and an effective angle of shearing resistance of 30°. The unit weight of water is 9.81 kN/m^3. A slope is to be made on this soil deposit in which the seepage occurs parallel to the slope up to the free surface. Under this seepage condition for a factor of safety of 1.5, the safe slope angle (in degree, round off to 1 decimal place) would be ______.

40. A portal frame shown in figure (not drawn to scale) has a hinge support at joint P and a roller support at joint R. A point load of 50 kN is acting at joint R in the horizontal direction. The flexural rigidity. EI, of each member is 10^6 kNm^2. Under the applied load, the horizontal displacement (in mm, round off to 1 decimal place) of joint R would be ______

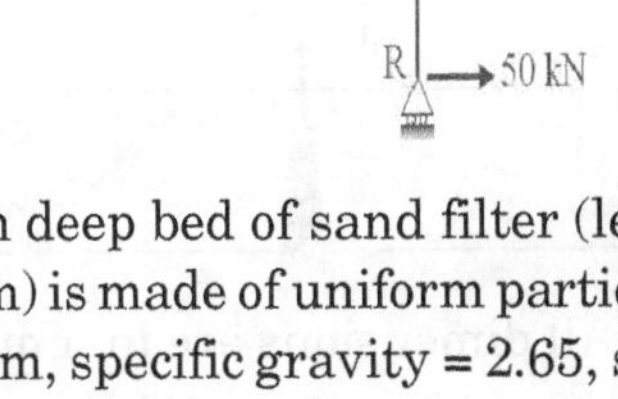

41. A 0.80 m deep bed of sand filter (length 4m and width 3m) is made of uniform particles (diameter = 0.40 mm, specific gravity = 2.65, shape factor = 0.85) with bed porosity of 0.4. the bed has to be backwashed at a flow rate of 3.60 m^3/min. During backwashing, if the terminal settling velocity of sand particles is 0.05 m/s, the expanded bed depth (in m, round off to 2 decimal places) is ______

42. A 3 m × 3 m square precast reinforced concrete segments to be installed by pushing them through an existing railway embankment for making an underpass as shown in the figure. A reaction arrangement using precast PCC blocks placed on the ground is to be made for the jacks.

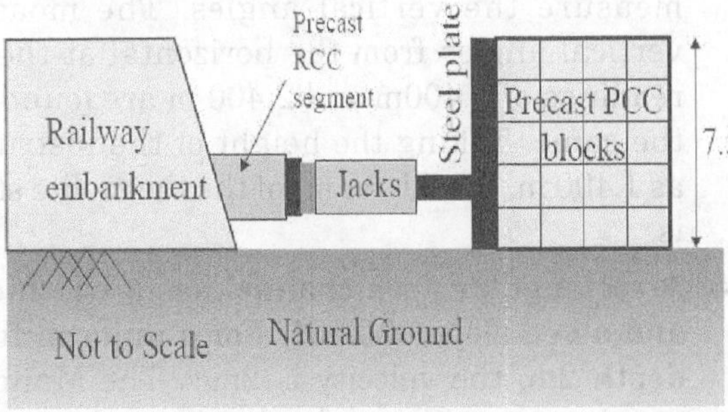

At each stage, the jacks are required to apply a force of 1875 kN to push the segment. The jacks will react against the rigid steel plate placed against the reaction arrangement. The footprint area of reaction arrangement on natural ground is 37.5 m^2. The unit weight of PCC block is 24 kN/m^3. The properties of the natural ground are: c = 17 kPaϕ = 25° and γ =18 kNm^3. Assuming that the reaction arrangement has rough interface and has the same properties that of soil, the factor of safety (round off to 1 decimal place) against shear failure is ______.

43. The hyetograph of a storm event of duration 140 minutes is shown in the figure.

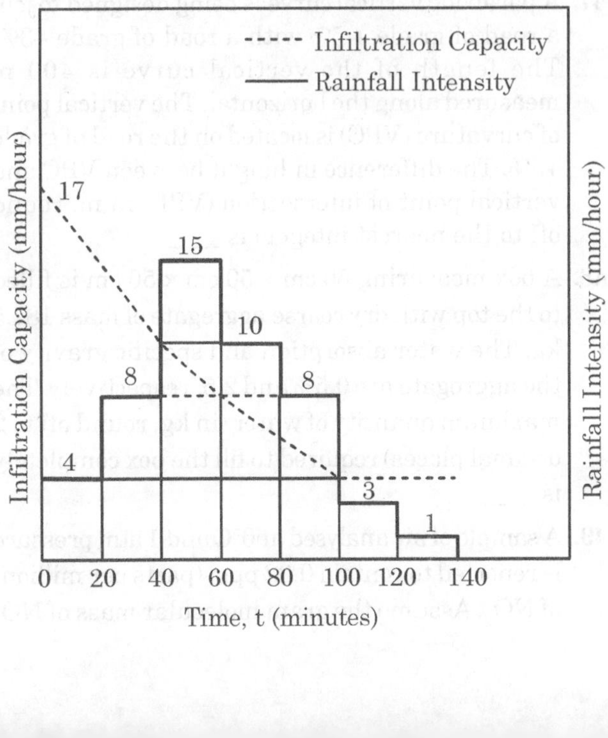

The infiltration capacity at the start of this event ($t = 0$) is 17mm/hour, which linearly decreases to 10 mm/hour after 40 minutes duration. As the event progresses, the infiltration rate further drops down linearly to attain a value of 4mm/hour at $t = 100$ minutes and remains constant thereafter till the end of the storm event. The value of the infiltration index, *f* Ö (in mm/hour, round off to 2 decimal places), is ______.

44. A staff is placed on a benchmark (BM) of reduced level (RL) 100.000 m and a theodolite is placed at a horizontal distance of 50m from the BM to measure the vertical angles. The measured vertical angles from the horizontal at the staff readings of 0.400m and 2.400 m are found to be the same. Taking the height of the instrument as 1.400 m, the RL (in m) of the theodolite station is ______.

45. A rectangular open channel has a width of 5m and a bed slope of 0.001. For a uniform flow of depth 2m, the velocity is 2m/s. The Manning's roughness coefficient for the channel is

(*a*) 0.017 (*b*) 0.050

(*c*) 0.033 (*d*) 0.002

46. Consider a laminar flow in the x-direction between two infinite parallel plates (Couette flow). The lower plate is stationary and the upper plate is moving with a velocity of 1 cm/s in the x-direction. The distance between the plates is 5mm and the dynamic viscosity of the fluid is 0.01 N–s/m^2. If the shear stress on the lower plate is zero, the pressure gradient, $\frac{\partial p}{\partial x}$ (in N/m^2 per m, round off to 1 decimal place) is______.

47. A parabolic vertical curve is being designed to join a road of grade + 5% with a road of grade –3%. The length of the vertical curve is 400 m measured along the horizontal. The vertical point of curvature (VPC) is located on the road of grade +5%. The difference in height between VPC and vertical point of intersection (VPI) (in m, round off to the nearest integer) is ___

48. A box measuring 50 cm × 50 cm × 50 cm is filled to the top with dry coarse aggregate of mass 187.5 kg. The water absorption and specific gravity of the aggregate are 0.5% and 2.5, respectively. The maximum quantity of water (in kg, round off to 2 decimal places) required to fill the box completely is _____.

49. A sample of air analysed at 0°C and 1 atm pressure is reported to contain 0.02 ppm (parts per million) of NO_2. Assume the gram molecular mass of NO_2 as 46 and its volume at 0°C and 1 atm pressure as 22.4 litres per mole. The equivalent NO_2 concentration (in microgram per cubic meter, round off to 2 decimal palces) would be ______.

50. A survey line was measured to be 285.5m with a tape having a nominal length of 30m. On checking, the true length of the tape was found to be 0.05 m too short. If the line lay on a slope of 1 in 10, the reduced length (horizontal length) of the line for plotting of survey work would be

(*a*) 285.6m (*b*) 283.6m

(*c*) 285.0m (*d*) 284.5m

51. A one-dimensional domain is discretized into N sub-domains of width Δx with node numbers $i = 0, 1, 2, 3, \ldots, N$. If the time scale is discretized in steps of Δt the forward-time and centered-space finite difference approximation at ith node and nth time step, for the partial differential equation $\frac{\partial v}{\partial t} = \beta \frac{\partial^2 v}{\partial x^2}$ is

(*a*) $\frac{v_{i+1}^{(n+1)} - v_i^{(n)}}{\Delta t} = \beta\left[\frac{v_{i+1}^{(n)} - 2v_i^{(n)} + v_{i-1}^{(n)}}{2\Delta x}\right]$

(*b*) $\frac{v_i^{(n)} - v_i^{(n-1)}}{2\Delta t} = \beta\left[\frac{v_{i+1}^{(n)} - 2v_i^{(n)} + v_{i-1}^{(n)}}{2\Delta x}\right]$

(*c*) $\frac{v_i^{(n+1)} - v_i^{(n)}}{\Delta t} = \beta\left[\frac{v_{i+1}^{(n)} - 2v_i^{(n)} + v_{i-1}^{(n)}}{(\Delta x)^2}\right]$

(*d*) $\frac{v_i^{(n)} - v_i^{(n-1)}}{\Delta t} = \beta\left[\frac{v_{i+1}^{(n)} - 2v_i^{(n)} + v_{i-1}^{(n)}}{(\Delta x)^2}\right]$

52.

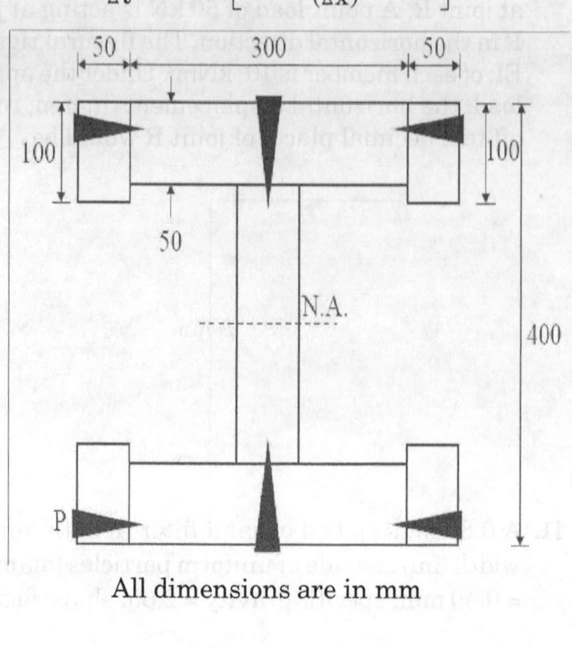

All dimensions are in mm

The cross-section of a built-up wooden beam as shown in the figure (not drawn to scale) is subjected to a vertical shear force of 8kN. The beam is symmetrical about the neutral axis (N.A.) shown, and the moment of inertia about N.A. is 1.5×10^9 mm^4. Considering that the nails at the location P are spaced longitudinally (along the length of the beam) at 60 mm, each of the nails at P will be subjected to the shear force of

(*a*) 240 N (*b*) 480 N

(*c*) 120 N (*d*) 60 N

53. A 16 mm thick gusset plate is connected to the 12mm thick flange plate of an I-section using fillet welds on both sides as shwon in the figure (not drawn to scale). The gusset plate is subjected to a point load of 350 kN acting at a distance of 100 mm from the flange plate. Size of fillet weld is 10 mm.

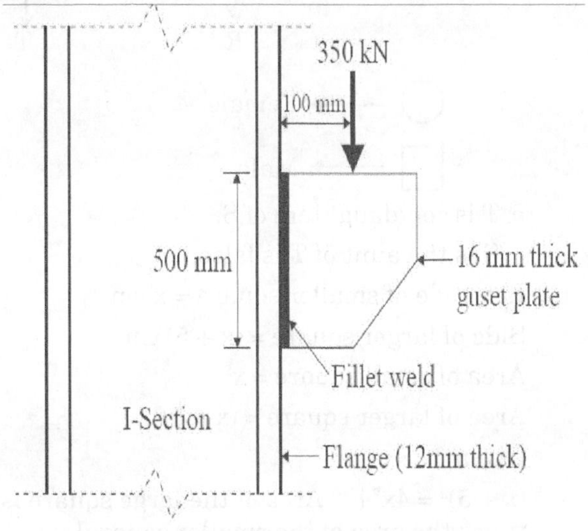

(Front view)

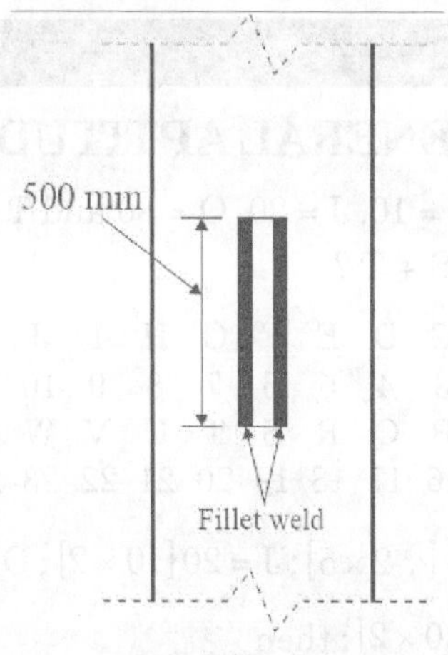

(Side view)

The maximum resultant stress (in MPa, round off to 1 decimal place) on the fillet weld along the vertical plane would be ______.

54. Consider two functions: $x = \psi \ell n \phi$ and $y = \phi \ell n \psi$. Which one of the following is the correct expression for $\dfrac{\partial \psi}{\partial x}$?

(*a*) $\dfrac{\ell n \psi}{\ell n \phi \ell n \psi - 1}$ (*b*) $\dfrac{\ell n \phi}{\ell n \phi \psi - 1}$

(*c*) $\dfrac{x \ell n \psi}{\ell n \phi \psi - 1}$ (*d*) $\dfrac{x \ell n \phi}{\ell n \phi \ell n \psi - 1}$

55. Tie bars of 12 mm diameter are to be provided in a concrete pavement slab. The working tensile stress of the tie bars is 230 MPa, the average bond strength between a tie bar and concrete is 2 MPa, and the joint gap between the slab is 10mm. Ignoring the loss of bond and the tolerance factor, the design length of the tie bars (in mm, round off to the nearest integer) is ______.

ANSWERS

General Aptitude

1. (*c*)	**2.** (*a*)	**3.** (*b*)	**4.** (*d*)	**5.** (*b*)	**6.** (*c*)	**7.** (*d*)	**8.** (*a*)
9. (*a*)	**10.** (*b*)						

Civil Engineering

1. (*c*)	**2.** (*d*)	**3.** (*b*)	**4.** (*b*)	**5.** (*a*)	**6.** (*c*)	**7.** (0.081)	**8.** (0.001)
9. (0.52)	**10.** (*a*)	**11.** (4)	**12.** (*a*)	**13.** (*a*)	**14.** (*a*)	**15.** (*c*)	**16.** (*c*)
17. (2094.9)	**18.** (*c*)	**19.** (*d*)	**20.** (*d*)	**21.** (*d*)	**22.** (17.9)	**23.** (100)	**24.** (*c*)
25. (*b*)	**26.** (150500)	**27.** (270)	**28.** (*b*)	**29.** (5.8)	**30.** (*b*)	**31.** (*a*)	**32.** (0.18)
33. (*c*)	**34.** (1800)	**35.** (6)	**36.** (*d*)	**37.** (8)	**38.** (1244.07)	**39.** (21.8)	**40.** (25)
41. (1.21)	**42.** (2.018)	**43.** (7)	**44.** (100)	**45.** (*a*)	**46.** (8)	**47.** (6.25)	**48.** (50.38)
49. **(41.07)**	**50.** (*b*)	**51.** (*a*)	**52.** (*a*)	**53.** (78.1)	**54.** (*a*)	**55.** (700)	

EXPLANATIONS

GENERAL APTITUDE

3. Given E = 10, J = 20, O = 30 and T = 40

P + E + S + T ?

A	B	C	D	E	F	G	H	I	J	K	L	M
1	2	3	4	5	6	7	8	9	10	11	12	13
N	O	P	Q	R	S	T	U	V	W	X	Y	Z
14	15	16	17	18	19	20	21	22	23	24	25	26

If $E = 10\ [\because 2 \times 5]$; $J = 20[10 \times 2]$; $O = 30[15 \times 2]$;

$T = 40[20 \times 2]$; then

$P = 32[16 \times 2]$; $S = 38[19 \times 2]$; $T = 40[20 \times 2]$

$\therefore P + E + S + T = 32 + 10 + 38 + 40 = 120.$

5.

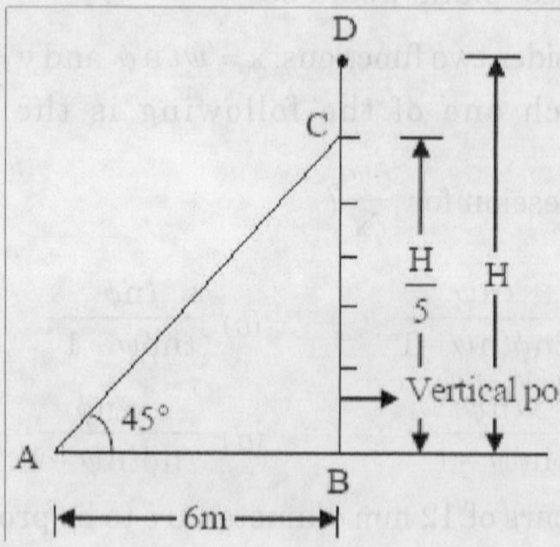

H denotes height of the pole from the bottom.

In ΔABC ;

$$\tan 45° = \frac{H/5}{6}$$

$$\Rightarrow 1 = \frac{H/5}{6} \qquad [\because \tan 45° = 1]$$

$$\Rightarrow H/5 = 6 \Rightarrow H = 30 \text{ m.}$$

7. Given,

Play only cricket → 105

Play only hockey → 70

Play only football → 50

Play both cricket and hockey → 25

Play both hockey and football → 15

Play both cricket and football → 30

Total number of people in a sports academy = 300

Number of people who can play only one sport = 105 + 70 + 50 = 225

Number of people who can play at least two sports = Total – (people who can play only one sport) = 300 – 225 = 75

$$\therefore \text{Percentage} = \left[\frac{75}{300} \times 100\right]\% = 25\%$$

8. Given,

P is the brother of S,

Q is the wife of P,

R and T are the children of the siblings P and S respectively.

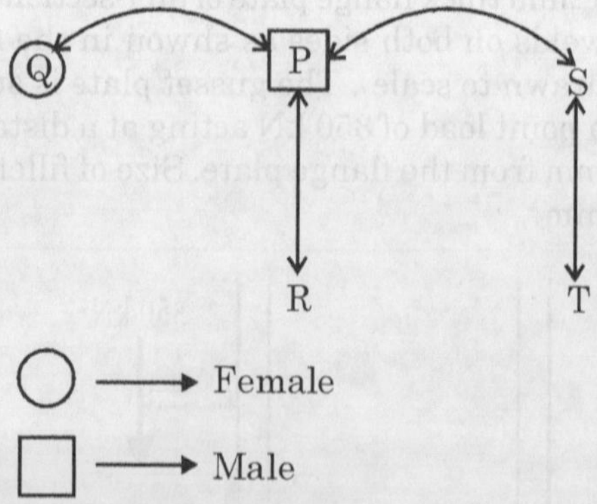

∴ T is son/daughters of S.

∴ S is the aunt of T is false.

10. Let, Side of smaller square = x cm

Side of larger square = (x + 5) cm

Area of small square = x^2

Area of larger square = $(x + 5)^2$

Given,

$(x + 5)^2 = 4x^2$ [∵ Area of the large square is four times the area of the smaller square]

$$\Rightarrow x^2 + 25 + 10x = 4x^2$$

$$\Rightarrow 3x^2 - 10x - 25 = 0$$

$$\Rightarrow x = 5; \ -\frac{5}{3} \rightarrow \text{which is not possible.}$$

∴ Side if larger square 10 cm.

CIVIL ENGINEERING

1. Soil mass is fully saturated.

By applying Darcy's law

$$V_x = K_x . i_x = K_x . \frac{dh}{dx}$$

$$V_z = K_z . i_z - K_z . \frac{dh}{dz}$$

From the continuity equation

$$\frac{\partial u}{\partial x} + \frac{\partial v}{\partial z} = 0$$

$$K_x.\frac{\partial^2 h}{\partial x^2}+K_z.\frac{\partial^2 h}{\partial z^2}=0$$

For isotropic soil $K_x = K_z$

$$\frac{\partial^2 h}{\partial x^2}+\frac{\partial^2 h}{\partial z^2}=0$$

2. 1st Stage

$\mu_1 = f_r$, $n_1 = +G\%$

$$\text{Braking distance } (d_1) = \frac{V^2}{2g(\mu+0.01n)}$$

$$=\frac{V^2}{2g(f_r+0.01G)}$$

2nd Stage

$\mu_2 = f_r$, $n_2 = +2G\%$

$$\text{Braking distance } (d_2) = \frac{V^2}{2g(\mu+0.01n)}$$

$$=\frac{V^2}{2g(f_r+0.01\times 2G)}=\frac{V^2}{2g(f_r+0.02G)}$$

$$\%\text{ change} = \frac{d_1-d_2}{d_1}\times 100$$

$$=\frac{\frac{V^2}{2g(f_r+0.01G)}-\frac{V^2}{2g(f_r+0.02G)}}{\frac{V^2}{2g(f_r+0.01G)}}\times 100$$

$$=\frac{\frac{1}{f_r+0.01G}-\frac{1}{f_r+0.02G}}{\frac{1}{f_r+0.01G}}$$

$$=\frac{(f_r+0.02G)-(f_r+0.01G)}{f_r+0.02G}=\frac{0.01G}{f_r+0.02G}\times 100$$

3.

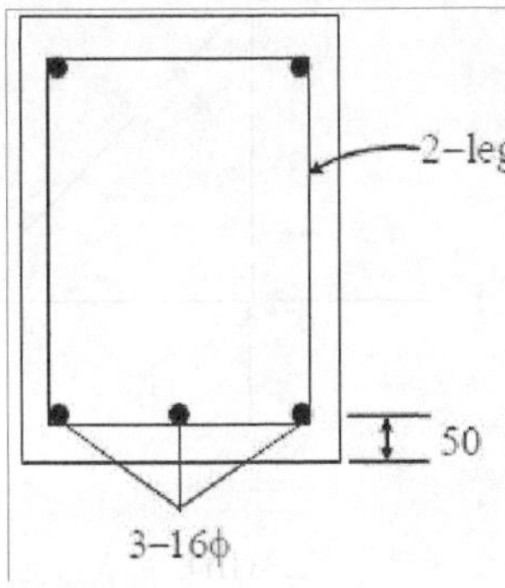

Nominal cover

$$= \text{Cover} - \frac{\text{diameter of bar}}{2} - \text{diameter of stirrup}$$

$$= 50\text{ mm} - \frac{16}{2}\text{ mm} - 12$$

$= 50 - 8 - 12 = 50 - 20 = 30$ mm

4. We know that, if three members are meeting at the joint, two of them are collinear and there is no point load acting on the joint, then the third member will carry zero force.

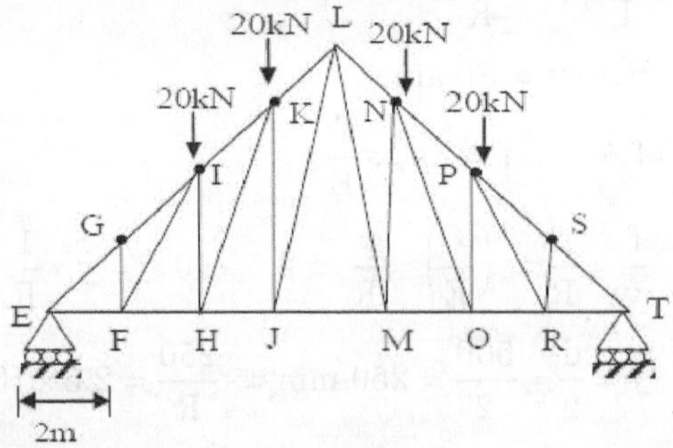

From the above statement,
We can say the FG, FI, RS, PR members will carry zero force.

5. Based on origin, air pollutants are classified into two categories

(i) Primary air pollutants: Particulates, CO, SO_x, NO_x, hydrocarbons ....etc.

(ii) Secondary air pollutants: Ozone, PAN, PBN, PPN, Smog.... etc.

7. Discharge $= Q = C_D\dfrac{\sqrt{2gH}A_1A_2}{\sqrt{A_1^2-A_2^2}}$

$$A_1=\frac{\pi}{4}(0.3)^2;\ A_2=\frac{\pi}{4}(0.15)^2$$

$$\begin{bmatrix} A_1 = \text{Area of C/S (1) – (1)} \\ A_2 = \text{Area of C/S (2) – (2)} \end{bmatrix}$$

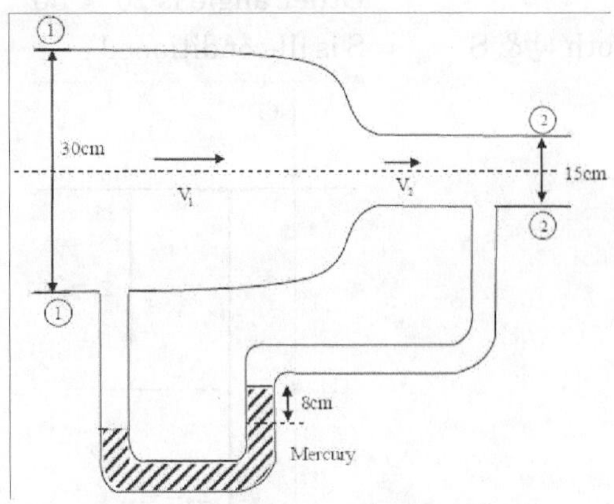

$$H = \text{Differential head} = \left(\frac{S_{Hg}}{\rho_w}-1\right)\times\text{sum}$$

$$=\left(\frac{13.6\times 1000}{1000}-1\right)\times 0.08 = 1.008\text{ m}$$

$$Q=\frac{\sqrt{2\times 9.81\times 1.008}\times \pi/4\times(0.3)^2\times\pi/4\times(0.15)^2}{[\pi/4\times(0.3)]^2-[\pi/4\times(0.15)]^2}$$

$[C_d \approx 1$; Frictionless flow$]$

$= 0.081$ m^3/s

8. Overall depth (d) = 500 mm

We know that, from bending equation

$$\frac{M}{I} = \frac{f}{y} = \frac{E}{R}$$

Stress $\propto$ Strain

$$f \propto \varepsilon \Rightarrow f = E\varepsilon \Rightarrow \frac{f}{E} = \varepsilon$$

$$\frac{f}{y} = \frac{E}{R} \Rightarrow \left(\frac{f}{E}\right) = \frac{y}{R} \Rightarrow 2.5 \times 10^{-4} = y \times \frac{1}{R}$$

$$y = \frac{d}{2} = \frac{500}{2} = 250 \text{ mm} \Rightarrow \frac{250}{R} = 2.5 \times 10^{-4}$$

$$\text{Curvature}\left(\frac{1}{R}\right) = \frac{2.5 \times 10^{-4}}{250} = 1 \times 10^{-6} \text{ mm}^{-1}$$

$$= 1 \times 10^{-3} \text{ m}^{-1} = 0.001 \text{ m}^{-1}$$

9. Given (maximum number of vehicle) in 5 minutes = 160 vehicles

Flow in the peak hour = 1000 vehicles

5 minute-Peak hour factor

$$= \frac{\text{Peak flow}}{12(5 \text{ minute traffic})} = \frac{1000}{12(160)} = 0.52$$

10. If any interior angle of a triangle is more than 120° or less than 30°; then the triangle is called ill-conditioned.

Triangle Q:	One angle is 25° < 30°
	Q is ill-conditioned
Triangle S:	One angle is 130° < 120°
	Other angle is 20° < 30°
Both Q & S	S is ill-conditioned

11.

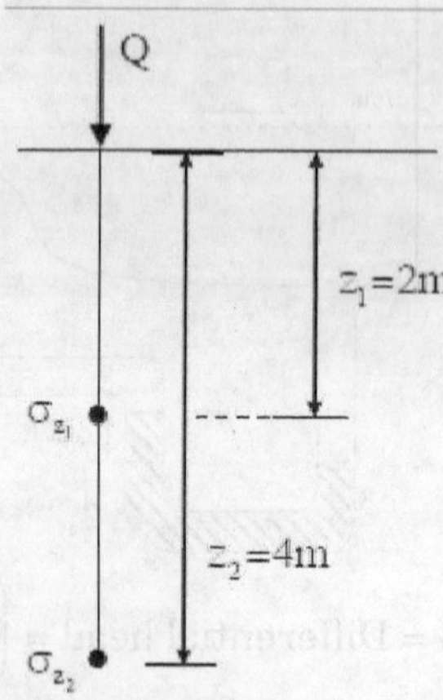

As per Boussinegq's theory

$$\text{Vertical stress, } (\sigma_Z) = \frac{Q}{z^2}\frac{3}{2\pi}\left[\frac{1}{1+\left(\frac{r}{z}\right)^2}\right]^{5/2}$$

$$\sigma_z = K.\frac{Q}{z^2} \Rightarrow \sigma_z \propto \frac{1}{z^2} \quad \frac{\sigma_{z1}}{\sigma_{z2}} = \left(\frac{z_2}{z_1}\right)^2 = \left(\frac{4}{2}\right)^2 = 4$$

12. In case of upward seepage, force becomes equal to the bouyant weight of the soil. The effective stress in the soil becomes zero ($\sigma = 0$)

$$H.\gamma_{sub} - i.H.\gamma_w = 0$$

$$H.\gamma_{sub} = iH.\gamma_w$$

$$i = \frac{\gamma_{sub}}{\gamma_w} = i_{cr} \Rightarrow i = i_{cr}$$

For quick sand condition $\sigma' = 0$ and $i = i_{cr}$

13.

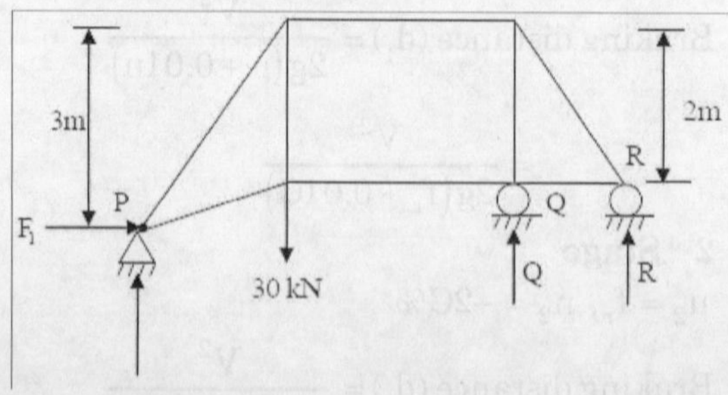

$$\Sigma F_H = 0$$

$$F_1 = 0 \qquad \ldots \text{(i)}$$

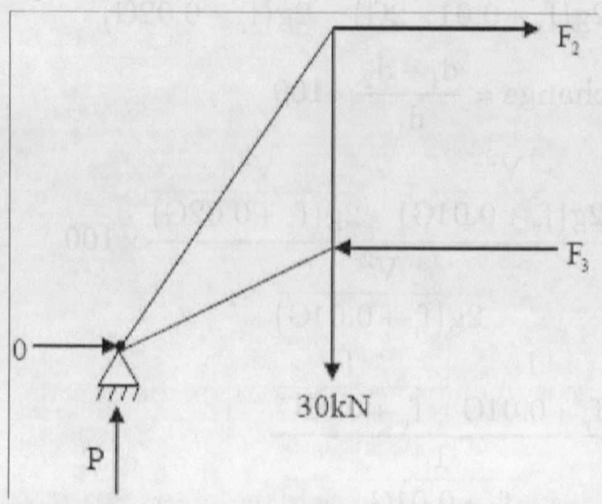

$$\Sigma F_v = 0$$

$$P = 30 \qquad \ldots \text{(ii)}$$

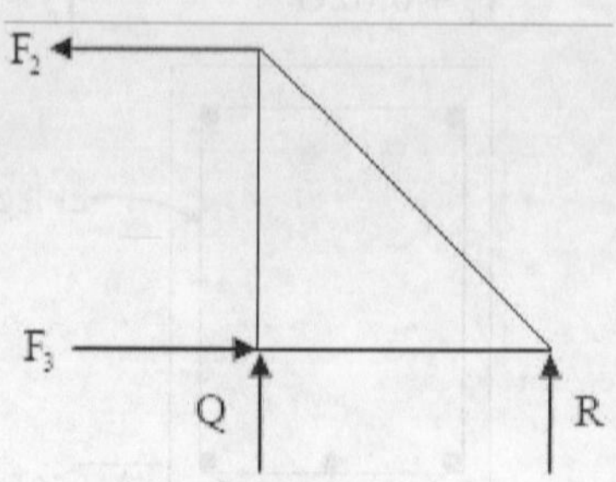

$$\Sigma F_H = 0$$

$$F_2 = F_2 \qquad \ldots \text{(iii)}$$

$$F_2 \times 3 + 30 \times 3 - F_3 \times 1 = 0$$

$$F_2 \times 3 - F_2 \times 1 + 90 = 0$$

$$2F_2 = -90$$

$$F_2 = -45$$

$$\Rightarrow F_2 = -45$$

$Q + R = 0$

$F_3 \times 2 + R \times 3 = 0$

$-45 \times 2 + R \times 3 = 0$

$R = 30 \Rightarrow Q = -30$

15. **Case I: Low shear**

$V < 0.6V_d$

$M_{d_1} = \frac{Z_e f_y}{\gamma_{m_0}}$ = Design bending strength of a semi-compact section in case of low shear.

Case II: High shear

$V > 0.6V_d$

$M_{d_2} = \frac{Z_e f_y}{\gamma_{m_0}}$ = Design bending strength of a semi-compact section in case of high shear.

$M_{d_1} = M_{d_2} \Rightarrow$ zero reduction

16. We know that velocity head $= \frac{V^2}{2g}$

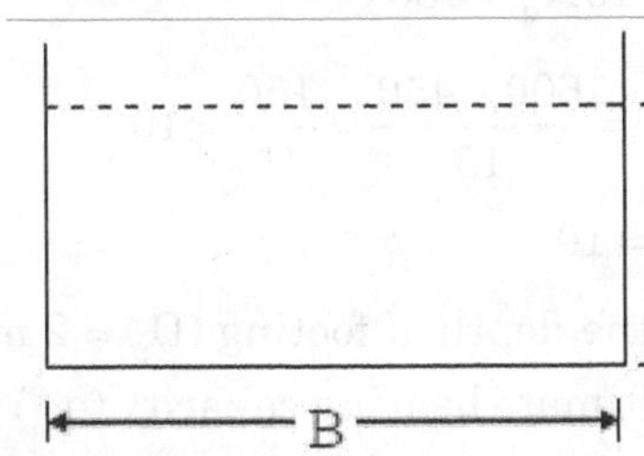

Area = B.y

Top surface width (T) = B

Q = A.V

Q = B.y.V

For critical flow conditin Froude number $(F_r) = 1$

$$\Rightarrow \frac{Q^2 T}{gA^3} = 1$$

$$Q^2 T = gA^3$$

$$(B.y.V)^2\, T = gA^3$$

$$B^2.y^2.V^2(B) = g(B.y)^3$$

$$B^3.y^2V^2 = g.B^3y^3 \Rightarrow \frac{V^2}{g} = y$$

By Multiplying with $\frac{1}{2}$ on both sides, we get

$$\frac{V^2}{2g} = \frac{y}{2} \Rightarrow \frac{\frac{V^2}{2g}}{y} = \frac{1}{2} \Rightarrow \frac{\text{Velocity head}}{\text{depth}} = \frac{1}{2}$$

17. Given that

Specific gravity (G) = 2.65

Mass density of water $(\gamma_w) = 1000\ kg/m^3$

Moisture content (w) = 10%

By the zero-air void line

$$\text{dry density } (\gamma_{dry}) = \frac{G_S \cdot \gamma_W}{1 + WG} = \frac{2.65 \times 1000}{1 + 0.10 \times 2.65}$$

$$= \frac{2650}{1.265} = 2094.86\ kg/m^3$$

18. As the base of slab is frictionless base

$\Rightarrow$ Thermal stress = 0

19. The Taylor's series expansion for f(x + h) is

$$f(x+h) = f(x) + hf'(x) + \frac{h^2}{2!} f''(x) + \frac{h^3}{3!} + f'''(x) + \ldots \infty$$

20. $\lim\limits_{x \to 0} \frac{\sin 4x}{\sin 2x} \quad \left(\frac{0}{0}\right)$

$$= \lim_{x \to 0} \frac{4.\cos 4x}{2.\cos 2x} \quad [\because \text{from L'hospital rule}]$$

$$= \frac{4}{2} = 2$$

$\lim\limits_{x \to 0} \frac{\tan x}{x} \quad \left(\frac{0}{0}\right)$

$$= \lim_{x \to 0} \frac{\sec^2 x}{1} = 1 \quad [\because \text{from L'hospital rule}]$$

21. Given $\varepsilon_x = 0.0030$; $\varepsilon_y = 0.0020$

Normal stress in the plane of maximum shear strain is the centre of Mohr circle.

$$\text{Centre of Mohr circle} = \frac{\varepsilon_x + \varepsilon_y}{2}$$

$$= \frac{0.0030 + 0.0020}{2} = \frac{0.0050}{2} = 0.0025$$

22. Probability of flood discharge exceeds at least once in next 5 years = 0.25

Risk = 0.25

$$1 - \left(1 - \frac{1}{T}\right)^n = 0.25 \Rightarrow 1 - 0.25 = \left(1 - \frac{1}{T}\right)^5$$

$$(0.75)^{\frac{1}{5}} = 1 - \frac{1}{T}$$

$$\frac{1}{T} = 1 - (0.75)^{\frac{1}{5}} \Rightarrow \frac{1}{T} = 1 - 0.944 = 0.559$$

$\Rightarrow$ T = 17.89 years

23. As we know that settling velocity for discrete particles is given by stokes law as

$$V_T = \frac{(G-1)\gamma d^2}{18\mu} \Rightarrow V_T \propto d^2$$

For 30 second duration if 0.35 mm particle size settles completely then % removal of particle size 0.45 mm and 0.50 mm will be100% respectively for each. As settling velocity of particle size 0.45 mm and 0.50 mm will be greater than settling velocity of size 0.35 mm $\left(V_T \propto d^2\right)$.

24. $m.\ddot{z} + kz = 0$

$$\ddot{z} + \frac{k}{m}.z = 0$$

Comparing with $\ddot{x} + \omega_n^2 x = 0 \Rightarrow \omega_n^2 = \frac{k}{m}$

Natural frequency, $(\omega_n) = \sqrt{\frac{k}{m}}$

26.

Activity	Normal duration	Crash duration	Normal cost	Crash cost	C/S slope = $\frac{C.C-N.C}{N.D-C.D}$
P	6	4	15000	25000	5000
Q	5	2	6000	12000	2000
R	5	3	8000	9500	750
S	6	3	7000	10000	1000
T	3	2	6000	9000	3000
U	2	1	4000	6000	2000
V	4	2	20000	28000	4000

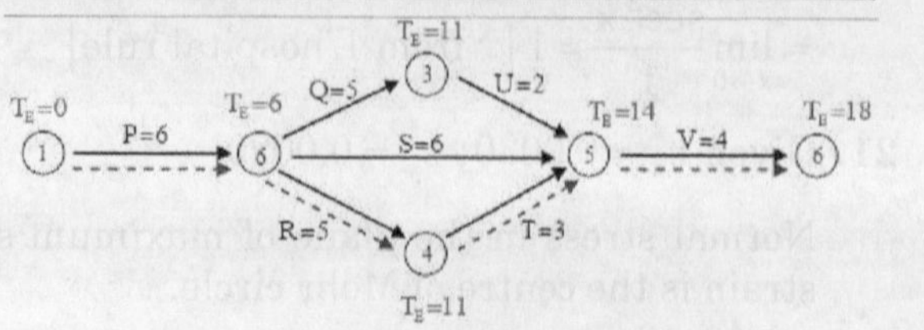

Critical path ① P ② R ④ T ⑤ V ⑥

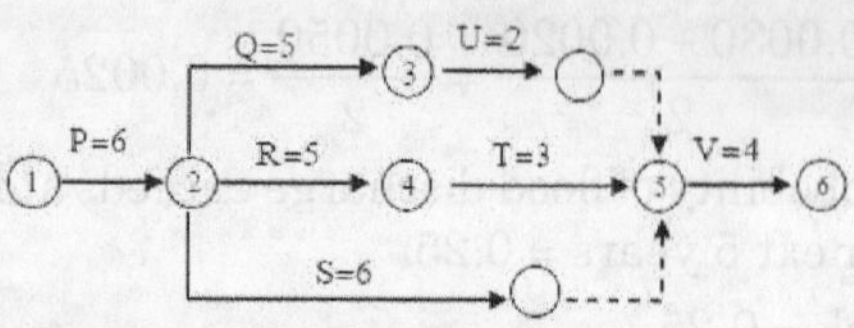

Path PRTV is critical path and corresponding normal duration is 18 days.

For 18 days:

Direct cost = 66000

Indirect cost = 18 × 5000 = 90000

Normal total cost = 156000

Crash activity R by 1 day

For 17 days,

Direct cost = 66000 + 750 = 66750

Indirect cost = 17 × 5000 = 85000

Total cost = 151750 Rs.

Now further crash activity Q, S and R by 1 day.

For 16 days, Direct cost = 66750 + 2000 + 1000 + 750 = 70500

Indirect cost = 16 × 5000 = 80,000

Total cost = 150500 Rs.

27. Given that soil is sand⇒ C = 0

Also given ultimate bearing capacity (q_u) = 600 kPa

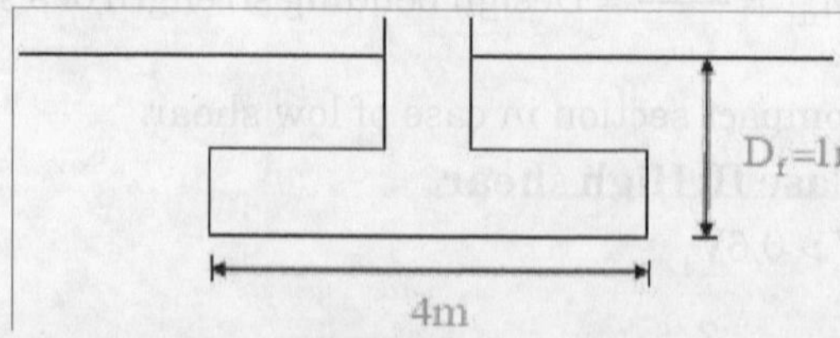

sand, $\gamma_d = 15$ kN/m^3

As per Terzaghi's theory, For square footing Ultimate bearing capacity (q_u)

$$= 1.3CN_C + \gamma D.N_q + 0.4\gamma B.N_\gamma$$

$$600kPa = 0 + 15 \times 1 \times N_q + 0.4 \times 15 \times 4 \times 18.75$$

$$600 = 15N_q + 450$$

$$\Rightarrow N_q = \frac{600-450}{15} = \frac{150}{15} = 10$$

$$\therefore N_q = 10$$

Now, the depth of footing (D_f) = 2 m

Net ultimate bearing capacity (q_{nu})

$$= 1.3 \times CN_C + \gamma D\left(N_q - 1\right) + 0.4\gamma B \times N_\gamma$$

$$= 0 + 15 \times 2 \times (10-1) + 0.4 \times 15 \times 4 \times 18.75$$

= 270 + 450 = 720 kPa

Safe bearing capacity (q_{safe})

$$= \frac{q_{nu}}{F.O.S.} + \gamma D_f = \frac{720}{3} + 15 \times 2 = 240 + 30$$

$$\Rightarrow q_{safe} = 270kPa$$

28. Let $y = x\sqrt{x} = x^{\frac{1}{x}} \Rightarrow \ell n\, y = \ell n\, x^{\frac{1}{x}} = \frac{\ell n\, x}{x}$

$$\Rightarrow y = e^{\frac{\ell n x}{x}}$$

$\therefore$ If $\frac{\ell n\, x}{x}$ is maximum then $y = \frac{\ell n\, x}{x}$ is maximum for same 'x'

If $\frac{\ell n\, x}{x}$ is minimum then $y = \frac{\ell n\, x}{x}$ is minimum for same 'x'

$\therefore$ Consider the function $f(x) = \dfrac{\ell n x}{x}$

$$\Rightarrow f'(x) = 0 \Rightarrow \frac{x\left(\frac{1}{x}\right) - \ell n x}{x^2} = 0$$

$$\Rightarrow \frac{1 - \ell n x}{x^2} = 0$$

$\Rightarrow x = e \rightarrow$ stationary point

$$f''(x) = \frac{x^2\left[0 - \frac{1}{x}\right] - [1 - \ell n x]2x}{x^4}$$

$$\Rightarrow f''(x) = \frac{-x - 2x + 2x\,\ell n x}{x^4} = \frac{-3x + 2x\ell n x}{x^4}$$

$$\therefore f''(e) = \frac{-3e + 2e(1)}{e^4} = \frac{-e}{e^4} = -e^{-3} < 0$$

$\therefore$ x = e is a point of maximum

$\therefore y = e^{\frac{\ell n x}{x}} \Rightarrow y = \sqrt[x]{x}$ has global maximum at x = e.

29.

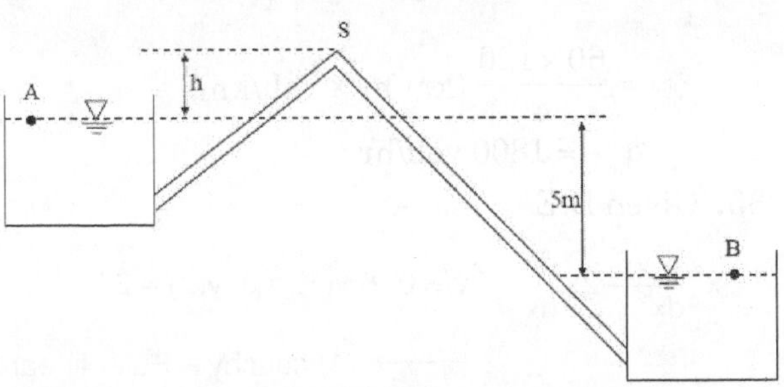

Applying Bernoulli's equation at A & B:

$$\frac{P_A}{\gamma_w} + \frac{V_A^2}{2g} + Z_A = \frac{P_B}{\gamma_w} + \frac{V_B^2}{2g} + Z_B + h_t$$

$$P_A = P_B = P_{atm};\ V_A = V_B = 0$$

$$\Rightarrow Z_A - Z_B = h_f = \frac{f\ell V^2}{2gd} \quad \left[\begin{array}{l} V = \text{Velocity in pipe} \\ \ell = 5000 \text{ m} \\ d = 0.1 \\ f = 0.02 \\ h_f = Z_A - Z_B = 5 \end{array}\right]$$

$$\Rightarrow 5 = \frac{0.02 \times 5000 \times V^2}{2 \times 9.81 \times 0.1}$$

$$\Rightarrow V = 0.313 \text{ m/s}$$

Applying Bernoulli's equation at A & S:

$$\frac{P_A}{\gamma_w} + \frac{V_A^2}{2g} + Z_A = \frac{P_S}{\gamma_w} + \frac{V_s^2}{2g} + Z_S + h_f$$

$$\frac{P_s}{\gamma_w} = 2.5 \text{ m} \quad \left[\begin{array}{l} V_S = V \\ P_A = P_{atm} \\ V_A = 0 \end{array}\right]$$

$$\frac{P_{Ar}}{\gamma_w} = \frac{P_{atm}}{\gamma_w} = 10.326 \text{ m}$$

[P_{atm} is equivalent to 10.326 m head of water]

$$10.326 + Z_A = 2.5 + \frac{V^2}{2g} + Z_s + h_f$$

$$\Rightarrow Z_s - Z_A = h;$$

h_f = friction head loss in pipe along inlet leg

$$= \frac{f\ell V^2}{2gd}\left[\begin{array}{l} \ell = 2000 \text{ m} \\ v = 0.313 \text{ m/s} \end{array}\right]$$

$$\Rightarrow 10.326 - 2.5 = \frac{(0.313)^2}{2 \times 9.81} + h + \frac{0.02 \times 2000 \times (0.313)^2}{2 \times 9.81 \times 0.1}$$

$$\Rightarrow h = 5.823 \text{ m} \approx 5.8 \text{ m}$$

30. Using slope deflection method.

$$M_{QR} = \frac{2EI}{\ell}\left(\theta_R - \frac{3\Delta}{\ell}\right)$$

$$M_{RQ} = \frac{2EI}{\ell}\left(2\theta_R - \frac{3\Delta}{\ell}\right)$$

$$M_{RS} = \frac{2EI}{\ell}(2\theta_R)$$

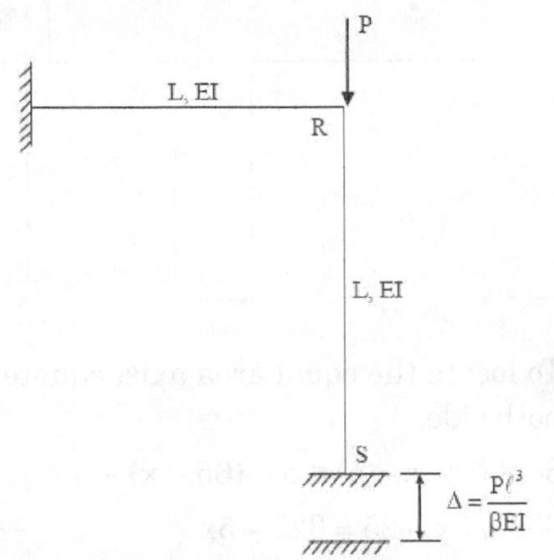

If reaction at S is equal to zero

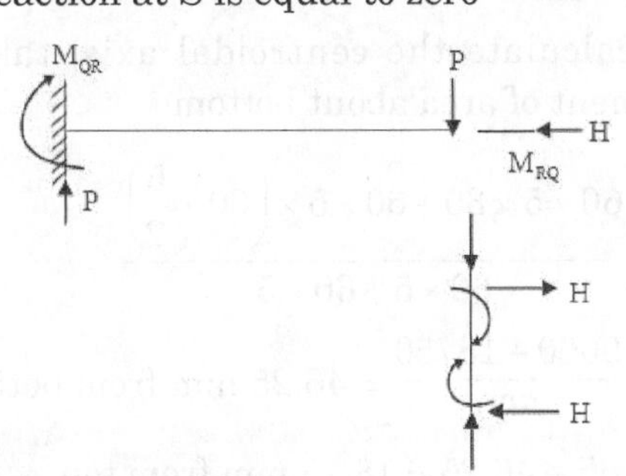

$$M_{RQ} + M_{QR} + Pl = 0$$

$$\frac{6EI\theta_R}{\ell} - \frac{12EI\Delta}{\ell^2} + P\ell = 0$$

$$\frac{6EI\theta_R}{\ell} - \frac{12EI}{\ell^2} \times \frac{P\ell^3}{\beta EI} + P\ell = 0$$

$$\frac{6EI\theta_R}{\ell} - \frac{12P\ell}{\beta} + P\ell = 0 \quad \text{...(i)}$$

From equilibrium of joint,

$$M_{RQ} + M_{RS} = 0$$

$$\frac{8EIQ_R}{\ell} - \frac{6EI\Delta}{\ell^2} = 0$$

$$\frac{6EI\theta_R}{\ell} = \frac{6}{8}\left(\frac{6EI}{\ell^2} \times \frac{Pl^3}{\beta EI}\right)$$

$$\frac{6EI\theta_R}{\ell} = \frac{36P\ell}{8\beta} \quad ...(ii)$$

$\Rightarrow$ From (i) & (ii),

$$\frac{36P\ell}{8\beta} - \frac{96P\ell}{8\beta} - P\ell = 0$$

$$\Rightarrow -\frac{60P\ell}{8\beta} + P\ell = 0 \Rightarrow 8\beta = 60$$

$$\beta = \frac{60}{8} = 7.5$$

31.

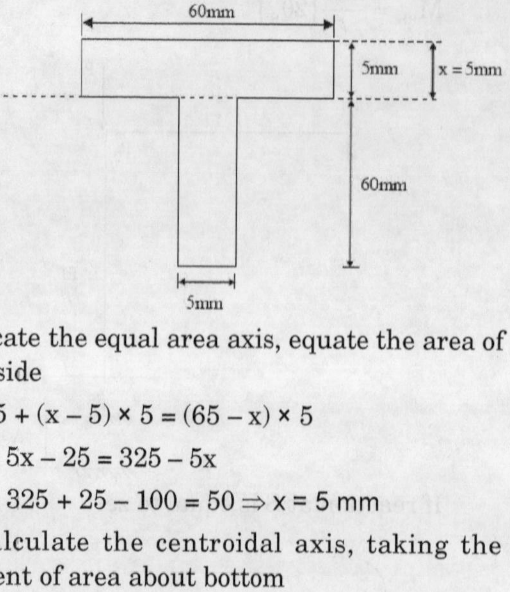

To locate the equal area axis, equate the area of both side

$60 \times 5 + (x - 5) \times 5 = (65 - x) \times 5$

$300 + 5x - 25 = 325 - 5x$

$10x = 325 + 25 - 100 = 50 \Rightarrow x = 5$ mm

To calculate the centroidal axis, taking the moment of area about bottom

$$\bar{x} = \frac{60 \times 5 \times 30 + 60 \times 5 \times \left(60 + \frac{5}{2}\right)}{60 \times 5 + 60 \times 5}$$

$$= \frac{9000 + 18750}{600} = 46.25 \text{ mm from bottom}$$

$\bar{x} = 65 - 46.25 = 18.75$ mm from top

For plastic state,

Area in compression = Area in tension

$60 \times 5 = 60 \times 5 \Rightarrow$ N.A N.A lies at the intersection web and flang E

$\bar{y} = 5$ mm from top

Difference in N.A = 18.75 – 5 = 13.75 mm

32. λ(rate) = 360 vehicle/hour = 0.1 vehicle/hour

Let 'X' denote the headway between two successive arrivals. [clearly exponential R.V.].

$\therefore f(n) = \lambda e^{-\lambda x}$ [pdf of x] $\rightarrow$ exponential distribution

$\Rightarrow f(n) = 0.1e^{-1.0x}$

$\therefore$ Required probability

$$P\left[6 < \underset{6<x<10}{x} < 10\right] = \int_6^{10} f(n)dx$$

$$= \int_6^{10} 0.1e^{-0.1x} = 0.1\frac{\left[e^{-0.1x}\right]}{(-0.1)}\Bigg|_6^{10}$$

$$= -\left[e^{-0.1(10)} - e^{-0.1(6)}\right] = -\left[e^{-1} - e^{-0.6}\right] = e^{-0.6} - e^{-1}$$

$$\Rightarrow P[6 < x < 10] = 0.18.$$

34. Free flow speed (V_f) = 60 km/hr

Jam density (K_J) = 120 veh/km

For a linear speed – density relationship

$$\text{Maximum flow}(q_{max}) = \frac{V_f \times K_J}{4}$$

$$= \frac{60 \times 120}{4} \text{(km/hr × veh/km)}$$

$$q_{max} = 1800 \text{ veh/hr}$$

35. Given D.E

$$x^2\frac{d^2y}{dx^2} - 2x\frac{dy}{dx} + 2y = 0 \quad \& \ y(1) = 0, \ y(2) = 2.$$

$\rightarrow$ (1) cauchy – Euler linear D.E.

$$\Rightarrow \left[x^2D^2 - 2xD + 2\right]y = 0$$

Let $xD = \theta$; $x^2D^2 = \theta(\theta - 1)$; where $\theta = \frac{d}{dz}$ & $x = e^z$

$[\theta(\theta - 1) - 2\theta + 2]y = 0$

$\Rightarrow \left[\theta^2 - 3\theta + 2\right]y = 0$

The A.E is $\theta^2 - 3\theta + 2 = 0$

$\Rightarrow (\theta - 2)(\theta - 1) = 0$

$\Rightarrow \theta = 1, 2 \rightarrow$ real & distinct

$\Rightarrow$ The solution of eq^n(1) is

$y = C_1e^z + C_2e^{2z}$

$\Rightarrow y = C_1x + C_2x^2 \left[\because x = e^z\right]$ ...(2)

Given $\therefore$ y = 0 at x = 1

(2) $\Rightarrow 0 = C_1 + C_2 \Rightarrow C_1 = -C_2$ & y = 2 at x = 2

(2) $\Rightarrow 2 = 2C_1 + 4C_2 \Rightarrow 2 = -2C_2 + 4C_2$

$C_2 = 1$ and $C_1 = -1$

From (2) ; the solution is

$y = -x + x^2 \Rightarrow y(3) = -3 + 3^2 \Rightarrow y(3) = 6.$

36. We know that

Settling velocity

$$(V_s) = \frac{g(\rho_s - \rho_\omega)d^2}{18\mu}$$

$$V_s = \frac{980 \times (2.51 -)(25 \times 10^{-4})^2}{18 \times 0.01} = 0.051 \text{ cm/sec}$$

Given flow rate(Q) = 0.2 m²/sec

We know that surface overflow rate (V_0)

$$= \frac{\text{Volume / Time}}{\text{Surface Area}}$$

$$V_0 = \frac{Q}{\text{Surface area}}$$

Surface area = B × L = 8 × 32 = 256 m²

$$V_0 = \frac{Q}{BL} = \frac{0.2 \text{ m}^3/\text{sec}}{256}$$

$= 7.8125 \times 10^{-4}$ m/sec

$$V_0 = 7.8125 \times 10^{-2} \text{ cm / sec}$$

$$\text{Particle removal efficiency } (\eta) = \frac{V_s}{V_0} \times 100$$

$$\eta = \frac{0.051}{0.078} \times 100 = 65.38\%$$

37. Waste water flow rate (Q) = 36 m³/hr

Model is $N_t = N_0.e^{-Kt}$

N_t = Number of micro-organism survived at time = t = 1% = 0.01

N_0 = Number of micro-organism at t = 0

$N_t = N_0.e^{-Kt}$

$0.01 = 1.e^{-0.23 \times t}$

$e^{-0.23t} = 0.01$

$-0.23t = \text{In. } 0.01$

$-0.23t = -4.605$

t = 20 minutes

Volume = Q × t_d = 36 m³/hr × 20 minutes

$$= 36 \frac{m^3}{hr} \times \frac{20}{60} hrs$$

Volume = 12 m³

BLH = 12

1 × L × 1.5 = 12

$$\text{Length} = \frac{12}{1.5} = 8\text{m}$$

38. For Clays,

$Q_u = Q_{pf}$

$= A_s.f_s = (\pi dL) \times (\propto C)$

$= \pi \times 0.6 \times 12 \times (0.5 \times 110) = 1244.07$ kN

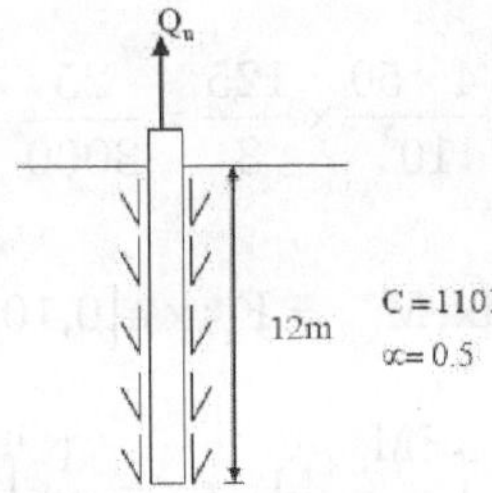

39. F.O.S = 1.5

If there is seepage parallel to the slope

$$\text{F.O.S} = \frac{C^1 + \gamma^1 z.\cos^2\beta.\tan\phi}{\gamma_{sat}.z.\cos\beta.\sin\beta}$$

$$\text{F.O.S} = \frac{\gamma^1 z.\cos^2\beta.\tan\phi}{\gamma_{sat}.z.\cos\beta.\sin\beta}$$

$$\text{F.O.S} = \frac{\gamma^1}{\gamma_{sat}} \frac{\tan\phi}{\tan\beta}$$

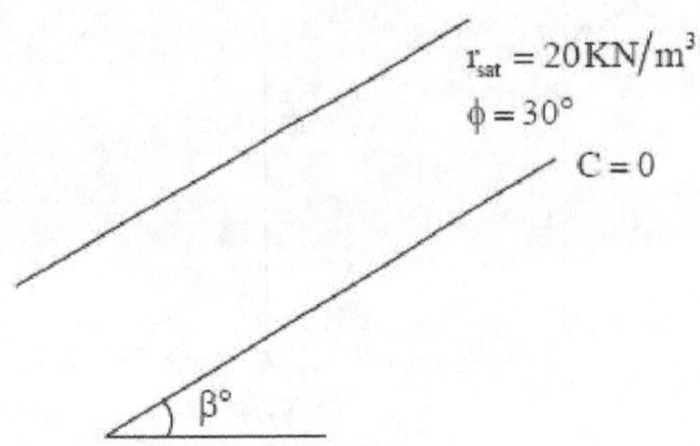

$$1.5 = \frac{(20 - 9.81)}{9.81} \frac{\tan 30}{\tan\beta}$$

$$\frac{\tan 30}{\tan\beta} = 1.444$$

$\tan\beta = 0.399 \Rightarrow \beta = 21.79°$

40. $\Sigma F_x = 0$

$\Rightarrow H_p = F = 0 \Rightarrow H_P = -F$

$\Sigma M_P = 0$

$\Rightarrow V_R \times 5 + F \times 10 = 0 \Rightarrow V_R = -2F$

$\Sigma F_y = 0$

$\Rightarrow V_P + V_R = 0 \Rightarrow V_p = 2F$

$$\Delta \text{ R, horizontal} = \frac{\partial U}{\partial F} = \frac{VU_{PQ}}{\partial F} + \frac{\partial V_{QR}}{\partial F}$$

(Castigliano's theorem)

$$U_{PQ} = \int \frac{M^2 PQ}{2EI} dx; \; M_{PQ} = V_P x = 2F_x ; x \in [0, 5]$$

$$\Rightarrow \frac{\partial U_{PQ}}{\partial F} = \frac{1}{EI} \int \frac{\partial M_{PQ}}{\partial F} \times M_{PQ} dx = \frac{1}{EI} \int_0^5 2Fx \times 2x \, dx$$

$$= \frac{4F}{EI}\left[\frac{x^3}{3}\right]_0^5 = \frac{4\times 50}{10^6}\times\frac{125}{3} = \frac{25}{3000}$$

$$U_{QR} = \int \frac{M_{QR}^2}{2EI}dx; M_{QR} = F_x; x \in [0,10]$$

$$\Rightarrow \frac{\partial U_{QR}}{\partial F} = \frac{1}{EI}\int \frac{\partial M_{QR}}{\partial F} M_{QR} dx = \frac{1}{EI}\int_0^{10} Fx \times x dx$$

$$= \frac{F}{EI}\left[\frac{x^3}{3}\right]_0^{10} = \frac{50}{10^6}\times\frac{1000}{3} = \frac{50}{3000}$$

$$\Rightarrow \frac{\partial U}{\partial F} = \frac{25}{3000} + \frac{50}{3000} = \frac{25}{1000} = 0.025 \text{ m} = 25 \text{ mm}$$

$$\therefore \Delta R \,]\, \text{Horizontal} = \frac{\partial U}{\partial F} = 25 \text{ mm}$$

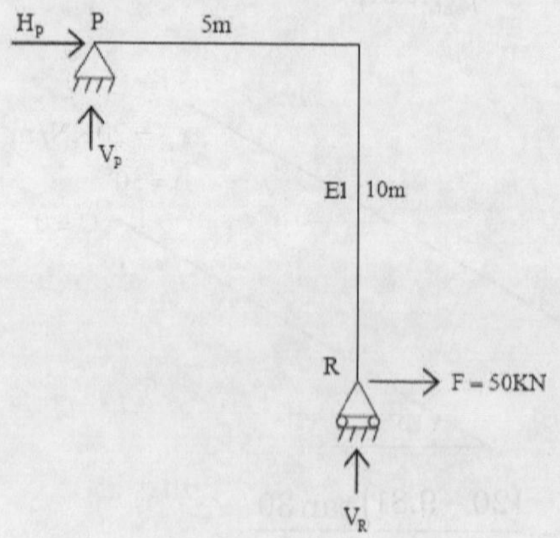

41. Expanded bed depth = De = $D\left[\frac{1-n}{1-n_e}\right]$

D = Depth of bed = 0.8 m;
n = Bed porosity = 0.4

n_e = porosity of expanded bed = $\left(\frac{V_b}{V_s}\right)^{0.22}$

V_S = terminal velocity of settling particles during back wasting = 0.05 m/s
V_b = backwash velocity

$$= \frac{Q}{BL}(Q = \text{Back wash flow rate})$$

$[Q = 3.6\,m^3/\text{min}\ B = 3m\ L = 4m]$

$$\Rightarrow V_b = \frac{3.6}{60\times 3\times 4} = 0.005 \text{ m/s}$$

$$\Rightarrow n_e = \left(\frac{0.005}{0.05}\right)^{0.22} = 0.60255$$

$$\therefore \text{ De} = 0.8\left[\frac{1-0.4}{1-0.60255}\right] = 1.21 \text{ m/s}$$

42. FOS against shear failure

$$= \frac{\text{strength}}{\text{Applied load}} = \frac{(c+\sigma\tan\phi)A}{P}$$

$$\sigma = \frac{N}{A} = \frac{24\frac{kN}{m^3}\times 37.5m^2\times 7.5m}{37.5m^2}$$

$$= 24\times 7.5 \text{ kN/m}^2$$

$$\Rightarrow \text{FOS} = \frac{(c+\sigma\tan\phi)A}{P}$$

$$= \frac{(17+24\times 7.5\times\tan 25°)\times 37.5}{1875}$$

FOS = 2.0187

43.

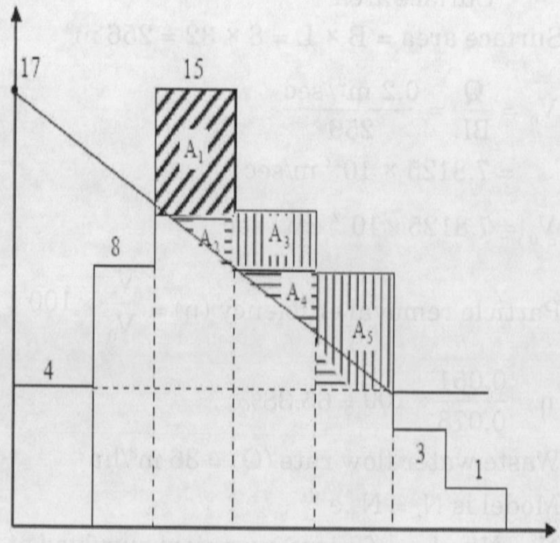

From the above diagram we can say that $A_6 = A_4$

$$A_4 + A_5 = (A_5 + A_6)$$

Infiltration capacity = 17 mm/hr

15 mm/hr, 10 mm/hr, and 8 mm/hr rainfall will effect the run off the, shaded area will shows the effective rainfall i.e. runoff

Total runoff = $A_1 + A_2 + A_3 + A_4 + A_5$

$$= A_1 + A_2 + A_3 + (A_5 + A_6)$$

$$= \left(\frac{5}{60}\times 20\right) + \frac{1}{2}\left(\frac{2}{60}\times 20\right) + \left(\frac{2}{60}\times 20\right) + \left(\frac{4}{60}\times 20\right)$$

= 4 mm

Precipitation = (15 + 10 + 8) mm/hr $\frac{20}{60}$ = 11mm

$$\phi = \text{index} = \frac{P-R}{t} = \frac{(11-4)}{3\times 20 \text{ min}} \text{mm}$$

= 0.116 mm/min = 7 mm/hr

44.

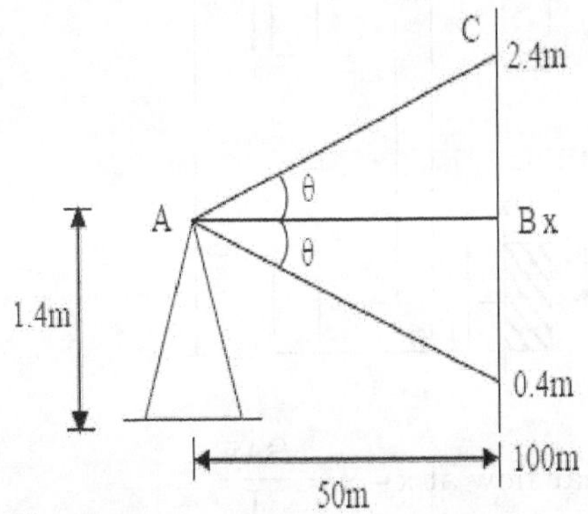

$$\tan\theta = \frac{2.4-x}{50} = \frac{x-0.4}{50}$$

$2x = 2.8$

$x = 1.4$ m

H.O.I = 100 + 1.4 = 101.4 m

RL of theodelite station = 101.4 – theodelite height

= 101.4 – 1.4 = 100 m

45.

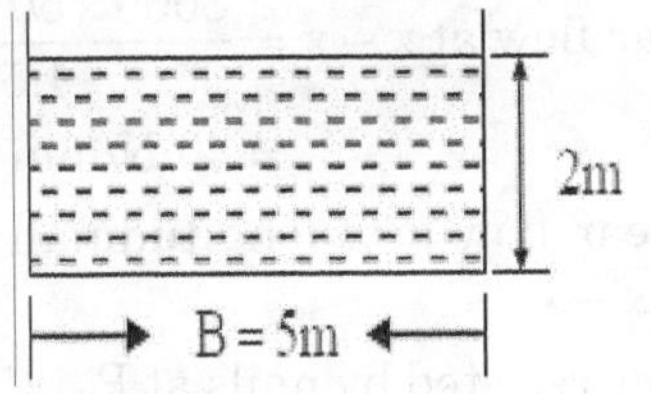

Width (B) = 5 m

Depth (y) = 2 m

Area (B × y) = 5 × 2 = 10 m^2

Perimeter (R) = B + 2y = 5 + 2 × 2 = 9 m

$$\text{Hydraulic radius (R)} = \frac{\text{Area}}{\text{Perimeter}}$$

$$R = \frac{10}{9} = 1.11$$

By the manning's equation,

$$\text{Velocity (V)} = \frac{1}{n}R^{2/3}S^{1/2}$$

$$2 = \frac{1}{n}(1.11)^{2/3}(0.001)^{1/2}$$

$$n = \frac{1.072 \times 0.03162}{2} = 0.0169 \cong 0.017$$

Manning's roughness coefficient (n) = 0.017

46. Given data;

Velocity of plate, V = 1 cm/sec

Distance between the late = 5 mm

Dynamic viscosity of fluid = 0.01 N – S/m^2

Shear stress at lower plate = 0

$$\text{Pressure gradient } \frac{\partial P}{\partial x} = ?$$

We know that, in case of coquette flow, shear stress (τ) is given by

$$\tau = \frac{\mu V}{B} + \left(-\frac{\partial P}{\partial x}\right)\left(\frac{B}{2} - y\right)$$

At lower plate, y = 0; τ = 0 [Given]

$$0 = \frac{0.01 \times 0.01}{0.005}\left(\frac{\partial P}{\partial x}\right)\left[\frac{0.005}{2} - 0\right]$$

$$\frac{\partial P}{\partial x} = 8\text{N/m}^2\text{per m}$$

47. $h_1 = 0.05$

$h_2 = -0.03$

L = 400 m

$M = h_1 - h_2 = 0.05 - (-0.03) = 0.08$

The difference in the height between VPC and VPI,

$$y = \frac{N}{2L}x^2 = \frac{0.08}{2 \times 400}x^2$$

∵ The highest point is given by,

$$x = \frac{n_1 L}{N} = \frac{0.05 \times 400}{0.08} = 250\text{m}$$

$$\text{So, } y = \frac{0.08}{2 \times 400} \times (250)^2 = 6.25\text{m}$$

48. Volume of box = 0.5 × 0.5 × 0.5 = 0.125 m^3

Volume (V) = $V_C + V_W$

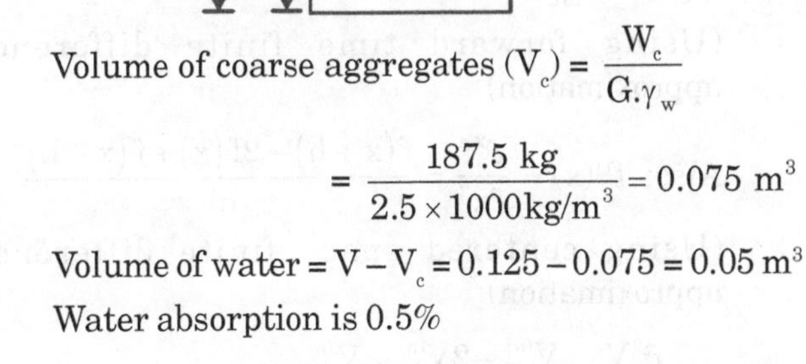

$$\text{Volume of coarse aggregates } (V_c) = \frac{W_c}{G.\gamma_w}$$

$$= \frac{187.5 \text{ kg}}{2.5 \times 1000\text{kg/m}^3} = 0.075 \text{ m}^3$$

Volume of water = $V - V_c$ = 0.125 – 0.075 = 0.05 m^3

Water absorption is 0.5%

$$\text{Water absorbed} = 0.5 \times \frac{0.075}{100} = 0.000375$$

Total volume of water = 0.05 + 0.000375

= 0.050375

Weight of water = Volume × γ_w

= 0.050375 × 1000 = 50.375 kg

49. 1 ppm $NO_2 = \dfrac{1 \text{ part of } NO_2}{10^6 \text{ parts of air}} = \dfrac{1m^3 \text{ of } NO_2}{10^6 m \text{ of air}}$

$\Rightarrow$ 1 mole of NO_2 occupies 22.4 litres of volume at STP.

$\Rightarrow$ 1 mole of NO_2 occupies 0.0224 m³ volume at STP.

$\Rightarrow$ lm³ of NO_2 contains $\dfrac{1}{0.0224}$ moles

$= 44.64$ moles

1 mole of NO_2 = 46 gms

$\Rightarrow$ 1m³ of NO_2 = 44.64 × 46 gms = 2053.57 gms

$\Rightarrow$ 0.02 m³ of NO_2 = 41.07 gms of NO_2

$\therefore$ 0.02 ppm of $NO_2 = \dfrac{0.02m^3 \text{ of } NO_2}{10^6 m^3 \text{ of air}}$

= 41.07 gms per 10^6 m³

= 41.07 × 10^{-6} gm/m³

= 41.07 µg/m³

50. Measured distance (ℓ) = 285.5 m

Length of tape = 30 m

True length of tape = 30 m – 0.05 m = 29.95

Correct length $(\ell^1) = \dfrac{29.95}{300} \times 28.5.5 = 285.024$ m

The line measured on 1/10 slope

Correction due to slope

$$= \frac{h^2}{2\ell} = \frac{\left(\frac{285.024}{10}\right)^2}{2 \times 285.024} = 1.425$$

Corrected length = 285.024 – 1.425 = 283.598 m

51. $\dfrac{\partial v}{\partial t} = \beta \dfrac{\partial^2 V}{\partial x^2}$

$\dfrac{\partial v}{\partial t} = \dfrac{V_i^{(n+1)} V_i^{(n)}}{\Delta t}$

(Using forward time finite difference approximation)

Also; $f^{11}(x) = \dfrac{\partial^2 f}{\partial x^2} = \dfrac{f(x+h) - 2f(x) + f(x-h)}{h^2}$

(Using centered space finite difference approximation)

$$\Rightarrow \frac{\partial^2 V}{\partial x^2} = \frac{V_{i+1}^{(n)} - 2V_i^{(n)} + V_{i-1}^{(n)}}{(\Delta x)^2}$$

$\therefore \dfrac{\partial V}{\partial t} = \beta \dfrac{\partial^2 v}{\partial x^2}$ can be represented as

$$\frac{V_i^{(n+1)} - V_i^{(n)}}{\Delta t} = \beta \left[\frac{V_{i+1}^{(n)} - 2V_i^{(n)} + V_{i-1}^{(n)}}{(\Delta x)^2}\right]$$

52.

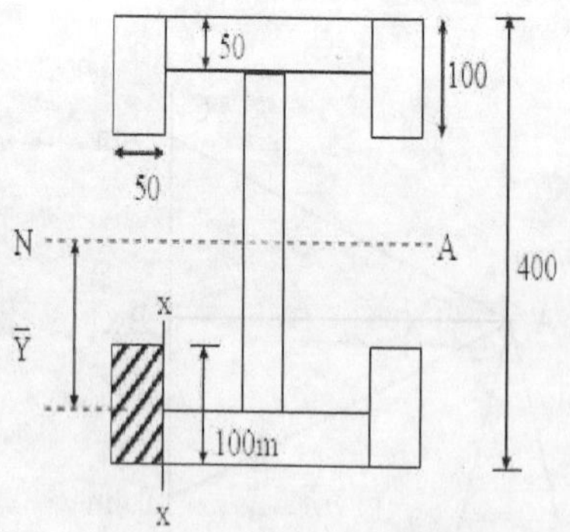

Shear flow at x – x = $\dfrac{SA\bar{y}}{I}$

S = shear force = 8 kN = 8000 N

A = Area of shaded portion = 50 × 100 mm²

$\bar{Y}$ = Distance of centroid of shaded portion from N.A

$= 200 - \dfrac{100}{2} = 150$ mm

I = 1.5 × 10^9 mm⁴

$\Rightarrow$ Shear flow at x – x = $\dfrac{8000 \times 50 \times 1000 \times 150}{1.5 \times 10^9}$

= 40 N/mm

This shear flow acts longitudinally throughout section x – x.

$\Rightarrow$ Shear resisted by nails at P

= shear flow × sparing of nails

= 40 × 60 = 240 N

53.

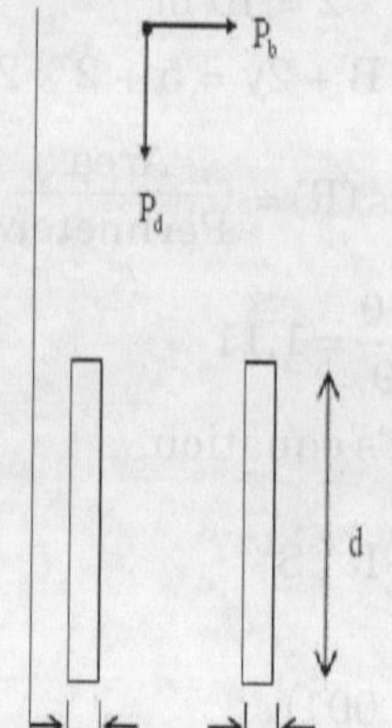

Maximum resultant stress

$= \sqrt{P_d^2 + P_b^2 + 2P_dP_b \cos\theta}$

P_d = Direct shear stress

P_b = Bending stress

θ = 90° (angle between P_d & P_b)

$\Rightarrow$ Maximum resultant stress = $\sqrt{P_d^2 + P_b^2}$

$$P_d = \frac{\text{Force}}{\text{Shear Area}} = \frac{350\text{KN}}{dt + dt}$$

(Shear Area = dt + dt)

d = 500 mm; t = KS = 0.7 × 10 = 7 mm

$$\begin{bmatrix} t = \text{effective throat thickness} \\ S = \text{size of weld} \end{bmatrix}$$

$$\Rightarrow P_d = \frac{350 \times 1000}{2 \times 500 \times 7} = 50 \text{ N/mm}^2$$

$$P_b = \frac{Pe}{z} = \frac{350 \times 1000 \times 100}{\frac{2 \times 7 \times 500^2}{6}} \left[z = \frac{(2t)d^2}{6} \right]$$

$$\Rightarrow \text{Maximum resultant stress} = \sqrt{50^2 + 60^2}$$

$$= 78.1 \text{ N/mm}^2 = 78.1 \text{ MPa}$$

54. Given, $x = \psi \ell n\phi$ & $y = \phi \ell n \psi$

$$\Rightarrow \psi = \frac{x}{\ell n\phi};$$

$$\Rightarrow \psi = \frac{x}{\ell n\left[\frac{y}{\ell n\psi}\right]} \left(\because y = \phi \ell n \psi \Rightarrow \phi = \frac{y}{\ell n\psi}\right)$$

$$\Rightarrow \psi = \frac{x}{\ell n[y] - \ell n[\ell n\psi]}$$

$$\left(\because \ell n\left[\frac{A}{B}\right] = \ell n(A) - \ell n(B)\right)$$

∴ Differentiating ψ partially w.r.t 'x', we have

$$\frac{\partial\psi}{\partial x} = \frac{\left[(\ell n(y) - \ell n[\ell n\psi])(1) - x\left[0 - \frac{1}{\ell n\psi}\frac{1}{\psi}\frac{\partial\psi}{\partial x}\right]\right]}{\left[\ell n(y) - \ell n(\ell n\psi)\right]^2};$$

since y is constant

$$= \frac{\frac{x}{\psi} + x\left[\frac{1}{\ell n\psi}\frac{1}{\psi}\frac{\partial\psi}{\partial x}\right]}{\left(\frac{x}{\psi}\right)2};$$

$$\text{since } \ell ny - \ell n(\ell n\psi) = \ell n\phi = \left[\frac{x}{\psi}\right]$$

$$\Rightarrow \frac{\partial\psi}{\partial x} = \frac{1 + \frac{1}{\ell n\psi}\frac{\partial\psi}{\partial x}}{\frac{x}{\psi}}$$

$$\Rightarrow \frac{x}{\psi}\frac{\partial\psi}{\partial x} = 1 + \frac{1}{\ell n\psi}\frac{\partial\psi}{\partial x}$$

$$\Rightarrow \frac{x}{\psi}\frac{\partial\psi}{\partial x} - \frac{1}{\ell n\psi}\frac{\partial\psi}{\partial x} = 1$$

$$\Rightarrow \frac{\partial\psi}{\partial x}\left[\frac{x}{\psi} - \frac{1}{\ell n\psi}\right] = 1$$

$$\Rightarrow \frac{\partial\psi}{\partial x}\left[\frac{x\ell n\psi - \psi}{\psi\ell n\psi}\right] = 1$$

$$\Rightarrow \frac{\partial\psi}{\partial x} = \frac{\psi\ell n\psi}{\psi\ell n\psi - \psi} = \frac{\ell n\psi}{\frac{x}{\psi}\ell n\psi - 1} = \frac{\ell n\psi}{\ell n\phi\ell n\psi - 1};$$

$$\text{Since } \frac{x}{\psi} = \ell n\phi \Rightarrow \frac{\partial\psi}{\partial x} = \frac{\ell n\psi}{\ell n\phi\ell n\psi - 1}$$

55.

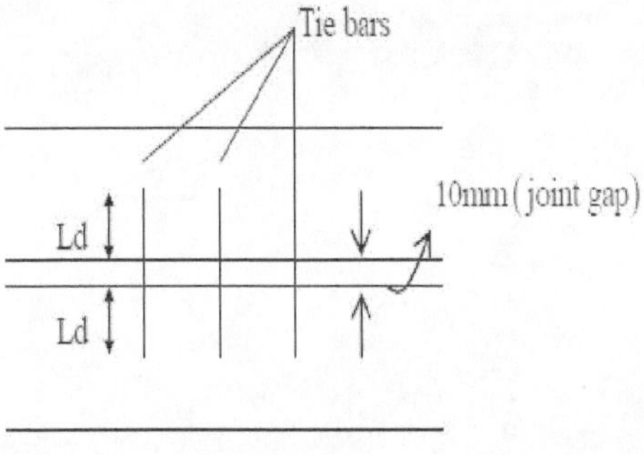

Length of Tie bars

$= 2L_d$ + joint gap

$$= 2L_d + 10 \quad \left[L_d = \frac{\phi\sigma_{st}}{4\tau_{bd}}\right]$$

$$= \frac{2\phi\sigma_{st}}{4\tau_{bd}} + 10 = \frac{2 \times 12 \times 230}{4 \times 2} + 10$$

$$= 700 \text{ mm}$$

SOLVED PAPER - 2020

SET 1

INSTRUCTIONS

1. Total of 65 questions carrying 100 marks, out of which 10 questions carrying a total of 15 marks are in General Aptitude (GA)
2. The Engineering Mathematics will carry around **15% of the total marks,** the General Aptitude section will carry **15% of the total marks** and the **remaining 70% of the total marks.**
3. **Types of Questions**
   (*a*) Multiple Choice Questions (MCQ) carrying 1 or 2 marks each in all papers and sections. These questions are objective in nature, and each will have a choice of four options, out of which the candidate has to mark the correct answer(s).
   (*b*) Numerical Answer Questions of 1 or 2 marks each in all papers and sections. For these questions the answer is a real number, to be entered by the candidate using the virtual keypad. No choices will be shown for these type of questions.
4. For **1-mark** multiple-choice questions, **1/3 marks** will be deducted for a wrong answer. Likewise, for **2-marks** multiple-choice questions, **2/3** marks will be deducted for a wrong answer. There is no negative marking for numerical answer type questions.

Chapter-Wise Analysis

#	Subject	Marks		
		1	2	Total
1	General Aptitude	5	5	15
2	Engineering Mathematics	5	4	13
3	Engineering Mechanics	1	0	1
4	Solid Mechanics	0	1	2
5	Structural Analysis	1	3	7
6	Construction Materials and Management	2	0	2
7	Concrete Structures	1	2	5

#	Subject	Marks		
		1	2	Total
8	Steel Structures	0	1	2
9	Soil Mechanics and Foundation Engineering	1	6	13
10	Fluid Mechanics and Hydraulics	5	2	9
11	Hydrology & Irrigation	2	2	6
12	Environmental Engineering	2	6	14
13	Transportation Engineering	4	2	8
14	Surveying	0	2	2

GENERAL APTITUDE

Q 1 – Q 5 carry one mark each

1. It is a common criticism that most of the academicians live in their _____., so, they are not aware of the real life challenges.

(a) homes (b) ivory towers
(c) glass palaces (d) big flats

2. His hunger for reading is insatiable. He reads indiscriminately. He is most certainly a/an _____ reader.

(a) all-round (b) precocious
(c) voracious (d) wise

3. Select the word that fits the analogy:
Fuse : Fusion :: Use: _______.

(a) Usage (b) User
(c) Uses (d) Usion

4. If 0, 1, 2, ..., 7, 8, 9 are coded as O, P, Q ....., V, W, X, then 45 will be coded as _______.

(a) TS (b) ST
(c) SS (d) SU

5. The sum of two positive numbers is 100. After subtracting 5 from each number, the product of the resulting numbers is 0. One of the original numbers is _______ .

(a) 80 (b) 85
(c) 90 (d) 95

Q 6 – Q 10 carry two marks each

6. The American psychologist Howard Gardener expounds that human intelligence can be subcategorised into multiple kinds, in such a way that individuals differ with respect to their relative competence in each kind. Based on this theory, modern educationists insist on prescribing multi-dimensional curriculum and evaluation parameters that enable development and assessment of multiple intelligences.

Which of the following statements can be inferred from the given text?

(a) Howard Gardner insists that the teaching curriculum and evaluation needs to be multi-dimensional
(b) Howard Gardner wants to develop and assess the theory of multiple intelligences
(c) Modern educationists want to develop and assess the theory of multiple intelligences.
(d) Modern educationists insist that the teaching curriculum and evaluation needs to be multidimensional

7. Five friends P, Q, R, S and T went camping. At night, they had to sleep in a row inside the tent. P, Q and T refused to sleep next to R since he snored loudly. P and S wanted to avoid Q as he usually hugged people in sleep.

Assuming everyone was satisfied with the sleeping arrangements. What is the order in which they slept?

(a) RSPTQ (b) SPRTQ
(c) QRSPT (d) QTSPR

8. Insert seven numbers between 2 and 34, such that the resulting sequence including 2 and 34 is an arithmatic progression. The sum of these inserted seven numbers is _____.

(a) 120 (b) 124
(c) 126 (d) 130

9. The unit's place in 26591749^{110016} is _______.

(a) 1 (b) 3
(c) 6 (d) 9

10. The total expenditure of a family, on different activities in a month, is shown in the pie-chart. The extra money spent on education as compared to transport (in percent) is _______ .

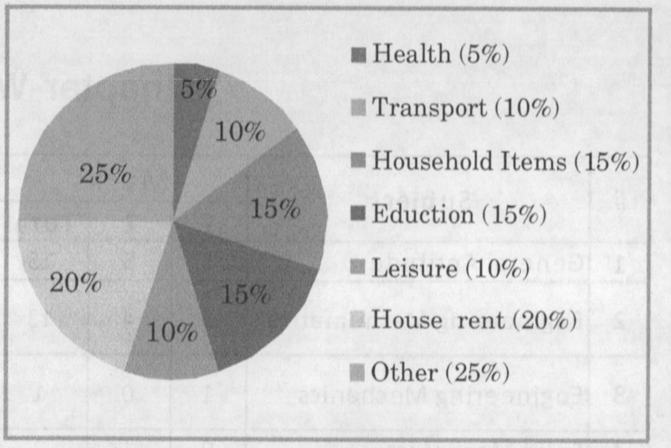

(a) 5 (b) 33.3
(c) 50 (d) 100

CIVIL ENGINEERING

Q 1 – Q 25 carry one mark each

1. In the following partial differential equation, θ is a function of t and z, and D and K are functions of θ

$$D(\theta)\frac{\partial^2\theta}{\partial z^2} + \frac{\partial K(\theta)}{\partial z} - \frac{\partial\theta}{\partial t} = 0$$

The above equation is

(a) a second order linear equation
(b) a second degree linear equation
(c) a second order non-linear equation
(d) a second degree non-linear equation

2. The value of $\lim_{x \to \infty} \frac{x^2 - 5x + 4}{4x^2 + 2x}$ is

(a) 0 (b) $\frac{1}{4}$

(c) $\frac{1}{2}$ (d) 1

3. The true value of $ln(2)$ is 0.69. If the value of $ln(2)$ is obtained by linear interpolation between $ln(1)$ and $ln(6)$, the percentage of absolute error (round off to the nearest integer), is

(a) 35 (b) 48

(c) 69 (d) 84

4. The area of an ellipse represented by an equation $\frac{x^2}{a^2} + \frac{y^2}{b^2} = 1$ is

(a) $\frac{\pi ab}{4}$ (b) $\frac{\pi ab}{2}$

(c) πab (d) $\frac{4\pi ab}{3}$

5. Consider the planner truss shown in the figure (not drawn to the scale)

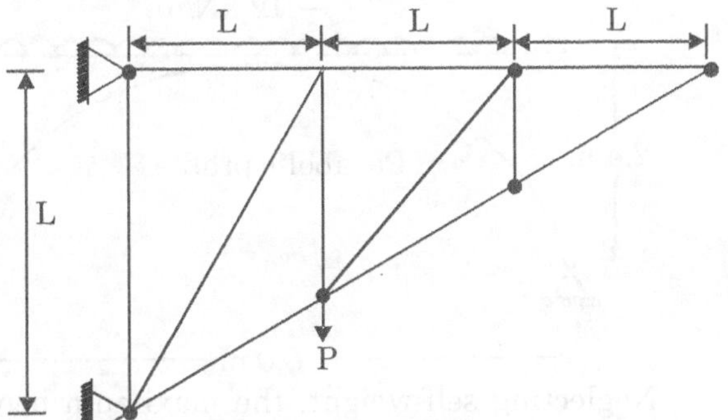

Neglecting self-weight of the members, the number of zero-force members in the truss under the action of the load P, is

(a) 6 (b) 7

(c) 8 (d) 9

6. A reinforcing steel bar, partially embedded in concrete, is subjected to a tensile force P. The figure that appropriately represents the distribution of the magnitude of bond stress (represented as hatched region), along the embedded length of the bar, is

(a) 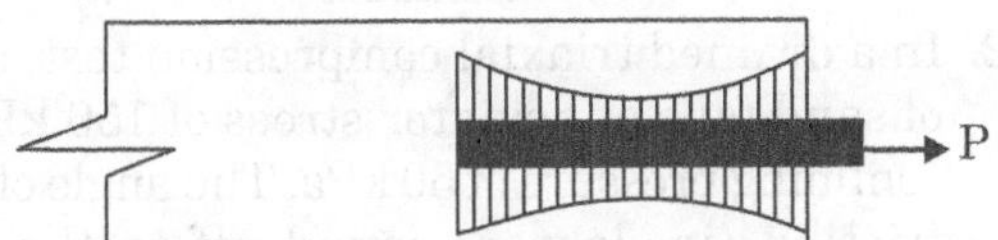

(b) 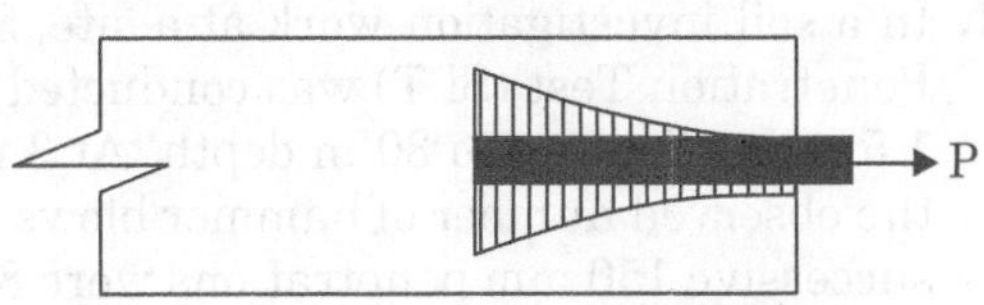

(c) 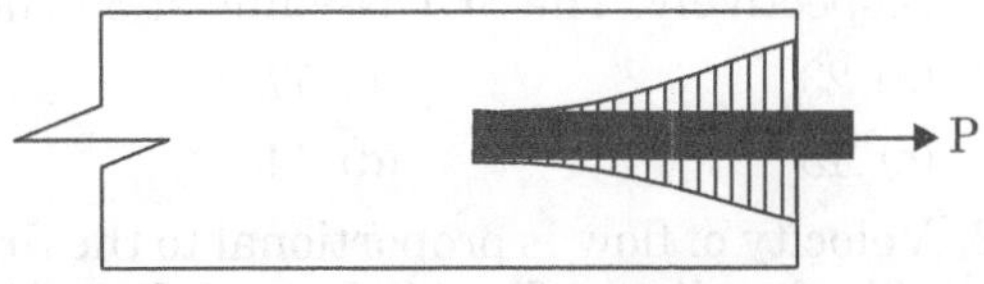

(d) 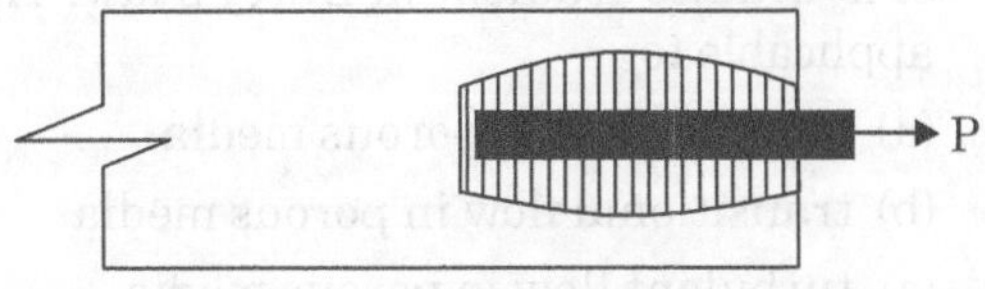

7. In a two-dimensional stress analysis, the state of stress at a point P is $[\sigma] = \begin{bmatrix} \sigma_{xx} & \tau_{xy} \\ \tau_{xy} & \sigma_{yy} \end{bmatrix}$

The necessary and sufficient condition for existence of the state of pure shear at the point P, is

(a) $\sigma_{xx}\sigma_{yy} - \tau_{xy}^2 = 0$ (b) $\tau_{xy} = 0$

(c) $\sigma_{xx} + \sigma_{yy} = 0$ (d) $(\sigma_{xx} - \sigma_{yy})^2 + \tau_{xy}^2 = 0$

8. During the process of hydration of cement, due to increase in Dicalcium Silicate (C_2S) content in cement clinker, the heat of hydration

(a) increases

(b) decreases

(c) intially decreases and then increases

(d) does not change

9. The Los Angeles test for stone aggregates is used to examine

(a) abrasion resistance (b) crushing strength

(c) soundness (d) specific gravity

10. Which one of the following statements is NOT correct?

(a) A clay deposit with a liquidity index greater than unity is in a state of plastic consistency.

(b) The cohesion of normally consolidated clay is zero when triaxial test is conducted under consolidated undrained condition.

(c) The ultimate bearing capacity of a strip foundation supported on the surface of sandy soil increases in direct proportion to the width of footing.

(d) In case of a point load, Boussinesq's equation predicts higher value of vertical stress at a point directly beneath the load as compared to Westergaard's equation.

11. In a soil investigation work at a site, Standard Penetration Test (SPT) was conducted at every 1.5 m interval up to 30 m depth. At 3 m depth, the observed number of hammer blows for three successive 150 mm penetrations were 8, 6 and 9 respectively. The SPT N-value at 3 m depth, is

(a) 23 (b) 17
(c) 15 (d) 14

12. Velocity of flow is proportional to the first power of hydraulic gradient in Darcy's law. This law is applicable to

(a) laminar flow in porous media
(b) transitional flow in porous media
(c) turbulent flow in porous media
(d) laminar as well as turbulent flow in porous media

13. A body floating in a liquid is in a stable state of equilibrium if its

(a) metacentre lies above its centre of gravity
(b) metacentre lies below its centre of gravity
(c) metacentre coincides with its centre of gravity
(d) centre of gravity is below its centre of bouyancy

14. Uniform flow with velocity U makes an angle θ with the y-axis, as shown in the figure.

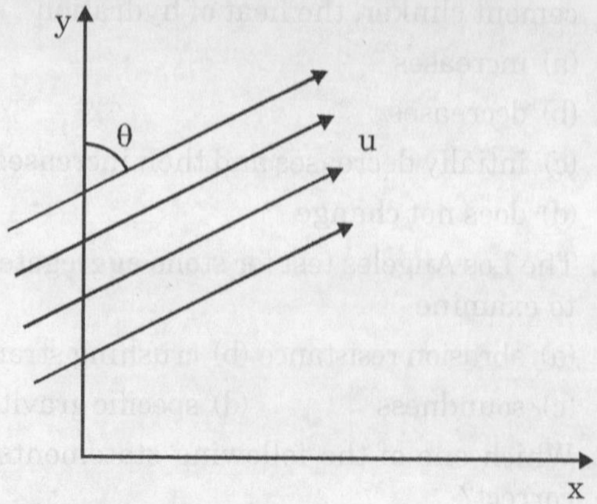

The velocity potential (ϕ), is

(a) $\pm U(x \sin\theta + y \cos\theta)$
(b) $\pm U(y \sin\theta - x \cos\theta)$
(c) $\pm U(x \sin\theta - y \cos\theta)$
(d) $\pm U(y \sin\theta + x \cos\theta)$

15. The data for an agricultural field for a specific month are given below:

Pan Evaporation = 100 mm

Effective Rainfall = 20 mm (after deducting losses due to runoff and deep percolation)

Crop Coefficient = 0.4

Irrigation Efficiency = 0.5

The amount of irrigation water (in mm) to be applied to the field in that month, is

(a) 0 (b) 20
(c) 40 (d) 80

16. During chlorination process, aqueous (aq) chlorine reacts rapidly with water to from Cl^-, HOCl and H^+ as shown below.

$$Cl_2(aq) + H_2O \rightleftharpoons HOCl + Cl^- + H^+$$

The most active disinfectant in the chlorination process from amongst the following is

(a) H^+ (b) HOCl
(c) Cl^- (d) H_2O

17. An amount of 35.67 mg HCl is added to distilled water and the total solution volume is made to one litre. The atomic weights of H and Cl are 1 and 35.5, respectively. Neglecting the dissociation of water, the pH of the solution, is

(a) 3.50 (b) 3.01
(c) 2.50 (d) 2.01

18. The probability that a 50 year flood may NOT occur at all during 25 years life of a project (round off to two decimal places), is _____.

19. A planar elastic structure is subjected to uniformly distributed load, as shown in the figure (not drawn to the scale)

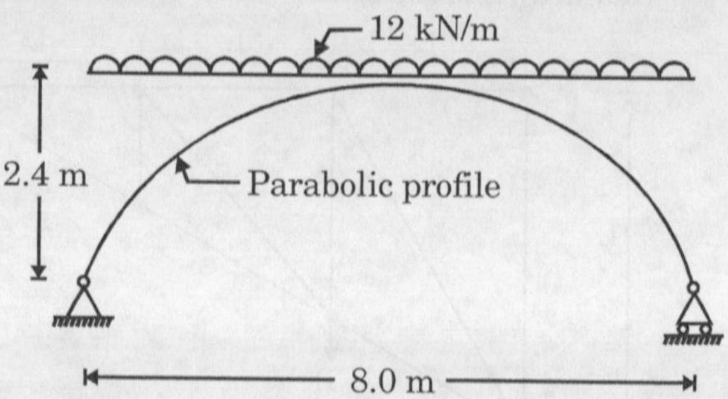

Neglecting self-weight, the maximum bending moment generated in the structure (in kN.m, round off to the nearest integer), is _____.

20. In an urban area, a median is provided to seperate the opposing streams of traffic. As per IRC : 86-1983, the desirable minimum width (in m, expressed as integer) of the median, is _____.

21. A road in a hilly terrain is to be laid at a gradient of 4.5%. A horizontal curve of radius 100 m is laid at a location on this road. Gradient needs to be eased due to combination of curved horizontal and vertical profiles of the road. As per IRC, the compensated gradient (in %, round off to one decimal place), is _______.

22. In a drained triaxial compression test, a sample of sand fails at deviator stress of 150 kPa under confining pressure of 50 kPa. The angle of internal friction (in degree, round off to the nearest integer) of the sample, is _____.

23. A fully submerged infinite sandy slope has an inclination of 30° with the horizontal. The saturated unit weight and effective angle of internal friction of sand are 18 kN/m^3 and 38° respectively. The unit weight of water is 10 kN/m^3. Assume that the seepage is parallel to the slope. Against shear failure of the slope, the factor of safety (round off to two decimal places) is ______.

24. A 4 m wide rectangular channel carries 6 m^3/s of water. The Manning's 'n' of the open channel is 0.02. Considering g = 9.81 m/s^2, the critical velocity of flow (in m/s, round off to two decimal places) in the channel, is ______.

25. A river has a flow of 1000 million litres per day (MLD), BOD_5 of 5 mg/litre and Dissolved Oxygen (DO) level of 8 mg/litre before receiving the wastewater discharge at a location. For the existing environmental conditions, the saturation DO level is 10 mg/litre in the river. Wastewater discharge of 100 MLD with the BOD_5 of 200 mg/litre and DO level of 2 mg/litre falls at that location. Assuming complete mixing of wastewater and river water, the immediate DO deficit (in mg/litre, round off to two decimal places), is ______.

Q 26 – Q 55 carry two marks each

26. For the Ordinary Differential Equation

$\frac{d^2x}{dt^2} - 5\frac{dx}{dt} + 6x = 0$, with initial conditions

$x(0) = 0$ and $\frac{dx}{dt}(0) = 10$, the solution is

(a) $-5e^{2t} + 6e^{3t}$ (b) $5e^{2t} + 6e^{3t}$

(c) $-10e^{2t} + 10e^{3t}$ (d) $10e^{2t} + 10e^{3t}$

27. A continuous function f(x) is defined. If the third derivative at x_i is to be computed by using the fourth order central finite-divided-difference scheme (with step length = h), the correct formula is

(a) $$f'''(x_i) = \frac{-f(x_{i+3}) - 8f(x_{i+2}) - 13f(x_{i+1}) + 13f(x_{i-1}) - 8f(x_{i-2}) + f(x_{i-3})}{8h^3}$$

(b) $$f'''(x_i) = \frac{f(x_{i+3}) - 8f(x_{i+2}) - 13f(x_{i+1}) + 13f(x_{i-1}) + 8f(x_{i-2}) + f(x_{i-3})}{8h^3}$$

(c) $$f'''(x_i) = \frac{-f(x_{i+3}) - 8f(x_{i+2}) - 13f(x_{i-1}) + 13f(x_{i-1}) + 8f(x_{i-2}) - f(x_{i-3})}{8h^3}$$

(d) $$f'''(x_i) = \frac{f(x_{i+3}) - 8f(x_{i+2}) - 13f(x_{i+1}) + 13f(x_{i-1}) + 8f(x_{i-2}) - f(x_{i-3})}{8h^3}$$

28. Distributed load(s) of 50 kN/m may occupy any position (s) (either continuously or in patches) on the girder PQRST as shown in the figure (not drawn to the scale)

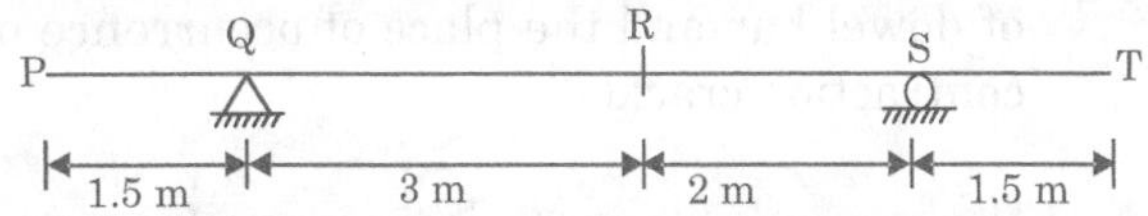

The maximum negative (hogging) bending moment (in kN.m) that occurs at point R, is

(a) 22.50 (b) 56.25

(c) 93.75 (d) 150.00

29. A rigid weightless platform PQRS shown in the figure (not drawn to the scale) can slide freely in the vertical direction. The platform is held in position by the weightless member OJ and four weightless, frictonless rollers. Points O and J are pin connections. A block of 90 kN rests on the platform as shown in the figure.

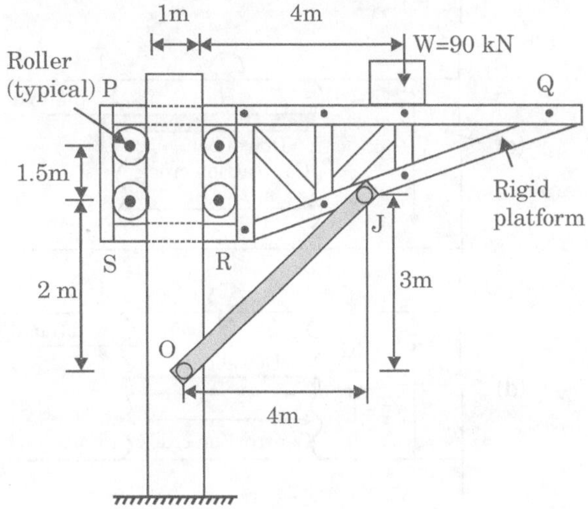

The magnitude of horizontal component of the reaction (in kN) at pin O, is

(a) 90 (b) 120

(c) 150 (d) 180

30. A cantilever beam PQ of uniform flexural rigidity (EI) is subjected to a concentrated moment M at R as shown in the figure.

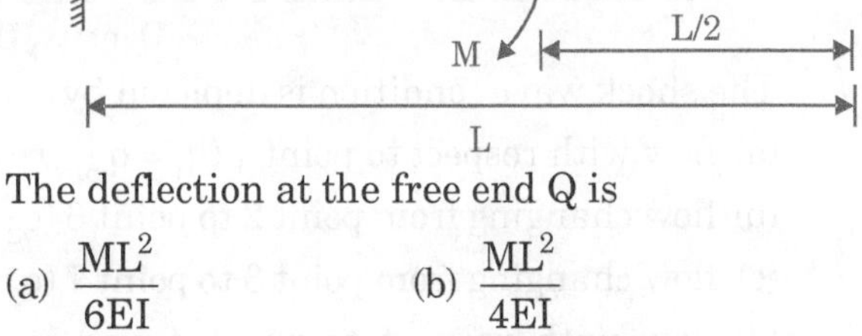

The deflection at the free end Q is

(a) $\frac{ML^2}{6EI}$ (b) $\frac{ML^2}{4EI}$

(c) $\frac{3ML^2}{8EI}$ (d) $\frac{3ML^2}{4EI}$

31. A dowel bar is placed at a contraction joint. When contraction occurs, the concrete slab cracks at predetermined location (s). Identify the arrangement, which shows the correct placement of dowel bar and the place of occurrence of the contraction crack(s).

(a) 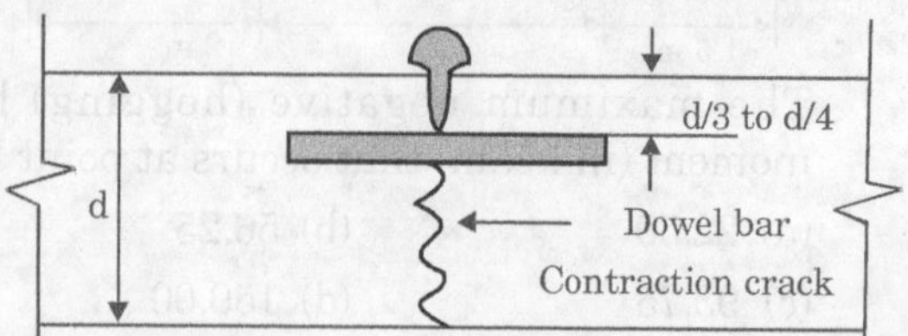

(b) 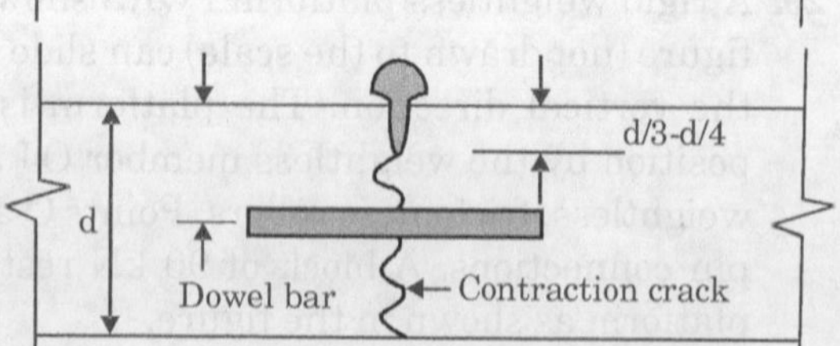

(c) 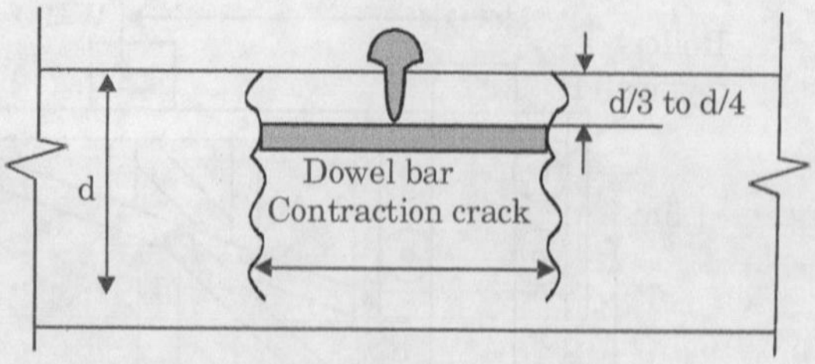

(d) 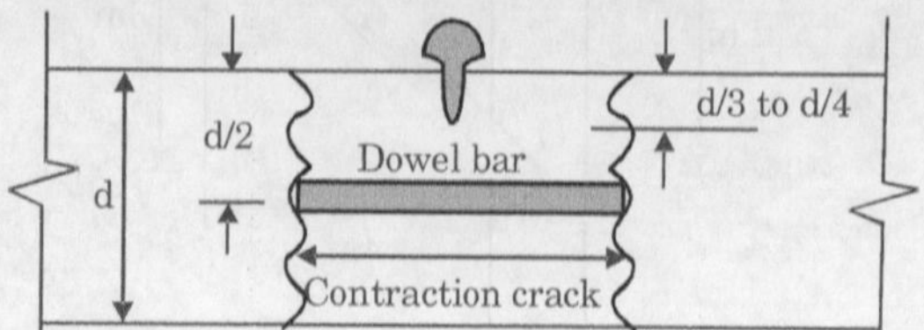

32. The relationship between traffic flow rate (q) and density (D) is shown in the figure.

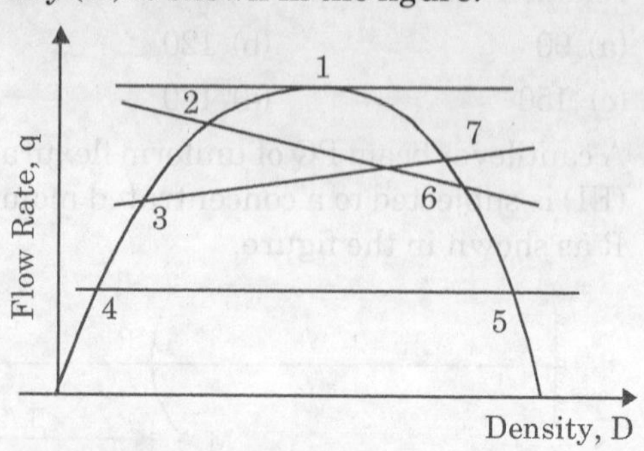

The shock wave condition is depicted by

(a) flow with respect to point 1 ($q_1 = q_{max}$)

(b) flow changing from point 2 to point 6 ($q_2 > q_6$)

(c) flow changing from point 3 to point 7 ($q_3 < q_7$)

(d) flow with respect to point 4 and point 5 ($q_4 = q_5$)

33. The appropriate design length of a clearway is calculated on the basis of 'Normal Take-off' condition. Which one of the following options correctly depicts the length of the clearway? (*Note: None of the options are drawn to scale*)

(a) 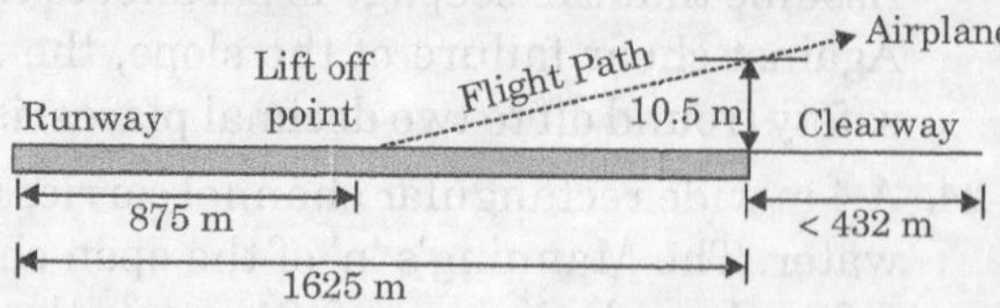

(b) 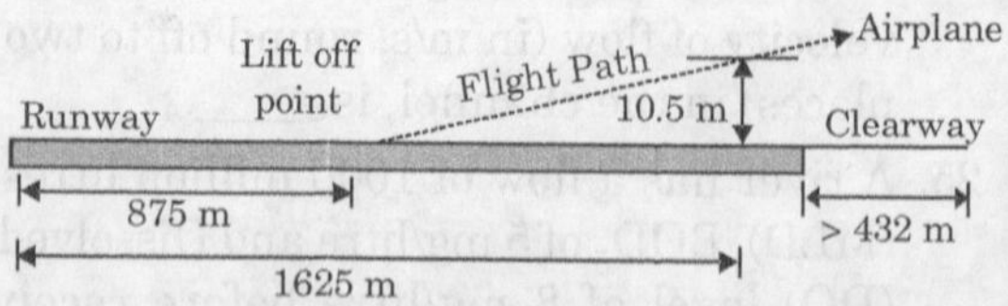

(c) 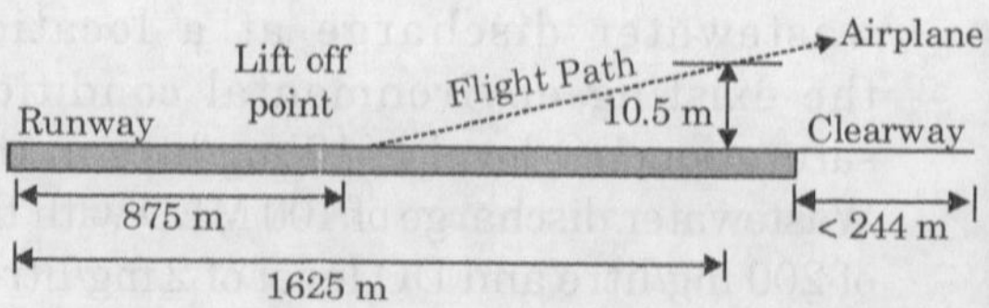

(d) 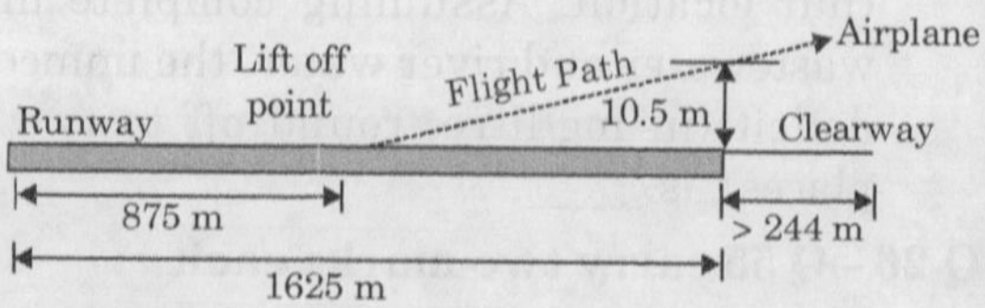

34. The total stress paths corresponding to different loading conditions, for a soil specimen under the isotropically consolidated stress state (O), are shown below.

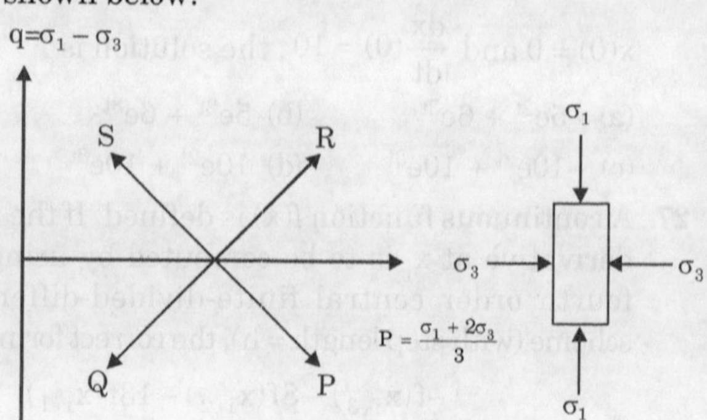

Stress Path	**Loading Condition**
OP	I-Compression loading (σ_1-increasing; σ_3-constant)
OQ	II-Compression unloading (σ_1-constant; σ_3-decreasing)
OR	III-Extension unloading (σ_1-decreasing; σ_3-constant)
OS	IV-Extension loading (σ_1-constant; σ_3-increasing)

The correct match between the stress paths and the listed loading conditions, is

(a) OP-I, OQ-II, OR-IV, OS-III

(b) OP-IV, OQ-III, OR-I, OS-II

(c) OP-III, OQ-II, OR-I, OS-IV

(d) OP-I, OQ-III, OR-II, OS-IV

35. The soil profile at a site up to a depth of 10 m is shown in the figure (not drawn to the scale). The soil is preloaded with a uniform surcharge (q) of 70 kN/m^2 at the ground level. The water table is at a depth of 3m below ground level. The soil unit weight of the respective layers is shown i the figure. Consider unit weight of water as 9.81 kN/m^3 and assume that the surcharge (q) is applied instantaneously.

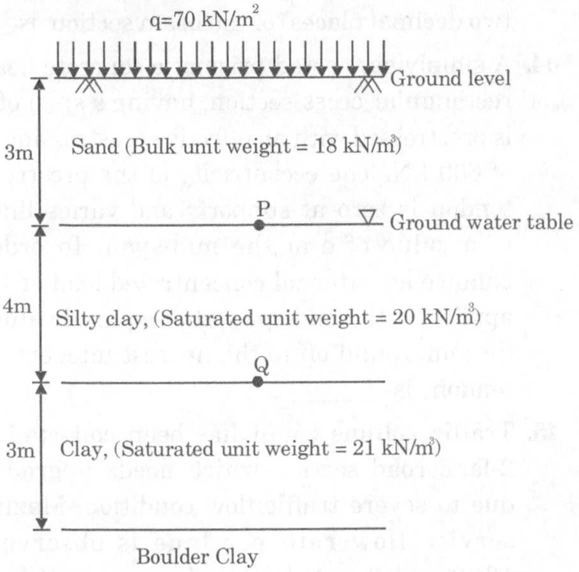

Immediately after preloading, the effective stresses (in kPa) at points P and Q, respectively, are

(a) 124 and 204 (b) 36 and 90

(c) 36 and 126 (d) 54 and 95

36. Water flows at the rate of 12 m^3/s in a 6 m wide rectangular channel. A hydraulic jump is formed in the channel at a point where the upstream depth is 30 cm (just before the jump). Considering acceleration due to gravity as 9.81 m/s^2 and density of water as 1000 kg/m^3, the energy loss in the jump is

(a) 114.2 kW (b) 114.2 MW

(c) 141.2 h.p. (d) 141.2 J/s

37. A water supply scheme transports 10 MLD (Million Litres per Day) water through a 450 mm diameter pipeline for a distance of 2.5 km. A chlorine dose of 3.50 mg/litre is applied at the starting point of the pipeline to attain a certain level of disinfection at the downstream end. It is decided to increase the flow rate from 10 MLD to 13 MLD in the pipeline. Assume exponent for concentration, n = 0.86. With this increased flow, in order to attain the same level of disinfection, the chlorine dose (in mg/litre) to be applied at the starting point should be

(a) 3.95 (b) 4.40

(c) 4.75 (d) 5.55

38. An open traverse PQRST is surveyed uisng theodolite and the consecutive coordinates obtained are given in the table

Line	Consecutive Coordinates			
	Northing (m)	Southing (m)	Easting (m)	Westing (m)
PQ	110.2	–	45.5	–
QR	80.6	–	–	60.1
RS	–	90.7	–	70.8
ST	–	105.4	55.5	–

If the independent coordinates (Northing, Easting) of station P are (400 m, 200 m), the independent coordinates (in m) of station T, are

(a) 194.7, 370.1 (b) 205.3, 429.9

(c) 394.7, 170.1 (d) 405.3, 229.9

39. If C represents a line segment between (0, 0, 0) and (1, 1, 1) in Cartesian coordinate system, the value (expressed as integer) of the line integral

$$\int_C [(y + z)dx + (x + z)dy + (x + y)dx]$$

is ________.

40. Consider the system of equations

$$\begin{vmatrix} 1 & 3 & 2 \\ 2 & 2 & -3 \\ 4 & 4 & -6 \\ 2 & 5 & 2 \end{vmatrix} \begin{vmatrix} x_1 \\ x_2 \\ x_3 \end{vmatrix} = \begin{vmatrix} 1 \\ 1 \\ 2 \\ 1 \end{vmatrix}$$

The value of x_3 (round off to the nearest integer), is ________.

41. A rigid, uniform, weightless, horizontal bar is connected to three vertical members P, Q and R as shown in the figure (not drawn to the scale). All three members have identical axial stiffness of 10 kN/mm. The lower ends of bars P and R rest on a rigid horizontal surface. When no load is applied, a gap of 2 mm exists between the lower end of the bar Q and the rigid horizontal surface. When a vertical load W is placed on the horizontal bar in the downward direction, the bar still remains horizontal and gets displaced by 5 mm in the vertically downward direction.

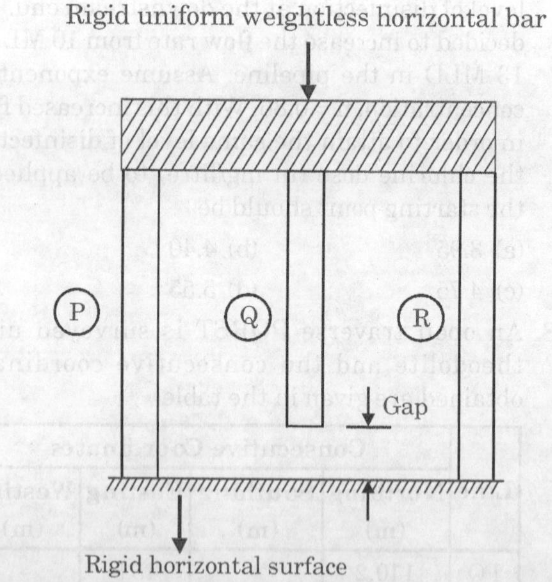

The magnitude of the load W (in kN, round off to the nearest interger), is ______.

42. The flange and web plates of the doubly symmetric built-up section are connected by continuous 10 mm thick fillet welds as shown in the figure (not drawn to the scale). The moment of inertia of the section about its principal axis X-X is 7.73×10^6 mm^4. The permissible shear stress in the fillet welds is 100 N/mm^2. The design shear strength of the section is governed by the capacity of the fillet welds.

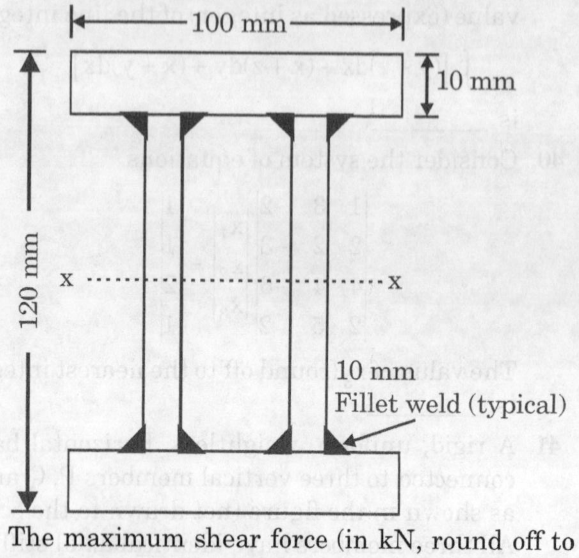

The maximum shear force (in kN, round off to one decimal place) that can be carried by the section, is ______

43. The singly reinforced concerete beam section shown in the figure (not drawn to the scale) is made of M25 grade concrete and Fe500 grade reinforcing steel. The total cross-sectional area of the tension steel is 942 mm^2.

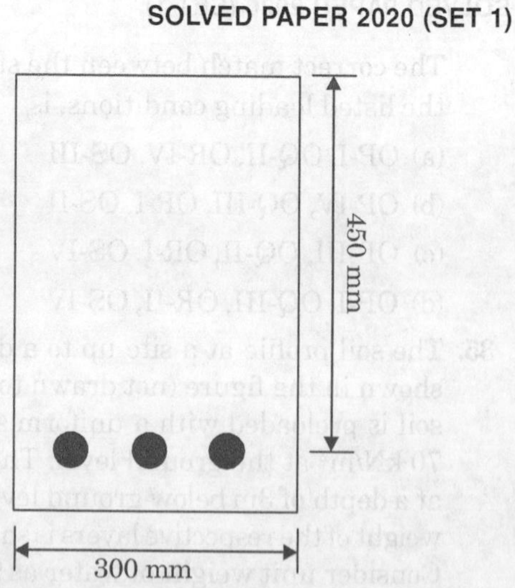

As per Limit State Design of IS 456 : 2000, the design moment capacity (in kN.m, round off to two decimal places) of the beam section, is ______.

44. A simply supported prismatic concrete beam of rectangular cross-section, having a span of 8 m, is prestressed with an effective prestressing force of 600 kN. The eccentricity of the prestressing tendon is zero at supports and varies linearly to a value of e at the mid-span. In order to balance an external concentrated load of 12 kN applied at the mid-span, the required value of e (in mm, round off to the nearest integer) of the tendon, is ______.

45. Traffic volume count has been collected on a 2-lane road section which needs upgradation due to severe traffic flow condition. Maximum service flow rate per lane is observed as 1280 veh/h at level of service 'C'. The Peak Hour Factor is reported as 0.78125. Historical traffic volume count provides Annual Average Daily Traffic as 12270 veh/day. Directional split of the traffic flow is observed to be 60 : 40. Assuming that traffic stream consists of 'All cars' and all drivers are 'Regular Commuters', the number of extra lane(s) (round off to the next higher integer) to be provided, is ______.

46. A vertical retaining wall of 5 heights has to support soil having unit weight of 18 kN/m^3, effective cohesion of 12 kN/m^2, and effective friction angle of 30°. As per Rankine's earth pressure theory and assuming that a tension crack has occured, the lateral active thrust on the wall per meter length (in kN/m, round off to two decimal places), is ______.

47. Water flows in the upward direction in a tank through 2.5 m thick sand layer as shown in the figure. The void ratio and specific gravity of sand are 0.58 and 2.7, respectively. The sand is fully saturated. Unit weight of water is 10 kN/m^3.

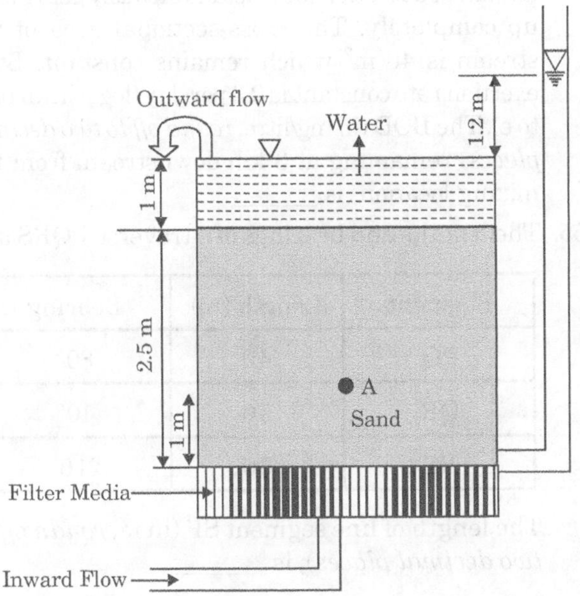

The effective stress (in kPa, round off to two decimal places) at point A, located 1 m above the base of tank, is ______.

48. A 10 m thick clay layer is resting over a 3 m thick sand layer and is submerged. A fill of 2 m thick sand with unit weight of 20 kN/m^3 is placed above the clay layer to accelerate the rate of consolidation of the clay layer. Coefficient of consolidation of clay is 9×10^{-2} m^2/year and coefficient of volume compressibility of clay is 2.2×10^{-4} m^2/kN. Assume Taylor's relation between time factor and average degree of consolidation.

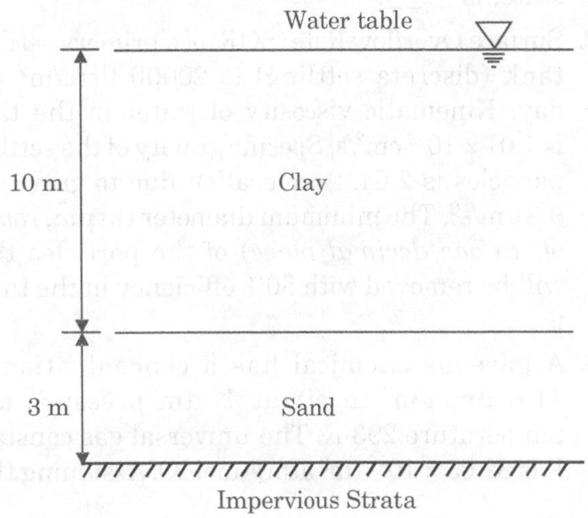

The settlement (in mm, round off to two decimal places) of the clay layer, 10 years after the construction of the fill, is ______.

49. Three reservoirs P, Q and R are interconnected by pipes as shown in the figure (not drawn to the scale). Piezometric head at the junction S of the pipes is 100 m. Assume acceleration due to gravity as 9.81 m/s^2 and density of water as 1000 kg/m^3. The length of the pipe from junction S to the inlet of reservoir R is 180 m.

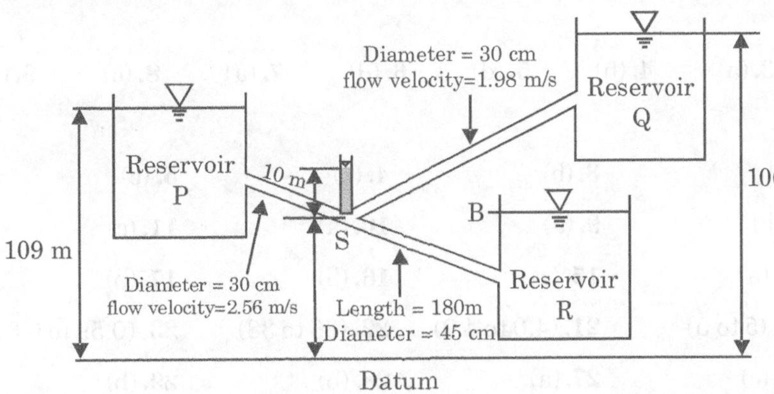

Considering head loss only due to friction (with friction factor of 0.03 for all the pipes), the height of water level in the lowermost reservoir R (in m, round off to one decimal place) with respect to the datum, is ______.

50. In a homogeneous unconfined aquifer of area 3.00 km^2, the water table was at an elevation of 102.00 m. After a natural recharge of volume 0.90 million cubic meter (Mm^3), the water table rose to 103.20 m. After this recharge, ground water pumping took place and the water table dropped down to 101.20 m. The volume of ground water pumped after the natural recharge, expressed (in Mm^3 and round off to two decimal places), is ____.

51. A circular water tank of 2 m diameter has a circular orifice of diameter 0.1 m at the bottom. Water enters the tank steadily at a flow rate of 20 litre/s and escapes through the orifice. The coefficient of discharge of the orifice is 0.8. Consider the acceleration due to gravity as 9.81 m/s^2 and neglect frictional losses. The height of the water level *(in m, round off to two decimal places)* in the tank at the steady state, is _____.

52. Surface Overflow Rate (SOR) of a primary settling tank (discrete settling) is 20000 litre/m^2 per day. Kinematic viscosity of water in the tank is 1.01×10^{-2} cm^2/s. Specific gravity of the settling particles is 2.64. Acceleration due to gravity is 9.81 m/s^2. The minimum diameter (in µm, *round off to one decimal place*) of the particles that will be removed with 80% efficiency in the tank, is _____.

53. A gaseous chemical has a concentration of 41.6 µmol/m^3 in air at 1 atm pressure and temperature 293 K. The universal gas constant R is 82.05×10^{-6} (m^3 atm)/(mol K). Assuming that ideal gas law is valid, the concentration of the gaseous chemical (in ppm, *round off to one decimal place*), is _____.

54. A stream with a flow rate of 5 m^3/s is having an ultimate BOD of 30 mg/litre. A wastewater discharge of 0.20 m^3/s having BOD_5 of 500 mg/litre joins the stream at a location and instantaneously gets mixed up completely. The cross-sectional area of the stream is 40 m^2 which remains constant. BOD exertion rate constant is 0.3 per day (logarithm base to e). The BOD (in mg/litre, *round off to two decimal places)* remaining at 3 km downstream from the mixing location, is _____.

55. The lengths and bearings of a traverse PQRS are:

Segment	Length (m)	Bearing
PQ	40	80°
QR	50	10°
RS	30	210°

The length of line segment SP (in m, *roudn off to two decimal places*), is _____

ANSWERS

General Aptitude

1. (b) **2.** (c) **3.** (a) **4.** (b) **5.** (d) **6.** (d) **7.** (a) **8.** (c) **9.** (a) **10.** (c)

Civil Engineering

1. (c) **2.** (b) **3.** (b) **4.** (c) **5.** (c) **6.** (c)
7. (c) **8.** (b) **9.** (a) **10.** (a) **11.** (c) **12.** (a)
13. (a) **14.** (a) **15.** (c) **16.** (b) **17.** (b) **18.** (0.59 to 0.61)
19. (97 to 97) **20.** (5 to 5) **21.** (4.0 to 4.0) **22.** (36 to 38) **23.** (0.59 to 0.61) **24.** (2.35 to 2.55)
25. (2.45 to 2.65) **26.** (c) **27.** (a) **28.** (b) **29.** (b) **30.** (c)
31. (b) **32.** (b) **33.** (c) **34.** (b) **35.** (d) **36.** (a)
37. (c) **38.** (c) **39.** (3 to 3) **40.** (3 to 3) **41.** (129 to 131) **42.** (393.5 to 399.1)
43. (158.00 to 158.60) **44.** (39 to 41) **45.** (1 to 1) **46.** (21.00 to 22.00)
47. (8.70 to 9.30) **48.** (18.50 to 19.50) **49.** (97.2 to 97.8) **50.** (1.40 to 1.60) **51.** (0.50 to 0.54)
52. (14.0 to 15.0) **53.** (0.9 to 1.1) **54.** (49.00 to 50.00) **55.** (44.00 to 45.00)

EXPLANATIONS

GENERAL APTITUDE

1. Ivory towers means lack of concern with or interest in practical matters.
2. Voracious means hungry for knowledge.
3. Verb and noun analogy.
4. O – 0, P – 1, Q – 2, R – 4, T – 5, U – 6, V – 7, W – 8, X – 9

   So, ST = 45
5. **As Given that :**

   a + b = 100

   and (a – 5) (b – 5) = 0

   So, either a = 5 or b = 5

   Since 5 is not in any of the option.

   If a = 5, then b = 95 or if b = 5, then a = 95

   Hence of the original number is 95.
6. The last two lines support the answer.
7. P, Q & T refused to sleep next to R

   So option (c, b, d) eliminated

   Only option (a) is satisfies the given conditions in the paragraph.
8. 2, –, –, –, –, –, –, –, 34

   $t_1 = a = 2$ $\quad$ $t_a = a + (n-1)d = 34$

   $= 2 + 8d = 34$

   $d = 4$

   So, 2, 6, 10, 14, 18, 22, 26, 30, 34

   Sum of inserted 7 AP numbers

   $$= \frac{n}{2}(\text{I}^{st}\text{ terms + last term}) = \frac{n}{2}[a + (a + 6d)]$$

   $$= \frac{7}{2}(6 + 30) = 18 \times 7 = 126$$
9. Units place in $26591749^{110016} = 9^{110016}$

   $= 9^{even} = 1 \qquad (\because 9^2 = 81)$

   Hence, answer will be 1.
10. Money spent on Education $= \frac{15}{100} \times 100 = 15\%$

    Money spent on transport $= \frac{10}{100} \times 100 = 10\%$

    So, required % of extra money spent on education compared to transport :

    $$= \frac{15-10}{10} \times 100$$

    $= 50\%$

CIVIL ENGINEERING

1. As given that, the partial differential equation :

   $$D(\theta)\frac{\delta^2\theta}{\delta z^2} + \frac{\delta k(\theta)}{\delta z} - \frac{\delta\theta}{\delta t} = 0$$

   The 1st term of given partial differential equation contains product of dependent variable with its derivative, So it is non-linear and also we have 2nd order derivative, Hence, its order is two i.e., 2nd order non-linear equation.
2. $\lim\limits_{x\to\infty} \frac{x^2\ 5x + 4}{4x^2 + 2x}$

   By L-hospital rule :

   $$= \lim_{x\to\infty} \frac{x^2\left(1 - \frac{5}{x} + \frac{4}{x^2}\right)}{x^2\left(4 + \frac{2}{x}\right)}$$

   $$= \frac{1 - 0 + 0}{4 + 0} = \frac{1}{4}$$
3. The linear interpolation for two data points (1, *l*n(1), (6, *l*n (6))

   $$\frac{y - \ln(1)}{\ln(6) - \ln(1)} = \frac{x-1}{5}$$

   $$y = \left(\frac{x-1}{5}\right)(\ln(6) - \ln(1))$$

   $$y = \left(\frac{x-1}{5}\right)\ln(6)$$

   $$y = \frac{\ln(6)}{5} = 0.358 = A \text{ (approximate value)}$$

   $T = ln(2) = 0.69$ (exact value)

   Absolute error $= |(T - A)| = |0.69 - 0.358| = 0.332$

   $$\% \text{ error} = \frac{T - A}{T} \times 100$$

   $$= \frac{0.332}{0.69} \times 100 = 48.11\% = 48 \text{ (approx)}$$
4. Area of the ellipse $\frac{x^2}{a^2} + \frac{y^2}{b^2} = 1$

   $$\text{Area} = \iint (1)dy\,dx$$

   $$= \int_{x=-a}^{a} \int_{y=-\frac{b}{a}\sqrt{-x^2+a^2}}^{+\frac{b}{a}\sqrt{-x^2+a^2}}$$

   $$= 4\int_{x=0}^{a} \int_{y=0}^{+\frac{b}{a}\sqrt{a^2-x^2}} (1)dydx = \pi ab$$

5.

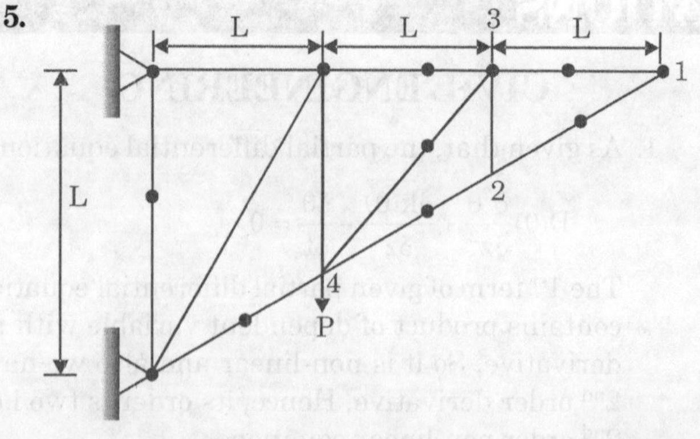

Supports at A & B are hinged supports, member AB will not undergo change in length, then strain is zero, stress is zero & force in AB also zero.

As $\Delta_{AB} = 0$, hence $F_{AB} = 0$

Total number of zero force member = 8

6. Probable variation of anchorage bond stress

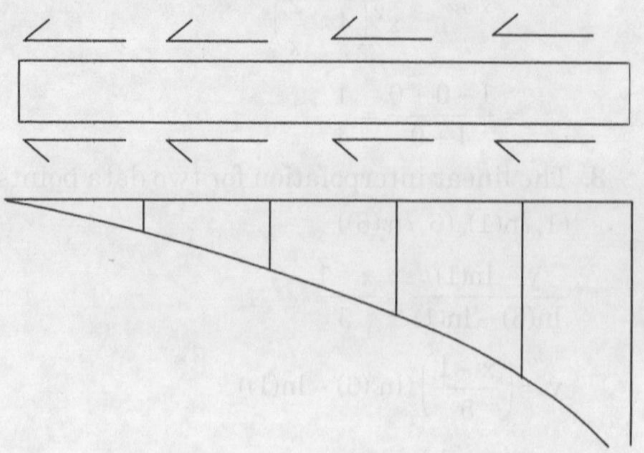

Note: Max at the junction of support and beam.

7.

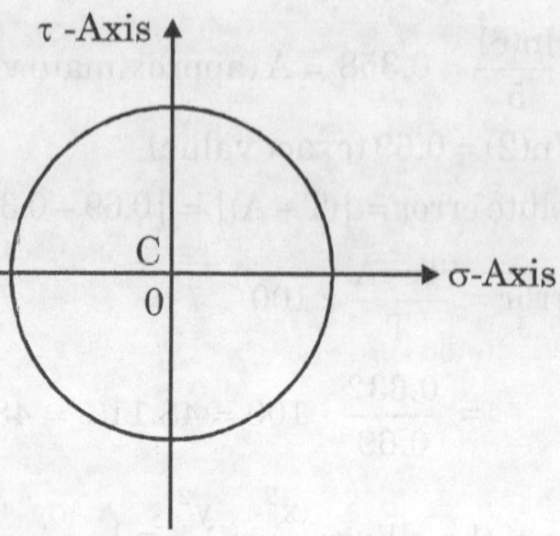

$\sigma_{xx} + \sigma_{yy} = 0$

Centre of Mohr circle coincide with origin.

8. Among all the Bogues compounds, C_2S has the least heat of hydration. So, the C_2S proportion increases, the heat of hydration of cement decreases.

9. Los Angeles for stone aggregate is used to examine the abrasion resistance.

10. By taking each options.

(a) Liquidity Index, $I_L = \dfrac{w_n - w_p}{w_L - w_P}$

If the natural w.c. of soil, $w_n > w_L$ then the I_L is > 1 and soil is in liquid state.

Hence, statement (a) is wrong

(b) In the case of NC clay in CU test, the failure envelope is a shown below.

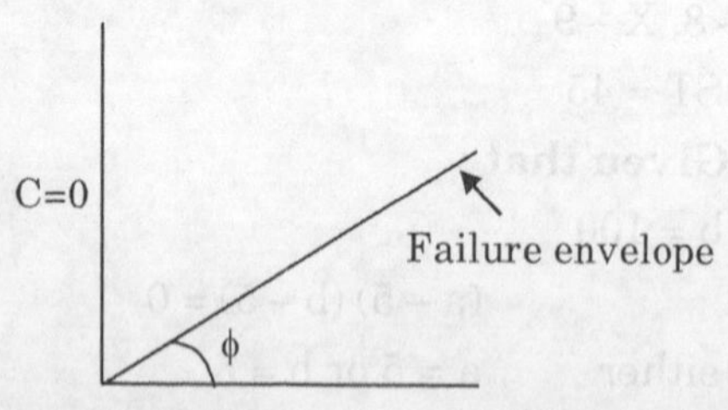

Hence statement (b) is correct

(c) For strip footing resting on surface of sandy soil, q_u is :

$q_u = 0.5\,\gamma B N_\gamma ...$

$\therefore$ Hence statement (c) is correct.

(d) Vertical stress at any depth directly beneath the load is as shown below.

Boussinesq's equation : $\sigma_z = \dfrac{Q}{z^2} \cdot \dfrac{3}{2\pi}$

Westergaard's equation : $\sigma_z = \dfrac{Q}{z^2} \cdot \dfrac{1}{\pi}$

Hence statement (d) is correct.

11. Given : At 3m depth

$N_1 = 8 \rightarrow$ 1st 15 cm

$N_2 = 6 \rightarrow$ 2nd 15 cm

$N_3 = 9 \rightarrow$ 3rd 15 cm

Standard penetration number is the total number of Blows required for last 30 cm by neglecting First 15 cm.

So, observed SPT number.

$N = N_2 + N_3$

$= 6 + 9 = 15$

12. From Darcy's law $V \propto i$

Assumptions of Darcy's law:

- Flow through soil is laminar
- The flow is continuous and steady.
- Soil is saturated

13. For stability of floating body, stable equilibrium is attained if GM > 0

$\Rightarrow$ Meta centre (M) lies above centre of gravity (G).

14.

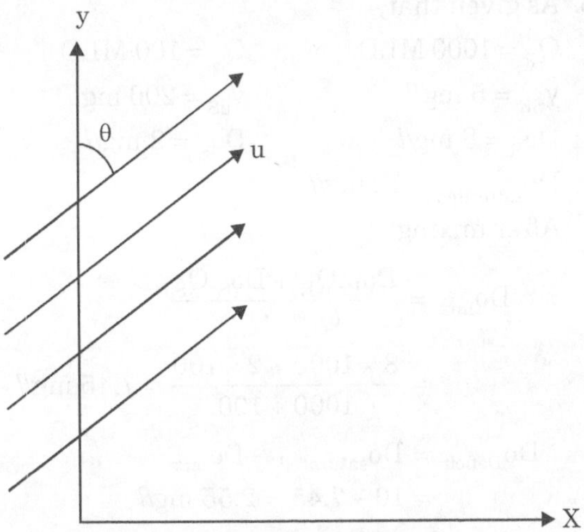

The velocity in x-depth:

$U_x = U \sin\theta$

The velocity in y-depth:

$U_y = U \cos\theta$

$U_x = \frac{-2\phi}{2x} \Rightarrow U \sin\theta = -\frac{2\phi}{2x}$

$\Rightarrow \phi = -U \times \sin\theta + \phi(y)$

$U_y = \frac{-2\phi}{2y} \Rightarrow U \cos\theta = -\frac{2\phi}{2y}$

$\Rightarrow \phi = -Uy \cos\theta + \phi(x)$

Hence, $\phi = \pm U (x \sin\theta + y \cos\theta)$

15. As given that, Pan evaporation = 100 mm

Crop coefficient = 0.4

Water required by crop :

$C_u = 0.4 \times (100) = 40$ mm

Additional water required

$P_e = 20$ mm

Effective rain fall

$CIR = C_u - P_e = 20$ mm

Irrigation efficiency $\eta = 0.5$

FIR = Depth of water to be supplied after accounting irrigation efficiency

$$= \frac{CIR}{\eta} = \frac{20}{0.5} = 40 \text{ mm}$$

16. Most powerful disinfectant is HOCL

17. $H^+ + Cl^- \rightarrow HCl$

Mol. wt of $H^+ = 1$, Mol wt of HCl : 36.5

36.5 parts of HCl = 1 Part

$$1 \text{ part of HCl} = \frac{1}{36.5} \text{ Part of } H^+$$

$$35.67 \text{ mg/l of HCl} = \frac{1}{36.5} \times 35.67 \text{ mg/l of } H^+$$

$$= 0.977 \text{ mg/l of } H^+$$

$$pH = \log_{10}[H^+]$$

$$pH = -\log_{10}[0.97726 \times 10^{-3}]$$

$$= -\log_{10}(0.977) + 3\log_{10}(10)$$

$$\approx 3.01$$

18. As given that

T = 50 years, n = 25 years

$P_{o,n} = ?$

$P_{o,n} = q^n, q = 1 - P, P = \frac{1}{T}$

$P = \frac{1}{T}$

$= \frac{1}{50} = 0.02$

$q = 1 - P = 1 - 0.02 = 0.98$

The probability of non-occurance of an event is given by

$(P_{0.25}) = q^n = 0.98^{25} = 0.60346 = 60.35\%$

19.

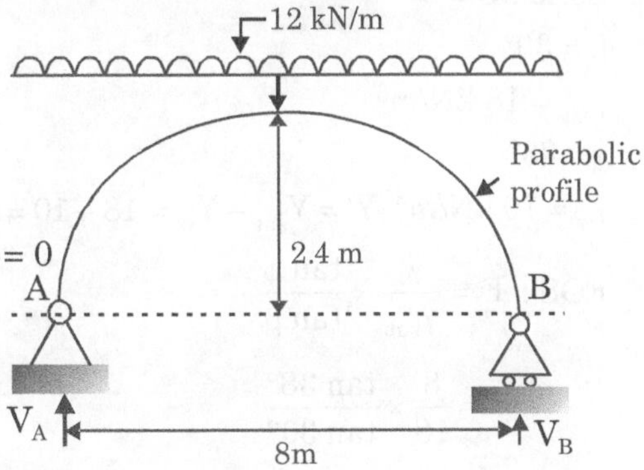

$$BM_{x-x} = 48x - \frac{12x^2}{2} = 48x - 6x^2 \quad ...(i)$$

As horizontal thrust is zero, so it behaves like a curved beam.

So, $\frac{\partial BM_{x-x}}{\partial x} = 0$

$\frac{\partial}{\partial x}(48x - 6x^2) = 0$

$48 - 12x = 0$

$x = 4$ m

By putting the value of x in equation (i).

$BM_{x-x} = 48 \times 4 - 6 \times 4^2 = 96$ kN-m

20. As per IRC: 86 – 1983:

Desirable minimum width of median in urban roads is 5 m and minimum width = 1.2 m.

21. Gradient G = 4.5%, R = 100 m

Grade compensation, $GC = \min\left\{\frac{30+R}{R}, \frac{75}{R}\right\}$

$G.C = \min\left\{\frac{130}{100}, \frac{75}{100}\right\} = 0.75$

Compensated gradient, (CG) = gradient = G – ac

$= 4.50 - 0.75 = 3.75$

As per IRC, CG $\not<$ 4%, Hence, CG = 4%

22. For Drained Triaxial Test:

Sand: $C' = 0$ $\quad \sigma'_d = 150$ kPA

$\sigma'_c = 50$ kPa $\quad f = ?$

$\sigma'_1 = \sigma'_c + \sigma'_d = 200$ kPA $\quad \sigma'_3 = \sigma'_c = 50$ kPa

$\therefore \phi = \sin^{-1}\left(\frac{\sigma'_1 - \sigma'_3}{\sigma'_1 + \sigma'_3}\right) \quad \phi = \sin^{-1}\left(\frac{200-50}{200+50}\right)$

$\phi = 36.87° \approx 37°$

Hence, the angle of internal friction to the nearest intger is 37.

23. For infinite slope:

Sand : C = 0

$\beta = 30°$

$\gamma_{sat} = 18$ kN/m^3

$\phi = 38°$

$\gamma_w = 10$ kN/m^3, $Y' = Y_{sat} - Y_w = 18 - 10 = 8$ kN/m^3

$$FOS : F = \frac{\gamma'}{\gamma_{sat}} \times \frac{\tan\phi}{\tan\beta}$$

$$= \frac{8}{18} \times \frac{\tan 38°}{\tan 30°}$$

$$= 0.6014 \approx 0.60$$

24. As given that,

B = 4 m $\quad$ Q = 6 m^3/s $\quad$ n = 0.02

$V_C = ?$

$$q = \frac{6}{4} = 1.5 \text{ m}^2/\text{s/m}$$

As we know that critical depth :

$$\therefore y_c = \left(\frac{q^2}{g}\right)^{\frac{1}{3}} = \left(\frac{1.5^2}{9.81}\right)^{\frac{1}{3}}$$

$\therefore y_c = 0.612$ m

$$F_r = \frac{V_c}{\sqrt{gy_c}} = 1$$

So, critical velocity (V_C) :

$$\therefore V_C = \sqrt{9.81 \times 0.612} = 2.45 \text{ m/s}$$

25. As given that,

$Q_R = 1000$ MLD $\quad Q_S = 100$ MLD

$y_{SR} = 5$ mg/l $\quad y_{uS} = 200$ mg/l

$Do_R = 8$ mg/l $\quad Do_S = 2$ mg/l

$Do_{saturated} = 10$ mg/l

After mixing:

$$Do_{mix} = \frac{Do_R.Q_R + Do_S.Q_S}{Q_R + Q_S}$$

$$= \frac{8 \times 1000 + 2 \times 100}{1000 + 100} = 7.45 \text{ mg}/l$$

$$Do_{Deficit} = Do_{saturated} - Do_{mix}$$

$$= 10 - 7.45 = 2.55 \text{ mg}/l$$

26. As given that, the ordinary differential equation :

$$\frac{d^2x}{dt^2} - 5\frac{dx}{dt} + 6x = 0$$

The auxiliary equation is

$D^2 - 5D + 6 = 0$

Roots are D = 2, 3

The general solution is

$$x = C_1e^{2t} + C_2\,e^{3t} \quad ...(1)$$

$$\frac{dx}{dt} = 2C_1e^{2t} + 3C_2e^{3t} \quad ...(2)$$

Now, by using initial condition;

$x(0) = 0$, (1)

$\Rightarrow 0 = C_1 + C_2$

$\Rightarrow C_2 = -C_1$

$\Rightarrow \frac{dx(0)}{dt} = 10$, (2)

$\Rightarrow 10 = 2C_1 + 3C_2$

$\Rightarrow 10 = 2C_1 - 3C_1 = -C_1$

$\Rightarrow C_1 = -10$ & $C_2 = 10$

$\therefore$ The particular solution is

$x = -10\,e^{2t} + 10\,e^{3t}$

Note: Among four options, only option (d) satisfies x(0) = 0. So option (d) is correct.

27. The six point central difference formula for third derivative is

$$\left.\frac{\delta^3 u}{\delta x^3}\right|_{x_i} = \left[\frac{-u_{i+3} + 8u_{i+2} - 13u_{i+1} + 13u_{i-1} - 8u_{i-2} + u_{i-3}}{8\Delta h^3}\right]$$

$$f''(x_i) = \left[\frac{[-f(x_{i-3}) - 8f(x_{i-2}) + 13f(x_{i-1}) - 13f(x_{i+1}) + 8f(x_{i+2}) - f(x_{i+3})]}{8h^3}\right]$$

28.

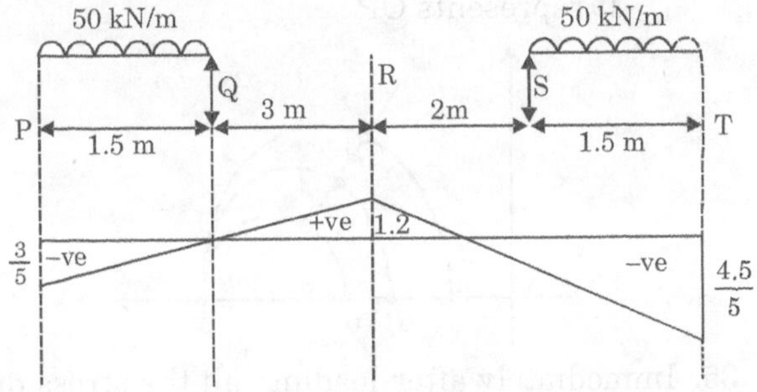

ILD for BM at R.

To get Maximum negative hogging BM at R = Intensity of ud$l$ × Area of ILD under the ud$l$ over PQ and ST

Maximum –ve BM at R

$$= -50 \times \frac{1}{2} \times 1.5 \times \frac{3}{5} - 50 \times \frac{1}{2} \times 1.5 \times \frac{4.5}{5}$$

$= -22.5 - 33.75$

$= -56.25$ kN-m

$= 56.25$ kN-m (hogging)

29. Vertical component of OJ $= F_{OJ} \sin\theta = W$

$$F_{OJ} = \frac{W}{\sin\theta} \text{ (comp)}$$

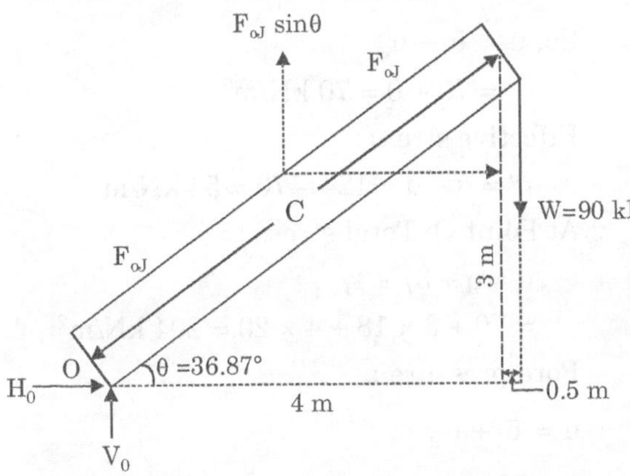

$W = 90$ kN

Horizontal component of reaction at pin O

$$= H_D = F_{DJ} \cos\theta = \frac{W}{\sin\theta} \times \cos\theta$$

$$= W \times \frac{4}{3} = \frac{90 \times 4}{3} = 120 \text{ kN}$$

30. Deflection at free end

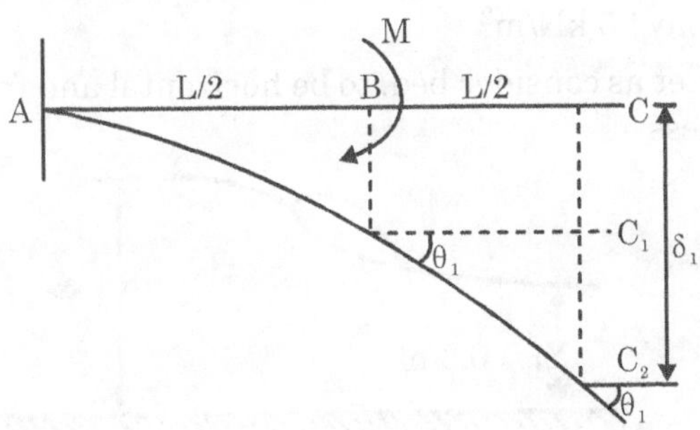

$$\delta_C = \delta_1 = CC_1 + C_1C_2$$

$$= BB_1 + C_1C_2$$

$$\delta_C = \frac{M\left(\frac{L}{2}\right)^2}{2EI} + \frac{M\left(\frac{1}{2}\right)}{EI}\left(\frac{L}{2}\right)$$

$$= \frac{ML^2}{EI}\left[\frac{1}{8} + \frac{1}{4}\right] = \frac{3ML^2}{8EI}$$

31. Dowel bar must be at mid depth and the crack at the middle of dowel.

32.

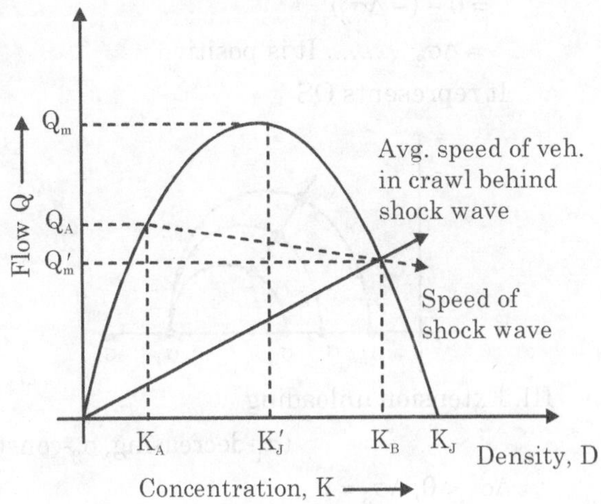

33. For normal take off condition

$$\text{Clearway} \not> \frac{1}{2}\left[\left(\frac{100+15}{100}\right) \times (\text{LOD} - \text{TOD})\right]$$

$$\not> \frac{1}{2}\{(1.15 \times 1625 - 875)\}$$

$$\not> \frac{1}{2}(750)$$

$$\not> 431.25 \text{ m}$$

Hence, clearway is less then 432 m.

34. I: Compression loading:

(σ_1 - increasing, σ_3 - constant)

$\Delta\sigma_1 > 0, \Delta\sigma_3 = 0$

$$P = \frac{\sigma_1 + 2\sigma_3}{3}$$

$$= \frac{\Delta\sigma_1 + 2\Delta\sigma_3}{3} \Rightarrow \frac{\Delta\sigma_1}{3} + 0$$

$q = \sigma_1 - \sigma_3 = \Delta\sigma_1 - \Delta\sigma_3$

Both p & q positive

∴ it represent OR

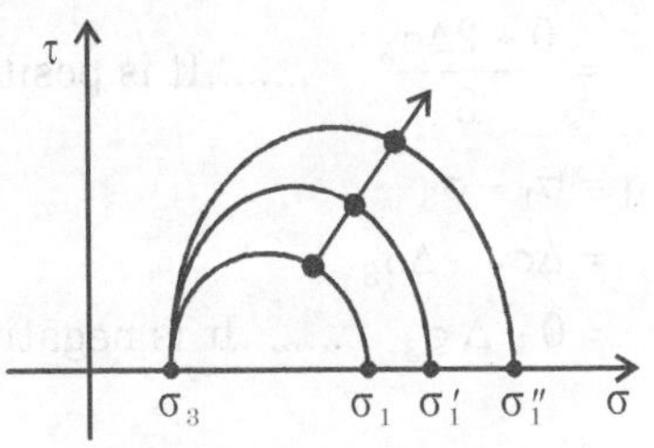

II. Compression unloading (σ_1 = constant, σ_3 - decreasing)

$\Delta\sigma_1 = 0 \qquad \Delta\sigma_3 < 0$

$$P = \frac{\sigma_1 + 2\sigma_3}{3}$$

$$P = \frac{\Delta\sigma_1 + 2(-\Delta\sigma_3)}{3}$$ .....it is negative

$$q = \sigma_1 - \sigma_3$$
$$= \Delta\sigma_1 - \Delta\sigma_3$$
$$= 0 - (-\Delta\sigma_3)$$
$$= \Delta\sigma_3$$ .... It is positive

$\therefore$ It represents OS

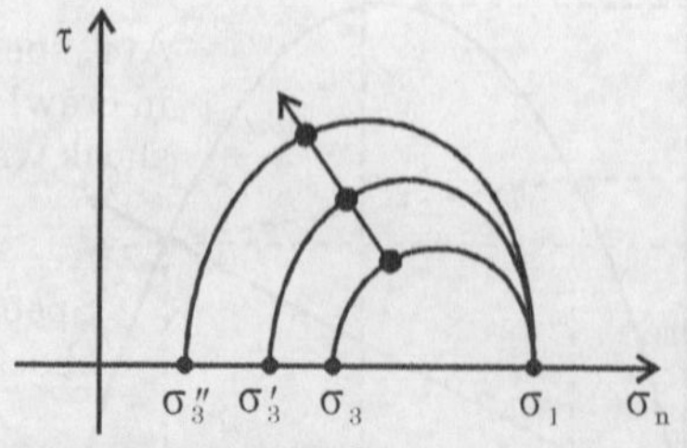

III. Extension unloading

(σ_1-decreasing, σ_3-constant)

$\therefore \ \Delta\sigma_1 < 0, \Delta\sigma_3 = 0$

$$P = \frac{\sigma_1 + 2\sigma_3}{3}$$
$$= \frac{\Delta\sigma_1 + 2\Delta\sigma_3}{3}$$
$$= \frac{-\Delta\sigma_1 + 0}{3}$$ ....It is negative

$\therefore$ It represent OQ

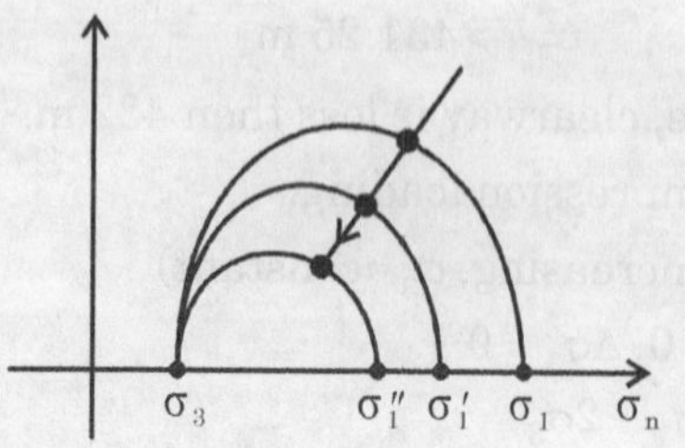

IV. Extension loading (σ_1-constant, σ_3-increases)

$-\Delta\sigma_1 = 0 \qquad \Delta\sigma_3 > 0$

$$P = \frac{\sigma_1 + 2\sigma_3}{3}$$
$$= \frac{\Delta\sigma_1 + 2\Delta\sigma_3}{3}$$
$$= \frac{0 + 2\Delta\sigma_3}{3}$$ ....It is positive

$$q = \sigma_1 - \sigma_3$$
$$= \Delta\sigma_1 - \Delta\sigma_3$$
$$= 0 - \Delta\sigma_3$$ ....It is negative

$\therefore$ It represents OP

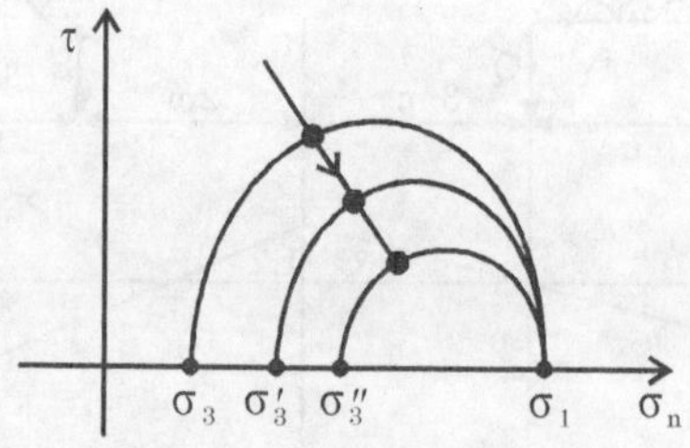

35. Immediately after loading, all the stress due to the preloading is taken by water only.

Initial excess pore water pressure,

$$\bar{u}_i = 70 \text{ kN/m}^2$$

At Point P: total stress

$$\sigma = q + 3\gamma$$
$$= 70 + 3 \times 18 = 124 \text{ kN/m}^2$$

Pore water pressure :

$$u = \bar{u}_i + u_{st}$$

Where $\bar{u}_i$ = Initial excess pore water pressure due to external loading

u_{st} = hydrostatic pore water pressure

$\therefore \ \bar{u}_i = 70 \text{ kN/m}^2, u_{st} = 0$

So, $u = \bar{u}_i + u_{st}$

$$= 70 + 0 = 70 \text{ kN/m}^2$$

Effective stress :

$\therefore \ \sigma' = \sigma - u = 124 - 70 = 54 \text{ kN/m}^2$

At Point Q: Total stress :

$$\sigma = q + 3\gamma + 4\gamma_{sat}$$
$$= 70 + 3 \times 18 + 4 \times 20 = 204 \text{ kN/m}^2$$

Pore pressure :

$$u = \bar{u}_i + u_{st}$$
$$u = 70 + 4\gamma_w$$
$$= 70 + 4 \times 9.81$$
$$= 109.24 \text{ kN/m}^2$$

Effective stress :

$$\sigma' = \sigma - u$$
$$= 204 - 109.24$$
$$= 94.76 \text{ kN/m}^2$$

Say 95 kN/m^2

36. Let as consider bed. to be horizontal and friction less

$Q = 12\ m^3/s,\ B = 6\ m$

$\therefore\ q = \frac{12}{6} = 2\ m^2/s/m$

$\therefore\ y_1 = 30\ cm \rightarrow$ pre jump depth

$= 0.3\ m$

$g = 9.81\ m/s^2$

$\delta_w = 100\ kg/m^3$

When specific force are same

$\therefore\ F_1 = F_2$

$$\frac{q^2}{g} = \frac{y_1 y_2 (y_1 + y_2)}{2}$$

$$\frac{2^2}{9.81} = \frac{0.3 y_2 (0.3 + y_2)}{2}$$

$$2.718 = 0.3 y_2 + y_2^2$$

$$y_2^2 + 0.3 y_2 - 2.718 = 0$$

By solving $y_2 = 1.505\ m$

Head loss in the jump (h_L)

$$\therefore\ h_L = \Delta E = \frac{(y_2 - y_1)^3}{4.y_1 y_2}$$

$$= \frac{(1.505 - 0.3)^3}{4 \times 1.505 \times 0.3} = 0.968\ m$$

Power loss in the jump $= \gamma_w \cdot Q \cdot h_L$

$= 9.81 \times 12 \times 0.968$

$= 114.05\ kW$

37. For $Q_1 = 10$ MLD; $C_1 = 3.5\ mg/l$

$Q_2 = 13$ MLD ; $C_2 = ?$

If disinfection is not first order or exponential, [as given n = 0.86] then the relationship between C & t are given by waterson law:

$t.C^n$ = constant, $t_1 . C_1^{\,n} = t_2 . C_2^{\,n}$

$$\left[\frac{V_1}{Q_1}\right] C_1^n = \left[\frac{V_2}{Q_2}\right] C_2^n$$

But same disinfection unit is used, so $V_1 = V_2$

$$\therefore\ \frac{C_1^n}{Q_1} = \frac{C_2^n}{Q_2}$$

$$C_2^n = C_1^n \cdot \frac{Q_2}{Q_1}$$

$$C_2^{0.86} = [3.5]^{0.86} \times \frac{13}{10}$$

$$C_2 = \left(\frac{13}{10}\right)^{\frac{1}{0.86}} \times 3.5$$

$\therefore\ C_2 = 4.75\ mg/L$

38.

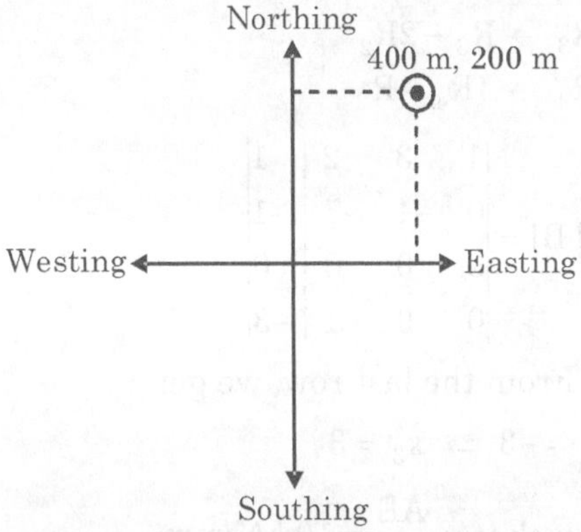

Where, $\Delta L = -5.3$, $\Delta D = -29.9$

So, T, Northing {400 + (–5.3)} = 394.7

T, Easting {200 + (–29.9)} = 170.1

Station	Consecutive coordinates				Independent coordinate	
	Northern(+)	Southern(–)	Eastern(+)	Western(–)	L	D
P	–	–	–	–	400	200
Q	110.2	–	45.5	–	510.2	245.5
R	80.6	–	–	60.1	590.8	185.4
S	–	90.7	–	70.8	500.1	114.6
T	–	105.4	55.5	–	394.7	170.1

39. $I = \int_c ((y+z)dx + (x+z)dy + (x+y)dz)$

$$= \int_c [ydx + zdx + xdy + zdy + xdz + ydz]$$

$$= \int_c [(ydx + xdy) + (zdx + xdz) + (zdy + ydz)]$$

$$= \int_{(0,0,0)}^{(1,1,1)} [d(xy) + d(xz) + d(yz)]$$

$$= [xy + xz + yz]_{(0,0,0)}^{(1,1,1)} = (1+1+1) - (0+0+0) = 3$$

40. The augmented matrix is

$$[A\,|\,B] = \left[\begin{array}{ccc|c} 1 & 3 & 2 & 1 \\ 2 & 2 & -3 & 1 \\ 4 & 4 & -6 & 2 \\ 2 & 5 & 2 & 1 \end{array}\right]$$

By converting into an echelon form :

$R_2 \rightarrow R_2 - 2R_1$

$R_3 \rightarrow R_3 - 4R_1$

$R_4 \rightarrow R_4 - 2R_1$

$$\text{So, } [A\,|\,B] = \left[\begin{array}{ccc|c} 1 & 3 & 2 & 1 \\ 0 & -4 & -7 & -1 \\ 0 & -8 & -14 & -2 \\ 0 & -1 & -2 & -1 \end{array}\right]$$

Now, $R_3 \rightarrow R_3 - 2R_2$

$R_4 \rightarrow 4R_4 - R_2$

$$\text{So, } [A \mid B] = \left|\begin{array}{ccc|c} 1 & 3 & 2 & 1 \\ 0 & -4 & -7 & -1 \\ 0 & 0 & 0 & 0 \\ 0 & 0 & -1 & -3 \end{array}\right|$$

Hence, From the last row, we get

$-x_3 = -3 \Rightarrow x_3 = 3$

41. As Given that : $\frac{AE}{l} = 10$ kN/mm

$\delta_P = \delta_R = 5$mm

$\delta_Q = 5 - \text{gap} = 5 - 2 = 3$ mm

$$W_P = W_R = \frac{(\delta)AE}{l} = 5 \times (10 \text{ kN/mm}) = 50 \text{ kN}$$

$$W_Q = \frac{(\delta)AE}{l} = (3 \text{ mm})\ (10 \text{ kN/m}) = 30 \text{ kN}$$

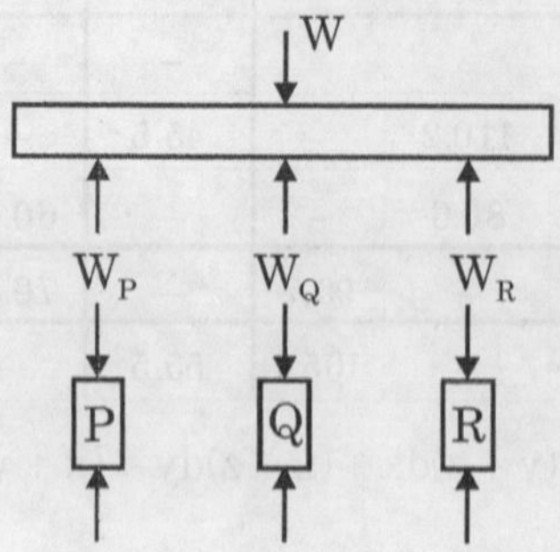

So, Total load = W = $W_P + W_Q + W_R$

= 50 + 30 + 50 = 130 kN

42. As given that,

Permissible shear stress in fillet weld

$q = 100$ N/mm^2

Size of fillet weld S = 10 mm

Effective throat thickness t_t

$= K \times S = 0.7 \times 10 = 7$ mm

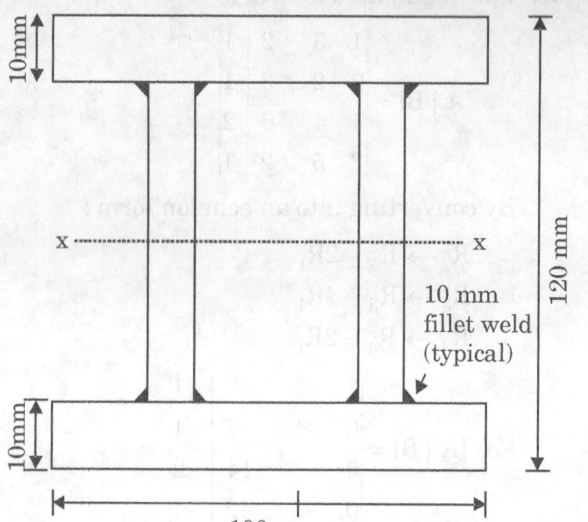

Moment of inertia of fillet weld about X-X axis

$I_{xx} = 7.73 \times 10^6$ mm^4

Horizontal shear per mm length of built-up section at junction between flange and web of built up section $q = \frac{FA\bar{Y}}{Ib}$

Where

F = Maximum shear force

$A\bar{Y}$ = Moment of area about X-X axis

$= 100 \times 10 \times \left(\frac{120}{2} - \frac{10}{2}\right)$

$= 55000$ mm^3

b = Effective width of weld section (4 welds)

$= 4 \times t_1 = 4 \times 7 = 28$ mm

$$q = \frac{FA\bar{Y}}{Ib}$$

$$100 = \frac{F \times 55000}{7.73 \times 10^6 \times 28}$$

$F = 393.5 \times 10^3$ N

= 393.5 kN

43. Limiting depth of N.A

(i) $X_{u\,max} = 0.46\, d = 0.46 \times 450 = 207$ mm

(ii) Actual dpeth of N.A

C = T

$0.36\, f_{ck}\, b\, x_u = 0.87\, f_y A_{st}$

$0.36 \times 25 \times 300 \times x_u = 0.87 \times 500 \times 942$

$$x_u = \frac{0.87 \times 500 \times 942}{0.36 \times 25 \times 300}$$

$x_u = 151.8$ mm

∴ U.R.S.

(iii) $X_u < X_{u\,max}$

Moment of resistance

(iv) MR = TZ

$= 0.87\, f_y\, A_{st}\, [d - 0.42\, x_u]$

$= 0.87 \times 500 \times 942\, [450 - 0.42 \times 151.8]$

= 158.28 kN-m

44.

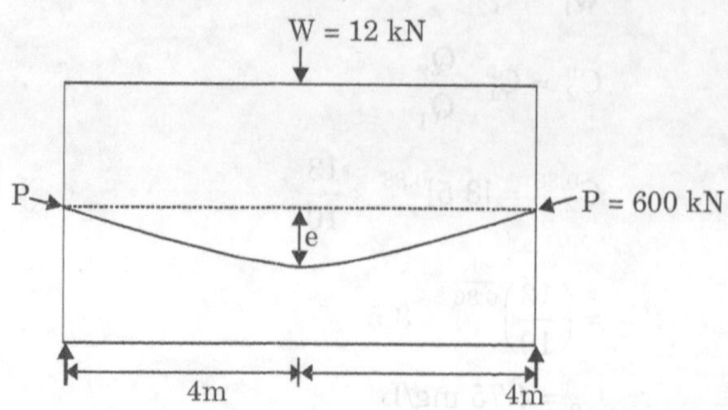

For load balancing condition equate moments due to prestressing force and external load

$\because$ Balancing load = Point load

$M_P = M_E$

$\Rightarrow$ $Pe = \frac{WL}{4}$

$e = \left[\frac{WL}{4P}\right]$

$e = \frac{12 \times 10^3 \times 8 \times 10^3}{4 \times 600 \times 10^3}$

= 40 mm

45. As given that

Maximum service flow rate observed per lane

= 1280 veh/hr

PHF = 0.78125

The peak hour service volume per lane

$= \frac{1280 \text{ Veh/hr}}{\text{PHF}}$

$= \frac{1280}{0.78125}$

= 1638.4 Veh/hr

Annual average daily traffic (AADT)

= 12270 veh/day.

Now, Average hourly traffic on the 2-lane road

$= \frac{12270}{24}$

= 511.25 veh/hr

Lane split = 60 : 40 (60% on critical lane)

So, Average hourly traffic on critical lane of a road

$= 511.25 \times \frac{60}{100}$

= 306.75 veh/hr

No. of lane required to accommodate the traffic

$N = \frac{\text{Peak hour service volume per lane}}{\text{average volume per lane}}$

$= \frac{1638.4}{306.75}$

= 5.34

≈ 6 hours

No. of additional lanes required = 6 – 1 = 5

On each side

Hence, the total number of additional lanes required

= 5 + 5 = 10 nos

46.

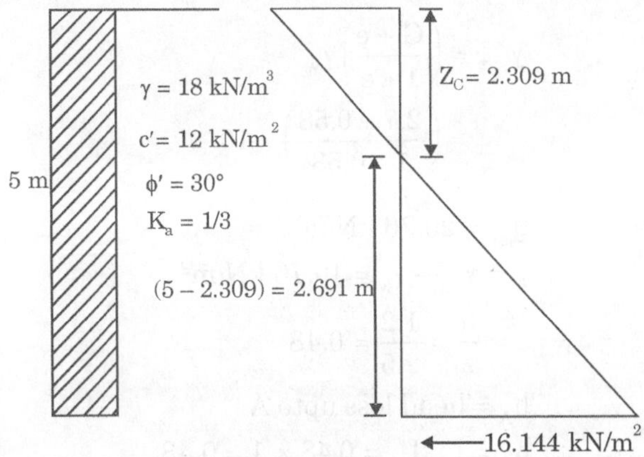

$\therefore$ As soil is homogenous (single layer)

$Z_c = \frac{2c}{\gamma\sqrt{K_a}} = \frac{2 \times 12}{18 \times \sqrt{\frac{1}{3}}}$

$Z_c = 2.309$ m

At base of retaining wall

$P_a = K_a\sigma_v - 2C\sqrt{K_a}$

$\sigma_v = 18 \times 5 = 90$ kPa

$\therefore$ $P_a = \frac{1}{3} \times 90 - 2 \times 12 \times \sqrt{\frac{1}{3}}$

= 16.144 Kpa

After formation of crack : Neglect tension zone

$\therefore$ $F_a = \frac{1}{2} \times 2.691 \times 16.144 = 21.714$ kN/m

47. As given that :

Z = 2.5 m G = 2.7 e = 0.58

$\gamma_w = 10$ kN/m³

At 'd'

$\therefore$ Z_A = depth of soil upto 'A'

$(h_S)_A$ = Seepage head at 'A'

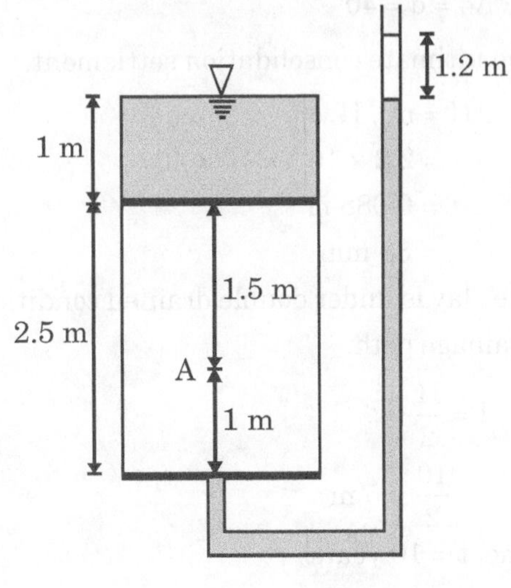

$$\gamma_{sat} = \left(\frac{G+e}{1+e}\right)\gamma_w$$

$$= \left(\frac{2.7+0.58}{1+0.58}\right) \times 10$$

$\therefore$ $g_{sat} = 20.76$ kN/m^3

$\gamma' = \gamma_{sat} - \gamma_w = 10.76$ kN/m^2

$\therefore$ $i = \frac{h}{z} = \frac{1.2}{2.5} = 0.48$

$\therefore$ h_A = head loss upto A

$\therefore$ $h_A = i \times L_A = 0.48 \times 1 = 0.48$

$(h_s)_A = 1.2 - 0.48 = 0.72$

$\therefore$ $\sigma' = \gamma'.Z_A - \gamma_w (h_s)_A = 10.76 \times 1.5 - 10 \times 0.72 =$ 8.94 kN/m^2

48.

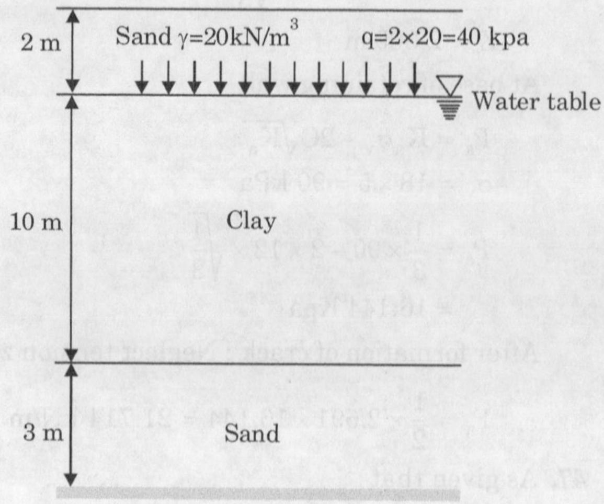

The fill weight acts as a surcharge of intensity,

$q = 2 \times \gamma = 2 \times 20 = 40$ kPa

$\therefore$ Increase in vertical stress at all depths below G.L

$\Delta\bar{\sigma} = q = 40$

The ultimate consolidation settlement,

$$\Delta H = m_v.\ H.\ \Delta\bar{\sigma}$$
$$= 2.2 \times 10^{-4} \times 10 \times 40$$
$$= 0.088 \text{ m}$$
$$= 88 \text{ mm}$$

The clay is under double drained condition

Drainage path,

$$d = \frac{H}{2}$$

$\therefore$ $= \frac{10}{2} = 5$ m

time, t = 10 years

$\therefore$

$$T_v = \frac{C_v.t}{d^2} = \frac{9\times 10^{-2} \times 10}{5^2} = 3.6 \times 10^{-2} = 0.036$$

$$T_v = \frac{\pi}{4}\left[\frac{U\%}{100}\right]^2$$

$$0.036 = \frac{\pi}{4}\left[\frac{U\%}{100}\right]^2$$

$$U = \sqrt{\frac{0.036 \times 4}{\pi}} \times 100$$

$$= 0.2141 \times 100$$

$\therefore$ U = 21.41%

$$U = \frac{\Delta h}{\Delta H} \times 100$$

$$21.41 = \frac{\Delta h}{88} \times 100$$

$\therefore$ Settlement Δh after 10 years = 18.84 mm

49. As given that

$h_p^* = 109$ m $\qquad h^*_Q = 106$ m

$h_s^* = 100$ m $\qquad g = 9.81$ m/s^2

$r = 1000$ kg/m^3

As we know that,

$$Q_1 = A_1V_1 = \frac{\pi}{4}(0.3)^2 \times 2.56$$

$$= 0.1809 \text{ m}^3/\text{s}$$

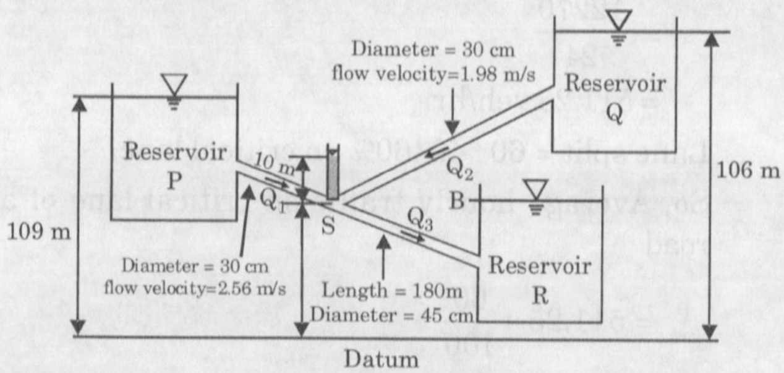

$$Q_2 = A_2V_2 = \frac{\pi}{4}(0.3)^2 \times 1.98$$

$$= 0.14 \text{ m}^3/\text{s}$$

Q@ Junction = 0

Now by apply conutinuity

$\therefore$ $Q_3 = Q_1 + Q_2 = 0.1809 + 0.14 = 0.3209$ m^3/s

$$h = \frac{fLQ^2}{12.1D^5}$$

$\Rightarrow$ $h = \frac{0.03 \times 180 \times 0.3209^2}{12.1 \times 0.45^5}$ $\quad h = 2.49$ m

$\therefore$ $h^*_s - h^*_R = 2.49$ m

$\Rightarrow$ $h^*_R = (100 - 2.49)$ m

$\Rightarrow$ 97.51 m

50. $S_y = \frac{V_T}{B} = \frac{V_{wy}}{\text{Area} \times Z}$

$= \frac{0.9 \times 10^6}{3 \times 10^6 \times (103.20 - 102.0)}$

$= 0.25$

$S_y = \frac{V_D}{V}$

$0.25 = \frac{V_D}{A \times Z}$

$0.25 = \frac{V_D}{3 \times 10^6 (103.20 - 101.20)}$

$V_D = 1.5 \times 10^6 \text{ m}^3 = 1.5 \text{ Mm}^3$

51. For steady state of tank

$Q_{orifice} = Cd. \frac{\pi}{4} d^2 V_{orifice} = 0.02$

$\Rightarrow V_{orifice} = \frac{4 \times 0.02}{0.8 \times \pi \times 0.01}$

$= \frac{10}{\pi}$ m/s 3.183 m/s

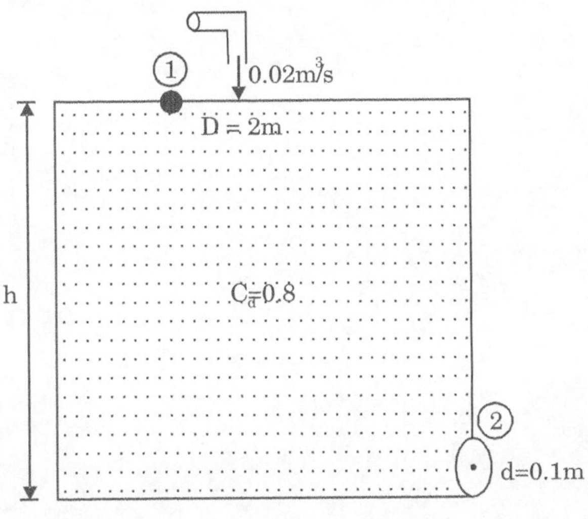

Assume H is the level of water in the tank in steady condition.

For steady water level in the tank :

By continuity equation :

D is charge through orifice = water enters in the tank.

$A_{orifice} V_{orifice} = A_{tank} \cdot V_{tank}$

$\Rightarrow V_{tank} = \frac{(0.1)^2}{2^2} \times \frac{10}{\pi}$

$= \frac{1}{40\pi}$

$= 0.00795$ m/s

By Bernoulli equation

$\frac{P_1}{\gamma} + \frac{V_1^2}{2g} + Z_1 = \frac{P_2}{\gamma} + \frac{V_2^2}{2g} + Z_2 + h$

$H = \frac{V_2^2 - V_1^2}{2g}$

$= 0.5163$ m

52. $V_s = 20000$ litre/m^2

$\eta = 80\%$

$U_s = V_0 \cdot \eta$

$= 20000 \times 0.8$

$= 16000$ Litre/m^2

$= 16$ m/D

$= 2[16/24 \times 60 \times 60]$

For estimating Diameter of particle

$U_s = \frac{gd^2}{18\upsilon}[G - 1]$

$\frac{16}{24 \times 60 \times 60} = \frac{9.81 d^2}{18 \times 1.01 \times 10^{-6}}[2.64 - 1]$

$d = 1.446 \times 10^{-5}$ m

$\therefore$ $d = 14.46$ µm

53. As we know that

$PV = nRT \rightarrow V = \frac{nRT}{P}$

$V = \frac{82.05 \times 10^{-6} \times 293}{1}$

$= 0.024 \text{ m}^3 / \text{mole}$

$= 24.04$ L/mole

Now, Chemical concentration in µg/m^3

$= \text{Concentration in PPM} \times \frac{\text{Mol. Wt}}{\text{Volume}} \times 10^3$

[L/mole]

$41.6 = \frac{X_{(PPM)} \times 1 \times 10^3}{24.04}$

So, $x_{[PPM]} = \frac{41.6 \times 24.04}{10^3}$

$= 1.000064$ ppm >> 1 ppm

54. As given that

$Q_R = 5 \text{ m}^3/\text{s}$ $y_{u_1} = 30$ mg/l $Q_S = 0.2 \text{ m}^3/\text{s}$

$y_{u_2} = 500 \text{ mg/l} \rightarrow y_{u_2} = \frac{500}{[1 - e^{-0.3 \times 5}]}$

$= 643.66$ mg/l

Area = 40 m^2

L = 3 km = 3000 m

$$BOD_{mix} = \left[\frac{Q_R y_{u_1} + Q_S y_{u_2}}{Q_R + Q_S}\right]$$

$$\therefore\ BOD_{mix} = y_{u\,mix} = \frac{(5 \times 30) + (0.2 \times 643.66)}{5 + 0.2}$$

$$= 53.6 \text{ mg/l}$$

$$\therefore\ t = \frac{3000 \times 40}{(5 + 0.2) \times 24 \times 60 \times 60} = 0.26 \text{ days}$$

BOD Remaining:

$$y_R = y_{u\,mix} e^{-kt}$$

$$= 53.66 \times e^{-0.3 \times 0.26}$$

$$= 49.57 \text{ mg/l}$$

55. In a closed traverse $\Sigma L = 0;\ \Sigma D = 0$

$\Sigma L = 0 \Rightarrow 40 \cos 80° + 50 \cos 10° + 30 \cos 210° + l \cos \theta = 0$

$\Rightarrow \Delta L = l \cos \theta = \Delta L = -30.2055$

$\Sigma D = 0 \Rightarrow 40 \sin 80° + 50 \sin 10° + 30 \sin 210° + l \sin \theta = 0$

$\Rightarrow l \sin \theta = \Delta D = -33.0747$

So, length, $SP = \sqrt{\Delta L^2 + \Delta D^2}$

$= \sqrt{2006.308}$

$SP = l = 44.792$ m

SOLVED PAPER - 2020

SET 2

INSTRUCTIONS

1. Total of 65 questions carrying 100 marks, out of which 10 questions carrying a total of 15 marks are in General Aptitude (GA)
2. The Engineering Mathematics will carry around **15% of the total marks,** the General Aptitude section will carry **15% of the total marks** and the **remaining 70% of the total marks.**
3. **Types of Questions**
   (*a*) Multiple Choice Questions (MCQ) carrying 1 or 2 marks each in all papers and sections. These questions are objective in nature, and each will have a choice of four options, out of which the candidate has to mark the correct answer(s).
   (*b*) Numerical Answer Questions of 1 or 2 marks each in all papers and sections. For these questions the answer is a real number, to be entered by the candidate using the virtual keypad. No choices will be shown for these type of questions.
4. For **1-mark** multiple-choice questions, **1/3 marks** will be deducted for a wrong answer. Likewise, for **2-marks** multiple-choice questions, **2/3** marks will be deducted for a wrong answer. There is no negative marking for numerical answer type questions.

Chapter-Wise Analysis

#	Subject	Marks		
		1	2	Total
1	General Aptitude	5	5	15
2	Engineering Mathematics	5	4	13
3	Engineering Mechanics	1	0	1
4	Solid Mechanics	0	1	2
5	Structural Analysis	1	3	7
6	Construction Materials and Management	2	0	2
7	Concrete Structures	0	2	4

#	Subject	Marks		
		1	2	Total
8	Steel Structures	1	1	3
9	Soil Mechanics and Foundation Engineering	4	5	14
10	Fluid Mechanics and Hydraulics	2	2	6
11	Hydrology & Irrigation	3	2	7
12	Environmental Engineering	3	4	11
13	Transportation Engineering	3	4	11
14	Surveying	0	2	4

GENERAL APTITUDE

(Q 1 – Q 5) carry one mark each

1. Rescue teams deployed _____ disaster hit areas combat ___ a lot of difficulties to save the people.

(a) with, at (b) in, with
(c) with, with (d) to, to

2. Select the most appropriate word that can replace the underlined word without changing the meaning of the sentence :
Now-a-days, most children have a tendency to belittle the legitimate concerns of their parents.

(a) disparage (b) applaud
(c) reduce (d) begrudge

3. Select the word that fits the analogy :
Partial : Impartial :: Popular : _____

(a) Impopular (b) Dispopular
(c) Mispopular (d) Unpopular

4. After the inauguration of the new building, the Head of the Department (HoD) collated faculty preferences for office space. P wanted a room adjacent to the lab. Q wanted to be close to the lift. R wanted a view of the playground and S wanted a corner office.

Assuming that everyone was satisfied, which among the following shows a possible allocation

(a)
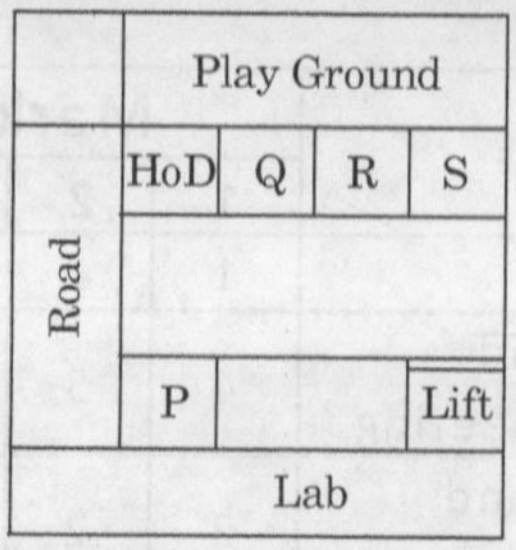

(b)
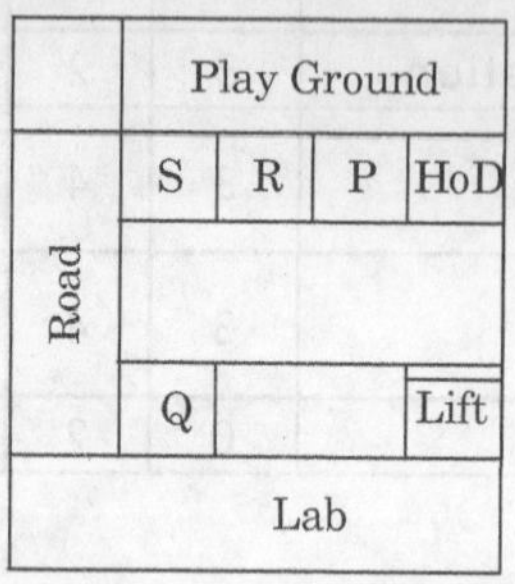

(c)
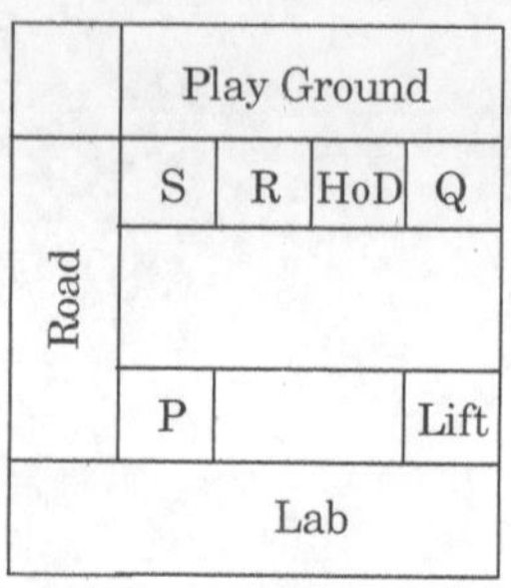

(d)
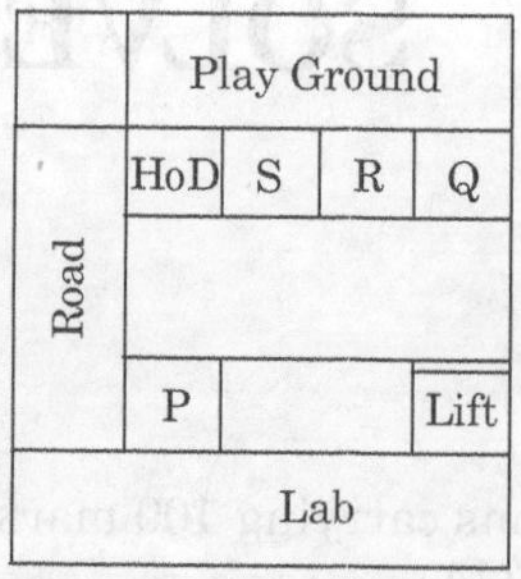

5. If $f(x) = x^2$ for each $x \in (-\infty, \infty)$, then $\dfrac{f(f(f(x)))}{f(x)}$ is equal to _____

(a) $f(x)$ (b) $(f(x))^2$
(c) $(f(x))^3$ (d) $(f(x))^4$

(Q 6 – Q 10) carry two marks each

6. Nominal interest rate is defined as the amount paid by the borrower to the lender for using the borrowed amount for a specific period of time. Real interest rate calculated on the basis of actual value (inflation-adjusted), is approximately equal to the difference between nominal rate and expected rate of inflation in the economy.

Which of the following assertions is best supported by the above information?

(a) Under high inflation, real interest rate is low and borrowers get benefited.
(b) Under low inflation, real interest rate is high and borrowers get benefited.
(c) Under high inflation, real interest rate is low and lenders get benefited.
(d) Under low inflation, real interest rate is low and borrowers get benefited.

7. For the year 2019, which of the previous year's calendar can be used?

(a) 2011 (b) 2012
(c) 2013 (d) 2014

8. The ratio of 'the sum of the odd positive integers from 1 to 100' to 'the sum of the even positive integers from 150 to 200' is _____ .

(a) 45 : 95 (b) 1 : 2
(c) 50 : 91 (d) 1 : 1

9. In a school of 1000 students, 300 students play chess and 600 students play football. If 50 students play both chess and football, the number of students who play neither is _____

(a) 200
(b) 150
(c) 100
(d) 50

10. The monthly distribution of 9 Watt LED bulbs sold by two firms X and Y from January to June 2018 is shown in the pie-chart and the corresponding table. If the total number of LED bulbs sold by two firms during April-June 2018 is 50000, then the number of LED bulbs sold by the firm Y during April-June 2018 is _____.

Percentage of 9 watt LED bulbs sold by the firms X & Y from January, 2018 to June, 2018.

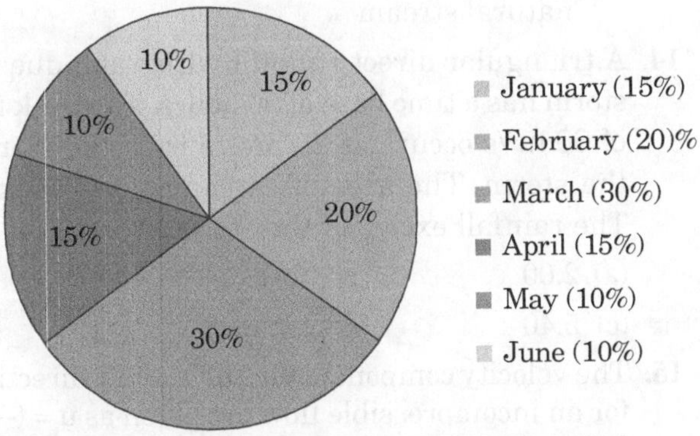

Month	Ratio of LED bulbs sold by two firms (X : Y)
January	7 : 8
February	2 : 3
March	2 : 1
April	3 : 2
May	1 : 4
June	9 : 11

(a) 11250 (b) 9750
(c) 8750 (d) 8250

CIVIL ENGINEERING

(Q 1 – Q 25) carry one mark each

1. The ordinary differential equation

$$\frac{d^2u}{dx^2} - 2x^2u + \sin x = 0 \text{ is}$$

(a) linear and homogeneous
(b) linear and non homogeneous
(c) nonlinear and homogeneous
(d) non linear and non homogeneous

2. The value of $\lim_{x\to\infty} \frac{\sqrt{9x^2 + 2020}}{x + 7}$ is

(a) $\frac{7}{9}$ (b) 1
(c) 3 (d) indeterminable

3. The integral

$$\int_0^1 (5x^3 + 4x^2 + 3x + 2)\, dx$$

is estimated numerically using three alternative methods namely the rectangular, trapezoidal and Simpson's rules with a common step size. In this context, which one of the following statements is **TRUE**?

(a) Simpson's rule as well as rectangular rule of estimation will give NON-zero error.
(b) Simpson's rule, rectangular rule as well as trapezoidal rule of estimation will give NON-zero error.
(c) Only the rectangular rule of estimation will give zero error.
(d) Only Simpson's rule of estimation will give zero error.

4. The following partial differential equation is defined for u : u (x, y)

$$\frac{\partial u}{\partial y} = \frac{\partial^2 u}{\partial x^2}; \quad y \geq 0;\ x_1 \leq x \leq x_2$$

The set of auxiliary conditions necessary to solve the equation uniquely, is

(a) three initial conditions
(b) three boundary conditions
(c) two initial condition and one boundary condition
(d) one initial condition and two boundary conditions

5. The ratio of the plastic moment capacity of a beam section to its yield moment capacity is termed as

(a) aspect ratio (b) load factor
(c) shape factor (d) slenderness ratio

6. The state of stress represented by Mohr's circle shown in the figure is

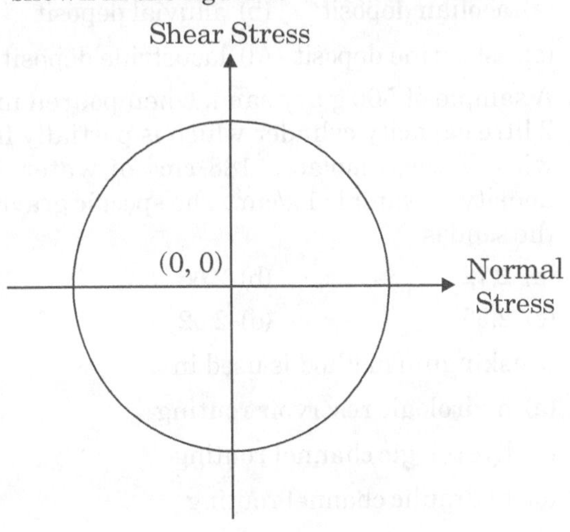

(a) uniaxial tension
(b) biaxial tension of equal magnitude
(c) hydrostatic stress
(d) pure shear

7. A weightless cantilever beam of span L is loaded as shown in the figure. For the entire span of the beam, the material properties are identical and the cross-section is rectangular with constant width.

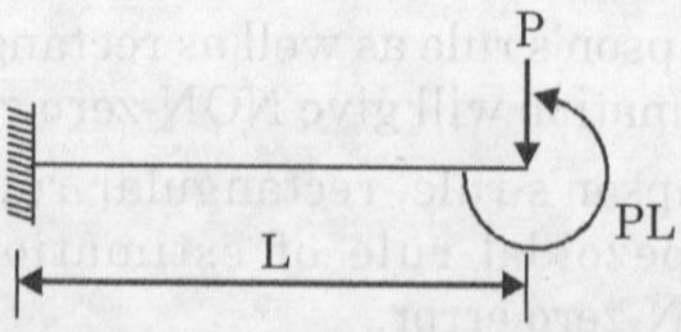

From the flexure-critical perspective, the most economical longitudinal profile of the beam to carry the given loads amongst the options given below, is

(a)  (b) 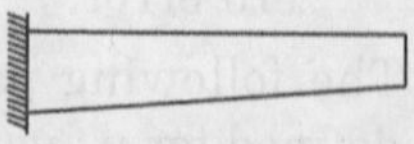

(c) 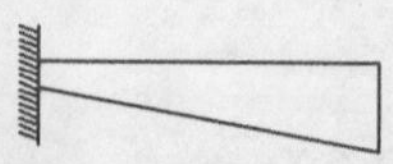 (d) 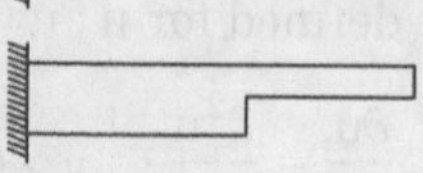

8. As per IS 456 : 2000, the pH value of water for concrete mix shall **NOT** be less than

(a) 4.5 (b) 5.0
(c) 5.5 (d) 6.0

9. The traffic starts discharging from an approach at an intersection with the signal turning green. The constant headway considered from the fourth or fifth headway position is referred to as

(a) discharge headway
(b) effective headway
(c) intersection headway
(d) saturation headway

10. Soil deposit formed due to transportation by wind is termed as

(a) aeolian deposit (b) alluvial deposit
(c) estuarine deposit (d) lacustrine deposit

11. A sample of 500 g dry sand, when poured into a 2 litre capacity cylinder which is partially filled with water, displaces 188 cm^3 of water. The density of water is 1 g/cm^3. The specific gravity of the sand is

(a) 2.72 (b) 2.66
(c) 2.55 (d) 2.52

12. Muskingum method is used in

(a) hydrologic reservoir routing
(b) hydrologic channel routing
(c) hydraulic channel routing
(d) hydraulic reservoir routing

13. Super passage is a canal cross-drainage structure in which

(a) natural stream water flows with free surface below a canal
(b) natural stream water flows under pressure below a canal
(c) canal water flows with free surface below a natural stream
(d) canal water flows under pressure below a natural stream

14. A triangular direct runoff hydrograph due to a storm has a time base of 90 hours. The peak flow of 60 m^3/s occurs at 20 hours from the start of the storm. The area of catchment is 300 km^2. The rainfall excess of the storm (in cm), is

(a) 2.00 (b) 3.24
(c) 5.40 (d) 6.48

15. The velocity components in the x and y directions for an incompressible flow are given as u = (– 5 + 6x) and v = – (9 + 6y), respectively. The equation of the streamline is

(a) $(-5 + 6x) - (9 + 6y) = \text{constant}$

(b) $\dfrac{-5 + 6x}{9 + 6y} = \text{constant}$

(c) $(-5 + 6x)(9 + 6y) = \text{constant}$

(d) $\dfrac{9 + 6y}{-5 + 6x} = \text{constant}$

16. The relationship between oxygen consumption and equivalent biodegradable organic removal (i.e., BOD) in a closed container with respect to time is shown in the figure.

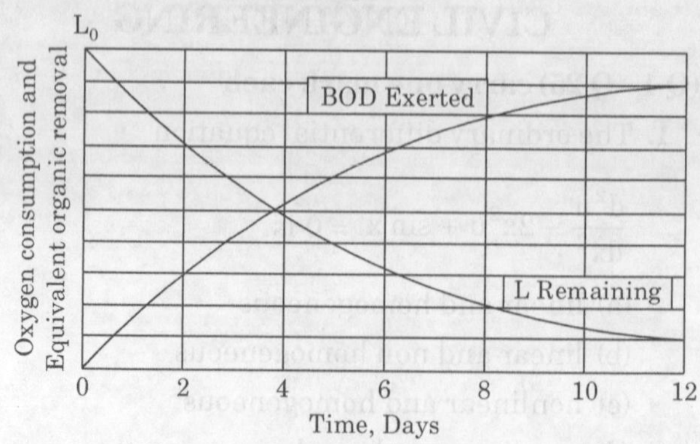

Assume that the rate of oxygen consumption is directly proportional to the amount of degradable organic matter and is expressed as $\dfrac{dL_t}{dt} = -kL_t$, where, L_t (in mg/litre) is the oxygen equivalent of the organics remaining at time t and k(in d^{-1}) is the degradation rate constant. L_o is the oxygen equivalent of organic matter at time, t = 0.

In the above context, the correct expression is

(a) $BOD_5 = L_5$ (b) $BOD_t = L_0 - L_t$

(c) $L_0 = L_t e^{-kt}$ (d) $L_t = L_0 (1 - e^{-kt})$

17. A gas contains two types of suspended particles having average sizes of 2 μm and 50 μm. Amongst the options given below, the most suitable pollution control strategy for removal of these particles is

(a) settling chamber followed by bag filter

(b) electrostatic precipitator followed by venturi scrubber

(c) electrostatic precipitator followed by cyclonic separator

(d) bag filter followed by electrostatic precipitator

18. A fair (unbiased) coin is tossed 15 times. The probability of getting exactly 8 Heads (round off to three decimal places), is _____.

19. The maximum applied load on a cylindrical concrete specimen of diameter 150 mm and length 300 mm tested as per the split tensile strength test guidelines of IS 5816 : 1999 is 157 kN. The split tensile strength (in MPa, round off to one decimal place) of the specimen is _____.

20. For an axle load of 15 tonne on a road, the Vehicle Damage Factor (round off to two decimal places), in terms of the standard axle load of 8 tonne, is _____.

21. 24-h traffic count at a road section was observed to be 1000 vehicles on a Tuesday in the month of July. If daily adjustment factor for Tuesday is 1.121 and monthly adjustment factor for July is 0.913, the Annual Average Daily Traffic (in veh/day, round off to the nearest integer) is _____.

22. A soil has dry unit weight of 15.5 kN/m^3, specific gravity of 2.65 and degree of saturation of 72%. Considering the unit weight of water as 10 kN/m^3, the water content of the soil (in %, round off to two decimal places) is _____.

23. A one-dimensional consolidation test is carried out on a standard 19 mm thick clay sample. The oedometer's deflection gauge indicates a reading of 2.1 mm. just before removal of the load, without allowing any swelling. The void ratio is 0.62 at this stage. The initial void ratio (round off to two decimal places) of the standard specimen is _____ .

24. Velocity distribution in a boundary layer is given by $\frac{u}{U_\infty} = \sin\left(\frac{\pi}{2}\frac{y}{\delta}\right)$, where u is the velocity at vertical co-ordinate y, U_∞ is the free stream velocity and δ is the boundary layer thickness. The values of U_∞ and δ are 0.3 m/s and 1.0 m, respectively. The velocity gradient $\left(\frac{\partial u}{\partial y}\right)$ (in s^{-1}, round off to two decimal places at y = 0, is _____.

25. Two identically sized primary settling tanks receive water for Type-I settling (discrete particles in dilute suspension) under laminar flow conditions. The Surface Overflow Rate (SOR) maintained in the two tanks are 30 m^3/m^2. d and 15 m^3/m^2.d. The lowest diameters of particles, which shall be settled out completely under SORs of 30m^3/m^2.d and 15 m^3/m^2.d are designated as d_{30} and d_{15}, respectively. The ratio, $\frac{d_{30}}{d_{15}}$ (round off to two decimal places), is _____.

(Q 26 – Q 55) carry two marks each

26. An ordinary differential equation is given below

$$6\frac{d^2y}{dx^2} + \frac{dy}{dx} - y = 0$$

The general solution of the above equation (with constants C_1 and C_2), is

(a) $y(x) = C_1 e^{-\frac{x}{3}} + C_2 e^{\frac{x}{2}}$

(b) $y(x) = C_1 e^{\frac{x}{3}} + C_2 e^{-\frac{x}{2}}$

(c) $y(x) = C_1 x e^{-\frac{x}{3}} + C_2 e^{\frac{x}{2}}$

(d) $y(x) = C_1 e^{-\frac{x}{3}} + C_2 x e^{\frac{x}{2}}$

27. A 4 × 4 matrix [P] is given below

$$[P] = \begin{bmatrix} 0 & 1 & 3 & 0 \\ -2 & 3 & 0 & 4 \\ 0 & 0 & 6 & 1 \\ 0 & 0 & 1 & 6 \end{bmatrix}$$

The eigen values of [P] are

(a) 0, 3, 6, 6

(b) 1, 2, 3, 4

(c) 3, 4, 5, 7

(d) 1, 2, 5, 7

28. A prismatic linearly elastic bar of length L, cross-sectional area A, and made up of a material with Young's modulus E, is subjected to axial tensile force as shown in the figures. When the bar is subjected to axial tensile forces P_1 and P_2, the strain energies stored in the bar are U_1 and U_2, respectively.

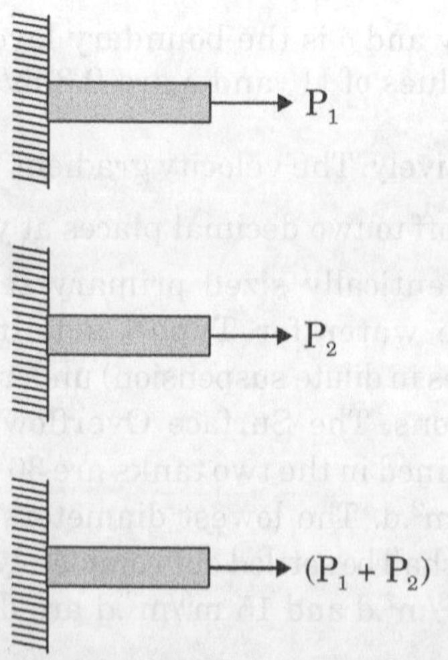

If U is the strain energy stored in the same bar when subjected to an axial tensile force $(P_1 + P_2)$, the correct relationship is

(a) $U = U_1 + U_2$ (b) $U = U_1 - U_2$

(c) $U < U_1 + U_2$ (d) $U > U_1 + U_2$

29. The planar structure RST shown in the figure is roller-supported at S and pin-supported at R. Members RS and ST have uniform flexural rigidity (EI) and S is a rigid joint. Consider only bending deformation and neglect effects of self-weight and axial stiffening.

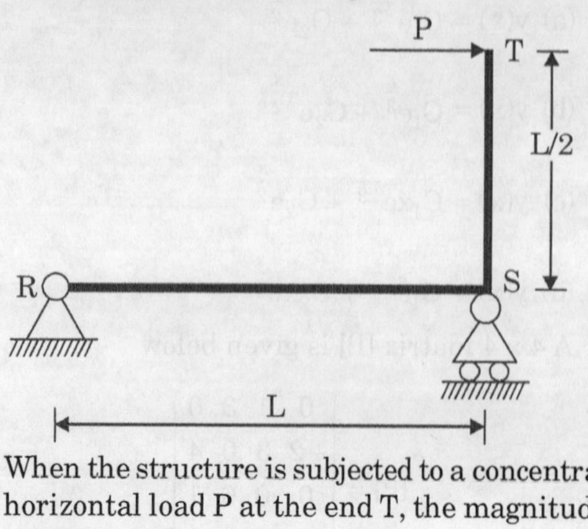

When the structure is subjected to a concentrated horizontal load P at the end T, the magnitude of rotation at the support R, is

(a) $\frac{PL^3}{12EI}$ (b) $\frac{PL^2}{12EI}$

(c) $\frac{PL^2}{6EI}$ (d) $\frac{PL}{6EI}$

30. Joints I, J, K, L, Q and M of the frame shown in the figure (not drawn to the scale) are pins. Continuous members IQ and LJ are connected through a pin at N. Continuous members JM and KQ are connected through a pin at P. The frame has hinge supports at joints R and S. The loads acting at joints, I, J and K are along the negative Y direction and the loads acting at joints L and M are along the positive X direction.

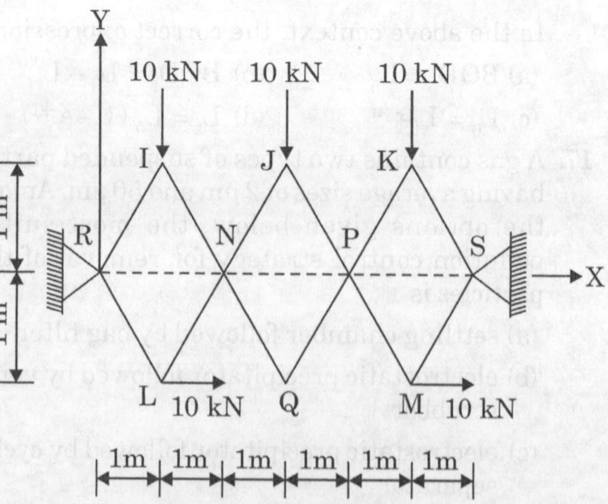

The magnitude of the horizontal component of reaction (in kN) at S, is

(a) 5 (b) 10

(c) 15 (d) 20

31. The flow-density relationship of traffic on a highway is shown in the figure

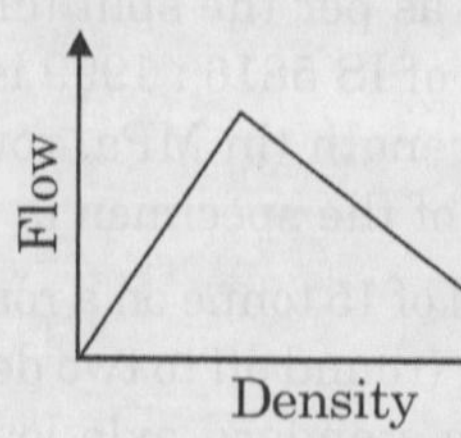

The correct representation of speed-density relationship of the traffic on this highway is

(a)

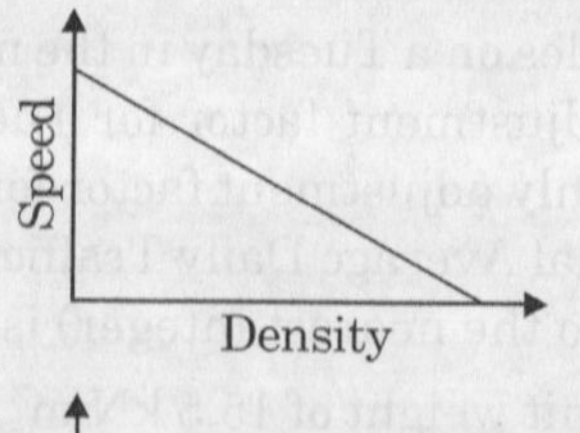

(b)

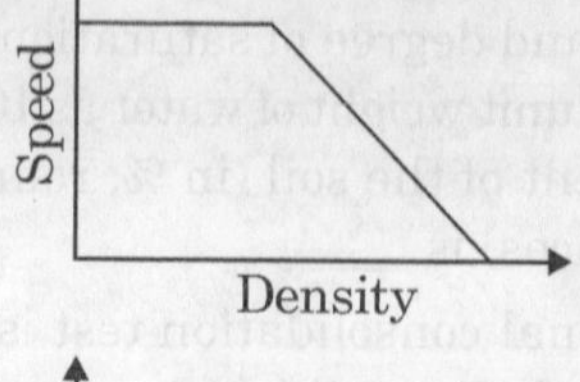

(c)

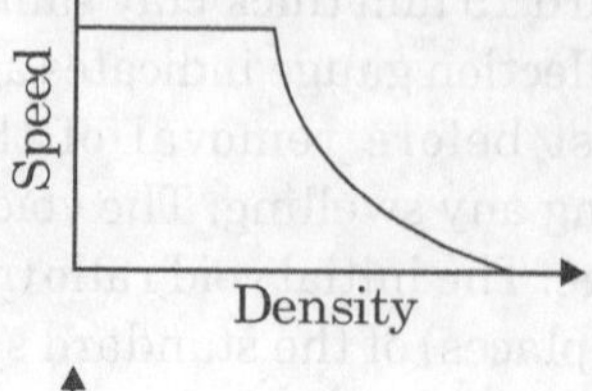

(d)

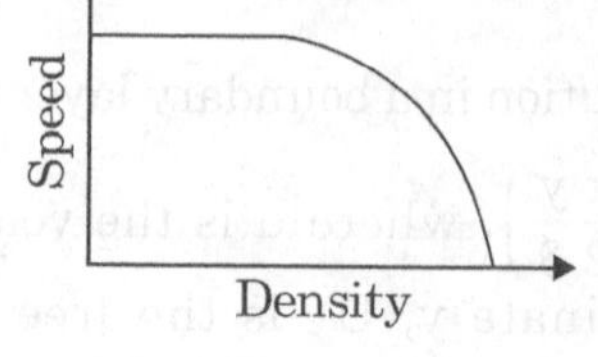

32. Group-I gives a list of test methods for evaluating properties of aggregates. Group-II gives the list of properties to be evaluated.

Group-I (Test Methods)

P. Soundness test

Q. Crushing test

R. Los Angeles abrasion test

S. Stripping value test

Group-II (Properties)

1. Strength
2. Resistance to weathering
3. Adhesion
4. Hardness

The correct match of test methods under Group-I to properties under Group-II, is

(a) P-4; Q-1; R-2; S-3

(b) P-2; Q-1; R-4; S-3

(c) P-3; Q-4; R-1; S-2

(d) P-2; Q-4; R-3; S-1

33. For the hottest month of the year at the proposed airport site, the monthly mean of the average daily temperature is 39°C. The monthly mean of the maximum daily temperature is 48°C for the same month of the year. From the given information, the calculated Airport Reference Temperature (in °C), is

(a) 36 (b) 39

(c) 42 (d) 48

34. Permeability tests were carried out on the samples collected from two different layers as shown in the figure (not drawn to the scale). The relevant horizontal (k_h) and vertical (k_v) coefficients of permeability are indicated for each layer.

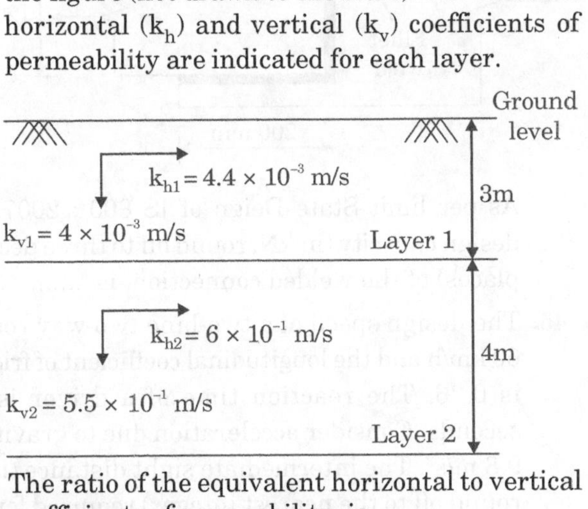

The ratio of the equivalent horizontal to vertical coefficients of permeability, is

(a) 37.29 (b) 80.20

(c) 68.25 (d) 0.03

35. A 10 m high slope of dry clay soil (unit weight = 20 kN/m^3), with a slope angle of 45° and the circular slip surface, is shown in the figure (not drawn to the scale). The weight of the slip wedge is denoted by W. The undrained unit cohesion (c_u) is 60 kPa.

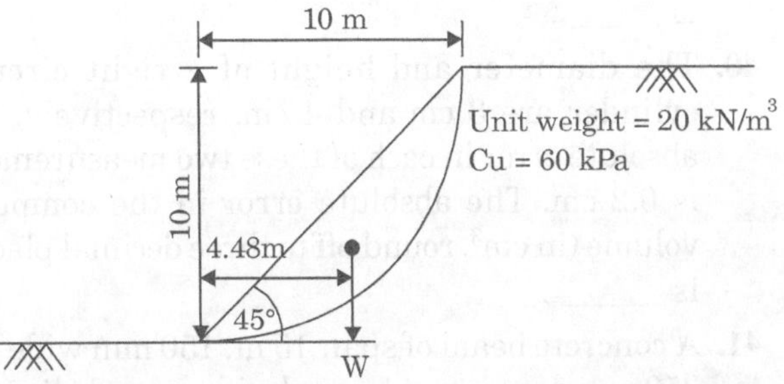

The factor of safety of the slope against slip failure, is

(a) 1.84 (b) 1.57

(c) 0.58 (d) 1.67

36. Crops are grown in a field having soil, which has field capacity of 30% and permanent wilting point of 13%. The effective depth of root zone is 80 cm. Irrigation water is supplied when the average soil moisture drops to 20%. Consider density of the soil as 1500 kg/m^3 and density of water as 1000 kg/m^3. If the daily consumptive use of water for the crops is 2 mm, the frequency of irrigating the crops (in days) is

(a) 7 (b) 10

(c) 11 (d) 13

37. Alkalinity of water, in equivalent/litre (eq/litre), is given by

$\{HCO_3^-\} + 2\{CO_3^{2-}\} + \{OH^-\} - \{H^+\}$

Where, {} represents concentration in mol/litre. For a water sample, the concentrations of HCO_3^- = 2×10^{-3} mol/litre, $CO_3^{2-} = 3.04 \times 10^{-4}$ mol/litre and the pH of water = 9.0. The atomic weights are : Ca = 40; C = 12; and O = 16. If the concentration of OH^- and H^+ are NEGLECTED, the alkalinity of water sample (in mg/litre as $CaCO_3$), is

(a) 130.4 (b) 100.0

(c) 50.0 (d) 65.2

38. A theodolite was set up at a station P. The angle of depression to a vane 2 m above the foot of a staff held at another station Q was 45°. The horizontal distance between stations P and Q is 20 m. The staff reading at a benchmark S of RL 433.050 m is 2.905 m. Neglecting the errors due to curvature and refraction, the RL of the station Q (in m), is

(a) 413.050 (b) 413.955

(c) 431.050 (d) 435.955

39. The Fourier series to represent $x - x^2$ for $-\pi \le x \le \pi$ is given by

$$x - x^2 = \frac{a_o}{2} + \sum_{n=1}^{\infty} a_n \cos nx + \sum_{n=1}^{\infty} b_n \sin nx$$

The value of a_o (round off to two decimal places), is ______ .

40. The diameter and height of a right circular cylinder are 3 cm and 4 cm, respectively. The absolute error in each of these two measurements is 0.2 cm. The absolute error in the computed volume (in cm^3, round off to three decimal places), is ______ .

41. A concrete beam of span 15 m, 150 mm wide and 350 mm deep is prestressed with a parabolic cable as shown in the figure (not drawn to the scale). Coefficient of friction for the cable is 0.35, and coefficient of wave effect is 0.0015 per meter.

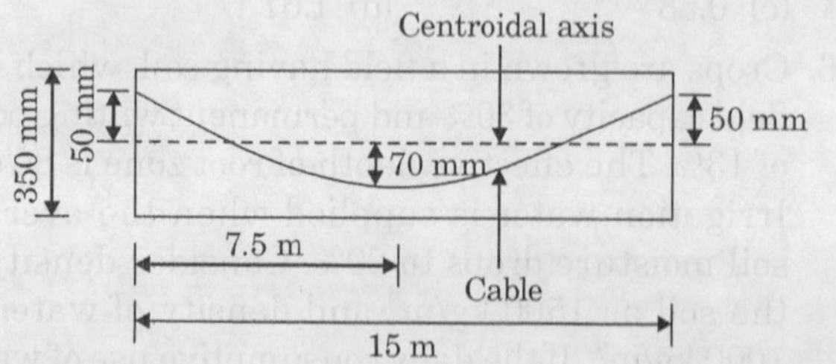

If the cable is tensioned from one end only, the percentage loss (round off to one decimal place) in the cable force due to friction, is ______ .

42. The plane truss has hinge supports at P and W and is subjected to the horizontal forces as shown in the figure (not drawn to the scale)

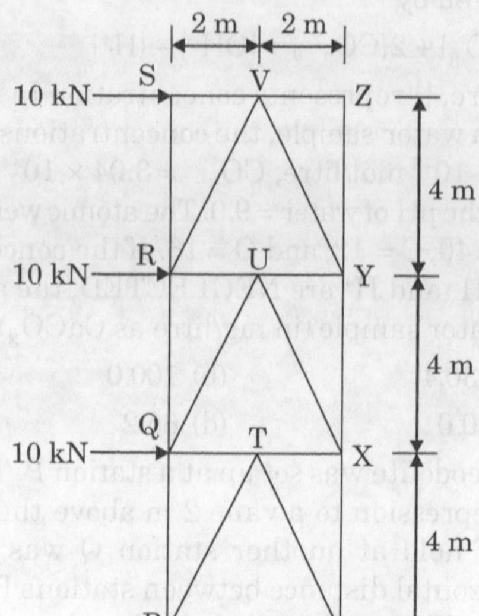

Representing the tensile force with '+' sign and the compressive force with '–' sign, the force in member XW (in kN, round off to the nearest integer), is ______ .

43. The cross-section of the reinforced concrete beam having an effective depth of 500 mm is shown in the figure (not drawn to the scale). The grades of concrete and steel used are M35 and Fe550, respectively. The area of tension reinforcement is 400 mm^2. It is given that corresponding to 0.2% proof stress, the material safety factor is 1.15 and the yield strain of Fe550 steel is 0.0044.

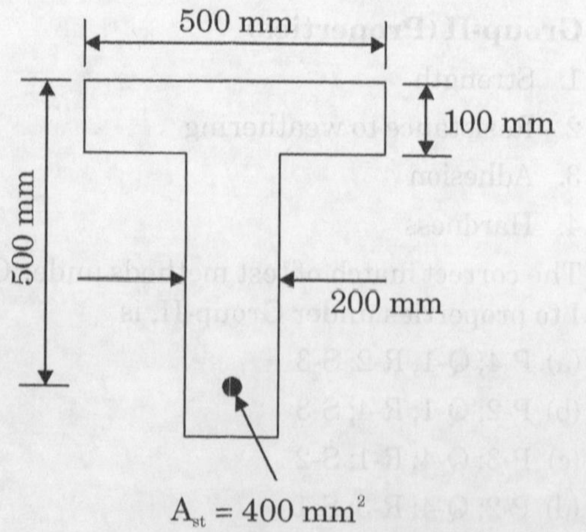

As per IS 456 : 2000, the limiting depth (in mm, round off to the nearest integer) of the neutral axis measured from the extreme compression fiber, is ______ .

44. Two steel plates are lap jointed in a workshop using 6 mm thick fillet weld as shown in the figure (not drawn to the scale). The ultimate strength of the weld is 410 MPa.

As per limit State Deign of IS 800 : 2007, the design capacity (in kN, round off to three decimal places) of the welded connection, is.............. .

45. The design speed of a two-lane two-way road is 60 km/h and the longitudinal coefficient of friction is 0.36. The reaction time of a driver is 2.5 seconds. Consider acceleration due to gravity as 9.8 m/s^2. The intermediate sight distance (in m, round off to the nearest integer) required for the road is ______ .

46. A footing of size 2 m × 2 m transferring a pressure of 200 kN/m², is placed at a depth of 1.5 m below the ground as shown in the figure (not drawn to the scale). The clay stratum is normally consolidated. The clay has specific gravity of 2.65 and compression index of 0.3.

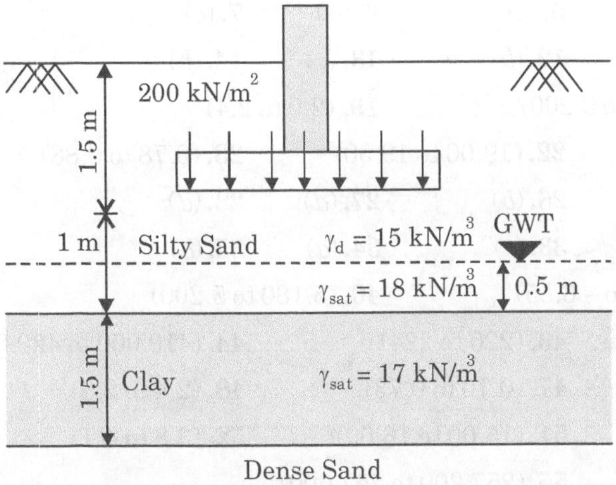

Considering 2 : 1 (vertical to horizontal) method of load distribution and γ_w = 10 kN/m³, the primary consolidation settlement (in mm, round off to two decimal places) of the clay stratum is _____.

47. A constant-head permeability test was conducted on a soil specimen under a hydraulic gradient of 2.5. The soil specimen has specific gravity of 2.65 and saturated water content of 20%. If the coefficient of permeability of the soil is 0.1 cm/s, the seepage velocity (in cm/s, round off to two decimal places) through the soil specimen is ____.

48. A 5 m high vertical wall has a saturated clay backfill. The saturated unit weight and conhesion of clay are 18 kN/m³ and 20 kPa, respectively. The angle of internal friction of clay is zero. In order to prevent development of tension zone, the height of the wall is required to be increased. Dry sand is used as backfill above the clay for the increased portion of the wall. The unit weight and angle of internal friction of sand are 16 kN/m³ and 30°, respectively. Assume that the back of the wall is smooth and top of the backfill is horizontal. To prevent the development of tension zone, the minimum height (in m, round off to one decimal place) by which the wall has to be raised, is _____ .

49. A cast iron pipe of diameter 600 mm and length 400 m carries water from a tank and discharges freely into air at a point 4.5 m below the water surface in the tank. The friction factor of the pipe is 0.018. Consider acceleration due to gravity as 9.81 m/s². The velocity of the flow in pipe (in m/s, round off to two decimal places) is _____.

50. A hydraulic jump occurs in a triangular (V-shaped) channel with side slopes 1 : 1 (vertical to horizontal). The sequent depths are 0.5 m and 1.5 m. The flow rate (in m³/s, round off to two decimal places) in the channel is _____ .

51. A concrete dam holds 10 m of static water as shown in the figure (not drawn to the scale). The uplift is assumed to vary linearly from full hydrostatic head at the heel, to zero at the toe of dam. The coefficient of friction between the dam and foundation soil is 0.45. Specific weights of concrete and water are 24 kN/m³ and 9.81 kN/m³, respectively.

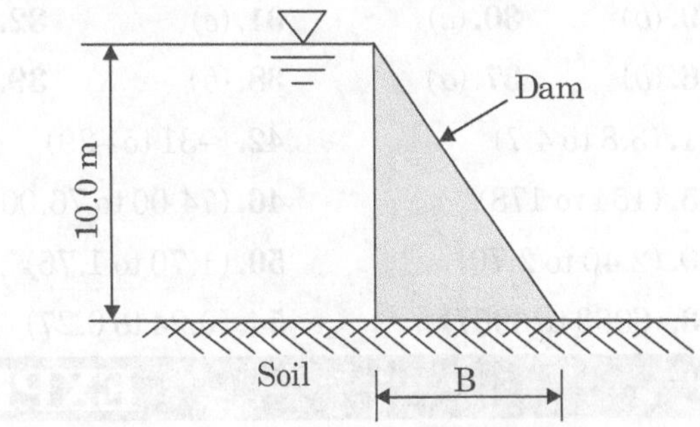

For NO sliding condition, the required minimum base width B (in m, round off to two decimal places) is _____ .

52. The ion product of water (pK_w) is 14. If a rain water sample has a pH of 5.6, the concentration of OH^- in the sample (in 10^{-9} mol/litre, round off to one decimal place), is _____ .

53. A waste to energy plant burns dry solid waste of composition : Carbon = 35%, Oxygen = 26%, Hydrogen = 10%, Sulphur = 6%, Nitrogen = 3% and Inerts = 20%. Burning rate is 1000 tonnes/d. Oxygen in air by weight is 23%. Assume complete conversion of Carbon to CO_2, Hydrogen to H_2O, Sulphur to SO_2 and nitrogen to NO_2.

Given Atomic weights : H = 1, C = 12, N = 14, O = 16, S = 32

The stoichiometric (theoretical) amount of air (in tonnes/d, round off to the nearest integer) required for complete burning of this waste, is _____ .

54. A sample of water contains an organic compound $C_8H_{16}O_8$ at a concentration of 10^{-3} mol/litre. Given that the atomic weight of C = 12 g/mol, H = 1 g/mol, and O = 16 g/mol, the theoretical oxygen demand of water (in g of O_2 per litre, round off to two decimal places), is _____ .

55. A theodolite is set up at station A. The RL of instrument axis is 212.250 m. The angle of elevation to the top of a 4 m long staff, held vertical at station B, is 7°. The horizontal distance between stations A and B is 400 m. Neglecting the errors due to curvature of earth and refraction, the RL (in m, round off to three decimal places) of station B is _____ .

ANSWERS

General Aptitude

1. (*b*) **2.** (*a*) **3.** (*d*) **4.** (c) **5.** (*c*) **6.** (*a*) **7.** (*c*) **8.** (*c*) **9.** (*b*) **10.** (*b*)

Civil Engineering

1. (*b*) **2.** (*c*) **3.** (*d*) **4.** (*d*) **5.** (*c*) **6.** (*d*) **7.** (*c*)
8. (*d*) **9.** (*d*) **10.** (*a*) **11.** (*b*) **12.** (*b*) **13.** (*c*) **14.** (*b*)
15. (*c*) **16.** (*b*) **17.** (*a*) **18.** (0.190 to 0.200) **19.** (2.0 to 2.4)
20. (12.30 to 12.40) **21.** (1021 to 1024) **22.** (19.00 to 19.50) **23.** (0.78 to 0.88)
24. (0.45 to 0.49) **25.** (1.40 to 1.42) **26.** (*b*) **27.** (*d*) **28.** (*d*)
29. (*b*) **30.** (*c*) **31.** (*c*) **32.** (*b*) **33.** (*c*) **34.** (*a*) **35.** (*a*)
36. (*b*) **37.** (*a*) **38.** (*b*) **39.** (–6.61 to – 6.55) **40.** (5.180 to 5.200)
41. (3.8 to 4.7) **42.** (–31 to –29) **43.** (220 to 224) **44.** (410.000 to 420.000)
45. (154 to 178) **46.** (74.00 to 76.00) **47.** (0.70 to 0.73) **48.** (2.4 to 2.6)
49. (2.40 to 2.70) **50.** (1.70 to 1.76) **51.** (15.00 to 16.00) **52.** (3.8 to 4.1)
53. (6963 to 6967) **54.** (0.24 to 0.27) **55.** (257.200 to 257.600)

EXPLANATIONS

GENERAL APTITUDE

1.
$$f(x) = x^2$$
$$f(f(x)) = f(x^2) = x^4$$
$$f(f(f(x))) = f(x^4) = x^8$$

So, $$\frac{f(f(f(x)))}{f(x)} = \frac{x^8}{x^2} = x^6 = (f(x))^3$$

2. Then term **'inflation-adjusted'** suggest option (c).

3. Belittle means disparage which means to describe someone as unimportant.

5. Deployed in : combat with

6. 1. P wanted a room adjacent to lab option 'c' eliminated.

2. S wanted a corner office option 'd' eliminated.

3. Q want to be close to the shift option 'b' eliminated.

Best possible option A, satisfying faculty preferences

7. Antonyms analogy is applied.

8. Non-leap year repeats after 6 (or) 11 years

2013 + 6 = 2019

2013 & 2019 calender is same

Method - 2

	2013	2014	2015	2016	2017	2018	2019
odd days	1	1	1	2	1	1	

odd days = 7= 0

So, 2013 & 2019 have same calendar

9. Sum of odd positive integers from 1 to 100

$$= 50\frac{(1+99)}{2} = 2500$$

Sum of even positive integers from 150 to 200

$$= 26\frac{(150+200)}{2} = 13 \times 350$$

$$\text{Ratio} = \frac{2500}{13 \times 350} = \frac{50}{91}$$

10.

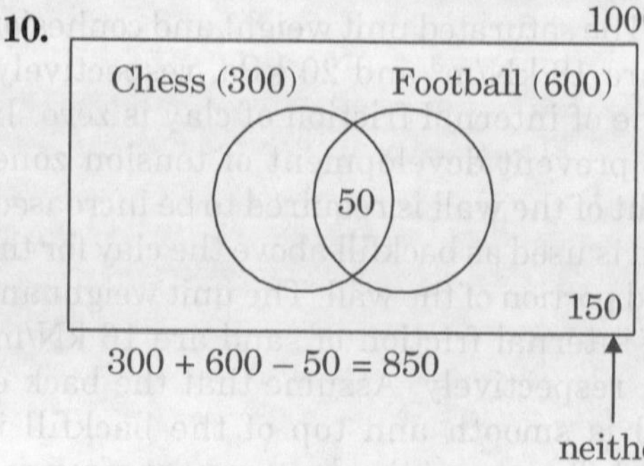

300 + 600 − 50 = 850

150 students play neither chess not football

CIVIL ENGINEERING

1. Hydrologic channel routing is done by Muskingum method.

2. $$\frac{d^2u}{dx^2} - 2x^2u + \sin x = 0$$

It is a linear non-homogeneous equation

3. Given : $$\frac{U}{U_\infty} = \sin\left(\frac{\pi}{2}\frac{y}{\delta}\right)$$

$$U_\infty = 0.3 \text{ m/s}$$
$$\delta = 1 \text{ m}$$

To find: $\left(\dfrac{du}{dy}\right)_{y=0}$

$$\frac{du}{dy} = U_\infty \cdot \frac{\pi}{2\delta} \cdot \cos\left(\frac{\pi}{2}\frac{y}{\delta}\right)$$

$$\cos\left(\frac{\pi}{2}\frac{y}{\delta}\right) = 1$$

$$\Rightarrow \left(\frac{du}{dy}\right)_{y=0} = \frac{0.3 \times \pi}{2 \times 1} = 0.471\ s^{-1}$$

4. As per clause 5.4.2 of IS 456 : 2000, the pH value of water should not be less than 6.
5. Saturation headways starts after fourth or fifth headway
6. In Super passage, canal runs below drain. FSL of canal is lower than bed of natural stream canal water runs under gravity.
7. ADT = Daily adjustment factor × 24 hr traffic

   ADT = 1.21 × 1000 = 1121 v/day

   AADT = Monthly exp factor × ADT

   = 0.913 × 1121

   = 1023.5 vehicles/day
8. $$VDF = \left(\frac{\text{actual load}}{\text{standard load}}\right)^4$$

   $$VDF = \left(\frac{15}{8}\right)^4 = 12.359 \approx 12.36$$
9. Given : $\dfrac{\partial u}{\partial y} = \dfrac{\partial^2 u}{\partial x^2}$; $y \geq 0$; $x_1 \leq x \leq x_2$

   It needs one initial and two boundary conditions to solve.
10. Under pure shear condition, centre of Mohr's circle coincides with the origin.
11. 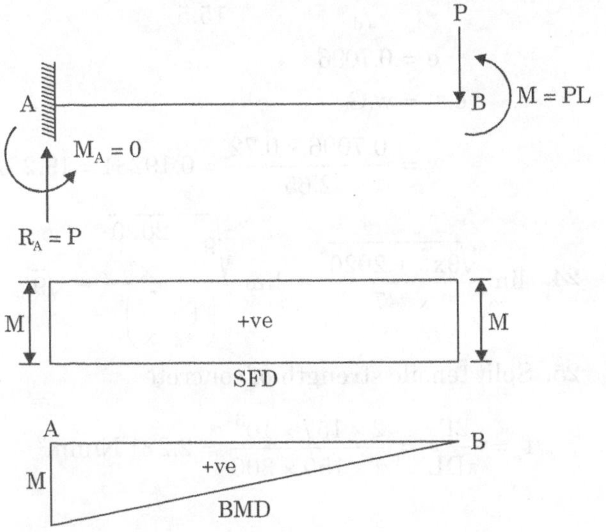

$$\Sigma F_y = 0$$
$$R_A - P = 0$$
$$R_A = P$$
$$+\curvearrowright \Sigma M_A = 0$$
$$P(L) - PL - M_A = 0$$
$$M_A = 0$$

For most economical (Beam of uniform strength) non prismatic beam should be used.

Maximum cross section is provided where the bending moment is maximum

i.e. 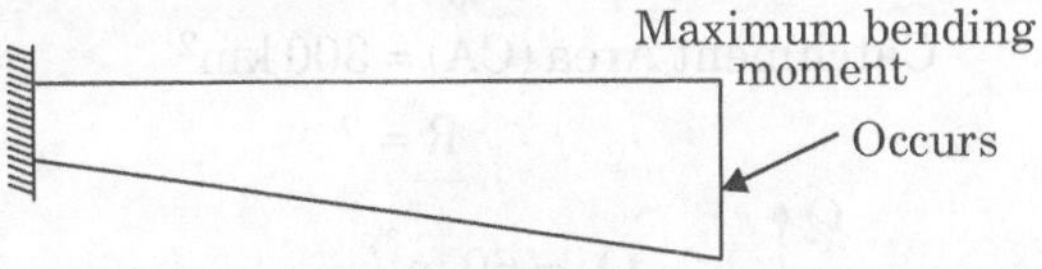

12. Particles of 50 μ separated by gravity settling chamber and 2 μ are separated by Fabric filters [bag filters]
13. **As per BOD Kinetics :**

   L_o = Initial organic matter concentration

   L_t = Remaining organic matter concentration at any time 't'

   $$\therefore \frac{dL}{dt} \propto L_t$$

   Consider exponential decline, $\dfrac{dL}{dt} = -K.L_t$

   $$\int \frac{dL}{dt} = \int -Kdt$$

   $$\therefore \ln(L_t) = -k(t) + C$$

   Applying Boundary Conditions :

   At $t = 0,\ L_t = L_o$

   $$\ln(L_o) = C$$

   $$\therefore \ln(L_t) = -Kt + \ln(L_o)$$

   $$\therefore \frac{L_t}{L_o} = e^{-kt}$$

   Organic matter remaining at time t,

   $$L_t = L_o \cdot e^{-Kt}$$

   y_t = BOD consumed durintg 't' days

   = Organic matter consumed

   $$y_t = L_o - L_t$$

   $$\therefore y_t = L_o - L_o \cdot e^{-kt}$$

   $$= L_o\,[1 - e^{-kt}]$$
14. Initial After loading

   H = 19 mm ΔH = 2.1 mm

   e_o = ? $e_f = 0.62$

   $$\frac{\Delta H}{H} = \frac{\Delta e}{1 + e_o}$$

$$\frac{2.1}{19} = \frac{e_o - 0.62}{1 + e_o}$$

$$0.11 = \frac{e_o - 0.62}{1 + e_o}$$

$$0.11 + 0.11\,e_o = e_o - 0.62$$

$$e_o = 0.820$$

$$e_o = 0.82 \quad \text{...Ans.}$$

15. Triangular Direct runoff hydrograph

Time base 90 hrs = T_B

$Q_p = 60\ m^3/s$

Catchment Area (CA) = 300 km^2

R = ?

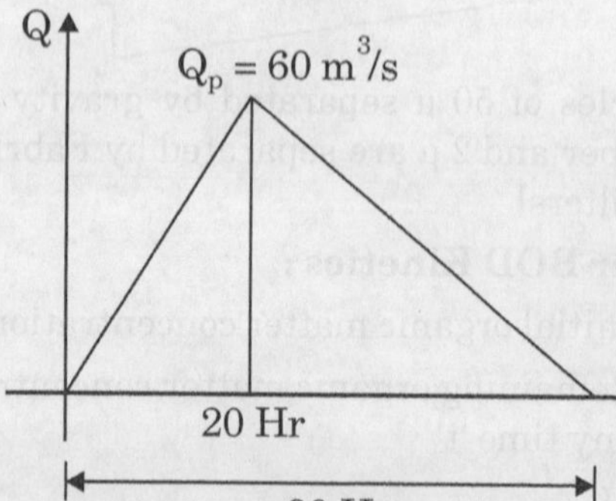

Area of DRH = C.A × R

$$\frac{1}{2} \times 90 \times 3600 \times 60 = 300 \times 10^6 \times R$$

$$R = 0.0324\ m$$

$$R = 3.24\ cm$$

16. Soil deposited by wind is known as Aeolian soil

Soil deposited by water on river bed is known as Alluvial soil

Soil deposited by water lake bed is known as Lacustrine soil.

17. Given : $n = 15,\ P = \frac{1}{2},\ q = \frac{1}{2}$

Let x = Number of heads

$$P(x = 8) = {}^nC_x p^x q^{n-x}$$

$$= {}^{15}C_8 \left(\frac{1}{2}\right)^8 \left(\frac{1}{2}\right)^7 = \frac{{}^{15}C_8}{(2)^{15}} = 0.196$$

18. The integrand is

$f(x) = 5x^3 + 4x^2 + 3x + 2$

It is a polynomial of degree 3

Simpson's rule gives the exact value of the integral

19. Given :

Weight of solid particles, W_S = 500 g

The volume of water replaced by dry soil, is equal to the volume of the soil solid particles.

$$\therefore \quad V_S = 188\ cm^3$$

$$\gamma_S = \frac{W_S}{V_S} = \frac{500}{188} = 2.66\ g/cc$$

The specific gravity of soil (soil solids),

$$G_S = \frac{\gamma_S}{\gamma_W} = \frac{2.66}{1}$$

$$\therefore \quad G_S = 2.66$$

20. For PST, $V_o = V_s$

$$V_s \propto d^2$$

$$\therefore \quad \frac{d_{30}^2}{d_{15}^2} = \frac{(V_s)_{30}}{(V_s)_{15}}$$

$$\therefore \quad \frac{d_{30}}{d_{15}} = \sqrt{\frac{30}{15}} = \sqrt{2}$$

21. $$\text{Shape factor} = \frac{M_p}{M_y} = \frac{\text{Plastic Moment Capacity}}{\text{Yield Moment Capacity}}$$

22.
$$u = (-5 + 6x)$$

$$v = -(9 + 6y)$$

$$\frac{dy}{dx} = \frac{-(9 + 6y)}{(-5 + 6x)}$$

$$\Rightarrow \quad \frac{dx}{(-5 + 6x)} = \frac{-dy}{(9 + 6y)}$$

By integration

$$\frac{1}{6}\ln(-5 + 6x) = -\frac{1}{6}\ln(9 + 6y) + \ln C$$

$$\Rightarrow (-5 + 6x)(9 + 6y) = \text{constant}$$

23. $\gamma = 15.5\ kN/m^3 \qquad G = 2.65$

$S_r = 72\% \qquad \gamma_w = 10\ kN/m^3$

w = ?

$$\gamma_d = \frac{G\gamma_w}{1 + e}$$

$$e = \frac{G\gamma_w}{\gamma_d} - 1 = \frac{2.65 \times 10}{15.5} - 1$$

$$\therefore \quad e = 0.7096$$

$$\therefore \quad e.s_r = w.G$$

$$w = \frac{0.7096 \times 0.72}{2.65} = 0.19281 = 19.28\%$$

24. $$\lim_{x \to \infty} \frac{\sqrt{9x^2 + 2020}}{x + 7} = \lim_{x \to \infty} \frac{\sqrt{9 + \frac{2020}{x^2}}}{\left(1 + \frac{7}{x}\right)} = \sqrt{9} = 3$$

25. Split tensile strength of concrete

$$f_s = \frac{2P}{\pi DL} = \frac{2 \times 157 \times 10^3}{\pi \times 150 \times 300} = 2.221\ N/mm^2$$

26. $CO_3^{-2} = 3.04 \times 10^{-4}$ mole/L

$= 3.04 \times 10^{-4} \times 60 \times 10^3$

$= 18.24$ mg/L

$HCO_3^- = 2 \times 10^{-3}$ mole/L

$= 2 \times 10^{-3} \times 61 \times 10^3$

$= 122$ mg/L

$$\therefore \text{Total Alkalinity} = [CO_3^{-2}] \times \frac{50}{30} + [HCO_3] \times \frac{50}{61}$$

$$\therefore \quad TA = 18.24 \times \frac{50}{30} + 122 \times \frac{50}{61}$$

$$= 30.4 + 100$$

$$= 130.4 \text{ mg/L as } CaCO_3$$

27.

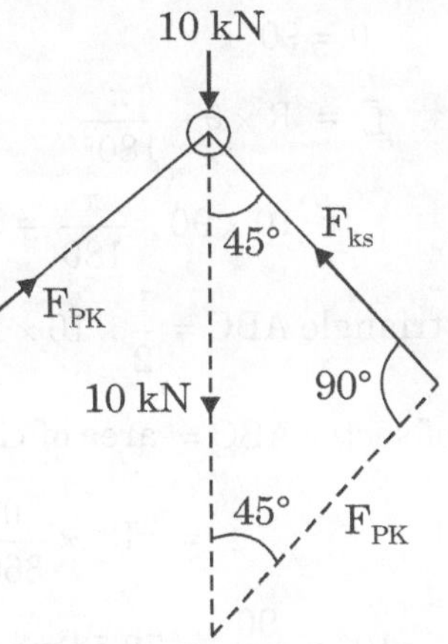

By applying Lami's Theorem

$$\frac{F_{KS}}{\sin 45} = \frac{10}{\sin 90}$$

$$F_{KS} = \frac{10}{\sqrt{2}} \text{ kN(compression)}$$

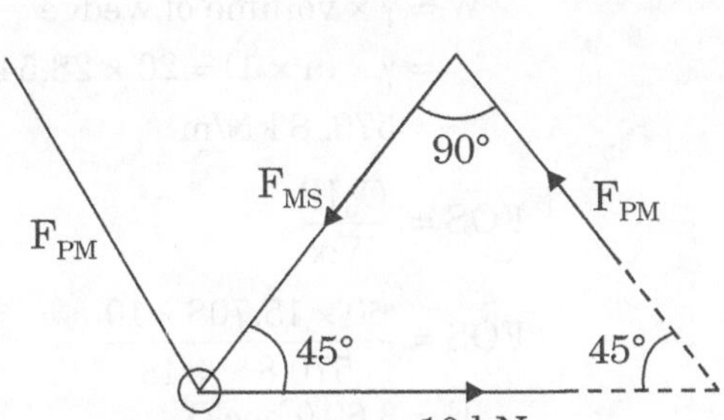

By applying Lami's theorem

$$\frac{F_{MS}}{\sin 45} = \frac{10}{\sin 90}$$

$$F_{MS} = \frac{10}{\sqrt{2}} \text{(compression)}$$

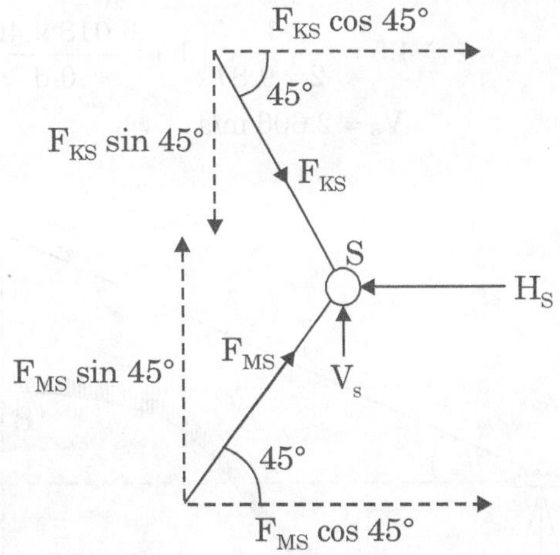

$V_s = 0$

The magnitude of the horizontal component of reaction at S

$$H_S = F_{KS} \cos 45° + F_{MS} \cos 45°$$

$$= \frac{10}{\sqrt{2}} \times \frac{1}{\sqrt{2}} + \frac{10}{\sqrt{2}} \times \frac{1}{\sqrt{2}} = 10 \text{ kN}$$

28.

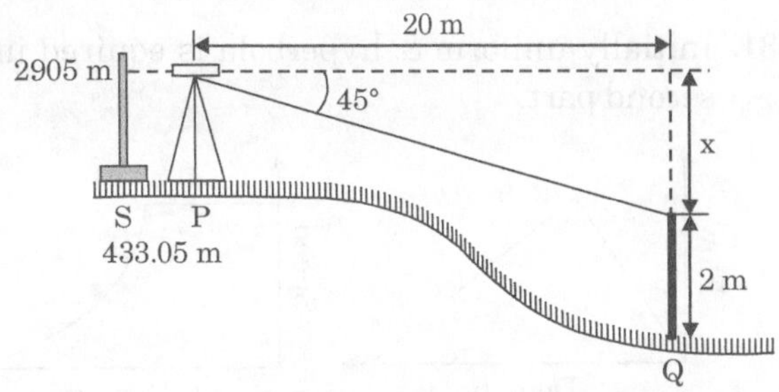

$$\tan 45° = \frac{x}{20} = x = 20 \text{ m}$$

$$\text{RL of Q} = 433.05 + 2.905 - 20 - 2$$

$$= 413.955 \text{ m}$$

29. Given :

L = 400 m

D = 0.6 m

f = 0.018

g = 9.81 m/s²

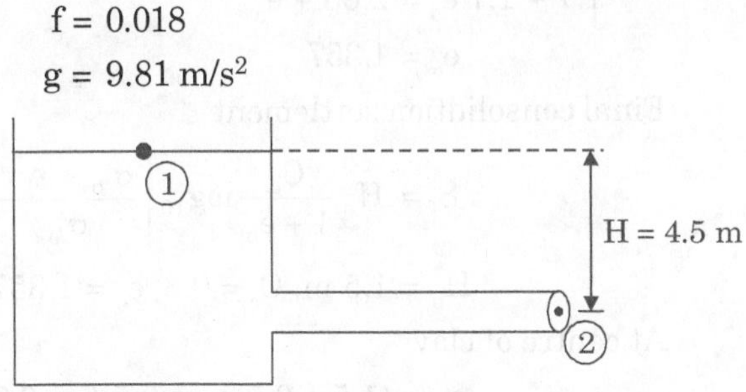

To find : V = ?

By applying Bernaulli's equation between (1) and (2)

$$\frac{P_1}{q} + \frac{V_1^2}{2g} + Z_1 = \frac{P_2}{q} + \frac{V_2^2}{2g} + Z_2 + h$$

$$4.5 = \frac{V_2^2}{2g} + \frac{fL}{D}\frac{V_2^2}{2g}$$

$$\Rightarrow \quad 4.5 = \frac{1}{2 \times 9.81}\left(1 + \frac{0.018 \times 400}{0.6}\right)V_2^2$$

$$\Rightarrow \quad V_2 = 2.606 \text{ m/s}$$

30.

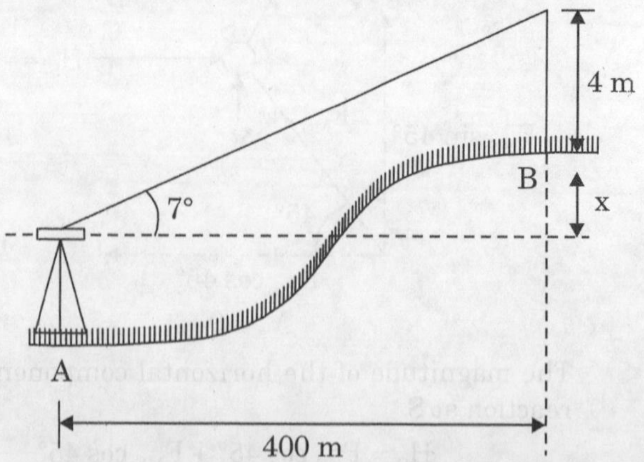

$(4 + x) = 400 \tan 7° \Rightarrow x = 45.113$ m

RL of B = 212.25 + 45.113 = 257.363 m

31. Initially uniform & hyperbola is equired in the second part.

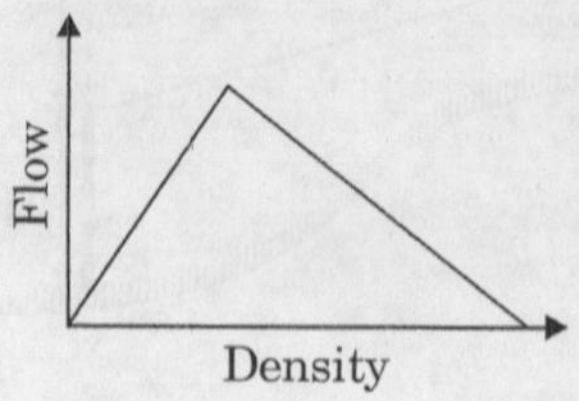

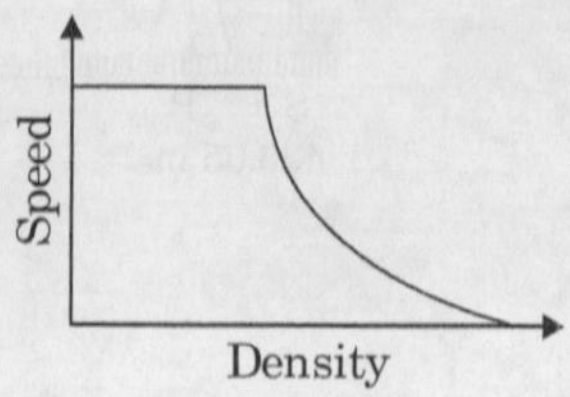

32. Footing size = 2m × 2m

Clay :

$$\gamma_{sat} = \frac{G + e_o}{1 + e_o} \cdot \gamma_w$$

$$\frac{17}{10} = \frac{2.65 + e_o}{1 + e_o}$$

$$1.7 + 1.7\, e_o = 2.65 + e_o$$

$$e_o = 1.357$$

Final consolidtion settlement

$$S_f = H_o \frac{C_c}{1 + e_o} \log_{10}\left[\frac{\sigma'_o + \Delta\sigma'}{\sigma'_o}\right]$$

$$H_o = 1.5 \text{ m}, C_c = 0.3, e_o = 1.357$$

At centre of clay

$$\sigma'_o = (1.5 + 0.5)\,\gamma_d + 0.5\gamma'_1 + 0.7\,\gamma'_2$$

$$\sigma'_o = 2 \times 15 + 0.5 \times 8 + 0.75 \times 7$$

$$\sigma'_o = 39.25 \text{ kPa}$$

Increase in stress at centre of clay layer due to footing load,

$$\Delta\sigma' = \frac{Q}{(B + Z)^2} \text{ ... for 2V and 1H load dipersion}$$

$$\Delta\sigma' = \frac{800}{(2 + 1.75)^2} = 56.89 \text{ kPa}$$

$$S_f = \frac{0.3}{1 + 1.357} \times 1.5 \times \log_{10}\left(\frac{39.25 + 56.89}{39.25}\right)$$

$$S_f = 0.7427 \text{ m} = 74.27 \text{ mm}$$

33.

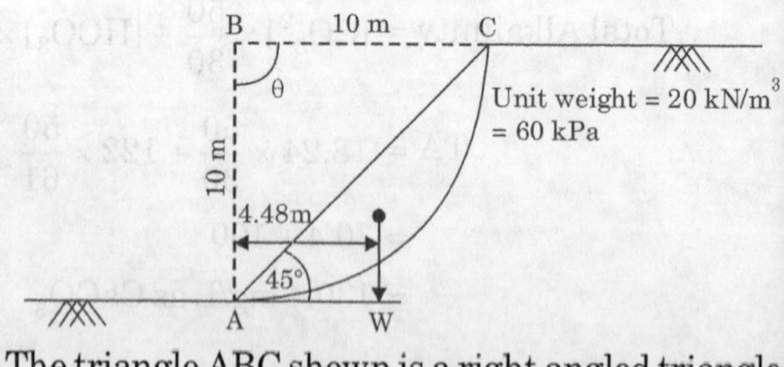

The triangle ABC shown is a right angled triangle

$\therefore \quad \theta = 90°$

$$\bar{L} = R \times \theta \times \frac{\pi}{180°}$$

$$= 10 \times 90 \times \frac{\pi}{180} = 15.708 \text{ m}$$

$$\text{Area of the triangle ABC} = \frac{1}{2} \times 10 \times 10 = 50 \text{ m}^2$$

$$\text{Area of sector ABC} = \text{area of circle} \times \frac{\theta°}{360°}$$

$$= \pi R^2 \times \frac{\theta}{360}$$

$$\pi 10^2 \times \frac{90}{360} = 78.54 \text{ m}^2$$

Area of the wedge

= area of the sector – area of triangle

a = 78.54 – 50

a = 28.54 m²

Weight of the wedge

W = γ × volume of wedge

= γ × (a × 1) = 20 × 28.54 × 1

= 570. 8 kN/m

$$\text{FOS} = \frac{C_u IR}{W \cdot x}$$

$$\text{FOS} = \frac{60 \times 15.708 \times 10}{570.8 \times 4.48}$$

F = 3.69 (check)

34.

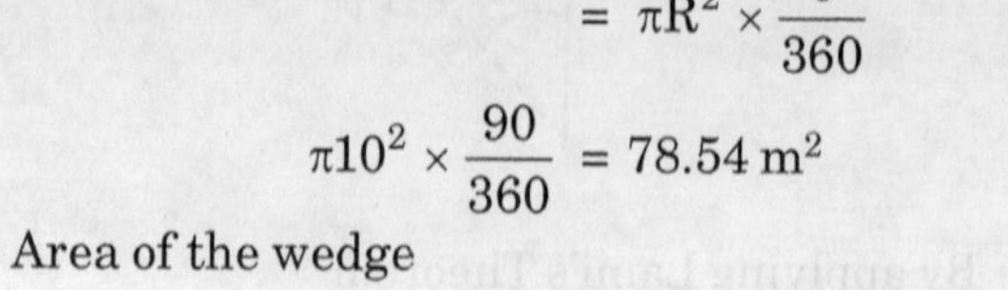

$$QR = \frac{ML}{6EI} = \frac{\left(\frac{PL}{2}\right) \times L}{6EI} \times \frac{PL^2}{12EI}$$

35.
$$ART = T_a + \left(\frac{T_m - T_a}{3}\right)$$
$$= 39 + \left(\frac{48 - 39}{3}\right) = 42°C$$

36.

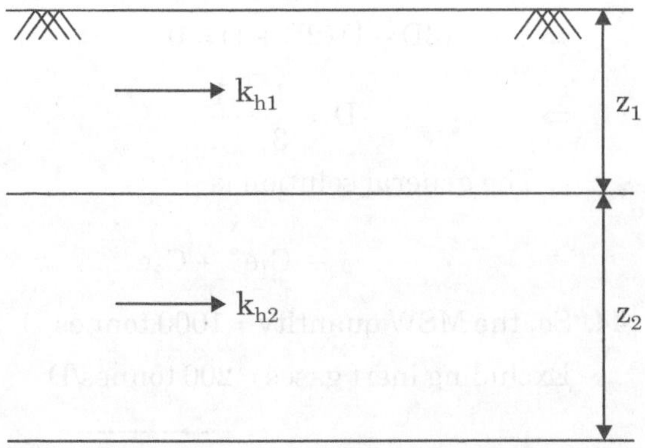

K_h :
$K_{h_1} = 4.4 \times 10^{-3}$ m/s
$K_{h_2} = 6 \times 10^{-1}$ m/s

K_v :
$k_{v_1} = 4 \times 10^{-3}$ m/s
$K_{v_2} = 5.5 \times 10^{-1}$

Equivalent Horizontal permeability,
$$(K_h)_e = \frac{K_{h_1} Z_1 + K_{h_2} Z_2}{Z_1 + Z_2}$$
$$= \frac{4.4 \times 10^{-3} \times 3 + 6 \times 10^{-1} \times 4}{3 + 4}$$
$$= 0.3447 \text{ m/s}$$

Equivalent vertical permeability,

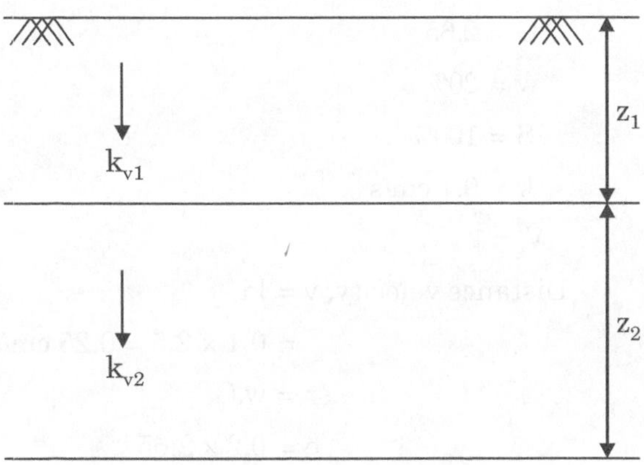

$$(K_v)_e = \frac{Z_1 + Z_2}{\frac{Z_1}{K_{v_1}} + \frac{Z_2}{K_{v_2}}}$$
$$(K_v)_e = \frac{3 + 4}{\frac{3}{4 \times 10^{-3}} + \frac{4}{5.5 \times 10^{-1}}} = \frac{7}{757.272}$$
$$= 9.2437 \times 10^{-3}$$
$$\left(\frac{K_h}{k_v}\right)_e = \frac{0.3447}{9.2437 \times 10^{-3}} = 37.29$$

37. $V = \frac{\pi d^2 h}{4}$
$$\delta V = \frac{2\pi dh\delta d}{4} + \frac{\pi d^2 \delta h}{4}$$
given $\delta d = 0.2$ cm, $\delta h = 0.2$ cm
$d = 3$ cm, $h = 4$ cm
$$\delta V = \frac{2\pi + 3 \times 4 \times 0.2}{4} + \frac{\pi \times 3^2 \times 0.2}{4}$$
$$\delta V = 1.65\pi$$
$$\delta V = 5.1836 \text{ cm}^3 = 5.184 \text{ cm}^3$$

38. Limiting depth of N.A
$$X_{u\,max} = \left[\frac{0.0035}{0.0035 + 0.0044}\right] d = 0.44 \text{ d}$$
$$= 0.44 \times 500 = 220 \text{ mm}$$
$X_{u\,max} > 100$ mm $\therefore$ N.A lies in web

39. Design Shear capacity of fillet weld (P_{dw}) = P
$$P_{dw} = P = \frac{L_w + t_t + f_u}{\sqrt{3} \times \gamma_{mw}}$$
$$L_w = 2 \times 200 + 120 = 520 \text{ mm}$$
$$t_t = K \times S = 0.7 \times 6 = 4.2 \text{ mm}$$
$\gamma_{mw} = 1.25$ (for workshop welding)
$= 1.50$ (for site welding)
$$P = P_{dw} = \frac{520 \times 4.2 \times 410}{\sqrt{3} \times 1.25}$$
$$= 413586.02 \text{ N}$$
$$= 413.586 \text{ kN}$$

40.

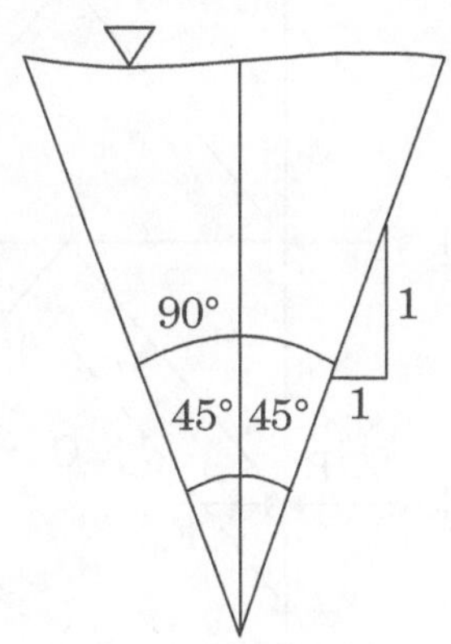

Given : Side slopes = 1 : 1
depths $y_1 = 0.5$ m
$y_2 = 1.5$ m
To find : Flow rate (m^3/s)
Specific force equation :
$$\frac{Q^2}{g.A_1} + z_1 = \frac{Q^2}{g.A_2} + z_2$$

For triangle channel $z_1 = \frac{y_1}{3}$,

$$z_2 = \frac{y_2}{3}$$

$\therefore \quad A_1 = y_1^2,$

$$A_2 = y_2^2$$

$$\Rightarrow \frac{Q^2}{g}\left[\frac{1}{y_1^2} - \frac{1}{y_2^2}\right] = \frac{1}{3}\left[y_2^3 - y_1^3\right]$$

$$\Rightarrow \frac{Q^2}{9.81}[3.556] = \frac{1}{3}[3.25]$$

$$\Rightarrow Q^2 = \frac{3.25 \times 9.81}{3 \times 3.556}$$

$$\Rightarrow Q^2 = 2.9886$$

$$\Rightarrow Q = \sqrt{2.9886}$$

$$\Rightarrow Q = 1.7287$$

$$\simeq 1.73 \text{ m}^3/\text{s}$$

41. $pH = 5.6$

$p^{OH} = 14 - 5.6$

$= 8.4$

$[OH^-] = 10^{-84}$ mole/L

$= 3.98 \times 10^{-9}$ mole/L

42.

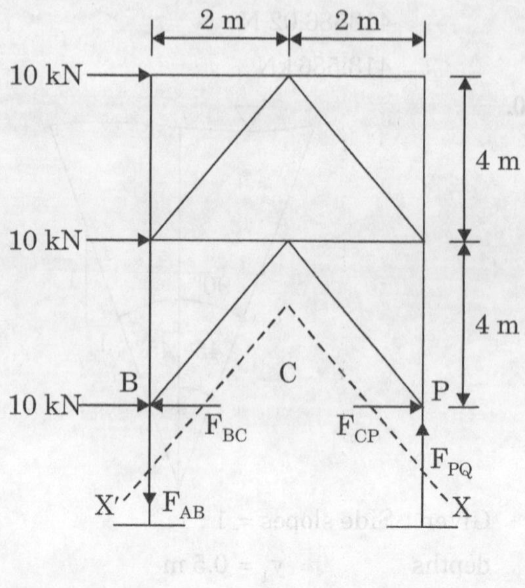

Adopting method of sections and taking section X-X as shown in above figure and selecting the upper part of section cut :

$\Sigma M_B = 0 \Rightarrow 10 \times 4 + 10 \times 8 - F_{PQ} \times 4 = 0$

$$F_{PQ} = \frac{10 \times 4 + 10 \times 8}{4} = 30 \text{ kN (compression)}$$

43. The auxiliary equation is

$$6D^2 + D - 1 = 0$$

$$\Rightarrow 6D^2 + 3D - 2D - 1 = 0$$

$$\Rightarrow 3D(2D + 1) - (2D + 1) = 0$$

$$\Rightarrow (3D - 1)(2D + 1) = 0$$

$$\Rightarrow D = \frac{1}{3}, -\frac{1}{2}$$

$\therefore$ The general solution is

$$y = C_1 e^{\frac{x}{3}} + C_2 e^{-\frac{x}{2}}$$

44. So, the MSW quantity = 1000 tonnes/D

Excluding inert gases = 200 tonnes/D

---

800 tonnes/D

80 gm MSW needs 160 g of Oxygen 800 tonnes MSW needs :-

$$O_2 = \frac{800 \times 160}{80} = 1600 \text{ ton/day}$$

Air contains only 0.23 [23%] of O_2,

So, $\quad \text{Air} = \frac{1600}{0.23} = 6956.5$ ton/D

45. Constant Head test :

$i = 2.5$

$G = 2.65$

$w = 20\%$

$S = 100\%$

$k = 0.1$ cm/s

$V_s = ?$

Distance velocity, $v = ki$

$= 0.1 \times 2.5 = 0.25$ cm/s

$\therefore \quad e.sr = w.G$

$e = 0.2 \times 2.65$

$e = 0.53$

$$n = \frac{e}{1 + e}$$

$= 0.3464$

Seepage velocity, $v_s = \frac{v}{n}$

$$\therefore \quad V_s = \frac{0.25}{0.3464}$$

$= 0.7216$ cm/s

$V_s = 0.72$ cm/s

46.

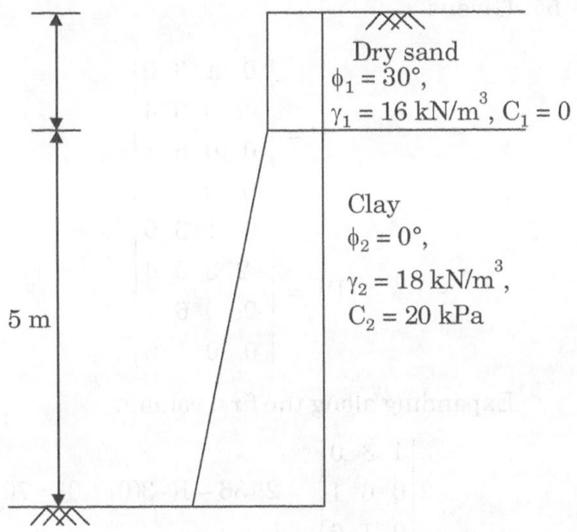

$$K_{a_1} = \frac{1 - \sin\phi_1}{1 + \sin\phi_1} = \frac{1 - \sin 30}{1 + \sin 30} = \frac{1}{3}$$

$$K_{a_2} = \frac{1 - \sin\phi_2}{1 + \sin\phi_2} = 1$$

For no tension crack condition, the active pressure at the top of clay shall be zero.

At the depth of 'x' from top, as shown in figure, the active pressure, P_a just below the sand is :

$$P_a = K_{a_2}\sigma_v - 2C_2\sqrt{K_{a_2}}$$

$$0 = 1 \times \gamma_1 x - 2 \times 20\sqrt{1}$$

$$0 = 1 \times 16x - 2 \times 20\sqrt{1}$$

$$x = 2.5\ m$$

47. Let $f(x) = x - x^2, -\pi \le x \le \pi$

$$a_o = \frac{1}{\pi}\int_{-\pi}^{\pi} f(x)\,dx$$

$$= \frac{1}{\pi}\int_{-\pi}^{\pi} (x - x^2)\,dx$$

$$= \frac{1}{\pi}\left\{\int_{-\pi}^{\pi} x\,dx - \int_{-\pi}^{\pi} x^2\,dx\right\}$$

$$= \frac{1}{\pi}\left[0 - 2\int_0^{\pi} x^2\,dx\right]$$

$$= \frac{1}{\pi}\left[-2\left(\frac{x^3}{3}\right)_0^{\pi}\right]$$

$$= \frac{1}{\pi}\left[-\frac{2\pi^3}{3}\right]$$

$$= -\frac{2}{3}\pi^2 = -6.57$$

48.

$$d_w = sd\,(EC - OMC)$$

$$= \frac{1500}{1000}\,80\,\frac{(30 - 20)}{100}$$

$$= 12\ cm = 120\ mm$$

$$c_u = 2\ mm/day$$

$$f = \frac{dw}{c_u} = \frac{120}{2} = 60\ days$$

49. Loss of stress due to friction

$$\Delta\sigma = \sigma_o[\mu\alpha + KL]$$

% Loss of stress due to friction when cable is tensioned from one end

$$\Rightarrow \quad h = 50 + 70\ mm$$

$$= 120\ mm = 0.12\ m$$

$$\Rightarrow \quad \alpha = \frac{8h}{L} = \frac{8 \times 0.12}{15} = 0.064$$

$$\Rightarrow \quad \frac{\Delta\sigma}{\sigma_o} \times 100 = \frac{\sigma_o[\mu\alpha + KL]}{\sigma_o} \times 100$$

$$= [0.35 \times 0.064 + 0.0015 \times 15] \times 100$$

$$= 4.49\% \approx 4.5\%$$

50.

$$U_{Axial\ load} = \frac{P^2L}{2AE}$$

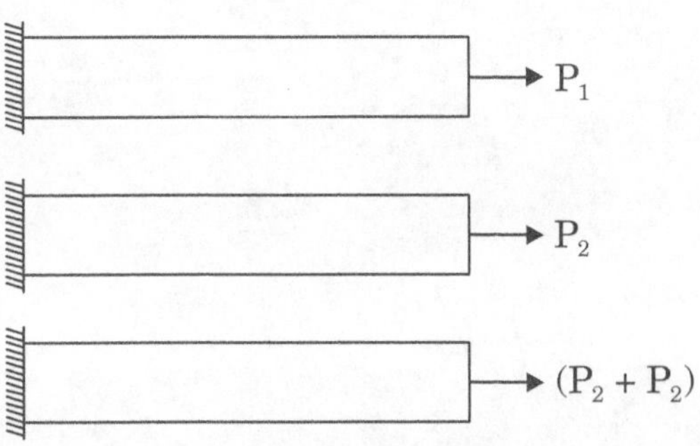

as $(P_1 + P_2)^2 > P_1{}^2 + P^2{}_2$

$$U > U_1 + U_2$$

51.
1. Soundness Test of aggregates is used to determine their resistance against weathering affects.
2. Crushing test is used to determine the strength of the aggregates.
3. Los Angeles Abrasion test is used to determine the harness of aggregates.
4. Stripping value test is used to determine the adhesion better aggregates and bitumen.

52. V = 60 kmph (16.67 m/s),

f = 0.36; t = 2.5 sec

g = 9.8 m/s

$$ISD = 2(SSD) = 2\left[vt + \frac{v^2}{2gf}\right]$$

$$= 2\left[16.67 \times 2.5 + \frac{16.67^2}{2 \times 9.8 \times 0.36}\right]$$

$$= 162 \text{ m}$$

53. $C_8H_{16}O_8 = 10^{-3}$ mole/L

Molecular weight of $C_8H_{16}O_8 = 240$ gm/mole

$\therefore \quad C_8H_{16}O_8 = 240 \times 10^{-3} = 0.24$ g/L

$$C_8H_{16}O_8 + 8O_2 \rightarrow 8CO_2 + 8H_2O$$

Mol. weight : 240 gm $\qquad$ 8 × 32 gm

weight : 0.24 g/L $\qquad$?

$$\text{Theoretical oxygen demand} = 0.24 \times \frac{8 \times 32}{240}$$

$$= 0.256 \text{ g/L}$$

54.

$$b = \frac{H}{\mu(S - C)} = \frac{10}{0.45\left(\frac{24}{9.81} - 1\right)}$$

$$= \frac{10}{0.45(1.446)} = 15.36 \text{ m}$$

$$b_{safe} = 15.36 \text{ m}$$

i.e., $\quad = 16$ m

55. Given :

$$P = \begin{vmatrix} 0 & 1 & 3 & 0 \\ -2 & 3 & 0 & 4 \\ 0 & 0 & 6 & 1 \\ 0 & 0 & 1 & 6 \end{vmatrix}$$

$$|P| = \begin{vmatrix} 0 & 1 & 3 & 0 \\ -2 & 3 & 0 & 4 \\ 0 & 0 & 6 & 1 \\ 0 & 0 & 1 & 6 \end{vmatrix}$$

Expanding along the first column

$$2\begin{vmatrix} 1 & 3 & 0 \\ 0 & 6 & 1 \\ 0 & 1 & 6 \end{vmatrix} = 2\{(36 - 1(-3(0) + 0\} = 70$$

We know that det (P) = Product of eigen values

SOLVED PAPER - 2021
SET 1

INSTRUCTIONS

1. Total of 65 questions carrying 100 marks, out of which 10 questions carrying a total of 15 marks are in General Aptitude (GA)
2. The Engineering Mathematics will carry around **15% of the total marks,** the General Aptitude section will carry **15% of the total marks** and the **remaining 70% of the total marks.**
3. **Types of Questions**
   (*a*) **Multiple Choice Questions (MCQ)** carrying 1 or 2 marks each in all papers and sections. These questions are objective in nature, and each will have a choice of four options, out of which the candidate has to mark the correct answer(s).
   (*b*) **Numerical Answer Questions** of 1 or 2 marks each in all papers and sections. For these questions the answer is a real number, to be entered by the candidate using the virtual keypad. No choices will be shown for these type of questions.
   (*c*) **Multiple Select Questions (MSQ)** carrying 1 or 2 marks each in all the papers and sections. These questions are objective in nature, and each will have choice of four answers, out of which ONE or MORE than ONE choice(s) is / are correct.
4. For **1-mark** multiple-choice questions, **1/3 marks** will be deducted for a wrong answer. Likewise, for **2-marks** multiple-choice questions, **2/3** marks will be deducted for a wrong answer. There is no negative marking for numerical answer type questions & multiple selection questions.

Chapter-Wise Analysis

#	Subject	Marks		
		1	2	Total
1	General Aptitude	5	5	15
2	Engineering Mathematics	5	4	13
3	Engineering Mechanics	0	1	2
4	Solid Mechanics	2	2	6
5	Structural Analysis	1	1	3
6	Construction Materials and Management	1	2	5
7	Concrete Structures	1	1	3

#	Subject	Marks		
		1	2	Total
8	Steel Structures	0	1	2
9	Soil Mechanics and Foundation Engineering	3	5	13
10	Fluid Mechanics and Hydraulics	2	3	8
11	Hydrology & Irrigation	2	2	6
12	Environmental Engineering	3	4	11
13	Transportation Engineering	4	3	10
14	Surveying	1 (MSQ)	1	3

GENERAL APTITUDE (GA)

Q. 1 to Q. 5 : Carry One Mark Each.

1. Getting to the top is __________ than staying on top.

(a) more easy (b) much easy

(c) easiest (d) easier

2.

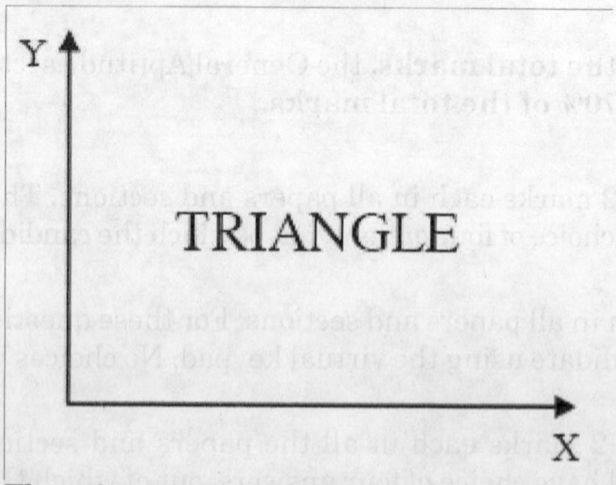

The mirror image of the above text about the x-axis is

(a)  (b) 

(c)  (d) 

3. In a company, 35% of the employees drink coffee, 40% of the employees drink tea and 10% of the employees drink both tea and coffee. What % of employees drink neither tea nor coffee?

(a) 15 (b) 25

(c) 35 (d) 40

4. $\oplus$ and $\odot$ are two operators on numbers p and q such that

$$p \oplus q = \frac{p^2 + q^2}{pq} \text{ and } p \odot q = \frac{p^2}{q};$$

If $x \oplus y = 2 \odot 2$, then x =

(a) $\frac{y}{2}$ (b) y

(c) $\frac{3y}{2}$ (d) 2y

5. Four persons P, Q, R and S are to be seated in a row, all facing the same direction, but not necessarily in the same order. P and R cannot sit adjacent to each other. S should be seated to the right of Q. The number of distinct seating arrangements possible is:

(a) 2 (b) 4

(c) 6 (d) 8

Q. 6 to Q. 10 : Carry Two Marks Each.

6. Statement: Either P marries Q or X marries Y

Among the options below, the logical NEGATION of the above statement is:

(a) P does not marry Q and X marries Y.

(b) Neither P marries Q nor X marries Y.

(c) X does not marry Y and P marries Q.

(d) P marries Q and X marries Y.

7. Consider two rectangular sheets, Sheet M and Sheet N of dimensions 6 cm x 4 cm each.

Folding operation 1: The sheet is folded into half by joining the short edges of the current shape.

Folding operation 2: The sheet is folded into half by joining the long edges of the current shape.

Folding operation 1 is carried out on Sheet M three times.

Folding operation 2 is carried out on Sheet N three times.

The ratio of perimeters of the final folded shape of Sheet N to the final folded shape of Sheet M is ______.

(a) 13 : 7 (b) 3 : 2

(c) 7 : 5 (d) 5 : 13

8.

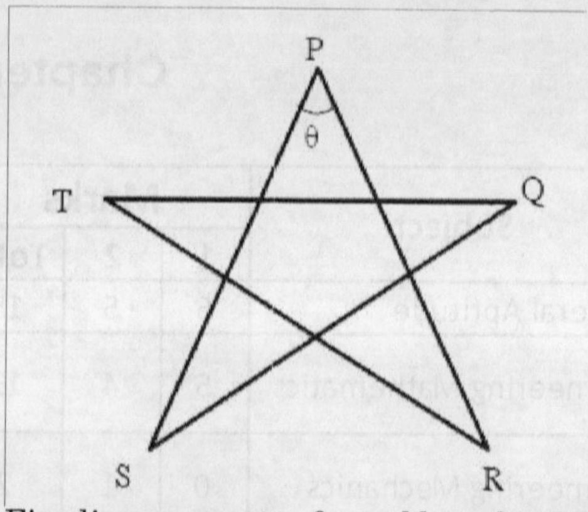

Five line segments of equal lengths, PR, PS, QS, QT and RT are used to form a star as shown in the figure above.

The value of θ, in degrees, is

(a) 36 (b) 45

(c) 72 (d) 108

9. A function, λ, is defined by

$$\lambda(p,q) = \begin{cases} (p-q)^2, & \text{if } p \geq q, \\ p+q, & \text{if } p < q. \end{cases}$$

The value of the expression

$$\frac{\lambda(-(-3+2),(-2+3))}{(-(-2+1))} \text{ is :}$$

(a) –1 (b) 0

(c) $\frac{16}{3}$ (d) 16

10. Humans have the ability to construct worlds entirely in their minds, which don't exist in the physical world. So far as we know, no other species possesses this ability. This skill is so important that we have different words to refer to its different flavors, such as imagination, invention and innovation.

Based on the above passage, which one of the following is TRUE?

(a) No species possess the ability to construct worlds in their minds.

(b) The terms imagination, invention and innovation refer to unrelated skills.

(c) We do not know of any species other than humans who possess the ability to construct mental worlds.

(d) Imagination, invention and innovation are unrelated to the ability to construct mental worlds.

CIVIL ENGINEERING

Q. 1 to Q. 25 : Carry One Mark Each.

1. The rank of matrix $\begin{bmatrix} 1 & 2 & 2 & 3 \\ 3 & 4 & 2 & 5 \\ 5 & 6 & 2 & 7 \\ 7 & 8 & 2 & 9 \end{bmatrix}$ is

(a) 1 (b) 2

(c) 3 (d) 4

2. If $P = \begin{bmatrix} 1 & 2 \\ 3 & 4 \end{bmatrix}$ and $Q = \begin{bmatrix} 0 & 1 \\ 1 & 0 \end{bmatrix}$ then $Q^T P^T$ is

(a) $\begin{bmatrix} 1 & 2 \\ 3 & 4 \end{bmatrix}$ (b) $\begin{bmatrix} 1 & 3 \\ 2 & 4 \end{bmatrix}$

(c) $\begin{bmatrix} 2 & 1 \\ 4 & 3 \end{bmatrix}$ (d) $\begin{bmatrix} 2 & 4 \\ 1 & 3 \end{bmatrix}$

3. The shape of the cumulative distribution function of Gaussian distribution is

(a) Horizontal line

(b) Straight line at 45 degree angle

(c) Bell-shaped

(d) S-shaped

4. A propped cantilever beam EF is subjected to a unit moving load as shown in the figure (not to scale). The sign convention for positive shear force at the left and right sides of any section is also shown:

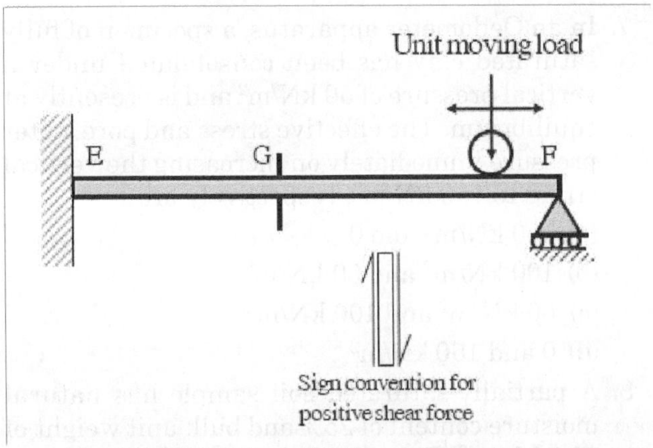

The CORRECT qualitative nature of the influence line diagram for shear force at G is

(a)

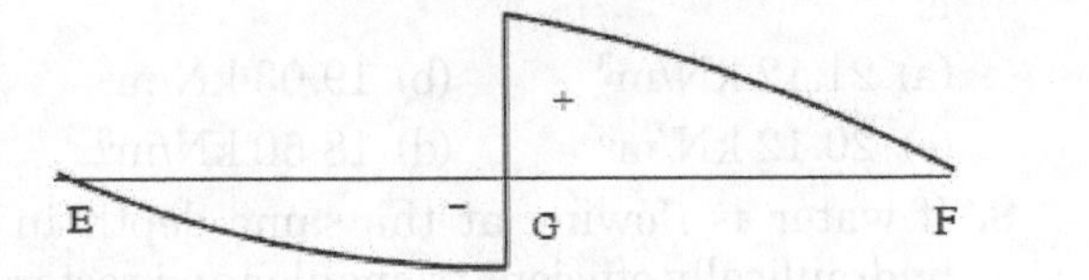

(b)

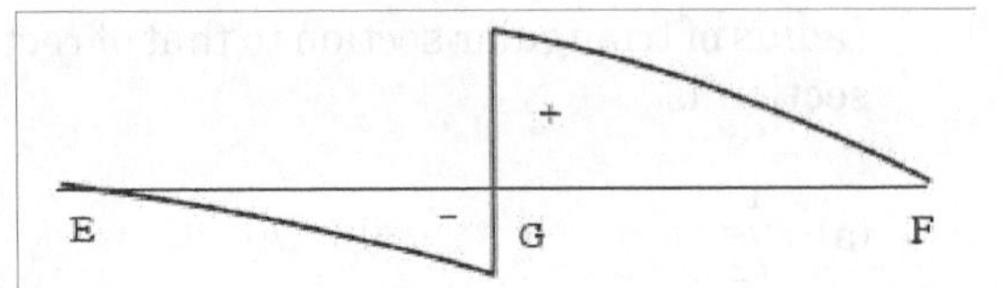

(c)

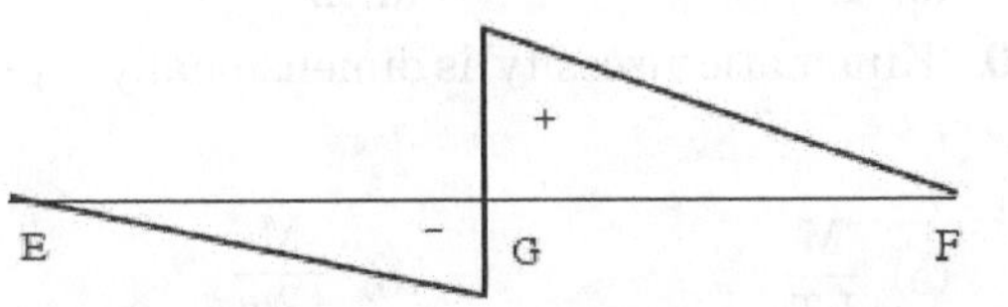

(d)

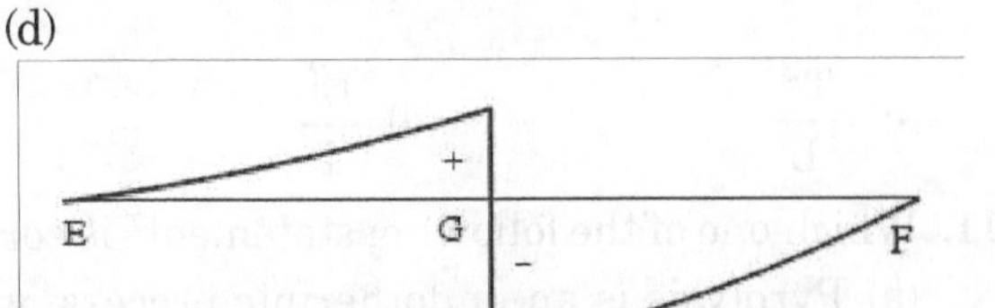

5. Gypsum is typically added in cement to

(a) prevent quick setting

(b) enhance hardening

(c) increase workability

(d) decrease heat of hydration

6. The direct and indirect costs estimated by a contractor for bidding a project is Rs.160000 and Rs.20000 respectively. If the mark up applied is 10% of the bid price, the quoted price (in Rs.) of the contractor is

(a) 200000 (b) 198000

(c) 196000 (d) 182000

7. In an Oedometer apparatus, a specimen of fully saturated clay has been consolidated under a vertical pressure of 50 kN/m^2 and is presently at equilibrium. The effective stress and pore water pressure immediately on increasing the vertical stress to 150 kN/m^2, respectively are

(a) 150 kN/m^2 and 0

(b) 100 kN/m^2 and 50 kN/m^2

(c) 50 kN/m^2 and 100 kN/m^2

(d) 0 and 150 kN/m^2

8. A partially-saturated soil sample has natural moisture content of 25% and bulk unit weight of 18.5 kN/m^3. The specific gravity of soil solids is 2.65 and unit weight of water is 9.81 kN/m^3. The unit weight of the soil sample on full saturation is

(a) 21.12 kN/m^3 (b) 19.03 kN/m^3

(c) 20.12 kN/m^3 (d) 18.50 kN/m^3

9. If water is flowing at the same depth in most hydraulically efficient triangular and rectangular channel sections then the ratio of hydraulic radius of triangular section to that of rectangular section is

(a) $\frac{1}{\sqrt{2}}$ (b) $\sqrt{2}$

(c) 1 (d) 2

10. 'Kinematic viscosity' is dimensionally represented as

(a) $\frac{M}{LT}$ (b) $\frac{M}{L^2T}$

(c) $\frac{T^2}{L}$ (d) $\frac{L^2}{T}$

11. Which one of the following statements is correct?

(a) Pyrolysis is an endothermic process, which takes place in the absence of oxygen.

(b) Pyrolysis is an exothermic process, which takes place in the absence of oxygen.

(c) Combustion is an endothermic process, which takes place in the abundance of oxygen.

(d) Combustion is an exothermic process, which takes place in the absence of oxygen.

12. Which one of the following is correct?

(a) The partially treated effluent from a food processing industry, containing high concentration of biodegradable organics, is being discharged into a flowing river at a point P. If the rate of degradation of the organics is higher than the rate of aeration, then dissolved oxygen of the river water will be lowest at point P.

(b) The most important type of species involved in the degradation of organic matter in the case of activated sludge process based wastewater treatment is *chemoheterotrophs*.

(c) For an effluent sample of a sewage treatment plant, the ratio $BOD_{5\text{-day},20°C}$ upon ultimate BOD is more than 1.

(d) A young lake characterized by low nutrient content and low plant productivity is called *eutrophic* lake.

13. The liquid forms of particulate air pollutants are

(a) dust and mist (b) mist and spray

(c) smoke and spray (d) fly ash and fumes

14. The shape of the most commonly designed highway vertical curve is

(a) circular (single radius)

(b) circular (multiple radii)

(c) parabolic

(d) spiral

15. A highway designed for 80 km/h speed has a horizontal curve section with radius 250 m. If the design lateral friction is assumed to develop fully, the required super elevation is

(a) 0.02 (b) 0.05

(c) 0.07 (d) 0.09

16. Which of the following is NOT a correct statement?

(a) The first reading from a level station is a 'Fore Sight'.

(b) Basic principle of surveying is to work from whole to parts.

(c) Contours of different elevations may intersect each other in case of an overhanging cliff.

(d) Planimeter is used for measuring 'area'.

17. Which of the following is/are correct statement(s)?

(a) Back Bearing of a line is equal to Fore Bearing ± 180°.

(b) If the whole circle bearing of a line is 270°, its reduced bearing is 90° NW.

(c) The boundary of water of a calm water pond will represent contour line.

(d) In the case of fixed hair stadia tachometry, the staff intercept will be larger, when the staff is held nearer to the observation point.

18. Consider the limit:

$$\lim_{x \to 1}\left(\frac{1}{\ln x} - \frac{1}{x-1}\right)$$

The limit (*correct up to one decimal place*) is_____

19. The volume determined from $\iiint_V 8\,xyz\,dV$ for $V = [2, 3] \times [1, 2] \times [0,1]$ will be *(in integer)* ________

20. The state of stress in a deformable body is shown in the figure. Consider transformation of the stress from the x-y coordinate system to the X-Y coordinate system. The angle θ, locating the X-axis, is assumed to be positive when measured from the x-axis in counter-clockwise direction.

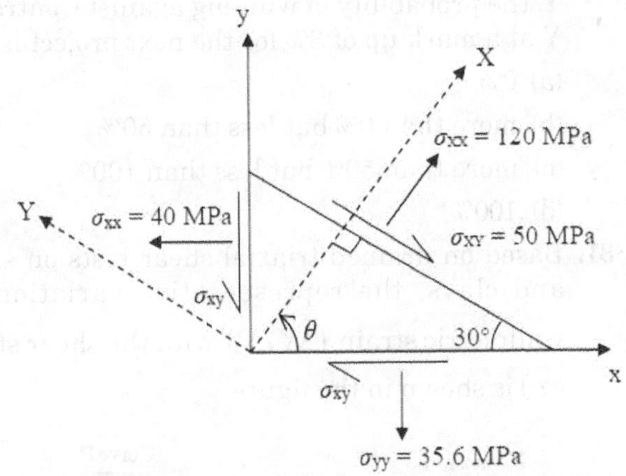

The absolute magnitude of the shear stress component σ_{xy} (in MPa, *round off to one decimal place*) in x-y coordinate system is ________

21. The equation of deformation is derived to be $y = x^2 - xL$ for a beam shown in the figure.

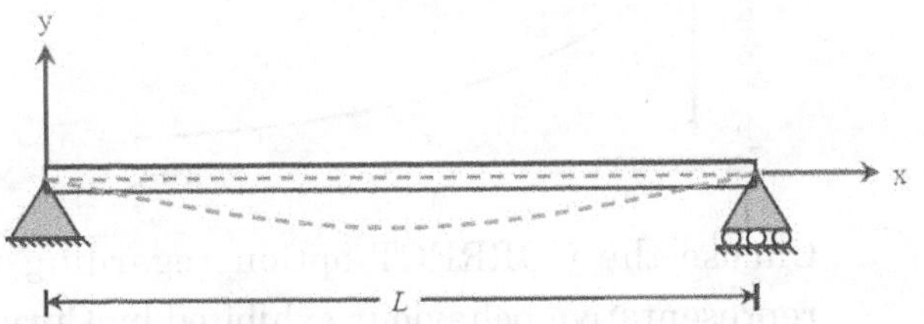

The curvature of the beam at the mid-span (in units, *in integer*) will be ________

22. A truss EFGH is shown in the figure, in which all the members have the same axial rigidity R. In the figure, P is the magnitude of external horizontal forces acting at joints F and G.

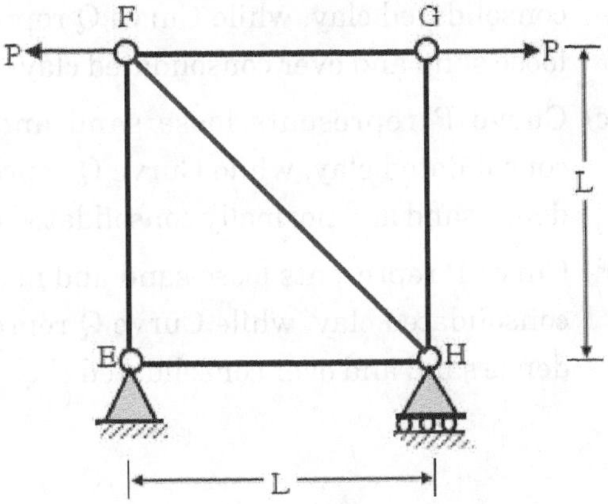

If $R = 500 \times 10^3$ kN, P = 150 kN and L = 3 m, the magnitude of the horizontal displacement of joint G (in mm, *round off to one decimal place*) is ________

23. The cohesion (c), angle of internal friction (ϕ) and unit weight (γ) of a soil are 15 kPa, 20° and 17.5 kN/m³, respectively. The maximum depth of unsupported excavation in the soil (in m, *round off to two decimal places*) is ________

24. Two reservoirs are connected through a homogeneous and isotropic aquifer having hydraulic conductivity (K) of 25 m/day and effective porosity (η) of 0.3 as shown in the figure (not to scale). Ground water is flowing in the aquifer at the steady state.

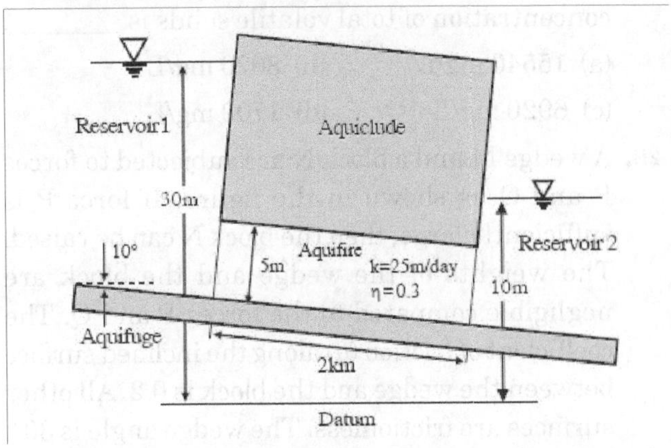

If water in Reservoir 1 is contaminated then the time (in days, *round off to one decimal place*) taken by the contaminated water to reach to Reservoir 2 will be ________

25. A signalized intersection operates in two phases. The lost time is 3 seconds per phase. The maximum ratios of approach flow to saturation flow for the two phases are 0.37 and 0.40. The optimum cycle length using the Webster's method (in seconds, *round off to one decimal place*) is ________

Q. 26 to Q. 35 : Carry Two Marks Each.

26. The solution of the second-order differential equation $\frac{d^2y}{dx^2} + 2\frac{dy}{dx} + y = 0$ with boundary conditions y(0) = 1 and y(1) = 3 is

(a) $e^{-x} + (3e - 1)xe^{-x}$

(b) $e^{-x} - (3e - 1)xe^{-x}$

(c) $e^{-x} + \left[3e\sin\left(\frac{\pi x}{2}\right) - 1\right]xe^{-x}$

(d) $e^{-x} - \left[3e\sin\left(\frac{\pi x}{2}\right) - 1\right]xe^{-x}$

27. The value of $\int_0^1 e^x dx$ using the trapezoidal rule with four equal subintervals is

(a) 1.718 (b) 1.727

(c) 2.192 (d) 2.718

28. A 50 mL sample of industrial wastewater is taken into a silica crucible. The empty weight of the crucible is 54.352 g. The crucible with the sample is dried in a hot air oven at 104 °C till a constant weight of 55.129 g. Thereafter, the crucible with the dried sample is fired at 600 °C for 1 h in a muffle furnace, and the weight of the crucible along with residue is determined as 54.783 g. The concentration of total volatile solids is ______.

(a) 15540 mg/L (b) 8620 mg/L

(c) 6920 mg/L (d) 1700 mg/L

29. A wedge M and a block N are subjected to forces P and Q as shown in the figure. If force P is sufficiently large, then the block N can be raised. The weights of the wedge and the block are negligible compared to the forces P and Q. The coefficient of friction (μ) along the inclined surface between the wedge and the block is 0.2. All other surfaces are frictionless. The wedge angle is 30°.

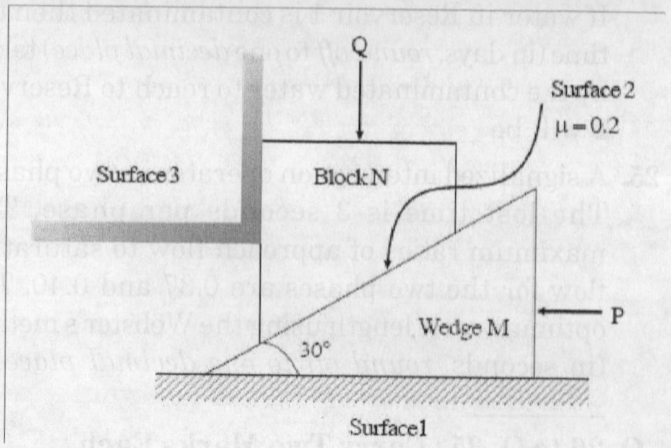

The limiting force P, in terms of Q, required for impending motion of block N to just move it in the upward direction is given as $P = \alpha Q$. The value of the coefficient 'α' (*round off to one decimal place*) is

(a) 0.6 (b) 0.5

(c) 2.0 (d) 0.9

30. Contractor X is developing his bidding strategy against Contractor Y. The ratio of Y's bid price to X's cost for the 30 previous bids in which Contractor X has competed against Contractor Y is given in the Table

Ratio of Y's bid price to X's cost	Number of bids
1.02	6
1.04	12
1.06	3
1.10	6
1.12	3

Based on the bidding behaviour of the Contractor Y, the probability of winning against Contractor Y at a mark up of 8% for the next project is

(a) 0%

(b) more than 0% but less than 50%

(c) more than 50% but less than 100%

(d) 100%

31. Based on drained triaxial shear tests on sands and clays, the representative variations of volumetric strain $(\Delta V/V)$ with the shear strain (γ) is shown in the figure.

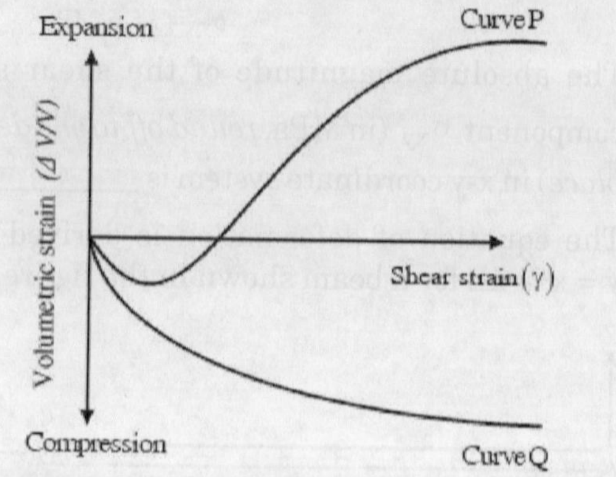

Choose the CORRECT option regarding the representative behaviour exhibited by Curve P and Curve Q.

(a) Curve P represents dense sand and over consolidated clay, while Curve Q represents loose sand and normally consolidated clay

(b) Curve P represents dense sand and normally consolidated clay, while Curve Q represents loose sand and over consolidated clay

(c) Curve P represents loose sand and over consolidated clay, while Curve Q represents dense sand and normally consolidated clay

(d) Curve P represents loose sand and normally consolidated clay, while Curve Q represents dense sand and over consolidated clay

32. A fluid flowing steadily in a circular pipe of radius R has a velocity that is everywhere parallel to the axis (centerline) of the pipe. The velocity distribution along the radial direction is $V_r = U\left(1 - \frac{r^2}{R^2}\right)$, where r is the radial distance as measured from the pipe axis and U is the maximum velocity at r = 0. The average velocity of the fluid in the pipe is

(a) $\frac{U}{2}$ (b) $\frac{U}{3}$

(c) $\frac{U}{4}$ (d) $\left(\frac{5}{6}\right)U$

33. A water sample is analyzed for coliform organisms by the multiple-tube fermentation method. The results of confirmed test are as follows:

Sample size (mL)	Number of positive results out of 5 tubes	Number of negative results out of 5 tubes
0.01	5	0
0.001	3	2
0.0001	1	4

The most probable number (MPN) of coliform organisms for the above results is to be obtained using the following MPN Index.

MPN Index for Various Combinations of Positive Results when Five Tubes used per Dilution of 10.0 mL, 1.0 mL and 0.1 mL	
Combination of positive tubes	MPN Index per 100 mL
0 - 2 - 4	11
1 - 3 - 5	19
4 - 2 - 0	22
5 - 3 - 1	110

The MPN of coliform organisms per 100 mL is

(a) 1100000 (b) 110000

(c) 1100 (d) 110

34. Ammonia nitrogen is present in a given wastewater sample as the ammonium ion $\left(NH_4^+\right)$ and ammonia $\left(NH_3\right)$. If pH is the only deciding factor for the proportion of these two constituents, which of the following is a correct statement?

(a) At pH above 9.25, only NH_4^+ will be present.

(b) At pH below 9.25, NH_3 will be predominant.

(c) At pH 7.0, NH_4^+ and NH_3 will be found in equal measures.

(d) At pH 7.0, NH_4^+ will be predominant.

35. On a road, the speed – density relationship of a traffic stream is given by u = 70 – 0.7k (where speed, u, is in km/h and density, k, is in veh/km). At the capacity condition, the average time headway will be

(a) 0.5 s (b) 1.0 s

(c) 1.6 s (d) 2.1 s

Q. 36 to Q. 55 : Carry Two Marks Each.

36. The values of abscissa (x) and ordinate (y) of a curve are as follows:

X	y
2.0	5.00
2.5	7.25
3.0	10.00
3.5	13.25
4.0	17.00

By Simpson's 1/3rd rule, the area under the curve (*round off to two decimal places*) is ________

37. Vehicular arrival at an isolated intersection follows the Poisson distribution. The mean vehicular arrival rate is 2 vehicle per minute. The probability (*round off to two decimal places*) that at least 2 vehicles will arrive in any given 1-minute interval is ________

38. Refer the truss as shown in the figure (not to scale).

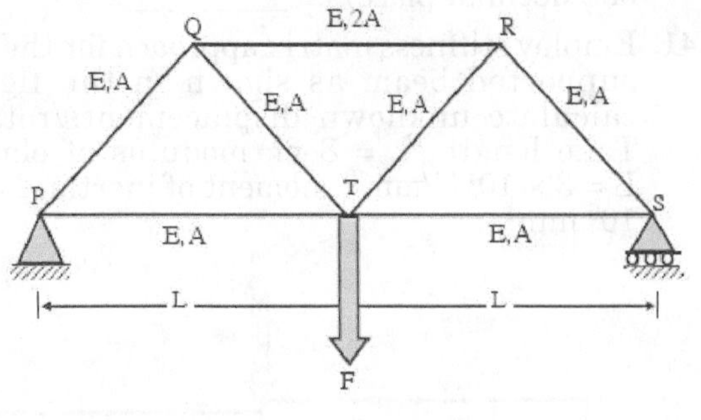

If load, F = $10\sqrt{3}$ kN, moment of inertia, I = 8.33 × 10^6 mm^4, area of cross-section, A = 10^4 mm^2, and length, L = 2 m for all the members of the truss, the compressive stress (in kN/m^2, *in integer*) carried by the member Q-R is ________

39. A prismatic cantilever prestressed concrete beam of span length, L = 1.5 m has one straight tendon placed in the cross-section as shown in the following figure (not to scale). The total prestressing force of 50 kN in the tendon is applied at d_c = 50 mm from the top in the cross-section of width, b = 200 mm and depth, d = 300 mm.

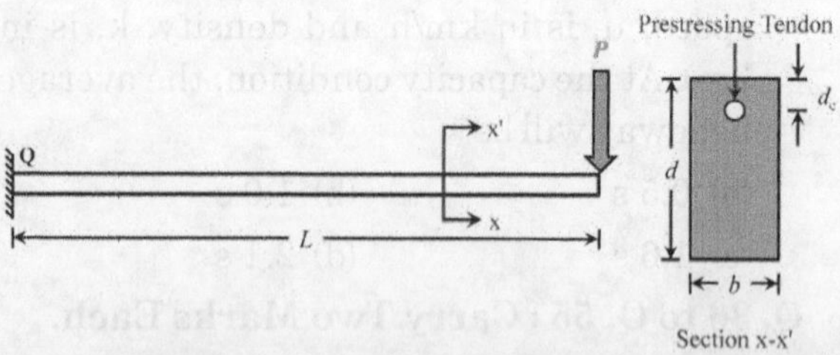

If the concentrated load, P = 5 kN, the resultant stress (in MPa, in integer) experienced at point 'Q' will be ________

40. A column is subjected to a total load (P) of 60 kN supported through a bracket connection, as shown in the figure (not to scale).

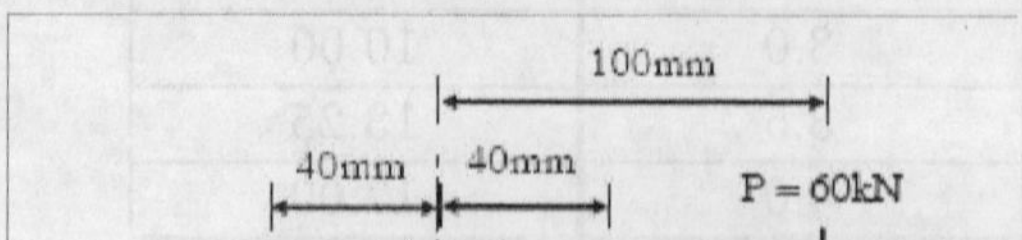

The resultant force in bolt R (in kN, round off to one decimal place) is ________.

41. Employ stiffness matrix approach for the simply supported beam as shown in the figure to calculate unknown displacements/rotations. Take length, L = 8 m; modulus of elasticity, E = 3 × 10^4 N/mm²; moment of inertia, I = 225 × 10^6 mm⁴.

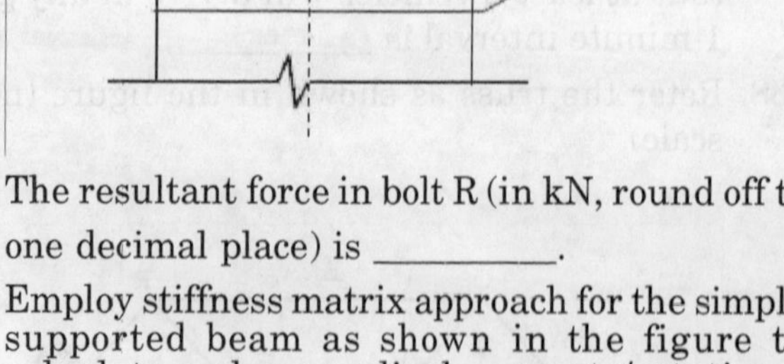

The mid-span deflection of the beam (in mm, *round off to integer*) under P = 100 kN in downward direction will be ________

42. A square plate O-P-Q-R of a linear elastic material with sides 1.0 m is loaded in a state of plane stress. Under a given stress condition, the plate deforms to a new configuration O-P'-Q'-R' as shown in the figure (not to scale). Under the given deformation, the edges of the plate remain straight.

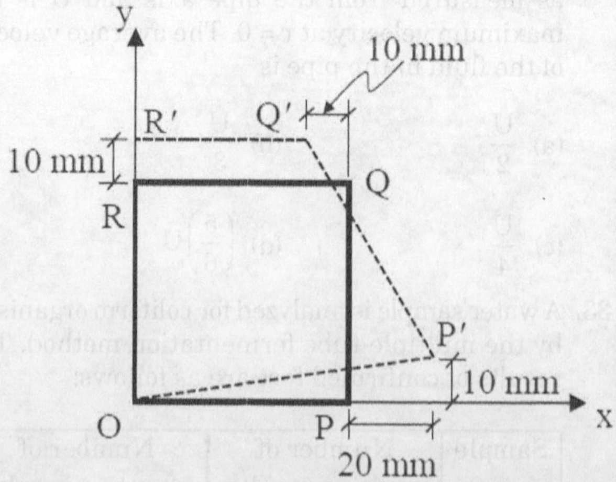

The horizontal displacement of the point (0.5 m, 0.5 m) in the plate O-P-Q-R (in mm, *round off to one decimal place*) is ________

43. A small project has 12 activities – N, P, Q, R, S, T, U, V, W, X, Y, and Z. The relationship among these activities and the duration of these activities are given in the Table.

Activity	Duration (in weeks)	Depends upon
N	2	-
P	5	N
Q	3	N
R	4	P
S	5	Q
T	8	R
U	7	R, S
V	2	U
W	3	U
X	5	T, V
Y	1	W
Z	3	X, Y

The total float of the activity "V" (in weeks, *in integer*) is ________

44. The soil profile at a construction site is shown in the figure (not to scale). Ground water table (GWT) is at 5 m below the ground level at present. An old well data shows that the ground water table was as low as 10 m below the ground level in the past. Take unit weight of water, γ_w = 9.81 kN/m³.

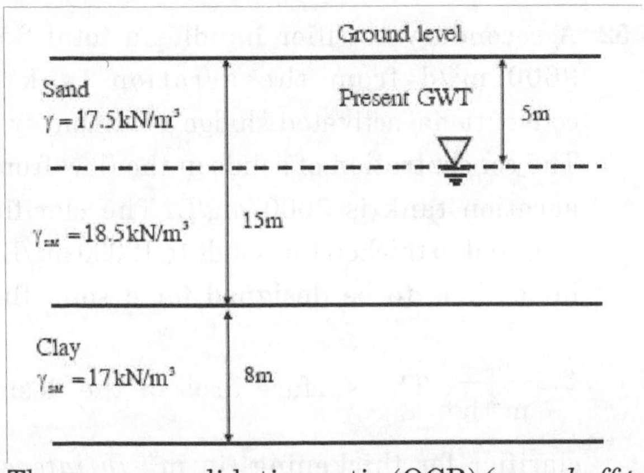

The over consolidation ratio (OCR) *(round off to two decimal places)* at the mid-point of the clay layer is ________

45. A retaining wall of height 10 m with clay backfill is shown in the figure (not to scale). Weight of the retaining wall is 5000 kN per m acting at 3.3 m from the toe of the retaining wall. The interface friction angle between base of the retaining wall and the base soil is 20°. The depth of clay in front of the retaining wall is 2.0 m. The properties of the clay backfill and the clay placed in front of the retaining wall are the same. Assume that the tension crack is filled with water. Use Rankine's earth pressure theory. Take unit weight of water, $\gamma_w = 9.81 kN/m^3$.

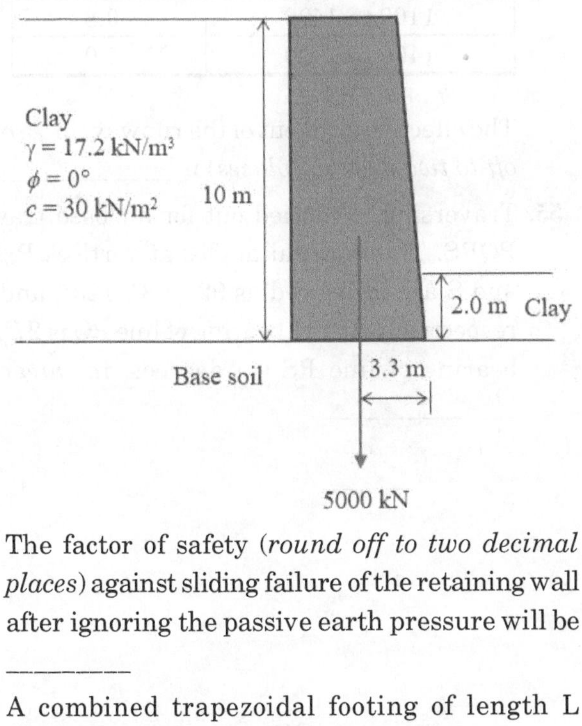

The factor of safety *(round off to two decimal places)* against sliding failure of the retaining wall after ignoring the passive earth pressure will be

____

46. A combined trapezoidal footing of length L supports two identical square columns (P_1 and P_2) of size 0.5 m × 0.5 m, as shown in the figure. The columns P_1 and P_2 carry loads of 2000 kN and 1500 kN, respectively.

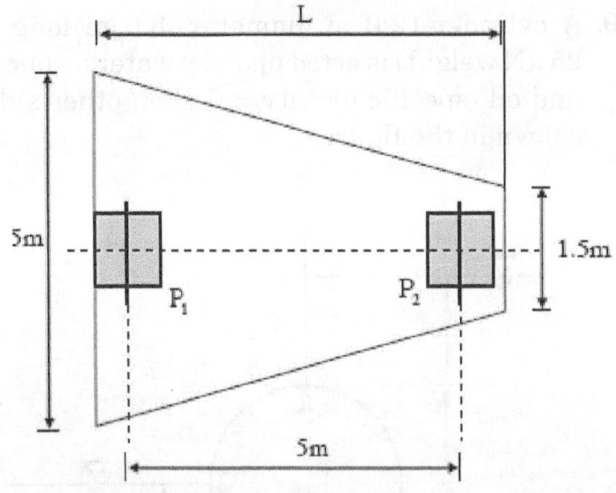

If the stress beneath the footing is uniform, the length of the combined footing L (in m, *round off to two decimal places*) is ________

47. An unsupported slope of height 15 m is shown in the figure (not to scale), in which the slope face makes an angle 50° with the horizontal. The slope material comprises purely cohesive soil having undrained cohesion 75 kPa. A trial slip circle KLM, with a radius 25 m, passes through the crest and toe of the slope and it subtends an angle 60° at its center O. The weight of the active soil mass (W, bounded by KLMN) is 2500 kN/m, which is acting at a horizontal distance of 10 m from the toe of the slope. Consider the water table to be present at a very large depth from the ground surface.

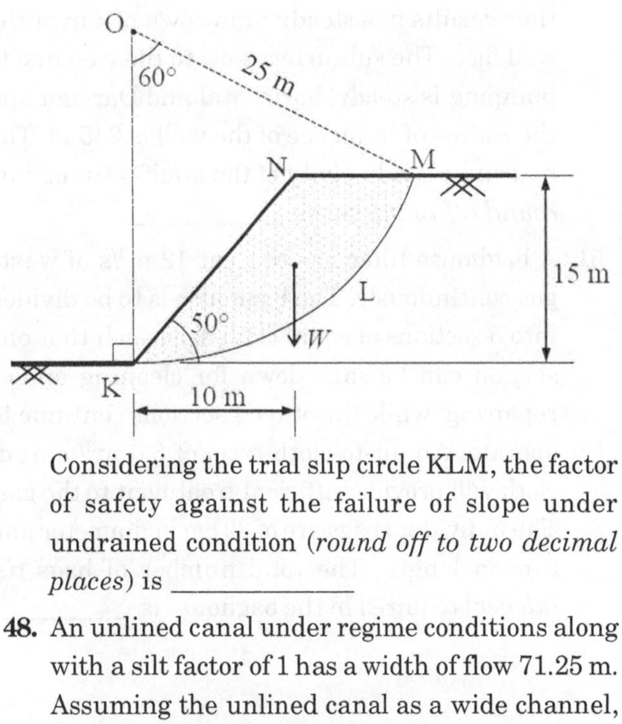

Considering the trial slip circle KLM, the factor of safety against the failure of slope under undrained condition *(round off to two decimal places)* is ________

48. An unlined canal under regime conditions along with a silt factor of 1 has a width of flow 71.25 m. Assuming the unlined canal as a wide channel, the corresponding average depth of flow (in m, *round off to two decimal places*) in the canal will be ________

49. A cylinder (2.0 m diameter, 3.0 m long and 25 kN weight) is acted upon by water on one side and oil (specific gravity = 0.8) on other side as shown in the figure.

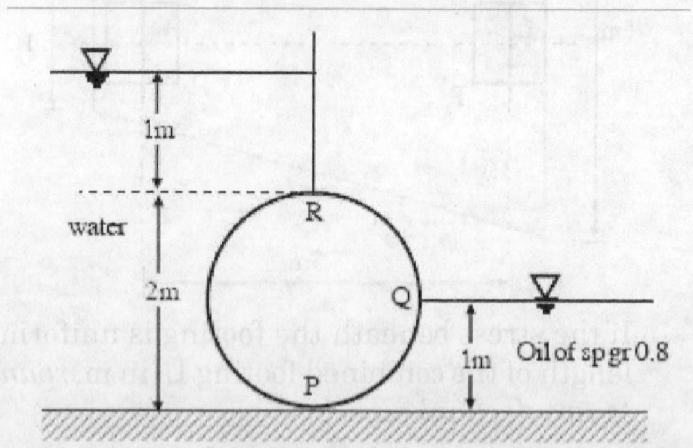

The absolute ratio of the net magnitude of vertical forces to the net magnitude of horizontal forces (*round off to two decimal places*) is ______

50. A tube-well of 20 cm diameter fully penetrates a horizontal, homogeneous and isotropic confined aquifer of infinite horizontal extent. The aquifer is of 30 m uniform thickness. A steady pumping at the rate of 40 litres/s from the well for a long time results in a steady drawdown of 4 m at the well face. The subsurface flow to the well due to pumping is steady, horizontal and Darcian and the radius of influence of the well is 245 m. The hydraulic conductivity of the aquifer (in m/day, *round off to integer*) is ______

51. A baghouse filter has to treat 12 m^3/s of waste gas continuously. The baghouse is to be divided into 5 sections of equal cloth area such that one section can be shut down for cleaning and/or repairing, while the other 4 sections continue to operate. An air-to-cloth ratio of 6.0 m^3/min-m^2 cloth will provide sufficient treatment to the gas. The individual bags are of 32 cm in diameter and 5 m in length. The total number of bags (*in integer*) required in the baghouse is ______

52. A secondary clarifier handles a total flow of 9600 m^3/d from the aeration tank of a conventional activated-sludge treatment system. The concentration of solids in the flow from the aeration tank is 3000 mg/L. The clarifier is required to thicken the solids to 12000 mg/L, and hence it is to be designed for a solid flux of $3.2 \frac{kg}{m^2.h}$. The surface area of the designed clarifier for thickening (in m^2, *in integer*) is ______

53. Spot speeds of vehicles observed at a point on a highway are 40, 55, 60, 65 and 80 km/h. The space-mean speed (in km/h, *round off to two decimal places*) of the observed vehicles is ______

54. The longitudinal section of a runway provides the following data:

End-to-end runway (m)	Gradient (%)
0 to 300	+1.2
300 to 600	–0.7
600 to 1100	+0.6
1100 to 1400	–0.8
1400 to 1700	–1.0

The effective gradient of the runway (in %, *round off to two decimal places*) is ______

55. Traversing is carried out for a closed traverse PQRS. The internal angles at vertices P, Q, R and S are measured as 92°, 68°, 123°, and 77°, respectively. If fore bearing of line PQ is 27°, fore bearing of line RS (in degrees, in *integer*) is ______

ANSWERS

General Aptitude (GA)

1. (d)	**2.** (b)	**3.** (c)	**4.** (b)	**5.** (c)	**6.** (b)
7. (a)	**8.** (a)	**9.** (b)	**10.** (c)		

Civil Engineering

1. (b)	**2.** (d)	**3.** (d)	**4.** (b)	**5.** (a)	**6.** (a)
7. (c)	**8.** (b)	**9.** (a)	**10.** (d)	**11.** (a)	**12.** (b)
13. (b)	**14.** (c)	**15.** (b)	**16.** (a)	**17.** (a;b;c)	**18.** (0.5 to 0.5)
19. (15 to 15)	**20.** (95 to 97)	**21.** (2 to 2)	**22.** (0.9 to 0.9)	**23.** (4.88 to 4.92)	**24.**(2400 to 2400)
25. (60.7 to 61.1)	**26.** (a)	**27.** (b)	**28.** (c)	**29.** (d)	**30.** (b)
31. (a)	**32.** (a)	**33.** (b)	**34.** (d)	**35.** (d)	**36.** (20.00 to 21.00)
37. (0.58 to 0.60)	**38.** (490 to 510)	**39.** (0 to 0)	**40.** (27.0 to 29.0)	**41.** (100 to 130)	**42.** (2.4 to 2.6)
43. (0 to 0)	**44.** (1.18 to 1.26)	**45.** (4.20 to 4.35)	**46.** (5.75 to 5.90)	**47.** (1.94 to 1.98)	**48.** (2.80 to 2.95)
49. (0.35 to 0.40)	**50.** (34 to 38)	**51.** (30 to 30)	**52.** (375 to 375)	**53.** (55.50 to 58.50)	**54.** (0.30 to 0.34)
55. (196 to 196)					

EXPLANATIONS

GENERAL APTITUDE (GA)

1. When the comparison is between two things we use the second degree of the adjective the degree from of easy are : easy → easier → easiest.

Hence, comparative degree is easier.

2. The mirror image of TRIANGLE about the X-axis is

∵ Top to Bottom will interchange (or) symmetrical about X-axis.

3.

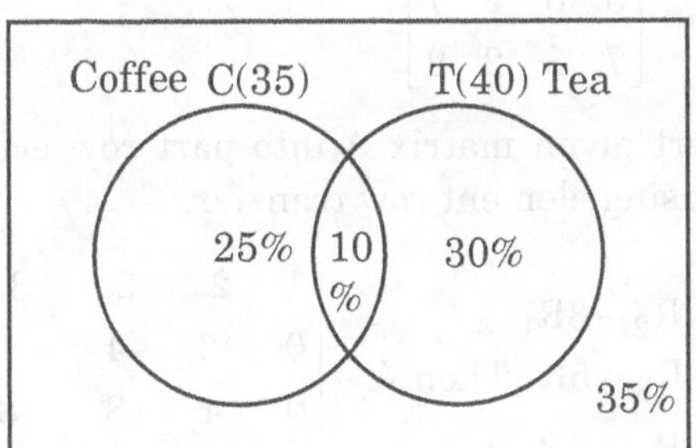

C → Coffee → Percentage of employees drink Coffee = 35%

T → Tea → Percentage of employees drink Tea = 40%

→ Percentage of employees drink both Tea and Coffee = 10%

→ Percentage of employees drink neither Tea and Coffee = 100% – n(TUE)

= 100% – (25% + 10% + 30%)

= 35%

4. As given that,

$$\left[\because p \oplus q = \frac{p^2 + q^2}{pq} \text{ and } p \odot q = \frac{p^2}{q}\right]$$

Now, L.H.S $\Rightarrow$ $x \oplus y = \dfrac{x^2 + y^2}{x.y}$

R.H.S $\Rightarrow$ $x \oplus y = 2 \odot 2 = \dfrac{2^2}{2}$

So, L.H.S = R.H.S

$$\Rightarrow \frac{x^2 + y^2}{xy} = \frac{2^2}{2}$$

$\Rightarrow x^2 + y^2 = 2xy \Rightarrow x^2 + y^2 - 2xy = 0$

$\Rightarrow (x - y)^2 = 0$

$\Rightarrow x - y = 0$

$\Rightarrow \boxed{x = y}$

5. According to the given information,

Possible way are

R Q S P

P Q S R

R Q P S

P Q R S

Q R S P

Q P S R

⇒ Total number of ways = 2 + 2 + 2 = 6

∴ The number of distinct seating arrangements possible is 6.

6. As Given statement:
Either P marries Q or X marries Y.
The logical negation of the Either P marries Q or X marries Y "above statement is Neither P marries Q nor X marries Y."
Hence option (b) is correct.

7. Shortest side of folding:

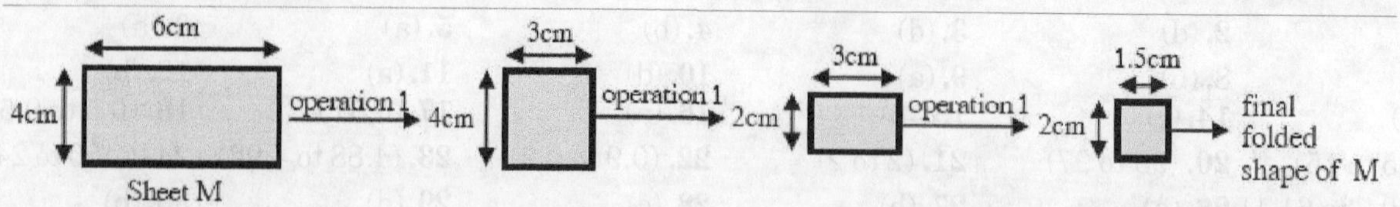

So, perimeter of final folded shape of M = 2 (2 + 1.5) = 7
Longest side in folded:

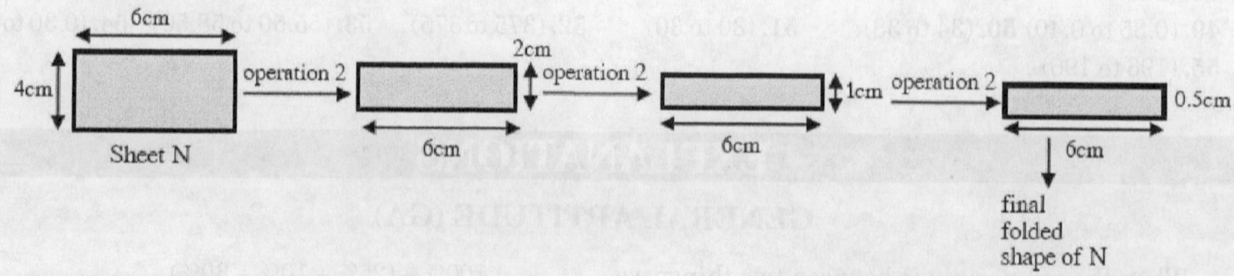

$\therefore$ Perimeter of final folded shape of N = 2(6 + 0.5) = 13
$\therefore$ Required ratio is (N:M) = 13:7

8.

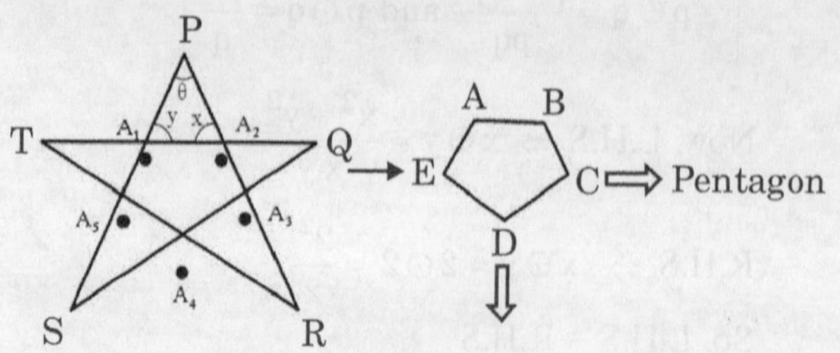

PR = PS = QS = QT = RT (Given)
Then ABCDE will be regular pentagon.
So, Sum of internal angles formed at pentagon = (5 – 2) × 180

$\Rightarrow$ Each internal angle = $\frac{3\times180}{5} = 108°$

So, $\angle\theta$ = ?; $\angle x$ = 180° – 108° = 72°; $\angle y$ = 180 – 108 = 72
$\therefore \angle x + \angle y + \theta = 180$
$\Rightarrow \theta = 180 - 72 - 72 = 36°$

9. According to the question

$$\frac{\lambda(-(-3+2),(-2+3)}{(-(-2+1)} = \frac{\lambda(-1(-1),1)}{1}$$

So, 1st definition will be applicable as p = q
[$\therefore \lambda(p, q) = (p - q)^2$ if p = q]
So, $\lambda(1, 1) = (1 - 1)^2 = 0$

10. Only humans have the ability to construct worlds entirely in their minds and no other species has this.
Hence, answer is option (c) which reflects the information given in the passage.

CIVIL ENGINEERING

1. Let $A = \begin{bmatrix} 1 & 2 & 2 & 3 \\ 3 & 4 & 2 & 5 \\ 5 & 6 & 2 & 7 \\ 7 & 8 & 2 & 9 \end{bmatrix}$

Convert given matrix A into part row echelon form using element row transfer.

$R_2 \to R_2 - 3R_1$
$R_3 \to R_3 - 5R_1$ Then $A \sim \begin{bmatrix} 1 & 2 & 2 & 3 \\ 0 & -2 & -4 & -4 \\ 0 & -4 & -8 & -8 \\ 0 & -6 & -12 & -12 \end{bmatrix}$
$R_4 \to R_4 - 7R_1$

$R_3 \to R_3 - 2R_2$
$R_4 \to R_4 - 3R_2$ $A \sim \begin{bmatrix} 1 & 2 & 2 & 3 \\ 0 & -2 & -4 & -4 \\ 0 & 0 & 0 & 0 \\ 0 & 0 & 0 & 0 \end{bmatrix}$ is row

Echelon form
Hence, rank (A) is the number of non zero rows = 2

2. As we know that,
$(PQ)^T = Q^T P^T$
Where $Q^T P^T = (PQ)^T$
$= (QP)^T$

$$Q^T = \begin{bmatrix} 0 & 1 \\ 1 & 0 \end{bmatrix}; P^T = \begin{bmatrix} 1 & 3 \\ 2 & 4 \end{bmatrix}$$

$$Q^T P^T = \left[\begin{bmatrix} 0 & 1 \\ 1 & 0 \end{bmatrix}\begin{bmatrix} 1 & 3 \\ 2 & 4 \end{bmatrix}\right]^T = \begin{bmatrix} 2 & 4 \\ 1 & 3 \end{bmatrix}$$

3. As we know that,
The Gaussian distribution also known as normal distribution is a bell-shaped curve. So, the cummulative gaussion distribution curve is S-shaped curve.

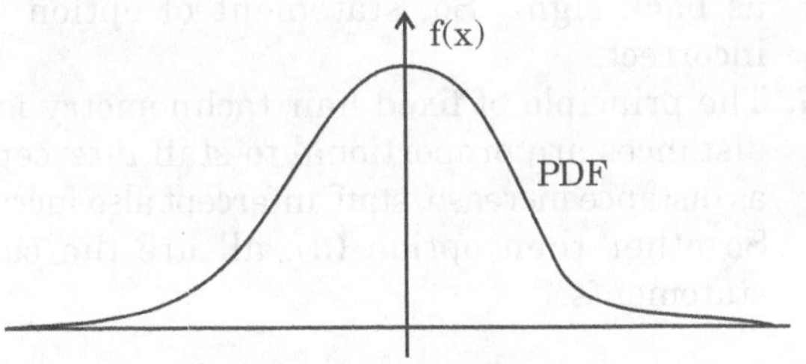

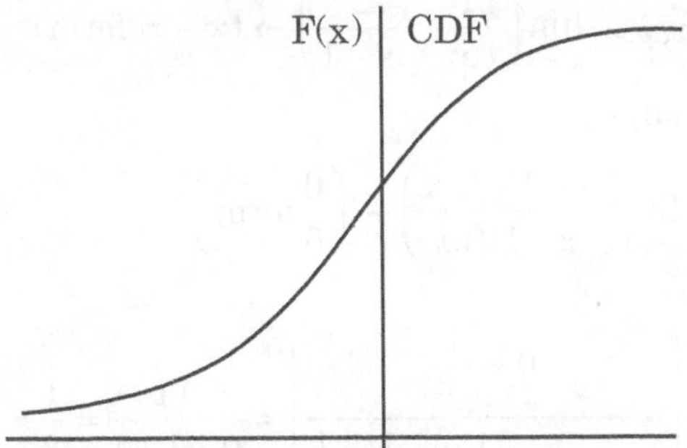

So, P.D.F: $f(x) = \frac{1}{\sigma\sqrt{2\pi}} e^{-(x-\mu)^2/(2\sigma^2)}$

C.D.F: $f(x) = \frac{1}{2}\left[1 + \text{erf}\left[\frac{(x-\mu)}{\sigma\sqrt{2}}\right]\right]$

4. From muller Breslauce principle to obtain I LD for any stress function, we release the restrain offered by that stress function and apply a directly related unit load. the deflected shape so obtained to some scale, is the I LD for the stress function.

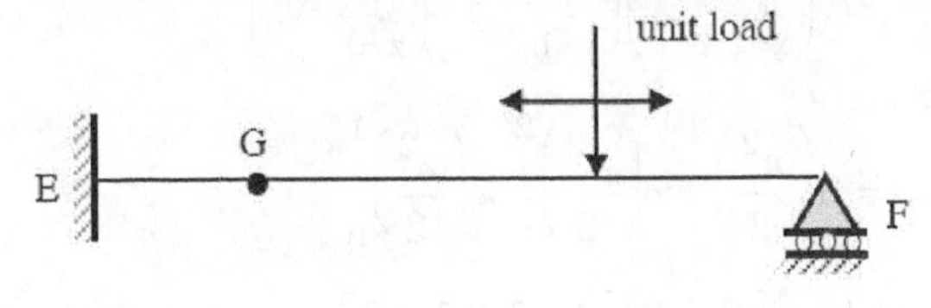

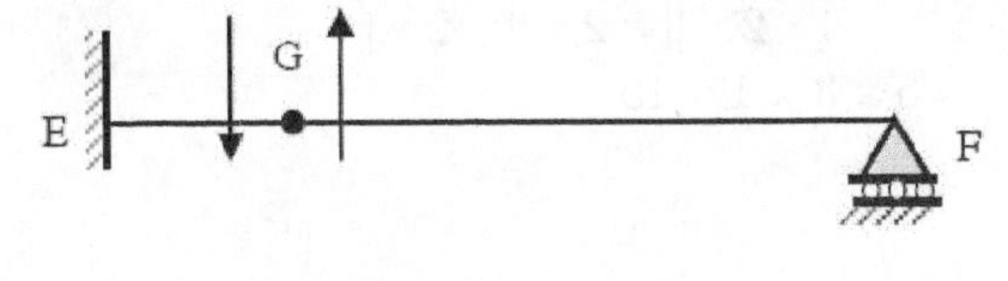

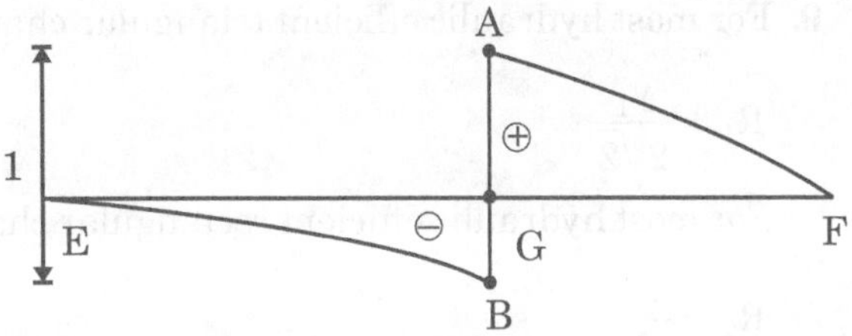

Hence, slope at A and B will be same.

5. The gypsum is added to cement at the end of grinding clinker it is added to prevent quick setting of concrete.

6. As given that,
Direct cost = 160000
Indirect cost = 20000
Total cost = [Direct cost + Indirect cost]
= 1,60,000 + 20,000 = 1,80,000
Mark up = 10%
Mark up cost = 0.10 × 1,80,000 = 18,000
Bid price = 1,80,000 + 18,000 = 1,98, 00

So, quoted bid price $= \frac{1,80,000}{0.9} = 2,00,000$

7. The sample is consolidated to 50 kN/m^2 and then suddenly the load has been increased to 150 kN/m^2.
So, Increase in vertical stress ($\Delta\sigma$) = 150 – 50
= 100 kN/m^2
The Clay soil is saturated and no shearing has been imposed, so
$\Delta\sigma$ = Pore water pressure increase
ΔP = 100 kN/m^2
The soil is in equilibrium at 50 kN/m^2,
The initial pore water pressure = 0
Final pore water pressure = 100 + 0 = 100 kN/m^2
Effective stress = Total stress – Pore water pressure = 150 – 100 = 50 kN/m^2

8. As given that,
Moisture content (w) = 25% = 0.25
Bulk unit weigh (γ) = 18.5 kN/m^3
Specific gravity (G) = 2.65

Saturated unit weight $(\gamma_{sat}) = ?$

As we know that full saturation
And,
e = wG = 0.25 × 2.65 = 0.6625

$$\gamma_{sat} = \left(\frac{G_s + e}{1 + e}\right)\gamma_w$$

$$\gamma_{sat} = \left(\frac{G_s + wG}{1 + wG}\right)\gamma_w = \left(\frac{2.65 + 0.25 \times 2.65}{1 + 0.25 \times 2.65}\right) \times 9.81$$

= 19.54 kN/m^3

9. For most hydraulic efficient triangular channel

$$R_1 = \frac{y_1}{2\sqrt{2}}$$

For most hydraulic efficient rectangular channel

$$R_2 = \frac{y_2}{2}$$

As given that $(y_1 = y_2 = y)$

$$\frac{R_1}{R_2} = \frac{\frac{y}{2\sqrt{2}}}{\frac{y}{2}} = \frac{1}{\sqrt{2}}$$

10. As we know that,
Kinematic Viscocity

$$(\nu) = \frac{\text{Dynamic viscosity}(\mu)}{\text{Density}(\rho)}$$

In S.I. Unit form:

$$\nu = \frac{kg/m.s}{kg/m^3} = m^2/S$$

$$[\nu] = \frac{m^2}{S} = \left[\frac{L^2}{T}\right]$$

11. Pyrolysis is an endothermic process as there is a substantial heat occur to raise the biomass to the reaction temperature.
So, pyrolysis occur in the absence of oxygen. Whereas combustion exothermic occur in the presence of air.

12. Chemoheterotrophs are the organisms which generate energy with the help of chemical reactions by consuming organic matter during their metabolism. Chemoheterotrophs can be either autotrophic or heterotropic. These are the species which help to degrade the organic matter in ASP but in ASP heterotrophs only involved in biodegradation of organics.

13. The liquid forms of particulate air pollutants are mist and spray.
Particulates sub classes are solid and liquid. And its typical numbers of sub classes are: Dust, Smoke, fumes, fly ash mist, spray.
Mist is a cloud made of very small drops of water in the air just above the ground which reduces the visibility.

14. The most preferred curve for vertical alignment is parabolic. So, the ideal vertical curve is 2° parabola.

15. As given that,
Design speed (v) = 80 kmph

$$= 80 \times \frac{5}{18} = 22.22 \text{ m/s}$$

Design lateral friction (f) = 0.15 (as per IRC)
Radius (R) = 250 m
As we know that,

$$e + f = \frac{v^2}{gR}$$

$$e + 0.15 = \frac{(22.22)^2}{9.81 \times 250} \quad (\because g = 9.81)$$

$$e = \frac{(22.22)^2}{9.81 \times 250} - 0.15$$

$$e = 0.201 - 0.15$$

$$\Rightarrow e = 0.0513 \approx 0.05$$

16. The first reading from a level station B is known as back sight. So, statement of option (a) is incorrect.

17. The principle of fixed hair tachnometry is that distances are proportional to staff intercept. So as distance increase, staff intercept also increases. So other then option (d), all are the correct statements.

18. Let, $l = \lim_{x\to 1}\left(\frac{1}{\ell nx} - \frac{1}{x-1}\right) \to (\infty - \infty \text{ form})$

Then,

$$l = \lim_{x\to 1}\left(\frac{x-1-\ell nx}{(x-1)\ell nx}\right) \to \left(\frac{0}{0}\text{ form}\right)$$

$$l = \lim_{x\to 1}\left[\frac{0+\frac{1}{x^2}}{(x-1)\left(-\frac{1}{x^2}\right)+\frac{1}{x}+\frac{1}{x}}\right] = \frac{+1}{0+1+1} = \frac{1}{2} = +0.5$$

$$\boxed{\therefore l = +0.5}$$

19. x limits : 2 to 3 ; y limits : 1 to 2 and z limits: 0 to 1

$$\therefore \text{Volume} = 8\int_{x=2}^{3}\int_{y=1}^{2}\int_{z=0}^{1} xyz\, dx\, dy\, dz$$

$$= 8\left(\int_{x=2}^{3} x dx\right)\left(\int_{y=1}^{2} y dy\right)\left(\int_{z=0}^{1} z dz\right)$$

$$= 8\left(\frac{x^2}{2}\right)_2^3 \times \left(\frac{y^2}{2}\right)_1^2 \times \left(\frac{z^2}{2}\right)_0^1$$

$$V = 8\left[\frac{9-4}{2}\right]\left[\frac{4-1}{2}\right]\left[\frac{1-0}{2}\right]$$

$$= 5 \times 3 \times 1 = 15$$

20.

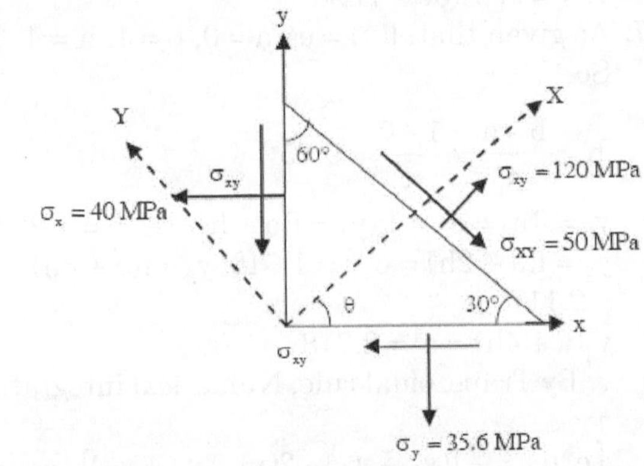

Resultant stress at angle θ is

$$\sigma_\theta = \left(\frac{\sigma_x + \sigma_y}{2}\right) + \left(\frac{\sigma_x - \sigma_y}{2}\right)\cos 2\theta + \tau_{xy}.\sin 2\theta$$

$$\sigma_x = 40\text{MPa}, \sigma_y = 35.6$$

$$\sigma_\theta = 120, \theta = 30°$$

$$120 = \left(\frac{40+35.6}{2}\right) + \left(\frac{40-35.6}{2}\right)\cos(2\times 30) + \tau_{xy}.\sin(2\times 30)$$

$$\tau_{xy} = 96.18 \text{ MPa}$$

21. Given equation

$y = x^2 - xL$

Slope, $\frac{dy}{dx} = 2x - L$

Curvature, $\boxed{\frac{d^2y}{dx^2} = 2}$

So, the curvature of $\left(x = \frac{L}{2}\right)$ mid span is 2m.

Hence, the beam is having constant curvature throughout the beam.

22.

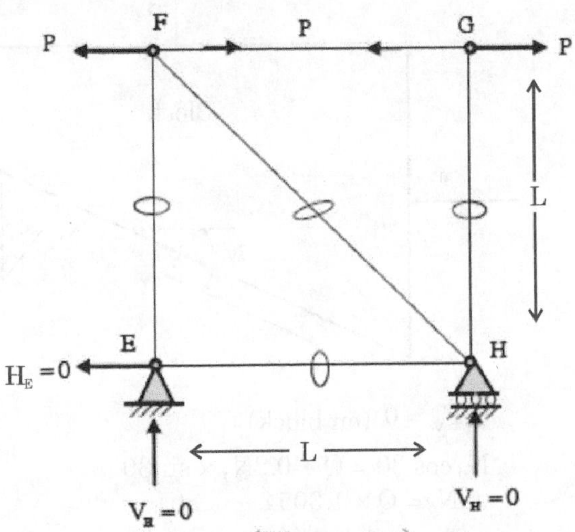

(P force system)

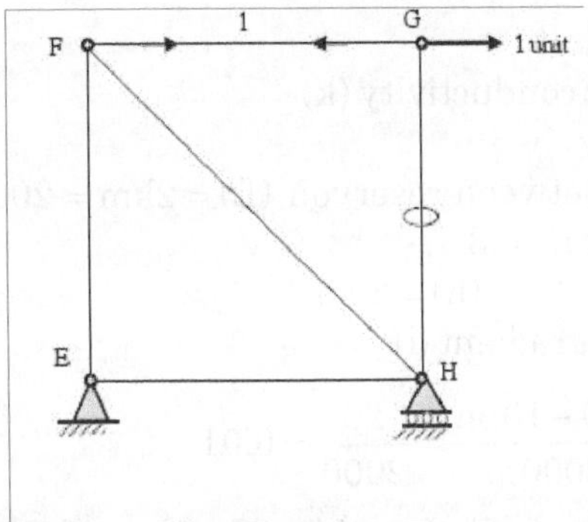

(k force system)

Taking moment about 'E'.

$\sum M_E = 0$ (– ve) – ve

$-R_H + P \times L - P \times L = 0$

$R_H = R_E = 0$

Here, No need to calculate 'k' force in all members because 'P' fore is zero force all members except FG By unit load method.

$$\delta_{HG} = \sum_{i=1}^{n} \frac{P_i k_i L_i}{A_i E_i}$$

δ_{HG} = Horizontal deflection at joint 'G'

$$\delta_{HG} = \underbrace{\frac{P_{FG} \times K_{FG} \times L_{FG}}{AE}}_{\text{(FG–member)}} \times \underbrace{'O'}_{\text{(For all other members)}}$$

($\because$ AE (Axial rigidity) = 500 × 10^3 kN)

$$\delta_{HG} = \frac{PL}{AE} = \left(\frac{+150 \times (+1) \times 3}{500 \times 10^3} \times 10^3\right) \text{mm} = 0.90 \text{ mm}$$

23. As given that,

Cohesion (c) = 15 kPa = 15kN/m²

Angle of internal friction (ϕ) = 20°

Unit weight of soil (γ) = 17.5 kN/m³

As we know that,

Maximum depth of unsupported excavation (H_C) = 2 × critical depth

$$= 2 \times \frac{2c}{\gamma\sqrt{k_a}} = \frac{4c}{\gamma\sqrt{k_a}}$$

Where,

$$k_a = \frac{1-\sin\phi}{1+\sin\phi} = \frac{1-\sin 20}{1+\sin 20} = 0.49$$

So,

$$(H_c) = \frac{4c}{\gamma\sqrt{k_a}} = \frac{4\times 15}{17.5\sqrt{0.49}} = 4.897\text{m}$$

24. As given that,
Hydraulic conductivity (k)
= 25 m/day
Distance between reservoir (L) = 2km = 2000 m
Porosity (n) = 0.3
Head difference (h) = 20m
Hydraulic gradient (i)

$$= \frac{\Delta h}{L} = \frac{(30-10)m}{2000m} = \frac{20}{2000} = 0.01$$

Discharge velocity (v) = ki = 25m/day × 0.01
= 0.25 m/day

$$\text{Actual velocity } (V_S) = \frac{V}{n} = \frac{0.25}{0.3} = 0.8333 \text{ m / day}$$

$$\therefore \text{ Velocity} = \frac{\text{distance}}{\text{time}}$$

So, number of days required by the contaminated water to reach reservoirs.

$$\text{time} = \frac{2000m}{\left(\frac{0.25}{0.3}\right) m / day} = 2400 \text{ days}$$

25. As given that
Number of phases (n) = 2
Lost time per phase (l) = 3 seconds
Total lost time per cycle (L) = 2 × 3 = 6 seconds

$$\text{Critical ratios } y_1 = \frac{q_1}{s_1} = 0.37, y_2 = \frac{q_2}{s_2} = 0.40$$

$y = y_1 + y_2 = 0.37 + 0.4 = 0.77$
As per Webster method
Optimum cycle time length (C_0)

$$= \frac{1.5L+5}{1-y} = \frac{1.5 \times 6 + 5}{1-0.77} = 60.869 \text{ seconds}$$

26. As given D.E is

$$\frac{d^2y}{dx^2} + \frac{2dy}{dx} + y = 0$$

A.E of given D.E is $m^2 + 2m + 1 = 0$
$\Rightarrow (m + 1)(m + 1) = 0$
$\Rightarrow m = -1, -1$
(two equal real roots), since R.H.S is 0
∴ D.E is homogenous equation
∴ Solution is y = C.F
$\Rightarrow y = (c_1 + c_2x)e^{-x}$ ...(1)
Using boundary condition y(0) =1 (i.e., y = 1, x = 0) and y (1) = 3 (i.e., y = 3, x = 1), equation(1) gives
$1 = c_1$ & $3 = (c_1 + c_2)e^{-1}$
$\Rightarrow 3e = 1 + c_2$
$\Rightarrow c_2 = 3e - 1$
Now, particular solution:

$\therefore y = (1 + (3e - 1)x)e^{-x}$

27. As given that, $f(x) = e^x$, a = 0, b = 1, n = 4
So,

$$h = \frac{b-a}{n} = \frac{1-0}{4} = 0.25$$

$y_0 = f(a) = e^0 = 1$; $y_1 = f(a + h) = e^{0.25} = 1.284$
$y_2 = f(a + 2h) = e^{0.5} = 1.648$; $y_3 = f(a + 3h) = e^{0.75}$
= 2.117
$y_4(a + 4h) = e^1 = 2.718$
∴ By Trapezoidal rule, Numerical integration is:

$$\int_0^1 e^x dx = \frac{h}{2}[(y_0 + y_4) + 2(y_1 + y_2 + y_3)]$$

$$= \frac{1}{8}[(1 + 2.718) + 2(1.284 + 1.648 + 2.117)]$$

$\simeq 1.727$

28. As given that
Volume of sample (V) = 50 mL.
Wt. of cruicide = 54.352 gm.

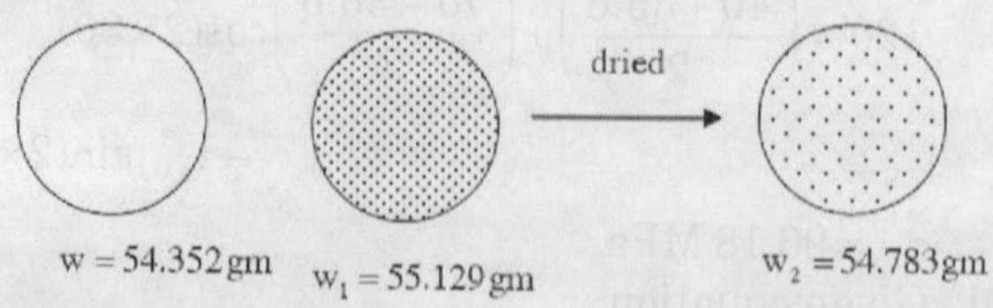

As we know that,

$$\text{Total volatile solids} = \frac{W_1 - W_2}{V}$$

$$= \frac{(55.129 - 54.783)}{50} \text{gm / mL}$$

$$= \left(\frac{55.129 - 54.783}{50}\right) \times \frac{10^3}{10^{-3}} \text{mg / L}$$

= 6920 mg/L

29. 1. FBD of Block N:

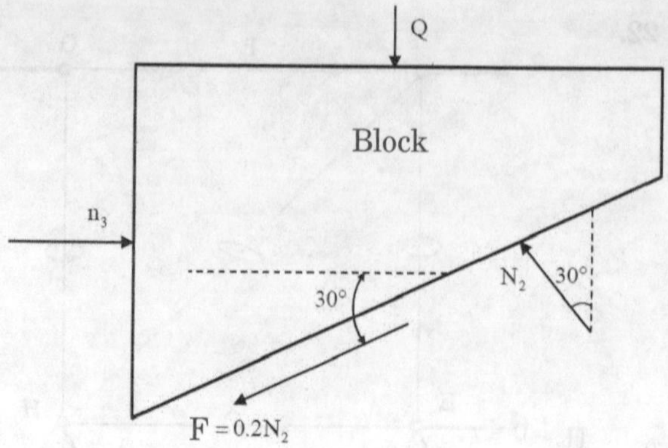

$\Sigma F_y = 0$ (on block)
$N_2 \cos 30 = Q + 0.2N_2 \times \sin 30°$
$\therefore N_2 = Q \times 1.3054$ ...(i)

2. FBD of Wedge M:

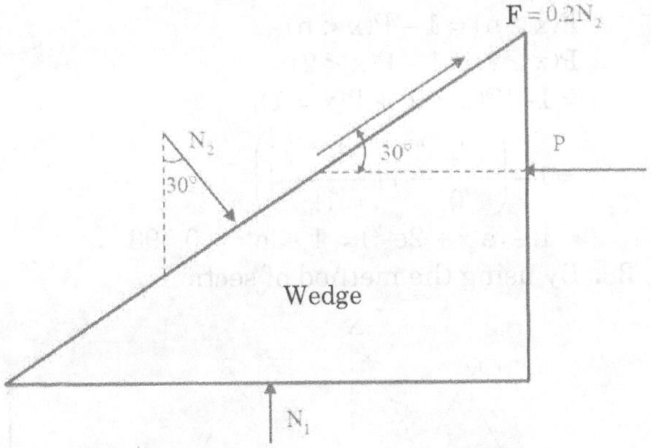

$\Sigma F_x = 0$

$P = N_2 \sin 30 + 0.2\ N_2 \cos 30$

$P = N_2 \times 0.6732$ ... (ii)

From equation (i)

$P = Q \times 1.3054 \times 0.6732$

$P = 0.8788 \times Q = \alpha \times Q$

$\therefore \alpha \simeq 0.9$.

30.

Ratio of Y's bid price to X's Cost	Number of bids
1.02	6
1.04	12
1.06	3
1.10	6
1.12	3
	Total = $\left(\sum n\right) = 30$

Mean = Value of ratio of y's bid price to x's cost

$$= \left[\frac{1.02 \times 6 + 1.04 \times 12 + 1.06 \times 3 + 1.10 \times 6 + 1.12 \times 3}{30}\right]$$

= 1.058

Mean = 1.058

Normal distribution of y's bidding

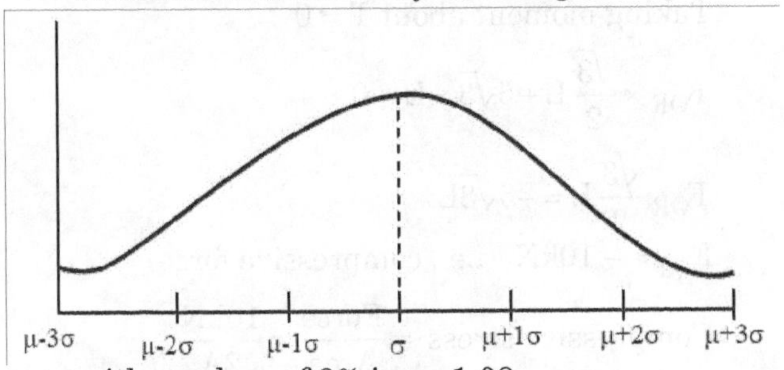

with markup of 8% i.e. = 1.08

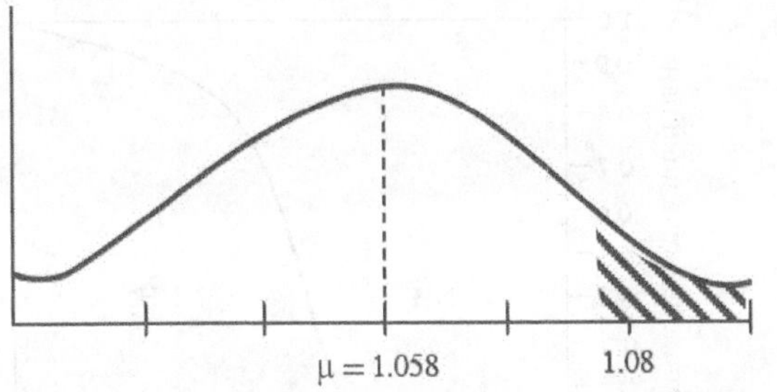

Shaded portion is probability of X contractor winning contract over y which is less than 50%.

31. In dense sand and OC (Over consolidated) clays volumetric strain initially decreases and then increases with the increase in shear strain.

In loose sand s NC clays, volumetric strain decreases with the increase in shear strain.

32. As, given that,

$$V_r = U\left(1 - \frac{r^2}{R^2}\right)$$

$$Q = \int V_r . dA = \int_0^R U\left(1 - \frac{r^2}{R^2}\right)(2\pi r)dr$$

$$Q = \frac{2\pi U}{R^2}\left(\frac{R^2 r^2}{2} - \frac{r^4}{4}\right)\Bigg|_0^R = \frac{2\pi U R^2}{4} = \frac{\pi}{2} U R^2$$

$$\left(\because A = \pi R^2\right)$$

$$\overline{V} = \frac{Q}{A} = \frac{\frac{\pi}{2} U R^2}{\pi R^2} = \frac{U}{2}$$

33. MPN against (5 – 3 –1) for (10.0 – 1–0.1) mL = 110

MPN against (5 – 3 – 1) for sample (0.01 – 0.001 – 0.0001) per 100 mL

$$= 110 \times \frac{10\text{mL}}{0.01} = 110 \times 1000$$

$= 110000$

34. Ammonia exist in the form of NH_4^+ & NH_3

- At PH < 8, all ammonia is in the form of NH_4^+.
- At PH = 9.5, there are 50% NH_3 & 50% NH_4.
- At PH > 11, all Ammonia is in the form of NH_3.

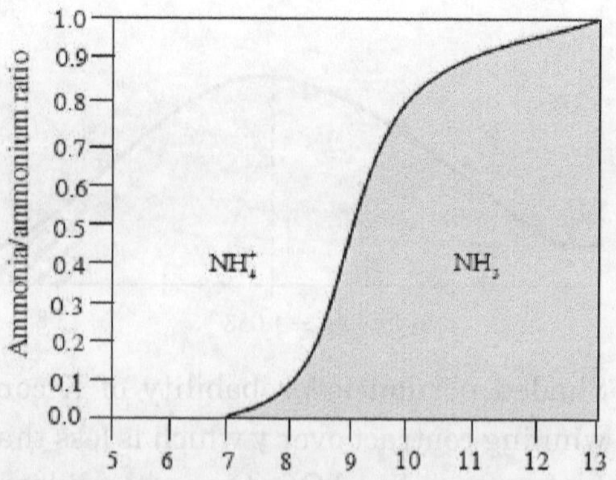

From the above curve, it is evident that at pH = 7.0, NH_4^+ will be predominant

35. As Given that, the speed density relationship:

$u = 70 - 0.7k$

$q = ku = k(70 - 0.7k)$

$q = 70k - 0.7k^2$

For maximum flow $\frac{dq}{dk} = 0$

$$\frac{d}{dk}(70k - 0.7k^2) = 0$$

$70 - 2 \times 0.7k = 0$

$$k = \frac{70}{1.4} = 50$$

Maximum flow $(q_{max}) = 70 \times 50 - 0.7 \times 50^2$

$= 3500 - 1750 = 1750$ veh/hr

$$\text{Average time headway} = \left(\frac{3600}{q_{max}}\right) = \left(\frac{3600}{1750}\right) \text{sec}$$

$= 2.06$ seconds $\simeq 2.1$ sec.

36. As given that,

$h = 0.5, y_0 = 5, y_1 = 7.25; y_2 = 10, y_3 = 13.25, y_4 = 17$

The formula of Simpson's $\frac{1}{3}$rd rule,

So, numerical integration of the area $= \int_2^4 y dx$

$$= \frac{h}{3}[(y_0 + y_4) + 4(y_1 + y_3) + 2(y_2)]$$

$$= \frac{0.5}{3}[(5+17) + 4(7.25+13.25) + 2(10)]$$

$$= \frac{0.5}{3}[22+82+20] = \frac{124}{6} = 20.66$$

37. As given that,

The mean vehicular arrival rate, $(\lambda) = 2$veh / min

Let x = Number of vehicles arrive in a minute.

As per Poisson's distribution

$P(x \geq n) = 1 - P(x < n)$

$P(x \geq 2) = 1 - P(x < 2)$

$= 1 - \{P(x = 0) + P(x = 1)\}$

$$= 1 - \left(\left(\frac{\lambda^0 e^{-\lambda}}{0!}\right) + \left(\frac{\lambda^1 e^{-\lambda}}{1!}\right)\right)$$

$= 1 - (e^{-2} + 2e^{-2}) = 1 - 3e^{-2} = 0.593$

38. By using the method of sections.

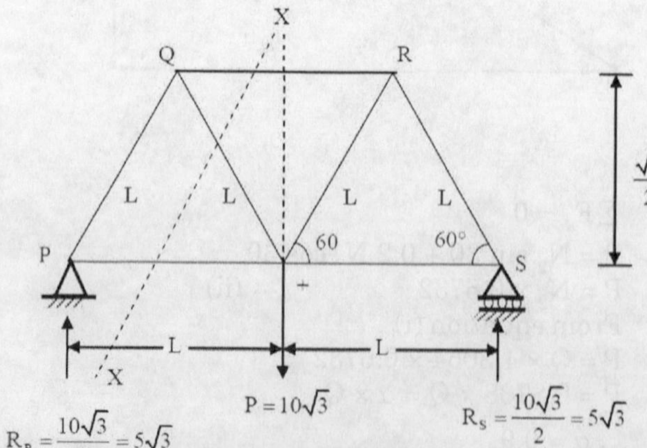

Here, all the members are of equal length sign convention.

Tension ⊕, Compression ⊖

Now, taking Cutting section X-X:

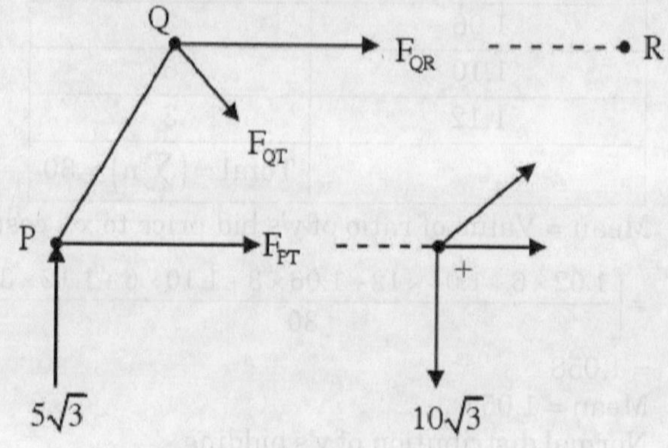

Taking moment about T = 0

$$F_{QR} \times \frac{\sqrt{3}}{2} L + 5\sqrt{3} \times L = 0$$

$$F_{QR} \frac{\sqrt{3}}{2} L = -5\sqrt{3}L$$

$F_{QR} = -10$kN (i.e., compressive force)

$$\text{Compressive stress} = \frac{\text{Force}}{\text{Area}} = \frac{10\text{kN}}{2A}$$

$$= \frac{10\text{kN}}{2 \times 10^4 \text{mm}^2} = 5 \times 10^{-4} \text{kN/mm}^2$$

$= 5 \times 10^{-4} \times 10^6$ kN/m^2 = 500 kN/m^2

39. As given that,

$L = 1.5$ m, $P_0 = 50$ kN, dc = 50mm.

b = 200 mm, d = 300 mm, P = 5 kN

Moment at Q = PL = 5 × 1.5

= 7.5 k.N.m

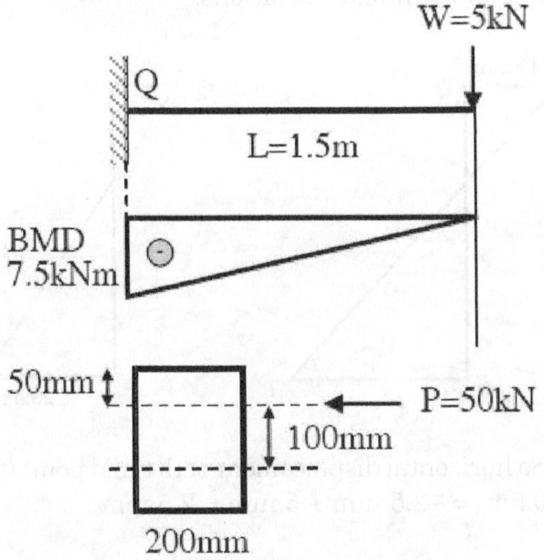

Let, T = top and B = Bottom

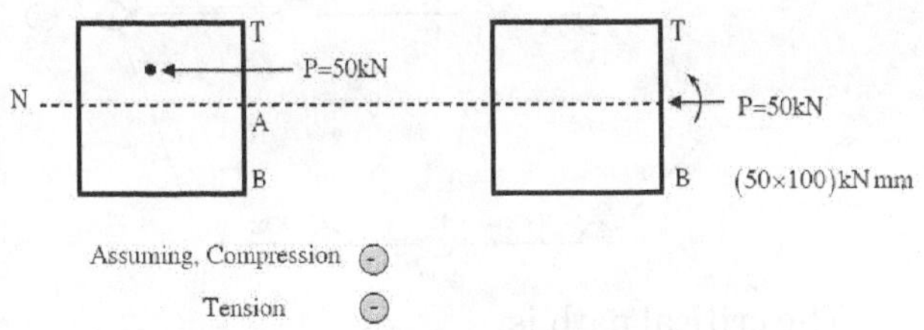

Consider a cross section element at Q.

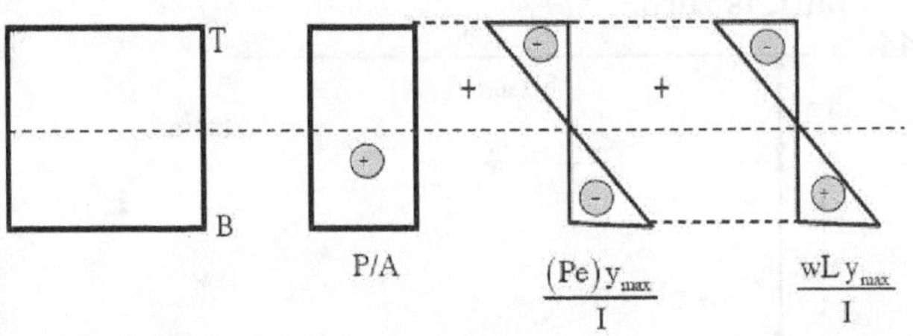

$$\text{MOI of C/S (1)} = \frac{1}{12}BD^3 = \frac{1}{12} \times 200 \times 300^3$$

$= 450 \times 10^6$ mm^4

$$\text{Section modulus (Z)} = \frac{1}{y} = \frac{\frac{BD^2}{12 \times D}}{2} = \frac{BD^2}{6}$$

$$= \frac{450 \times 10^6 \text{ mm}}{150} = 3 \times 10^6 \text{ mm}^4$$

Note: For rectangular cross-section

$$\text{Section Modulus(Z)} = \frac{BD^2}{6} = \frac{200 \times 300^2}{6}$$

$= 3 \times 10^6$ mm^4

'Q' is the top point

Stress at top, (σ_T)

$$(\sigma_T) = \frac{50 \times 10^3 \text{N}}{200 \times 300} + \frac{(50 \times 10^3) \times 100}{3 \times 10^6}$$

$$-\frac{(5 \times 10^3) \times (1.5 \times 10^3)}{3 \times 10^6}$$

= 0.833 + 1.667 − 2.5

$= 3 \times 10^{-3}$ N/mm^2

Resultant stress at 'Q' rounded of in integer = 0

40. Bolt (R) will be subjected to direct shear force (P_1) and torsional shear force (P_2).

Resultant (R) = $P_1 + P_2$

$$\text{Shear force } (P_1) = \frac{P}{n} = \frac{60\text{kN}}{6} = 10\text{kN}$$

$$\text{Torsional force } (P_2) = \frac{Pe.r}{\sum r^2}$$

$$= \frac{60 \times 100 \times 40}{(4 \times 50^2 + 2 \times 40^2)}$$

So, resultant force in bolt 'R'

$$= \sqrt{P_1^2 + P_2^2 + 2P_1P_2\text{Cos}\theta}$$

$$= \sqrt{(P_1 + P_2)^2}$$

$(\because \theta = 0)$

$= P_1 + P_2$

= 10 + 18.18

= 28.18 kN.

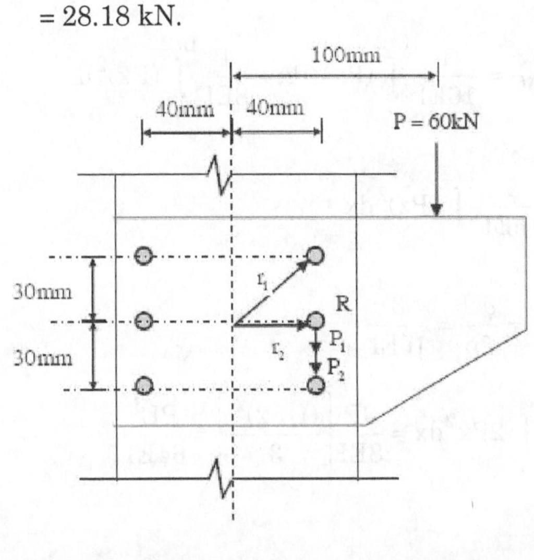

41. As Given that P = 100kN, E = 3 × 10^4, I = 225 × 10^6 mm^4, L = 8 m.

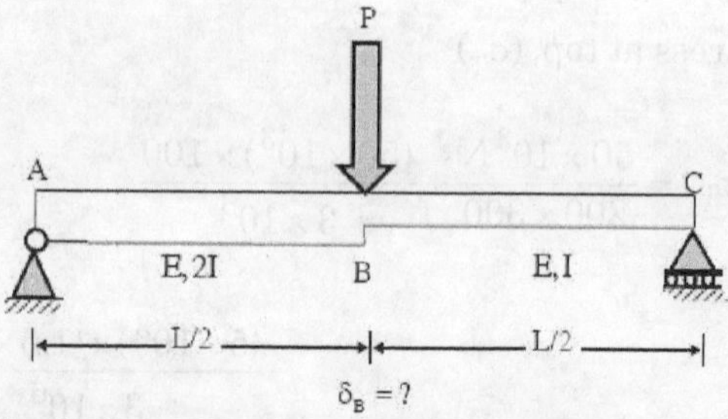

As deflection to be find out at mid span (point load location) strain energy method is the easiest to find.

$U_{AC} = U_{AB} + U_{BC}$

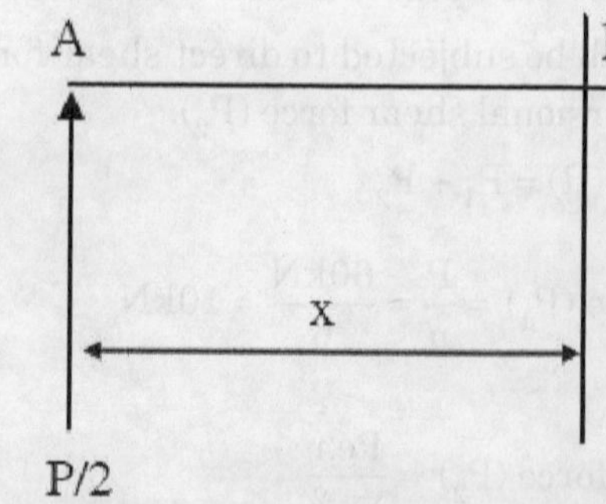

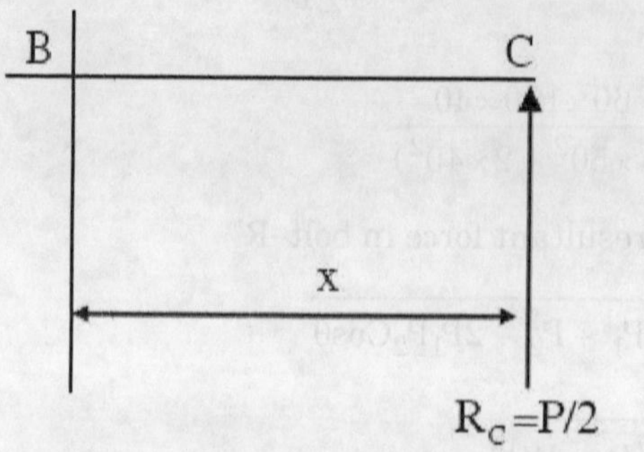

$$U_{AB} = \int_0^{L/2} \frac{\left(\frac{P}{2}x\right)^2}{4EI} dx, \quad U_{BC} = \int_0^{L/2} \frac{\left(\frac{P}{2}x\right)^2}{2EI} dx$$

$$U_{AC} = U_{AB} + U_{BC}$$

$$U_{AC} = \frac{1}{16EI}\int_0^{L/2} (Px)^2 dx + \frac{1}{8EI}\int_0^{L/2} (Px)^2 dx$$

$$= \frac{3}{16EI}\int_0^{L/2} (Px)^2 dx$$

$$\Delta_B = \frac{\partial U}{\partial p} = \frac{3}{16EI}$$

$$= \int_0^{L/2} 2Px^2 dx = \frac{3P}{8EI}\left[\frac{(L/2)^2}{3}\right] = \frac{PL^3}{64EI}$$

$$\Delta_B = \frac{100 \times 10^3 \times (8000)^3}{64 \times 3 \times 10^4 \times 225 \times 10^6}$$

$= 118.5$ mm ≈ 119mm

42. Assuming linear variations.

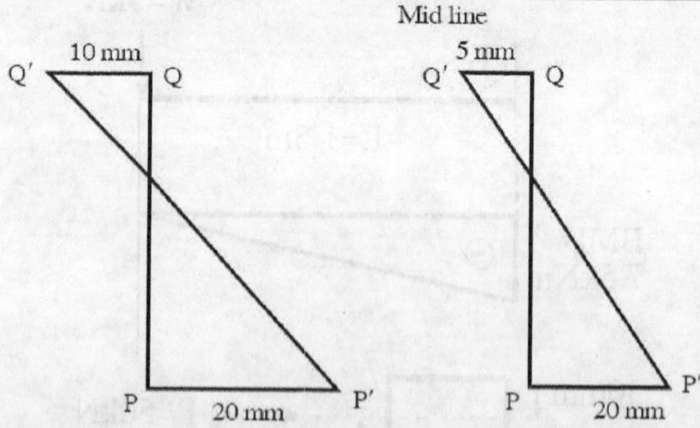

So horizontal displacement of the mid point (0.5m, 0.5m) = –2.5 mm + 5 mm = 2.5 mm

43.

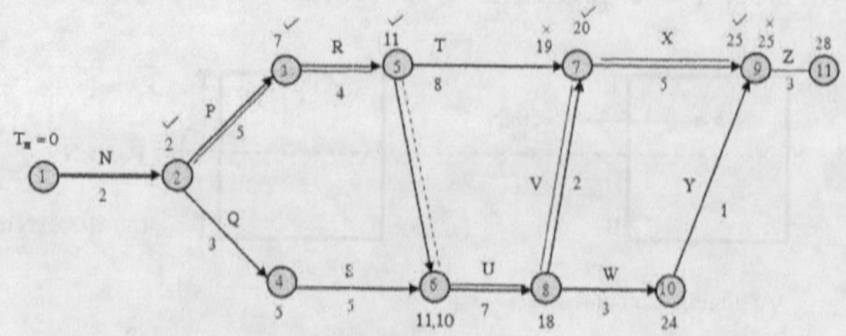

The critical path is

N – P – R – U – V – X – Z

And the total float on activity which is in critical path is zero.

44.

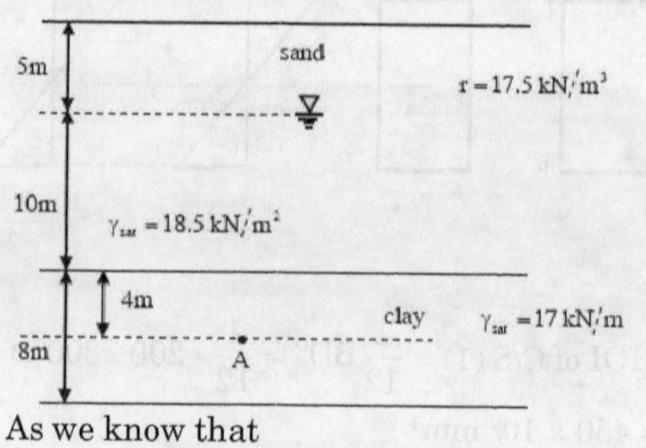

As we know that

Over consolidation ratio (OCR) = $\dfrac{\text{Pre consolidation status}(\bar{\sigma}_x)}{\text{Present stress}(\bar{\sigma}_A)}$

When water table at 10m depth from the ground

Effective stress at

A = 10 × 17.5 + 5 × (18.5 – 9.81) + 4 (17 – 9.81)

= 247.21 kN m^2

$(\bar{\sigma}) = 247.21 kN/m^2$

Presently water table at a depth of 5m from the ground level

Effective stress at A = 5 × 17.5 + 10 (18.5 – 9.81) + 4(17 – 9.81)

$(\bar{\sigma}) = 203.16 \text{kN/m}$

$$\text{OCR} \frac{\bar{\sigma}_C}{\bar{\sigma}} = \frac{247.21}{203.16} = 1.22$$

45. According to the question:

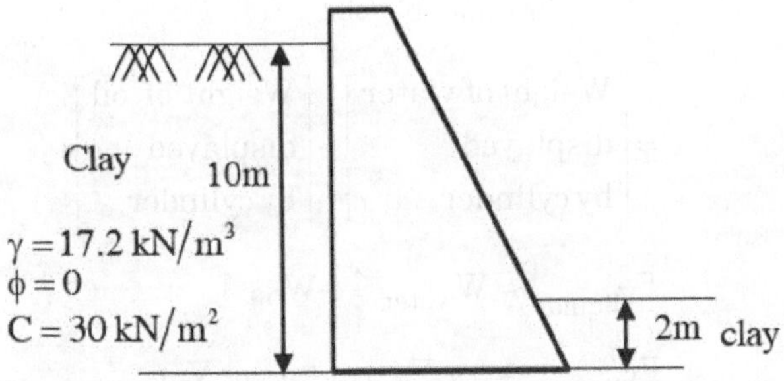

As the soil clay

$$\text{Critical depth } (Z_z) = \frac{2c}{\gamma\sqrt{k_a}}$$

$$\text{Where, } k_a = \left[\frac{1-\sin\phi}{1+\sin\phi}\right]$$

For $\phi = 0$

$$k_a = \frac{1-\sin 0}{1+\sin 0} = 1$$

$$z_c = \frac{2\times 30}{17.2\sqrt{1}} = 3.49\text{m}$$

$$P_a = k_a.\lambda_{sub}.z - 2c\sqrt{k_a}$$

$$= 1 \times (17.2 - 9.81) \times 10 - 2 \times 30\sqrt{1}$$

$P_a = 13.9 \text{ kN/m}^2$

$P_w = k_a.\gamma_w z = 1 \times 9.81 \times 10 = 98.1 \text{ kN/m}^2$

Total active thrust

$$= K_a.\frac{1}{2}\gamma H^2 - 2C\sqrt{K_a}H + \frac{2C^2}{\gamma} + \frac{\gamma_w / z_C{}^2}{2}$$

$$= 1\times\frac{1}{2}\times 17.2\times 10^2 - 2\times 30\sqrt{1}\times 10 + \frac{2\times 30^2}{17.2} + \frac{9.81\times 3.488^2}{2}$$

$F_a = 424.326$ kN

Frictional force = N tanδ = w tanδ

$\Rightarrow$ 5000 tan 20°

= 1819.85 kN

$$\text{F.O.S} = \frac{\text{Frictional force}}{\text{Active thurst}}$$

$$= \frac{1819.851}{424.326} = 4.29$$

46.

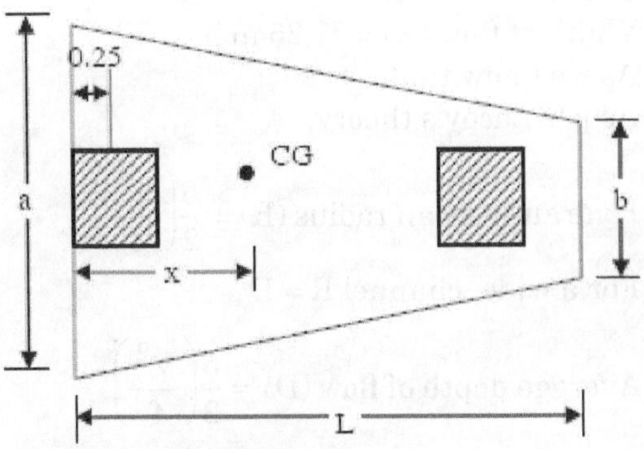

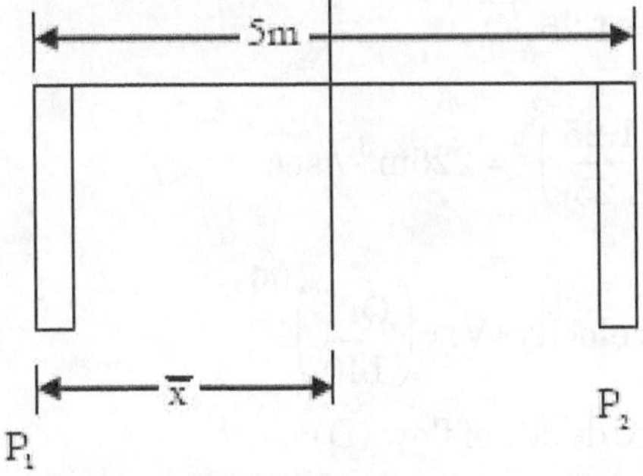

Taking moment about centre of column

$$\bar{x} = \frac{P_1\times 0 + P_2\times 5}{P_1 + P_2} = \frac{2000\times 0 + 1500\times 5}{2000 + 1500} = 2.143\text{m}$$

Footing is proportioned such that the centre of gravity of the footing lies on the line of action of the resultant of the column load.

$$x = \left(\frac{a+2b}{a+b}\right)\times\frac{L}{3} = \left(\frac{5+2\times 1.5}{5+15}\right)\frac{L}{3}$$

$$x = \bar{x} + 0.25$$

$$\left(\frac{5+2\times 1.5}{5+1.5}\right)\frac{L}{3} = 2.143 + 0.25$$

L = 5.833 m

47. As we know that,

$$\text{FOS(F)} = \frac{\text{Resisting moment}}{\text{Actuating moment}}$$

$$\text{FOS(F)} = \frac{C_u\hat{L}R}{w\bar{x}}$$

$\hat{L}$ = length of ac KLM

$\bar{x}$ = Distance of 'w' from toe

$$\Rightarrow FOS(F) = \frac{75 \times 2\pi \times 25 \times \frac{60}{36} \times 25}{2500 \times 100}$$

$$\Rightarrow FOS\,(F) = 1.96$$

48. As given that,
Slit factor (f) = 1,
Width of flow (B) = 71.25 m
As we know that,
As per Lacey's theory,

$$\text{Hydraulic mean radius (R)} = \frac{5}{2}\left(\frac{V^2}{f}\right)$$

For a wide, channel R = D

$$\text{Average depth of flow (D)} = \frac{5}{2}\left(\frac{V^2}{f}\right)$$

$$\text{Perimeter (P)} = \text{Width (B)} = 4.25\sqrt{Q}$$

$$71.25 = 4.25\sqrt{Q}$$

$$Q = \left(\frac{71.25}{4.25}\right)^2 = 225\text{m}^3/\text{sec}$$

$$\text{Mean velocity (V)} = \left(\frac{Qf^2}{140}\right)^{1/6}$$

Average depth of flow (D)

$$= \frac{5}{2}\left(\frac{\left(\left(\frac{Qf^2}{140}\right)^2\right)}{f}\right) = \frac{5Q^2f^3}{2140^2} = \frac{5}{2} \times \frac{225^2 \times 1^3}{140^2} = 2.938\text{m}$$

49. According to the question:

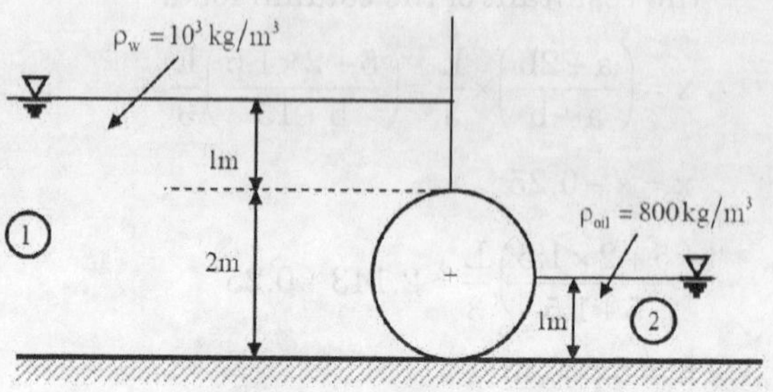

Net horizontal force due to liquid.

a. $\vec{F}_{H_{NET}} = \vec{F}_{H_{Water}} + \vec{F}_{H_{Oil}}$

$$F_{H_{Net}} = \rho_{water}.g.\bar{h}_{water}.A_{projected\,cylinder}$$

$$-\rho_{oil}.g.\bar{h}_{oil}.\text{Aprojected half cylinder}$$

$$F_{H_{net}} = 1000 \times 9.81 \times \left(1 + \frac{2}{2}(2 \times 3)\right)$$

$$-800 \times 9.81 \times \left(\frac{1}{2}\right)(1 \times 3)$$

$$F_{Net} = 117720 - 11772$$

$$|F_{H_{Net}}| = 105948\text{N} = 105.948\text{kN} \quad \ldots (1)$$

b. Net vertical force due to liquid.

$$F_v \uparrow = F^{\uparrow}_{v_{water}} + F_{V_{oil}}$$

$$= \begin{bmatrix}\text{Weight of water}\\ \text{displayed}\\ \text{by cylinder}\end{bmatrix} + \begin{bmatrix}\text{Weight of oil}\\ \text{displayed}\\ \text{by cylinder}\end{bmatrix}$$

$$F_{V_{liquids}} = W_{water} \uparrow + W_{Oil} \uparrow$$

$$F_{V_{liquids}} = \rho.g.V_{water} + \rho_{Oil}g.V_{oil}$$

$$= \rho_w.g.\frac{\pi R^2.L}{2} + \rho_{oil}.g.\frac{\pi R^2.L}{4}$$

$$F_{V_{liquids}} = 100 \times 9.81 \times \frac{\pi(1)^2(3)}{2} + 800 \times 9.81 \times \frac{\pi(1)^2}{4} \times 3$$

$$= 46228.536 + 18491.414$$

$$F_{V_{liquids}} = 64719.95\text{N} \cong 64.72\text{kN} \uparrow \sqrt{a^2 + b^2}$$

$$|F_{V_{Net}}|_{liquids} = (F_V)_{liquids} \uparrow - W_{cylinder} \downarrow$$

$$|F_{V_{liquids}}| = 64.72 \uparrow -25 \downarrow$$

$$\therefore |F_{V_{NET}}| = 39.72\text{kN} \uparrow \quad \ldots(2)$$

Ratio:

$$\frac{F_{V_{Net}}}{F_{H_{Net}}} = \frac{39.720(\text{kN})}{108.948(\text{kN})} = 0.3749 \cong 0.375$$

$$= 0.38$$

50. For unconfined aquifer, by using Thiem's equation,

$$Q = \frac{2\pi TS_w}{\ell n\left(\frac{R}{r_w}\right)} \quad \cdots (i)$$

$$Q = 40 \text{ lit/sec} = \left[\frac{40 \times 10^{-3}}{\frac{1}{24 \times 60 \times 60}}\right] = 3456$$

$\because$ 1 day = 24 × 60 × 60 sec

$$1\,\text{sec} = \left(\frac{1}{24 \times 60 \times 60}\right)\text{day}$$

$Q = 3456\ m^3/day$

From (i) equation;

$$3456m^3/\text{day} = \frac{2 \times \pi \times T \times 4}{\ell n\left(\frac{245m}{0.1m}\right)}$$

$T = 1073.11\ m^2/day$

$\therefore$ Transmissibility, (T) = kB.

$$\text{So, k} = \frac{T}{B} = \frac{1073.11}{30} \quad (\because B = 30)$$

= 35.76 m/day

51. As given that,

Discharge (Q) = $12m^3/sec$

So, total area of filter required

$$= \frac{\text{Discharge}}{\text{velocity}} = \frac{12m^3/\text{sec}}{\frac{6}{60}m/\text{sec}} = \frac{720}{6} = 120m^2$$

Surface area of each bag = $(\pi D)L = \pi \times 0.32 \times 65$

= $5.626\ m^2$

As we know that,

$$\text{Number of bags} = \frac{\text{Total are required}}{\text{Area of each bag}}$$

$$= \frac{120}{5.626} = 23.81 = 24\text{bags}$$

$$\text{Number of bags per section} = \frac{24}{4} = 6$$

So, number of bags for five sections = 6 × 5 = 30 bags.

52. As given that,

Total flow (Q) = $9600\ m^3/day$

Concentration of solids = 3000 mg/L

Solid flux = $3.2\ kg/m^2.h$

As we know that

$$\text{Surface area} = \frac{\text{Total solids}}{\text{Solid flux rate}} = \frac{\text{MLSS} \times Q}{\text{Solid flux rate}}$$

$$= \frac{9600(m^3/\text{day})Q}{3.2\ kg/m^2.hr} = \frac{9600 \times \frac{3}{24}}{3.2} = 375m^2$$

53. As we know that,

$$\text{Space mean speed } (V_s) = \frac{n}{\sum_{i=1}^{n} \frac{1}{V_i}}$$

$$V_s = \frac{5}{\frac{1}{40} + \frac{1}{55} + \frac{1}{60} + \frac{1}{65} + \frac{1}{85}}$$

Hence, Space mean speed = 56.99 kmph.

54. Assuming RL of start of runway as datum i.e., RL = 0m

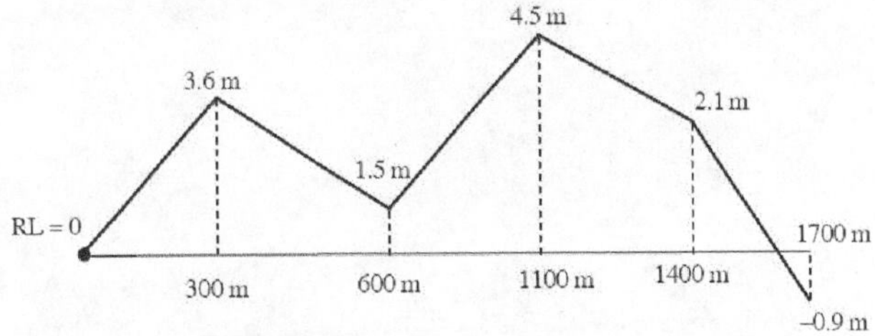

As we know that,

Effective gradient =

$$\left[\frac{\text{Minimum difference in reduced level}}{\text{Total runway length}}\right]$$

$$= \left[\frac{4.5 - (-0.9)}{1700} \times 100\right]\%$$

$= 0.3176\% \simeq 0.32\%$

55.

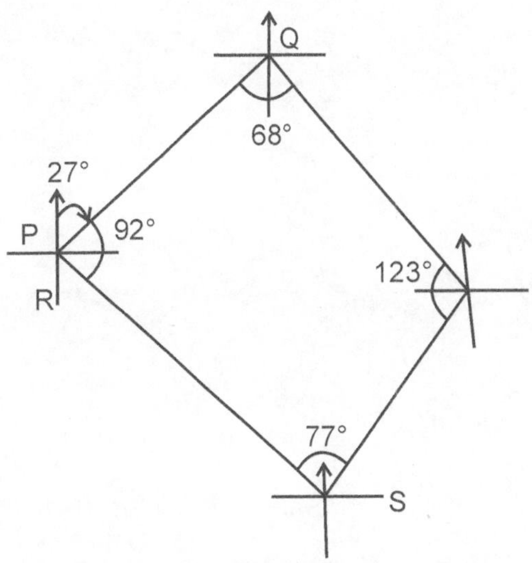

Back bearing:

$QR = BB_{PQ} - \angle Q = (27 + 180) - 68° = 139°$

Back bearing RS = Back bearing of QR – $\angle R$

= (139 + 180) – 123

$FB_{RS} = 196°$

SOLVED PAPER - 2021

SET 2

INSTRUCTIONS

1. Total of 65 questions carrying 100 marks, out of which 10 questions carrying a total of 15 marks are in General Aptitude (GA)
2. The Engineering Mathematics will carry around **15% of the total marks,** the General Aptitude section will carry **15% of the total marks** and the **remaining 70% of the total marks.**
3. **Types of Questions**
   (*a*) **Multiple Choice Questions (MCQ)** carrying 1 or 2 marks each in all papers and sections. These questions are objective in nature, and each will have a choice of four options, out of which the candidate has to mark the correct answer(s).
   (*b*) **Numerical Answer Questions** of 1 or 2 marks each in all papers and sections. For these questions the answer is a real number, to be entered by the candidate using the virtual keypad. No choices will be shown for these type of questions.
   (*c*) **Multiple Select Questions (MSQ)** carrying 1 or 2 marks each in all the papers and sections. These questions are objective in nature, and each will have choice of four answers, out of which ONE or MORE than ONE choice(s) is / are correct.
4. For **1-mark** multiple-choice questions, **1/3 marks** will be deducted for a wrong answer. Likewise, for **2-marks** multiple-choice questions, **2/3** marks will be deducted for a wrong answer. There is no negative marking for numerical answer type questions & multiple selection questions.

Chapter-Wise Analysis

#	Subject	Marks		
		1	2	Total
1	General Aptitude	5	5	15
2	Engineering Mathematics	4	5	14
3	Engineering Mechanics	1	2	5
4	Solid Mechanics	2	1	4
5	Structural Analysis	2	1	4
6	Construction Materials and Management	3	0	3
7	Concrete Structures	3	3	9

#	Subject	Marks		
		1	2	Total
8	Steel Structures	1	0	1
9	Soil Mechanics and Foundation Engineering	4	5	14
10	Fluid Mechanics and Hydraulics	0	3	6
11	Hydrology & Irrigation	2	2	6
12	Environmental Engineering	2 + (1 MSQ)	1	5
13	Transportation Engineering	4	3	10
14	Surveying	0	2	4

GENERAL APTITUDE

Q. 1 – Q. 5 carry one mark each

1. (i) Arun and Aparna are here.

(ii) Arun and Aparna is here.

(iii) Arun's families is here.

(iv) Arun's family is here.

Which of the above sentences are grammatically CORRECT?

(a) (i) and (ii) (b) (i) and (iv)

(c) (ii) and (iv) (d) (iii) and (iv)

2.

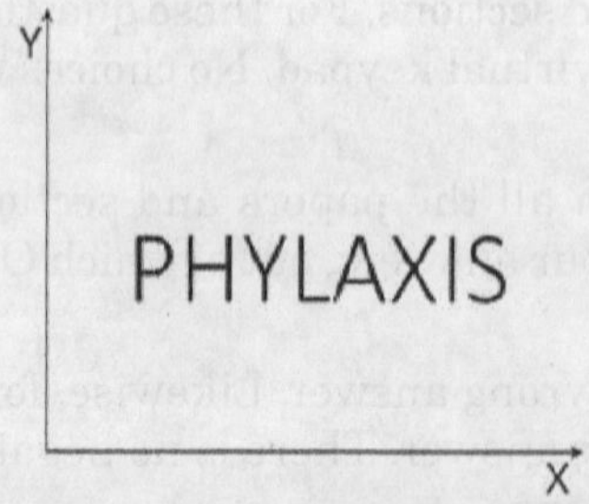

The mirror image of the above text about the x-axis is

(a) 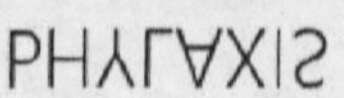 (b) 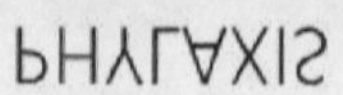

(c) 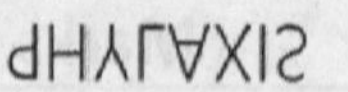 (d) 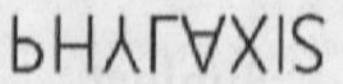

3. Two identical cube shaped dice each with faces numbered 1 to 6 are rolled simultaneously. The probability that an even number is rolled out on each dice is:

(a) 1/36 (b) 1/12

(c) 1/8 (d) 1/4

4. $\oplus$ and $\odot$ are two operators on numbers p and q such that $p \odot q = p - q$, and $p \oplus q = p \times q$

Then, $(9 \odot (6 \oplus 7)) \odot (7 \oplus (6 \odot 5)) =$

(a) 40 (b) –26

(c) –33 (d) –40

5. Four persons P, Q, R and S are to be seated in a row. R should not be seated at the second position from the left end of the row. The number of distinct seating arrangements possible is:

(a) 6 (b) 9

(c) 18 (d) 24

Q. 6 – Q. 10 carry two marks each

6. On a planar field, you travelled 3 units East from a point O. Next you travelled 4 units South to arrive at point P. Then you travelled from P in the North-East direction such that you arrive at a point that is 6 units East of point O. Next, you travelled in the North-West direction, so that you arrive at point Q that is 8 units North of point P. The distance of point Q to point O, in the same units, should be ______

(a) 3

(b) 4

(c) 5

(d) 6

7. The author said, "Musicians rehearse before their concerts. Actors rehearse their roles before the opening of a new play. On the other hand, I find it strange that many public speakers think they can just walk on to the stage and start speaking. In my opinion, it is no less important for public speakers to rehearse their talks."

Based on the above passage, which one of the following is TRUE?

(a) The author is of the opinion that rehearsing is important for musicians, actors and public speakers.

(b) The author is of the opinion that rehearsing is less important for public speakers than for musicians and actors.

(c) The author is of the opinion that rehearsing is more important only for musicians than public speakers.

(d) The author is of the opinion that rehearsal is more important for actors than musicians.

8. 1. Some football players play cricket.

2. All cricket players play hockey.

Among the options given below, the statement that logically follows from the two statements 1 and 2 above, is:

(a) No football player plays hockey.

(b) Some football players play hockey.

(c) All football players play hockey.

(d) All hockey players play football.

9. 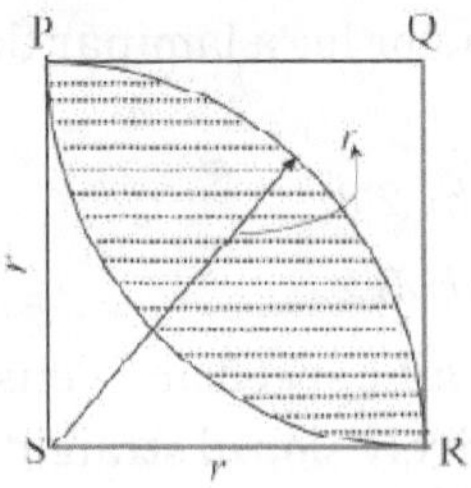

In the figure shown above, PQRS is a square. The shaded portion is formed by the intersection of sectors of circles with radius equal to the side of the square and centers at S and Q.

The probability that any point picked randomly within the square falls in the shaded area is ______

(*a*) $4-\dfrac{\pi}{2}$ (*b*) 1/2

(*c*) $\dfrac{\pi}{2}-1$ (*d*) $\dfrac{\pi}{4}$

10. In an equilateral triangle PQR, side PQ is divided into four equal parts, side QR is divided into six equal parts and side PR is divided into eight equal parts. The length of each subdivided part in cm is an integer.

The minimum area of the triangle PQR possible, in cm², is

(*a*) 18 (*b*) 24

(*c*) $48\sqrt{3}$ (*d*) $144\sqrt{3}$

CIVIL ENGINEERING (CE) - 2

Q. 1 – Q. 16 carry one mark each

1. The value of $\lim\limits_{x\to\infty}\dfrac{x\ln(x)}{1+x^2}$ is

(*a*) 0 (*b*) 1.0

(*c*) 0.5 (*d*) ∞

2. The rank of the matrix $\begin{bmatrix} 5 & 0 & -5 & 0 \\ 0 & 2 & 0 & 1 \\ -5 & 0 & 5 & 0 \\ 0 & 1 & 0 & 2 \end{bmatrix}$ is

(*a*) 1 (*b*) 2

(*c*) 3 (*d*) 4

3. The unit normal vector to the surface $X^2+Y^2+Z^2-48=0$ at the point (4, 4, 4) is

(*a*) $\dfrac{1}{\sqrt{2}},\dfrac{1}{\sqrt{2}},\dfrac{1}{\sqrt{2}}$ (*b*) $\dfrac{1}{\sqrt{3}},\dfrac{1}{\sqrt{3}},\dfrac{1}{\sqrt{3}}$

(*c*) $\dfrac{2}{\sqrt{2}},\dfrac{2}{\sqrt{2}},\dfrac{2}{\sqrt{2}}$ (*d*) $\dfrac{1}{\sqrt{5}},\dfrac{1}{\sqrt{5}},\dfrac{1}{\sqrt{5}}$

4. If A is a square matrix then orthogonality property mandates

(*a*) $AA^T = I$

(*b*) $AA^T = 0$

(*c*) $AA^T = A^{-1}$

(*d*) $AA^T = A^2$

5. In general, the CORRECT sequence of surveying operations is

(a) Field observations → Reconnaissance → Data analysis → Map making

(b) Data analysis → Reconnaissance → Field observations → Map making

(c) Reconnaissance → Field observations → Data analysis → Map making

(d) Reconnaissance → Data analysis → Field observations → Map making

6. Strain hardening of structural steel means

(a) experiencing higher stress than yield stress with increased deformation

(b) strengthening steel member externally for reducing strain experienced

(c) strain occurring before plastic flow of steel material

(d) decrease in the stress experienced with increasing strain

7. A single story building model is shown in the figure. The rigid bar of mass 'm' is supported by three massless elastic columns whose ends are fixed against rotation. For each of the columns, the applied lateral force (P) and corresponding moment (M) are also shown in the figure. The lateral deflection (δ) of the bar is given by $\delta = \dfrac{PL^3}{12EI}$, where L is the effective length of the column, E is the Young's modulus of elasticity and I is the area moment of inertia of the column cross-section with respect to its neutral axis.

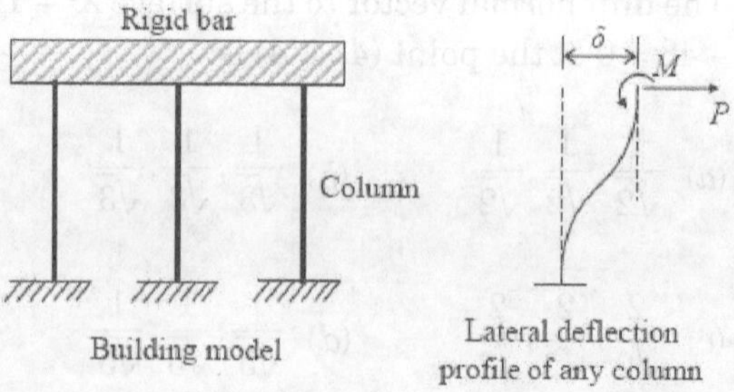

For the lateral deflection profile of the columns as shown in the figure, the natural frequency of the system for horizontal oscillation is

(*a*) $6\sqrt{\frac{EI}{mL^3}}$ rad / s (*b*) $\frac{1}{L}\sqrt{\frac{2EI}{m}}$ rad / s

(*c*) $2\sqrt{\frac{6EI}{mL^3}}$ rad / s (*d*) $\frac{2}{L}\sqrt{\frac{EI}{m}}$ rad / s

8. Seasoning of timber for use in construction is done essentially to

(a) increase strength and durability

(b) smoothen timber surfaces

(c) remove knots from timber logs

(d) cut timber in right season and geometry

9. In case of bids in Two-Envelop System, the correct option is

(a) Technical bid is opened first

(b) Financial bid is opened first

(c) Both (Technical and Financial) bids are opened simultaneously

(d) Either of the two (Technical and Financial) bids can be opened first

10. The most appropriate triaxial test to assess the long-term stability of an excavated clay slope is

(a) consolidated drained test

(b) unconsolidated undrained test

(c) consolidated undrained test

(d) unconfined compression test

11. As per the Unified Soil Classification System (USCS), the type of soil represented by 'MH' is

(a) Inorganic silts of high plasticity with liquid limit more than 50%

(b) Inorganic silts of low plasticity with liquid limit less than 50%

(c) Inorganic clays of high plasticity with liquid limit less than 50%

(d) Inorganic clays of low plasticity with liquid limit more than 50%

12. The ratio of the momentum correction factor to the energy correction factor for a laminar flow in a pipe is

(*a*) 1/2 (*b*) 2/3

(*c*) 1 (*d*) 3/2

13. Relationship between traffic speed and density is described using a negatively sloped straight line. If v_f is the free-flow speed then the speed at which the maximum flow occurs is

(*a*) 0 (*b*) $\frac{v_f}{4}$

(*c*) $\frac{v_f}{2}$ (*d*) v_f

14. Determine the correctness or otherwise of the following Assertion [a] and the Reason [r].

Assertion [a]: One of the best ways to reduce the amount of solid wastes is to reduce the consumption of raw materials.

Reason [r]: Solid wastes are seldom generated when raw materials are converted to goods for consumption.

(a) Both [a] and [r] are true and [r] is the correct reason for [a]

(b) Both [a] and [r] are true but [r] is not the correct reason for [a]

(c) Both [a] and [r] are false

(d) [a] is true but [r] is false

15. The hardness of a water sample is measured directly by titration with 0.01 M solution of ethylenediamine tetraacetic acid (EDTA) using eriochrome black T (EBT) as an indicator. The EBT reacts and forms complexes with divalent metallic cations present in the water. During titration, the EDTA replaces the EBT in the complex. When the replacement of EBT is complete at the end point of the titration, the colour of the solution changes from

(a) blue-green to reddish brown

(b) blue to colourless

(c) reddish brown to pinkish yellow

(d) wine red to blue

16. The softening point of bitumen has the same unit as that of

(a) distance (b) temperature

(c) time (d) viscosity

17. Which of the following statement(s) is/are correct?

(a) Increased levels of carbon monoxide in the indoor environment result in the formation of carboxyhemoglobin and the long term exposure becomes a cause of cardiovascular diseases.

(b) Volatile organic compounds act as one of the precursors to the formation of photochemical smog in the presence of sunlight.

(c) Long term exposure to the increased level of photochemical smog becomes a cause of chest constriction and irritation of the mucous membrane.

(d) Increased levels of volatile organic compounds in the indoor environment will result in the formation of photochemical smog which is a cause of cardiovascular diseases.

Q. 18 – Q. 25 carry one mark each

18. The value (round off to one decimal place) of $\int_{-1}^{1} xe^{|x|}dx$ is ________

19. A solid circular torsional member OPQ is subjected to torsional moments as shown in the figure (not to scale). The yield shear strength of the constituent material is 160 MPa.

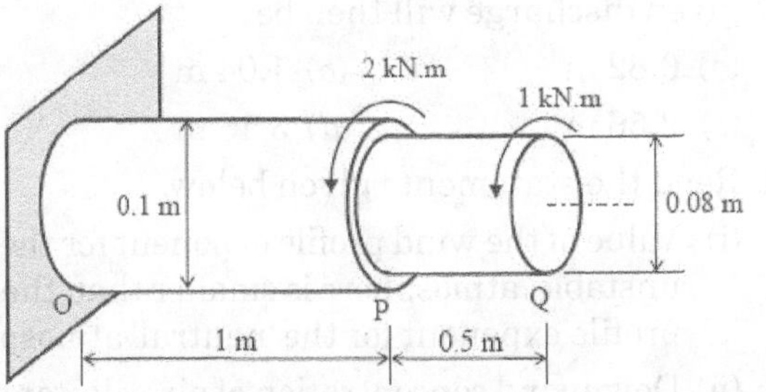

The absolute maximum shear stress in the member (in MPa, round off to one decimal place) is ________

20. A propped cantilever beam XY, with an internal hinge at the middle, is carrying a uniformly distributed load of 10 kN/m, as shown in the figure.

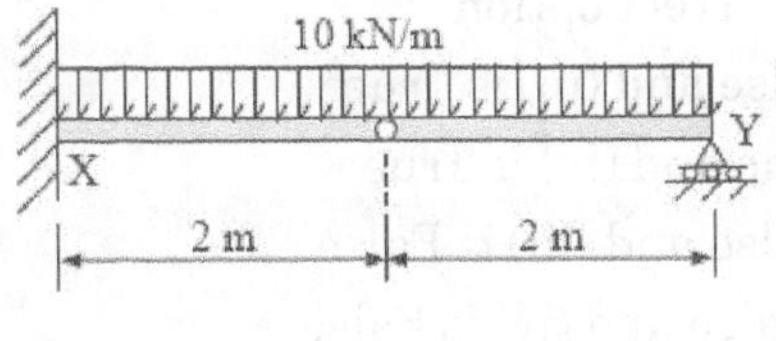

The vertical reaction at support X (in kN, in integer) is ________

21. The internal (d_i) and external (d_o) diameters of a Shelby sampler are 48 mm and 52 mm, respectively. The area ratio (A_r) of the sampler (in %, round off to two decimal places) is ________

22. A 12-hour unit hydrograph (of 1 cm excess rainfall) of a catchment is of a triangular shape with a base width of 144 hour and a peak discharge of 23 m^3/s. The area of the catchment (in km^2, round off to the nearest integer) is ________

23. A lake has a maximum depth of 60 m. If the mean atmospheric pressure in the lake region is 91 kPa and the unit weight of the lake water is 9790 N/m^3, the absolute pressure (in kPa, round off to two decimal places) at the maximum depth of the lake is ________

24. In a three-phase signal system design for a four-leg intersection, the critical flow ratios for each phase are 0.18, 0.32, and 0.22. The total loss time in each of the phases is 2 s. As per Webster's formula, the optimal cycle length (in s, round off to the nearest integer) is ________

25. A horizontal angle θ is measured by four different surveyors multiple times and the values reported are given below.

Surveyor	Angle θ	Number of observations
1	36°30'	4
2	36°00'	3
3	35°30'	8
4	36°30'	4

The most probable value of the angle θ (in degree, round off to two decimal places) is ________

Q. 26 – Q. 35 carry two mark each

26. If k is a constant, the general solution of $\frac{dy}{dx} - \frac{y}{x} = 1$ will be in the form of

(a) $y = x\ln(kx)$ (b) $y = k\ln(kx)$
(c) $y = x\ln(x)$ (d) $y = xk\ln(k)$

27. The smallest eigenvalue and the corresponding eigenvector of the matrix $\begin{bmatrix} 2 & -2 \\ -1 & 6 \end{bmatrix}$, respectively, are

(*a*) 1.55 and $\begin{Bmatrix} 2.00 \\ 0.45 \end{Bmatrix}$ (*b*) 2.00 and $\begin{Bmatrix} 1.00 \\ 1.00 \end{Bmatrix}$

(*c*) 1.55 and $\begin{Bmatrix} -2.55 \\ -0.45 \end{Bmatrix}$ (*d*) 1.55 and $\begin{Bmatrix} 2.00 \\ -0.45 \end{Bmatrix}$

28. A prismatic steel beam is shown in the figure.

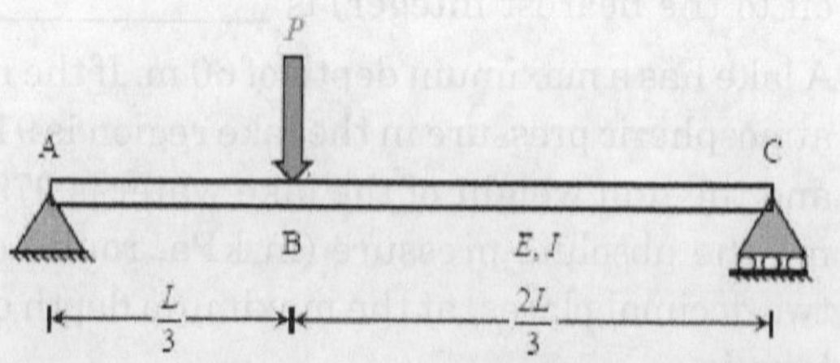

The plastic moment, Mp calculated for the collapse mechanism using static method and kinematic method is

(*a*) $M_{p,static} > \frac{2PL}{9} = M_{p,kinematic}$

(*b*) $M_{p,static} = \frac{2PL}{9} \neq M_{p,kinematic}$

(*c*) $M_{p,static} = \frac{2PL}{9} = M_{p,kinematic}$

(*d*) $M_{p,static} < \frac{2PL}{9} = M_{p,kinematic}$

29. A frame EFG is shown in the figure. All members are prismatic and have equal flexural rigidity. The member FG carries a uniformly distributed load w per unit length. Axial deformation of any member is neglected.

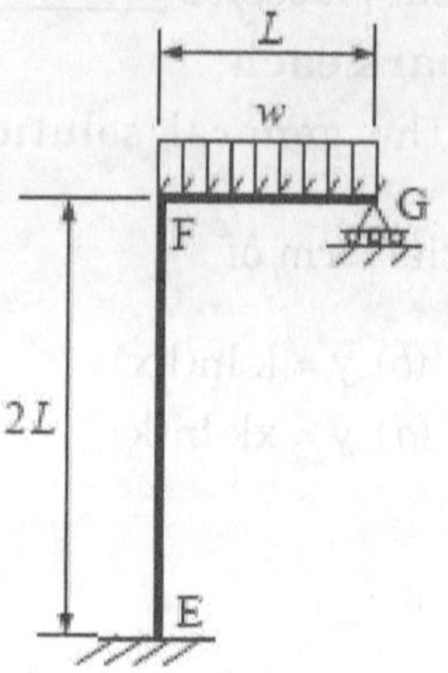

Considering the joint F being rigid, the support reaction at G is

(*a*) 0.375 wL (*b*) 0.453 wL

(*c*) 0.482 wL (*d*) 0.500 wL

30. A clay layer of thickness H has a preconsolidation pressure p_c and an initial void ratio e_0. The initial effective overburden stress at the mid-height of the layer is p_0. At the same location, the increment in effective stress due to applied external load is Δp. The compression and swelling indices of the clay are C_c and C_s, respectively. If $p_0 < p_c < (p_0 + \Delta p)$, then the correct expression to estimate the consolidation settlement (s_c) of the clay layer is

(*a*) $s_c = \frac{H}{1+e_0}\left[C_c \log\frac{p_c}{p_0} + C_s \log\frac{p_0+\Delta p}{p_c}\right]$

(*b*) $s_c = \frac{H}{1+e_0}\left[C_s \log\frac{p_c}{p_0} + C_c \log\frac{p_0+\Delta p}{p_c}\right]$

(*c*) $s_c = \frac{H}{1+e_0}\left[C_c \log\frac{p_0}{p_c} + C_s \log\frac{p_0+\Delta p}{p_c}\right]$

(*d*) $s_c = \frac{H}{1+e_0}\left[C_s \log\frac{p_0}{p_c} + C_c \log\frac{p_0+\Delta p}{p_c}\right]$

31. A rectangular open channel of 6 m width is carrying a discharge of 20 m^3/s. Consider the acceleration due to gravity as 9.81 m/s^2 and assume water as incompressible and inviscid. The depth of flow in the channel at which the specific energy of the flowing water is minimum for the given discharge will then be

(*a*) 0.82 m (*b*) 1.04 m

(*c*) 2.56 m (*d*) 3.18 m

32. Read the statements given below.

(i) Value of the wind profile exponent for the 'very unstable' atmosphere is smaller than the wind profile exponent for the 'neutral' atmosphere.

(ii) Downwind concentration of air pollutants due to an elevated point source will be inversely proportional to the wind speed.

(iii) Value of the wind profile exponent for the 'neutral' atmosphere is smaller than the wind profile exponent for the 'very unstable' atmosphere.

(iv) Downwind concentration of air pollutants due to an elevated point source will be directly proportional to the wind speed.

Select the correct option.

(a) (i) is False and (iii) is True

(b) (i) is True and (iv) is True

(c) (ii) is False and (iii) is False

(d) (iii) is False and (iv) is False

33. A water filtration unit is made of uniform-size sand particles of 0.4 mm diameter with a shape factor of 0.84 and specific gravity of 2.55. The depth of the filter bed is 0.70 m and the porosity is 0.35. The filter bed is to be expanded to a porosity of 0.65 by hydraulic backwash. If the terminal settling velocity of sand particles during backwash is 4.5 cm/s, the required backwash velocity is

(*a*) 5.79×10^{-3} m/s (*b*) 6.35×10^{-3} m/s

(*c*) 0.69 cm/s (*d*) 0.75 cm/s

34. For a given traverse, latitudes and departures are calculated and it is found that sum of latitudes is equal to +2.1 m and the sum of departures is equal to –2.8 m. The length and bearing of the closing error, respectively, are

(a) 3.50 m and 53°7'48" NW

(b) 2.45 m and 53°7'48" NW

(c) 0.35 m and 53.13° SE

(d) 3.50 m and 53.13° SE

35. From laboratory investigations, the liquid limit, plastic limit, natural moisture content and flow index of a soil specimen are obtained as 60%, 27%, 32% and 27, respectively. The corresponding toughness index and liquidity index of the soil specimen, respectively, are

(a) 0.15 and 1.22 (b) 0.19 and 6.60

(c) 1.22 and 0.15 (d) 6.60 and 0.19

Q. 36 – Q. 55 carry two mark each

36. A function is defined in Cartesian coordinate system as $f(x, y) = xe^y$. The value of the directional derivative of the function (in integer) at the point (2, 0) along the direction of the straight line segment from point (2, 0) to point $\left(\frac{1}{2}, 2\right)$ is ________

37. An elevated cylindrical water storage tank is shown in the figure. The tank has inner diameter of 1.5 m. It is supported on a solid steel circular column of diameter 75 mm and total height (L) of 4 m. Take, water density = 1000 kg/m³ and acceleration due to gravity = 10 m/s².

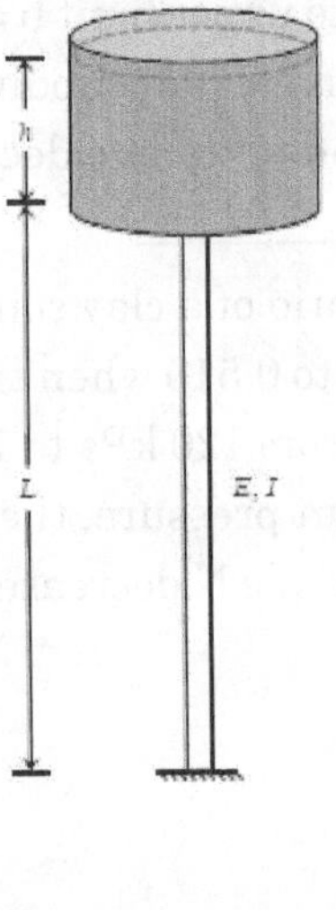

If elastic modulus (E) of steel is 200 GPa, ignoring self-weight of the tank, for the supporting steel column to remain unbuckled, the maximum depth (h) of the water permissible (in m, round off to one decimal place) is ________

38. A prismatic fixed-fixed beam, modelled with a total lumped-mass of 10 kg as a single degree of freedom (SDOF) system is shown in the figure.

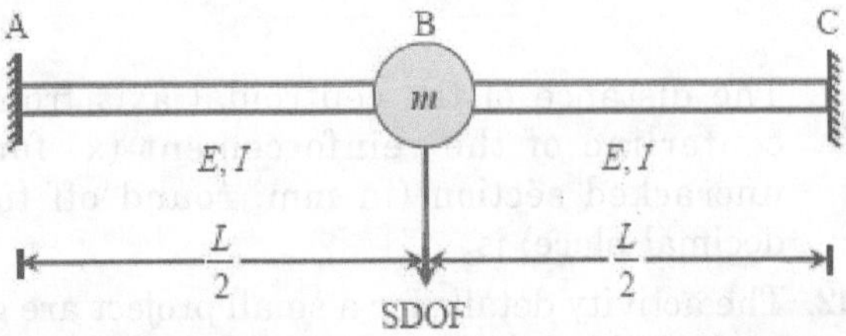

If the flexural stiffness of the beam is $4\pi^2$ kN/m, its natural frequency of vibration (in Hz, in integer) in the flexural mode will be ________

39. A perfectly flexible and inextensible cable is shown in the figure (not to scale). The external loads at F and G are acting vertically.

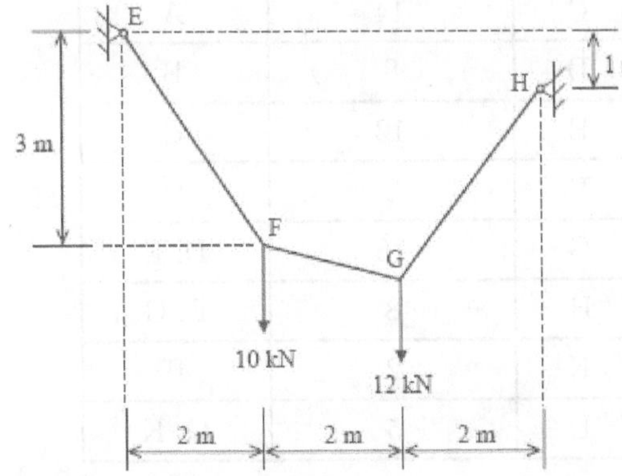

The magnitude of tension in the cable segment FG (in kN, round off to two decimal places) is ________

40. A fire hose nozzle directs a steady stream of water of velocity 50 m/s at an angle of 45° above the horizontal. The stream rises initially but then eventually falls to the ground. Assume water as incompressible and inviscid. Consider the density of air and the air friction as negligible, and assume the acceleration due to gravity as 9.81 m/s². The maximum height (in m, round off to two decimal places) reached by the stream above the hose nozzle will then be ________

41. A rectangular cross-section of a reinforced concrete beam is shown in the figure. The diameter of each reinforcing bar is 16 mm. The values of modulus of elasticity of concrete and steel are 2.0×10^4 MPa and 2.1×10^5 MPa, respectively.

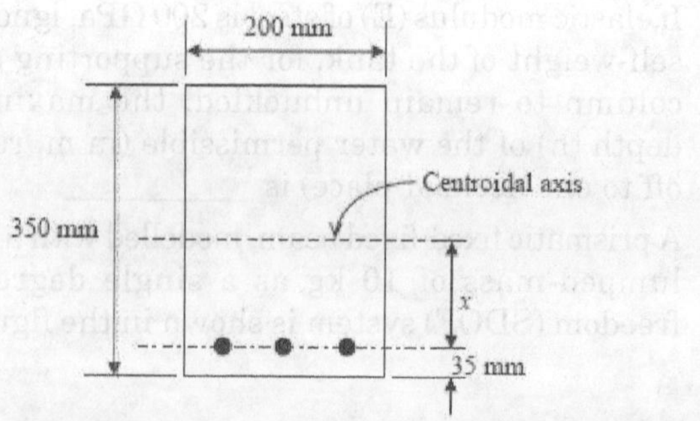

The distance of the centroidal axis from the centerline of the reinforcement (x) for the uncracked section (in mm, round off to one decimal place) is ________

42. The activity details for a small project are given in the Table.

Activity	Duration (days)	Depends on
A	6	-
B	10	A
C	14	A
D	8	B
E	12	C
F	8	C
G	16	D, E
H	8	F, G
K	2	B
L	5	G, K

The total time (in days, in integer) for project completion is ________

43. An equipment has been purchased at an initial cost of Rs. 160000 and has an estimated salvage value of Rs. 10000. The equipment has an estimated life of 5 years. The difference between the book values (in Rs., in integer) obtained at the end of 4th year using straight line method and sum of years digit method of depreciation is ________

44. A rectangular footing of size 2.8 m × 3.5 m is embedded in a clay layer and a vertical load is placed with an eccentricity of 0.8 m as shown in the figure (not to scale). Take Bearing capacity factors: N_c = 5.14, N_q = 1.0, and $N\gamma$ = 0.0; Shape factors: s_c = 1.16, s_q = 1.0 and s_γ = 1.0; Depth factors: d_c = 1.1, d_q = 1.0 and d_γ = 1.0; and Inclination factors: i_c = 1.0 and i_q = 1.0 and i_γ = 1.0.

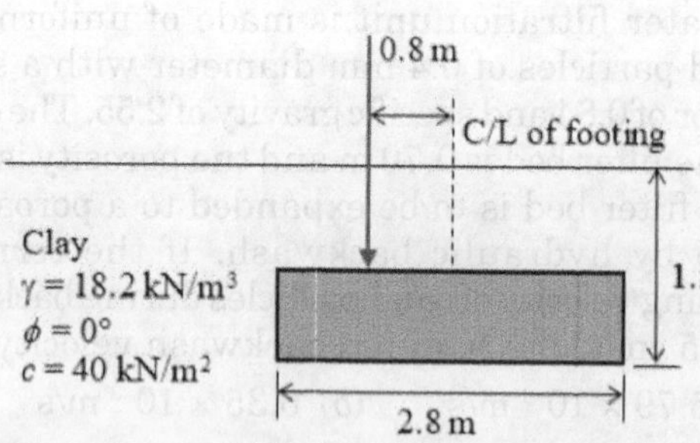

Using Meyerhoff's method, the load (in kN, round off to two decimal places) that can be applied on the footing with a factor of safety of 2.5 is ________

45. The soil profile at a road construction site is as shown in figure (not to scale). A large embankment is to be constructed at the site. The ground water table (GWT) is located at the surface of the clay layer, and the capillary rise in the sandy soil is negligible. The effective stress at the middle of the clay layer after the application of the embankment loading is 180 kN/m^2. Take unit weight of water, γ_w = 9.81 kN/m^3.

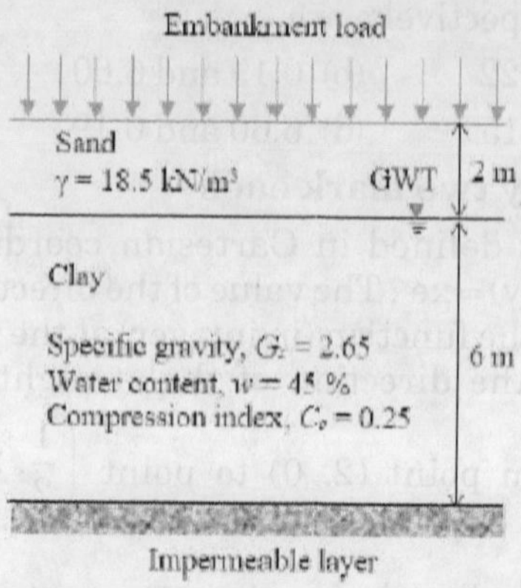

The primary consolidation settlement (in m, round off to two decimal places) of the clay layer resulting from this loading will be ________

46. Numerically integrate, f(x) = 10x – 20x^2 from lower limit a = 0 to upper limit b = 0.5. Use Trapezoidal rule with five equal subdivisions. The value (in units, round off to two decimal places) obtained is ________

47. The void ratio of a clay soil sample M decreased from 0.575 to 0.510 when the applied pressure is increased from 120 kPa to 180 kPa. For the same increment in pressure, the void ratio of another clay soil sample N decreases from 0.600 to 0.550.

If the ratio of hydraulic conductivity of sample M to sample N is 0.125, then the ratio of coefficient of consolidation of sample M to sample N (round off to three decimal places) is ________

48. The hyetograph in the figure corresponds to a rainfall event of 3 cm.

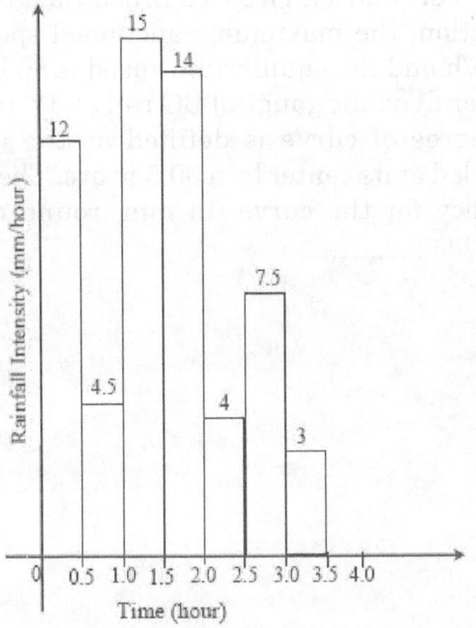

If the rainfall event has produced a direct runoff of 1.6 cm, the ϕ-index of the event (in mm/hour, round off to one decimal place) would be ________

49. A venturimeter as shown in the figure (not to scale) is connected to measure the flow of water in a vertical pipe of 20 cm diameter.

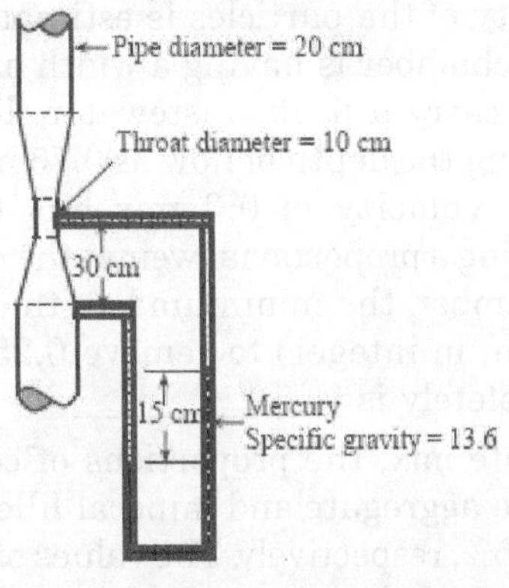

Assume $g = 9.8\ m/s^2$. When the deflection in the mercury manometer is 15 cm, the flow rate (in lps, round off to two decimal places) considering no loss in the venturimeter is ________

50. A reservoir with a live storage of 300 million cubic metre irrigates 40000 hectares (1 hectare = $10^4\ m^2$) of a crop with two fillings of the reservoir. If the base period of the crop is 120 days, the duty for this crop (in hectares per cumec, round off to integer) will then be ________

51. An activated sludge process (ASP) is designed for secondary treatment of 7500 m^3/day of municipal wastewater. After primary clarifier, the ultimate BOD of the influent, which enters into ASP reactor is 200 mg/L. Treated effluent after secondary clarifier is required to have an ultimate BOD of 20 mg/L. Mix liquor volatile suspended solids (MLVSS) concentration in the reactor and the underflow is maintained as 3000 mg/L and 12000 mg/L, respectively. The hydraulic retention time and mean cell residence time are 0.2 day and 10 days, respectively. A representative flow diagram of the ASP is shown below.

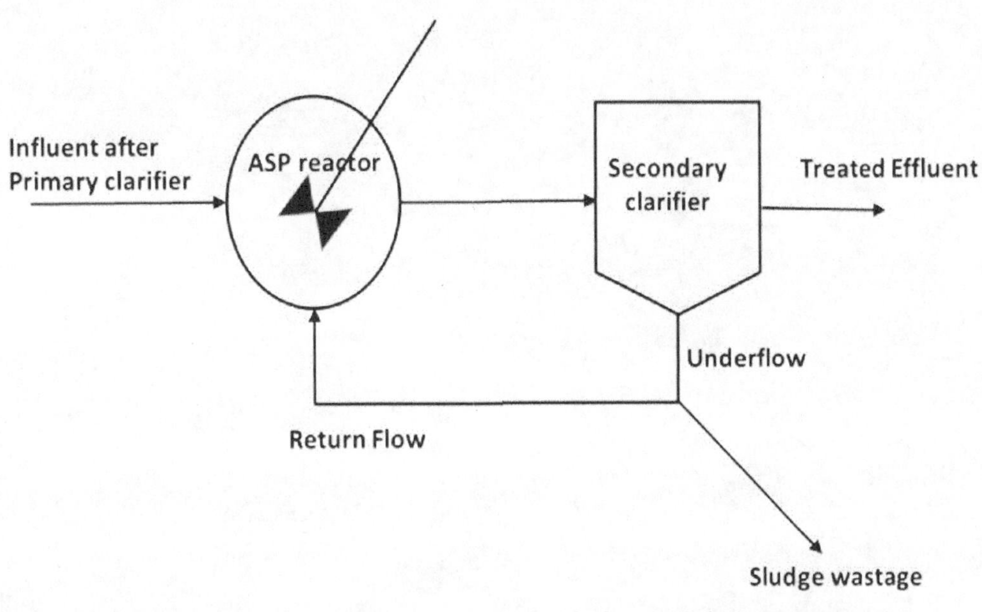

The underflow volume (in m^3/day, round off to one decimal place) of sludge wastage is ________

52. A grit chamber of rectangular cross-section is to be designed to remove particles with diameter of 0.25 mm and specific gravity of 2.70. The terminal settling velocity of the particles is estimated as 2.5 cm/s. The chamber is having a width of 0.50 m and has to carry a peak wastewater flow of 9720 m^3/d giving the depth of flow as 0.75 m. If a flow-through velocity of 0.3 m/s has to be maintained using a proportional weir at the outlet end of the chamber, the minimum length of the chamber (in m, in integer) to remove 0.25 mm particles completely is ______________

53. In an aggregate mix, the proportions of coarse aggregate, fine aggregate and mineral filler are 55%, 40% and 5%, respectively. The values of bulk specific gravity of the coarse aggregate, fine aggregate and mineral filler are 2.55, 2.65 and 2.70, respectively. The bulk specific gravity of the aggregate mix (round off to two decimal places) is ______________

54. The stopping sight distance (SSD) for a level highway is 140 m for the design speed of 90 km/h. The acceleration due to gravity and deceleration rate are 9.81 m/s^2 and 3.5 m/s^2, respectively. The perception/reaction time (in s, round off to two decimal places) used in the SSD calculation is ______________

55. For a 2° curve on a high speed Broad Gauge (BG) rail section, the maximum sanctioned speed is 100 km/h and the equilibrium speed is 80 km/h. Consider dynamic gauge of BG rail as 1750 mm. The degree of curve is defined as the angle subtended at its center by a 30.5 m arc. The cant deficiency for the curve (in mm, round off to integer) is ______________

ANSWERS

General Aptitude

1. (*b*) **2.** (*b*) **3.** (*d*) **4.** (*d*) **5.** (*c*) **6.** (*c*) **7.** (*a*) **8.** (*b*) **9.** (*c*) **10.** (*d*)

Civil Engineering (CE)

1. (*a*)	**2.** (*c*)	**3.** (*b*)	**4.** (*a*)	**5.** (*c*)	**6.** (*a*)
7. (*a*)	**8.** (*a*)	**9.** (*a*)	**10.** (*a*)	**11.** (*a*)	**12.** (*b*)
13. (*c*)	**14.** (*d*)	**15.** (*d*)	**16.** (*b*)	**17.** (*a; b; c*)	
18. (0.0 to 0.0)	**19.** (14 to 16)	**20.** (30 to 30)	**21.** (17.25 to 17.45)		
22. (595 to 598)		**23.** (677.50 to 679.50)		**24.** (48 to 52)	
25. (36 to 36)	**26.** (*a*)	**27.** (*a*)	**28.** (*c*)	**29.** (*c*)	**30.** (*b*)
31. (*b*)	**32.** (*d*)	**33.** (*b*)	**34.** (*a*)	**35.** (*c*)	
36. (*1 to 1*)	**37.** (2.5 to 2.9)	**38.** (10 to 10)	**39.** (8.10 to 8.40)	**40.** (63.50 to 63.90)	
41. (129.0 to 130.0)		**42.** (56 to 56)		**43.** (20000 to 20000)	
44. (439.00 to 442.00)		**45.** (0.32 to 0.34)		**46.** (0.38 to 0.42)	
47. (0.090 to 0.100)		**48.** (4.2 to 4.2)		**49.** (49.0 to 50.0)	
50. (689 to 693)		**51.** (37.0 to 38.0)		**52.** (9 to 9)	
53. (2.58 to 2.61)		**54.** (1.90 to 2.10)		**55.** (55 to 59)	

EXPLANATIONS

GENERAL APTITUDE

1. When two or more singular nouns joined by 'and' the verb should be plural.

2.

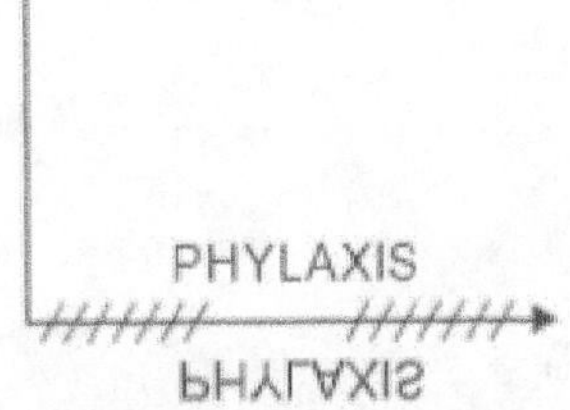

With respect 'x' – axis.

3. Let, $E_1 \rightarrow$ occurrence of even number on first dice

$E_2 \rightarrow$ occurrence of even number on second dice

Probability of even number on a first dice is:

$$P(E_1) = \frac{3}{6} = \frac{1}{2}$$

And probability of even number on a second dice is:

$$P(E_2) = \frac{3}{6} = \frac{1}{2}$$

So, the probability that an even number is rolled out on each dice is:

$$P(E_1 \cap E_2) = P(E_1) \times P(E_2)$$

$$= \frac{1}{2} \times \frac{1}{2} = \frac{1}{4}$$

4. As we know that

$p \odot q = p - q$

$p \oplus q = p \times q$

$\Rightarrow 9 \odot (6 \oplus 7) \odot (7 \oplus (6 \odot 5))$

$= 9 - (6 \times 7) - (7 \times (6 - 5))$

$= 9 - 42 - (7 \times 1)$

$= -40$

5. Possible arrangents:

$\because$ 'R' should not be seated at second position from the left end:

(–, R, –, –)

So, possible arrangments:

	P			= 3!
	Q			= 3!
	S			= 3!

Hence, total number of ways = 3 × 3! = 18 chances.

6.

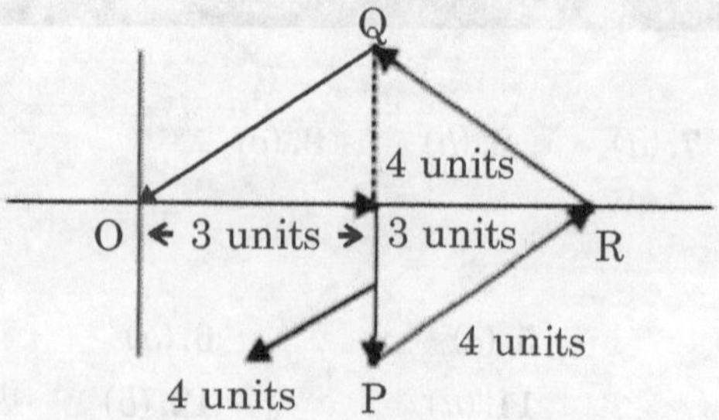

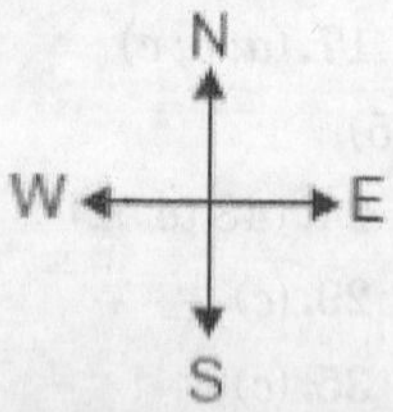

$\therefore\ OQ = \sqrt{(3)^2 + (4)^2}$

$= 5$ units

7. The author states musicians rechearse, actors rehearse and like wise he states public speakers also should rechearse and this is stated in option (a).

8. (i) Some foot ball player play cricket:

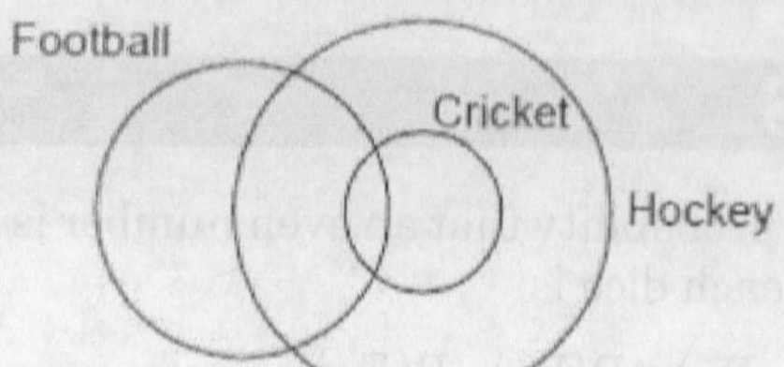

(ii) All, cricket players play hockey:

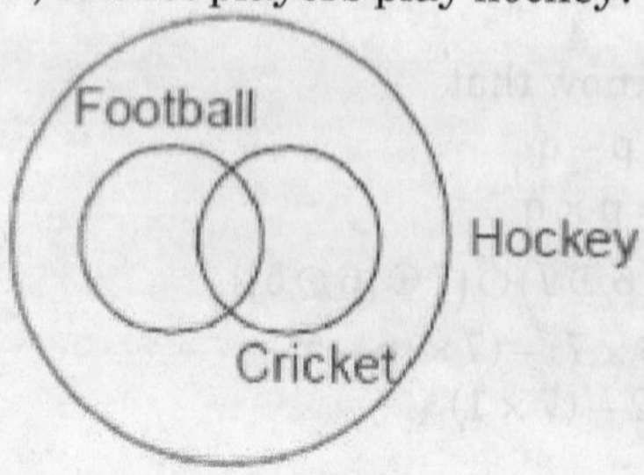

Hence, some football players play hockey is satisfied.

9. According to question:

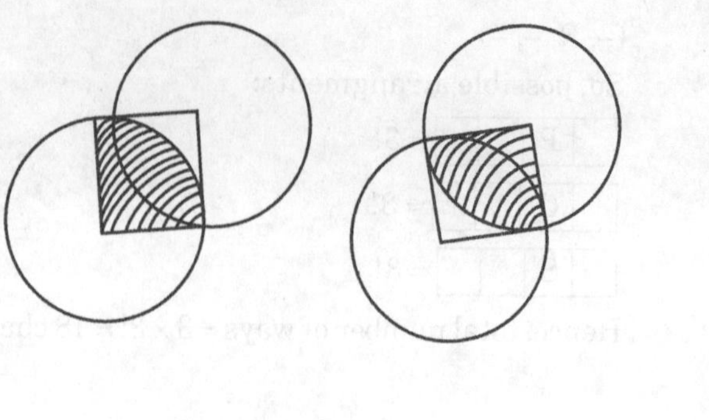

Area of shaded region from the figure is $\frac{\pi r^2}{4}$.

Shaded area from the figure:

$$= 2\times\left[\frac{\pi r^2}{4} - \frac{1}{2}\times r\times r\right]$$

$$= r^2\left[\frac{\pi}{2} - 1\right]$$

Total area $= r^2$

As we know that,

$$\therefore\ \text{Required probability} = \left[\frac{\text{Shaded area}}{\text{Total area}}\right]$$

$$= \frac{r^2\left[\frac{\pi}{2}-1\right]}{r^2} = \left(\frac{\pi}{2} - 1\right)$$

10.

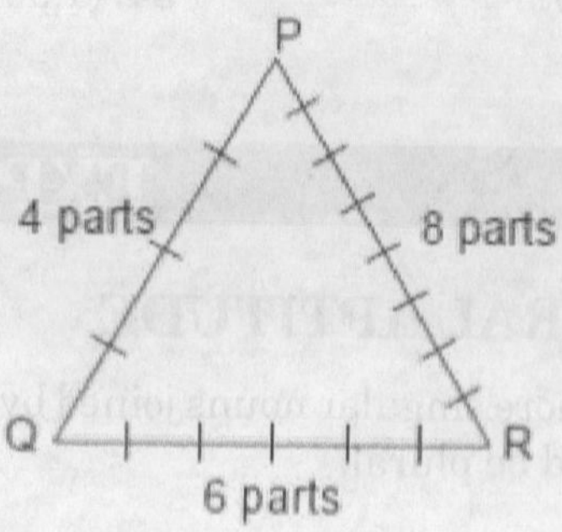

Let the dimension of equilateral

$\Delta PQR = a$ units

As per question,

$\left(\frac{a}{4}, \frac{a}{6}, \frac{a}{8}\right)$ is integer,

Minimum area of triangle

(A) = LCM (4, 6, 8) = 24 units

$$\therefore\ \text{Area of } \Delta PQR = \frac{\sqrt{3}}{4}\times a^2$$

$$= \left[\frac{\sqrt{3}}{4}\times 24\times 24\right] = 144\sqrt{3}\ \text{sq. units}$$

CIVIL ENGINEERING

1. $\lim_{x\to\infty}\left[\frac{x\times \ln x}{1+x^2}\right] = \left[\frac{\infty}{\infty}\text{ form}\right]$ (Indeterminate form)

$$= \lim_{x\to\infty}\left[\frac{1\times\ln x + x\times\left(\frac{1}{x}\right)}{0+2x}\right] = \lim_{x\to\infty}\left[\frac{1\times \ln x}{2x}\right]$$

$$= \left[\text{Again}\left(\frac{\infty}{\infty}\right)\text{form}\right]$$

Apply 'L' Hospital rule again, we get,

$$= \lim_{x\to\infty}\left[\frac{0+\left(\frac{1}{x}\right)}{2}\right]$$

$$= \lim_{x\to\infty}\left(\frac{1}{2x}\right) = \left(\frac{1}{2\times\infty}\right) = 0$$

$$\therefore \lim_{x\to\infty}\frac{x\ln(x)}{(1+x^2)} = 0$$

2. As given that,

$$A = \begin{bmatrix} 5 & 0 & -5 & 0 \\ 0 & 2 & 0 & 1 \\ -5 & 0 & 5 & 0 \\ 0 & 1 & 0 & 2 \end{bmatrix}$$

$$\sim \begin{bmatrix} 5 & 0 & -5 & 0 \\ 0 & 2 & 0 & 1 \\ 0 & 0 & 0 & 0 \\ 0 & 1 & 0 & 2 \end{bmatrix} (R_3 \to R_1 + R_3)$$

$$\sim \begin{bmatrix} 5 & 0 & -5 & 0 \\ 0 & 2 & 0 & 1 \\ 0 & 0 & 0 & 0 \\ 0 & 0 & 0 & 1.5 \end{bmatrix} \left(R_4 \to R_4 - \frac{R_2}{2}\right)$$

which is the form of Echelon matrix.

So, the rank of matrix is the number of non-zero

Hence, the rank of the matrix, i.e. $\rho(A) = 3$.

3. The unit normal vector to the sphere

$x^2 + y^2 + z^2 = r^2$

at a point (x_1, y, z_1) is given by

$$\because \nabla\phi = 2x\hat{i} + 2y\hat{j} + 2z\hat{k}$$

$$\nabla\phi_{(4,4,4)} = 2\times 4\hat{i} + 2\times 4\hat{j} + 2\times 4\hat{k}$$

$$= 8\hat{i} + 8\hat{j} + 8\hat{k}$$

$$\hat{n} = \frac{\nabla\phi}{|\nabla\phi|}$$

$$= \frac{8\hat{i}+8\hat{j}+8\hat{k}}{\sqrt{8^2+8^2+8^2}}$$

$$= \frac{\hat{i}+\hat{j}+\hat{k}}{\sqrt{3}}$$

Hence, option (b) is correct.

4. A square matrix A is called an orthogonal matrix if the product of matrix A with its transpose matrix

A^T is an identity matrix, i.e.,

$AA^T = A^TA = I.$

or $A^T = A^{-1}$

5. Reconnaissance → Site visit

Field observation → Preliminary/Detailed survey

6. Strain hardening means increased resistance (stress) against further deformation due to change in crystalline structure of steel.

In figure AB = Strain hardening zone

OA = Linear elastic zone.

Stress corrosponding to point 'A' is yield stress.

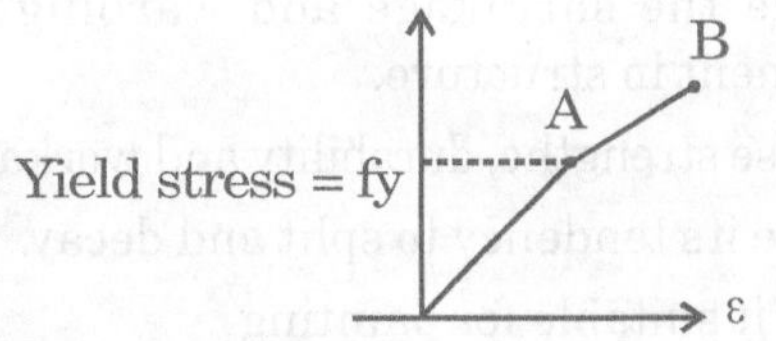

7. As we know that

For one column:

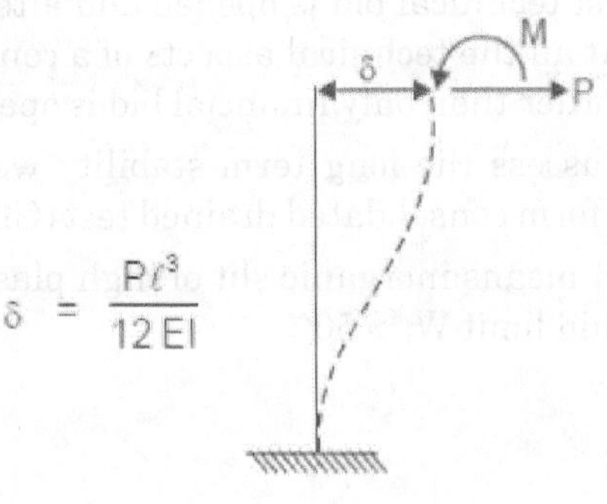

We also know that:

For combined system

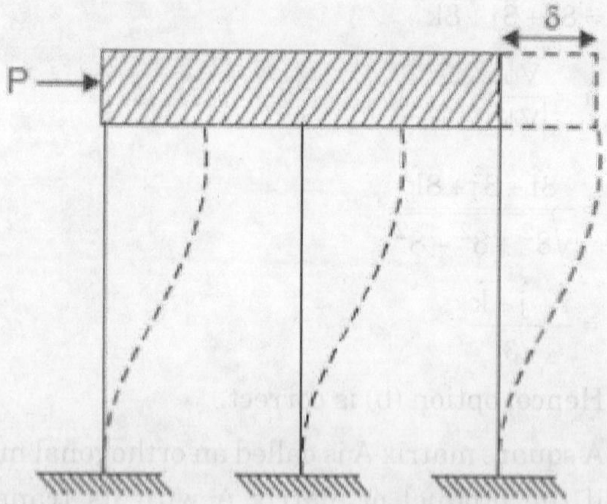

$$\delta = \frac{P\ell^3}{36EI} = \frac{P}{K}$$

$$\Rightarrow K = \frac{36EI}{\ell^3}$$

$$w = \sqrt{\frac{K}{m}} = 6\sqrt{\frac{EI}{m\ell^3}} \text{ rad/sec}$$

8. Seasoning is the process of reducing the moisture content (drying) of timber in order to prevent the timber from possible fermentation and making it suitable for use.

Some of the objectives of seasoning wood are as follows:

(i) Reduce the shrinkage and warping after placement in structure.

(ii) Increase strengthc, durability and workability.

(iii) Reduce its tendency to split and decay.

(iv) Make it suitable for painting.

(v) Reduce its weight.

9. In case of bids in two envelop system technical bid is opened first.

First technical bid is opened and after ensuring that all the technical aspects of a contractor are in order than only financial bid is opened.

10. To assess the long term stability we generally perform consolidated drained test (CD test).

11. MH means inorganic slit of high plasticity with liquid limit $W_L > 50\%$

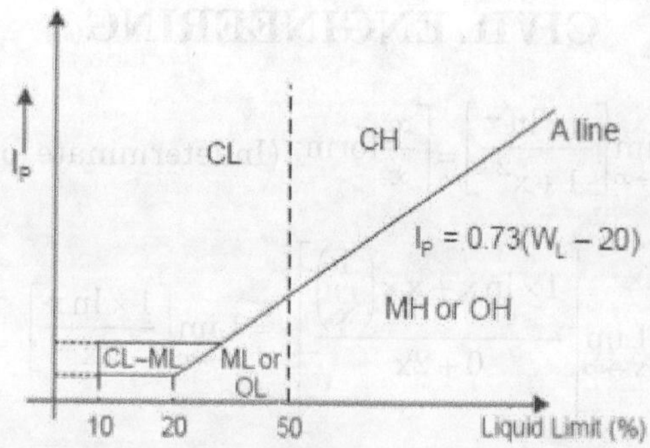

12. According to the question

In case of laminar flow for circular pipe,

$$\text{Momentum correction factor } (\beta) = \frac{4}{3}$$

$$\text{Energy correction factor } (\alpha) = 2$$

$$\therefore \frac{\beta}{\alpha} = \frac{\frac{4}{3}}{2} = \frac{2}{3}$$

13. As per greenshield, model, the relationship between the speed and density should be assumed linear.

As we know that,

Relationship between speed and density is given by

$$V_s = V_f - \frac{V_f}{K_j} \times K \quad ...(i)$$

As we know that,

Traffic volume (q) = Space mean speed (V_s) × Density (K)

$$\Rightarrow q = V_s \times K$$

$$= V_s \left[(V_f - V_s) \times \frac{K_j}{V_f} \right] \text{ (from eq. (i))}$$

$$\boxed{q = K_j \times V_s - \left(\frac{K_j}{V_f}\right) V_s^2}$$

So,

$$\text{For maximum flow, } \frac{dq}{dV_s} = 0$$

$$\Rightarrow \frac{d\left[K_j V_s - \frac{K_j V_s^2}{V_f}\right]}{dV_s} = 0$$

$$\Rightarrow K_j - \frac{2K_jV_s}{V_f} = 0$$

$$\Rightarrow \boxed{V_s = \left(\frac{V_f}{2}\right)}$$

Hence, for maximum flow, speed should be half of free mean speed.

14. Solid waste generated by the use of raw material processing and finished goods.

15. EBT addition to water make wine red colour and at hardness end point of titration, colour changes red to blue.

16. The softening point of bitumen or tar is the temperature at which the substance attains particular degree of softening.

As per IS 334:2002, it is the temperature in °C at which a standard ball passes through a sample of bitumen in a mould and falls through a height of 2.5 cm, when heated under water or glycerine at specified conditions of test.

17. **Photochemical Smog:**

- This type of smog is also called as Los Angeles type smog (observed in Los Angeles in 1950s)
- It does not require any smoke or fog. The word "smog" is misnoere here.
- Photochemical smog is the chemical reaction of sunlight, nitrogen oxides and volatile organic compounds in the atmosphere, which leaves airborne particles and ground-level ozone.
- It is formed in the month of summer during afternoon when there is bright sunlight so that photochemical reactions takes place.
- Indoor environment photo chemical smog does not form.

Increased levels of carbonmonoxide in the indoor environment result in the formation of carboxyhemoglobin and the long term exposure becomes a cause of cardiovascular diseases and even lead to death.

18. As given expression

$$\int_{-1}^{1} xe^{|x|}dx = \sum_{-1}^{0} xe^{-x}dx + \int_{0}^{1} xe^{x}dx$$

$$= \left[x\left(-e^{-x}\right) - \int\left(-e^{-x}\right)dx\right]_{-1}^{0} + \left[x \times e^{x} - \int e^{x}dx\right]_{0}^{1}$$

$$= \left[-e^{-x}(x+1)\right]_{-1}^{0} + \left[e^{x}(x-1)\right]_{0}^{1}$$

$$= \begin{bmatrix} -e^{-0}(0+1) - \\ (-e \times (-1+1)) \end{bmatrix} + \begin{bmatrix} e(1-1) \\ -e^{0}(0-1) \end{bmatrix}$$

Here, $f(x) = xe^{|x|}$

$f(-x) = -xe^{|-x|} = -xe^{|x|} = -f(x)$

So, the given function is odd.

$$\therefore \int_{-1}^{1} f(x)dx = \int xe^{|x|}dx = 0$$

19.

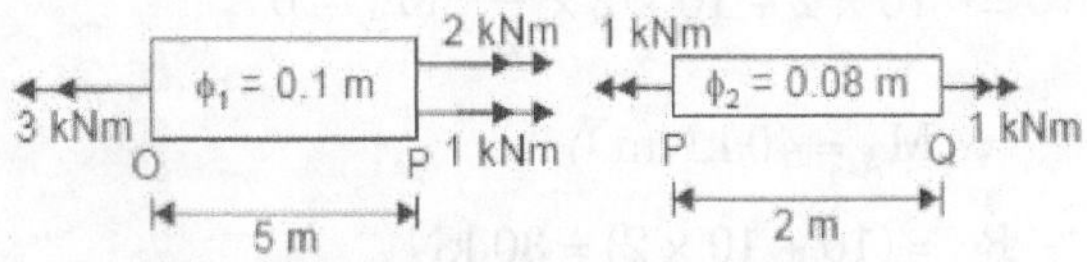

According to the question

FBD of OP:

$$(\tau_{max})_{OP} = \frac{16\, T_{OP}}{\pi \times \phi_1^3}$$

$$= \frac{16 \times 3 \times 10^6}{\pi \times (100)^3}$$

$= 15.279$ MPa

FBD of PQ

$$(\tau_{max})_{PQ} = \frac{16 \times T_{PQ}}{\pi \times \phi_2^3}$$

$$= \frac{16 \times 1 \times 10^6}{\pi \times (80)^3}$$

$= 9.947$ MPa

$\therefore$ Absolute maxima shear stress in member

$= \max [(t_{max})_{OP}, (\tau_{max})_{PQ}]$

$= 15.279$ MPa

20. According to the question.

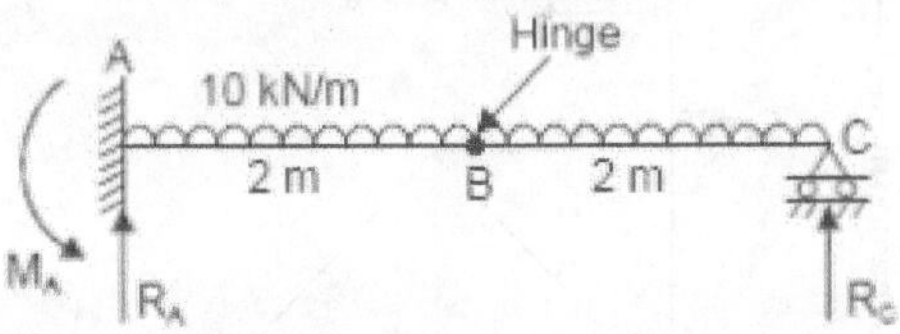

At internal hinge B, $\sum M_B = 0$

$$\Rightarrow R_C \times 2 = 10 \times 2 \times \frac{2}{2}$$

$$\Rightarrow \boxed{R_C = 10\text{kN}}$$

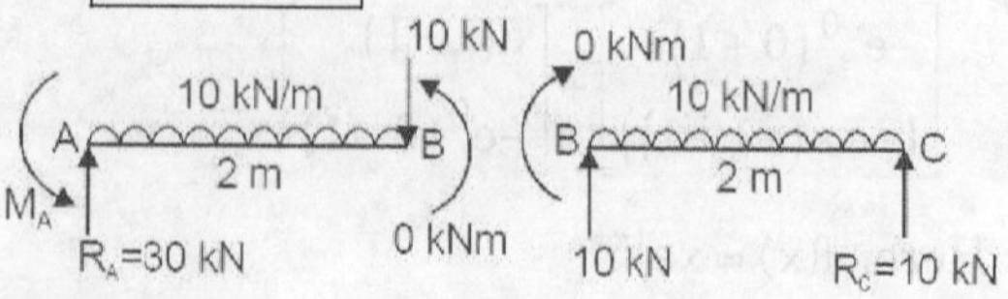

In span AB,

So, the vertical reaction at support(t).

$$\sum F_y = 0$$

$$\sum M_A = 0$$

$$\Rightarrow 10 \times 2 + 10 \times 2 \times \frac{2}{2} - M_A = 0$$

$$\Rightarrow M_A = 40 \text{ kNm}$$

$$R_A = (10 + 10 \times 2) = 30 \text{ kN}$$

21. As given that,

Internal diameter (D_i) = 48 mm,

Outer diameter (D_0) = 52 mm

As we know that

Area ratio (A_r):

$$A_r = \left[\frac{A_0 - A_i}{A_i} \times 100\right]\%$$

$$A_r = \left[\frac{\frac{\pi}{4} \times (D_0)^2 - \frac{\pi}{4} \times (D_i)^2}{\frac{\pi}{4} \times (D_i)^2} \times 100\right]\%$$

$$A_r = \left[\frac{(D_0)^2 - (D_i)^2}{(D_i)^2} \times 100\right]\%$$

$$A_r = \left[\frac{(52)^2 - (48)^2}{(48)^2} \times 100\right]\%$$

$$A_r = 17.361\%$$

22. Let consider catchment area = A km^2

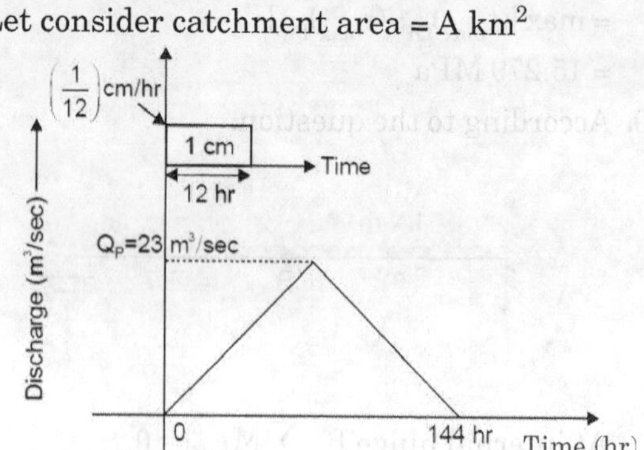

As, we know that,

Area of hydrograph

= [(Area of 12 hr hyetograph) × Catchment area]

$$\Rightarrow \frac{1}{2} \times 144 \times 3600 \times 23$$

$$= \left(12 \times \frac{1}{12} \times \frac{1}{100}\right) A \times 10^6 \text{ m}^2 = 596.16\text{k}^2$$

Hence,

$$\Rightarrow \boxed{A = 596.16 \text{ km}^2}$$

23. As we know that,

Absolute pressure at maximum depth of the lake (P_{abs}) = P_{atm} + ρgh

$$= \left[91 + \frac{9790 \times 60}{1000}\right] \text{kPa}$$

= 678.4 kPa

24. As we know that,

According to Webster method, the optimum signal cycle time is given by

$$C_0 = \frac{(1.5L + 5)}{(1 - y)}$$

where, L = Total lost time per cycle, in seconds

= 2 × (Number of phase)

= (2 × 3) = 6 sec

$Y = \sum y_i = (y_1 + y_2 + y_3) = (0.18 + 0.32 + 0.22) = 0.72$

Cycle time as per webster is:

$$\therefore \quad C_0 = \left[\frac{1.5 \times 6 + 5}{1 - 0.72}\right] = 50 \text{ sec}$$

25. As we know that,

Treating number of observations as weight

$$\text{Most prabable value (MPV)} = \frac{\sum W_i \times x_i}{\sum W_i}$$

$$= \left[\frac{(36°30' \times 4) + (36° \times 3) + (35°30' \times 8) + (36°30' \times 4)}{4 + 3 + 8 + 4} = 36°\right]$$

26. As given that,

$$\frac{dy}{dx} - \frac{y}{x} = 1$$

Compaire D. E is in the form of $\frac{dy}{dx} + Py = Q$

$$\Rightarrow \frac{dy}{dx} + \underbrace{\left(\frac{-1}{x}\right)}_{P} y = \underbrace{1}_{Q} \quad \text{...(i)}$$

$\therefore$ Integrating factor (I.F.)

$$= e^{\int P dx} = e^{\int -\frac{1}{x} dx}$$

$$= e^{-1 \times \ln x}$$

$$= x^{-1} = \left(\frac{1}{x}\right)$$

Solution of the D. E is given by:

$$y \times (\text{I.F.}) = \int Q(\text{I.F.}) dx + C$$

$$\Rightarrow y \times \left(\frac{1}{x}\right) = \int 1 \times \frac{1}{x} dx + C$$

$$\Rightarrow y \times \left(\frac{1}{x}\right) = \ln x + \ln k = \ln(kx)$$

$$\Rightarrow y = x\ln(kx)$$

27. As we know that,

The question the characteristic equation

$$|A - \lambda| = 0 \Rightarrow \begin{vmatrix} 2-\lambda & -2 \\ -1 & 6-\lambda \end{vmatrix} = 0$$

$\Rightarrow (2 - \lambda)(6 - \lambda) - 2 = 0$

$\Rightarrow 12 - 2\lambda - 6\lambda + \lambda^2 - 2 = 0$

$\lambda^2 - 8\lambda + 10 = 0$

$\lambda = 4 \pm \sqrt{6}$

Smallest eigen value $= 4 - \sqrt{6} = 1.55$

$$\begin{bmatrix} 2-\lambda & -2 \\ -1 & 6-\lambda \end{bmatrix} \begin{bmatrix} x_1 \\ x_2 \end{bmatrix} = \begin{bmatrix} 0 \\ 0 \end{bmatrix}$$

$$\begin{bmatrix} 2-(4-\sqrt{6}) & -2 \\ -1 & 6-(4-\sqrt{6}) \end{bmatrix} \begin{bmatrix} x_1 \\ x_2 \end{bmatrix} = \begin{bmatrix} 0 \\ 0 \end{bmatrix}$$

$$\begin{bmatrix} -2+\sqrt{6} & -2 \\ -1 & 2+\sqrt{6} \end{bmatrix} \begin{bmatrix} x_1 \\ x_2 \end{bmatrix} = \begin{bmatrix} 0 \\ 0 \end{bmatrix}$$

$$-2x_1 + \sqrt{6}x_1 - 2x_2 = 0$$

$$\left(-2 + \sqrt{6}\right)x_1 = 2x_2$$

$$x_1 = \frac{2x_2}{-2 + \sqrt{6}}$$

$$\frac{x_1}{2} = \frac{x_2}{-2 + \sqrt{6}}$$

$$\frac{x_1}{2} = \frac{x_2}{0.45}$$

Hence, $\begin{bmatrix} x_1 \\ x_2 \end{bmatrix} = \begin{bmatrix} 2 \\ 0.45 \end{bmatrix}$

28.

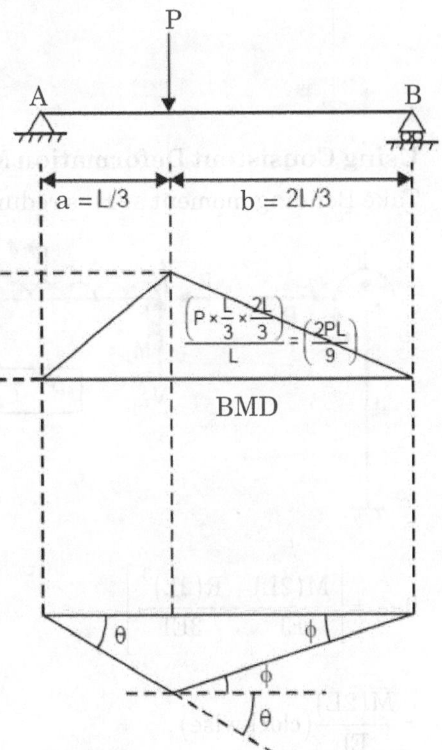

(i) Static Method :

$$M_P, \text{static} = \left(\frac{2PL}{9}\right)$$

(ii) Kinematic Method :

Here, $\frac{2L}{3} \times \phi = \frac{L}{3}\theta \Rightarrow \boxed{\theta = 2\phi}$

Now, $P\times\left(\frac{L}{3}\right)\times\theta = M_P\times(\theta+\phi)$

$$\Rightarrow P\times\frac{L}{3}\times 2\phi = M_P\times 3\phi$$

$$\Rightarrow M_{P,Kinematic} = \left(\frac{2PL}{9}\right)$$

$$\therefore M_{P,\,Static} = \left(\frac{2PL}{9}\right) = M_{P,Kinematic}$$

29. According to the question:

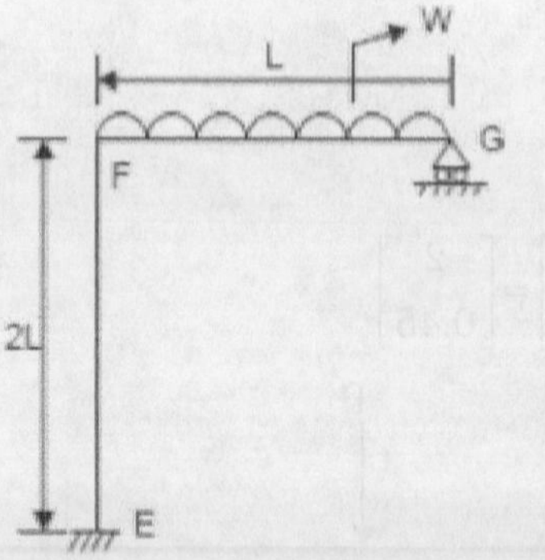

Using Consistent Deformation Method:

Take Bending moment at F as redundant

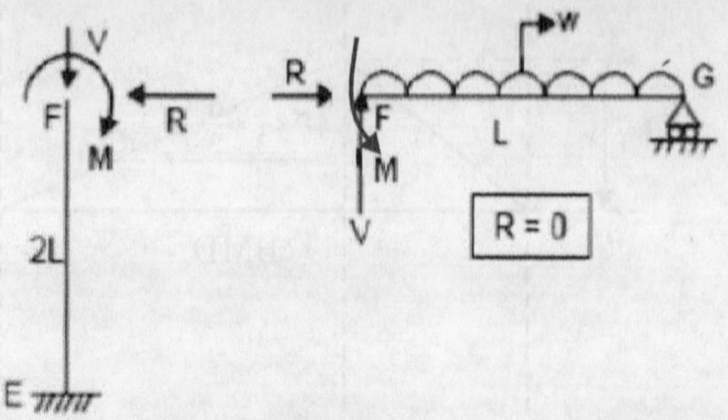

$$\theta_{FE} = \left[\frac{M(2L)}{EI} - \frac{R(2L)^3}{3EI}\right]$$

$$= \frac{M(2L)}{EI}\text{(clockwise)}$$

$$\theta_{FG} = \frac{wL^3}{24EI} - \frac{ML}{3EI}\ \text{(clockwise)}$$

$$\theta_{FE}\curvearrowright = \theta_{FG}\curvearrowright$$

$$\frac{M(2L)}{EI} = \frac{wL^3}{24EI} - \frac{ML}{3EI}$$

$$\Rightarrow M = \frac{3}{7}\left(\frac{WL^3}{24EI}\right)\times\frac{EI}{L} = \frac{wL^2}{56}$$

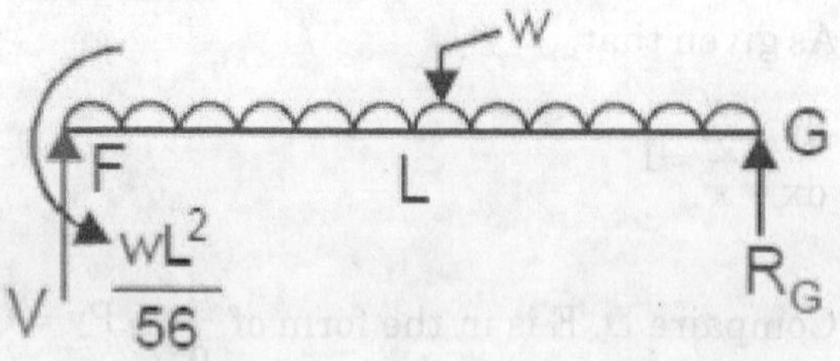

$$\sum M_F \curvearrowleft = 0$$

$$R_G\times L - \frac{w\times L^2}{2} + \frac{wL^2}{56} = 0$$

$$R_G = \left(\frac{-wL^2}{56} + \frac{wL^2}{2}\right)\times\frac{1}{L}$$

$$R_G = 0.482\ wL$$

30. As the soil is initally in over consolidate state ($P_O < P_C$). due to the external applied load (ΔP), the soil initially undergoes re-compression upto P_C and then changes to virgin compression from P_C to P_F.

The re-compression index (Cr) is also called swelling index, (C_S).

$$S_C = \frac{H_0}{1+e_0}\left[C_s\log\left[\frac{P_C}{P_O}\right] + C_C\log\left[\frac{P_O+\Delta P}{P_C}\right]\right]$$

31. As given that,

Rectangle: B = 6m

$Q = 20m^3/s$

$g = 9.81\ m/s^2$

We know that,

For a given discharge, specific energy of the flowing water is minimum at critical depth.

For a rectangular channel,

$$\text{Critical depth } (y_c) = \left[\frac{q^2}{g}\right]^{\frac{1}{3}}$$

where, q = (Discharge per unit width)

$$= \left[\frac{20}{6}\right] m^3/sec/m$$

$$\Rightarrow y_c = \left[\frac{\left(\frac{20}{6}\right)^2}{9.81}\right]^{\frac{1}{3}}$$

= 1.0424 m

32. It is assumed that the plume has a Gaussain concentration distribution in both the z(vertical) and y (horizontal) direction.

$$\text{Concentration} \propto \frac{1}{\text{Wind Speed}}$$

Concentration (C) of a gas or aerosol (< 20μ) calculated at group level for a distance downwind (x) is given by

$$C_{x,y} = \frac{Q}{\mu u \sigma_z \sigma_y} e^{\left[-\frac{1}{2}\left(\frac{H}{\sigma_z}\right)\right]} e^{\left[-\frac{1}{2}\left(\frac{y}{\sigma_y}\right)^2\right]}$$

where, C = pollutant concentration, g/m^3

Q = pollutant emission rate, g/s

u = mean wind speed, m/s

σ_y = standard deviation of horizontal plume concentration, evaluated in terms of downwind distance, x meter

σ_x = standard deviation of vertical plume concentration evaluated in terms of downwind distance, x meter

H = effective stack height, m

x = downwind distance along plume mean centerline from point source, m

y = crosswind distance from the centerline of the plume, m

33. As given that,

d = 0.4 mm, Z = 0.7 m

Q = 0.84, n = 0.35

S = 2.55

n_{ex} = 0.65

V_S = 4.5 cm/sec

We Know that,

$$\text{Porosity of expanded bed } (n_{ex}) = \left(\frac{V_B}{V_s}\right)^{0.22}$$

$$\Rightarrow 0.65 = \left[\frac{V_B}{4.5\text{cm/s}}\right]^{0.22}$$

$$\Rightarrow V_B = \left[(0.65)^{\frac{1}{0.22}} \times 4.5\right] \text{cm/sec}$$

= 0.63506 cm/sec

= 6.3506 × 10^{-3} m/sec

34. The x and y components of the closing error can be computed as

$$e_x = \sum \text{Departure}(D) = -2.8 \text{ m}$$

$$e_y = \sum \text{Latitude}(L) = +2.1 \text{ m}$$

Total length of closing error (e) is given as

$$e = \sqrt{\left(\sum D\right)^2 + \left(\sum L\right)^2}$$

$$= \sqrt{(-2.8)^2 + (2.1)^2}$$

= 3.5 m

The bearing of the closing error is given by

$$\tan\theta = \left(\frac{\sum D}{\sum L}\right) = \left(\frac{-2.8}{+2.1}\right)$$

$$\Rightarrow \theta = 53.13° = 53°7'48.37''$$

$$\text{As, } \tan\theta = \frac{\sum D}{\sum L} = \frac{\ominus}{\oplus},$$

closing error lies in IIIrd quadrant.

∴ Bearing of closing error = 53°7′48.37″ NW

35. As given that

w_{LL} = 60%

$w_{P.L.}$ = 27%

w = 32%

I_f = 27%

As we know that,

$$\text{Toughnes Index} = \left[\frac{\text{Plasticity index}}{\text{Flow index}}\right]$$

$$= \frac{w_{LL} - w_{P.L.}}{I_f}$$

$$I_t = \frac{60 - 27}{27} = 1.22$$

$$\text{Liquidity Index} = \frac{w - w_{PL}}{w_{LL} - w_{PL}}$$

$$= \frac{32 - 27}{60 - 27} = 0.15$$

36. As given that,

f(x, y) = xe^y

$\nabla\phi = \nabla f$

$$\nabla\phi = \frac{\partial f}{\partial x}\hat{i} + \frac{\partial f}{\partial y}\hat{j} + \frac{\partial f}{\partial z}\hat{k}$$

$$\nabla\phi = e^y\hat{i} + xe^y\hat{j} + 0\hat{k}$$

$$\nabla\phi_{(2,0)} = \hat{i} + 2\hat{j}$$

$$\vec{a} = (x_2 - x_1)\hat{i} + (y_2 - y_1)\hat{j}$$

$$= -\frac{3}{2}\hat{i} + 2\hat{j}$$

$$\hat{n} = \frac{\vec{a}}{|\vec{a}|} = \frac{-\frac{3}{2}\hat{i} + 2\hat{j}}{\sqrt{\left(-\frac{3}{2}\right)^2 + (2)^2}}$$

$$= \frac{2}{5}\left(-\frac{3}{2}\hat{i} + 2\hat{i}\right)$$

$$= -\frac{3\hat{i}}{5} + \frac{4}{5}\hat{j}$$

$$DD = \nabla f . \hat{n}$$

$$= -\frac{3}{5} + 2 \times \frac{4}{5} = 1$$

37. As we know that $P_{cr} = \frac{\pi^2 EI}{l_{eff}^2}$

= Buckling load = Weight of water

For no buckling of column, maximum weight of water should be less than buckling load of column.

(maximum volume of water) $\rho g < P_{cr}$

$$\left(\frac{\pi D^2}{4} \times h_{max}\right) \times \rho g < \frac{\pi^2 EI}{(l_{eff})^2}$$

$$\frac{\pi \times (1.5)^2}{4} \times h_{max} \times 1000 \times 10$$

$$< \frac{\pi^2 \times 2 \times 10^5 \times \frac{\pi}{64} \times 75^4}{(2 \times 4 \times 1000)^2}$$

$h_{max} < 2.71$ m

$\Rightarrow$ $\boxed{h_{max} = 2.71 \text{ m}}$

38. As given that,

m = 10 kg

$k = 4\pi^2 \times 10^3$

As we know that

$$\text{Natural frequency (f)} = \frac{\omega}{2\pi}$$

$$= \frac{1}{2\pi}\sqrt{\frac{k}{m}}$$

$$= \frac{1}{2\pi}\sqrt{\frac{4\pi^2 \times 1000}{10}}$$

= 10 Hz

39.

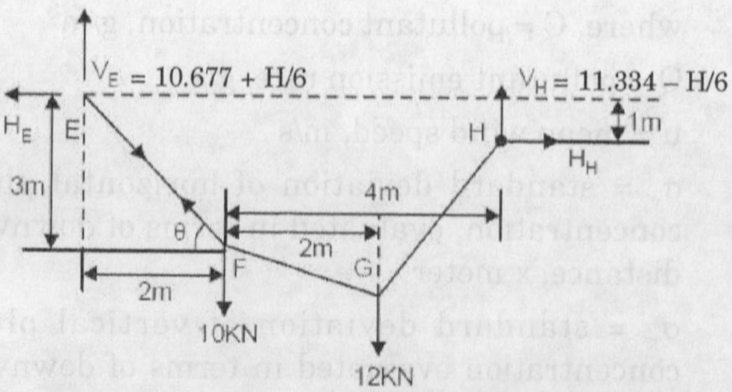

left of F, $\sum M_F = 0$,

$$H_E \times 3 + V_E \times 2 = 0$$

$$H_E = \frac{2}{3} V_E \quad ...(i)$$

At support H, $\sum M_H = 0$

$$V_E \times 6 - H_E \times 1 - 10 \times 4 - 12 \times 2 = 0$$

$$6V_E - H_E = 64 \quad ...(ii)$$

Put value of H_E in equation (ii)

$$6V_E - \frac{2}{3} V_E = 64$$

$$V_E = \frac{3}{16} \times 64 = 12KN$$

$$\boxed{V_E = 12KN}$$

$$\Rightarrow \quad H_E = \frac{2}{3} \times 12 = 8KN$$

$$\boxed{H_E = 8KN}$$

Tension, in EF,

$$F_{EF} = \sqrt{V_E^2 + H_E^2} = \sqrt{(12)^2 + (8)^2} \text{ KN}$$

$$\boxed{E_{EF} = 4\sqrt{13} \text{ KN}}$$

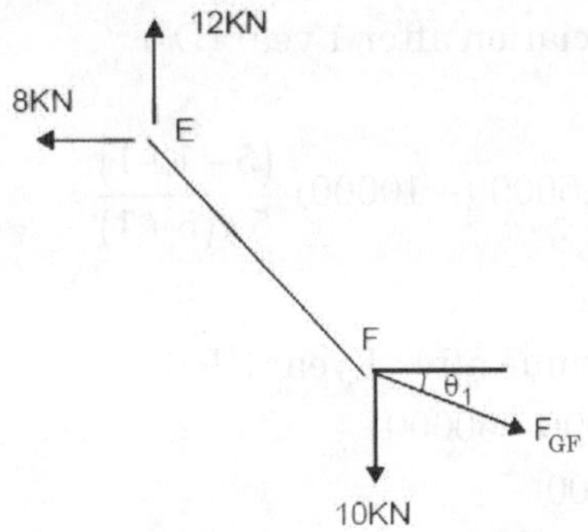

$\sum F_H = 0$

$F_{GF} \cos \theta_1 = 8$ ...(iii)

$\sum F_V = 0$

$F_{GF} \sin \theta_1 + 10 = 12$

$F_{GF} \sin \theta_1 = 2$ ...(iv)

$(F_{FG})^2 = (8)^2 + (2)^2$

$F_{GF} = 8.2462$ KN

$\boxed{F_{GF} \simeq 8.25 KN}$

Hence, the tension in segment GF is 825 kN

40. According to the question.

At maximum height, vertical component of velocity will become zero.

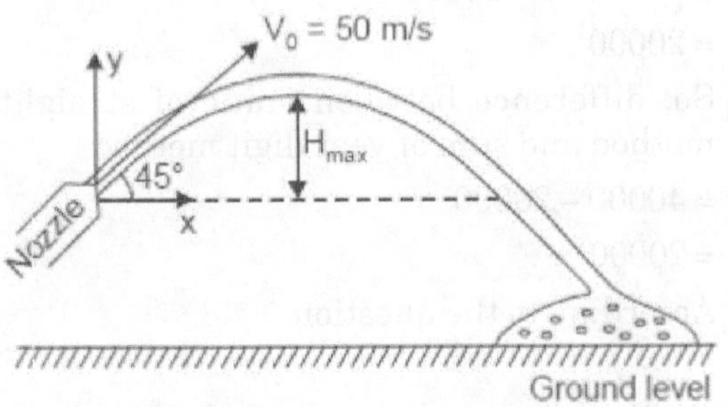

Equation of motion in y-direction

$$V_{fy}^2 = V_{iy}^2 + 2as$$

Where,

V_{fy} = Final velocity in y-direction

V_{iy} = Initial velocity in y-direction

At $s = H_{max}$, $V_{fy} = 0$, $a = -g$ m/s^2

$\Rightarrow V_{iy} = V_0 \sin 45°$

$\Rightarrow 0 = (V_0 \sin 45°)^2 - 2gH_{max}$

$$H_{max} = \frac{(V_0 \sin 45°)^2}{2g}$$

$$= \frac{\left(50 \times \frac{1}{\sqrt{2}}\right)^2}{2 \times 9.81} = 63.71 \text{ m}$$

Maximum height above fire hose nozzle.

$\boxed{H_{max} = 63.71 m}$

41. According to the question.

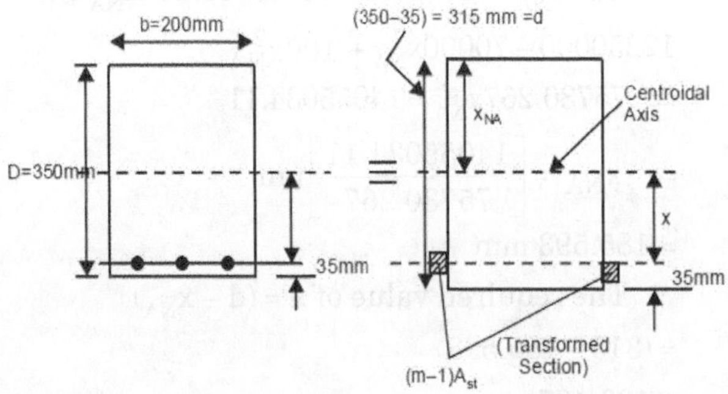

As given that,

A_{st} = 3 nos 16 mm diameter bars

$$= 3 \times \frac{\pi}{4} \times \left(16^2\right) = 603.186 \text{mm}^2$$

Modulus of elasticity of concrete

$= E_C = 2.0 \times 10^4$ MPa

Modulus of Elasticity of Steel

$= E_{st} = 2.1 \times 10^5$ MPa

As we know that,

$$\therefore \text{ Modular ratio (m)} = \left[\frac{E_{st}}{E_C}\right]$$

$$= \left[\frac{2.1 \times 10^5}{2 \times 10^4}\right]$$

= 10.5

Calculation of Neutral Axis :

$$\begin{pmatrix}\text{Moment of} \\ \text{Compression area} \\ \text{in transformed} \\ \text{section} \\ \text{about Neutral axis}\end{pmatrix} = \begin{pmatrix}\text{Moment of} \\ \text{tension area} \\ \text{in transformed} \\ \text{section about} \\ \text{neutral axis}\end{pmatrix}$$

$$\Rightarrow \left[b \times x_{NA} \times \frac{x_{NA}}{2}\right]$$

$$= \begin{bmatrix}(m-1) A_{st} \times (d - x_{NA}) + \\ b \times (D = x_{NA}) \times \dfrac{(D - x_{NA})}{2}\end{bmatrix}$$

$$\Rightarrow \frac{200 \times x_{NA}^2}{2}$$

$$= \begin{bmatrix}(10.5 - 1) \times 603.186 \times (315 - x_{NA}) \\ + \dfrac{200}{2} \times (350 - x_{NA})^2\end{bmatrix}$$

$\Rightarrow\ 100x_{NA}^2 = 1805034.105 - 5730.267x_{NA} + 100 \times \left[(350)^2 - 2\times350\times x_{NA} + x_{NA}^2\right]$

$\Rightarrow\ 100x_{NA}^2 = 1805034.105 - 5730.267x_{NA} + 12250000 - 70000x_{NA} + 100x_{NA}^2$

$\Rightarrow\ 75730.267x_{NA} = 14055034.11$

$\Rightarrow\ x_{NA} = \left[\frac{14055034.11}{75730.267}\right]$ mm

= 185.593 mm

$\therefore$ The required value of $x = (d - x_{NA})$

= (315 – 185.593)

= 129.407 mm

42. According to the question:

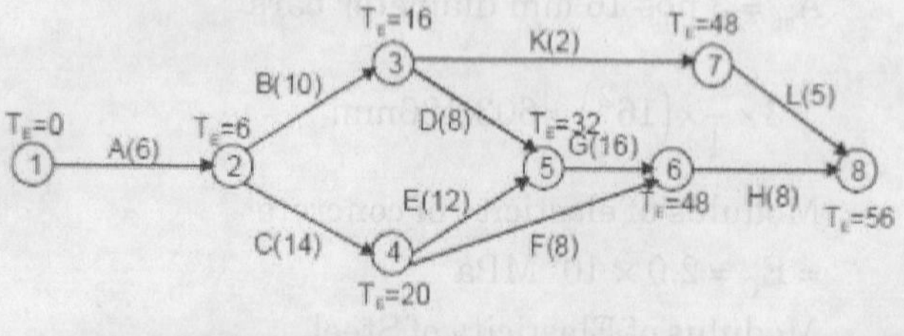

Hence, total project time = 56 days

43. As given that,

Initial Cost (CP) = 160000

Salvage Value (SV) = 10000

n = life of asset

= 5 year

As we know that,

The annual depreciation (D)

$= \left(\frac{P - SV}{5}\right)$

Now, Book value after 4th year by straight line method

= Initial cost (P) – 4 Annual depreciation (I)

$= 1{,}60{,}000 - 4\left(\frac{160000 - 10000}{5}\right)$

= 40000

Depreciation after mth year by sum of year digit method

$= (C_i - C_s)\left[\frac{n - m + 1}{\frac{n(n+1)}{2}}\right]$

Depreciation after 1 year (D_1)

$D_1 = (160000 - 10000)\ \frac{(5-1+1)}{\frac{5\times(5+1)}{2}}$

= 50000

Book value after 1 year (B_1)

= 160000 – 50000

= 110000

$D_2 = (160000 - 10000) \times \frac{4}{15}$

= 40000

$B_2 = 110000 - 40000$

= 70000

$D_3 = (160000 - (1000) \times \frac{3}{15}$

= 30000

$B_3 = 70000 - 30000$

= 40000

$D_4 = (160000 - 10000) \times \frac{2}{15}$

= 20000

$B_4 = 40000 - 20000$

= 20000

So, difference between value of straight line method and sum of year digit method

= 40000 – 20000

= 20000

44. According to the question.

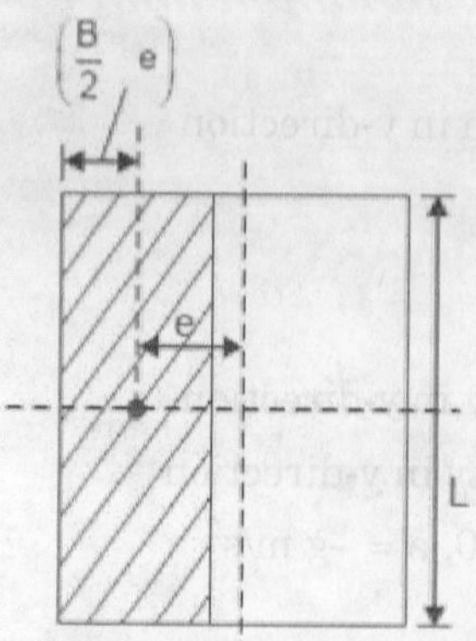

So,

$B' = 2\left(\frac{B}{2} - e\right)$

= (B – 2e) = 2.8 – 2 × 0.8 = 1.2 m

$L' = L = 3.5$ m

$q_u = CN_c.s_cd_ci_c + qN_qs_qd_qi_q + (0.5)B'\gamma O_\gamma s_\gamma d_\gamma i_\gamma$

$q_u = CN_c.s_cd_ci_c + {}_qN_qs_qd_qi_q$ $[q = \gamma D_f]$

Load that can be applied on the footing is the (gross safe bearing capacity) × L'B'

As we know that

Gross safe bearing capacity

$$= \frac{q_{nu}}{F.O.S} = \gamma D_f$$

$$= \frac{(q_u - \gamma D_f)}{F.O.S} + \gamma D_f$$

$q_u = (40 \times 5.14)\,(1.16 \times 1.1 \times 1) + (18.2 \times 1.5)\,(1) \times (1 \times 1 \times 1)$

$q_{nu} = q_u - \gamma D_f = (40 \times 5.14)\,(1.16 \times 1 \times 1)$

⇒ Load that can be applied on the footing

$$= \left[\frac{40 \times 5.14 \times 1.16 \times 1.1}{2.5} + 18.2 \times 1.5\right] \times (3.5 \times 1.2) \text{ kN}$$

= 555.4 kN

When net load is asked, load that can be applied

$= 555.4 - (18.2 \times 1.5) \times 3.5 \times 1.2$

= 440.74 kN

45. As we know that,

1° -consolidation settlement

$$= \frac{C_C H}{1+e_0}\log_{10}\left(\frac{\bar{\sigma}_0 + \Delta\bar{\sigma}}{\bar{\sigma}_0}\right)$$

Where,

$\bar{\sigma}_0 + \Delta\bar{\sigma}$ = Eff. stress at the centre of clay layer after embankment loading = 180 kN/m^2

$\bar{\sigma}_0$ = Eff. stress at the centre of clay layer before embankment loading.

We also know that,

$$\gamma_{sub} \text{ of clay layer } = \left[\frac{G-1}{1+e}\right]\gamma_w$$

$$= \frac{(2.65 - 1)}{1 + \frac{wG}{1}} \times 9.81$$

$$= \frac{1.65 \times 9.81}{1 + 0.45 \times 2.65} = 7.383 \text{ KN/m}^3$$

$\bar{\sigma}_0 = (18.5 \times 2) + (7.383) \times 3$

= 59.149 KN/m^2

$$e_0 = \frac{WG}{1} = 0.45 \times 2.65$$

= 1.1925

$$\Rightarrow \quad \Delta H = \frac{0.25 \times 6}{1 + 1.1925}\log_{10}\left(\frac{180}{59.149}\right)$$

$= 0.3307 \simeq 0.33$ m.

46. As given that,

a = 0, b = 0.5, n = 5

As we know that,

$$h = \left(\frac{b-a}{n}\right)$$

$$= \frac{0.5 - 0}{5} = 0.1$$

$x_0 = 0, y_0 = 0$

$x_1 = 0.1, y_1 = 0.8$

$x_2 = 0.2, y_2 = 1.2$

$x_3 = 0.3, y_3 = 1.2$

$x_4 = 0.4, y_4 = 0.8$

$x_5 = 0.5, y_5 = 0$

So, Numerical intergration

$$= \frac{h}{2}[(y_0 + y_5) + 2(y_1 + y_2 + y_3 + y_4)]$$

$$= \frac{0.1}{2}[(0 + 0) + 2(0.8 + 1.2 + 1.2 + 0.8)]$$

$$= \frac{0.1 \times 8}{2} = 0.4$$

47. As we know that

$\Delta H = m_v\left(\Delta\bar{\sigma}\right).H$

And,

$$\therefore m_v = \frac{k_v}{C_v \gamma_w}$$

$$\Rightarrow \quad \Delta H = \frac{k_v}{C_v \gamma_w}\left(\Delta\bar{\sigma}\right)H$$

Now ΔH for M & N

$$\frac{(\Delta H)_M}{(\Delta H)_N} = \frac{(k_v)_M}{(k_v)_N} \times \frac{(C_v)_N}{(C_v)_M}\cdot\frac{(\Delta\bar{\sigma})_M}{(\Delta\bar{\sigma})_N}\frac{(H)_M}{(H)_N}$$

$$\frac{\left(\frac{\Delta H}{H}\right)_M}{\left(\frac{\Delta H}{H}\right)_N} = (0.125)\frac{(C_v)_N}{(C_v)_M} \times 1$$

$$\frac{\left(\frac{\Delta e}{1+e_0}\right)_M}{\left(\frac{\Delta e}{1+e_0}\right)_N} = (0.125)\left(\frac{(C_v)_N}{(C_v)_M}\right)$$

$$\frac{\left(\frac{0.575-0.51}{1+0.575}\right)}{\left(\frac{0.6-0.55}{1+0.6}\right)} = 0.125\left(\frac{(C_v)_N}{(C_v)_M}\right)$$

$$\Rightarrow \left(\frac{(C_v)_M}{(C_v)_N}\right) = 0.095$$

48. As given that,

Total Rainfall (P) = 3 cm = 30 cm

Direct runoff (R) = 1.6 cm = 16 mm

Infiltration = (30 – 16) = 14 mm

Time of rainfall (t) = 3.5 hr

Considering the whole period of rainfall to be the period of rainfall excess.

As we know that $\phi \text{In dex} = \left[\frac{P-R}{t}\right]$

$$\therefore \quad \phi\text{-index} = \left[\frac{30-16}{3.5}\right] = 4\text{mm/hr}$$

Since, time period (2.0 to 2.5 hr) and (3.0 to 3.5 hr) are not the period of rainfall excess, these periods are not considered in calculation of ϕ-index.

So,

$$\phi\text{-index} = \left[\frac{30-0.5\times 3-0.5\times 4-16}{3.5-2\times 0.5}\right] \text{mm/hr}$$

= 4.2 mm/hr

49. As we know that,

For Venturimeter

flow rate,

$$Q = C_d\left(\frac{a_1 a_2}{\sqrt{a_1^2-a_2^2}}\right)\times\sqrt{2gh}$$

[where g = 9.8 m/s²]

Piezometric head difference,

$$h = \left(\frac{G_{Hg}}{G_{water}}-1\right)x$$

$$h = \left(\frac{13.6}{1}-1\right)\times 0.15$$

= 1.89 m

Now, For no loss, coefficient of discharge, ($C_d = 1$)

Area of pipe, $a_1 = \frac{\pi}{4}\times d_1^2$

Diameter of pipe, d_1 = 20 cm = 0.2 m

$$\Rightarrow a_1 = \frac{\pi}{4}\times(0.2)^2$$

Area of throat, $a_2 = \frac{\pi}{4}d_2^2$

Diameter of throat, d_2 = 10 cm = 0.1 m

$$\Rightarrow a_2 = \frac{\pi}{4}\times(0.1)^2$$

$$Q = \left[\frac{\frac{\pi}{4}\times(0.2)^2\times\frac{\pi}{4}\times(0.1)^2}{\sqrt{\left(\frac{\pi}{4}\times(0.2)^2\right)^2-\left(\frac{\pi}{4}\times(0.1)^2\right)^2}}\right]\times\sqrt{2\times 9.8\times 1.89}$$

Q = 0.04937 m³/s

Hence, Q = 49.37 litres/second

50. As given that,

Live storage = 300 mm³

Crop Area (A) = 40000 hectares

Since, the tank gets filled twice in a year.

$\therefore$ Volume of water required for the crop area (V) = (2 × 300) = 600 Mm³

$$\Rightarrow \text{Delta } (\Delta) = \left(\frac{V}{A}\right) = \left[\frac{600\times 10^6 \text{ m}^3}{40000\times 10^4 \text{ m}^2}\right]$$

= 1.5 m

As we know that

$$\text{Flow Duty (D)} = \left(\frac{6.64B}{\Delta}\right)$$

$$= \left(\frac{8.64\times 120}{1.5}\right)\text{ha/cumec}$$

= 691.2 ha/cumec

51. As given that,

Q_C = 7500 m³/d

S_0 = 200 mg/l

S = 20 mg/l

$X = 3000$ mg/l

$X_u = 12000$ mg/l

$$\frac{V}{Q_0} = 0.2 \text{ day}$$

As we know that

$$\theta_C = \frac{VX}{Q_w X_u + (Q_0 - Q_w)X_e}$$

= 10 days

Neglecting, Xe, $\theta_C = \dfrac{VX}{Q_W X_u}$ ($\because x_e$ is not given)

$$\Rightarrow Q_w = \frac{Vx}{\theta_C X_u}$$

$\because V = Q \times (D.T)$

$= 7500 \times 0.2$

$= 1500 \text{ m}^3$

$$Q_w = \frac{(0.2 \times 7500)\text{m}^3 \times 3000 \text{ mg/l}}{10 \text{days} \times 12000 \text{ mg/l}}$$

$Q_W = 37.5 \text{ m}^3/\text{day}$

52. As given that,

$d = 0.25$ mm

$S = 2.7$

$V_S = 2.5$ cm/sec

$B = 0.5$ m

$Q = 9720 \text{ m}^3/\text{day}$

$H = 0.75$ m

As we know that

$L_{min} = V_{flow} \times t_d$

$V_{flow} = 0.3$ m/s

$$t_d = \frac{H}{V_{setting}}$$

$$= \frac{0.75}{2.5 \times 10^{-2} \text{m/s}}$$

= 30 sec.

Hence, $L_{min} = 0.3\dfrac{\text{m}}{\text{s}} \times 30 \text{sec} = 9\text{m}$

53. As we know that,

Bulk specific gravity (G_{th})

$$= \left[\frac{W}{\left(\frac{0.55W}{2.55} + \frac{0.40W}{2.65} + \frac{0.05W}{2.70}\right)}\right] = 2.5964 \simeq 2.6$$

54. We know that,

Stopping sight distance (SSD)

$$= vt_r + \frac{v^2}{2\mu g}$$

As given that

$a = \mu g = 3.5 \text{ m/sec}^2$

$$\Rightarrow 140 = \left(\frac{5}{18} \times 90 \times t_r\right) + \frac{\left(\frac{5}{18} \times 90\right)^2}{2 \times 3.5}$$

$\Rightarrow t_r = 2.028 \text{ sec} \simeq 2.03 \text{ sec}$

55. As we know that,

$$\text{Radius of Curve} = \left(\frac{1750}{D}\right)$$

$$= \left(\frac{1750}{2}\right) = 875 \text{ m}$$

Equilibrium superelevation for 80 kmph

$$= \left(\frac{GV^2}{127R}\right)$$

$$= \left[\frac{1750 \times (80)^2}{127 \times 875}\right] = 100.79 \text{ mm}$$

Equilibrium superelevation for maximum sanctioned speed (100 kmph)

$$= \left[\frac{GV^2}{127R}\right] = \left[\frac{1750 \times (100)^2}{127 \times 875}\right]$$

$= 157.48$ mm

Hence, Cant deficiency = (157.48 – 100.79) mm = 56.69 mm $\simeq$ 57 mm